# Fundamentals of Fabrication and Welding Engineering

**F.J.M. Smith Ho ISME**

Formerly Senior Lecturer responsible for Fabrication and Welding Engineering, Tottenham College of Technology.

*This book is dedicated to the memory of my dear wife Ivy Eileen*

# Acknowledgements

Longman Scientific & Technical
Longman Group UK Limited,
Longman House, Burnt Mill, Harlow,
Essex, CM20 2JE, England
*and Associated Companies throughout the world.*

© Longman Group UK Limited 1993

All rights reserved: no part of this publication may be reproduced, stored in a retrieval system, or transmitted in any form or by any means, electronic, mechanical, photocopying, recording, or otherwise without either the prior written permission of the Publishers, or a licence permitting restricted copying in the United Kingdom issued by the Copyright Licensing Agency Ltd, 90 Tottenham Court Road, London, W1P 9HE.

First published 1993

British Library Cataloguing-in-Publication Data

ISBN 0-582-09799-1

Set by 4 in 10/12 pt Century
Printed in Hong Kong
AP/01

We are grateful to the following for permission to reproduce copyright photographs and drawings:

AGA (UK) Ltd
Douglas Barnes (Machinery) Ltd
The British Oxygen Company Ltd
British Steel Technical
British Standards Institution
Carver & Co. (Engineering) Ltd
Duplex Electric Tools Ltd
F.J. Edwards Ltd
Elliot Machine Equipment Ltd
GKN Lincoln Electric Ltd
Hatfield Machine Tool Co. Ltd
L.J. Hydleman & Co. Ltd
I.P.C. Business Press Ltd
The Kingsland Engineering Co. Ltd
Moore & Wright (Sheffield) Ltd
A.J. Morgan & Son (Lye) Ltd
The National Federation of Building Trade Employers (Constructional Safety)
Neldco Ltd
James North & Sons Ltd
Philpott & Cowlin Ltd
William Press & Son Ltd
Rockweld Ltd
Rushton & Co. (Sowerby Bridge) Ltd
Tin Research Institute
The Welding Institute
Wolf Electric Tools Ltd

# Contents

|   |   |   |
|---|---|---|
| Preface | | viii |
| **1** | Safety | 1 |
| | Part A Safety on site | 1 |
| | Part B Arc welding | 8 |
| | Part C Gas welding and cutting | 16 |
| | Part D Miscellaneous | 20 |
| **2** | Communications | 33 |
| **3** | Metal joining — mechanical fastenings | 69 |
| **4** | Science | 97 |
| | Part A Basic science — thermal operations | 97 |
| | Part B Engineering science — fundamentals and applications | 115 |
| **5** | Measurement | 139 |
| **6** | Materials | 176 |
| **7** | Material removal | 231 |
| **8** | Restraint and location | 278 |
| **9** | Workshop operations | 298 |
| **10** | Fabrication processes | 339 |
| **11** | Metal joining — thermal and adhesive processes | 393 |
| **12** | Welding | 426 |
| | Part A Oxy-acetylene welding | 426 |
| | Part B Metal-arc welding | 440 |
| | Part C Welding defects | 451 |
| Appendix 1 | | 457 |
| Appendix 2 | | 459 |

# Preface

This new book supersedes *Basic Fabrication and Welding Engineering* which was originally written by the author as a specialised complement for the very successful 'Longman Craft Series' on basic engineering, edited by R.L. Timings, former Head of Engineering at Henley College of Further Education, Coventry. The aim of this series was to provide a primary source of information for students following the Engineering Craft Studies Course Part 1 (CGLI). However, for full coverage of fabrication and welding students' needs, it was necessary to study the contents of *Basic Fabrication and Welding Engineering* in conjunction with *Basic Engineering* (R.L. Timings) which contained, in addition to two chapters on 'Metal Joining', several cross-references to other essentially fabrication topics.

Most of the original contents of *Basic Fabrication and Welding Engineering* have been embodied in this new work together with all the relevant topics contained in *Basic Engineering*. The chapter on 'Calculations' has been deliberately omitted and numerous 'worked examples' have been interspersed throughout the book. These worked examples have been carefully selected to support associated topics and their solutions are presented in an easy to follow format.

Many of the chapters have been greatly enhanced with additional essential topics, in particular, the chapters on 'Materials' (which now includes some elementary metallurgy), 'Materials Removal' and 'Basic Science'. The overall text has been expanded to provide not only more than adequate coverage for 'Basic Engineering Craft Studies' (CGLI Part 1) but also a solid foundation and introduction to Craft Studies in Sheet Metal and Thin Plate, Structural and Thick Plate, and Welding (CGLI Part II). In addition the text should also prove to be a valuable source of information for Technicians attending colleges on BTEC courses in Engineering which contain Fabrication and Welding units of study. It is also the author's intention that this book be used as a 'terms of reference' by anyone employed in or associated with the fabrication industry.

This comprehensive new work presents each topic logically and with continuity — any cross-references are clearly indicated. Wherever possible the text has been kept to a minimum and, in order to provide interesting reading, extensive use has been made of illustrations, diagrams and tables throughout. Each chapter contains sufficient description and explanation to provide readers with adequate material to satisfy their basic requirements, and to provide a firm foundation for future studies.

As we approach the twenty-first century, far-reaching changes will inevitably occur in technical education. However, we must not forget that the essential fundamentals of engineering will always remain the starting platform for any future technological developments.

The author particularly wishes to thank the publishers for their encouragement, advice and assistance freely given during the preparation of this manuscript.

May 1991　　　　　　　　　　　　　　　　　　　　　　　　F.J.M. SMITH

# 1 Safety

## Part A Safety on site

The welding and fabrication engineer is often expected to work *on site*. This entails working in partly-completed buildings or on equipment being erected out of doors which introduces additional hazards. When working under such conditions additional precautions must be taken. The erector must not only be mindful of the safety codes of his own trade, but also of the *codes of safety for building sites* as well. Sections 1.1 to 1.6 indicate some precautions that should be observed when working *on site*.

## 1.1 Protective clothing

Popular types of protective clothing are shown in Figure 1.1. For general workshop purposes the boiler suit is the most practical and safest form of body protection. The bib and brace type of overall is much cooler to wear but does not offer as much protection as the boiler suit. The coat overall or dust-coat is very comfortable to wear. It is normally worn by supervisors, inspectors and stores persons. It is preferred by those skilled in detail or precision sheet-metal work.

In recent years, tremendous advances have been made in the design and range of protective clothing. Such very necessary clothing is manufactured to the highest standards with regard to materials, construction, and resistance to wear or to damage.

Examples of protective clothing in everyday use are Wellington boots, overalls, aprons, leggings, headgear, and gloves. Because of adverse weather conditions when working *on site*, protective clothing generally includes jackets and overtrousers made from oilskin, plastic, rubber, or other waterproof materials. Figure 1.2 shows a few of the numerous types of protective garments available.

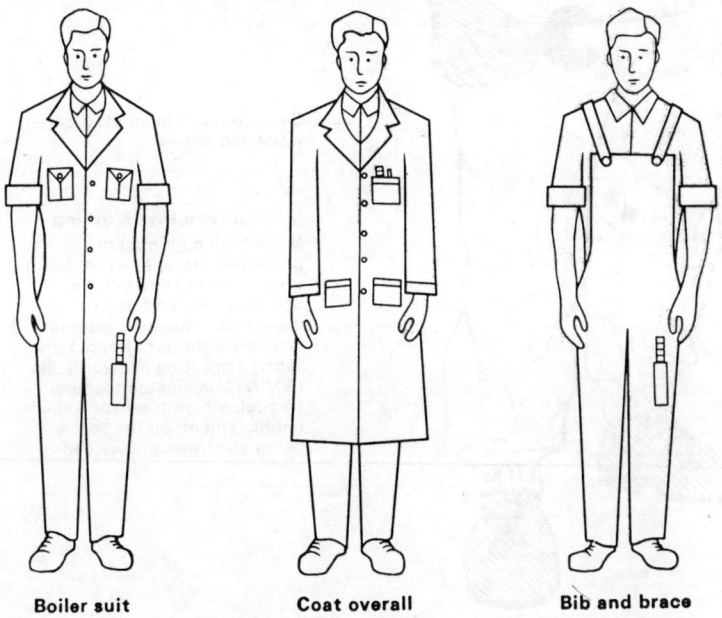

Boiler suit     Coat overall     Bib and brace

**Fig. 1.1 Types of overall**

*(a)* **'Foul-weather clothing'**

Made from high quality protective material which has a great resistance to tearing, abrasion, oils, and most chemicals. One such material is a very light but strong 3-ply fabric consisting of two sheets of VINYL laminated together through a high-tenacity NYLON reinforcement, all the seams being electronically welded

*(b)* **'Donkey Jacket'**

Generally made from hard wearing melton cloth reinforced with P.V.C. on shoulders, back and cuffs

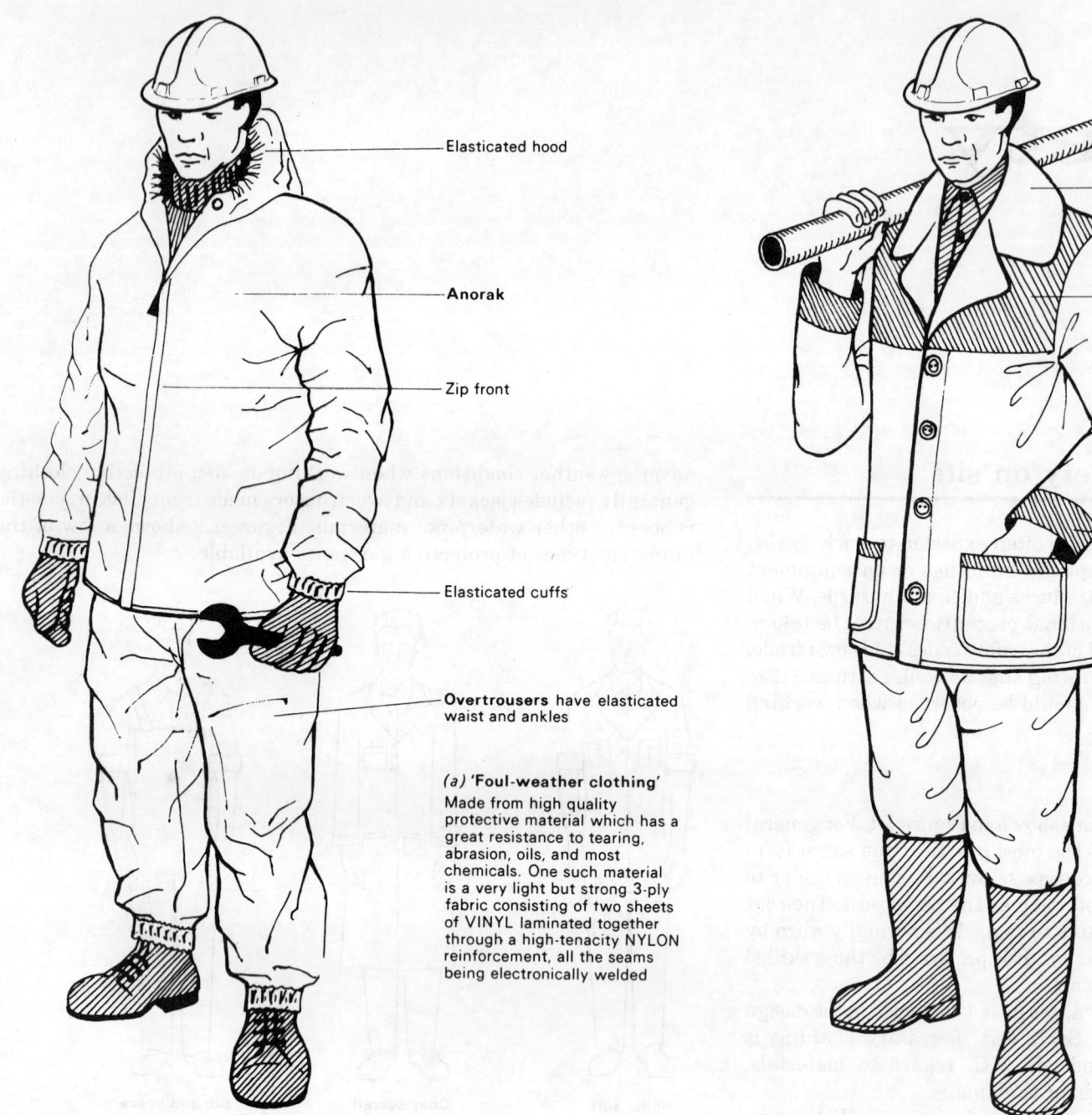

**Fig. 1.2 Protective clothing — site work**

*Because a workman's hands are in constant use they are always at risk.* He has to handle dirty, oily, greasy, rough, sharp, brittle, hot, and maybe corrosive materials. Gloves and 'palms' of a variety of styles and types of materials are available to protect the hands, whatever the nature of the work. Some hand-protection methods are illustrated in Figure 1.3.

## 1.2 Head protection (BS 2826)

When working *on site*, all persons should wear safety helmets because of the ever present danger from falling objects. *Even small objects such as nuts or bolts can cause very serious head injuries when dropped from heights.* When entering on to a site THE FIRST PRIORITY IS ADEQUATE HEAD PROTECTION. Figure 1.4 shows a typical safety helmet.

Safety helmets are made to BS 2826 from moulded plastic or fibre-glass reinforced polyester. Colour coded for personnel identification, they are light and comfortable to wear, yet despite their lightness have a high resistance to impact and penetration.

To eliminate the possibility of electric shock, safety helmets have *no metal parts*. The materials used to manufacture the outer shell have to be non-flammable and their electrical insulation must withstand 35 000 volts.

Figure 1.5 shows the harness inside the safety helmet. This provides ventilation and a *fixed safety clearance* between the helmet and the wearer's skull.

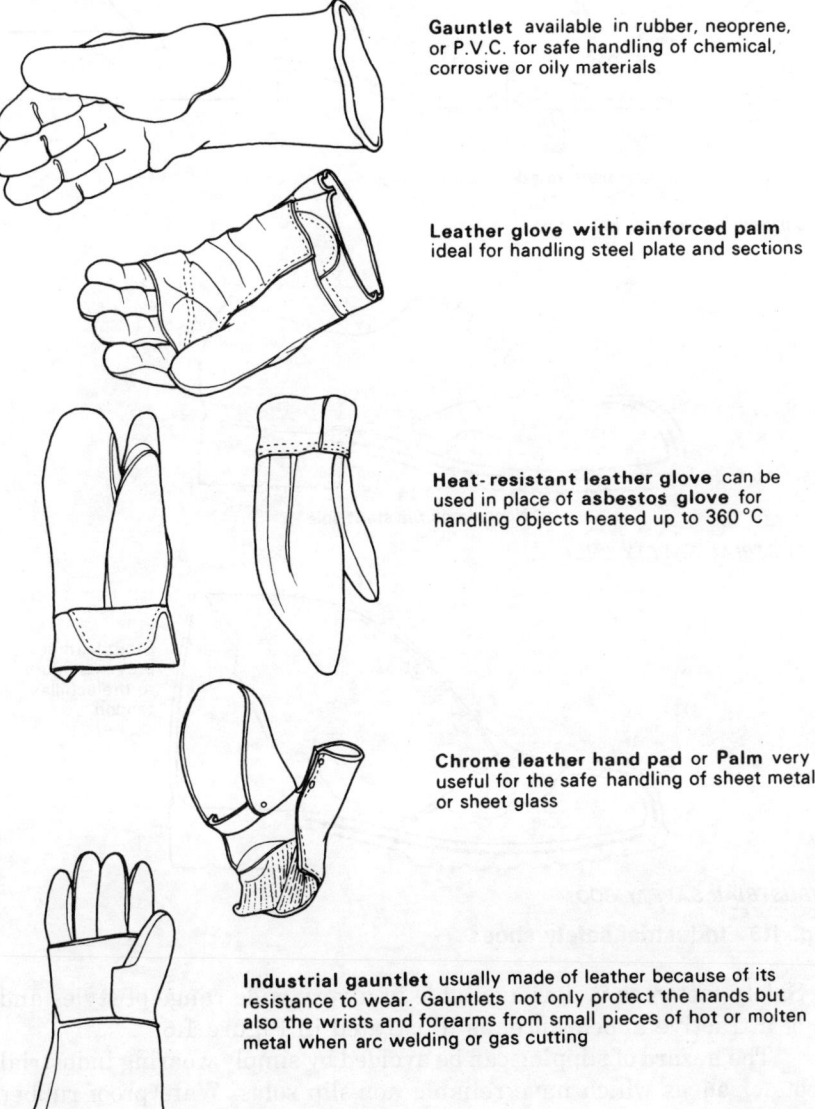

**Gauntlet** available in rubber, neoprene, or P.V.C. for safe handling of chemical, corrosive or oily materials

**Leather glove with reinforced palm** ideal for handling steel plate and sections

**Heat-resistant leather glove** can be used in place of **asbestos glove** for handling objects heated up to 360 °C

**Chrome leather hand pad** or **Palm** very useful for the safe handling of sheet metal or sheet glass

**Industrial gauntlet** usually made of leather because of its resistance to wear. Gauntlets not only protect the hands but also the wrists and forearms from small pieces of hot or molten metal when arc welding or gas cutting

**Fig. 1.3 Hand protection**

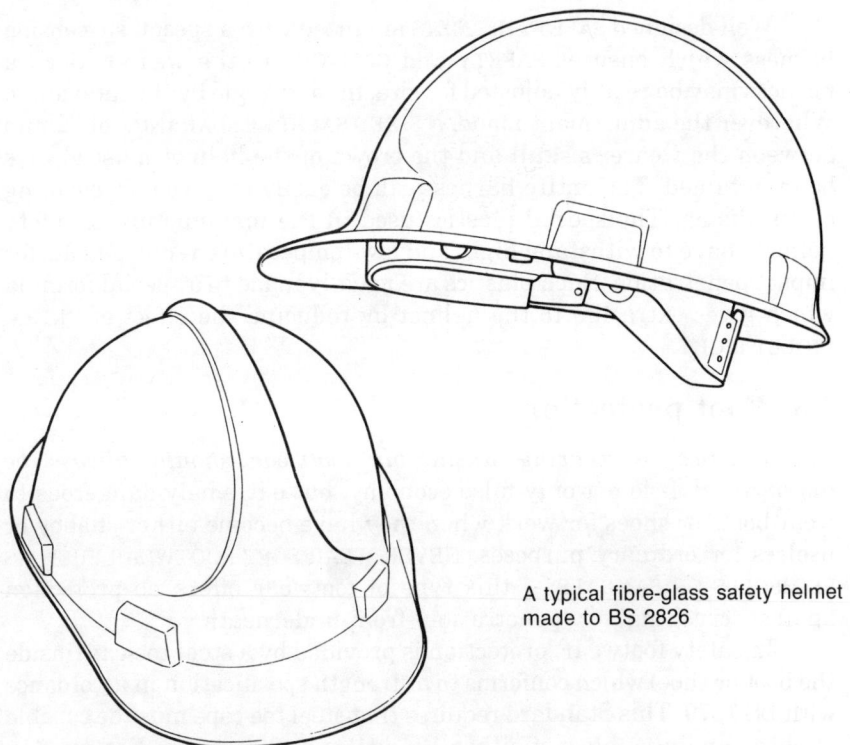

A typical fibre-glass safety helmet made to BS 2826

**Fig. 1.4 Safety helmets**

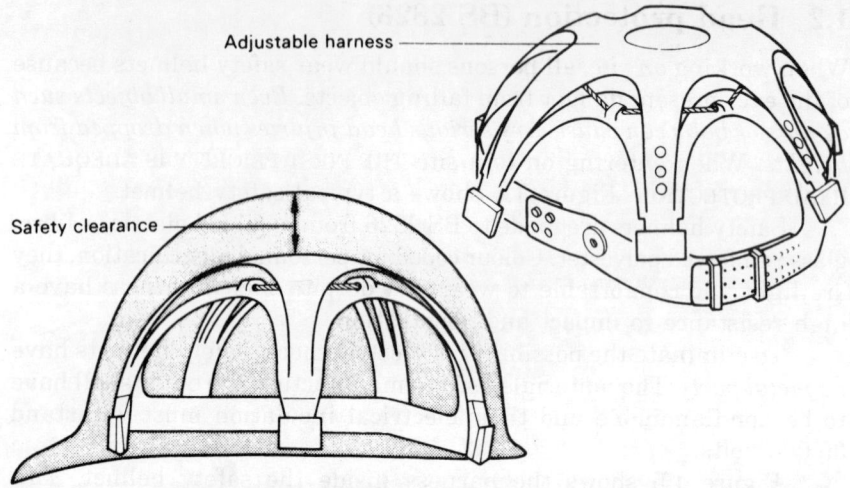

**Fig. 1.5 Safety helmet harness**

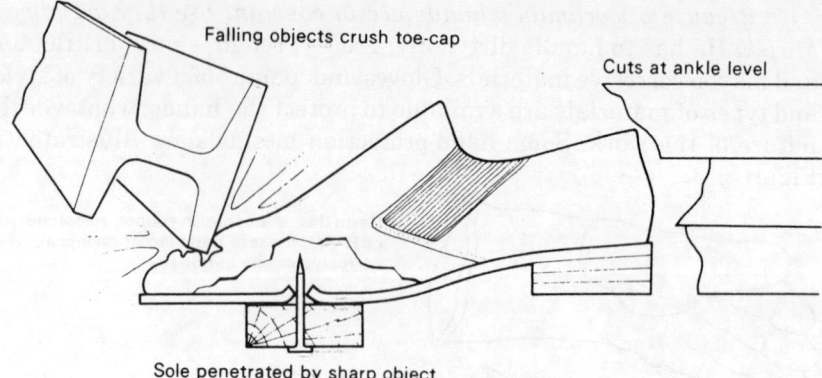

*Lightweight shoes offer NO protection*

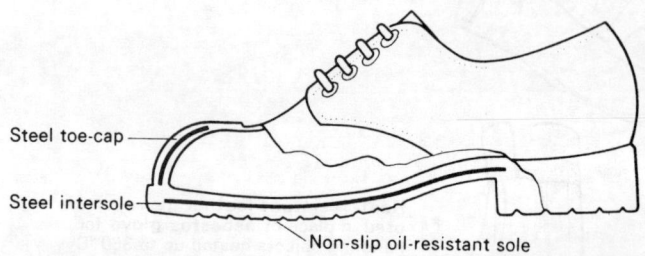

*INDUSTRIAL SAFETY SHOE*

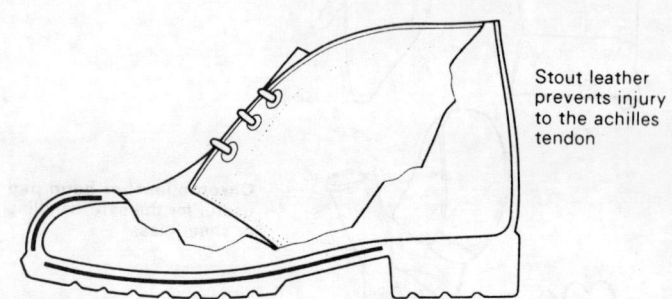

*INDUSTRIAL SAFETY BOOT*

**Fig. 1.6 Industrial safety shoes**

Well-designed SAFETY HELMETS are fitted with a special suspension harness which ensures SAFETY and COMFORT to the wearer. Such a harness may be readily adjusted for size, fit, and angle by the individual. Whatever the adjustment made, a FIXED SAFETY CLEARANCE of 32 mm between the wearer's skull and the crown of the helmet must always be maintained. The entire harness can be easily removed for cleaning or sterilising. The special plastics used in the manufacture of safety helmets have to withstand high- and low-temperature requirements for impact penetration. Such plastics are usually made to a special formula which gives extra life to the helmet by reducing the effect of ULTRA-VIOLET LIGHT.

## 1.3 Foot protection

*The practice of wearing unsuitable footwear should always be discouraged.* It is not only false economy, but extremely dangerous to wear boots or shoes for work when they have become either shabby or useless for ordinary purposes. NEVER WEAR SOFT FOOTWEAR SUCH AS PLIMSOLLS OR SANDALS — this type of footwear offers no protection against 'crushing', or 'penetration' from underneath.

In safety footwear, protection is provided by a steel toe-cap (inside the boot or shoe) which conforms to a strength specification in accordance with BS 1870. This Standard requires that steel toe-caps must be capable of withstanding a blow of either 134 joules or 200 joules. Footwear in the former category has to be marked GRADE 2, and in the latter GRADE

1. Safety footwear is now available in a very wide range of styles and is of attractive appearance, as illustrated in Figure 1.6.

The hazard of slipping can be avoided by simply wearing industrial boots or shoes which have reliable non-slip soles. Waterproof rubber ANKLE BOOTS or KNEE BOOTS are also available with special cleated heels and soles.

## 1.4 Precautions when working aloft

Whenever persons are required to ascend to a height in order to reach the workplace, the following are a few of the many rules or regulations which should be observed for safety:

1. *Always have some breakfast, or at least a hot drink such as a cup of tea or coffee before you go to work* — AN EMPTY STOMACH CAN OFTEN RESULT IN A SUDDEN ATTACK OF FAINTNESS in even the healthiest person.
2. SAFETY BELTS OR HARNESS should be worn whenever possible, and these must be securely anchored. Figure 1.7 illustrates a safety device which can be fitted permanently to any structure where personnel are moving in a vertical plane.

    This type of 'SAFETY BLOCK' has many applications in construction work, maintenance, and other engineering operations where workers are moving up or down shafts, chimneys, masts, ladders, cranes, or on the outside of structures. THE BLOCK GIVES COMPLETE SECURITY AGAINST FALLS IN THESE KINDS OF SITUATIONS.

    It operates in both directions and the worker is protected when travelling both up and down. Should the worker slip, his fall is stopped within 300 mm with a gradual deceleration that causes no discomfort. The block unlocks automatically when the pull on the rope is released.
3. It is often desirable for a worker wearing a safety harness to have a limited safe radius of operation, this can be achieved by use of a special safety block which is illustrated in Figure 1.8.

    This block is completely self-contained. It holds approximately 5 metres of STEEL WIRE ROPE mounted on a springloaded drum. With normal movement the rope pulls out and winds up automatically, keeping the rope taut and giving the user freedom of movement with absolute safety.

    The desired limited safe radius of operation can be obtained with the safety block by pulling out the required length of rope and locking it with a ratchet pawl as shown. A sudden pull engages a locking device and the rope is stopped. Because of a friction brake on the rope drum in the block there is no jolt on the worker in the event of a fall, yet the fall is stopped after about 300 mm. Figure 1.9 shows how the self-contained block is used in practice.
4. All stairways and walkways should be provided with handrails. On stairways which would otherwise be open-sided, a lower rail

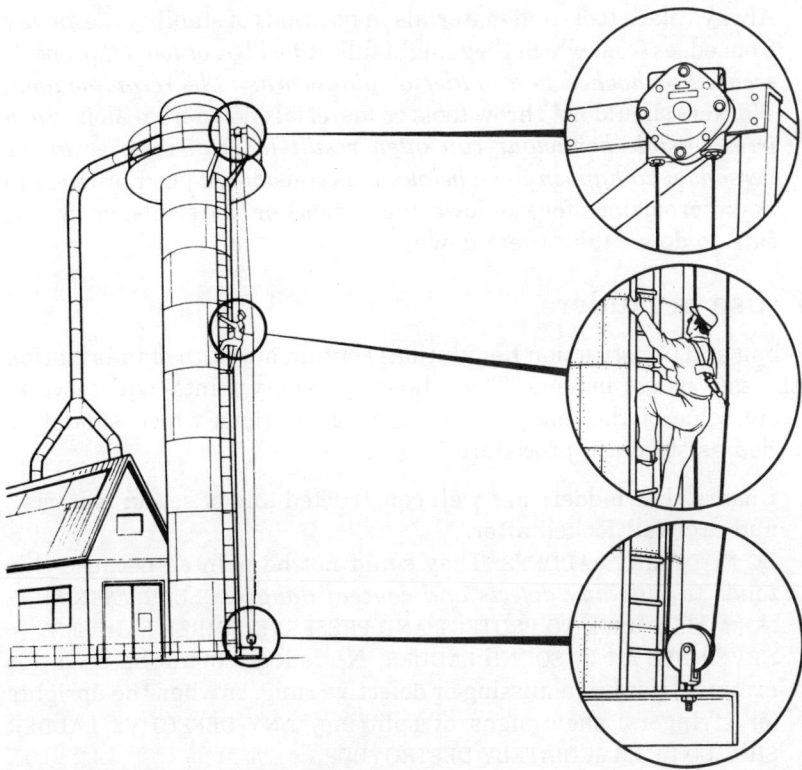

**Fig. 1.7 Use of a static safety block**

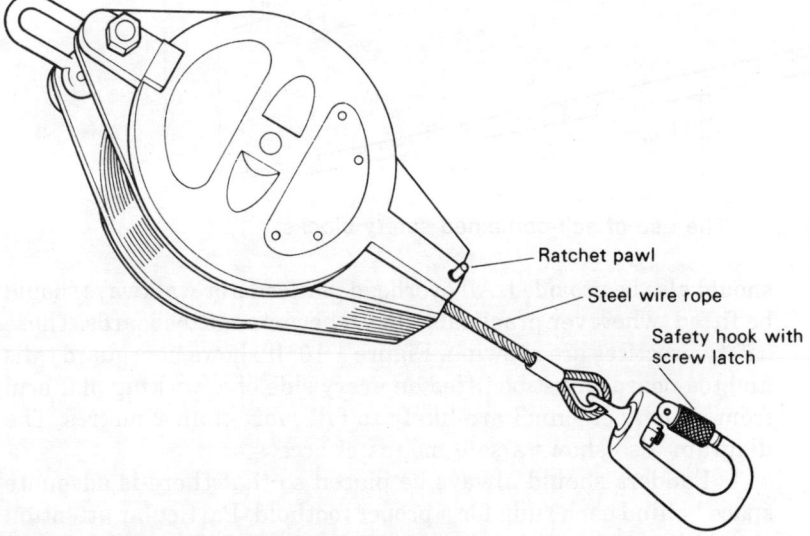

Fig. 1.8 **The self-contained safety block**

**Fig. 1.9 The use of self-contained safety blocks**

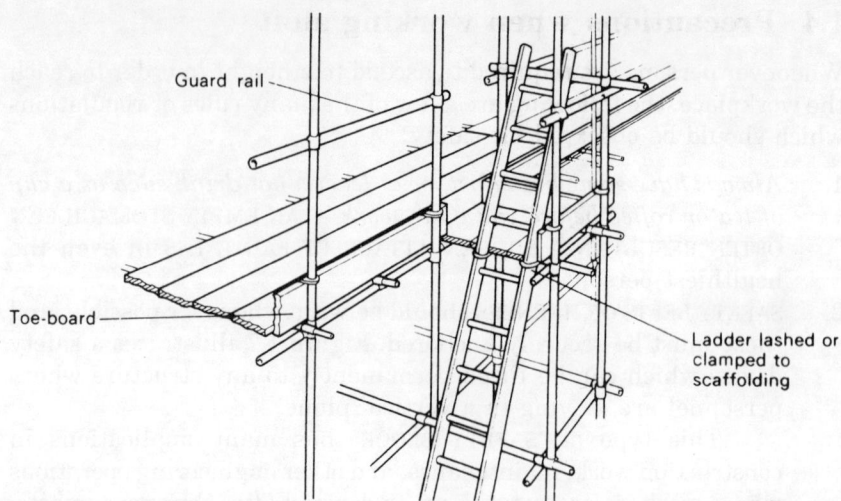

**Fig. 1.10 Overhead platforms**

5. Always place tools and materials in positions of stability, i.e., away from edges from which they might fall. *A tool box or tool kit properly secured or hooked to a ladder or platform is to be recommended.*
6. Workers should not throw tools or materials down from aloft. *Such irresponsible behaviour can often result in fatal injuries to the persons who happen to be below.* It is considered good practice to make provisions for the lowering of tools or materials, or, if it is safe to do so, take them down.

## 1.5 Use of ladders

Building and Construction Regulations contain a wealth of information on the safe use of ladders. The following 'Safety Hints' will serve to indicate some of the many elementary precautions which should be regarded as standard procedure:

1. Ensure that ladders are well constructed and of sound material, and are well looked after.
2. DO NOT PAINT LADDERS. They sould not be painted because this tends to *hide any defects and conceal danger.* CLEAR VARNISH IS NORMALLY USED TO PROTECT AND PRESERVE TIMBER LADDERS.
3. NEVER USE AN UNSOUND LADDER. No ladder should be used, for example, if it has a missing or defective rung, or when the uprights (or stringers) show signs of splitting. ANY DEFECTIVE LADDER SHOULD BE IMMEDIATELY DESTROYED.
4. *The correct pitch of a ladder must always be observed.* THE

should also be provided. All overhead platforms or walkways should be fitted, wherever practicable, with permanent toeboards. These safety practices are shown in Figure 1.10. It shows how guard rails and toe-boards must be fitted on every side of a working platform from which personnel are likely to fall more than 2 metres. The diagram also shows a safe means of access.

Ladders should always be placed so that there is adequate space behind each rung for a proper foothold. Particular attention should be paid to this point at the landing platform.

VERTICAL HEIGHT FROM THE GROUND OR BASE TO THE POINT OF REST SHOULD BE FOUR TIMES THE DISTANCE BETWEEN THE BASE OF THE VERTICAL HEIGHT AND THE BASE OF THE LADDER, as illustrated in Figure 1.11.

5. Make certain that the ladder reaches at least 1 metre above the landing platform. *This is to provide a handhold while stepping from the ladder*, as shown in Figure 1.11.
6. Use the correct length of ladder for the job. NEVER LASH TWO SHORT LADDERS TO MAKE A LONGER ONE.
7. MAKE SURE THAT THE LADDER IS SET ON A FIRM LEVEL BASE. Care must be taken on sloping surfaces — *the packing of ladder feet is a dangerous habit which must be discouraged.* Figure 1.12 shows a few of the safe practices observed when setting ladders.
8. Lash the ladder as soon as possible (see Figure 1.11). Until this is done, ensure that a workmate is standing on the bottom or 'footing it'.
9. Always face the ladder when climbing or descending, and *beware of wet, greasy, or icy rungs.*
10. INSPECT LADDERS BEFORE USE AND REGULARLY WHEN STORED. Store

1. Ladders should be set 0·3m out for each 1·2m of height, i.e., use the 'FOUR TO ONE RULE' so that the ladder is at an angle of about 75° to the horizontal

2. Stringers should be lashed or clamped securely to some convenient anchorage to prevent slipping sideways

3. Ladders must extend 1·07m above the stepping-off point to ensure adequate handhold. If this is not possible then a nearby handhold of equivalent height must be provided, as shown below

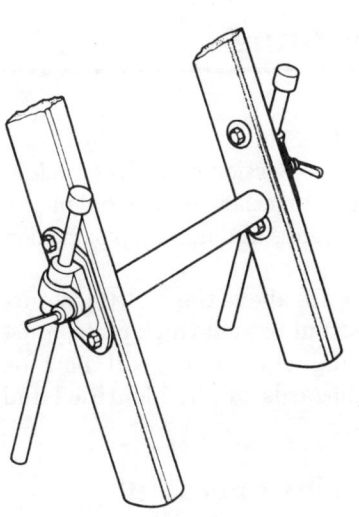

Adjustable safety device to suit any uneven contour of surface. Can easily be fitted to most ladders

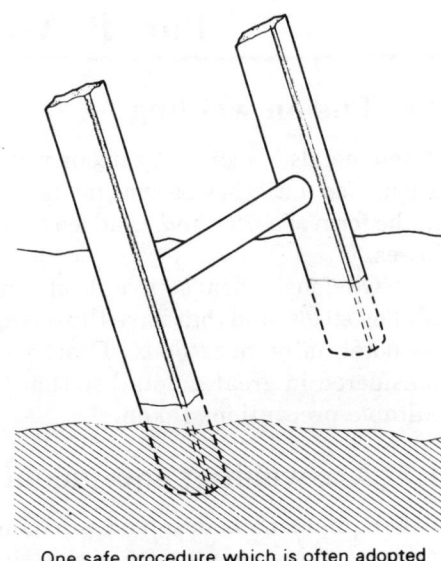

One safe procedure which is often adopted on site is to bury the foot of the ladder in the ground up to the first rung

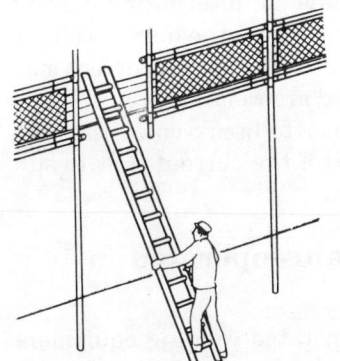

If in doubt have a person 'footing the ladder'

**Fig. 1.11 Setting a ladder**

**Fig. 1.12 Safety precautions — ladders**

7

ladders where they will not suffer damage either by the weather or by mechanical means. Always rest ladders, DO NOT HANG THEM FROM THE STILES OR RUNGS.

(Note: The *uprights* (or *stringers*) are also known as *stiles* in some parts of the country.)

The author is indebted to the Royal Society for the Prevention of Accidents for much of the information used in this section of the chapter.

# Part B  Arc welding

## 1.6  Fusion welding

When metals are joined by fusion welding (see Section 11.29) their edges are heated until they become molten and run together. Additional metal, in the form of a *filler rod*, is added to fill any gaps and make up oxidation losses.

Obviously, heat sources that can operate above the melting points of alloy steels, and that have the energy output to melt thick plate, must be potentially hazardous. Fusion welding equipment will now be considered in greater detail so that the hazards can be identified and suitable precautions taken.

## 1.7  Arc-welding equipment (mains operated)

This equipment is designed to change the high voltage alternating-current mains supply into a safe, low-voltage, heavy-current supply suitable for arc welding. Figure 1.13 shows examples of some typical arc-welding sets.

It will be seen that the output can have an alternating current waveform or a direct current waveform. For safety, the output voltage is limited to between 50 and 100 volts; however, the output current may be as high as 500 amps. Figure 1.14, a typical arc-welding circuit, shows that a welding set is basically a transformer to break down the high mains voltage, and a tapped choke to control the current flow to suit the gauge of electrode used.

## 1.8  Hazards associated with mains-operated arc-welding equipment

The hazards that may arise from mains-operated welding equipment are set out in Table 1.1. To understand these more fully, constant reference should be made to Figure 1.15 which shows a schematic

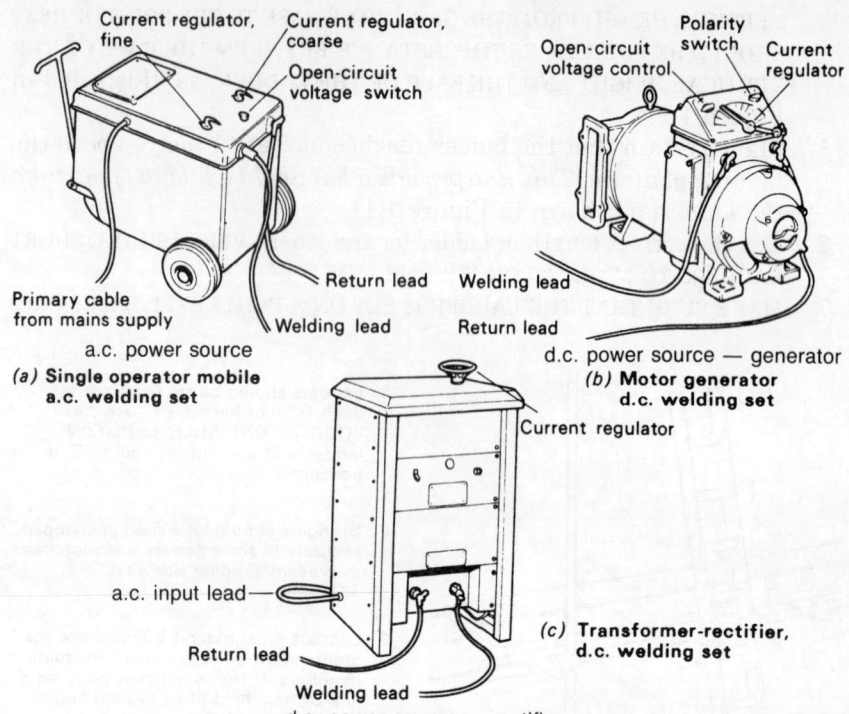

Fig. 1.13  Mains-operated arc welding equipment

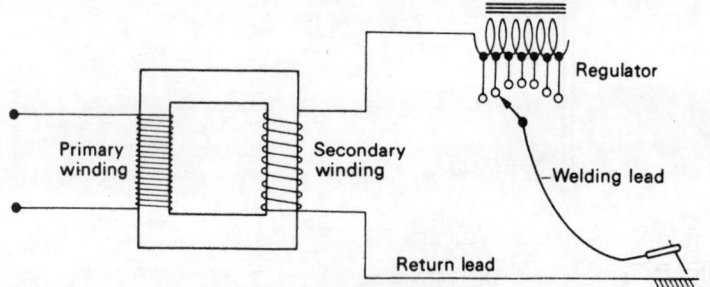

Fig. 1.14  Circuit diagram of an alternating-current arc-welding set

diagram of a typical manual, metallic-arc welding circuit. It will be seen that the circuit conveniently divides into two parts:

1. The primary (high voltage) circuit, which should be installed and maintained by a skilled electrician.
2. The secondary (low voltage) external welding circuit, which is normally set up by the welder to suit the job in hand.

**TABLE 1.1  Arc-welding hazards**

CIRCUIT — HIGH VOLTAGE — Primary

| *Fault:* | *Hazard:* |
|---|---|
| 1. Damaged insulation | Fire — loss of life and damage to property<br>Shock — severe burns and loss of life |
| 2. Oversize fuses | Overheating — damage to equipment and fire |
| 3. Lack of adequate earthing | Shock — if fault develops — severe burns and loss of life |

CIRCUIT — LOW VOLTAGE — Secondary (very heavy current)

| *Fault:* | *Hazard:* |
|---|---|
| 1. Lack of welding earth | Shock — if a fault develops — severe burns and loss of life |
| 2. Welding cable — damaged insulation | Local arcing between cable and any adjacent metalwork at earth potential causing fire |
| 3. Welding cable — inadequate capacity | Overheating leading to damaged insulation and fire |
| 4. Inadequate connections | Overheating — severe burns — fire |
| 5. Inadequate return path | Current leakage through surrounding metalwork — overheating — fire |

To eliminate these hazards as far as possible, the following precautions should be taken. These are only the *basic* precautions and a check should always be made as to whether the equipment and working conditions require special, additional precautions:

1. Make sure that the equipment is fed from a switch-fuse so that it can be isolated from the mains supply. Easy access to this switch must be provided at all times.
2. Make sure that the trailing primary cable is armoured against mechanical damage as well as being heavily insulated against the high supply voltage (415 volts).
3. Make sure that all cable insulation is undamaged and all terminations are secure and undamaged. *If in doubt, do not operate the equipment until it has been checked by a skilled electrician.*
4. Make sure that all the equipment is adequately earthed with conductors capable of carrying the heavy currents used in welding.
5. Make sure the current regulator has an 'OFF' position so that in the event of an accident the welding current can be stopped without having to follow the primary cable back to the isolating switch.

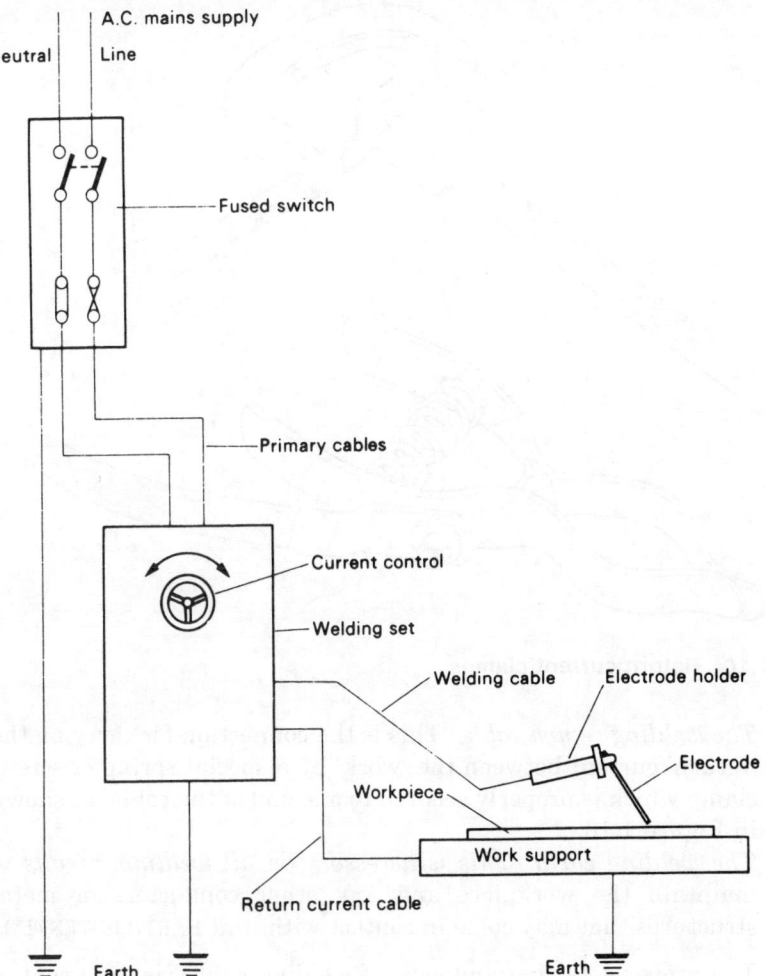

**Fig. 1.15  Manual metal-arc welding circuit diagram**

6. Make sure that the 'external welding circuit' is adequate for the heavy currents it has to carry.

## 1.9  The external arc-welding circuit

This is normally set up by the welder himself to suit the job in hand. There are three important connections for every welding circuit:

1. *The welding lead*  This is used for carrying the welding current from the power source to the ELECTRODE HOLDER.

9

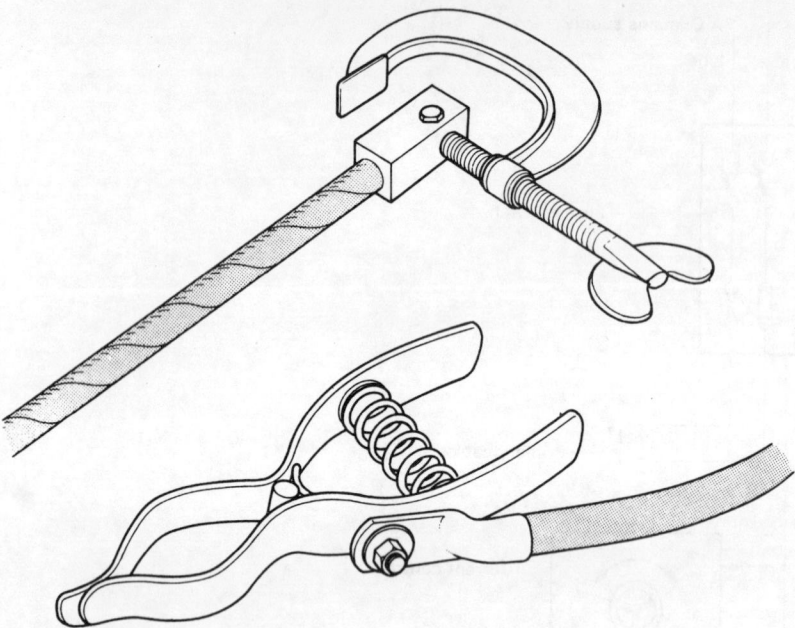

**Fig. 1.16 Return current clamps**

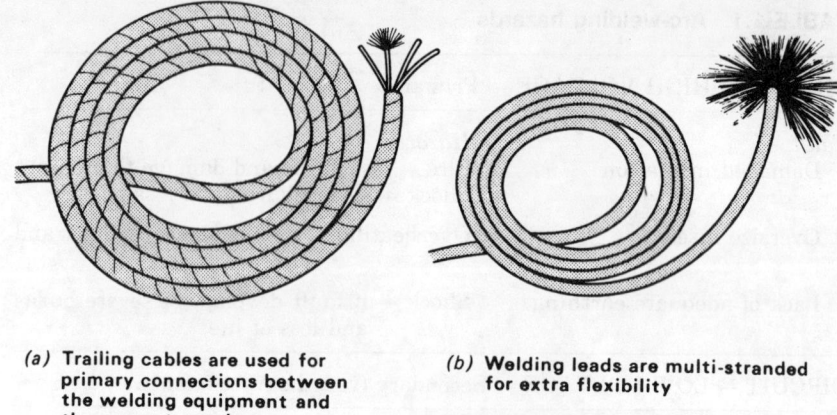

(a) Trailing cables are used for primary connections between the welding equipment and the current supply

(b) Welding leads are multi-stranded for extra flexibility

**Fig. 1.17 Welding cables**

2. *The welding return cable* This is the connection for carrying the 'return' current between the 'work' by a special spring or screw-clamp which is properly secured to one end of the cable, as shown in Figure 1.16.
3. *The welding earth This is necessary on all welding circuits* to maintain the workpiece and any other conductors or metal structures that may come in contact with it at EARTH POTENTIAL.

The proper selection and care of welding cables is an essential safety requirement, for it is surprising how many faults occur in practice through neglecting to take elementary precautions. The most common faults are:

1. Bad connections;
2. The use of longer cables than necessary (causing excessive 'voltdrop');
3. Defective insulation;
4. The use of secondary cables of incorrect current carrying capacity.

Figure 1.17 shows the fundamental difference between the primary trailing cable and the welding lead. The former has three or four cores of relatively thick stranded wire, whereas the lattter derives its flexibility and heavy current-carrying capacity from hundreds of strands of very fine wire. This fine wire is contained within a paper wrapping, to allow them to slip readily when the cable is bent. Wear resistance is usually provided by a tough outer sheath of insulating material, such as tough vulcanised rubber (TVR).

The main hazards associated with the external welding circuit are given in Table 1.1, and are overcome by the use of cables of adequate capacity and terminating them correctly. Table 1.2 gives details of 'secondary welding cables'. Figure 1.18 shows typical connectors for welding and return cables.

## 1.10 The welding return and the welding earth

With arc-welding processes ALL ELECTRIC CURRENT FED TO THE WELDING ARC MUST FIND ITS WAY BACK TO ITS SOURCE OF SUPPLY. As arc welding using 300 amps is common, and since some work involves the operator using 600 amps, it is imperative that an efficient welding return be provided.

The diagrams shown in Figures 1.19, 1.20 and 1.21 serve to illustrate the importance of the welding return and the welding earth. The possible dangers that arise, should the electric current be allowed to establish other return paths through adjacent metalwork, can provide a number of possibilities which may cause trouble and accidents, for example:

1. PORTABLE POWER TOOLS — the earthing on the metal case can be burnt out. Should the tool become faulty before the earth is repaired, a serious or even fatal accident may arise.

**Table 1.2 Secondary welding cables**

Constructional details and current ratings of rubber insulated/rubber sheathed and rubber insulated/P.C.P. sheathed cable with copper conductors.

| CONDUCTOR DETAILS | | CURRENT RATINGS (Amps) at a maximum duty cycle of | | | | |
|---|---|---|---|---|---|---|
| Nominal area (mm$^2$) | Nominal number and nominal diameter of wires (mm) | 100% | 85% | 60% | 30% | 20% |
| 16 | 513/0.20 | 105 | 115 | 135 | 190 | 235 |
| 25 | 783/0.20 | 135 | 145 | 175 | 245 | 300 |
| 35 | 1107/0.20 | 170 | 185 | 220 | 310 | 380 |
| 50 | 1566/0.20 | 220 | 240 | 285 | 400 | 490 |
| 70 | 2214/0.20 | 270 | 295 | 350 | 495 | 600 |
| 95 | 2997/0.20 | 330 | 360 | 425 | 600 | 740 |
| 120 | 608/0.50 | 380 | 410 | 490 | 690 | 850 |
| 185 | 925/0.50 | 500 | 540 | 650 | 910 | 1120 |

The cables connecting the welding set to the electrode holder and the EARTH RETURN connecting the workpiece to the welding set *must be of the appropriate size for the maximum* WELDING CURRENT *of the set.*

Duty cycle — A POWER SOURCE for manual arc welding is never loaded continuously because of the periods spent on such operations as electrode changing, slag removal, etc.
The ratio of the actual welding or arcing time and the total working time expressed as a percentage is called the 'duty cycle'

$$\text{DUTY CYCLE} = \frac{\text{ARCING TIME}}{\text{TOTAL WORKING TIME}} \times 100\%$$

(where the total working time = arcing time + handling time).

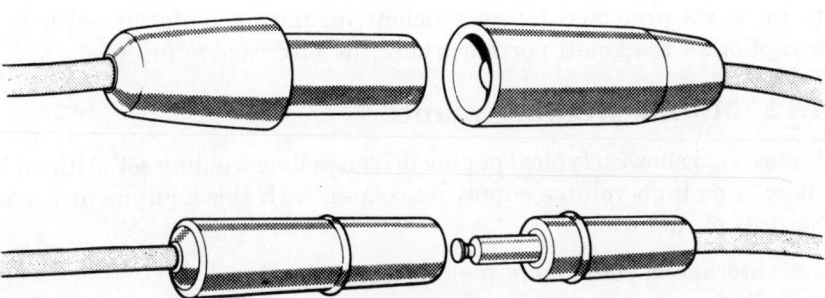

**Fig. 1.18 Fully insulated quick-acting cable connectors**

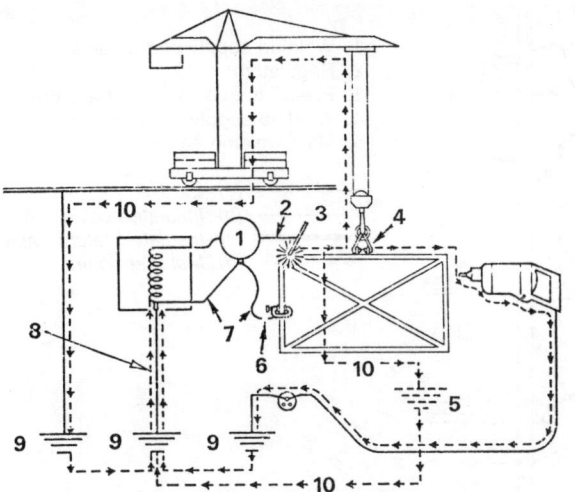

Key:
1. Regulator
2. Welding lead
3. Electrode
4. Chain slings
5. Chance earth
6. Break in the welding return
7. Welding return back to power source
8. Connection which may be burnt out or damaged
9. Earth
10. Possible return paths for current (dashed line, arrowed)

This diagram shows how a high impedance or break in the welding return cable permits electric current to establish a return path without interfering with the operation, but at the same time *creating a number of hazards.*

**Fig. 1.19 An unsafe welding circuit**

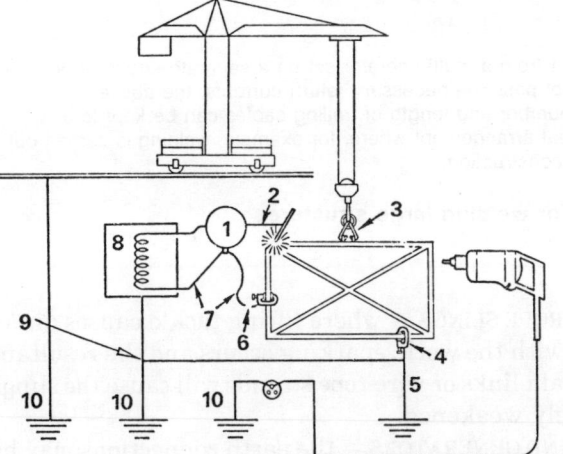

Key:
1. Regulator
2. Welding lead
3. Chain slings
4. Earth clamp
5. Welding earth at workpiece bonded to the earth continuity system
6. High resistance or break in welding return cable
7. Welding return
8. Power source (transformer tank)
9. N.B. — *This connection does not carry any welding current*
10. Earth

This is a similar circuit to that shown in Figure 1.19, but in this case the welding earth is connected to the workpiece and no alternative return paths become possible. Should a break or a high resistance occur in the welding return cable welding must cease, thus making the welding circuit safe.

**Fig. 1.20 The importance of the welding earth**

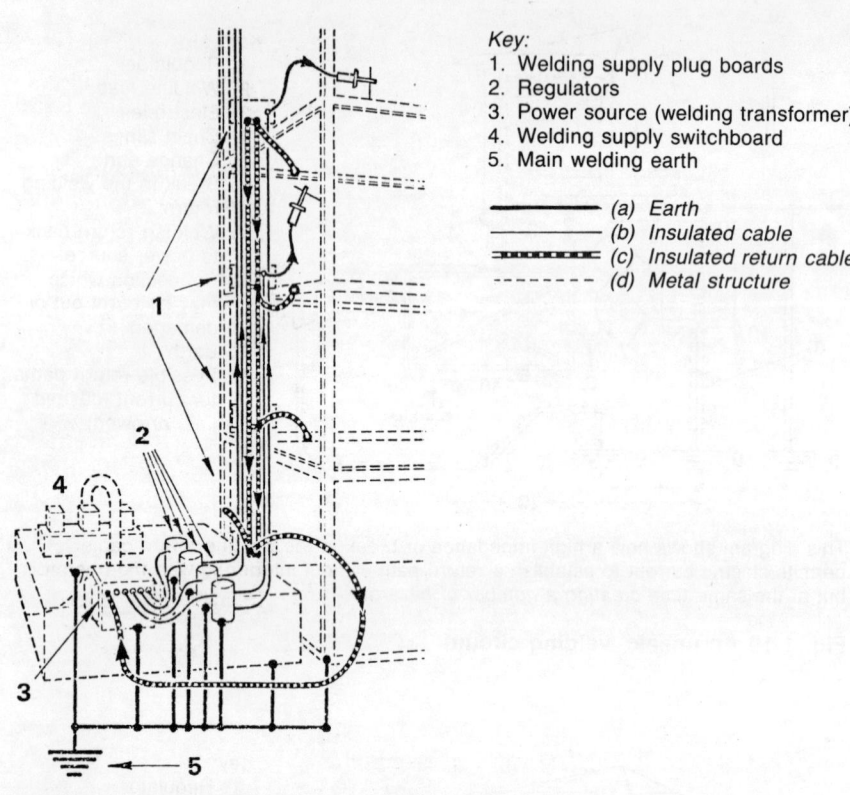

Key:
1. Welding supply plug boards
2. Regulators
3. Power source (welding transformer)
4. Welding supply switchboard
5. Main welding earth

──────── (a) Earth
──────── (b) Insulated cable
▬▬▬▬▬▬ (c) Insulated return cable
-------- (d) Metal structure

Where welding is carried out from a multi-operator set on a separate structure or on a single structure which cannot pass the necessary return currents, the above arrangement is used. The number and length of trailing cables can be kept to a minimum, and this is an ideal arrangement where, for example, welding is carried out on the hull of a ship under construction.

**Fig. 1.21 Cable layout for welding large structures**

2. CHAIN AND WIRE ROPE SLINGS — where lifting tackle causes direct metallic contact with the work, sparking occurs and the resultant burning of the chain links or wire rope strands will cause the slings to become severely weakened.
3. TRANSFORMERS AND GENERATORS — the earth connections may be damaged or burnt out.
4. METAL CONDUITS AND PIPEWORK — conduits, gas, air and steam pipes or hydraulic mains can all be damaged. *Where pipework contains an inflammable liquid or gas there is the added danger of the welding current arcing and causing a fire or an explosion!*

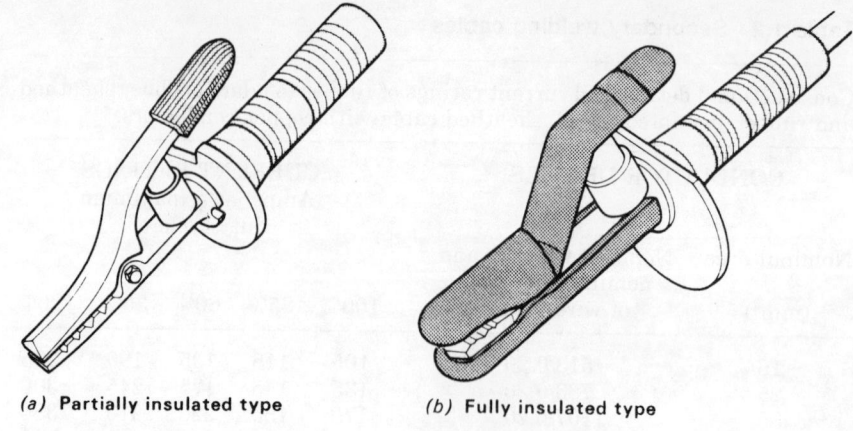

*(a)* Partially insulated type     *(b)* Fully insulated type

**Fig. 1.22 Electrode holders**

## 1.11 Electrode holders

These should be soundly connected to the welding lead. They should be of adequate rating for the maximum welding current to prevent them heating up and becoming too hot to handle. Many types of electrode holders are available; some are partially-insulated and others fully-insulated, as illustrated in Figure 1.22.

Electrode holders of the partially-insulated type have, in addition to a handle made of a heat-resisting non-ignitable insulating material, a protective guard in the form of a disc between the handle and the exposed metal parts. This guard has a dual purpose — it affords protection to the operator's hand by preventing it slipping down on to the exposed metal parts, which are both electrically alive and physically hot; it also enables the holder to be placed on the workpiece or supporting structure without the exposed parts 'shorting'.

The fully-insulated electrode holder, as the name implies, has all metal parts protected by an efficient insulating material, with the exception of the small portion where the electrode is inserted.

## 1.12 Mobile welding plant

Figure 1.23 shows a typical engine-driven mobile welding set. Although there is no high voltage supply associated with this equipment it has hazards of its own:

1. Storage of flammable fuel oil for the engine.
2. Toxic exhaust gases. (Adequate ventilation must be provided to disperse these if the equipment is used indoors.)

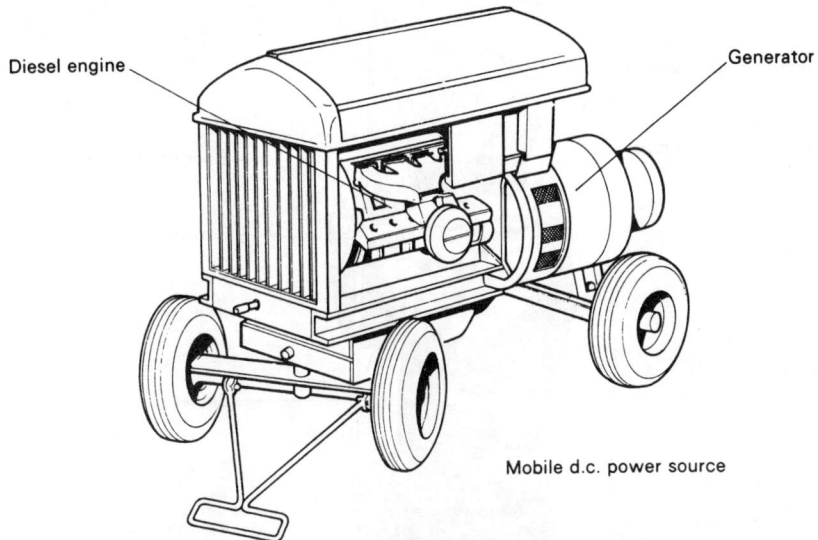

**Fig. 1.23  A mobile engine-driven direct-current welding set**

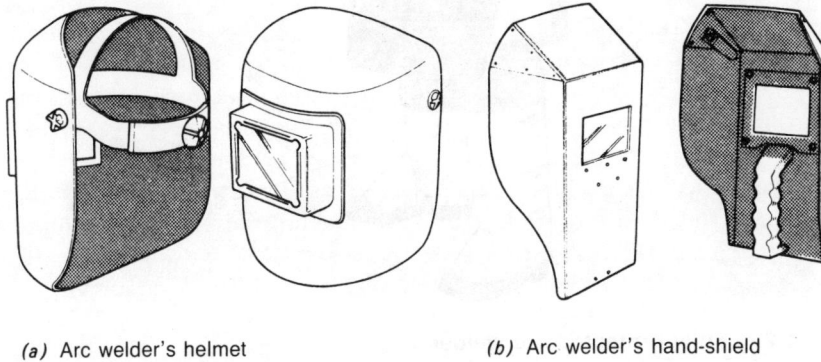

(a) Arc welder's helmet         (b) Arc welder's hand-shield

**Fig. 1.24  Eye and head protection**

## 1.13  Protective clothing for the eyes and head

For all arc-welding operations it is essential to protect the welder's head from radiation, spatter, and hot slag, and for this purpose either a helmet or a hand shield must be worn. Examples are shown in Figure 1.24.

An arc welder's HAND-SHIELD protects one hand as well as the face. It is fitted with a handle which is made of material which insulates against heat and electricity. The handle may either be fixed inside the shield to protect the hand from the heat and rays of the arc, or fixed outside and provided with an effective guard for the same purpose.

Welder's HEAD-SHIELDS are usually fitted with an adjustable band to fit the wearer's head. *This band, and the means of adjustment, should be thoroughly insulated from the wearer's head, the insulation being non-absorbent as a precaution against dampness such as perspiration which tends to make it conductive.* Head-shields are designed to pivot so as to provide two definitely located positions:

1. Lowered in front of the face — the welding position — for protection; and,
2. Raised in a horizontal position to enable the welder to see when not striking the arc.

Some welders prefer to use a hand-shield, rather than a head-shield because it is the least tiring protection to use. *However, head-shields provide better protection and allow the welder the free use of both hands.*

The injurious effect of the radiations emitted by an electric arc are similar whether a.c. or d.c. is used for welding. Exposing the eyes and face to INFRA-RED rays would lead to the face becoming uncomfortably hot and might induce serious eye troubles. If too much ULTRA-VIOLET radiation is received by the welder or anyone in the vicinity, it can cause an effect similar to sunburn on the skin and a condition known as ARC EYE. In addition, too much visible light will dazzle the operator, and too little can cause eyestrain and headaches.

The obvious precaution is to prevent the harmful radiations from the welding arc and the molten weld pool from damaging the eyes. *This is achieved by the use of special glass filters of suitable colour and density, which also reduce the intensity of the visible light rays.* All helmets and hand-shields are provided with a filter glass and a less expensive protective cover glass on the outside. Table 1.3 gives some examples of the filter glasses recommended in BS 679.

The removal of slag by 'chipping' can create an accident hazard to the face and eyes. Although welders are often tempted to chip away without protection, the use of a shield or goggles having clear glass windows is essential in all such operations. Where a screen is to be used solely for protection when de-slagging it is advisable to use a shatterproof glass or plastic cover. *When using a hand-shield or helmet of the dual-purpose type, the change-over device for raising or lowering the dark filter glass should fail to 'safe'.*

## 1.14  Body protection

Figure 1.25 shows a welder fully equipped with protective clothing. The welder's body and clothing must be protected from radiation and burns

**Table 1.3 Filters for manual arc welding**

| GRADE OF FILTER REQUIRED | APPROXIMATE RANGE OF WELDING CURRENT (AMPS) |
|---|---|
| 8/EW | Up to 100 |
| 9/EW | 100 to 200 |
| 10/EW | Up to 200 |
| 11/EW | Up to 300 |
| 12/EW<br>13/EW<br>14/EW | For use with currents over 300 |

Each filter purporting to comply with BS 679 should be permanently marked as follows:

1. BS 679
2. The certification mark of the British Standards Institute
3. The manufacturer's name, symbol or licence number
4. The figures and letters denoting its shade and type of welding process (gas or electric)

Filter glasses are expensive, therefore they should be used with a clear plain cover glass on the outside in order to protect them from damage by spatter and fumes.
These cover glasses are relatively cheap and easily replaceable.

caused by flying globules of molten metal. It may be necessary for a welder to wear an apron, usually of asbestos or thick leather, to protect his trunk and thighs whilst seated at a bench welding. An apron should also be worn if the welder's clothing is made of flammable material.

When deep gouging or cutting is carried out using metal-arc processes, the amount of 'spatter' is considerably greater than that experienced with normal arc welding, and therefore *it is necessary to protect the feet and legs in the same way as the hands and forearms.* Suitable leather leggings and spats are available and should be used to prevent burns to the legs, feet, and ankles. Further details are given in Section 1.18.

## 1.15 Screens

People working in the vicinity of a welding arc, including other welders, can be exposed to stray radiation from the arc, and can be caused considerable discomfort. Looking at an unscreened welding arc, even from a distance of several metres and for a few seconds only, can cause

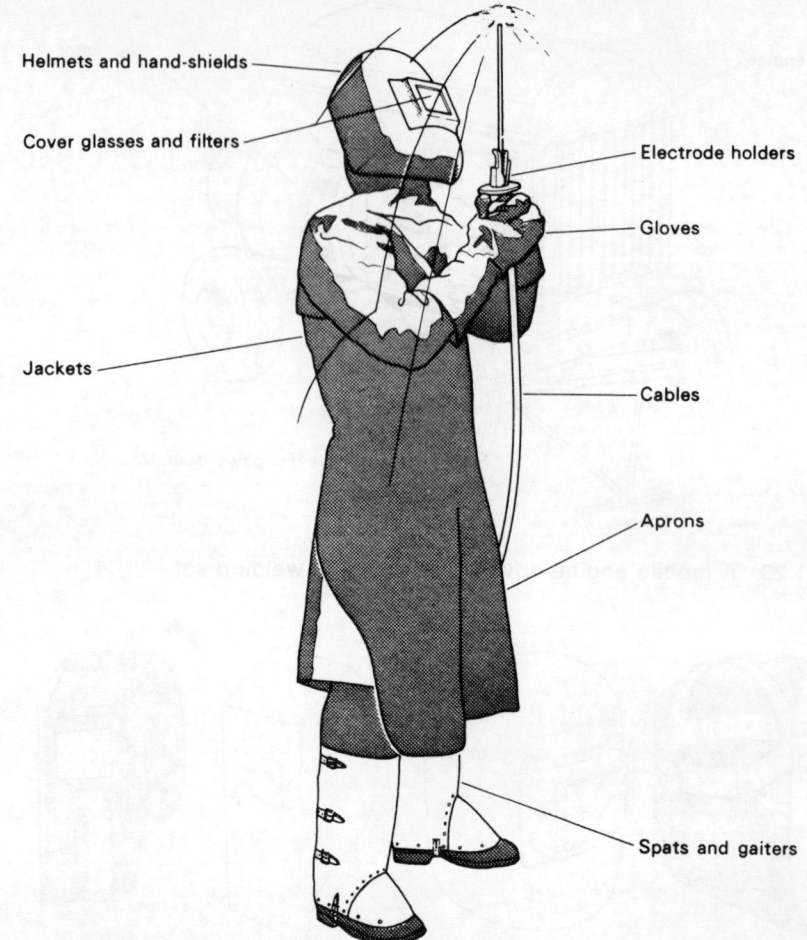

**Fig. 1.25 Fully protected arc welder**

'Arc Eye'. The painful effects of exposure will not be felt until between four and twelve hours later, and it is very likely that the affected worker wakes up at night with the characteristic pains of arc eye. *Persons affected usually complain of a feeling of 'sand in the eyes' which become sore, burn, and water.* Where possible, each arc should be screened in such a way that stray radiation is kept to a minimum. *The walls of welding shops or individual welding booths should be painted with a matt, absorbent type of paint with a very low reflective quality.* The colour of the paint does not have to be black, as experiments have proved that matt pastel shades of grey, blue, or green are equally efficient.

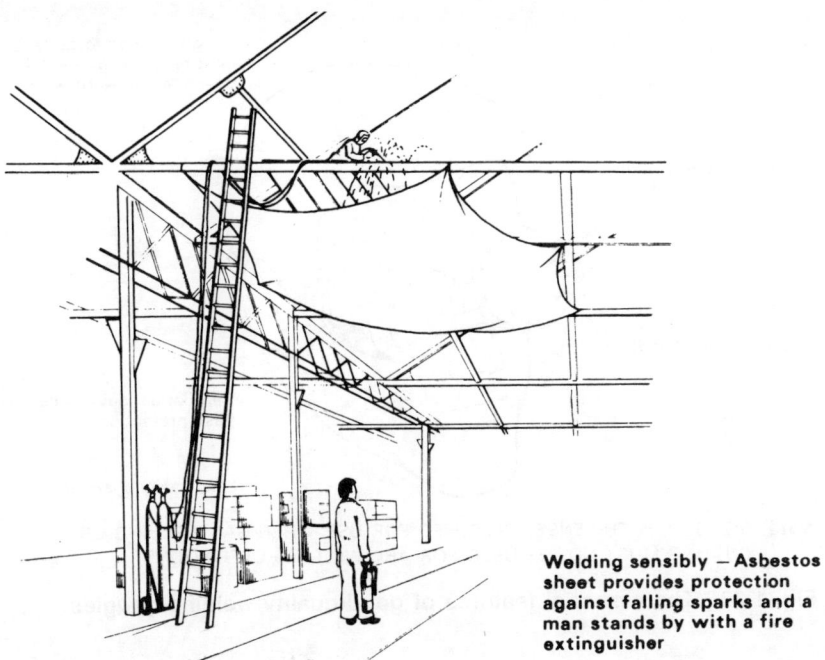

**Fig. 1.26  Overhead protection**

*Welding sensibly — Asbestos sheet provides protection against falling sparks and a man stands by with a fire extinguisher*

If a person is exposed to a flash, the effects of 'Arc Eye' *can* be minimised by the immediate use of a SPECIAL EYE LOTION which *should* be available in the FIRST AID BOX.

*One experience of the distressing results which follow exposure to stray radiation is usually sufficient to make the sufferer more careful in the future!*

## 1.16  Fire hazards

Attention must be drawn to one particular hazard which is very often ignored, that is the danger of burns from freshly welded metal. Such pieces should be clearly marked HOT in chalk or special marking crayon. This simple safety precaution should also be adopted when FLAME CUTTING operations are carried out. In addition to the precautions taken to protect the operator himself from burns, it must be realised that *sparks, molten metal, and hot slag can cause a fire to start if flammable material is left in the welding area.* All old rags, cotton waste, sacks, paper, etc., should be removed. Metal bins, surrounded by sand or asbestos sheet, should be provided for spent electrode stubs, for hot electrode stubs can burn and penetrate through the soles of footwear and are a hazard when thrown on to the workshop floor. *Treading on an electrode stub can have the same disastrous effect as, for example, inadvertently stepping on a roller skate.* General cleanliness, not always very apparent in welding workshops, is essential.

Wherever gas or electric welding or cutting operations take place the work area should be screened off with asbestos blankets or metal screens. Wooden flooring should be covered with sand, or overlapping metal sheets; sparks or molten globules of metal must not be allowed to fall into gaps between boards. Very often welders have to work overhead and Figure 1.26 illustrates the safety precautions which are necessary.

## 1.17  Ventilation

When using the majority of types of electrodes *the welder is not likely to suffer any ill effects from welding fumes provided that reasonable ventilation is available.* Localised exhaust ventilation can be provided by a good fume extractor, which not only dilutes and removes fumes but also assists in keeping down the temperature and adds to the comfort and efficiency of the welder. Figure 1.27 illustrates a typical portable fume extractor.

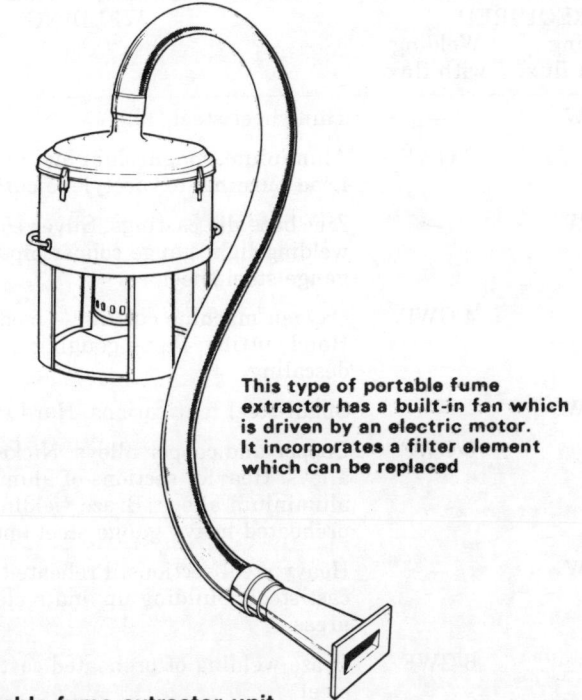

*This type of portable fume extractor has a built-in fan which is driven by an electric motor. It incorporates a filter element which can be replaced*

**Fig. 1.27  Portable fume extractor unit**

15

# Part C  Gas welding and cutting

## 1.18  Personal protection

The need for protective equipment and clothing should be the first consideration before commencing gas welding and cutting operations.

### GOGGLES

These are essential and *must be worn to protect the eyes from heat and glare, and from particles of hot metal and scale.* Goggles used for welding and cutting are fitted with approved filter lenses, as listed in Table 1.4. Figure 1.28 shows the main features of good quality welding goggles.

### PROTECTIVE CLOTHING

Garments made of wool are generally considered not to be readily flammable. However, a high percentage of outer clothing, normally worn

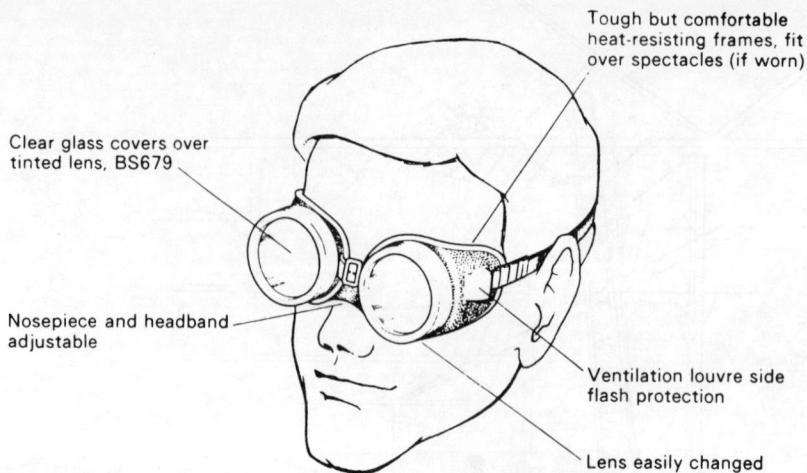

*Note:* GOGGLES WITH LENSES SPECIFIED FOR USE WHEN GAS WELDING OR CUTTING MUST NOT BE USED FOR ARC WELDING OPERATIONS

**Fig. 1.28  The essential features of good quality welding goggles**

**Table 1.4  Filters for gas welding and cutting**

| GRADE OF FILTER REQUIRED | | RECOMMENDED FOR USE WHEN WELDING |
|---|---|---|
| Welding without flux | Welding with flux | |
| 3/GW | — | Thin sheet steel |
| — | 3/GWF | Aluminium, magnesium and aluminium alloys. Lead burning, oxyacetylene cutting |
| 4/GW | — | Zinc-base die castings. Silver soldering. Braze welding light gauge copper pipes and light gauge steel sheet. |
| — | 4/GWF | Oxygen machine cutting — medium sections. Hand cutting, flame gouging and flame descaling |
| 5/GW | — | Small steel fabrications. Hard surfacing |
| — | 5/GWF | Copper and copper alloys. Nickel and nickel alloys. Heavier sections of aluminium and aluminium alloys. Braze welding of un-preheated heavy gauge steel and cast iron |
| 5/GW | — | Heavy steel sections. Preheated cast iron and cast steel. Building up and reclaiming large areas |
| — | 6/GWF | Braze welding of preheated cast iron and cast steel |

by workers, is usually made from flammable materials. *Cuffs on overalls, or turn-ups on trousers are potential fire traps; hot slag, sparks, and globules of hot metal can so easily lodge in them.*

The protective clothing worn will depend upon the nature of the work, and the following suggestions are offered as a general guide:

(a) Asbestos or leather gloves should be worn for all cutting operations which involve the handling of hot metal.
(b) Safety boots should be worn to protect the feet from hot slag and, in particular, from falling off-cuts.
(c) The wearing of asbestos spats is strongly advised for most cutting operations.
(d) The wearing of an asbestos or leather apron will help to prevent sparks and hot metal globules reaching and burning the sensitive parts of the operator's body. *Many welders have experienced the folly of working with their shirts open to the waist, thus presenting a ready-made receptacle for sparks and hot metal spatter.*
(e) Leather gauntlets should be worn when welding or cutting on overhead or vertical structures.
(f) In situations where welding and cutting operations are carried out overhead it is advisable for those persons working or standing below to protect themselves against falling sparks. For this purpose it is recommended that safety helmets or leather skullcaps should

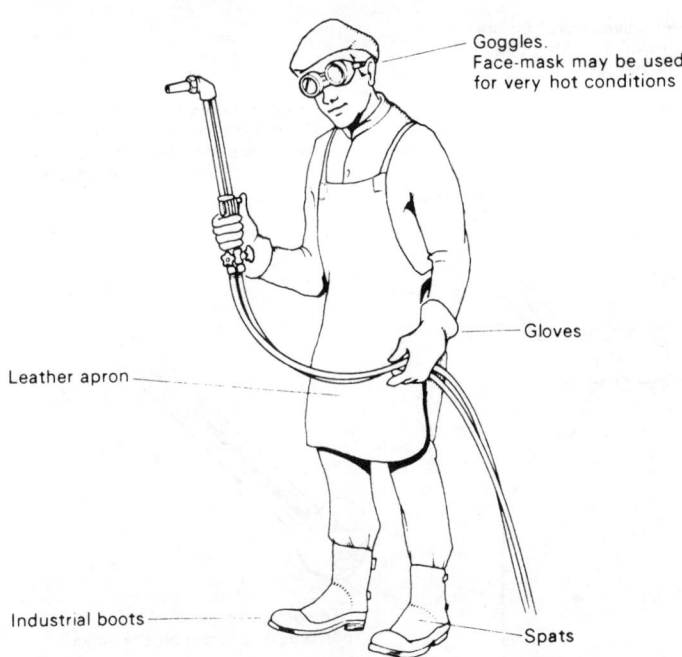

**Fig. 1.29** Protective clothing for cutting operations

be worn by those in the vicinity. Figure 1.29 shows typical protective clothing worn for cutting operations.

## 1.19 Fire hazards

The following important precautions should be rigidly observed:

1. Do not position gas cylinders and hoses where sparks or slag can fall on them.
2. Wooden floors should be kept thoroughly wetted with water, or completely covered, for example with sand.
3. Wooden structures should be adequately protected by sheet metal or asbestos.
4. All combustible materials should be removed to a safe position, or, if this is not possible, should be properly protected against flying sparks. *It must be realised that sparks from cutting can travel up to 9 metres along a floor.*
5. Suitable safety measures must be taken in the case of cracks or openings in walls or floors.
6. KEEP FIRE FIGHTING EQUIPMENT READY TO HAND. A responsible person should keep the site under observation for at least half an hour after the completion of the work in order to watch for, and deal with, any outbreak of fire. *There is always the danger of material smouldering for hours before a fire breaks out.*

## 1.20 Explosion risks

Basically, the heat required for gas welding and cutting operations is generated by the combustion of a suitable fuel gas with oxygen. The vast majority of gas welding processes employ an *Oxygen/Acetylene* gas mixture, and Table 1.5 lists a number of efficient welding gas mixtures in common use.

Explosions can occur when ACETYLENE gas is present in AIR in any proportions between 2 per cent and 82 per cent. This gas is also liable to explode when under unduly high pressure, even in the absence of air, therefore THE WORKING PRESSURE OF ACETYLENE SHOULD NOT EXCEED 0·62 BARS.

When using gas welding processes the first essential requirements are:

(*a*) ENSURE THAT THERE IS ADEQUATE AND PROPER VENTILATION.
(*b*) EXAMINE THE EQUIPMENT AND SEE THAT IT IS FREE FROM LEAKS.

Explosions in the equipment itself may be caused by FLASH-BACK *Flash-backs occur because of faulty equipment or incorrect usage.* Approximately 80 per cent of these occur when lighting welding or cutting torches, the other 20 per cent when they are in use. *As long as the flow of gas equals the burning speed a stable flame will be maintained at the torch nozzle, otherwise mixed gases will arise in one of the hoses, resulting in a flash-back.*

**Table 1.5** Welding gas mixtures

| FUEL GAS | MAXIMUM FLAME TEMPERATURE | |
| --- | --- | --- |
| | with air (°C) | with oxygen (°C) |
| Acetylene | 1755 | 3200 |
| Butane | 1750 | 2730 |
| Coal Gas | 1600 | 2000 |
| Hydrogen* | 1700 | 2300 |
| Propane | 1750 | 2500 |

*The oxy-hydrogen flame has an important application in under-water cutting processes.

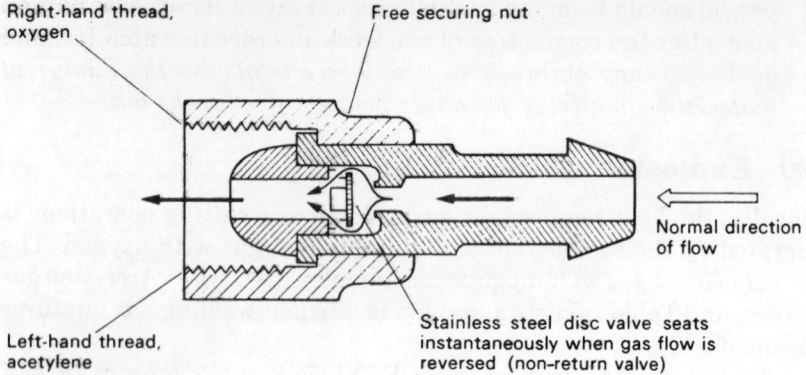

**Fig. 1.30 Hose protector**

Flash-backs may also occur in the following circumstances:

(a) By dipping the nozzle-tip into the molten weld pool.
(b) By putting the nozzle-tip against the work and stopping the flow.
(c) By allowing mud, paint, or scale to cause a stoppage at the nozzle.

In every case the obstruction will cause the OXYGEN to flow back into the ACETYLENE supply pipe and communicate ignition back towards the cylinder or source.

The British Oxygen Company issues many booklets on the safe use and storage of welding gases and these should be consulted before using gas welding equipment.

Figure 1.30 shows a suitable 'hose protector' which will arrest a flash-back; it is built into the hose union.

COPPER tube or fittings made of copper must never be used with ACETYLENE, and alloys used in the construction of pipes, valves, or fittings should not contain more than 70 per cent copper — THE ONLY EXCEPTION IS THE WELDING OR CUTTING NOZZLE. This is because copper, when exposed to the action of acetylene, forms *a highly explosive compound called COPPER ACETYLIDE, which is readily detonated by HEAT or FRICTION*.

Explosions can occur in the regulators on oxygen cylinders as a result of dust, grit, oil, or grease getting into the socket of the cylinder; dust (especially COAL DUST) is highly flammable. The outlet sockets of cylinder valves should be examined for cleanliness before fitting regulators otherwise, if not removed, foreign matter will be projected on to the regulator valve-seating when the cylinder valve is opened. The

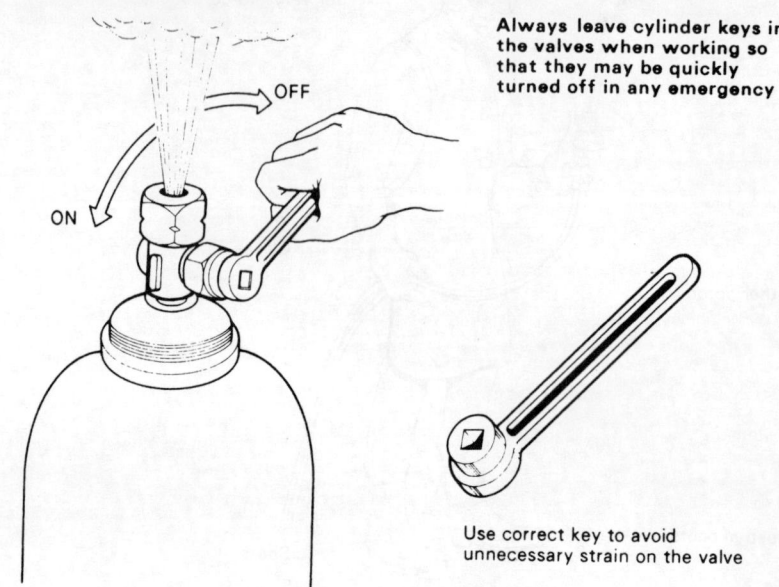

**Fig. 1.31 Cracking the cylinder**

outlet socket can usually be cleaned by turning on the cylinder valve for a brief moment and closing it soundly. This is commonly known as CRACKING THE CYLINDER, and is shown in Figure 1.31.

## 1.21 Testing for leaks

It is extremely dangerous to search for gas leaks with a naked flame. OXYGEN is *odourless* and, whilst it does not burn, it readily supports and speeds up combustion. ACETYLENE has an *unmistakable smell*, rather like garlic, and can be instantly ignited by a spark or even a piece of red-hot metal.

Before leak testing, it is considered good practice to 'pressurise' the system, and the procedure is as follows:

1. Open the control valves on the torch.
2. Release the pressure-adjusting control on the regulators.
3. Open the cylinder valves to turn on the gas.
4. Set the working pressures by adjusting the regulator control.
5. Having established the correct pressure for each gas, close the control valves on the torch.

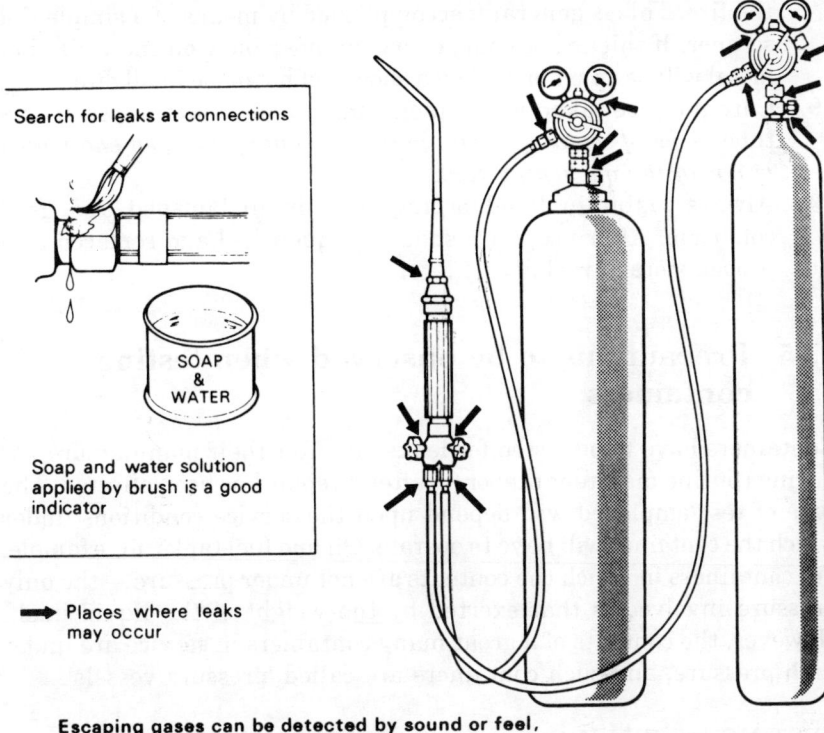

**Fig. 1.32 Testing for leaks**

The system is now ready for leak testing, for which the procedure is clearly indicated in Figure 1.32.

## 1.22 Safety in the use of gas cylinders

Cylinders for compressed gases are not themselves dangerous. They must comply with rigid government standards and should be regularly inspected and tested.

The fact that gas cylinders are a familiar sight in factories or on sites where welding and cutting is carried out is often the reason why ordinary safeguards are neglected.

For general identification purposes, narrow cylinders contain gases of HIGH PRESSURE whilst broad cylinders contain LOW PRESSURE gases.

A great deal of information is available, usually in the form of safety booklets issued by manufacturers, on the use, handling, and storage of gas cylinders. A few of the many safety precautions will now be considered.

1. Cylinders must be protected from mechanical damage during storage, transportation, and use. Acetylene cylinders must *always* be kept upright.
2. Cylinders must be kept cool. On no account should the welding flame, or any other naked light, be allowed to play on the cylinders or regulators. They must also be shielded from direct sunlight, wet, and frost on an open site.
3. Cylinders must always be stored in well-ventilated surroundings to prevent the build-up of pockets of explosive mixtures of gases should any leaks occur. **DO NOT SMOKE IN A GAS CYLINDER STORE.**
4. Correct automatic pressure regulators must be fitted to all cylinders prior to use. The cylinder valve must always be closed when the cylinder is not in use or whilst changing cylinders or equipment.
5. Keep cylinders free from contamination. Oils and greases ignite violently in the presence of oxygen; similarly, do not wear dirty or greasy clothes in the presence of compressed gases.

## 1.23 Welding in confined spaces

Normally AIR contains 21 per cent OXYGEN. Only a small increase in this percentage results in the air becoming 'oxygen enriched'. One danger to a welder working in a confined space, should the air become enriched with oxygen, is that his clothing could spontaneously ignite causing severe burns. ANY SITUATION IN WHICH THE NATURAL VENTILATION IS INADEQUATE SHOULD ALWAYS BE REGARDED AS A 'CONFINED SPACE'.

Safety regulations require specific measures to be taken:

(a) Prevent the use of oxygen to ventilate any confined space; and
(b) Ensure that adequate provision is made for ventilating, and that it is used.

*Ventilation is of vital importance when welding or cutting is carried out in confined or enclosed spaces, and neglect in this respect has often resulted in loss of life.*

(c) Any worker in a confined space must be kept under constant observation by an assistant outside. This assistant should be competent to control the supply of gases and carry out emergency procedures.
(d) The assistant outside the vessel or confined area should have immediately to hand a pail of water and a suitable fire extinguisher.

(e) When welding or cutting is carried out inside boilers or vessels, gas cylinders should always be kept OUTSIDE.
(f) ALWAYS PASS THE WELDING OR CUTTING TORCH TO THE OPERATOR WORKING IN A CONFINED SPACE ALREADY LIGHTED.
(g) Never leave torches and hoses in a confined space when they are not in use, as, for example, during a meal break or overnight. *A very small leak of either the oxygen or the fuel gas*, over such a period, *can produce a dangerous atmosphere if contained in a confined space.*

# Part D  Miscellaneous

## 1.24  Precausions to be observed when forging

The general safety precautions to be observed where forging operations are involved can be summarised as follows:

1. The floor area where a smith's hearth is installed should be of fireproof material.
2. It is essential that workers employed in forging operations should wear adequate protective clothing, such as safety footwear, leather gloves, and leather aprons.
3. Ensure that the fuel used in the hearth is free from foreign material such as stones.
4. A properly designed flue must be fitted to conduct any fumes to the outside atmosphere. DUST AND FUMES ARE A POTENTIAL SOURCE OF DANGER. *Carbon monoxide* is a dangerous gas produced by combustion, especially if the fuel in the hearth is allowed to burn without an adequate air supply. *The effect of this gas, which cannot be detected by smell, is to diminish the oxygen-carrying capacity of the blood. This could render the worker dizzy and weak, and it may lead to a state of unconsciousness.*
5. In respect of item 4, it is essential that the electricity supply to any extractor fan should be arranged independently of any emergency control ('stop button') to other equipment.
6. Anvils and quenching tanks should be situated as close as possible to the hearth. THE DANGERS OF CARRYING HOT METAL AROUND CANNOT BE OVER-EMPHASISED.
7. The smith's anvil should be mounted on a stable base, preferably a cast iron stand, and maintained in good condition. A damaged anvil face can be dangerous and should not be used.
8. Scale should be removed from the workpiece on withdrawal from the fire. This is generally accomplished by means of a simple flat scraper. If this is not done, every hammer blow on the white hot metal will be accompanied by a shower of hot oxide in all directions.
9. Care must be taken when quenching a hollow section, such as a tube. *A jet of scalding steam may shoot out of the open end which is not in the quenching tank.*
10. Always maintain hand-forging tools in undamaged and good condition. After use, tools should be quenched and replaced in a proper storage rack.

## 1.25  Precautions to be observed when testing containers

Containers have to be tested for leaks: (*a*) after their manufacture; (*b*) during routine maintenance; or (*c*) after a repair has been effected. The type of test employed will depend upon the service conditions under which the container will have to operate. Oil and fuel tanks, for example, are containers in which the contents are not under pressure — the only pressure involved is that exerted by the weight of the liquid itself. However, the contents of a great many containers in service are under high pressure, and such containers are called 'pressure vessels'.

### TESTING BY FILLING

*Open-topped containers*, such as water tanks or large vats, are usually tested by filling with water and being allowed to stand for a period, during which time any leaks will be indicated. *Precautions should be taken when very large containers are being tested because the very large volume of water required will be extremely heavy.*

It is recommended that large open-topped containers should be well supported on a strong base before commencing the test, and the test should preferably be carried out in the open.

### IMMERSION TESTS

An immersion test can only be applied to small tanks. It must be appreciated that for every $0 \cdot 028$ m$^3$ capacity the water would press upwards with a force of approximately 276 N. *It follows, therefore, that for a container with a capacity of several cubic metres, a considerable force would be required to hold it under the water.* Coupled with that, the container would need to be revolved in order to be sure of locating leaks on the under-side. Thus only relatively small containers can be tested easily by this method. In this test the tank or container is pressurised and completely immersed in water. Any leaks will be indicated by bubbles and their positions marked by an indelible pencil.

## PRESSURE TEST — PNEUMATIC

The tank should be rendered safe by the recommended methods laid down by the Factories Act. It is then partly filled (about one-tenth of its capacity) with paraffin and rocked so that all internal surfaces become covered with paraffin. The external seams and all other joints (including rivet and bolt heads) must then be coated with a paste prepared from METHYLATED-SPIRIT and WHITENING, applied with a brush. *The spirit will evaporate and the whitening surface remain.* It is general practice to seal off all outlets except one which is provided with a pump connection. Air is pumped into the tank to the correct test pressure, as indicated on a pressure gauge. *Any leakage will at first be indicated by a drop in pressure on the gauge.* The exterior of the container should be visually examined and *any cracks, faulty seams and joints, or faulty bolts and rivets will be clearly indicated by the pressurised paraffin discolouring the whitening.* The leaks are then marked with an indelible pencil, and the whitening removed with a suitable solvent. The container is drained and all volatile fumes expelled. The leaks should then be repaired and the same procedure adopted for any subsequent pressure tests, which are repeated until the container withstands the test.

There is always the risk of an explosion when testing a pressure vessel with compressed air. Such 'Pneumatic Tests' must be carried out under close supervision by the inspecting authority. ADEQUATE PRECAUTIONS, SUCH AS BLAST WALLS OR PITS AND MEANS OF REMOTE OBSERVATION ARE ESSENTIAL.

## PRESSURE TEST — HYDRAULIC

The safest method for testing vessels which operate at medium or high pressures in service is the 'HYDRAULIC TEST'. Any failure due to the internal pressure escaping by leakage will only result in a slight spillage. *Unlike air or gas under pressure, water cannot be compressed and is only a means of transmitting high pressures.*

With the hydraulic test, the vessel being tested is subjected to an internal test pressure of 1·5 times the safe working pressure for not less than 30 minutes. This is considered sufficient minimum length of time to permit a thorough examination to be made of all seams and joints. By comparison, when a vessel is subjected to a pneumatic test, the test pressure should not exceed the design pressure. *In the interest of safety, it is important that any vessel subjected to a hydraulic test should be properly vented so as to prevent the formation of 'air pockets' before

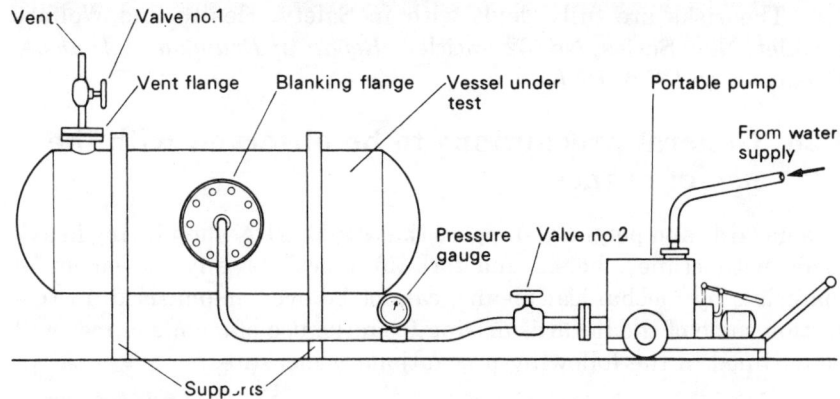

*Testing procedure*
1. Fill vessel with water by removing vent flange
2. Reconnect vent flange leaving valve No.1 open
3. Open valve No 2 and start pumping
4. When water is flowing out of vent pipe - INDICATING THAT ALL THE AIR HAS BEEN FORCED OUT - close valve No.1 ; PRESSURE WILL NOW BEGIN TO BUILD UP INSIDE THE VESSEL
5. When the desired pressure has been reached on the pressure gauge, turn off valve No.2 and stop pumping. ANY DROP IN PRESSURE ON THE GAUGE INDICATES LEAKAGE

**Fig. 1.33 Pressure testing**

*the test pressure is applied.* It is recommended that during the test the temperature of the water should not be below 7°C. Figure 1.33 illustrates the correct procedure for hydraulic pressure testing.

## REPAIRS TO CONTAINERS

Leaks in containers usually require repairing by some process which involves heat, such as welding, brazing, or soldering, therefore at this stage reference will be made to the requirements of the Factories Act. They specify that no plant, tank, or vessel which has contained any explosive or flammable substance should be subjected to:

(*a*) Any welding, brazing, or soldering operations;
(*b*) Any cutting operation which involves the application of heat; or
(*c*) Any operation involving application of heat for the purpose of taking apart or removing the plant, tank, or vessel or any part of it until all practicable steps have been taken to remove the substances and any fumes arising from them or render them non-explosive or non-flammable.

The risks are fully dealt with in Safety, Health and Welfare Booklet, New Series, No. 32 entitled: *Repair of Drums and Tanks — Explosion and Fire Risk.*

## 1.26 General precautions to be observed with the use of cranes

The hazards and potential dangers that could arise when lifting heavy loads with cranes, hoists and fork-lift trucks or any movement of materials by mechanical means, cannot be over emphasised. In this section some of the hazards arising from the use of mobile cranes will be outlined in the following precautionary measures:

1. Loads should only be lifted vertically. It is a hazard to swing loads out manually to gain additional radius, for in doing so the effect is to extend the length of the jib and throw stresses on the crane for which it was not designed. This effect is shown in Figure 1.34.
2. Loads should always be kept directly and vertically under the lifting point of the jib. Severe overstressing of the jib can be caused by dragging loads inwards or sideways or by moving loads out of the vertical. This effect is indicated in Figure 1.35.
3. NO PERSON SHOULD STAND UNDERNEATH A LOAD SUSPENDED FROM A LIFTING DEVICE; NEITHER SHOULD A LOAD BE TRAVERSED OVER ANY PERSON.
4. Tyres of mobile cranes need frequent checking. If they are faulty

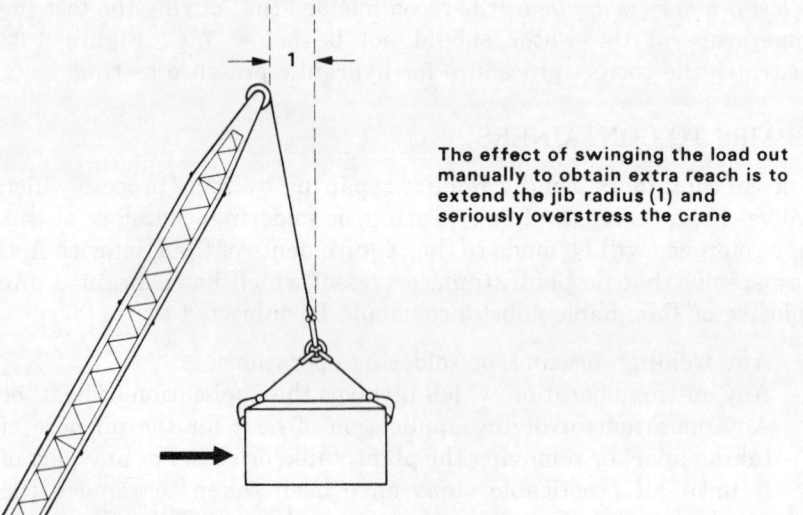

Fig. 1.34 Effect of over-reaching the jib

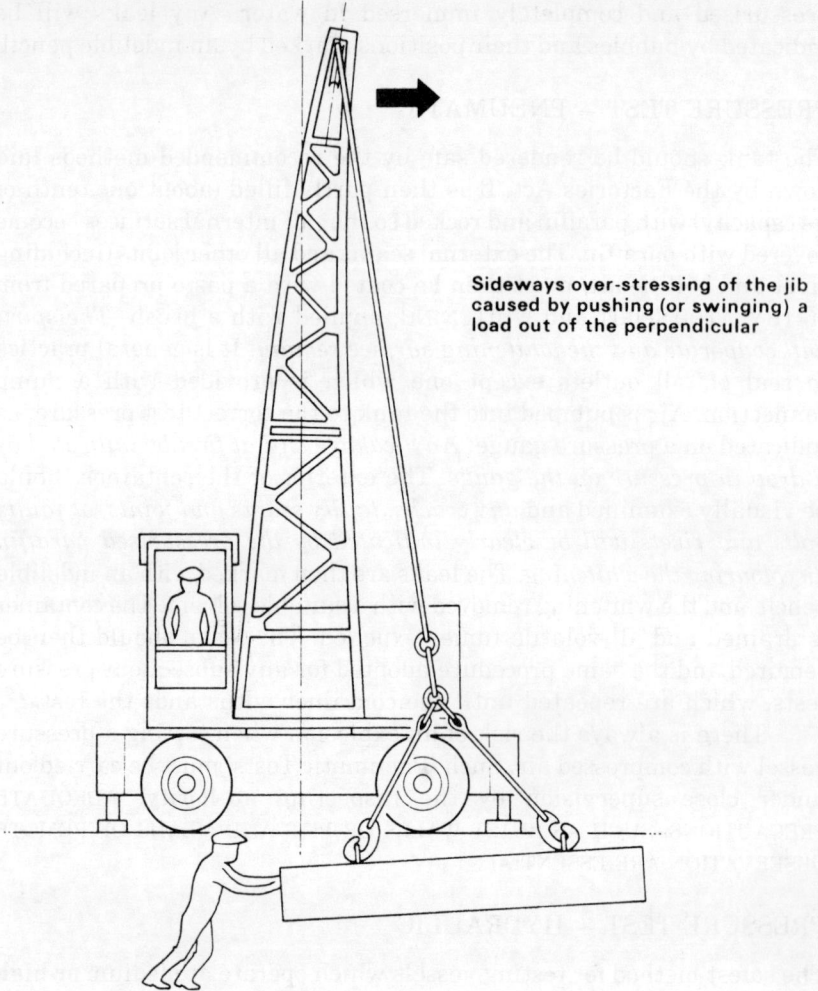

Fig. 1.35 Effect of swinging the load

or inflated at the wrong pressure, the result will be crane instability, as shown in Figure 1.36.

5. Mobile cranes should not be moved with the jib in near minimum radius. Figure 1.37 illustrates the danger of the jib whipping back and causing the crane to overturn.
6. Mobile cranes should be fitted with 'Outriggers'. These are auxiliary equipment for extending the effective bases of cranes thus increasing their stability. The crane shown in Figure 1.35 is fitted with outriggers. On soft ground it is recommended that strong

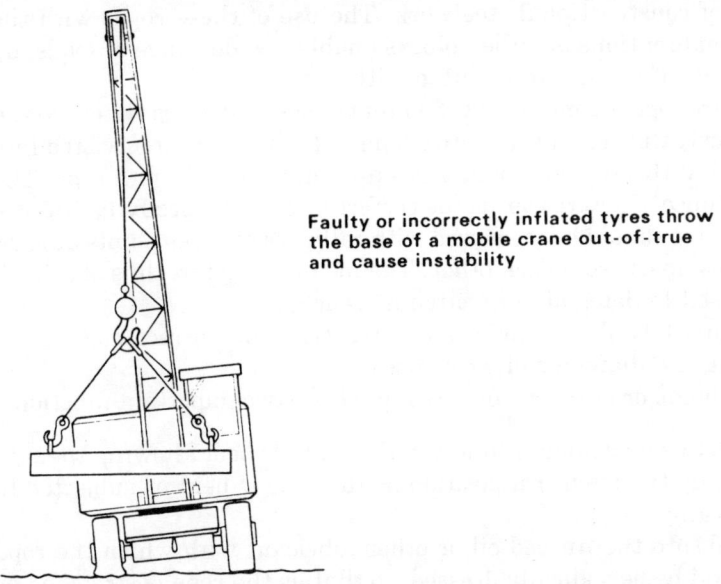

Fig. 1.36 Importance of stability

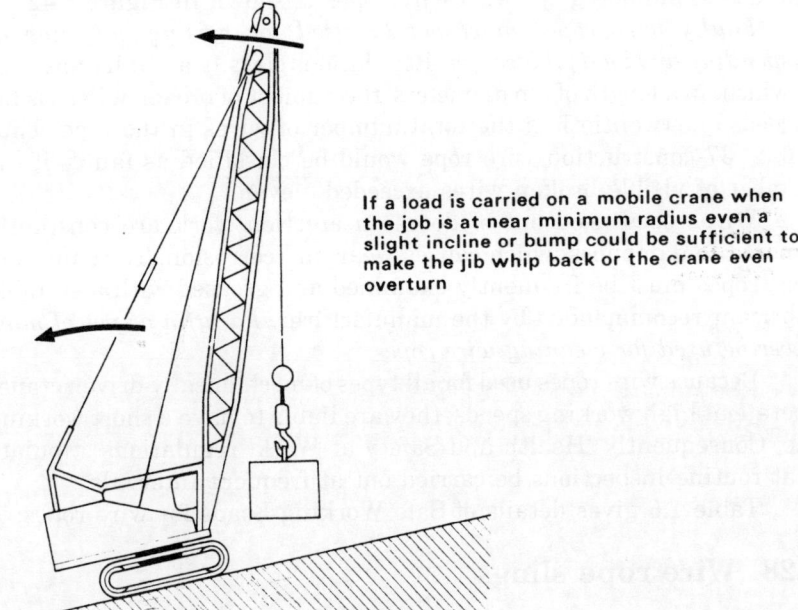

Fig. 1.37 Care when moving the load

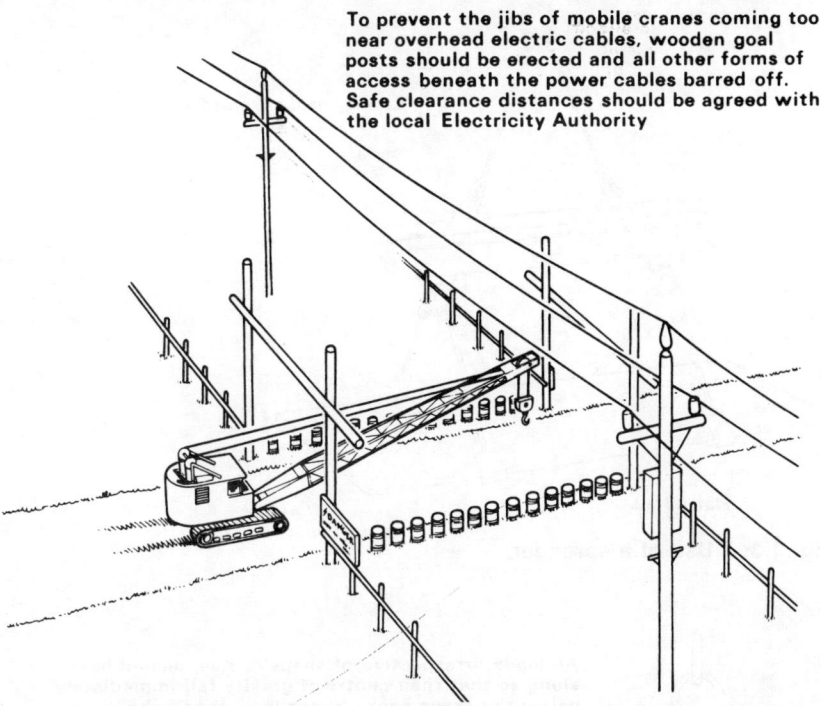

Fig. 1.38 Care when near overhead cables

timber baulks (for example, railway sleepers), be placed under the outriggers in order to spread the load and increase stability.

7. Where jibs of mobile cranes have to pass under or operate near to overhead power lines, there is a very real danger that high voltage can arc between the power cable and the jib. Figure 1.38 indicates one of the precautions taken on site to ensure that no part of site equipment can approach too near any live overhead cable.

8. Spreaders should be used where the load is a long one, and where it is necessary to distribute the loading to avoid excessive stresses in the object being lifted. Spreaders are frequently used when lifting long plates or rods which would be liable to buckle. Figure 1.39 shows use of a spreader with two pairs of plate lifters when handling long plates.

9. When a load is raised on a multi-leg sling, the legs should be evenly disposed about the *Centre Of Gravity* of the load. Failure to observe this precaution will result in the load tilting until the centre of gravity is vertically below the crane hook, and in an extreme case,

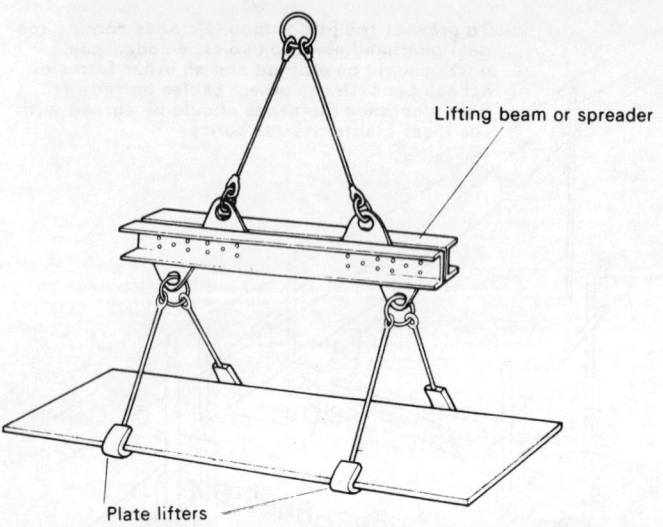

**Fig. 1.39 Use of a spreader**

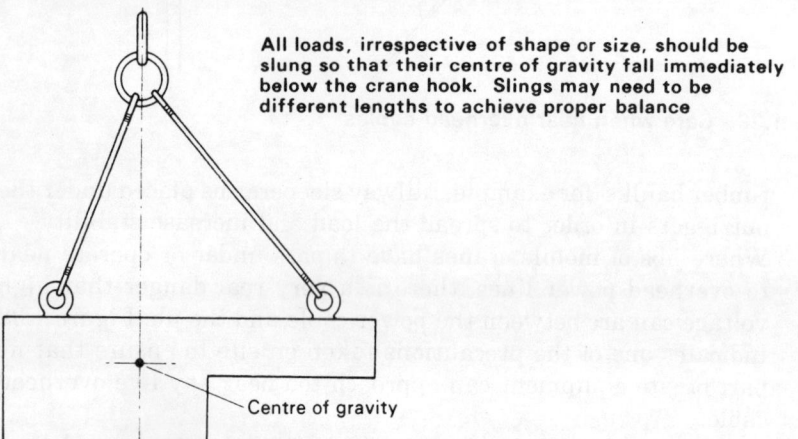

**Fig. 1.40 The use of slings**

one leg could hang vertically and therefore take the full load. This very important precaution is shown in Figure 1.40.

## 1.27 Wire ropes

Because of their large selection of size, strength, good wearing and handling capacities, wire ropes are virtually indispensable for the erection of constructional steelwork. The use of these ropes with the correct combinations of pulley blocks enable a wide range of loads, up to hundreds of tonnes, to be lifted with ease.

A wire rope is constructed of a number of small wires which extend continuously throughout its entire length; these wires are twisted into strands, and the strands themselves are 'laid up' to form a rope. The circumference of a wire rope is the correct method of specifying its size. The diameter should be measured with callipers at three points at least 1·5 metres apart from each other. The mean diameter, thus obtained, is then used to determine its circumference.

Figure 1.41 illustrates the construction and correct method for measuring the diameter of wire ropes.

The hemp or jute cores of wire ropes have two important functions:

1. To act as a cushion into which the strands bed, allowing them to take up their natural position as the rope is bent or subjected to a strain;
2. To absorb the linseed oil or other lubricant with which the rope should be periodically dressed, so that as the rope is stretched or flexed the oil is squeezed between the wires, thus lubricating and reducing the friction between them.

The use of bulldog grips with wire ropes is shown in Figure 1.42.

*Faulty wire ropes must not be used for raising, lowering or suspending any load whatsoever*. Regulations classify as faulty any rope in which, in a length of ten diameters, the number of broken wires visible exceeds one-twentieth of the total number of wires in the rope. Thus a $6 \times 37$ construction wire rope would be classified as faulty if the number of visible broken wires exceeded eleven.

Wire ropes used out of doors on erection work are constantly exposed to dampness which causes wear and corrosion. To counteract this, ropes must be frequently examined and greased with a suitable lubricant recommended by the manufacturer. *Paraffin or petrol must never be used for cleaning wire ropes.*

Because wire ropes used for all types of mechanically-driven cranes operate at high working speeds, they are liable to have a short working life. Consequently 'Health and Safety at Work' regulations stipulate that routine inspections be carried out at frequent intervals.

Table 1.6 gives details of Safe Working Loads for wire ropes.

## 1.28 Wire rope slings

Compared with chain slings, wire rope slings are much lighter, more flexible and easier to handle. *Thimbles* are used to terminate the ends

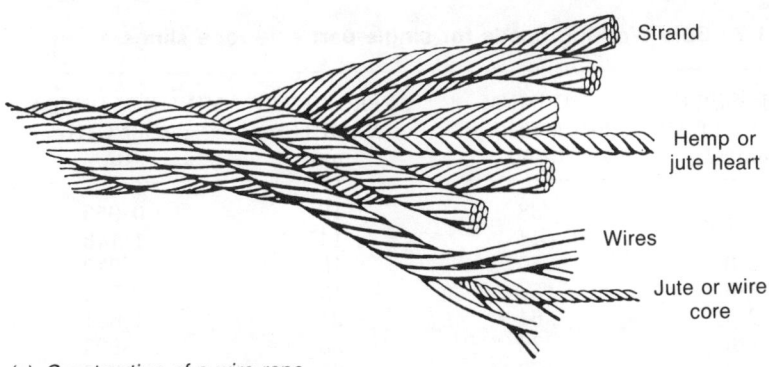

*(a) Construction of a wire rope*
All Extra Flexible Steel Wire Ropes consist of six strands.
The wires forming a strand are twisted left-handed around a jute or wire core.
The strands forming the rope are laid up right-handed around a hemp or jute heart (core).

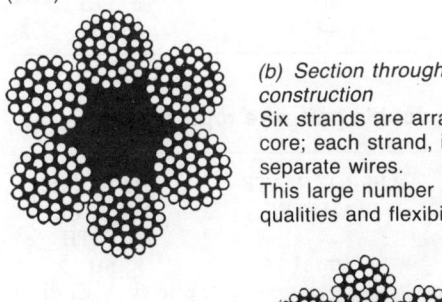

*(b) Section through a wire rope of 6 × 37 construction*
Six strands are arranged over a central fibre core; each strand, in turn, is composed of 37 separate wires.
This large number of wires imparts wearing qualities and flexibility.

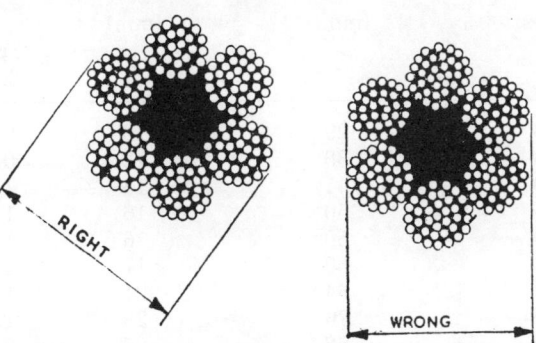

*(c) Right and wrong way to measure the diameter of a wire rope.*

**Fig. 1.41 Wire ropes**

of wire ropes used for slings, as shown in Figure 1.43. *Reeving thimbles are almost heart shaped and large enough to allow the ordinary thimble to pass through. It is essential when using brother slings that the included angle between the legs is kept to the minimum possible.* The Safe Working Loads for the various types of wire rope slings shown in Figure 1.43 are given in Tables 1.7, 1.8 and 1.9.

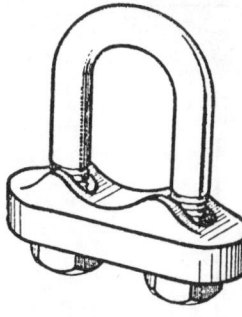

*(a) Bulldog grip.*
A screwed clamp for holding two parts of a wire rope together.
Available to fit wire ropes ranging from 6 mm to 51 mm diameter.
The U-bolt part of the grip should always be positioned over the tail end of the rope with the bridge on the standing or working part.
Three grips should be used on all ropes up to 76 mm circumference; four grips on ropes over 76 mm and up to 102 mm circumference, and five or more on ropes over 102 mm circumference.
The spacing between grips is three times the circumference of the rope.
MUST NOT BE USED TO JOIN TWO ROPES TOGETHER.

*(b) A temporary eye made with bulldog grips.*

**Fig. 1.42 Wire ropes — use of bulldog grips**

Table 1.6  Safe working loads for 6 × 37 construction wire ropes

| APPROXIMATE DIAMETER (mm) | CIRCUMFERENCE (mm) | WEIGHT (kg per 30m) | BREAKING LOAD (tonnes) | S.W.L.* (tonnes) |
|---|---|---|---|---|
| 8  | 25  | 8   | 3·5  | 0·6 |
| 10 | 30  | 9   | 4·2  | 0·7 |
| 12 | 40  | 19  | 8·4  | 1·4 |
| 16 | 50  | 29  | 13·2 | 2·2 |
| 19 | 60  | 41  | 18·0 | 3·0 |
| 22 | 70  | 55  | 24·7 | 4·1 |
| 25 | 80  | 71  | 32·4 | 5·4 |
| 29 | 90  | 88  | 39·7 | 6·6 |
| 32 | 100 | 108 | 49·4 | 8·2 |
| 35 | 110 | 139 | 62·0 | 10·3 |
| 38 | 120 | 165 | 74·7 | 12·4 |

* A safety factor of 6 (e.g. breaking load 18 tonnes; S.W.L. 3 tonnes).

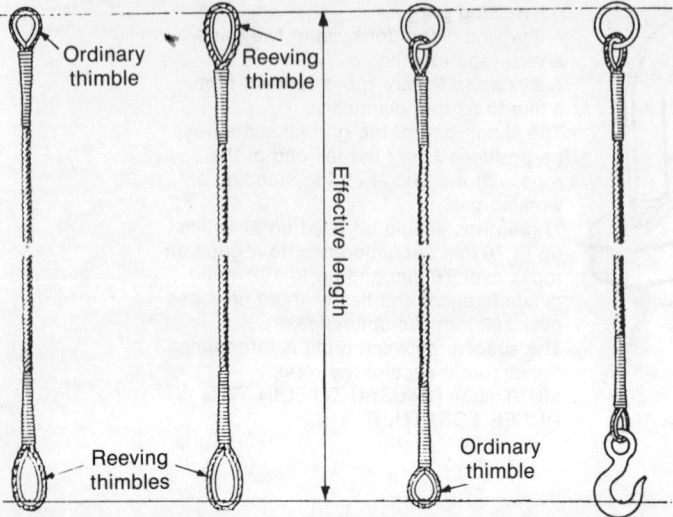

(a) *Single part spliced slings.*
A reeving thimble allows the rope to pass through.
The sling, far right, is useful for tackling small beams.

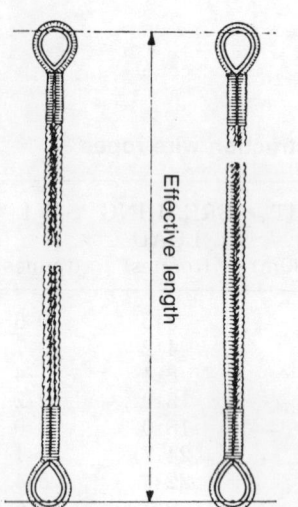

(b) *Double part spliced endless sling and double part endless grommet sling.*

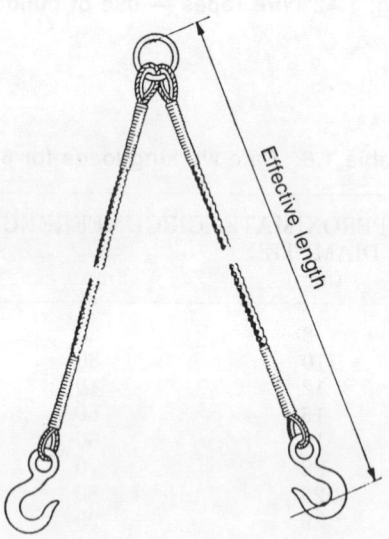

(c) *Two-legged or brother sling.*
Useful for slinging beams which have to be placed horizontally.

**Fig. 1.43 Wire rope slings**

Table 1.7 Safe Working Loads for single part wire rope slings

| PERMISSIBLE WORKING LOAD (tonnes) | MINIMUM CIRCUMFERENCE (mm) | APPROXIMATE DIAMETER (mm) | MINIMUM OVERALL LENGTH (m) |
|---|---|---|---|
| 1·0 | 38 | 12 | 0·953 |
| 1·5 | 44 | 14 | 1·143 |
| 2·0 | 50 | 16 | 1·270 |
| 2·5 | 57 | 19 | 1·534 |
| 3·0 | 64 | 21 | 1·651 |
| 4·0 | 76 | 23 | 1·905 |
| 5·0 | 79 | 25 | 2·032 |
| 6·0 | 86 | 27 | 2·159 |
| 8·0 | 102 | 33 | 2·667 |
| 10·0 | 111 | 35 | 2·794 |
| 12·5 | 128 | 40 | 3·175 |
| 15·0 | 140 | 44 | 3·556 |

Table 1.8 Safe Working Loads for double part wire rope slings

| PERMISSIBLE WORKING LOAD (tonnes) | MINIMUM CIRCUMFERENCE (mm) | APPROXIMATE DIAMETER (mm) | MINIMUM OVERALL LENGTH (m) Spliced Endless | MINIMUM OVERALL LENGTH (m) Endless Grommet |
|---|---|---|---|---|
| 1·0 | 29 | 10 | 0·762 | 0·241 |
| 1·5 | 38 | 12 | 0·953 | 0·292 |
| 2·0 | 41 | 13 | 1·016 | 0·318 |
| 2·5 | 50 | 16 | 1·270 | 0·394 |
| 3·0 | 50 | 16 | 1·270 | 0·394 |
| 4·0 | 60 | 17 | 1·524 | 0·483 |
| 5·0 | 64 | 21 | 1·651 | 0·508 |
| 6·0 | 76 | 23 | 1·905 | 0·597 |
| 8·0 | 83 | 26 | 2·096 | 0·660 |
| 10·0 | 95 | 30 | 2·375 | 0·749 |
| 12·5 | 102 | 33 | 2·667 | 0·838 |
| 15·0 | 111 | 35 | 2·784 | 0·876 |

## 1.29 Chain slings

Chain slings are provided with a ring large enough in diameter to enable it to be used over a hook intended for a larger safe working load. This facilitates, for example, the use of a 1-tonne chain sling on the hook of a 5-tonne crane. They are manufactured from either standard or special

**Table 1.9  Safe Working Loads for two-legged wire rope slings**

| PERMISSIBLE WORKING LOAD OF ONE LEG (tonnes) | PERMISSIBLE WORKING LOAD OF TWO-LEGGED WIRE SLING IF INCLUDED ANGLE OF LEGS IS: |  |  |  |  |
|---|---|---|---|---|---|
|  | 0° (tonnes) | 30° (tonnes) | 60° (tonnes) | 90° (tonnes) | 120° (tonnes) |
| 1·0 | 2·0 | 1·95 | 1·73 | 1·41 | 1·0 |
| 1·5 | 3·0 | 2·90 | 2·60 | 2·13 | 1·5 |
| 2·0 | 4·0 | 3·86 | 3·46 | 2·83 | 2·0 |
| 2·5 | 5·0 | 4·83 | 4·33 | 3·53 | 2·5 |
| 3·0 | 6·0 | 5·80 | 5·20 | 4·24 | 3·0 |
| 4·0 | 8·0 | 7·73 | 6·92 | 5·65 | 4·0 |
| 5·0 | 10·0 | 9·65 | 8·65 | 7·05 | 5·0 |
| 6·0 | 12·0 | 11·59 | 10·39 | 8·50 | 6·0 |
| 7·5 | 15·0 | 14·48 | 13·00 | 10·60 | 7·5 |

*As the angle between the legs of the sling is increased, note the rapid decreases in the Safe Working Load*

**Table 1.10  Safe Working Loads for single-leg chain slings**

| NOMINAL DIAMETER OF CHAIN (mm) | PERMISSIBLE WORKING LOAD (tonnes) |
|---|---|
| 12·7 | 1·50 |
| 14·0 | 1·85 |
| 16·0 | 2·30 |
| 17·5 | 2·30 |
| 19·0 | 3·35 |
| 21·0 | 4·00 |
| 22·0 | 4·55 |
| 24·0 | 5·25 |
| 27·0 | 6·00 |
| 29·0 | 7·55 |
| 30·0 | 8·50 |

grades of wrought iron, and the strength of the chain is measured from the diameter of the individual links. The effective length of the chain sling is measured similarly to that of wire rope slings.

Tables 1.10 and 1.11 give details of Safe Working Loads for chain slings.

Hooks, rings and shackles, as well as chain and wire rope slings are termed 'lifting tackle'. *All lifting tackle must be tested before use and given an identification mark after testing with the S.W.L. clearly shown.*

Regulations stipulate that an examination of all lifting tackle must

**Table 1.11  Safe Working Loads for two-legged chain slings**

| NOMINAL DIAMETER OF CHAIN (mm) | PERMISSIBLE WORKING LOAD OF TWO-LEGGED CHAIN SLING IF INCLUDED ANGLE OF LEGS IS: |  |  |  |  |
|---|---|---|---|---|---|
|  | 0° (tonnes) | 30° (tonnes) | 60° (tonnes) | 90° (tonnes) | 120° (tonnes) |
| 12·7 | 3·00 | 2·90 | 2·60 | 2·10 | 1·50 |
| 14·0 | 3·75 | 3·60 | 3·25 | 2·65 | 1·85 |
| 16·0 | 4·60 | 4·45 | 4·00 | 3·25 | 2·30 |
| 17·5 | 5·60 | 5·45 | 4·85 | 4·00 | 2·80 |
| 19·0 | 6·75 | 6·50 | 5·85 | 4·75 | 3·35 |
| 21·0 | 8·00 | 7·70 | 6·90 | 5·65 | 4·00 |
| 22·0 | 9·10 | 8·80 | 7·90 | 6·45 | 4·55 |
| 24·0 | 10·50 | 10·15 | 9·10 | 7·40 | 5·25 |
| 25·4 | 12·00 | 11·60 | 10·40 | 8·50 | 6·00 |
| 27·0 | 13·50 | 13·05 | 11·70 | 9·55 | 6·75 |
| 29·0 | 15·10 | 14·60 | 13·10 | 10·70 | 7·55 |
| 30·0 | 17·00 | 16·40 | 14·70 | 12·00 | 8·50 |

be carried out by a competant examiner every six months. Annealing of all wrought iron lifting tackle must be carried out at regular intervals of fourteen months, with the exception of tackle constructed from bar of 12 mm diameter or less, in this case, the period between annealings is six months.

## 1.30  Hooks

A hook is a vital part of any lifting tackle and a variety of types are available for use with chains and slings. Some have trunnions which allow them to swivel and others are fitted with a safety catch.

The following terminology is used when refering to hooks:

The BILL is the point of a hook;
The SHANK is the body;
The CROWN is the bottom of a hook;
The BACK is that part of the shank opposite the bill;
The JAW is the space between the bill and the top of the shank;
The CLEAR is the inside diameter of the crown.

An ordinary tackle hook, if necessary, can be MOUSED to prevent unhooking as shown in Figure 1.44

## 1.31  Shackles

Shackles are couplings used for joining wire ropes and/or chains together or to some fitting. They are usually made of wrought iron or mild steel.

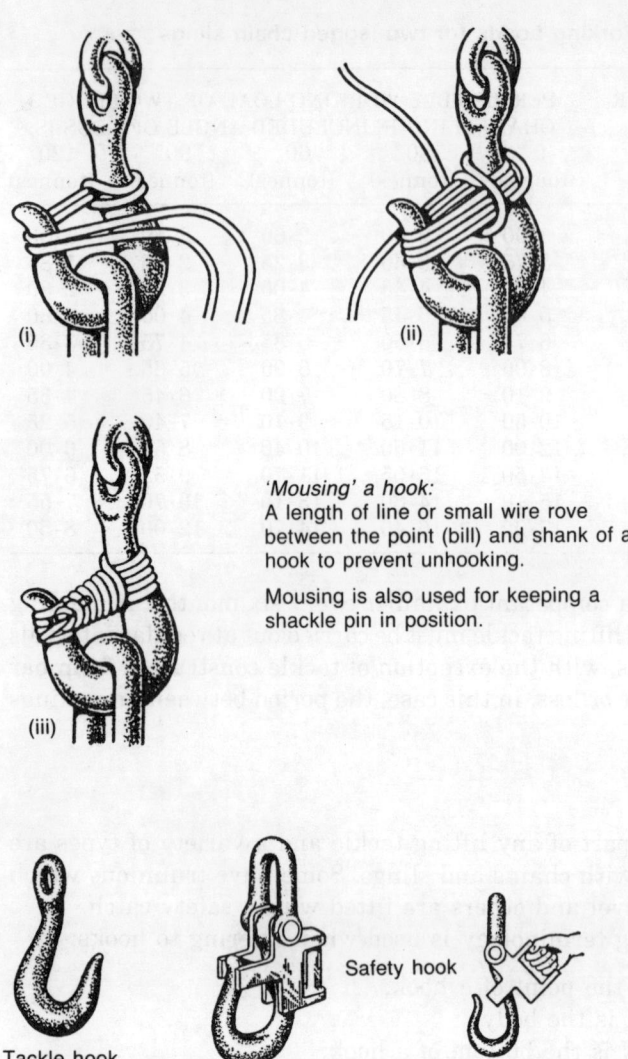

'Mousing' a hook:
A length of line or small wire rove between the point (bill) and shank of a hook to prevent unhooking.

Mousing is also used for keeping a shackle pin in position.

**Fig. 1.44 Lifting tackle — hooks**

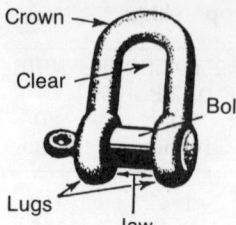

*Screw Shackle*
The end of the bolt (shackle pin) is screwed into one of the lugs.
The head of the bolt may be flanged or countersunk.
It should be 'moused'.

*Forelock Shackle*
A flat tapered pin called a 'forelock' is passed through the slot in the end of the bolt, projecting beyond one of the lugs. The forelock may be attached to the shackle by a keep chain.

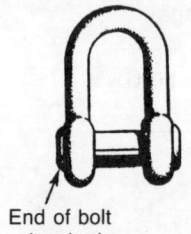

*Clenched Shackle*
The end of the bolt is heated and then hammered over so that it cannot be removed, thus closing it permanently.

**Fig. 1.45 Lifting tackle — shackles**

having a screw thread at one end and an eye at the other which enables it to be tightened or slackened with a toggle spanner. Other types of pin may have countersunk heads which require the use of a screwdriver. A general purpose shackle is usually named by the manner in which its bolt is secured in place, as illustrated in Figure 1.45.

## 1.32 Eye bolts, eye plates and ring bolts

These are widely used as an aid to lifting and transporting.

*EYE PLATES* are stamped out of mild steel or wrought iron and are used for securing an eye to a metal structure. They are either riveted or welded in place.

*EYE BOLTS* are made of mild steel or wrought iron. They may be secured by screwing into a tapped hole or with a nut and washer.

DEE SHACKLES (which are U-shaped and sometimes called 'straight shackles') and BOW SHACKLES (which have curved sides and are weaker than straight ones) are the two common patterns in general use. Both types are available with either large or small jaws and may be described as 'long in the clear' or 'wide in the clear'. The jaw is closed by a removable bolt, called a 'shackle pin', which passes through a hole in each lug. Shackle pins may be of various patterns, the most common

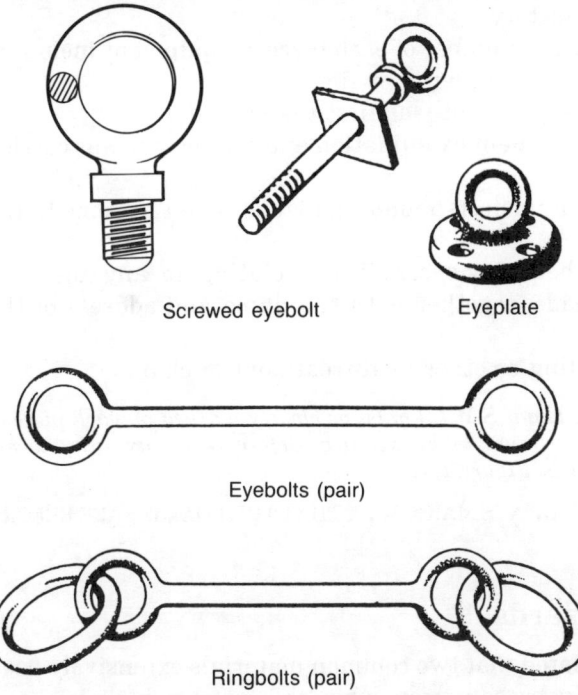

**Fig. 1.46 Lifting tackle — eye bolts, eye plates and ring bolts**

Some types are supplied in pairs which can be cut to length and threaded as and when required.

*RING BOLTS* are screwed eye bolts with a ring or link attached. Like eye bolts, they are available in pairs as shown in Figure 1.46.

## 1.33 Safety hints — lifting tackle

### SLINGS

1. Use the correct type of sling for the job.
2. Check the S.W.L. marked on the sling and against the load to be lifted.
3. Ensure that the sling is in good condition — check splices, rings and thimbles.
4. Check for any broken ends in wires.
5. Make sure that the slings are properly adjusted on the load.
6. Use timber packing or sacking when slinging loads with sharp edges.
7. Do not use wire rope slings for hot loads.
8. Do not stand under loads.
9. Safeguard your fellow workers; use proper signals.
10. Return the sling to stores after use.
11. WHEN USING TWO-LEGGED SLINGS THE HOOKS MUST BE TURNED SO THAT THEY POINT OUTWARDS, AWAY FROM EACH OTHER.

### CHAINS

1. Make sure that the chain is marked with its S.W.L. and select the right chain for the job.
2. Check all chains before using. Report immediately any chain with corroded, deformed, cracked or cut links.
3. Make sure that the chain is not kinked or twisted.
4. Use packing for chain slings when lifting any load with sharp edges.
5. Return chains to stores where they should be properly racked immediately after use.

DO NOT:

6. attempt to lengthen a chain by joining pieces together, or shorten it by knotting it;
7. drop chains on hard surfaces;
8. leave chains lying around where they can be run over or be otherwise ill-treated;
9. use chains or chain slings with a ring too small for the hook;
10. expose chains to acids or other corrosive substances;
11. lubricate chain slings, nor hoist chains if the lubricant is liable to pick up sand or grit.

REMEMBER — A CHAIN IS ONLY AS STRONG AS ITS WEAKEST LINK.

### SHACKLES

1. Check the S.W.L. of the shackle before use and select the right type for the job in hand.
2. Never use any shackle that is not marked with its S.W.L.
3. Examine for damage or distortion; destroy if doubtful.
4. Check the lugs and pin for excessive wear; destroy if wear is one-tenth or more of the original diameter.
5. Make sure the pin is free, but not loose in the tapped hole of the lug.
6. Check that threads are undamaged and without flats or appreciable wear.
7. Check the alignment of holes in lugs. The untapped hole should not be enlarged or worn.

8. Sound shackles should have a clear ring when tested by suspending and lightly tapping with a hammer.
9. When using a shackle with a 'nut and bolt' pin, the pin should be free to rotate when the nut is tight.
10. Never use a shackle where the pin can unscrew by 'rolling' under the load.
11. To prevent shackle pins from unscrewing, secure with a split pin or forelock if possible. Alternatively, 'mouse' with spun yarn.

NEVER USE HOME-MADE SHACKLES.

## HOOKS

1. Check for distortion. If in doubt consult drawings or standard tables reference dimensions.
2. Check the 'clear' and the 'jaw'. If the hook has opened by more than one-fifth of its original dimension, destroy it.
3. Carefully examine for evidence of cracks, cuts, dents or surface pitting caused by corrosion.
4. Swivel hooks should rotate freely. Make sure that the nut securing the hook to the trunnion is split-pinned or otherwise secured.
5. Where a swivel hook is welded in its trunnion, check the 'shank' for excessive wear and the weld for any deterioration.
6. Always 'mouse' hooks unless fitted with a safety catch.
7. When a hook is fitted with a safety catch, make sure that the catch operates freely.

## EYE BOLTS

1. Check that the centre line of the eye is central with the threaded portion of the bolt.
2. The collar or shoulder of the bolt should be flat, free from damage, and at right angles to the threaded portion.
3. Examine for damaged threads — check with a thread gauge.
4. Fit is important — check the thread in standard tapped holes.
5. Carefully examine for cracks, cuts, dents and evidence of surface corrosion by pitting.
6. Check eye for wear; if the wear is one-tenth or more of the original diameter, destroy it.

Health and Safety at Work legislation requires that a register be kept on all premises where slings and lifting tackle are used. This register must state:

1. Names of the occupier of the factory.
2. Address of the factory.
3. Identifying mark or number of each piece of equipment including ropes and slings (chain, wire, or fibre).
4. Date when the equipment was first taken into use.
5. Date of each subsequent examination and by whom it was carried out.
6. Particulars of any defects found and steps taken to remedy the defect.
7. Date and number of the certificate relating to any test and examination made, together with the name and address of the issuing officer.
8. Dates of annealing or other heat-treatment of chains.

*Note: the Safe Working Load (S.W.L.) must be clearly marked on each piece of equipment. Obviously this must not be exceeded. Before using any lifting tackle make the checks shown in Figure 1.47.*

WHEN LOADING, ALWAYS MAKE SURE THAT THE LOAD IS SAFE BEFORE TAKING A LIFT.

## 1.34 Health hazards

It may not be appreciated that two common materials extensively used in the fabrication industry can create a health hazard, namely lead and zinc.

*ZINC FUMES.* The zinc coating on steel produces a health hazard when the coated material is welded.

The metal zinc has a very low melting point (419°C) and boiling point (950°C). In common with any other liquid, once it starts to boil it quickly evaporates. The vapour given off combines with the oxygen in the ambient atmosphere to form *zinc oxide* dust. This dust is so fine that it remains suspended in the air as 'fume'.

Since the welding flame is operating at a temperature in excess of 1530°C when welding steel, it follows that the zinc vapourises long before the base metal fuses (melts) to form the weld.

Breathing in the zinc 'fumes' causes a temporary sickness called *metal fume fever*, commonly referred to as 'welders ague' or 'zinc chills'. The symptoms are nausea and vomiting. However, the effect is not cumulative as in the case of lead, and after about 24 hours it clears up. To prevent this particular health hazard, the following precautions should be observed.

1. In areas where adequate ventilation is not provided, wear an approved respirator (Figure 1.48).
2. Drink milk.

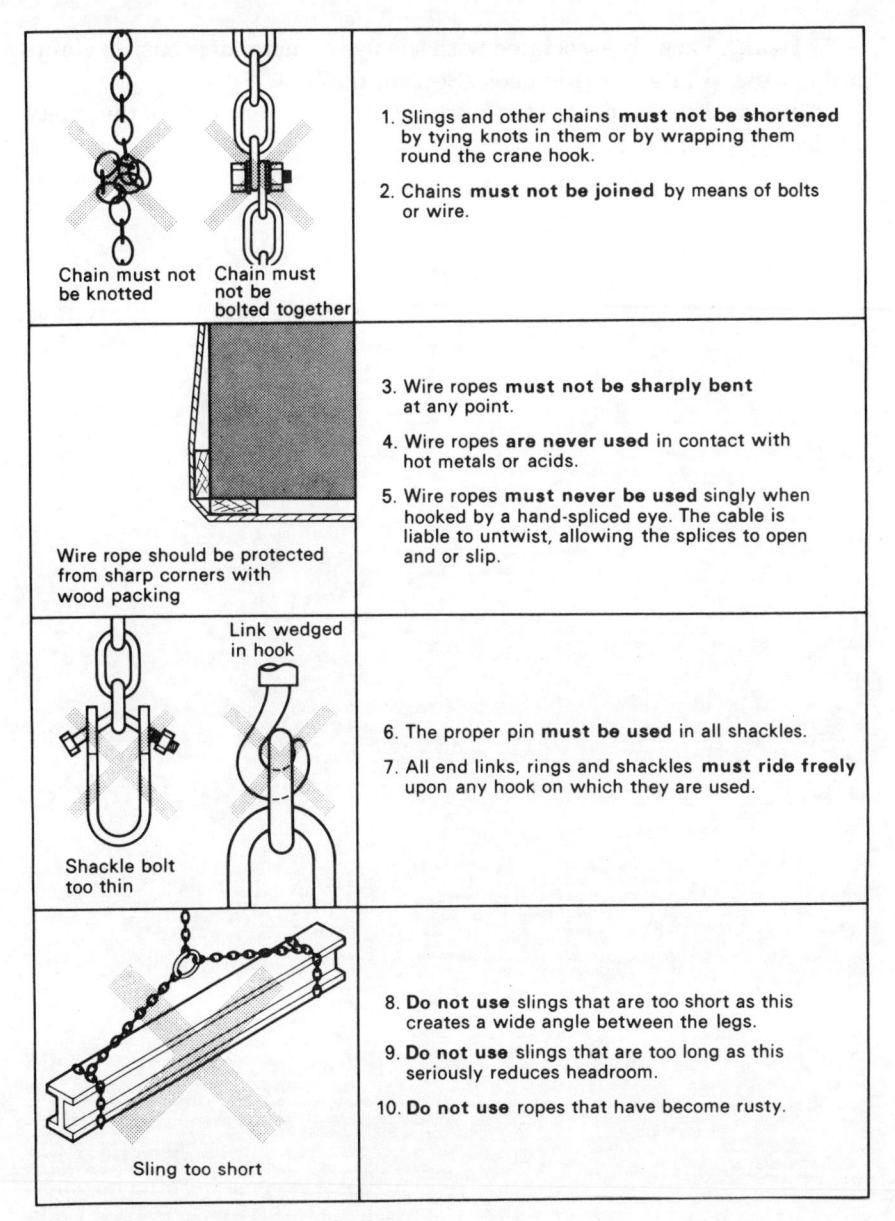

| | |
|---|---|
| *(chain illustrations)* Chain must not be knotted / Chain must not be bolted together | 1. Slings and other chains **must not be shortened** by tying knots in them or by wrapping them round the crane hook.<br>2. Chains **must not be joined** by means of bolts or wire. |
| *(wire rope illustration)* Wire rope should be protected from sharp corners with wood packing | 3. Wire ropes **must not be sharply bent** at any point.<br>4. Wire ropes **are never used** in contact with hot metals or acids.<br>5. Wire ropes **must never be used** singly when hooked by a hand-spliced eye. The cable is liable to untwist, allowing the splices to open and or slip. |
| *(shackle illustrations)* Link wedged in hook / Shackle bolt too thin | 6. The proper pin **must be used** in all shackles.<br>7. All end links, rings and shackles **must ride freely** upon any hook on which they are used. |
| *(sling illustration)* Sling too short | 8. **Do not use** slings that are too short as this creates a wide angle between the legs.<br>9. **Do not use** slings that are too long as this seriously reduces headroom.<br>10. **Do not use** ropes that have become rusty. |

**Fig. 1.47 Care and use of slings**

3. Provide adequate ventilation.
4. Use a portable fume extractor (Figure 1.27).
5. By removing the zinc coating in the vicinity of the joint as shown in Figure 1.48(a). This can be achieved by careful application of

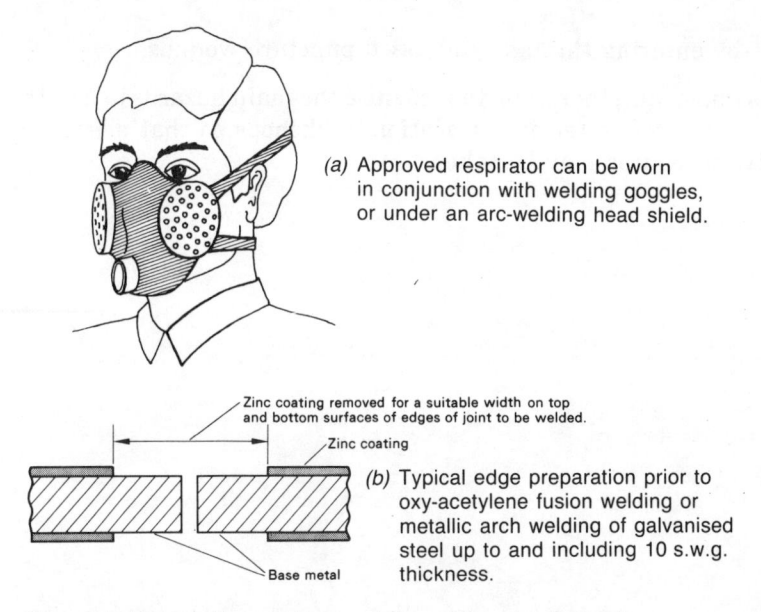

*(a)* Approved respirator can be worn in conjunction with welding goggles, or under an arc-welding head shield.

*(b)* Typical edge preparation prior to oxy-acetylene fusion welding or metallic arch welding of galvanised steel up to and including 10 s.w.g. thickness.

*Zinc coating removed for a suitable width on top and bottom surfaces of edges of joint to be welded.*
*Zinc coating*
*Base metal*

**Fig. 1.48 Precautions when welding galvanised steel**

HYDROCHLORIC ACID which dissolves the zinc; the residue, which is ZINC CHLORIDE, must be removed by rinsing with water because of its corrosive nature.

*LEAD.* The effects of lead poisoning are much more serious than for zinc. When lead is absorbed into the body it it stored in the system so that although individual doses may be harmless in themselves, they accumulate to produce harmful or even fatal effects.

As a precautionary measure it is important to wash thoroughly after handling this toxic material so as to avoid any contaminating particles being carried to the mouth on food or cigarettes, and thus transferred to the digestive system.

Aluminium, chromium, copper and arsenic are other metals that store up in the body and produce harmful or even fatal results.

In a workshop environment, toxic substances may enter the body in various ways other than handling food or cigarettes with dirty hands.

1. Dust, vapours and gases can be breathed in through the nose.
2. Liquids, powders, dust, vapours may all enter the body through the skin:
   (a) directly through the pores,
   (b) by destroying the outer horny layers of the skin and attacking the sensitive layers underneath,

(c) by entering through undressed puncture wounds.

It is important, therefore, to recognise the main hazards to health that may be encountered in fabrication workshops so that adequate preventive measures may be taken.

Health hazards associated with the use of metal degreasing plants are discussed later in this book (Section 6.27).

The author is indebted to *Construction Safety* for much of the safety advice offered in this section.

# 2 Communications

## 2.1 Orthographic drawing

In order to interpret orthographic drawings correctly it is very important for the craftsman to be able to distinguish between first- and third-angle projections. Any misinterpretation, even of a relatively simple component, will result in some details being fabricated in the opposite hand. In Figure 2.1 the same component is shown drawn (a) in first-angle (English) projection, and (b) in third-angle (American) projection, for comparison of the two methods.

## 2.2. Orthographic drawing (first-angle projection)

1 To draw the *End view*

Look in this direction — Draw what you see here

ELEVATION (side view)    (end view)

2 To draw the *Plan view*

Look in this direction

ELEVATION (side view)    (end view)

Draw what you see here

First-angle projection

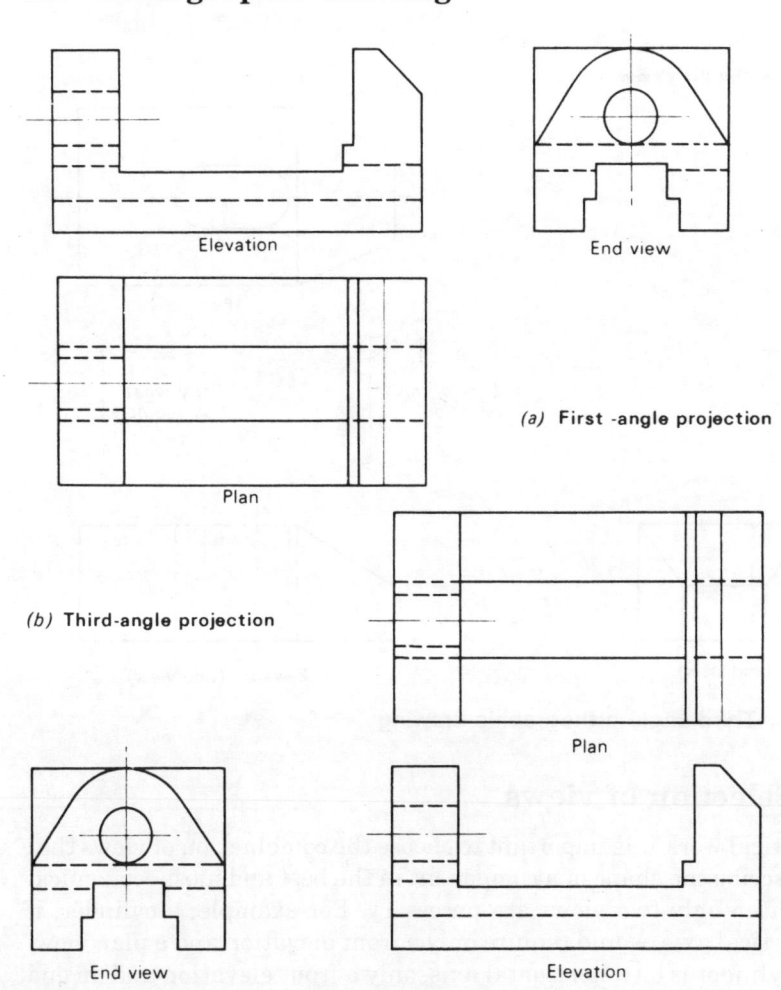

PLAN VIEW

Fig. 2.1(a) First- and (b) third-angle projection

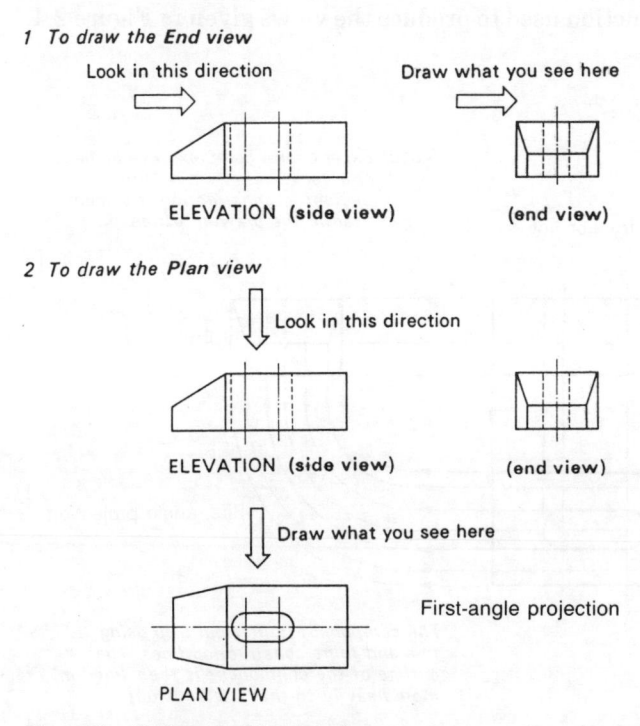

Fig. 2.2 Orthographic drawing

The production or the reading of an orthographic drawing is quite simple, providing a few rules are learnt.

Basically three views of the component are given:

(a) the elevation (side view)
(b) the plan view
(c) the end view

Figure 2.2 shows how these are conventionally positioned in a first-angle drawing. In practice, an end view can be at either end or even at both ends at the same time for clarity.

THIRD-ANGLE PROJECTION

It will be noticed in Figure 2.1 that a drawing in third-angle projection has exactly the same views as the first-angle projection, but the views are positioned in a different relationship to each other. Figure 2.2 shows how the views are conventionally positioned in a third-angle drawing. Figure 2.3 illustrates the geometrical construction employed to produce the views given in Figure 2.2. To avoid confusion *the projection used must always be clearly stated on the drawing*. Figure 2.5 shows the geometrical construction used to produce the views given in Figure 2.4.

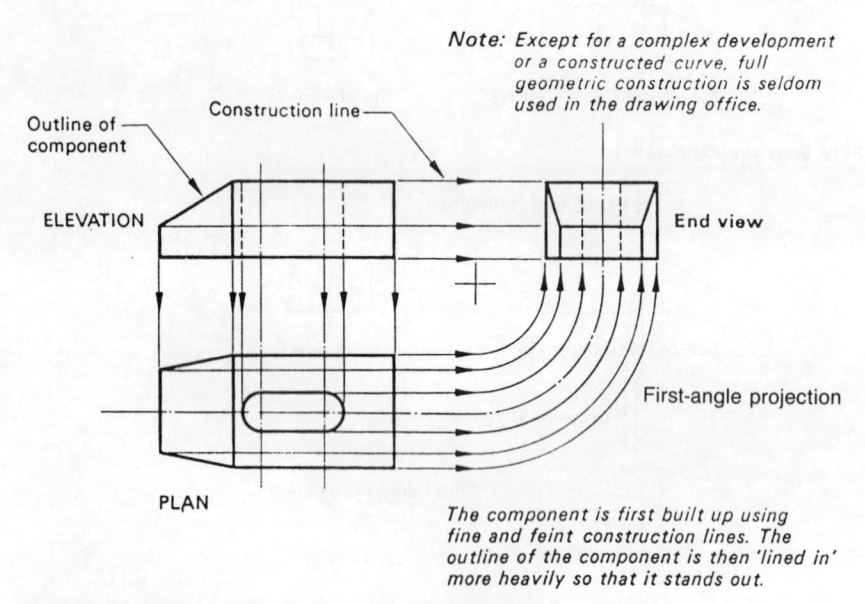

**Fig. 2.3 Orthographic drawing — geometric construction**

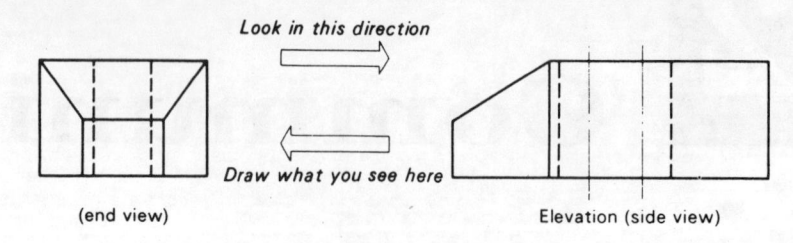

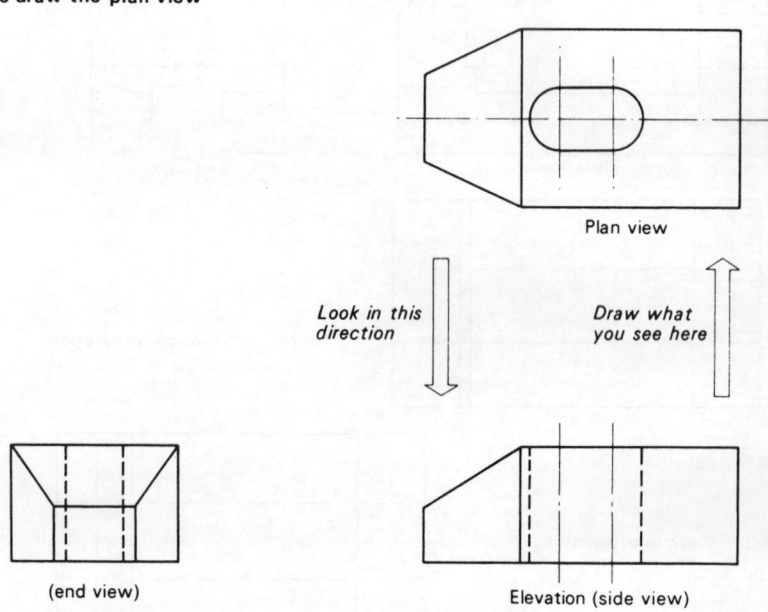

**Fig. 2.4 Third-angle orthographic drawing**

## 2.3 Selection of views

In practical work it is important to choose the combination of views that will describe the shape of a component in the best and most economical way. Often only two views are necessary. For example, a cylinder, if on a vertical axis, would require only a front elevation and a plan view. If the cylinder is on a horizontal axis, only a front elevation and an end view is required. Conical and pyramidal shapes can also be described

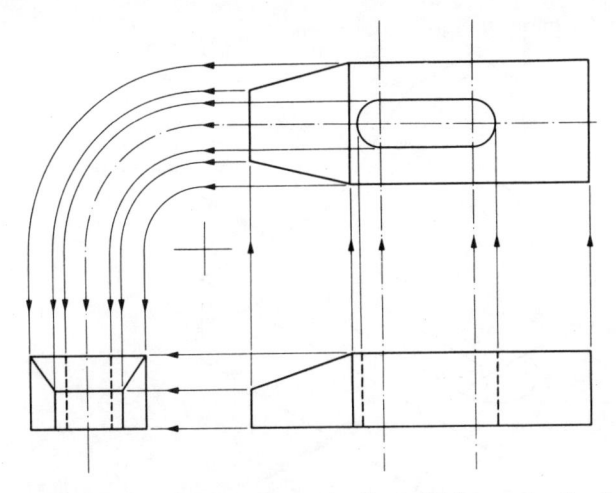

The component is first built up using fine and feint construction lines. The outline of the component is then 'lined in' more heavily so that it stands out.

**Note:** Except for a complex development or a constructed curve, full geometric construction is seldom used in the drawing office.

**Fig. 2.5 Third-angle — geometric construction**

by two views. Figure 2.6 illustrates two-view shapes drawn in third-angle projection.

Sometimes two views are used to describe an object or component on the assumption that the contour in the third direction is of a shape that can naturally be expected. Figure 2.7 explains why in many cases two views do not fully describe the shape of a component.

The two views shown at (*a*) would suggest that component 'A' is rectangular and of square cross-section. However, these two views may be the front elevation and plan views of the tapering rectangular component 'A' as shown in the two views at (*b*). The two views of a component shown at (*c*) do not describe the component at all. It might be assumed to be of square cross-section, but it could as easily be of round, triangular, quarter-round, or of any other shaped cross-section, *which can only be indicated by a side view*. The figure at (*d*) illustrates how the component shown at 'B' is clearly described by two views. Figure 2.8 illustrates the six principal views of an object and indicates which views need to be selected in order that the object be fully represented. In practice, any simple component would be visualised mentally and the necessary views selected without a pictorial sketch as shown in Figure 2.8 (*b*). However, in more complicated work, a pictorial sketch

*(a)* **Cylinder**

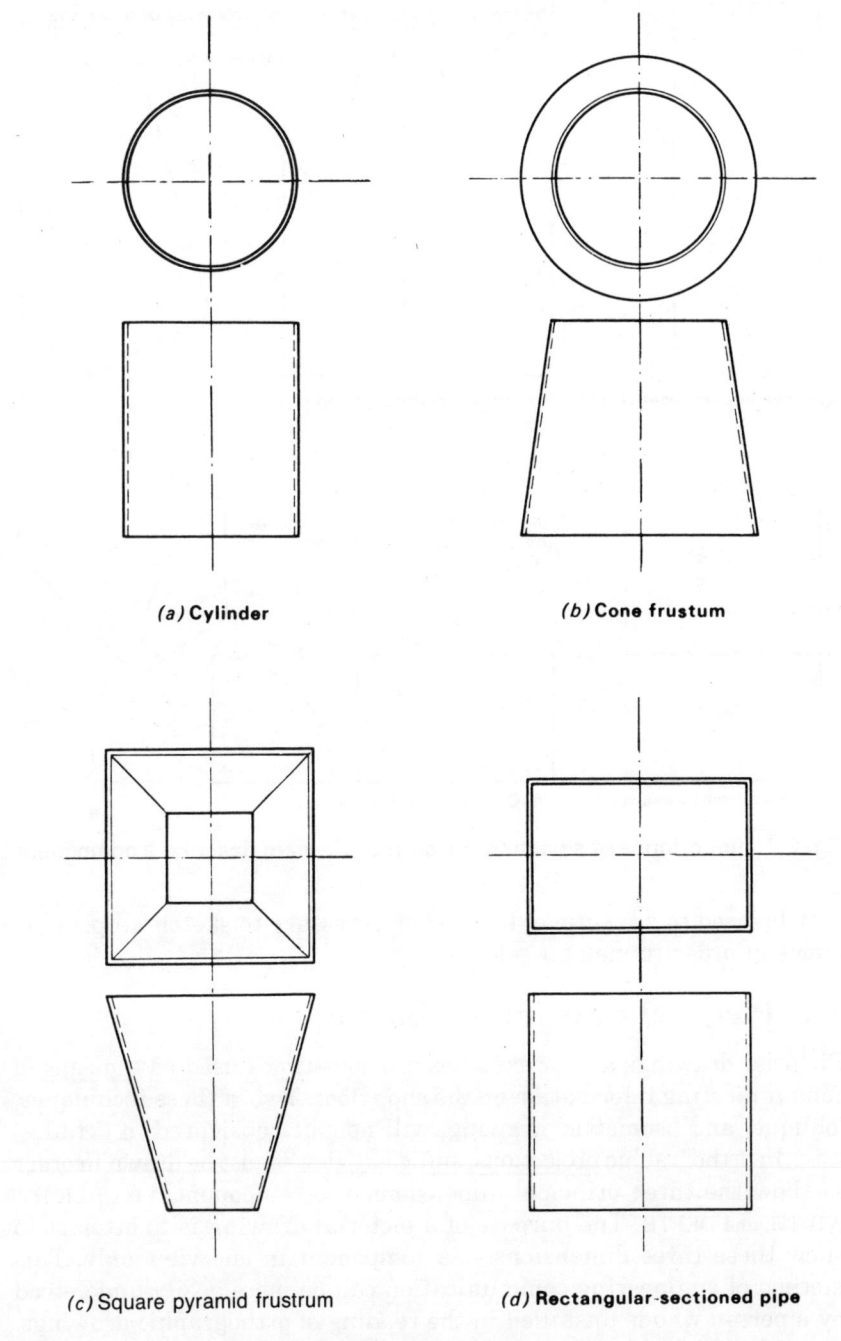

*(b)* **Cone frustum**

*(c)* **Square pyramid frustrum**

*(d)* **Rectangular-sectioned pipe**

**Fig. 2.6 Two-view shapes (third-angle projection)**

35

The two views shown at (a) do not correctly describe the component 'A'
The two views shown at (b) communicate the correct shape of 'A'

(a) Incorrect representation   (b) Correct representation

The two views at (d) communicate the correct shape of the component shown at 'B'

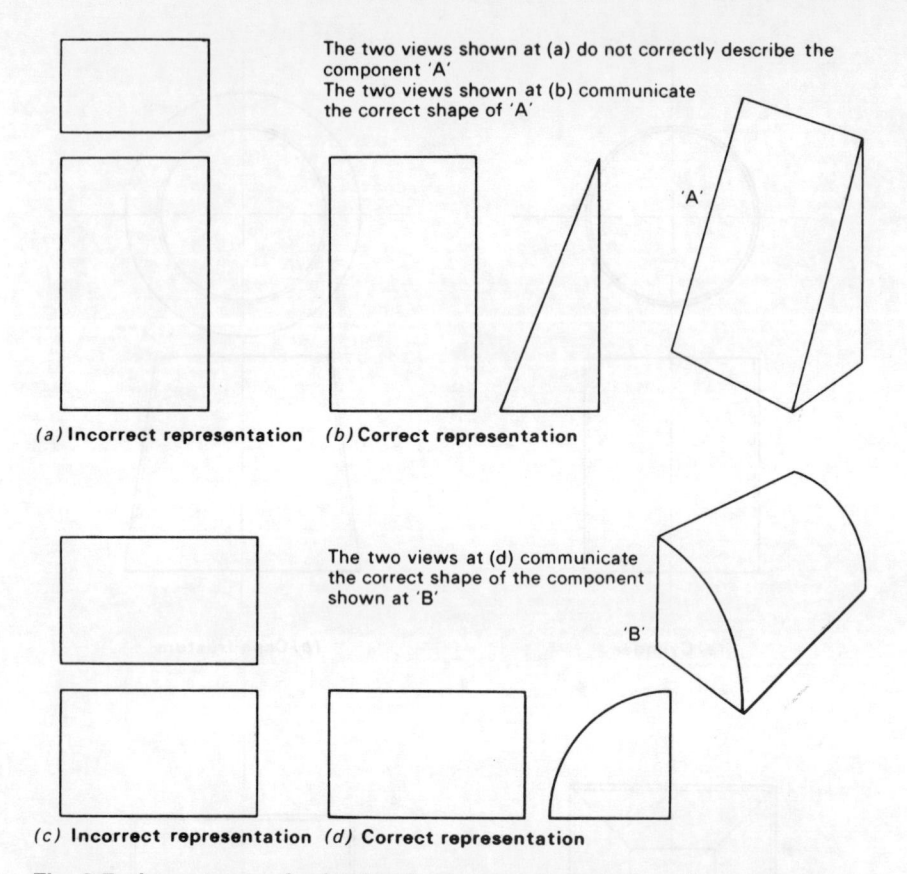

(c) Incorrect representation   (d) Correct representation

**Fig. 2.7  Importance of selecting the correct views to describe a component**

## 2.4  Pictorial drawing — oblique

Pictorial drawings and/or sketches are sometimes used as a means of communicating information on the shop floor. Two of these techniques, 'oblique' and 'isometric' drawing, will now be considered in detail.

In orthographic projections, sufficient views must be drawn in order to show the three principal dimensions of a component, i.e., HEIGHT, WIDTH, and DEPTH. The purpose of a pictorial drawing is to attempt to show these three dimensions of a component in one view only. This method of engineering communication can be more easily understood by a person who is unskilled in the reading of orthographic drawings.

Lines that disappear into the drawing are termed *'receding lines'*

Top (needed)
Rear (not needed)
Right side (needed)
Left side (not needed)
Front (needed)
Bottom (not needed)

(a)

*A view must be drawn in each direction (top, front, side) needed to communicate a true description of every feature of the component but unnecessary views must not be drawn*

ORTHOGRAPHIC THIRD-ANGLE

(b)

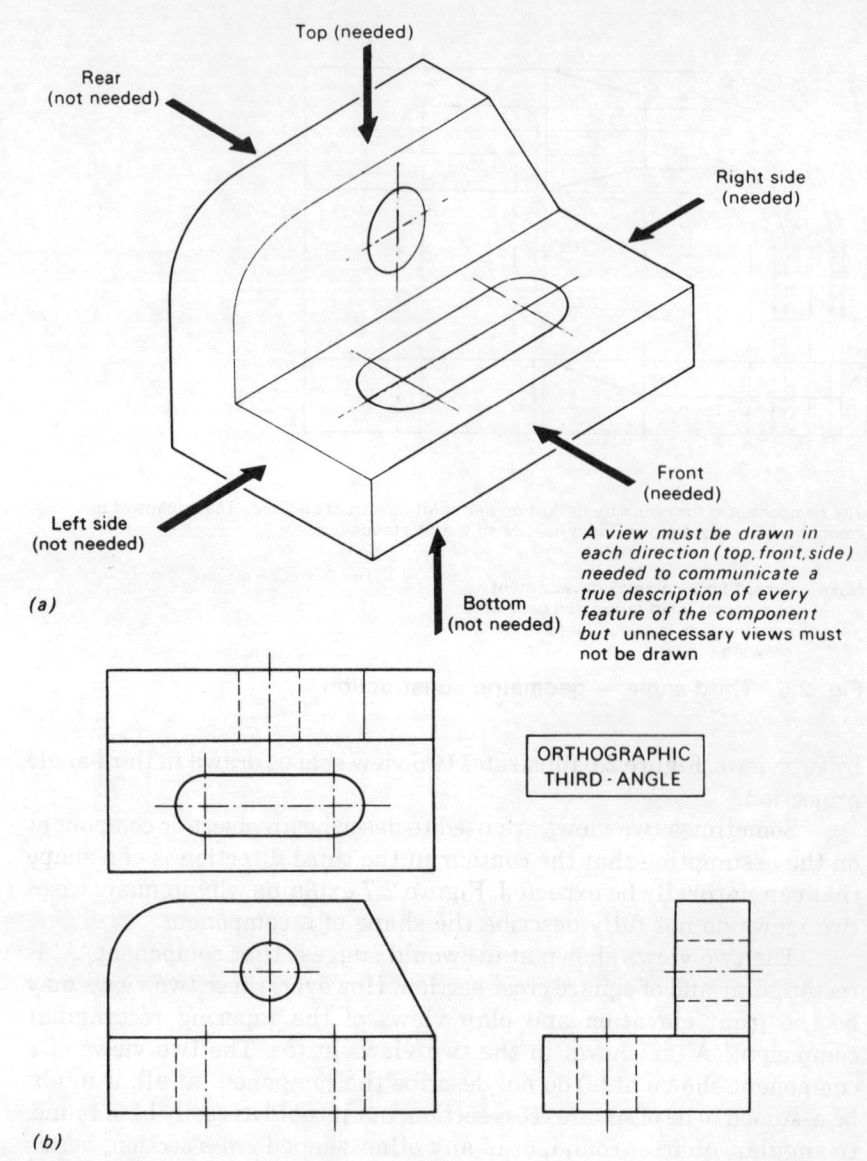

**Fig. 2.8  Selection of views**

or *'receders'*. These may have to be distorted in length and position in order to give the drawing 'realism'. The simplest technique is called OBLIQUE DRAWING. The procedure for making such a drawing is shown in Figure 2.9.

Receding lines are drawn HALF TRUE LENGTH and at 45° to the horizontal base line

45°

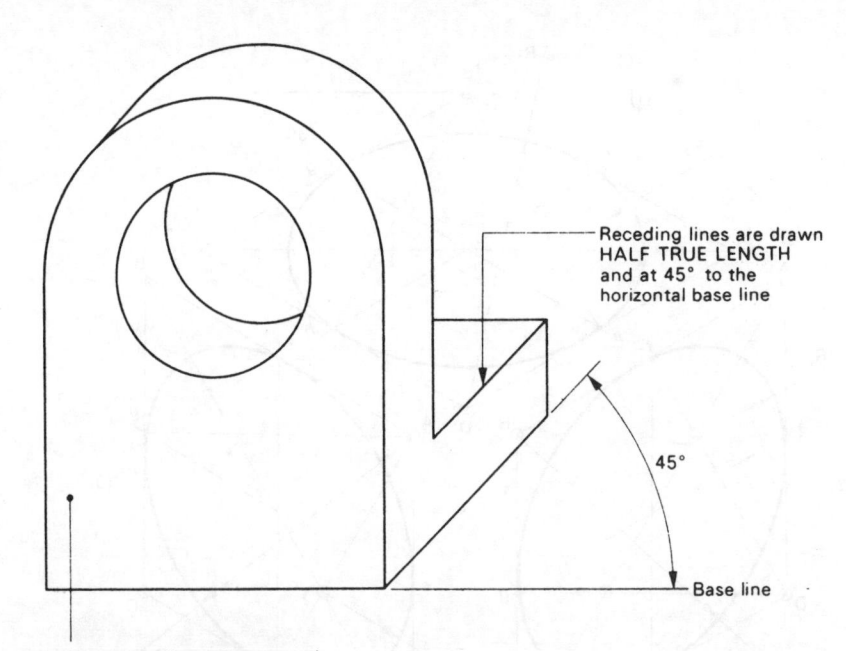

Base line

The elevation is drawn 'square on' as in orthographic drawing. All circles and arcs can be drawn with compasses. All lines are TRUE LENGTH

**Fig. 2.9  Oblique drawing**

## 2.5 Pictorial drawing — isometric

An ISOMETRIC DRAWING is rather more difficult to produce, but it has the advantage of showing horizontal surfaces more clearly. With this technique all the curved surfaces have to be constructed. Figure 2.10 illustrates the steps in making a simple isometric drawing, whilst Figure 2.11 shows alternative methods of how curved surfaces may be constructed.

### CURVE-GRID METHOD

Figure 2.11 (*a*) shows how a curve may be constructed using a *grid*. The advantage of this method is that complex, multi-radii curves may be easily plotted.

### ISOMETRIC CIRCLE — EIGHT-POINT METHOD

Figure 2.11 (*b*) shows the '*eight-point method*' of constructing an isometric circle.

All lines are true length

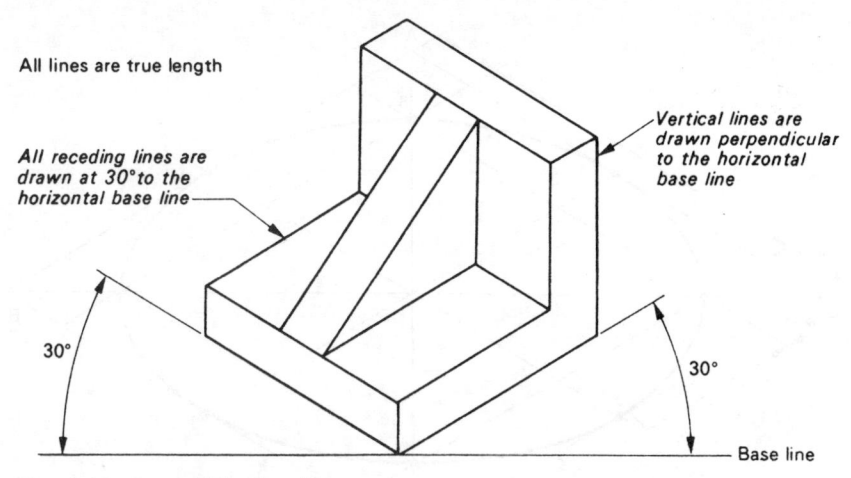

All receding lines are drawn at 30° to the horizontal base line

Vertical lines are drawn perpendicular to the horizontal base line

30°   30°

Base line

**Fig. 2.10  Isometric drawing**

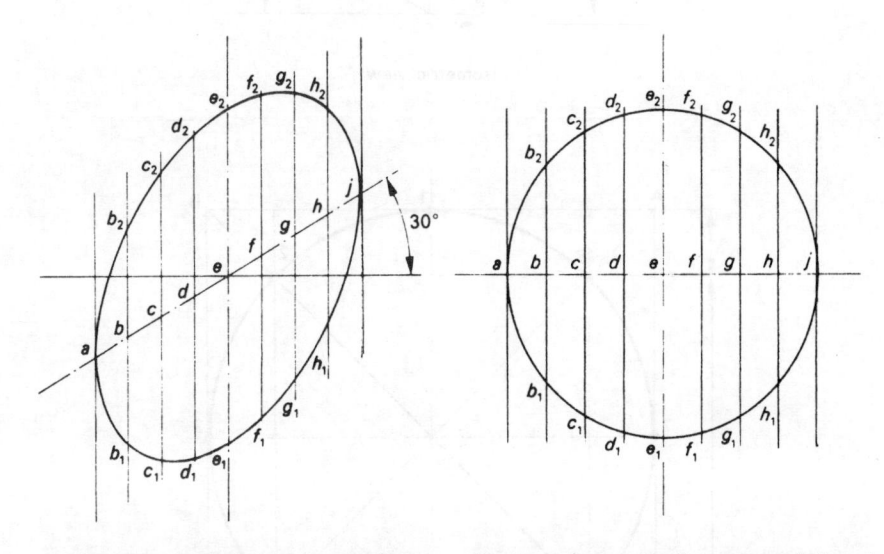

Circle - isometric. Becomes an ellipse      Circle - orthographic. True size and shape

1. *Construct a grid over the true circle by dividing its centre line into an equal number of parts, a b c ……j, and erecting a perpendicular at each point*

2. *Construct a similar grid on the isometric centre line*

3. *Step off distances $b_1$–$b$–$b_2$, $c_1$–$c$–$c_2$ etc. on the isometric grid by transferring the corresponding distances from the true circle*

4. *Draw a fair curve through the points plotted*

**(a)** **The construction of isometric curves**

**Fig. 2.11  Pictorial drawing (continued overleaf)**

37

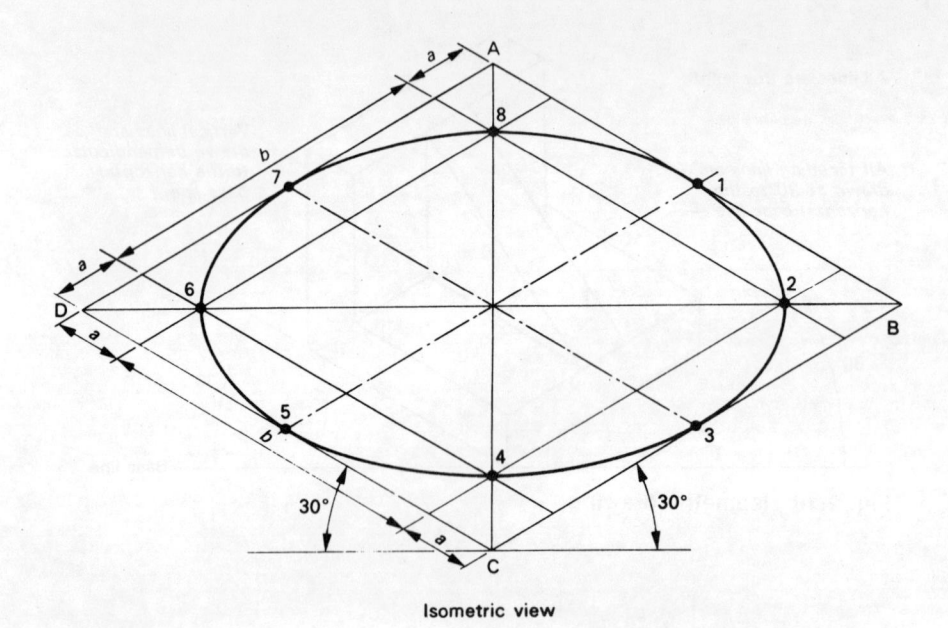

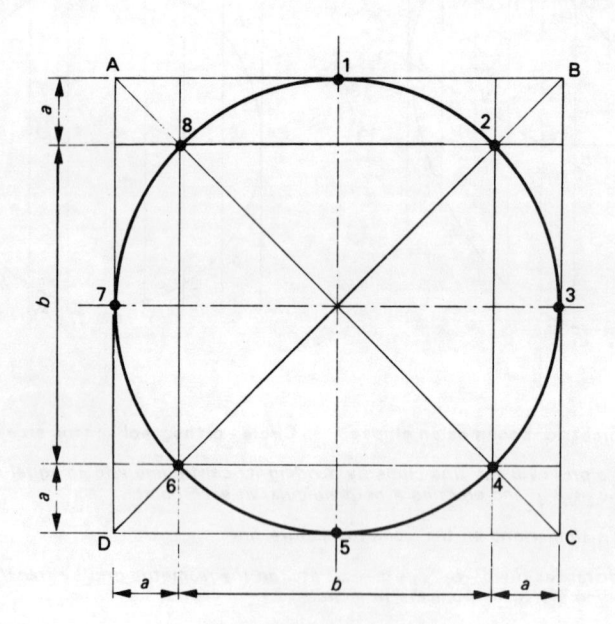

Isometric view

True view of circle in 'box'

(b) Construction of isometric circle — eight-point method

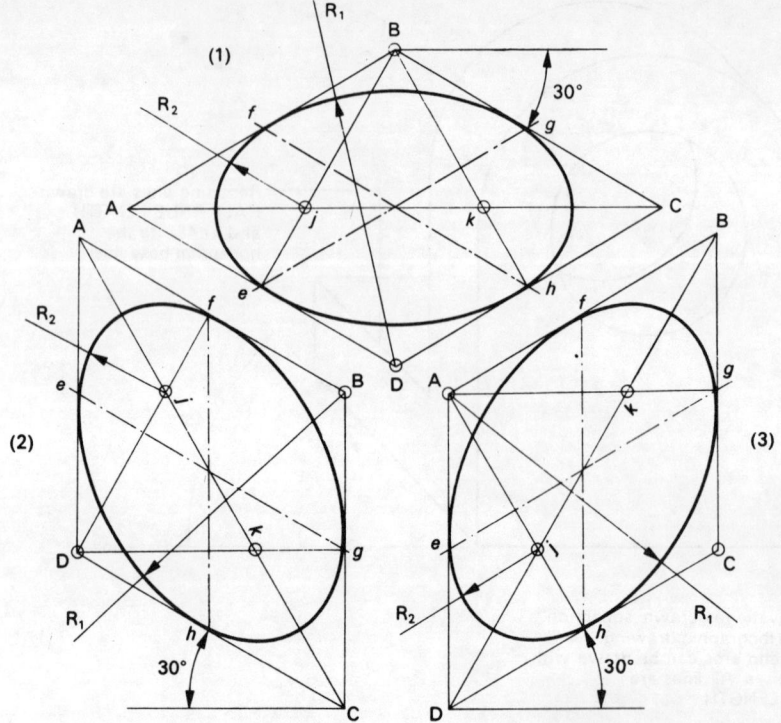

Three-dimensional isometric views of circle

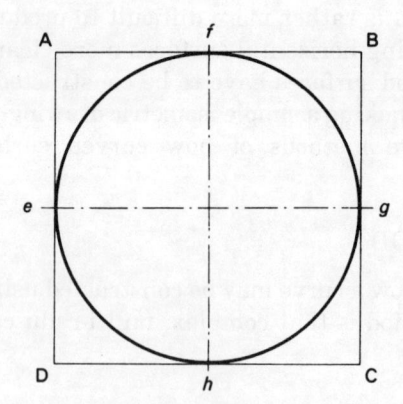

True view of circle in 'box'

(c) Construction of an isometric circle — circular arc method

**Fig. 2.11 (continued)**

38

*Method of construction*

1. Draw the isometric square and transfer the correct location of the centre lines from the true view.
2. The long diagonal AC or BD is drawn on the isometric square.
3. From B or D (in views (1) and (2)) or from A or C (in view (3)) draw lines to the mid-points of the opposite sides. In view (1) the mid-points opposite B are *e* and *h*. In view (2) the mid-points opposite D are *f* and *g*. In view (3) the mid-points opposite A are *g* and *h*. Where these two lines cross the long diagonal at *j* and *k*, are two centres for circular arcs to draw the approximate ellipse; the other two centres are the corners B and D (in views (1) and (2)) and the corners A and C (in view (3)).
4. Using these centres and radii $R_1$ and $R_2$, draw the arcs with compasses to complete the ellipse.

In the examples of isometric drawing shown in Figures 2.10 and 2.11, both the vertical lines and the receders have been drawn 'true length'. To be strictly correct, the receders should be drawn to *isometric scale* and only the vertical lines should be drawn true scale. Figure 2.12 shows the previous example drawn correctly to isometric scale. However, it is common practice to draw all the lines true length.

## 2.6 Use of square and isometric paper

Figure 2.13 shows how an oblique drawing of a simple component may be drawn on squared paper. For convenience, the component shown in Figure 2.9 has been dimensioned and redrawn on squared paper.

Squared paper is very useful for making oblique drawings for the following reasons:

1. It assists in the drawing of straight lines.
2. It helps to maintain the oblique proportion. In this example twenty of the 5 mm squares have been chosen for the 100 mm height of the front face of the component. All the other dimensions on the front face are drawn to this scale. The oblique faces are drawn to only one-half of this scale. For example, the 70 mm length is drawn to an equivalent length of seven squares.
3. The 45° lines for the oblique faces are drawn diagonally through the squares. Squared paper is also an aid to quick and accurate free-hand orthographic sketching as shown in Figure 2.14. Projections are much easier to make than on plain paper. To transfer dimensions from the top view (plan) to the side view, the squares are counted. In this example each square is 5 mm.

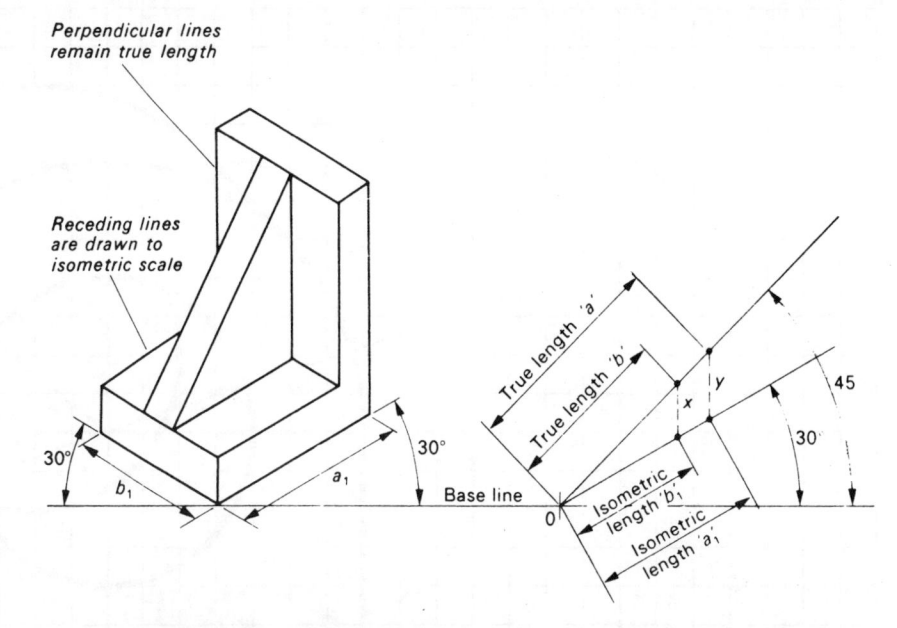

1. The isometric scale is constructed by drawing two lines at 30° and 45° respectively through the same point 'O' on the base line

2. True lengths are stepped off along the 45° line and perpendiculars (x, y) are dropped to cut the 30° line

3. The corresponding lengths, cut off the 30° line, represent the isometric scale lengths for the receding lines of the drawing

**Fig. 2.12  Isometric scale**

Figure 2.15 shows an isometric drawing of a square duct elbow produced on isometric paper. *This example shows that all truly isometric lines are either at 30° or 90° to the horizontal.*

When making isometric or oblique drawings, it is good practice to adopt the method of 'boxing' in which an imaginary 'box' is drawn to just contain the component. Additional boxes may be required to simplify the construction of parts of the component. Figure 2.16 illustrates the use of 'boxing' on isometric drawings of (*a*) a square hopper, and (*b*) a right cone. Both these examples have been drawn on isometric paper.

## 2.7 Dimensioning

The sections in this chapter so far have dealt basically with orthographic drawing techniques of describing the shape of the object or component.

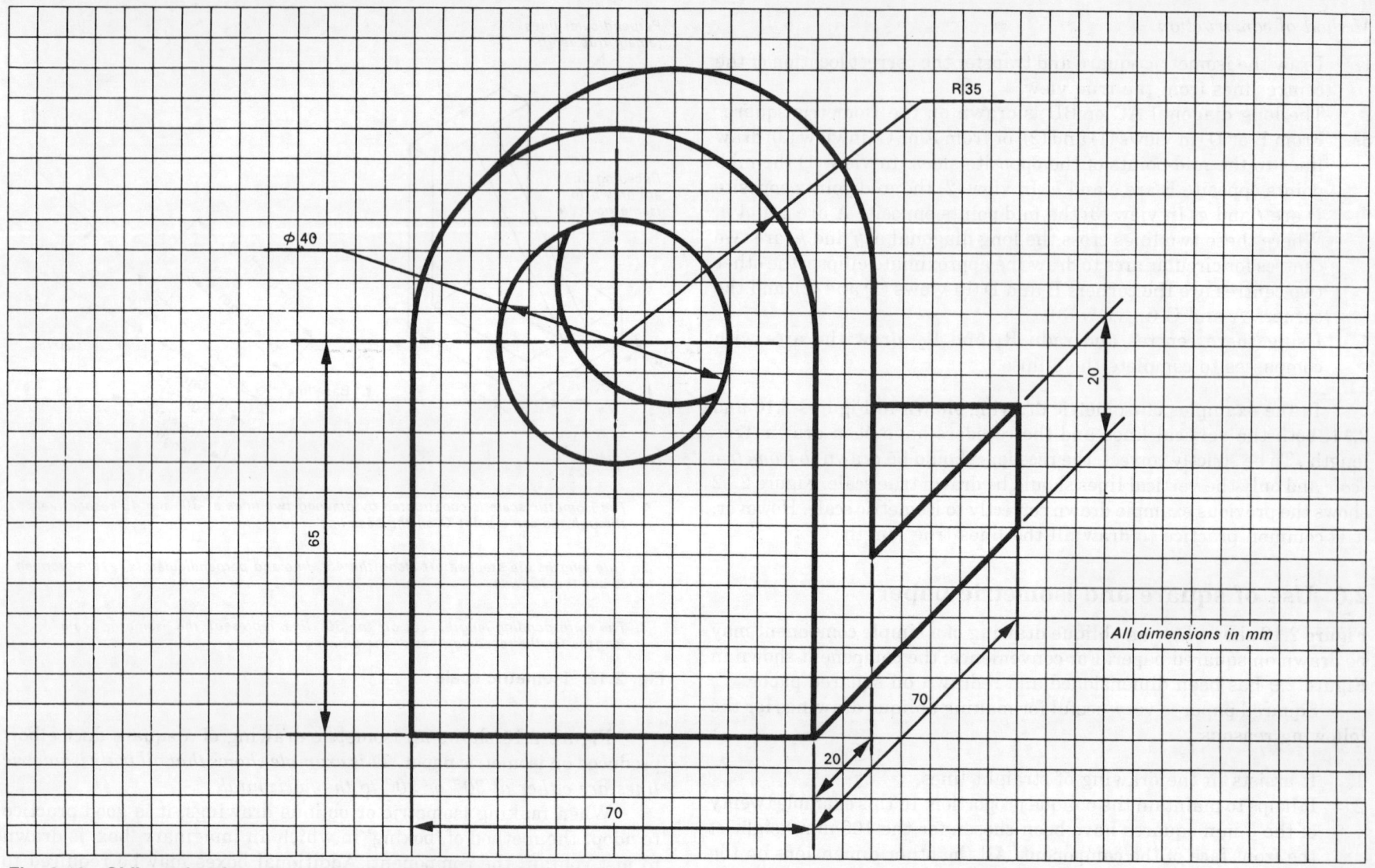

**Fig. 2.13 Oblique drawing on square or co-ordinate paper**

A two-view drawing may communicate the complete shape description of a sheet metal cylinder, as shown at (a) in Figure 2.17, but *neither communicates to the craftsman any size descriptions nor material specifications*. However, when size and material specifications are added, as shown at (b) in Figure 2.17, drawing becomes a 'working drawing', which is suitable for use in the fabrication shop.

In order to be proficient in communicating the required sizes of drawn shapes by 'dimensioning', three basic skills are required.

1. A knowledge of the techniques used in dimensioning.
2. The ability to select the essential dimensions which are needed in order to make the component.
3. The ability to position the selected dimensions according to accepted Engineering Drawing Practice.

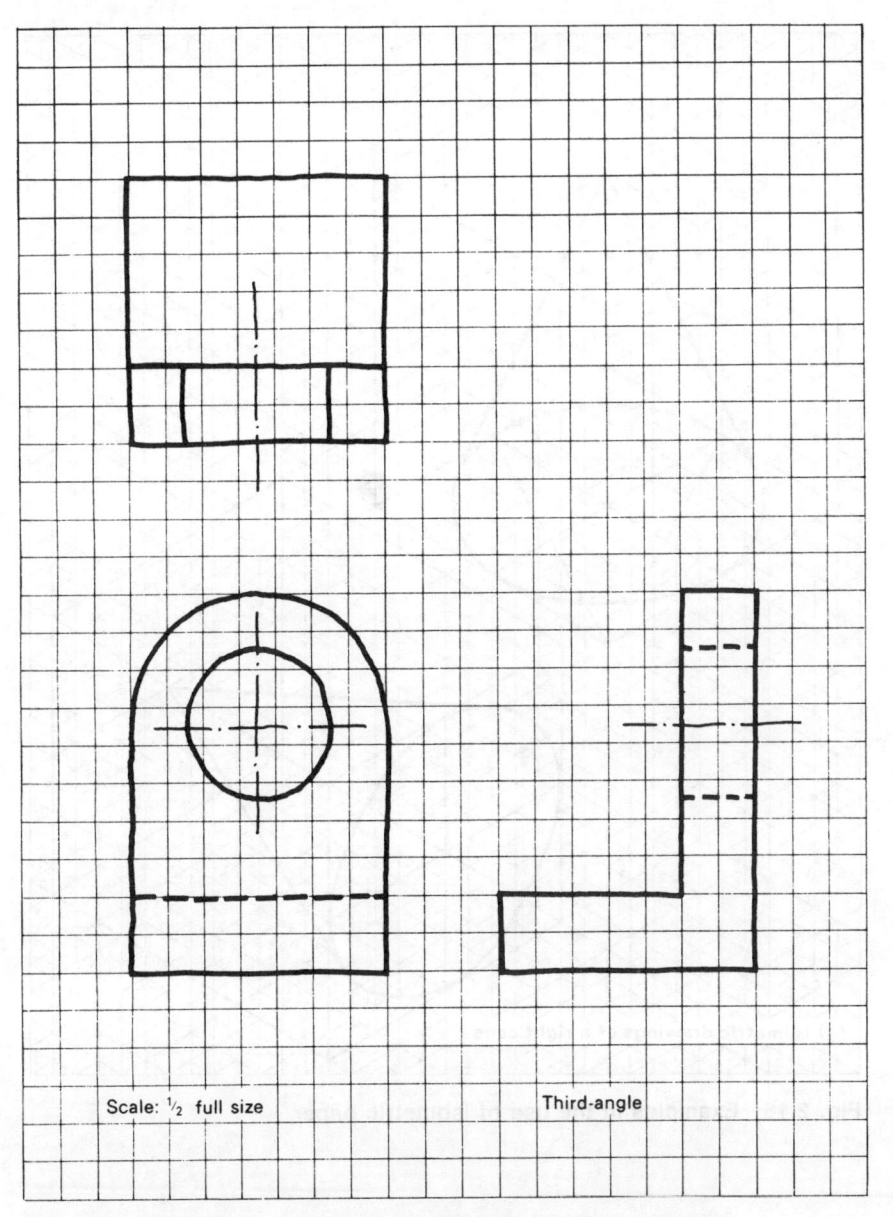

**Fig. 2.14  Free-hand orthographic sketching on square paper**

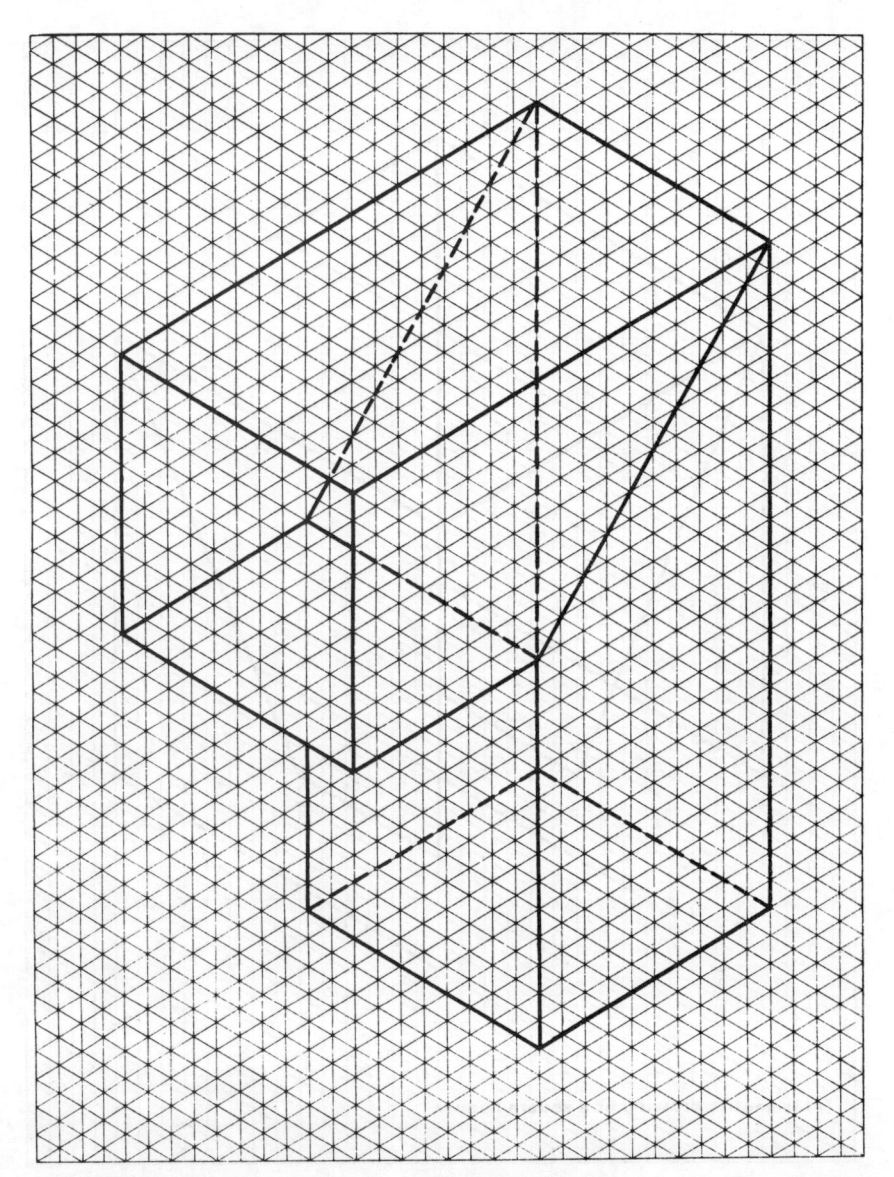

Scale: ½ full size    Third-angle

**Fig. 2.15  Isometric drawing on isometric paper**

The drawings in this chapter are in accordance with the standards of practice laid down in BS 308:1972.

Some of the basic methods of dimensioning, which should be adopted as standard practice, are illustrated in Figure 2.18.

Four types of lines are used in dimensioning an engineering drawing:

1. CENTRE LINES
2. PROJECTION LINES
3. DIMENSION LINES

41

*(a)* **Isometric drawing of a square hopper**

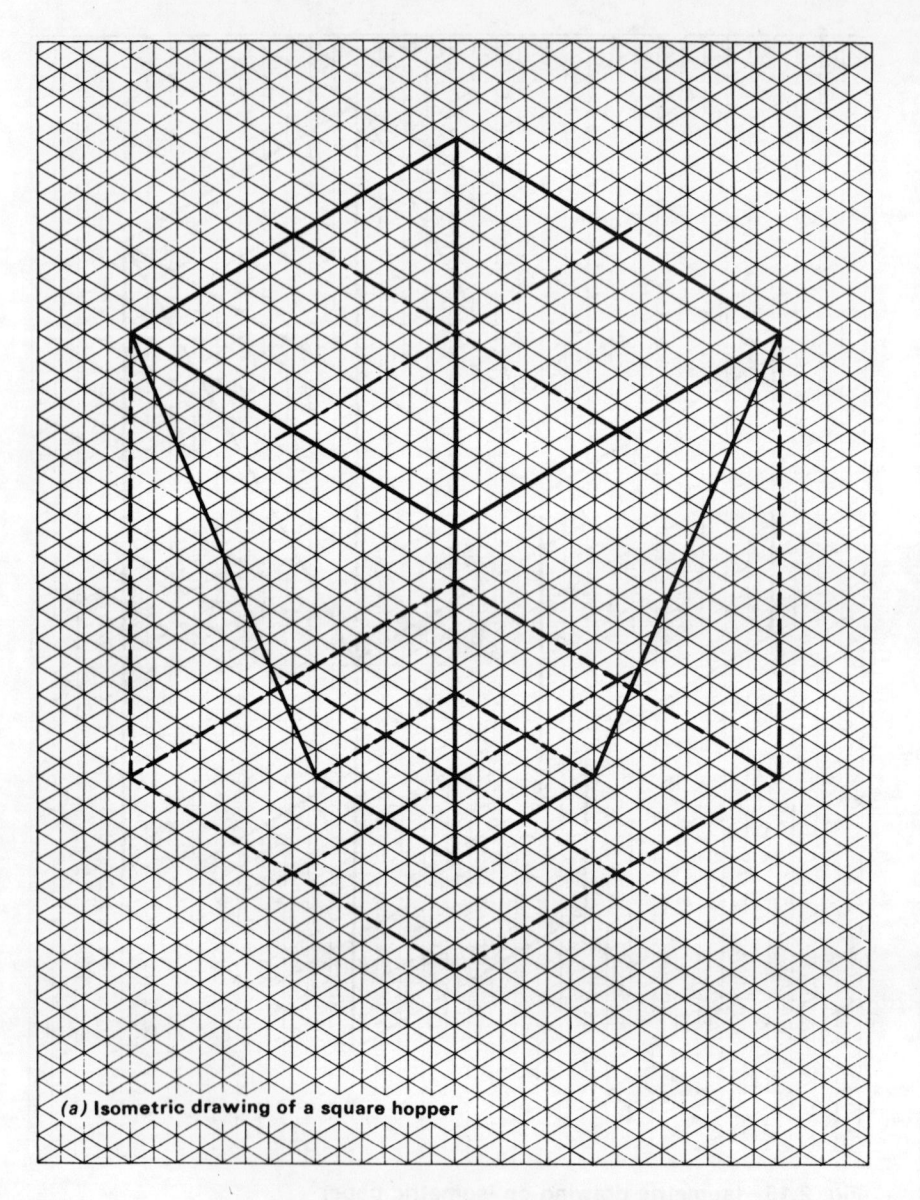

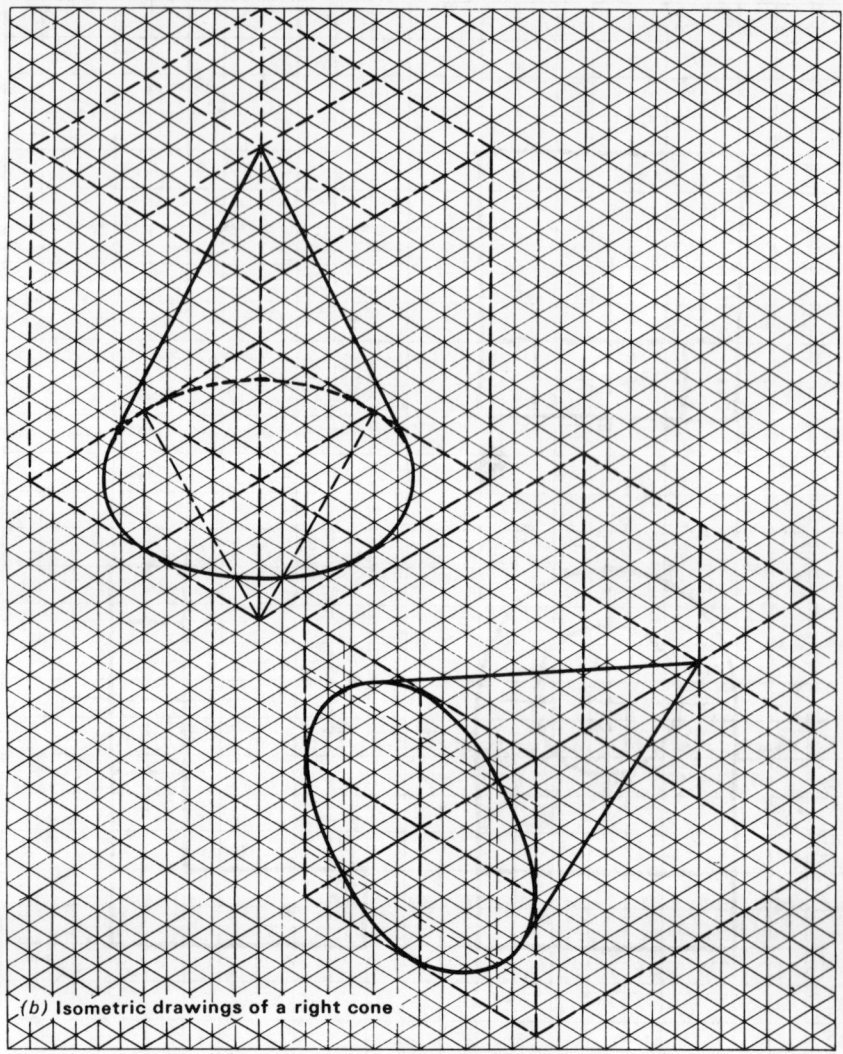

*(b)* **Isometric drawings of a right cone**

**Fig. 2.16 Examples in the use of isometric paper**

4. LEADER LINES

CENTRE LINES These are necesary to indicate the existence and location of circular features on drawings. They are thin, chain lines which are a valuable aid in reading a drawing, and must be included for all circular features in all views where appropriate.

PROJECTION LINES These are thin lines, sometimes termed 'Extension Lines', which extend away from the outline of the component in order to remove the dimension from the view. *In this manner the view remains unobstructed and easier to read.* Such dimension extension lines should start just clear of the drawing outline and extend a little beyond the arrowhead on the dimension line.

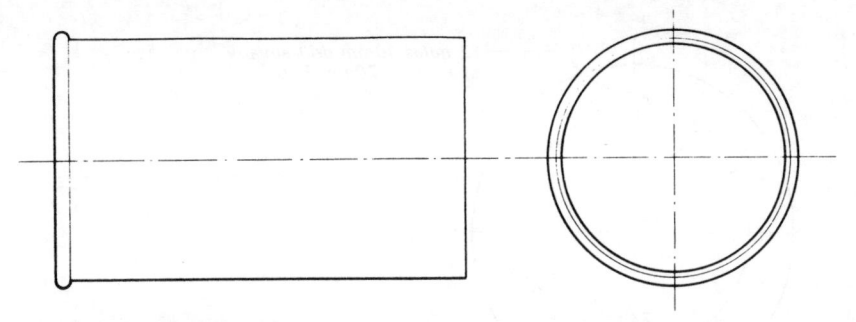

(a) These two views clearly indicate the exact shape of the component but do not communicate any dimensions or material specifications, and therefore cannot be used for a working drawing

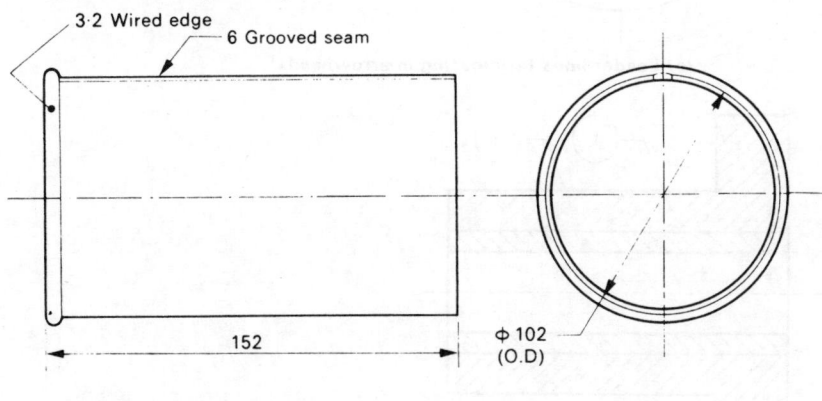

Dimensions in millimetres
Material: 22 swg galvanised steel
6-off

NOT TO SCALE   SLEEVE

(b) Required information added to the two views provides a working drawing which can be used by the sheet metal worker or fabricator

**Fig. 2.17  The need for dimensioning**

DIMENSION LINES  These should also be thin lines terminating in arrowheads. The arrowheads should be easily readable and normally not less than 3 mm long. *The point of an arrowhead should just touch the projection line.* Centre lines or any portion of the drawing outline must never be used as a dimension line, as doing so will lead to confusion.

LEADER LINES  While size descriptions are communicated by means of dimensions, many features are better described by means of a note. Leader lines are used for this purpose. Such lines are continuous thin lines terminating in arrowheads or dots. Arrowheads are always used where the leader line *touches the outline* of the feature of component.

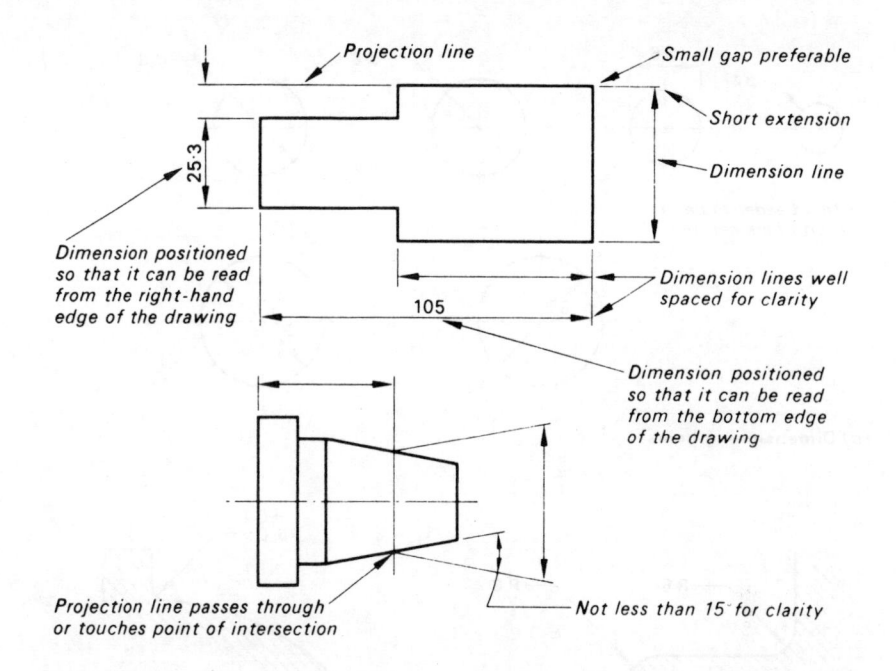

(a) Projection and dimension lines

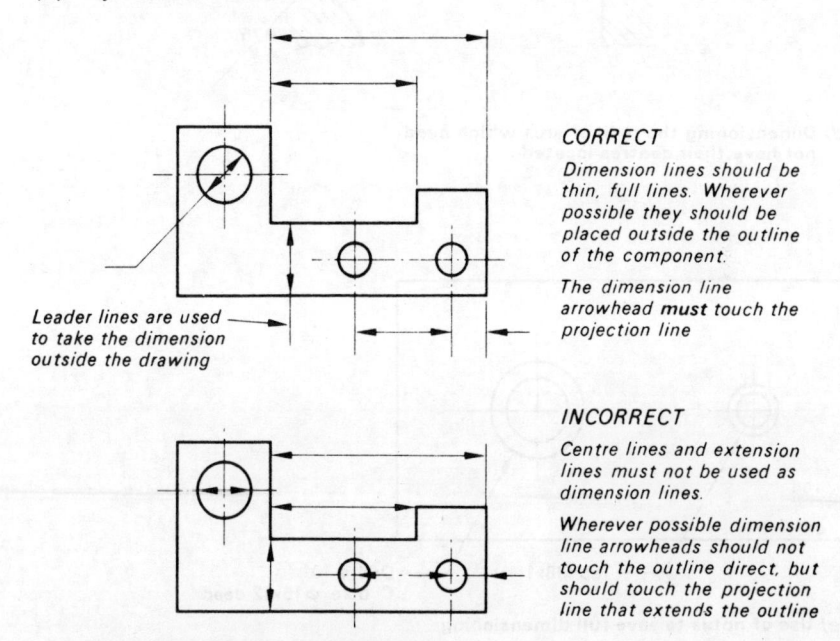

*CORRECT*

Dimension lines should be thin, full lines. Wherever possible they should be placed outside the outline of the component.

The dimension line arrowhead **must** touch the projection line

*INCORRECT*

Centre lines and extension lines must not be used as dimension lines.

Wherever possible dimension line arrowheads should not touch the outline direct, but should touch the projection line that extends the outline

(b) Correct and incorrect dimensioning

**Fig. 2.18  Dimensioning (continued overleaf)**

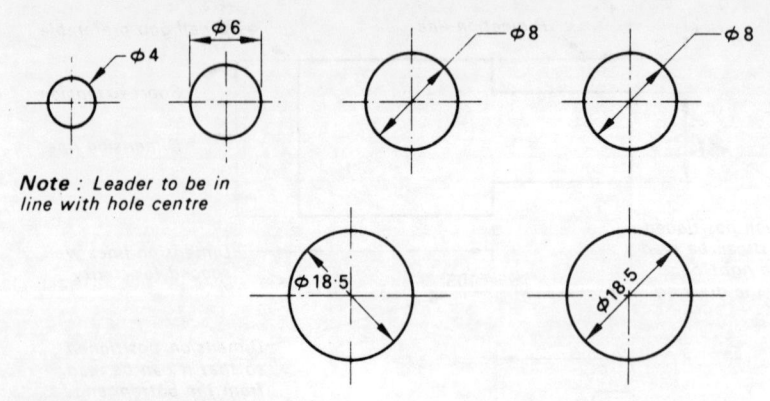

(c) Dimensioning holes

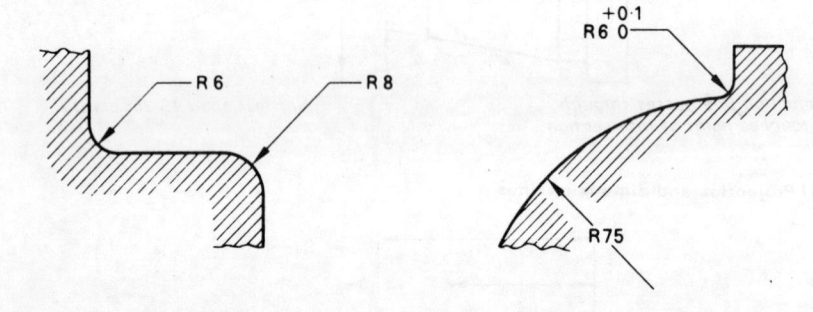

*Note: Leader to be in line with hole centre*

(d) Dimensioning the radii of arcs which need not have their centres located

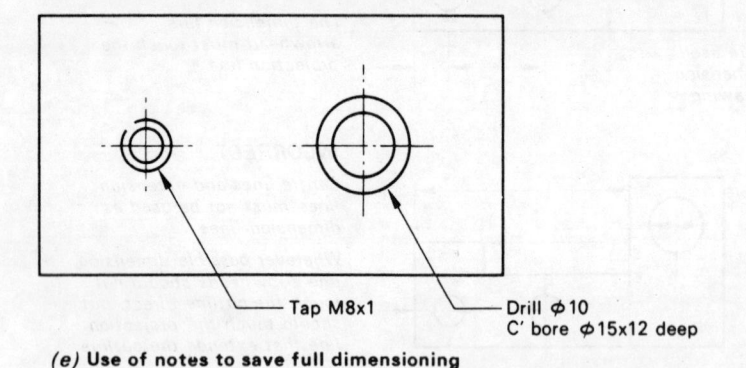

(e) Use of notes to save full dimensioning

Tap M8x1    Drill φ10 C'bore φ15x12 deep

**Fig. 2.18 (continued)**

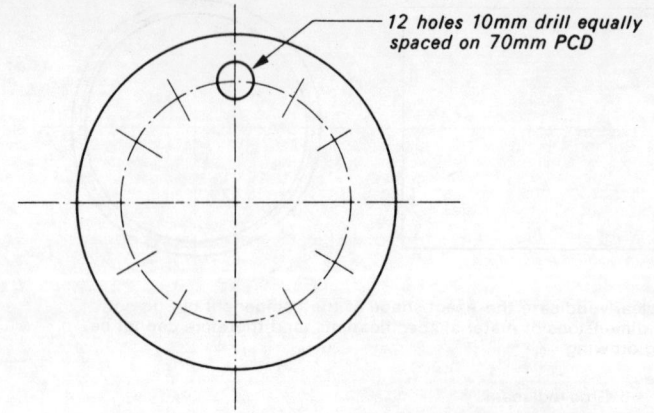

(a) Leader lines terminating in arrowheads

12 holes 10mm drill equally spaced on 70mm PCD

(b) Leader lines terminating in dots

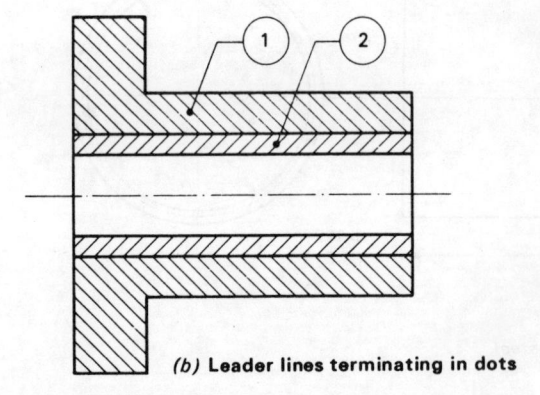

**Fig. 2.19 Leader lines**

*Dots* are used where the leader line terminates *within the outline* of the feature or component.

Leader lines which touch lines should be nearly normal to the surface. *They should not be parallel to adjacent dimension or projection lines where confusion may arise*. Typical examples of the use of 'leader lines' are shown in Figure 2.19.

## 2.8 Selection of dimensions

Certain basic principles apply to the selection of dimensions for various shapes. In the same way as the number of views in a drawing are kept to a minimum, so too must dimensions be selected so that the fabricator will not be confused by unnecessary details. Duplication of dimensions should also be avoided.

**Fig. 2.20 Examples of the application of dimensions**

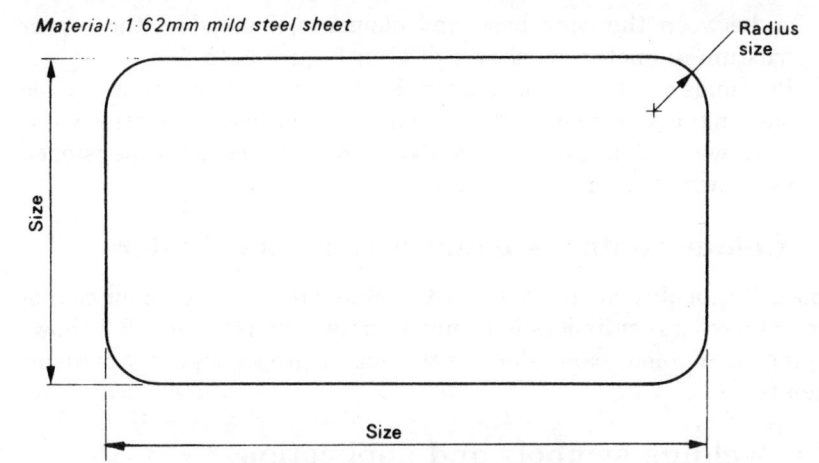

(a) Rectangular sheet metal blank with radiused corners
Material: 1.62mm mild steel sheet

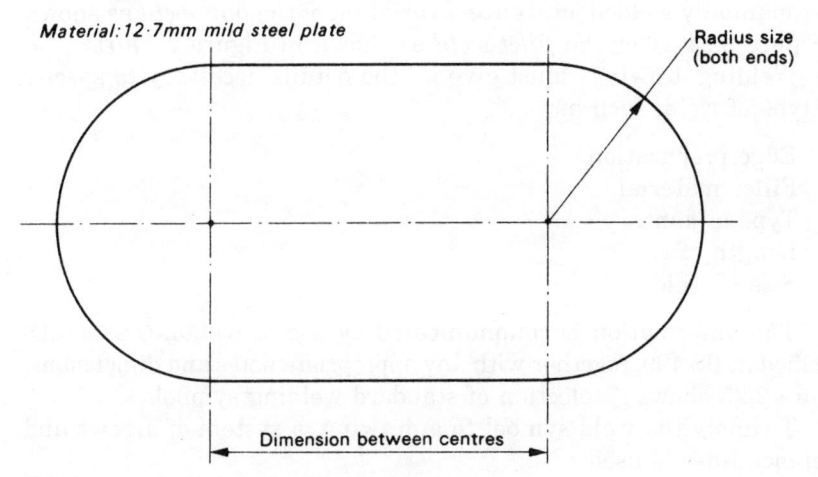

(b) Rectangular flat stock with rounded corners
Material: 12.7mm mild steel plate

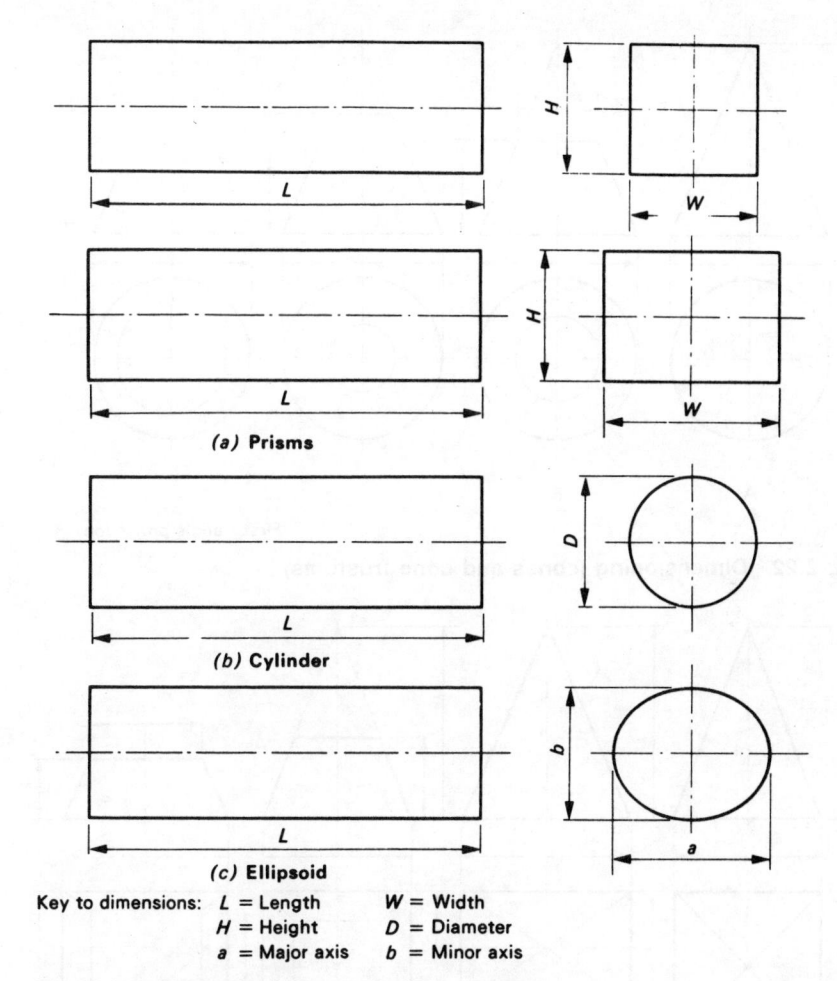

**Fig. 2.21 Dimensioning — selection of views**

(a) Prisms
(b) Cylinder
(c) Ellipsoid

Key to dimensions:  $L$ = Length    $W$ = Width
$H$ = Height    $D$ = Diameter
$a$ = Major axis    $b$ = Minor axis

In this section a few selected examples will be used to illustrate how the basic principles of dimensioning are applied.

1. Square and rectangular pieces of flat stock have the two basic dimensions of width and length. The thickness is usually given by means of a note or other specification.
2. Flat discs have only one basic dimension which is the diameter. The thickness is specified as in the previous example.
3. If the flat stock is square or rectangular but has rounded corners, the radius of the rounded corner is given. Figure 2.20 (a), shows a rectangular sheet metal blank with four rounded corners. In this case the overall dimensions of the blank are given and the corner radius is shown.
4. Figure 2.20 (b) shows how a metal blank with rounded ends is dimensioned. It will be noticed that only the centre-line distance (centre to centre) and the radius for the ends are given because these are the only dimensions required for marking-out and cutting the blank.
5. Prisms are dimensioned by indicating the width and length measurements on one view and the thickness or height dimension in a related view, as shown in Figure 2.21(a).

45

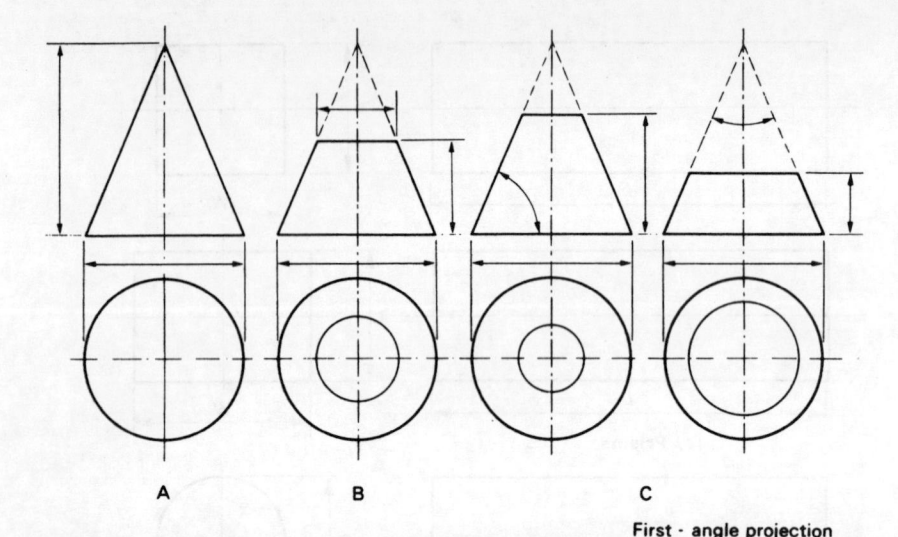

**Fig. 2.22 Dimensioning (cones and cone frustums)**

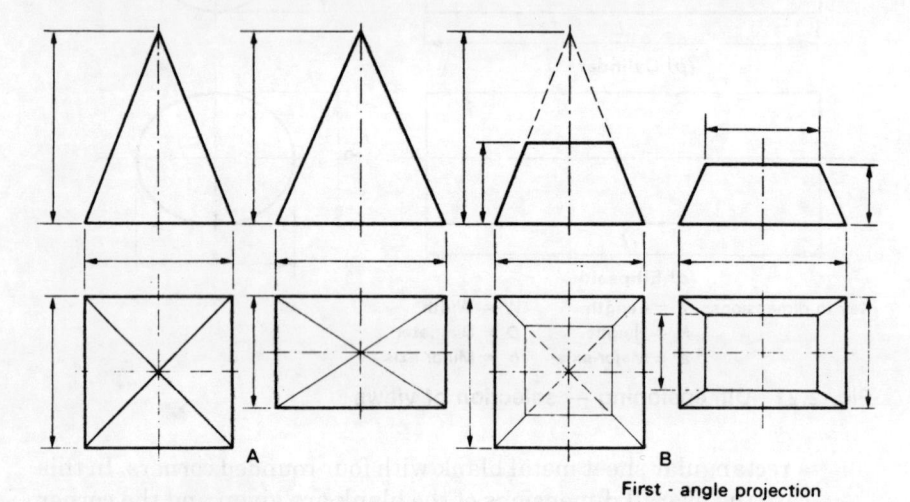

**Fig. 2.23 Dimensioning (pyramids and pyramid frustums)**

6. Cylinders are dimensioned as shown in Figure 2.21(b).
7. Cones are dimensioned by giving the diameter of the base and the true height in the elevation view, as shown at A in Figure 2.22. A cone frustum is dimensioned by giving the top and bottom diameters and the vertical distance between them in the elevation view, as shown at B in Figure 2.23. An alternative method of dimensioning is to give an angular value either between elements, or between the cone base and elements, instead of the top or frustum diameter, as shown at C in Figure 2.22.
8. Pyramids are dimensioned by indicating the measurements for the base in the plan view, and the vertical height in the elevation view, as shown at A in Figure 2.23. Pyramid frustums are dimensioned as shown at B in Figure 2.23.

## 2.9 Colour coding — compressed gas cylinders

Table 2.1 is an abstract from the British Standards for the colour coding of compressed gas cylinders in common industrial use. Table 2.2 shows how the use of colour bands denotes the hazard properties of the cylinder contents.

## 2.10 Welding symbols and application

Most manually welded joints are a variation of the *butt weld* as shown in Figure 2.24(a), or the *fillet weld* as shown in Figure 2.24(b).

Welding drawings must give all the details necessary to specify the type of weld, such as:

1. Edge preparation.
2. Filler material.
3. Type of joint.
4. Length of run.
5. Size of weld.

This information is communicated by use of WELDING SYMBOLS specified in BS 499, together with any appropriate notes and dimensions. Figure 2.25 shows a selection of standard welding symbols.

To apply the weld symbol to a drawing, a system of arrows and reference lines is used.

(a) When the weld symbol is *above* the reference line the weld is made on the side of the joint *opposite* the arrowhead, i.e., the OTHER SIDE.
(b) When the weld symbol is *below* the reference line the weld is made on the *same side* of the joint as the arrowhead, i.e., the ARROW SIDE.
(c) When the weld symbol is on *both sides* of the reference line, the welding is to be carried out on *both sides* of the joint.

The reference line that is connected to the arrow at an angle (as shown in the examples) is the *datum line* for determining the position of the weld symbol in order that the latter may indicate the location of the weld. *The welding symbol is inverted and suspended from this datum line for welds made from the arrow side of the joint. For welds to be made on the other side of the joint, the weld symbol is based on*

Table 2.1 Compressed gases in common industrial use

| NAME OF GAS | CHEMICAL FORMULA | BASE COLOUR OF CONTAINER | COLOUR OF BANDS | REMARKS |
|---|---|---|---|---|
| Acetylene | $C_2H_2$ | Maroon | None | Only acetylene regulators should be used. Cylinders must be stood upright. |
| Air | — | French grey | None | — |
| Ammonia | $NH_3$ | Black | Red and Yellow | The red band is adjacent to the valve fitting. |
| Argon | Ar | Peacock blue | None | Main use for gas-shielded arc welding processes. |
| Carbon dioxide | $CO_2$ | Black | None | |
| Coal gas | — | Signal red | None | Used for heating and cutting. Only coal gas regulator should be used. |
| Helium | He | Middle brown | None | Used for gas-shielded arc welding processes. |
| Hydrogen | $H_2$ | Signal red | None | Main use for heating and cutting. Only hydrogen regulator should be used. |
| Methane | $CH_4$ | Signal red | None | — |
| Nitrogen | $N_2$ | French grey | Black | — |
| Oxygen | $O_2$ | Black | None | Only oxygen regulator should be used |
| Propane (commercial) | — | Signal red | None | An acetylene regulator may be used with this gas. |

Note:
For the purpose of identifation the container should be marked with the name of the gas and the chemical formula in accordance with I.S.O. Recommendation R448 — 'Marking of Industrial Gas Cylinders for the identification of their content'.

Table 2.2 Colour bands to denote hazard properties of contents of compressed gas cylinders

| NATURE OF GAS OR MIXTURE | COLOUR OF BANDS | CONTAINER NECK |
|---|---|---|
| Non-flammable and non-poisonous | None | 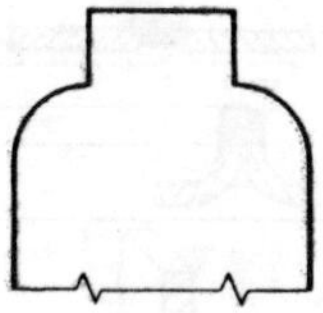 |
| Non-flammable and poisonous | Yellow | 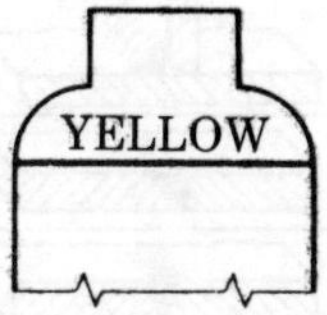 YELLOW |
| Flammable and non-poisonous | Red | 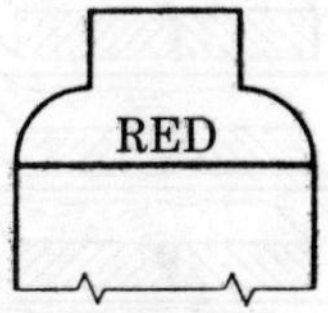 RED |
| Flammable and poisonous | Red and Yellow | 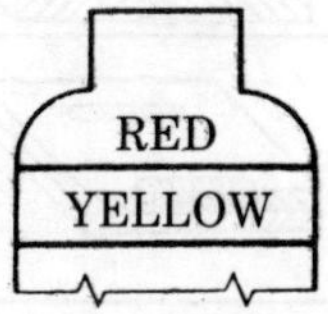 RED / YELLOW |

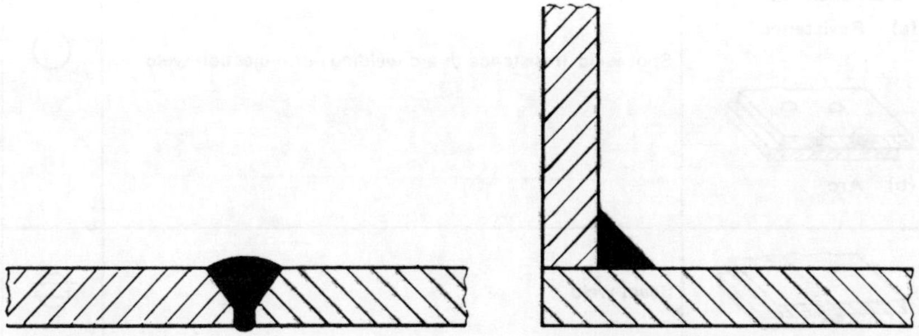

(a) Butt weld   (b) Fillet weld

Fig. 2.24 Basic welded joints

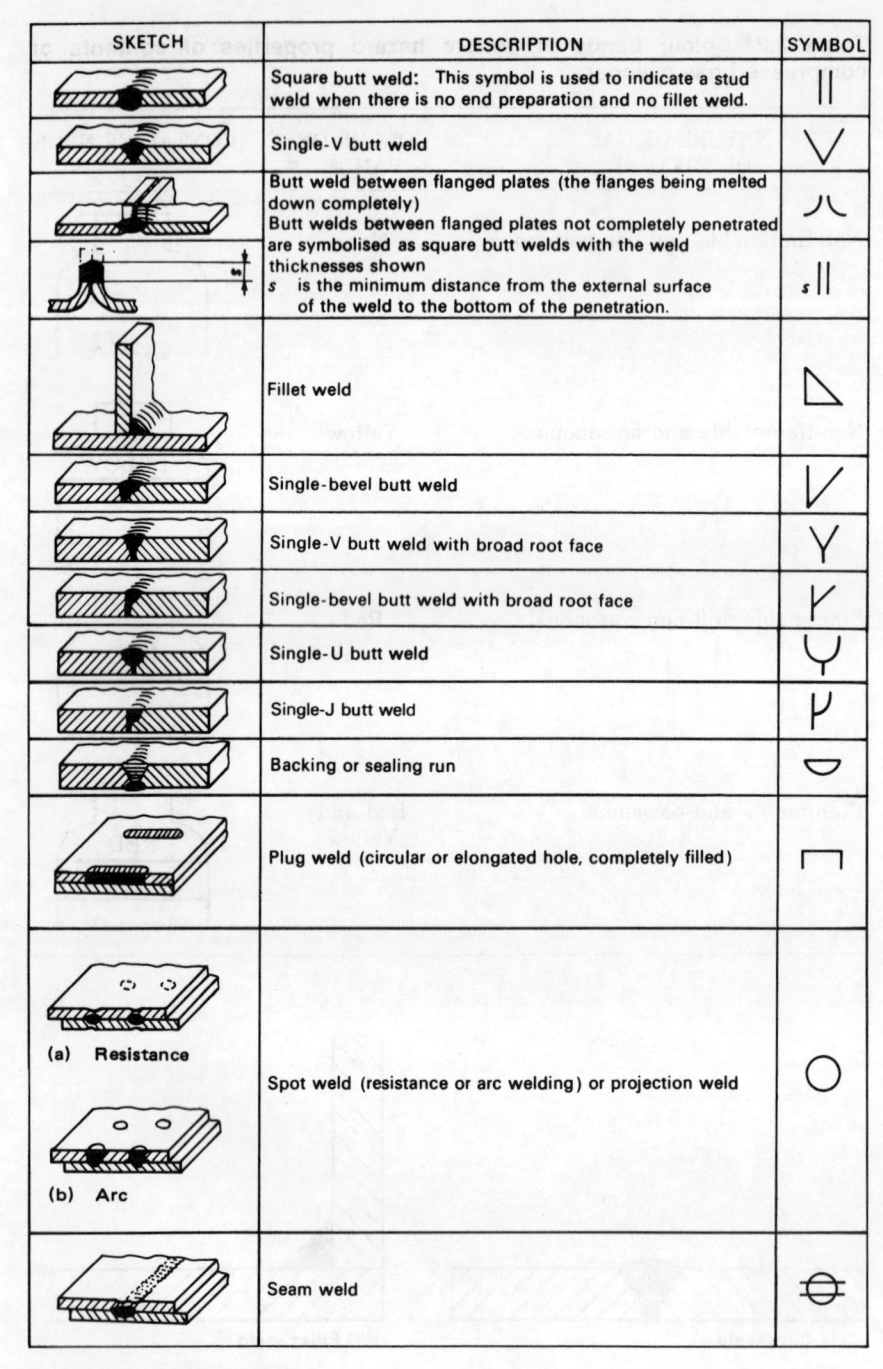

| SKETCH | DESCRIPTION | SYMBOL |
|---|---|---|
| | Square butt weld: This symbol is used to indicate a stud weld when there is no end preparation and no fillet weld. | $\|\|$ |
| | Single-V butt weld | $V$ |
| | Butt weld between flanged plates (the flanges being melted down completely) | $\curlywedge$ |
| | Butt welds between flanged plates not completely penetrated are symbolised as square butt welds with the weld thicknesses shown<br>$s$ is the minimum distance from the external surface of the weld to the bottom of the penetration. | $s\|\|$ |
| | Fillet weld | $\triangle$ |
| | Single-bevel butt weld | $V$ |
| | Single-V butt weld with broad root face | $Y$ |
| | Single-bevel butt weld with broad root face | $Y$ |
| | Single-U butt weld | $U$ |
| | Single-J butt weld | $P$ |
| | Backing or sealing run | $\smile$ |
| | Plug weld (circular or elongated hole, completely filled) | $\sqcap$ |
| (a) Resistance<br>(b) Arc | Spot weld (resistance or arc welding) or projection weld | $O$ |
| | Seam weld | $\ominus$ |

**Fig. 2.25** Weld symbols

*the datum line.* Applications of the use of welding symbols are illustrated in Figure 2.26.

In addition to the use of symbols, dimensional information is required to specify the size of the weld and the length of the run. Written notes can also be added to the drawing to specify the filler material and

SKETCH OF WELD          SYMBOLIC REPRESENTATION

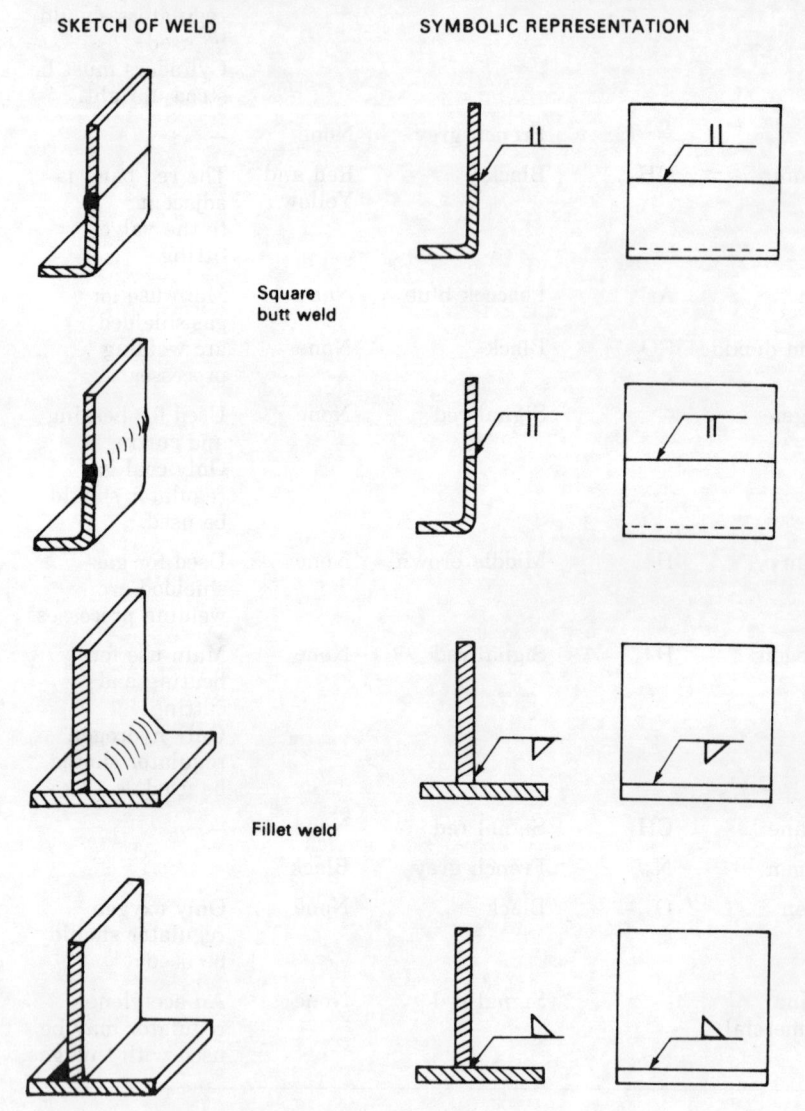

Square butt weld

Fillet weld

**Fig. 2.26** Some examples showing significance of the arrow and position of the weld symbol in relation to the reference line (continued)

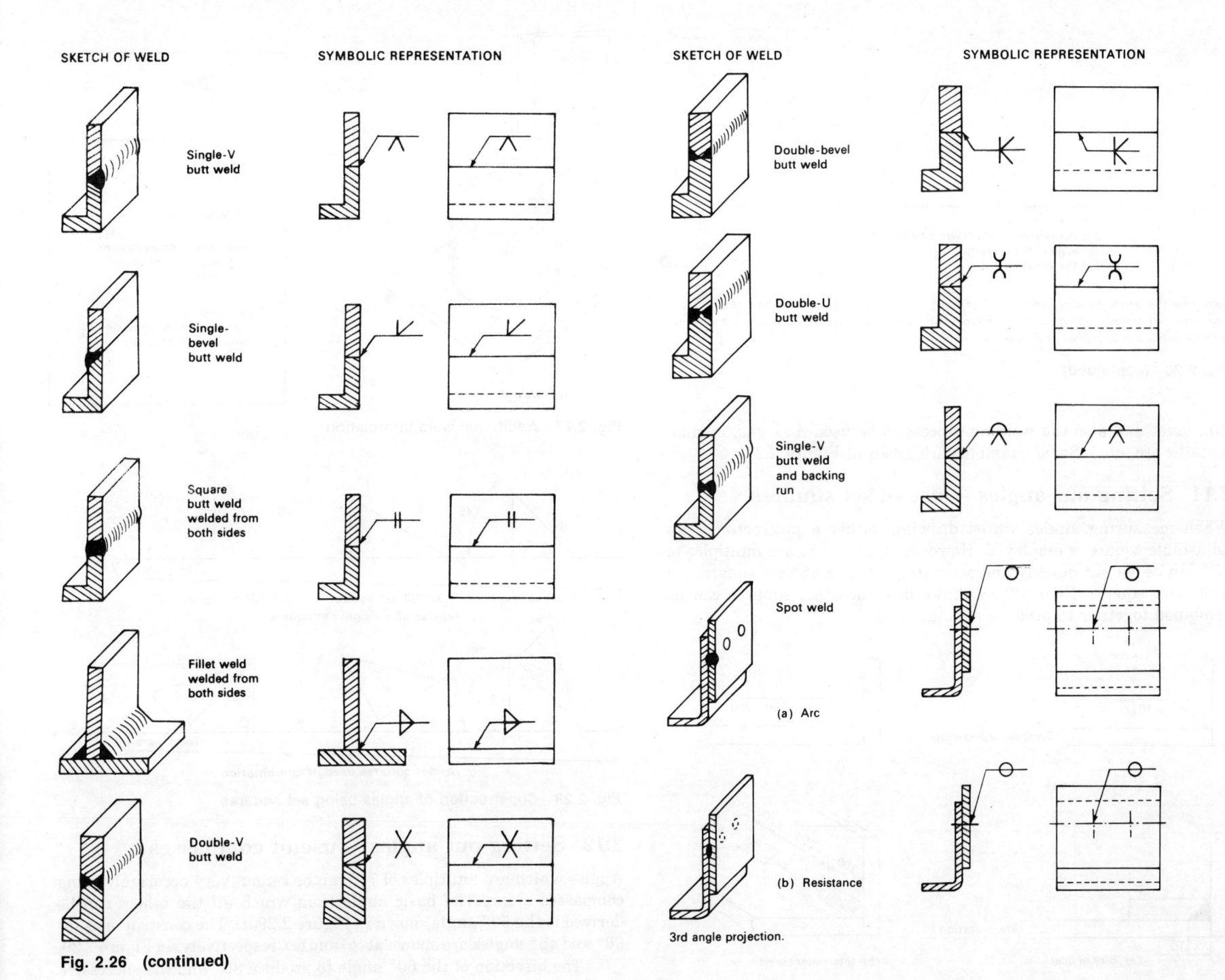

| SKETCH OF WELD | | SYMBOLIC REPRESENTATION | SKETCH OF WELD | | SYMBOLIC REPRESENTATION |

| | Single-V butt weld | | | Double-bevel butt weld | |
| | Single-bevel butt weld | | | Double-U butt weld | |
| | Square butt weld welded from both sides | | | Single-V butt weld and backing run | |
| | Fillet weld welded from both sides | | | Spot weld (a) Arc | |
| | Double-V butt weld | | | (b) Resistance | |

3rd angle projection.

**Fig. 2.26** (continued)

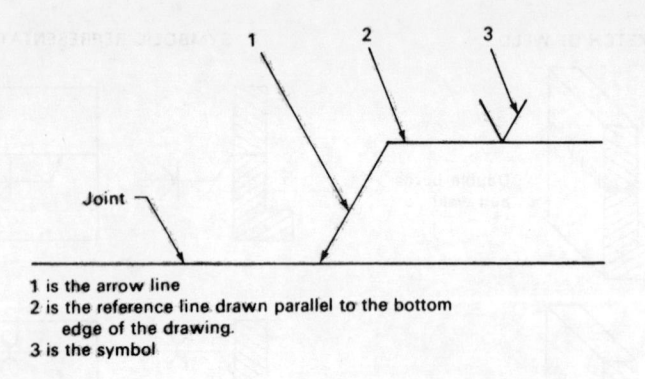

1 is the arrow line
2 is the reference line drawn parallel to the bottom edge of the drawing.
3 is the symbol

*Note how the arrow head is used to indicate which plate is prepared and how the disposition of the symbol about the reference line indicates the side from which the weld is to be made and therefore the wide part of the preparation*

**Fig. 2.26** (continued)

flux used, and also the welding process to be used (i.e., gas, manual metallic-arc, etc.). Some examples are given in Figure 2.27.

## 2.11 Setting out angles — use of set squares

When measuring angles whilst drawing, either a protractor or an adjustable square is employed. However, angles that are multiples of 15° can be set out quickly and accurately using a 45° set square and a 60° set square. Figure 2.28 shows how these set squares can be combined together to produce a range of angles.

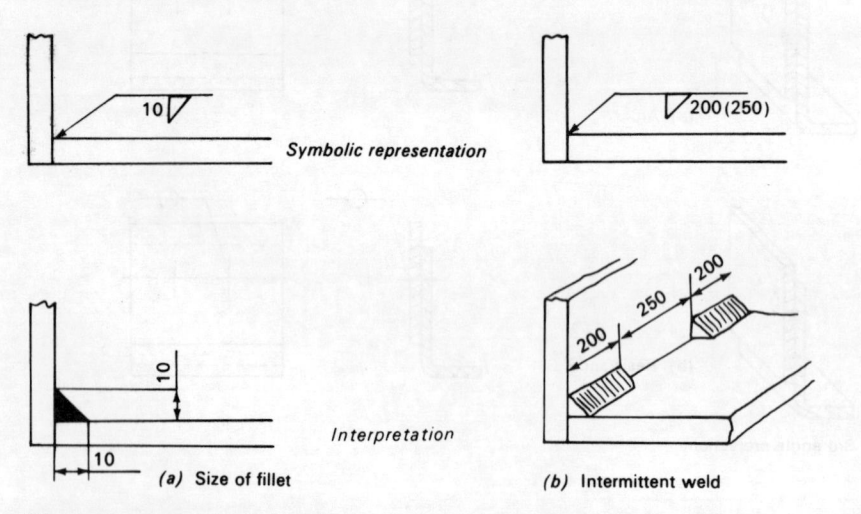

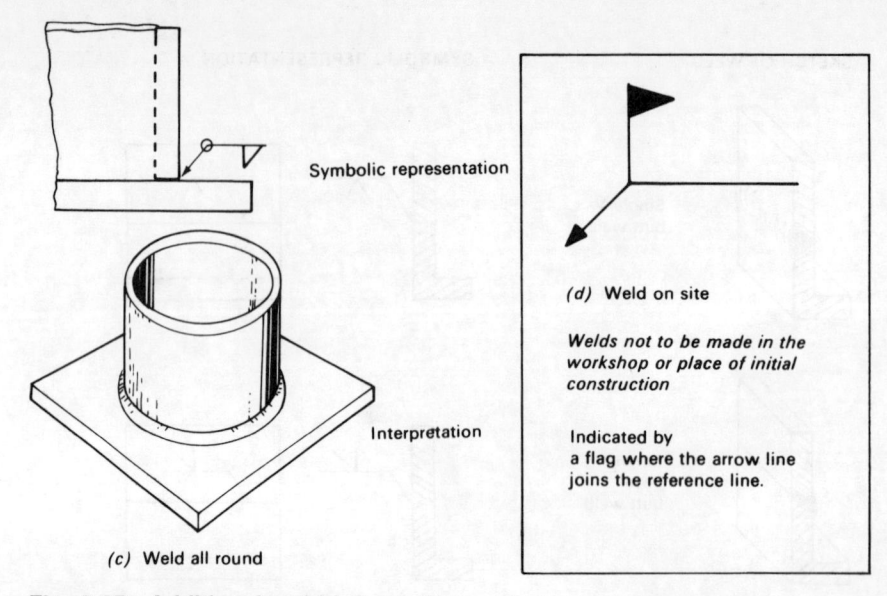

(a) Size of fillet

(b) Intermittent weld

Symbolic representation

Interpretation

(c) Weld all round

(d) Weld on site

*Welds not to be made in the workshop or place of initial construction*

Indicated by a flag where the arrow line joins the reference line.

**Fig. 2.27** Additional weld information

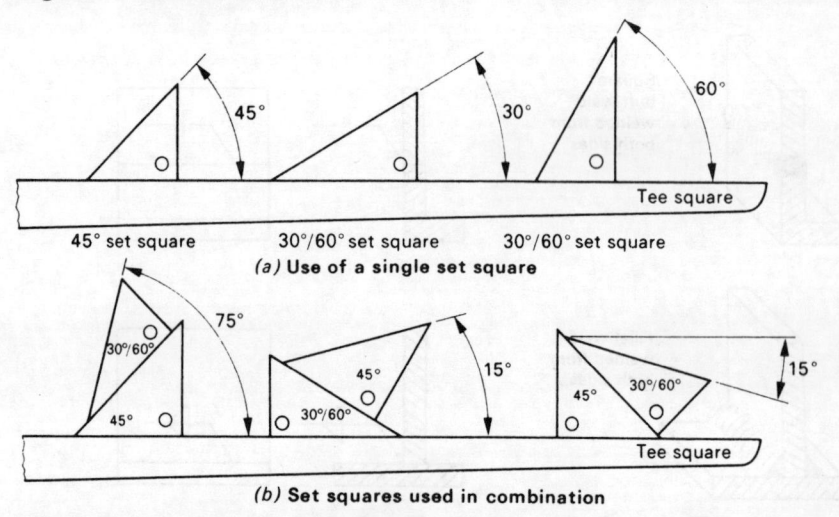

(a) Use of a single set square

(b) Set squares used in combination

**Fig. 2.28** Construction of angles using set squares

## 2.12 Setting out angles — use of compasses

Angles which are multiples of $7\frac{1}{2}°$ can be set out very accurately using compasses alone. The basic angle from which all the others can be derived is the 60° angle shown in Figure 2.29(a). The constructions for 90° and 45° angles are shown at (b) and (c), respectively, in Figure 2.29.

The bisection of the 60° angle to produce 30° and the successive

1. Let A be the apex of the angle
2. With centre A draw an arc BC of large radius
3. Step off BD equal in radius to AB
4. Draw a line AE through D
5. The angle EAB is 60°

(a) Construction of a 60° angle

1. Let A be the apex of the angle
2. With centre A draw an arc BC of large radius
3. Step off BD and DE equal in radius to AB
4. With centre D draw any arc F
5. With centre E draw an arc equal in radius to DF
6. Join AF with a straight line
7. Angle BAF is 90°

(b) Construction of a 90° angle

1. Draw AB and AC at right angles (90°) to each other as described in Fig. 2.29 (b)
2. With centre A, and with large radius, draw an arc to cut AB at D and AC at E
3. With centres E and D draw arcs of equal radius to intersect at F
4. Draw a straight line from A through F
5. Angle BAF is 45°
   (AF bisects – halves ∠BAF)

(c) Construction of a 45° angle

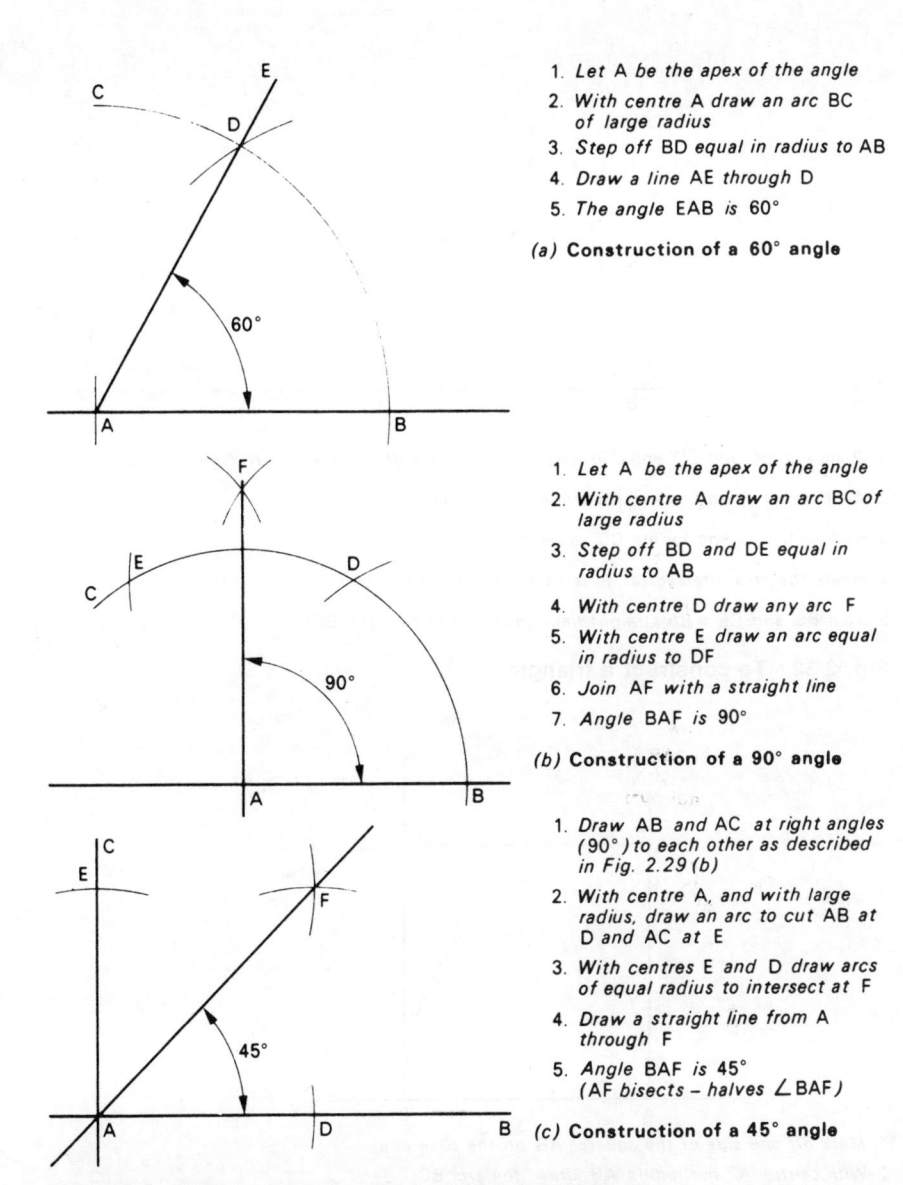

**Fig. 2.29 Construction of angles using compasses**

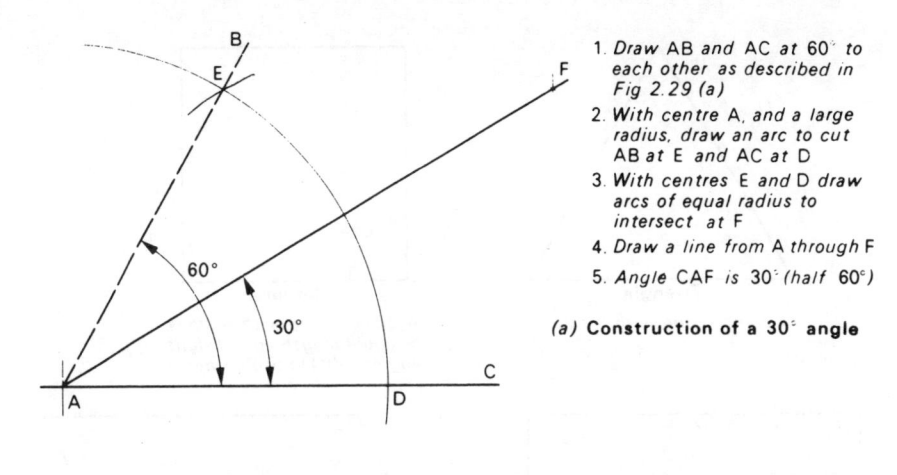

1. Draw AB and AC at 60° to each other as described in Fig 2.29 (a)
2. With centre A, and a large radius, draw an arc to cut AB at E and AC at D
3. With centres E and D draw arcs of equal radius to intersect at F
4. Draw a line from A through F
5. Angle CAF is 30° (half 60°)

(a) Construction of a 30° angle

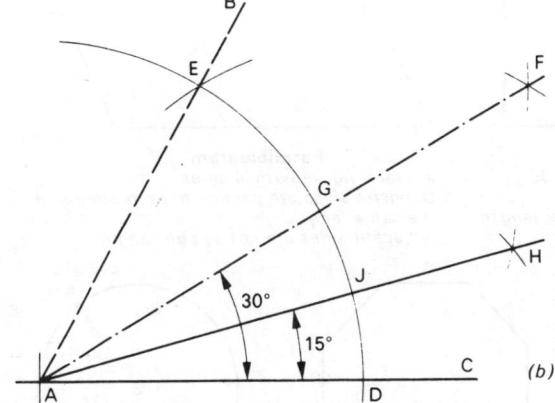

1. Draw AC and AF at 30° to each other as described in Fig. 2.30 (a)
2. With centres G and D draw arcs of equal radius to intersect at H
3. Draw a line from A through H
4. Angle CAH is 15° (half 30°)
5. With centres J and D a further bisection can be made as described in 2, 3 and 4 above. This would give angle of 7½°

(b) Construction of angles of 15° and 7½°

**Fig. 2.30 Bisection of angles**

Some of these figures, which are to be constructed in this section of the chapter, are described in Figure 2.31.

The basic constructions associated with these figures are given in the following illustrations:

1. Triangle — Figure 2.32
2. Square — Figure 2.33
3. Rectangle — Figure 2.34
4. Parallelogram — Figure 2.35
5. Pentagon — Figure 2.36
6. Hexagon — Figure 2.37
7. Octagon — Figure 2.38
8. Circle — Figure 2.39

bisection of this angle to produce 15° which is bisected to produce 7½° is shown in Figure 2.30.

## 2.13 Construction of plane figures

All the regular plane figures met within fabrication engineering can be set out by geometrical constructions, using compasses and set squares.

**Triangle**
A plane figure with 3 sides

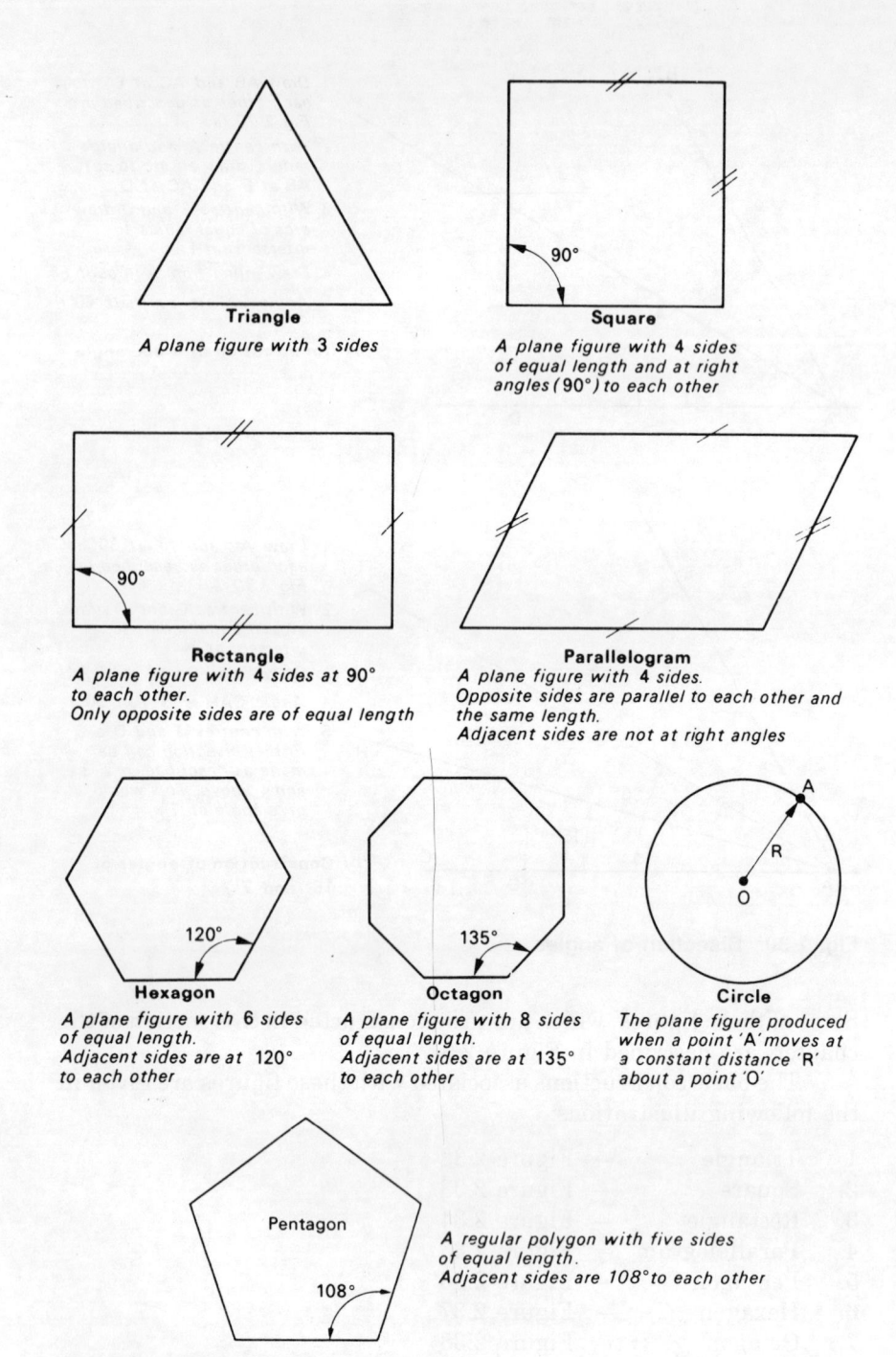

**Square**
A plane figure with 4 sides of equal length and at right angles (90°) to each other

**Rectangle**
A plane figure with 4 sides at 90° to each other.
Only opposite sides are of equal length

**Parallelogram**
A plane figure with 4 sides.
Opposite sides are parallel to each other and the same length.
Adjacent sides are not at right angles

**Hexagon**
A plane figure with 6 sides of equal length.
Adjacent sides are at 120° to each other

**Octagon**
A plane figure with 8 sides of equal length.
Adjacent sides are at 135° to each other

**Circle**
The plane figure produced when a point 'A' moves at a constant distance 'R' about a point 'O'

**Pentagon**
A regular polygon with five sides of equal length.
Adjacent sides are 108° to each other

**Fig. 2.31  Some regular plane figures**

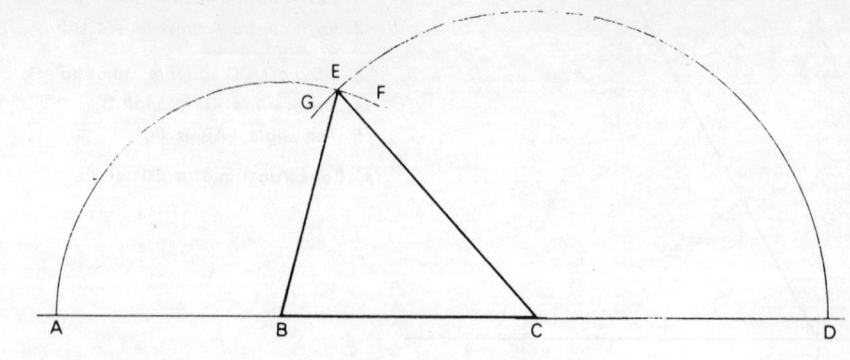

1. Draw AB, BC and CD equal in length to the sides of the required triangle
2. With centre B and radius AB draw the arc AF
3. With centre C and radius CD draw the arc DG
4. Where the arcs intersect at E is an apex of the triangle
5. Join BE and CE with straight lines to form the triangle BCE

**Fig. 2.32  To construct a triangle**

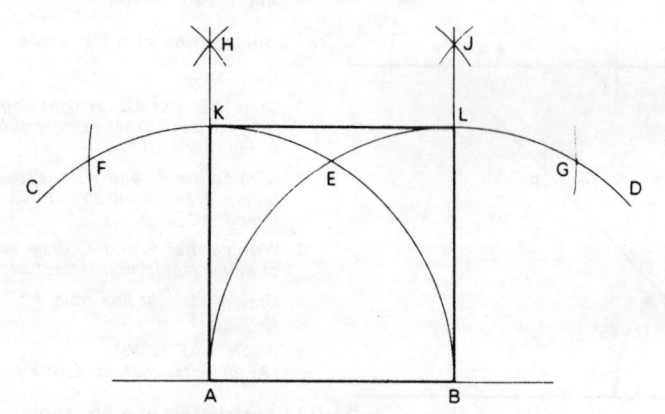

1. Mark off one side of the square AB on the base line
2. With centre 'A' and radius AB draw the arc BC
3. With centre 'B' and radius AB draw the arc AD
4. With centre 'E' and radius AB step off 'F' and 'G' on arcs BC and AD respectively
5. With centres 'E' and 'F' draw arcs of equal radius to intersect at H
6. With centres 'E' and 'G' draw arcs of equal radius to intersect at J
7. Erect perpendiculars AH and BJ
8. The arcs BC and AD cut the perpendiculars AH and BJ at 'K' and 'L' respectively
9. To complete the square join 'K' and 'L'

**Fig. 2.33  To construct a square**

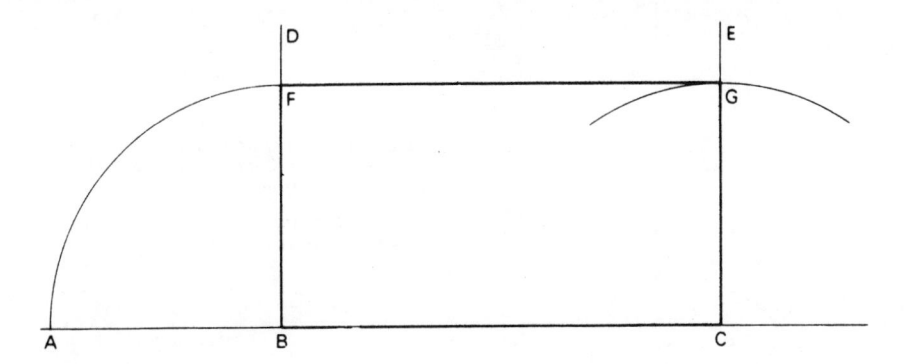

1. Mark off the lengths of a short side AB and a long side BC on the base line
2. Using the constructions demonstrated in the previous examples, or using a set square for simplicity, erect perpendiculars BD and CE at 'B' and 'C'
3. With centre 'B' and radius AB draw an arc to cut BD at 'F'
4. With centre 'C' and radius AB draw an arc to cut CE at 'G'
5. To complete the rectangle join F and G with a straight line

**Fig. 2.34  To construct a rectangle**

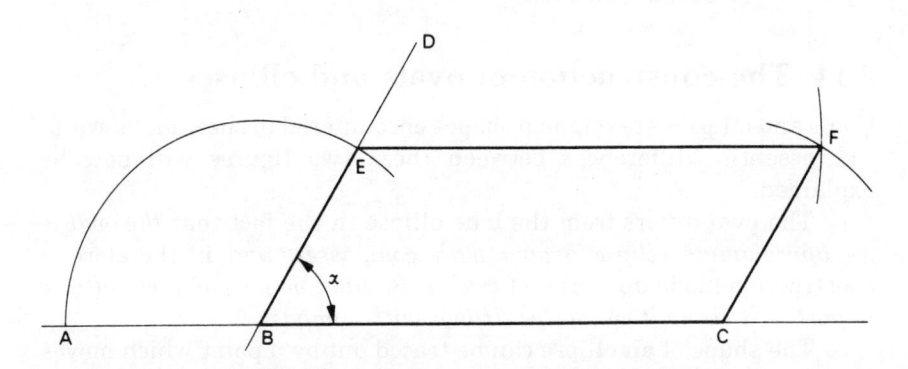

Method of construction
1. With centre O and compasses set to required radius draw the given circle
2. Draw a vertical diameter AB and a tangent at A.
3. With centre A and any suitable radius draw a semi-circle
4. With the compass divide the semi-circle into five equal parts (ie. same number as sides in the pentagon) and number 0–5
5. From tangent point A draw lines through the points 1, 2, 3 and 4 on the semi-circle to cut the given circle at C, D E and F
6. B Complete the pentagon by joining CDEF

**Note:** This method of construction may be used to construct any regular within a circle. The semi-circle is divided into the same number of equal parts as there are sides in the polygon

**(a) To construct a pentagon in a given circle**

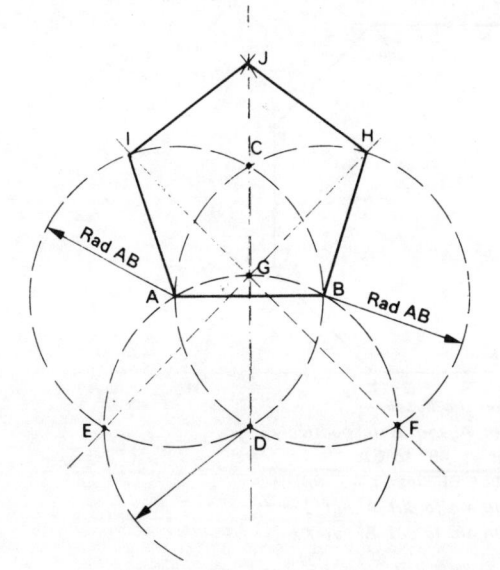

1. Mark off one short side AB and one long side BC on the base line
2. Draw BD at the required angle α to the base line using an adjustable set square or protractor
3. With centre 'B' and radius AB draw an arc to cut BD at 'E'
4. With centre 'C' and radius AB draw an arc
5. With centre 'E' and radius BC draw an arc to cut the arc drawn in (4) at F
6. To complete the parallelogram join EF and CF with straight lines

**Fig. 2.35  To construct a parallelogram**

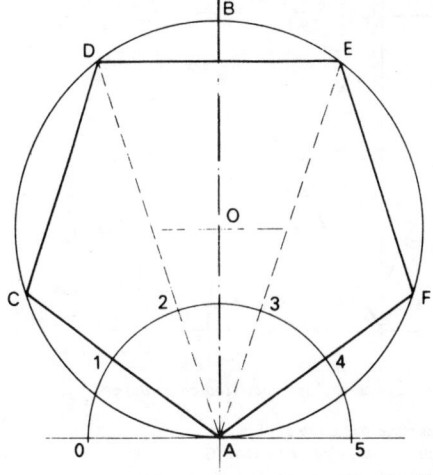

Method of construction
1. Draw the given side AB
2. With centres A and B and compasses set to radius AB, draw two circles to intersect at C and D
3. Join CD
4. With centre D and radius AB, draw an arc to cut the two previously drawn circles at E and F, and CD at G
5. From E and F draw lines through point G to cut the two circles at H and I
6. With radius AB and centres H and I draw arcs to interest at J on DC produced
7. Join AIJHB to complete the pentagon

**(b) To construct a pentagon given the length of one side**

**Fig. 2.36  Two methods of constucting a pentagon**

53

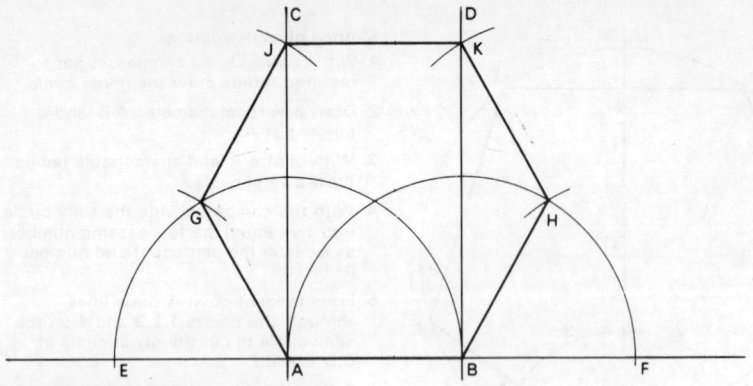

1. Mark off one side AB on the baseline
2. Erect perpendiculars AC and BD using one of the previously demonstrated constructions or set square
3. With centre 'A' and radius AB draw an arc to cut the base line at 'E'
4. With centre 'E' and radius AB draw an arc to cut the previous arc (3) at 'G'
5. With centre 'B' and radius AB draw an arc to cut the base line at 'F'
6. With centre 'F' and radius AB draw an arc to cut the previous arc (5) at 'H'
7. With centre 'G' and radius AB draw an arc to cut the perpendicular AC at 'J'
8. With centre 'H' and radius AB draw an arc to cut the perpendicular BD at 'K'
9. To complete the hexagon join AG, GJ, JK, KH and HB with straight lines

**Fig. 2.37 To construct a hexagon**

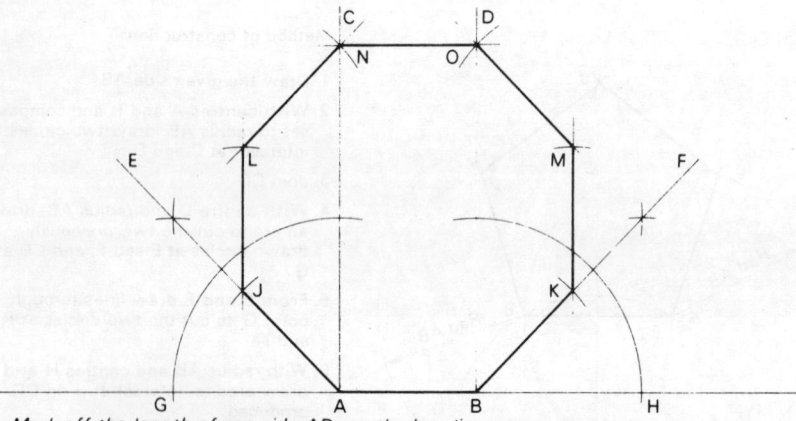

1. Mark off the length of one side AB on the baseline
2. Erect perpendiculars AC and BD from 'A' and 'B' respectively
3. Bisect the angle CAG so that EA lies at 45° to GA
4. Similarly, bisect the angle DBH so that BF lies at 45° to BH
5. With centre 'A' and radius AB draw an arc to cut AE at J
6. With centre 'B' and radius AB draw an arc to cut BF at K
7. Erect perpendiculars at 'J' and 'K'
8. With centres 'J' and 'K' and radius AB draw arcs to cut the perpendiculars at L and M respectively
9. With centres 'L' and 'M' and radius AB draw arcs to cut AC and BD at 'N' and 'O' respectively
10. To complete the octagon join AJ, JL, LN, NO, OM, MK and KB with straight lines

**Fig. 2.38 To construct an octagon**

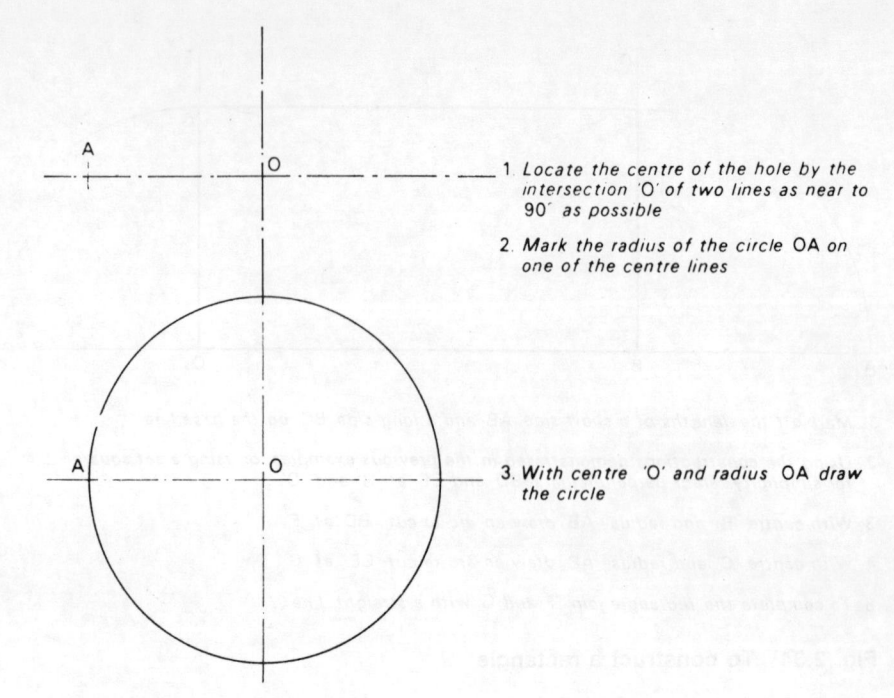

1. Locate the centre of the hole by the intersection 'O' of two lines as near to 90° as possible
2. Mark the radius of the circle OA on one of the centre lines
3. With centre 'O' and radius OA draw the circle

**Fig. 2.39 To construct a circle**

## 2.14 The construction of ovals and ellipses

Ovals and ellipses are common shapes encountered in sheet metal work. The essential differences between these two figures will now be explained.

The oval differs from the true ellipse in the fact that *the oval is an approximate ellipse drawn with compasses*, and is therefore a construction made up of arcs of circles. *Because no part of a true ellipse is part of a circle it cannot be drawn with compasses.*

The shape of an ellipse can be traced out by a point which moves so that the sum of its distances from two fixed points on the major axis is always constant. These fixed points are called the 'foci' or 'focal points' of the ellipse. This fundamental principle is illustrated in Figure 2.40.

The ELLIPSE is symmetrical about two AXES which bisect each other at 90°. The longer 'axis' is termed the MAJOR AXIS (or major diameter), and the shorter 'axis' is termed the MINOR AXIS (or minor diameter).

The path of an ellipse may be traced out by a point which moves from two fixed points on the major axis (termed FOCI) in such a way that the sum of its distances from the foci is always constant.

**Fig. 2.40 Principle of the ellipse**

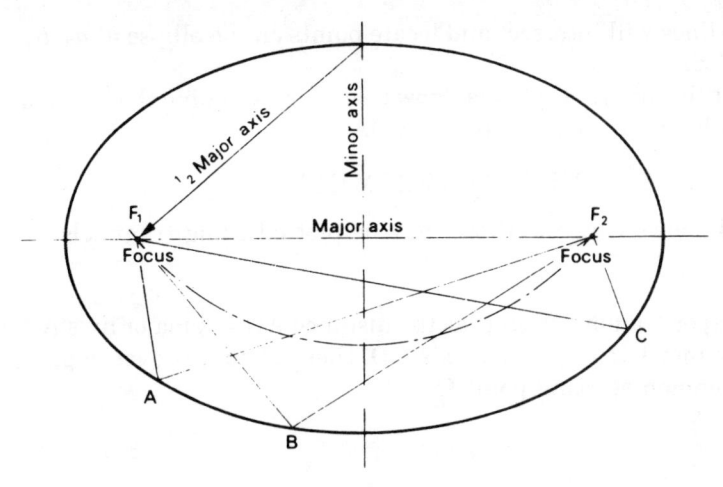

Note: Lines $F_1AF_2$, $F_1BF_2$ and $F_1CF_2$ are all the same length

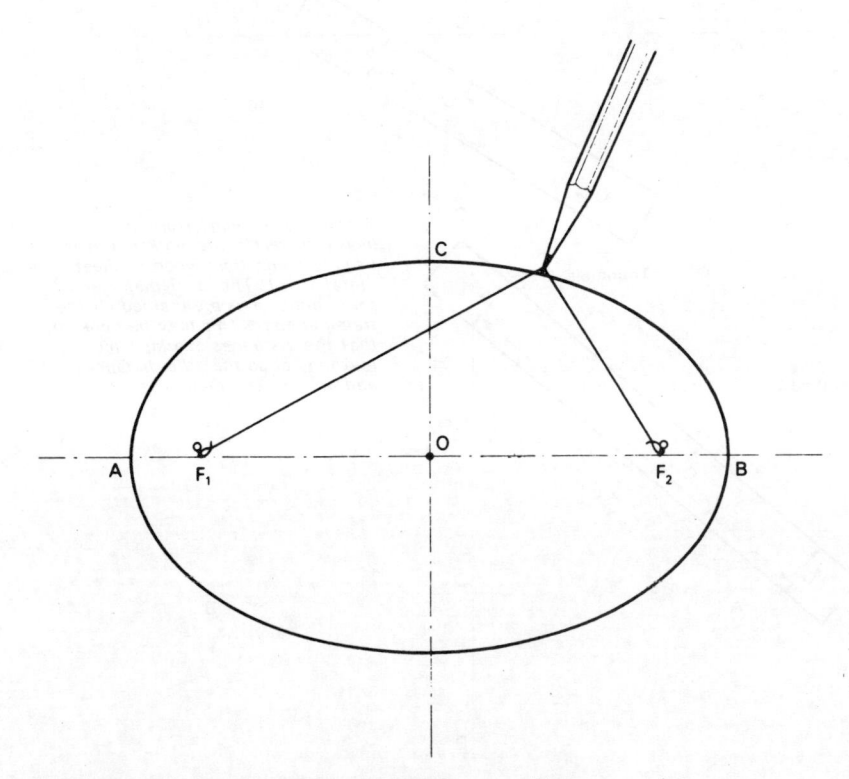

**Fig. 2.41 Practical method of drawing an ellipse**

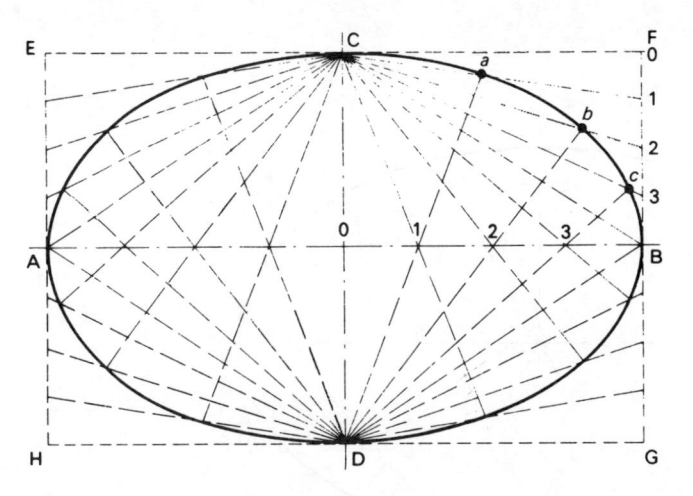

**Fig. 2.42 Construction of an ellipse — rectangle method**

*Method of construction*

1. Draw the major and minor axes.
2. With centre C and radius equal to AO ($\frac{1}{2}$ major axis) draw arcs to locate the focal points $F_1$ and $F_2$ on the major axis AB.
3. Fix pins in the focus points. Make two loops in a piece of cotton such that its length (including loops) is exactly equal to the major axis. Slip the loops over the locating pins.
4. Place the pencil point so that the cotton is stretched taut, as shown in the diagram. The path of the ellipse may now be traced by moving the pencil point whilst keeping the cotton taut.

*Method of construction*

1. Draw the major and minor axis AB and CD.
2. Draw the rectangle EFGH whose sides are equal to the length of the axis.
3. Divide the major axis AB into any equal number of parts, and divide the ends of the rectangle (EH and FG) into the same number of equal parts.

Note: For clarity only one-quarter of the rectangle has been numbered.

4. From each end of the minor axis CD, draw radial lines (as shown opposite), into points 1, 2 and 3.
5. Draw radial lines from C and D through the points 1, 2 and 3 on the major axis AB; these will intersect the previous radial lines at the points *a*, *b* and *c*.

55

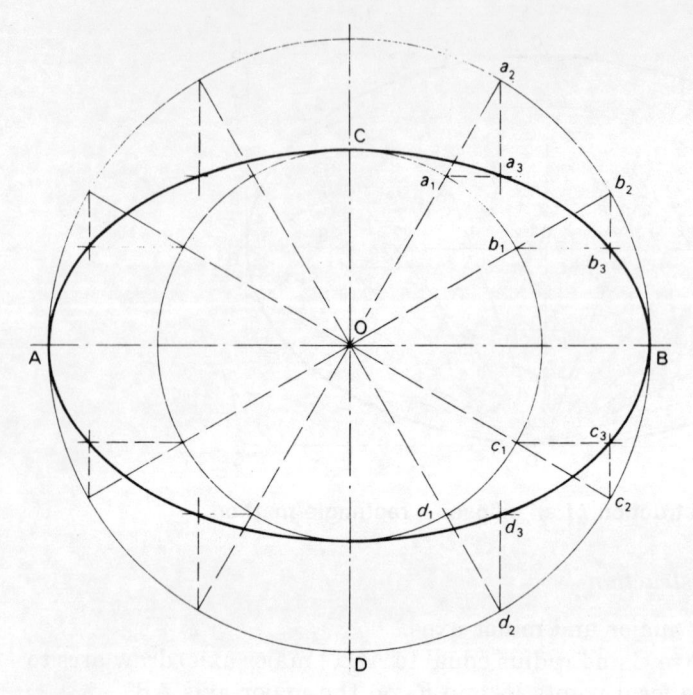

**Fig. 2.43 Construction of an ellipse — concentric circle method**

6. The outline of the ellipse may now be drawn by joining the location points between AC, CB, BD and DA.

## ELLIPSE CONSTRUCTION — CONCENTRIC CIRCLE METHOD

Figure 2.43 shows how an ellipse may be constructed by the concentric circle method.

*Method of construction*

1. Draw the major and minor axes, AB and CD. These intersect at O.
2. Using each axis as a diameter, with centre O, draw two CONCENTRIC CIRCLES as shown.
3. Using a 60°/30° set square in conjuction with a Tee square draw four diameters as indicated by $a$, $b$ $c$ and $d$, on the minor circle and $a_2$, $b_2$ and $d_2$, on the major circle.
4. From $a_2$, $b_2$, $c_2$ and $d_2$, draw vertical lines towards the major axis AB.
5. From $a$, $b$, $c$ and $d$, draw horizontal lines towards the minor axis CD.

6. These lines will intersect and locate points on the ellipse at $a_3$, $b_3$, $c_3$ and $d_3$.
7. The outline of the ellipse may now be drawn by joining the location points between AC, CB, BD and DA.

## PLOTTING AN ELLIPSE USING TRAMMELS

Figure 2.44 shows how an ellipse may be plotted using trammels.

*1st method*

Mark the paper trammel such that the distance AO = $\frac{1}{2}$ major axis AB and the distance CO = $\frac{1}{2}$ minor axis CD, these distances overlapping having a common starting point O.

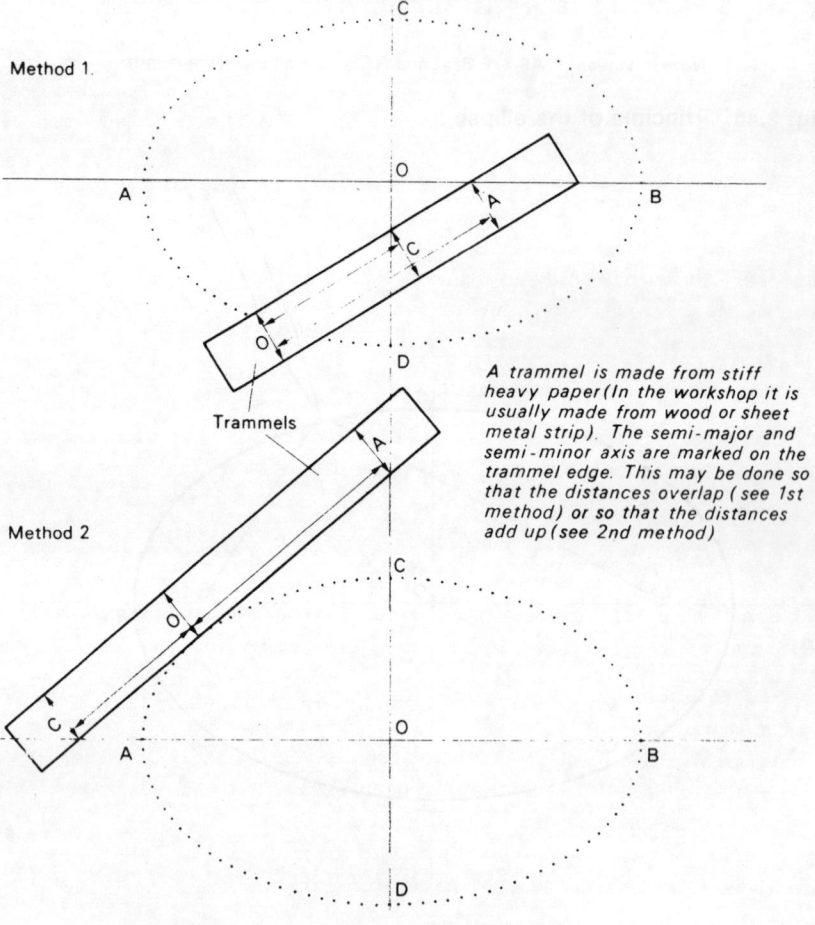

A trammel is made from stiff heavy paper (In the workshop it is usually made from wood or sheet metal strip). The semi-major and semi-minor axis are marked on the trammel edge. This may be done so that the distances overlap (see 1st method) or so that the distances add up (see 2nd method)

**Fig. 2.44 The plotting of an ellipse using trammels**

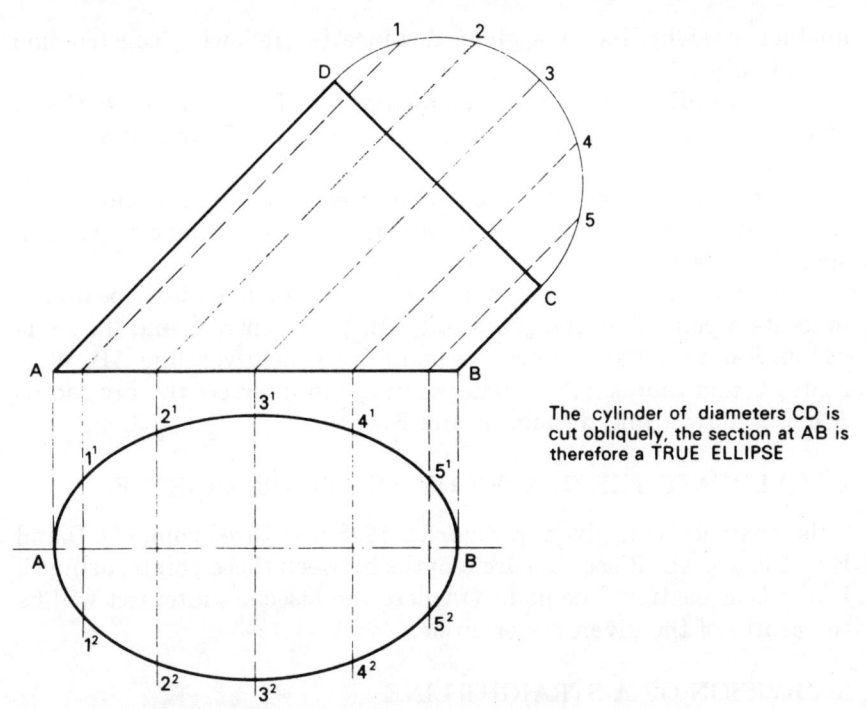

**Fig. 2.45** Construction of an ellipse by projection

1. Draw the major and minor axes AB and CD.
2. The semi-major point A is placed on the major axis and the semi-minor point C is placed on the minor axis. The point O on the trammel will locate one point on the ellipse (as shown). Additional points are located by moving the trammel keeping A and C on the axis.

*2nd method*

Mark the paper trammel such that the distances AO and CO representing the $\frac{1}{2}$ major and $\frac{1}{2}$ minor axes add up.

1. Draw the major and minor axes AB and CD and extend them.
2. The points A and C are positioned on the axis as shown above, and the point O on the trammel locates one point on the ellipse. Additional points for plotting the ellipse are obtained by moving the trammel keeping A and C on the axis.

ELLIPSE CONSTRUCTION — PROJECTION

Figure 2.45 shows how an ellipse may be projected.

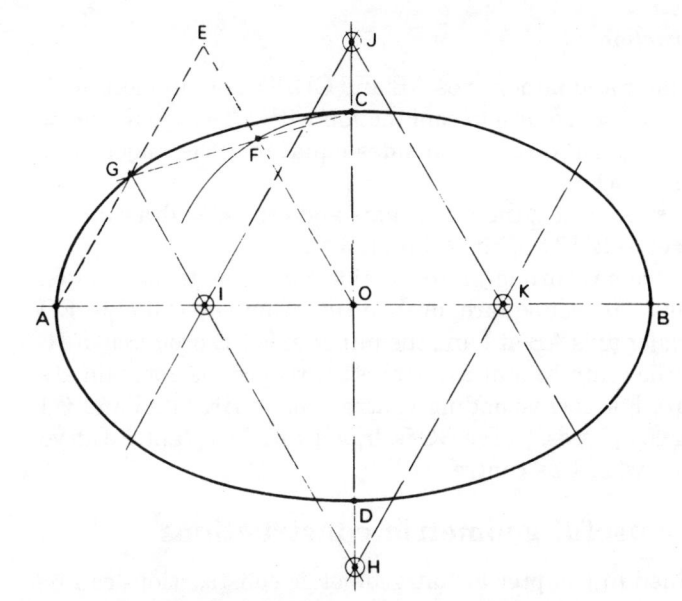

**Fig. 2.46** Construction of an oval (approximate ellipse)

*Method of projecting the ellipse*

1. Draw a semi-circle on the diameter CD and divide it into six equal parts.
2. From the points on the semi-circle project lines square to CD and through it back to the baseline AB.
3. In any convenient position below AB, draw the major axis of the ellipse $A^1B^1$ parallel to it.
4. Project lines from the points on the baseline AB and square to it through the major axis for plotting the ellipse as shown.
5. Mark off the various widths at 1, 2, 3, 4 and 5 on the semi-circle and transfer them above and below the major axis as points for plotting the ellipse.
6. The curve for the ellipse may now be drawn through points $A_1$, $1_1$, $2_1$, $3_1$, $4_1$, $5_1$, $B_1$, $5_2$, $4_2$, $3_2$, $2_2$, $1_2$ and $A_1$.

This method of constructing a true ellipse is used in pattern development.

OVAL CONSTRUCTION

Figure 2.46 shows how an oval (approximate ellipse) may be constructed.

57

*Method of construction*

1. Draw the major and minor axes AB and CD. These intersect at O.
2. Using a 60°/30° set square in conjunction with a Tee square, draw an EQUILATERAL TRIANGLE with sides equal to $\frac{1}{2}$ the major axis, as shown at AEO.
3. With compasses set to $\frac{1}{2}$ the minor axis and centre O, draw an arc from C to cut side EO of the triangle at F.
4. Draw a line from C through F to cut side EA of the triangle at G.
5. Using the 60°/30° set square, draw a line from G parallel to EO to cut the major axis AB at I and the minor axis CD produced at H.
6. H and I are the centres for drawing arcs to produce the approximate ellipse (oval). For corresponding centres, mark BK = AI and OJ = OH using compasses. Curve GC is drawn with H as centre. Curve GA is drawn with I as centre.

## 2.15 Further useful geometric constructions

It will be explained in Chapter 5 that geometric constructions can be used to simplify the marking out of geometric components. In addition to the plane figures just described (Sections 2.13 and 2.14) the following constructions are also of great use to the craftsman.

### 1 PARALLEL LINES

Figure 2.47 shows how parallel lines may be drawn.

It can be seen from Figure 2.47 (*a*) that to draw a line parallel to another straight line at a given distance the following construction should be used.

With radius $R$ (=to required distance) and any two points C and D on the given line AB, draw arcs. Draw the line EF tangential to the two arcs.

It can be seen from Figure 2.47 (*b*) that to draw a line parallel to a given straight line from a point outside it, the following construction should be used.

With the centre at the point P, and any radius R, describe an arc to locate a point C on the given line AB. With centre C and the same radius $R$ draw an arc to locate a point D on the given line AB. With centre C and radius $r$ (= DP) draw an arc to intersect the arc radius $R$ at E. Draw a line through P and E.

### 2 TO LOCATE THE CENTRE OF AN ARC OR A CIRCLE

This construction is given in Figure 2.48. Select three points C, D and E on the arc AB. Bisect the arc lengths between these points, using C, D and E as centres. The point O where the bisectors intersect will be the centre of the given arc or circle.

### 3 DIVISION OF A STRAIGHT LINE

When drawing, it is often more convenient and accurate to divide lines by geometrical constructions rather than by direct measurement. Figure

*(a)* To draw a line parallel to another straight line at a given distance

*(b)* To draw a line parallel to a given straight line from a point outside it

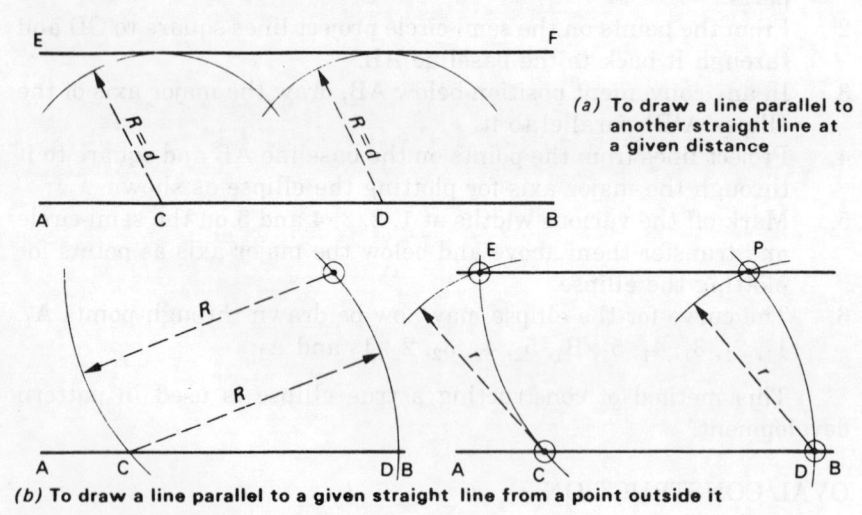

**Fig. 2.47 Construction of parallel lines**

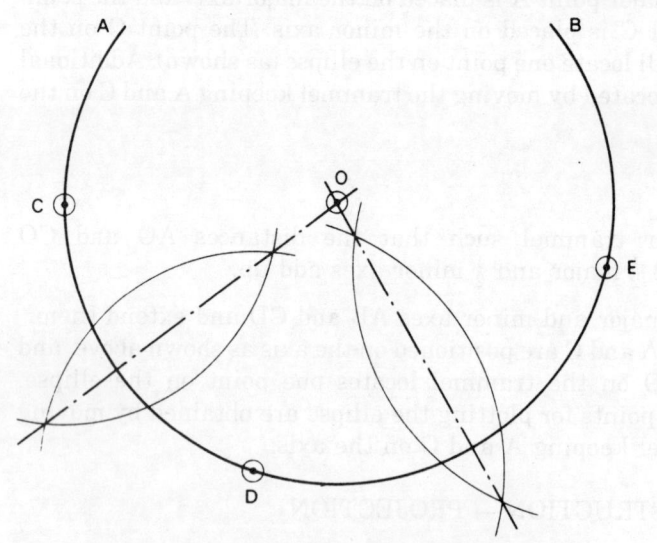

**Fig. 2.48 To locate the centre of a given arc or circle**

2.49(a) shows how a straight line may be bisected with the use of compasses. The method of dividing a straight line into any number of equal parts using a Tee square, set square and compasses, is illustrated in Figure 2.49(b).

(1) To bisect a line means to cut it into two equal parts
(2) Set your compasses to a radius that is greater than half the length of the line AC (i.e. greater than AB or BC)
(3) With centre 'A' draw an arc. With centre 'C' draw another arc. The arcs should cross (intersect) at 'D' and 'E'.
(4) Join 'D' to 'E' and where DE cuts the line AC at 'B' is the centre of AC (i.e. AB=BC)
(5) DE will be at RIGHT ANGLES to AC
   This construction can be used for drawing mutually perpendicular lines

(a) Bisecting a line

(b) Dividing a line into any number of equal parts

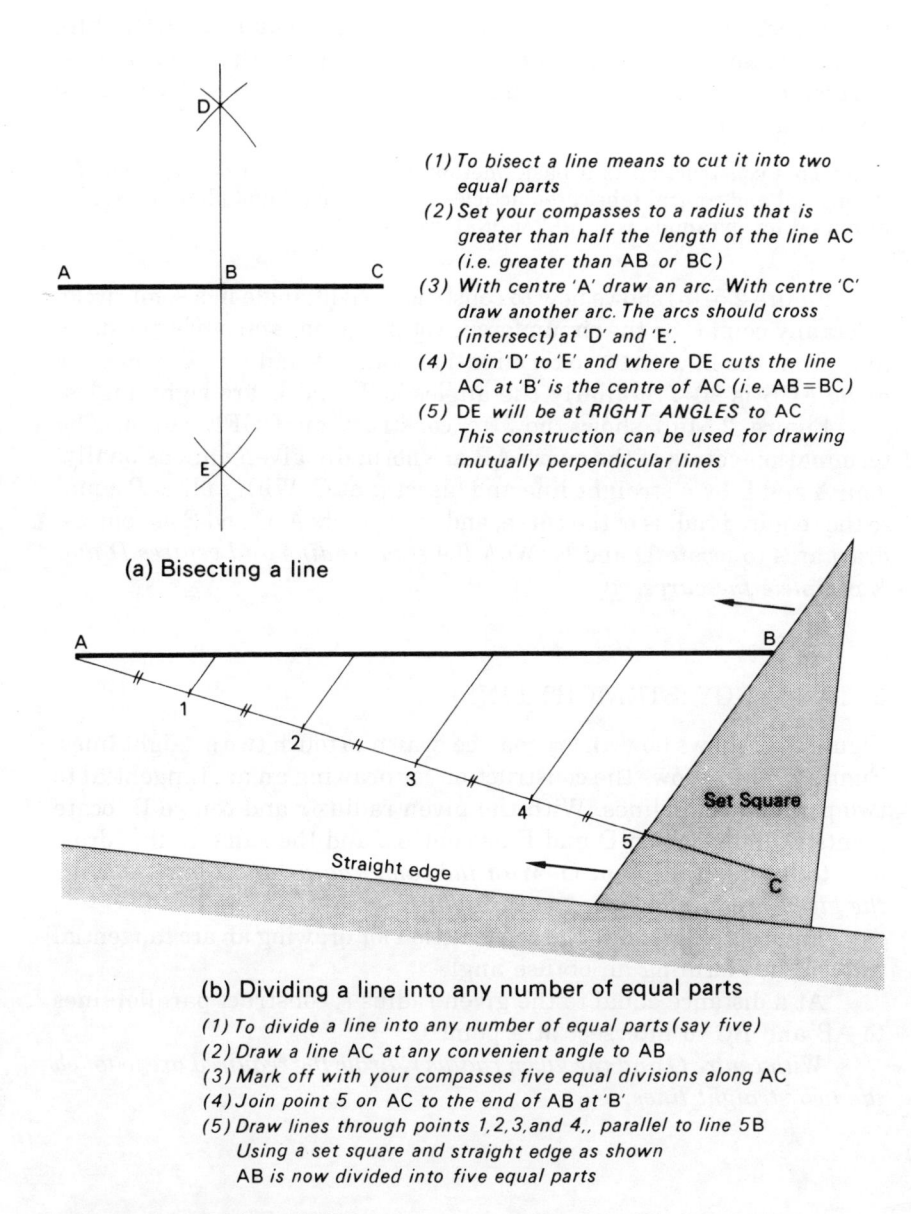

(1) To divide a line into any number of equal parts (say five)
(2) Draw a line AC at any convenient angle to AB
(3) Mark off with your compasses five equal divisions along AC
(4) Join point 5 on AC to the end of AB at 'B'
(5) Draw lines through points 1,2,3, and 4,, parallel to line 5B
   Using a set square and straight edge as shown
   AB is now divided into five equal parts

**Fig. 2.49 Division of lines**

## 4 TO CONSTRUCT A CIRCLE INSIDE OR OUTSIDE A GIVEN TRIANGLE

Figure 2.50(a) shows how a circle is drawn inside (*inscribed*) a triangle. The method is to bisect any two angles (Figure 2.30). The point O where the bisectors intersect is the centre of the circle which is tangential to all three sides of the given triangle.

Figure 2.50(b) shows how a circle is drawn outside (*circumscribed*) a triangle. The method is to bisect any two sides of the given triangle. The point O where the two bisectors intersect is the centre of the circle which may be drawn to past the points A, B and C.

(a) To construct a circle within a given triangle

(b) To construct a circle to circumscribe a given triangle

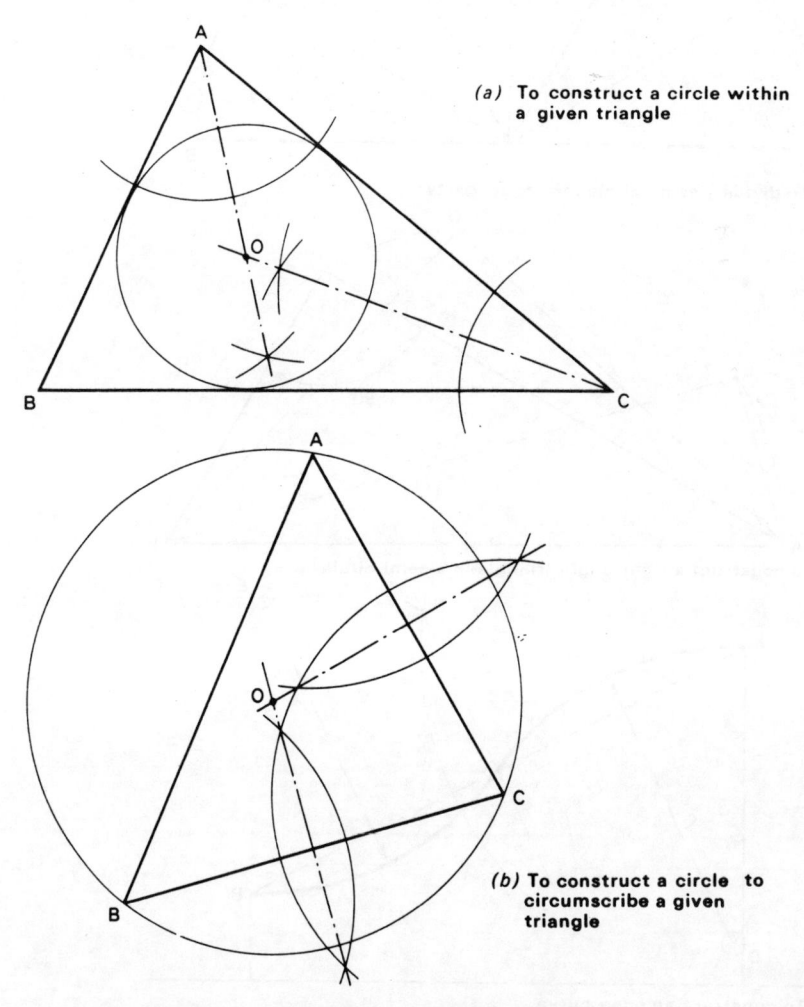

**Fig. 2.50 Inscribed and circumscribed triangles**

## 5 ARCS OF CIRCLES

Figure 2.51 shows various constructions associated with arcs of circles. Figure 2.51(a) shows how to divide a semi-circle into a number of equal parts.

Bisect the DIAMETER AC. With a radius equal to HALF THE DIAMETER, and centres A, D and C draw arcs to cut the semi-circular circumference at points 1, 2, 3 and 4. The semi-circle is now divided into SIX EQUAL PARTS.

Note: This construction is a basic method used in pattern development for cylindrical and conical fabricated articles in sheet metal and plate work. It is also used for pyramids with HEXAGONAL bases.

Figure 2.51(b) shows how to construct a right-angle in a semi-circle. Select any point C on the circumference of the given semi-circle and draw lines from the DIAMETER to termination points A and B to C. *Then the angle ACB is 90°.* Similarly the angles at E and D are right-angles.

Figure 2.51(c) shows how to construct an OGEE curve. The termination points of the curve AB are normally given dimensionally. Join A and B by a straight line and bisect it at C. With radius $R$ equal to the required radius of the curve, and with points A, C and B as centres, draw arcs to locate D and E. *With the same radius and centres D and E complete the curve.*

## 6 TANGENCY (STRAIGHT LINE)

Figure 2.52 shows how an arc may be drawn to touch two straight lines. Figure 2.52(a) shows the construction for drawing an arc tangential to two perpendicular lines. With the given radius $r$ and centre B locate points E and D. With D and E as centres, and the same radius draw arcs to intersect at point O. *With the same radius and centre O draw the given arc to touch the two straight lines.*

Figure 2.52(c) shows the construction for drawing an arc tangential to two lines forming an obtuse angle.

At a distance equal to the given radius $r$, construct parallel lines to AB and BC to intersect at a point O.

*With centre O and the given radius $r$, draw the required arc to touch the two straight lines.*

*(a) To divide a semi-circle into equal parts*

*(b) To construct a right-angle triangle in a semi-circle*

*(c) To construct an ogee curve*

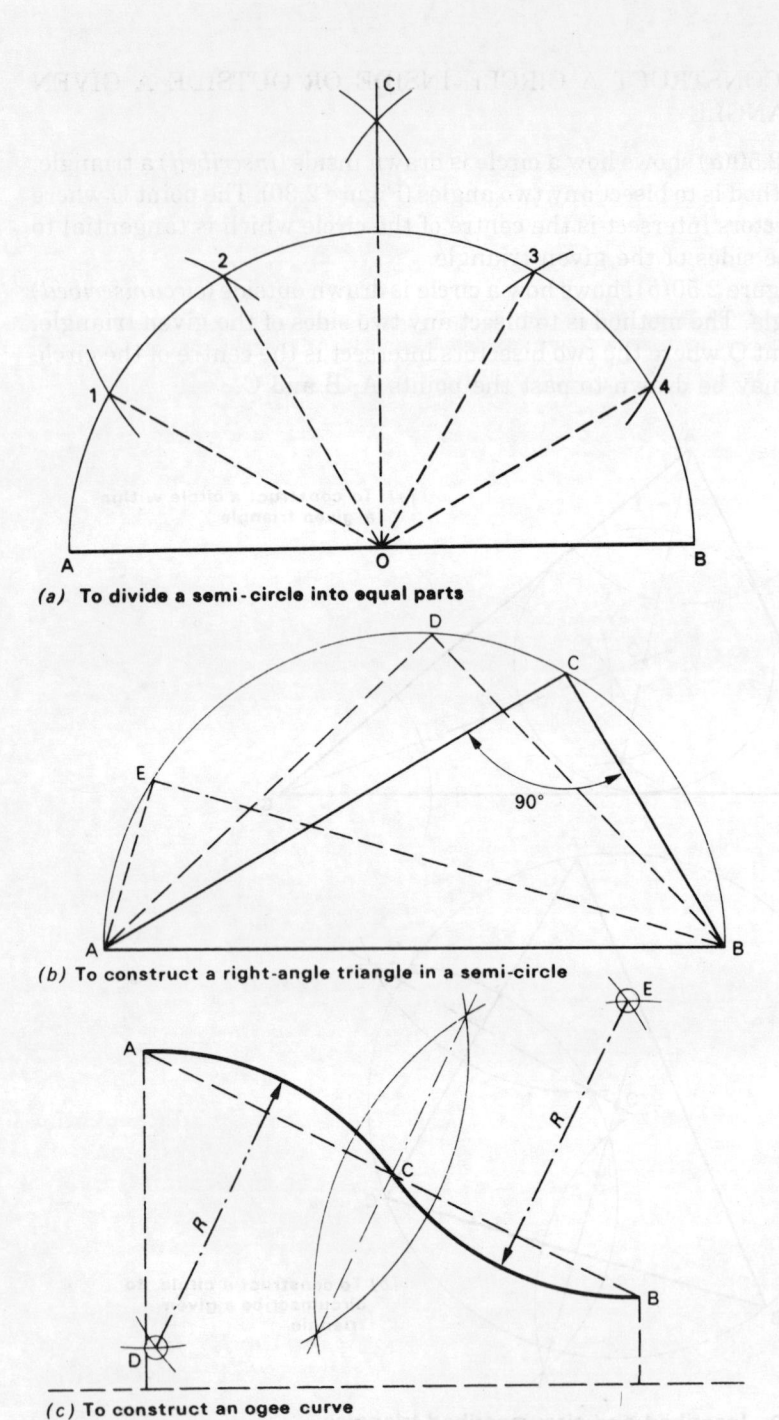

**Fig. 2.51 Constructions associated with arcs of circles**

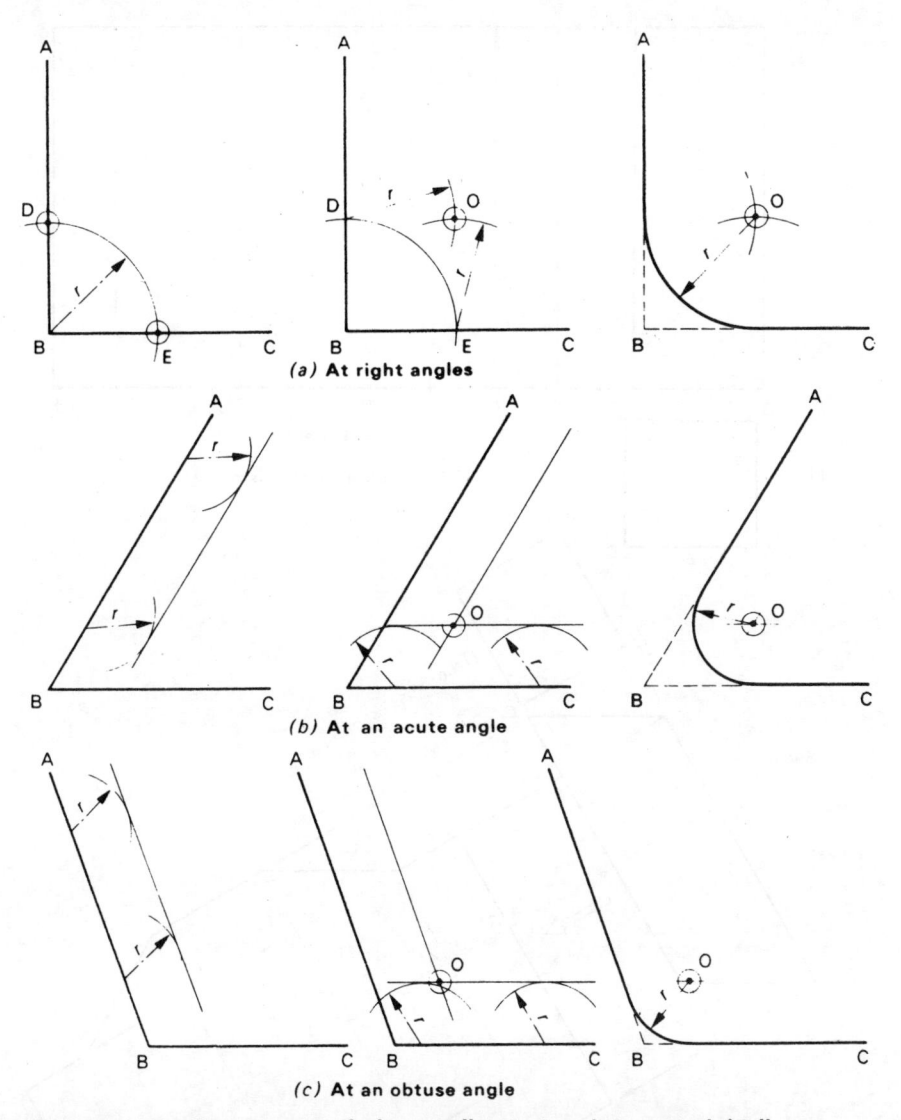

*(a)* At right angles

*(b)* At an acute angle

*(c)* At an obtuse angle

**Fig. 2.52 To draw an arc of given radius to touch two straight lines**

## 7 TANGENTIAL ARCS

Figure 2.53 shows the constructions associated with tangential arcs. Figure 2.53(a) shows the construction for an arc that is tangential to a circle and a straight line. Construct a parallel line, at a distance equal to the given radius $R_1$, to the given straight line AB. Add $R_1$ to the radius $R$ of the given circle and from the centre of the circle draw an

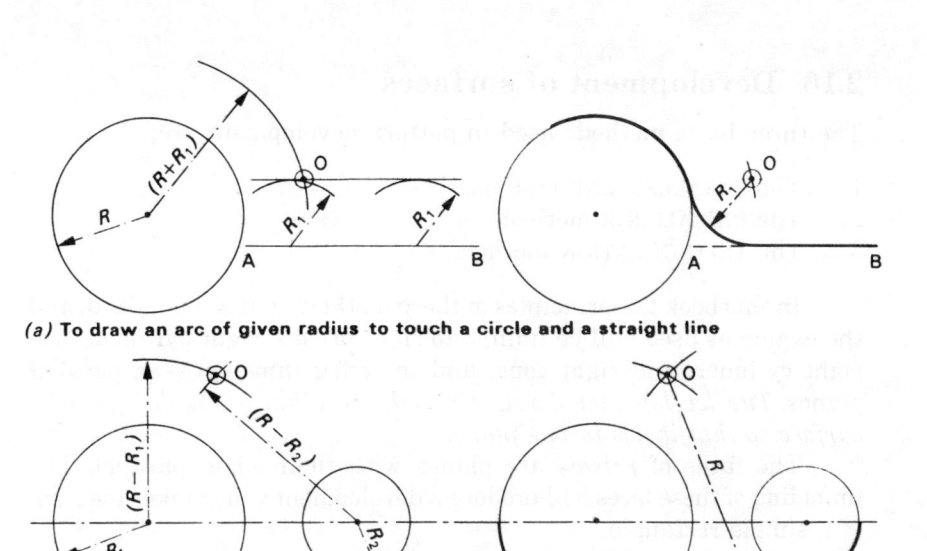

*(a)* To draw an arc of given radius to touch a circle and a straight line

*(b)* To draw an arc of given radius to touch two circles externally

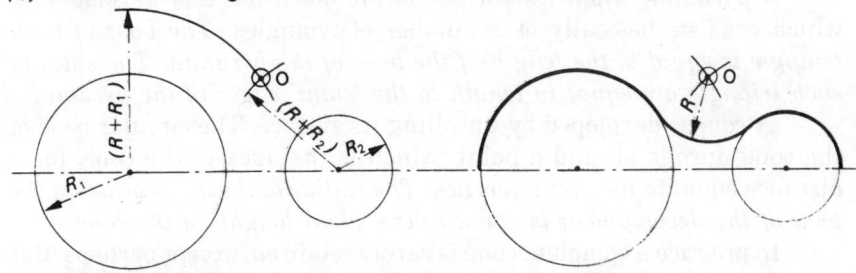

*(c)* To draw an arc of given radius to touch two circles internally

**Fig. 2.53 Tangential arcs**

arc to locate point O. *With centre O and radius $R_1$ draw the required arc to touch the straight line and the circle.*

Figure 2.53(b) shows the construction of an external, tangential arc. Let $R$ represent the radius of the given arc, and $R_1$ and $R_2$ the radii of the circles it is required to touch. Subtract the circle radii from the given radius of the arc, in turn, and from the respective centres of the circles draw arcs to intersect at point O. *With O as centre and radius R draw the required arc to touch the circles externally.*

Figure 2.53(c) shows the construction of an internal, tangential arc. Add the circle radii to the given radius of the arc $R$. From the respective centres of the circles draw arcs to intersect at a point O. *With O as centre and radius R, draw the required arc to touch the circles internally.*

61

## 2.16 Development of surfaces

The three basic methods used in pattern development are:

1. The PARALLEL LINE method.
2. The RADIAL LINE method.
3. The TRIANGULATION method.

In this book the principles of these methods will be explained, and the examples used will be limited to right prisms, right pyramids, the right cylinder, and right cone, and their frustums between parallel planes. *The development of a surface is the unrolling or unfolding of that surface so that it lies in one plane.*

The faces of *prisms* are planes with their edges parallel. The unfolding of these faces will produce a development which takes the form of a simple rectangle.

*A cylinder* is developed by unrolling its surface, thus producing a *rectangle* having *one side equal in length to the 'circumference' of the cylinder*, the other side being equal to the *length or height of the cylinder*.

*A pyramid*, when its surface is unfolded, forms a development which consists basically of a number of triangles. *The base of each triangle is equal to the length of the base of the pyramid. The sides of each triangle are equal in length to the 'slant' edges of the pyramid.*

*A cone* is developed by unrolling its surface. The circular base of the cone unrolls around a point, which is the apex of the cone, for a distance equal to its circumference. *The radius of the arc producing the base of the development is equal to the 'slant height' of the cone.*

In practice a complete cone is rarely required, except perhaps 'flat cones' which are used as 'caps', for example, on stove pipes. However, in the fabrication industry conical sections are constantly required to be manufactured. These components are part cones, often referred to as TRUNCATED CONES. When the cone is cut off parallel with its base, i.e., the top portion removed, the remaining portion is called the FRUSTUM.

## 2.17 Parallel line developments

The 'parallel line method' of pattern development depends upon a principle of locating the shape of the pattern on a series of parallel lines. All articles or components which belong to the class of PRISMS, *which have a constant 'cross-section' throughout their length, may be developed by the parallel line method*.

Elementary examples of parallel line development are illustrated in Figures 2.54 to 2.56 inclusive.

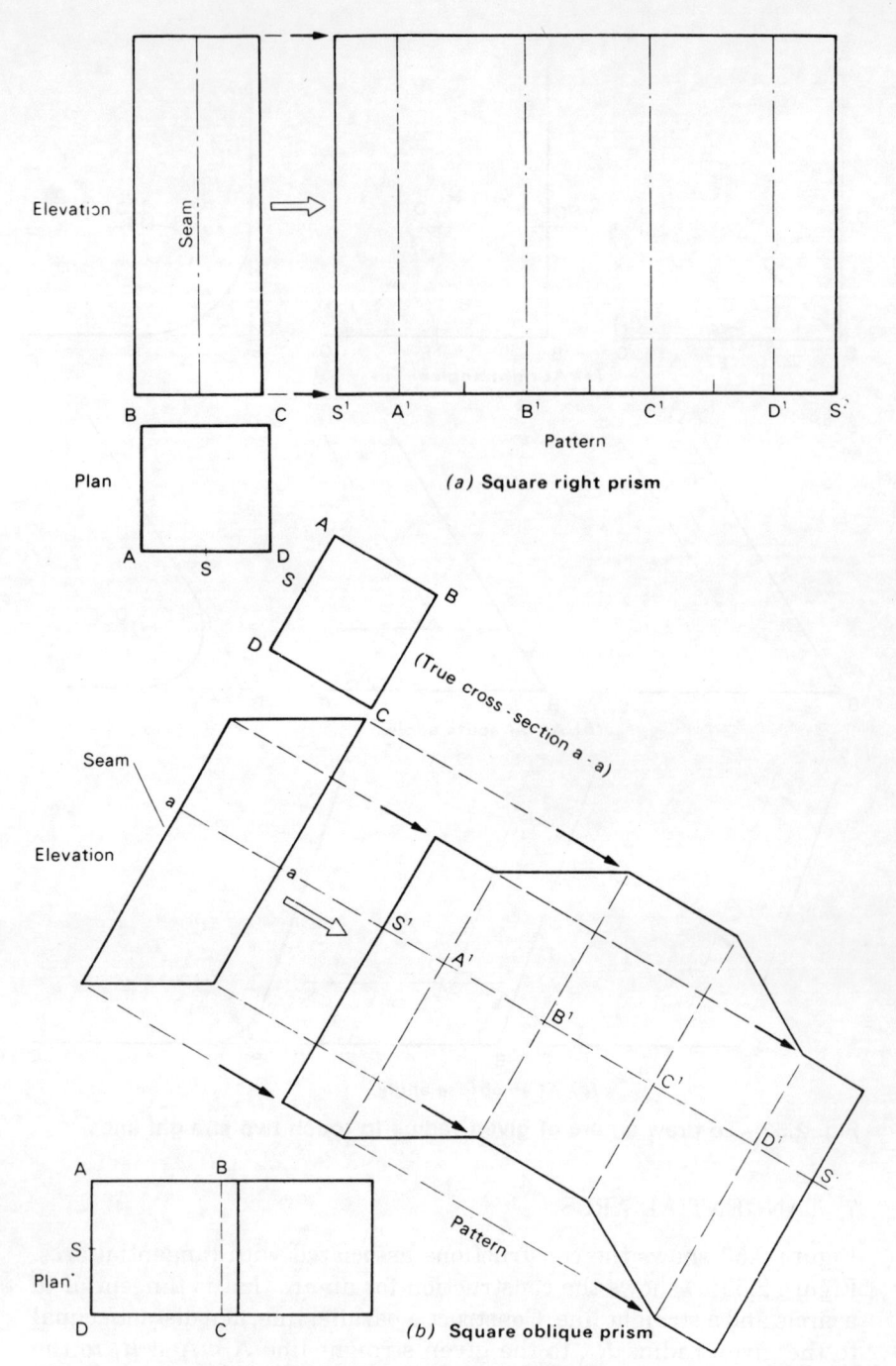

**Fig. 2.54 Developments of square prisms (parallel line)**

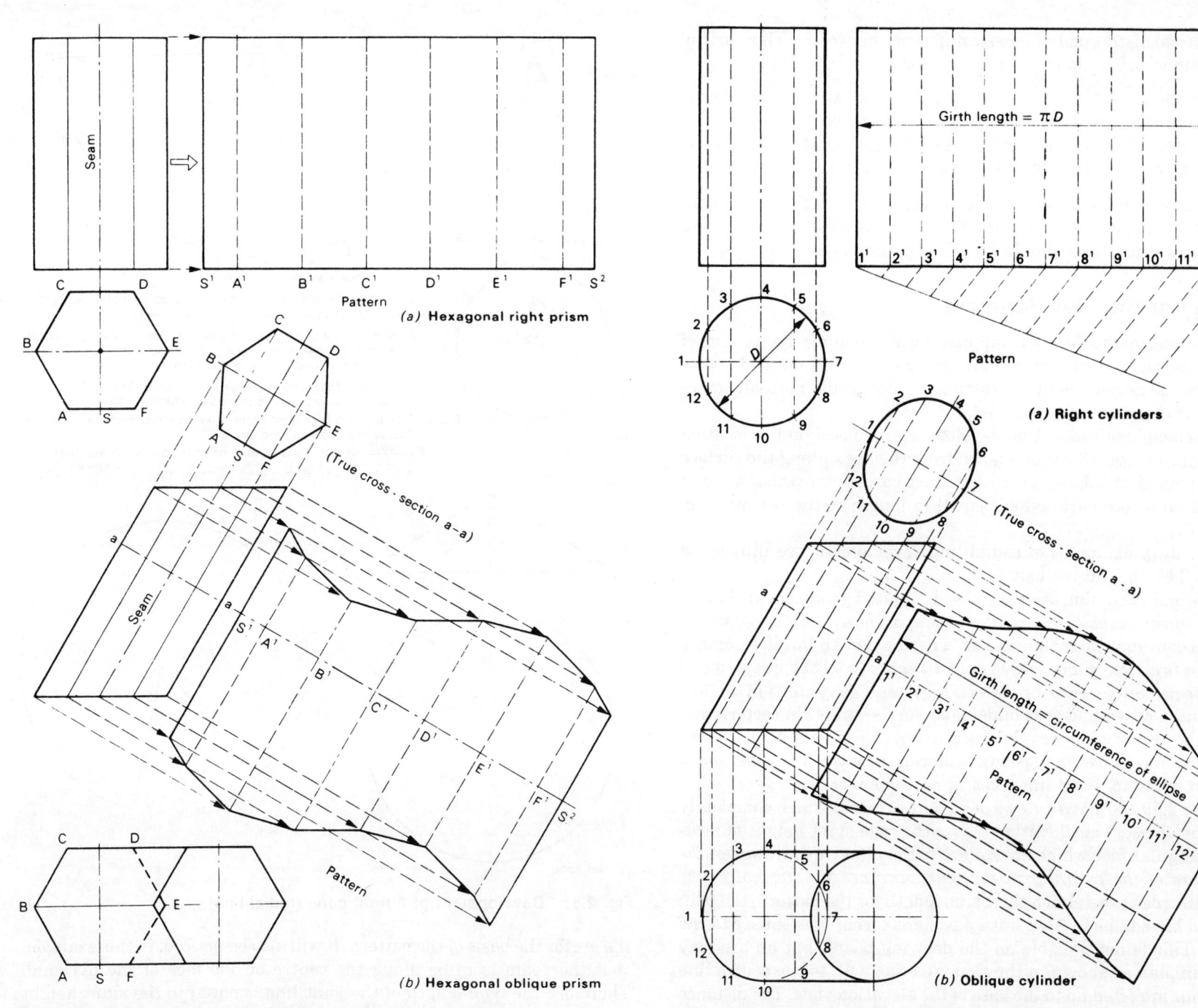

**Fig. 2.55** Developments of hexagonal prisms (parallel line)

*(a) Hexagonal right prism*

*(b) Hexagonal oblique prism*

**Fig. 2.56** Developments of cylinders (parallel line)

*(a) Right cylinders*

*(b) Oblique cylinder*

63

In order to distinguish the basic difference between a 'right prism' and an 'oblique prism' it is essential to apply the basic rule:

IF THE CROSS-SECTION OF THE ENDS BETWEEN PARALLEL PLANES IS AT 90° TO THE AXIS, THE COMPONENT IS A RIGHT PRISM.

IF THE CROSS-SECTION OF THE ENDS BETWEEN PARALLEL PLANES IS NOT NORMAL TO THE AXIS, THE COMPONENT IS AN OBLIQUE PRISM.

*Therefore in order to develop the correct 'stretch-out' for the length of the pattern the distance around the* TRUE CROSS-SECTION *must be used.* The examples illustrated should be self-explanatory.

## 2.18 Radial line development

The 'radial line method' may be applied for developing the pattern of any article or component which *tapers to an apex*. This method is also adaptable to the development of 'frustums' which would normally *taper to an apex if the sides are produced.*

The principle of radial line development is based on the location of a series of lines which radiate down from the apex along the surface of the component to a base, or an assumed base, from which a curve may be drawn whose perimeter is equal in length to the perimeter of the base.

Elementary examples of radial line development are illustrated in Figures 2.57 to 2.60 inclusive.

Cones and pyramids are very closely related geometrical shapes. *A pyramid may be considered as 'a cone with a limited number of sides'.* Similarly a cone may be considered as 'a pyramid of an infinite number of sides'. *In practice, many large conical shapes in heavy gauge metal are often formed on the press brake as if they were many-sided pyramids.*

Although cones and pyramids have very similar characteristics, care must be taken when developing patterns for pyramids. It is very important to recognise one specific difference between a cone and a pyramid in order to avoid mistakes in development.

Figure 2.59 shows two views of a right pyramid which completely describe the object. The elevation shows the true slant height of faces of each triangular face which are square to the plan view. However, *the slant corners of the pyramid in the plan view are not normal to the elevation.* In order to establish their true length for the pattern, the plan view would have to be rotated until one slant corner was square to the elevation. This is not possible on the drawing board, but an arc may be drawn in plan as shown in the figure (radius OB) to the centre line and the point projected up to the base of the elevation view. The distance from the apex to this point will provide the true radius for swinging

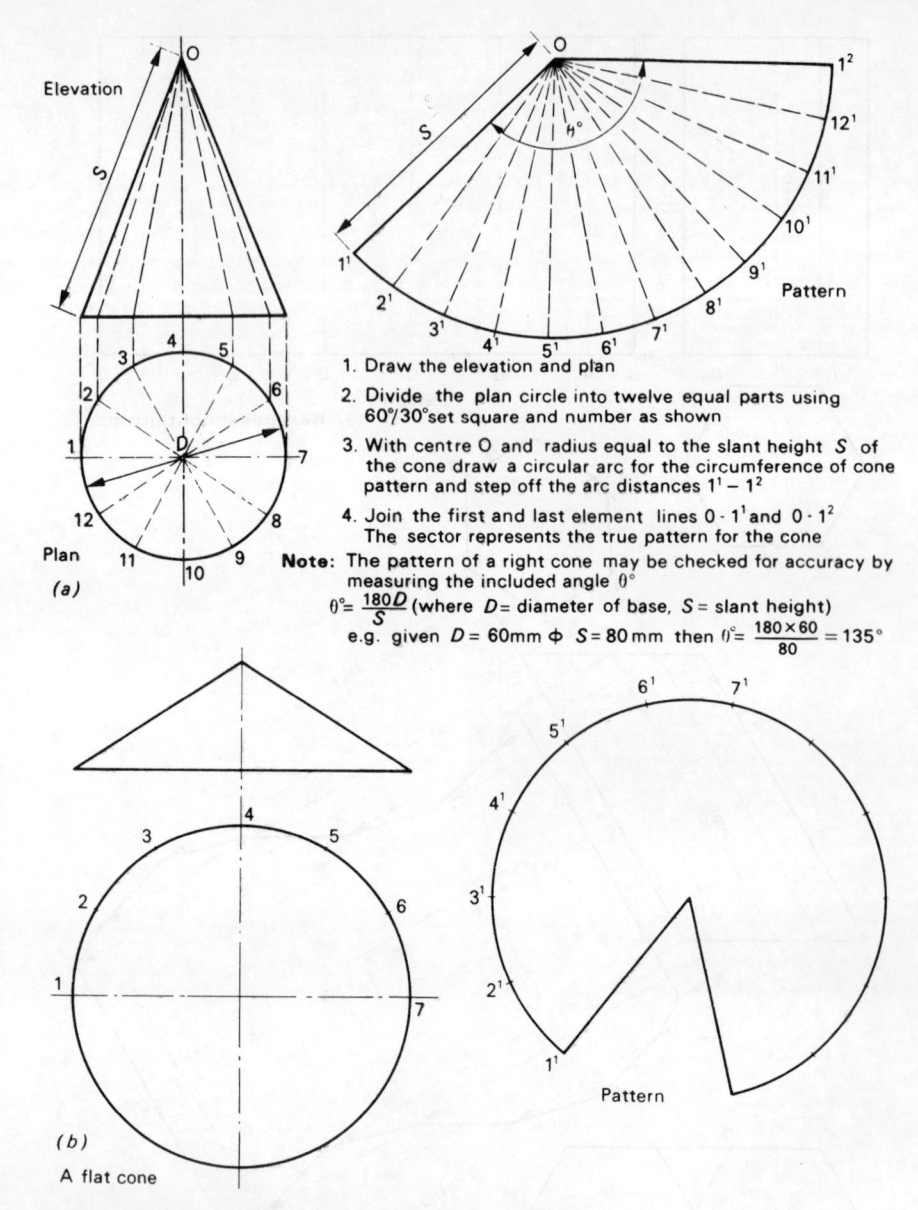

1. Draw the elevation and plan
2. Divide the plan circle into twelve equal parts using 60°/30° set square and number as shown
3. With centre O and radius equal to the slant height S of the cone draw a circular arc for the circumference of cone pattern and step off the arc distances 1¹ – 1²
4. Join the first and last element lines 0 - 1¹ and 0 - 1² The sector represents the true pattern for the cone

Note: The pattern of a right cone may be checked for accuracy by measuring the included angle $\theta°$

$\theta° = \frac{180 D}{S}$ (where $D$ = diameter of base, $S$ = slant height)

e.g. given $D = 60 mm \phi$  $S = 80 mm$ then $\theta° = \frac{180 \times 60}{80} = 135°$

(a)

(b) A flat cone

**Fig. 2.57 Development of a right cone (radial line)**

the arc for the basis of the pattern. It will also be noticed, in this example, that the seam is to be along the centre of one face of the pyramid. Therefore the true length of the joint line is equal to the slant height shown in the elevation. The three full sides are marked off along the

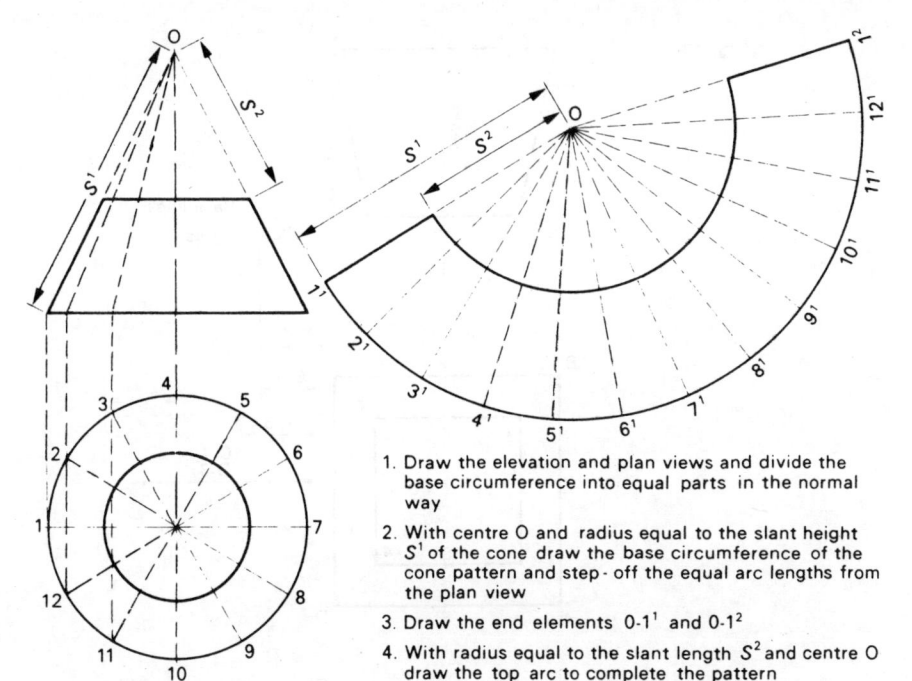

**Fig. 2.58 Development of a right cone frustum (radial line)**

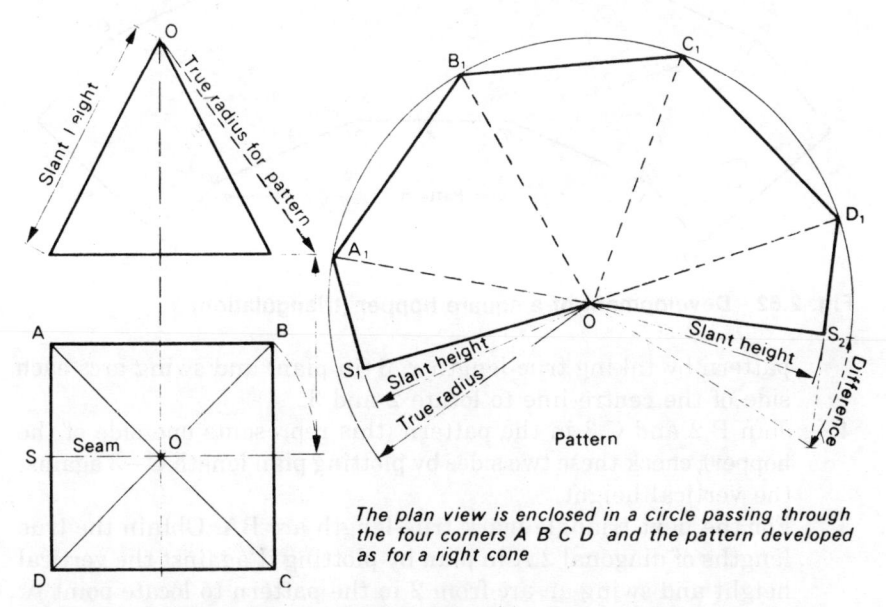

1. Draw the elevation and plan views and divide the base circumference into equal parts in the normal way
2. With centre O and radius equal to the slant height $S^1$ of the cone draw the base circumference of the cone pattern and step-off the equal arc lengths from the plan view
3. Draw the end elements 0-1¹ and 0-1²
4. With radius equal to the slant length $S^2$ and centre O draw the top arc to complete the pattern

**Fig. 2.59 Development of a square-based pyramid (radial line)**

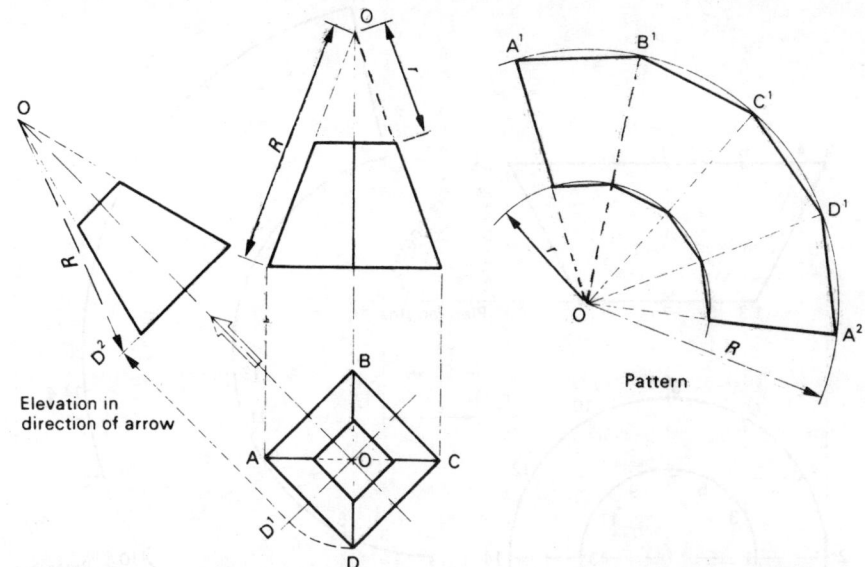

The plan view is enclosed in a circle passing through the four corners A B C D and the pattern developed as for a right cone

**Fig. 2.60 Development of a square-based pyramid frustum (radial line)**

basis curve in the pattern, in the same manner as for a right cone, an arc is swung each end using centre O and radius equal to the slant height, and the last two triangles are completed by swinging arcs from A and D with a radius equal to half the length of one side in the plan. The fundamental difference between the cone and pyramid is clarified in Figure 2.60.

## 2.19 Development by triangulation

Triangulation is by far the most important method of pattern development since a great number of fabricated components transform from one cross-section to another. A typical 'square-to-round' transformer is illustrated in Chapter 5. The basic principle of triangulation is to develop a pattern by dividing the surface of the component into a number of triangles, determine the true size and shape of each, and then lay them down side by side in the correct order to produce a pattern.

*To obtain the true size of each triangle, the true length of each side must be determined and then placed in the correct relationship to the other sides.*

THE GOLDEN RULE OF TRIANGULATION:

'PLACE THE PLAN LENGTH OF A LINE AT RIGHT ANGLES TO ITS VERTICAL HEIGHT; THE DIAGONAL WILL REPRESENT ITS TRUE LENGTH.'

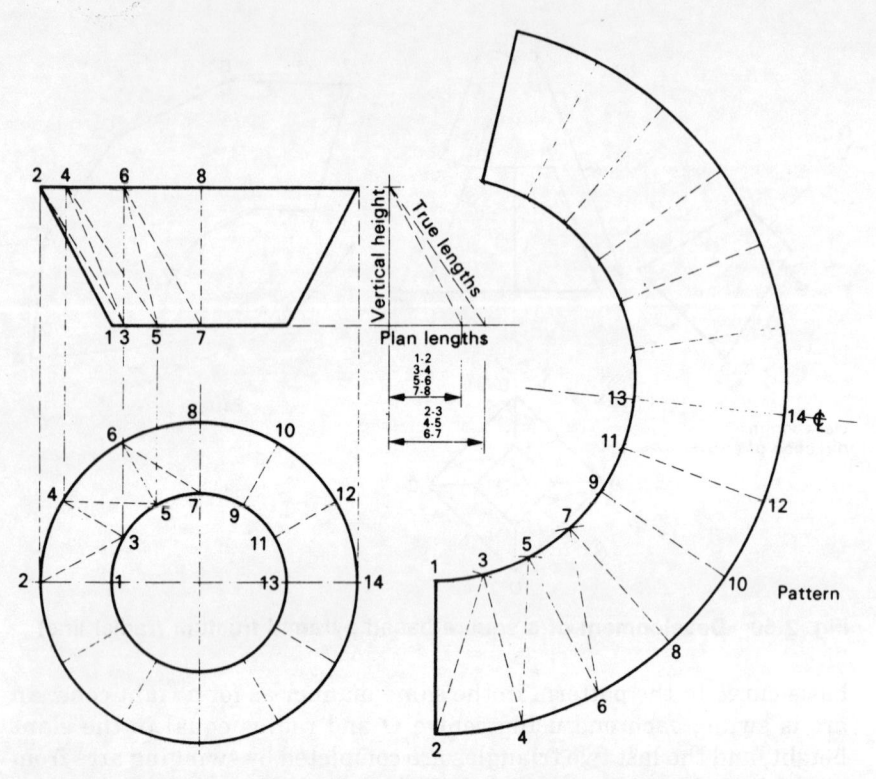

**Fig. 2.61 Development of a truncated cone (triangulation)**

An elementary example of the method of triangulation is shown in Figure 2.61.

A more complex example of triangulation is shown in Figure 2.62.

1. Draw the elevation and plan views. The corner points in plan are lettered A, B, C, D and numbered 1, 2, 3, 4 and 5 (5 denotes the seam).
2. It will be seen that the lengths AB, BC, CD and DA of the large square, and lengths 1,2, 2,3, 3,4 and 4,1 of the small square are TRUE LENGTHS in plan since they lie in the same horizontal plane, and therefore have no vertical height.
3. For the first triangle in the pattern take the true length distance BC (in plan) and mark it in the pattern. Draw a vertical centre line $x,x$. Mark the plan length $x,x$ along the baseline at 90° to the vertical height, and obtain its TRUE LENGTH and mark it on the pattern. Obtain the true length of diagonal $x,3$ in plan and swing arcs from $x$ (on BC) in the pattern. Complete the triangle in the pattern by taking true length $x, 3$ (in plan) and swing arcs each side of the centre line to locate 2 and 3.
4. Join B,2 and C,3 in the pattern (this represents one side of the hopper); check these two sides by plotting plan length C−3 against the vertical height.
5. For the next triangle mark true length arc BA. Obtain the true lengths of diagonal 2,A in plan by plotting it against the vertical height and swing an arc from 2 in the pattern to locate point A. Join BA in the pattern.

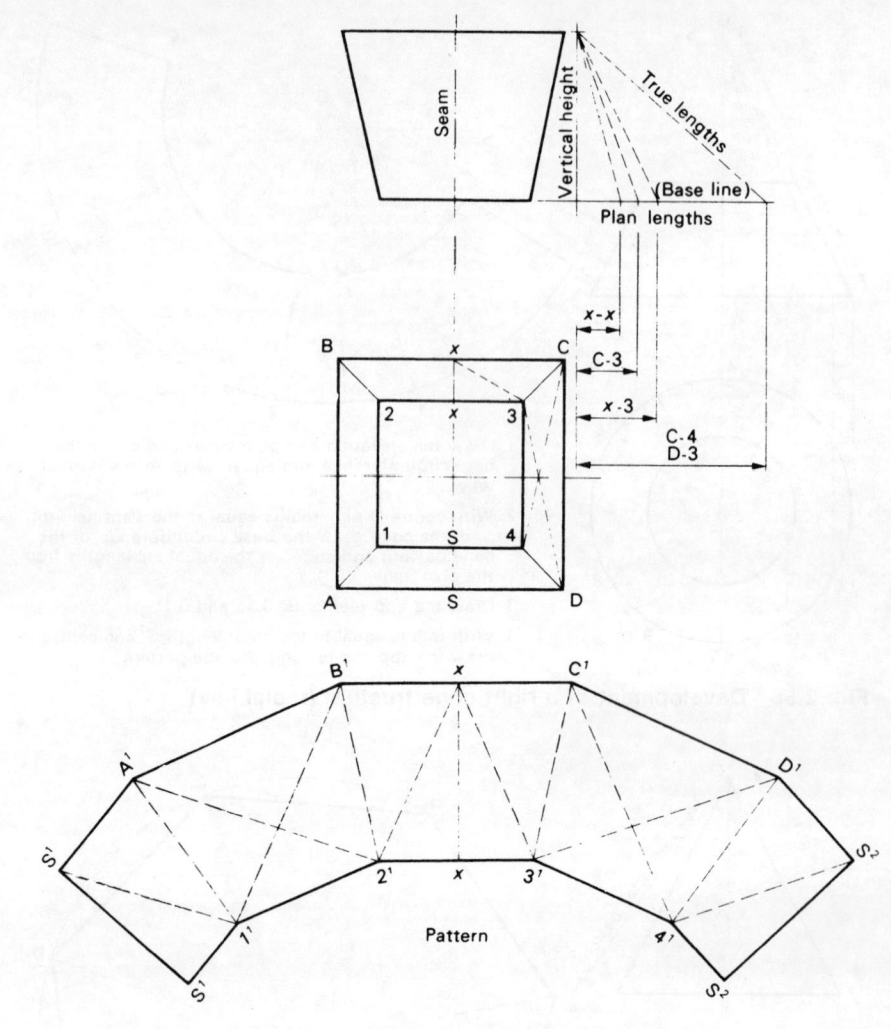

**Fig. 2.62 Development of a square hopper (triangulation)**

**Note:**

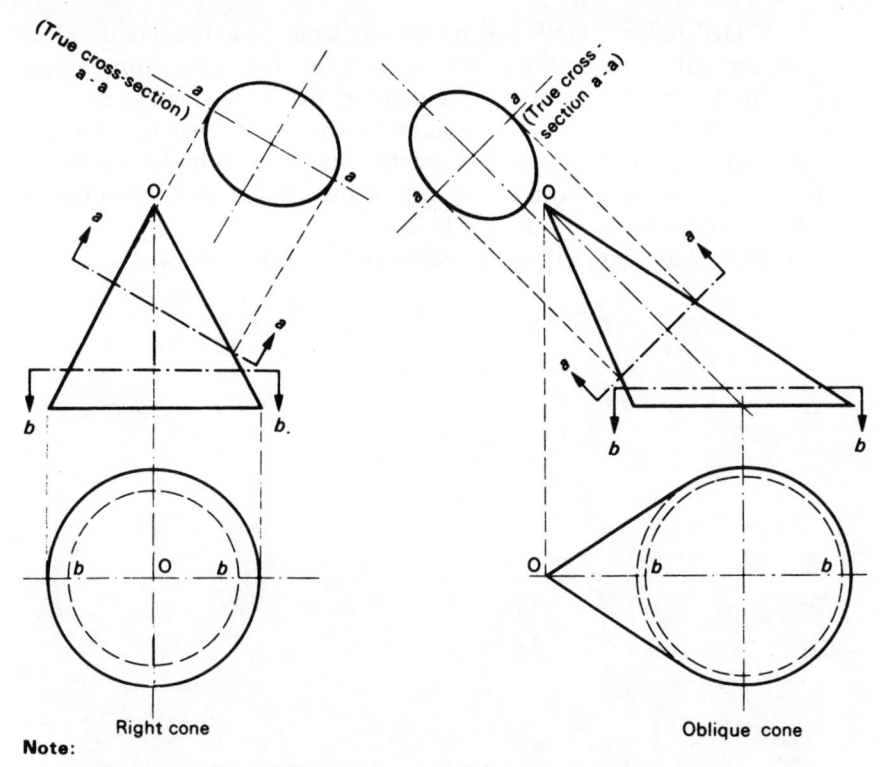

When an oblique cone is cut by a plane normal to its axis the cross-section produced is a true ellipse.

Whereas, in the case of the right cone, the distance from the apex to the base is the same at all positions round the surface, the oblique cone varies from point to point.

**Fig. 2.63 Comparison of right and oblique cones**

6. For the next triangle swing true length arc 2,1 and true length arc B,1 these will intersect to locate point 1 in the pattern. Join A,1 and 2,1 to complete a second side of the hopper.
7. Take 3,1 in plan and swing an arc from 1 in the pattern. Take true length AS from the plan view and swing an arc from A in the pattern to obtain true length 1,S by plotting its plan length against the vertical height and swing an arc from 1 in the pattern to locate points S. The last triangle S,1,S is completed by swinging an arc from S equal to the true length of the front line. Join AS, 1,S and S,S.

Note: By commencing the pattern in the middle at $x-x$ (i.e., opposite the seam). The whole pattern to be obtained by repeating the marking out procedure each side of the centre line.

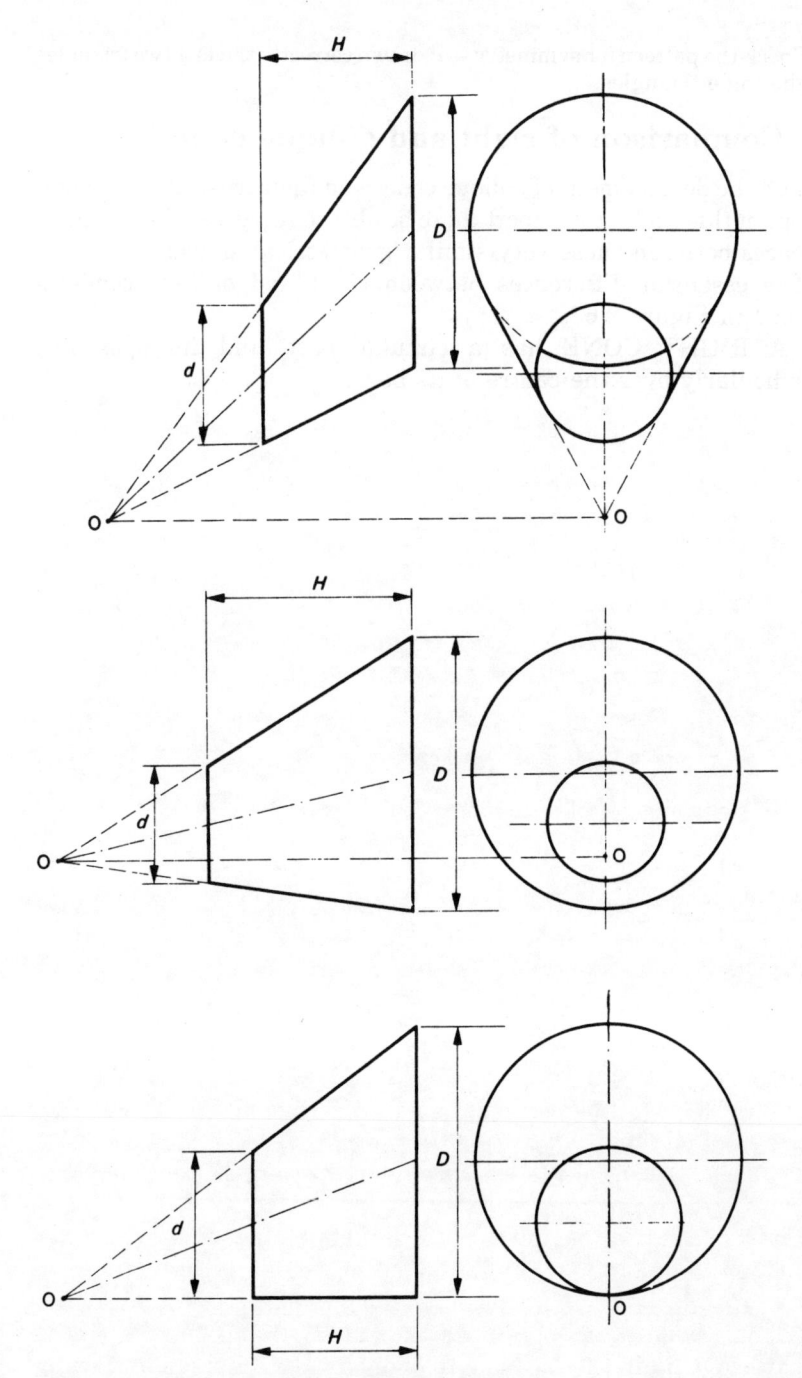

**Fig. 2.64 Typical oblique cone frustums**

67

Check the pattern for symmetry — if drawn correctly the last two triangles are right-angle triangles.

## 2.20 Comparison of right and oblique cones

Although the development of oblique cones and their frustums is beyond the scope of this book, it is important to be able to recognise the essential differences between these very similar geometrical shapes.

The essential differences between right and oblique cones is explained in Figure 2.63.

A RIGHT CONE has a circular base and its apex lies perpendicularly over the centre of its base.

AN OBLIQUE CONE has a circular base, but its apex does not lie perpendicularly over the centre of its base. The axis of an oblique cone leans to one side of the perpendicular. Both cones when cut by a plane parallel to the base (as shown at $b,b$) present a true circle at the plane of cutting. In the case of the right cone a cutting plane parallel to its base is normal to its axis. A right cone cut obliquely presents an elliptical cross-section (as shown at $a,a$).

Figure 2.64 shows typical frustums of oblique cones.

# 3 Metal joining — mechanical fastenings

Engineering components and fabrications can be joined together in a variety of ways. The type of joint may be either temporary or permanent. Table 3.1 indicates the different types of joint employed.

Thermal methods of joining are described in detail in Chapter 11.

## 3.1 Screwed fastenings

Various types of screwed fastenings are used where components must be assembled and dismantled regularly. Figure 3.1 shows a range of such fastenings.

The proportions of the fastenings shown are not given as these will vary with the screw thread system to which the fastening has been manufactured. Figure 3.2 shows some typical applications of these fastenings.

In order to prevent screwed fastenings slacking off due to vibration, various locking devices are employed. A selection of these are shown in Figure 3.3.

Spanners are used to tighten screwed fastenings. They are carefully proportioned so that their length enables a man of normal strength to tighten the fastening correctly. A spanner that is too long will provide too much 'leverage' and the fastening will break. This is why a spanner should not be lengthened with a piece of tube. Figure 3.4 shows a selection of spanners.

## 3.2 Screw thread elements

Figure 3.5 shows a section through a male screw thread, illustating the basic elements essential to the understanding of the following topics on screw thread forms and systems.

**Table 3.1 Types of joint**

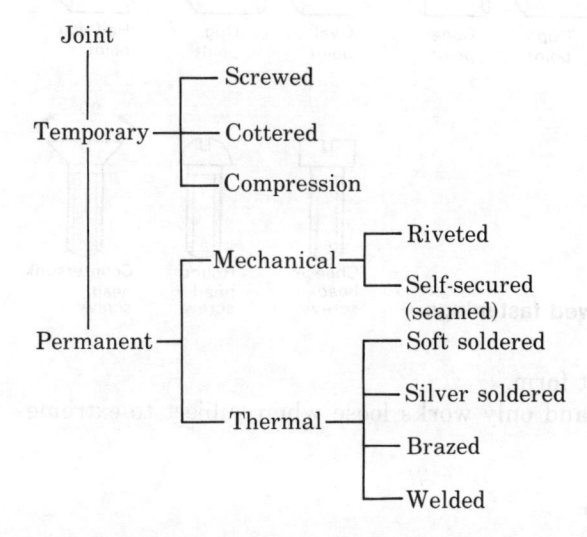

- Joint
  - Temporary
    - Screwed
    - Cottered
    - Compression
  - Permanent
    - Mechanical
      - Riveted
      - Self-secured (seamed)
    - Thermal
      - Soft soldered
      - Silver soldered
      - Brazed
      - Welded

## 3.3 Screw thread forms

Figure 3.6 shows a range of screw threads that are found on machines and other pieces of engineering equipment.

### 1. VEE THREAD

This thread form has many advantages which is why it is so widely used.

(a) It is the easiest and cheapest to manufacture.
(b) It is easily cut with taps and dies.

69

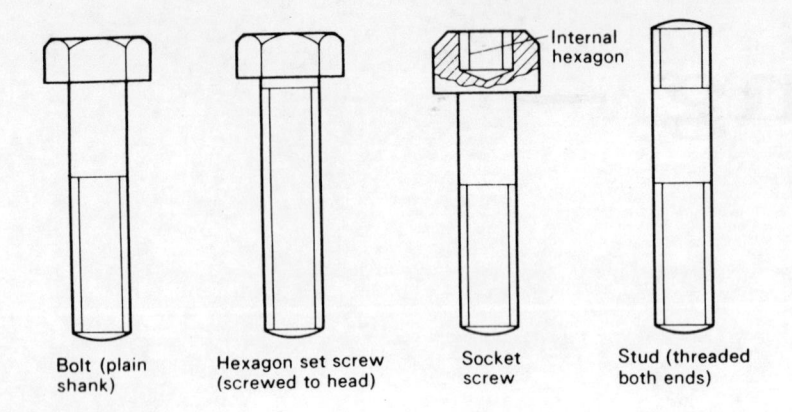

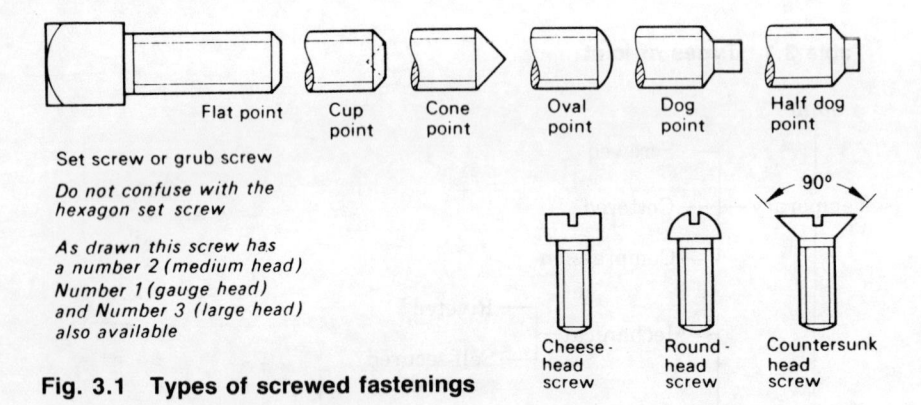

**Fig. 3.1 Types of screwed fastenings**

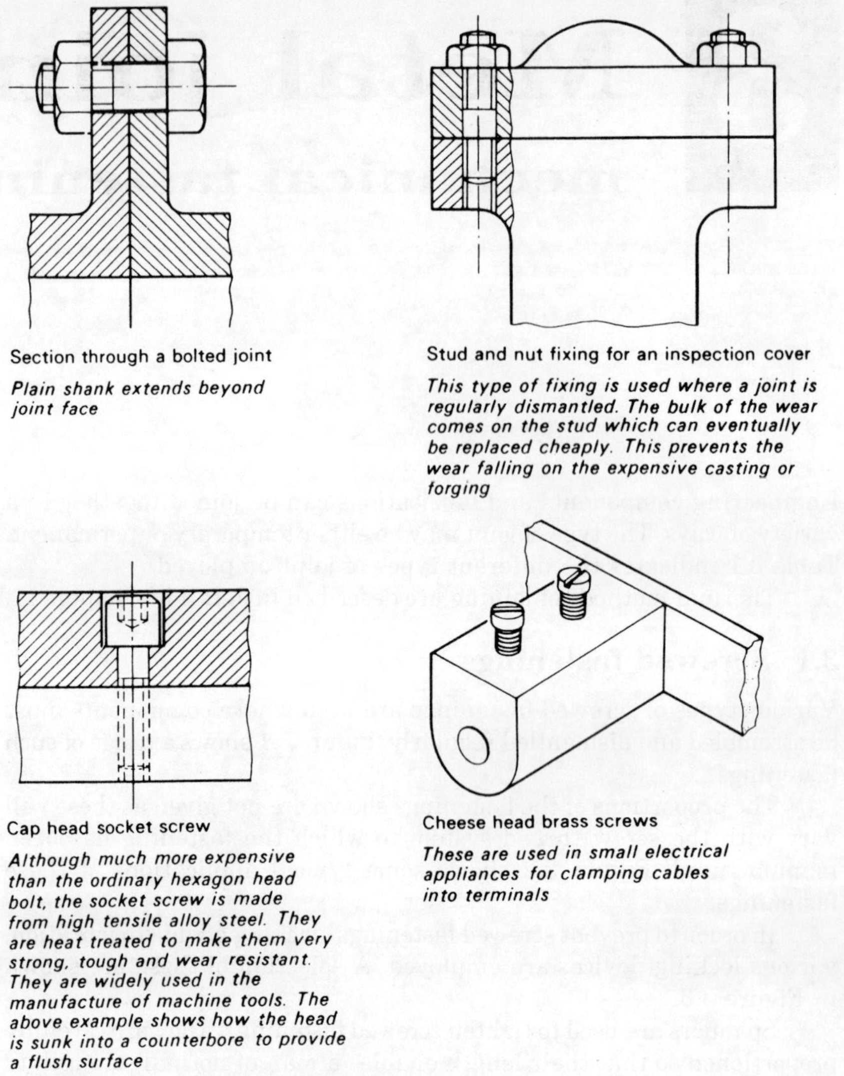

- Bolt (plain shank)
- Hexagon set screw (screwed to head)
- Socket screw (Internal hexagon)
- Stud (threaded both ends)

Set screw or grub screw
- Flat point
- Cup point
- Cone point
- Oval point
- Dog point
- Half dog point

*Do not confuse with the hexagon set screw*

*As drawn this screw has a number 2 (medium head) Number 1 (gauge head) and Number 3 (large head) also available*

- Cheese head screw
- Round head screw
- Countersunk head screw (90°)

**Fig. 3.2 Use of screwed fastenings**

Section through a bolted joint
*Plain shank extends beyond joint face*

Stud and nut fixing for an inspection cover
*This type of fixing is used where a joint is regularly dismantled. The bulk of the wear comes on the stud which can eventually be replaced cheaply. This prevents the wear falling on the expensive casting or forging*

Cap head socket screw
*Although much more expensive than the ordinary hexagon head bolt, the socket screw is made from high tensile alloy steel. They are heat treated to make them very strong, tough and wear resistant. They are widely used in the manufacture of machine tools. The above example shows how the head is sunk into a counterbore to provide a flush surface*

Cheese head brass screws
*These are used in small electrical appliances for clamping cables into terminals*

(c) It is the strongest form.
(d) It is self-locking and only works loose when subject to extreme vibration.

## 2. SQUARE THREAD

This thread form is used where rotary motion is to be transformed into linear motion; for example, a machine lead screw.

(a) It has less friction as there is no locking action.
(b) It is weaker than a vee thread.
(c) It is more difficult to cut than a vee thread.

## 3. KNUCKLE THREAD

This is a modified square thread used where rough usage and heavy wear would damage the corners of a true square form.

(a) It is used for railway couplings and fire hydrants.
(b) It is easier to cut than a square thread.

## 4. ACME THREAD

This is a modified square thread and is used for lead screws operating in conjuction with split nut.

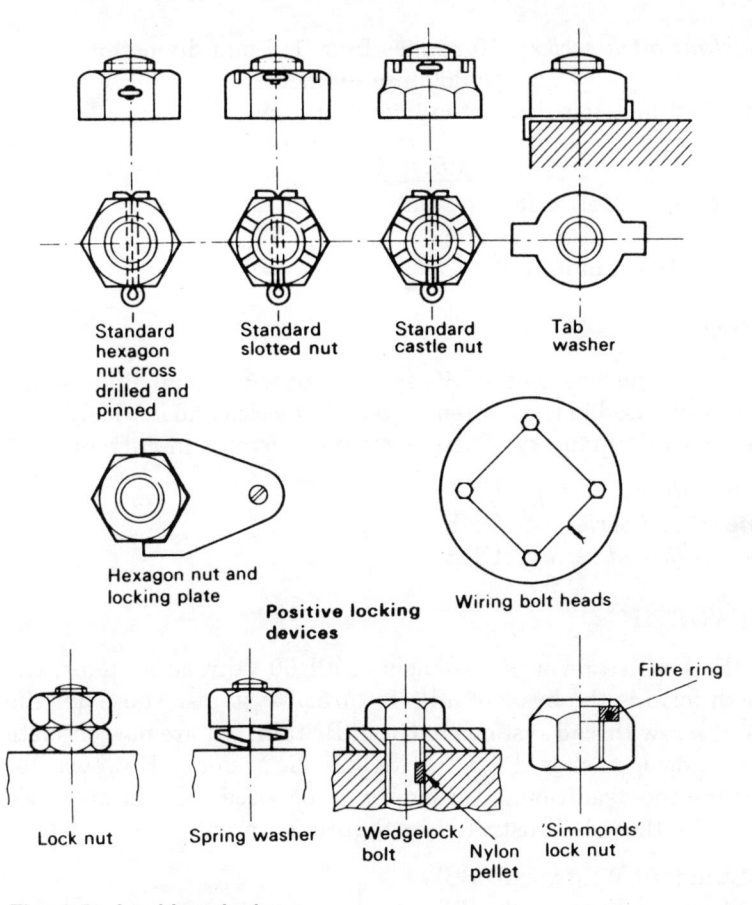

Fig. 3.3 Locking devices

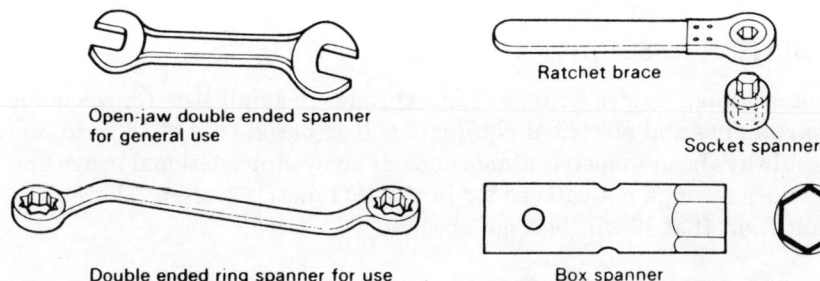

Fig. 3.4 Spanners

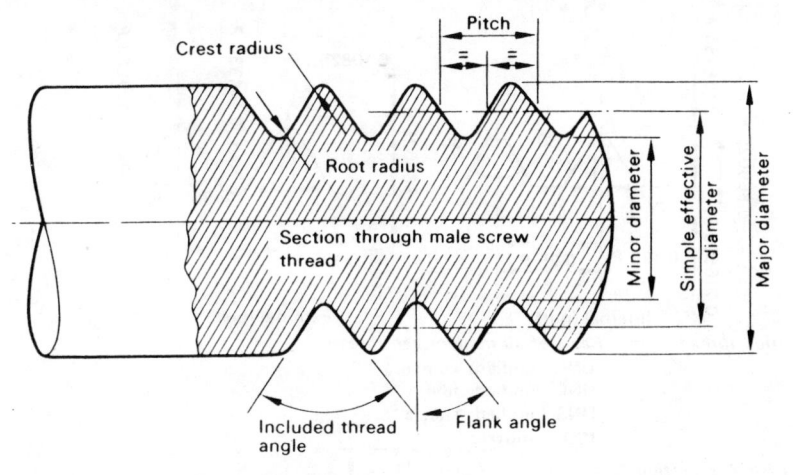

Fig. 3.5 Screw thread elements

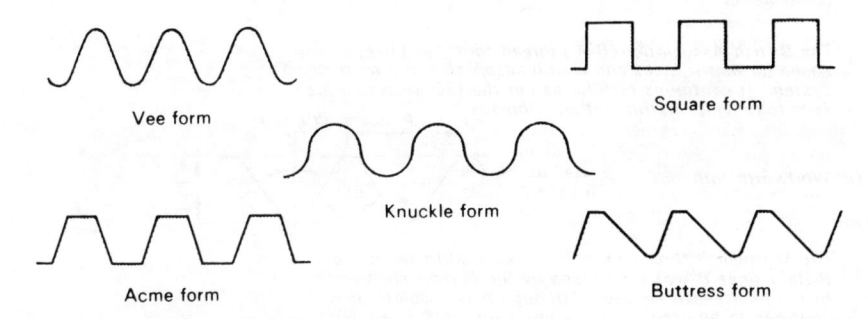

Fig. 3.6 Types of screw thread

(a) There is less friction when the nut is being engaged and disengaged.
(b) The taper gives the nut a lead when the nut is being engaged.
(c) Wear can be compensated for by moving the nut deeper into engagement.

5. BUTTRESS THREAD

As shown in Figure 3.6, this form is used where the axial force operates in one direction only.

(a) It is twice as strong as the square form.
(b) It is used for quick release vices and gun breach mechanisms.

71

**Fig. 3.7 Screw thread forms**

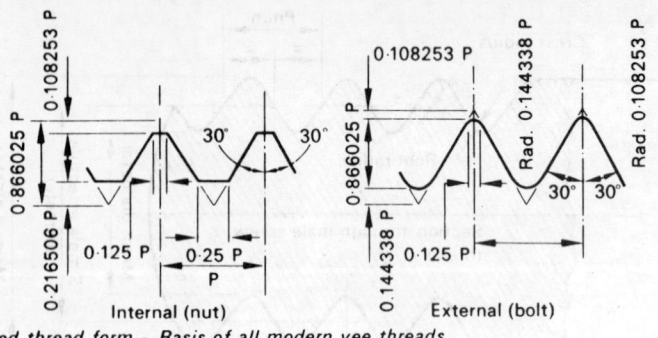

(a) Unified thread form - Basis of all modern vee threads
UNC - unified: course
UNF - unified: fine
UNS - unified: special
ISO - metric

(b) British Association form. Used for the screws in scientific instruments and small electronic components

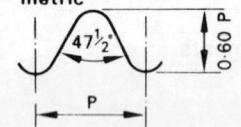

The British Association (B.A.) thread form has always been based on metric sizes and is still acceptable in a metricated system. It continues to fill a gap in the ISO system which fails to provide instrument size threads

(c) Whitworth form

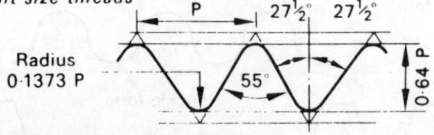

The Whitworth thread form has been traditional in Great Britain since it was introduced by Sir Joseph Whitworth in the nineteenth century. Although now obsolete, it will continue to be used for many years for replacement and maintainance purposes

## 3.4 Screw thread systems

There are many screw thread systems currently in use in Great Britain. Figure 3.7 shows a selection of the ones in more frequent use.

### 1. ISO-METRIC

This is the metric system adopted by Great Britain and will, in time, supersede all other screw thread systems for new designs and products. It is based on the form of the Unified thread.

*Coarse thread series*: 1·6 mm to 68 mm diameter.
*Fine thread series*: 1·8 mm to 68 mm diameter.
*Constant pitch series*: 10 ranges from 1·6 mm diameter to 300 mm diameter.

A typical metric screw thread would be specified as:

$$\underline{M8 \times 1}$$

where: M = Metric ISO system
8 = 8 mm diameter
1 = 1 mm pitch

### 2. UNIFIED

This uses the same form as the ISO system but with 'inch' dimensions. It is universally used in the United States of America and is widely used in the automobile industry. There are three systems in current use:

*Coarse thread series*: UNC
*Fine thread series*: UNF
*Special thread series*: UNS

### 3. WHITWORTH

Just as the Unified form is the basis of all 60° thread systems, the Whitworth form is the basis of all 55° thread systems. These are the traditional screw thread systems in Great Britain, but are now obsolete and all new designs should use the ISO-metric systems. However, for many years the traditional threads will be used for maintenance purposes. The threads illustrated in Figure 3.7 are:

British Standard Whitworth: BSW
British Standard Fine: BSF
British Standard Pipe: BSP
(Parallel and tapered)

### 4. BRITISH ASSOCIATION

This system provides a range of fine threads in small sizes for scientific instruments and electrical equipment. It is based on a $47\frac{1}{2}°$ form and has always been in metric dimensions. It enjoys international usage and provides a range not catered for in the ISO-metric system. There is no indication that it will become obsolete.

### 5. CONDUIT (ELECTRICAL)

As with all modern thread systems, the ISO-metric conduit thread has a 60° form.

Nominal diameters: 16 mm; 20 mm; 25 mm; 35 mm.

Irrespective of the diameter, the pitch is constant at 1·5 mm. The depth of thread on the conduit is constant at 0·920 mm.

## 3.5 Pitch/diameter relationship

### 1. COARSE THREAD SYSTEMS

Stronger and less inclined to 'strip'. They tend to vibrate loose. Usually used on soft and/or low strength materials.

### 2. FINE THREAD SYSTEMS

Not so strong but lock up tighter and are less inclined to work loose. Usually used on harder tough materials.

For example, the studs used for fixing the cylinder head to a light alloy engine block would have a coarse thread on one end for fixing into the light alloy block and a fine thread on the other end to take the steel nut.

## 3.6 Taps and dies

### 1. TAPS

These are used for cutting internal threads. There are three taps to a set and these are shown in Figure 3.8.

The tap is rotated by a tap wrench. This should be selected to suit the size of the tap. Too small a wrench results in excessive force being used to turn the tap. Too large a wrench results in lack of 'feel'. In either case there is lack of proper control that will result in a broken tap. Figure 3.9 shows a selection of tap wrenches.

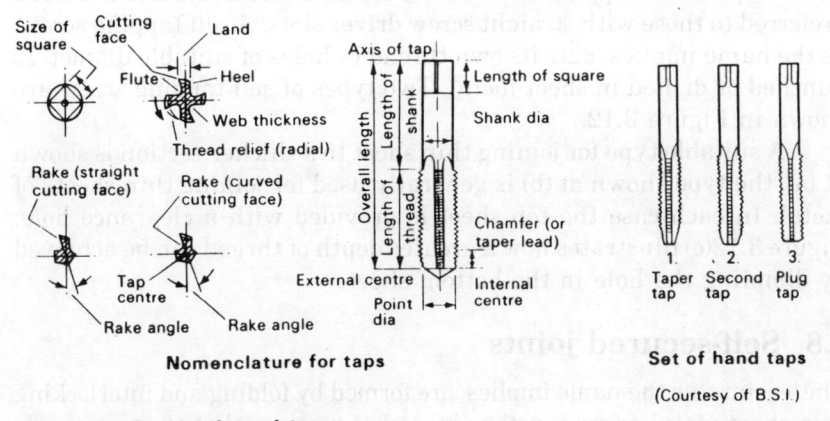

**Fig. 3.8 Screw thread taps**

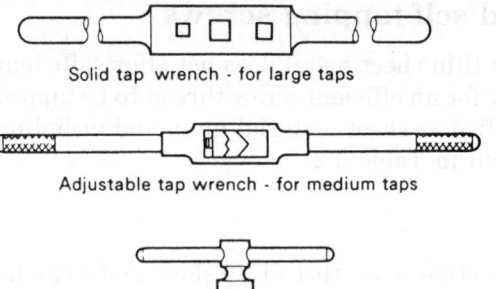

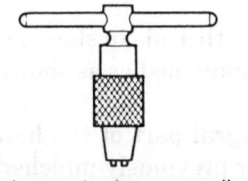

**Fig. 3.9 Tap wrenches**

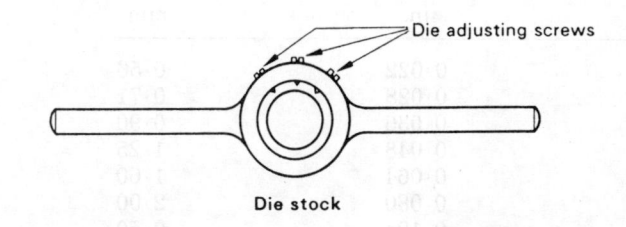

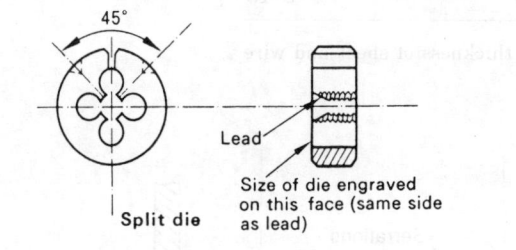

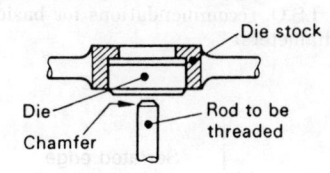

**Fig. 3.10 Dies and die-stock**

### 2. DIES

These are used for cutting external threads. Figure 3.10 shows a typical die and die stock.

It will be seen that the die has three adjusting screws. The centre screw spreads the die and reduces the depth of cut, while the outer screws close the die and increase the depth of cut.

73

## 3.7 Hank bushes and self-tapping screws

A hole punched or drilled in thin sheet metal does not offer sufficient thickness or depth necessary for an efficient screw thread to be tapped into it. Sheet metal is classified as sheet material up to and including 0·32 mm, thickness as shown in Table 3.2.

### 1. HANK BUSHES

The problem of producing effective screw threads in sheet metal can be effectively solved by the use of rivet or hank bushes as shown in Figure 3.11.

The rivet shank, which is an integral part of the hank bush, is inserted into a hole of suitable diameter previously punched or drilled

**Table 3.2 Thickness of sheet metal in general use**

| S.W.G. | in. | mm* |
|---|---|---|
| 24 | 0·022 | 0·56 |
| 22 | 0·028 | 0·71 |
| 20 | 0·036 | 0·90 |
| 18 | 0·048 | 1·25 |
| 16 | 0·064 | 1·60 |
| 14 | 0·080 | 2·00 |
| 12 | 0·104 | 2·50 |
| 10 | 0·128 | 3·15 |

(*I.S.O. recommendations for basic thickness of sheet and wire diameters)

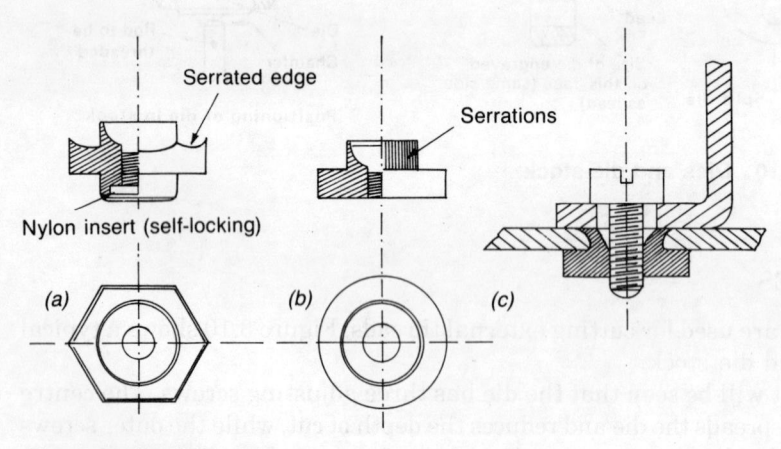

**Fig. 3.11 Hank bushes**

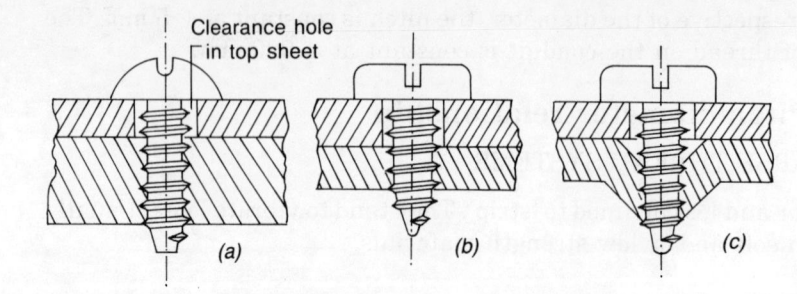

**Fig. 3.12 Self-tapping screws**

in the sheet metal. With the bush supported on a solid flat surface or anvil, the protruding part of the shank is hammered down flush with the top surface of the sheet (Figure 3.11(c)). The best results are obtained by using a suitable ball-faced punch to spread or flare the rivet shank before finally hammering down. When secured by riveting, the serrations provided on the bush itself (Figure 3.11(a)) or round the rivet shank (Figure 3.11(b)) effectively grip the metal and prevent the hank bush revolving. Some hank bushes are manufactured with a fibre or nylon annular insert to provide self-locking (Figure 3.11(a)). Hank bushes with a range of BRITISH ASSOCIATION thread sizes are extensively used on electronic components and instrument panels.

### 2. SELF-TAPPING SCREWS

Sheet metal fabrications are often held together by means of self-tapping screws.

A range of sizes are available and are supplied with a variety of head shapes. Self-tapping screws with Phillips recessed heads are often preferred to those with straight screw driver slots. A self-tapping screw, as the name implies, cuts its own thread in holes of suitable diameters punched or drilled in sheet metal. Two types of self-tapping screw are shown in Figure 3.12.

A suitable type for joining thin sheet to a thicker section is shown at (a), the type shown at (b) is generally used for joining thin sheets of metal. In each case the top sheet is provided with a clearance hole. Figure 3.12(c) illustrates how a greater depth of thread can be achieved by dimpling the hole in the bottom sheet.

## 3.8 Self-secured joints

These joints, as the name implies, are formed by folding and interlocking thin sheet metal edges together in such a manner that they are made

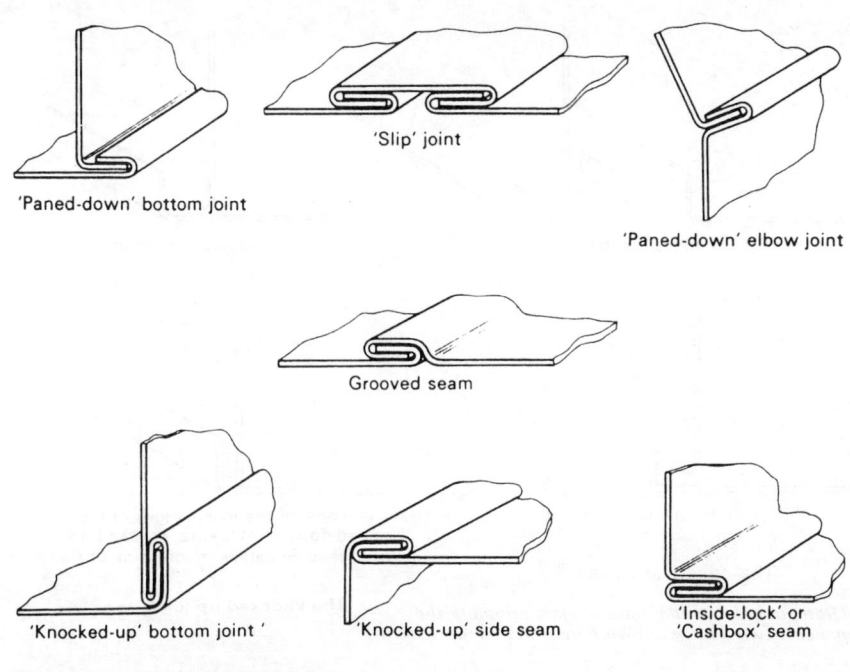

**Fig. 3.13 Self-secured joints**

secure without the aid of any additional jointing process. Their use is confined to fabrications or components constructed with light gauge sheet metal less than 1·6 mm thick.

A selection of these joints is shown in Figure 3.13. Of these, the following are the most widely used:

(a) The grooved seam.
(b) The paned-down joint.
(c) The knocked-up joint.

## 3.9 The grooved seam

This seam consists of two folded edges, called 'locks' (shown in Figure 3.14(a)). The two edges (one up and the other down) are hooked together (Figure 3.14(b)) and finally 'locked' together (Figure 3.14(c)) by means of a tool called a 'hand groover', or with a 'grooving machine'.

## 3.10 The internal grooved seam

This is really a COUNTERSUNK GROOVED SEAM. It is constructed in exactly the same manner as a normal grooved joint. The only difference is that where a 'grooving tool' cannot be used on the inside of the seam,

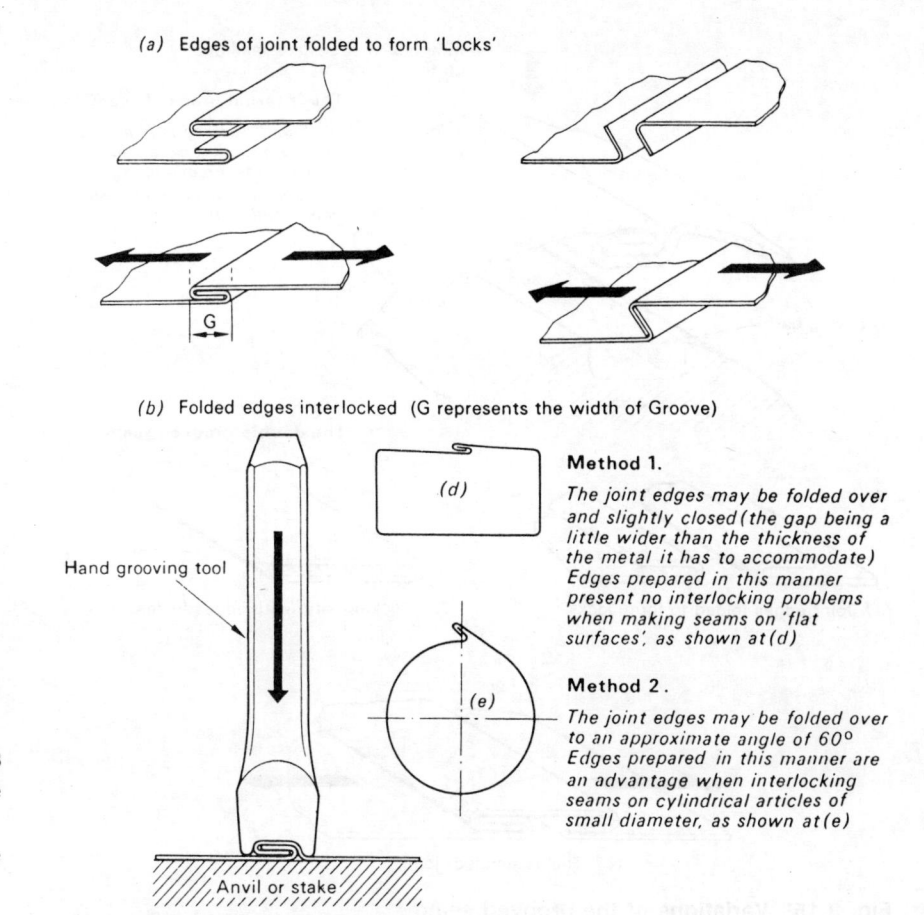

**Fig. 3.14 The grooved seam**

then the interlocking edges are placed over a special 'grooving bar and the groove is sunk with a mallet or hammer, as shown in Figure 3.15.

## 3.11 The double grooved seam

Use is made of a detachable 'locking strip' to hold the folded edges of the joint firmly together. This strip or cap combines strength with good appearance.

The seam is made by folding 'locks' to the desired size, as shown in (a), Figure 3.15. A metal strip of the proper width, with the edges folded (b) is then 'slipped' over the locks of the two pieces to be joined,

75

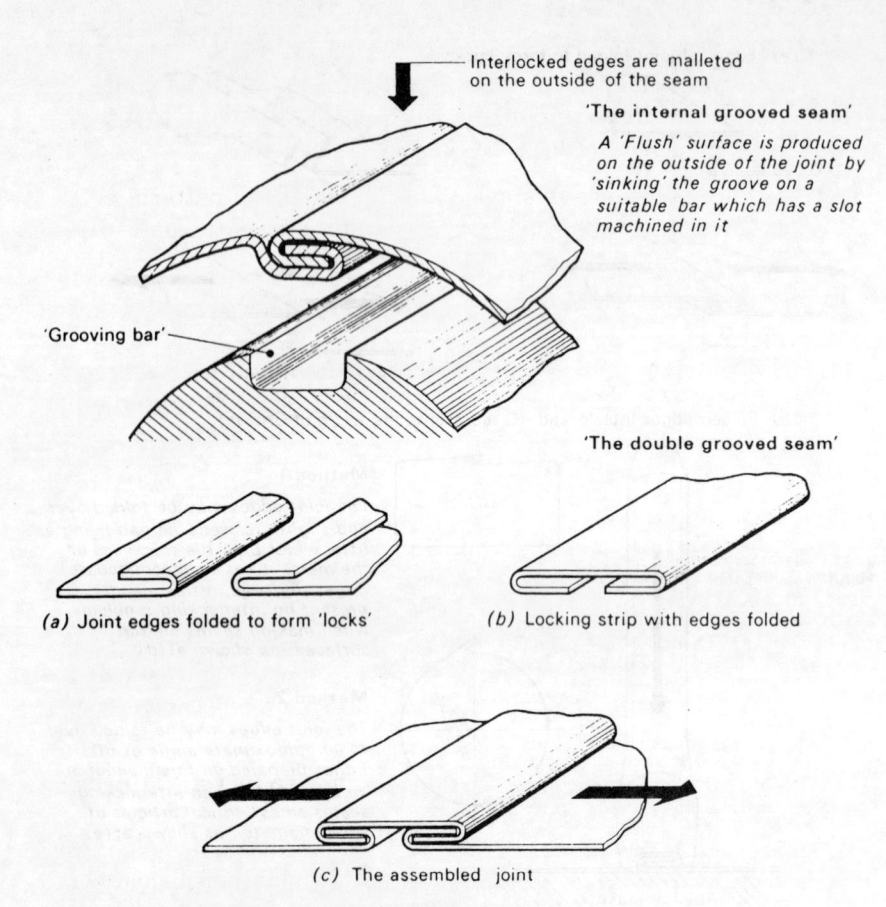

**Fig. 3.15** Variations of the grooved seam

as at (*c*). Usually, in practice, the cap or locking strip has to be driven by lightly striking one end with a hammer.

## 3.12 The paned-down joint

The simple construction of a paned-down joint is illustrated in Figure 3.16.

The two edges of the sheet metal to be joined are each folded at 90°, and to the desired width, as in (*a*). These 'flanges' are placed in position as at (*b*), and one flange is closed over the other, i.e. PANED-DOWN as shown at (*c*), with a special hammer called a PANING HAMMER. Where a large number of these seams are produced a PANING-DOWN MACHINE is employed.

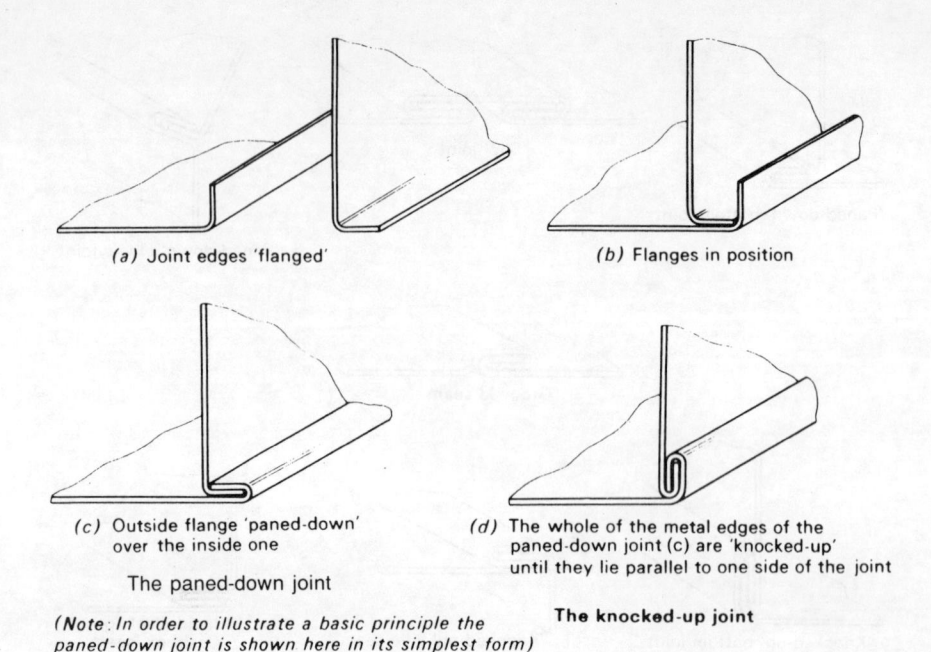

**Fig. 3.16** The paned-down and knocked-up joint

## 3.13 The knocked-up joint

This very useful joint is very much stronger than the paned-down joint. Basically it is a paned-down joint that has been 'knocked-up'.

Figure 3.16 illustrates how *the knocked-up joint is simply an extension of the paned-down joint*. The whole of the metal edges in the paned-down joint (*c*) are turned over — i.e. 'knocked-up' — until they lie parallel to one side of the joint, as shown at (*d*).

Strong and efficient self-secured joints will result if the following basic procedure is adopted:

1. The metal to be joined is folded along the joint edges.
2. The folded edges are interlocked.
3. The assembled joint is secured in such a way that it cannot be easily pulled apart.

## 3.14 Folding and jointing allowances

When making self-secured joints or seams, it is necessary to make allowance for the amount of material that is to be added for the folds or locks. This allowance depends largely upon two basic factors which should be considered:

1. The **width of the folded edge**.
2. The **thickness of the metal**.

Consider the enlarged cross-sectional views of the common self-secured joints illustrated in Figure 3.17.

ALLOWANCE FOR A GROOVED JOINT

If we fold over the edges to a width W and form the joint shown, the final completed width of the joint G will be clearly seen to be somewhat greater than W. In fact it can be seen that the final width of the GROOVE will have a minimum value of W+3T, where T represents the metal thickness.

Key:
W = Width of folded edge
G = Width of grooved seam. (or size of groove)
T = Thickness of metal
C = Width of capping strip
P = Width of paned-down seam
K = Width of knocked-up seam

Allowances for self-secured joints are governed by the width of the folded edge and the metal thickness

| TYPE OF JOINT | ALLOWANCE |
|---|---|
| Grooved seam | Total allowance = 3G<br>This is shared by the two joint edges:<br>(i) Add 1½G to each edge;<br>(ii) Add G to one edge and 2G to the other. |
| Double grooved seam | (a) Add W-T to each edge of sheet to be jointed;<br>(b) Total allowance for capping strip = 4W + 4T |
| Paned-down and knocked-up seams | Allowance for the single edge = W.<br>Allowance for the double edge = 2W + T |

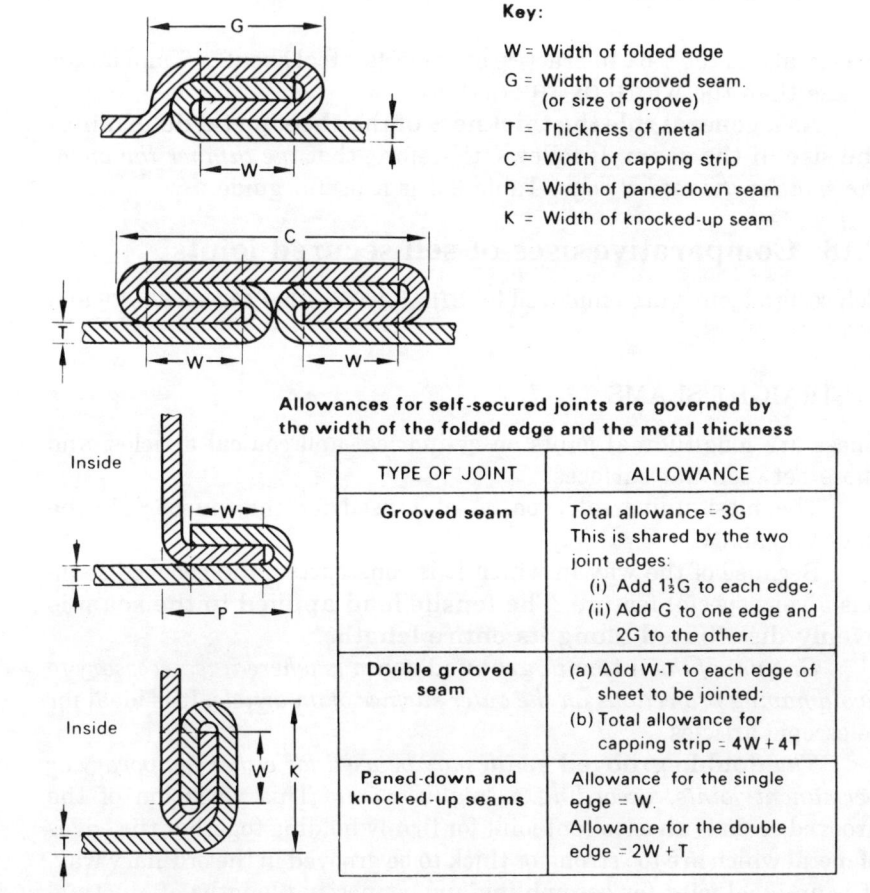

**Fig. 3.17 Allowances for self-secured joints**

The popular allowance for a grooved seam is three times the specified width of the seam (or the width of the grooving tool).

This allowance may be used in one of two ways:

1. Half this allowance is made on each side of the pattern, or
2. One-third of the allowance is made on one side of the pattern and the remaining two-thirds made on the other.

*The thickness of the metal must be taken into consideration when selecting a grooving tool to close or set the joint.* For example: a joint is required to be made on 0·6 mm (24 SWG) sheet metal with a finished groove width of 6·35 mm. It follows that the **width of the folded edge**, i.e., the distance in from the edge to the 'fold line,' must be 6·35 mm minus 1½ times the thickness of the metal.

ALLOWANCE FOR FOLDING = **Width of groove minus one and a half times the thickness of metal.**
= 6·35 − 1·5 × 0·6 mm
= 6·35 − 0·9 mm
WIDTH OF FOLDED EDGE = 5·45 mm (round off downwards to 5·4 mm to ensure clearance is maintained)

**Remember it is the size of edge that has to be folded; not the size of the groove**.

If the metal had been folded over 6·35 mm from the edge, then reference to Figure 3.17 will clearly indicate that the finished width of groove would be somewhat greater.

WIDTH OF GROOVE = **Width of folded edge plus three thicknesses of metal**
= 6·35 + 3 × 0·6 mm
= 6·35 + 1·8 mm
SIZE OF GROOVER = 8·15 (minimum)(round off upwards to 8·2 mm to ensure clearance is maintained)

## 3.15 Allowance for the double grooved seam

The allowance for the edges of the sheet is **the width of the folded edge minus a thickness of metal**. If an allowance were not made for the metal thickness, the dimension across the seam would be increased by two metal thicknesses. It will be seen in Figure 3.17 that the width of the capping strip is equivalent to twice the width of the folded edge plus four thicknesses of metal, i.e., C = 2W + 4T. Thus the complete allowance for the capping strip will be four times the width of the folded edge plus four thicknesses of metal.

## 3.16 Allowance for paned-down and knocked-up joints

As previously stated the knocked-up joint is an extension of the paned-down joint. Therefore, the allowances will be identical.

In the views given in Figure 3.17, P represents the size of the paned-down joint, and K is similarly used for the size of the knocked-up joint.

**The size of paned-down and knocked-up joints is determined by the width of the single folded edge.** On examination of the respective sectional views it will be seen that the size of the paned-down seam, 'P', will be equivalent to 2W + 2T, and 'K', which represents the size of the knocked-up joint, is equivalent to 2W + 3T.

The allowances for making both these joints is **the width of the folded edge for the single fold and twice the width of the folded edge plus one thickness of metal for the double edge.**

## 3.17 The hand-grooving tool

The simple tool used for closing grooved seams by hand is made of medium carbon steel. It has a half-round slot or groove machined into its rectangular working face.

**The width of the groove denotes the size of the tool**, and hand-grooving tools are available in sizes varying from 3·2 mm to 19 mm. The usual variation between sizes in this range is about 1·6 mm. Reference to Figure 3.14 will show that a completed grooved seam will have a width equal to the size of the single folded edge plus three thicknesses of metal, and a total thickness of four times that of the metal. Thus when making a grooved seam one has a choice of approach:

1. To determine the correct width of the single folded edge, i.e., the FOLDING ALLOWANCE, for a specific size of GROOVE by deducting one and a half times the metal thickness;
2. To make the interlock with a specific width of single folded edge, and then select a suitable size of GROOVING TOOL to close the seam.

Very often, in practice, the second approach is adopted, and reference to Figure 3.17 will show why. *The sectional view of the seam shows that the edges of the material do not fit accurately into the radiused corners of the locks.* This obviously affects the tolerance between the GROOVE ALLOWANCE and the size of the GROOVING TOOL. Hence, in the initial stages of closing the seam, a further clearance is essential between

**Table 3.3 Relationship between metal thickness and width of seam**

| METAL THICKNESS MM | S.W.G. | MINIMUM WIDTH FOR SINGLE FOLDED EDGE (mm) |
|---|---|---|
| 0·315 | 30 | 3 |
| 0·400 | 28 | 3·5 |
| 0·500 | 26 | 3·5 |
| 0·600 | 24 | 4·0 |
| 0·800 | 22 | 4·0 |
| 1·000 | 20 | 5·0 |
| 1·250 | 18 | 6·0 |
| 1·600 | 16 | 6·0 |

the metal and the tool. In practice one selects a tool about 1·6 mm larger in size than the WIDTH OF THE GROOVE.

**As a general rule the thickness of the sheet metal determines the size of the seam**. It follows, therefore, that *the thinner the metal the smaller the folded edge*. Table 3.3 is a useful guide.

## 3.18 Comparative uses of self-secured joints

Self-secured joints are employed for STRAIGHT SEAMS, CORNER SEAMS and BOTTOM JOINTS.

### 1. STRAIGHT SEAMS

These are longitudinal joints on cylindrical and conical articles, and those between flat surfaces.

The most universal type of joint used for this purpose is the GROOVED SEAM.

Because of the way in which it is constructed, the grooved seam has a very special feature. **The tensile load applied to the seam is evenly distributed along its entire length.**

*One use of the* **internal grooved seam** *is where it is necessary to avoid having projections on the outer surface of an article.* It is ideal for telescopic articles.

*The* **double grooved seam** *may be used for either temporary or permanent joints, according to requirements.* This variation of the grooved seam is an excellent joint for firmly holding together the edges of metal which are too strong or thick to be grooved in the ordinary way. It is an ideal joint for assembling and connecting lengths of 'ducting' together, especially on site. Thermal insulation covers, which have to be periodically dismantled, incorporate this type of joint. The capping strip is easily removable when the need arises. It is often referred to as a 'SLIP JOINT'.

## 2. CORNER SEAMS

These seams are either straight or curved and are usually employed on the corners of sections of trunking, square and rectangular duct elbows, boxes and tanks with flat surfaces. When used for joining the corners of square duct elbows, *the single folded edge is always made on the flat sides or cheeks*. The double edge is placed on the 'throat' and 'back' which are the curved sides.

The types of joints used for securing corners are the PANED-DOWN and the KNOCKED-UP JOINTS.

**The knocked-up joint is the strongest of all the self-secured joints**.

## 3. BOTTOM JOINTS

These can be made on articles of any cross-section provided it is possible to fold up an edge.

The type of self-secured joints employed for making strong bottom joints are the PANED-DOWN and the KNOCKED-UP joints.

Figure 3.18 illustrates practical examples of the use of self-secured joints.

## 4. SEALING

Very often self-secured joints are employed in the manufacture of containers which have to hold foodstuffs and liquids.

In the case it is necessary to seal the joints and seams. Use is made of LATEX RUBBER inserts, and SOFT SOLDER with a high TIN content.

## 3.19 Joints — not 'self-secured'

Figure 3.19 shows a selection of joints requiring additional connections such as bolting or riveting. These techniques will now be dealt with in the following sections of this chapter. Thermal and adhesive bonding processes would also be appropriate and these are considered in Chapter 11.

## 3.20 Riveted and bolted joints

Riveting and bolting are two of the most common methods used for fastening sheet metal and metal plate work.

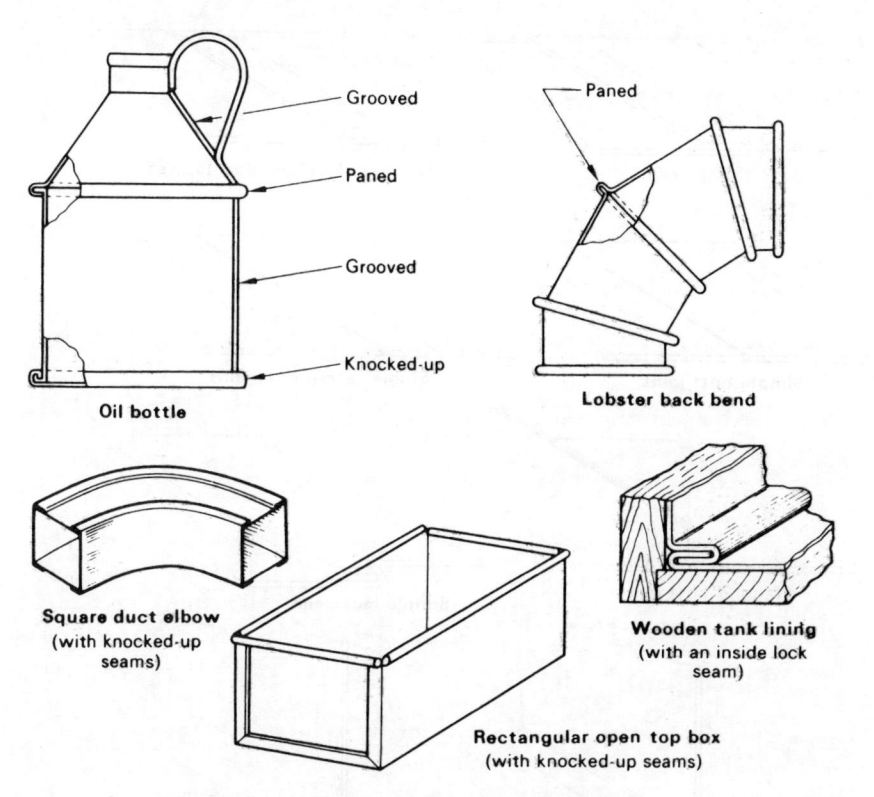

**Fig. 3.18 Practical examples of self-secured joints**

1. **Riveting** *is a method of making permanent joints*. The process consists of drilling or punching the plates to be riveted, inserting the rivet, then closing it by an applied compression force so that *it completely fills the hole* and forms a rigid joint.

    Rivets may be worked either HOT or COLD, depending on the type of structure. Smaller diameter rivets are used for light fabrications and are worked cold. As no allowance is required for thermal expansion, the 'cold' rivet, when initially inserted, is a close fit in the hole. When 'hot riveting' is employed, for example, in boiler and structural steel work, *the shank of the rivet should be somewhat smaller than the hole in order that the rivet may be easily inserted when made 'red hot'*. One advantage of 'hot riveting' as compared to 'cold riveting' is that it produces a stronger joint. *Upon cooling, the metal in the rivet* **contracts** *and increases the tightening effect produced by the shrinking or* **compression force** *between the two heads*, as shown in Figure 3.20.

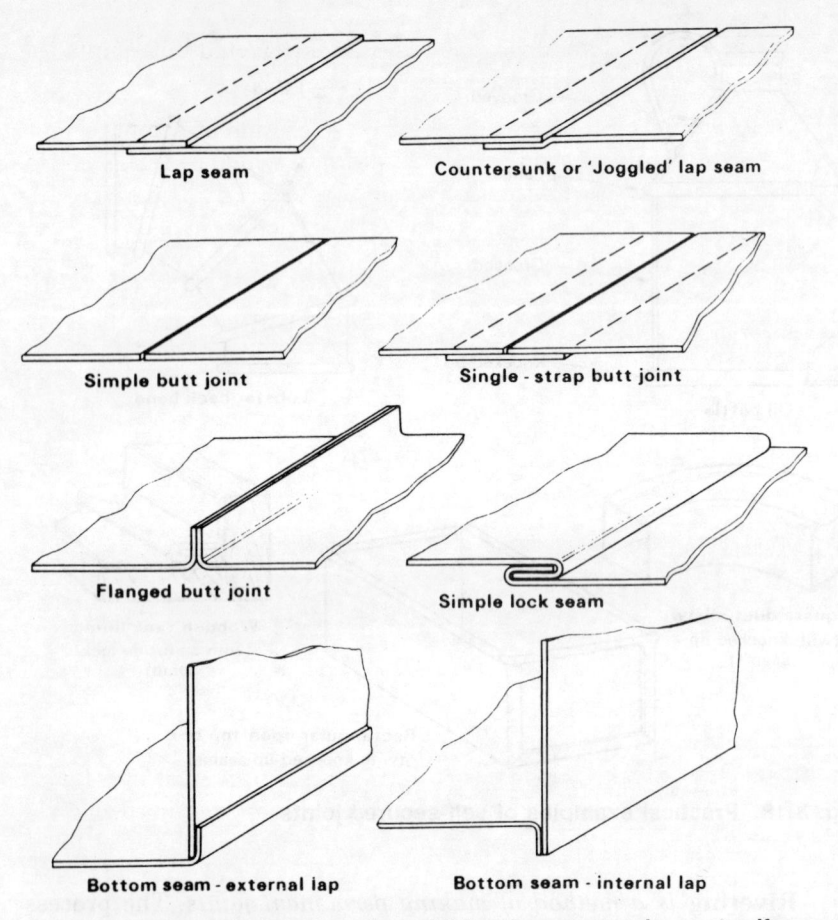

**Fig. 3.19 Types of joints requiring additional processes, i.e., not self-secured**

2. **Bolting** *is a method of making temporary joints*. The process consists of inserting a bolt of the correct length into the matching clearance holes in the plates or components to be fastened and securing it by means of a nut. It is common practice to insert a suitable washer under the nut, as shown in Figure 3.20.

## 3.21 Types of riveted joint

A variety of riveted joints is used for constructional and fabrication work, those most commonly employed being:

(i) Single riveted lap joint
(ii) Double riveted lap joint

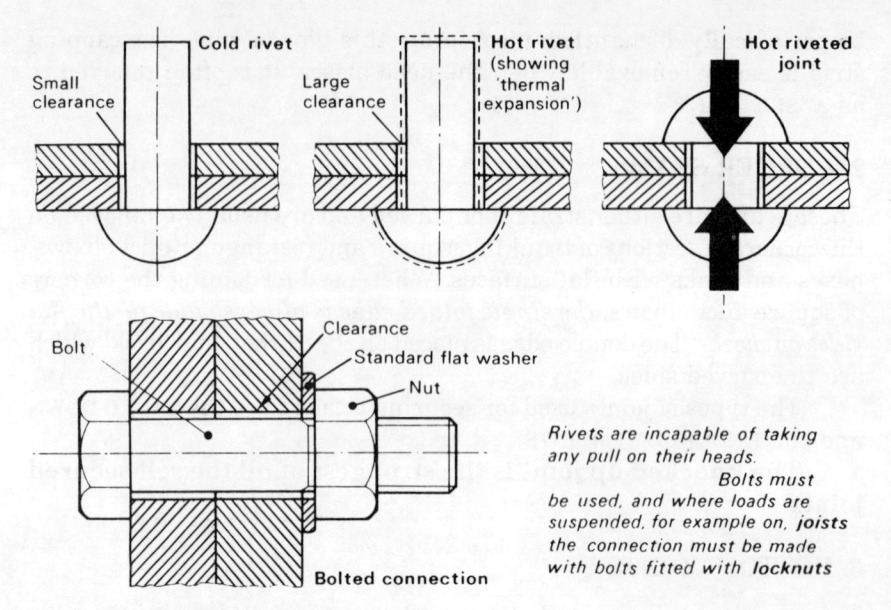

**Fig. 3.20 Riveted and bolted joints**

(iii) Single-strap butt joint
(iv) Double-strap butt joint

These are illustrated in Figure 3.21.

### SINGLE RIVETED LAP JOINT

This is the simplest of all riveted joints and is extensively used for joining both thick and thin plates. The plates to be joined are overlapped a short distance and a single row of rivets, conveniently spaced along the middle of the lap, completes the joint.

### DOUBLE RIVETED LAP JOINT

A lap joint having two rows of rivets is known as a 'double riveted lap joint'. Sufficient overlap must be provided in order to accommodate a double row of rivets. *This type of joint may have the two rows of rivets arranged in a square formation, known as* **'chain riveting'**, *or the rivets may be arranged diagonally to form triangles, called* **'zig-zag riveting'**.

### SINGLE-STRAP BUTT JOINT

When two cover plates are riveted one each side of a butt joint, the joint is known as a 'double-strap butt joint'.

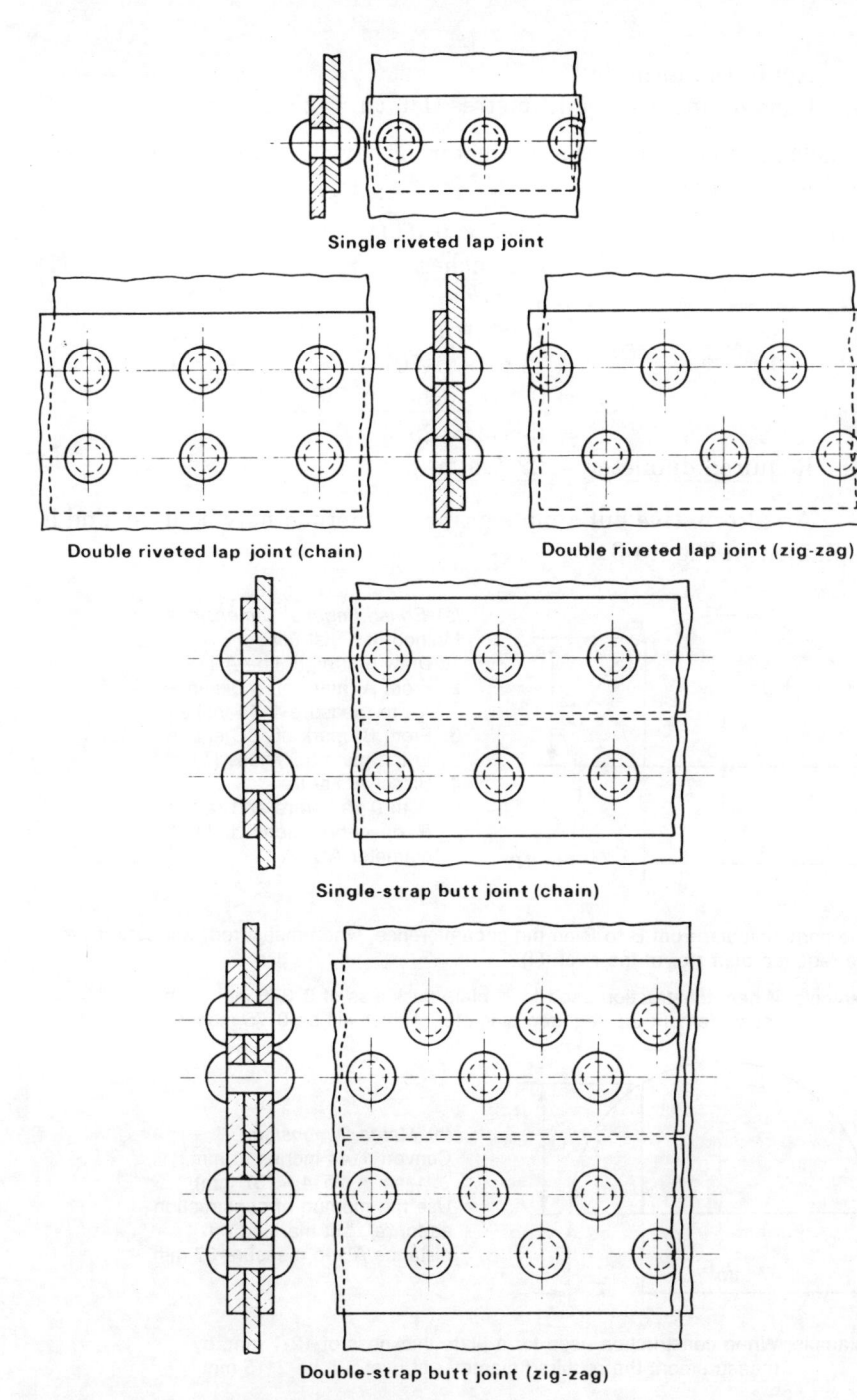

**Fig. 3.21 Types of riveted joint**

- Single riveted lap joint
- Double riveted lap joint (chain)
- Double riveted lap joint (zig-zag)
- Single-strap butt joint (chain)
- Double-strap butt joint (zig-zag)

When single or double straps are used for riveted butt joints the arrangement of the rivets may be:

(a) **Single riveted:** one row of rivets on each side of the butt; *or*
(b) **Double, triple or quadruple riveted:** in which case the 'chain' or the 'zig-zag' formation may be employed.

## 3.22 The strength of riveted joints

A riveted joint is only as strong as its weakest part, and it must be borne in mind that it may fail in one of four ways:

(i) Shearing of the rivet:
(ii) Crushing of the metal:
(iii) Splitting of the metal:
(iv) Rupture *or* tearing of the plate

These four undesirable effects are illustrated in Figure 3.22.

*Note:* For design purposes the rivet should only be loaded in shear and its tensile strength is assumed to be zero.

## 3.23 Selecting the correct size of rivet

To obtain the full strength of a riveted joint, a rivet of the correct diameter and length must be used. For example, if a rivet of a larger diameter were inserted in a thin sheet, the pressure required to drive the rivet would cause bulging of the thin metal around the rivet head, as illustrated in Figure 3.27.

The diameters of rivets for metal plate work may be determined by use of the following formula:

$$D = 1 \cdot 25\sqrt{T} \quad \text{(Urwin's formula)}$$

where D represents the required diameter of the rivet in inches, and T represents the plate thickness in inches.

EXAMPLE: Determine the diameter of rivet required for a plate thickness of $0 \cdot 25$ inches ($\frac{1}{4}$ inch)

*Solution*:
$$D = 1 \cdot 25\sqrt{T}$$
$$= 1 \cdot 25\sqrt{0 \cdot 25}$$
$$= 1 \cdot 25 \times 0 \cdot 5$$

Required diameter = $0 \cdot 625$ inches or 5/8 inch

*Note*: WHEN THE THICKNESS OF THE PLATE IS GIVEN IN MILLIMETRES:

1. Convert millimetres to inches (1 mm = $0 \cdot 0394$ inches)

*(a)* **Shearing of rivet**

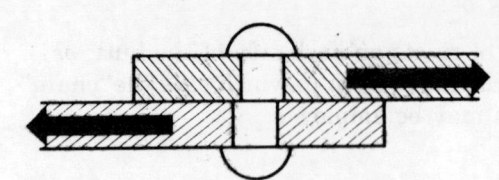

**Cause** Diameter of rivet too small compared with thickness of plate. The diameter of the rivet must be greater than the thickness of the plate in which it is to be inserted.

**Prevention** Select the correct diameter rivet for the thickness of the plate.

*(b)* **Crushing of the metal**

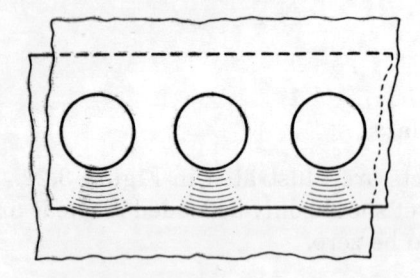

**Cause** Diameter of rivet too large compared with thickness of plate. The rivets when driven tend to bulge and crush the metal in front of them.

**Prevention** Select the correct diameter rivet for the thickness of the metal plate.

*(c)* **Splitting of the metal**

**Cause** Rivet holes punched or drilled too near edge of plate. Metal is likely to fail by splitting in front of the rivets.

**Prevention** Drill or punch the rivet holes at the correct edge distance, and use the correct lap allowance for the diameter of rivet selected.

*(d)* **Tearing of the plate**

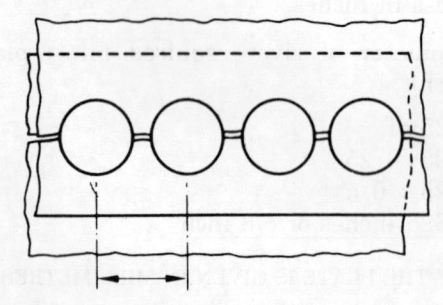

**Cause** Plates weakened by rivet holes being too close together. Plates tend to rupture along the centre line of the rivets.

**Prevention** Punch or drill rivet holes at the correct spacing or 'pitch'. In addition remove all burrs from the holes before final assembly.

**Fig. 3.22 Common causes of failure in riveted joints**

2. Apply the formula
3. Convert inches to millimetres (1 inch = 25·4 mm)

EXAMPLE: Determine the diameter of rivet required for a plate thickness of 12·7 mm.

*Solution*:

$$T = 12 \cdot 7 \times 0 \cdot 0394$$
$$= \underline{0 \cdot 5 \text{ inches}} \quad (1)$$
$$D = 1 \cdot 25\sqrt{T} \quad (2)$$
$$= 1 \cdot 25\sqrt{0 \cdot 5}$$
$$= 1 \cdot 25 \times 0 \cdot 707$$
$$= \underline{0 \cdot 844 \text{ inches or } 7/8 \text{ inches}}$$
$$= 0 \cdot 844 \times 25 \cdot 4$$

Required diameter = $\underline{22 \cdot 5 \text{ mm}}$ (3)

An alternative but simple graphical method may be used, and is explained in Figure 3.23.

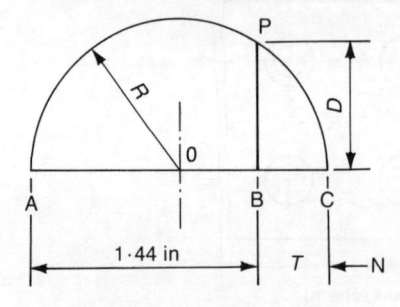

*(a) British Imperial Dimensions*
Method of construction:
1. Draw a straight line AN
2. From A, mark off a distance AB to measure 1·44 inches
3. From B, mark off BC equal to the plate thickness (*T*)
4. Bisect AC at 0
5. With 0 as centre and radius *R*, describe a semi-circle of diameter AC

The perpendicular from B to P on the circumference, when measured, will determine the required diameter of the rivet (*D*).

Example: When construction used for a plate thickness of 0·375 inches, by measurement the required diameter of rivet will be 0·75 inches.

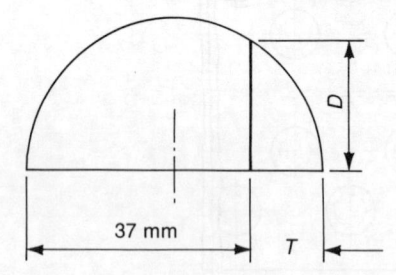

*(b) Metric Dimensions*
Convert 1·44 inches to mm:
(1·44 × 25·4 = 37 mm)
Use the method of construction as for *(a)*, but marking off distance AB to measure 37 mm.

Example: When construction used for a plate thickness of 12·7 mm, by measurement the required diameter of rivet will be 21·5 mm.

**Fig. 3.23 Graphical method to determine rivet diameters**

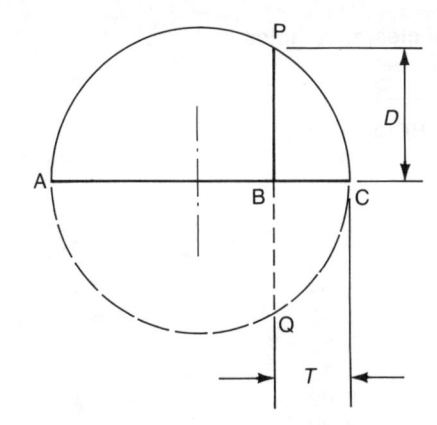

By Intersecting Chords:

$PB \times BQ = AB \times BC$
$PB^2 = AB \times BC$  (PB = BQ)
$PB = \sqrt{AB \times BC}$

PB represents the required rivet diameter (*D*), and BC the thickness of the plate (*T*)

$D = \sqrt{AB \times T}$

(where AB represents a length of 1·44 inches or 37 mm).

(a) *British Imperial Dimensions:*
To calculate required diameter of rivet (*D*) for a given thickness of plate (*T*), use the formula: $D = \sqrt{1.44T}$ (where *T* is given in inches).

Example: Calculate the required diameter of rivet where the plate thickness is 0·375 inches.

Solution:  $D = \sqrt{1.44 \times 0.375}$
$= \sqrt{0.54}$
Diameter of rivet = 0·7348 in.  (nearest stock size ¾ in.)

(b) *Metric Dimensions:*
To calculate required diameter of rivet where the thickness of plate (*T*) is given in mm, use the formula: $D = \sqrt{37T}$

Example: Calculate the required diameter of rivet where the plate thickness is 6·35 mm.

Solution:  $D = \sqrt{37 \times 6.35} = \sqrt{234.95}$
Diameter of rivet = 15·32 mm.

**Fig. 3.24 Intersecting chords**

Figure 3.24 illustrates how, by the application of INTERSECTING CHORDS, alternative formulae can be derived from the information contained in Figure 3.23.

Table 3.4 gives the diameter of rivets commonly used for assembling two sheets or plates of equal thickness.

It is also important to use a rivet of the correct length. A rivet which is too long has a tendency to bend when headed. On the other hand, if a rivet is too short, it will prove too difficult to form a satisfactory head. **The correct length of the rivet should be equal to the total thickness of the sheets through which it is inserted plus the necessary allowance for forming the head.**

**Table 3.4  Rivet sizes**

| METAL THICKNESS | | DIAMETER OF RIVET |
|---|---|---|
| mm | S.W.G. | (mm) |
| 0·80 U | 22 | 1·587 |
| 1·00 U | 20 | 2·381 |
| 1·25 U | 18 | 3·175 |
| 1·60 U | 16 | 3·969 |
| 2·50 U | 14 | 4·763 |
| 2·80 U | 12 | 4·763 or 6·350 |
| 3·55 U | 10 | 6·350 or 7·938 |
| 4·76 | | 9·525 |
| 6·35 | | 11·113 or 12·700 |
| 7·94 | | 12·700 or 15·875 |
| 9·53 | | 15·875 or 19·050 |

Code letter U denotes I.S.O. metric preferred series.

The length of 'shank' required to form the 'head' of the rivet (i.e., the length standing proud of the sheet surface when the rivet is inserted in the hole and *held up tight*) depends upon the form of the head and the 'clearance' between the rivet and the rivet hole. For ROUNDHEAD or SNAPHEAD forms the length of shank required is $1\frac{1}{2}$ to $1\frac{3}{4}$ times the diameter of the rivet, the TOTAL LENGTH of shank required is this length plus the total thickness of the plates through which it is inserted.

The total length of shank required is equal to the total thickness of the metal to be joined plus the allowance for making the head, i.e.,
$L = T + T + 1.5D$

D = Diameter of rivet
T = Thickness of metal
L = Length of shank

EXAMPLE: What length of 4 mm diameter rivet is required to form a snaphead and join two pieces of 1·6 mm sheet metal together?

*Solution*:  $L = T + T + 1.5D$
$= 1.6 + 1.6 + 1.5 \times 4$ mm
**Rivet length** = 3·2 + 6 mm = **9·2 mm**

For the COUNTERSUNK HEAD form, the length of shank required is equal to the diameter of the rivet.

The total length of shank required is equal to the total thickness of the metal to be joined plus the allowance for making the head, i.e.
$L = T + T + D$

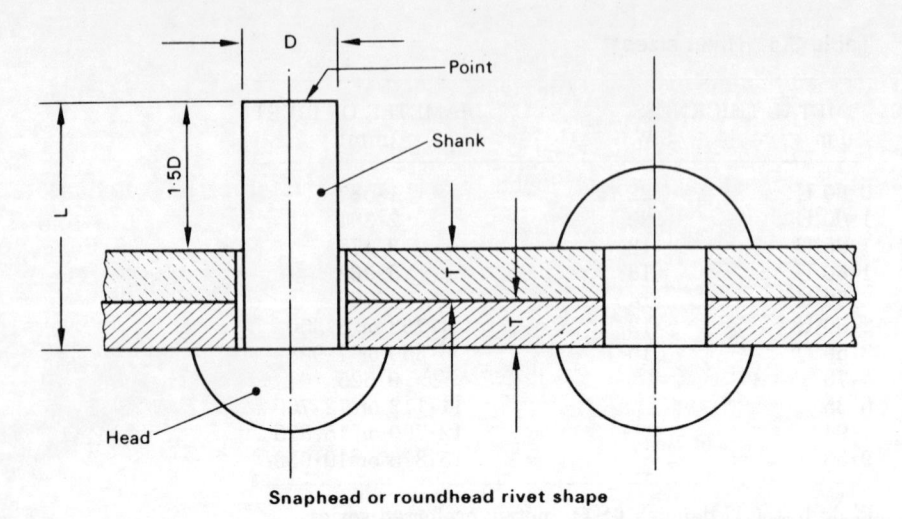

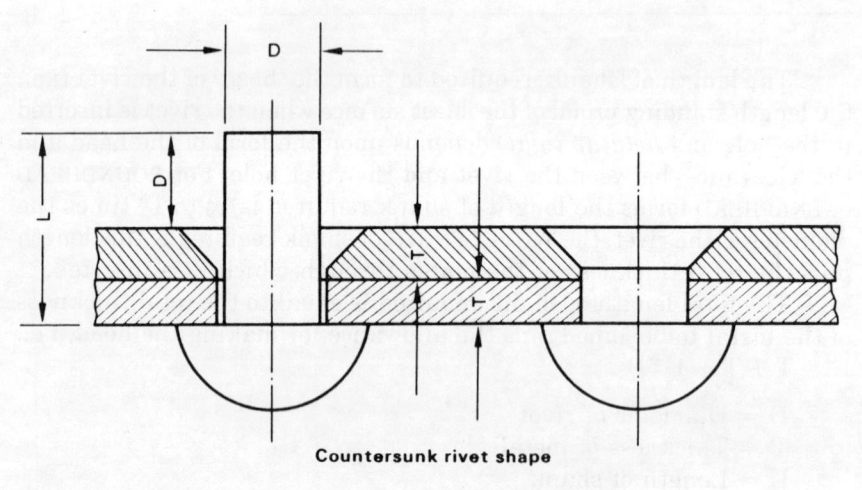

**Fig. 3.25** Factors influencing choice of rivet lengths

EXAMPLE: What length of 9·5 mm diameter rivet is required to form a countersunk head and join two pieces of 4·76 mm plate?

Solution:
$$L = T + T + D$$
$$= 4\cdot76 + 4\cdot76 + 9\cdot5 \text{ mm}$$
$$= 19\cdot02 \text{ mm}$$

**Rivet length** = 19 mm

Figure 3.25 illustrates both forms of rivet head.

Hole clearance is very important and should be kept to the absolute minimum. Table 3.5 shows the recommended hole clearances for various diameters of rivet.

**Table 3.5 Recommended diameters of clearance holes for rivets**

| RIVET DIAMETER (mm) | HOLE DIAMETER (mm) |
|---|---|
| 1·59 | 1·63 |
| 2·38 | 2·43 |
| 3·18 | 3·25 |
| 3·97 | 4·03 |
| 4·76 | 4·85 |
| 5·56 | 5·61 |
| 6·35 | 6·52 |
| 7·94 | 8·02 |
| 9·53 | 9·80 |
| 11·11 | 11·40 |
| 12·70 | 13·10 |

## 3.24 Rivet spacing

Rivet holes should be spaced according to the specification of the job.

*The space or distance from the edge of the metal to the centre of any rivet should be at least twice the diameter of the rivet*, to prevent the rivets from tearing out.

A useful rule is to make the edge distance equal to $1\frac{1}{2}$ diameters plus 9·5 mm. The maximum distance from the edge is governed by the necessity of preventing the sheets from 'gaping' and should be limited to 10 times the thickness of the sheet metal or plate.

*The minimum distance between rivets (known as the '***pitch***') should be sufficient to allow the rivets to be driven without interference* (see Section 3.26) *or about three times the rivet diameter.*

The maximum distance between rivets should never be such that the material is allowed to buckle between the rivets and in practice should never exceed 24 times the thickness of the sheet. Rivet spacing is illustrated in Figure 3.26.

## 3.25 Defects in riveted and bolted joints

Most of the faults or defects associated with riveted and bolted connections are caused by bad workmanship. Numerous defects may be avoided by adopting an intelligent approach to the job in hand, especially in the preparation stage, and by careful use of tools.

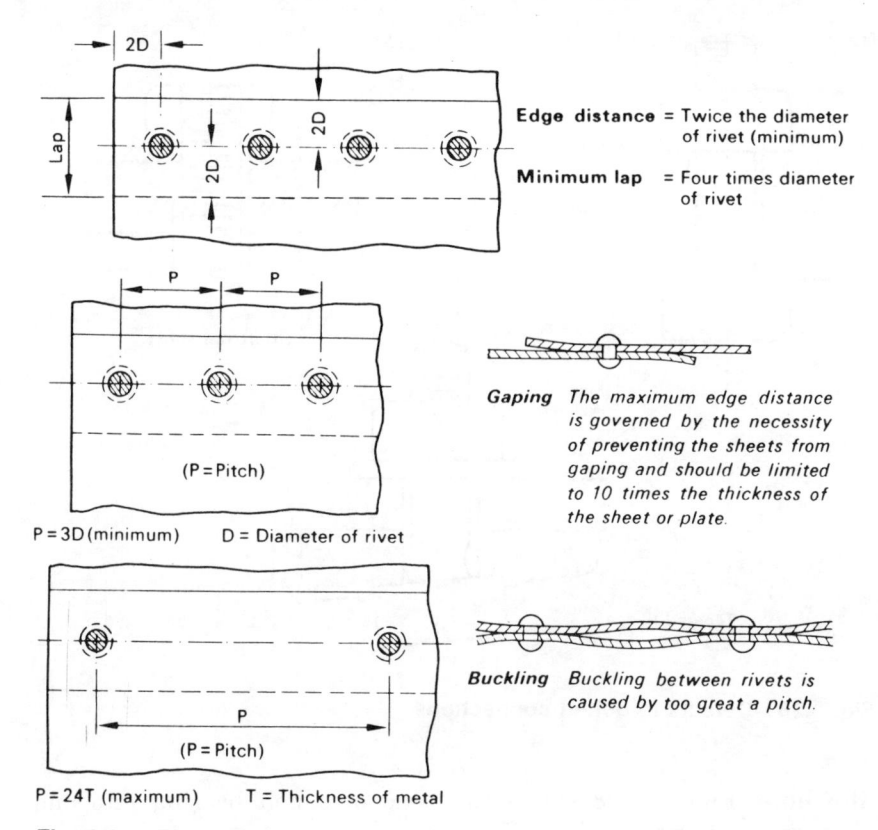

**Fig. 3.26 Rivet spacings**

Edge distance = Twice the diameter of rivet (minimum)
Minimum lap = Four times diameter of rivet

P = 3D (minimum)   D = Diameter of rivet
P = 24T (maximum)  T = Thickness of metal

*Gaping* The maximum edge distance is governed by the necessity of preventing the sheets from gaping and should be limited to 10 times the thickness of the sheet or plate.

*Buckling* Buckling between rivets is caused by too great a pitch.

When making connections which require the use of rivets or bolts, the following are a few of the basic details which should be considered in order that many common defects may be prevented:

(i) When marking out use the correct allowance for edge clearance and pitch;
(ii) All drilled or punched holes should be made to the correct clearance size to suit the bolt or rivet diameter, or as specified on the drawing;
(iii) Remove any 'burrs' from around the edges of all holes before final assembly of the parts to be joined;
(iv) Ensure that holes are correctly lined and matched before inserting the bolt or rivet;
(v) Use the proper type of rivet or bolt as specified on the drawing;
(vi) Use rivets or bolts of the correct length;
(vii) Where bolts are employed always use a suitable washer under the nut;
(viii) When using bolts do not overtighten the nut;
(ix) When inserting bolts or rivets, do not attempt to force or drive them into the hole;
(x) Always use the correct tools for the job.

Some of the common forms of defects associated with riveted and bolted connections are illustrated in Figure 3.27. Figure 3.28(a), (b) and (c) also illustrates associated defects:

**(a)** Unless a suitable washer is fitted under the nut, damage will be caused to the contact surface of the plate or structural member.

As the nut is tightened it will tend to bite into the metal and spin up an annular bead as it penetrates the bearing surface. The bolt will always tend to become slack, and any subsequent tightening of the nut will cause it to bite still further into the metal. Eventually, when the whole of the threaded portion of the bolt has completely penetrated the nut, no further tightening can take place and the joint becomes weak.

**(b)** With any bolted joint at least two screw threads on the bolt should protrude through the nut. Therefore, a bolt of the correct length should be used. If a bolt is too short the nut cannot obtain sufficient

| CAUSE OF RIVETING DEFECT | RESULTANT EFFECT |
|---|---|
| Insufficient hole clearance | Rivet not completely 'drawn through'. Not enough shank protruding to form head<br><br>Original head of rivet 'stands proud', the formed head is weak and mis-shapened |
| Hole too large for rivet | Hole not filled<br><br>Rivet tends to bend and deform. Head weak and poorly shaped |
| Rivet too short | Not enough shank protruding to produce a correctly shaped head<br><br>Plate surface damaged — Countersinking not completely filled |

**Fig. 3.27 Common defects in riveting (continued overleaf)**

| CAUSE OF RIVETING DEFECT | RESULTANT EFFECT |
|---|---|
| Rivet too long | Too much shank protruding to form required head<br>'Flash' formed around head (Jockey cap) — Countersinking over-filled |
| Rivet set or dolly not struck square | Badly shaped head - off centre<br>Sheet damaged by riveting tool |
| Drilling burrs not removed | Not enough shank protruding to form the correct size head<br>Plates or sheets not closed together. Unequal heads |
| Sheets not closed together — rivet not drawn up sufficiently | Weak joint. Rivet shank swells between the plates<br>Not enough shank protruding to form correctly shaped head |
| Rivet holes not matched | Weak mis-shapened head<br>Rivet deformed and does not completely fill the hole |

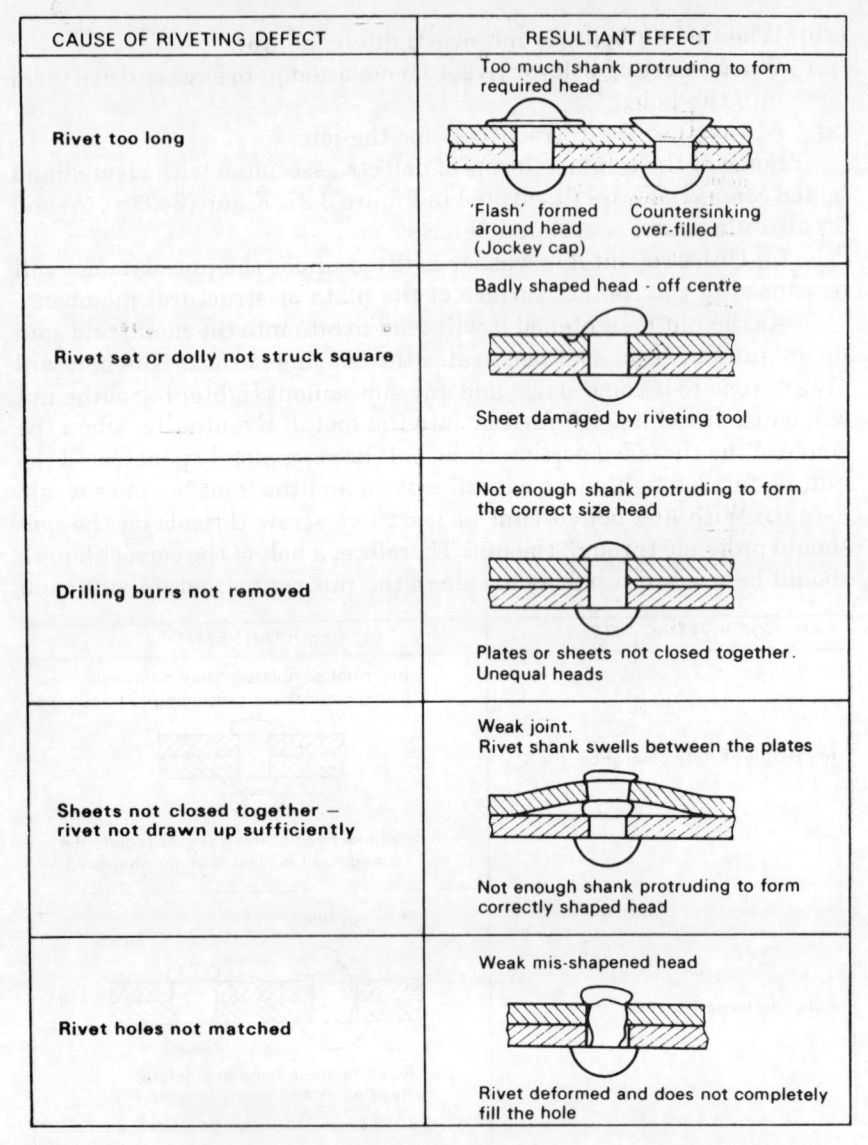

**Fig. 3.27 (continued)**

grip on the screw thread of the bolt, and there is always a tendency for it to work loose. Any attempt to tighten the nut generally results in stripping the thread, especially when a bolt with a fine thread is used.

**No part of the thread on a bolt should be in the hole.**

**(c) The contact surfaces of a nut or bolt must be square to the hole.** Figure 3.28 shows the head of a bolt bearing into and damaging a flange.

A '*tapered washer*' must be used, otherwise the tapering surfaces of British Standard Beams (BSB), Channels or Tee-sections will cause the shank of bolt to bend. *If only one washer is used when bolting sections with tapered flanges, then the bolt must be arranged so that a tapered washer is fitted under the nut.*

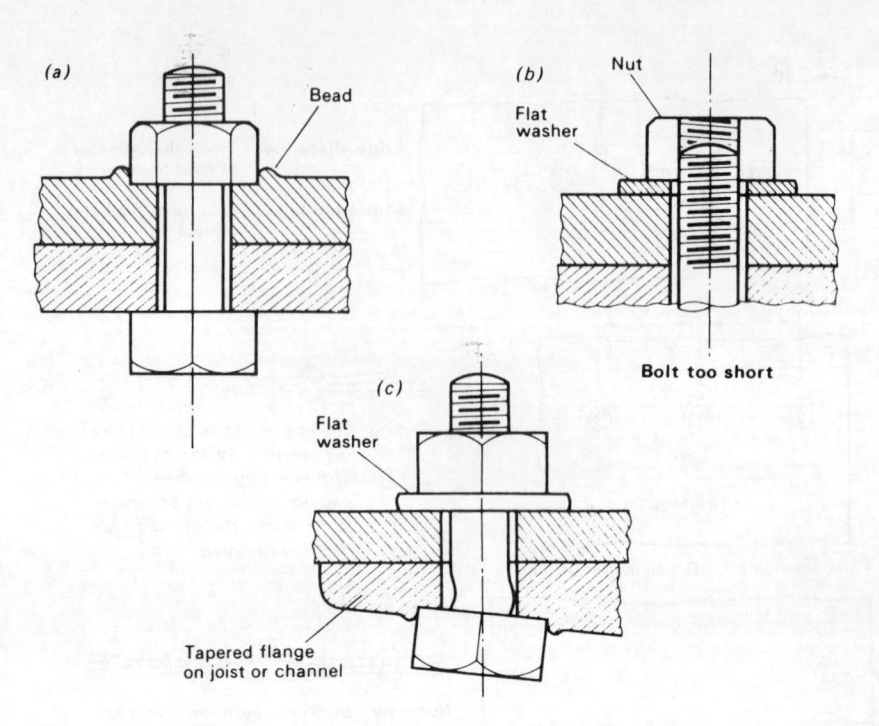

**Fig. 3.28 Defects in bolted connections**

## 3.26 Clearances

The diameter of the holes and the clearances allowed for bolts are usually specified on the drawing, and in the case of structural steelwork, for example, standard clearance dimensions are used.

*The minimum distance between centres of bolts is $2\frac{1}{2}$ times the nominal diameter of the bolt.* This dimension may have to be increased to ensure sufficient 'working clearance', and also with the size of washer.

### WORKING CLEARANCE

In all bolted or riveted connections sufficient clearance must be provided to allow for turning of a bolt-head or nut with a spanner, or driving a rivet with a rivet set or 'dolly'.

Reference to some standard clearances will be made in Section 3.29.

## 3.27 Applications — types of rivet and rivet head

The solid rivet has been in use for many hundreds of years. Today rivets in general use are made from SOFT IRON, MILD STEEL, COPPER, BRASS, ALUMINIUM and a range of ALUMINIUM ALLOYS.

The standard types of rivet heads are illustrated in Figure 3.29 together with the method employed when countersunk rivets are used for joining thin material to thick material.

*Tinners* and *flat head* rivets are used in most general sheet metal fabrications, where the metal is very thin and little strength is required. The *countersunk head* is used when a flush surface is required, and the *roundhead* or *snaphead* is most widely used where the joint must be as strong as possible. *Mushroom head* or *'knobbled rivets'* as they are termed in the steel construction industry, are used where it is necessary to curtail the height of the rivet head above the metal surface, as, for example, on the outer fuselage 'skins' of aircraft in order to decrease 'drag', or in the case of steel chutes and bunkers, to reduce obstruction on the inside surfaces. *Pan head* rivets are very strong, and are, therefore, widely used for girders and heavy constructional engineering.

'Tinners' are similar to flat head rivets. They are made of soft iron and are usually coated with tin to prevent corrosion and to make them easier to soft solder — hence their name 'tinners'.

**Always use the correct rivet for a particular metal to be riveted.** When riveting aluminium, for example, use aluminium rivets; and when riveting copper use copper rivets.

## 3.28 Applications — pop riveting

Pop rivets, unlike solid rivets, are tubular and therefore much lighter in weight. They are manufactured from either aluminium alloy for lightness or nickel alloy for additional strength and corrosion resistance. They were originally designed for blind or one-sided riveting, by which rivets can be set in places otherwise inaccessible. Only one operator is required, the rivets being set or clinched with the aid of special hand-held 'lazy tongs' or pliers. Although their main application has been in aircraft construction and motor vehicle body building, where frequently it is necessary to join thin material to thicker supporting members and lightness is important, they are often used in place of solid rivets for general riveting. They are available in diameters 2.4 mm, 3.2 mm, 4 mm and 4.8 mm for joining thicknesses up to 12.7 mm. With regard to speed of application, on a reasonably straight run it is possible to set pop rivets at a rate of twenty rivets per minute. Figure 3.30 illustrates the setting of pop rivets.

Pop rivets are supplied with either a 'break-stem' or a 'break-head'

**Fig. 3.29 Rivet heads and applications**

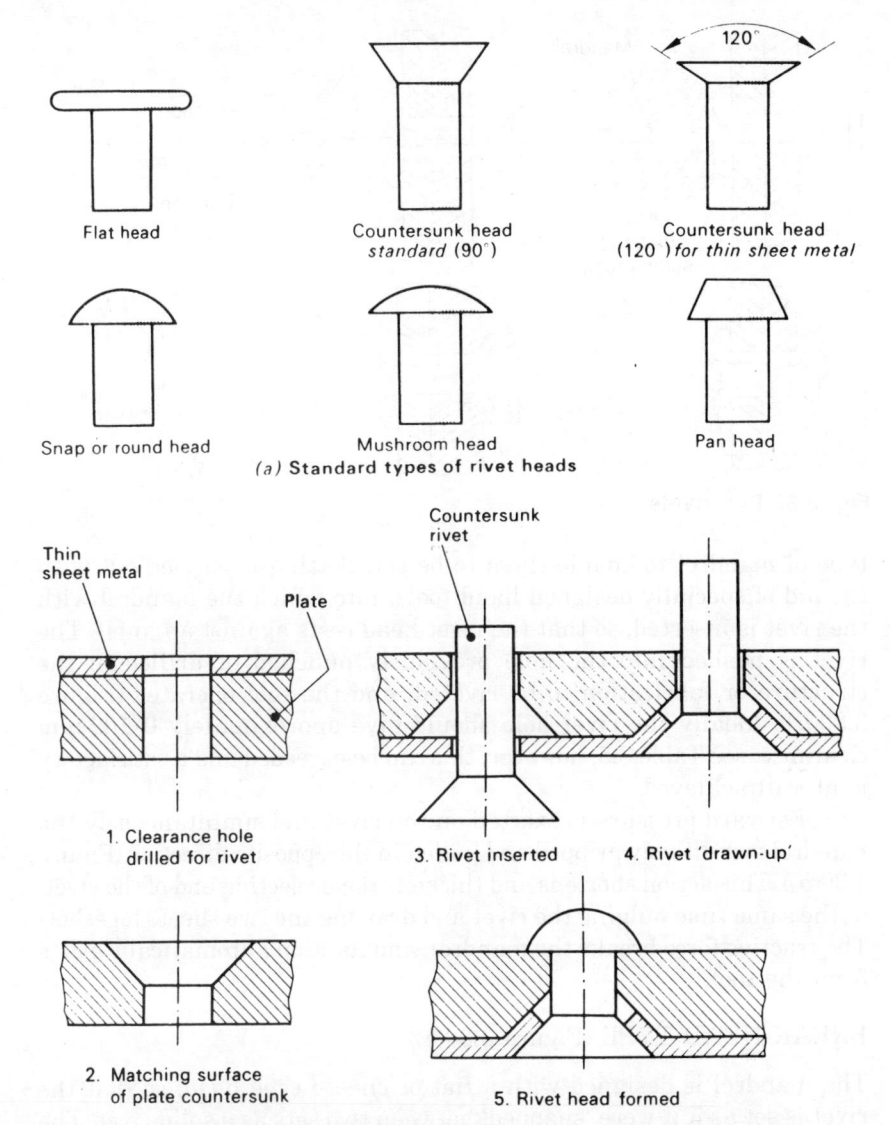

(a) Standard types of rivet heads: Flat head, Countersunk head *standard* (90°), Countersunk head (120°) *for thin sheet metal*, Snap or round head, Mushroom head, Pan head

(b) Countersunk riveting of thin material to thick material
1. Clearance hole drilled for rivet
2. Matching surface of plate countersunk
3. Rivet inserted
4. Rivet 'drawn-up'
5. Rivet head formed

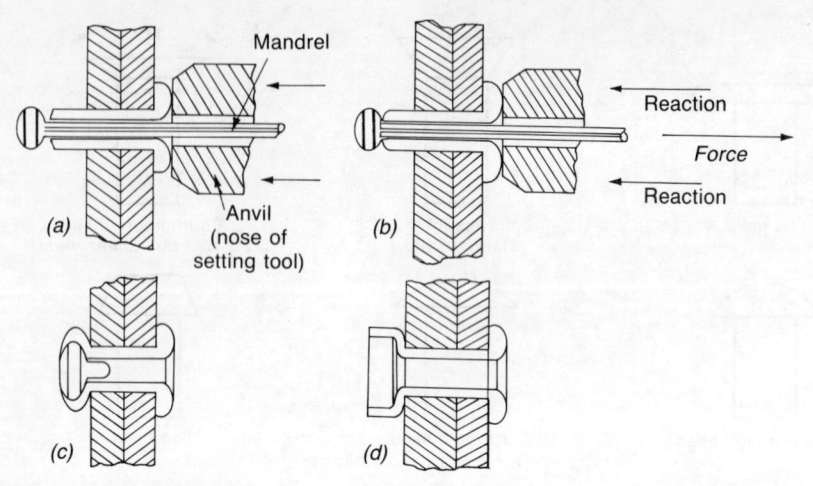

**Fig. 3.30 Pop rivets**

type of mandrel to enable them to be set. Setting is carried out with the aid of specially designed hand tools, into which the mandrel with the rivet is inserted, so that the rivet head rests against an anvil. The rivet is pushed into the hole previously punched or drilled in the structure or fabrication to be riveted, and the tool operated (Figure 3.30(a)). Ideally the rivet hole should give approximately 0.102 mm clearance; see Table 3.5; however, this can be exceeded and a satisfactory joint still achieved.

Forward pressure is exerted on the rivet and simultaneously the mandrel is positively gripped and pulled in the opposite direction (Figure 3.30(b)). This action shortens and thickens the projecting end of the rivet, at the same time bulging the rivet and drawing the two sheets together. The tractive force breaks the mandrel, and its stem automatically ejects from the tool.

BREAK-HEAD TYPE (Figure 3.30(c))

The mandrel is designed with a flat or cheese type head, so that the rivet is set as if it were 'snapped' between two sets as a solid rivet. The breaking point is immediately beneath the head of the mandrel — hence its name. On completion of the rivet setting, the mandrel head breaks off at its designed load (135 N to 360 N, according to the size of the rivet used).

BREAK-STEM TYPE (Figure 3.30(d))

The head on this mandrel is very similar to that on the break-head type, but its stem breaks approximately 1.3 mm from the shoulder of its head.

Break-stem pop rivets are used when it is necessary to retain the head, and therefore are ideally suitable when riveting is carried out on a totally enclosed member.

## 3.29 Applications — black and turned bolts

In most bolted assemblies 'hexagon bolts and nuts' are preferred. *This is because a bolt head or nut of hexagon shape offords more spanner positions* and can be used in places where the 'square nut' would be difficult to tighten because of lack of spanner-turning space, thus ensuring maximum tightening.

The cheapest type of bolt is the 'BLACK BOLT' — so-called from its appearance — it is forged from round steel stock, the head being 'upset' by machine, and the only machining on it is the cut screw thread (this also applies to the nut). These bolts are supplied with a unified coarse thread (UNC).

Black bolts are used in 'clearance holes' which are normally 1.6 mm larger in diameter and transmit their loads by direct bearing or shear at relatively low stresses. This means that in certain connections which are to be highly stressed, black bolts cannot be used. In such cases use is made of the 'turned bolt' which is capable of carrying much higher loads.

TURNED BOLTS, also known as 'precision', 'fitted' or 'close tolerance bolts', are more expensive because they are machined on the 'shank' or 'barrel' and under the head the nuts are 'faced' on one side. *They are used where greater dimensional accuracy is required and in holes with a very small clearance.* Clearance holes for these bolts are generally 0.127 mm larger than the nominal diameter of the bolt — if a larger clearance is used it should not exceed 0.397 mm. *When making connections 'on site' greater attention has to be given to the alignment or matching of the holes, and care must be taken when inserting turned bolts in order that no damage is caused either to the bolts themselves or to the metal surrounding them.* Turned bolts are supplied with either a unified coarse thread (UNC) or a unified fine thread (UNF). Figure 3.31 illustrates the essential features of 'black' and 'turned' bolts.

## 3.30 High strength friction grip bolts

When using turned or fitted bolts for connections that are highly stressed the major problem is the insertion of the bolts. *To overcome this problem it is standard practice to have the holes drilled undersize in the workshop, make the connection by assembling the sections with 'temporary tacking bolts', and then reaming the holes to the correct size before the bolts are inserted.* Such a procedure is time consuming and expensive and a

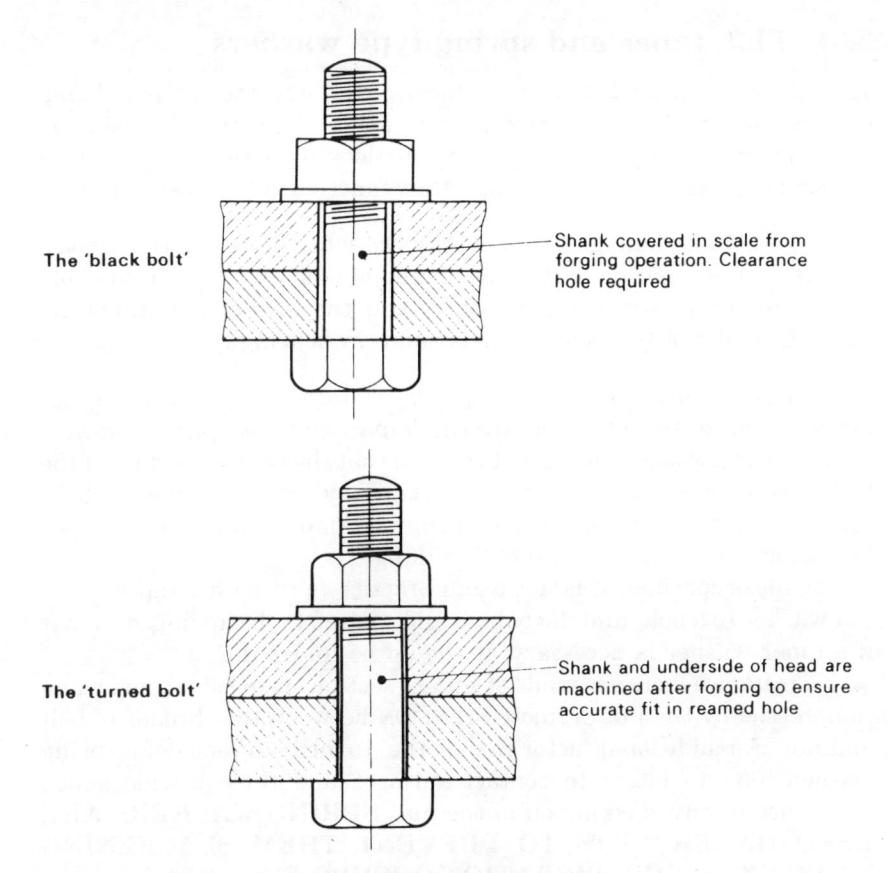

**Fig. 3.31 Comparison of black and turned bolts**

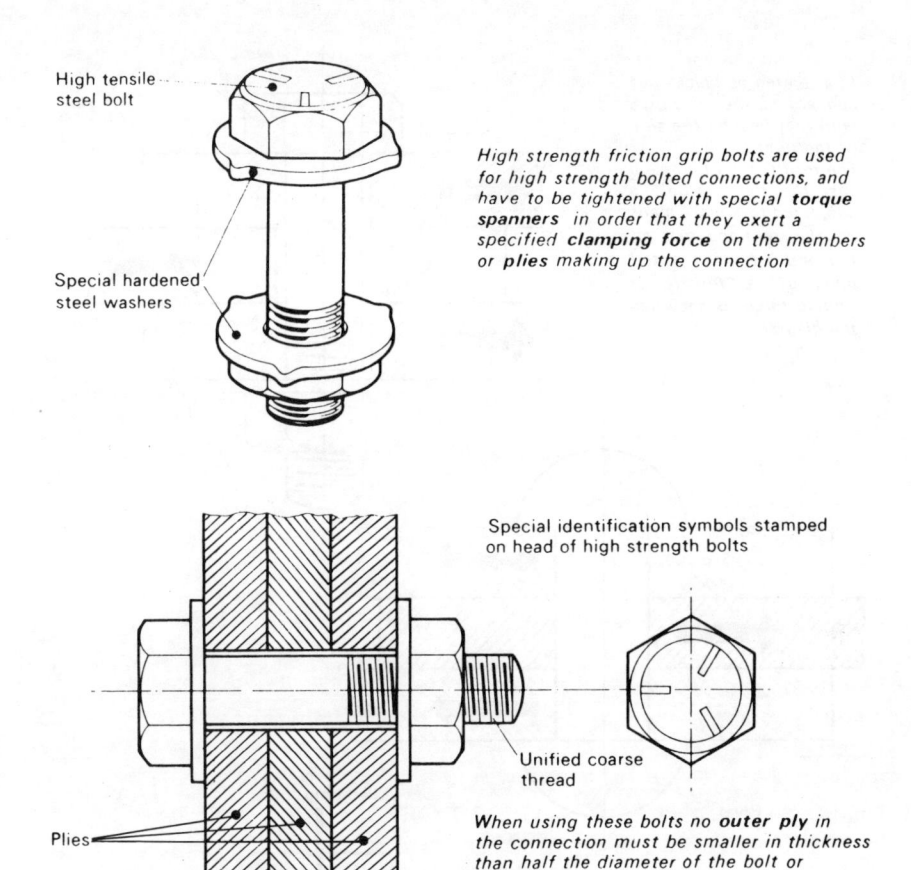

**Fig. 3.32 High strength friction grip bolts**

method of bolting known as 'HIGH STRENGTH BOLTING' is now quite frequently used.

*High strength friction grip bolts* are forged from HIGH TENSILE STEEL and the underside of the head and the bearing surface of the nut are semi-finished — *special identification symbols are stamped on the head*. Such bolts can be inserted in normal clearance holes (usually 1·6 mm) as if the connection were to be made with standard black bolts. *Each bolt is provided with two hardened washers which must be fitted under the head and the nut*, as shown in Figure 3.32. These washers spread the load, preventing the head and the nut becoming embedded in the relatively softer steel of the structural members.

*When high strength bolting is employed the bolts are tightened to such an extent that they carry the shear and bearing loads imposed on them by FRICTION between the faces of the connected parts.*

High strength bolts have to be tightened until they exert a specified CLAMPING FORCE on the members making up the connection. *Thus the function of the bolt is to hold the 'plies' clamped so that frictional forces are produced at their contact surfaces.*

The essential differences between riveted or black bolt connections and connections made with high strength bolts is illustrated in Figure 3.33.

When steel stuctures are made up by bolting, it is common practice to descale or remove rust, and then paint the contact surfaces with 'red lead'. **When using high strength bolts no paint should be applied to contact surfaces adjacent to the bolts.** *The bolts must be square to the bearing surfaces under the bolt head and nut, hardened 'taper washers' being used where necessary.*

*In a riveted or black bolt connection, the plies give until they bear on the side of the rivet or bolt.*
*The result is that in each case the rivet or bolt is subjected to 'shear' loads on the sections marked A-A and B-B when acted upon by a 'tensile' or pulling force, as shown in the diagrams.*

*The **clamping force** of the bolt produces **frictional forces** on the contact surfaces of the plies which resist any shear loads which tend to occur at sections A-A and B-B when the connection is subjected to a tensile force.*

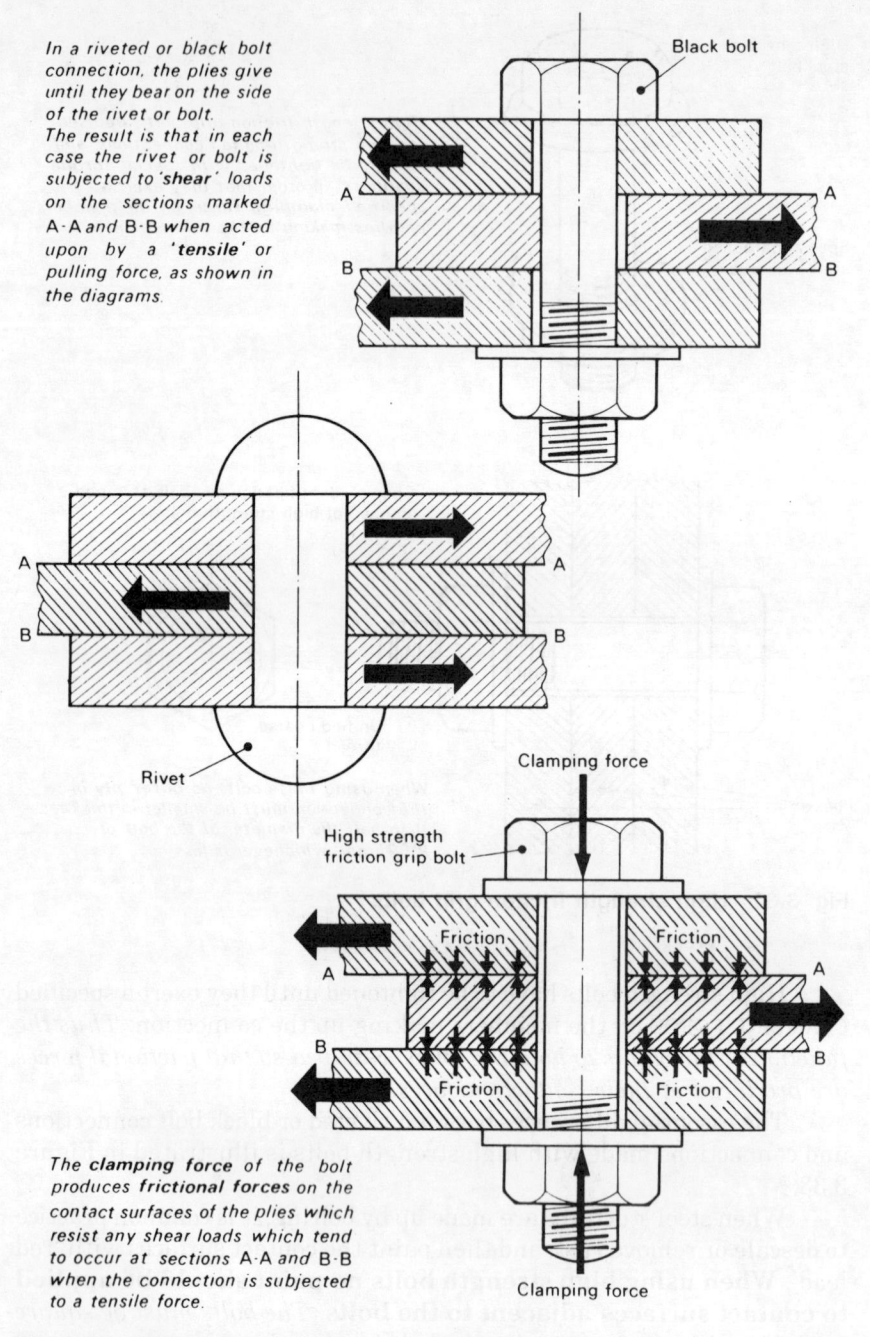

**Fig. 3.33  The essential difference between riveted or black bolt connections and high-strength boltings**

## 3.31 Flat, taper and spring type washers

The use of a washer between the faces of the nut and the part being fastened or connected increases the FRICTIONAL GRIP, thereby reducing the tendency to 'slip' and improving the locking action of the nut on the screw thread. Two other important functions of a washer are:

(i) It prevents damage to the workpiece when nuts are tightened;
(ii) It ensures that no part of the thread of a bolt shall be in the hole in which it is inserted. By doing this the washer maintains the full value of the bolt 'in shear' and 'in bearing'.

In addition to the common *'plain' or 'flat' washer* various other types are used, two of which are the *'taper' and the 'spring' washer*.

Tapered washers are used to ensure that the contact surface of the bolt head or nut are square to the hole. They are obtainable with 5° taper for use on the inside of joist or channel flanges, and with 8° taper for use on the inside surface of Tee flanges.

Unless specified it is not usual practice to fit both a taper and a flat washer to a bolt, and the bolt should, therefore, be arranged so that if a taper washer is necessary, it should be under the nut.

Spring washers are made of square section spring steel, and when compressed by the nut promotes TENSION between the threads of bolt and nut. An additional factor is that the 'turned-out' ends of a spring washer tend to bite into contact surfaces and thus provide added resistance to any slacking-off of the nut. **SPRING WASHERS ARE USED UNDER NUTS TO PREVENT THEM SLACKENING WHEN EXCESSIVE VIBRATION OCCURS**. Figure 3.34 illustrates applications of washers.

## 3.32 The alignment of holes

'DRIFTS' are employed when assembling connections for riveting and bolting, especially with large steel fabrications.

They are made of hardened and tempered steel and are available to suit all sizes of hole. The three basic types are described below and each has a specific function when used for aligning parts coming together so that holes will be 'faired' to admit bolts or rivets, as shown in Figure 3.35.

### TAPER DRIFTS

Used for 'fairing' or aligning rivet and bolt holes. The holes in the plates move into the correct position as the drift is hammered into them.

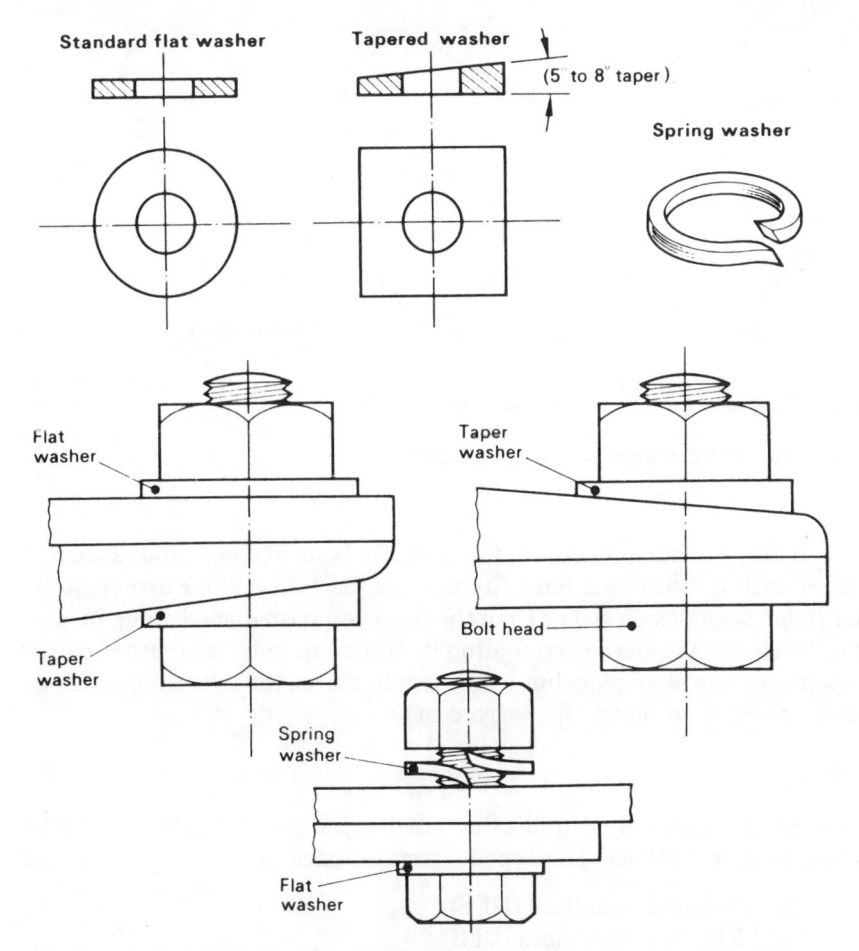

**Fig. 3.34 Flat, taper and spring type washers**

## BARREL DRIFTS

Very useful when aligning holes in confined spaces. The drift is hammered until it passes through the holes.

## PARALLEL DRIFT

Used mainly for holding and locating holes in position during riveting and bolting operations. (*Parallel drifts are generally 0·08 mm smaller in diameter than the size of the hole.*)

'PODGER SPANNERS' are used when making bolted connections. These special spanners are available in two patterns — 'straight' or 'joggled' — and in a range of bolt sizes. The 'joggled' pattern has a 'crank' in the shank to allow for clearance space between the user's hand and the work. The end of the shank is tapered for use in 'podgering' or 'fairing' holes. Both patterns are illustrated in Figure 3.35.

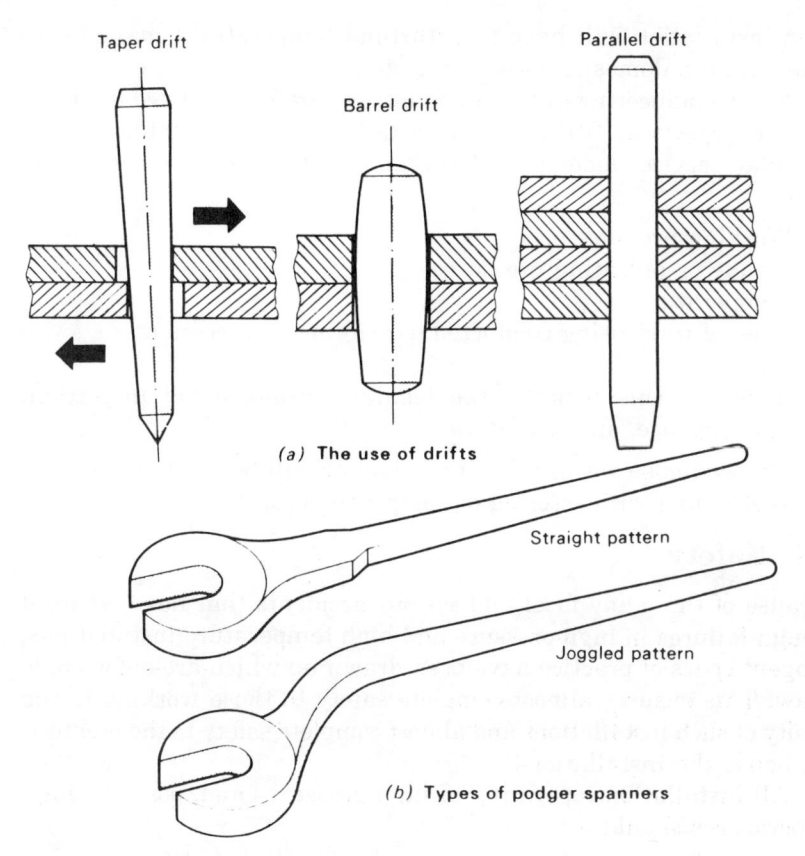

**Fig. 3.35 Fairing or aligning holes**

## 3.33 Pipe fitting

Except for coolant, fuel and lubrication services, engineering pipework has to withstand much higher temperatures and pressures than those met in domestic plumbing. The light gauge copper pipe, compression jointed, found in modern houses is satisfactory up to pressures of $1·0$ $MN/m^2$ ($10·0$ bar) providing its temperature does not exceed boiling point.

An industrial hydraulic system may be called upon to withstand pressures up to $41·4$ $MN/m^2$ (414 bar) and a high pressure super-heated steam installation will have to withstand temperatures many times greater than a domestic hot-water system.

Every engineering installation must be carefully planned to ensure that the pipework, fittings, and installations are suitable for the particular service under consideration. The factors to be considered include:

1. Working pressure
2. Steady or pulsating pressures
3. Temperature
4. Type of fluid being conducted (possibility of corrosion)
5. Cost
6. Safety — this includes the legal requirements of inspection, maintenance, and insurance.

The systems considered in this chapter will be limited to copper and steel piping with screwed or compression joints.

### 3.34 Safety

Because of the many fatal and serious accidents that have occurred through failures in high-pressure and high-temperature installations, stringent codes of practice have been drawn up which are enforceable at law. This ensures almost complete safety to those working in the vicinity of such installations and almost complete safety to the premises that house the installations.

All installations operating at high pressures and, possibly, high temperatures should be:

1. Designed by experts to meet the requirements of the Factory Acts.
2. Constructed only from equipment that has been tested and certified.
3. Constructed from equipment that follows the recommendations of the British Standards Institution.
4. Inspected by the insurance surveyor upon completion.
5. Inspected at regular intervals thereafter, usually annually, by the insurance surveyor.
6. Colour coded to indicate the content of the pipework (see Section 3.40).

### 3.35 Types of pipe

#### COPPER

For pressures up to 0·7 MN/m$^2$ (7·0 bar), and temperatures up to 100°C, copper pipe is usually chosen. This will be solid drawn, that is, it will not contain a seam running along its length (Figure 3.36).

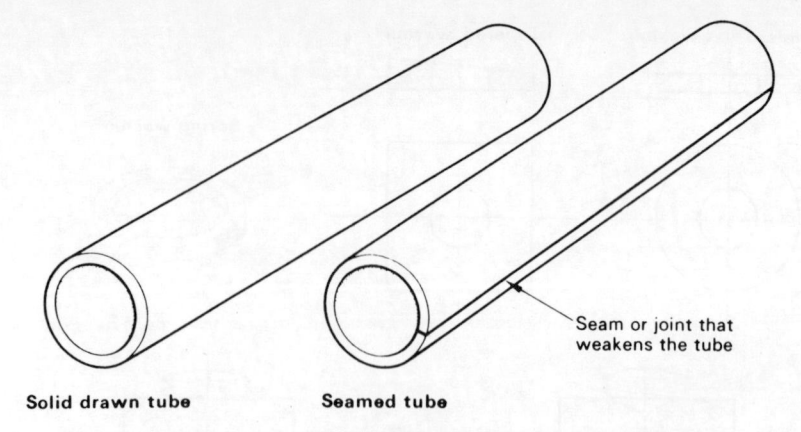

**Fig. 3.36 Solid drawn and seamed tube**

It has the advantages that it is easily bent to shape and is easily joined with compression joint fittings (see Section 3.38). Further, it is not liable to corrosion either from the fluid being conducted along inside it or by the atmosphere surrounding it. Unfortunately, it is rather more expensive than steel pipe, but in the smaller sizes the advantages listed above more than offset the extra cost.

#### STEEL

Both plain carbon steel and alloy steel pipes are available for high pressure installations. The types normally used are:

Hot Finished Seamless (HFS)
Cold Finished Seamless (CFS)
Electric Resistance Welded (ERW)

**Hot finished seamless** pipe can be used at the highest pressures and temperatures. It has a rough finish and cannot be used with compression jointed fittings. It is used with screwed or welded joints.

**Cold finished seamless** pipe is used for the same purposes as hot finished, but as it has a smooth surface it can be used with compression joint fittings. It is more costly than hot finished pipe.

**Electric resistance welded** pipe is very much cheaper than solid drawn pipe. It has a good finish, and is accurate enough to be used with compression joint fittings. It is only available in sizes up to 50 mm outside diameter.

**Note:** the wall thickness for pipes carrying water are increased for any given pressure to allow for corrosion.

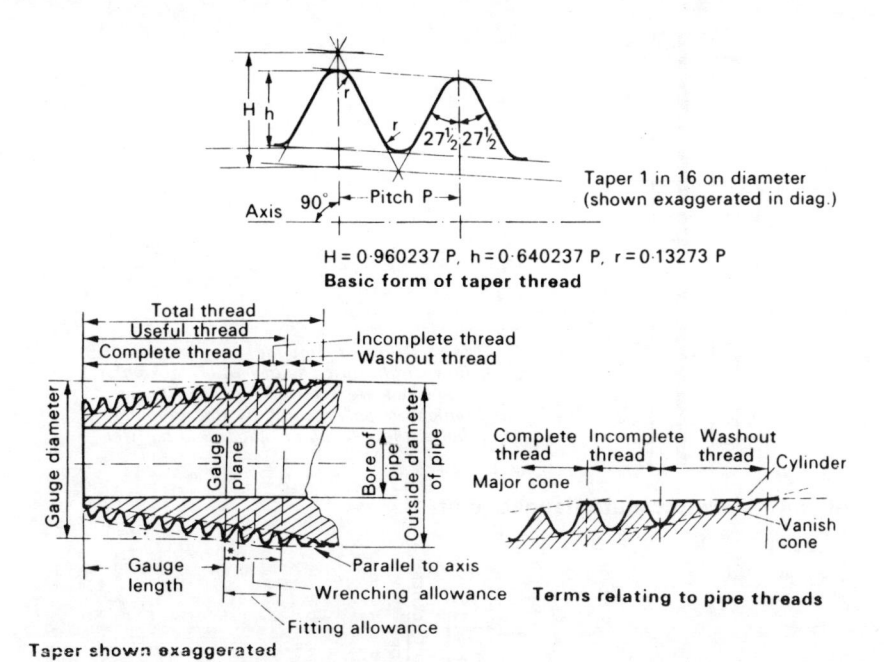

**Fig. 3.37 British Standard pipe threads**

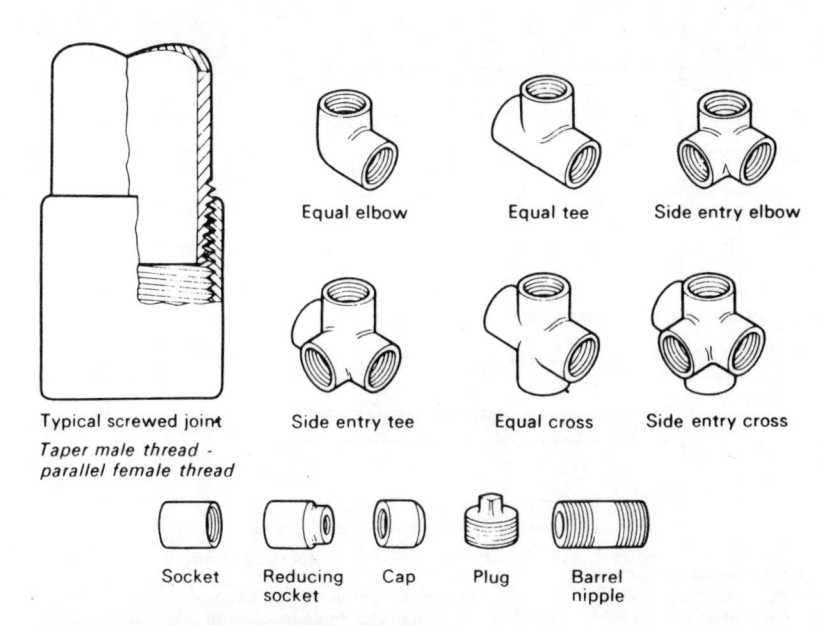

**Fig. 3.38 Typical screwed fittings**

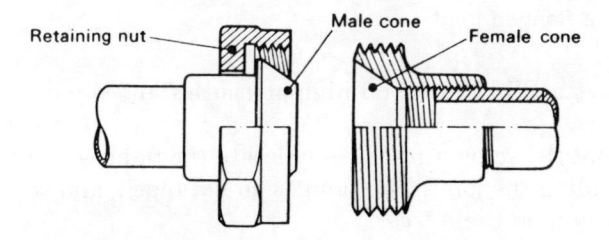

**Fig. 3.39 Union joint**

## 3.36 Screwed joints

At present, British Standard Pipe Threads are used for all screwed joints, although in time ISO-metric threads will be used.

The thread is of Whitworth form and may be tapered or parallel, as shown in Figure 3.37.

Tapered threads are self-seal when pulled up tight and are used with fittings threaded parallel internally. Some examples are shown in Figure 3.38.

To ensure the pressure tightness, the joint is usually sealed by brushing a 'jointing compound' on to the threads before screwing the joint together. There are many proprietary makes of jointing compound on the market. Linseed-oil based, putty type compounds are suitable at low temperatures and pressures. They have the advantage that the joint may be easily broken out if a fitting has to be altered or replaced. At high pressures and temperatures, a hard setting synthetic resin-based compound should be used. At very high temperatures, such as super-heated steam, an all-metal joint should be used (see Section 3.37).

Although the types of joint shown are simple to make and neat in appearance, they have some fundamental limitations.

1. Valves and other fittings cannot be readily broken out of the pipe run for maintenance or replacement.
2. If the pipe is cranked, a large amount of room is required to swing it round as it is screwed home.

To overcome these limitations a number of unions (Figure 3.39) have to be left in the pipework and these are a source of weakness.

## 3.37 Flanged joints

These overcome the disadvantages of the plain screwed joint, but are:

1. More costly to produce.
2. More bulky.

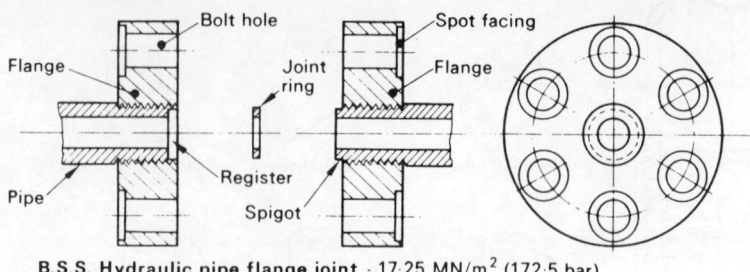

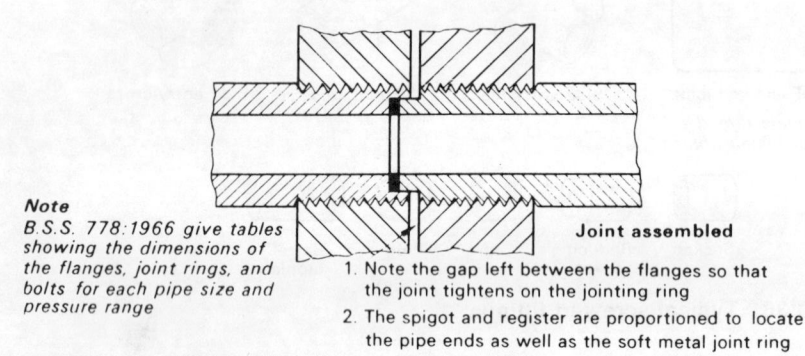

**Fig. 3.40 High-pressure flanged joint**

Figure 3.40 shows a BSS 778 : 1966 high-pressure flanged joint with an all metal seal.

It will be seen that the screw threads only locate the flanges and in no way assist in sealing the joint. The flanges do not touch, and so do not require machining on their faces.

It is usual to spot face the nut and bolt seatings. This type of joint is not specified for pipe work above 6 inches outside diameter. Above this size, the flange is usually welded to the pipe.

Figure 3.41 shows a simpler type of flange joint used at low pressures and temperatures.

## 3.38 Compression joints

These are becoming increasingly used in place of screwed joints for the smaller size pipes. Not only do they save time in installation (the pipe does not have to be threaded), but lighter gauge pipe can be used since it is not weakened by the screw thread (Figure 3.42).

Initially, compression joints were made from brass and limited to copper pipework and pressures of 1 MN/m² (10 bar). Nowadays, they are being made from steel for pressures up to 41·4 MN/m² (414 bar). Figure 3.43 shows some typical joints.

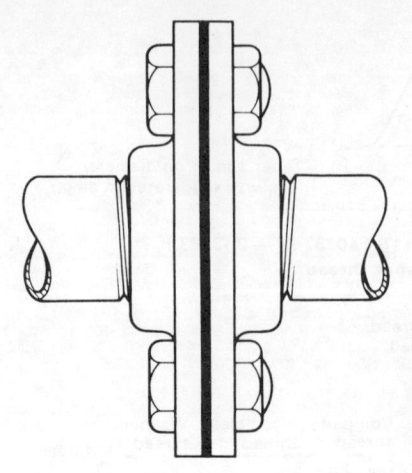

**Fig. 3.41 Low-pressure flanged joint**

*In this type of joint the flanges are pulled down tight onto the gasket or packing. The pipes are screwed into the flanges which are only located by the retaining bolts, there being no spigot and register*

**Fig. 3.42 Effect of thread on wall strength**

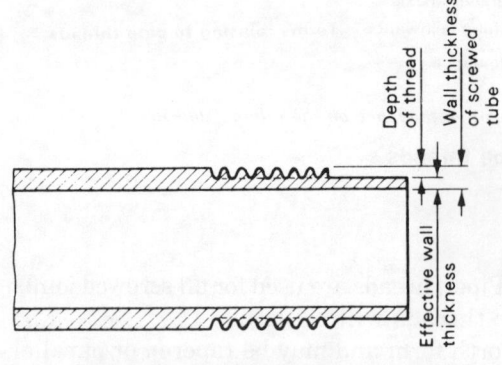

1. The tube is only as strong as the **effective** wall thickness left after screwing. Therefore except where the pipe is to be threaded, it is unnecessarily thick. This is a waste of material

2. A pipe used with compression joint fittings need only have effective wall thickness throughout its length. This not only saves material, the pipe is easier to bend

## 3.39 Screwing equipment

Pipework may be screwed by hand on the site, or machine screwed in the workshop if the pipework is pre-fabricated. The solid button dies used for electrical conduit are largely unsuitable for pressure-tight joints as the thread cut cannot be adjusted and the finish is too rough.

Adjustable dieheads are used, as shown in Figure 3.44(a). The thread may be produced in a number of cuts. This prevents tearing and allows the thread to be adjusted in size so that the pipe will tighten up in the fitting to which it is assembled.

Considerable force is required to rotate the diehead when threading large diameter tubes.

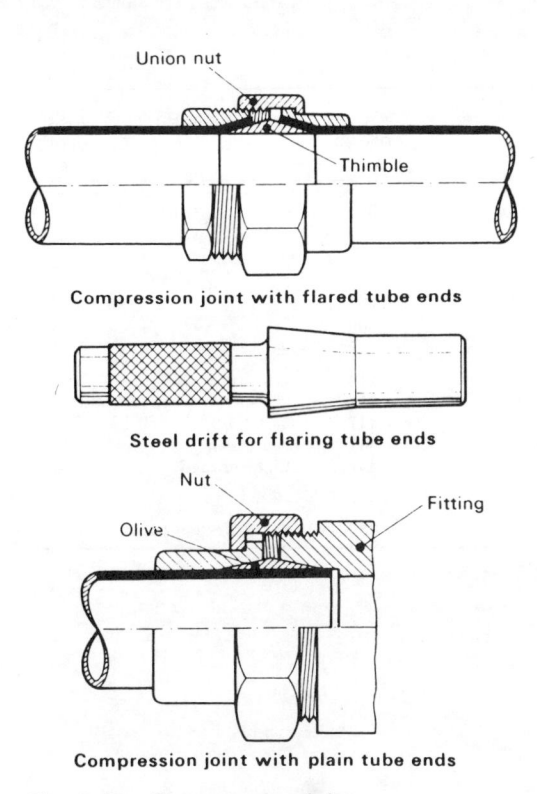

**Fig. 3.43 Compression joint**

For this reason, a long tommy bar is fitted to the diehead so that the fitter can obtain sufficient leverage. It would not be convenient to swing this long tommy bar round the tube, so it is attached to the diehead by means of a ratchet. Thus the bar need only be swung back and forth through a small arc.

When working with a portable pipe vice and stand, the torque of the diehead must be balanced. The fitter's mate will hold the pipe from turning with chain grips as shown in Figure 3.44(b).

Hand- and motor-operated screwing machines are available both in the workshop and as portable units. They usually have opening dieheads so that they do not have to be reversed while the pipe is removed. A typical machine is shown in Figure 3.45.

## 3.40 Colour coding

For safety reasons all pipework installations should be colour coded in accordance with BSS 1710 1960. The recommendations are given in Table 3.6.

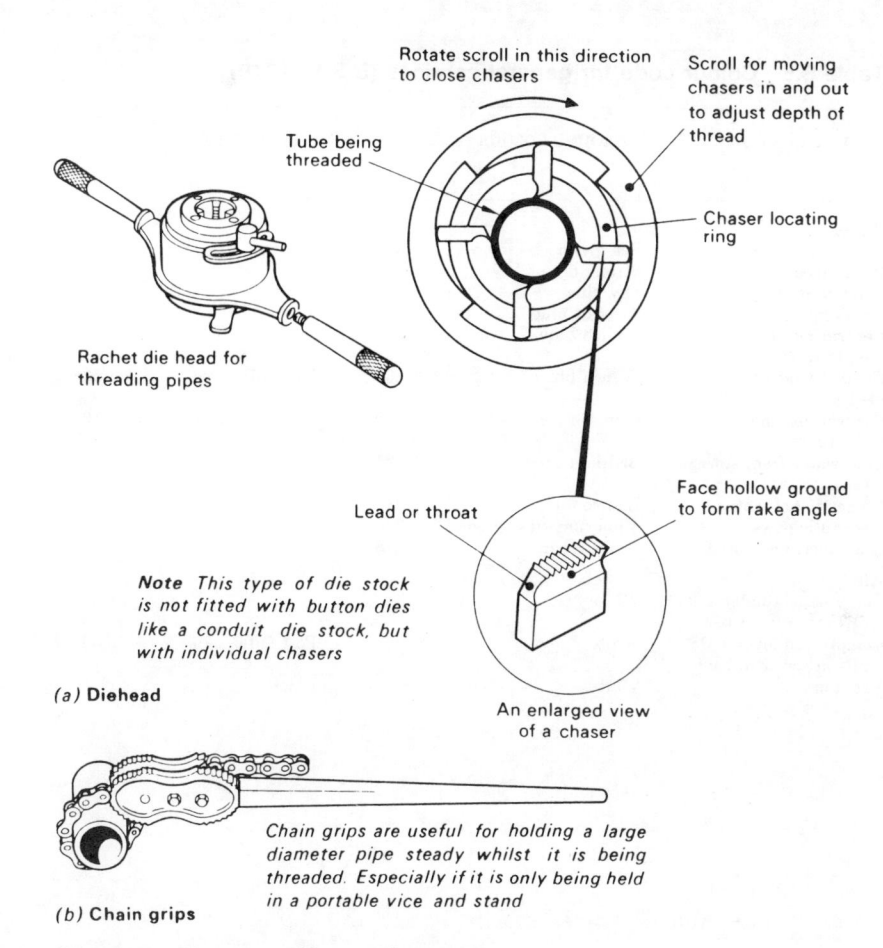

**Fig. 3.44 Pipe threading equipment**

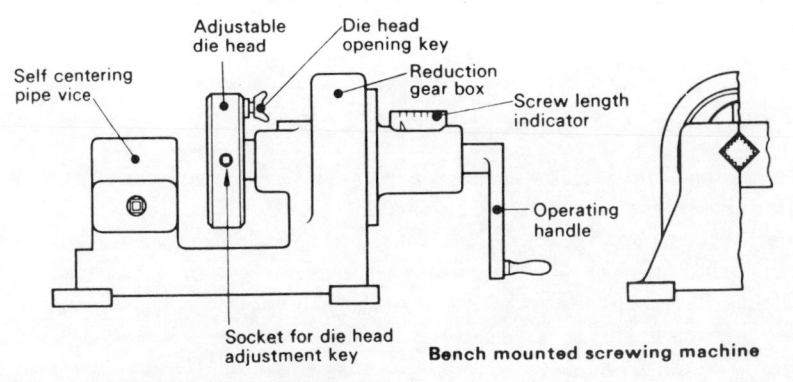

**Fig. 3.45 Screwing machine**

95

## Table 3.6 Colour code for general services (B.S.S. 1710)

| PIPE CONTENTS | GROUND COLOUR | B.S.S. COLOUR No* | COLOUR BAND | B.S.S. COLOUR No* |
|---|---|---|---|---|
| **Water** | | | | |
| Cooling (primary) | Sea green | 217 | — | — |
| Boiler feed | Strong blue | 107 | — | — |
| Condensate | Sky blue | 101 | — | — |
| Drinking | Aircraft blue | 108 | — | — |
| Central heating (less than 40°C) | French blue | 166 | — | — |
| Central heating (0–100°C) | French blue | 166 | Post Office red | 538 |
| Central heating (over 100°C) | Crimson | 540 | French blue | 166 |
| Cold water from storage tank | Brilliant green | 221 | — | — |
| Domestic hot water | Eau-de-nil | 216 | — | — |
| Hydraulic power | Mid-brunswick green | 226 | — | — |
| Sea, river, untreated | Grass green | 218 | — | — |
| **Air** | | | | |
| Compressed (under 1·38 MN/m$^2$ or 13·8 bar) | White | — | — | — |
| Compressed (over 1·38 MN/m$^2$ or 13·8 bar) | White | — | Post Office red | 538 |
| Vacuum | White | — | Black | — |

| PIPE CONTENTS | GROUND COLOUR | B.S.S. COLOUR No* | COLOUR BAND | B.S.S. COLOUR No* |
|---|---|---|---|---|
| **Steam** | Aluminium or crimson | 540 | — | — |
| **Drainage** | Black | — | — | — |
| **Electrical services** | Light orange | 557 | — | — |
| **Town gas** | Canary yellow | 309 | — | — |
| **Oils** | | | | |
| Diesel fuel | Light brown | 410 | — | — |
| Furnace fuel | Dark brown | 412 | — | — |
| Lubricating | Salmon pink | 447 | — | — |
| Hydraulic power | Salmon pink | 447 | Sea green | 217 |
| Transformer | Salmon pink | 447 | Light orange | 557 |
| **Fire installations** | Signal red | 537 | — | — |
| **Chemicals** | Dark grey | 632 | See B.S.S.349:1319 | — |

*See B.S.S. 381C — *Ready mixed paints*

# 4 Science

## Part A  Basic science — thermal operations

### 4.1  Sources of heat energy

The three most commonly used heat sources that are found in the fabrication engineering workshop are as follows:

1. THE FLAME

Solids such as coke or charcoal burn without a flame, only gases and the vapours of liquid fuels burn with a flame. Flames may be *luminous* or *non-luminous* depending upon the fuel being burnt and the efficiency and completeness of combustion. A smoky flame means that particles of carbon are escaping into the atmosphere without being burnt. This is both dirty and uneconomical, and in many instances objectionable.

It is essential when using an oxy-acetylene welding torch that, once the acetylene control valve has been opened and the gas ignited, the correct flame condition be achieved without delay. A clean flame condition results when the oxygen control valve is opened and adjusted for complete combustion. A very dirty smoky flame indicates incomplete combustion, causing objectionable soot deposits on work surfaces and anyone in the close vicinity.

Figure 4.1 shows the basic structure of some typical flames.

2. THE ELECTRIC ARC

Cool dry air is a good electrical insulator. However, when it is heated it becomes *ionised* and capable of conducting electricity. (This is the main reason why lightning invariably strikes the chimney of a building.) In the electric arc a heavy electric current flows through a path of heated air (see Figure 4.2).

The electric arc radiates ultra-violet rays that harm the eyes and skin. That is why protective clothing must be worn when arc welding.

3. THE ELECTRIC RESISTANCE

The work done by an electric current in overcoming the resistance of a wire causes the wire to become hot. This is wasteful and dangerous in wires transmitting electricity from one place to another. However, use can be made of this effect in such devices as heat-treatment furnaces (as shown in Figure 4.3) and soldering irons (Section 11.7). The wire from which the heating elements are wound is an alloy of nickel and chromium (nichrome) which has a high resistance and can withstand high temperatures without oxidising and burning away.

### 4.2  Chemical changes in metals

Chemical changes in metals occur as a result of using different oxy-acetylene *flame settings*. These basic flame settings are fully explained in Part A of Chapter 12. When a metal is heated at high temperatures, it rapidly becomes oxidised by oxygen which is ever present in the ambient atmosphere. Therefore, where thermal processes for joining are employed, it is essential that adequate provision be made to exclude the possibility of oxidation. However, some thermal processes use oxidation to advantage. Oxy-fuel gas flame cutting of steel depends upon oxidation (see Chapter 7).

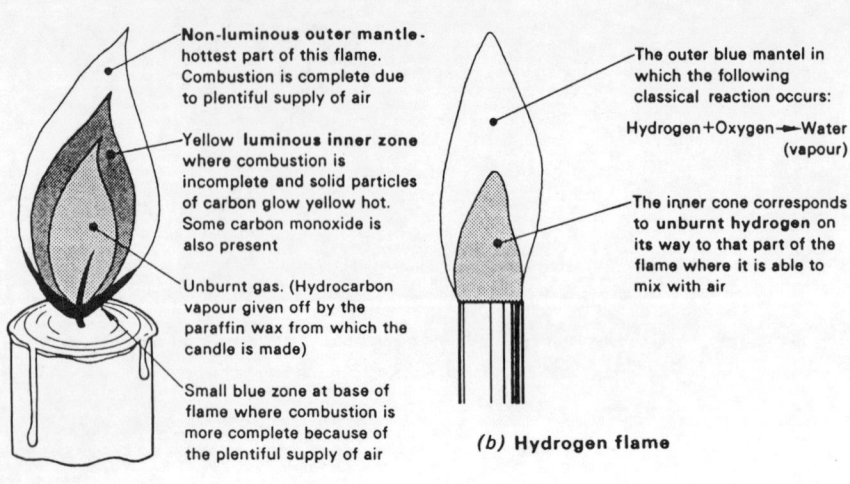

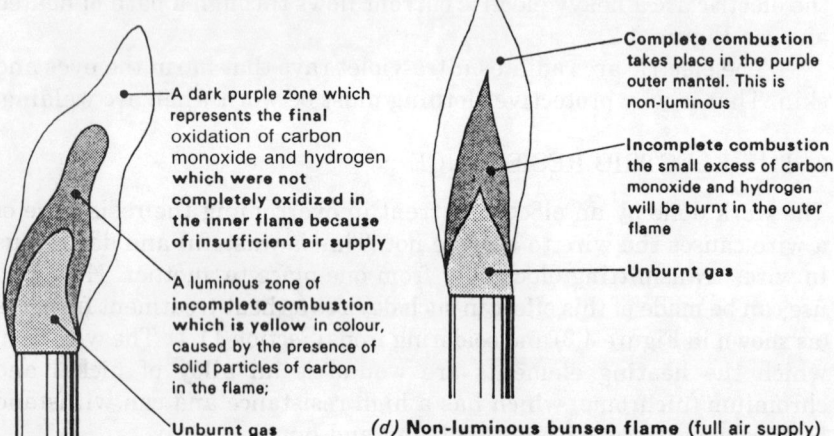

### Fig. 4.1 The structure of flames

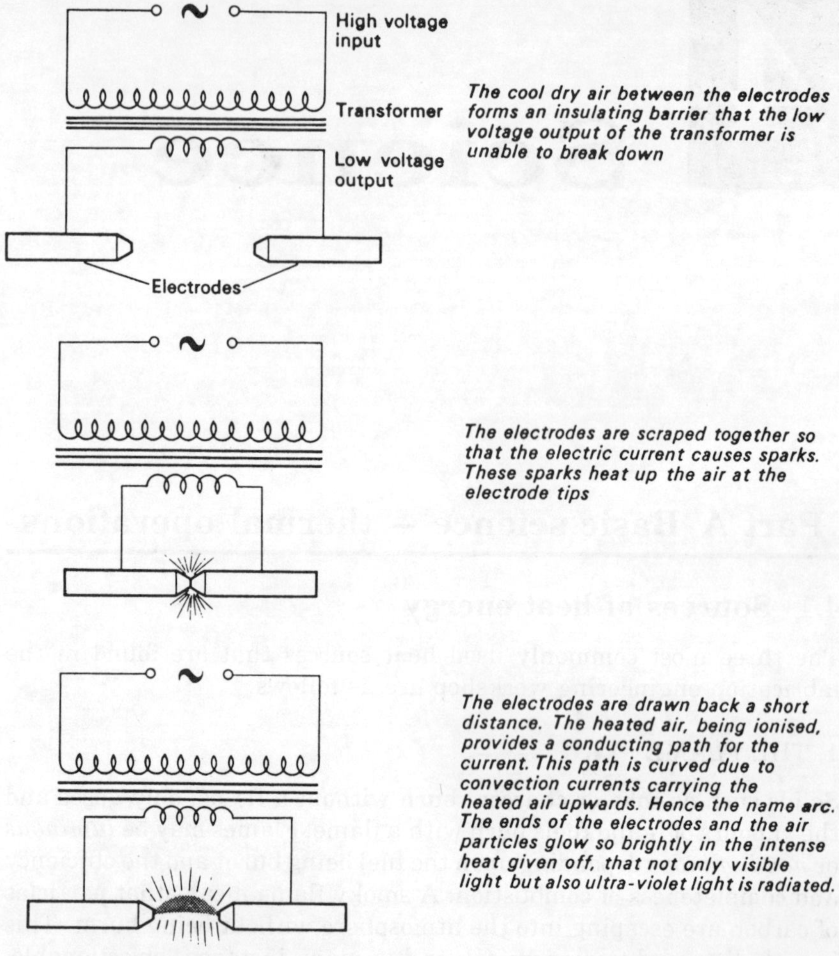

**(a) Candle flame**
- Non-luminous outer mantle - hottest part of this flame. Combustion is complete due to plentiful supply of air
- Yellow luminous inner zone where combustion is incomplete and solid particles of carbon glow yellow hot. Some carbon monoxide is also present
- Unburnt gas. (Hydrocarbon vapour given off by the paraffin wax from which the candle is made)
- Small blue zone at base of flame where combustion is more complete because of the plentiful supply of air

**(b) Hydrogen flame**
- The outer blue mantel in which the following classical reaction occurs:
  Hydrogen + Oxygen → Water (vapour)
- The inner cone corresponds to unburnt hydrogen on its way to that part of the flame where it is able to mix with air

**(c) Luminous bunsen flame (air shut off)**
- A dark purple zone which represents the final oxidation of carbon monoxide and hydrogen which were not completely oxidized in the inner flame because of insufficient air supply
- A luminous zone of incomplete combustion which is yellow in colour, caused by the presence of solid particles of carbon in the flame
- Unburnt gas

This is a dirty smoky flame which will deposit soot (Carbon) on any cold surface with which it comes in contact. A similar type of dirty flame results when burning Acetylene with insufficient air supply

**(d) Non-luminous bunsen flame (full air supply)**
- Complete combustion takes place in the purple outer mantel. This is non-luminous
- Incomplete combustion the small excess of carbon monoxide and hydrogen will be burnt in the outer flame
- Unburnt gas

The coal gas flame with full air supply. Again there are three distinct zones with the difference this time that the central zone of incomplete combustion is bright blue, the flame is hotter and free from soot. With sufficient oxygen it is possible to obtain almost perfect combustion, this produces a clean and much hotter flame. All flames contain a zone of unburnt gas, and in all flames in which hydrocarbons burn, the actual oxidation occurs in two or more stages, for example:
1. Little air, combustion products - soot and water vapour,
2. More air, combustion products - carbon monoxide, water vapour,
3. Sufficient air, where the final products of combustion are carbon dioxide and water vapour

### Fig. 4.2 The electric arc

*The cool dry air between the electrodes forms an insulating barrier that the low voltage output of the transformer is unable to break down*

*The electrodes are scraped together so that the electric current causes sparks. These sparks heat up the air at the electrode tips*

*The electrodes are drawn back a short distance. The heated air, being ionised, provides a conducting path for the current. This path is curved due to convection currents carrying the heated air upwards. Hence the name* **arc**. *The ends of the electrodes and the air particles glow so brightly in the intense heat given off, that not only visible light but also ultra-violet light is radiated.*

*Thermit welding* is a unique thermal joining process in which tremendous heat energy is released as a direct result of a very violent oxidation reaction. The fundamentals of the thermit welding process are illustrated in Figure 4.4.

The thermit reaction can be expressed as:

$$Fe_2O_3 + 2Al \rightarrow Al_2O_3 + 2Fe$$

Iron oxide powder, aluminium, aluminium oxide, molten iron

To set off the reaction an easily ignited priming charge is placed in the top of the thermit mixture.

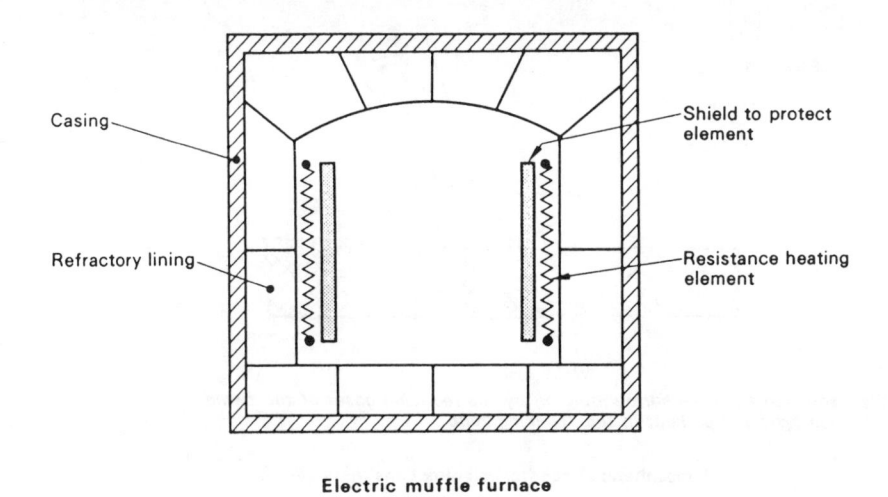

**Fig. 4.3 Electric resistance heating**

The thermit welding process is used extensively for joining and repairing railway lines. The equipment is portable and can be set up quickly *in situ* wherever required on the track.

Other chemical reactions also take place when metals are joined by thermal processes.

## 4.3 Chemical reactions during welding

Whether in the solid or molten state, heated materials tend to absorb and/or react with gases in the surrounding atmosphere. *The higher the temperature is raised, the more rapid becomes the rate of absorption and reaction, and these usually show a marked increase as the material becomes molten.*

The fuel gases used for gas-welding flames disassociate when ignited in the presence of oxygen or air. They break down into CARBON MONOXIDE, CARBON DIOXIDE, HYDROGEN, and WATER VAPOUR.

It is these new gases which react with materials rather than the original fuel gases.

If metal is welded without protection from the atmosphere, the OXYGEN and NITROGEN in the air will chemically combine with the molten metal to form OXIDES and NITRIDES. If these are allowed to dissolve in the WELD POOL the result will be the production of poor and brittle welds.

OXYGEN is generally harmful and a metal that oxidises readily is likely to be difficult to weld. Oxidised welds are undesirable and can

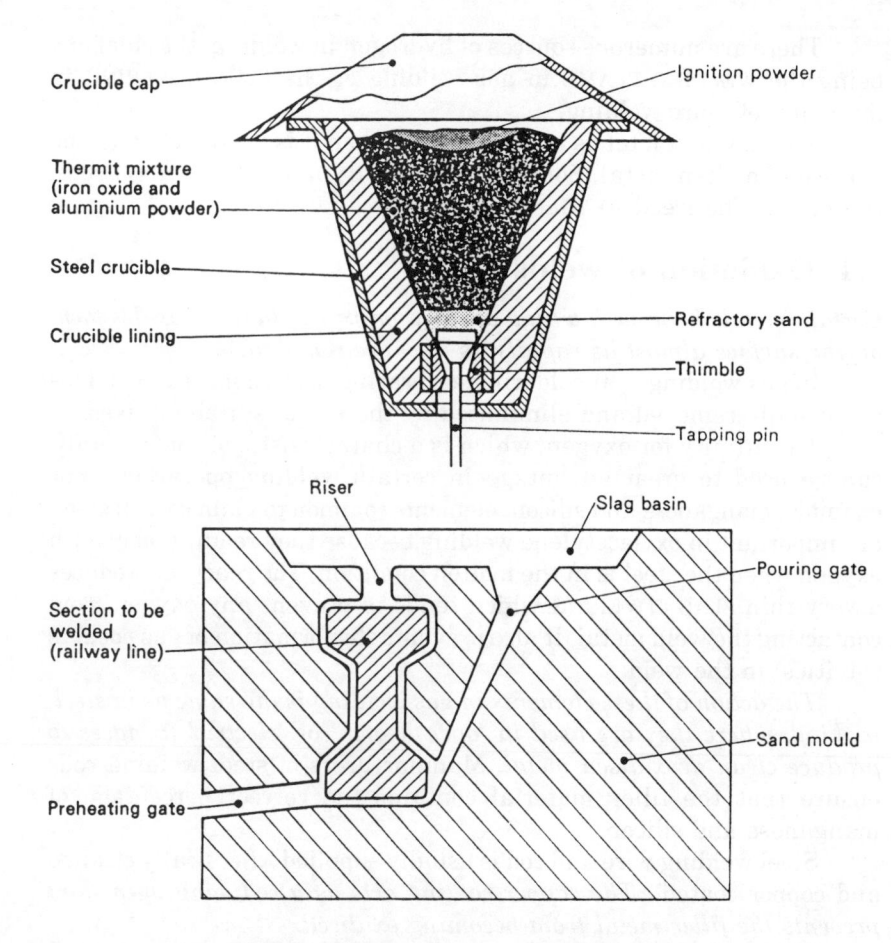

*1000g of Thermit mixture produces 524g of iron and 476g of slag.*

**Fig. 4.4 The Thermit welding process**

easily be identified by the appearance of their surface which is irregular and pitted. Such welds lack STRENGTH and DUCTILITY.

NITROGEN dissolves in many liquid materials and may react with some of their constituents. If, when welding certain steels, nitrogen is allowed to enter into the molten weld pool, the resultant weld will be brittle, porous, and low in ductility.

HYDROGEN can cause undesirable problems with many molten metals, in particular steel, aluminium, and 'tough-pitch' copper. The chief cause of GAS POROSITY is the presence of hydrogen in weld metal, or the reaction of hydrogen with oxides present in the molten parent metal which results in the production of STEAM.

There are numerous sources of hydrogen in welding, the chief one being the WELDING FLAME in gas welding or the FLUX COVERING in manual metal-arc welding.

WATER VAPOUR tends to disassociate when it is in contact with the surface of molten metal. Some of the freed HYDROGEN will *dissolve* in the metal. The freed OXYGEN tends to *react* with it.

## 4.4 Oxidation of welds

*Certain metals have such a high affinity for oxygen that the oxides form on the surface almost as rapidly as they are removed.*

In gas welding, soft soldering, or brazing operations, these oxides are usually removed and eliminated by the use of suitable fluxes.

The affinity for oxygen, which is a characteristic of some metals, can be used to great advantage in certain welding operations. For example, manganese and silicon, elements common to plain carbon steel, are important in oxy-acetylene welding because they readily react with oxygen when the steel is in the molten condition. The reaction produces a very thin SLAG COVERING which tends to prevent any oxygen from contacting the weld metal. It also prevents the formation of gas pockets (cavities) in the weld.

*The action of these elements in gas-welding is the same as in steel-making, where they are used in 'open hearth' or 'electric' furnaces to produce clean deoxidised metal.* Manufacturers of steel welding rods ensure that the filler material contains the correct percentage of manganese and silicon.

Steel welding wire and rod is usually supplied 'chemically cleaned and copper coated'. *The copper coating acts as a reducing agent and prevents the filler metal from becoming oxidised.*

## 4.5 Protecting the 'weld pool' from the atmosphere

The type of flame used in welding various materials plays an important role in securing the most desirable weld deposit.

*The proper type of flame condition with the correct welding technique will provide a shielding medium which will protect the molten weld metal from the harmful effects of oxygen and nitrogen in the atmosphere.* When welding plain carbon steels, use of the correct flame condition has the effect of stabilising the molten weld metal and preventing the 'burning-out' of CARBON, MANGANESE, and other alloying elements.

Figure 4.5(*a*) illustrates how the molten metal or 'weld pool' is protected from the harmful gases of the surrounding atmosphere by the REDUCING GASES present in the oxy-acetylene flame.

In manual metal-arc welding the flux covering on the electrode

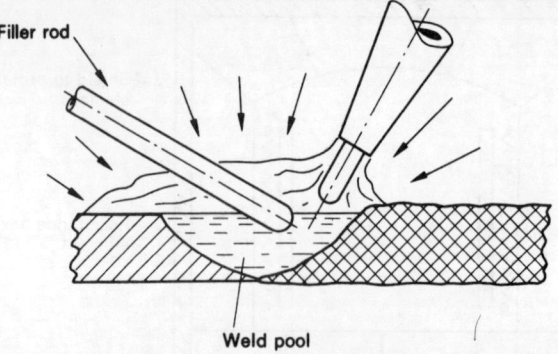

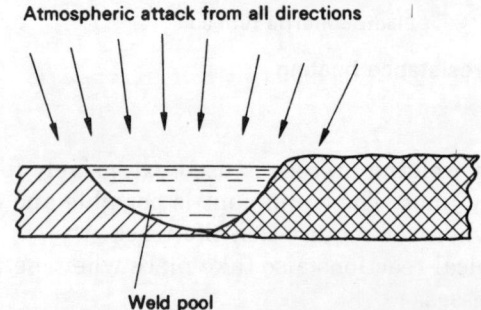

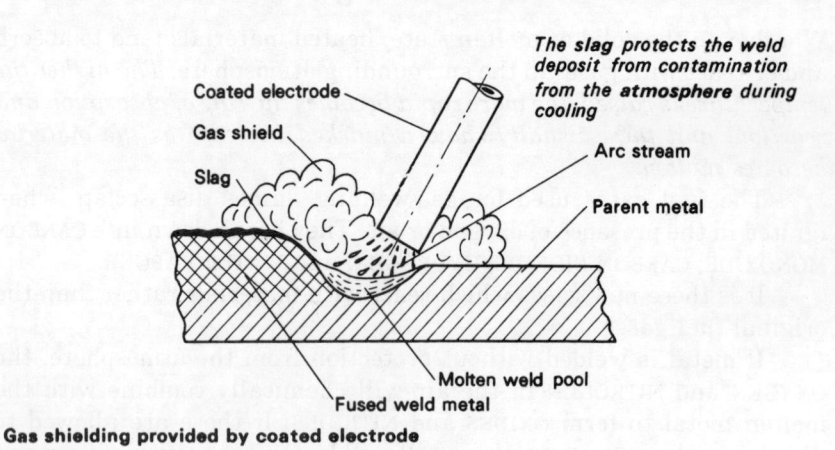

**Fig. 4.5** Protection of the weld (a) by the oxy-acetylene flame and (b) by manual metal-arc welding

protects the weld from atmospheric contamination, as shown in Figure 4.5(b).

## 4.6 Shielding gases in arc welding (specific)

Two arc-welding processes which are rapidly gaining popularity in most modern fabrication and welding workshops, and in shipyards, are the TUNGSTEN INERT GAS METAL-ARC WELDING PROCESS, commonly referred to as TIG, and the METAL INERT GAS WELDING PROCESS, known also as MIG, Although the mechanics of these two important arc-welding processes are beyond the scope of this book, a brief description of the functions of the shielding gases is given below.

TUNGSTEN INERT GAS WELDING

The TIG welding process employs a tungsten electrode which is non-consumable. Tungsten has a very high melting point (in excess of 3000°C), and therefore is not consumed by the heat of the arc. The atmosphere is excluded from the weld pool by a supply of inert gas. *An inert gas is one which does not chemically combine with any other element, and therefore cannot affect the weld metal.* ARGON and HELIUM are inert gases which are used with this welding process to shield the molten weld pool and parent metal from the atmosphere. When using this process, care must be taken to see that the gas flow is correct. Should it be too low, or should a draught interfere with the shielding of weld, atmospheric contamination will result. With the TIG welding process the arc is maintained between the tip of the tungsten electrode and the parent metal, and unlike other metal-arc welding processes, the filler material is added as in gas welding. Figure 4.6 shows how the atmosphere is excluded from the weld pool.

METAL INERT GAS WELDING

The MIG welding processes are basically automatic or semi-automatic. The electrode is a bare wire which is continuously fed into the welding gun by means of an electrode wire drive unit as the metal is transferred across the arc. Some electrode holders are termed 'welding torches' because the design of the curved neckpiece resembles an oxy-acetylene welding torch.

The shielding gases used are ARGON, HELIUM, and CARBON DIOXIDE. When carbon dioxide is used, the process is referred to as '$CO_2$ welding'. Carbon dioxide is not an inert gas, and when it passes through the arc it tends to break down into carbon monoxide and oxygen. To ensure that the liberated oxygen does not contaminate the weld metal,

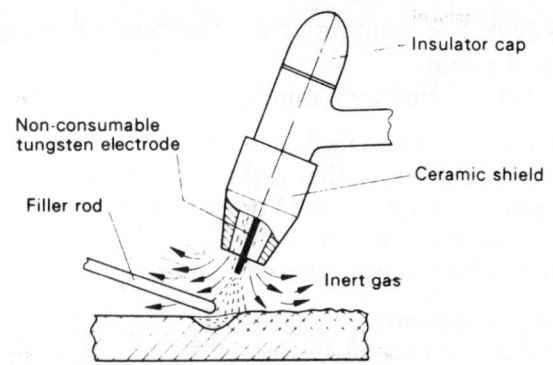

*The atmosphere is excluded from the weld by shielding with an inert gas. No chemical reactions take place*

**Fig. 4.6 Protection of the weld — TIG process**

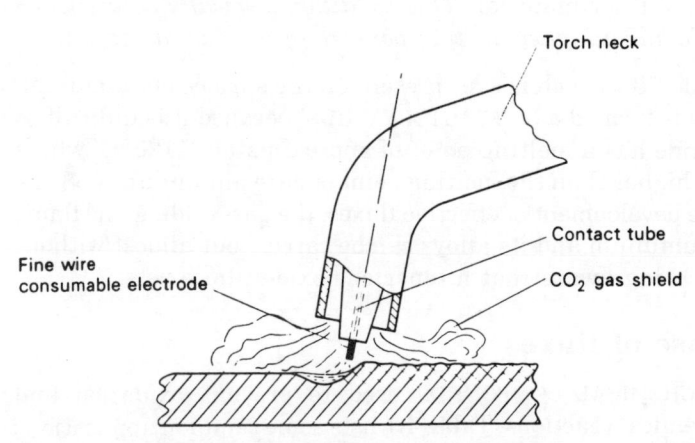

**Fig. 4.7 Protection of the weld — MIG process**

'DEOXIDISERS' are included in the welding wire which is specially manufactured for these welding processes. These deoxidisers combine with the oxygen to form a very thin slag on the surface of the completed weld. Figure 4.7 illustrates the method of shielding the weld in $CO_2$ welding.

## 4.7 The problem of oxides in welding

The removal and dispersion of the oxide film which forms when metals are heated is one of the traditional problems associated with welding.
Practical difficulties in welding occur:

1. When the surface oxide forms a tenacious film.

2. When the oxide has a MELTING POINT very much higher than that of the parent metal.
3. When the oxide forms very rapidly.

When welding or brazing aluminium, all three factors operate, but it is the tenacity of the oxide film that is the main source of trouble. One of the most important factors in weld quality is the removal of oxides from the surface of the metal to be welded. Unless the oxides are removed, the following undesirable conditions may result:

1. Fusion may be difficult.
2. Inclusions may be present in the weld metal.
3. The joint will be weakened.
4. The oxides will not flow from the welding zone but remain to become entrapped in the solidifying metal, interfering with the addition of filler material. *This condition generally occurs when the oxide has a higher melting point than the parent metal.*

The oxide film which is ever present on the surface of aluminium and its alloys is termed a 'REFRACTORY OXIDE' because it is difficult to melt. This oxide has a melting point of approximately 1120°C, which is very much higher than the melting point of pure aluminium (658°C). Thanks to the development of effective fluxes, the gas-welding and flame brazing of aluminium and its alloys can be carried out almost without the operator being aware that a tenacious oxide film exists.

## 4.8 The use of fluxes

Fluxes are CHEMICAL COMPOUNDS used to prevent oxidation and unwanted chemical reactions. Table 4.1 lists some common applications of fluxes for gas-welding.

**Table 4.1 Common applications of fluxes for gas-welding**

| METAL or ALLOY | FLUX | REMARKS |
| --- | --- | --- |
| Mild steel | — | No flux is required because the oxide produced has a lower melting point than the parent metal. Being less dense it floats to the surface of the molten weld metal as scale which is easily removed after welding. Use a neutral flame. |
| Copper | Borax base with other compounds | If Borax alone is used, a hard scale of copper borate is formed on the surface of the weld which is difficult to remove. Use a neutral flame. |
| Aluminium and aluminium alloys | Contains chlorides of lithium and potassium | Aluminium fluxes absorb moisture from the atmosphere — i.e., they are hygroscopic. Always replace the lid firmly on the container when the flux is not in use. The flux residue is very corrosive. On completion of the weld it is essential to remove all traces of this residue. This can be accomplished by scrubbing the joint area with a 5% nitric acid solution or hot soapy water. Use carburising flame. |
| Brasses and bronzes | Borax type containing sodium borate with other chemicals | The flux residue is a hard glass-like compound which can be removed by chipping and wire brushing. Use an oxidising flame. |
| Cast iron | Contains borates, carbonates, and bicarbonates plus other slag-forming compounds. | Oxidation is rapid at red heat, and melting point of the oxide is higher than that of the parent metal. For this reason it is important that the flux combines with the oxides to form a slag which floats to the surface of the weld pool and prevents further oxidation. |

*FLUX REMOVAL:* Many types of fluxes are corrosive to the metals or alloys with which they are used. Therefore it is important that residual flux be removed from the surface immediately after the welding operation. *Methods of removal generally employed include mechanical methods such as chipping and scratch brushing, rinsing or scrubbing with water, and use of acids or other chemicals.*

## THE ESSENTIAL FUNCTIONS OF A GOOD FLUX

It is important that the basic functions and characteristics of fluxes used in gas-welding and allied processes be clearly understood. A flux should:

1. Assist in removing the oxide film present on the surface of the metal by attacking and dissolving it.
2. Assist in removing any oxides which may occur during welding by forming 'fusible slags' which float to the surface of the molten weld pool, and not interfere with the deposition and fusion of filler material.
3. Protect the weld pool from atmospheric oxygen and *prevent the absorption and reaction of other gases in the welding flame*, without obscuring the welder's vision, or hampering his manipulation of the molten pool.
4. Have a LOWER MELTING POINT than the metal being welded and the filler material used.

Electrode flux coverings will be fully discussed in Chapter 12.

Most methods of preventing absorption of OXYGEN, HYDROGEN, and NITROGEN by the molten weld metal, aim at *providing a barrier between the molten metal and the surrounding atmosphere*. These can be summarised as:

1. The addition of *Reducing Agents* ('Deoxidisers') to the filler material.
2. Providing an *Inert Atmosphere*, such as *Argon* or *Helium*.
3. Providing a *Shielding Gas*, such as *Carbon Dioxide*.
4. Covering with an *Inert Layer Of Slag*.
5. Covering with a *Slag Capable Of Removing Oxides*.

## 4.9 Chemical changes in metal affected by the oxy-acetylene flame

### THE *NEUTRAL* FLAME CONDITION

A NEUTRAL flame condition is produced when equal volumes of oxygen and acetylene are supplied to the mixing chamber and the chemical reactions in the flame take place in two stages. The primary reaction produces CARBON MONOXIDE and HYDROGEN which are both REDUCING GASES. The secondary reaction for complete combustion takes extra oxygen from the surrounding atmosphere which means that there is little chance of oxygen combining with the weld metal. With complete combustion there is no free CARBON to be picked up by the weld metal, and, therefore, there is no change in the structure of the weld metal when using a neutral flame, as shown in Figure 4.8.

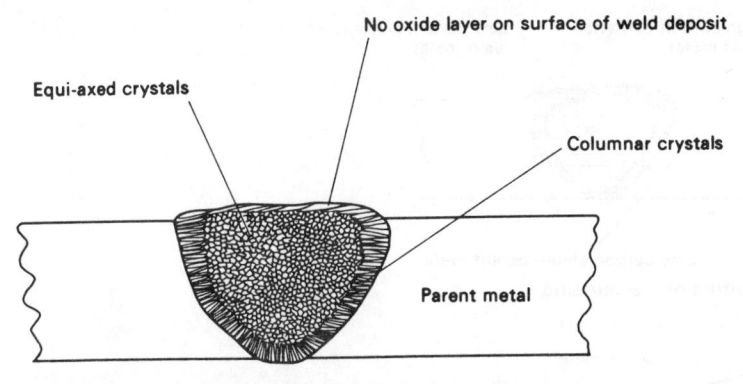

(Oxy-acetylene weld in low carbon steel — neutral flame)

**Fig. 4.8 Structure of weld metal when using a neutral flame**

The weld metal is in the normal 'cast' condition, consisting of a uniform equi-axed crystal formation surrounded by long columnar crystals caused by a faster rate of cooling. No oxides are present, and no addition of carbon to the weld occurs because of complete combustion of the acetylene. The effect on the structure of the parent metal is explained in Chapter 6.

### THE *CARBURISING* FLAME CONDITION

When the ratio of oxygen to acetylene becomes less than 1, the flame condition is said to be 'carburising'. *Excess acetylene results in incomplete combustion which produces free CARBON.* The weld metal becomes heavily carburised which results in a major change in the HARDNESS of the weld structure. Figure 4.9 illustrates the effect of the carbon from the flame:

1. The carbon content of molten weld metal is increased by DIFFUSION, as shown in Figure 4.9(a). *This means that there is a higher carbon content in the weld metal than in the parent metal.* This gives rise to the possibility of HARDENING OF THE WELD AREA during cooling if the carbon content exceeds 0·4 per cent (see Section 6.28).
2. Iron and steel will dissolve small amounts of CARBON into the structure in the same way as sugar is dissolved in water. The effect of solutions of metals and gases or non-metals (e.g., carbon) is to alter the MELTING TEMPERATURE 'UP' or 'DOWN'.

*When CARBON is dissolved in IRON the melting temperature is always lowered, as shown in Figure 4.9(b).* This effect is advantageous

103

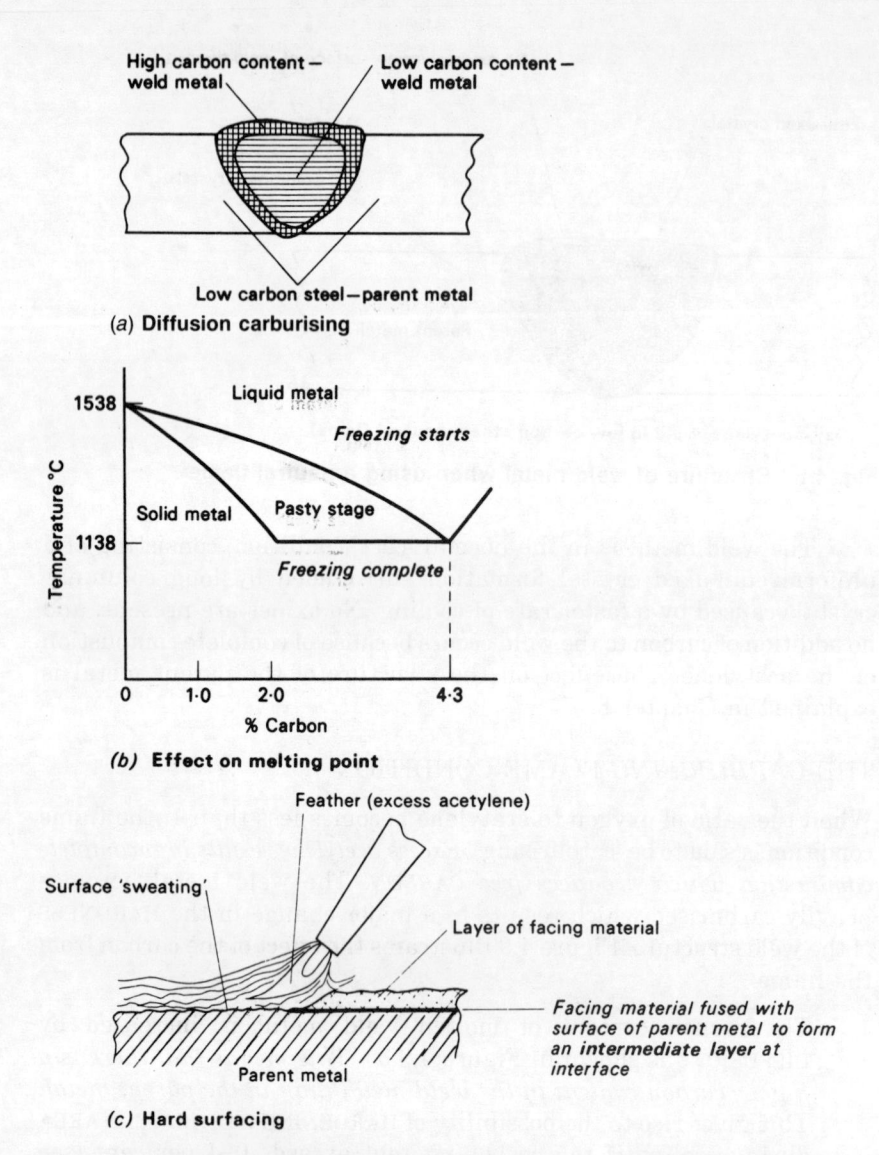

**Fig. 4.9 Effects of using a carburising flame**

(a) Diffusion carburising
(b) Effect on melting point
(c) Hard surfacing

when welding low-carbon steel; as the welding flame becomes increasingly carburising, the amount of 'CARBON PICK-UP' by the surface layers of the weld pool increases, *thus reducing the melting point and speeding up the welding action.*

When low-carbon steels are being hard surfaced the excess acetylene is kept to a minimum. The surface of the steel is made to melt at a lower temperature by the excess carbon in the flame. This is known as surface sweating which allows a layer of hard surfacing to be 'semi-fused' on to it, as shown in Figure 4.9.

### THE *OXIDISING* FLAME CONDITION

For complete combustion, 1 volume of acetylene requires $2\frac{1}{2}$ volumes of oxygen, thus with the neutral flame the extra $1\frac{1}{2}$ volumes is obtained from the surrounding atmosphere. When an oxidising flame is used the oxygen is not taken from the atmosphere since there is an excess of oxygen supplied from the cylinder. This means that the weld metal is in contact with an atmosphere which is rich in oxygen. OXYGEN IS A VERY REACTIVE GAS WHICH WILL CHEMICALLY COMBINE WITH MOST METALS TO FORM OXIDES, AND THE CHEMICAL REACTION IS SPEEDED UP BY THE APPLICATION OF HEAT. The effects of using an oxidising flame condition are illustrated in Figure 4.10.

When welding low-carbon steels with an OXIDISING FLAME condition the excess OXYGEN tends to combine with the CARBON in the

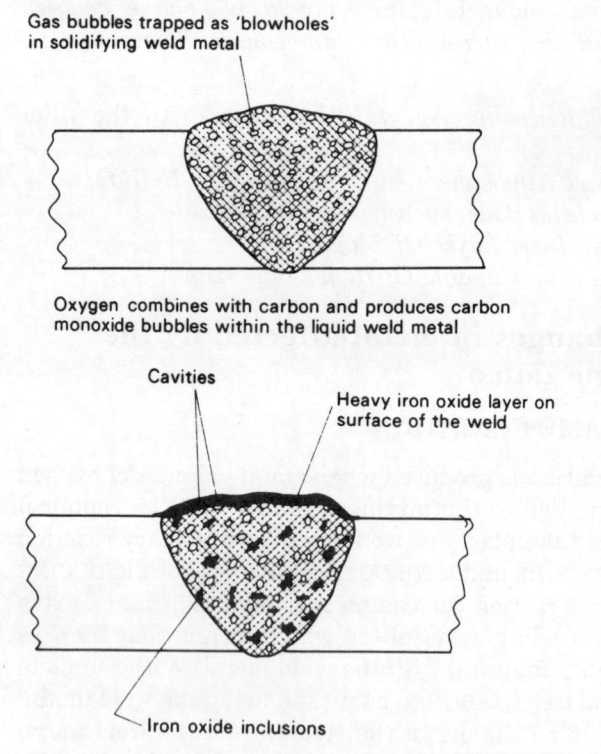

**Fig. 4.10 Effects of using an oxidising flame**

steel which is a 'REDUCING AGENT'. This results in CARBON MONOXIDE bubbles forming within the liquid weld metal. As the liquid weld metal commences to freeze, some of these bubbles will become trapped as 'blowholes'. The structure of the weld metal will be very porous, and it will have less carbon content near the surface than in the body of the weld structure.

Once the carbon has been oxidised, the *oxygen* will tend to chemically combine with the IRON in the steel to form IRON OXIDES. Iron oxides will be present in the weld structure as undesirable 'INCLUSIONS'. They are harmful to the weld structure because they tend to break up an otherwise normal homogenous weld; *they weaken the cohesive strength of the weld.*

THE HIGHEST FLAME TEMPERATURE IS ATTAINED WITH A HIGHLY OXIDISED FLAME CONDITION. This means that the molten weld pool will tend to boil vigorously. *When steel is heated to a high temperature, this may result in a condition which cannot be remedied by HEAT TREATMENT, and the steel is said to be 'BURNT'.* This condition is due to the fact that the grain boundaries become OXIDISED as a result of absorption of OXYGEN at high temperature, and hence the steel is weakened. The presence of brittle IRON OXIDE films at the grain boundaries renders the steel unfit for service, except as scrap for remelting. *Therefore, an oxidised weld can only be removed (by gouging or machining) and remade with a NEUTRAL FLAME condition.* OXIDES IN WELDS CAN ALSO LEAD TO CORROSION PROBLEMS. Table 4.2 indicates the effect of varying flame conditions when welding low-carbon steel.

## 4.10 Linear expansion

If a bar of steel is HEATED UNIFORMLY in a furnace it will EXPAND naturally in ALL DIRECTIONS if it is NOT RESTRAINED in any way. If allowed to COOL EVENLY and WITHOUT RESTRAINT it CONTRACTS to its original shape and size WITHOUT DISTORTION, as shown in Figure 4.11.

If the original dimensions, the rise in temperature, and the expansion of the metal block (Figure 4.11) are measured, it will be found that they are related to each other. Thus, for a given metal:

(i) The expansion is proportional to the rise in temperature.
(ii) The expansion is proportional to the original size of the component.

Expressed mathematically this gives the formula:

$$x = l \alpha t$$

where

$x$ = increase in size
$l$ = original length

**Table 4.2** Effects of varying flame conditions when welding low-carbon steel (oxy-acetylene process)

| BEFORE TEST | | GAS RATIO $(R) = \dfrac{\text{OXYGEN}}{\text{ACETYLENE}}$ | | | | | |
|---|---|---|---|---|---|---|---|
| | | OXIDISING FLAMES | | | | NEUTRAL FLAME | CARBURISING FLAME |
| ELEMENT | PER CENT | $R=1.14$ | $R=1.33$ | $R=2$ | $R=2.37$ | $R=1$ | $R=0.82$ |
| Carbon | 0.15 | 0.054 | 0.054 | 0.058 | 0.048 | 0.15 | 1.56 |
| Oxygen | — | 0.04 | 0.07 | 0.09 | — | 0.02 | 0.01 |
| Nitrogen | — | 0.015 | 0.023 | 0.03 | — | 0.012 | 0.023 |

The above are the results of an experiment carried out with various flame conditions when making welds on low carbon steel.
Both the filler rod and the parent metal used had the following analysis:

Carbon     0.15%
Manganese  0.56%
Silicon    0.03%
Sulphur    0.03%
Phosphorus 0.018%

The results clearly show that with a NEUTRAL FLAME there is no decarburation of the weld when $R = 1$.
With an increase of 14 per cent OXYGEN ($R = 1.14$) the carbon content of the weld drops to 0.05 per cent — SLIGHTLY OXIDISING FLAME.
Use of a CARBURISING FLAME, where $R = 0.8$, causes the weld to absorb free CARBON from the flame, and the final carbon content in the weld is 1.56. These results clearly indicate the effects of the three basic oxy-acetylene flame conditions on the weld metal.

$\alpha$ = coefficient of linear expansion
$t$ = rise in temperature

Table 4.3 gives the values of $\alpha$ for some common engineering materials ($\alpha$ is the Greek letter 'alpha').

Expansion in long straight runs of pipeline can cause problems, as the following example will illustrate:

EXAMPLE: A straight length of copper pipeline measures 30 metres at room temperature (15°C). After hot water has been passing through it for some considerable time, its temperature is raised to 60°C. Calculate its increase in length.

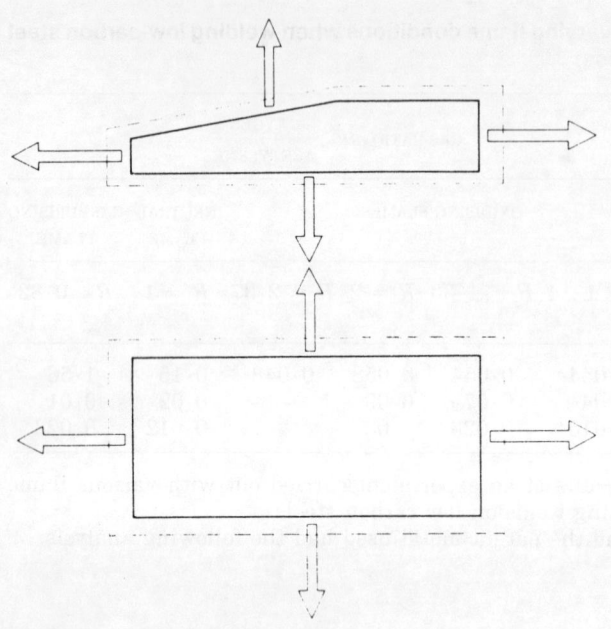

**Fig. 4.11 Linear expansion**

**Table 4.3 Coefficients of linear expansion**

| MATERIAL | COEFFICIENT OF LINEAR EXPANSION |
|---|---|
| Lead | 0·000 028/°C |
| Aluminium | 0·000 023/°C |
| Brass | 0·000 02/°C |
| Copper | 0·000 017/°C |
| Mild steel | 0·000 011/°C |
| Wrought iron | 0·000 011/°C |

Solution:
$$x = l \alpha t$$
$$= 30 \times 0.000\ 017 \times (60-15)$$
$$= 30 \times 0.000\ 017 \times 45$$
Increased length $= 0.022\ 95$ metres $= \underline{22 \cdot 95 \text{ mm}}$

An increase in length of 23 mm is quite substantial, therefore pipelines for carrying steam or hot liquids require special provisions to compensate for expansion. Figure 4.12 shows typical examples of expansion bends which can be introduced into pipeline systems.

There are some engineering operations where the expansion of metals can be used to advantage. When a metal ring or hoop is heated and placed over another part, the resultant contraction, upon cooling,

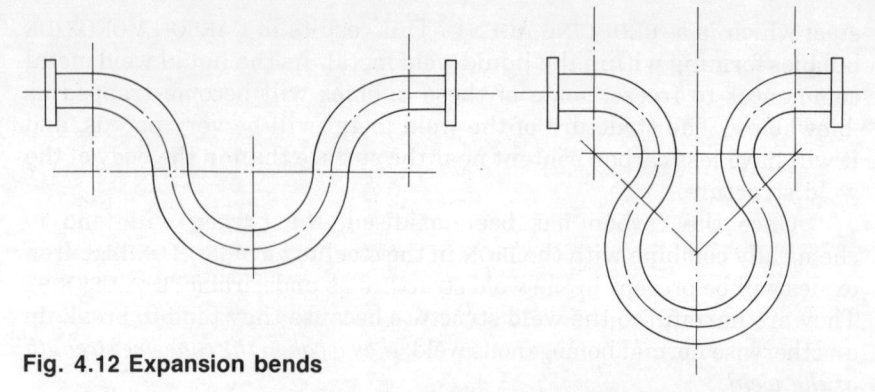

**Fig. 4.12 Expansion bends**

will ensure an extremely tight fit with the encircled part. This is known as 'shrink fit' and is used, for example, when fitting a steel tyre to the rim of a locomotive wheel.

## 4.11 Expansion and contraction in welding and cutting processes

It has just been explained how metals become larger when heated, and become smaller upon cooling. During welding and cutting operations the flame or the arc heats the metal causing it to become larger, or *expand*. When the heat source is removed, the surrounding metal and air exerts a cooling effect upon the heated area, resulting in the metal becoming smaller, or *contracting*. *If this expansion and contraction is not controlled, excessive* DISTORTION *is likely to occur*. There is another problem associated with welding, for if expansion and contraction is *restrained* or controlled too rigidly, severe *stresses* may occur and seriously impair the strength of the weld. *Summarising: metals* EXPAND WHEN HEATED *and* CONTRACT ON COOLING, *it is this effect which is responsible for the introduction of undesirable stresses in the weld and distortion in the work.*

## 4.12 Theory of distortion

Two factors are necessary to avoid distortion, which in this case are:

1. UNIFORM HEATING AND COOLING OF THE ENTIRE METAL BAR.
2. FREEDOM FROM RESTRAINT, so that EXPANSION and CONSTRACTION can take place unhindered.

*Unfortunately, when making a weld, unhindered expansion and contraction in all directions is not possible.* Thus distortion will result unless suitable precautions are taken.

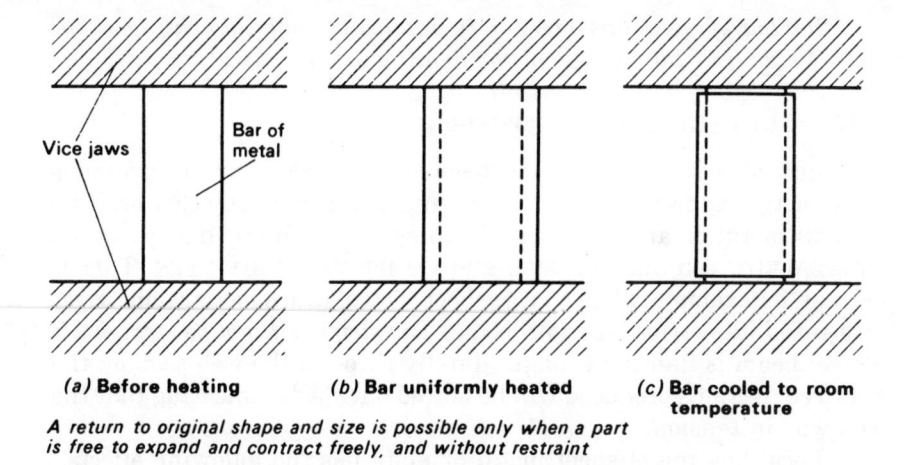

**Fig. 4.13 Effect of heating a steel bar under restraint**

*If a metal bar is restrained in any way during heating, it will not be able to expand in the direction of the restraint.* The effect of heating a metal bar under restraint is shown in Figure 4.13.

A metal bar is placed in a vice so that the jaws close against the two ends, as shown in Figure 4.13(a). The bar is in restraint and therefore cannot expand in the direction of the restraint.

When the bar is heated uniformly it can expand in all directions except in the direction where it is held by the vice jaws. The expansion is indicated by the full lines. The dotted lines represent the original shape and size (Figure 4.13(b)).

When it contracts upon cooling, however, there is no restraint in any direction and the bar is free to contract in all directions. It does not return to its original shape and size, as shown opposite, but becomes shorter in length, thicker in cross-section, and drops out of the vice jaws. The dotted lines represent the original shape and size (Figure 4.13(c)).

*Should a bar be heated over a small area, the expansion will be local and uneven, as is the case in welding.* The mass of the surrounding and relatively cool metal will not expand and tends to restrain the expansion of the heated area in all directions except upon the surface. Figure 4.14 illustrates the effects of heating a spot on a steel plate.

When a spot is heated on a smooth steel plate, the metal which is locally heated expands and exerts a force around it. The relatively cooler metal which surrounds the heated spot acts as a restraint and causes the hot plastic metal to be forced upwards into the air.

During the cooling period CONTRACTION takes place and the local area of metal continues to shrink until room temperature is reached.

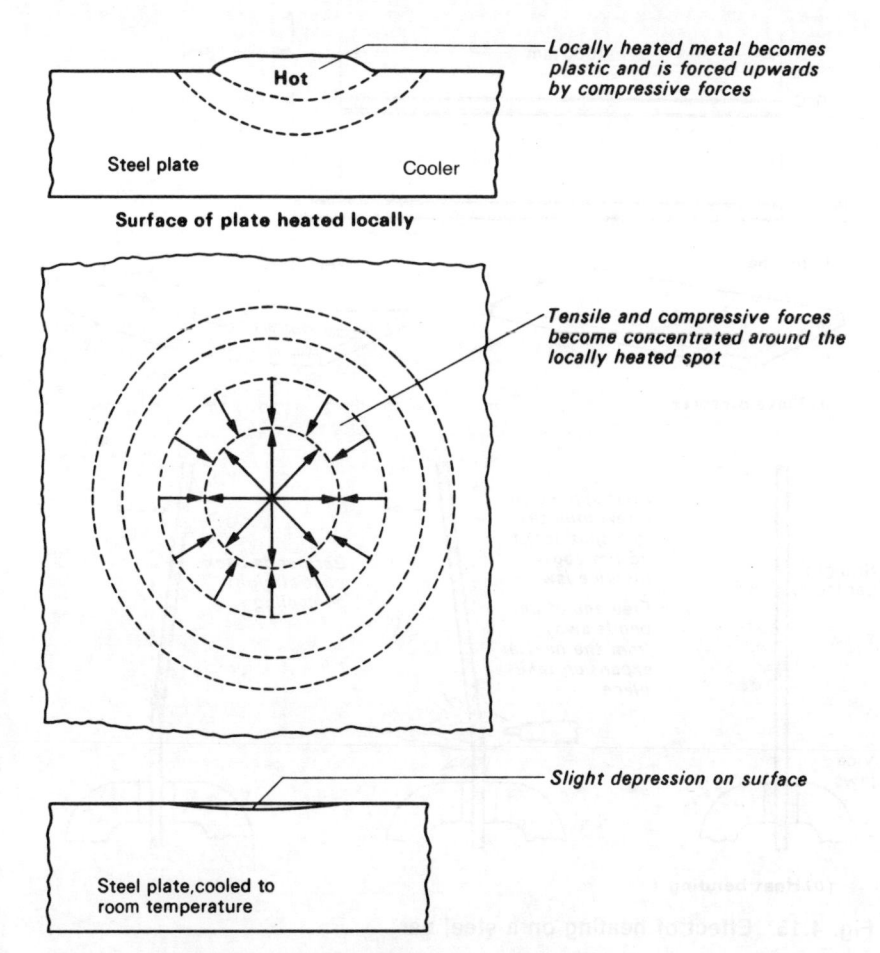

**Fig. 4.14 Effects of local heating on a steel plate**

While the metal is shrinking, TENSILE stresses develop around the spot. These tensile stresses in the vertical direction tend to draw the metal downwards, resulting in a slight depression on the surface of the plate where it was locally heated.

Two effects of heating on a steel bar are shown in Figure 4.15 one of which can be used to great advantage as explained in Section 4.20.

When a steel rod, 1 metre long, is heated from 0°C to 600°C, it expands approximately 9 mm in length (LINEAR EXPANSION). When a weld bead is laid on the surface of a flat steel bar, the weld bead itself can be likened to the steel rod heated well above 600°C. The heated rod contracts and returns to its original length upon cooling (Figure 4.15(a)). The weld bead (heated rod) also contracts in its length upon

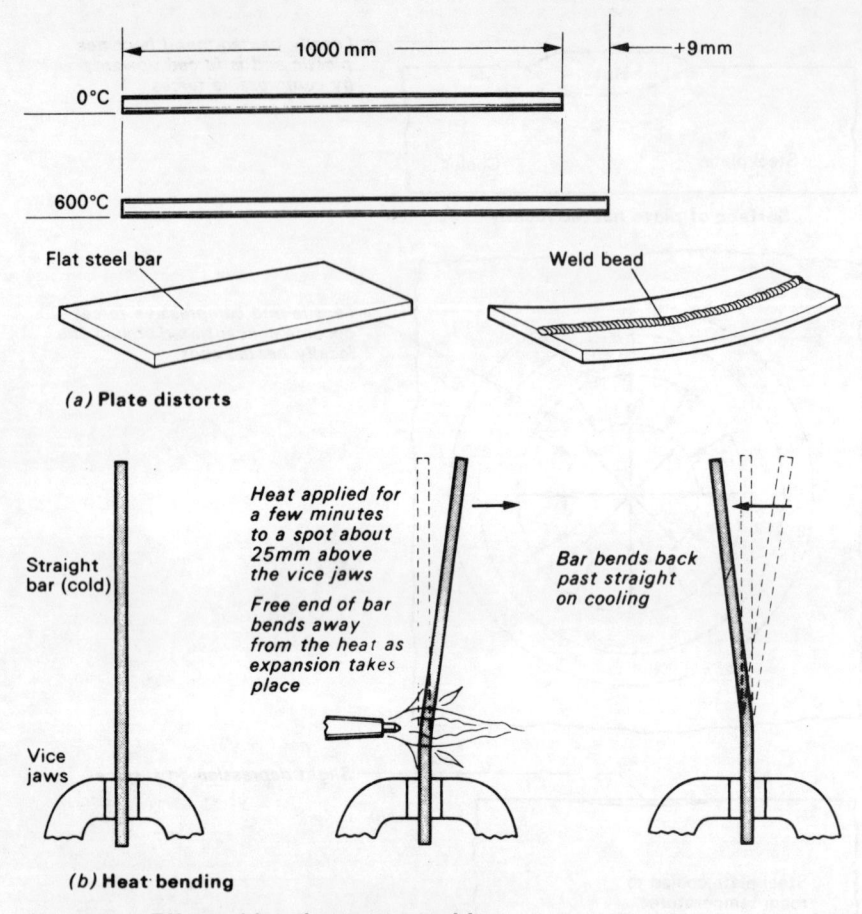

Fig. 4.15  Effect of heating on a steel bar

cooling, but is restrained by the steel plate. The flat steel bar bends as the weld bead cools. This is because the weld bead is not free to move in the direction of its length when contracting.

The locally heated spot of metal expands in all directions. In expanding, some of the force of expansion is expended by bending the bar. Some of the expansion force is relieved by expanding away from the surface of the metal, but some of the expansion force remains in the hot metal as a compressive force. The heated metal has expanded beyond its ELASTIC LIMIT in compression. when the metal cools, the spot shrinks and causes the bar to bend back, but since the metal has taken a permanent set in compression it contracts (shrinks) to a length shorter than the original. The final result is that the bar is no longer straight (Figure 4.15(b)).

*If a bar of metal has a bow in it, application of local heating on the convex side (i.e., the outside of the bend) will straighten it.*

## 4.13 Manufacturing stresses

Internal stresses may have been introduced during previous manufacturing processes such as rolling or forging. Any HOT or COLD ROLLED metal or any CAST metal, unless it has been HEAT TREATED ('ANNEALED'), contains TENSILE and COMPRESSION STRESSES. This is evident, for example, when a large sheet of steel which has straight edges is sheared into strips on a guillotine. *The strips will not be straight.* If a steel beam is flame cut longitudinally through the web section, the two 'Tee' sections produced will be bowed outwards indicating that the web was in tension.

Local heating of sheet metal or plate has the following effects:

1. Sheet metal or thin plate becomes buckled.
2. Thick plate acquires stresses upon cooling.
3. A sheet or plate which is heated at the edge warps upon cooling.

## 4.14 The problem of distortion in welding

During all welding operations the weld metal and the heated parent metal undergo considerable contraction on being cooled to room temperature.

In oxy-acetylene or manual metal-arc welding, the heating is not uniform because the deposited weld metal is far hotter than the adjacent parent metal in the joint, which in turn is hotter than the parent metal remote from the weld area. This means that the weld will contract more than the parent metal unless prevented by restraint. The volume of the weld metal will contract about 1 per cent in all dimensions. This contraction results in considerable stress in the weld itself. In addition, because of uneven heating, stress is imposed in the parent metal, and these stresses combine in causing the structure or welded fabrication to distort and shrink, unless the component parts are prevented from moving.

Under such conditions, RESIDUAL STRESSES up to the YIELD POINT of the metal may be expected. Should the plastic flow required exceed the metal's capacity to flow, CRACKING may result.

## 4.15 Types of distortion

Figure 4.16 illustrates the spread of heat from the weld through the parent metal (*a*) and the effects of single-side welding of an unrestrained butt joint (*b*).

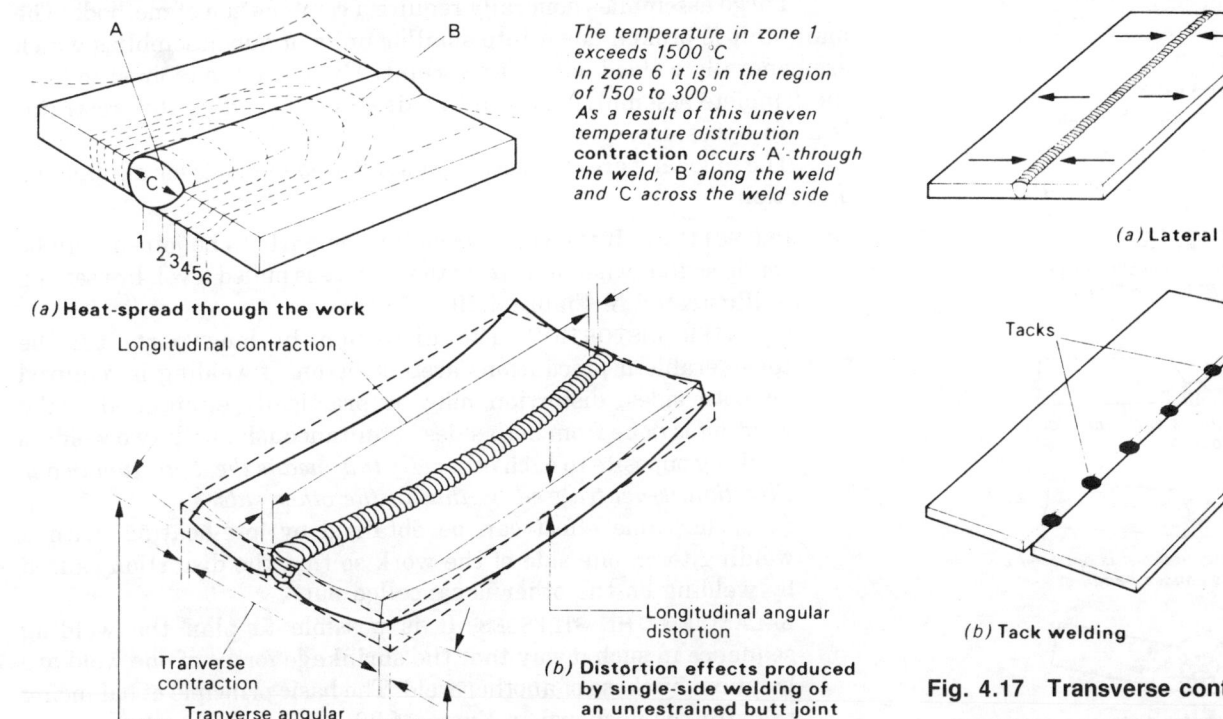

**Fig. 4.16** Effects of welding on a metal plate

(a) Heat-spread through the work

The temperature in zone 1 exceeds 1500 °C
In zone 6 it is in the region of 150° to 300°
As a result of this uneven temperature distribution contraction occurs 'A' through the weld, 'B' along the weld and 'C' across the weld side

(b) Distortion effects produced by single-side welding of an unrestrained butt joint

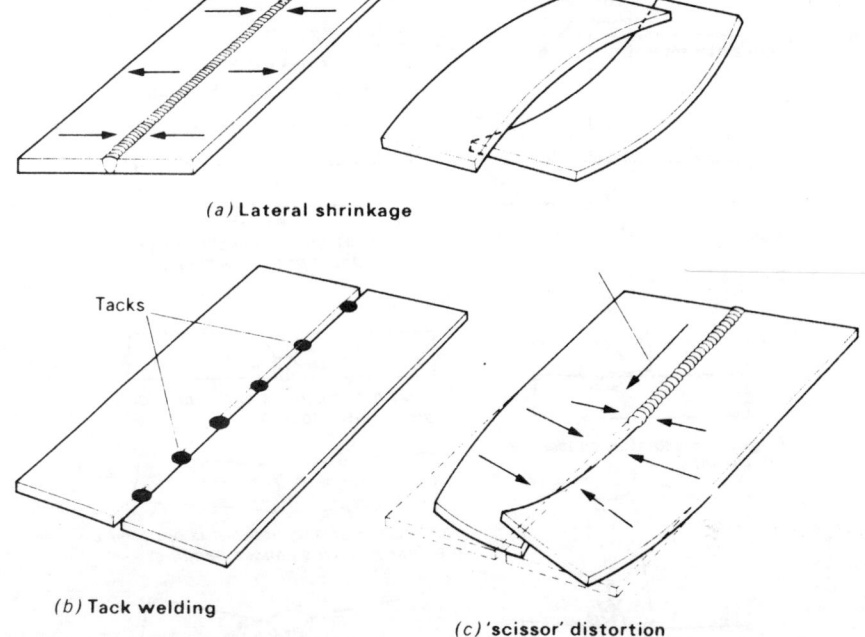

**Fig. 4.17** Transverse contraction

(a) Lateral shrinkage
(b) Tack welding
(c) 'scissor' distortion

*Longitudinal and transverse distortion.* Longitudinal distortion refers to bending which occurs along the length of the joint, and *transverse distortion* to that in the direction at right angles to the joint.

*Transverse contraction.* If two plates are spaced slightly apart and not 'tack-welded', the gap will close up in advance of the flame or the arc as welding proceeds from one end to the other. This is referred to as *transverse contraction*, and is illustrated in Figure 4.17.

When two plates are butt-joined together by welding there are COMPRESSION STRESSES at the ends and TENSILE STRESSES at the middle of the joint. *This results in lateral shrinkage*, as shown in Figure 4.17(a). If the same amount of heat that was applied to the joint during welding, is applied to one edge of each of the plates (not held together) they would acquire the shape shown.

Sheet metal or plates to be welded can be held in restraint by tack-welding as shown in Figure 4.17(b). This prevents the pre-set gap between the butt edges from closing up in advance of the welding arc or flame due to transverse contraction of the weld metal.

If the plates are spaced slightly apart and not tack-welded, and welding commenced at one end, the effect shown in Figure 4.17(c) will occur. The gap between the plates has narrowed and closed up. *If the plates are thin and long enough they eventually 'scissor' or overlap each other, so great is the contractional force.*

EFFECTIVE TACK-WELDING IS ESSENTIAL TO MAINTAIN THE ROOT GAP IN A JOINT SO THAT ADEQUATE PENETRATION CAN BE OBTAINED.

*Angular distortion.* When two pieces of sheet metal are butt welded together with an oxy-acetylene flame, it will be noticed the completed joint has 'bird-winged' because of greater contractional forces on the top surface of the weld. This is known as ANGULAR DISTORTION, and is illustrated in Figure 4.18.

## 4.16 Methods of minimising distortion

In spite of the hazards of internal stressing, restrained welding is the generally-employed process; the parts may be held in fixtures (jigs), secured by clamps, or merely tack-welded.

The alternative to the restrained method of assembly is to allow almost complete freedom of the parts during welding, this having the

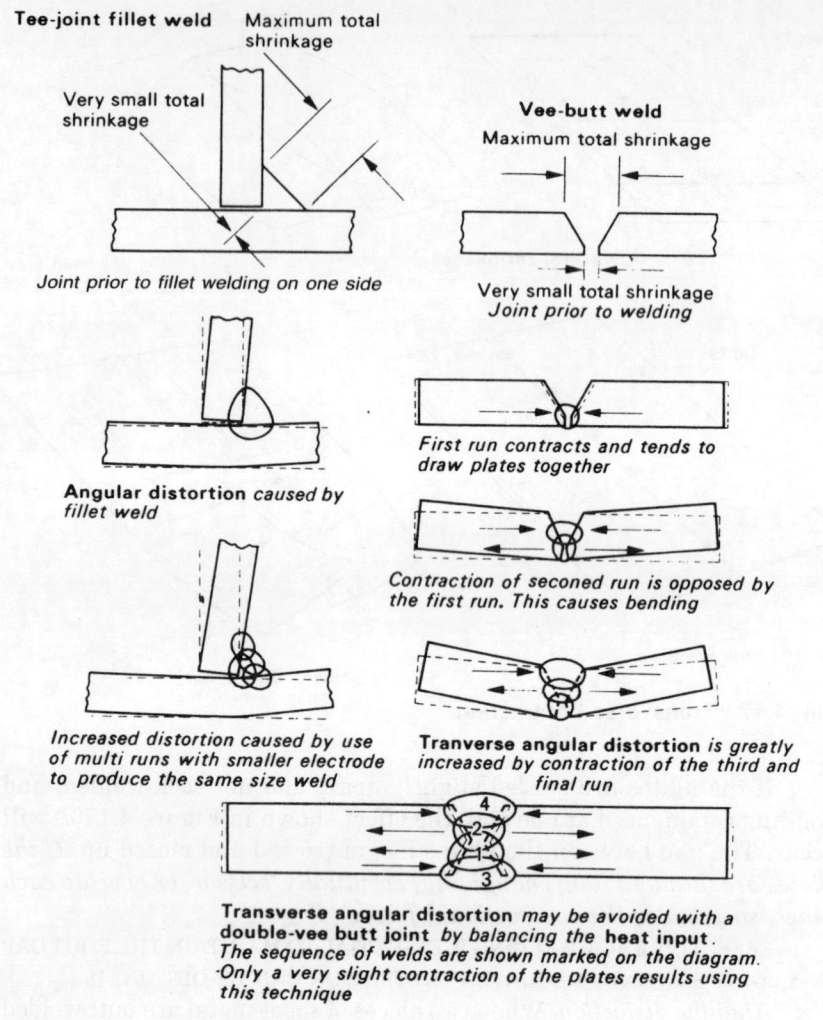

**Fig. 4.18 Angular distortion**

*Figure labels:*
- Tee-joint fillet weld — Maximum total shrinkage; Very small total shrinkage. Joint prior to fillet welding on one side.
- Vee-butt weld — Maximum total shrinkage; Very small total shrinkage. Joint prior to welding.
- Angular distortion caused by fillet weld
- First run contracts and tends to draw plates together
- Contraction of second run is opposed by the first run. This causes bending
- Increased distortion caused by use of multi runs with smaller electrode to produce the same size weld
- Transverse angular distortion is greatly increased by contraction of the third and final run
- Transverse angular distortion may be avoided with a double-vee butt joint by balancing the heat input. The sequence of welds are shown marked on the diagram. Only a very slight contraction of the plates results using this technique

advantage of ensuring the joints are practically free from locked-up stresses.

*Every effort must be made to anticipate distortion and make arrangements to correct it before it happens.* It may be possible to 'pre-set' the parts so that they distort to their correct positions.

The stresses can often be reduced by using a carefully planned welding sequence, or by pre-heating.

It is not always necessary to prevent movement in all directions; sometimes it is an advantage to allow movement in one direction.

Large assemblies generally require a combination of methods. The main unit is broken down into smaller units or sub-assemblies which are generally welded without restraint. These are then joined to form the complete assembly, and it is at this stage of welding that restraint is often necessary.

The following are the most commonly-used methods for minimising distortion:

1. PRE-SETTING If the shape permits, the parts to be joined may be set up so that when it distorts the surface is pulled level. Pre-setting is illustrated in Figure 4.19.

2. COUNTER DISTORTION The distortion in long parts can be considerable if precautions are not taken. If welding is required on both sides, distortion may be practically eliminated if the welding is done from both sides simultaneously, with two welders working opposite to each other. *By this means the distortion in one direction is neutralised by that of the other side.*

   The same effect can be obtained by 'pre-heating' with a welding torch one side of the work so that the distortion caused by welding on the other is cancelled out.

3. BALANCING THE STRESSES It is possible to plan the welding sequence in such a way that the shrinkage forces of one weld are balanced by those of another weld. The basic principle of balancing the stresses is shown in Figure 4.20.

4. BACK-STEPPING AND SKIP-WELDING Long joints are particularly liable to distortion if welded continuously from end to end, and for this reason one or both of the following methods should be adopted:

   (a) Whenever possible, the weld should start at the middle of the joint, working towards the free ends.
   (b) Either STEP-BACK or SKIP-WELDING should be employed in order to distribute the welding heat over a large area.

The *step-back* method entails making a series of short welds in the opposite direction to the general direction of progression, as illustrated in Figure 4.21. *By this means one section of the weld is fairly cool before the next is joined to it.*

*Skip-welding* or *planned wandering* is a method employed to distribute the heat by making a series of short welds in different places along the joint. The first run is made in the centre and thereafter, portions on each side of the centre are welded in turn, run by run, as shown in Figure 4.22. The runs are usually deposited in the same direction as the general direction of progression. This technique distributes the welding heat over a much wider area than the *step-back* method.

Pre-setting the two plates prior to making the butt weld to compensate for estimated **angular distortion**

Pre-setting the vertical member of a Tee-joint fillet weld, to be welded on one side, prior to welding. The members should be square after the **angular distortion** has taken place

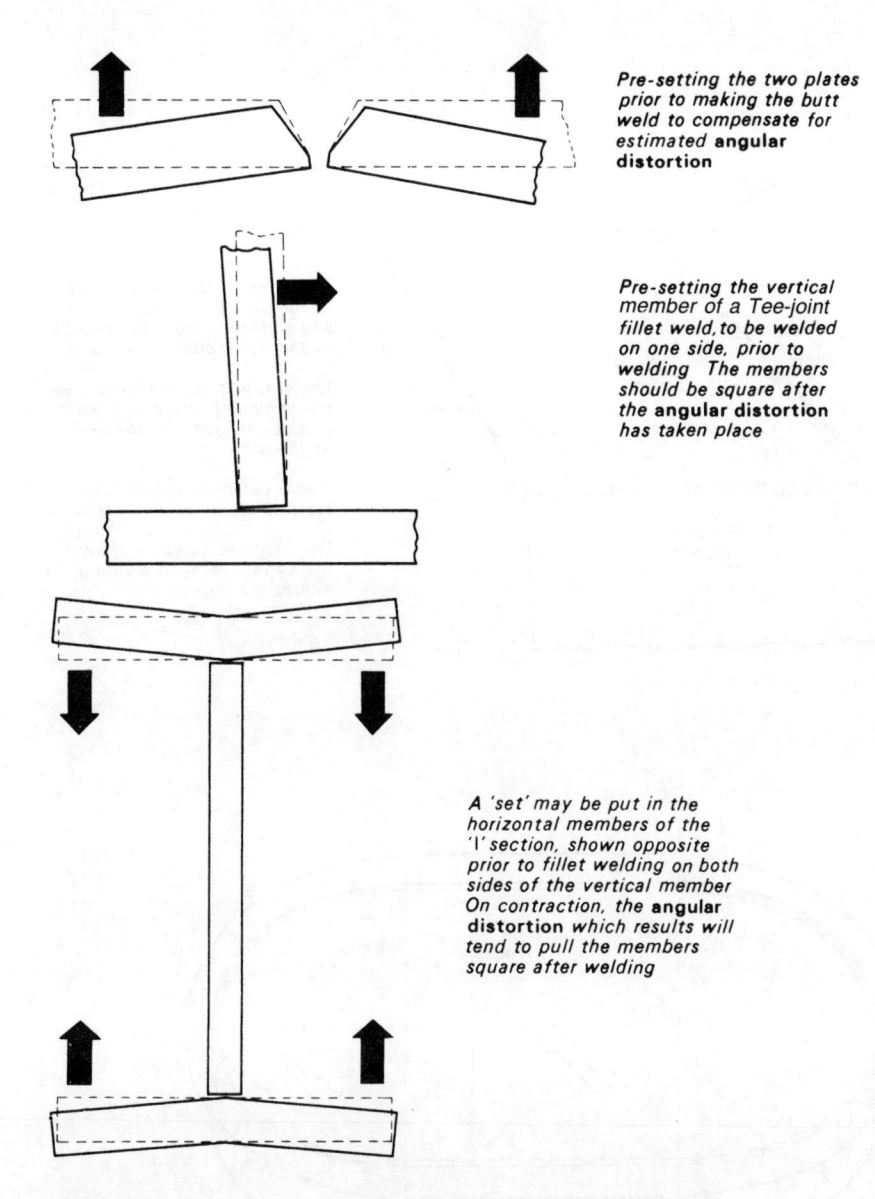

A 'set' may be put in the horizontal members of the 'I' section, shown opposite prior to fillet welding on both sides of the vertical member. On contraction, the **angular distortion** which results will tend to pull the members square after welding

**Fig. 4.19  Pre-setting to compensate for angular distortion**

Fillet weld both sides of Tee-joint

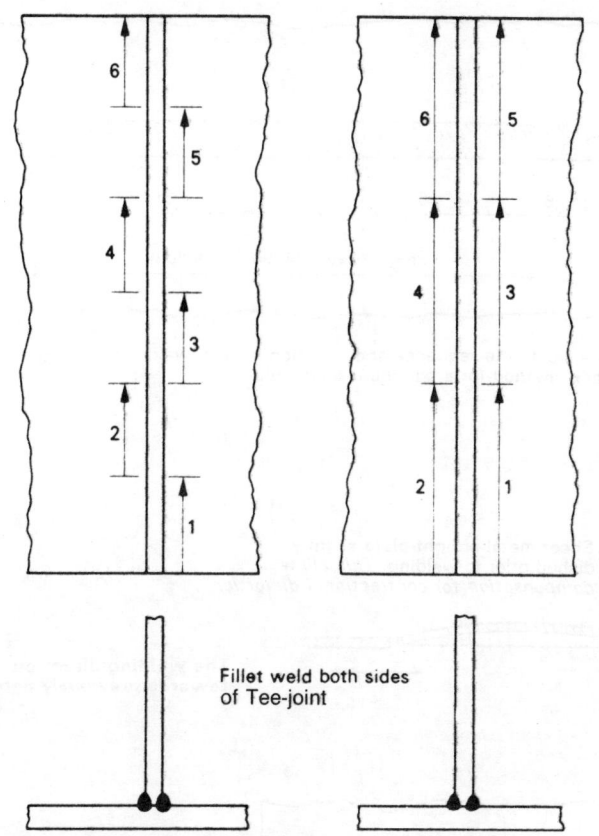

The arrows indicate the sequence of welding; on very long welds, the length of each run is the length of metal deposited by each electrode

**Fig. 4.20  Principle of balancing contractional stresses**

5.  LOCAL HEATING  Localised heat causes distortion, and this effect may often be put to good use to correct distortion arising from welding heat. If carefully chosen areas are heated with a welding torch, the contraction stresses that occur during cooling can sometimes be used to correct distortion due to welding. Angular distortion on a long welded joint may often be corrected by applying heat in straight lines on the reverse side, following the path of the joint. This is particularly useful in the case of distortion arising from fillet welds.

IN ALL WELDING OPERATIONS WHERE DISTORTION IS A FACTOR TO BE AVOIDED, THE DIRECTION OF WELDING SHOULD BE AWAY FROM THE POINT OF RESTRAINT AND TOWARDS THE POINT OF MAXIMUM FREEDOM. The importance of this statement is illustrated in Figure 4.23.

Consider three cast iron plates as shown in Figure 4.23(*a*):

1.  If the first weld is made along the seam A to B, *no cracking* takes place because the plates either side of the welded joint are free to expand and contract.

111

The arrows indicate the sequence and direction of each weld when using the 'step-back' method for a continous long weld

Sheet metal or light plate slightly dished prior to welding This allows compensation for contractional distortion

The welding direction of each run is towards previously deposited metal

Welding in a circular patch by the 'step-back' method This method gives a balanced heat distribution to compensate for the fact that the circular patch becomes much hotter than the surrounding plate during welding On cooling the weld is subject to stresses in all directions

*For short welds always start in the middle and weld out to one end, then from the middle to the other end*

*The greatest stresses occur at the commencement of a weld and at the crater at the end of the run*

*Avoid commencing and restarting welds at sharp corners*

*The diagram opposite shows the correct method welding around a sharp corner*

Sequence of welding to avoid distortion

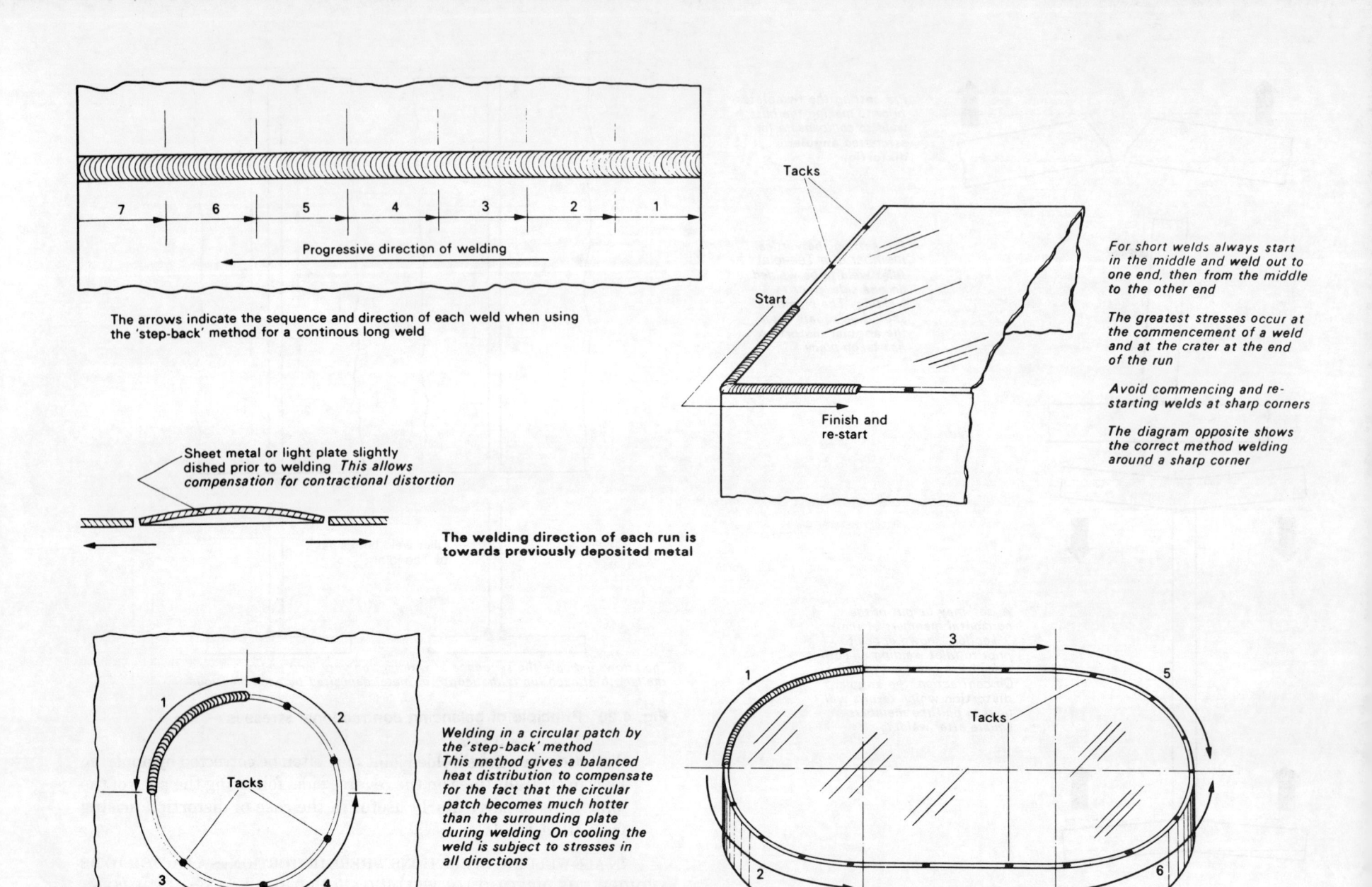

**Fig. 4.21 Principles of 'step-back' welding**

**Fig. 4.22 Skip-welding (planned wandering) (continued opposite)**

⊄ of joint length

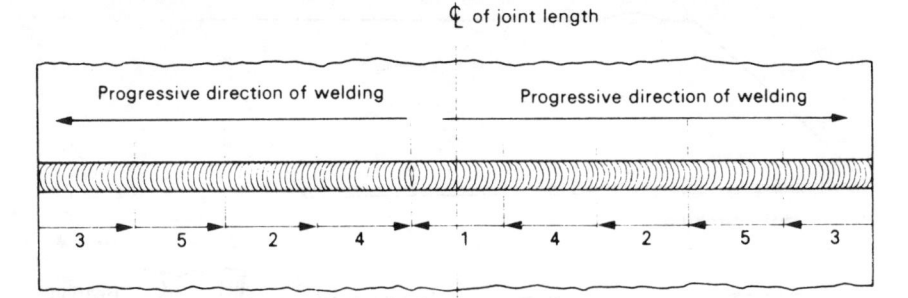

*'Down-hand' or 'flat' positional welding on butt-joint*

The diagram shows the sequence of welding a long butt-weld with two operators welding simultaneously after the first run has been made.
The arrows indicate the direction and sequence of each weld relative to the centre line of the joint.

**Fig. 4.22 (continued)**

2. If the next weld is made from D to C, *cracking* will take place because of the direction of welding — from a free end to a fixed (restrained) one.
3. When welding from C to D, *no cracking* takes place, as this is from a fixed end to a free (unrestrained) one.

## 4.17 Distortion caused by flame cutting

Figure 4.24 illustrates the possible types of plate deformation (distortion) which may result through flame cutting.

When a narrow strip is flame cut from a plate, the cut strip, on cooling will bow inwards towards the cut edge, as shown in Figure 4.24(*a*). This effect results from the hot edge contracting. The other edge of the strip is relatively cold and acts as a restraint to thermal expansion. Figure 4.24(*b*) shows the remaining plate deformed. The effect of flame cutting has released *residual stresses* (locked-up stresses) which were left in as a result of hot rolling during manufacture.

When flame cutting on a profile machine, the cutting head is adjustable up and down. This is to allow the operator to maintain the correct distance between the cutting flame and the top surface of the plate which tends to buckle during cutting.

When a hole is flame cut out of a plate, the cut edge will contract. Other areas of the plate, being longer, will bend and distort, as shown in Figure 4.24(*c*).

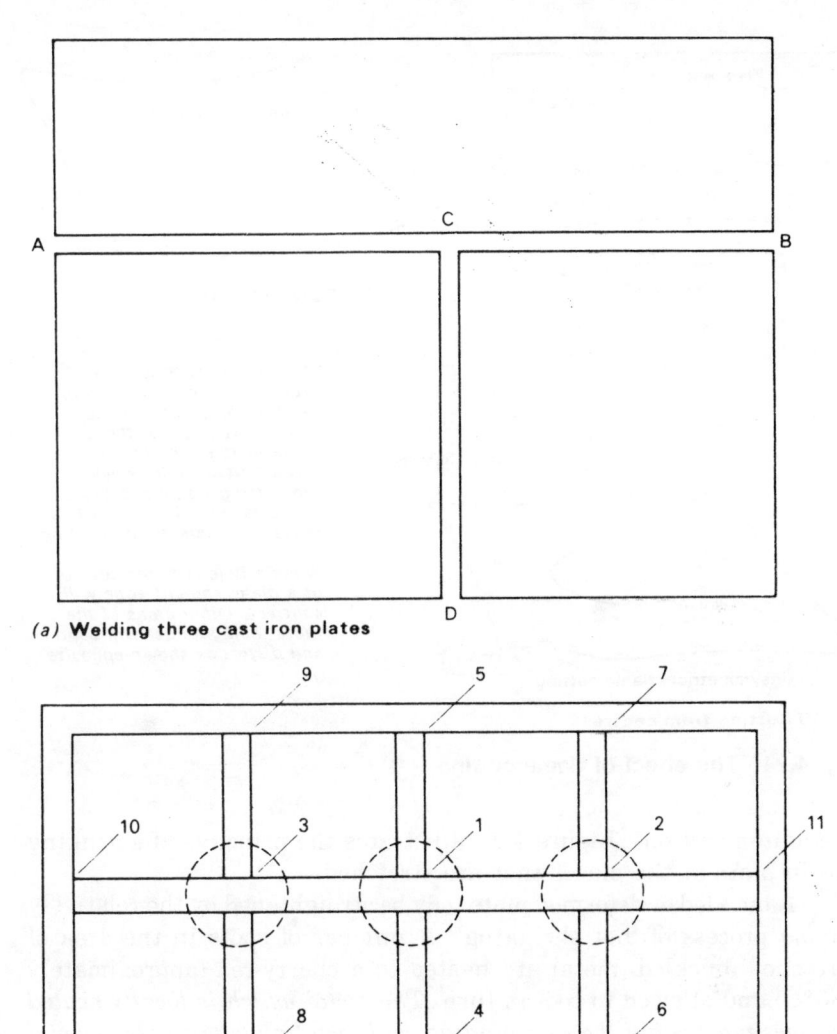

*(a)* **Welding three cast iron plates**

*(b)* **A suitable welding sequence for a steel frame structure**

**Fig. 4.23 The importance of a correct welding sequence**

## 4.18 The use of heat for straightening plate and sections

It has been seen that the application of heat can produce distortion. Heat can be used to advantage, for those same forces of expansion and contraction can be harnessed to remove distortion in plates or to

113

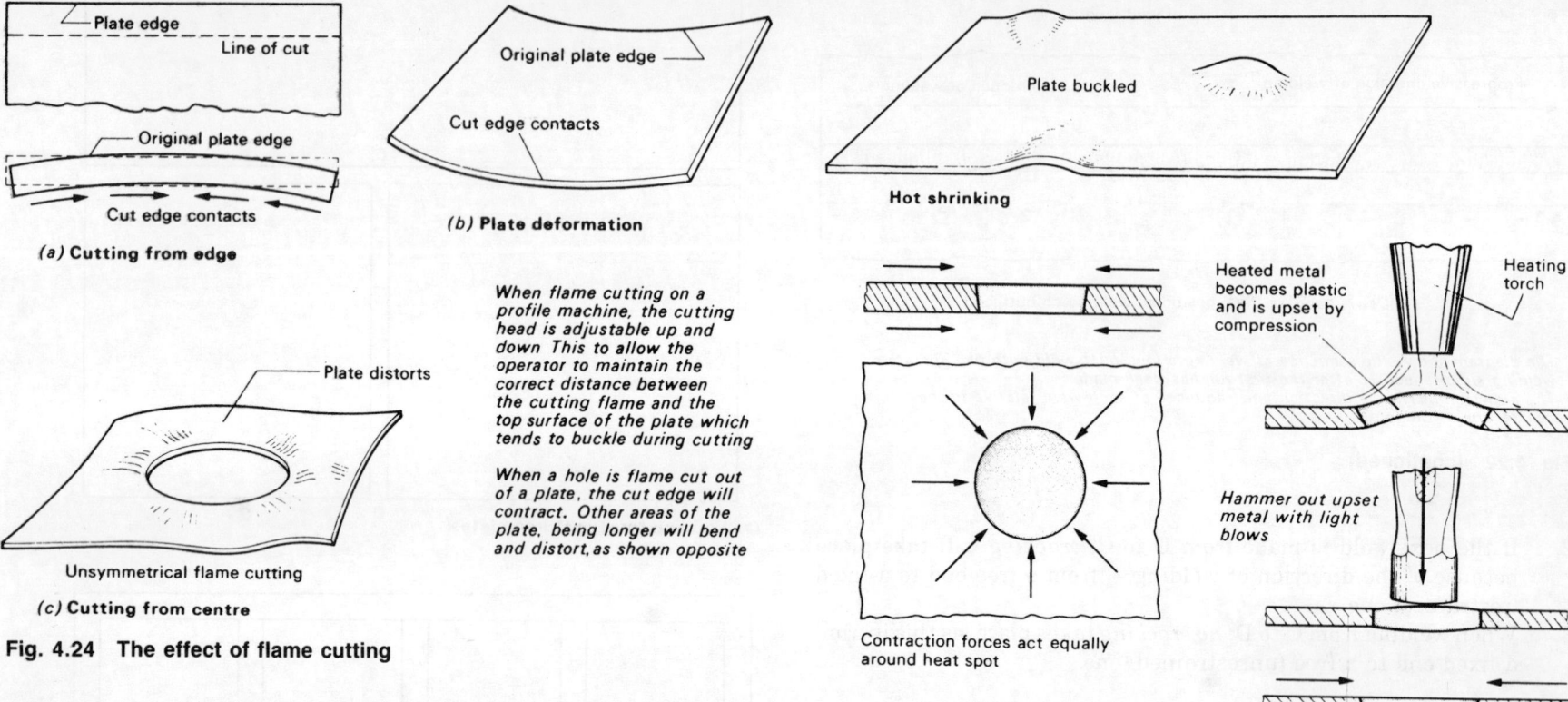

**Fig. 4.24 The effect of flame cutting**

(a) Cutting from edge
(b) Plate deformation
(c) Cutting from centre — Unsymmetrical flame cutting

*When flame cutting on a profile machine, the cutting head is adjustable up and down This to allow the operator to maintain the correct distance between the cutting flame and the top surface of the plate which tends to buckle during cutting*

*When a hole is flame cut out of a plate, the cut edge will contract. Other areas of the plate, being longer will bend and distort, as shown opposite*

**Fig. 4.25 Principle of hot shrinking**

straighten sections. Figure 4.25 illustrates the principle of shrinking a thin plate at the places that are stretched.

A buckled or deformed plate may be straightened by the relatively simple process of 'hot shrinking'. A number of spots in the area of stretched (buckled) metal are heated to a cherry-red (approximately 750°C) and allowed to cool in turn. *The metal which is locally heated becomes plastic, but the surrounding cold metal plate prevents thermal expansion.* The plastic area becomes upset by compressive forces. When a heated spot is allowed to cool, the metal will tend to contract, and it is during this shrinkage that contractional stresses will occur.

The process is repeated until the stretched areas of metal are compressed, and the plate is restored to a straight and flat condition. THIS PROCESS IS WIDELY USED IN LIGHT VEHICLE CRASH REPAIR AND PANEL-BEATING WORKSHOPS.

## 4.19 Use of heat strips

Figure 4.26 shows the use of *heat strips* for the 'hot straightening' and 'hot shrinking' of plate and wide sections.

The shrinking forces will be approximately equal for both sides of the plate. Figure 4.26 shows the application of a heat strip which, upon cooling, causes the metal to become compressed, because the contraction forces come in at right angles to the strip.

The length and width of a particular heat strip can be determined by the thickness of the plate. As a general guide: for thicknesses from about 10 mm to 30 mm, the width of the heat strip should be between 20 and 30 mm and the length of the heat strip should be between 130 and 200 mm.

Heating is commenced at one end of the strip, making sure that the correct heat goes right through the plate (cherry-red, 750°C). The whole heating operation is a continuous one, employing a zig-zag movement of the heating torch towards the opposite end. On cooling, the plate will be shorter in length in the locally heated area.

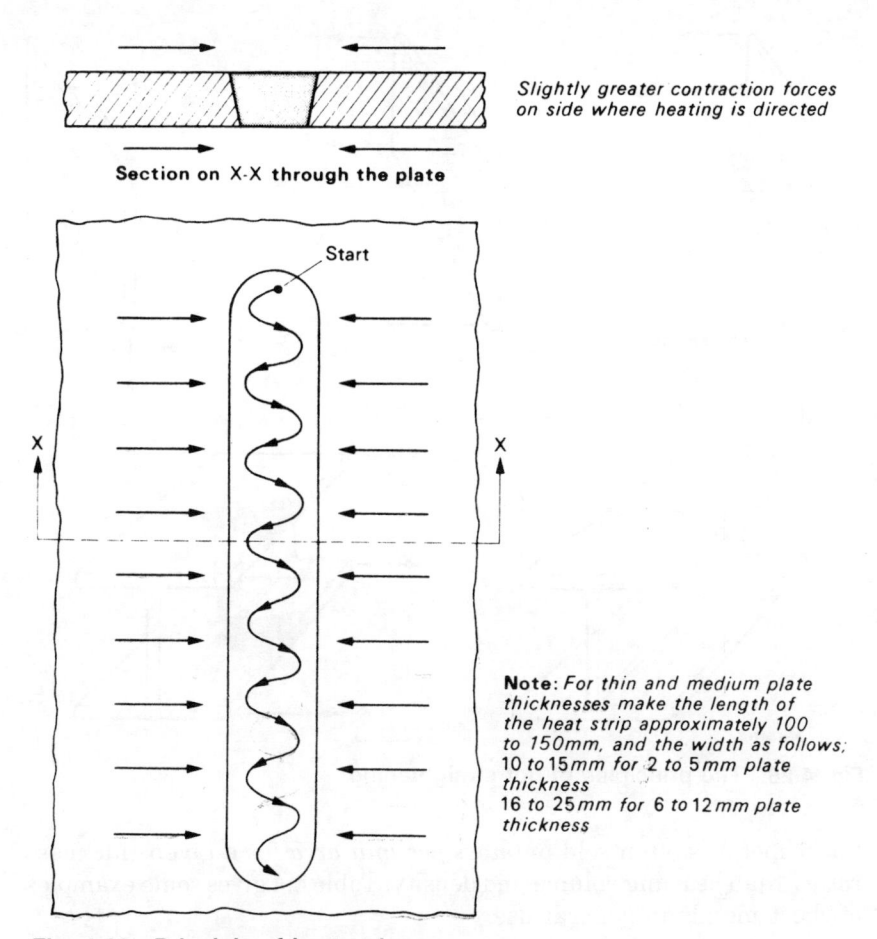

*Slightly greater contraction forces on side where heating is directed*

Section on X-X through the plate

**Note:** *For thin and medium plate thicknesses make the length of the heat strip approximately 100 to 150 mm, and the width as follows:*
*10 to 15 mm for 2 to 5 mm plate thickness*
*16 to 25 mm for 6 to 12 mm plate thickness*

**Fig. 4.26 Principle of heat strips**

## 4.20 Use of heat triangles

The use of *heat triangles* for straightening thin angle and flat sections, and the use of 'triangles' of heat strips for the bending and straightening of plate and wide sections are shown in Figure 4.27.

Simple heat triangles may be used as shown in Figure 4.27(a). This entails starting with the heating torch at the apex of the triangle and working towards the base with a gradually widening zig-zag movement. When allowed to cool, the base of the heat triangle will start to contract the most, and the contracting forces tend to cause the plate to bend, as shown in Figure 4.27(b).

The resultant effects of using triangles of heat strips are exactly the same as for the simple heat triangles. Simple heat triangles are used

(a) Simple heat triangle

(b) Effects of contracting forces

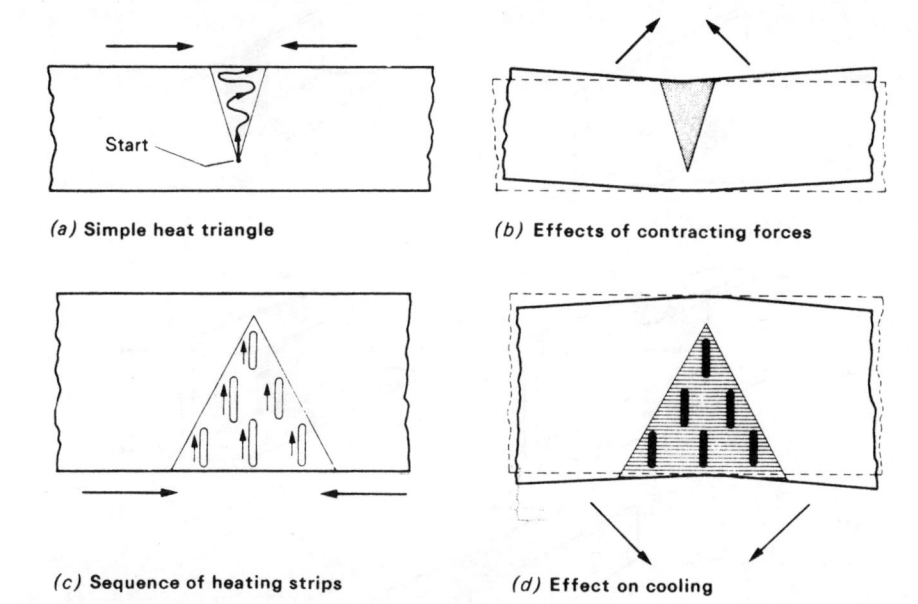

(c) Sequence of heating strips

(d) Effect on cooling

**Fig. 4.27 Use of heat triangles**

for straightening of thin plate and light sections. Triangles of heat strips are preferred when bending or straightening thick plate and heavy sections.

The order in which the heat strips are applied, in the triangle, is shown in Figure 4.27(c). Heating with the torch is commenced a short distance in from the edge of the plate, progressively heating from the outside inwards.

## 4.21 Straightening simple sections

The basic principles of straightening simple sections are explained in Figure 4.28, shown overleaf.

# Part B Engineering science — fundamentals and applications

## 4.22 Force and mass

Engineering is concerned with the production of a force which may be exerted in a particular manner and a certain direction in order to fulfil a specific purpose. The presence of a force can usually be detected by

115

**Angle section**

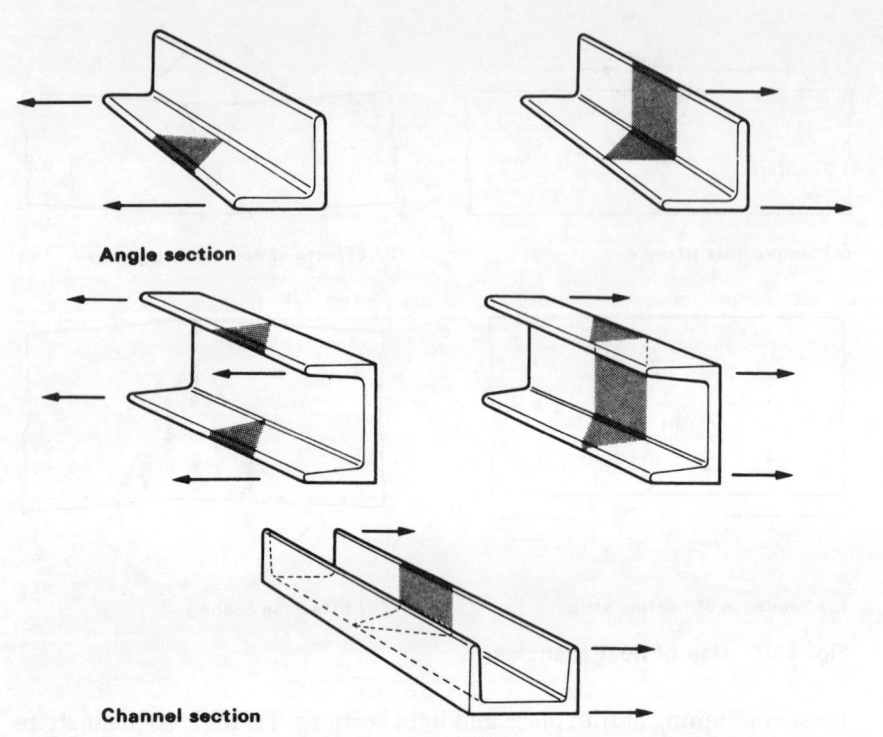

**Channel section**

*The arrows indicate the direction in which the ends of the section will move on cooling, after the application of localised heating — 'heat triangles'*

**Tee section**

**Beams**

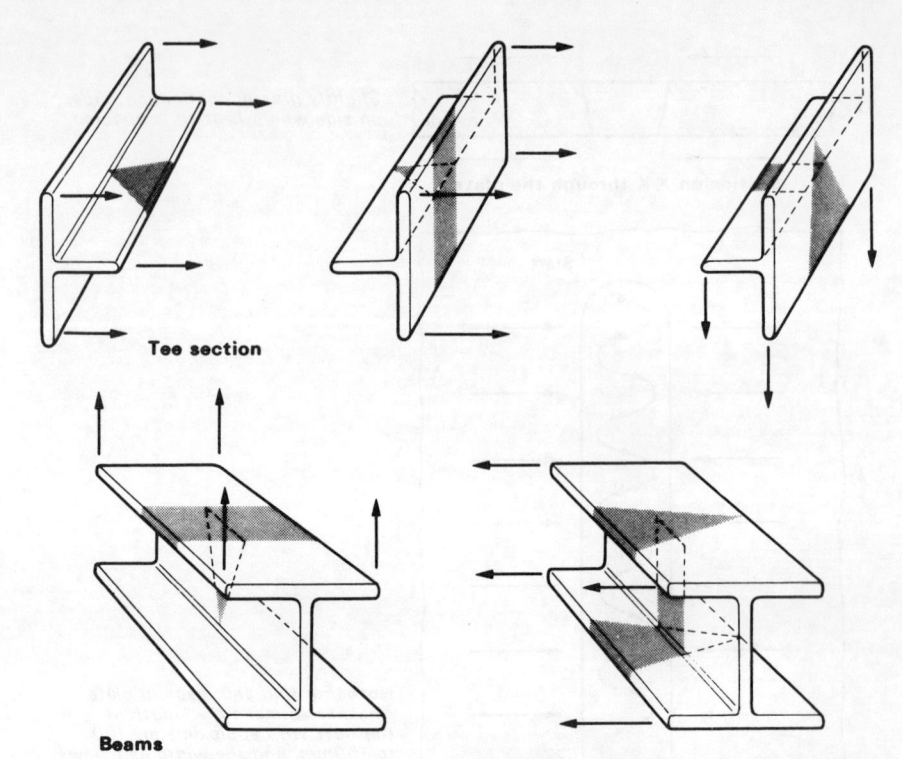

**Fig. 4.28   The principles of hot straightening**

means of its effects. A force can be exerted to lift something or start or stop it from moving. A moving object can be made to move faster or slower by means of a force. Compressing, stretching, bending or even breaking an object can be achieved by a force. By definition, *a force is that which changes, or tends to change, the state of rest or uniform motion of a body. A force can also deform a body if the motion of the body is resisted.*

## 1. MASS

Mass is the quantity of 'matter' or material in a component. Providing additional material is not added to the component, or any material cut off the component, its MASS IS CONSTANT. This mass will remain constant on Earth; and on the moon; and anywhere in space.

The S.I. unit for mass is the kilogramme (kg).

1000 grammes = 1 kilogramme
1000 kilogrammes = 1 tonne
MASS = VOLUME × DENSITY

Sheet metal is often sold by *mass per unit area* for a given thickness rather than be using volume and density. Table 4.4 gives some examples of sheet metals in general use.

## 2. WEIGHT

The effect of the force of 'gravity' acting on a mass of a component is called its 'weight'. The gravitational force acting upon a body or component is the measure of the *force of attraction exerted on its mass by the Earth* (Figure 4.29).

The S.I. unit for weight is the Newton (N), and *the gravitational force on a mass of 1 kg is 9·81 N.*

Weight is a variable quantity because the gravitational force exerted by the Earth on a given mass is not constant at all points on the Earth's surface. The Earth is not a true sphere, the Poles are nearer the centre of the Earth than a point on the Equator. Therefore a component of *constant mass* will weigh less at the Equator than it does at the North and South Poles.

**Table 4.4 Mass/unit area for sheet metal**
(The values given are for mild steel. For other metals see notes.)

| THICKNESS (mm) ISO R388 | | MASS/UNIT AREA | |
|---|---|---|---|
| 1st choice | 2nd choice | g/cm$^2$ | kg/m$^2$ |
| 0·020 | | 0·015 5 | 0·155 |
| | 0·022 | 0·017 0 | 0·170 |
| 0·025 | | 0·019 3 | 0·193 |
| | 0·028 | 0·021 6 | 0·216 |
| 0·032 | | 0·024 7 | 0·247 |
| | 0·036 | 0·027 8 | 0·278 |
| 0·040 | | 0·030 9 | 0·309 |
| | 0·045 | 0·034 8 | 0·348 |
| 0·050 | | 0·038 7 | 0·387 |
| | 0·056 | 0·043 3 | 0·433 |
| 0·063 | | 0·048 7 | 0·487 |
| | 0·071 | 0·054 9 | 0·549 |
| 0·080 | | 0·061 8 | 0·618 |
| | 0·090 | 0·069 6 | 0·696 |
| 0·100 | | 0·077 3 | 0·773 |
| | 0·112 | 0·086 5 | 0·865 |
| 0·125 | | 0·096 6 | 0·966 |
| | 0·140 | 0·108 2 | 1·082 |
| 0·160 | | 0·123 7 | 1·237 |
| | 0·180 | 0·139 1 | 1·391 |
| 0·200 | | 0·154 6 | 1·546 |
| | 0·224 | 0·173 2 | 1·732 |
| 0·250 | | 0·193 3 | 1·933 |
| | 0·280 | 0·216 4 | 2·164 |
| 0·315 | | 0·243 5 | 2·435 |
| | 0·355 | 0·274 4 | 2·744 |
| 0·400 | | 0·309 2 | 3·092 |
| | 0·450 | 0·347 9 | 3·479 |
| 0·500 | | 0·386 5 | 3·865 |
| | 0·560 | 0·432 9 | 4·329 |
| 0·630 | | 0·487 0 | 4·870 |
| | 0·710 | 0·548 8 | 5·488 |
| 0·800 | | 0·618 4 | 6·184 |
| | 0·900 | 0·695 7 | 6·957 |
| 1·00 | | 0·773 0 | 7·730 |
| | 1·120 | 0·865 0 | 8·650 |
| 1·25 | | 0·996 2 | 9·962 |
| | 1·40 | 1·082 2 | 10·822 |
| 1·60 | | 1·236 8 | 12·368 |
| | 1·80 | 1·391 4 | 13·914 |
| 2·00 | | 1·546 0 | 15·460 |
| | 2·24 | 1·731 5 | 17·315 |
| 2·50 | | 1·932 5 | 19·325 |
| | 2·80 | 2·164 4 | 21·644 |
| 3·15 | | 2·434 9 | 24·349 |
| | 3·5 | 2·744 1 | 27·441 |
| 4·00 | | 3·092 0 | 30·920 |

*Notes on use of table*:

1. The mass/unit area given in the tables can be converted to the values for metals other than steel by use of the following multiplying factors.

   Aluminium × 0·331 1
   Brass (70/20) × 1·062
   Bronze × 1·102
   Copper × 1·119
   Lead × 1·475
   Tin × 0·944
   Zinc × 0·906

2. To calculate the mass/unit area of 1 mm thick copper sheet. Mass/unit area for 1 mm thick steel is 7·73 kg/m$^2$ multiplying factor for copper is 1·119. Therefore mass/unit area for copper 1 mm thick will be.

   7·73 × 1·119 = 8·65 kg/m$^2$

## 3. CENTRE OF GRAVITY

The centre of gravity (c.g.), or 'centre of mass' of a body or component, is the single point at which the whole 'weight' of a body will be concentrated whatever the position in which that body might be. A thin flat sheet is termed a LAMINA. The centres of gravity for symmetrical shapes of laminae are shown in Figure 4.30(a).

The surface area of a lamina has no mass and, therefore, cannot experience a gravitational force. When only a shape or area is considered, the term CENTROID or 'centre of area', is used where the centre of gravity of a lamina of that shape would be. The centre of gravity of the lamina would be in the middle of its thickness.

It is rather more difficult to determine the position of the centre of gravity of a solid object or component not in the form of a lamina. Examples of some solid shapes which can be determined by inspection are illustrated in Figure 4.30(b).

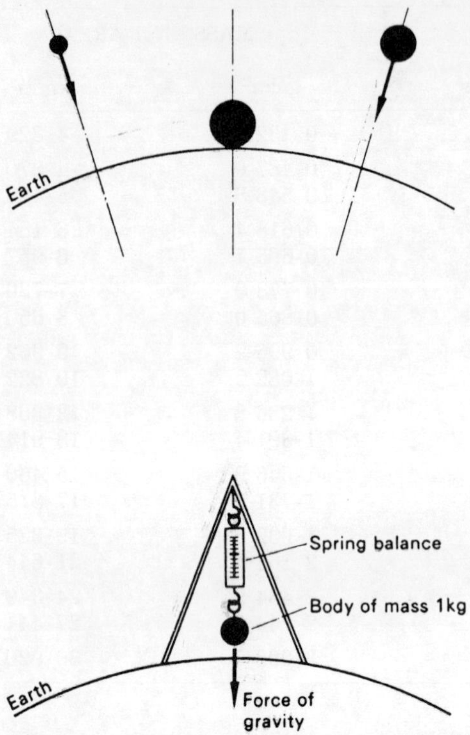

1. All objects within the earth's gravitational field are pulled towards the centre of the earth by the **Force of gravity**

2. Unsupported objects must, therefore, fall downwards towards the centre of the earth. In practice they fall until they reach the ground or some other solid object

3. At the surface of the earth the force of gravity exerts a pull of approximately 9·81N on each kilogram mass of the body. In the example shown, the pull on the body of mass 1 kg would be 9·81 N
Thus we would say that its **weight is 9·81 N**

4. If its mass had been 3 kg then its weight would have been
3 × 9·81 = 29·43 N

**Fig. 4.29 Weight**

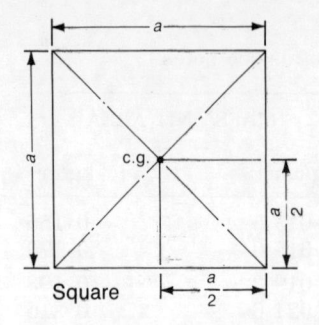

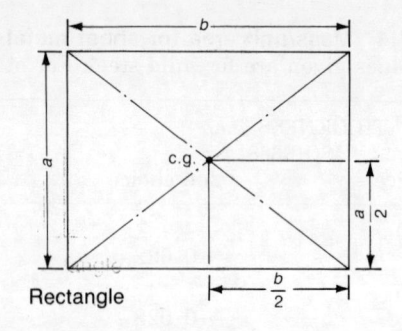

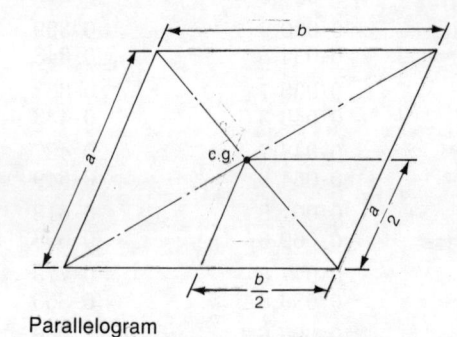

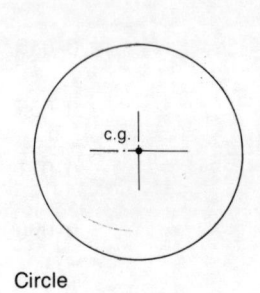

The centre of gravity of a square, rectangle or parallelogram is the intersection of the diagonals, or lines drawn down the centre and parallel to any side. The centre of gravity of a triangular lamina lies at the point of intersection of lines (medians) drawn from a corner (apex) to the mid-point of the side opposite. *Note: with triangular laminae the centre of gravity always lies on a point one-third of the way up from the base along any median.* The centre of gravity of a circular lamina is its geometric centre.

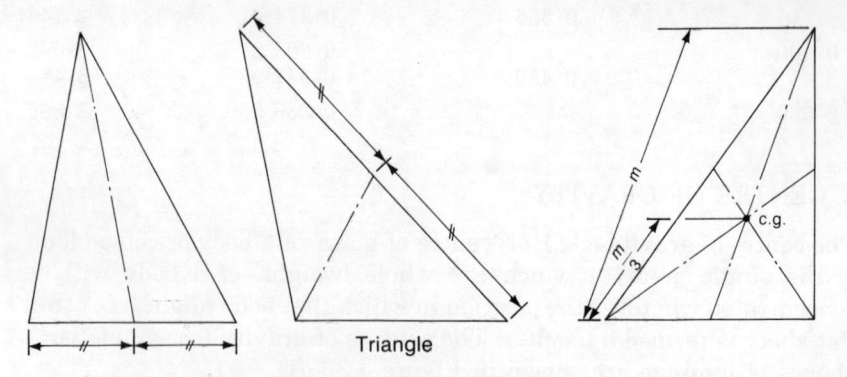

## 4. CENTRE OF GRAVITY — COMPOSITE FIGURES

The centroid of a composite area made up from geometric shapes, or the centre of gravity of a composite solid built up in this way, can be determined by the application of the principle of moments. The moment of an area about an axis can be calculated as if the area were a force acting through its centroid. Symbols $x$ and $y$ may be used to indicate the horizontal and vertical distances, respectively, of either the centroid or the centre of gravity from some datum line or point.

*It is important when lifting heavy loads or machines by slinging or hoisting, that the hoisting position is vertically above the centre of gravity of the load* (see Chapter 1).

The principle of moments adopted to determine the centroid of a composite area, or the centre of gravity of a composite solid, is explained using numerical examples, in Table 4.5.

**Fig. 4.30(a) Centre of gravity — laminar geometrical shapes**

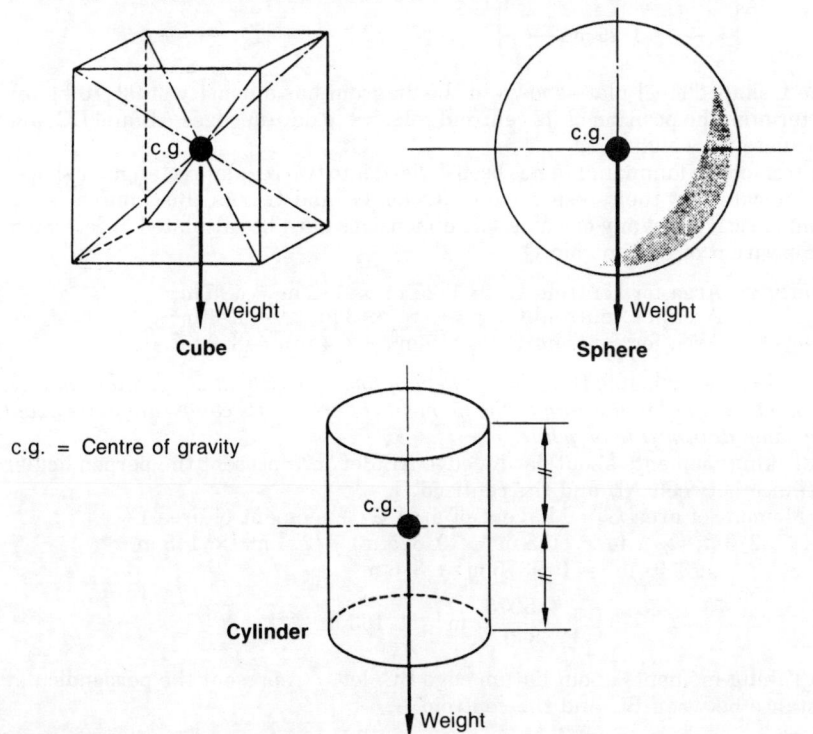

**Fig. 4.30(b) Centre of gravity — solid shapes**

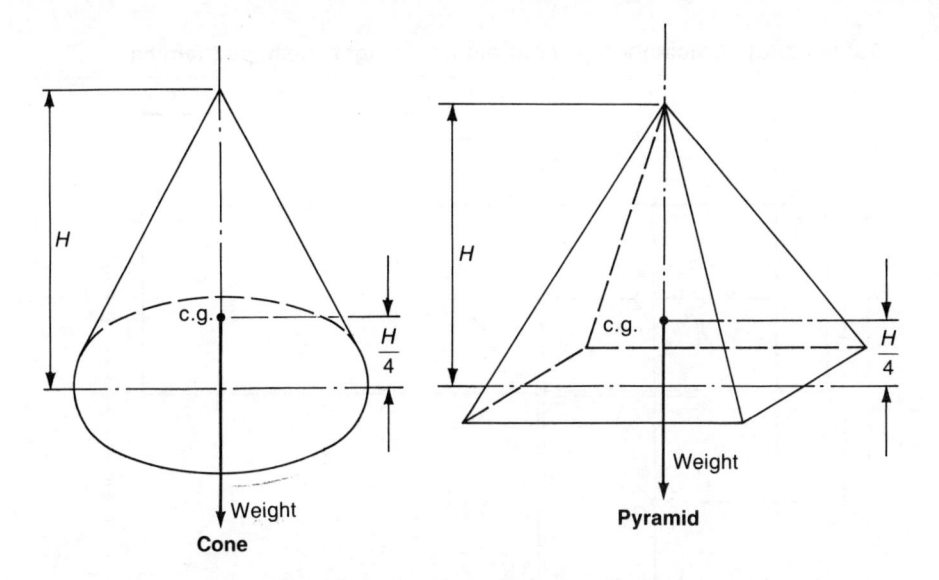

*For a body of regular shape the centre of gravity is at its geometric centre*

c.g. = Centre of gravity

**Table 4.5(a)  Calculation — centroid, circular composite lamina**

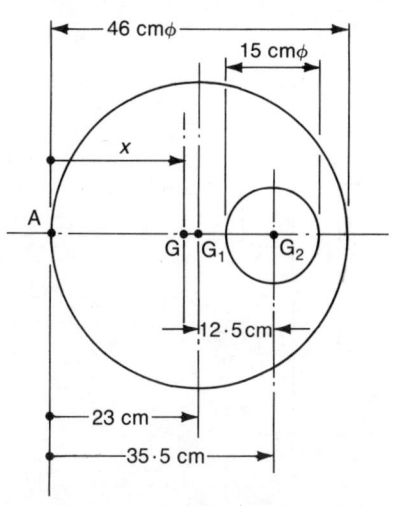

Details are given of a circular metal plate of 35·5 cm diameter in which a hole of 15 cm diameter has been cut. Calculate the position of its centroid. This composite lamina is symmetrical about a centre line 'A', therefore only one dimension '$x$' is required to be determined.
The centroids of the composite geometrical shapes are shown as $G_1$ and $G_2$.

*Solution*:  Area for centroid $G_1 = \pi R^2$ where $R = 23$ cm
   $= \pi \times 23^2 = 529\pi \text{cm}^2$.
Area for centroid $G_2 = \pi r^2$ where $r = 7.5$ cm
   $= \pi \times 7.5^2 = 56.25\pi \text{cm}^2$.
Area for centroid $G = \pi(R^2 - r^2)$
   $= \pi(529 - 56.25) = 472.75\pi \text{cm}^2$.

*Moment of area = Area × perpendicular distance of centroid from a given datum.*

Taking moments about datum 'A', let '$x$' represent the distance between centroid G and the datum:
   Moment of area G $= 472.75\pi \text{cm}^2 \times x$ cm $= 472.75\pi x$ cm$^3$.
   Moment of area $G_1 = 529\pi \text{cm}^2 \times 23$ cm $= 12167\pi$ cm$^3$.
   Moment of area $G_2 = 56.25\pi \text{cm}^2 \times 35.5$ cm $= 1996.88\pi$ cm$^3$.

Moment of area G = Total moments of area of composite parts $G_1$ and $G_2$.
   $472.75x$ cm$^3$ $= 12167$ cm$^3$ $- 1996.88$ cm$^3$
   $472.75x$ cm$^3$ $= 10170.12$ cm$^3$.

$$x = \frac{10170.12}{472.75} \text{ cm} = 21.51 \text{ cm}.$$

The centroid of this composite is 14·9 mm to the left of $G_1$ on the central axis.

## 5. DENSITY

Equal masses of different materials do not occupy equal volumes. For example, although 1 kg of water and 1 kg of mercury have an identical mass, by contrast, their volumes are extremely different. *The volume of mercury would exceed $13\frac{1}{2}$ times the volume of an equal mass of water!*

**Table 4.5(b) Calculation — Centroid, rectangle/square composite lamina**

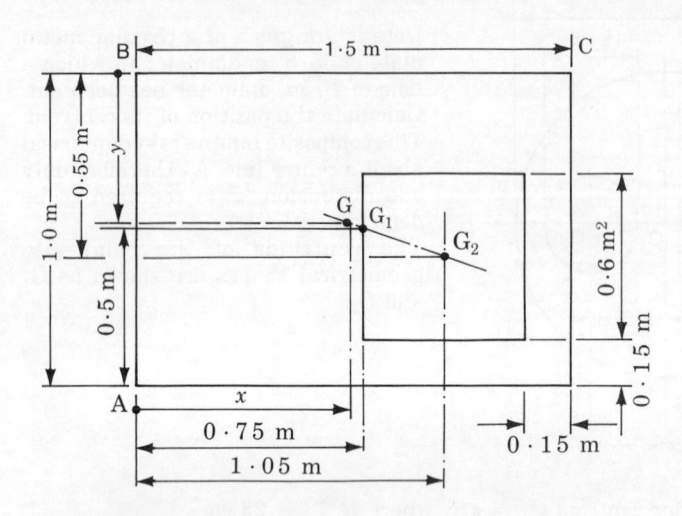

The diagram shows a 1·5 m by 1·0 m rectangular steel plate of uniform thickness with a 0·6 m square apperture. Determine the position of its centroid relative to the datum edges AB and BC.
    The centroids for the rectangle and the square composites are marked at $G_1$ and $G_2$ respectively. The composite shape is not symmetrical, therefore two dimensions '$x$' and '$y$' will have to be determined.

*Solution*: Area for centroid $G_1$ = 1·5 m × 1·0 m    = 1·5 m$^2$.
           Area for centroid $G_2$ = 0·6 m × 0·6 m    = 0·36 m$^2$.
           Area for centroid G   = 1·5 m$^2$ − 0·36 m$^2$ = 1·14 m$^2$.

Moment of area G = Moment of area of $G_1$ − Moment of area of $G_2$
Taking moments about datum AB, let '$x$' represent the perpendicular distance between centroid G and datum AB:

$1·14 \times x^3 = (1·5 \text{ mm}^2 \times 0·75 \text{ m}) - (0·36 \text{ m}^2 \times 1·05 \text{ m})$
$1·14x \text{ m}^3 = 1·125 \text{ m}^3 - 0·378 \text{ m}^3$

$$x = \frac{0·747}{1·14} \text{ m} = 0·655 \text{ m}.$$

Taking moments about datum BC, let '$y$' represent the perpendicular distance between centroid G and datum BC:

$1·14 \times y \text{ m}^3 = (1·5 \text{ m}^2 \times 0·5 \text{ m}) - (0·36 \text{ m}^2 \times 0·55 \text{ m})$
$1·14y \text{ m}^3 = 0·75 \text{ m}^3 - 0·198 \text{ m}^3$

$$y = \frac{0·552}{1·14} \text{ m} = 0·484 \text{ m}.$$

The centroid G is located on a straight line drawn through $G_1$ and $G_2$. Its position is 0·655 m from datum edge AB and 0·484 m from datum edge BC.

---

**Table 4.5(c) Calculation — centroid and weight, L-shaped lamina**

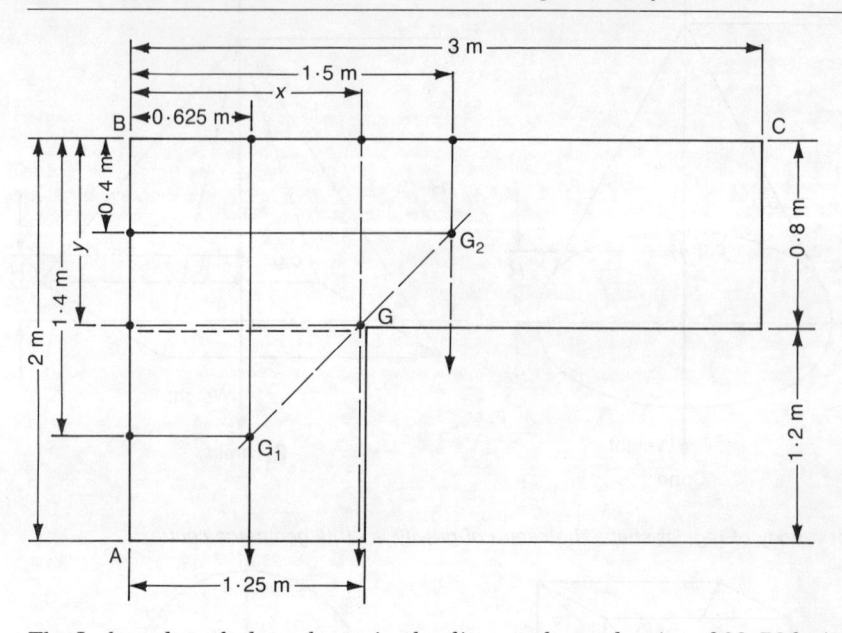

The L-shaped steel plate shown in the diagram has a density of 92·76 kg/m$^2$. Determine the position of its centroid relative to datum edges AB and BC, and calculate its weight.
The composite laminar area has been divided into two regular rectangular shapes as indicated by their respective centroids $G_1$ and $G_2$. As this lamina is not symmetrical about any one axis, two dimensions must be calculated to determine the position of its centroid G.

*Solution*: Area for centroid $G_1$ = 1·25 m × 1·2 m = 1·5 m$^2$
           Area for centroid $G_2$ = 0·8 m × 3 m   = 2·4 m$^2$
           Area for centroid G   = 1·5 m$^2$ + 2·4 m$^2$ = 3·9 m$^2$

Applying the principle that *the moment of total area about a given datum line, or point, is equal to the sum of the moments of area of its composite parts about the same datum line or point*:
(a) Taking moments about datum edge AB, let '$x$' represent the perpendicular distance between AB and the centroid G:
    Moment of area G = Moment of area $G_1$ + moment of area $G_2$
$3·9 \text{ m}^2 \times x \text{ m} = (1·5 \text{ m}^2 \times 0·625 \text{ m}) + (2·4 \text{ m}^2 \times 1·5 \text{ m})$
$3·9x \text{ m}^3 = 0·9375 \text{ m}^3 + 3·6 \text{ m}^3$

$$x = \frac{4·5375}{3·9} \text{ m} = 1·163 \text{ m}.$$

(b) Taking moments about datum edge BC, let '$y$' represent the perpendicular distance between BC and the centroid G:

$3·9 \text{ m}^2 \times y \text{ m} = (1·5 \text{ m}^2 \times 1·4 \text{ m}) + (2·4 \text{ m}^2 \times 0·4 \text{ m}^2)$
$3·9y \text{ m}^3 = 2·1 \text{ m}^3 + 0·96 \text{ m}^3$

$$y = \frac{3 \cdot 06}{3 \cdot 9} \text{ m} = 0 \cdot 785 \text{ m}.$$

The centroid G lies on a straight line between $G_1$ and $G_2$ at a position $1 \cdot 163$ m from datum edge AB and $0 \cdot 785$ m from datum edge BC.

$$\begin{aligned}\text{Mass of plate} &= \text{Area (m}^2) \times \text{Density (kg/m}^2)\\ &= 3 \cdot 9 \text{ m}^2 \times 92 \cdot 76 \text{ kg/m}^2 = 361 \cdot 76 \text{ kg}.\\ \text{Weight} &= 361 \cdot 76 \text{ kg} \times 9 \cdot 81 \text{ N/kg}\\ &= 3548 \cdot 87 \text{ N} \qquad = 3 \cdot 55 \text{ kN}.\end{aligned}$$

### Table 4.5(d)  Calculations — centre of gravity, weight composite solid

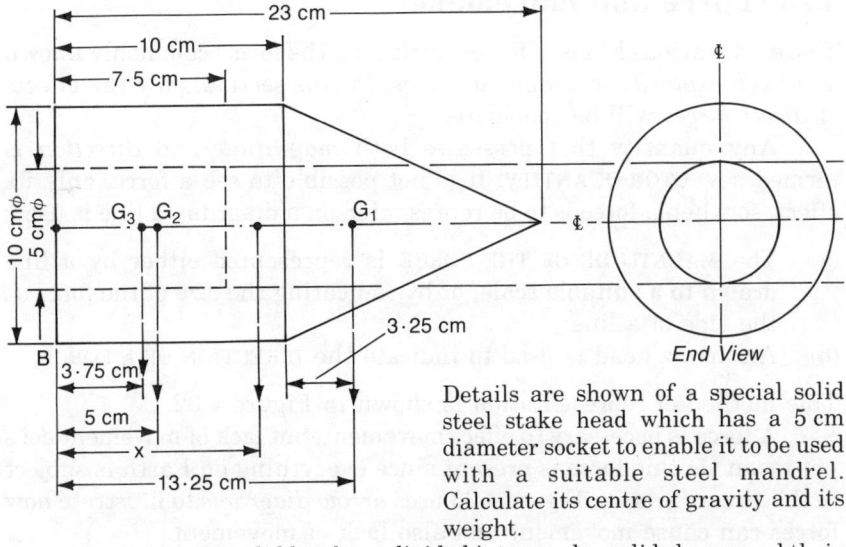

Details are shown of a special solid steel stake head which has a 5 cm diameter socket to enable it to be used with a suitable steel mandrel. Calculate its centre of gravity and its weight.

This composite solid has been divided into regular solid shapes and their centre of gravity is shown as $G_1$, $G_2$ and $G_3$. As it is symmetrical about a central axis only one dimension is required to be determined.

The material has uniform density, so masses are proportional to volumes. In this case, therefore, 'moments of volumes' can be taken about the datum axis AB.

*Solution:*

$$\begin{aligned}\text{Volume of cone about } G_1 &= \frac{\pi R^2 H}{3} \text{ where } R = 5 \text{ cm and } H = 13 \text{ cm}\\ &= \frac{\pi(25 \times 13)}{3} \text{ cm}^3 = 108 \cdot 33\pi \text{cm}^3\\ \text{Volume of cylinder about } G_2 &= \pi R^2 H \text{ where } R = 5 \text{ cm}, H = 10 \text{ cm}\\ &= \pi(5 \times 5 \times 10) \text{ cm}^3 = 250\pi \text{cm}^3\end{aligned}$$

$$\begin{aligned}\text{Volume of cylinder about } G_3 &= \pi r^2 n \text{ where } r = 2 \cdot 5 \text{ cm}, h = 7 \cdot 5 \text{ cm}\\ &= \pi(2 \cdot 5 \times 2 \cdot 5 \times 7 \cdot 5) \text{ cm}^3 = 46 \cdot 875\pi \text{cm}^3\\ \text{Volume of stake about } G &= 108 \cdot 33\pi \text{cm}^3 + 250\pi \text{cm}^3 - 46 \cdot 875\pi \text{cm}^3\\ &= 311 \cdot 455\pi \text{cm}^3\end{aligned}$$

Taking moments about datum axis AB, let '$x$' represent the perpendicular distance between AB and the centre of gravity G:

Moment of volume G = Moment of volume $G_1$ + moment of volume $G_2$ − moment of volume $G_3$.

$$311 \cdot 455\pi \text{cm}^3 \times x \text{ cm} = (108 \cdot 33\pi \text{cm}^3 \times 13 \cdot 25 \text{ cm}) + (250\pi \text{cm}^3 \times 5 \text{ cm}) -$$
$$(46 \cdot 875\pi \text{cm}^3 \times 3 \cdot 75 \text{ cm}^3)$$
$$311 \cdot 455x \text{ cm}^4 = 1435 \cdot 3725 \text{ cm}^4 + 1250 \text{ cm}^4 - 175 \cdot 78125 \text{ cm}^4$$
$$x = \frac{2509 \cdot 5913}{311 \cdot 455} \text{ cm} = 8 \cdot 0576 \text{ cm}.$$

The centre of gravity for this composite solid lies on its central axis at a distance of $8 \cdot 058$ cm from the datum AB.

$$\begin{aligned}\text{Volume of stake} &= 311 \cdot 455 \times 3 \cdot 142 \text{ cm}^3 = 978 \cdot 592 \text{ cm}^3\\ \text{Mass} &= \text{Volume (cm}^3) \times \text{Density (g/cm}^3)\\ &= 978 \cdot 592 \text{ cm}^3 \times 7 \cdot 83 \text{ k/cm}^3 \text{ (see Table 4.6)} = 7662 \cdot 375 \text{ g}\\ &= 7 \cdot 662 \text{ kg}\\ \text{Weight} &= 7 \cdot 662 \times 9 \cdot 81 \text{ N/kg} = 75 \cdot 16 \text{ N}.\end{aligned}$$

The relationship between mass and volume is known as 'density', and is defined as *mass per unit volume*.

$$\text{DENSITY} = \frac{\text{MASS OF MATERIAL}}{\text{VOLUME OCCUPIED BY THE MATERIAL}}$$

The S.I. units of measurement for density are *kilogrammes per cubic metre* (kg/m$^3$). Table 4.6 gives densities of common engineering materials.

## 6. RELATIVE DENSITY

ALUMINIUM is much lighter (less dense) than STEEL which, in turn, is lighter than LEAD, when comparing equal masses (Table 4.6). *One gramme of pure water at a temperature of 4°C occupies one cubic centimetre.* Hence the density of water is 1 g/cm$^3$ or 1000 kg/m$^3$. Therefore it is usual to compare the density of any substance with the density of water expressed in the same units. This is known as 'relative density':

$$\text{RELATIVE DENSITY OF A SUBSTANCE} =$$
$$\frac{\text{DENSITY OF SUBSTANCE}}{\text{DENSITY OF WATER (in same units)}}$$

**Table 4.6  Densities of common engineering materials**

| MATERIAL | DENSITY | | |
|---|---|---|---|
| | g/mm³ | g/cm³ | kg/m³ |
| Aluminium | 0·002 56 | 2·56 | 2 560 |
| Brass (70/30) | 0·008 21 | 8·21 | 8 210 |
| Chromium | 0·007 03 | 7·03 | 7 030 |
| Copper | 0·008 65 | 8·65 | 8 650 |
| Iron (cast) | 0·007 20 | 7·20 | 7 200 |
| Iron (wrought) | 0·007 75 | 7·75 | 7 750 |
| Lead | 0·011 4 | 11·4 | 11 400 |
| Nickel | 0·008 73 | 8·73 | 8 730 |
| Steel | 0·007 83 | 7·33 | 7 330 |
| Tin | 0·007 3 | 7·30 | 7 300 |
| Titanium | 0·003 54 | 3·54 | 3 540 |
| Zinc | 0·007 | 7·0 | 7 000 |

*The relative density of a substance has no units and is expressed numerically indicating the number of times that a given volume of a substance is heavier, or lighter, than an equal volume of water.*

EXAMPLES:

1. Given the density of aluminium is 2560 kg/m³, determine its relative density.

*Solution*:

$$\text{Relative density of aluminium} = \frac{\text{Density of aluminium}}{\text{Density of water}}$$

$$= \frac{2560 \text{ kg/m}^3}{1000 \text{ kg/m}^3}$$

$$= \underline{2 \cdot 56}$$

2. Given that the relative density of copper is 8·65, determine its density.

*Solution*:

$$\text{Relative density of copper} = \frac{\text{Density of copper}}{\text{Density of water}}$$

Density of copper = R.D of copper × density of water
= 8·65 × 1000 kg/m³
= $\underline{8650 \text{ kg/m}^3}$

3. A brass bar is 0·76 m in length and has a cross-section of 15 cm × 4 cm. Given that the relative density of brass is 8·48, determine its weight.

*Solution*:

Density of brass = 8·48 × 1000 kg/m³ = $\underline{8480 \text{ kg/m}^3}$
Volume of brass = 0·76 m × 0·15 m × 0·04 m = $\underline{0 \cdot 004\ 56 \text{ m}^3}$
Mass of brass = Density × volume
= 8480 × 0·004 56 kg = $\underline{38 \cdot 67 \text{ kg}}$
Weight of brass = 38·67 × 9·81 N = $\underline{379 \cdot 4 \text{N}}$

Further examples are given in Figure 4.31.

## 4.23 Force and movement

There are various kinds of forces, including the forms commonly known as *direct, explosive* or *magnetic* forces. In this section, only the effects of *direct forces* will be considered.

Any quantity that possesses both *magnitude* and *direction* is termed a VECTOR QUANTITY. It is not possible to see a force, only its effect, so when a force is to be represented on a diagram, a line is used:

(a) The MAGNITUDE OF THE FORCE is represented either by a line drawn to a suitable scale, or by indicating the size of the force at the side of a line;

(b) An arrow head is used to indicate the DIRECTION OF A FORCE.

This method of representation is shown in Figure 4.32.

A force is necessary to effect movement, but lack of movement does not mean that no force is present since everything on Earth is subject to the *force of gravity*. Figure 4.33 uses *arrow diagrams* to illustrate how forces can cause movement and also lack of movement.

## 4.24 Force — vector diagrams

A body may be acted upon by a number of forces. The gravitational force of a body, which is known as its *weight*, acts towards the centre of the Earth (Figure 4.29), and for most purposes is considered to be acting vertically downwards. When a body is lifted upwards, the applied force must act vertically upwards to overcome the gravitational force. For a body to be moved sideways, the force must be exerted in a sideways direction.

*When more than one force acts upon a body, the final result will depend upon the magnitude and direction of the forces being applied to that body.* The body will move as if a RESULTANT FORCE were being

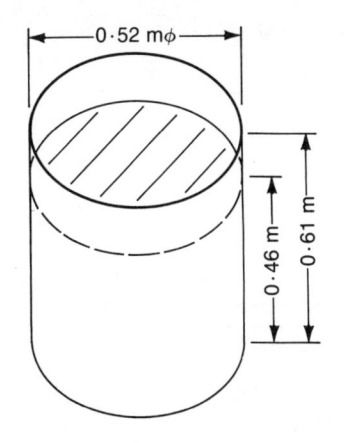

An open top cylindrical container fabricated from brass of 3·15 mm thickness has an outside diameter of 0·52 m and a height of 0·61 m. It contains paraffin to a depth of 0·46 m. Given that the density of paraffin is 800 kg/m³, calculate:
(a) the total mass of the vessel and its contents;
(b) the total weight.

Area of Sheet Metal:

Cylindrical body = $\pi DH$ (where $D$ = mean diameter and metal thickness $T$ = 0·003 15 m)
= 3·142 × (0·52 − 0·003 15) × 0·61
= 3·142 × 0·517 × 0·61 = 0·991 m²

Circular bottom = $R^2$ (where $D$ = 0·52 − 2T) ($R$ = 1/2 × 0·514 = 0·257 m)
= 3·142 × 0·257² = 0·208 m²

Total area = 0·991 + 0·208 = 1·199 m²

Mass per unit area for brass of 3·15 mm thickness (see Table 4.4)
= 24·35 × 1·072 kg/m² = 26·1 kg/m²

*Mass of container* = 1·199 × 26·1 = 31·29 kg

Volume of Paraffin:
= $\pi R^2 H$ (where $R$ = 0·257 and depth $H$ = 0·46 m)
= 3·142 × 0·257² × 0·46 = 0·095 m³

*Mass of paraffin* = Volume × Density
= 0·095 × 800 kg/m² = 76 kg

(a) Total mass of container and contents = 31·29 + 76 = 107·29 kg
(b) Total weight = 107·29 × 9·81 N = 1052·5 N

**Fig. 4.31(a) Mass and weight — cylindrical vessel and contents**

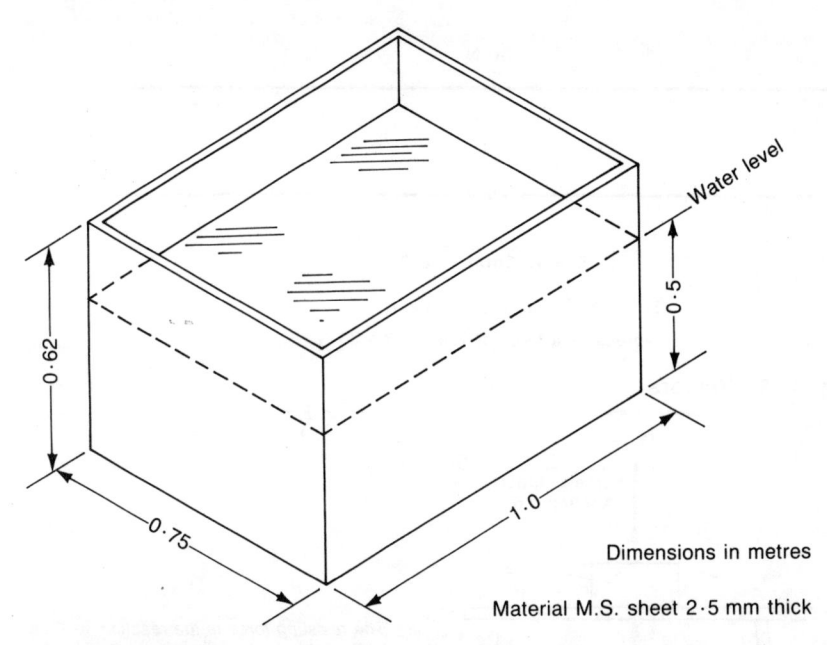

To find the mass and weight of the open top container shown together with water filled to a depth of 0·5 m.

Area of Sheet Metal:

Ends = 2[(0·62 − T) × (0·75 − 2T)] (where $T$ = 0·0025 m)
= 2 × 0·618 × 0·745 = 0·921 m²

Sides = 2(0·62 − T) × 1·0
= 2 × 0·618 × 1·0 = 1·236 m²

Bottom = (0·75 − 2T) × (1·0 − 2T)
= 0·745 × 0·995 = 0·741 m²

Total area = 0·921 + 1·236 + 0·741 = 2·898 m²

Mass per unit area for M.S. sheet of 2·5 mm thickness = 19·325 kg/m² (Table 4.4)

*Mass of container* = 2·898 × 19·325 = 56 kg

Volume of Water = Cross-section × Depth = 0·741 × 0·5  0·371 m³
*Mass of water* = Volume × Density = 0·371 × 1000 kg/m³ = 371 kg

(a) Total mass of container and contents = 56 + 371 = 427 kg
(b) Total weight = 427 × 9·81 N = 4189 N

**Fig. 4.31(b) Mass and weight — rectangular tank and contents**

applied. Three special cases for any two forces acting on the same body are graphically represented by vectors in Figure 4.34.

## 4.25 Applications — slings and cranes

Vector diagrams can often be used to determine the forces acting on a component without having to resort to the use of complicated mathematics. The example shown in Figure 4.35(a) gives the forces acting on a sling when a heavy load is being lifted in the workshop, whilst Figure 4.35(b) indicates the effect of increasing the sling angle.

Further examples of the use of vector diagrams are illustrated in Figure 4.36.

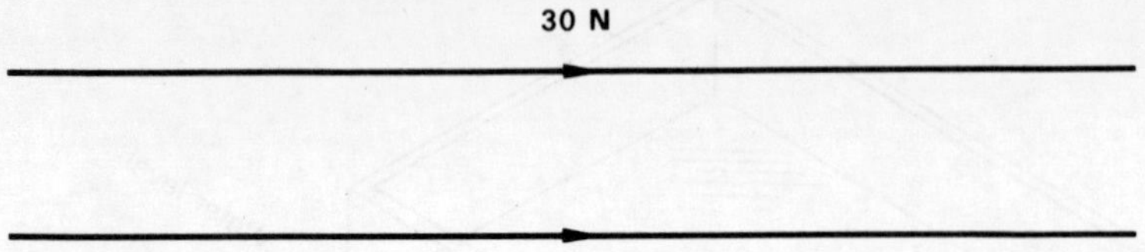

30 N

Scale: 1mm = 0·3 N

Since the line is 100mm long it represents a force of 100 x 0·3 = <u>30 N</u>

**Fig. 4.32 Vectors**

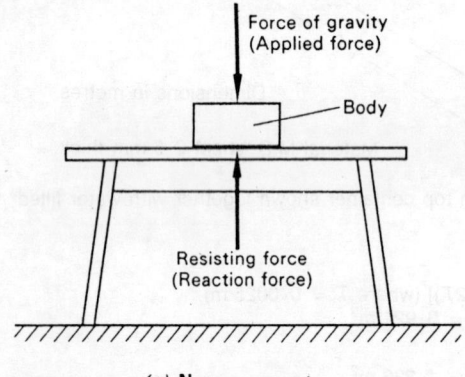

*(a) No movement*

The resisting force is the reaction of the table to the force of gravity acting on the body. Unless the table collapses the forces will balance and no movement will take place.

50 N → □ ← 50 N

*(b) No movement*

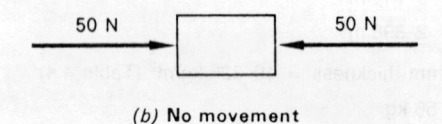

The applied forces are equal in size but acting in opposite directions. Therefore they cancel out and there is no movement.

75 N → □ ← 50 N

75 − 50 = 25 N → □

Direction of movement

*(c) Movement*

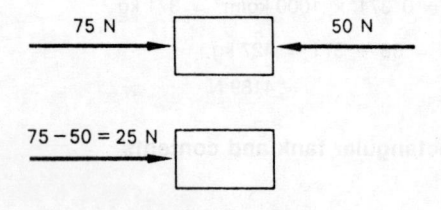

When the force system acting on a body is unbalanced, the system can be replaced by a single equivalent (resultant) force. Since it is not opposed, movement takes place in the direction of that force.

**Fig. 4.33 Force and movement**

Scale: 1 cm = 1 N

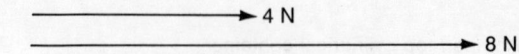

→ 4 N
→ 8 N

*(a) Force in the same direction along the same line of action*

*Note:* The vectors are drawn slightly apart and parallel for clarity.

Magnitude of Resultant Force = 4 + 8 = 12 N

4 N ← • → 8 N

*(b) Forces in opposite directions along the same line of action*

Magnitude of Resultant Force = 8 − 4 = 4 N

In this case the direction of the resultant force is from left to right.

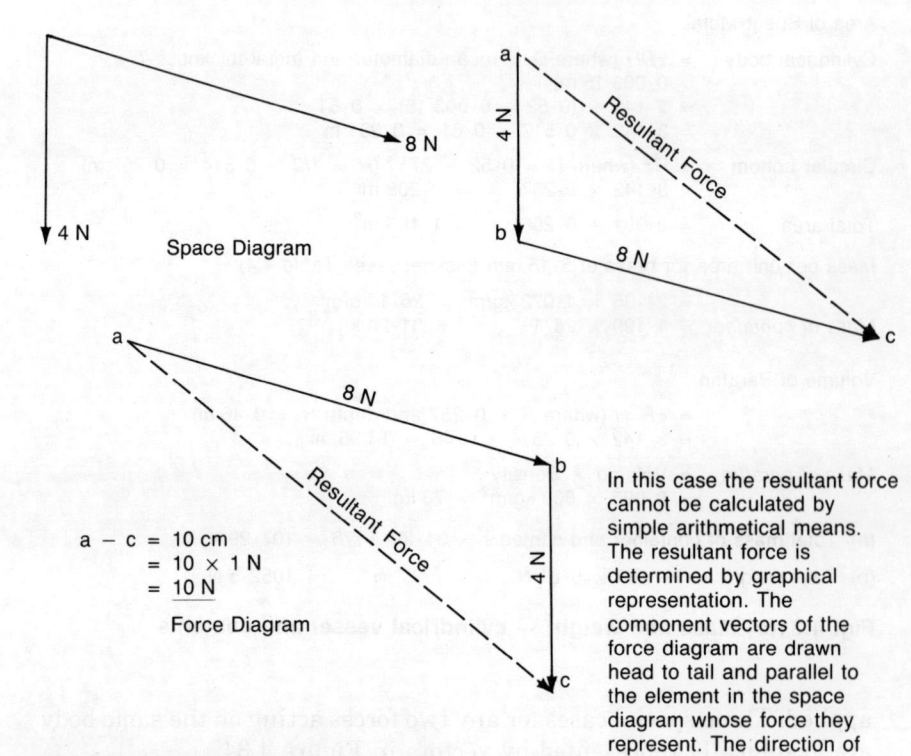

Space Diagram

4 N ↓    8 N

Force Diagram

a − c = 10 cm
     = 10 × 1 N
     = <u>10 N</u>

In this case the resultant force cannot be calculated by simple arithmetical means. The resultant force is determined by graphical representation. The component vectors of the force diagram are drawn head to tail and parallel to the element in the space diagram whose force they represent. The direction of the resultant vector is a to c, i.e. from start to finish of the component. The same result is obtained whichever force element is taken to start with. By measurement the resultant force a − c = 10 N.

*(c) Forces not having the same line of action*

**Fig. 4.34 Two forces acting on the same body**

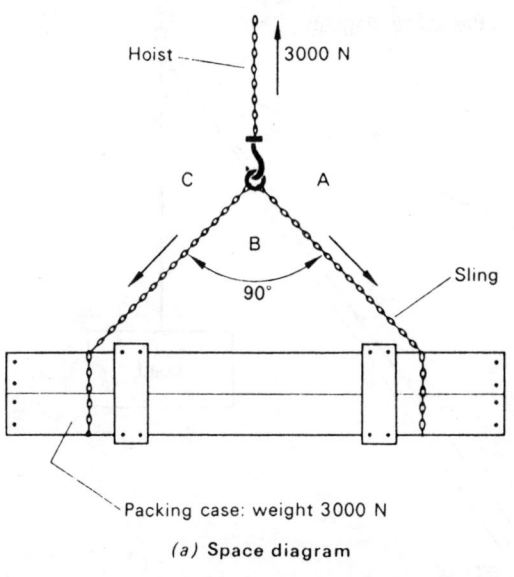

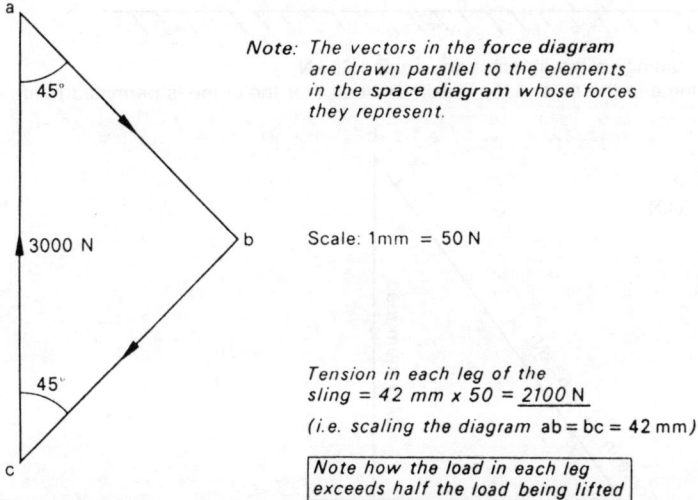

**Fig. 4.35(a)  Forces on a sling**

Hoist — 3000 N
Sling
B 90°
C    A
Packing case: weight 3000 N

*(a) Space diagram*

Note: The vectors in the *force diagram* are drawn parallel to the elements in the *space diagram* whose forces they represent.

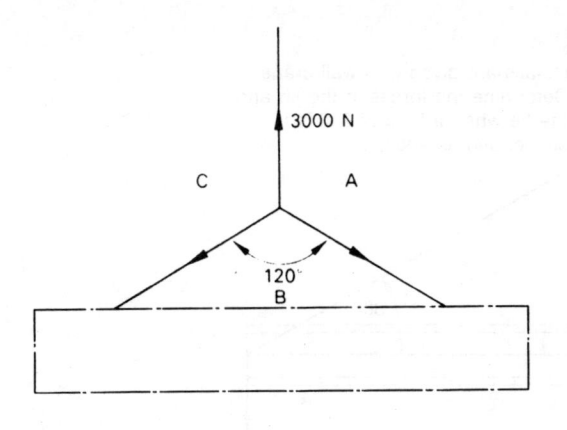

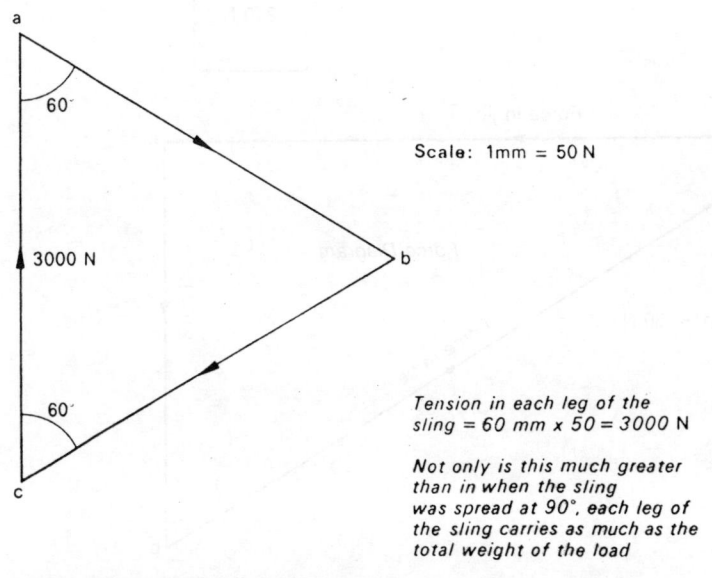

Scale: 1mm = 50 N

45°
3000 N
45°

Tension in each leg of the sling = 42 mm × 50 = **2100 N**
(i.e. scaling the diagram ab = bc = 42 mm)

Note how the load in each leg exceeds half the load being lifted

**Fig. 4.35(b)  Effect of sling spread**

3000 N
C    A
120°
B

60°
3000 N
60°

Scale: 1mm = 50 N

Tension in each leg of the sling = 60 mm × 50 = 3000 N

Not only is this much greater than in when the sling was spread at 90°, each leg of the sling carries as much as the total weight of the load

## 4.26 Moment of a force

A force will have a *turning effect* about any point not directly in its line of action. This can be demonstrated when opening or shutting a door. When a door is pushed, or pulled, at the handle side, the line of action does not pass through the hinges of the door. The hinge fastenings prevent the door moving bodily in the direction of the applied force and the door *turns* on its hinges. One only has to exert a relatively small effort to produce a turning effect when holding the door handle. However, as the distance of the applied force decreases toward the door hinges, the effort to move or turn the door greatly increases. On the other hand, if the edge of the door is pushed, the line of action passes through the hinge and no turning effect takes place. These forces are illustrated in Figure 4.37.

Details are given of a wall crane. Determine the forces in the jib and the tie when a load of 210 N is suspended as shown.

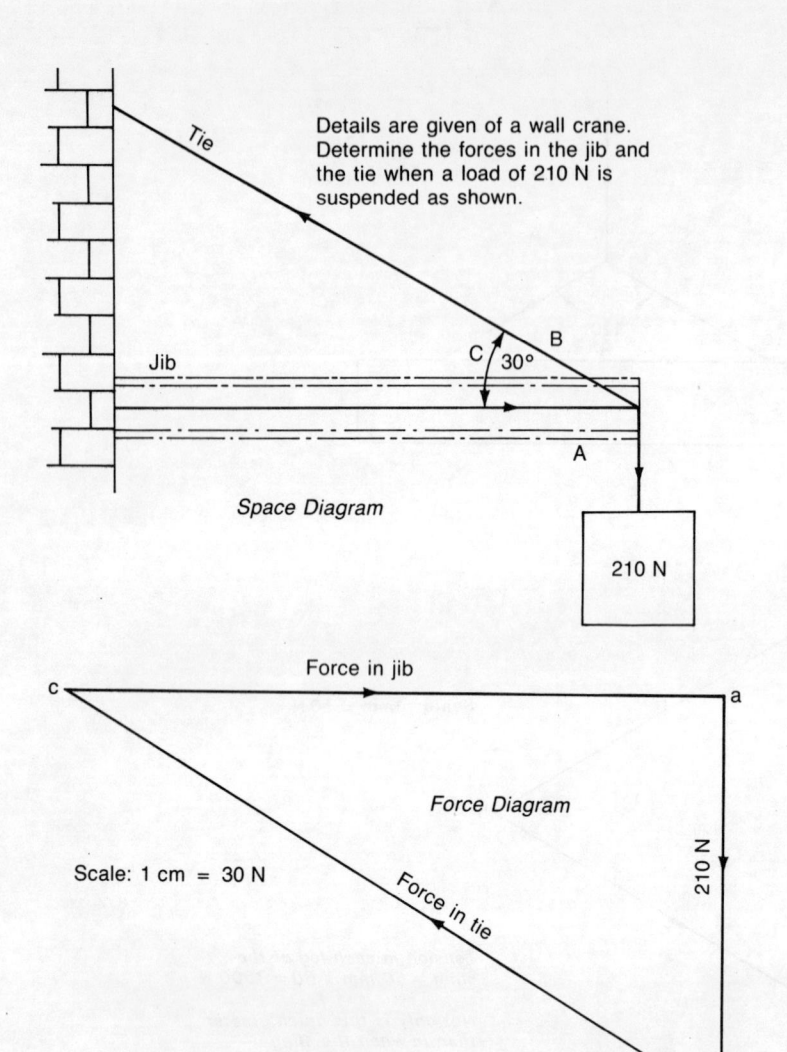

*Space Diagram*

*Force Diagram*

Scale: 1 cm = 30 N

The component vectors of the force diagram are drawn head to tail and parallel to the element in the space diagram whose magnitude of force they represent.

*By measurement:* Force in the jib (c − a) = 12·1 cm
= 12·1 × 30 = 363 N (Tension)

Force in tie (b − c) = 14 cm
= 14 × 30 = 420 N (Compression)

*Note:* In each case the force in the jib and the tie are both greater than the force applied by the load. This illustrates the necessity for a Maximum Safe Working Load.

**Fig. 4.36(a) Vector diagrams — wall crane**

Details of a jib crane are shown in the space diagram.

*Space Diagram*

Scale: 1 cm = 2 m

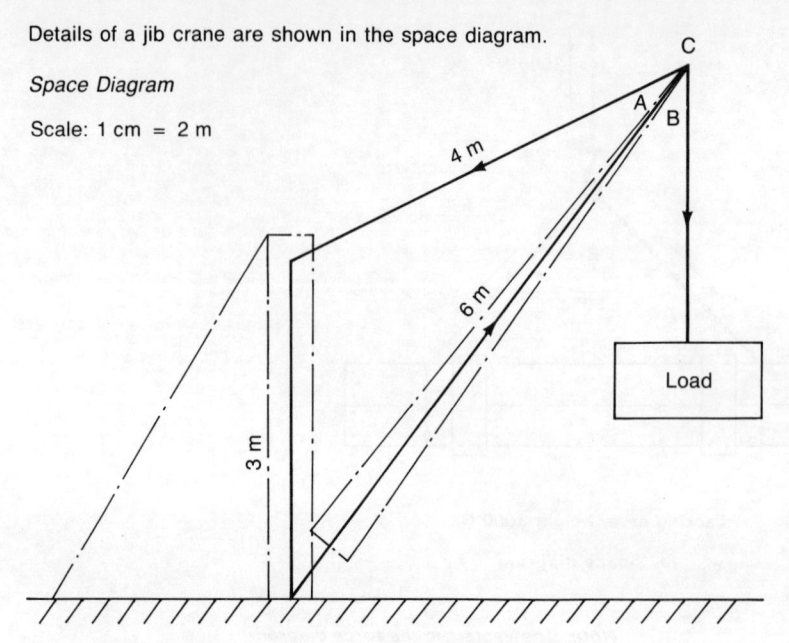

For safety, the loading on the jib must not exceed 65 kN.
Determine the force in the tie and the maximum load that the crane is permitted to lift.

*Force Diagram*

Scale: 1 cm = 5 kN

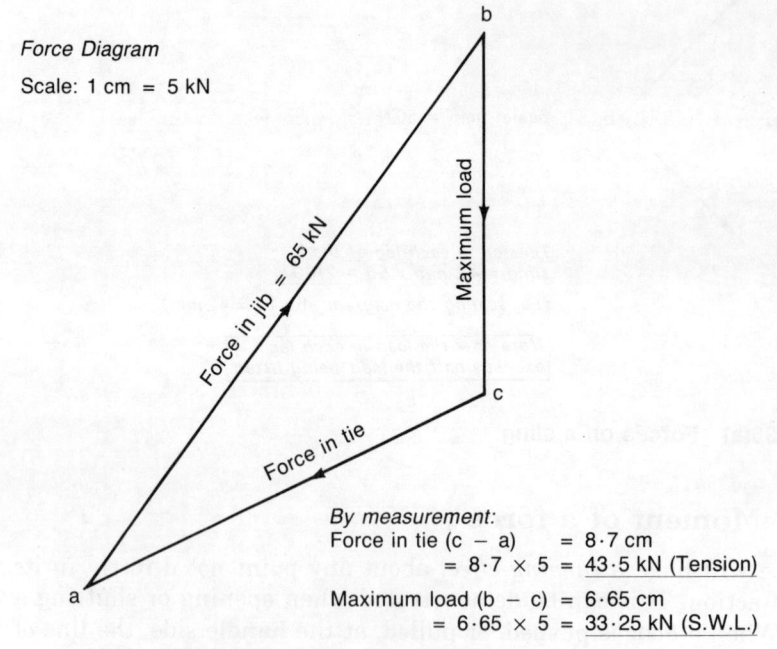

*By measurement:*
Force in tie (c − a) = 8·7 cm
= 8·7 × 5 = 43·5 kN (Tension)

Maximum load (b − c) = 6·65 cm
= 6·65 × 5 = 33·25 kN (S.W.L.)

**Fig. 4.36(b) Vector diagrams — jib crane**

*(a)* Turning effect of a force at A is greater than at B

*(b)* No turning effect when the line of action of a force passes through the hinge of the door

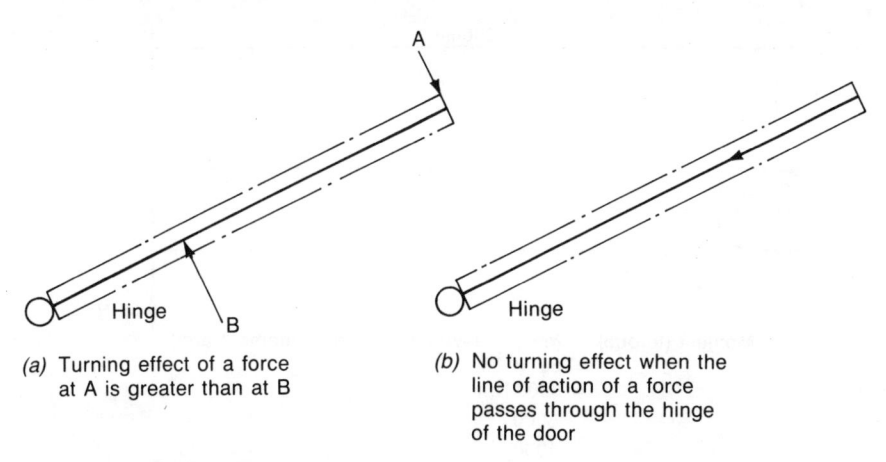

**Fig. 4.37 Turning effect of a force**

This turning effect is called the MOMENT of the force about a point, and may be measured as *the product of the force and the perpendicular distance between its line of action and the point considered*. The hinges of the door, or point about which rotation may take place is called the FULCRUM.

The units in which the moment of a force is measured are:

(Force units) × (Length units), expressed in NEWTON METRES (Nm)

Figure 4.38 shows the turning moment or TORQUE of a force. Some examples of the many applications of the turning effect of a force are shown in Figure 4.39.

The moment of a force can be easily calculated as follows:

MOMENT (TORQUE) = MAGNITUDE OF FORCE × LEVERAGE DISTANCE

An example is shown in Figure 4.40.

## 4.27 Principle of moments

A lever is a device by which an effort can overcome a load. Whether the effort is greater or smaller than the load will depend upon the position of the pivot (fulcrum). Some examples of levers are shown in Figure 4.41.

It can be seen from Figure 4.41 that for a lever to remain stationary (in a state of balance or EQUILIBRIUM), the load and effort must be acting upon it in opposite directions. In order to establish a state of equilibrium:

THE TOTAL CLOCKWISE MOMENTS MUST EQUAL
THE TOTAL ANTI-CLOCKWISE MOMENTS

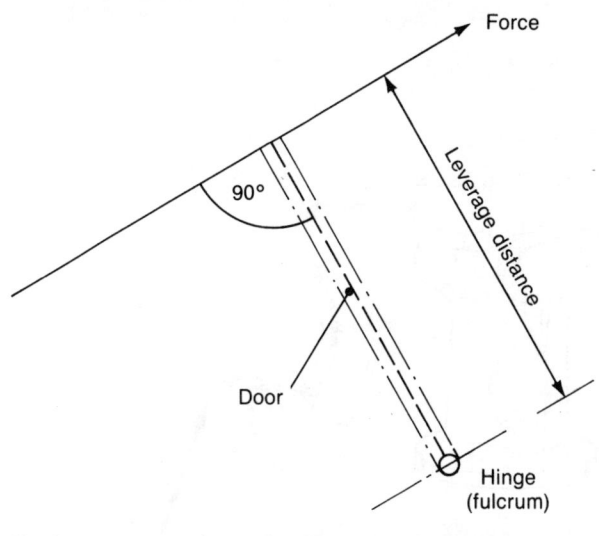

Turning moment = Leverage distance × Applied force

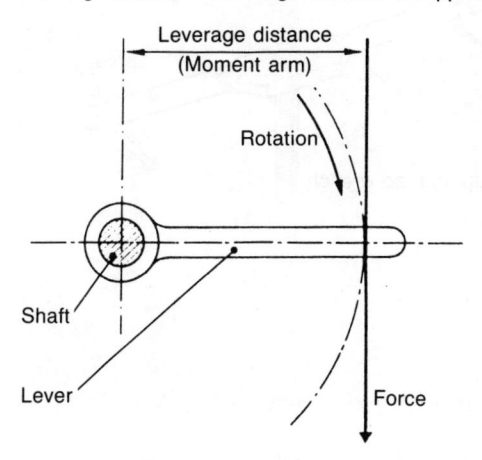

**Fig. 4.38 Moment (turning effect) of a force**

This is called the '*Principle of Moments*', and is used when calculating the forces acting on a lever. An example is shown in Figure 4.42.

Figure 4.43 shows how the principle of moments is applied to a clamp. (Further examples are given in Figures 7.34 and 7.38.)

## 4.28 Stress and strain

Figure 4.44 shows a block of metal being supported on a block of rubber. The force of gravity acts on the mass of the metal block causing it to

127

**Swaging machine**

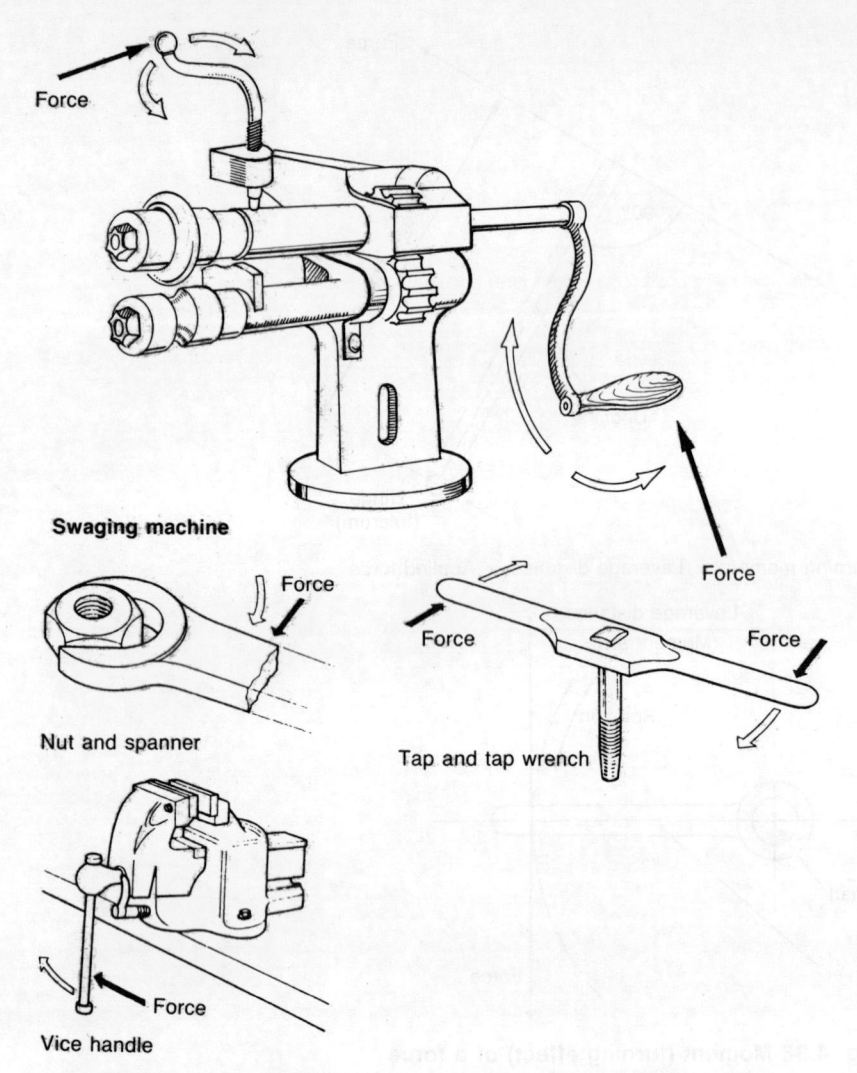

**Nut and spanner**

**Tap and tap wrench**

**Vice handle**

**Fig. 4.39 Some workshop examples of the turning effect of a force**

Moment (torque) = force × leverage distance (moment arm)
= 30 × 200
= 6000 Nmm
= $\frac{6000}{1000}$ Nm
= 6 Nm (Newton metres — the Unit of torque)

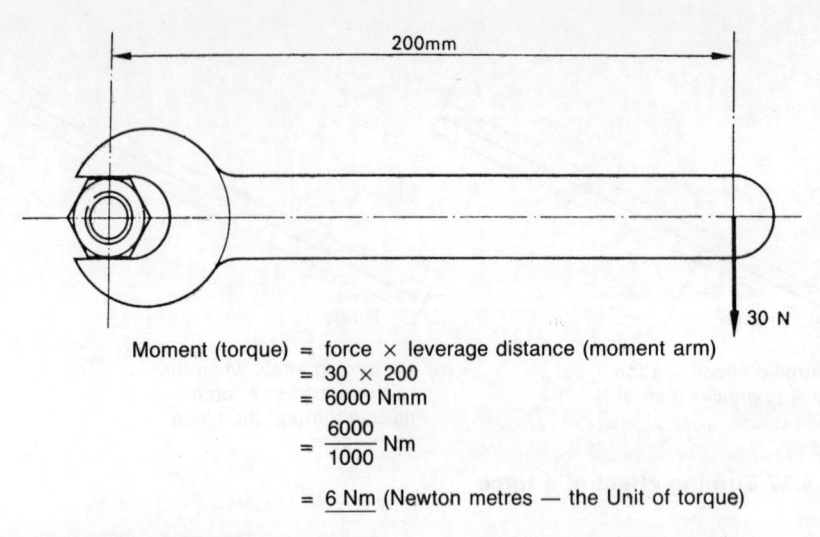

**Fig. 4.40 Calculation of torque**

apply a downward force to the rubber, which in turn, is supported by something else. As a result, the downward force of the metal distorts the rubber as shown. This distortion is termed STRAIN. The downward force of the metal is resisted by an *internal reaction* force, distributed over the cross-sectional area of the rubber block. This internal reaction force is called STRESS. As long as the metal block rests on the rubber block, the rubber is said to be in a *state of stress*. *Every external force acting upon a body has an opposite and equal reaction force.*

The more the rubber is compressed and distorted by the load, the greater will be the stress. Eventually the stress will balance the load and no further distortion (increase in strain) will occur.

### 4.29 Types of stress and strain

When a material is subjected to forces that tend to cause distortion (STRAIN), the material is said to be in a state of STRESS. The *intensity* of this stress depends upon the *magnitude* and *direction* of the applied force, and the *cross-sectional area* of the material withstanding the stress.

For direct TENSILE and COMPRESSIVE stresses, this cross-sectional area is measured at right angles to the direction of the force. For SHEAR stress, the area will be that of the adjacent surfaces which the stress is tending to slide over one another.

The S.I. unit of intensity of stress is the same as for PRESSURE. *The term pressure means the same as the term intensity of compressive stress at the surface of the material under pressure* (see Section 4.37).

The intensity of stress is measured in $N/m^2$ or $N/mm^2$.

Note: the units of length used in calculating strain must be the same for the original length and for the change in length. Therefore strain is expressed at a RATIO, that is it has no units — only a numerical value. For a material under a constant stress, a strain of 0·0003 would mean that a length of 1 mm would change by 0·0003 mm and, likewise, a length of 1 metre would change by 0·0003 metres.

**Fig. 4.42 Principle of moments**

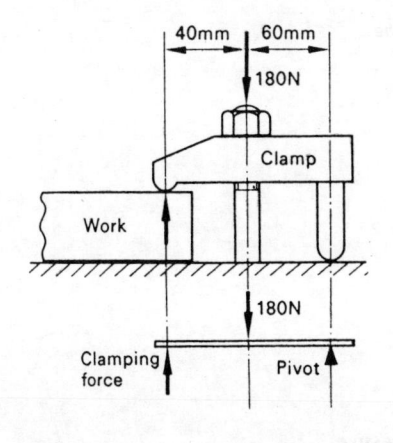

**Example**
To calculate the required **effort** to balance a load of 210N

Clockwise moments = Anti-clockwise moments
Effort × Distance = Load × Distance
Effort × 450 = 210 × 150
Effort = $\frac{210 \times 150}{450}$
Effort = 70N

*Note:* If the effort is less than 70 N the lever will rotate in an *anti-clockwise* direction.
If the effort is greater than 70 N the lever will rotate in a *clockwise* direction.

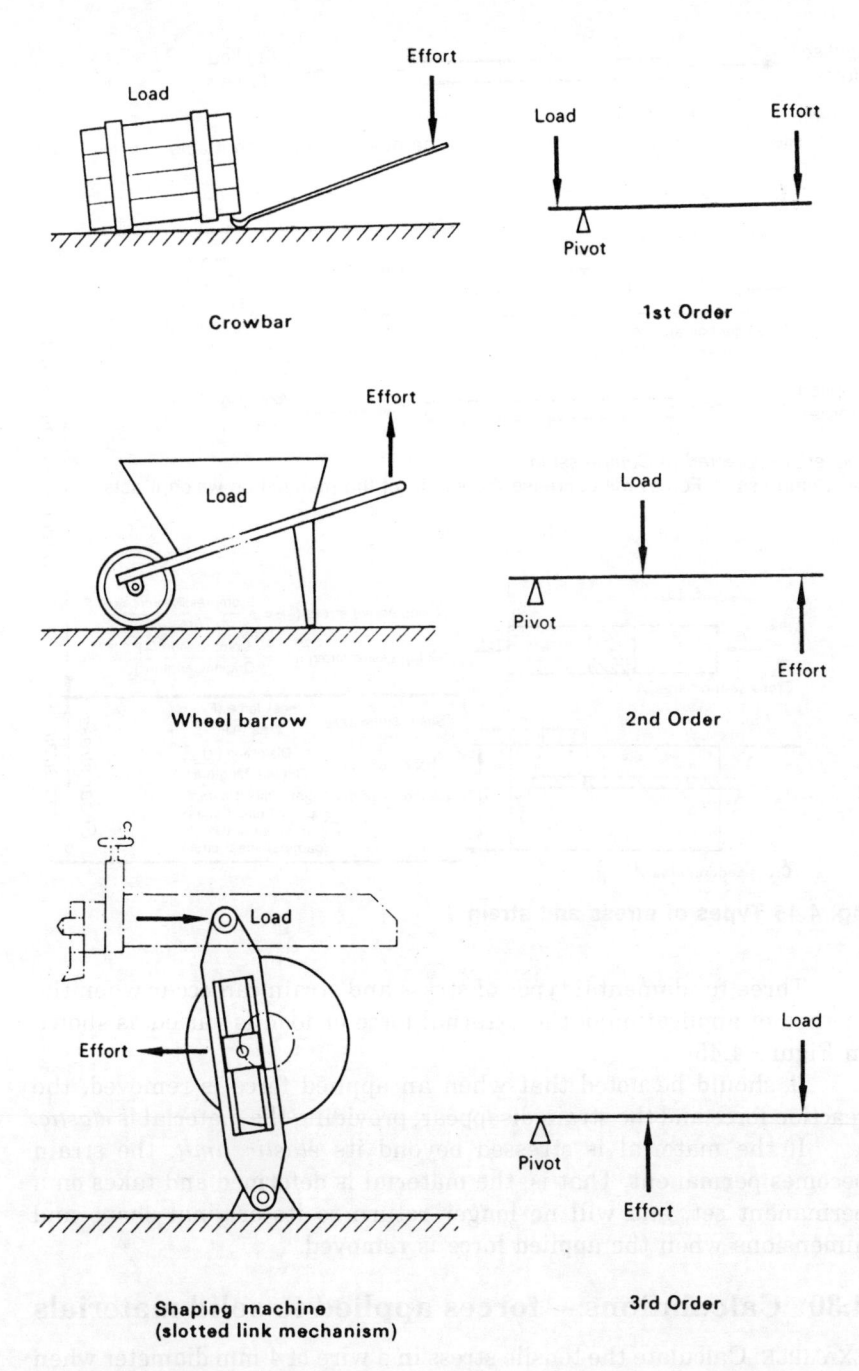

**Fig. 4.41 Levers**

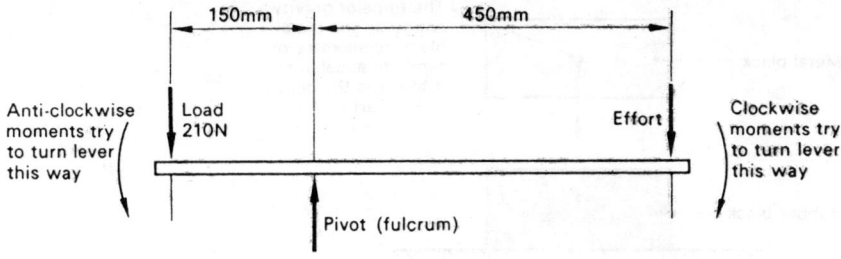

**Fig. 4.43 Clamping force**

**Example**
To calculate the clamping force on the work when the nut is tightened until it exerts a force of 180N on the clamp

Clockwise moments = Anti-clockwise moments
Force × distance = Force × distance
Clamp force × (40 + 60) = 180 × 60
Clamp force × 100 = 180 × 60
Clamp force = $\frac{180 \times 60}{100}$
= **108N**

*Note:*
1. The balance of the force exerted by the nut (180 − 108 = 72 N) is exerted on the packing and is wasted
2. By working this example again for different positions of the nut and bolt, it will be found that the nearer the bolt is to the work; the greater will be the proportion of the force exerted on the work.

129

## Fig. 4.44 Stress and strain

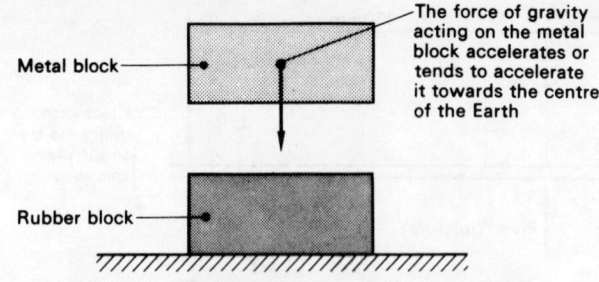

**(a)** Since the metal block is not yet in contact with the rubber block, it is not yet applying an external force to the rubber. Therefore the rubber is not yet providing an internal reaction force in opposition to the applied force.

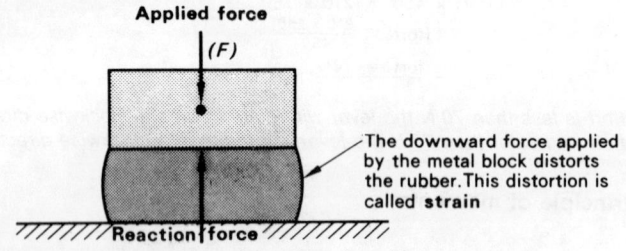

**(b)** As soon as the metal block touches the rubber, the rubber provides an internal reaction force in opposition to the applied force. At first this reaction force will be insufficient to balance the applied force and the metal block continues to move downwards.

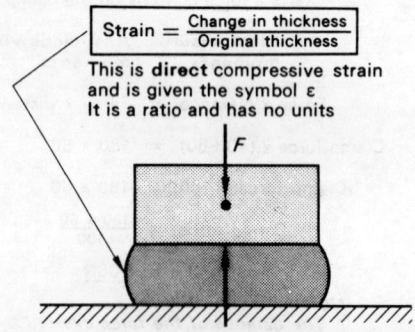

$$\text{Strain} = \frac{\text{Change in thickness}}{\text{Original thickness}}$$

This is **direct** compressive strain and is given the symbol $\varepsilon$. It is a ratio and has no units.

**(c)** Providing the downward force (F) is not excessive the rubber will continue to distort until the **internal reaction force balances the applied force (F)**. At this moment the downward movement of the metal block will cease and there will be no further increase in distortion (strain) of the rubber block. The rubber block is in a **state of stress**.

$$\text{Stress } (\sigma) = \frac{\text{Force } (F)}{\text{Area of rubber block}}$$

---

### Material subjected to Tension
A 'Tensile Force' will increase the length of the material on which it acts.

$$\text{Tensile stress } (\sigma) = \frac{\text{Tensile force } (F)}{\text{Area } (A)}$$

$$\text{Tensile strain } (\varepsilon) = \frac{\text{Extension } (x)}{\text{Original length } (l)}$$

### Material subjected to Compression
A 'Compressive Force' will decrease the length of the material on which it acts.

$$\text{Compressive stress } (\sigma) = \frac{\text{Compressive force } (F)}{\text{Area } (A)}$$

$$\text{Compressive strain } (\varepsilon) = \frac{\text{Compression } (x)}{\text{Original length } (l)}$$

$$\text{Shear stress } (\tau) = \frac{\text{Shear force } (F)}{\text{Area } (A)}$$

$$\text{Shear strain } (\gamma) = \frac{\text{Distortion } (x)}{\text{Original length } (l)}$$

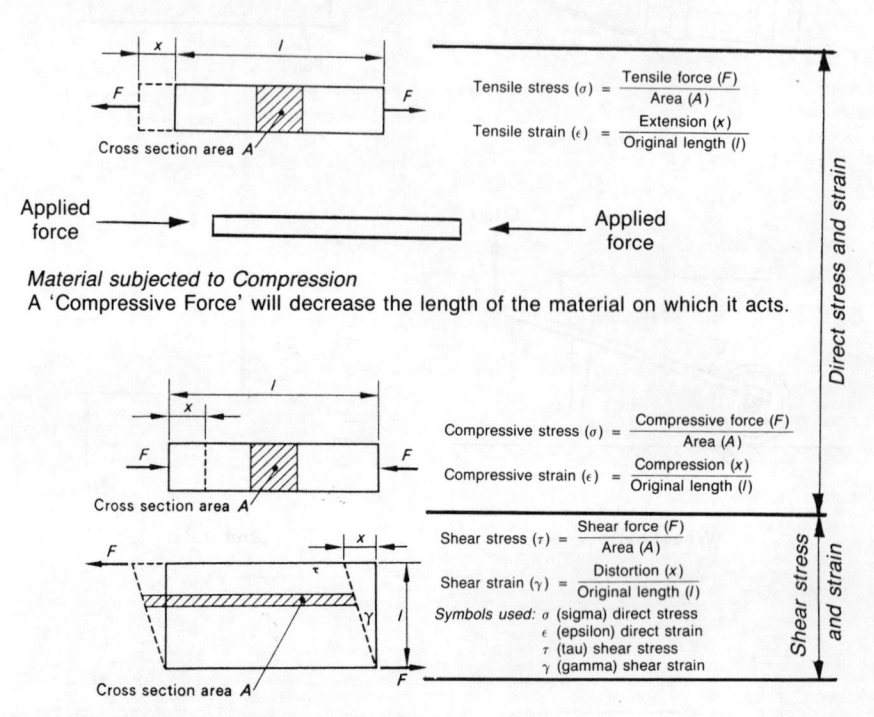

Symbols used: $\sigma$ (sigma) direct stress
$\varepsilon$ (epsilon) direct strain
$\tau$ (tau) shear stress
$\gamma$ (gamma) shear strain

**Fig. 4.45 Types of stress and strain**

Three fundamental types of stress and strain can occur when the method of application of the external force or load is varied as shown in Figure 4.45.

It should be noted that when an applied force is removed, the reaction force and the strain disappear, providing the material is *elastic*.

If the material is stressed beyond its *elastic limit*, the strain becomes permanent. That is, the material is deformed and takes on a permanent set, and will no longer return to its original shape and dimensions when the applied force is removed.

## 4.30 Calculations — forces applied to solid materials

EXAMPLE: Calculate the tensile stress in a wire of 4 mm diameter when a direct tensile force of 200 N is applied.

*Solution*:

$$\text{TENSILE STRESS} = \frac{\text{LOAD}}{\text{CROSS-SECTIONAL AREA}}$$

$$\text{Cross-sectional area of wire} = \frac{\pi D^2}{4} = \frac{3 \cdot 142 \times 4 \times 4}{4} \text{ mm}^2$$
$$= 12 \cdot 568 \text{ mm}^2$$

$$\text{Tensile stress} = \frac{200 \text{ N}}{12 \cdot 568} \text{ mm}^2 = 15 \cdot 91 \text{ N/mm}^2$$

EXAMPLE: A solid steel anvil has a mass of 90 kg. It is supported on a four-legged stand such that each leg provides an equal reactive force. Calculate the compressive stress in each leg whose cross-sectional area is 600 mm².

*Solution*: A mass of 90 kg represents a load of 90 kg × 9·81 N/kg:

$$\text{LOAD} = 882 \cdot 9 \text{ N}$$

$$\text{COMPRESSIVE STRESS} = \frac{\text{LOAD}}{\text{CROSS-SECTIONAL AREA}}$$

Since each leg produces the same reaction force, the compressive force (load) acting on each leg:

$$\text{LOAD ACTING ON EACH LEG} = \frac{882 \cdot 9}{4} \text{ N} = 220 \cdot 73 \text{ N}$$

$$\text{Compressive stress (each leg)} = \frac{220 \cdot 73 \text{ N}}{600 \text{ mm}^2} = 0 \cdot 368 \text{ N/mm}^2$$

Further numerical examples are given in Table 4.7.

## 4.31 Practical effects of stressing material

Engineering materials are constantly subject to the types of force illustrated in Figure 4.46 either singly or in combination. In service these materials are stressed so that they behave in an *elastic manner*. Figure 4.45 shows the effects of various types of load applied under practical working conditions.

Sometimes engineering materials are deliberately over-stressed so that they behave in a *plastic manner* and take a permanent set. This enables the materials to be formed and cut as shown in Figure 4.47.

## 4.32 The deflection of beams

Figure 4.48 shows a metal bar supported as a single beam. It also shows what happens when the beam is bent under a load.

**Table 4.7  Calculations, forces acting on solid materials**

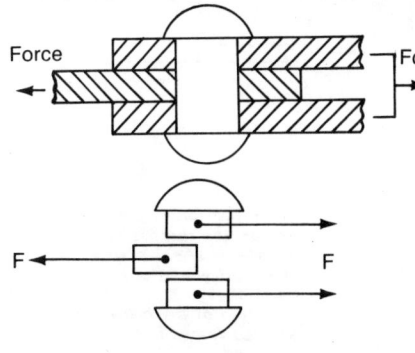

Three steel plates are joined together with two rivets of 6·3 mm diameter. If the shear stress must not exceed 45 N/mm², calculate the applied tensile force required to shear the rivets.

As seen in the diagram both rivets would have to shear twice (across the two faces of each).

*Solution*: Total area resisting shear (2 rivets in double shear):

$$A = 2 \times 2 \times \frac{\pi D^2}{4} \text{ mm}^2 = \pi D^2 \text{ mm}^2$$
$$= 3 \cdot 142 \times 6 \cdot 3 \times 6 \cdot 3 \text{ mm}^2 = 124 \cdot 706 \text{ mm}^2$$

$$\text{Shear Stress} = \frac{\text{Shear Force } (F)}{\text{Area resisting Shear } (A)}$$

Shear Force $(F)$ = Shear Stress × Area resisting shear $(A)$
$$F = 45 \text{ /Nmm}^2 \times 124 \cdot 706 \text{ mm}^2$$
$$= 5611 \cdot 77 \text{ N} = 5 \cdot 6 \text{ kN}$$

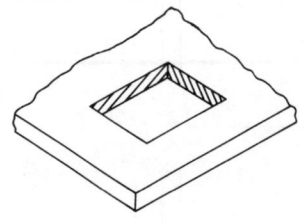

A rectangular hole is to be punched out of sheet metal of 1·6 mm thickness. Given that the dimensions for the hole are 25 mm by 32 mm and the shear stress required to cause fracture is 375 N/mm², calculate the compressive force applied to, and compressive stress in the punch.

*Solution*:
Area of metal to be sheared = Perimeter of hole × metal thickness
$A = 2(32 \text{ mm} + 25 \text{ mm}) \times 1 \cdot 6 \text{ mm} = 114 \text{ mm} \times 1 \cdot 6 \text{ mm} = 182 \cdot 4 \text{ mm}^2$

$$\text{Shear Stress} = \frac{\text{Force on Punch}}{\text{Area resisting Shear } (A)} \text{ N/mm}^2$$

Force on Punch = Shear Stress × Area resisting Shear
$$= 357 \text{ N/mm}^2 \times 182 \cdot 4 \text{ mm}^2 = 68\,400 \text{ N} = 68 \cdot 4 \text{ kN}.$$

Cross-sectional area of punch = $25 \times 32 \text{ mm}^2 = 800 \text{ mm}^2$

$$\text{Compressive Stress in Punch} = \frac{\text{Load}}{\text{Cross-sectional Area}}$$
$$= \frac{68 \cdot 4 \text{ kN}}{800 \text{ mm}^2} = 0 \cdot 0855 \text{ kN/mm}^2 = 85 \cdot 5 \text{ N/mm}^2$$

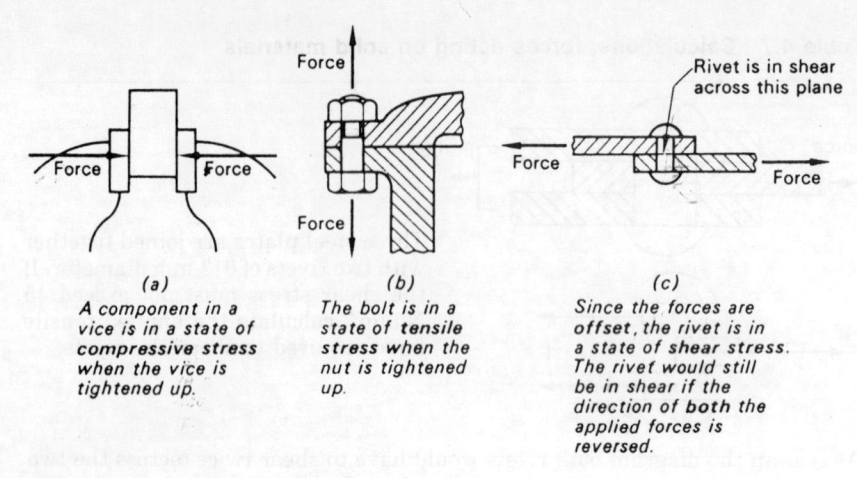

**Fig. 4.46 Practical examples of stress**

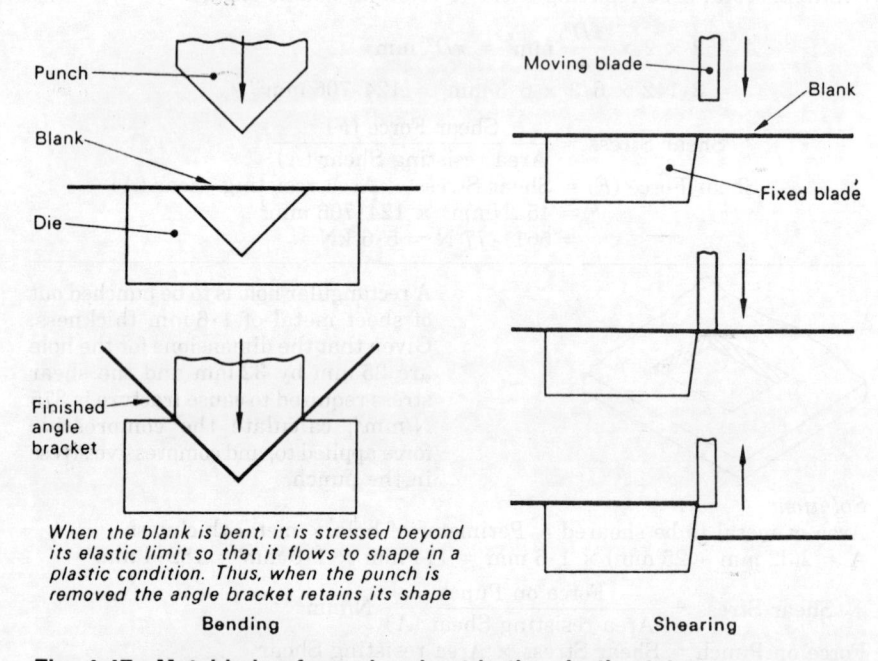

(a) A component in a vice is in a state of **compressive** stress when the vice is tightened up.

(b) The bolt is in a state of **tensile** stress when the nut is tightened up.

(c) Since the forces are offset, the rivet is in a state of **shear** stress. The rivet would still be in shear if the direction of **both** the applied forces is reversed.

**Fig. 4.47 Metal being formed and cut in the plastic state**

Bending: When the blank is bent, it is stressed beyond its elastic limit so that it flows to shape in a plastic condition. Thus, when the punch is removed the angle bracket retains its shape.

Shearing.

**Fig. 4.48 A simple beam**

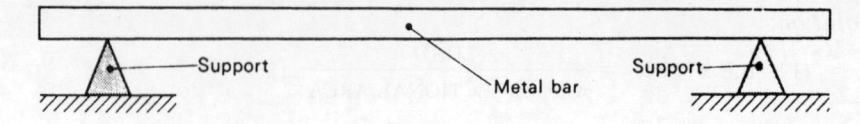

(a) As shown, the metal bar is supported as a simple beam.

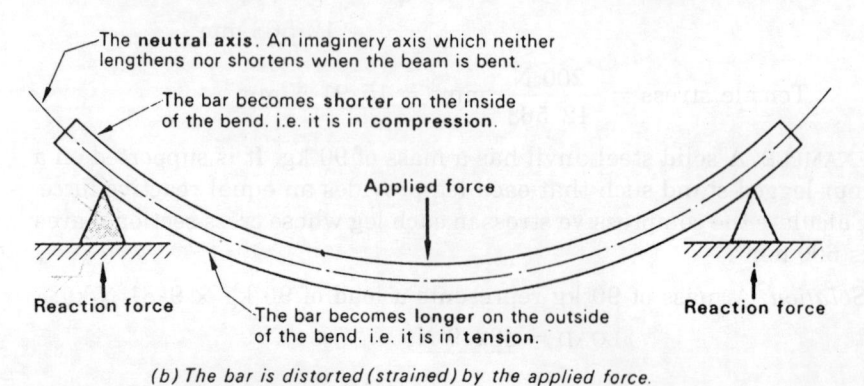

The **neutral axis**. An imaginery axis which neither lengthens nor shortens when the beam is bent.

The bar becomes **shorter** on the inside of the bend. i.e. it is in **compression**.

The bar becomes **longer** on the outside of the bend. i.e. it is in **tension**.

(b) The bar is distorted (strained) by the applied force. Under this condition it is stressed as shown above.

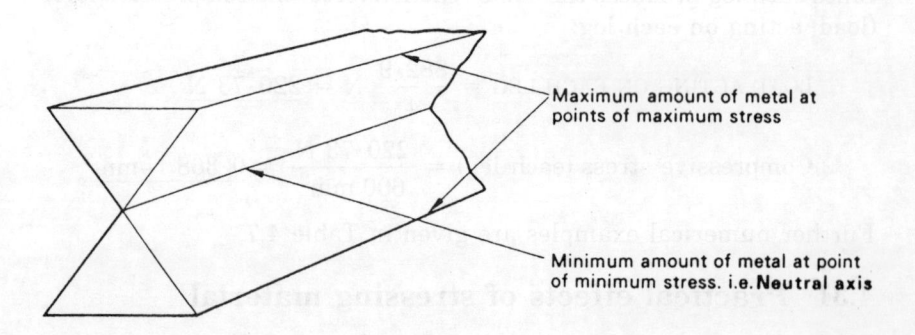

**Fig. 4.49 A theoretical beam section**

Maximum amount of metal at points of maximum stress

Minimum amount of metal at point of minimum stress. i.e. **Neutral axis**

It will be seen that the length of the *neutral axis* remains unchanged when the beam is bent, which means no distortion has taken place; therefore *no strain* and *no stress*. It also shown that the greatest distortion and, therefore, the greatest strain and stress occurs in those parts of the beam furthest from the neutral axis. To be most efficient, ideally, the material from which the beam is made should be concentrated at the points of greatest stress as shown in Figure 4.49.

The beam illustrated in Figure 4.49 is purely a theoretical one; however, some very practical sections are shown in Figure 4.50.

Most of the load is carried by the *flanges*, the purpose of the *web* being to keep the flanges in place, and to carry any shear loading to which the beam is subjected. In fact if weight is important, as it is in long unsupported spans, the web is pierced to make it lighter. Fabricated lattice beams are also used. Examples are shown in Figure 4.51.

**Fig. 4.50 Practical beams**

*British standard beam* / *British standard channel* — Hot rolled from the solid

*Fabricated beam using plate and standard angle section riveted together*

**Fig. 4.51 Reducing the weight of beams**

When weight is important the mass, and therefore the weight of the beam can be reduced by piercing 'lightening' holes in the web. This does not appreciably weaken the beam

A lightweight beam fabricated from mild steel flats for the flanges and heavy gauge tube for the web. The flanges and web are welded together

## 4.33 Holes and notches in beams

When a framed building or structure is designed, the structural engineer estimates the applied loads and selects beams, stanchions, stringers and other members of a size capable of carrying the loads. He allows a considerable factor of safety in case of unexpected gale-force winds or of excessive loads being placed on upper floors. Often this factor of safety is abused when holes and notches are cut in the structure so that additional pipe runs or pieces of equipment can be attached to it. For safety, a qualified engineer should be consulted before any holes or

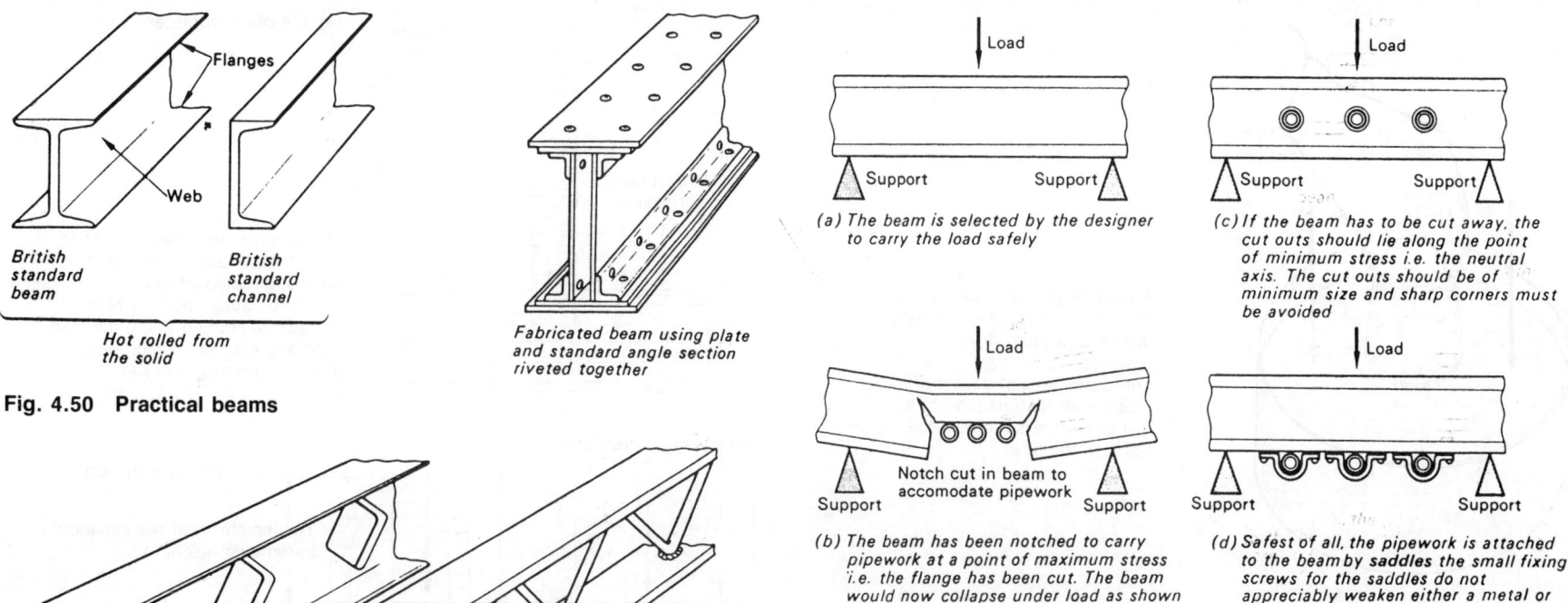

*(a) The beam is selected by the designer to carry the load safely*

*(b) The beam has been notched to carry pipework at a point of maximum stress i.e. the flange has been cut. The beam would now collapse under load as shown*

*(c) If the beam has to be cut away, the cut outs should lie along the point of minimum stress i.e. the neutral axis. The cut outs should be of minimum size and sharp corners must be avoided*

*(d) Safest of all, the pipework is attached to the beam by **saddles** the small fixing screws for the saddles do not appreciably weaken either a metal or concrete beam*

**Fig. 4.52 Notches in beams**

notches are cut. On no account should holes or notches be cut in concrete beams, but holes are permissible in metal beams providing they are small and placed at a point of low stress. Some examples are given in Figure 4.52.

## 4.34 Pressure of the atmosphere

Details of two very simple, but interesting experiments which show that air has weight are illustrated in Figure 4.53.

At sea level the density of air is $1 \cdot 225$ kg/m$^3$. However, since there is an effective height of several kilometres of air above ground level, the resultant pressure is measured in kN/m$^2$.

Due to climatic conditions ATMOSPHERIC PRESSURE, caused by the great mass of air above the ground, usually varies to within 10% of $101 \cdot 3$ kN/m$^2$ ($1 \cdot 103$ bar). Figure 4.54(a) shows a simple barometer consisting of a glass tube closed at one end, filled with a liquid, and then inverted in a trough of the liquid. *If WATER were the liquid used, normal atmospheric pressure would support a column $10 \cdot 4$ m high.* MERCURY is the liquid used for barometers, since it has a RELATIVE DENSITY of

133

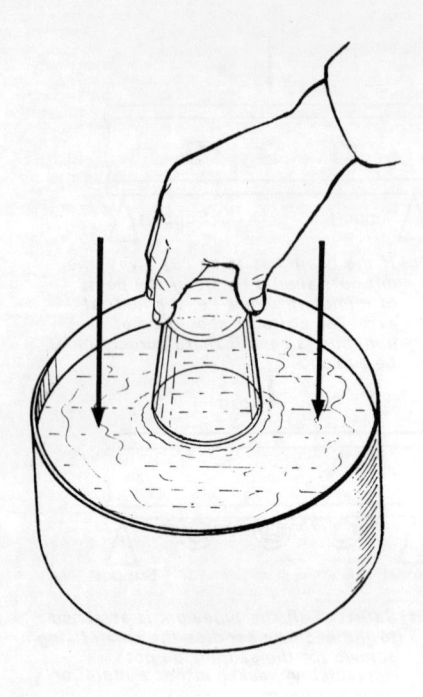

A tumbler placed under water in a washing-up bowl is turned upside down and drawn upwards.
The water will not descend inside it so long as the rim is just below the surface of the water in the bowl

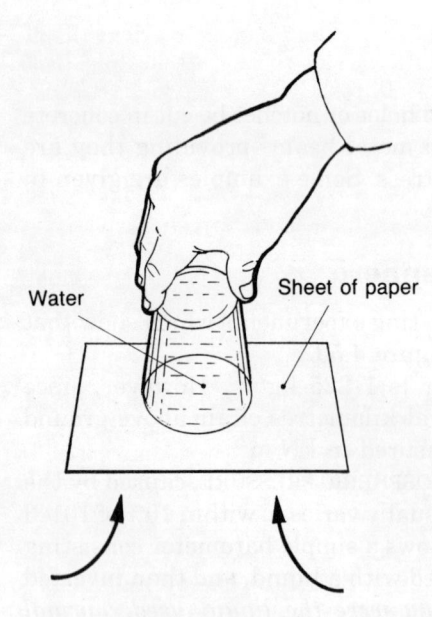

When a piece of paper is placed on a tumbler brimming over with water, it can be turned upside down in air without emptying it.
The paper remains as if it were glued to the rim of the tumbler.

**Fig. 4.53 Simple experiments to show that air has weight**

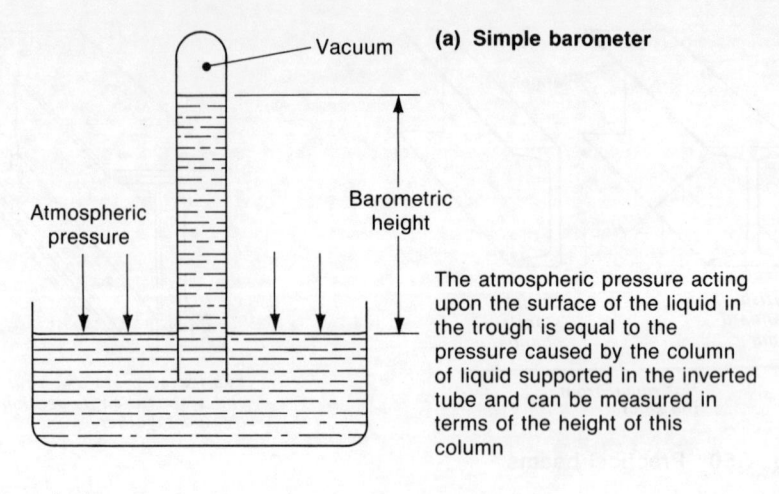

**(a) Simple barometer**

The atmospheric pressure acting upon the surface of the liquid in the trough is equal to the pressure caused by the column of liquid supported in the inverted tube and can be measured in terms of the height of this column

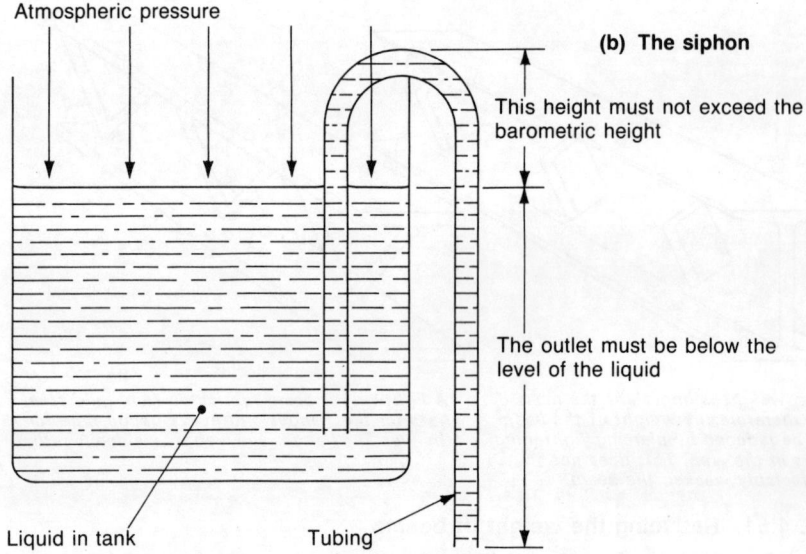

**(b) The siphon**

This height must not exceed the barometric height

The outlet must be below the level of the liquid

Atmospheric pressure may be used to cause a liquid to flow up a tube and out of a container into which the tube is placed. This action is called siphoning

**Fig. 4.54 Applications of atmospheric pressure**

13·6. Thus the height of the mercury is only 1/13·6 times the height of a corresponding column of water.

Atmospheric pressure is often quoted as a *'head of mercury'*, a normal value being 760 mm. *Accurate mercury barometers enable barometric heights to be measured to within a tenth of a millimetre.*

Atmospheric pressure can be used to practical advantage when syphoning liquid from a tank or container, as shown in Figure 4.54(b).

## 4.35 Use of compressed air — pneumatics

All the wide variety of operations that can be effected using electrical tools may also be performed with the use of PNEUMATIC TOOLS.

TYPES OF PNEUMATIC TOOLS

1. *Rotary Type*: Compressed air produces a rotary motion via a suitably geared turbine rotor or a power vane. This arrangement affords perfect balance and complete absence of vibration. A governing device prevents 'racing' and waste of air when 'running light'.
2. *Hammer Type*: There are diverse arrangements of valves employed in the design and construction of hammer type pneumatic tools. However, in general, the principle of operation is a sliding valve driven to and fro by the action of the piston itself. As the piston travels forward it uncovers a port which allows air, at working pressure, to travel to the opposite side of the valve, and hence to the front of the piston, causing it to return to its original position. A taper or 'ball-seat' valve provides very sensitive control, allowing a gradual starting up and enabling the force of the blows to be varied from light taps to full power.
3. *Compression Type*: The operation of this type of pneumatic tool is effected by a single stroke through the medium of a piston actuating a pivoted beam which forces down a ram. The 'squeeze pressure' can be adjusted by means of a regulator, and the movement or 'thrust' of the piston controlled by pedal, leaving the operator's hands completely free. Squeeze riveters and resistance welding machines are operated by this pneumatic system.

Several portable pneumatic tools are listed in Table 4.8.

## 4.36 Fluid pressure

The pressure of the fingers on a punctured rubber ball filled with water will force water out of the pores at all angles as shown in Figure 4.55(a). *When a fluid completely fills a vessel, and pressure is applied to it at any part of the surface, that pressure is transmitted throughout the whole of the enclosed fluid.*

This very important principle, known a 'Pascal's Principle', may be demonstrated by the simple apparatus shown in Figure 4.55(b). It

**Table 4.8 Pneumatically operated tools/machines**

*ROTARY ACTION*:

PORTABLE DRILLS: The cases or bodies of small rotary drills are either a small cylindrical hand-grip type or a pistol-grip type. Heavy service models are available with either a spade handle, straight side handle, a breast plate or with a screwfeed.

TAPPERS: Engagement and reverse on a 'push and pull' principle is employed in the hand-grip and pistol-grip tappers, the reverse speed being double that for tapping.

GRINDERS: Light grinders driven by a turbine rotor attain very high speeds and have a long fluted stem to hold in the hand. Heavier machines are available, having either pistol-grips, extended shanks or straight handles. Rotary wire brushes are substituted for grinding wheels when cleaning or removing paint and scale.

SCREWDRIVERS or NUT RUNNERS: These have an adjustable clutch mechanism to suit different screws or nuts, setting them all to the same degree of tightness.

SHEARS: A mechanism converts the rotary action to a reciprocating motion for a small nibbling blade which is opposed to a fixed blade on a helical anvil (see Figure 7.14).

*HAMMER ACTION*:

CHIPPING and RIVETING HAMMERS: Pneumatic hammers are available with long or short stroke designed to be used with interchangeable tooling. Used extensively in the fabrication industry for 'chipping', 'caulking' and 'riveting'. Light weight short-stroke riveters are used in situations where space is limited, such is often the case in aircraft and automobile construction.

*COMPRESSION TYPE*:

YOKE or SQUEEZE RIVETERS: The yoke can be comparatively deep or very shallow, depending on the shape of the structure to be riveted. Stationary or column-mounted yoke riveters are fitted with a pedal control, leaving the hands free. The hammer cylinder closes on the work pneumatically, and riveting does not commence until the rivet is clamped between the 'snaps'. On releasing the pedal, hammering ceases and the cylinder lifts clear.

RESISTANCE WELDING MACHINES: When the pedal control is operated, a piston activates the upper movable electrode arm which descends, clamping the workpiece between the electrodes during the welding cycle (see Section 11.3). On release of the pedal, the pressure is removed and the electrode arm ascends to its original position.

consists of a glass cylinder with a plunger (piston); the cylinder has a bulb-shaped end pierced with a number of holes of uniform size. It may be filled by immersing the bulb end in water and slowly raising the plunger.

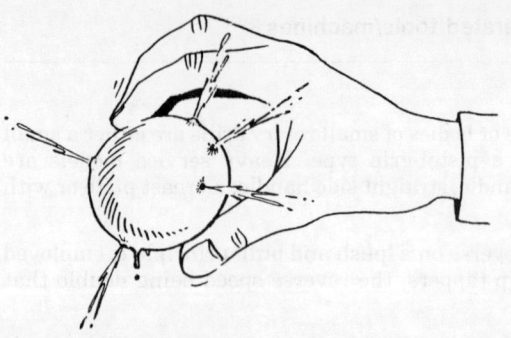

**(a) Pressure applied to a punctured rubber ball filled with water**

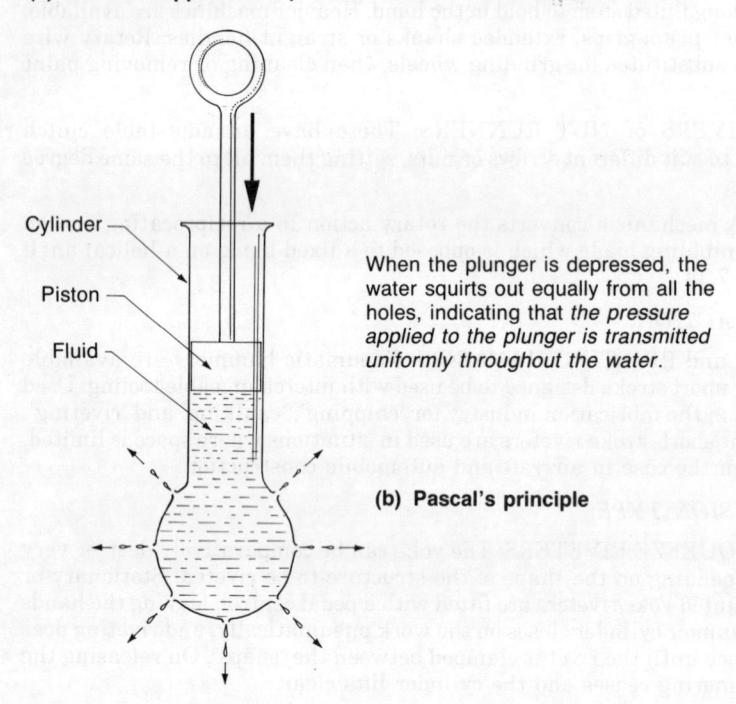

When the plunger is depressed, the water squirts out equally from all the holes, indicating that *the pressure applied to the plunger is transmitted uniformly throughout the water.*

**(b) Pascal's principle**

**Fig. 4.55 Transmission of pressure in fluids**

## 4.37 Transmission of pressure in fluids — hydraulics

A pressure change in one part of a fluid at rest in a closed container, is transmitted equally, without loss, throughout the fluid in all directions and to the walls of the container.

*The pressure on the fluid's surface is measured by the FORCE PER UNIT AREA of the surface, with the area measured at right angles to the direction of the force.*

Thus, if a force of 500 N acts at right angles to a surface having an area of 5 m², the pressure on that surface is given by:

$$\text{PRESSURE} = \frac{\text{FORCE}}{\text{AREA AT RIGHT ANGLES TO THE FORCE}}$$

$$= \frac{500 \text{ N}}{5 \text{ m}^2} = 100 \text{ N/m}^2$$

To understand how very large forces may easily be produced by much smaller forces, by the transmission of pressure through fluids, study the diagram shown in Figure 4.56.

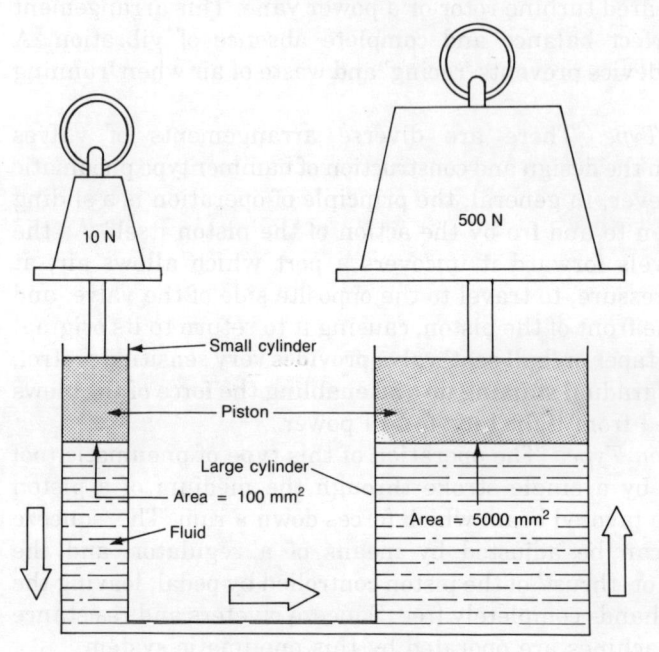

If the force acting down on the piston in the small cylinder is 10 N, and the contact area of the piston is 100 mm², then the pressure produced:

$$\text{Pressure} = \frac{\text{Force}}{\text{Area}} = \frac{10}{100} = 0 \cdot 1 \text{ N/mm}^2$$

This pressure is transmitted equally throughout the whole of the fluid to act upon the piston in the large cylinder.
If the contact area of the large piston is 5000 mm², then the total thrust exerted against it, forcing it outwards can be shown as:

$$\text{Thrust} = \text{Pressure} \times \text{Area} = 0 \cdot 1 \times 5000 = 500 \text{ N}$$

The force ratio may be determined by comparing the load and effort or the cross-sectional areas, as follows:

$$\frac{\text{Load}}{\text{Effort}} = \frac{500 \text{ N}}{100 \text{ N}} \quad \text{or} \quad \frac{\text{Cross-sectional area of load cylinder}}{\text{Cross-sectional area of effort cylinder}} = \frac{5000 \text{ mm}^2}{100 \text{ mm}^2}$$

F.R. = 5 : 1   F.R. = 5 : 1

**Fig. 4.56 Transmission of pressure in fluids — hydraulics**

In a hydraulic system, a pressure exerted on a piston produces an equal increase in pressure on another piston in the system. *If the second piston, in the system, has an area ten times greater than that of the first piston, the force on the second piston is ten times greater, although the pressure is the same as that on the first piston.*

## THE HYDRAULIC PRESS

A hydraulic press is a device consisting of a cylinder fitted with a sliding piston that exerts force upon a confined fluid, which, in turn, produces a *compressive force* upon a stationary anvil or base plate, the fluid is forced into the cylinder by a pump. Small portable machines have a hand operated pump, whilst with heavy duty machines, the pump is electrically operated. The fluid used is a special hydraulic oil.

For heavy lifts, the hydraulic jack is most suitable. Standard models can lift loads of up to 200 tonnes. A hydraulic jack consists of a ram and pump chamber, forming an integral unit, as shown in Figure 4.57. Hydraulic jacks available with the pump chamber separate; this makes them very adaptable in confined spaces, the two units being connected by flexible piping.

The formula used in Figure 4.57 to calculate the thrust ($T$) on the ram by the pressure transmitted through the fluid by the load ($L$) on the plunger, was derived as follows:

Let $L$ represent the load on the plunger, and $T$ represent the thrust on the ram. The cross-sectional area of the plunger is $\pi d^2/4$, whilst $\pi D^2/4$ represents the cross-sectional area of the ram. Because the pressure is transmitted equally, then:

$$\frac{\text{THRUST}}{\text{CROSS-SECTIONAL AREA OF RAM}} = \frac{\text{LOAD}}{\text{CROSS-SECTIONAL AREA OF PLUNGER}}$$

or, THRUST × CROSS-SECTIONAL AREA OF PLUNGER = LOAD × CROSS-SECTIONAL AREA OF RAM

$$T \times \frac{\pi d^2}{4} = L \times \frac{\pi D^2}{4}$$

$$T \times d^2 = L \times D^2$$

$$T = \frac{D^2}{d^2} \times L$$

In its simplest form the HYDRAULIC PRESS consists of a cylinder and piston (RAM) of large diameter which is connected by pipe to a

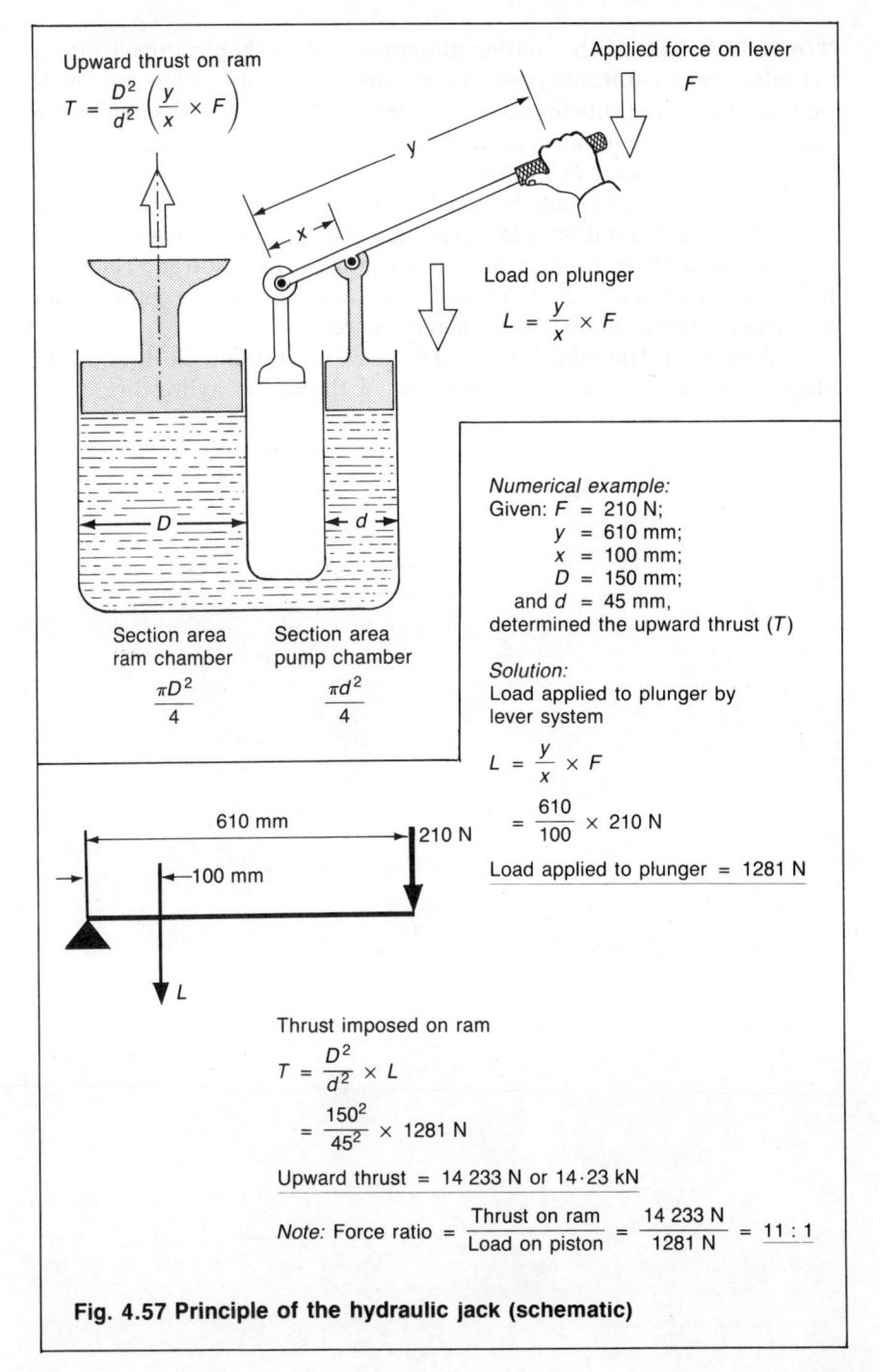

---

Upward thrust on ram
$$T = \frac{D^2}{d^2}\left(\frac{y}{x} \times F\right)$$

Applied force on lever $F$

Load on plunger
$$L = \frac{y}{x} \times F$$

Section area ram chamber
$$\frac{\pi D^2}{4}$$

Section area pump chamber
$$\frac{\pi d^2}{4}$$

*Numerical example:*
Given: $F$ = 210 N;
$y$ = 610 mm;
$x$ = 100 mm;
$D$ = 150 mm;
and $d$ = 45 mm,
determined the upward thrust ($T$)

*Solution:*
Load applied to plunger by lever system

$$L = \frac{y}{x} \times F$$

$$= \frac{610}{100} \times 210 \text{ N}$$

Load applied to plunger = 1281 N

Thrust imposed on ram

$$T = \frac{D^2}{d^2} \times L$$

$$= \frac{150^2}{45^2} \times 1281 \text{ N}$$

Upward thrust = 14 233 N or 14·23 kN

*Note:* Force ratio = $\dfrac{\text{Thrust on ram}}{\text{Load on piston}}$ = $\dfrac{14\ 233 \text{ N}}{1281 \text{ N}}$ = 11 : 1

**Fig. 4.57 Principle of the hydraulic jack (schematic)**

'FORCE PUMP' of much smaller diameter. Oil is then pumped into a cylinder, from a suitable reservoir, causing the RAM to move outwards and exert considerable force. A 'RELEASE VALVE' is incorporated in the system which, when opened, allows the oil to return to the source of supply as illustrated in Figure 4.58.

Hydraulic presses are designed for a multiplicity of operations, and are widely used in industry for forming metals, bending pipes and steel sections, and other tanks where a large force is required. They are available in a variety of styles and sizes, and capacities ranging from less than 1 tonne to 10 000 tonnes or more.

Aircraft retractable undercarriages and automobile brake and clutch systems are everyday examples of the use of hydraulics.

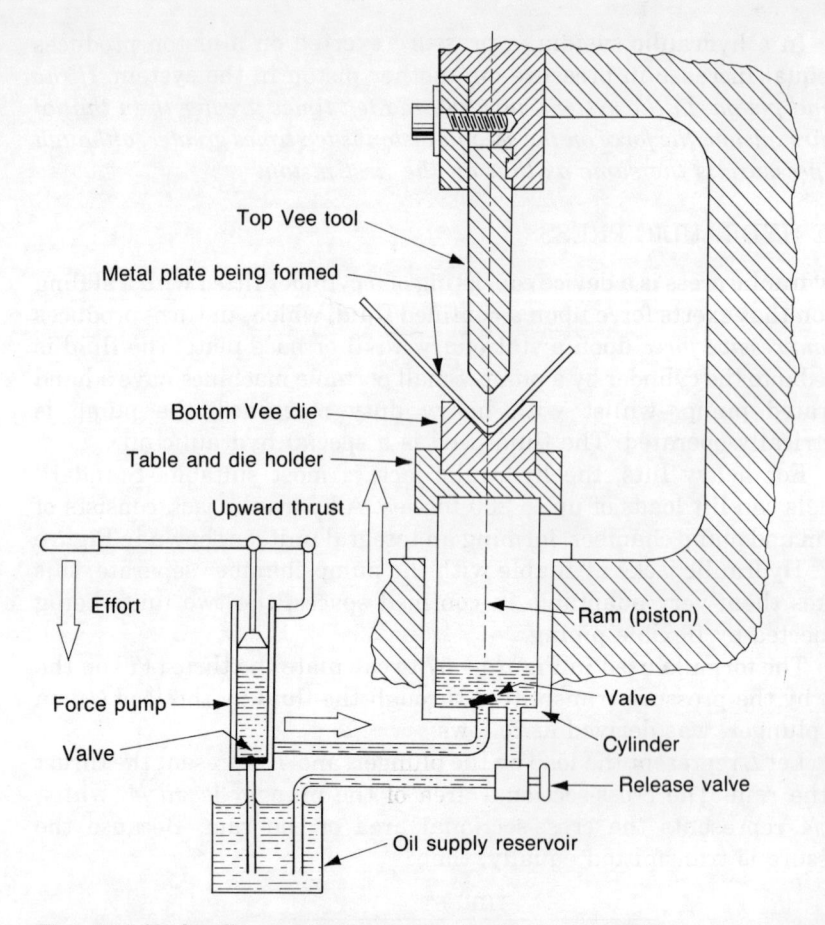

**Fig. 4.58 Hydraulic up-stroking press brake (schematic)**

# 5 Measurement

Since the United Kingdom has now adopted the international system (S.I.) of weights and measures as its legal standards, the international standard yard has become obsolete except for the manufacture of replacement components.

*The international metre has become the legal standard of length for the United Kingdom.*

## 5.1 Types of measurement

Accurate measurement is the basis of good engineering practice, and two principal types of measurement are employed.

### 1. DIRECT EYE MEASUREMENT METHOD

This method is exemplified by the use of the steel rule as a standard of measurement. It is placed alongside the work and the corresponding measurement is read off on the standard scale. Because of the difficulty of sighting the graduations in line with the feature being measured, all measurements made with a rule are of limited accuracy. Use of a low-power magnifying lens, such as an instrument maker's eye glass, increases the accuracy of reading. The sheet metal worker or fabricator uses the rule for those least important measurements whose accuracy is less than 0·5 mm. Flexible steel tapes are used when making measurements which extend beyond the normal range of the steel rule. Figure 5.1(a) shows a typical steel rule, whilst Figure 5.1(b) shows a typical steel tape.

### 2. CONTACT SURFACE MEASUREMENT METHOD

This is a two-surface contact method in which the object to be measured is arranged to make contact at its two ends between the surfaces of a

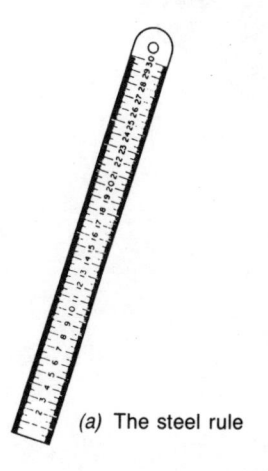

(a) The steel rule

Standard rules used in the fabrication industry are made from hardened and tempered spring steel. They have a finely divided precision engraved scale, and are available with a satin chrome finish which reduces glare, making them easier to read.
Their edges are ground which enables them to be used as *straight edges* when scribing lines or for testing the flatness of a surface.
Stainless steel rules are very popular because standard steel rules are very prone to rusting when used or left in damp or corrosive surroundings.

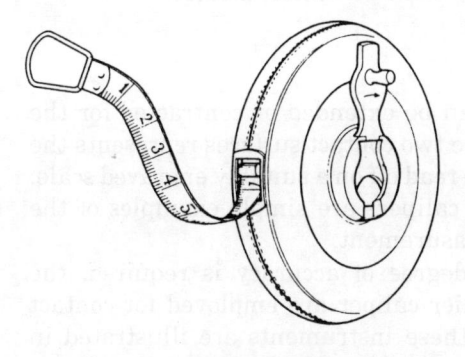

(b) The steel tape

In cases where long measurements have to be made, the flexible steel tape is used. Woven or fabric tapes are unsuitable for use in the engineering environment.
The graduations may be the usual black lines on a steel surface or, black lines on a white surface.
Flexible steel tapes are available in sizes ranging from the handy pocket size (150 mm) to the larger ones used for measurements up to and exceeding 30 m.

**Fig. 5.1 Standards of measurement — the rule and steel tape**

139

*(a) The vernier caliper*

*(b) The micrometer caliper*

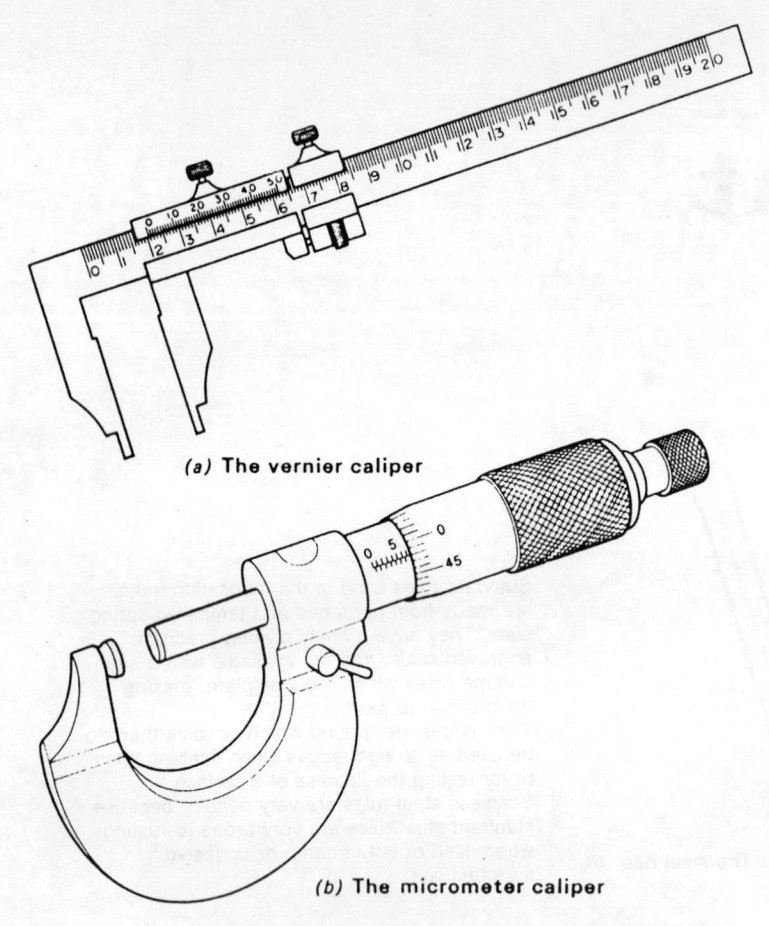

**Fig. 5.2** Precision measuring instruments — contact method

**Table 5.1** Workshop standards of length

| NAME | RANGE | READING ACCURACY |
| --- | --- | --- |
| Steel rule | 150 mm to 1000 mm (1 m) | 0·05 mm |
| Vernier caliper | 0/150 mm to 0/2000 mm (2 m) | 0·02 mm |
| Micrometer caliper | 0/25 mm to 1775/1800 mm (1·8 m) | 0·01 mm |

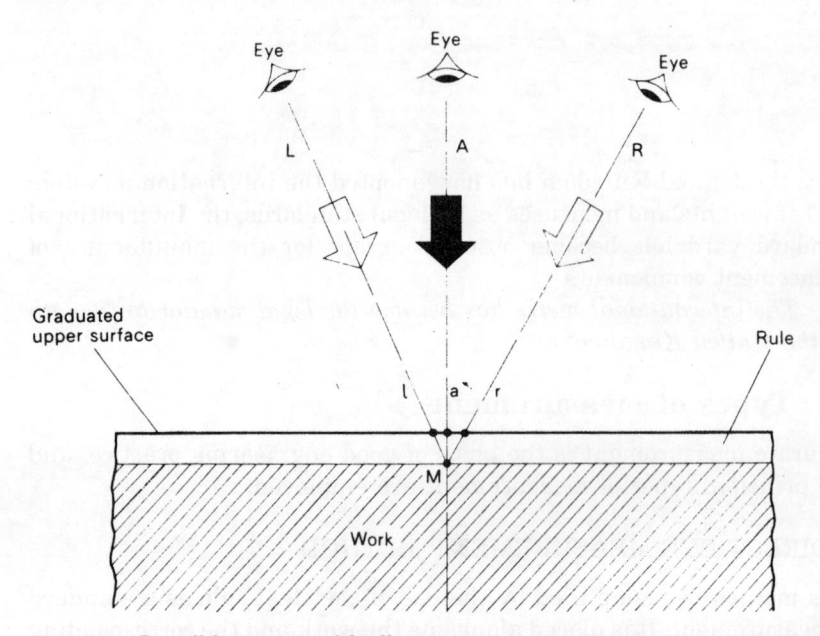

**Fig. 5.3** Possible errors with direct eye measurement

measuring device. This device can be extended or contracted for the purpose. The distance between the two contact surfaces represents the dimension of the object and can be read off on a suitably engraved scale. The ordinary inside and outside calipers are simple examples of the application of this method of measurement.

In cases where a greater degree of accuracy is required, the micrometer caliper and the vernier caliper are employed for contact measurement, and examples of these instruments are illustrated in Figure 5.2.

Table 5.1 compares the degree of accuracy in reading measurements.

## 5.2 Direct eye measurement

The ENGINEER'S RULE, used for making direct measurements, depends upon VISUAL ALIGNMENT of a mark or surface on the work to be measured against the nearest division on its scale. This may appear to be a relatively simple exercise, but in practice errors can very easily occur, as shown in Figure 5.3.

These errors can be minimised by using a rule whose thickness is as small as possible — this emphasises the importance of using a thin steel rule.

It is important when making measurements with an Engineer's rule to have the eye directly opposite and at 90° to the mark on the work, otherwise there will be an error — known as 'PARALLAX' — which is the result of any sideways positioning of the direction of sighting.

Reference to Figure 5.3 will show that:

1. 'M' represents the mark on the work whose position is required to be measured by means of a rule laid alongside it. The graduations of measurement are on the upper face of the rule, as indicated.
2. If the eye is placed along the sighting line A–M, which is at 90° to the work surface, a TRUE READING will be obtained at 'a', for it is then directly opposite 'M'.
3. If, however, the eye is not on this sighting line, but displaced to the right, as at 'R', the division 'r' on the graduated scale will appear to be opposite 'M' and an INCORRECT READING will be obtained. Similarly, if the eye is displaced to the left, as at 'L', an incorrect reading on the opposite side, as at 'l' will result.

## 5.3 Possible error when using scribers

Care must be taken marking a straight line or marking around a template with a scriber. To reduce the possibility of an error, the scriber must be held against the straight edge or the periphery (in the case of a template) as illustrated in Figure 5.4(b).

## 5.4 Use of calipers

The application and reading of the rule can be extended by the use of calipers. They are used as a method of transferring the distances between two contact surfaces to a rule for reading, thus reducing possible sighting errors. Their main use is for measuring inside and outside dimensions of workpieces.

### FRICTION-JOINT CALIPERS

The simplest form of caliper is the friction-joint type illustrated in Figure 5.5. *When using this type of caliper for small movements, they should not be forced open by hand or by any instrument.* The correct procedure for adjusting the caliper legs involves the use of a block of wood. Lightly tapping the back of the friction joint on the block will gradually open the caliper. Holding one leg of the caliper in the fingers and tapping the other one on the wooden block has the opposite effect, closing the caliper.

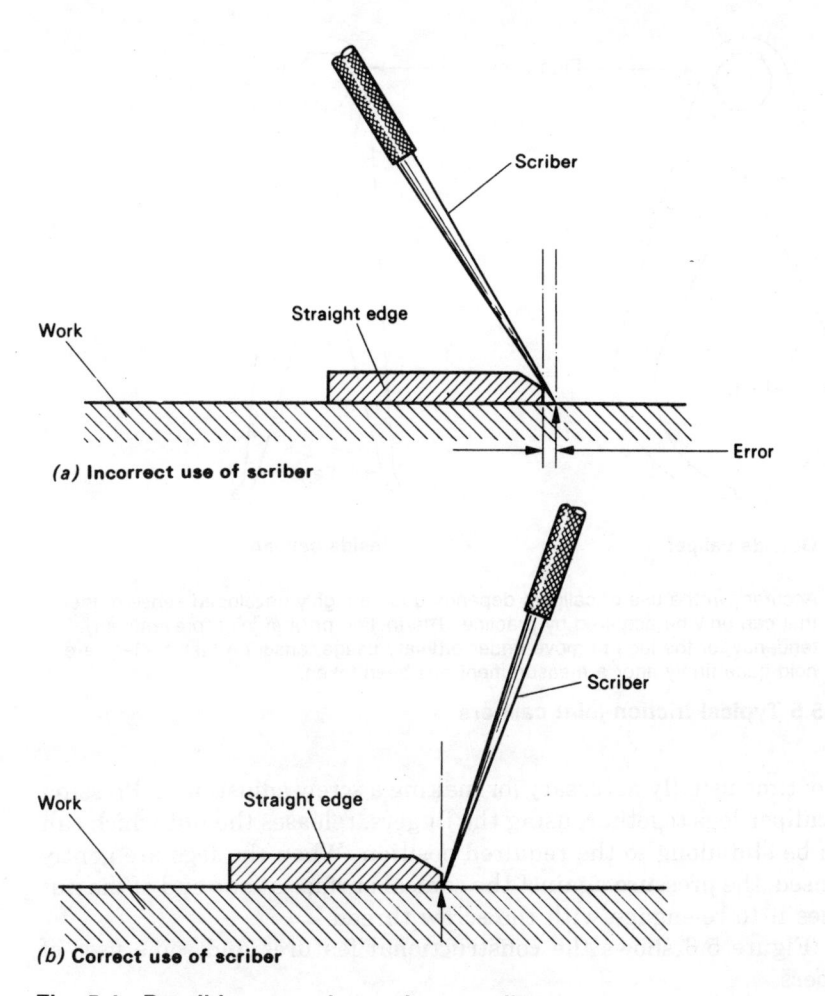

**Fig. 5.4 Possible error when using a scriber**

### SPRING CALIPERS

These are a marked improvement on the friction-joint type. The upper end of each leg hinges on a common flanged pin, as shown in Figure 5.6. At a slight distance above the axis of the pin, notches in the ends of the legs engage with a curved spring. The effect of the spring's pressure is to force the lower ends of the caliper legs apart such that movement of the adjustment nut enables them to either open with, or close against the pressure of the spring.

Spring calipers are available with either a fixed nut or a quick-release split nut for adjustment purposes. Use of the latter avoids the

**Outside caliper**     **Inside caliper**

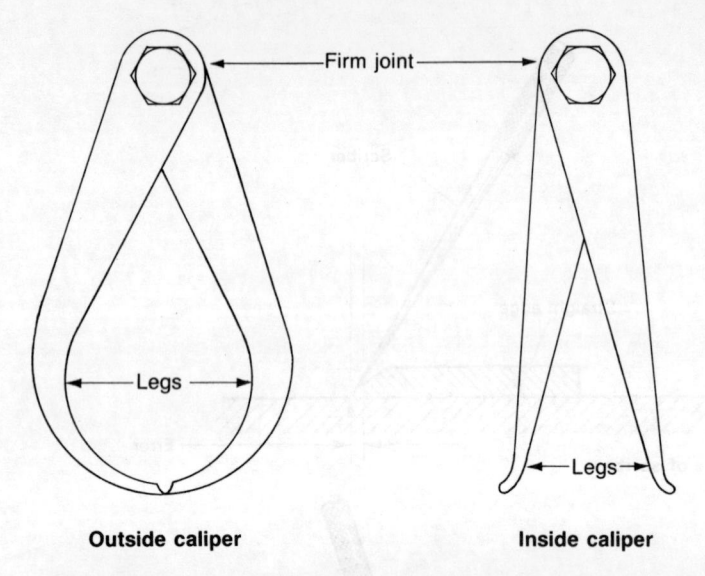

*Note:* Accuracy in the use of calipers depends upon a highly developed sense of feel that can only be acquired by practice. The friction or 'firm joint' prevents any tendency for the legs to move under ordinary usage, ensuring that the legs are held quite firmly after a measurement has been taken.

**Fig. 5.5 Typical friction-joint calipers**

loss of time usually necessary for making a screw adjustment. Pressing the caliper legs together, using the fingers, releases the nut which can then be slid along to the required position. When the legs are gently released, the pressure against the conical surface at one end of the nut causes it to re-engage with the screw thread.

Figure 5.6 shows the constructional features and some uses of calipers.

## 5.5 Micrometer caliper

The simple contact type measuring devices so far described have only a very limited degree of accuracy. Where precise measurement is essential, contact type measuring instruments possessing far greater accuracy and sensitivity must be employed. Figure 5.7 gives details of one of the most familiar precision measuring instruments found in the workshop, namely the MICROMETER CALIPER.

ADJUSTMENT OF THE MICROMETER

In the box supplied with the instrument will be found a small double-ended spanner to be used when making two basic adjustments:

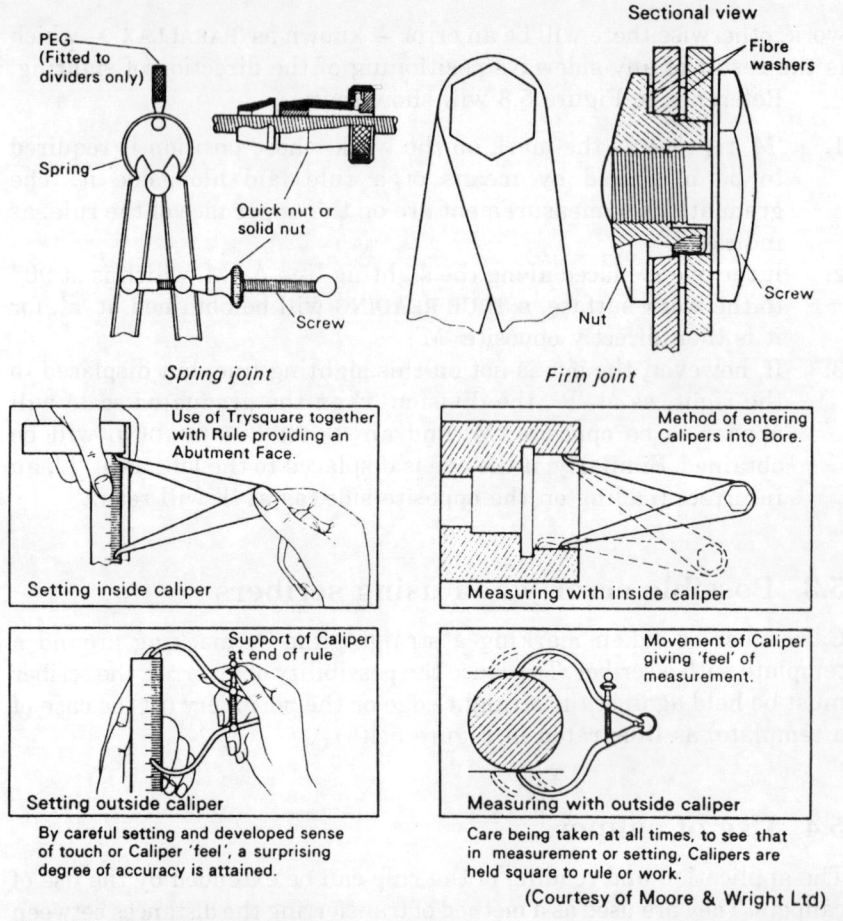

**Fig. 5.6 Construction and use of calipers**

1. *Looseness in the screw.* This is taken up by a slight turn of the screw adjustment nut — item (6) in Figure 5.7.
2. *Zero error.* Periodically the anvil faces (contact surfaces) should be cleaned and closed using the ratchet to give the correct measuring pressure. Two 'clicks' of the ratchet are sufficient. If the zero line of the thimble does not coincide with the datum line on the barrel, turn the barrel (item (4) in Figure 5.7) with the 'C' spanner until the datum line corresponds at zero.

## 5.6 Reading the metric micrometer

The metric micrometer has a screw of 0·5 mm pitch. The barrel is graduated in 'whole' and 'half' millimetres. The thimble is graduated

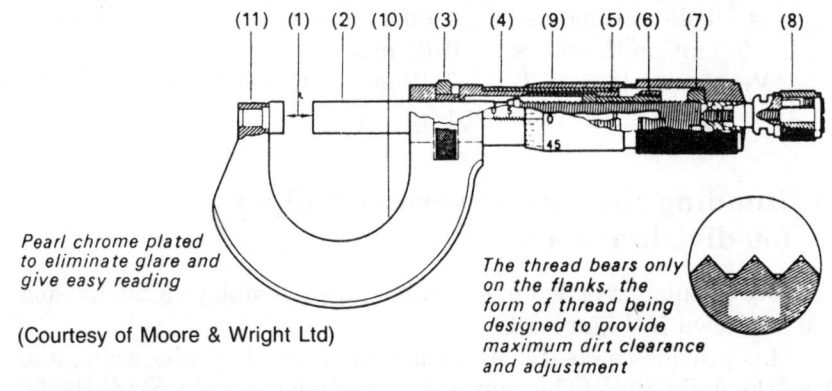

**Pearl chrome plated to eliminate glare and give easy reading**

*(Courtesy of Moore & Wright Ltd)*

**The thread bears only on the flanks, the form of thread being designed to provide maximum dirt clearance and adjustment**

(1) **Spindle and anvil faces** — Glass hard and optically flat, also available with **Tungsten carbide** faces
(2) **Spindle** — Thread ground, and made from alloy steel, hardened throughout, and stabilised
(3) **Locknut** — effective at any position. Spindle retained in perfect alignment
(4) **Barrel** — Adjustable for zero setting. Accurately divided and clearly marked. Pearl chrome plated
(5) **Main nut** — Length of thread ensures long working life
(6) **Screw adjusting nut** — For effective adjustment of main nut effective
(7) **Thimble adjusting nut** — Controls position of thimble
(8) **Ratchet** — Ensures a constant measuring pressure
(9) **Thimble** — Accurately divided and every graduation clearly numbered
(10) **Steel frame** — Drop forged. Marked with useful decimal equivalents
(11) **Anvil end** — Cutaway frame facilitates usage in narrow slots

**Fig. 5.7 Construction of the micrometer caliper**

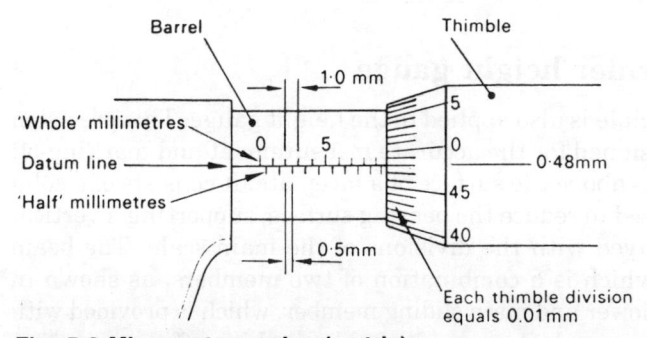

**Fig. 5.8 Micrometer scales (metric)**

with 50 divisions equally spaced around its circumference. Figure 5.8 shows the micrometer scales.

Since the pitch of the screw is 0·5 mm and the barrel divisions are 0·5 mm apart; one revolution of the thimble (and therefore the screw) moves the thimble along a distance of one barrel division. (The barrel divisions are placed on alternate sides of the datum line for clarity.)

Therefore, since the thimble has 50 equal divisions and one revolution of the thimble equals 0·5 mm, then a movement of one thimble division equals:

$$\frac{0 \cdot 5}{50} = 0 \cdot 01 \text{ mm}$$

*The metric micrometer reading equals:*

The largest visible 'whole' millimetre
+
The largest visible 'half' millimetre
+
The thimble division in line with the datum line.

The reading in Figure 5.8 is as follows:

9 'whole' millimetre = 9·00
1 'half' millimetre = 0·50
48 hundredths of a mm = 0·48
9·98 mm

## 5.7 The vernier caliper

Figure 5.9 gives details of the vernier caliper which, unlike the micrometer caliper, can make inside and outside measurements on the

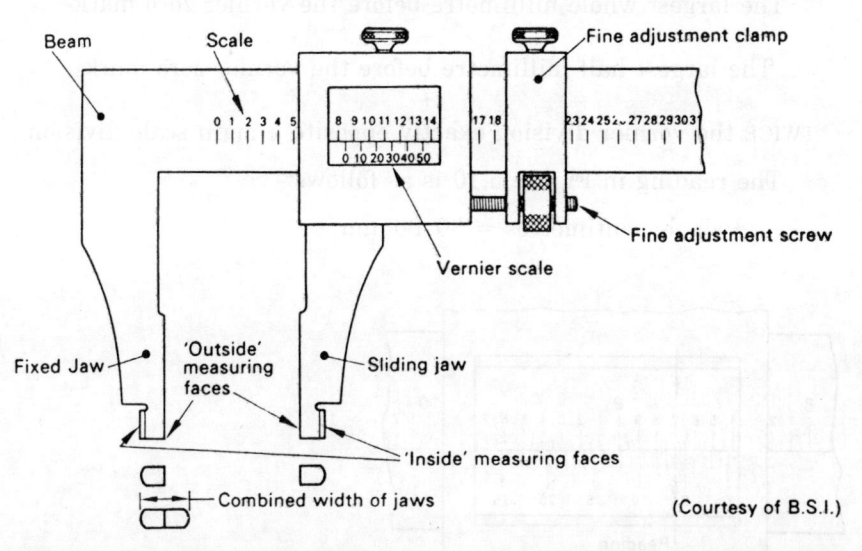

*(Courtesy of B.S.I.)*

*Note:* For inside readings, the thickness of the jaws must be added to the scale reading.

**Fig. 5.9 The vernier caliper**

143

one instrument. The vernier caliper reads from zero to the full length of its beam scale, whereas the micrometer only reads over a narrow range of 25 mm.

## 5.8 Reading the metric vernier caliper (25-division scale)

Figure 5.10 shows the scales of a metric vernier caliper. The main scale is divided into centimetres, and sub-divided into millimetres and half-millimetres. The vernier scale is divided into 25 equal divisions and it will be seen that these occupy 24 divisions (12 mm) on the main scale. Thus the length of each vernier division equals:

$$\frac{1}{25} \text{ of } 12 \text{ mm} \quad or \quad 0 \cdot 48 \text{ mm}$$

But *each* main scale division equals $\frac{1}{2}$ or 0·5 mm. Therefore, the difference between one fixed scale division and one vernier scale division is:

$$0 \cdot 50 - 0 \cdot 48 = \underline{0 \cdot 02 \text{ mm}}$$

*The metric vernier reading equals*:

The largest whole centimetre before the vernier zero mark
+
The largest whole millimetre before the vernier zero mark
+
The largest half millimetre before the vernier zero mark
+
*TWICE the vernier division exactly opposite a main scale division

The reading in Figure 5.10 is as follows:

```
  8 whole centimetres   =  80·00 mm
  4 whole millimetres   =   4·00 mm
  0 'half' millimetres  =   0·00 mm
 *Vernier scale 6( × 2) =   0·12 mm   (*6 × 0·02 mm)
                           ─────────
                            84·12 mm
```

## 5.9 Reading the metric vernier caliper (50-division scale)

The more popular 'metric only' vernier caliper employs a 50-division scale as shown in Figure 5.11.

This provides a clearer and more easily read vernier scale, and allows the main scale to be engraved in millimetres only. Since the 50 divisions of the vernier scale occupy 49 divisions (49 mm) on the main scale, the reading accuracy is still 0·02 mm.

*The metric vernier reading equals*:

The largest whole millimetre before the vernier zero mark
+
*TWICE the vernier division exactly opposite a main scale division

Thus the reading shown in Figure 5.11 is as follows:

```
 32 whole millimetres   = 32·00 mm
 *Vernier scale 11 (×2) =  0·22 mm   (*11 × 0·02 mm)
                          ─────────
                          32·22 mm
```

## 5.10 The vernier height gauge

The vernier principle is also applied to the height gauge. This precision instrument is designed for the accurate measurement and marking-off of vertical heights above the surface of a faceplate. It consists of a solid block base, recessed to reduce the bearing surface, supporting a vertical steel beam engraved with the divisions of the main scale. The beam carries a slider which is a combination of two members, as shown in Figure 5.12. The lower and main sliding member, which is provided with an engraved vernier scale, has an adjustable marking-off or measuring knife-edged pointer and a clamping screw. The upper sliding member is a fine adjustment device, also fitted with a clamping screw.

In use the instrument is placed upon a faceplate or the machined surface of a marking-off table and the pointer adjusted to suit any point to be measured. It can be used to measure the difference between two heights by taking vernier scale readings and subtracting the lesser from the greater reading. For scribing lines, when marking out, the pointer

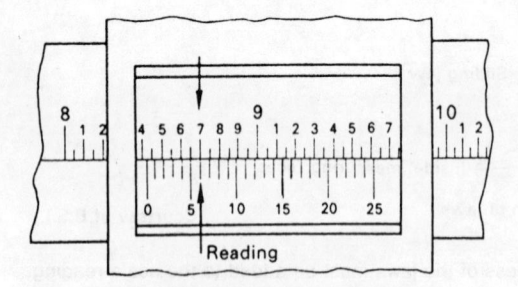

**Fig. 5.10 25-Division vernier scale**

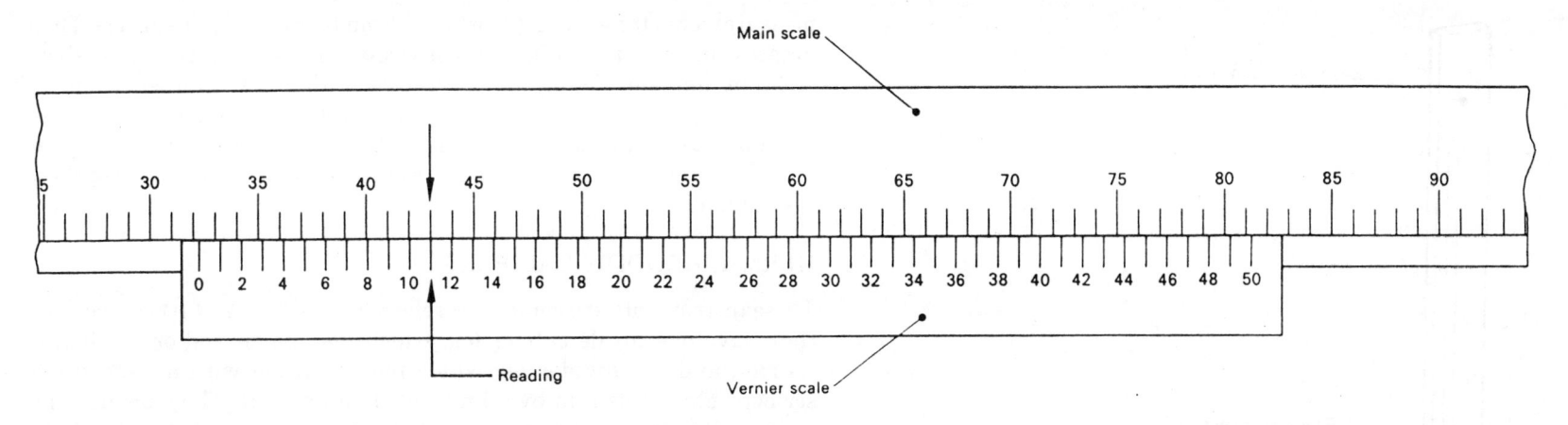

*Note:* Each division on the main scale represents one whole millimetre. Each division on the vernier scale represents one-hundredth of a millimetre multiplied by two.
  For example, division 16 represents 16 × 2 = 0·32 mm.

**Fig. 5.11 50-Division vernier scale**

is set with the fine adjustment device to the required height and clamped in position. Its use for marking out precision sheet metal work will be discussed in Section 5.43.

## 5.11 Marking-out (introduction)

The term MARKING-OUT, in general, means the scribing of lines on a metal surface to indicate, for example, a required outline or profile; the outline of any holes or apertures required to be cut; the position of any hole centres; the position of bending or folding lines, and allowances for edge preparations.

Scribing tools, unlike pencils and pens used for drawing, cut into the metal surface producing lines or marks which are reasonably permanent. *In order to avoid cutting through the plated surfaces of coated materials such as tinplate, terneplate and galvanised steel, exposing the mild steel base to possible corrosion, a brass scriber or a hard pencil should be used.* Prior to marking-out, it may be necessary to prepare the surface of the metal by the application of coating in a contrasting colour so that any scribed line will show up clearly.

### WHITEWASH

The surfaces of steel plate are often covered with a black scale (oxidation) as a result of the hot-rolling process during manufacture. On such surfaces it is not easy to distinguish a scribed line, as is the case with the rough surfaces of castings. To overcome this problem the surface to be marked-out is either 'chalked' or coated with 'whitewash'.

### CELLULOSE LACQUER

The once-popular *copper sulphate solution*, used to deposit a thin coating of copper on a clean steel surface, has been superseded by the use of cellulose lacquer which is available in a variety of colours including blue. Compared with lacquer, copper sulphate solution has certain disadvantages:

(*a*)  It stains or corrodes any marking out instruments it comes into contact with.
(*b*)  Its application is limited to steel surfaces only.
(*c*)  Once applied, the copper coating is difficult to remove.

Cellulose lacquer can be applied to any bright surface. When applied thinly it quickly dries by evaporation and, if required, the original metal surface may be restored by an application of liquid lacquer remover.

## 5.12 The scribed line

The choice of tool or instrument and the method of using it for measuring or marking-out is influenced by the degree of accuracy required. In order to produce accurately-dimensioned finely-defined scribed lines, it is

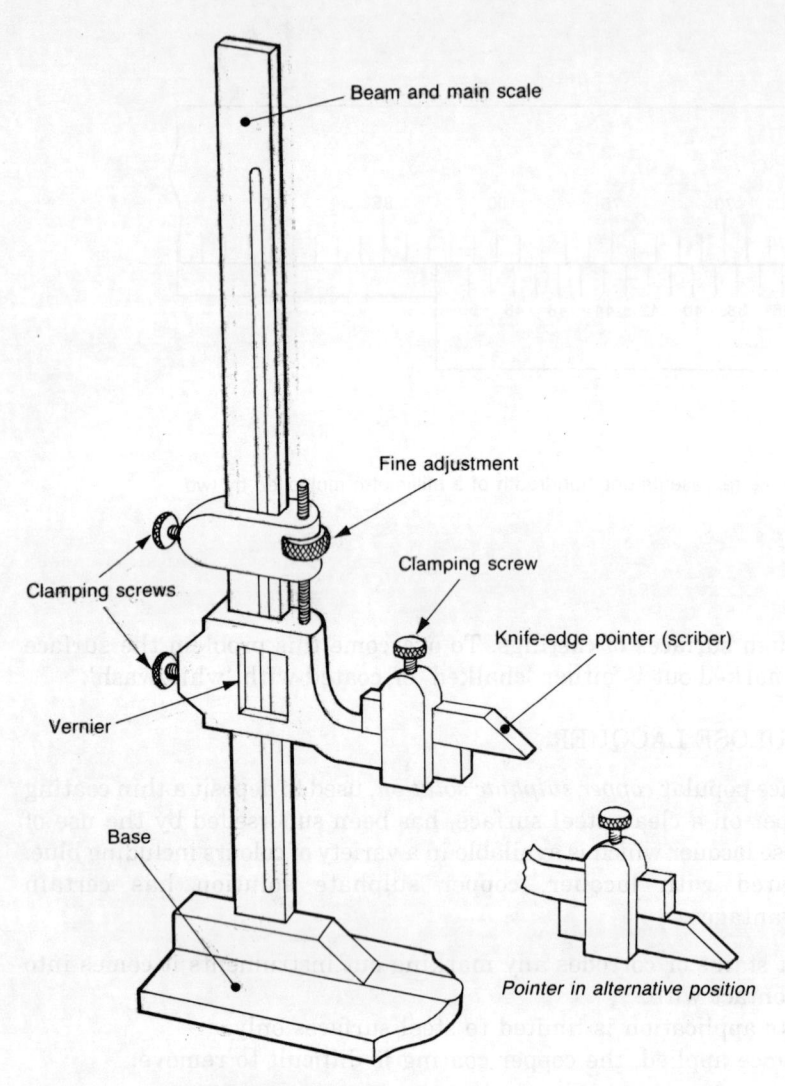

**Fig. 5.12 The vernier height gauge**

essential that the scribing points of scribers, dividers or compasses, jenny odd-legs, trammels and surface gauges be maintained in needle-sharp condition.

## DIVIDERS AND TRAMMELS

Dividers are a type of calipers used for measuring distances between two points, or parallel lines, on a flat surface and for transferring measurements taken direct from a scale on to another flat surface. Their greatest use is for scribing arcs or circles on flat surfaces up to about 150 mm diameter. Wing compasses are available with a leg length of 300 mm capable of scribing a circle of about 900 mm diameter. However, the limited range afforded by these instruments can be extended by the use of trammels (see Section 5.17). Figure 5.13 gives examples of these instruments.

## HERMAPHRODITE CALIPERS

These instruments are commonly referred to as JENNY ODD-LEGS because they have one inside caliper leg, which may be curved or notched at its foot, and one divider leg which may be fitted with an adjustable scriber held in position by a knurled clamping nut. They are used for scribing lines parallel to any straight or curved edge (see Section 5.22), as shown in Figure 5.14.

## THE SURFACE GAUGE

The marking-out of work often includes the scribing of lines at a given height from some face of the work or the construction of lines around its several surfaces. To do this an instrument called a *'surface gauge'* has been devised for holding the scriber. The surface gauge, commonly known as the *'scribing block'* is mostly used for scribing lines parallel to a datum surface (see Section 5.15) and for checking parallelism. Typical scribing blocks are illustrated in Figure 5.15.

## 5.13 Setting scribing instruments

To ensure accuracy when setting dividers or odd-leg calipers, never rely on the eyes alone. Figure 5.16 shows how use is made of the engraved divisions on the rule into which the scribing point can be made to 'click' into the required position.

## 5.14 Datum lines and datum edges

*A DATUM can be defined as a fixed point, line or surface that can be used as a 'foundation' from which measurements may be taken.*

Figure 5.17 explains how errors can be avoided by marking each dimension of a component individually from a chosen datum.

*Datum lines* are frequently used in plate work where the production of a suitable datum edge(s) would be a major operation, (see Section 5.17). However, most components require two *datum edges* which must be accurately cut, machined or filed at right angles to each other, as in Figure 5.18.

(a) Dividers

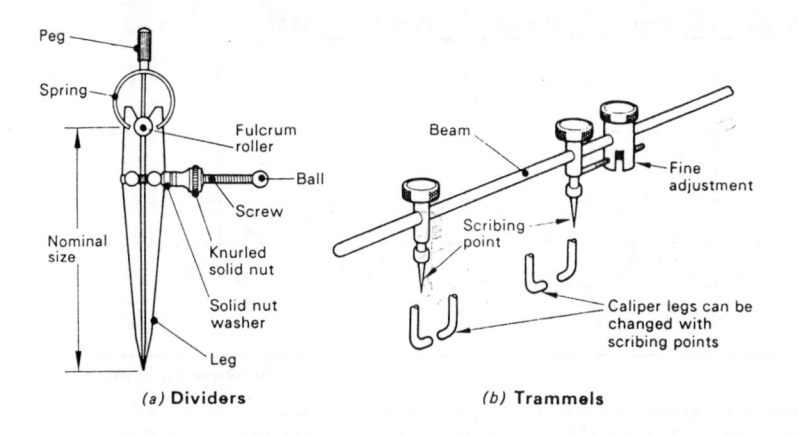

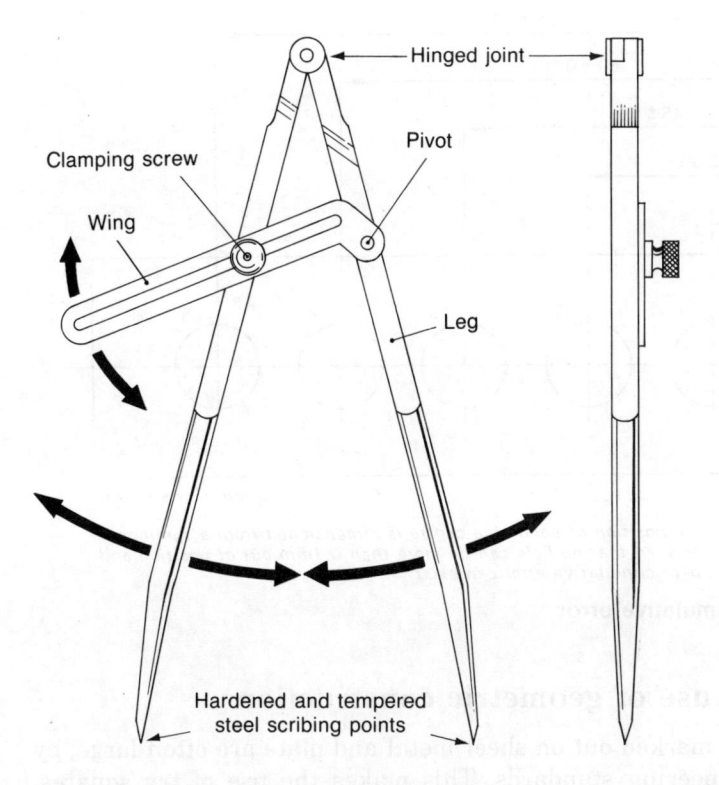

(b) Trammels

(c) Wing compasses (swivelling-link pattern)

**Fig. 5.13 Scribing instruments for radial lines**

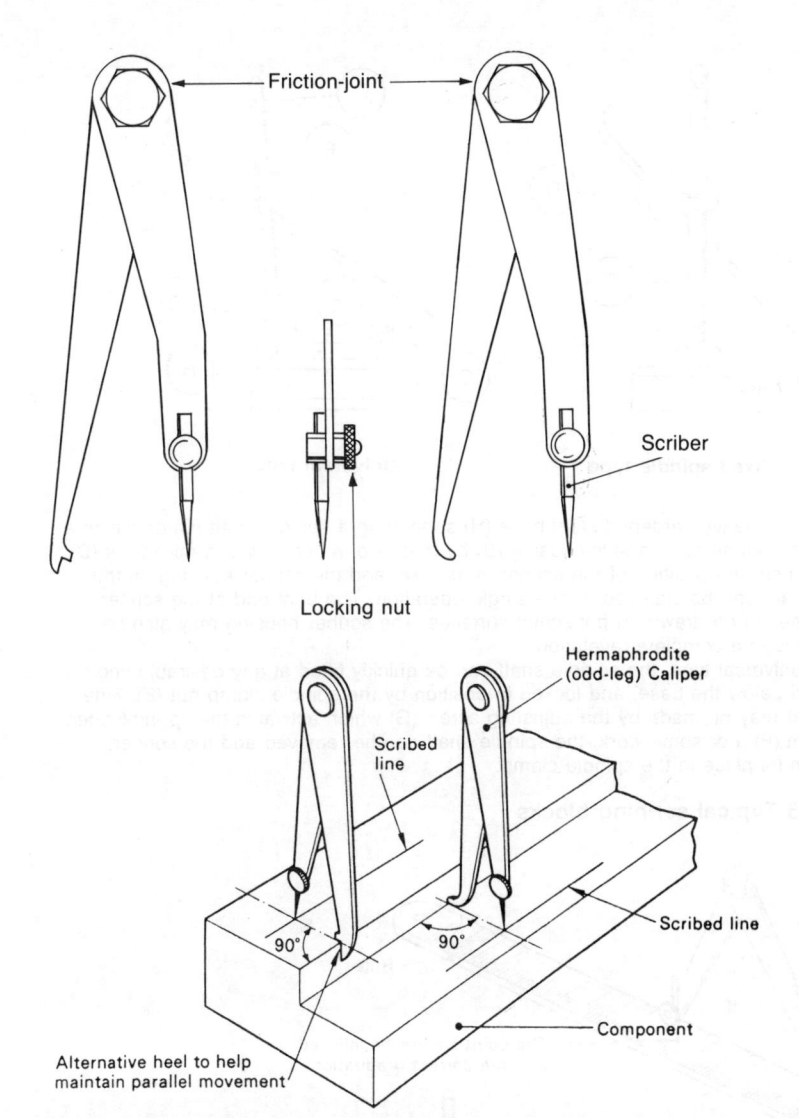

**Fig. 5.14 Hermaphrodite calipers**

## 5.15 The datum surface

The most accurate method of marking-out is by use of a *datum surface*. This is provided by a faceplate or the machined surface of a marking-out table together with the necessary equipment shown in Figure 5.19.

For precision marking-out, the scribing block would be replaced by the vernier height gauge (see Section 5.10). Figure 5.20 explains the marking-out procedure.

147

**Fixed spindle type**     **Universal type**

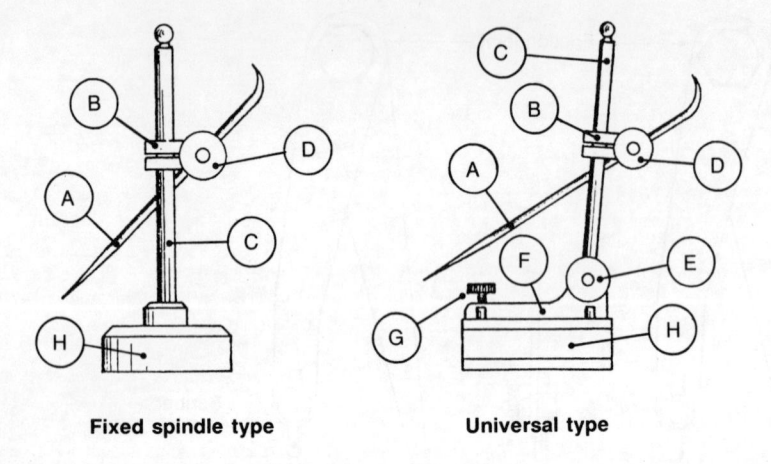

Consists of a heavy hardened steel base (H) supporting a spindle shaft (C) on which a scriber (A) is clamped in a split-housing (B) by means of a screw with a milled nut (D), such that both the position of the scriber in its hole, and the scriber housing on the spindle shaft, can be clamped in one single operation. The bent end of the scriber permits lines to be drawn on horizontal surfaces. The scriber housing may also be turned through a complete revolution.

With the universal type, the spindle shaft can be quickly tilted at any desirable angle, above and below the base, and locked in position by the spindle clamp nut (E). Fine adjustment may be made by the adjusting screw (G) which activates the spring-loaded rocker arm (F). For some work, the spindle shaft can be removed and the scriber inserted in its place in the spindle clamp.

**Fig. 5.15 Typical scribing blocks**

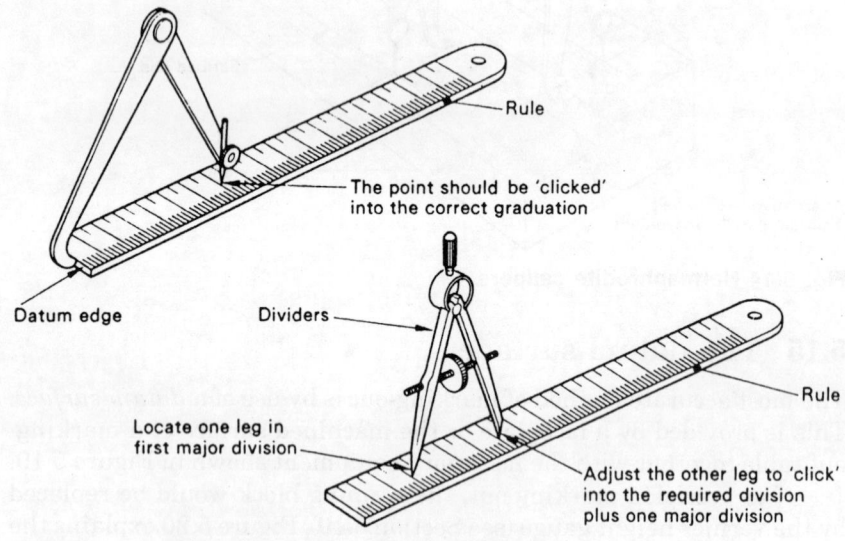

**Fig. 5.16 Setting dividers and odd-leg calipers**

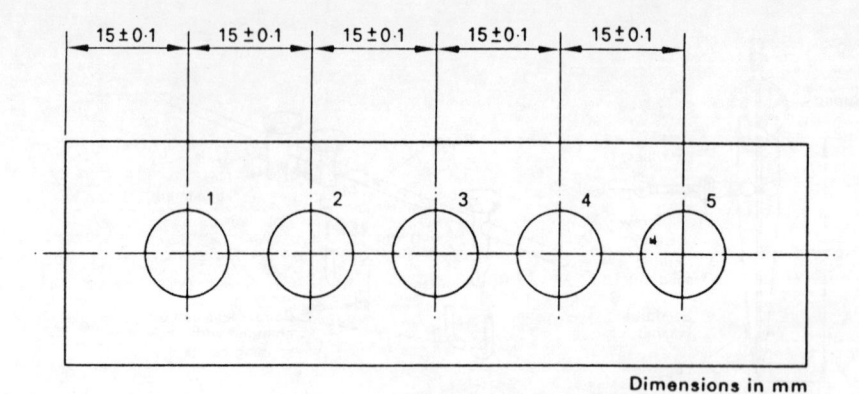

(a) In example (a) each hole centre is dimensioned from the one next to it. If each dimension is on top limit, then the build up in error between holes 1 and 5 is 4 ± 0.1 = 0.4mm. This is known as a cumulative or 'built up' error

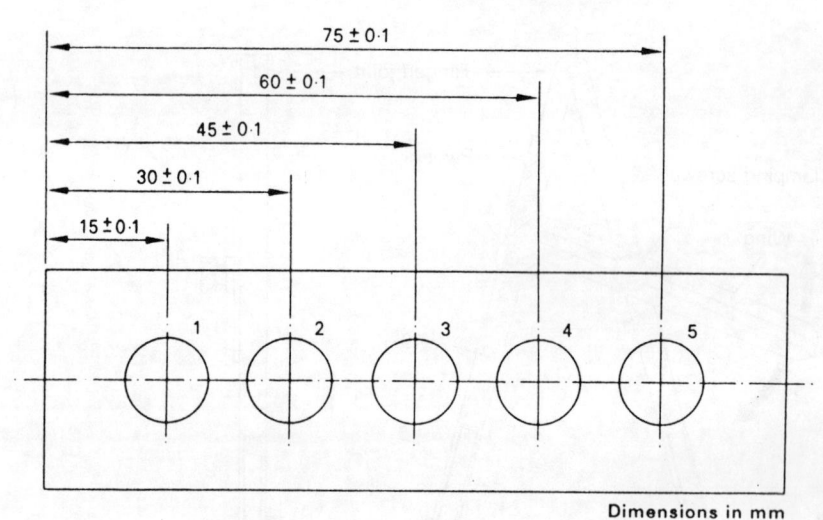

(b) In example (b) the position of each hole centre is dimensioned from a common datum. It will be seen that no hole can be more than 0.1mm out of position and that no 'built up' or cumulative error can exist

**Fig. 5.17 Cumulative error**

## 5.16 The use of geometric constructions

Components marked out on sheet metal and plate are often large, by normal engineering standards. This makes the use of try squares, straight edges, protractors, etc., difficult and sometimes impossible. It is often easier to construct the shape required geometrically, using dividers and trammels (beam compasses). Section 2 gave a wide variety

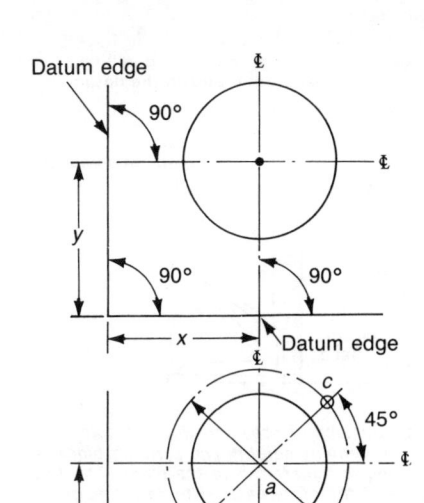

The hole is positioned from the datum edges by the dimensions 'x' and 'y'. These dimensions are the co-ordinates for the hole centre. Examples 'A' and 'B' show hole centres positioned on a pitch circle diameter and this information has to be translated into co-ordinate dimensions.

Given: 'x' = 42·86 mm, 'y' = 38 mm
and pitch circle diameter = 32 mm.

*By ratio of sides:*

$$ac = \frac{32}{2} = \underline{16 \text{ mm}}$$

$$am = cm = \frac{ac}{\sqrt{2}}$$

$$= \frac{16}{1·414} = \underline{11·32 \text{ mm}}.$$

| Hole centre | Ordinates (mm) | |
|---|---|---|
| b | x − 11·32 = 31·54 | y − 11·32 = 26·68 |
| c | x + 11·32 = 54·18 | y + 11·32 = 49·32 |

*By ratio of sides:*

$$ac = \underline{16 \text{ mm}}$$

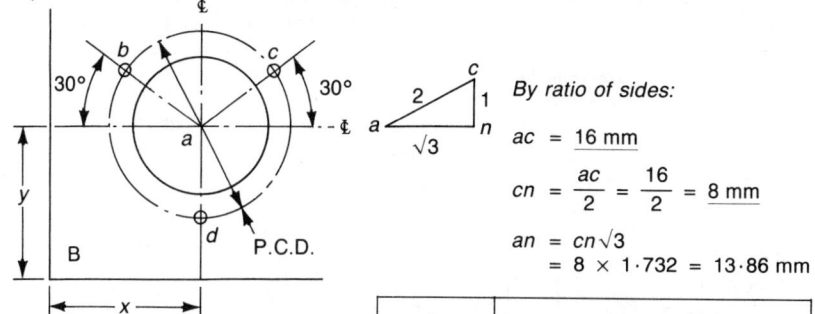

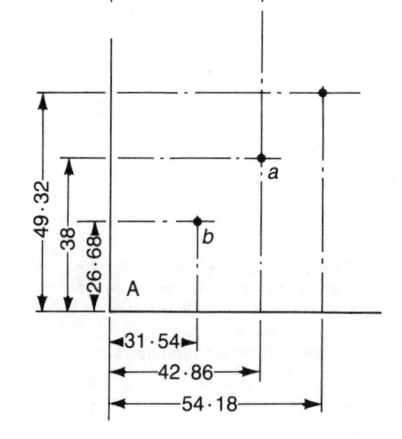

$$cn = \frac{ac}{2} = \frac{16}{2} = \underline{8 \text{ mm}}$$

$$an = cn\sqrt{3}$$
$$= 8 \times 1·732 = \underline{13·86 \text{ mm}}$$

| Hole centre | Ordinates (mm) | |
|---|---|---|
| b | x − 13·86 = 29 | y + 8 = 46 |
| c | x + 13·86 = 56·72 | y + 8 = 46 |
| d | x = 42·86 | y − 16 = 22 |

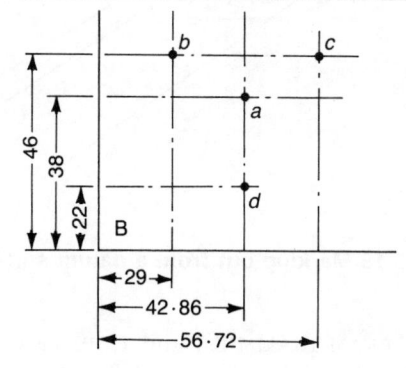

**Fig. 5.18 Use of co-ordinates**

## 5.17 Marking-off a large template or workpiece (general)

Large sheet metal fabrications and platework jobs have to be marked out 'in the flat'. Often it is convenient to cut a sheet or plate to the correct shape and dimensions before marking the position of rivet or bolt holes if these are around the edge only, otherwise the entire job must be laid out.

If the required sheet or plate is not too large, a 'DATUM LINE' may be scribed adjacent to one edge with the aid of a straight edge and scriber.

When marking-out large steel plates a CHALKLINE is employed for producing long straight lines. The use of a chalkline is illustrated in Figure 5.21.

1. The chalkline may be positioned at one end by means of a magnet;
2. The line is then thoroughly 'chalked', either with ordinary chalk or French chalk;
3. The line is stretched and firmly held against the work in line with 'witness marks' previously made with a centre punch;
4. The line is then 'flicked' by lifting it vertically with finger and thumb and releasing it;
5. On release, the line strikes the surface of the steel plate producing a truly straight chalk line in the required position.

NOTE: *When marking out very long lines, it is good practice to place a try-square against the line near its centre. The chalkline is then pulled straight up against the edge of the square and released.*

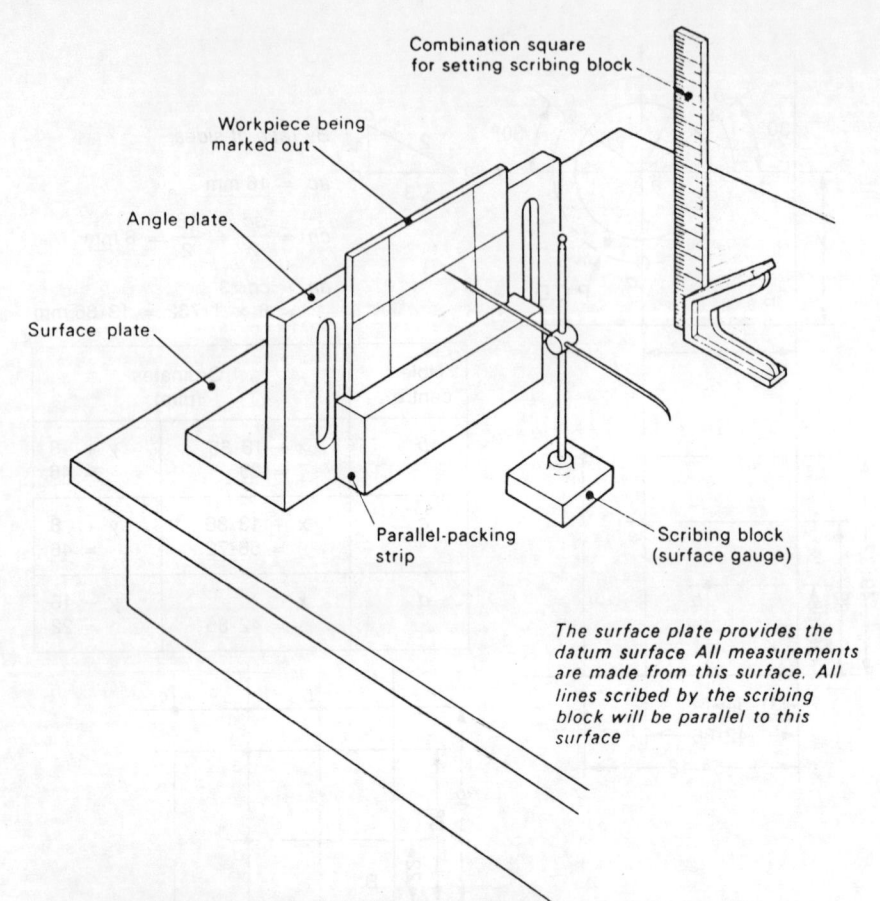

**Fig. 5.19 Marking-out from a datum surface**

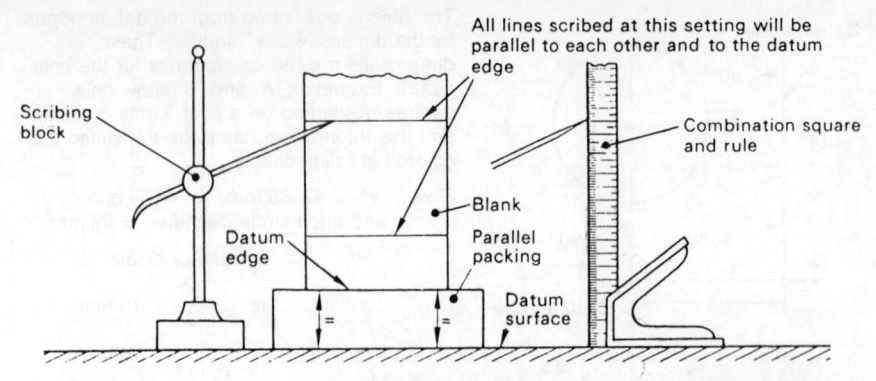

(a) *Blank is placed on datum edge on surface plate (datum surface). In this example parallel packing is used to raise the blank to a convenient height. The scribing block is set to the combination square and rule. The setting is transferred to the blank. The line so scribed will be parallel to the datum surface and therefore parallel to datum edge of the blank*

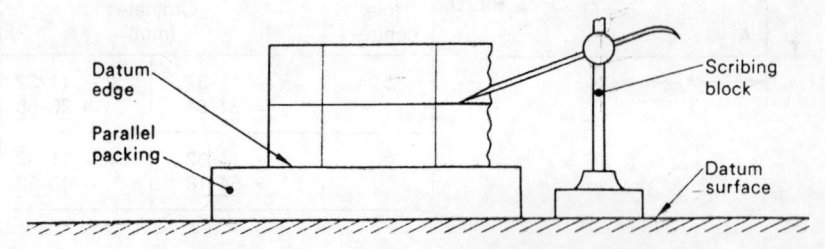

(b) *Blank is turned through 90° so that it rests on the other datum edge. This enables the centre line to be scribed in at right angles to the first two.*

**Fig. 5.20 Marking-out procedure when using a datum surface**

BEAM TRAMMELS and TAPE MEASURES are used for striking lines at 90° to each other, and for measuring distances accurately. It is common practice for the craftsman to use a pair of trammel heads or 'trams' and any convenient beam such as a length of wooden batten. Figure 5.22 shows the arrangement of the trammel and the simple means of fine adjustment for accurate marking-out.

Lines making angles of 90°, i.e., lines square with each other, may be set out with the aid of beam trammels or a steel tape, as shown in Figure 5.23(a).

A KNOWLEDGE OF GEOMETRIC CONSTRUCTIONS AND ARITHMETIC IS ESSENTIAL FOR MARKING-OUT.

NOTE: Where possible existing straight edges and square corners on the plate to be marked should be used.

The normal accuracy obtainable when marking-out with DIVIDERS, ODD-LEGS and TRAMMELS is within 0·15 mm of the TRUE DIMENSION.

Figure 5.23(b) shows how the properties of a right-angled triangle can be used to set out perpendicular lines.

*If the ratio of the sides of a triangle is 3:4:5 then it is a RIGHT-ANGLE TRIANGLE.*

Scribe a datum line on the plate and mark off a length AB representing 4 units of measurement with a steel tape or beam trammels. Centre-punch mark points A and B. With centre A and the trammels set at a distance equal to 3 units, scribe an arc. With centre B and the trammels set at a distance equal to 5 units scribe another arc. The two arcs intersect at a point C which is marked with a centre punch. With a chalkline, mark a line passing through the witness marks A and C.

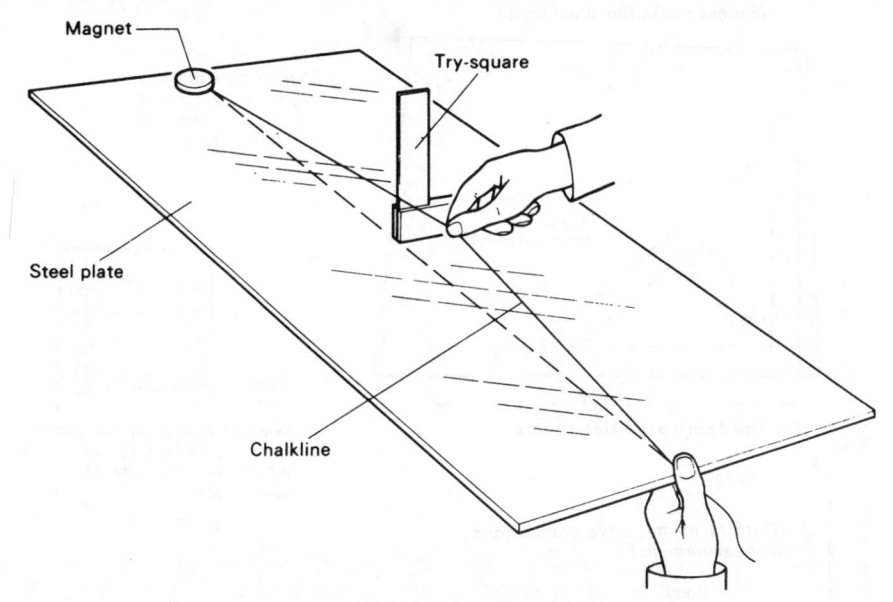

**Fig. 5.21** The use of a chalkline for marking plate

## 5.18 Basic methods used for marking-off large-size plates

For economic reasons advantage should always be taken of as many good and straight edges as possible before commencing marking off large-size plates for cutting. *Unnecessary shearing or flame-cutting can be avoided if the edges of the steel plates are examined before marking-off.*

Any one or a combination of THREE BASIC METHODS may be used for obtaining parallel and squared lines to enable a plate to be cut to the required dimensions:

### METHOD 1 – USE OF SQUARE AND STEEL TAPE

A FLAT SQUARE is used for marking out on large flat surfaces. The flat square differs from an Engineer's try-square in that it is laid on the flat surface of the sheet metal or plate to be marked out. It is larger than the try-square and is made in one piece, consisting of a long arm termed the 'body' and a short arm termed the 'tongue'. The body and tongue are of uniform thickness and form a 90° angle, as illustrated in Figure 5.24. In many fabrication workshops, use is made of a simple made-up square of either wood or light gauge steel.

A suitable steel tape is used in conjunction with the flat square.

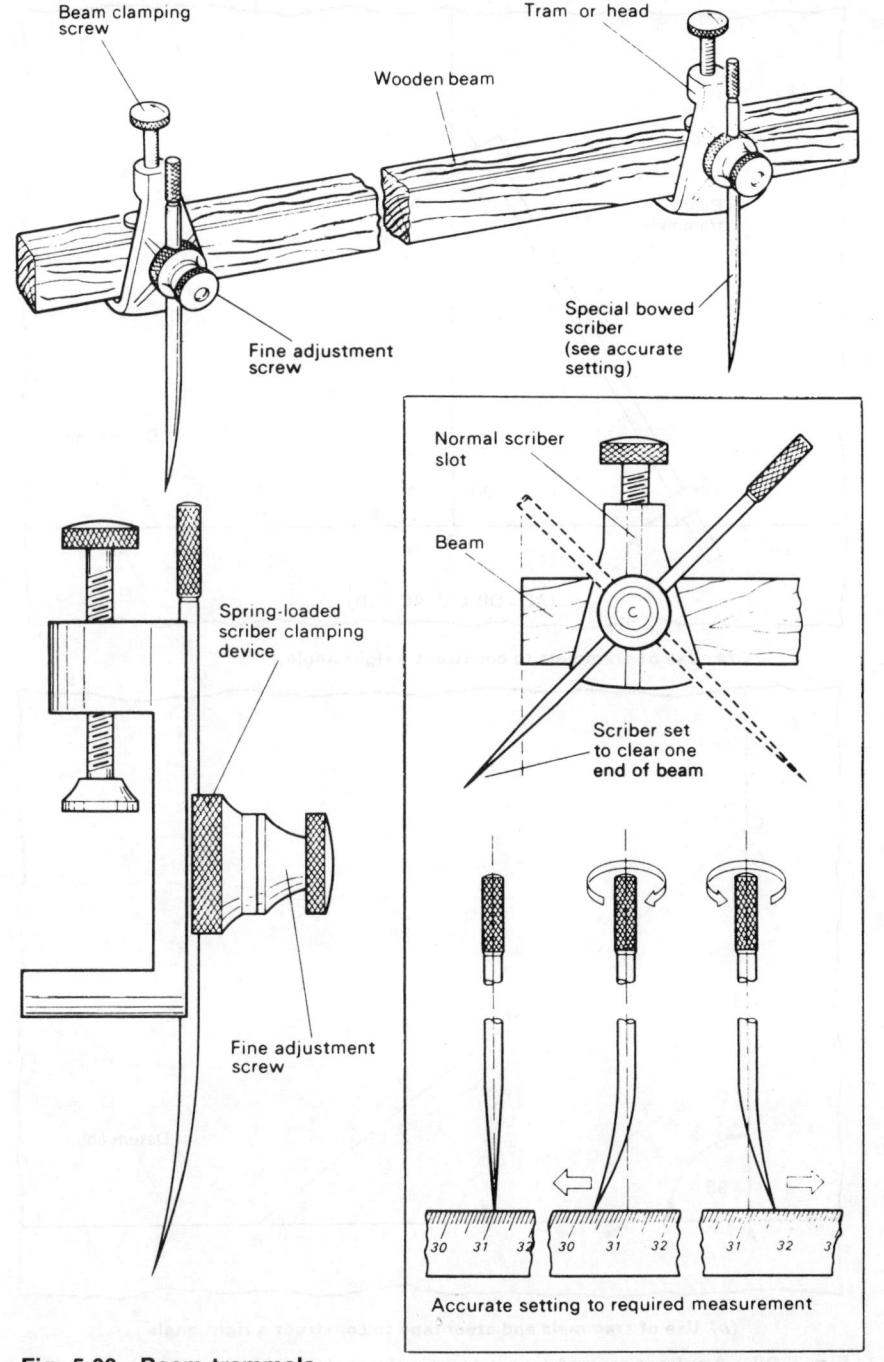

**Fig. 5.22** Beam trammels

151

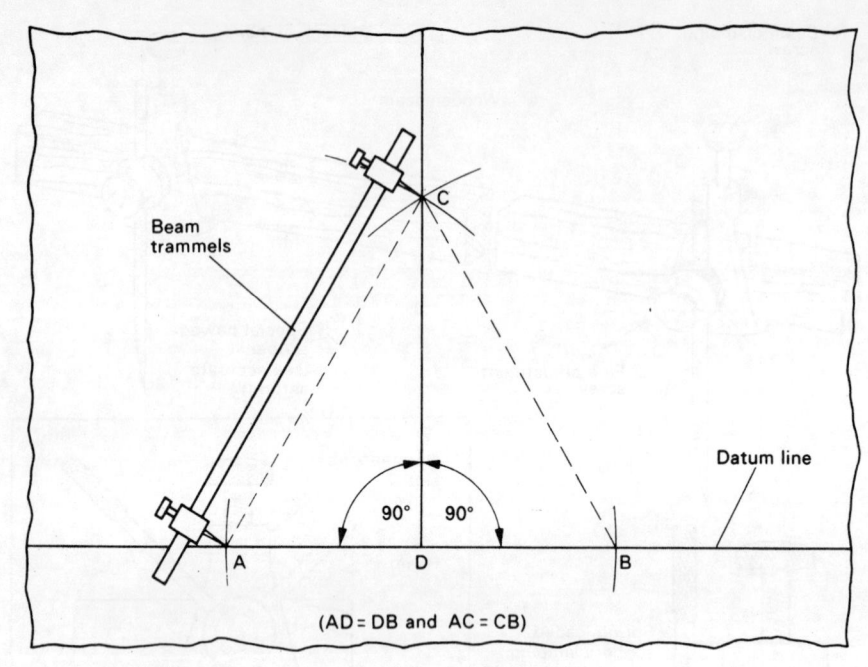

## (a) Use of trammels to construct a right angle

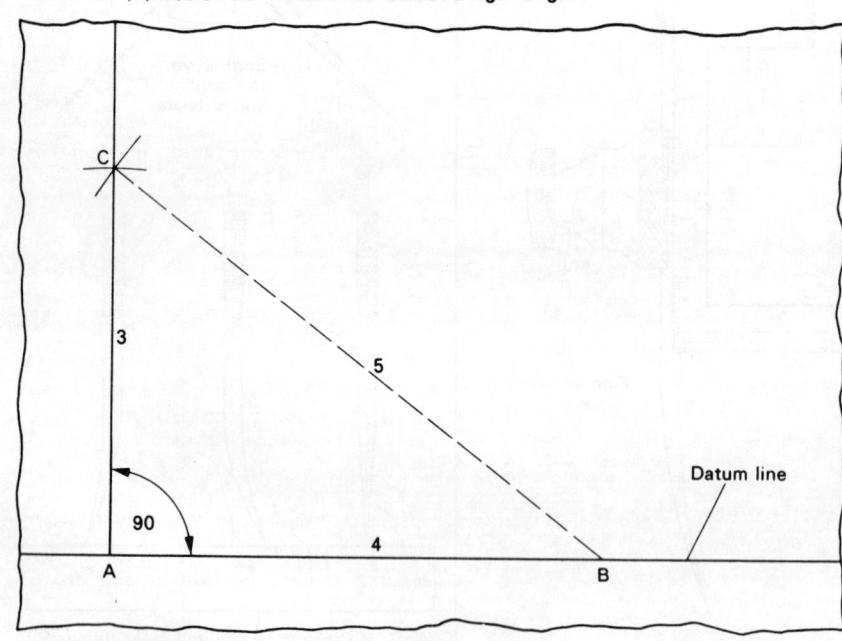

(AD = DB and AC = CB)

## (b) Use of trammels and steel tape to construct a right angle

**Fig. 5.23** Applications of beam trammels and steel tape (marking-out)

## (a) The fabricated flat square

## (b) The steel flat square (one piece)

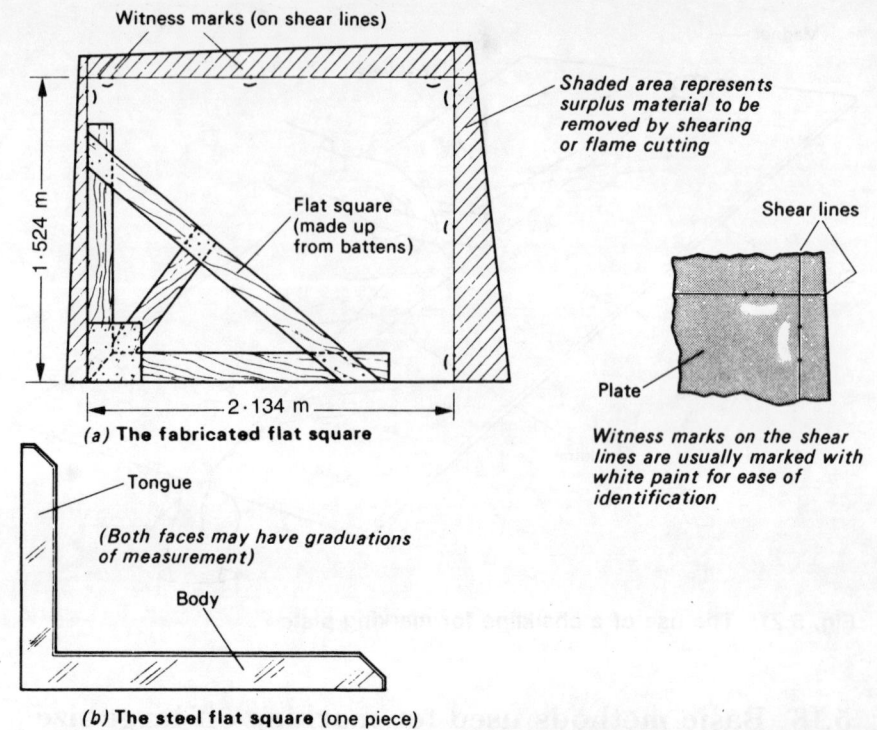

**Fig. 5.24** The flat square

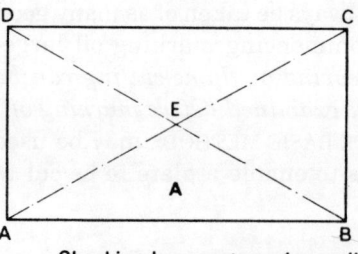

Witness marks on the shear lines are usually marked with white paint for ease of identification

## Checking long rectangular outlines

TRUE SQUARENESS of marking may be easily *checked by measuring diagonal corner distances*

**A** — In an OBLONG or RECTANGULAR figure, the diagonals bisect each other such that
A-E = C-E = B-E = D-E.
The diagonals A-C and B-D are equal, therefore the angles in a rectangle, or square are each 90°

**B** — In a PARALLELOGRAM the diagonals bisect each other such that:
A-E = E-C and B-E = E-D
The diagonals A-C and B-D are NOT EQUAL therefore none of the angles are 90°

**This principle is used to check large rectangular outlines for squareness when marking out**

**Fig. 5.25** Checking large rectangular outlines

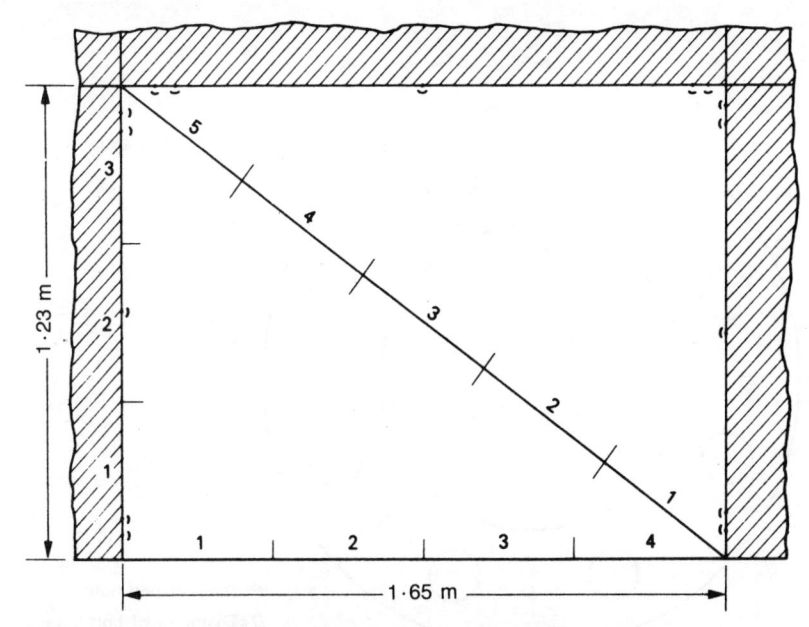

**Fig. 5.26** Use of the steel tape for marking-off

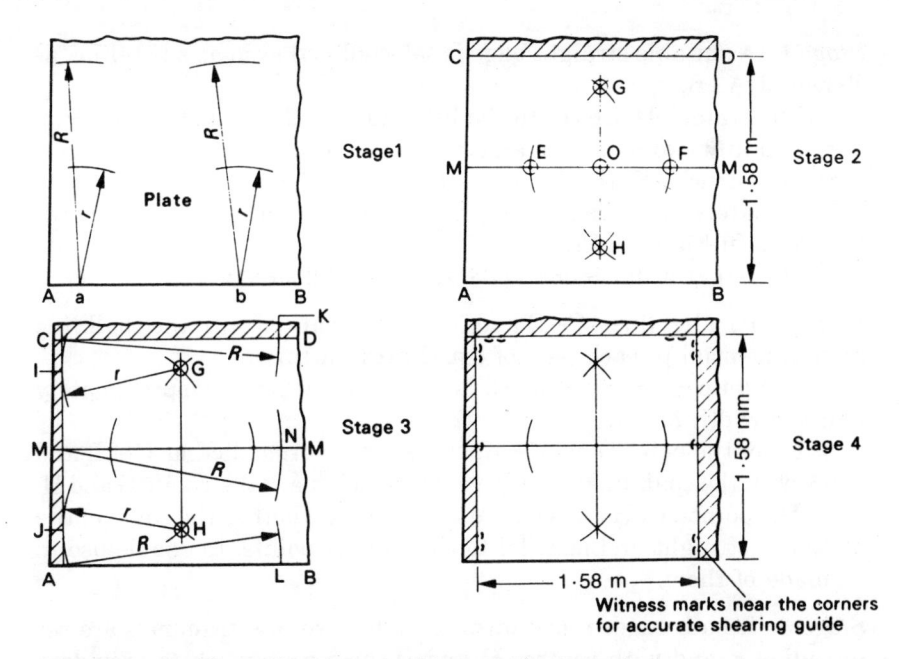

**Fig. 5.27** Marking-off with a steel tape and trammels

Select one straight edge on the plate, and with the aid of a flat square and a stick of French chalk, mark a line at 90° to this datum. Extend this line using a chalkline. From these two datums the required dimensions are marked off with French chalk. A steel tape is used for measuring all dimensions. The shear lines are completed with the aid of a chalkline, and witness marks are made on them with a centre punch.

*Before commencing to mark out a large plate*:
1. Always check for squareness
2. Where possible, select one straight edge and use as a base datum. Figure 5.24 shows how a square and steel tape are used for marking-off steel plate for cutting. Figure 5.25 shows how squareness may be checked.

## METHOD 2 — USE OF STEEL TAPE

A plate of any size may be marked-off with square corners by measuring with a steel tape, units of length in the proportion of 3, 4 and 5 to produce the datum lines at right angles to each other.

Figure 5.26 illustrates the use of a steel tape for marking-off a plate to measure 1·65 m by 1·23 m.

Select one straight edge on the plate for straightness and use as a baseline, otherwise mark a datum line with the aid of a chalkline.

In this example the plate is required to be marked out 1·23 m by 1·65 m, using a steel tape only.

The method employed has been explained in Figure 5.24.

In this case the most suitable measurement to be used for the 3:4:5 ratio of the sides of a 90° triangle will be 410 mm, giving the following dimensions to be used for the steel tape:

1230 mm (3 × 410) : 1640 mm (4 × 410) : 2050 mm (5 × 410)

Once a line has been constructed at 90° to the base datum, the dimensions of the sides are measured with the steel tape, the outlines made with a chalkline and witness marked.

*The outline is checked for true squareness as explained in Figure 5.25.*

Arcs may be swung with a steel tape by holding the French chalk in the hook at the zero end of the tape.

## METHOD 3 — USE OF STEEL TAPE AND TRAMMELS

Figure 5.27 illustrates the method of marking-off a steel plate which is required to be 1·58 m × 1·58 m with square corners, using a steel tape and trammels.

*Stage 1* A suitable straight edge is selected and used as a baseline, as shown at A–B.

The trammels are set to the full width of the plate ($R = 1 \cdot 58$ m) and with any two points 'a' and 'b' (on the baseline A–B) as centres, arcs are struck. With the same centres and the trammels set to approximately half this dimension (radius $r$) two other arcs are shown struck as in Figure 5.27.

THE STEEL TAPE IS USED FOR ALL MEASUREMENTS.

*Stage 2* Parallel lines, C–D and M–M are marked with the chalkline held tangential to each pair of equal arcs, in turn.

A light centre punch mark is made at O which is approximately half the width M–M.

From the point O on M–M, construct a perpendicular G–H, and mark with the chalkline. Lightly centre-punch mark the points G and H.

The points, G, H and O are used to check whether the edges of the plate are straight and parallel to this line of points, to enable use to be made of them.

*Stage 3* If both edges prove unsuitable for use, the trammels are set to radius $r$, and with centres G and H, arcs are struck to provide a suitable shearing margin at points I and J.

The end shear line is made with the chalkline held at a tangent to these arcs.

The plate edge measurements for the length of the plate are made from this line (through I and J). The trammels are set to $R = 1 \cdot 58$ m, and a chalkline is made at a tangent to the arcs at points K, N and L, as shown in Figure 5.27.

*Stage 4* The shear lines are witness marked with a centre punch, and white paint marks are made near them.

The finished outline is checked for SQUARENESS by measuring the diagonal lengths.

## 5.19 Method of marking-out bolt holes for flanges

Many fabrications such as boilers, chemical plant and pressure vessels incorporate the use of flanged inlets and outlets. Manholes and inspection covers are also bolted to flanges. Pipes of various diameters are connected by means of flanges.

The flanges are welded to the fabrications and on the ends of pipes, and the connections are made by bolting. Figure 5.28 illustrates the method of laying out the bolt holes on flanges. In practice the standard size of the required flange, the PITCH CIRCLE DIAMETER and the number and size of the bolts is specified in BS 1560.

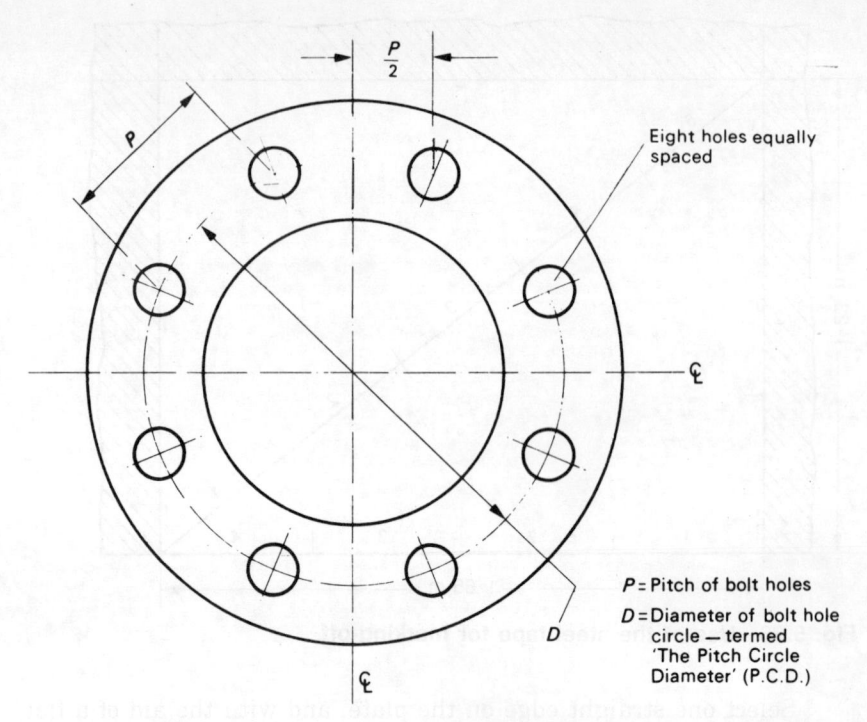

**Fig. 5.28 Marking-out bolt holes for flanges**

The centre of the flange is plugged with a suitable piece of wood or piece of flat bar, which is 'tack-welded' in position, to enable the centre of the flange to be located. On flanges up to about 460 mm in diameter horizontal and vertical centre lines (℄) may be marked with the aid of a height gauge in conjunction with an angle plate on a marking-out table, and a pair of trammels. On very large diameter flanges, use is made of a large centre-square to locate the centre for the bolt hole circle.

*For any specific size of flange for a particular class of work, details such as the BOLT CIRCLE DIAMETER, NUMBER OF BOLT HOLES and DIAMETER OF BOLTS are obtained form the appropriate standard table (see Tables 5.2 and 5.3).*

Having located the centre and marked the horizontal and vertical centre lines, the appropriate bolt circle is marked by means of trammels.

The pitch is constant and is usually obtained from tables of 'bolt hole locations' which *provide a constant that has to be multiplied by the diameter of the bolt hole circle to obtain the required pitch.*

To obtain the position of the first hole, divide the pitch by 2, set the dividers to this dimension and mark off from the intersection of the vertical centre line and the bolt circle, and centre punch.

**Table 5.2 Data for marking-out pipe flanges**

| FLANGE SIZE DESIGNATION (Nominal bore of pipe) (mm) | APPROXIMATE O.D. OF STEEL PIPE (mm) | DIAMETER OF FLANGE (mm) | BOLT CIRCLE DIAMETER (mm) | NUMBER OF BOLTS | DIAMETER OF BOLT (mm) |
|---|---|---|---|---|---|
| 152 | 165 | 279 | 235 | 4 | 16 |
| 254 | 273 | 406 | 356 | 8 | 19 |
| 305 | 324 | 457 | 406 | 8 | 22 |
| 432 | 457 | 610 | 523 | 12 | 25 |
| 584 | 610 | 787 | 724 | 12 | 25 |
| 686 | 711 | 870 | 857 | 16 | 25 |
| 889 | 914 | 1073 | 1016 | 24 | 25 |
| 1067 | 1092 | 1251 | 1194 | 28 | 25 |
| 1524 | 1549 | 1784 | 1702 | 32 | 25 |
| 1829 | 1861 | 2108 | 2019 | 36 | 35 |

The above data is a selection of recommended flange sizes for steel pipes from BS 1560 Class 150 to withstand working pressures up to 1034 kN/m$^2$ (10·5 kgf/cm$^2$).
Bolt-holes in flanges are marked out in accordance with Standard Tables, and the holes drilled to suit the correct diameter of bolt.

The remainder of the bolt hole centres may now be located with the dividers set at the correct pitch (CHORD LENGTH) and centre-punched in readiness for drilling to required size.

The procedure for drilling bolt holes in very large flanges is outlined in Chapter 9.

## 5.20 The bevel

A useful marking-off tool is the BEVEL which is frequently used in fabrication work for the marking-off of angles or mitres on steel sections. This simple tool is illustrated in Figure 5.29(*a*).

It consists of a blade and base which are set in position by a locking screw.

The bevel is set to the required angle and is used for checking, transferring and marking-out angles.

## 5.21 The pipe square

The PIPE SQUARE is used to check the correct alignment of pipe flanges on assembly, and is illustrated in Figure 5.29(*b*).

Care must be taken when using the pipe square for assembly work to ensure that it does not come into contact with the arc-welding electrode or the earthing cable clamp.

**Table 5.3 Constants for bolt hole location (flanges)**

| NUMBER OF BOLT HOLES | CONSTANT (to be multiplied) by Bolt Circle Diameter or P.C.D.) |
|---|---|
| 4 | 0·4071 |
| 8 | 0·3827 |
| 12 | 0·2588 |
| 16 | 0·1951 |
| 20 | 0·1564 |
| 24 | 0·1305 |
| 28 | 0·1120 |
| 32 | 0·0980 |
| 36 | 0·0872 |

Use of tables: Example
To mark out a 305 mm Class 150 Standard Flange.
From Table 5.2 the P.C.D. = 406 mm and the number of bolts = 8.
(The diameter of the flange = 457 mm)
The pitch (chord length of bolt holes centres) = the 'P.C.D.' multiplied by the 'constant' from Table 5.3.
The 'constant' for 8 holes = 0·3827
Therefore, 'pitch' = 406 × 0·3827
P = 155·76 mm
For any number of holes not shown in the table, multiply the SINE of HALF THE SUBTENDED ANGLE between a pair of holes by the 'P.C.D.'

The subtended angle = $\dfrac{360}{\text{Number of holes}}$

Example:
Number of holes = 15

Subtended Angle = $\dfrac{360}{15}$ = 24°

The constant = Sine 12° = 0·2079 (for 15 holes)

## 5.22 The scratch gauge

This useful tool is similar in principle to the carpenter's tool used for marking parallel lines to a given plane surface.

Such a gauge is usually made of steel with a hardened high carbon steel head. The head has a split bushing through it, against which a set screw acts to hold it firm. The steel beam is usually graduated in millimetres. The small scribing point consists of a thin square piece of suitably hardened and tempered steel, which is held firmly against one end of the beam. The scratch gauge is illustrated in Figure 5.29(*c*), together with a simple but effective scratch gauge which can readily be produced from a small piece of scrap sheet metal.

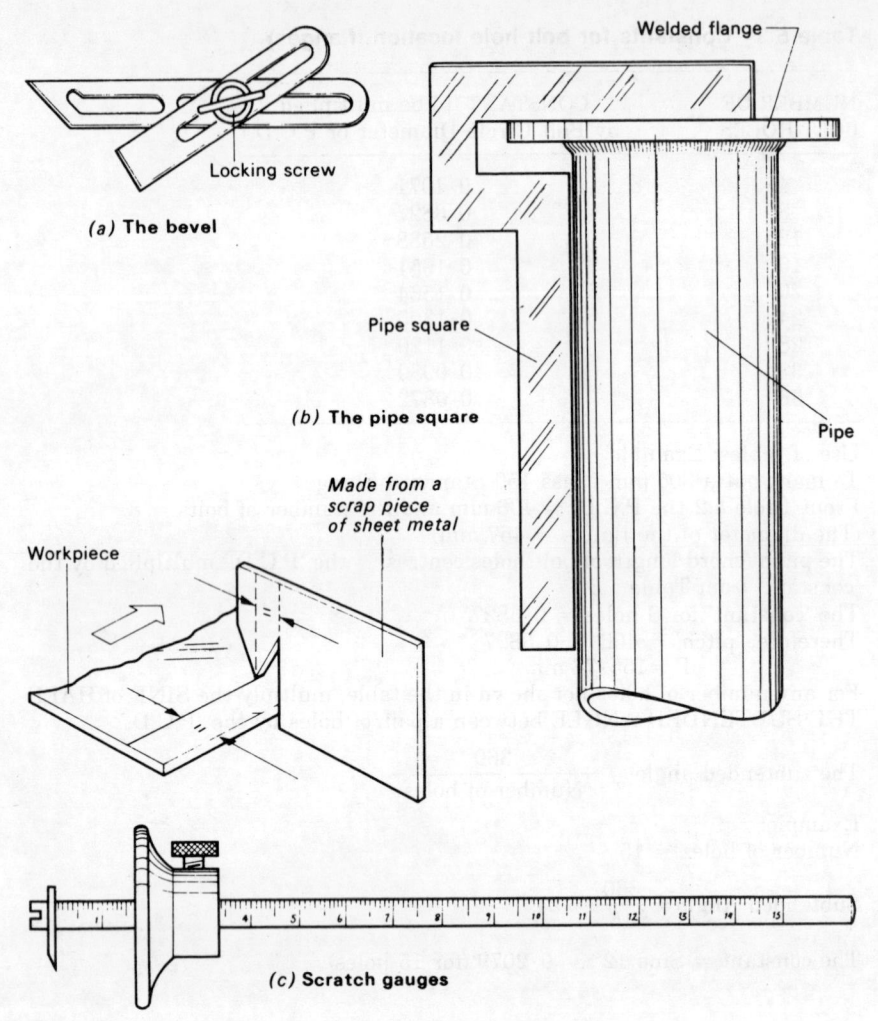

**Fig. 5.29** Useful marking-out and measuring tools

(a) The bevel — Locking screw
(b) The pipe square — Welded flange, Pipe square, Pipe, Made from a scrap piece of sheet metal, Workpiece
(c) Scratch gauges

## 5.23 Centre punches

These tools are made from high carbon steel, usually of hexagonal or octagonal section for easy grip, although a cylindrical section with a knurled grip is also common.

The centre punch is used for making circular indentations in the surface of metal for location, marking, drilling and general machining purposes. If it is intended to be used on soft metals, the tapered point is generally of smaller dimensions than when the punch is designed for use on tough steels. According to the purpose for which it is to be used, the diameter at the base of the conical point ranges from about 3 to 6·35 mm.

Centre punches for marking centres of holes to be drilled are made from 9·5 mm octagonal or cylindrical high carbon steel from about 127 mm long, and the tapered point is ground to an included angle of 90°. *The indentation the punch makes in the surface of the metal will not only locate the drill point, but will prevent the drill from wandering.*

Centre punches used for marking positions of lines and centres of circles and arcs to be drawn by dividers or trammels, are the smaller type of punch, made from about 6·35 mm diameter high carbon steel with a sharper point, usually about 60°.

'Dot' or 'nipple' punches are especially useful for repetition work when a TEMPLATE is used. The template has accurately positioned holes, all of which are of the same small diameter. The 'dot' or 'nipple' punch is designed with a parallel point, the end of which has a 'pip' ground on it — such punches are accurately machined, concentric to the parallel point, on a lathe.

The parallel point is made in a number of precise diameters to locate accurately in the template holes which are pre-determined by the diameter of the punch to be used. Thus by locating the punch in the holes in the template, the hole centres on each workpiece can be easily marked out precisely and identically.

In a similar manner, the indentations made by a 'dot' or 'nipple' punch can be used to great advantage to locate any shape of punch used for piercing sheet metal, provided it has a centre pip.

Figure 5.30 gives details of centre punches commonly used in the fabrication industry.

## 5.24 The use of plumb-lines

A typical application of the use of the plumb-line in steel fabrication work is shown in Figure 5.31.

The bottom vessel is securely bolted into the correct position and the top vessel is aligned to it, as shown, with the aid of a plumb-line before being bolted into the required position to receive the connection pipe.

## 5.25 The use of the tensioned wire

On large fabricated components, a tensioned wire may be used to check straightness and alignment. Piano wire or stainless steel wire of about 0·55 mm diameter is used for this purpose, and when not in use should be kept on a suitable reel.

Knurled grip

(a) The centre punch

Steel plate (to be flame cut)

60°  90°

'One-off' parts may be cut by hand with an oxygen-cutting torch after scribing the required profile and centre-punching it with 'witness marks'

Hammer blow

Template

Nipple punch

Workpiece

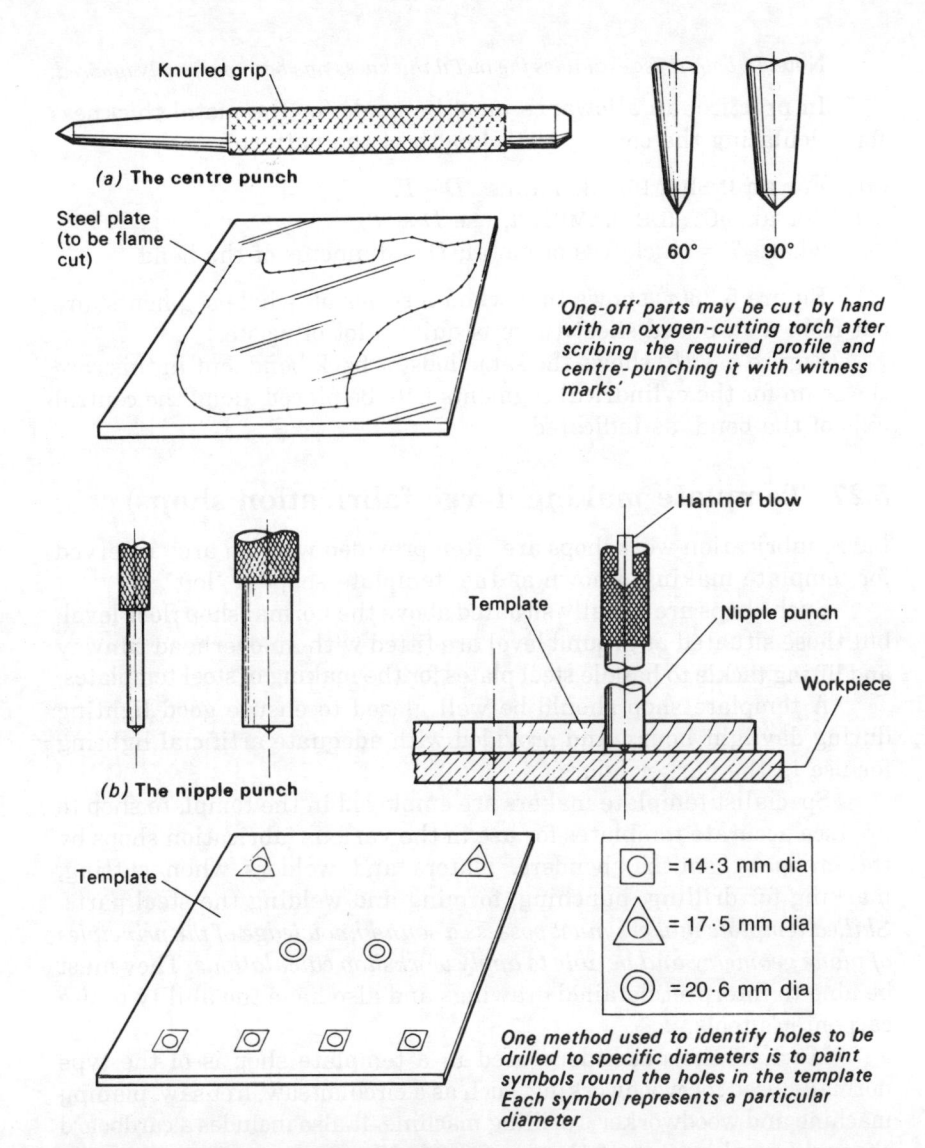

(b) The nipple punch

Template

○ = 14·3 mm dia
△ = 17·5 mm dia
◎ = 20·6 mm dia

One method used to identify holes to be drilled to specific diameters is to paint symbols round the holes in the template. Each symbol represents a particular diameter

**Fig. 5.30** Centre punches (applications)

Timber support for hanging the plumb-line

Pipe flanges

Connecting pipe

Top vessel

Top vessel support base

Bottom vessel

Plumb line

Bottom vessel support base

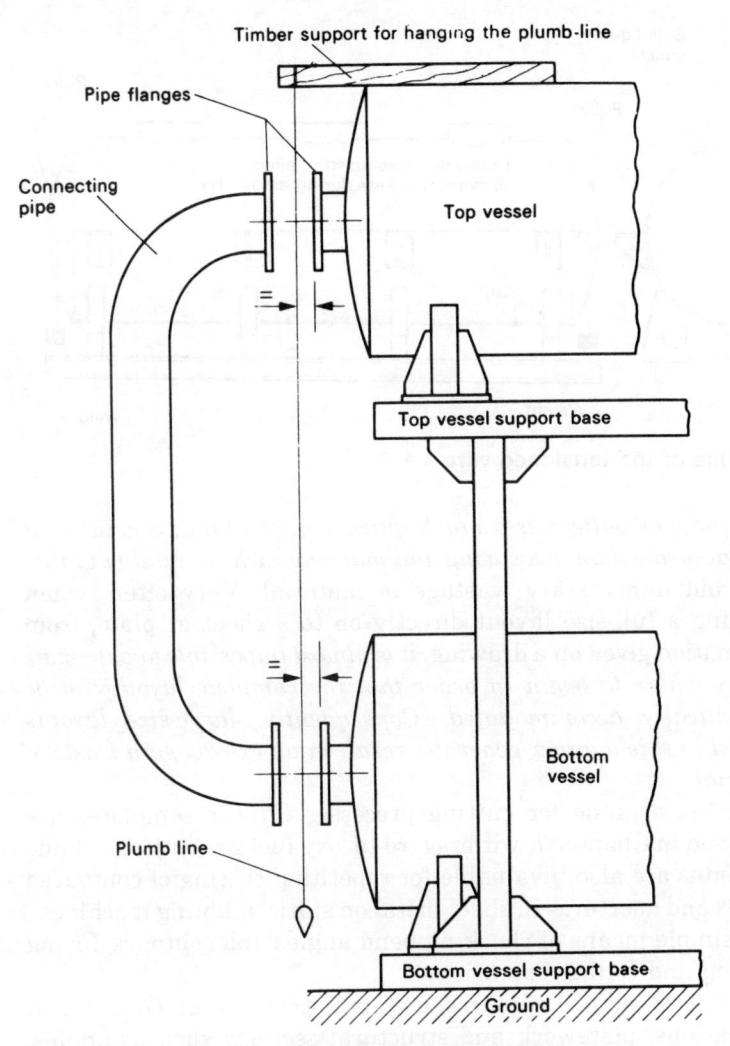

Ground

**Fig. 5.31** The use of a plumb-line (fabrication)

## 5.26 The need for templates

There are several reasons for the use of templates or patterns in the sheet metal and plate fabrication industries. For example:

1. To avoid repetitive measuring and marking-off of the same dimensions, where a number of identical parts or articles are required. *Marking-off large numbers of exactly the same type from*

When in use for measuring or checking, both ends of the wire are hung over supports which are rounded, such as round bar section or pulleys, and weighted sufficiently to keep the wire in TENSION. Alternatively the wire may be secured by means of adjustable clamping devices. Figure 5.32 illustrates the use of a tensioned wire.

157

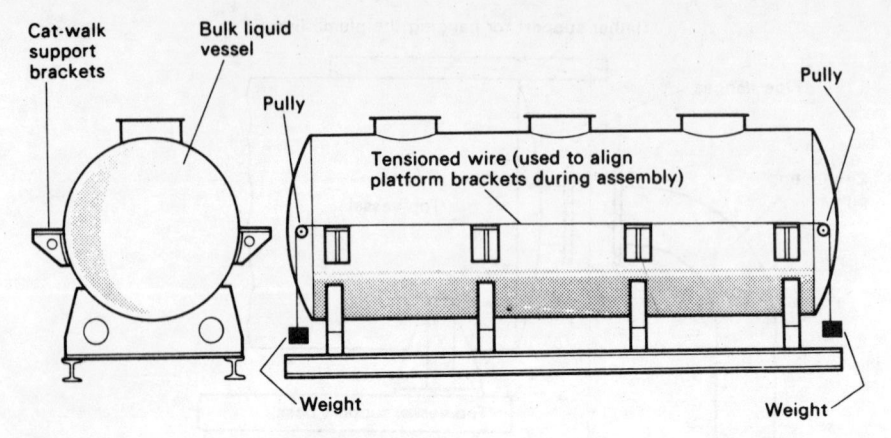

**Fig. 5.32** Use of the tensioned wire

*a template or pattern is a much quicker method and a great deal more accurate than measuring and marking each part individually.*

2. To avoid unnecessary wastage of material. Very often, when marking a full-size layout directly on to a sheet or plate, from information given on a drawing, *it is almost impossible to anticipate exactly where to begin in order that the complete layout can be economically accommodated. Consequently, large-size layouts tackled in this manner generally result in an extravagant waste of material.*

3. To act as a guide for cutting processes. Profile templates are discussed in Chapter 7, with regard to 'oxy-fuel gas cutting'. Guide templates are also invaluable for repetition cutting of contoured shapes and apertures in sheet metal on static 'nibbling machines'.

4. As a simple means of checking bend angles and contours during forming and rolling operations.

5. As a precise method of marking-off hole positions on sheet metal fabrications, platework and structural sections such as angles, channels, columns and beams, gusset plates and angle cleats.

Figure 5.33 illustrates the economical arrangement of patterns on a standard dimensioned sheet of metal.

The example shown in Figure 5.33 is a 'LOBSTER BACK BEND' consisting of 5 segments and 2 half-segments.

It is to be made from 1·6 mm-thick mild steel, and all joints are to be butt welded. The seam for the cylindrical segments is to be placed in the throat, as indicated.

It is general practice to make the throat radius $R$ equal to the diameter $D$ of the bend.

Note: *In both these examples the metal thickness has been purposely ignored.*

In practice, an allowance must be made for the metal thickness for calculating the correct girth length of the pattern:

(a) For an INSIDE DIAMETER, use $D - T$,
(b) For an OUTSIDE DIAMETER, use $D + T$,
where $T$ = thickness of metal; $D$ = diameter of the bend.

Figure 5.33(a) shows that when a series of 'whole segments' are cut from a standard sheet there is quite a lot of waste.

Figure 5.33(b) shows the same lobster back bend, but in this case the seam for the cylindrical segments is to be placed along the central axis of the bend, as indicated.

## 5.27 Template making (large fabrication shops)

Large fabrication workshops are often provided with an area reserved for template making, known as the 'template shop' or 'loft'.

Such shops are usually situated above the normal shop floor level, but those situated at ground level are fitted with an overhead runway and lifting tackle to handle steel plates for the making of steel templates.

A template shop should be well glazed to ensure good lighting during daylight hours, and provided with adequate artificial lighting for use in the darker hours.

Specialist template makers are employed in the template shop to produce accurate templates for use in the various fabrication shops by the croppers, smiths, benders, platers and welders when cutting, marking for drilling, punching, forming and welding the steel parts. *Skilled template makers must possess a sound knowledge of the principles of plane geometry and be able to apply workshop calculations.* They must be able to interpret detailed drawings and also have the ability to use carpenter's tools.

Much of the machinery used in a template shop is of the type normally used for woodworking, such as a circular saw, fret-saw, planing machine and woodworker's drilling machine. It also includes a cardboard shearing machine to cut the special template paper.

## 5.28 The setting-out floor

It is essential that the floor used for full-size laying-out consists of floor boarding placed diagonally across the floor joists. *If the floor boards were laid in the conventional manner (lengthways) or square across the shop, the joints between the boards (which tend to shrink) would offer a serious handicap, as most lines are marked on the floor in these directions.* The

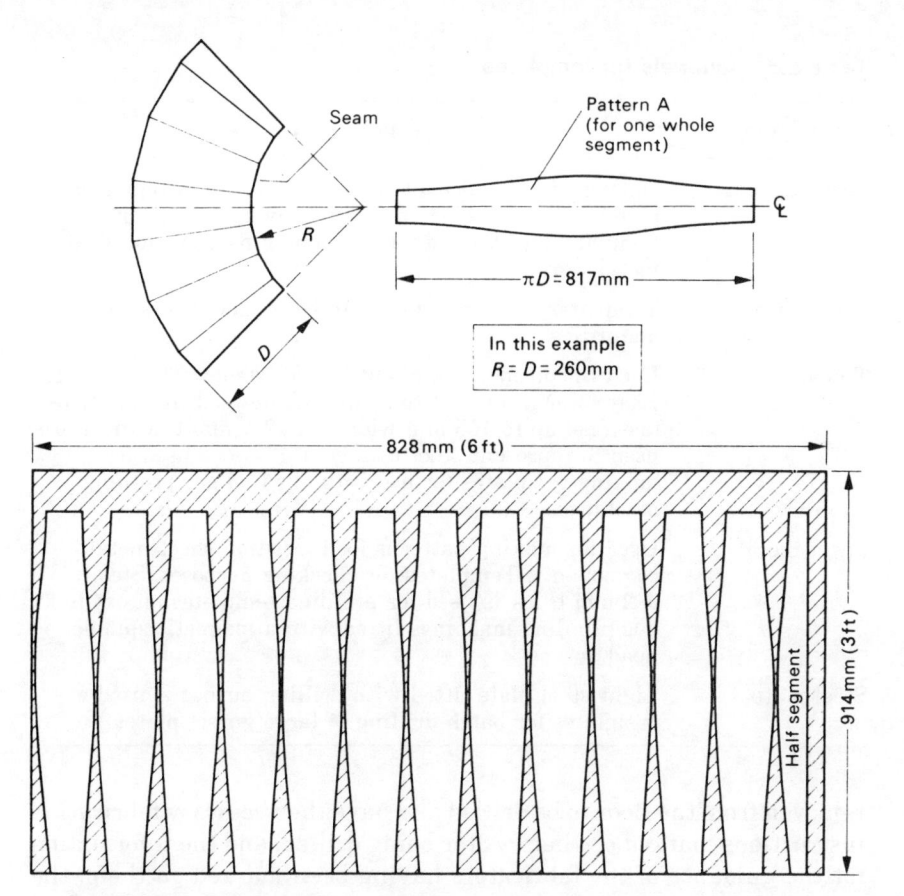

**In this example**
$R = D = 260$ mm

$\pi D = 817$ mm

828 mm (6 ft)

914 mm (3 ft)

Using Pattern A, The maximum of segments which can be cut from a standard sheet is 12½ as shown in the arrangement above.
(The shaded portion represents the amount of wasted material)

*(a) Use of whole segment*

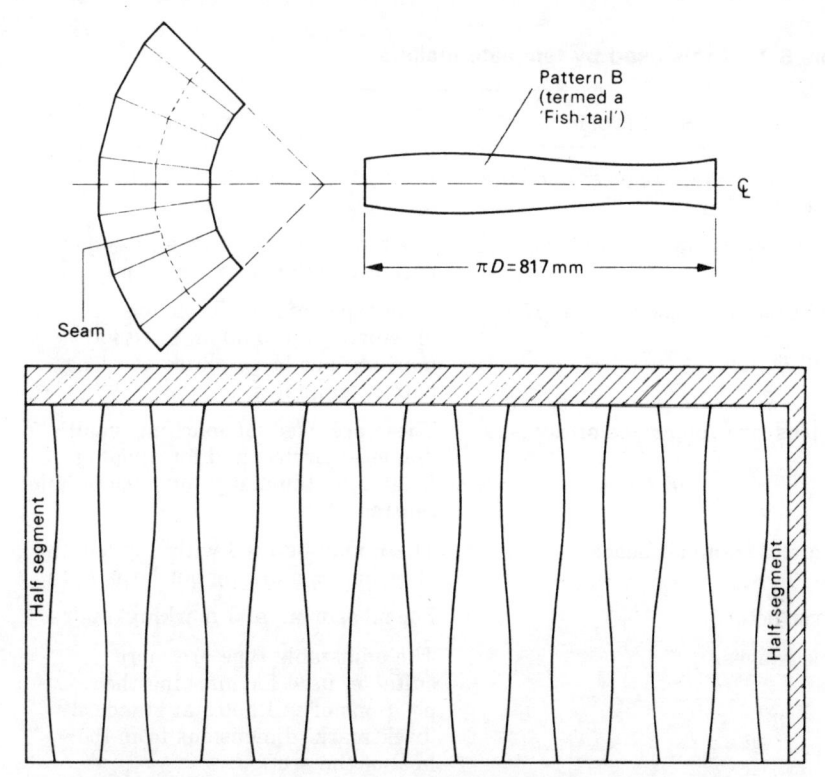

$\pi D = 817$ mm

Using Pattern B, the maximum of segments which can be cut from the same standard sheet is 15, as shown in the arrangement above.
When lobster back bends are to be produced in quantity, the adoption of a 'Fish-Tail pattern' will result in a greater economy in the use of material than the pattern A'
  A study of the two methods of marking the segments will clearly indicate that the use of a Fish-tail pattern will reduce the shearing operations by approximately 50%

*(b) Use of 'fish-tail' segments*

**Fig. 5.33 The economic arrangement of patterns**

joints between the boards may easily be mistaken for lines, or some portion of a line may coincide with an open joint. *Such problems are eliminated when the boards are laid diagonally.*

  The floor is given a coat of 'lamp black' and 'size' to ensure that the lines (made with a 'chalkline') can be clearly seen. Working from scale drawings, the template maker marks out full-size sets of steelwork on this black surface. The laying out of the drawing full size on the template shop floor is called 'lofting'.

## 5.29 Basic tools used by the template maker

The basic tools used by the template maker are listed in Table 5.4.

## 5.30 Materials used for templates

Table 5.5 gives details of the materials used for templates together with some of their applications.

  In the making of templates considerable use is made of timber: it is easy to drill and cut to shape, relatively light in weight, and fabrication instructions can be pencilled on it.

  Suitable wooden battens of various convenient widths and usually 10 mm or 12 mm thickness are cut to represent the steel members outlined on the template floor. These battens are then laid on the appropriate lines on the floor together with the paper or hardboard

**Table 5.4  Tools used by template makers**

| TOOLS OR ITEMS | REMARKS |
| --- | --- |
| Carpenter's saws, planes, hand-brace and bits. | For making wooden templates. |
| Joiner's marking gauge. | Used for scribing scrieve lines on batten templates for steelwork. |
| Steel tape to measure about 15 m. | Steel tapes are available for measuring up to 50 m. Used for marking out large plates and long batten templates. |
| Various size compasses or dividers. | These are used for marking small-diameter circles, and for dividing lengths on templates for pitch of hole centres. |
| A pair of trammel heads. | These may be used with any length of beam for marking out large radii. |
| A protractor. | For measuring and marking angles. |
| Back gauges. | The adjustable type are more suitable, used for marking the positions of tail holes at standard 'back mark' dimensions from the heel of the section. |
| Engineer's squares and flat squares. | For checking the squareness of two planes or marking a line square to another. |
| A steel straight edge. | For marking straight lines up to 2·5 m in length. |
| Hammers, centre and nipple punches. | For marking hole centres, and making witness marks. |
| Chalk line and soft chalk. French chalk. | For marking long straight lines. French chalk is generally supplied in sticks about 10 mm square and 100 mm long. |
| Coloured and indelible pencils. Crayons. | Used for marking instructions and information on templates. |

patterns representing gusset plates and cleat angle connections. All are temporarily nailed to the floor in their exact positions to represent the particular steel structure.

The centres of holes required for making the connections to be bolted or riveted are marked on the assembled templates, which are then

**Table 5.5  Materials for templates**

| MATERIAL | APPLICATIONS |
| --- | --- |
| Template paper | Outlines for small bent shapes, such as brackets, small pipe bends and bevelled cleats, may be set out on template paper. Used for developing patterns for sheet metal work. |
| Hardboard | Templates for gusset plates to be produced in small quantities. |
| Timber | Used in considerable quantities for steel-work templates. Easy to drill and cut to shape. Whitewood timber strips (battens) up to 153 mm wide and 12·7 mm thickness are used to represent steel members. Plywood used for making templates for use with oxy-fuel gas profiling machines. |
| Sheet metal | Used for making patterns for repetition sheet metal components. Templates for checking purposes. Steel, 3·2 mm thick is used for profiling templates on oxy-fuel gas profiling machines fitted with a magnetic spindle head. |
| Steel plate | Light steel plate fitted with drilling bushes is used as templates for batch drilling of large gusset plates. |

removed from the floor to be drilled and have the necessary fabrication instructions marked on them. After being drilled, and the information for the guidance of the fabricators having been marked on them, the whole assembly is replaced in the correct position on the template floor and checked for accuracy. They are then carefully stored until required on the workshop floor.

For economic reasons template battens may be used again after they have served their purpose in the fabrication shop. Long templates in which details are concentrated at several points may be cut and the drilled lengths can be re-used, after planing off the written instructions, for shorter templates. Very often a wooden template may be used again after the holes have been plugged and it has been planed.

## 5.31  Information given on templates

Information for the guidance of the various craftsmen employed in the fabrication shop and those employed on assembly work on site is marked on the templates and on the finished parts.

On wooden or hardboard templates the necessary information is best marked with an indelible pencil. Coloured pencils are also used for marking information. On sheet metal templates, for example, which are to be used for the marking of various diameter holes, it is common practice to mark rings, triangles or squares around the holes required to be of the same diameter with a distinguishing colour. On steel templates, whitewash or white paint is often used for marking the information.

Typical information 'written-up' on templates may be as follows:

1. Job or contract number,
2. Size and thickness of the plate,
3. Steel section and length,
4. Quantity required,
5. Bending or folding instructions,
6. 'This side up', 'left hand' or 'right hand',
7. Drilling requirements,
8. Cutting instructions,
9. Assembly reference mark.

## 5.32 The use of templates

Many detailed parts of a structure are so simple that they do not require to be set out on the template floor, instead they are marked out direct from drawings at the bench in the fabrication shop. However, templates are made for these simple details where a number of identical parts are required.

A few selected examples of the use of templates will now be considered in more detail.

## 5.33 Templates as a means of checking

These are usually made of sheet metal or wood, although for some applications template-making paper may be used. Figure 5.34 illustrates examples of the use of templates as a means of checking.

## 5.34 Templates for setting out sheet-metal fabrications

Tinplate or light-gauge sheet steel and template-making paper are the materials most commonly used when making templates or PATTERNS for sheet-metal fabrications. For economy reasons, many patterns are developed half-full-size or to scale from the drawing and then the information contained on them is transferred to full-size dimensions when the craftsman marks it off on the job 'in the flat'. Very often, on precision sheet-metal details, the job is marked off from a scale drawing which provides co-ordinates with precise dimensions marked on them. With many sheet-metal developments it is only necessary to use part patterns which are lined up with DATUM LINES.

Figure 5.35 shows examples of the use of 'Patterns' for marking out sheet metal prior to cutting and forming operations.

Figure 5.35(a) shows a smoke-cowl which is to be made out of 1·2 mm-thick mild steel. The edges of the open ends are to be wired with 3·2 mm-diameter wire. The connection flanges are 12 mm for spot welding, and the side seams are to be 6 mm grooved. The completed assembly is to be hot dipped galvanised.

Basically, this component is a combination of 'Tee'-pieces between cylinders of equal diameter, and only requires a part template which may be made from template paper or light-gauge sheet metal. This is then used to mark-out the contours of the intersection joint lines for the parts 'A', 'B' and 'C' whose developed sizes are marked-out in the flat with the appropriate DATUM LINES.

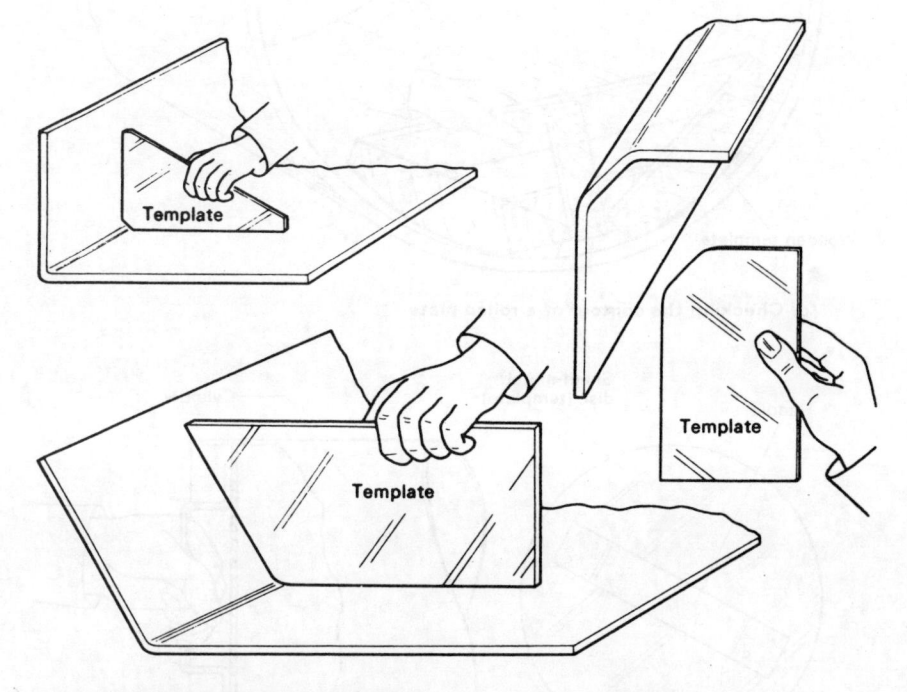

**(a) Checking angles with a template**

It is often necessary to make simple bending templates especially if the sheet or plate material *requires bending in several places to definite angles*
These templates are generally made from sheet metal

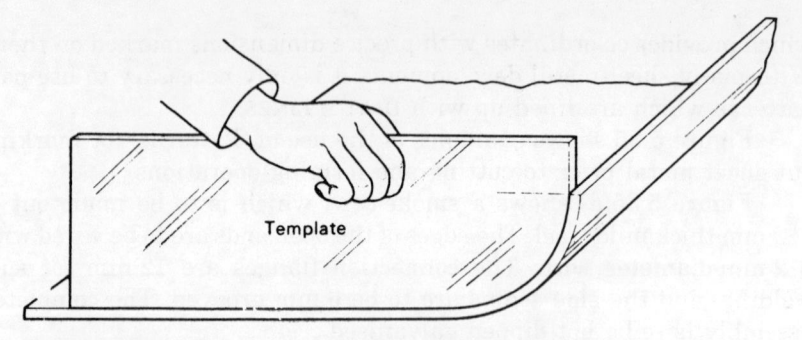

*(b) Checking the contour of a radiused corner*

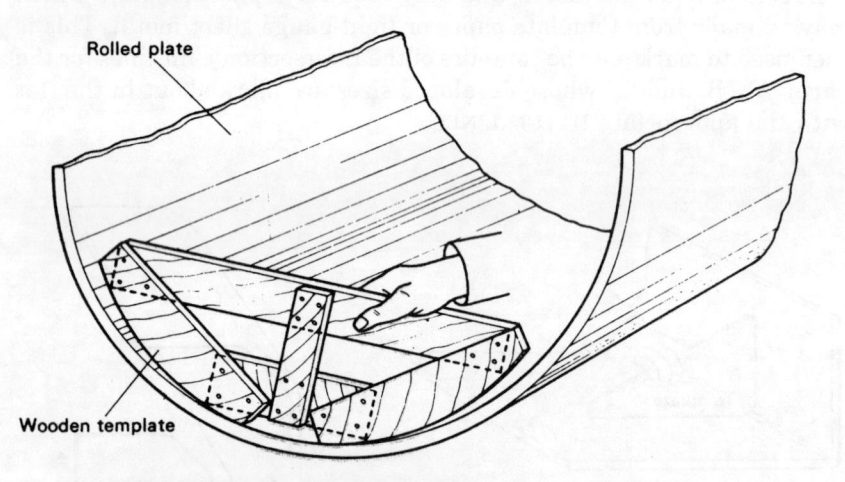

*(c) Checking the contour of a rolled plate*

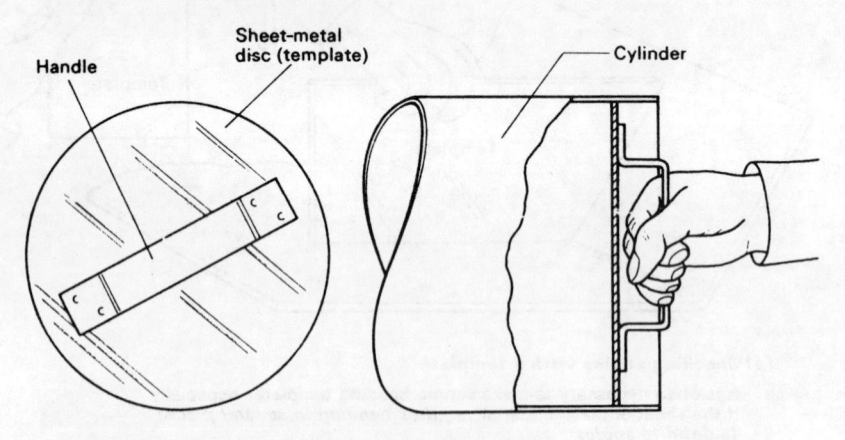

*(d) Template used for checking contour of cylindrical work such as ductwork*

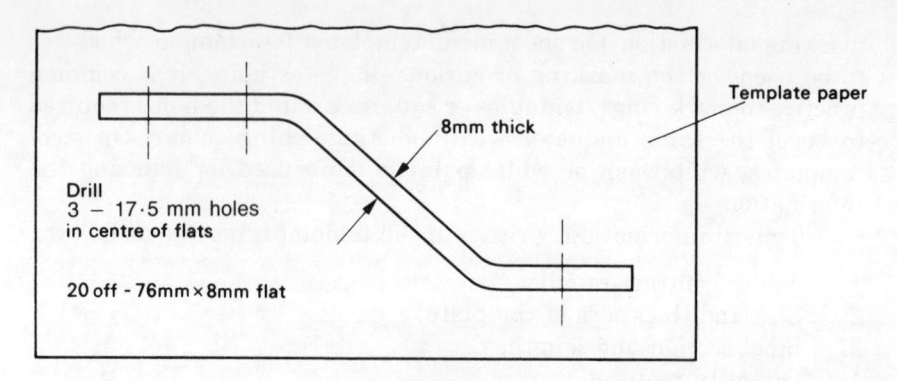

*(e) Use of template paper for checking bent flats*

**Fig. 5.34 The use of templates as a means of checking**

PROCEDURE FOR USING THE TEMPLATE
(See Figures 5.35(b) and 5.35(c))

1. On the developed part, datum lines are marked to represent quarter circumference lines, and the depth of the mitre. IN THIS EXAMPLE THE CONTOUR OF THE MITRE LINE IS IDENTICAL FOR EACH QUARTER BECAUSE THE INTERSECTION IS BETWEEN EQUAL DIAMETERS.

2. Place the mitre line template in alignment with the datum lines, as illustrated, and mark the contour for the mitre for each quarter in turn, reversing the template where necessary. An alternative method of marking the flange allowance is to move the template up the width of the flange and mark the contour, using a scriber.

Figure 5.36 shows a square to round transformer. Figure 5.36(a) shows an ISOMETRIC VIEW of the sheet-metal transforming piece to be used for connecting a circular duct to a square duct of equal cross-sectional area.

In this example, the diameter of the circular duct is 860 mm, the length of one side of the square duct is 762 mm, and the distance between the two ducts is 458 mm. The transformer is to be made from galvanised steel 1·2 mm thick.

Figure 5.36(b) represents a scale development pattern on which are marked the full-size ordinate dimensions. Such drawings are supplied by the drawing office for use by the craftsman for marking-out purposes. Any necessary allowances for seams and joints must be added to the layout (two off required).

(a) Chimney smoke cowl

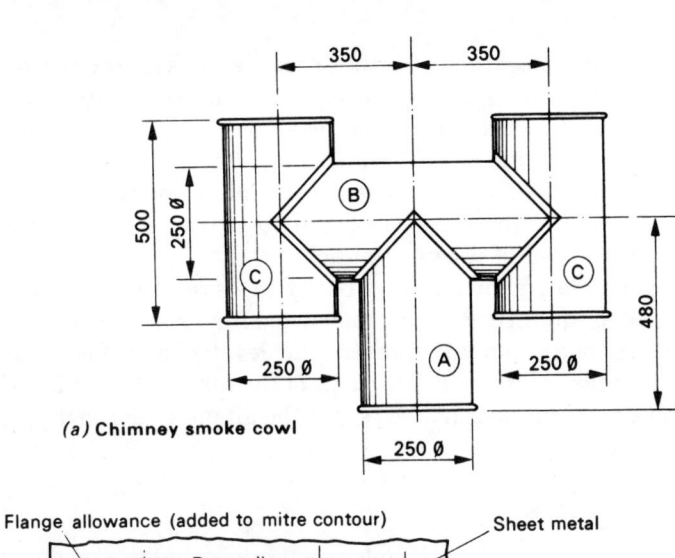

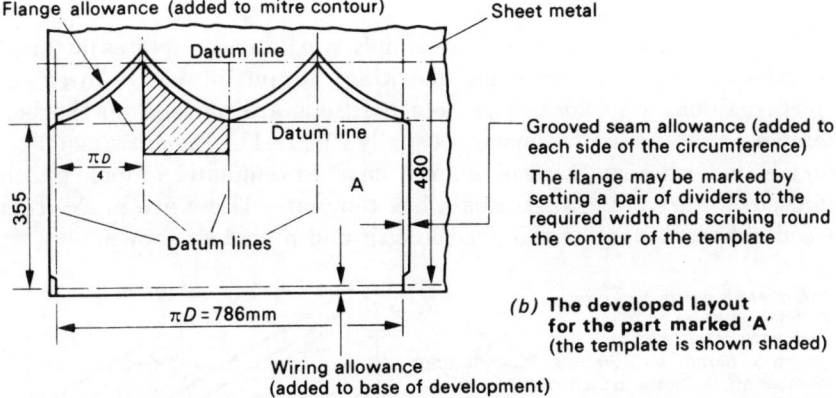

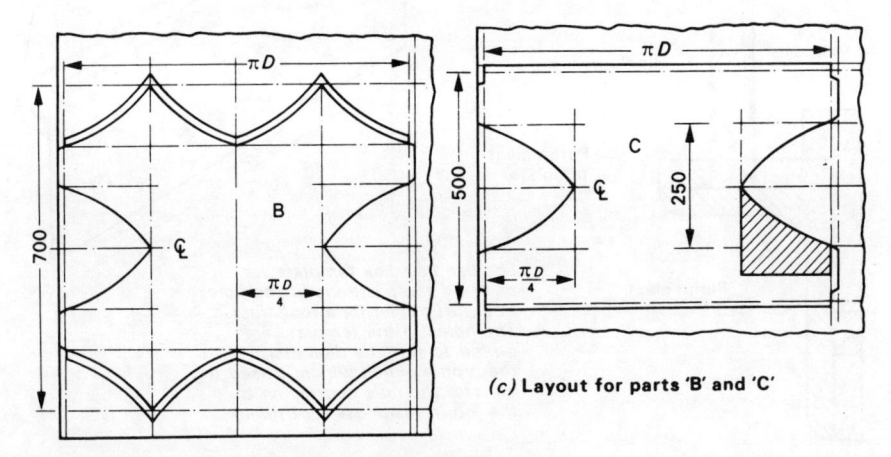

(b) The developed layout for the part marked 'A' (the template is shown shaded)

(c) Layout for parts 'B' and 'C'

**Fig. 5.35 An example of the use of templates**

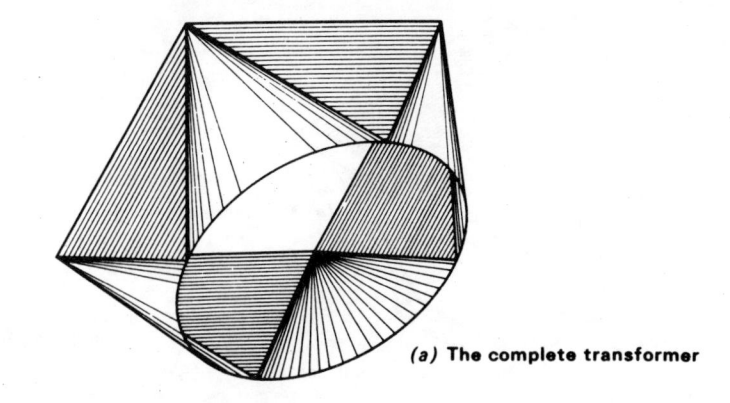

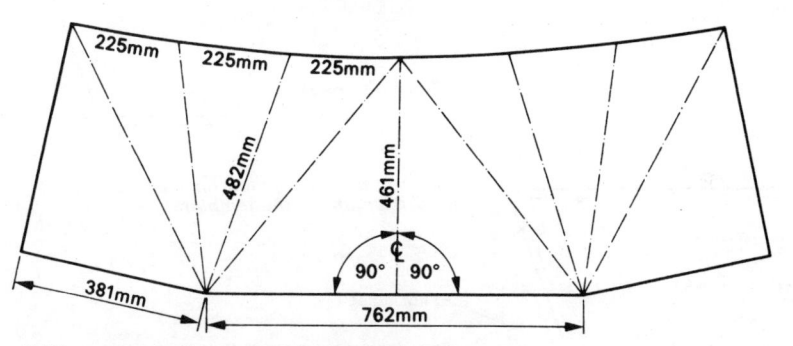

(a) The complete transformer

(b) Pattern or half-template without joint allowances

Fig. 5.36 Square to round transformer

## 5.35 Templates for hopper plates

Large steel hoppers are usually of riveted or welded construction made up of four tapered steel plates. The templates for these hopper plates may be made from wooden battens, sheet metal or template paper. The template is laid on the plate and the outline marked with French chalk and 'witness marks' are centre-punched at suitable positions.

Rivet holes (where applicable) are marked through the template with a nipple punch. When paper templates are used, the holes are not provided in the template, as is the case with wooden and metal templates. The centres of the hole positions are marked on a paper template and may be transferred on to the plate by centre-punching through the template. Figure 5.37 illustrates typical templates for marking off hopper plates.

Figure 5.37(a) shows the basic construction of a SQUARE HOPPER between parallel planes symmetrical to the centre line, made from steel

163

marked out on the paper, together with the exact positions for the centres of the holes. The template is then sheared to size on the guillotine.

Light-gauge sheet metal is ideal material for the template where a quantity of identical hoppers are required (see Figure 5.37(c)).

Small pilot holes are either punched or drilled in the template in the correct positions. The hole positions are then transferred to the hopper plate with the aid of a centre punch.

The batten template shown in Figure 5.37(d) is relatively light in weight and is used for quantities. It is constructed from suitable wooden battens, in the form of a frame to represent the developed profile of the required hopper plates. All the outside edges of the template are planed straight. The hole positions are drilled to suit the diameter of the nipple punch.

## 5.36 Box templates

These are made from wood and are simply two flange templates fastened together. They are used for marking up purlin cleats used in constructional steelwork. The hole positions are marked on the box template to standard dimensions, usually supplied by the drawing office, and drilled. when marking off holes from a box template, a nipple punch is used. Figure 5.38 illustrates a box template. These are made from wooden battens, cut to required length and nailed as shown.

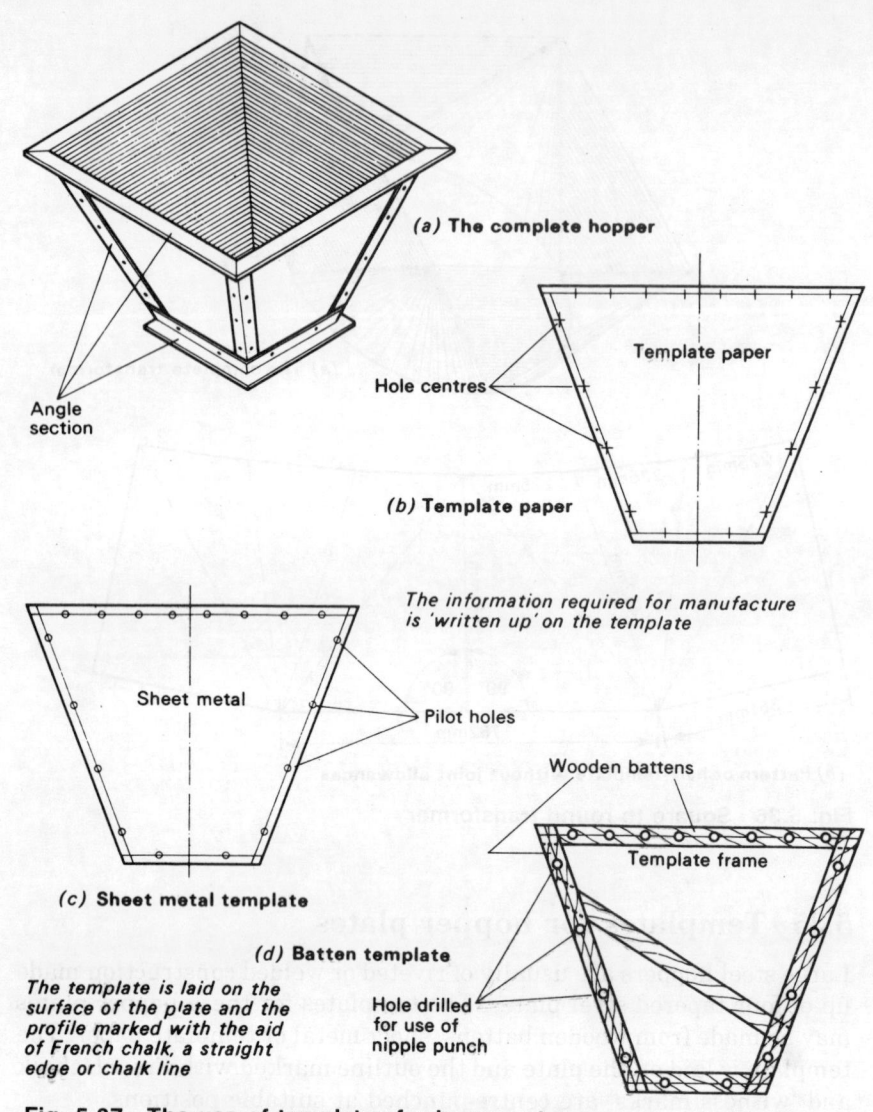

**Fig. 5.37** The use of templates for hopper plates

plates. Although many such hoppers are of welded construction, the example shown in of riveted construction, using angle sections.

This simple type of hopper is made up of four identical steel plates for which only one template is required. The choice of template material is as follows:

When only one or two off are required, template paper may be used, as shown in Figure 5.37(b). The profile of one side is developed and

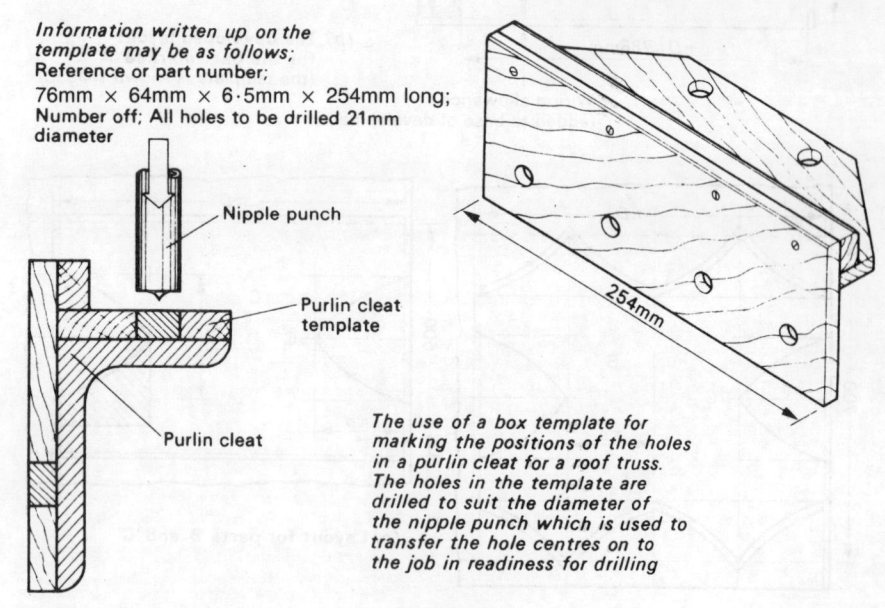

*Information written up on the template may be as follows;*
*Reference or part number;*
*76mm × 64mm × 6·5mm × 254mm long;*
*Number off; All holes to be drilled 21mm diameter*

*The use of a box template for marking the positions of the holes in a purlin cleat for a roof truss. The holes in the template are drilled to suit the diameter of the nipple punch which is used to transfer the hole centres on to the job in readiness for drilling*

**Fig. 5.38** Box templates for purlin cleats

164

## 5.37 Steel templates (ordinary and bushed)

For economy reasons, steel templates are made for part-drilling a batch of plates which may be separated for any additional holes to be drilled. *On the completion of the drilling the template itself becomes part of the fabrication.*

Bushed steel templates are employed for batch drilling over a long period of time, or where a high standard of accuracy is required. They are steel plates with hardened steel bushes tightly press-fitted into them. Hardened steel bushes, termed DRILL BUSHES are available with centre-holes to match a whole range of drill diameters.

When drilling large steel plates, for example GUSSET PLATES for large constructional steelwork, one plate is accurately marked out, carefully aligned as the top plate of a pile of three or four identical plates, the whole pile firmly clamped together and the assembly drilled as one.

Such a pile of plates are usually supported on a simple gantry and positioned under a radial drilling machine for the drilling of the holes, as shown in Figure 5.39.

On completion of the drilling, the top plate (TEMPLATE) is used as a component part of the steel assembly.

## 5.38 Marking-off holes in angle sections

Angle sections are usually cut to length and mitred, where applicable, before the hole positions are marked. Figure 5.40 illustrates the method employed for marking hole positions on angle sections.

First, a batten template is made as shown in Figure 5.40(a).

Second, the template is laid on the surface of the larger flange with the heel line of the template on the heel of the angle as shown in Figure 5.40(b). The holes marked 'A' on the template are marked through on to the surface of the flange with a nipple punch.

Third, the angle section is then turned over in the gantry and the template (bottom face up) is laid on the surface of the smaller flange with the heel line of the template in line with the heel of the angle as shown in Figure 5.40(c), and the 'TAIL HOLES' marked 'B' on the template marked through.

Alternatively when the tail holes are not drilled in the template, their positions are marked off with the aid of a set square, French chalk and a 'BACK GAUGE', as shown in Figure 5.40(d) and (e), and their centres marked with a centre punch.

The back gauge illustrated is of the adjustable type which may be set to a standard back mark dimension from the heel of the angle. These standard dimensions are usually supplied by the drawing office.

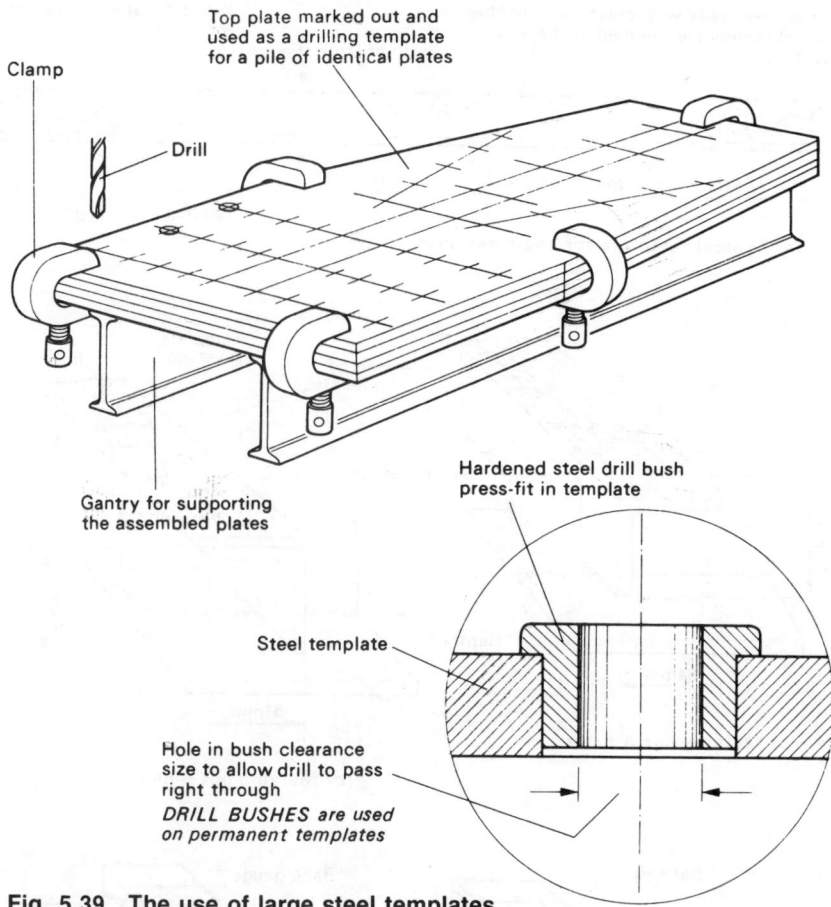

**Fig. 5.39 The use of large steel templates**

## 5.39 Marking-off holes in channel sections

Channel sections, cut to required lengths, are placed on a simple gantry with the web horizontal. The wooden template is placed in position, with the information uppermost, and clamped into position. Figure 5.41(a) shows the method of marking-off holes in channel section.

The hole positions in the web are marked through the template with a nipple punch, as shown in Figure 5.41(b). Whilst the template is clamped in position the positions of the tail holes are marked with the aid of a set square and French chalk, i.e., 'square-off' on the faces of both flanges.

When the template has been removed, a back gauge and French chalk are used to mark the position of the tail holes from the heel of

165

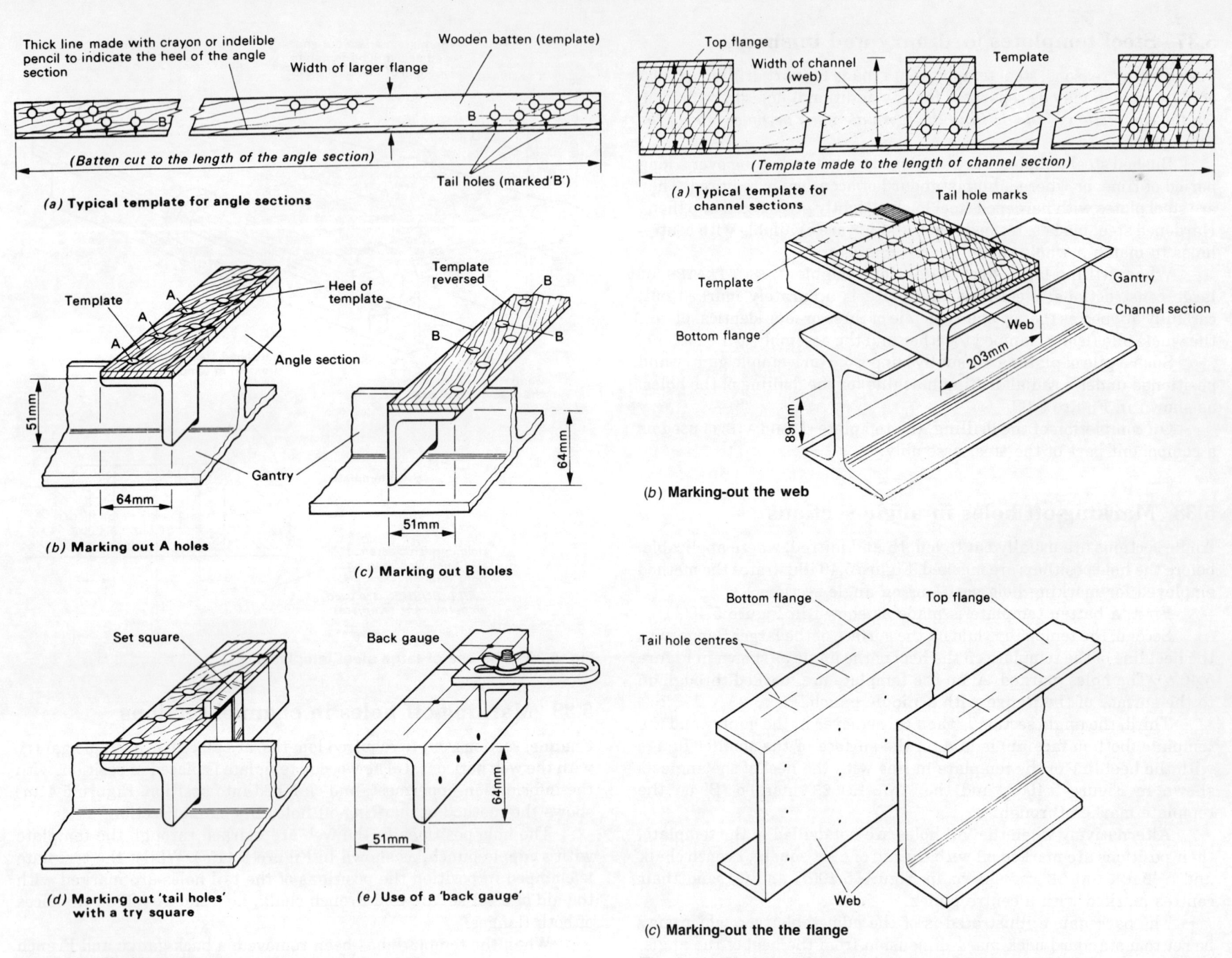

**Fig. 5.40 Marking-off hole positions in angle sections**

(a) Typical template for angle sections
(b) Marking out A holes
(c) Marking out B holes
(d) Marking out 'tail holes' with a try square
(e) Use of a 'back gauge'

**Fig. 5.41 Marking-off hole positions in channel sections**

(a) Typical template for channel sections
(b) Marking-out the web
(c) Marking-out the the flange

166

the flanges and their centres marked with a centre punch, as shown in Figure 5.41(c).

## 5.40 Marking-off holes in 'Tee' sections

One bottom template is generally used to mark off the hole positions on both the flange and the web or stalk, as illustrated in Figure 5.42(a).

Before applying the template, a centre line representing half the thickness of the stalk is marked with French chalk on both ends of the 'Tee' section.

The template (with the instructions uppermost) is laid on the surface of the flange with the centre line aligned with the centre lines marked each end of the 'Tee', and clamped in position, as shown in Figure 5.41(b). Except for the tail holes (marked 'B'), the holes are marked through to the face of the flange with a nipple punch.

The 'Tee' is then turned on the gantry with the stalk horizontal. A back-mark line is marked with French chalk on the face of the stalk at a standard dimension from the flange, as shown in Figure 5.42(c). The template is laid on the face of the stalk with the centre-line of the tail holes in line with the back-mark line, and clamped in position. The tail holes are marked through to the face of the stalk with a nipple punch.

## 5.41 Marking-off holes in columns or beams

The procedure for marking-off hole positions in columns and beams is illustrated in Figure 5.43. For reference purposes a beam has a 'top flange' and a 'bottom flange', whilst a column has an 'outside flange' and an 'inside flange'.

Figure 5.43(a) shows a typical template for marking-out columns and universal beams.

First, centre lines are marked on the web at both ends of the section for the purpose of locating with the centre line marked on the template.

Second, the template is laid on the surface of the top flange with the respective centre lines aligned as shown in Figure 5.43(b) clamped in position, and the plain and 'A' holes marked through with a nipple punch.

This procedure is repeated for the bottom flange where the plain and 'B' holes are marked through.

Finally, the beam is laid with the web horizontal on the gantry, and the web holes 'C' marked through in the same manner, as shown in Figure 5.43(c).

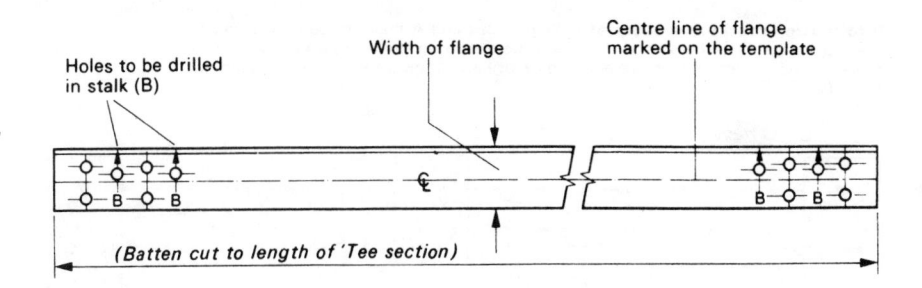

(a) Typical template for Tee sections

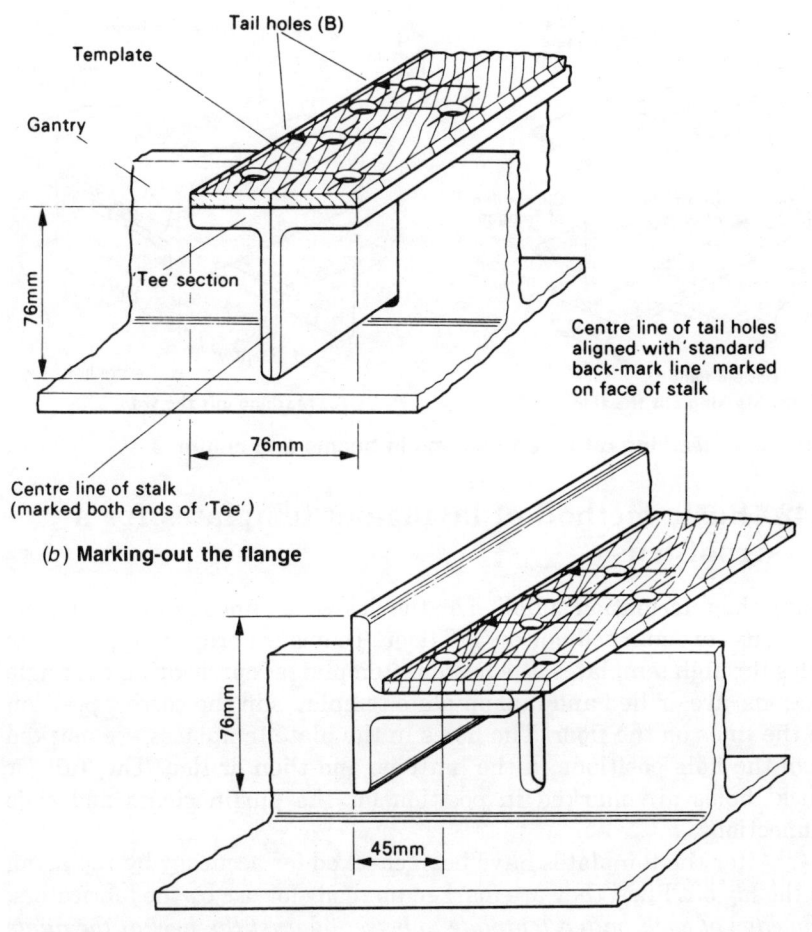

(b) Marking-out the flange

(c) Marking-out the web (stalk)

Fig. 5.42 Marking-off hole positions in 'Tee' sections

167

**Information marked on template**: Plain holes and A holes to be drilled on the top flange. Plain holes and B holes to be drilled on the bottom flange. C holes to be drilled in web. All flange holes to be drilled 14mm diameter. Web holes to be drilled 17.5mm diameter

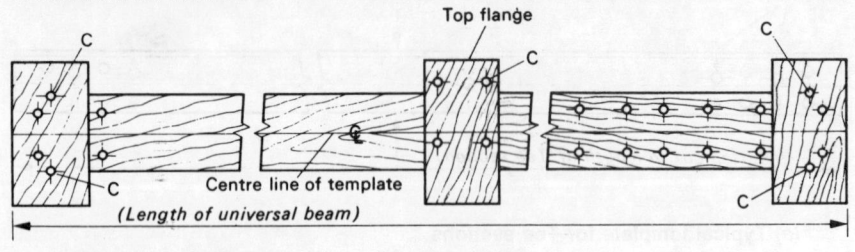

*(a)* Typical template for columns and universal beams

*(b)* Marking out the flange

*(c)* Marking out the web

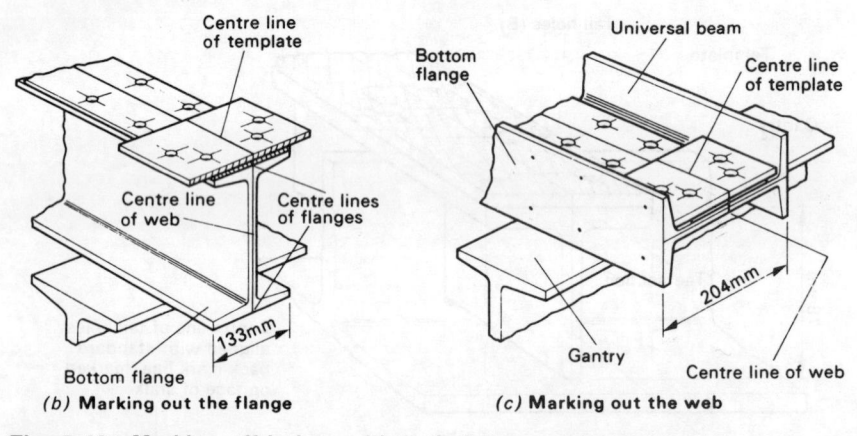

**Fig. 5.43** Marking-off hole positions in beams and columns

## 5.42 Basic method of laying-out templates for a roof truss

Using the information supplied on the drawings, lines representing the roof truss are marked out on the floor. To ensure proper alignment of holes through templates, the battens (templates representing the angle sections) are drilled and laid on plate templates in the correct position on the lines on the floor. The holes in the plate templates are marked from the hole positions in the battens, and then drilled. The 'tail' or 'back' holes are marked in position for the purlin cleats and shoe connections.

After the templates have been checked for accuracy by replacing on the lay-out lines, they are marked up ready for use by the fabricators. *The edge of each batten template to be set against the heel of the angle section is marked with a line close to that edge.*

Figure 5.44 shows a layout for a simple roof truss, together with information on how the various templates employed are marked up. Box templates are used for the purlin cleats as described in Figure 5.38.

Further information on the use of templates can be found in Chapter 9.

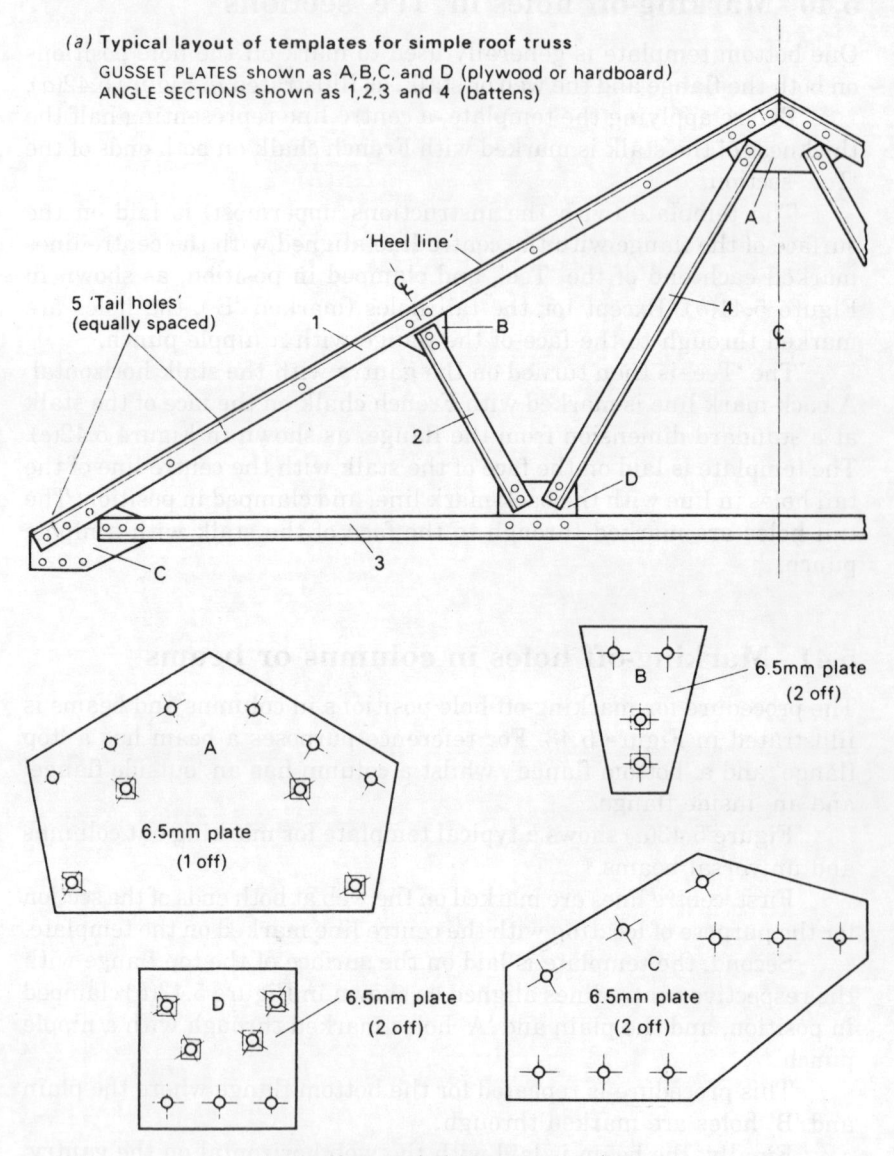

*(a)* Typical layout of templates for simple roof truss
GUSSET PLATES shown as A, B, C, and D (plywood or hardboard)
ANGLE SECTIONS shown as 1, 2, 3 and 4 (battens)

5 'Tail holes' (equally spaced)

A — 6.5mm plate (1 off)
B — 6.5mm plate (2 off)
C — 6.5mm plate (2 off)
D — 6.5mm plate (2 off)

**Instructions marked on gusset templates**
All plain holes to be drilled 21 diameter
Holes marked ⌀ to be drilled 17.5mm diameter

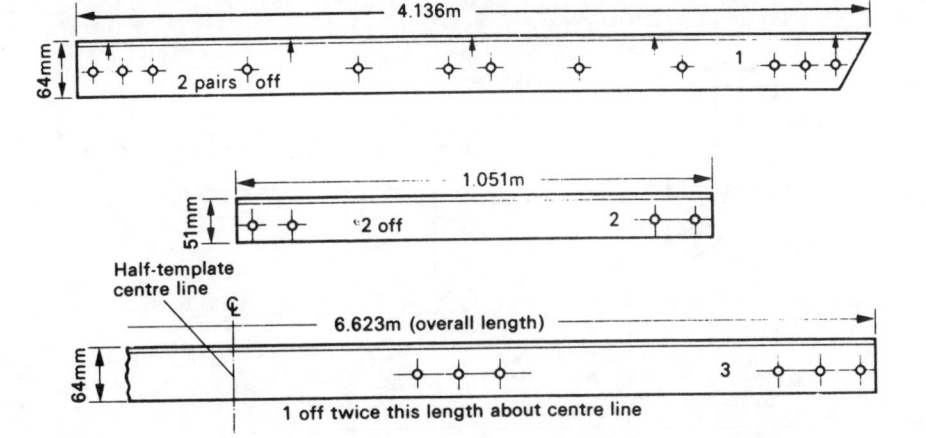

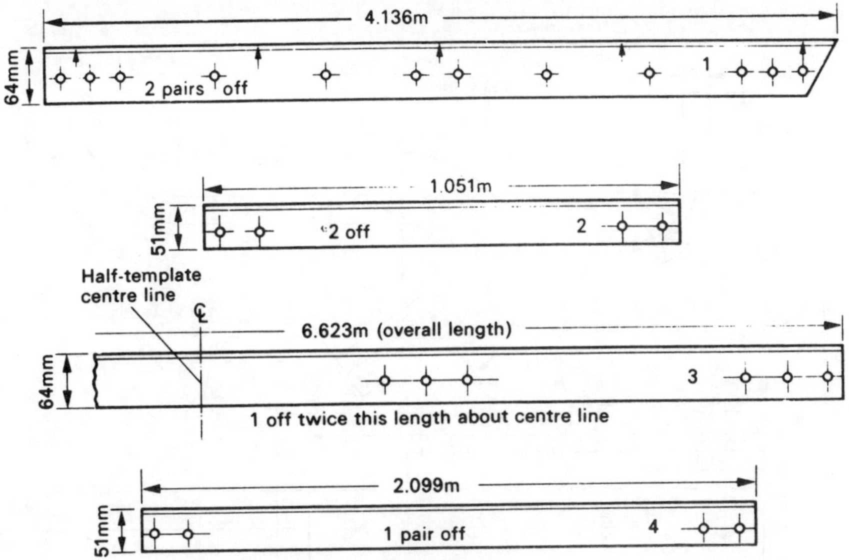

*(b)* **Batten templates for the angle sections**

*Note: For clarity these are not shown to scale*

*Instructions marked up on templates (generally written in indelible pencil)*

**Template 1.** 64mm × 51mm × 6.35mm × 4.136m long – 2 pairs off.
Tail holes to 29mm Back Gauge mark, drill 17.5mm diameter
Plain holes drill 21mm diameter

**Template 2.** 51mm × 51mm × 6.35mm × 1.051m long – 2 off.
Drill holes 17.5mm diameter

**Template 3.** 64mm × 51mm × 6.35mm × 6.623m long – 1 off
Drill holes 21mm diameter
THIS IS A HALF TEMPLATE TO THE CENTRE LINE
OF THE SECTION'S OVERALL LENGTH

**Template 4.** 51mm × 51mm × 6.35mm × 2.099m long – 1 pair off
Drill holes 17.5mm diameter.

**Fig. 5.44 Use of templates for structural steelwork**

## 5.43 Precision measurements and marking-out (sheet metal)

The precise marking-out of blanks and templates for sheet-metal components is performed from a DATUM SURFACE using a vernier height gauge in conjunction with an angle plate and parallels (see Section 5.15). However, before marking-out procedures can commence, dimensions for any hole centres, given on working drawings, have to be translated into CO-ORDINATES (see Section 5.14). *Metal thickness, inside bend radii, length of flats and bend allowances* must all be considered when calculating the true developed length of the blank (cutting sizes), as explained in Sections 10.27 and 10.28.

A study of Figure 5.45 will explain how the positions of the hole *centre lines* indicated on the sheet-metal section shown are translated into *ordinates* data from one datum edge only.

Figure 5.46 illustrates how the blank or template would be marked-out from the dimensions given on the sub-assembly shown.

The purpose of 'boxing' hole centres when marking out templates for precision sheet-metal work is explained in Figure 5.47.

In the aircraft industry, detail sheet-metal work is rigidly controlled by stringent specifications which must be adhered to. Figure 5.48 and the accompanying Table 5.6 give details of a typical specification in respect of sheet-metal corners. To avoid stress cracking

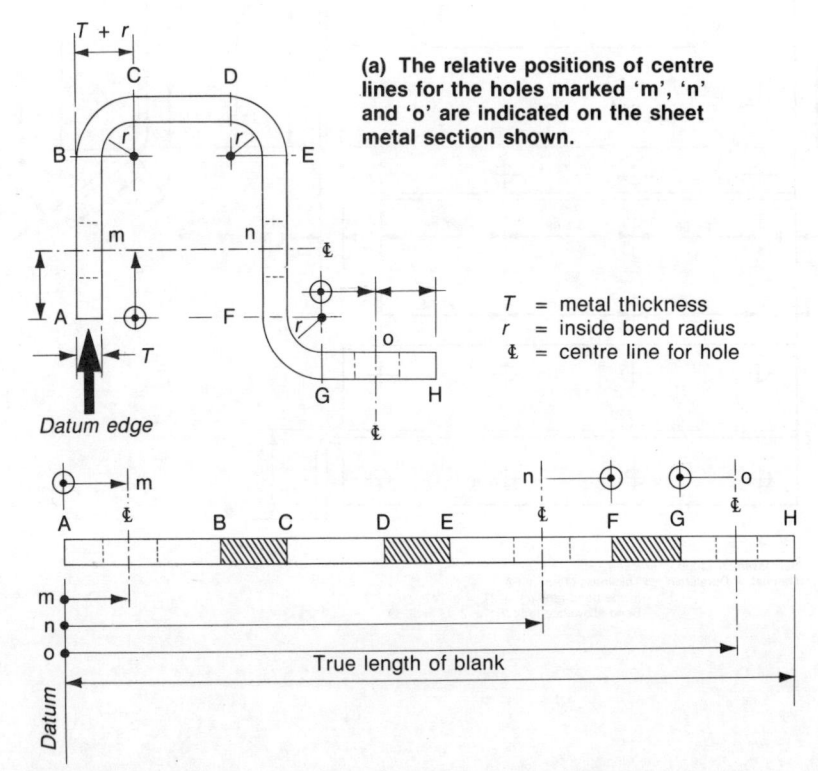

*(a)* The relative positions of centre lines for the holes marked 'm', 'n' and 'o' are indicated on the sheet metal section shown.

$T$ = metal thickness
$r$ = inside bend radius
₵ = centre line for hole

*(b)* The true position of each hole centre line is indicated by an ordinate from one datum edge as shown.
(The bend allowance in respect of each bend is indicated by the shaded areas.)

**Fig. 5.45 Use of ordinate data**

169

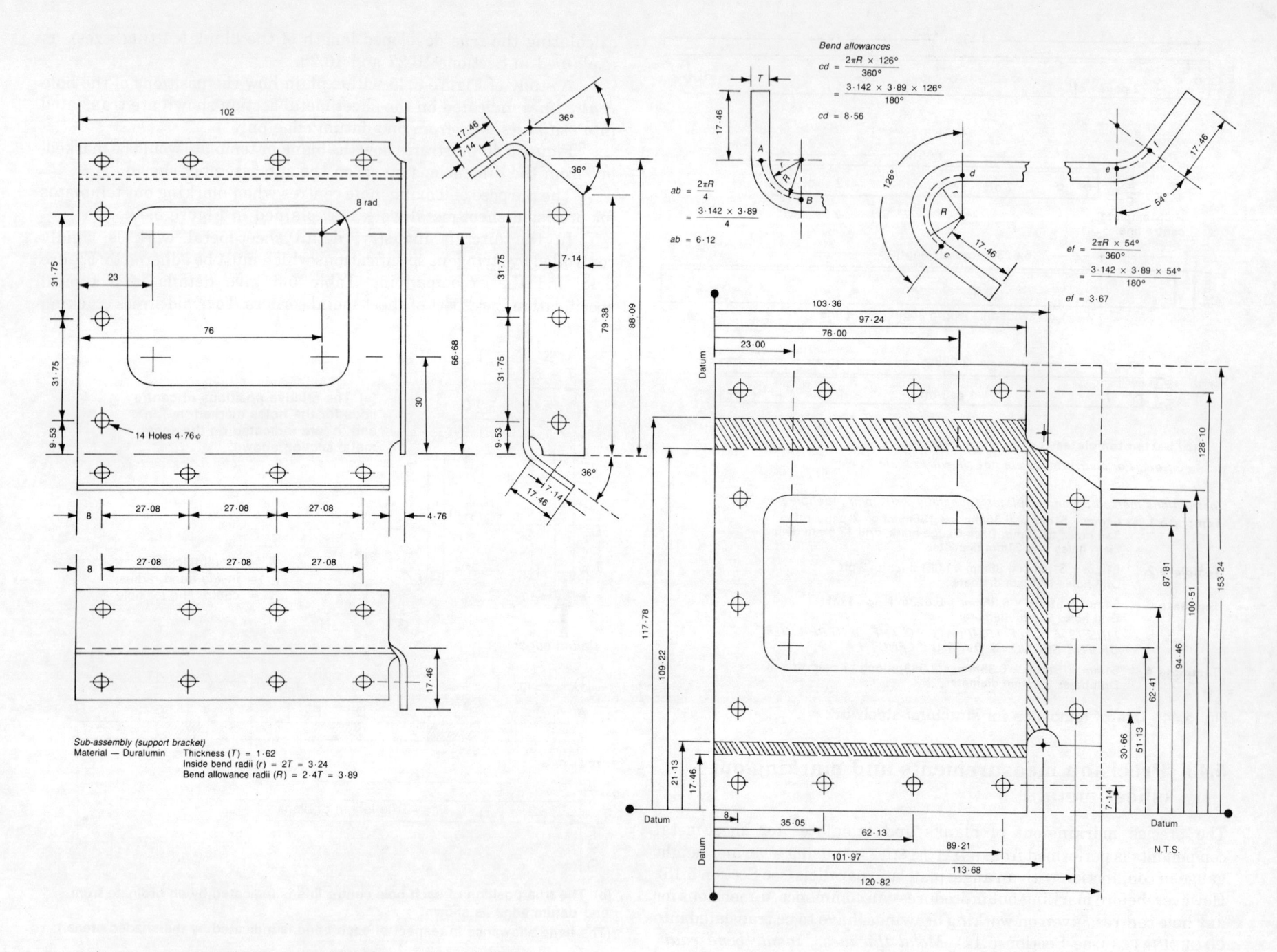

**Fig. 5.46 Development of sheet-metal blank — support bracket**

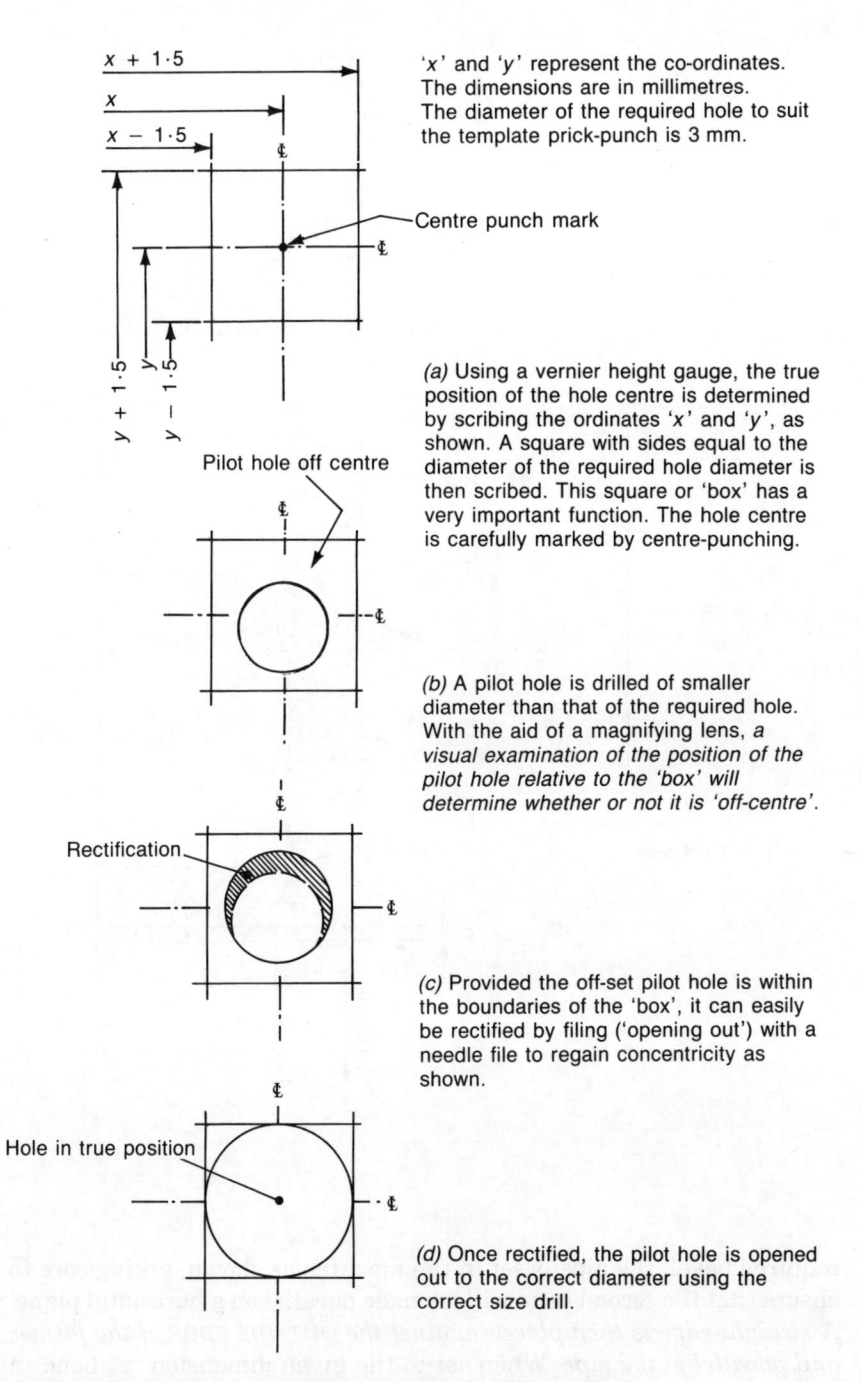

**Fig. 5.47 Principle of 'boxing' co-ordinates (templates)**

'x' and 'y' represent the co-ordinates. The dimensions are in millimetres. The diameter of the required hole to suit the template prick-punch is 3 mm.

(a) Using a vernier height gauge, the true position of the hole centre is determined by scribing the ordinates 'x' and 'y', as shown. A square with sides equal to the diameter of the required hole diameter is then scribed. This square or 'box' has a very important function. The hole centre is carefully marked by centre-punching.

(b) A pilot hole is drilled of smaller diameter than that of the required hole. With the aid of a magnifying lens, *a visual examination of the position of the pilot hole relative to the 'box' will determine whether or not it is 'off-centre'.*

(c) Provided the off-set pilot hole is within the boundaries of the 'box', it can easily be rectified by filing ('opening out') with a needle file to regain concentricity as shown.

(d) Once rectified, the pilot hole is opened out to the correct diameter using the correct size drill.

in the bends during service, sharp corners are not permissible; they must be radiused as indicated on the drawing.

## 5.44 Setting out bends (pipework)

On working drawings pipe bends are represented by centre lines only. In practice the actual setting out can be accomplished by either using a stick of chalk for marking on the workshop floor or a scriber for marking on a piece of steel sheet. A length of pipe may be used as a straight edge and for marking lines on each side to represent the diameter.

When two pipes running adjacent are required to change direction together, it is usual to make the outer bend with a standard machine former. This normally has a throat radius approximately equal to three times the diameter of the pipe. The inner, and necessarily sharper bend is made by hand. *It is important that the space between them is sufficient to allow for compression joints and the operation of tightening them up, and for screwing up fixing clips.*

### RIGHT-ANGLE BENDS

Figure 5.49(a) explains the method used for setting out two square bends running parallel with each other.

### OFFSET BENDS

Very often pipes have to change direction, for example, to pass an obstacle, such as a girder. This is achieved by 'offsetting' the pipe. The method adopted for setting out an offset is illustrated in Figure 5.49(b).

## 5.45 Setting out pipe bends (machine bending)

Different methods of specifying measurements for pipe bending are shown in Figure 5.50(a), and the setting up procedures for machine bending to these specific measurements are indicated in Fig. 5.50(b), (c) and (d).

*To bend to inside measurements* mark the pipe as shown by dimension 'x' in Figure 5.50(b). Place the set-square on the mark and insert the pipe in the machine such that *the square touches the INSIDE of the groove in the former.* Bend as usual.

*To bend to outside measurements* mark the pipe as shown by dimension 'y' in Figure 5.50(c). Place the set-square on the mark and carefully insert the pipe in the machine such that *the square touches the OUTSIDE EDGE of the former.* Bend as usual.

*To bend to measurements between centre lines* mark the pipe as shown by dimension 'z' in Figure 5.50(d). Place the set-square on the

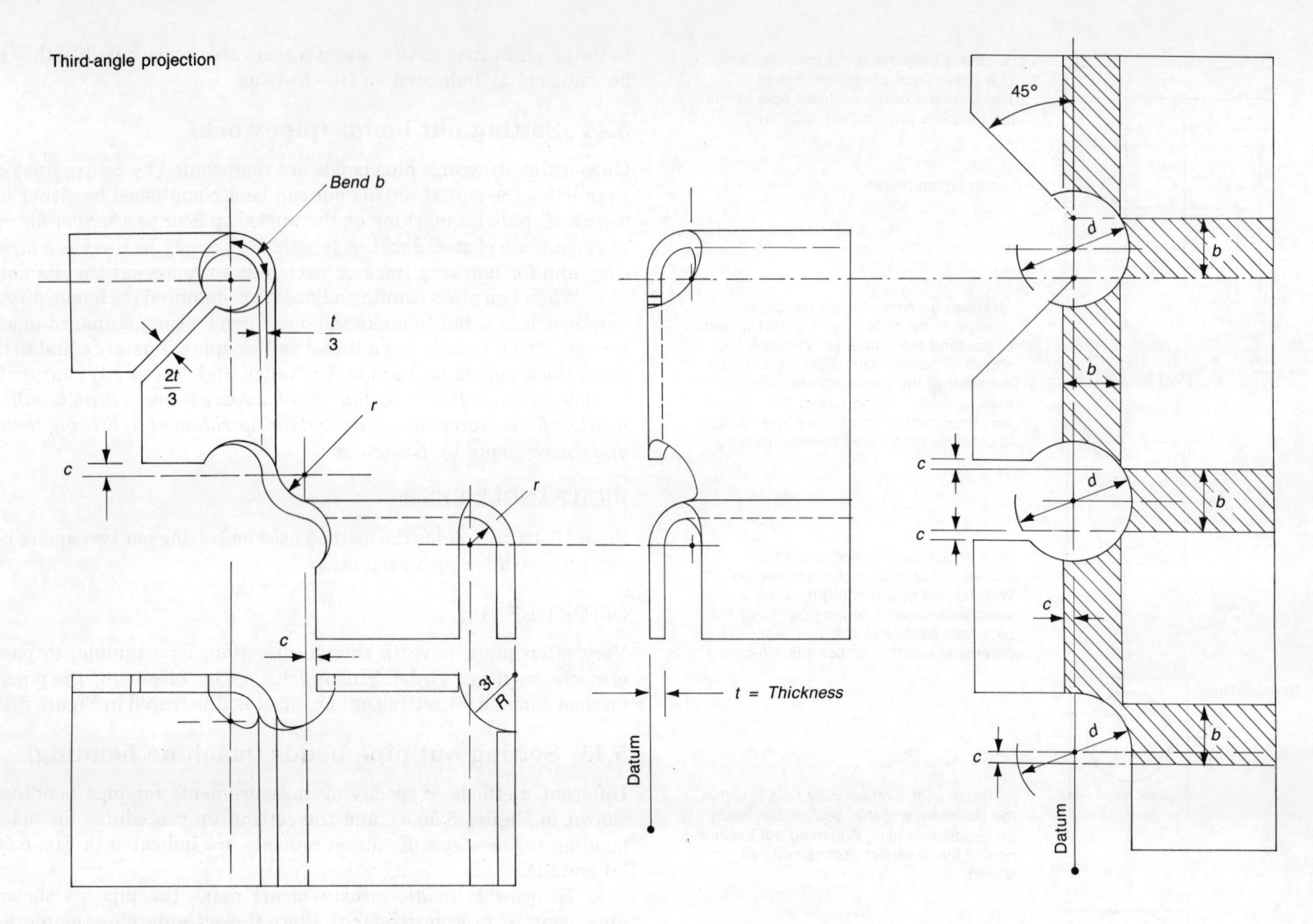

**Fig. 5.48 Sheet-metal corners (aircraft specification)**

mark and insert the pipe in the machine such that *the square touches the* OUTSIDE EDGE *of the former*, as when bending to outside measurements. Bend as usual.

Figure 5.50(e) shows *the set-up for making an 'offset' or 'double bend' to required measurements*. After making the first bend to the required bevel, the pipe is set in the machine as shown, taking care to ensure that the second bend will be made parallel on a horizontal plane. *A straight-edge is then placed against the* OUTSIDE EDGE *of the former and parallel to the pipe*. When set to the given dimension 'x', bend in the usual way.

172

**Table 5.6 Sheet-metal corners — specifications**

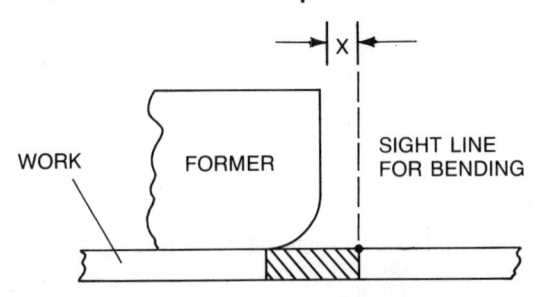

FOR MATERIAL WHERE MINIMUM BEND RADIUS = $1\frac{1}{2}t$.

| MATERIAL s.w.g. | BEND RADIUS r | BEND WIDTH b | OFFSET c | HOLE DIAMETER d | SIGHT LINE X |
|---|---|---|---|---|---|
| 26 | 0·794 | 1·588 | 0·397 | 3·175 | 0·794 |
| 24 | 0·794 | 1·588 | 0·397 | 3·175 | 1·191 |
| 22 | 1·191 | 2·382 | 0·397 | 4·763 | 1·588 |
| 20 | 1·588 | 3·175 | 0·794 | 6·350 | 1·984 |
| 18 | 1·984 | 3·969 | 0·794 | 7·938 | 2·381 |
| 16 | 2·381 | 4·763 | 0·794 | 9·525 | 2·778 |
| 14 | 3·175 | 6·350 | 1·191 | 12·700 | 3·572 |
| 12 | 3·969 | 7·938 | 1·191 | 15·875 | 4·763 |
| 10 | 4·673 | 9·525 | 1·588 | 19·050 | 5·556 |

All dimensions are in millimetres

## 5.46 Complex angle bends

A knowledge of geometry is very useful when setting out pipework bends in the workshop or on site. *Remember bends incorrectly made as a result of employing trial and error methods are very difficult if not impossible, to rectify.*

A graphical method for setting out a bend in a sloping pipe to enable it to pass around an external corner is illustrated in Figure 5.51. *Since the pipe is inclined, the true angle of the bend will be greater than 90°.*

The angle of the bend is shown by the triangle ABC in the three-dimensional sketch. This triangle can be drawn accurately to a suitable scale on the workshop floor, on a piece of sheet steel or on a sheet of paper, adopting the following procedure:

1. Set up the two side elevations for the pipe at the given angle of inclination ($\theta°$) as shown as 'Elevation 1 and Elevation 3'.

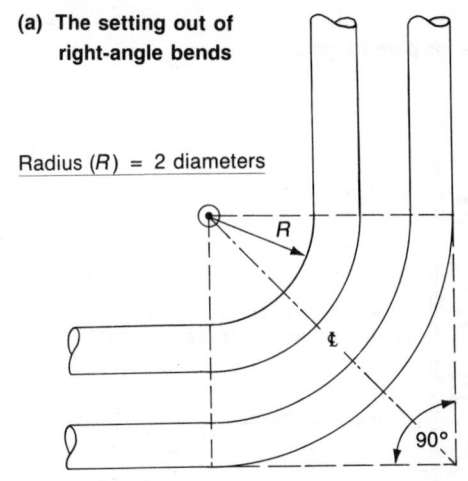

**(a) The setting out of right-angle bends**

Radius (R) = 2 diameters

The inner pipe is drawn first using a 'square' to make sure it is at right angles. The outer pipe is then drawn leaving sufficient distance between to allow for screwing up any fixing clips. The inner bend shown is drawn on a radius equal to two diameters to the throat or two and a half to the centre of the pipe. The arcs for each bend are struck from a common point (marked with a small circle) on the line bisecting the angle.

The dotted lines forming a right angle indicate the use of the square to obtain the right-angle bend.

**(b) The setting out of offset bends**

Radius (R) = 1 diameter

The pipes are drawn with the required amount of offset as shown. The angles are then bisected — the offset shown has an angle of 135°, but this angle will depend upon circumstances. The inner bends are made on a radius of one diameter to the throat or one and a half to the centre of the pipe. The arcs for each bend are struck from a point on the line bisecting the angle.

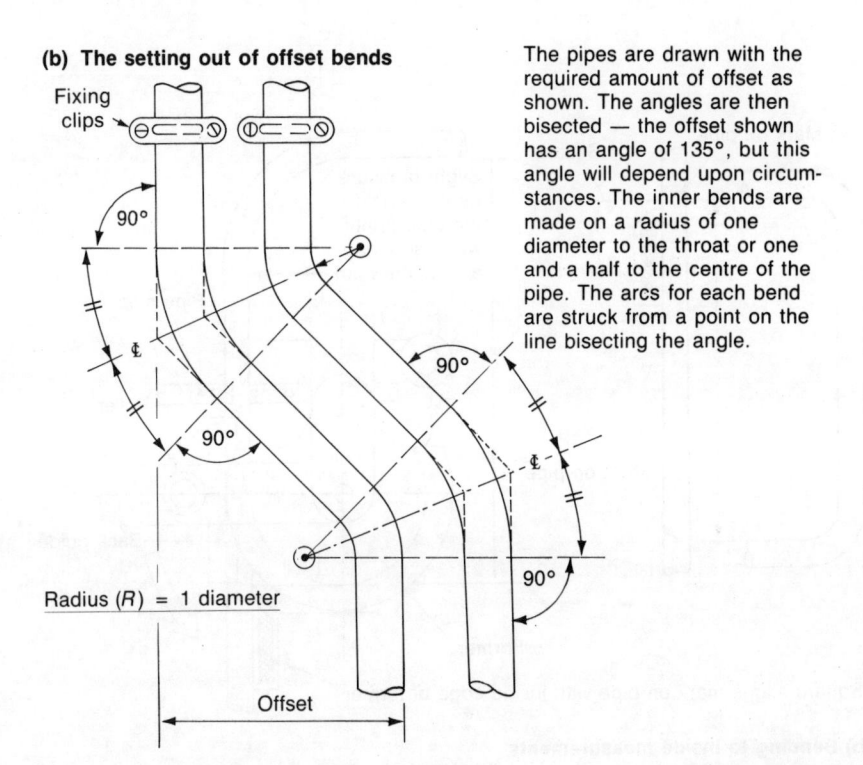

**Fig. 5.49 Setting out of pipe bends**

**(a) Methods of taking measurements for pipe bending**

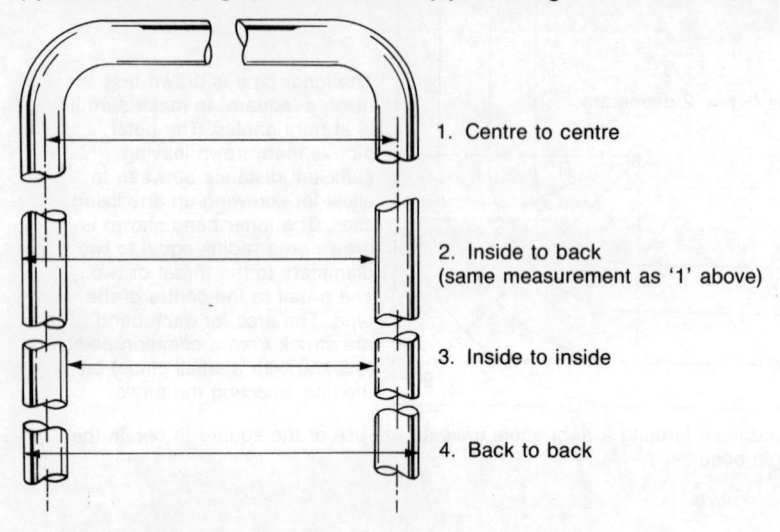

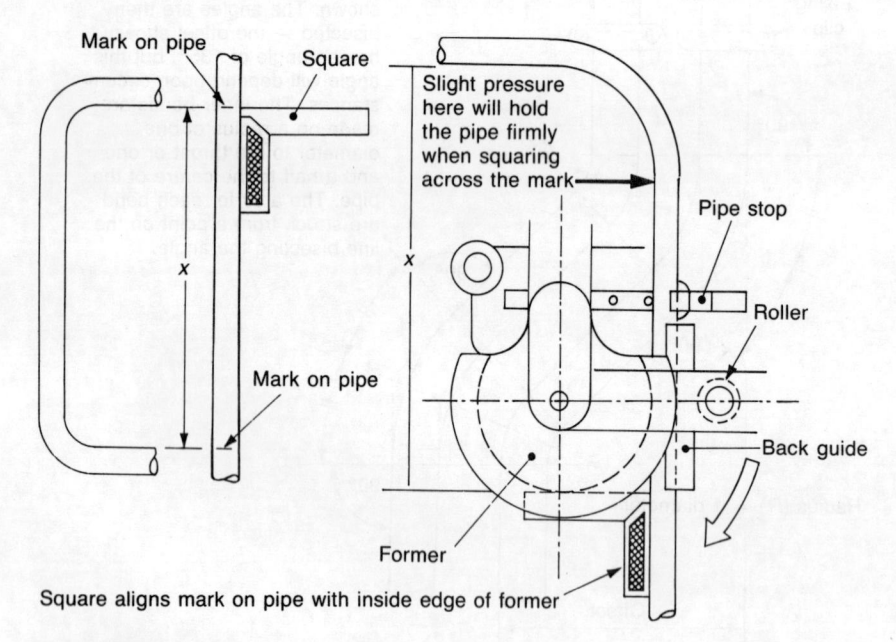

1. Centre to centre
2. Inside to back (same measurement as '1' above)
3. Inside to inside
4. Back to back

**(b) Bending to inside measurements**

Square aligns mark on pipe with inside edge of former

Labels: Mark on pipe, Square, Slight pressure here will hold the pipe firmly when squaring across the mark, Pipe stop, Roller, Back guide, Former, Mark on pipe

**(c) Bending to outside measurements**

(Square aligns mark on pipe with outside edge of former)

**(d) Bending to centre line measurements between bends**

(Square aligns mark on pipe with outside edge of former)

**(e) Making an offset or double bend**

Labels: Straight edge, Rule

**Fig. 5.50 Methods of making offset and double bends by machine**

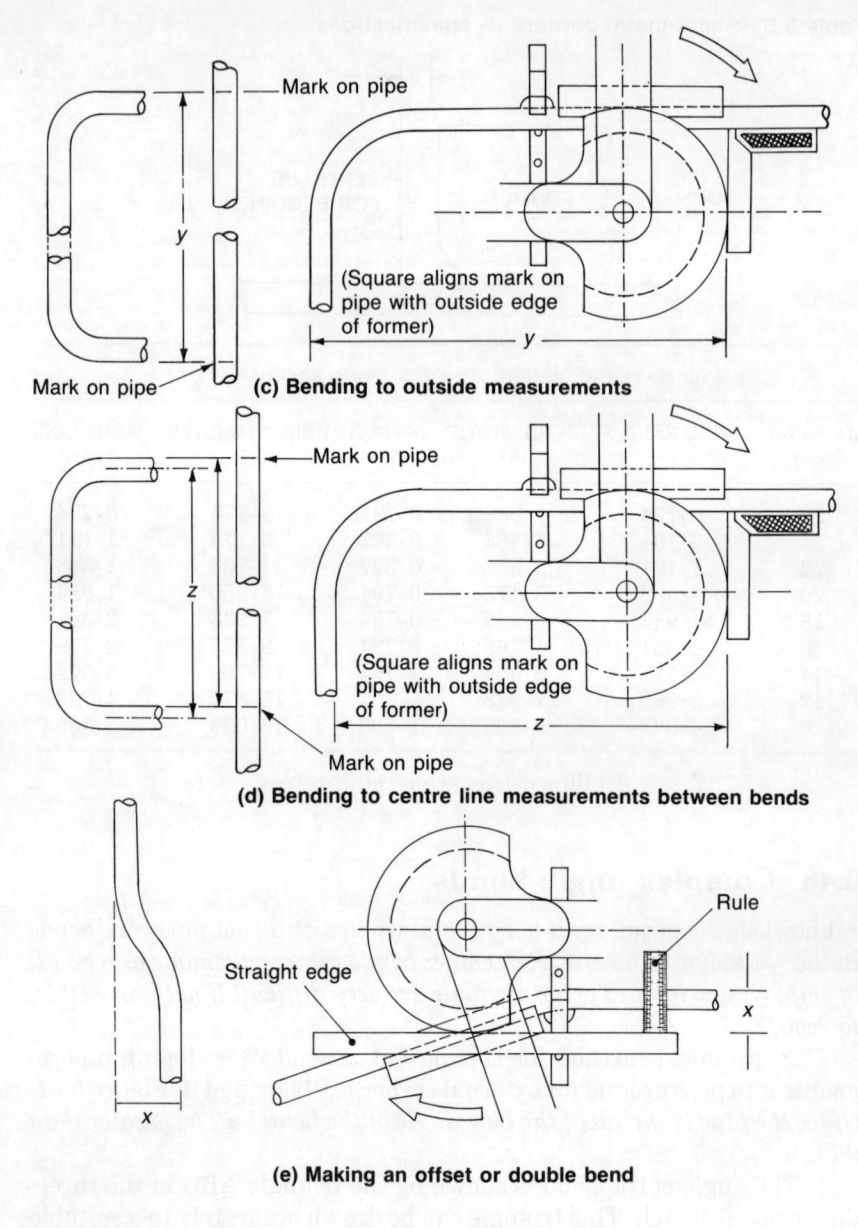

174

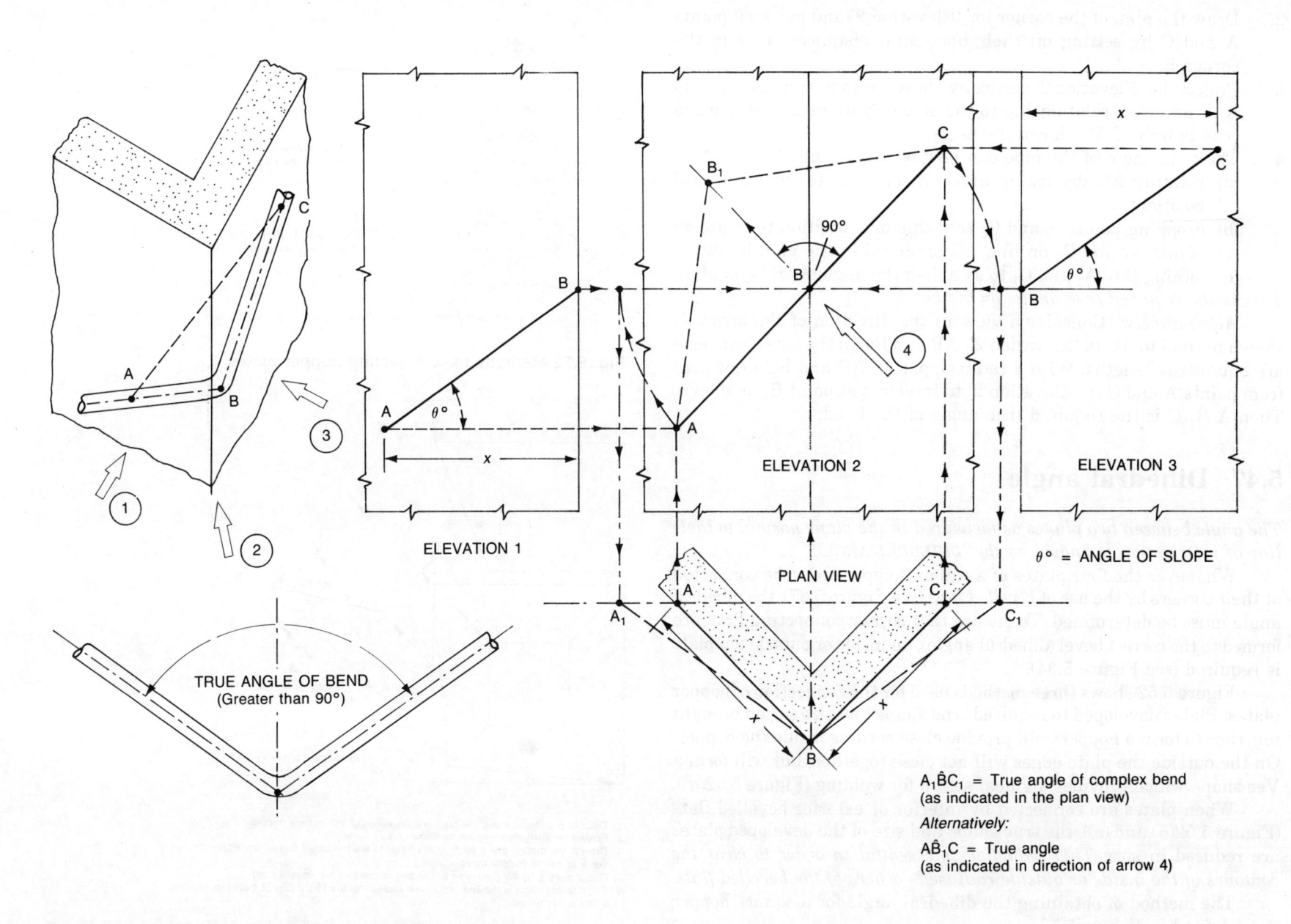

**Fig. 5.51 Setting out complex bend — graphical method**

$A_1\hat{B}C_1$ = True angle of complex bend
(as indicated in the plan view)

*Alternatively:*

$A\hat{B}_1C$ = True angle
(as indicated in direction of arrow 4)

$\theta°$ = ANGLE OF SLOPE

175

2. Draw the plan of the corner for 'Elevation 2' and mark off points A and C by setting off their horizontal distances 'x' from the corner B.
3. Project the 'Elevation 2' vertically above the plan and locate points A, B and C in relationship to the side elevations and obtain the true length of AC shown in the sketch.
4. The true angle of the pipe can now be determined by:
   (a) rotating AC (by use of arcs) on centre B to the horizontal position;
   (b) dropping points A and C from this new position to locate as points $A_1$ and $C_1$ on line AC produced in the plan below;
   (c) joining B to $A_1$ and $C_1$ to establish the angle $A_1 \hat{B} C_1$ in plan.

*This is the required true angle for the bend.*

*Alternatively*: Consider a view in the direction of the arrow 4, shown normal to AC in 'Elevation 2'. AB and BC in the side elevations are equal true lengths. With a radius equal to AB and BC draw arcs from points A and C in 'Elevation 2' to intersect at point $B_1$ as shown. Then $A \hat{B}_1 C$ is the required true angle of the bend.

## 5.47 Dihedral angle

*The angle between two planes as measured in the plane normal to their line of intersection is known as the 'DIHEDRAL ANGLE'.*

Whenever the four plates of a square hopper are to be connected at their corners by the use of *bevelled flats* (see Figure 5.37), the dihedral angle must be determined. To ensure that corner connecting flats are formed to the correct bevel (dihedral angle), an inside or outside template is required (see Figure 5.34).

Figure 5.52 shows three methods used for the connection of hopper plates. Plates developed to required true shape and size (when brought together to form a hopper) will provide close corners inside the hopper. On the outside the plate edges will not close together, but will form a Vee-shape which provides an ideal set-up for welding (Figure 5.52(*a*)).

When plates are connected by interior or exterior bevelled flats (Figure 5.52(*b*) and (*c*)), the true shape and size of the developed plates are reduced in size. *This reduction is essential in order to clear the contours of the inside or outside radiused corners of the bevelled flats.*

The method of obtaining the dihedral angle for a square hopper is explained in Figure 5.53.

(*a*) Welded. (No reduction in plate size)

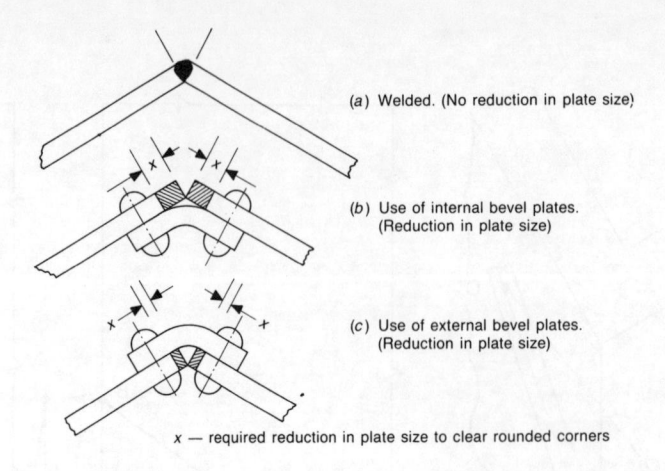

(*b*) Use of internal bevel plates. (Reduction in plate size)

(*c*) Use of external bevel plates. (Reduction in plate size)

*x* — required reduction in plate size to clear rounded corners

**Fig. 5.52 Methods for connecting hopper plates**

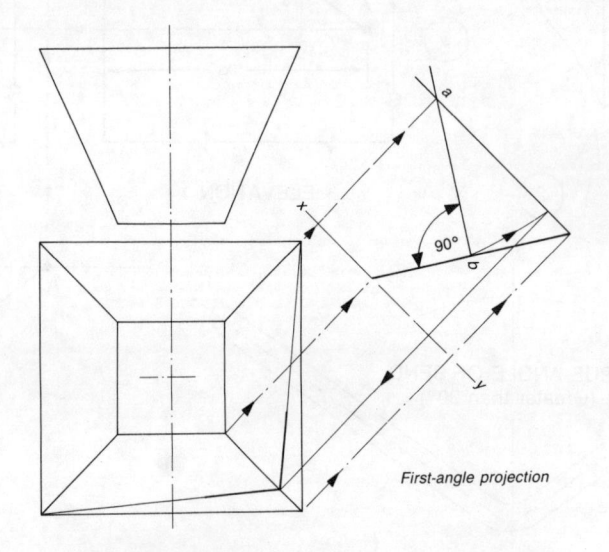

*First-angle projection*

1. Draw to a suitable scale the plan and elevation of the square hopper
2. Project an auxiliary elevation as shown on *xy*
3. Draw in the cutting plane *ab* at right angles to the projected corner line where the two surfaces of the hopper meet
4. Rabat point *b* to the plan, join to the two corners of the square (top)
5. The dihedral angle is shown lined in, and represents the TRUE BEVEL between adjacent sides of the hopper

**Fig. 5.53 Dihedral angle — Square hopper**

# 6 Materials

The topics discussed in this chapter are confined to those materials most frequently used in fabrication engineering — namely metallic materials.

Metallic materials can be classified *ferrous metals, non-ferrous metals and alloys*. Because pure or 'true' metals rarely possess the required properties, they are more commonly used in the form of alloys. However, before considering the properties and use of these materials, it is important to have a basic understanding of their structures.

## 6.1 Fundamentals

### METALS

A true metal is composed of many crystals or grains of a single metallic element, such as pure aluminium, pure copper, pure iron, pure tin, pure lead and pure zinc, etc. These crystals or grains are formed when the molten metal cools after a refining process.

*If the cooling is slow, the crystals or grains are relatively large, whilst rapid cooling produces correspondingly smaller crystals.* In either case the grains are packed so tightly together that they are held in a strong compact mass by their attraction to each other.

TRUE METALS have two distinctive characteristics:

1. They can be deformed to a considerable extent without breaking;
2. They are good conductors of electricity.

### CRYSTALS (GENERAL)

Many chemical substances are naturally *crystalline*, perhaps the most common examples in everyday life being table salt and sugar. Unlike tiny lumps of chalk which are of any shape, each grain of salt or sugar is bounded by flat surfaces and they are all of one shape. in the case of salt the shape is a perfect cube. In college or school laboratories, a common exercise is the growing of beautiful blue-green pyramid-shaped crystals of COPPER SULPHATE from a saturated solution of the compound. Crystals of salt and sugar have been formed in the same way by the evaporation of a solution of salt or sugar in water causing the solid to 'crystallise out'.

### METALLIC CRYSTALS

A metallic crystal is a systematic arrangement of the atoms of a metallic element in a definite geometric pattern called a '*lattice structure*'. There are several different lattice structures, among which are the CUBIC and HEXAGONAL structures.

When a single atom occupies each corner of an imaginary cube, the structure is known as 'cubic'.

If an additional atom is present in the centre of the cube, it is called a BODY-CENTRED-CUBIC structure and contains 9 atoms as shown in Figure 6.1(*a*).

If, however, the centre is vacant, and a single atom is contained in the centre of each face of the cube, the crystal structure is known as a FACE-CENTRED-CUBIC and contains 14 atoms as shown in Figure 6.1(*b*).

Figure 6.1(*c*) shows a more complicated crystal structure known as a CLOSE-PACKED-HEXAGONAL. This structure comprises a hexagon prism, having an atom at each corner, an atom in the centre of the top

177

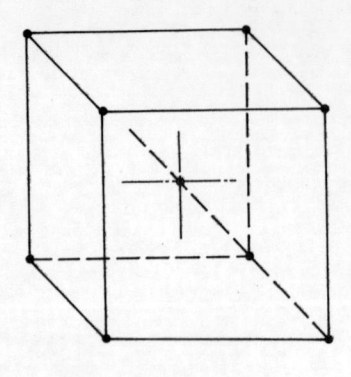

and bottom hexagons, and three atoms equally spaced in the centre of the prism — a total of 17 atoms.

Crystals are made up of many of these elementary cubes or hexagons. Even these crystals, however, are not visible except under a high-powered microscope. Crystals of irregular shape, in contact at all points with similar irregular crystals, are known as GRAINS.

## ALLOYS

Many so-called metals are not true metals, for example brass and stainless steel. They are either *solid solutions* of other metallic elements in true metals, or they are *aggregates* of true metals, solid solutions and *metallic compounds*.

## SOLID SOLUTIONS

In the liquid state, true metals can dissolve certain amounts of other metallic elements. Upon cooling and solidification, the dissolved elements sometimes remain in solution and are said to be solid solutions. *The existence of such dissolved elements, however, cannot be detected by any physical means.*

## AGGREGATES

An aggregate is a mechanical mixture which may be composed of a true metal, a solid solution, and a metallic compound. Any combination of two or more of these components may be present. *They can be readily distinguished from each other under a high-powered microscope.* Therefore an aggregate is quite different from a solid solution.

## 6.2 The crystalline structure of metals

When a mass of molten metal cools down through the solidification temperature it will start to crystallise out. Many crystals begin to grow and take shape throughout the mass. However, the growth of each individual crystal is impeded by that of its neighbours. The resultant solid mass is composed of irregularly shaped crystals known as '*grains*'. Their contact planes form a boundary between adjoining crystal grains. These contact surfaces are referred to as '*grain boundaries*' as shown in Figure 6.2.

## SOLIDIFICATION OF A PURE METAL

Crystal growth develops by a three-dimensional system of branches, rather like the branches and twigs of a fir tree. These needle-like

---

**(a) Body-centred-cubic structure**
Common to crystals of IRON at ordinary temperature and under ordinary conditions. Crystals of CHROMIUM and MOLYBDENUM have the same structure.

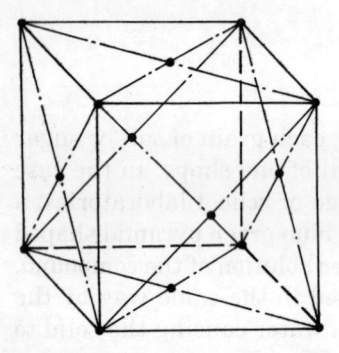

**(b) Face-centred-cubic structure**
Common to crystals of IRON at high temperatures
Crystals of ALUMINIUM, COPPER and NICKEL have the same structure.

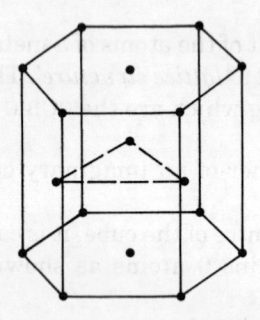

**(c) Close-packed-hexagonal structure**
Common to crystals of ZINC, MAGNESIUM and COBALT.

*Note*: The size of any elementary cube or hexagon is extremely small. Some conception of its size may be appreciated by considering that each side measures only 2·9 to 4·9 Angstrom units, and that each *single Angstrom unit is only one-ten-millionth of a millimeter.*

**Fig. 6.1 Common metal structures**

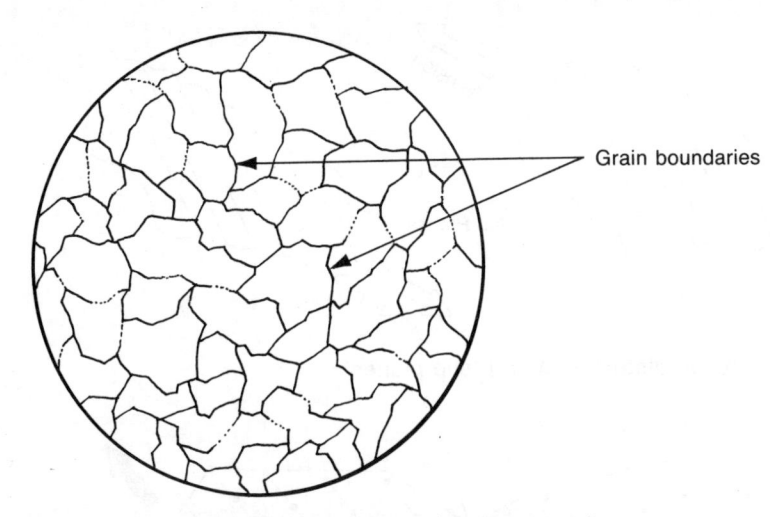

**Fig. 6.2  The crystalline structure of a pure metal**

At periodic intervals during crystal growth, secondary needle-like spurs radiate at right angles to each other.

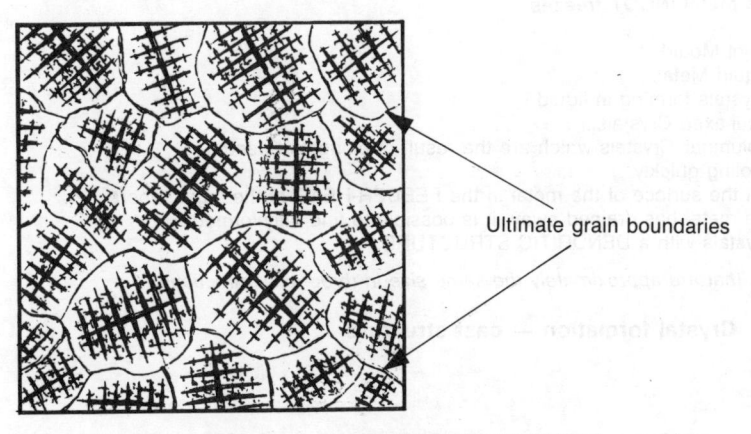

branches or spurs are called DENDRITES and the crystals are said to have a *dendritic structure* as shown in Figure 6.3.

As solidification continues throughout the melt, these dendrites increase in size until they come into contact with one another, thus restricting their direction of growth, causing solidification to occur in the interstices between the needle-like arms themselves. On complete solidification the crystals are eventually formed without a trace of their original dendritic structure, unless severe shrinking has occurred or impurities have been trapped between the needles of the dendrite.

## CRYSTAL FORMATION (CAST STRUCTURE)

When a mass of molten metal cools down, the parts of the mass to solidify first will, naturally, be those that are coolest. In the diagram shown in Figure 6.4, representing a mass of molten steel cooling in an ingot mould, crystallisation will commence along the bottom and sides where the heat is being withdrawn from the metal most rapidly. The initial direction of growth for the crystals will be predominately inwards towards the centre of the casting, i.e., *in the opposite direction to which the heat extraction is taking place.* This will produce a series of elongated crystals around the edge of the ingot. These are called COLUMNAR CRYSTALS and are often found in some types of weld deposits.

The continuation of the heat loss will cause the molten metal in the centre of the ingot to solidify at a number of points simultaneously. At each solidification point a crystal will start to grow outwards in all

*Note*: Although the illustration shows DENDRITIC GROWTH in one place, the actual phenomenon is three dimensional.

**Fig. 6.3  Crystal growth — dendrites**

directions, and will continue to do so, as long as molten metal is available to 'feed' it, until it meets the surfaces of neighbouring crystals growing out towards it. Crystal growth then stops.

The crystals in the centre of the solidifying mass show no preference to directional growth and are said to be EQUI-AXED. On completion of cooling, the result is a solid mass of irregularly shaped polygonal crystals separated by grain boundaries.

(a) How a metal INGOT freezes

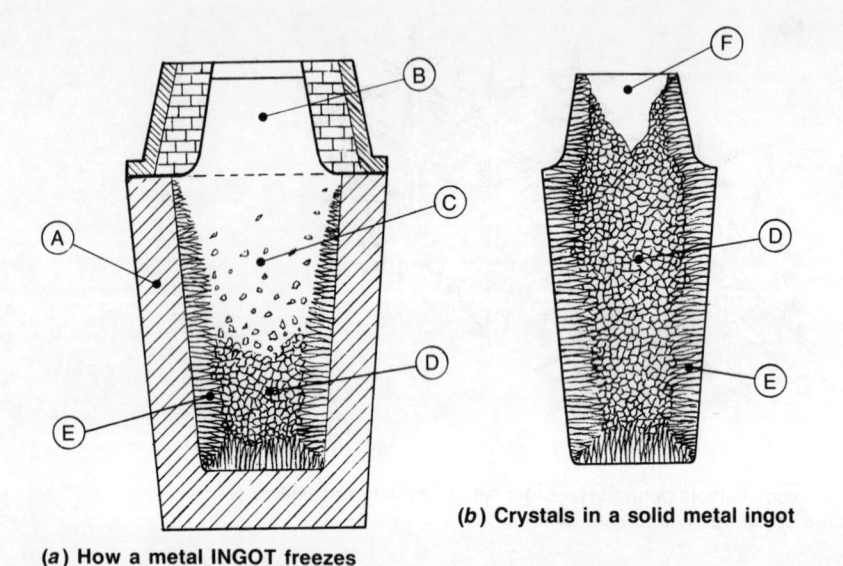

(b) Crystals in a solid metal ingot

A — Ingot Mould.
B — Liquid Metal.
C — Crystals forming in liquid.
* D — Equi-axed Crystals.
E — Columnar Crystals which are the result of elongated dendrites cooling quickly.
F — On the surface of the metal in the FEEDER-HEAD of the ingot, where the metal has drained away, it is possible to find fully-formed crystals with a DENDRITIC STRUCTURE.

* Crystals that are approximately the same size in three axial directions.

**Fig. 6.4 Crystal formation — cast structure**

## 6.3 Crystal structure — the effect of slip planes

Research has established that the deformation of metals takes place along a series of SLIP PLANES in the individual crystals. These slip planes are planes along which the atomic lattice structures can glide over each other, and for every individual crystal there are certain directions along which it can be more easily distorted than in other directions, as shown in Figure 6.5. If a piece of metal were to be made up entirely of a single crystal, these slip planes would be a very serious source of weakness. It would only require a comparatively low stress in the direction of the slip planes to cause the metal to distort and finally fail altogether. However, because metals are normally made up of a very large number of *randomly orientated* crystals this is not the case.

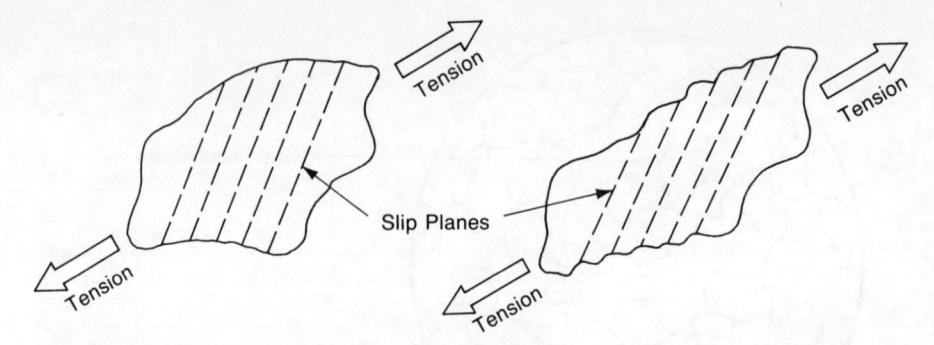

**Fig. 6.5 Crystal distortion along 'slip planes'**

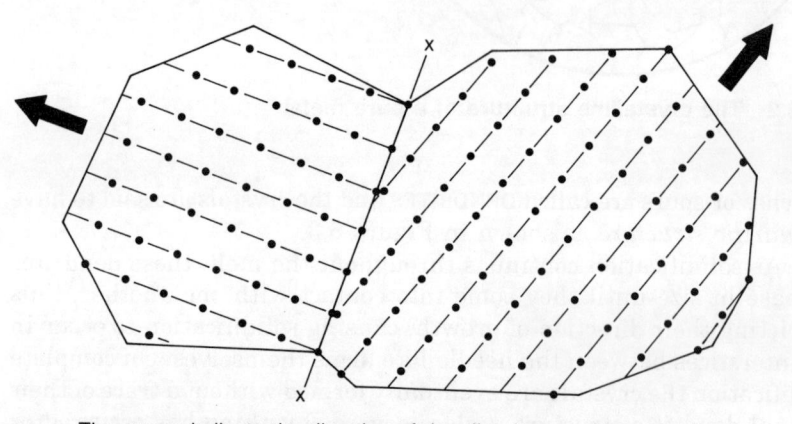

The arrows indicate the direction of the slip planes, whilst x–x shows the grain boundary between two adjoining crystals.
Within the irregular shape of the crystal grain, the atoms are aligned in their regular crystalline lattice structure. The direction of alignment, however, differs from crystal to crystal.

**Fig. 6.6 Disorientation of 'slip planes' between adjoining crystals**

Figure 6.6 illustrates how within the irregular shape of the crystal grain the atoms are aligned in their regular crystalline pattern. It also shows that the direction of the slip planes of one crystal is different from those of its immediate neighbour.

When a metal is stressed, whole planes of atoms within individual crystals are able to glide over each other. This slipping movement provides the inherent characteristic of a metal to be able to suffer deformation without fracture.

We may now appreciate the importance of crystal size in determining the strength and toughness of a metal.

## CRYSTAL SIZE

The crystal size should be as small as possible so that any 'slip' that does take place is quickly brought to a halt by meeting a change of direction in the slip plane at the grain boundary.

*The larger size of the individual crystals, the more damage any crystal that slips can do to the whole since the length of slip is all the greater.*

One of the ways of obtaining 'toughness' in a metal is to *refine the grain* or reduce the size of the crystals.

## 6.4 Recrystallisation

All metals possess the useful property of being able to recrystallise without necessarily having to be made molten again, and some metals can recrystallise more easily than others. Lead, for example, when bent, rolled or generally subjected to severe distortion will recrystallise under the action of 'cold work' alone. Because lead remains 'soft' when cold worked, it can be cold extruded into tubes and pipes with comparatively little force. However, the cold extrusion of most metals requires very much more force to accomplish since they tend to 'work-harden'.

Figure 6.7 illustrates the difference between lead and other metals when subjected to cold work.

Due to the differential rates of cooling that take place when a metal is cast into an ingot (Figure 6.4), the resultant crystalline structure is not a good one with regard to mechanical properties. However, this undesirable structure can be easily rectified. By reheating the ingot and passing it through the rolling mill, the original structure is broken down and replaced by a series of smaller uniform crystals. Since these crystals do not work-harden during the hot rolling process, they are left in the softened condition.

*The hot rolling and forging of metals, apart from being a convenient method of forming metals into desirable shapes, will reduce and regularise the size of the crystals, in other words 'refine the grain'.*

## 6.5 Properties

The properties possessed by metals will not only influence their suitability for a particular component or structure and its performance under service conditions, but also, to a great extent, the choice of manufacturing or fabrication process required to produce it.

Figure 6.8 illustrates those properties that allow materials to withstand loads.

Properties that permit a material to be worked and deformed are

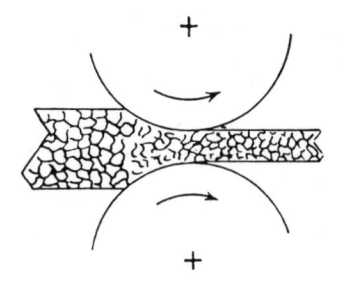

(a) COLD-ROLLING LEAD (RECRYSTALLISATION)

As lead passes between the rolls the act of squeezing alone, causes the original crystal structure to break down and be replaced by a fresh one, which shows no effect of distortion due to the action of the rolls.

*Note: Lead will recrystallise below room temperature.*

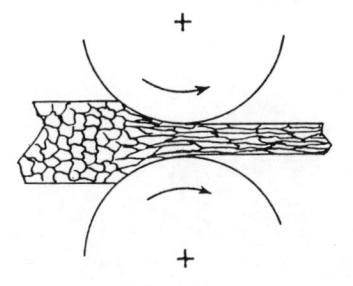

(b) COLD-ROLLING STEEL (CRYSTAL DEFORMATION)

Steel and other metals will not recrystallise in the cold state. The action of rolling is to cause 'slip' along the slip planes and the metal emerging from the rolls consists of the same crystals which are now elongated. *An elongated and distorted crystal structure of this kind gives the metal greater hardness and tensile strength and reduces the ductility, making it more resistant to further cold work.*

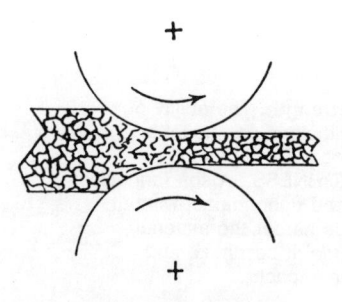

(c) HOT-ROLLING STEEL (RECRYSTALLISATION)

When metals are heated to a greater or lesser extent they come to resemble lead in that they are made softer and lose their ability to work-harden and now recrystallise under the action of work alone.
Hot rolling of metals has the same effect on the crystal structure as has the cold rolling of lead. Recrystallisation takes place continuously, the individual crystals becoming smaller with each rolling 'pass'.

**Fig. 6.7 The effect of cold and hot rolling on metal structure**

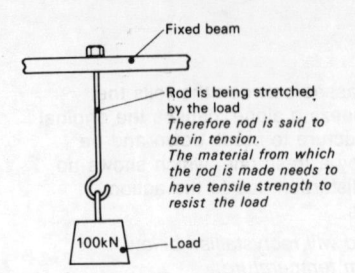

(a) **Tensile Strength.** The ability of a material to withstand a stretching load without fracturing.

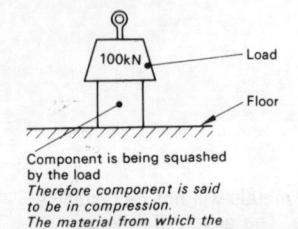

(b) **Compressive Strength.** The ability of a material to withstand a squeezing load without being crushed and fractured.

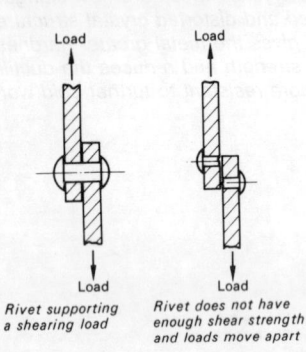

(c) **Shear Strength.** The ability of a material to withstand offset loads, or cutting actions without fracturing. Although the load shown is a tensile one, the same effect can be produced by a compressive load.

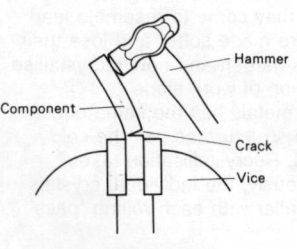

(d) **Impact Strength.** The ability of a material to withstand an impact or hammering load. This is often termed 'TOUGHNESS'. A soft (annealed) metal with bend upon impact without fracturing. The harder the material the more '*brittle*' it becomes, and will shatter on impact.

**Fig. 6.8 Material properties (strength)**

shown in Figure 6.9. Other important material properties are illustrated in Figure 6.10.

Mechanical tests are carried out to determine the properties of a material, to ascertain whether or not a raw material is up to required standard, and as a check upon manufacturing processes, for example heat treatment.

One of the principle mechanical tests carried out is the *tensile test* in which a specimen to be tested is *loaded in tension until fracture occurs*.

Figure 6.11 shows a typical stress/strain curve for a test specimen of mild steel, which is a *ductile* metal.

Not all materials when tensile tested, have a *yield point*, and in such cases a *yield stress* cannot be established. However, a *proof stress* can be determined by tensile testing, as shown in Figure 6.12.

Figure 6.13 compares characteristic tensile test curves for various common metals.

For further information refer to BSS 18 'Tensile Testing of Metals'.

## 6.6 Ferrous metals

The Latin name for iron is '*ferrum*' and the word '*ferrous*' means appertaining to iron. Thus *'ferrous' metals are those in which iron is the main constituent*. The word '*ferrite*' is used to describe *almost pure iron*, which outside the laboratory is rarely found in the pure state; it does in fact *hold a minute quantity of carbon in solution*.

Slight variations in the amount of carbon present can greatly influence the properties of the metal, and Table 6.1 shows how the addition of varying amounts of carbon, which is a *non-metal*, can produce a wide range of ferrous metals.

The effect of *carbon content* upon the properties of plain carbon steel is illustrated in Figure 6.14. On inspection, the diagram clearly indicates that *any increase in hardness and tensile strength is at the expense of a reduction in ductility*.

### PURE IRON

Commercially-pure iron is a soft ductile material which has a very limited use. However, *with the addition of silicon it has magnetic properties* which make it a most suitable material for transformer cores.

### WROUGHT IRON

Wrought iron is produced from high quality pig iron. It has a *characteristic fibrous structure* as a result of entrapped slag having been rolled and forged into long fibres throughout the entire length of the

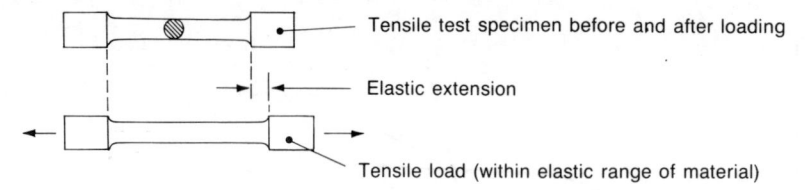

**(a) Elasticity** The ability of a material to deform under a load and return to its original shape when the load is removed. If the load exceeds the elastic limit of the material permanent deformation results.

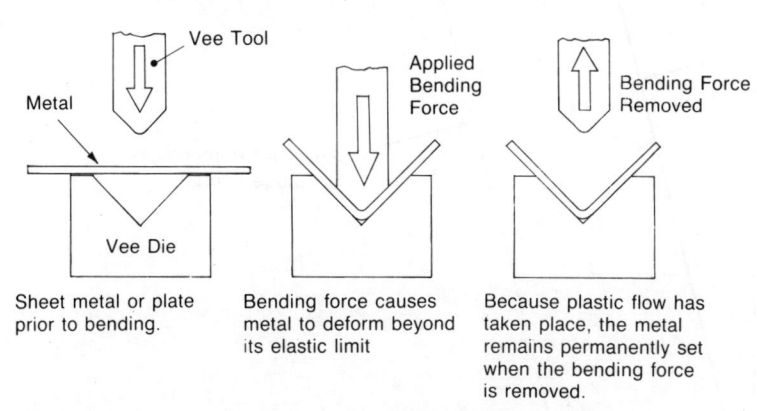

**(b) Plasticity** The ability of a material to deform under a load and retain its new shape when the load is removed.

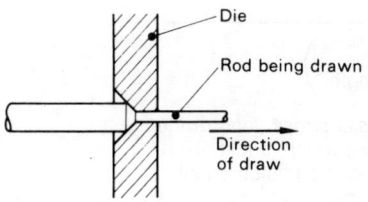

*A rod being drawn through a die to reduce its diameter requires the property of ductility.*

**(c) Ductility** This term is used when *plastic deformation* results from the application of a *tensile load*. Ductile metals are essential for such processes wire drawing, metal spinning (10.22) and producing panels by presswork (10.29).

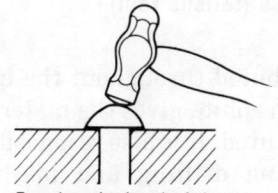

*Forming the head of rivet by hammering. The rivet needs to be made from a malleable material to withstand this treatment*

**(d) Malleability** This term is used when *plastic deformation* results from the application of a *compressive load*. Malleable metals can be cold formed by hollowing and raising (10.18) or by use of the wheeling machine (10.21).

**Fig. 6.9 Material properties (flow)**

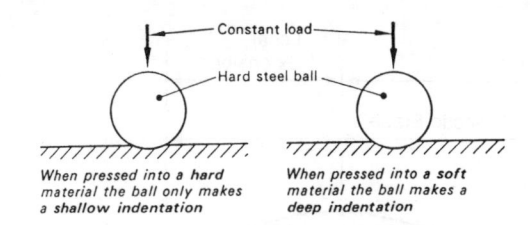

*When pressed into a hard material the ball only makes a shallow indentation*

*When pressed into a soft material the ball makes a deep indentation*

**(a) Hardness** The ability of a material to withstand scratching or indentation by another hard body.

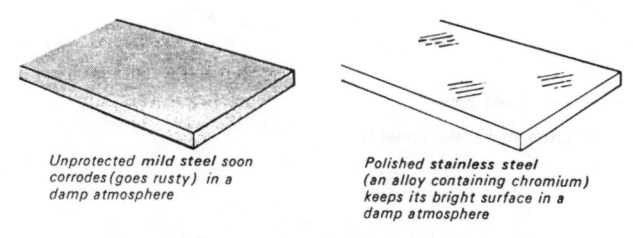

*Unprotected mild steel soon corrodes (goes rusty) in a damp atmosphere*

*Polished stainless steel (an alloy containing chromium) keeps its bright surface in a damp atmosphere*

**(b) Corrosion resistance** The ability to withstand chemical and electro-chemical attack.

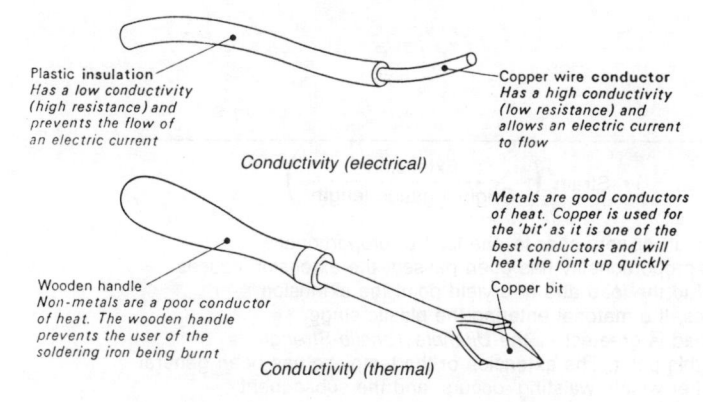

*Plastic insulation Has a low conductivity (high resistance) and prevents the flow of an electric current*

*Copper wire conductor Has a high conductivity (low resistance) and allows an electric current to flow*

*Conductivity (electrical)*

*Wooden handle Non-metals are a poor conductor of heat. The wooden handle prevents the user of the soldering iron being burnt*

*Metals are good conductors of heat. Copper is used for the 'bit' as it is one of the best conductors and will heat the joint up quickly*

*Copper bit*

*Conductivity (thermal)*

**(c) Conductivity** (Electrical or thermal) refers to the ability of a material to allow the passage of electricity or heat.

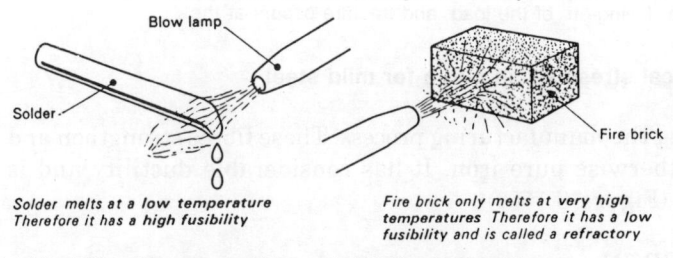

*Solder melts at a low temperature Therefore it has a high fusibility*

*Fire brick only melts at very high temperatures Therefore it has a low fusibility and is called a refractory*

**(d) Fusibility** The ease at which a material will melt. Materials that will only melt at very high temperatures are termed 'REFRACTORIES' (see 4.7)

**Fig. 6.10 Material properties (miscellaneous)**

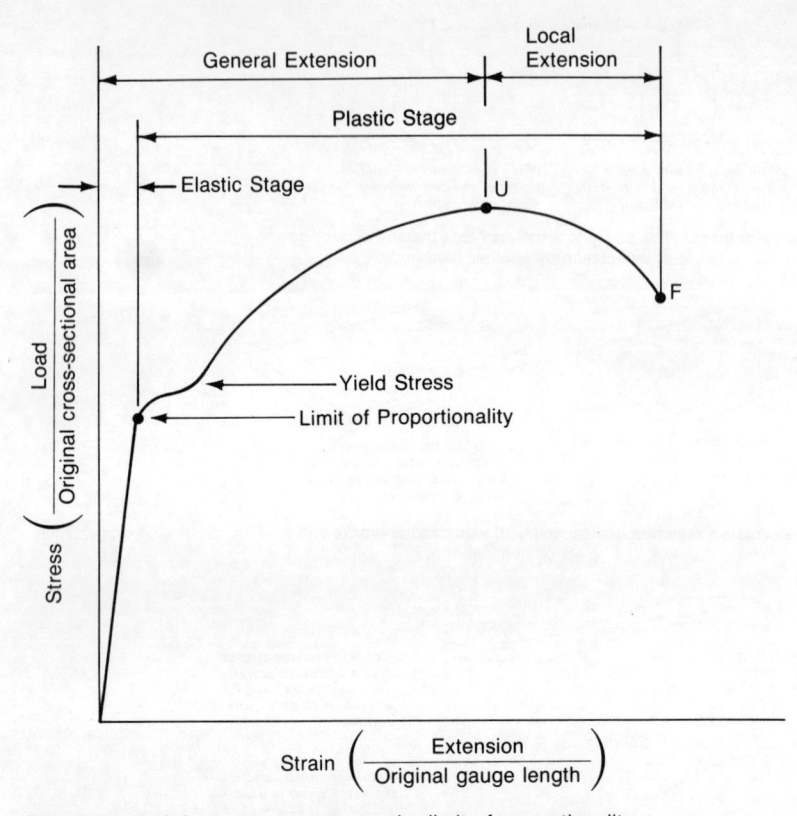

The elastic limit is at, or very near to, the limit of proportionality. When the limit of proportionality has been passed, the extension ceases to be proportional to the load and at a yield point the extension suddenly increases, the material entering the plastic stage.
At point 'U' the load is greatest — the *Ultimate Tensile Strength* is determined from this point. The extension of the testpiece has been general up to point 'U', after which 'waisting' occurs, and the subsequent extension is local. Since the cross-sectional area of the testpiece is considerably reduced at the 'waist', the actual stress increases, accompanied by a 'falling-off' of the load, and fracture occurs at the point 'F'.

**Fig. 6.11 Typical stress/strain curve for mild steel**

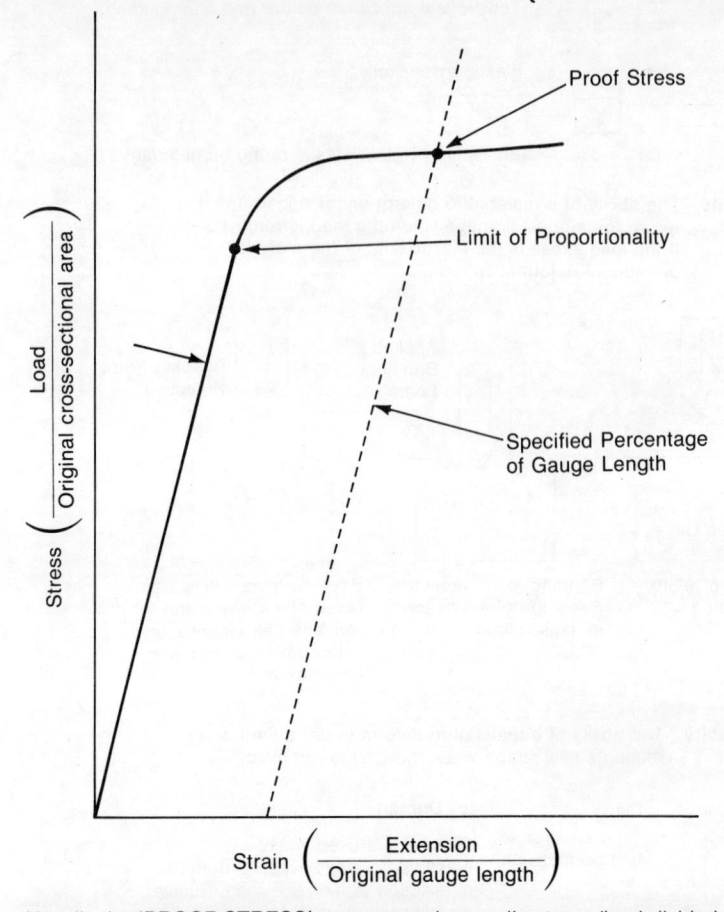

Usually the 'PROOF STRESS' represents the *tensile stress* (load divided by the original cross-sectional area of the testpiece) which is just sufficient to produce, under load, a non-proportional elongation equal to a specified percentage of the original gauge length.
Sometimes a proof stress is determined in *compression* or in *shear*.

**Fig. 6.12 Stress/strain curve — proof stress (tensile test)**

material during the manufacturing process. These fibres strengthen and toughen an otherwise pure iron. It has considerable ductility and is easily welded (Figure 6.15(a)).

## GREY CAST IRON

The carbon content of grey cast iron is far in excess of the upper limit of 1·7% carbon contained by plain carbon steels. Thus considerable uncombined or 'free' carbon exists distributed throughout the iron as fine flakes of GRAPHITE. The presence of graphite gives the material its characteristic grey appearance when fractured. It is also responsible for the dirty black dust experienced during drilling and machining operations. However, the graphite is important in that it not only acts as a self-lubricant when the material is drilled or machined, but also produces a cushioning effect between the iron grains which tends to dampen vibration. Cast iron has a high compressive strength but a low

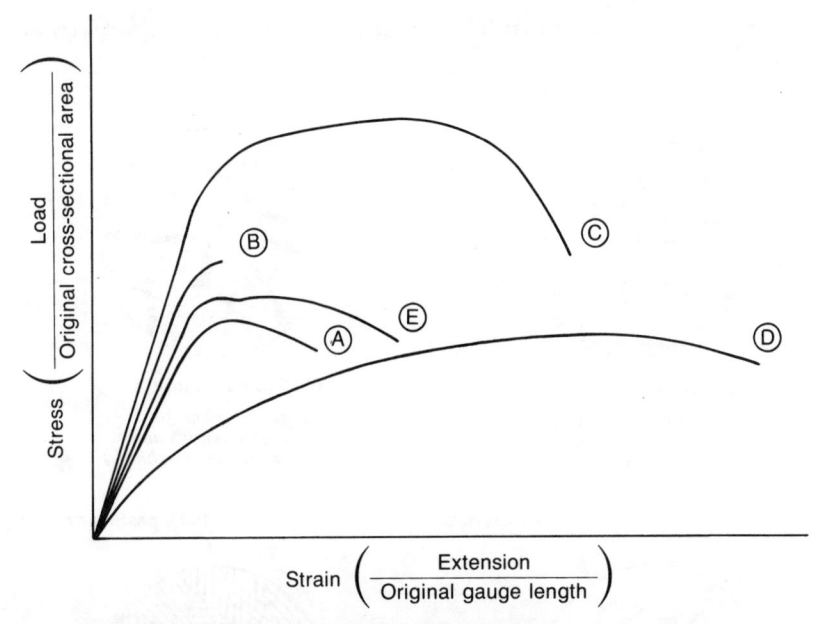

An inspection of the curves in the diagram shows that:
A — An ALUMINIUM ALLOY is fairly strong and ductile
B — The brittleness of CAST IRON is clearly indicated. Cast iron performs much better under *compression* and is rarely used where the load is a *tensile* one
C — HARD DRAWN BRASS is both strong and ductile
D — ANNEALED COPPER is very ductile
E — HARD DRAWN COPPER, on the other hand, is much stronger but less ductile than 'D'

**Fig. 6.13 Comparison of tensile test curves for various metals**

tensile strength; in compression it is about four times stronger than in tension. Because of its somewhat brittle structure it is not suitable for shaping by hot or cold working. However, *it is extremely fluid when molten and is therefore suitable for making intricate castings*. It can be both fusion and braze welded (Figure 6.15(*b*)).

## DEAD MILD STEEL

Because the carbon content of dead mild steel is very low, it possesses high ductility and malleability enabling it to be deep drawn and pressed into complicated shapes by cold working.

## MILD STEEL

This low carbon steel contains more *iron carbide* than dead mild steel making it much harder but slightly less ductile and malleable. It is

**Table 6.1 Ferrous metals**

| | CARBON-CONTENT % | TENSILE STRENGTH MN/m² | SOME APPLICATIONS |
|---|---|---|---|
| WROUGHT IRON | Less than 0·05 | 330 | Chains for lifting tackle, crane hooks, ornamental architectural ironwork. |
| GREY CAST IRON | 3·2 to 3·5 | 225 | General castings, machine beds, frames and sliding surfaces. |

Plain carbon steels are alloys of IRON and CARBON in which the iron and carbon are CHEMICALLY COMBINED at all times. The maximum solubility of carbon in iron is 1·7%, but the range of carbon content for carbon steel is from 0·1% to 1·2%.

| | CARBON-CONTENT % | TENSILE STRENGTH MN/m² | SOME APPLICATIONS |
|---|---|---|---|
| DEAD MILD STEEL | 0·1 to 0·15 | 360 | Sheets for deep-drawing and press forming of panels (aircraft and vehicle bodies), thin wire, rod and drawn tubing. |
| MILD STEEL | 0·15 to 0·3 | 460 to 500 | A general purpose material available in bars, plate, rods, sheet, wire and various structural sections. |
| MEDIUM CARBON STEEL | 0·3 to 0·5 | 700 to 800 | High tensile tubing, crankshafts, forgings and axles, hammer heads, screwdrivers, spanners, wood saws, wire ropes. |
| | 0·5 to 0·8 | | Cold chisels, high tensile tubing, leaf springs. |
| HIGH CARBON STEEL | 0·8 to 1·0 | 900 | Coil springs, shear blades, wood chisels, high tensile wire (piano wire). |
| | 1·0 to 1·2 | | Files, drills, taps and dies. |
| | 1·2 to 1·4 | | Ball bearings, files, metal turning tools, fine-edged tools, knives, etc. |

Medium carbon steels respond to heat-treatment to further increase their toughness and hardness, and are used for highly stressed components.

High carbon steels are mainly used where their properties — after suitable heat-treatment — of hardness and wear resistance can be exploited.

*Note*: 1 Ton/in² = 15·44 MN/m²; 1 lb/in² = 6·895 kN/m².

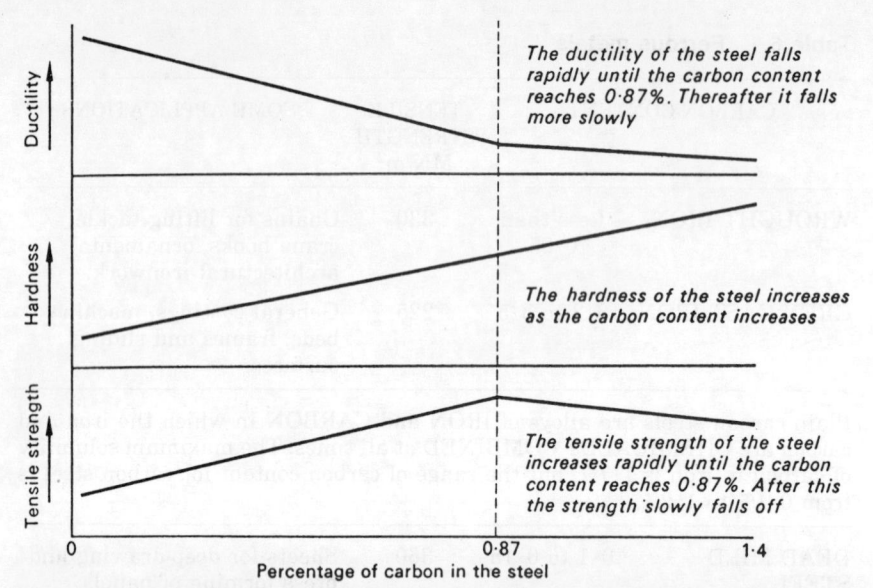

**Fig. 6.14** The effect of carbon content on the properties of carbon steels (annealed)

probably the most widely used fabrication material, being easily cold-worked or hot-worked (Figure 6.15(c)).

## MEDIUM CARBON STEEL

These steels, when they are suitably heat-treated, provide the tough materials needed for components that require hardness combined with strength and ductility. Medium carbon steel is less malleable than the low carbon steels and cannot be cold-worked to any great extent without risk of fracture (Figure 16.5(d)).

## HIGH CARBON STEEL

These steels combine hardness with high strength. They are less ductile and less tough than the medium carbon steels. Cold forming is not recommended, but they can be hot forged within a carefully controlled temperature range, and may be satisfactorily machined when heat-treated to a *normalised condition*. They can also be welded (Figure 6.15(e)).

Typical applications of the above materials are listed in Table 6.1.

## 6.7 Alloy steels

The required properties in respect of plain carbon steels depend, to a great extent, on correct heat-treatment. A relatively thin section of

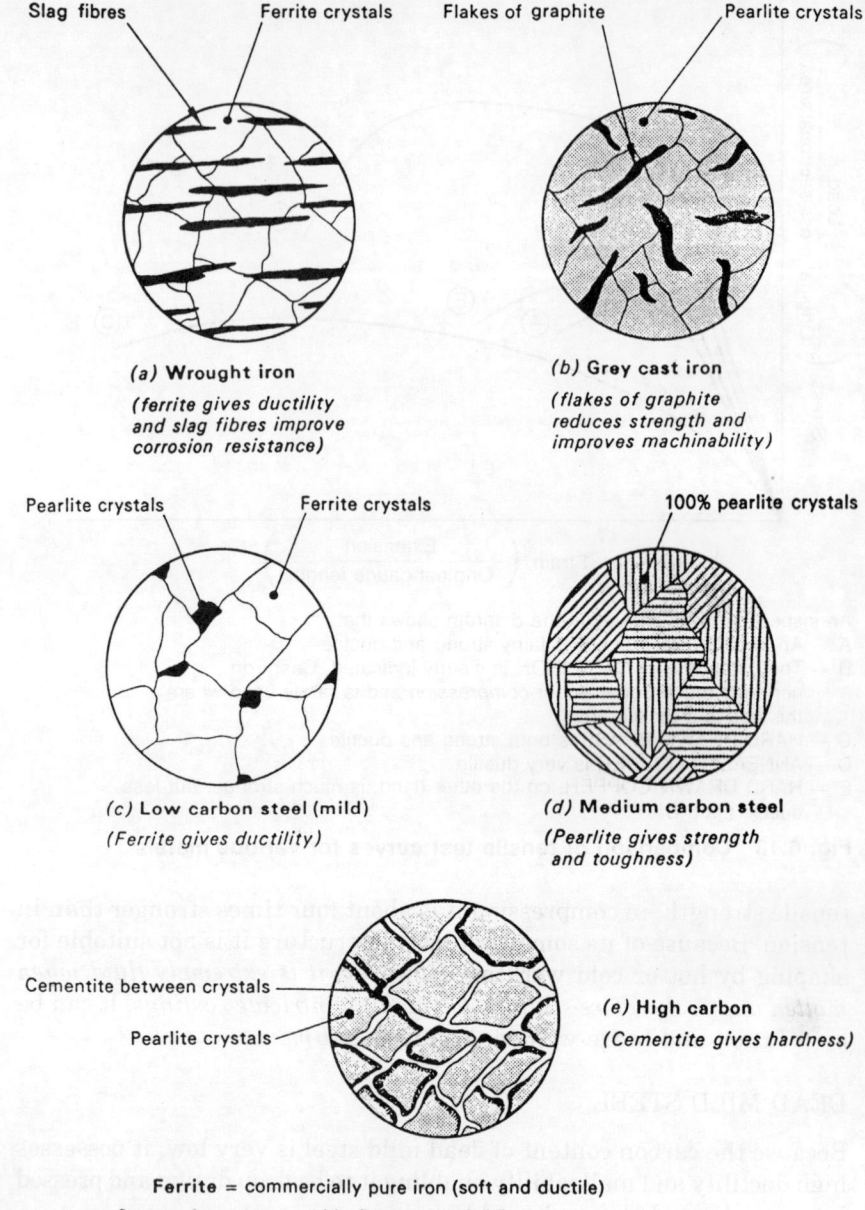

Ferrite = commercially pure iron (soft and ductile)
Cementite = iron carbide (hard and brittle)
Pearlite = laminated (like plywood) crystals of ferrite and cementite (maximum strength)

Magnification x 100

**Fig. 6.15** Ferrous metals micro-structures

Table 6.2  Metallic elements — alloy steels

| METAL | FUNCTION |
|---|---|
| Nickel | — increase the toughness. |
| Chromium | — improves the hardness, increases resistance to corrosion and imparts STAINLESS qualities. |
| Tungsten | — gives the steel the ability to cut at high temperatures without softening (e.g. HSS). |
| Manganese | — provides additional strength and resistance to wear. |
| Colbalt | — reduces the hardness. |
| Molybdenum Vanadium | — increases the hardness. |

The non-metal CARBON is an essential constituent of alloy steels, for without it they could not attain their useful mechanical properties. The carbon makes hardening and tempering possible, whilst the above alloying elements modify the effect of the heat-treatment (see Section 6.33).

Table 6.3  Composition of a typical high-speed tool steel

| ELEMENT | % | REMARKS |
|---|---|---|
| Carbon | 0·75 | High-speed steels are widely used for all types of cutting tools, lathe and shaper bits, milling cutters, routers, twist drills, reamers and saw blades. Used for components subjected to wear and abrasion in service. |
| Chromium | 4·7 | |
| Iron | 71·15 | |
| Tungsten | 22·0 | |
| Vanadium | 1·4 | |

carbon steel when heated and quenched will give uniform properties throughout the section, but when a large mass of carbon steel is subjected to the same treatment, the result is not so straightforward.

It is not possible to fully harden a plain carbon steel having a section in excess of approximately 30 mm because the quenching effect does not penetrate right to the inside. The result is that the interior of the section will cool fairly slowly, despite the rapid cooling on the outside, producing a gradual decrease in hardness from the outside to the inside of the quenched steel. Furthermore, unequal dimensions of a heavy block of high carbon steel can set up internal stresses causing the material to crack when quenched.

Although these factors place limitations on the uses of carbon steels, the difficulties have been overcome, and other advantages conferred, by the use of alloy steels, in which uniform hardness can be produced through the whole section or block when heated and quenched.

Alloy steels may be regarded as materials which contain in addition to iron and carbon, small additions of other metallic elements in order to produce the special qualities listed in Table 6.2.

The advantages of alloy steels over plain carbon steels are:

1. The additional elements (Table 6.2), enable special qualities to be imparted to the metal, such as improved strength and hardness, resistance to wear, elasticity or resistance to corrosion.
2. A combination of mechanical properties can be obtained by less drastic heat-treatment than would be necessary for a plain carbon steel.
3. Heat-treatment has a uniform effect in large masses of steel without the outer skin behaving differently from the interior, as would be the case if bulky sections of 'straight carbon' steel were treated.

HIGH-SPEED TOOL STEEL (HSS)

Plain carbon steels *temper* and lose hardness at comparatively low temperatures. The degree of softening depends upon the temperature to which the steel is heated, but at temperatures as low as 250°C, a high carbon tool steel will begin to soften and lose its keen cutting edge.

High-speed steels, of which there are a variety of grades, do not suffer from the low temperature tempering effect referred to; they will retain their hardness at high temperatures (600°C) and consequently will cut at high speed.

With all high-speed tool steels the main alloying element is TUNGSTEN which varies between 14% and 22%, depending on the grade. Table 6.3 lists the alloying elements of a top grade HSS together with applications.

## 6.8 Non-ferrous metals

The term 'non-ferrous metals' refers to those metals that contain no appreciable quantities of iron. Any iron that is present only exists as an impurity.

There are some 38 metals, apart from iron, which are known to man but very few of them are widely used in engineering. The non-ferrous metals most commonly used in fabrication engineering are listed

187

**Table 6.4  Common non-ferrous metals**

| METAL | DENSITY KG/M$^3$ | MELTING °C | PROPERTIES | TYPICAL USES |
|---|---|---|---|---|
| Aluminium | 2700 | 660 | High electrical and thermal conductivity soft, ductile, and low tensile strength 93 MN/m$^2$ | The base of many engineering alloys. Lightweight electrical conductors. |
| Copper | 8900 | 1083 | Soft, ductile, and low tensile strength 232 MN/m$^2$ Second only to silver in conductivity, it is much easier to joint by soldering and brazing than aluminium. Corrosion resistant. | The base of brass and bronze alloys. It is used extensively for electrical conductors and heat exchangers, such as motor-car radiators. |
| Lead | 11 300 | 328 | Soft, ductile and very low tensile strength. High corrosion resistance. | Electric cable sheaths. The base of 'solder' alloys. The grids for 'accumulator' plates. Lining chemical plant. Added to other metals to make them 'free-cutting'. |
| Magnesium | 1740 | 650 | The lightest of the commercially used metals low tensile strength. Burns with a brilliant white flame. The easiest of metals to machine. | The constituent and basis for many high strength lightweight alloys. Extensively used in the aircraft industry and for portable machine casings. |
| Tin | 7300 | 232 | Resists corrosion | Coats sheet mild steel to give 'tin plate'. Used in soft solders. One of the bases of 'white metal' bearings. An alloying element in bronzes. |
| Zinc | 7100 | 420 | Soft, ductile, and low tensile strength. Corrosion resistant. | Used extensively to coat sheet steel to give 'galvanised iron'. The base of die-casting alloys. An alloying element in brass. |

**Table 6.5  Further important non-ferrous metals**

| METAL | DENSITY (kg/m$^3$) | MELTING POINT (°C) | PROPERTIES | TYPICAL USES |
|---|---|---|---|---|
| CHROMIUM | 7500 | 1890 | Resists corrosion. Raises strength but lowers ductility of steels. Improves heat-treatment properties. | Used as an alloying element in high-strength and corrosion-resistant steels. Used for electro-plating. |
| COBALT | 8900 | 1495 | Improves wear resistance and 'hot hardness' of high-speed steels. | Used as an alloying element in 'super' high-speed steels and in permanent magnet alloys. |
| MANGANESE | 7200 | 1260 | High affinity for oxygen and sulphur. Soft and ductile. | Used to deoxidise steels and to offset the ill-effects of the impurity sulphur. Larger amounts improve wear resistance. |
| MOLYBDENUM | 9500 | 2620 | A heavy, heat-resistant metal that readily alloys with other metals. | Used as an alloying element in high-strength nickel/chrome steels to improve mechanical and heat-treatment properties. It reduces mass effect and temper-brittleness. |
| NICKEL | 8900 | 1458 | A strong, tough, corrosion-resistant metal widely used as an alloying element. | Used as an alloying element to improve the strength and mechanical properties of steel. Used for electroplating. |
| TUNGSTEN | 1930 | 3410 | A refractory metal, very hard and highly erosion-resistant at elevated temperatures. | The main alloying element for high-speed tool steels. The hard compound tungsten carbide is used for 'tipped' high-speed cutting tools. |
| VANADIUM | 7190 | 1890 | A hard corrosion-resistant that readily combines with nearly all non-metals, and is a 'scavenging agent' for removing non-metallic impurities. Produced commercially in the form of an iron alloy. | Major commercial use as alloying element in steels and cast irons to improve ductility and shock resistance. |

in Table 6.4, together with their applications. Other less common but important non-ferrous metals are given in Table 6.5.

Some very special non-ferrous metals, listed in Table 6.6, are relatively new to engineering although they have been known for many years. It is only since the Second World War that it has been possible to produce them in bulk. Due to advances in modern engineering technology there is an increase in the commercial demand for their use. However, these metals are very expensive compared with more common engineering materials and therefore are only used where their special properties can be best exploited.

In this modern 'jet age', TITANIUM has become a very important fabrication material. Commercially-pure titanium has a low density (4507 kg/m$^3$) and a relatively high melting point (1668°C), a very high STRENGTH/WEIGHT ratio, corrosion resistance and reasonable strength at elevated temperatures. All these properties are exploited to advantage in the aircraft production industry. Titanium is extensively used for

**Table 6.6 Modern non-ferrous metals**

| METAL | MELTING POINT (°C) | REMARKS |
|---|---|---|
| BERYLLIUM | 1283 | Has a low density, lighter than aluminium and nearly as light as magnesium but differs in that its melting point is some 600°C higher. When alloyed with copper and heat-treated, the alloy develops a strength two or three times that of mild steel and six times that of copper. Beryllium copper alloys are used to make instrument springs, and can be hardened to provide 'non-sparking' tools for use in oil fields and on gas rigs. |
| NIOBIUM | 2468 | Has excellent corrosion resistance; used as an alloying element in tools, dies and super conductive magnets. It is a major alloying element in nickel-based high-temperature alloys, and as an important additive to high-strength structural steels. Used for nuclear reactor components. |
| TANTULUM | 2996 | Has an unusually high resistance to corrosion, *hydrofluoric acid* being the only chemical agent that will attack it at room temperature. Used extensively in chemical equipment where heat has to be transferred under intensive corrosive conditions. Used in the form of carbides combined with other carbides for 'tipped' cutting tools. Its properties are exploited in nuclear reactors. |
| ZIRCONIUM | 1850 | The most important use of zirconium is in nuclear reactors. It has good strength at elevated temperatures and a high resistance to corrosion. Like magnesium, when in finely divided powder form, is reactive and ignites at low temperatures (photo-flash bulbs). When alloyed with magnesium it drastically reduces the grain size, thus greatly improving its mechanical properties. |
| TELLURIUM | 450 | Small amounts as an alloying element are used to increase the ductility of aluminium, the hardness and strength of tin alloys, and the machineability of stainless steel and copper. Used instead of lead in free-cutting alloys. |
| TITANIUM | 1668 | Used in 'supersonic' aircraft and rockets as it has very much higher STRENGTH/WEIGHT ratio than aluminium and retains its strength at elevated temperatures. Grades of commercially pure titanium are available with tensile strengths from 402 $MN/m^2$ to about 773 $MN/m^2$. |

'skinning' the fuselages of jet aircraft where friction with the air, during flight, raises the temperature of the 'skin' above the safe limit for aluminium and its alloys.

Two of the most important non-ferrous metals, ALUMINIUM and COPPER, apart from forming the basis for a wide range of alloys, are extensively used as commercially-pure metals (see Tables 6.7 and 6.8).

## 6.9 Aluminium alloys

Although a vast range of aluminium alloys are available as engineering materials and have various trade names, they may all be classified into one of four groups:

(*a*) wrought alloys (heat-treatable);
(*b*) cast alloys (heat-treatable);
(*c*) wrought alloys (non-heat-treatable);
(*d*) cast alloys (alloys (non-heat-treatable).

For further details refer to British Standards publications BS 1470–77 for wrought metals and BS 1940 for cast metals. Aircraft materials are covered by the 'L' series of British Standards and specifications published by the Directorate of Technical Development (Ministry of Supply) — commonly known as DTD specifications.

**Table 6.7 Aluminium**

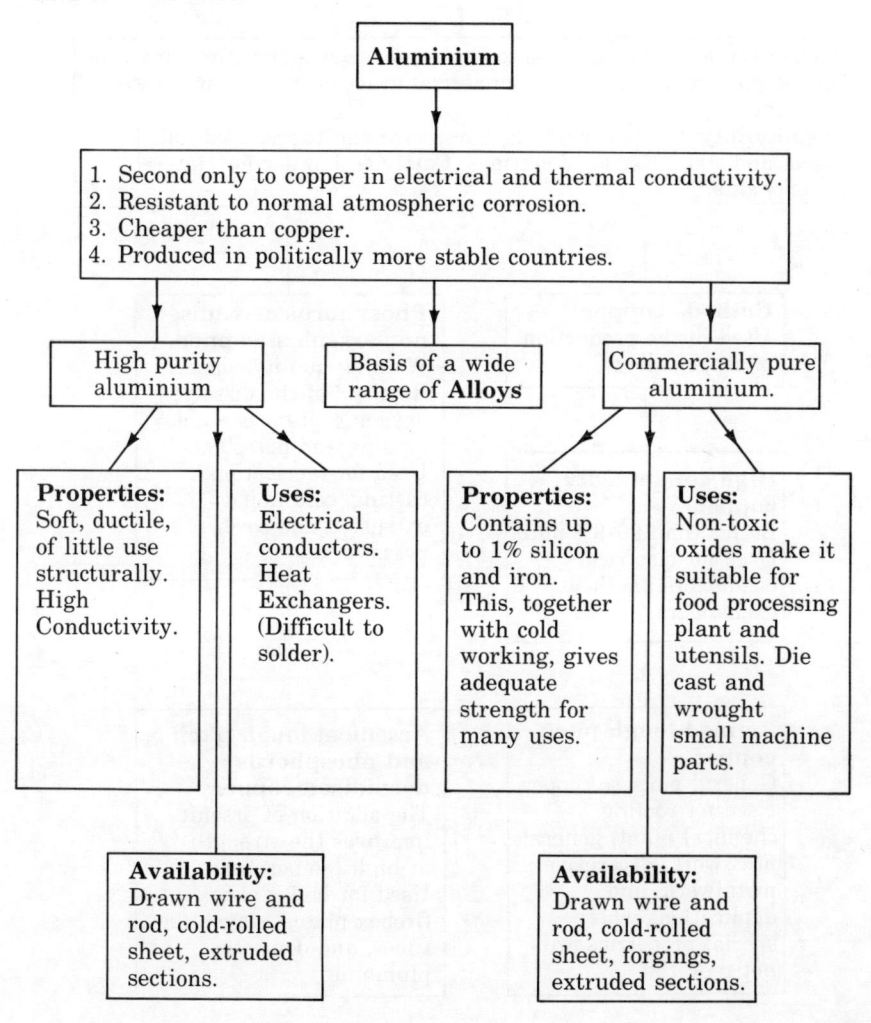

**Aluminium**

1. Second only to copper in electrical and thermal conductivity.
2. Resistant to normal atmospheric corrosion.
3. Cheaper than copper.
4. Produced in politically more stable countries.

**High purity aluminium**

**Properties:** Soft, ductile, of little use structurally. High Conductivity.

**Uses:** Electrical conductors. Heat Exchangers. (Difficult to solder).

**Availability:** Drawn wire and rod, cold-rolled sheet, extruded sections.

**Basis of a wide range of Alloys**

**Commercially pure aluminium.**

**Properties:** Contains up to 1% silicon and iron. This, together with cold working, gives adequate strength for many uses.

**Uses:** Non-toxic oxides make it suitable for food processing plant and utensils. Die cast and wrought small machine parts.

**Availability:** Drawn wire and rod, cold-rolled sheet, forgings, extruded sections.

189

**Table 6.8  Copper**

```
                    ┌──────────┐
                    │  Copper  │
                    └────┬─────┘
                         │
                         ▼
```

**Properties:**
1. Relatively high strength,
2. Very ductile (easily cold-worked).
3. Corrosion resistant.
4. Second only to silver as a conductor of heat and electricity
5. Easily joined by soldering, brazing and welding.

One of the few metals of use to the engineer as a structural material in the pure state, although commercial grades contain some trace elements.
**Availability:** Cold drawn rods, bars, wire and tubes. Cold-rolled sheet and strip. Extruded sections. Castings. Powder for sintered components.

**Cathode copper**
Used in the production of copper alloys.

**Phosphorus-deoxidised non-arsenical copper**
Welding quality copper. Removal of the dissolved oxygen content prevents gassing and porosity. Used for fabrication, casting, cold impact extrusion and severe presswork.

**High conductivity copper**
Better than 99·9% pure. Used for electrical conductors and heat exchangers.

**Refined tough pitch copper**
General purpose copper. Used for roofing, chemical plant, general presswork, decorative metalwork, and applications where special properties are not required.

**Arsenical tough pitch and phosphorus deoxidised copper**
The addition of arsenic improves the strength at high temperatures. Used for boiler and firebox plates, stays, flue tubes, and domestic plumbing.

---

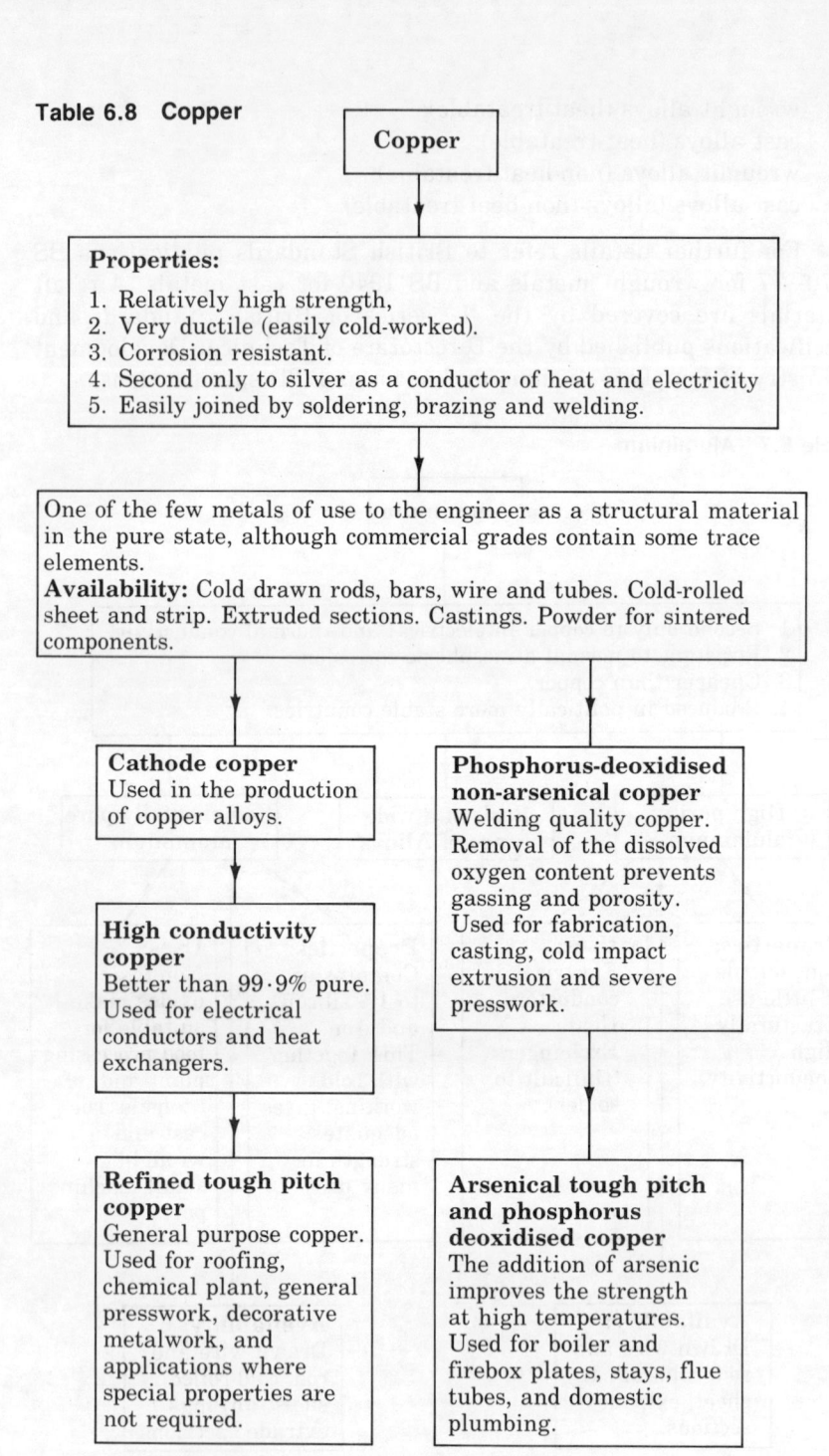

Table 6.9 gives details of a few typical aluminium alloys together with some applications.

DURALUMIN

This aluminium alloy has a tensile strength in the region of 450 MN/m$^2$ and although it is only about a third of the weight of steel, it is approximately as strong. Because it possesses a very high STRENGTH/WEIGHT ratio, it is employed extensively for airframe construction where lightness coupled with high strength is of paramount importance.

Duralumin alloys rely on a process called '*solution heat-treatment*' (discussed later in this chapter) for their excellent mechanical properties.

ALCLAD

One disadvantage of duralumin is that it corrodes more rapidly than other aluminium alloys, especially when in contact with salt water. To overcome this problem a manufacturing process known as 'cladding' was developed to provide the duralumin with a protective coating of aluminium. The efficiency of the resultant 'three-ply composite', in which the duralumin is sandwiched between two layers of aluminium, is such that a sheet of alclad of 1·6 mm in thickness will withstand exposure to the continuous action of a salt spray for a period in excess of 5 years without deterioration by corrosion.

The process of cladding is explained in Figure 6.16.

Alclad sheet is heat-treated in a similar manner to duralumin. It is heated rapidly to 495 °C and quenched, bringing it to the soft condition to enable the metal to be beaten, pressed, or otherwise cold-worked to the required shape. After cold working it is allowed to 'age' either at room temperature or in a controlled furnace. As a result, the duralumin core acquires the increased hardness and strength produced by 'age hardening'.

## 6.10  Magnesium alloys

Because of its silvery-white colour and lightness, magnesium is often mistaken for aluminium. However, the identity of each metal may be easily determined by a very simple workshop test. When a sample piece of each metal is dropped, in turn, on to a concrete floor or an anvil, aluminium will give a very dull sound whereas the magnesuim will emit a distinct metallic ring, i.e., magnesium is '*sonorous*'.

One of the greatest advantages of using magnesium is the readiness with which it can be machined. *No metal can be cut, drilled, filed, or shaped so easily or so fast.* It lends itself readily to welding and

Table 6.9 Some typical aluminium alloys

| TYPE | COPPER | IRON | MAGNESIUM | MANGANESE | SILICON | OTHERS | APPLICATIONS |
|---|---|---|---|---|---|---|---|
| WROUGHT NON-HEAT-TREATABLE | 0.1 max | 0.7 max | — | 0.1 max | 0.5 max | — | Fabricated assemblies; Electrical conductors; Food and brewing processing equipment; Ornamental architecture. |
| | 0.15 max | 0.75 max | 4.5 to 5.5 | 1.0 max | 0.6 max | 0.5 Chromium | High-strength shipbuilding and engineering products; Excellent corrosion resistance. |
| CAST NON-HEAT-TREATABLE | 1.6 | — | — | — | 10.0 | — | General purpose alloy for moderately stressed pressure die-castings. |
| | — | — | — | — | 10.0 to 13.0 | — | A widely used alloy; Ideal for sand, gravity and pressure die-castings; extensively used for large marine, automotive and general engineering castings. |
| WROUGHT HEAT-TREATABLE | 4.2 | 0.7 | 0.7 | 0.7 | 0.7 | 0.3 Titanium | High STRENGTH/WEIGHT ratio Duralumin alloy. Widely used for stressed sections and components, especially in the aircraft industry. |
| | — | — | 0.6 | — | 0.5 | — | Corrosion-resistant alloy used for lightly stressed components such as motor vehicle body parts, window sections and glazing bars. |
| CAST HEAT-TREATABLE | 1.8 | 1.0 | 0.2 | — | 2.5 | 0.15 Titanium 1.2 Nickel | High rigidity combined with moderate strength and shock resistance. Suitable for sand and gravity die-castings; excellent general purpose alloy. |
| | — | — | 10.5 | — | — | 0.2 Titanium | A strong, ductle and highly corrosion-resistant alloy. Used for a wide range of aircraft and marine castings. |

riveting. Although magnesium has no great strength, when suitably alloyed and worked its strength can be doubled and even trebled.

Magnesium and its alloys can be subjected to casting, die casting, extruding, forging, pressing and rolling, and are extensively used in the

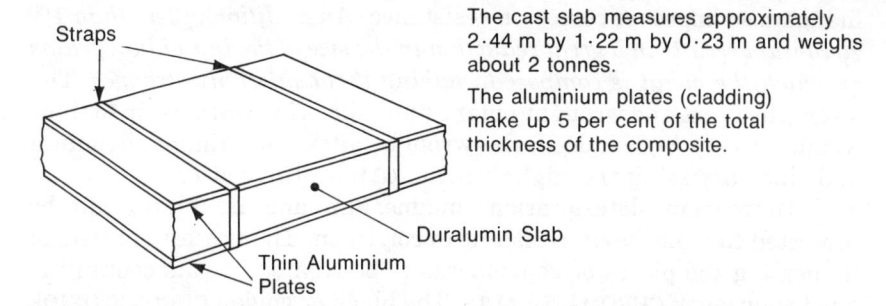

The cast slab measures approximately 2.44 m by 1.22 m by 0.23 m and weighs about 2 tonnes.

The aluminium plates (cladding) make up 5 per cent of the total thickness of the composite.

(a) Slab with cladding strapped into place. The straps are prised off after the first pass through the rolling mill. During the first passes the oxide films on both materials are disintegrated and dispersed by pressure. Recrystallisation takes place with crystal growth across the adjoining surfaces to give a perfect metallurgical bonding at the interface.

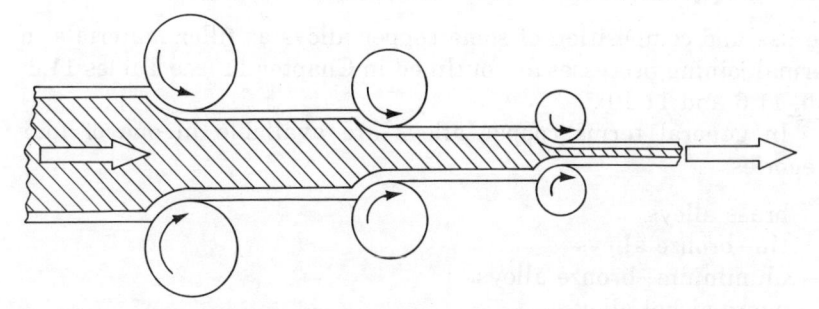

(b) Reduction by rolling, the cladding thickness remains of the same proportion to the core. The thickness from start to finish is usually within 5%.

Rapid cooling during the casting process produces brittle oxides and metallic compounds, formed by the alloying elements, which tend to segregate at the outer faces of the slab. These unwanted impurities are removed by milling or 'scalping' on a specially designed scalping machine. It is not necessary to mill the edges of the slab because the resultant sheet or strip material is usually edge trimmed during manufacture. A sandwich is prepared by placing the machined slab of duralumin between two thin plates of aluminium. The composite metal is then preheated to render it capable of the severe deformation involved in the rolling process. The alclad is then reduced to the required thickness by hot rolling. If required cladding can be added on to one side only of the slab.

Fig. 6.16 Cladding

cast form. In the wrought state (plate, rod, sheet, extruded sections, etc.), the alloying additions consist of MANGANESE; ALUMINIUM, ZINC and MANGANESE; or ZIRCONIUM and ZINC. The alloys with manganese contain about 2% of the element, while those containing zirconium and zinc have approximately 0.7% zirconium and 3% zinc. Aluminium and zinc act as hardeners to raise the mechanical properties whilst the presence of manganese improves corrosion resistance. *An addition of less than 1% zirconium results in a great reduction in the size of the individual grains of which the metal is composed, making it considerably stronger.* The strength of cast and wrought magnesium alloys in relationship to their weight is very high indeed; the wrought alloys containing zirconium and zinc possessing the highest STRENGTH/WEIGHT ratio.

To prevent deterioration, magnesium and its alloys can be protected to make them resistant to corrosion. The process consists of immersing the pieces or components to be treated in vats containing a hot solution of CHROMATE SALTS. The black or golden CHROMIC OXIDE film deposited on the surface of the metal forms a protective layer for the application of paint for further protection.

## 6.11 Copper, chromium and nickel alloys

The use and composition of some copper alloys as filler materials in thermal joining processes are outlined in Chapter 11 (see Tables 11.4, 11.5, 11.6 and 11.10).

In general terms copper alloys are available in one of four categories:

(1) brass alloys
(2) tin–bronze alloys
(3) aluminium–bronze alloys
(4) cupro-nickel alloys.

### BRASS ALLOYS

These are the most well-known copper alloys and contain up to about 40% zinc. Brasses which contain less than 36% zinc are ductile at room temperature and can be extensively cold-worked without the necessity for frequent annealing. The most ductile brass containing 30% zinc is familiarly known as 70/30 or '*cartridge brass*'. *The strength and hardness of brass gradually increases with rising zinc contents.* These properties are sharply altered at the 36% composition and further additions of zinc bring about a more rapid increase in hardness than before.

The atoms of pure zinc have a hexagonal lattice structure whereas those of pure copper are arranged in a face-centred cubic lattice. In addition, the space occupied by the copper and zinc atoms differs in that the zinc atoms are about 13% larger than the copper atoms. Therefore when these two metals are alloyed to form brass, considerable disruption of the parent metal's lattice structure occurs. When only small amounts of zinc are present in brass, the prevailing atomic arrangement is similar to that of copper, which means that zinc atoms, in solid solution, have to adapt to the face-centred cubic pattern. Each zinc atom takes the place of one copper atom, but because the space occupied by the zinc atom is greater than that of copper, the face-centred cubic lattice is slightly distorted and the whole of the lattice becomes slightly larger. This distortion of the lattice structure leads to a greater resistance to deformation than does the regular atomic pattern of the pure metal, thus when progressively increasing amounts of zinc are alloyed with copper, the brass becomes increasingly harder.

The effects of increasing amounts of zinc alloyed with copper are summarised in Table 6.10.

Some common brasses are listed in Table 6.11.

### TIN–BRONZE ALLOYS

Copper–tin alloys are termed 'bronzes' and contain a de-oxidiser. The purpose of the de-oxidiser is to prevent the tin content oxidising during casting and hot-working, rendering the alloy brittle. These alloys have a high resistance to wear and corrosion and are extensively used in the form of castings.

Two de-oxidisers are commonly used:

(a) PHOSPHORUS in the 'phosphor–bronze' alloys; and
(b) ZINC in the 'gun metal' alloys.

Some typical tin–bronze alloys are detailed in Table 6.12.

### ALUMINIUM–BRONZE ALLOYS

The metals copper and aluminium are both relatively weak, but when alloyed together can produce an alloy that is strong and has strikingly different properties to the parent metals. Copper with the addition of 5% aluminium produces an alloy twice as strong as copper, and with a 10% aluminium content, the resultant 'aluminium bronze' is as strong as mild steel, and very tough.

Figure 6.17 shows a graph representing the increase in strength of copper alloys containing aluminium. It will be seen that somewhat higher strength is obtained when the alloy is rapidly cooled, or 'chill cast', in metal moulds.

Aluminium–bronzes combine relatively high strength with good ductility. They are resistant to corrosion by chemicals and salt water, especially at high temperature. Unlike other copper alloys, they can be subjected to heat-treatments almost identical to plain carbon steel. Some typical aluminium–bronzes are listed in Table 6.13.

### Table 6.10 Effect of alloying zinc with copper

| COMPOSITION OF ALLOY | ATOMIC LATTICE STRUCTURE | CRYSTAL STRUCTURE AS SEEN UNDER A MICROSCOPE | ZINC % | DIAMOND PYRAMID HARDNESS NUMBER | TENSILE STRENGTH MN/m$^2$ | DUCTILITY %/50mm |
|---|---|---|---|---|---|---|
| Pure copper | Face-centred cubic. | Grains of pure copper | | 53 | 232 | 45 |
| Copper alloyed with up to 36% zinc | Face-centred cubic pattern which increases gradually in size. | Grains of solid solution of zinc in copper (known as 'alpha phase') | 10<br>20<br>30 | 60<br>62<br>65 | 278<br>309<br>324 | 55<br>65<br>70 |
| Copper with about 36% zinc | A new structure appears (body-centred cubic) in addition to the original face-centred cubic structure. | Small quantities of a new constituent appear (known as 'beta phase') | 36 | 70 | 340 | 60 |
| Copper with 36 to 42% zinc | With increasing zinc more and more of the alloy consists of the new constituent having a body-centred cubic structure. | The beta constituent increases in quantity while the alpha constituent diminishes | 40 | 85 | 417 | 45 |
| Copper with 42 to 52% zinc | The lattice is entirely body-centred cubic. | The structure is entirely beta | 45 | 90 | 432 | 20 |

*Note*:
1. When one metal exists in solid solution in another, the atoms of the added element usually take their place on the atomic lattice of the parent metal, in spite of the fact that the atomic size of the added metal is different and its lattice structure may be different. Such intrusions cause the lattice to become distorted, making the alloy harder and stronger than the parent metal.
2. When an alloy is made from two metals which possess a markedly different size of atom and which crystallise in different atomic patterns, the range of alloys divides into a number of 'phases'. The occurrence of a new phase is characterised by a change of the lattice arrangement and is also evident when the alloy is examined under a microscope.

## NICKEL–CHROMIUM ALLOYS

The 80/20 nickel–chromium alloys possess a high resistance to the passage of electricity and scaling at elevated temperatures. They are widely used for resistance wires (elements) in electric ovens and furnaces. The higher grades of resistance wire have the 80/20 composition, but lower grades are available in which 10% to 20% of iron replaces some of the nickel and chromium and these are suitable for operating at a dull red heat (500°C).

### Table 6.11 Brass alloys

| TYPE | COPPER | ZINC | OTHER ELEMENTS | APPLICATIONS |
|---|---|---|---|---|
| BASIS BRASS | 63 | 37 | — | The cheapest of the cold-working brasses. It has poor ductility and good strength. Has limited cold-working ability. |
| STANDARD BRASS | 65 | 35 | — | Soft and ductile, suitable for most engineering processes. Can be cast but not suitable for hot-working. |
| CARTRIDGE BRASS | 70 | 30 | — | Most ductile and malleable of the brasses. Will withstand severe deformation without cracking. Widely used for sheet-metal pressings and deep-drawn components. When work-hardened has a tenacity of 600NM/m$^2$ which is double its strength in the annealed state. Originally developed for the manufacture of cartridge cases, hence its name. |
| MUNTZ METAL | 60 | 40 | — | Not suitable for cold-working but works well hot. Relatively cheap due to its high zinc content. Widely used for castings and forgings, extrusions and hot stamping processes. |
| ADMIRALTY BRASS | 70 | 29 | 1·0 Tin | Similar to cartridge brass but its corrosion resistance is improved by the addition of tin. Resists attack by sea water and is used largely for condenser tubes. |
| NAVAL BRASS | 62 | 37 | 1·0 Tin | This is virtually muntz metal with a small amount of tin added to prevent corrosion by sea water. Used for marine pump castings, propellers, shafts and valve stems. |
| FREE-CUTTING BRASS | 58 | 39 | 3·0 Lead | Not suitable for cold-working, but can be machined at much higher speeds than would be possible in the absence of lead. The amount of lead content is reduced to between 1·0 and 2·5% (here the alloy has to be shaped by hot-stamping and then machined rapidly). Extensively used for making screw-threaded products. |

*Note*:
*HIGH TENSILE BRASS:* The addition of approximately 1·0% iron, 0·5% manganese and 1·0% tin to a muntz metal produces a hot-working brass of considerable strength and corrosion resistance. It is hard and tough and possesses a tenacity of over 600MN/m$^2$. It is misnamed 'MANGANESE BRONZE' but is essentially a high tensile brass used for castings.

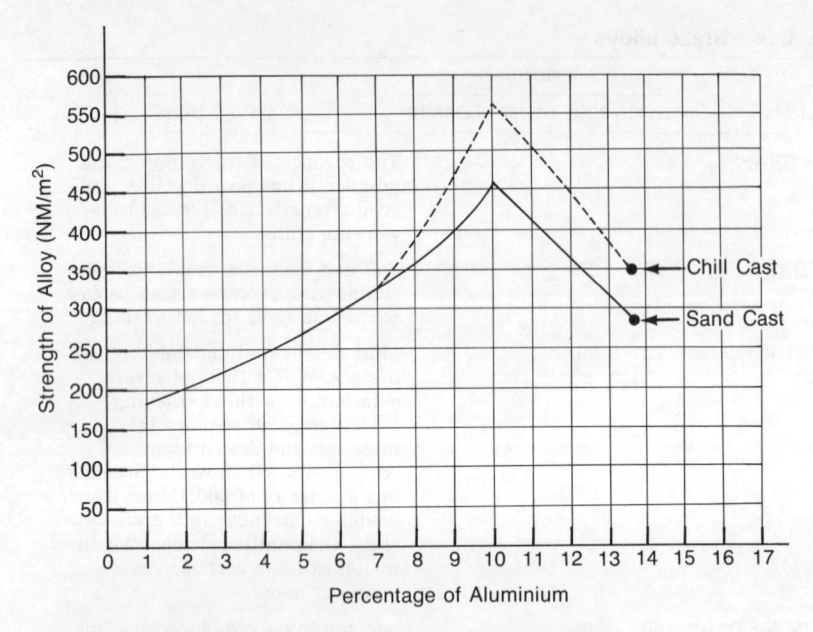

**Fig. 6.17 The effect of aluminium on the strength of copper–aluminium alloys**

'*Inconel*' is an alloy of 76% nickel and 15% chromium with the balance mainly iron. It resists corrosion by many organic and inorganic compounds, in addition to its ability to resist attack by oxidising atmospheres at high temperatures. It can be hot- or cold-worked, has a tensile strength in the region of 1082 MN/m$^2$ and can be joined by normal methods. It is used in food, chemical and textile-processing plant, and also for heat-treatment equipment and for steam turbine components.

## THE NIMONIC SERIES OF NICKEL–CHROMIUM ALLOYS

In service jet turbine aero-engines operate under a most severe combination of high stresses and elevated temperatures. In addition, corrosion by the products of combustion tends to weaken any metallic material that is employed.

When a material is loaded for long periods, it may gradually deform and eventually fracture at a stress that is well below its ultimate tensile stress. This continuous gradual extension under a steady load or stress is known as '*creep*' and must be considered when a material is to be used at temperatures approaching its recrystallisation temperature.

Because the high operational temperatures and rotational speeds

**Table 6.12 Tin–Bronze alloys**

| TYPE | COPPER | LEAD | PHOSPHORUS | TIN | ZINC | APPLICATIONS |
|---|---|---|---|---|---|---|
| LOW-TIN BRONZE | 96 | — | 0.1 to 0.25 | 3.9 to 3.75 | — | Can be hardened by severe cold-working. Good elasticity, corrosion and fatigue resistance. Is resilient, non-magnetic and a good conductor of electricity. Used extensively for springs and electrical contacting mechanisms. |
| DRAWN PHOSPHOR-BRONZE | 94 | — | 0.1 to 0.5 | 5.9 to 5.5 | — | This alloy is used in the work-hardened condition for machined components requiring strength and corrosion resistance, such as valve spindles. |
| CAST PHOSPHOR-BRONZE | Rem. | — | 0.03 to 0.25 | 10 | — | Has great fluidity, usually cast into rods and tubes for making bearing bushes and worm wheels. Has good anti-friction properties. |
| ADMIRALTY GUNMETAL | 88 | — | — | 10 | 2 | This alloy is suitable for sand castings where fine-grained, pressure-tight components are required, such as pump and valve bodies used in high-pressure steam plants. |
| LEADED GUNMETAL | 85 | 5 | — | 5 | 5 | This is a free-cutting bronze, used for the same purpose as standard Admiralty gunmetal, but although not quite as strong, has improved pressure tightness and machining properties. |
| LEADED BRONZE | 74 | 24 | — | 2 | — | An important material for lightly loaded bearings. The high lead content provides this bearing bronze with important anti-friction properties. Can contain up to 5% of other elements, such as nickel, tin and zinc. The copper provides high thermal conductivity, which is essential in avoiding over-heating, whilst the minor elements improve the lead distribution and increase the mechanical strength of the bearing. |

**Table 6.13 Aluminium–bronze alloys**

| COMPOSITION (Aluminium % only) | APPLICATIONS |
| --- | --- |
| 5 to 7·5 | Available in rod, strip, tube and wire form. More ductile than the tin–bronze alloys and can be readily cold-worked. Has excellent corrosion resistance and is used for pickling crates, condenser tubes and marine fittings. Sometimes used as a substitute for steel where non-magnetic properties and strength is desirable. |
| 7 | This alloy is used extensively for decorative architectural work — its rich golden colour is unaffected when exposed to the atmosphere. |
| 8 to 11·5 | Used in die-cast, sand-cast and wrought forms, and can be hot-worked. Typical uses are: gear selector forks, valve seat inserts, gears, locking rings and electrical switch mechanisms. Extensively used for a variety of marine applications. |

*Note*: Just as the commercial copper–zinc alloys may be classified as 'alpha' or 'alpha-beta', according to whether a single solid solution exists, or a pair of phases is present, so the copper–aluminium alloys may be classified. With up to about 7·5% aluminium the alloy consists of a single solid solution, while from 7·5 to 12%, two constituents are present.

produce creep, and the precise clearance between the tip of each blade and the turbine casing is critical, gas turbine blades must be made from a creep resisting material.

The '*Nimonic series*', capable of withstanding higher temperatures at higher stresses, was developed to meet the operating requirements for aircraft jet engines. Three of the most widely used wrought alloys in this series are 'Nimonic 75', 'Nimonic 80' and 'Nimonic 90' which operate satisfactorily at temperatures of 750°C, 800°C and 900°C, respectively. These alloys are very tough and work-harden easily, but are readily formed by spinning, rolling, pressing and wheeling. They can be joined by TIG, MIG and electric resistance welding processes.

'Nimonic 75' is a work-hardening alloy with a tensile strength of about 773 MN/m$^2$ at normal temperature in the cold-rolled condition. This 80/20 binary alloy is toughened by the precipitation of *titanium carbide* caused by the presence of approximately 0·3% of titanium and 0·1% of carbon. It is used for gas-turbine flame tubes and for furnace parts, and is suitable for operation at 750°C.

'Nimonic 80' is a heat-treatable alloy containing 1% aluminium which enables its mechanical properties to be maintained at even higher temperatures than Nimonic 75. After heat-treatment, Nimonic 80 has a tensile strength, at room temperature, of approximately 1066 MN/m$^2$, and it resists creep and fatigue at temperatures up to 850°C.

'Nimonic 90' contains 15% chromium, 20% cobalt and small amounts of other elements including aluminium and titanium, and the balance of nickel. It is heat-treatable and after treatment its tensile strength is in the region of 1236 MN/m$^2$ at room temperature. It is used for highly stressed components that operate at temperatures up to 900°C.

The manufacturer's instructions must be followed regarding the heat-treatment of Nimonic alloys to ensure that the optimum properties are obtained. In general, the heat-treatment consists of a lengthy *solution treatment* (up to 8 hours at about 1150°C) followed by slow cooling, and then a lengthy *precipitation treatment* (up to 16 hours at about 700°C) also followed by slow cooling.

## CUPRO-NICKEL ALLOYS

These are an important group of copper alloys, and because copper and nickel readily mix in the molten state, the useful range of alloys produced is not confined within any definite limits, as indicated in Figure 6.18.

Some typical copper–nickel alloys are summarised in Table 6.14.

## 6.12 Tin–lead alloys

Tin melts at 232°C and lead at 327°C; however, if lead is added to molten tin and the alloy is then slowly cooled, the 'freezing point' of the alloy is found to be lower than the freezing points of both tin and lead, as shown in Figure 6.19. For example, when a molten alloy containing 90% tin and 10% lead is cooled, a temperature of 217°C is reached before it begins to solidify. As it cools further, it gradually changes from a complete liquid state, through a 'pasty range', when it is like thick porridge, until finally it becomes completely solid at a temperature of 183°C. Similarly, an alloy containing 80% tin starts solidifying at 203°C, and also becomes completely solid at 183°C.

When tin is added to lead, the effect on the freezing point of the alloy is likewise lowered. For example, an alloy with only 20% tin starts to freeze at 279°C and is completely solid at the familiar temperature of 183°C.

*The whole range of tin–lead alloys finish freezing at the same temperature of 183°C.* However, one particular alloy, containing 63% tin and 37% lead, melts and solidifies entirely at 183°C, i.e., it is completely liquid at 184°C and completely solid at 182°C.

**Fig. 6.18 The melting ranges of the copper–nickel alloys**

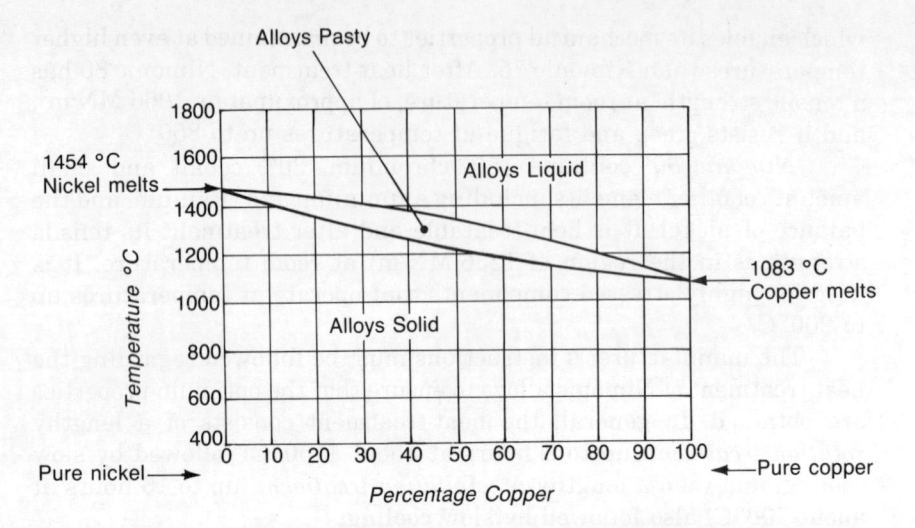

When one metal is alloyed with another, the melting point is always affected. A copper–nickel alloy does not melt or 'freeze' at one fixed and definite temperature, but progressively solidifies over a range of temperatures. Thus a 50/50% copper–nickel alloy, if liquified and then gradually cooled, will commence freezing at 1312 °C, and as the temperature falls, more and more of the alloy becomes solid until finally at 1248 °C it has become completely solidified. Except in certain special cases — 'eutectics' — this 'freezing range' occurs in all alloys, but is not found in pure metals, metallic, or chemical compounds, all of which melt and freeze at one definite temperature.

Note: A liquid metal is said to 'freeze' when it becomes solid. Thus copper freezes at 1083 °C, just as water freezes at 0 °C.

Similar effects occur in many other alloy systems and the composition which has the lowest freezing point of the series and which entirely freezes at that temperature has been given a special name. The particular alloy is known as a *'eutectic alloy'* and the freezing temperature (183 °C in the case of tin–lead alloys) is called the *'eutectic temperature'*.

For further information on soft solders and fusible alloys, refer to Chapter 11 (Tables 11.1, 11.11 and 11.36).

## 6.13 Hot and cold-working

The terms 'hot-working' and 'cold-working' are used from time to time in this chapter and elsewhere in the book. When subjected to these processes, the properties of a metal will change, for example, the condition of the grain structure can be changed during hot-working

**Table 6.14 Copper–nickel alloys**

| COMPOSITION (Nickel % only) | APPLICATIONS |
|---|---|
| 20 | One of the most ductile of the commercial wrought alloys. May be subjected to severe cold-working without the need of intermediate annealing. Can be readily forged and rolled at a temperature above 800 °C. Suitable alloy for drop forgings, cold stamping and pressing. Widely used in the construction of motor vehicle radiators. Other uses include bullet sheathing and resistance wires. |
| 30 | These alloys usually contain 0.5 to 1.5% of iron and are very resistant to corrosion by sea water. Extensively used for heat interchangers, steam condensers and desalination plants. |
| 45 | Because of its high electrical resistivity, is mainly used in resistors, thermo-couples and rheostats for control switchgear and other electrical apparatus. |
| 65 to 70 | This group of alloys are familiarly known as *'Monel metals'*. They are stronger than pure nickel, are non-magnetic and highly corrosion resistant. There are various grades available with tenacities up to 1000 MN/m$^2$. Readily fabricated by hot- and cold-working, machining and welding. Can be strengthened by small additions of carbon, iron, manganese and silicon. Possess exceptionally high strength at both room and elevated temperatures. Highly resistant to corrosion from acids, alkalis, food stuffs and salt water, and have many chemical plant and marine uses. The addition of a small percentage of aluminium and titanium renders them precipitation hardenable, these high-strength alloys are widely used for propeller shafts. |

*Note*: The addition of 2 to 45% of nickel to copper provides a series of alloys that are stronger and more resistant to oxidation at high temperature than is pure copper.

Ordinary wrought and cast nickel–copper alloys are not heat-treatable, but with the addition of between 2% and 4% aluminium respond to solution treatment followed by precipitation treatment. Such alloys available under the trade name *'K-Monel'*, fully heat-treated possess a tensile strength as high as 1545 MN/m$^2$.

during heating and cooling, and some of the effects are summarised in Figure 6.20.

Metals may be shaped by making them molten and CASTING them in moulds or they may be cut or MACHINED to shape.

However, there is the alternative of WORKING them to shape. This means shaping a piece of metal by stretching or squeezing it to shape.

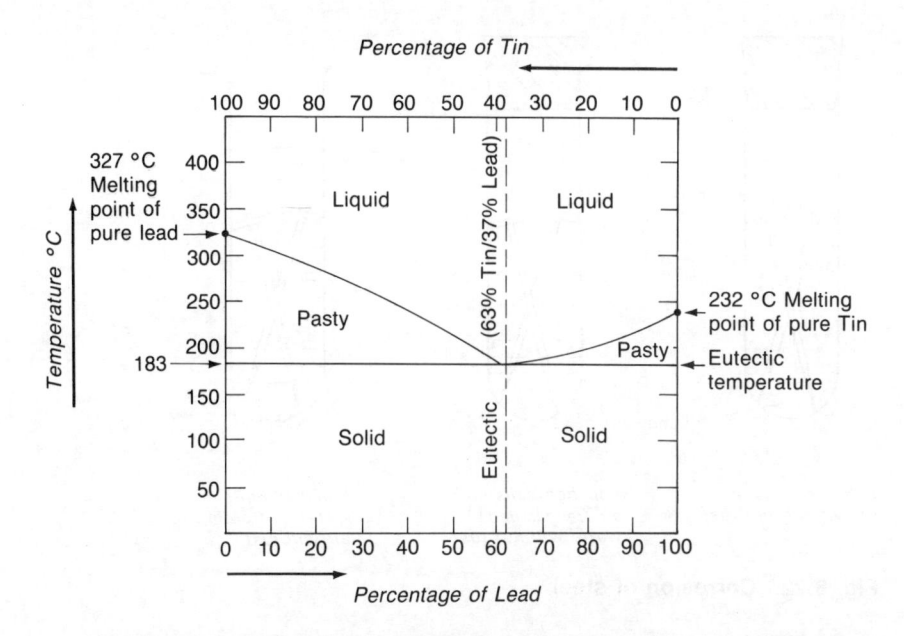

*Note:* The presence of eutectic in all ordinary tin–lead solders means that solders are no longer truly solid at 183 °C or higher, even though they may not be obviously 'pasty'. In the pasty range they all have no strength. *At 30 to 40 °C below the eutectic temperature, soft solders are only about half as strong as at room temperature, therefore it is important to avoid straining soldered joints before they have had time to cool.*

Reference to the diagram indicates that, in practice, *eutectic solder* (63% tin/37% lead) melts more quickly and freezes solid more quickly which makes it the fastest solder to work with.

**Fig. 6.19 The melting ranges of tin–lead alloys**

Metal may be HOT-WORKED or COLD-WORKED. The difference in temperature varies from metal to metal. It controls the changes that take place in the internal structure of the metal and is the subject of a fairly complicated metallurgical theory.

It is sufficient for the craftsman to know that when COLD-WORKING a metal it WORK-HARDENS and will crack if he does not stop working it and re-soften it by heat treatment.

When HOT-WORKING a metal it does not work-harden and if it were possible to prevent the metal cooling there would be no limit to the amount it could be worked. Further, metals require less force to make them flow to shape when heated.

Some examples of hot- and cold-working are illustrated in Figure 6.21.

**CRYSTAL STRUCTURE**

Cold working makes the crystals become elongated (work hardened)

*Crystal structure as it would appear through a microscope*

**EFFECT ON PROPERTIES**

Metal becomes stiffer (less ductile) and tends to crack. Its strength also increases.

Raising the temperature of the work-hardened metal into the region of 200-300°C produces no visible effect, but there will be considerable re-arrangement of the atoms in the crystals.

There is little change in the stiffness and strength of the metal, but it will have become stress relieved and is less likely to crack.

Raising the temperature of the metal further results in 're-crystallisation'. The 'normal' crystal structure of the metal reforms providing the metal is allowed to cool down slowly.

The crystals become 'equi-axed' instead of being elongated. The metal is now soft and ductile but there is an appreciable loss of strength.

Prolonged heating at high temperatures produce grain growth (excessively large crystals)

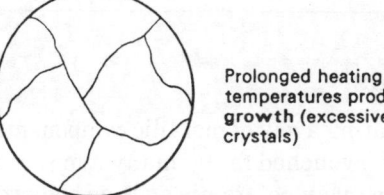

Large crystals make the metal weak and very soft. If the grain growth is excessive there is also a fall in ductility.

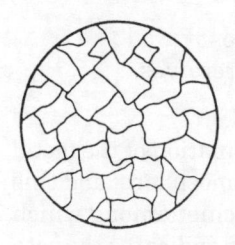

The crystal structure can be restored to normal by reheating the metal to the correct temperature and cooling it at the correct speed to give the required structure and, therefore, properties.

It is assumed that the overheating which produced the grain growth did not 'burn' the metal. That is, oxidise the crystal boundaries. This cannot be corrected and leaves the metal so weak and brittle that it has to be scrapped.

**Fig. 6.20 The changes of state in solid metals**

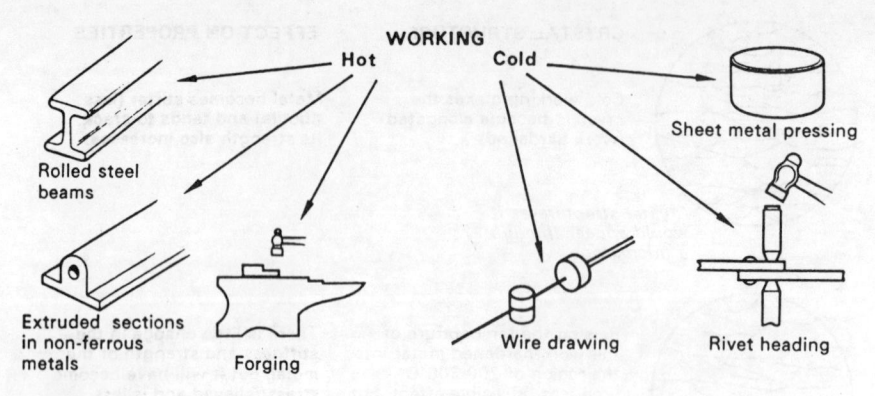

**Fig. 6.21 Examples of hot- and cold-working**

**Table 6.15 Rate of corrosion**

| TYPE OF ENVIRONMENT | TYPICAL RATE OF RUSTING FOR MILD STEEL IN TEMPERATE CLIMATES (mm/year) |
|---|---|
| Rural | 0·025–0·050 |
| Urban | 0·050–0·100 |
| Industrial | 0·100–0·200 |
| Chemical | 0·200–0·375 |
| Marine | 0·025–0·150 |

Comparable rates of rusting in different atmospheric environments.

## 6.14 The corrosion of metals

Corrosion is the slow but continuous eating away of metallic components. That it is costly and destructive will be vouched for by many a motorist who has seen his 'pride and joy' eaten away before his eyes as body rot sets in. Three factors govern the rate of corrosion.

(a) The metal from which the component is made.
(b) The treatment the surface of the component receives.
(c) The environment in which it is kept.

In the initial stages corrosion involves the combination of the metal surface with oxygen or other gases present in the atmosphere, and this process can be accelerated by a number of factors, chief among which are heat, moisture and electrolytic action. Table 6.15 indicates the rate of corrosion of unprotected steelwork.

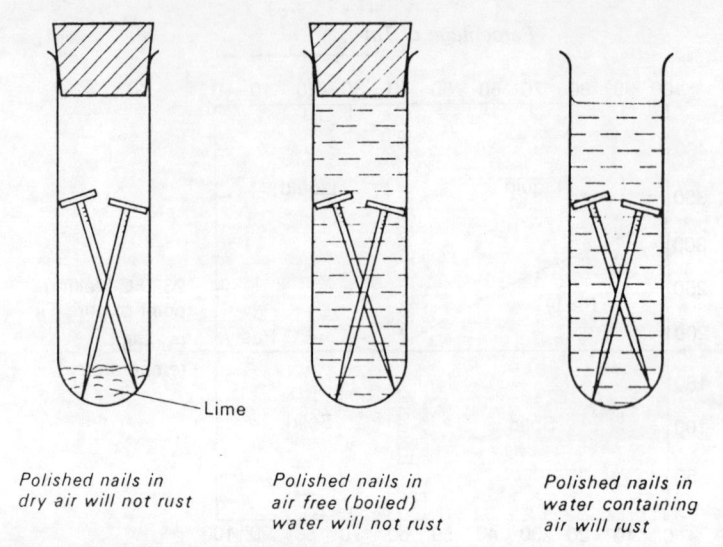

**Fig. 6.22 Corrosion of steel**

Polished nails in dry air will not rust

Polished nails in air free (boiled) water will not rust

Polished nails in water containing air will rust

## 6.15 Direct chemical corrosion (dry)

This type of corrosion involves direct combination between the metal and dry gases such as oxygen, carbon dioxide, sulphur dioxide and chlorine, usually at high temperatures. For example the 'scale' on hot rolled steel is an oxide of iron. If the corrosion film is continuous and 'self-healing' as in stainless steel and aluminium, then the corrosion will cease completely.

## 6.16 Direct chemical corrosion (wet)

The most common example of corrosion due to moisture is the rapid surface formation of a red rust resulting from the exposure of iron or steel to the atmosphere. This red rust is an oxide of iron, but a different oxide from the blue-black oxide formed under the action of heat, and is the direct result of a corrosive action due to the presence of moisture.

*It is shown in Figure 6.22 that iron or steel does not rust in dry air; neither does rusting occur in pure water that does not contain dissolved air or oxygen.*

Once 'rusting' commences, the action is self-generating. That is, it will continue even after the supply of moisture and air is removed. The rate of rusting slows down as the rust deepens or thickens, but as rain water washes off the surface rust the rate increases again. This cycle is continuous and once started is very difficult to control.

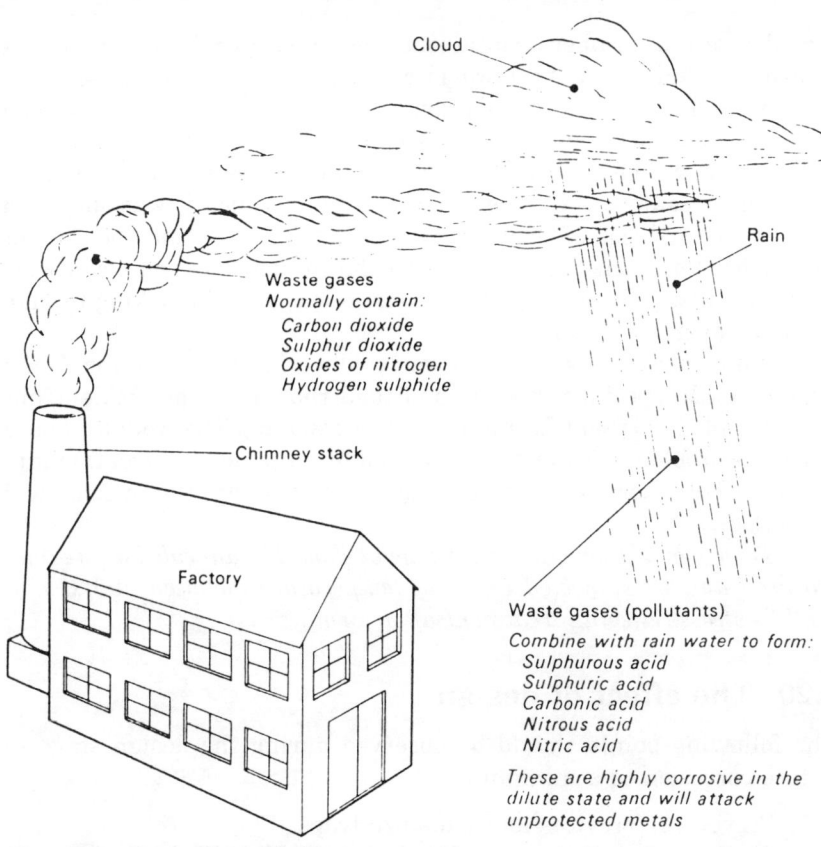

**Fig. 6.23 Corrosive pollution**

## 6.17 Direct chemical corrosion (pollution)

Atmospheric pollution rapidly increases the rate of rusting of iron and steel. It also attacks, but much more slowly, copper and zinc. Lead is virtually unaffected. How the gases given off by the burning of fuels and by industrial processes are converted into corrosive acids is shown in Figure 6.23. Gases that become acids when dissolved in water are called *acid anhydrides*.

In coastal areas the problem is increased by the presence of salt spray in the atmosphere.

## 6.18 Electro-chemical corrosion

When two disimilar metals come into contact in a moist or wet corrosive environment they behave like a simple electric cell. The current generated causes corrosion.

**Table 6.16 Electro-chemical series**

CORRODED (ANODIC)

| METAL | ELECTRODE POTENTIAL (VOLTS) |
|---|---|
| Sodium | −2·71 |
| Magnesium | −2·40 |
| Aluminium | −1·70 |
| Zinc | −0·76 |
| Chromium | −0·56 |
| Iron | −0·44 |
| Cadmium | −0·40 |
| Nickel | −0·23 |
| Tin | −0·14 |
| Lead | −0·12 |
| Hydrogen (reference potential) | 0·00 |
| Copper | +0·35 |
| Silver | +0·80 |
| Platinum | +1·20 |
| Gold | +1·50 |

PROTECTED (CATHODIC)

Metals can be arranged in a special order called the electro-chemical series. This series is given in Table 6.16. Note that in this context the gas hydrogen behaves like a metal.

If any two metals from the table come into contact in moist surroundings, the more negative one will corrode the most rapidly. For example:

(1) In galvanised iron (zinc coated steel), the zinc corrodes away whilst protecting the mild steel. The zinc is said to be SACRIFICIAL.
(2) In tin plate, the mild steel is corroded if the tin coating is broken. This is why lines should not be scribed on tin plate when marking-out.

Figure 6.24 shows what happens during electrolytic corrosion.

## 6.19 Stress corrosion

Chemical and electro-chemical corrosion is intensified when a metal is under stress. The stresses involved are usually internal stresses caused by some previous treatment such as cold-working. The corrosive attack is along the crystal boundaries and this considerably weakens the metal. Stress relieving usually prevents this form of corrosion.

**Fig. 6.24 Electrolytic corrosion**

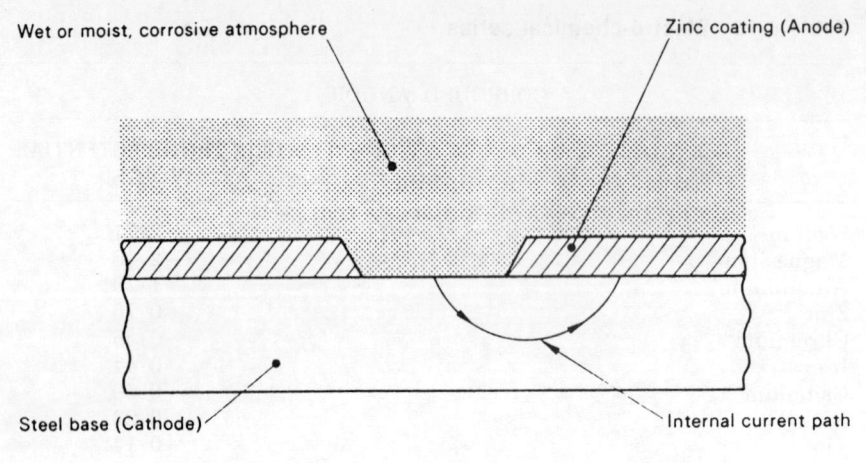

*(a) Protection by a sacrificial coating*
Coating is eaten away whilst protecting the base

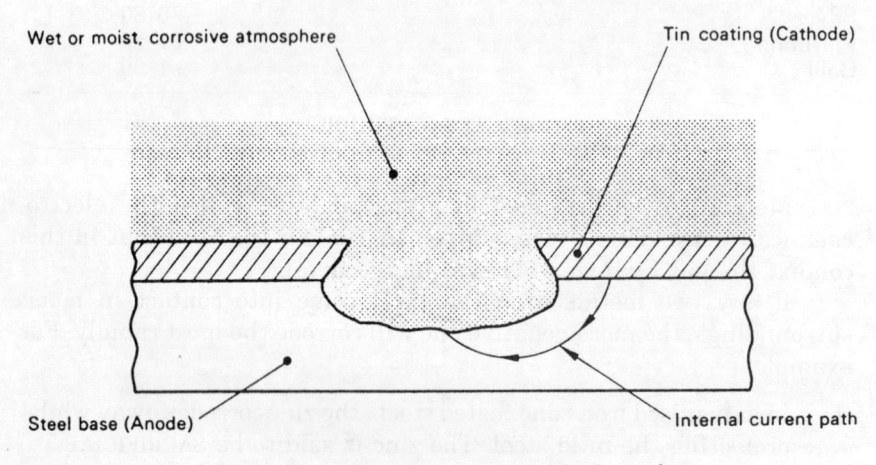

*(b) Protection by a purely mechanical coating*
Coating only protects the base if intact.
If coating is damaged, base is eaten away quicker than if coating were not present.

## SEASON CRACKING

Many years ago it was noticed that cartridge cases stored in India cracked spontaneously during the monsoon season and the name used to describe this metallurgical problem was *'season cracking'*.

Cold-worked (alpha) brasses undergo intergranular cracking when stored in an atmosphere containing *ammonium carbonate*, as is not unusual in India and other warm climates. It was also realised that season cracking only occurred some time after the metal had been severely cold-worked during fabrication and then later subjected to corrosive conditions. The unevenly distributed stresses remaining in the metal as a result of the severe deformation to which it was subjected — as is the condition of 'deep drawn' cartridge cases — may be present for a period of months or years until the influence of some corrosive medium causes the pent-up stress to be released and cracking to take place between the grains.

It may not be possible to prevent cold-worked material from being exposed to the special corrosive conditions that cause the problem, but it is possible to prevent failure by season cracking. The trouble — once a serious problem — can now be avoided by suitable heat-treatment to remove the internal stresses without destroying the mechanical properties.

*To prevent season cracking the metal should be annealed before cold-working and be subjected to a low-temperature (between 200°C and 300°C) stress-relieving treatment after forming.*

## 6.20 The effect of design

The following points should be observed during the design stage to reduce corrosion to a minimum.

1. Preventing crevices and moisture traps.
2. Sealing joints if they are not continuously welded.
3. Adequate ventilation and drainage.
4. Contact with corrosive fluids to be kept to a minimum.
5. Ease of washing down and cleaning.
6. Provision of protective (and decorative) coatings.

## 6.21 Metals that resist corrosion

### ALUMINIUM

The reaction of aluminium and its alloys with the atmosphere is somewhat different to that of copper and zinc, but the effect of the reaction is similar, insomuch that it results in retarding further action.

As previously stated, the metal has a great affinity for oxygen, and even highly polished aluminium surfaces quickly develop a thin transparent film of alumina, which if the metal is kept indoors, prevents further oxidation and retains its bright appearance. However, when aluminium is exposed to the atmosphere, this transparent oxide film

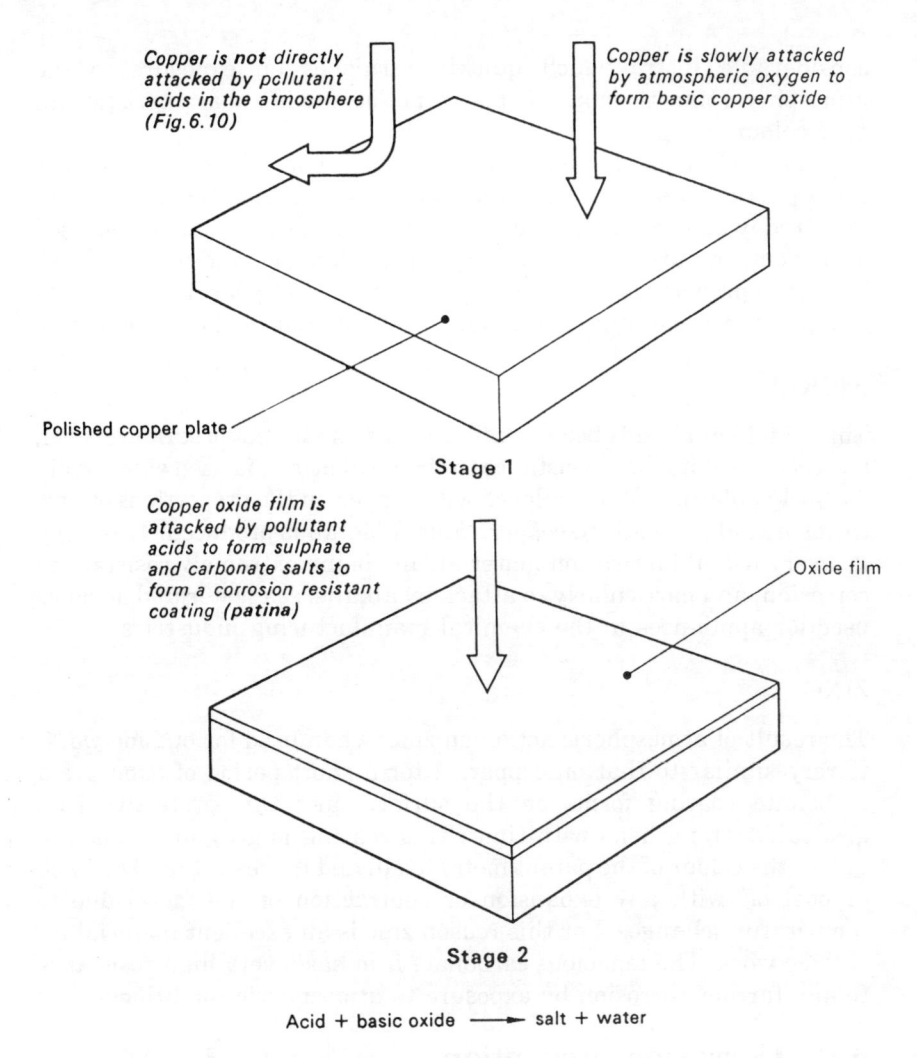

**Fig. 6.25 Formation of 'patina' on copper**

Figure labels:
- Copper is not directly attacked by pollutant acids in the atmosphere (Fig. 6.10)
- Copper is slowly attacked by atmospheric oxygen to form basic copper oxide
- Polished copper plate
- Stage 1
- Copper oxide film is attacked by pollutant acids to form sulphate and carbonate salts to form a corrosion resistant coating (patina)
- Oxide film
- Stage 2
- Acid + basic oxide ⟶ salt + water

thickens, becomes grey in colour and when sufficiently developed, protects the parent metal from further action.

## COPPER

When copper is exposed to atmospheric pollution for a long time, as for example when used as a roofing material, the surface develops a green coating or 'patina'. This coating or 'patina' is caused by the acids in the atmosphere (Figure 6.23) attacking the oxide coating of the copper to form sulphate and carbonate salts as shown in Figure 6.25.

The 'patina' which forms on copper giving it a protective skin must not be confused with the green compound 'verdigris' which forms on the surface of copper by the action of organic matter. Verdigris corrosion will completely destroy the copper in time.

Copper vessels used for industrial processing and in the preparation of food products are tinned to prevent verdigris forming.

## CHROMIUM

Although this metal is an important constituent of stainless steel, one of its more important uses is for plating articles against corrosion. It is highly resistant to corrosive influences and retains its high polish and colour for long periods.

Methods of preventing or retarding corrosion are largely applied to iron and steel, since, as already explained, most of the common non-ferrous metals and alloys form their own protective coating when exposed to the atmosphere.

## STAINLESS STEEL

Stainless steel is one of the most outstanding materials that is resistant to corrosion, which is remarkable considering the main constituent of the alloy is iron, an element that possesses no resistance when exposed to corrosive influences.

Stainless steels can be described under the general heading of *corrosion and heat-resisting steels.* They are a group of iron-based alloys which owe their resistance to corrosion at elevated temperatures to the presence of CHROMIUM or to a combination of CHROMIUM and NICKEL.

When more than 12% chromium is present in the steel it forms a tough continuous layer of CHROMIC OXIDE on the surface and protects the metal against attack by corrosion. Under oxidising conditions, if the film becomes damaged, it immediately reforms and continues its protective function. *Exposure to high temperature atmospheres also causes the oxide film to thicken.*

Nickel in amounts of 6% or more improves the corrosion and heat resistance of a stainless steel and makes fabrication much easier.

The various types of stainless steel can be grouped under the headings shown in Table 6.17.

## LEAD

When the surface of a lead sheet is scratched, cut or scraped, the newly exposed metal or fresh surface will have a silvery — or bluish white — appearance. However, the fresh surfaces of lead soon lose their

**Table 6.17 Stainless steels**

| GROUP | REMARKS |
|---|---|
| *Martensitic stainless steels* | Contain from 12% to a maximum of 16% of CHROMIUM as the main alloying element, and up to 0·3% of CARBON. These materials are capable of being hardened and, consequently, their welding presents difficulties unless special precautions are taken (pre-heating and post-heating treatment).<br>Used for cutlery, spindles, shafts and applications requiring good resistance to corrosion and scaling at elevated temperatures up to about 800°C. |
| *Ferritic stainless steels* | The CHROMIUM content varies from 16% to 30% and the CARBON content is usually low — about 0·1%.<br>These materials are not hardened by heat-treatment, but are liable to *brittleness* caused by excessive *grain growth* at high temperatures. This results in welds that are brittle at ordinary temperatures but which may be quite *tough* at 'red heat'.<br>Used when very high temperature scaling resistance is required, for example, in furnace parts and oil burners. |
| *Austenitic stainless steels* | By far the most important and largest group in the stainless steel range.<br>Contain a minimum of 17% CHROMIUM and 8% NICKEL. One familiar stainless steel is known as '18/8' because it contains 18% chromium and 8% nickel.<br>Many special alloys contain a greater percentage of both elements to provide greater corrosion resistance under specific conditions. The presence of 2% to 3% of MOLYBDENUM increases resistance to chemical attack — particularly SULPHURIC ACID.<br>The term 'AUSTENITIC' implies that *these steels cannot be hardened by heat-treatment* — drastic *quenching* from a high temperature actually soften them.<br>These steels '*work-harden*' rapidly when subjected to cold working. They are nearly as soft as MILD STEEL and are capable of as much cold working if frequently annealed.<br>Unlike other steels they are NON-MAGNETIC.<br>They are much more amenable to welding than the martensitic and ferritic types.<br>Used for a great variety of purposes, particularly in the chemical, brewing, dairy and food industries. |

*Note*: The many applications of these corrosion and heat-resisting steels range from chemical vessels to kitchen sinks; from petroleum plants to food processing equipment; from furnaces to freezers, from jet engines to storage tanks, and from brewery pipes to nuclear power plants, etc. Service conditions vary considerably in many of these applications, particularly in regard to resistance to corrosion, stress, and very high or low temperatures.

considerable lustre which quickly tarnishes on exposure to the atmosphere, thus the normal appearance of lead is a characteristic dull grey colour.

The white oxide film which is the result of exposure to the atmosphere is very tenacious and prevents further attack.

Today approximately one-third of the world's output of lead is used as a sheathing material for underground telephone and power cables. It has the property of being incorrodible year after year. It is also used as a protective metallic coating for sheet iron and steel ('Terne-plate').

NICKEL

This metal has already been mentioned as a constituent of stainless steel, but because of its high resistance to chemical attack is used extensively for 'nickel plating'. When alloyed with copper, in the proportions of two-thirds nickel to one-third copper, 'Monel Metal' is produced. This alloy is a very useful fabrication material, and being extremely resistant to corrosion, and particularly to attack by alkalis, sea water and acids, is used for appliances in the chemical manufacturing industries.

ZINC

The result of atmospheric action on zinc, when used for outside work, is very similar to that on copper. After a short period of exposure a carbonate coating forms on the surface, and this protective film gradually strengthens with time. This coating is grey in colour, not unlike the colour of the parent metal itself, and it does not tend to crack or peel off with any expansion or contraction of the metal due to temperature changes. For this reason zinc is an excellent material for outside work. The tenacious carbonate film has a very high resistance to any further corrosion by exposure to atmospheric conditions.

## 6.22 Corrosion prevention

For short-term protection, oiling or greasing up a component may be adequate. However, in the long term this is not satisfactory for the following reasons:

(1) The protective film will dry up and no longer seal the surface from the corrosive environment.
(2) Oils and greases absorb moisture from the atmosphere and corrosion can take place under the oil or grease film.
(3) Cheap oils and greases often contain active sulphur and acid impurities that will themselves attack and corrode the very metal surface that they are supposed to be protecting.

**Table 6.18 Corrosion-resistant coatings**

| | TYPE OF COATING | METHOD OF APPLICATION |
|---|---|---|
| *METALLIC* | **Hot-dipping** | Mild steel sheet is dipped into molten tin (*tinplate*), or into molten lead/tin alloy (*terneplate*), or into molten zinc (galvanised sheet). |
| | **Electro-plating** | Corrosion-resistant and decorative metals are electrolytically deposited upon a base metal. Generally the coatings are thinner and less protective than those produced by hot dipping. |
| | **Cladding** | Produced by hot-rolling a composite slab of base metal and a corrosion-resistant metal, e.g. ALCLAD (Figure 6.16). |
| | **Spraying** | Coatings of aluminium, brass, bronze, copper, lead, tin and zinc may be applied to the surfaces of a base metal by spraying. The coating metal in wire form is melted in a pistol (spray gun) and is impinged onto the required surfaces by a jet of compressed air. |
| | **Cementation** | *Calorising* — Components are baked in aluminium dust at 850–1000°C for several hours. <br> *Chromising* — Components are baked in aluminium and chromium dust in an atmosphere of hydrogen at 1300°C for 3 to 4 hours. <br> *Sheradising* — Components are baked in zinc dust at 350°C–370°C for up to twelve hours. |
| *NON-METALLIC* | **Anodising** | A hard oxide film is produced on the surface of aluminium by electrolysis in a chromic acid solution. For decorative purposes the film may be coloured by dyeing. |
| | **Chromating** | A hard oxide film is formed on magnesium by dipping in a solution of potassium dichromate and sealing with zinc chromate paint. |
| | **Phosphating** | Metallic phosphates are built up on the surfaces of a metal by immersion in suitable solutions. Phosphating is used extensively to prepare the steel panels of motor vehicles where it not only provides protection against corrosion but also provides a 'key surface' for painting. *Bonderising, Granodising, Parkerising* and *Walterising* are all phosphating processes. |
| | **Painting** | First a '*primer*' (containing a rust inhibitor such as lead oxide) is used on the metal surface to be protected and the protective coating is built up by one or more sealing coats. Is the only process that can readily be replaced in service. |
| | **Plastic** | A protective coating may be built up by dipping the component into molten plastic, or by shrinking an envelope of plastic around the component. |

*Note*: Protective coatings are often applied to metal components after fabrication, especially 'hot-dip galvanising'. This ensures that the protective coating is continuous, i.e., not broken at any point.

Table 6.18 gives details of more permanent methods of corrosion prevention.

## 6.23 Surface preparation

*It has been firmly established that the essential and most important factor of any efficient anti-corrosion treatment is surface preparation.*

However carefully selected the protective process may be, it cannot fulfill its purpose if it is applied to an ineffectively prepared surface. Surfaces carrying dirt, grease, corrosion or millscale are unsuitable for the direct application of an anticorrosion treatment. Table 6.19 shows how the surface may be prepared for treatment.

## 6.24 Pickling

This is a chemical method of cleaning in which dilute sulphuric acid or mixed acids are used to clean the surface of metal as shown in Figure 6.26.

## 6.25 Blast cleaning

This is a mechanical method of cleaning in which the scale and corrosion are dislodged by a high-velocity blast of abrasive particles as shown in Figure 6.27.

## 6.26 Flame descaling

The effectiveness of this process depends upon the difference in expansion between the scale and base metal when subject to local heat.

The steel surface that is to be descaled is heated with an oxy-fuel gas torch fitted with specially designed nozzles producing broad fan-shaped high-intensity flames. The rapid local thermal expansion of the

**Table 6.19 Surface preparation**

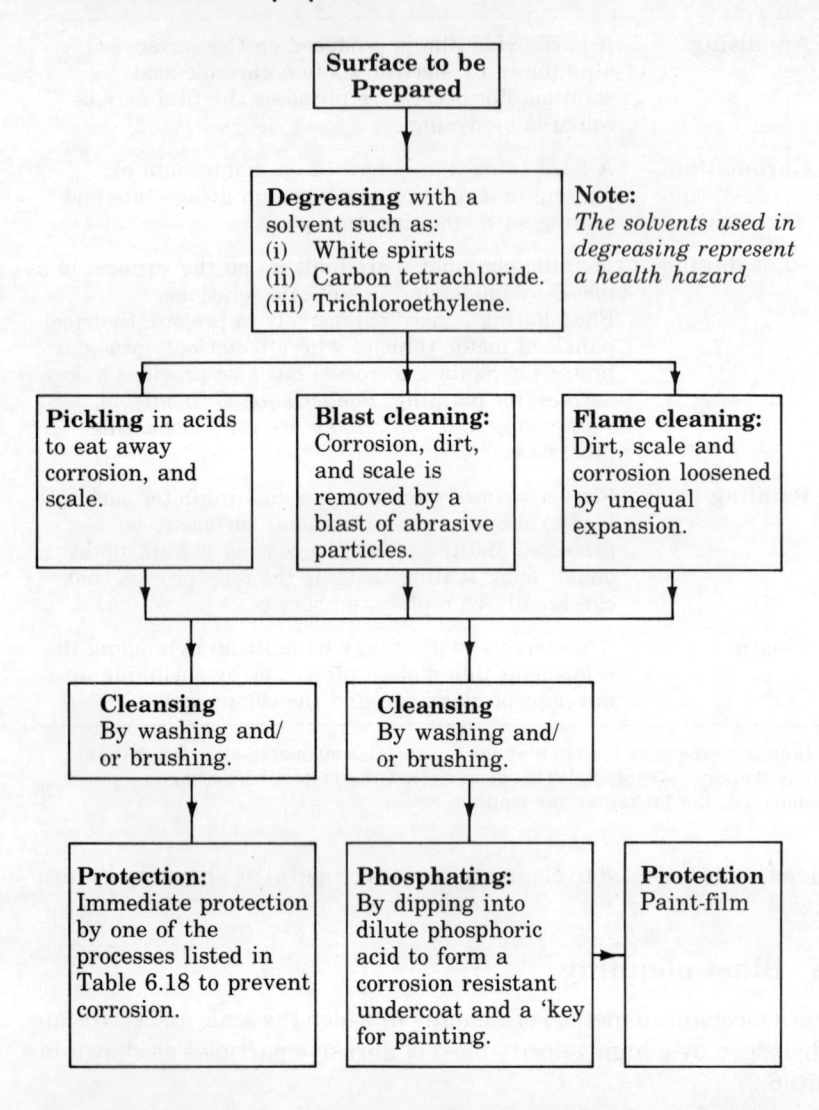

```
┌─────────────────────┐
│ Surface to be       │
│ Prepared            │
└──────────┬──────────┘
           ▼
┌─────────────────────────┐    ┌──────────────────────┐
│ **Degreasing** with a   │    │ Note:                │
│ solvent such as:        │    │ *The solvents used in│
│ (i)  White spirits      │    │ degreasing represent │
│ (ii) Carbon tetrachloride│   │ a health hazard*     │
│ (iii) Trichloroethylene.│    │                      │
└─────────────────────────┘    └──────────────────────┘
```

- **Pickling** in acids to eat away corrosion, and scale.
- **Blast cleaning:** Corrosion, dirt, and scale is removed by a blast of abrasive particles.
- **Flame cleaning:** Dirt, scale and corrosion loosened by unequal expansion.

- **Cleansing** By washing and/or brushing.
- **Cleansing** By washing and/or brushing.

- **Protection:** Immediate protection by one of the processes listed in Table 6.18 to prevent corrosion.
- **Phosphating:** By dipping into dilute phosphoric acid to form a corrosion resistant undercoat and a 'key' for painting.
- **Protection** Paint-film

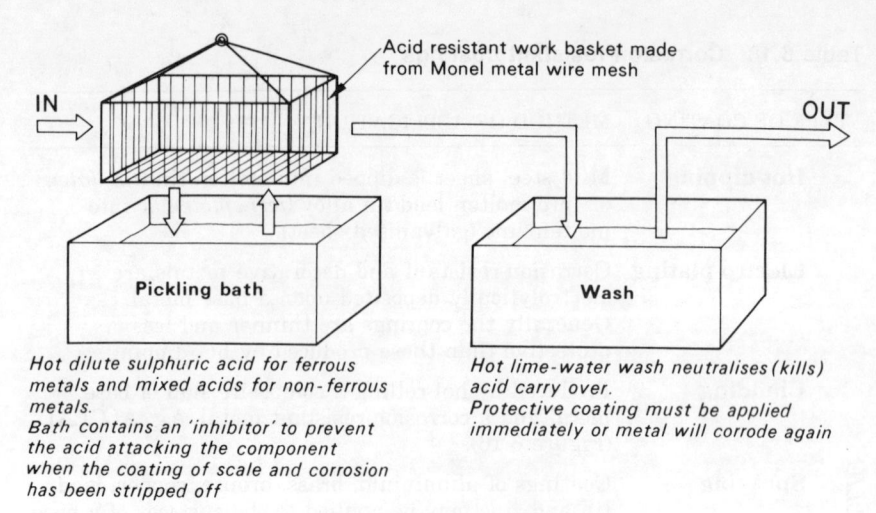

**Fig. 6.26 Pickling**

Acid resistant work basket made from Monel metal wire mesh

Pickling bath — Wash

Hot dilute sulphuric acid for ferrous metals and mixed acids for non-ferrous metals.
Bath contains an 'inhibitor' to prevent the acid attacking the component when the coating of scale and corrosion has been stripped off

Hot lime-water wash neutralises (kills) acid carry over.
Protective coating must be applied immediately or metal will corrode again

**Fig. 6.27 Blast cleaning**

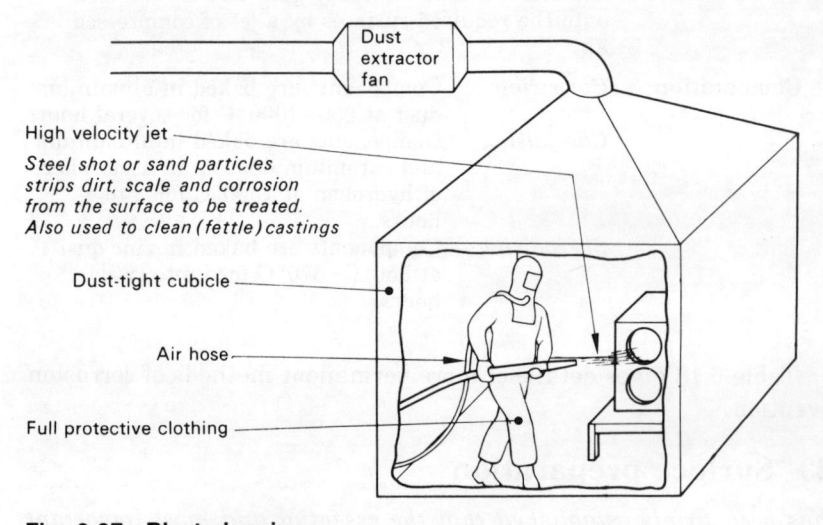

- Dust extractor fan
- High velocity jet
- *Steel shot or sand particles strips dirt, scale and corrosion from the surface to be treated. Also used to clean (fettle) castings*
- Dust-tight cubicle
- Air hose
- Full protective clothing

loosely adhering scale against the relatively unheated base metal causes the former to flake off. The process is also assisted by the generation of steam by any moisture that may be trapped under or absorbed in the scale itself.

With this process rust removal is excellent — any rust particles are converted to powder which can easily be brushed away before commencing to paint, and a dry surface is ensured for 'priming' while the metal is still warm. *For best results the primer should be applied when the steel is approximately 45°C, that is the temperature at which the hand can be held comfortably on the steel.*

Generally, flat nozzles which give a brush-like flame pattern are used for descaling flat surfaces and large areas which are free from projections, whilst round nozzles are used on all riveted and bolted areas, plate edges, corners, recesses and other places where it is not possible to operate flat nozzles effectively.

This method is often specified for the cleaning of heavily rusted steelwork *in situ*, prior to maintenance painting. However, it is not suitable for light steelwork which may buckle and distort due to the intense localised heat.

## 6.27 Metal degreasing

All surface treatments of metals such as *galvanising, nickel plating, plastic coating, painting*, etc., require surfaces free from grease, dirt or any other kind of contaminant before the actual treatment is carried out.

BASIC PRINCIPLES OF VAPOUR DEGREASING

By far the biggest bulk of degreasing is carried out by immersing the metal or metal component in a simple vapour plant. The whole cleaning operation is completed in one plant, from which the work emerges in a dry, neutral condition ready for subsequent processes such as heat, chemical or preservation treatment, inspection, repair, further fabrication or machining operations.

Vapour degreasing systems use inflammable chlorinated hydrocarbon solvents and the one most commonly used is TRICHLOROETHYLENE.

The basis of all degreasing systems is the solvent, and trichloroethylene is the most powerful grease solvent known. Its low boiling point, low specific heat and low latent heat of vaporisation make only moderate demands on fuel and cooling water. *The heating required to convert 1 kg of water to steam will vaporise 8·5 kg of trichloroethylene.*

The basic principle of vapour degreasing may be summarised as follows:

1. The contaminated component is suspended in the vapour rising from boiling trichloroethylene in a degreasing plant.
2. *The vapour immediately condenses on contact with the relatively cold metal surfaces, dissolves the contaminant* (usually oil or grease) *and runs back into the plant.*
3. Within a short space of time the component reaches the vapour temperature and condensation ceases. Degreasing is now complete.
4. The component is then slowly withdrawn from the plant in a perfectly clean condition.

Methods of degreasing vary according to the type of work and the nature of the contamination, and plants designed to cater for cleaning by:

(1) immersion in vapour
(2) immersion in liquor, or

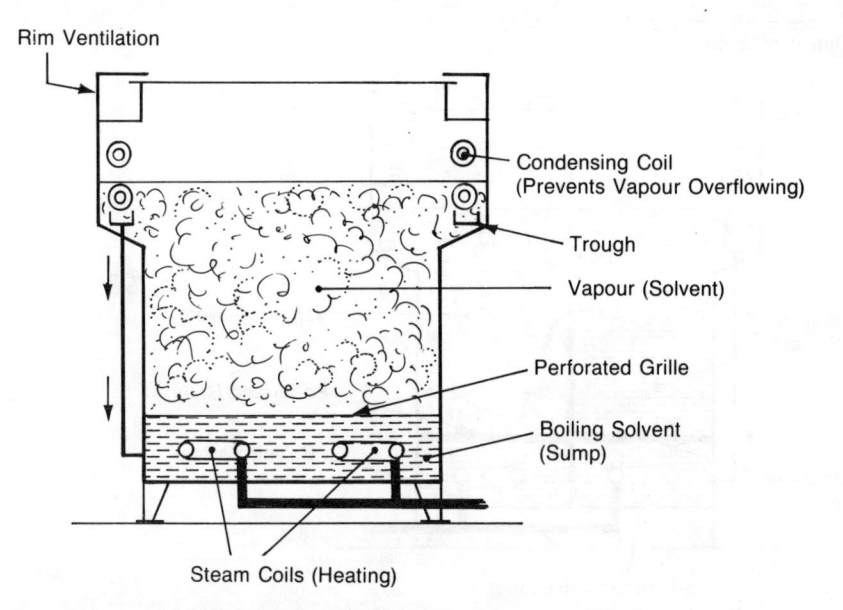

**Fig. 6.28 Standard vapour degreasing plant**

(3) a combination of both

are available.

STANDARD VAPOUR PLANTS

A standard vapour plant (Figure 6.28) consists essentially of a welded tank (of mild steel galvanised after fabrication, or of stainless steel) fitted with a '*condensing coil*' through which cold water flows. Solvent is boiled in the '*sump*', filling the plant with vapour up to the level of the condensing coil; here it condenses and is collected in a '*trough*', from which it may be diverted outside the plant to a drum or stock tank, or allowed to return to the sump.

LIQUOR DEGREASING

If dirt as well as oil or grease has to be removed, immersion of the work in boiling liquid solvent is more effective than vapour treatment. It is often preferred for the removal of tenacious fine polishing or lapping compounds.

STANDARD LIQUID–VAPOUR PLANTS

A standard liquid–vapour plant is similar in construction to the vapour

**Fig. 6.29 Standard liquid–vapour degreasing plant**

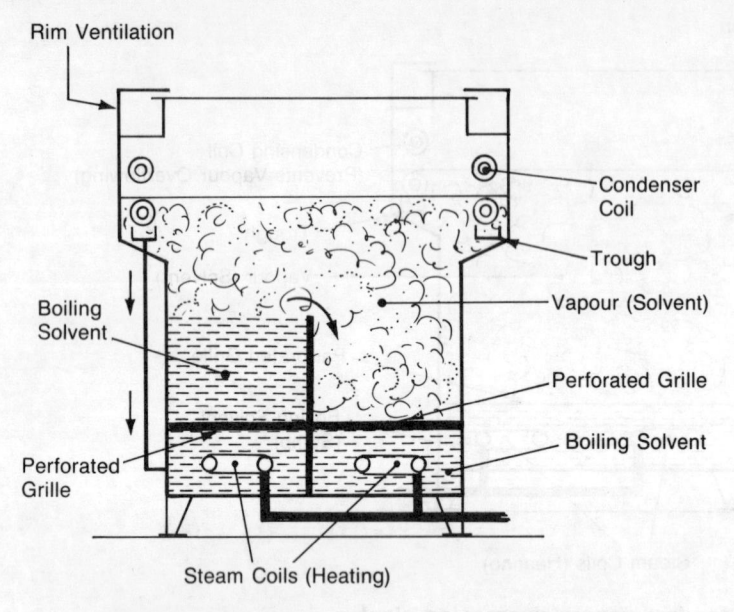

Work dipped into the boiling liquor is largely freed from dirt, grease, oil, polishing compound and swarf.

If the contamination of the liquor is low, the work will emerge in a sufficiently grease-free state for most following processes.

Perfect freedom from oil and grease, in any case, can be obtained by a subsequent immersion in the vapour compartment if the work is first allowed to cool slightly.

plant except that it is divided into two compartments, each fitted with heating facilities, as illustrated in Figure 6.29.

## METHODS OF HEATING

There are several ways to heat solvent to its boiling point as listed in Table 6.20.

## HANDLING OF WORK

To ensure that degreasing is effective, and to avoid 'drag-out' of solvent either as liquid or vapour, it is essential to load all work in such a position that it will drain freely. Some components, such as heavy machinery parts, are conveniently degreased while directly suspended from a hoist. At the other extreme, small items that have to be jigged or wired for plating may be suspended from light rods resting on top of the plant.

**Table 6.20 Methods of heating — decreasing plants**

| TYPE OF HEATING | REMARKS |
|---|---|
| **Steam heating** | Steam is recommended as the best heating medium for degreasing plants. A steam coil (located in the liquid) is fitted through a removable cleaning door cover to facilitate cleaning of both coil and sump. |
| **High-pressure hot-water heating** | Regulating valves are provided for correct adjustment of the heat input to each compartment. Maximum temperature 177°C and pressure of 7 to 17 bar (707 kN/m$^2$ to 1·717 MN/m$^2$). |
| **Electric heating** | Ring immersion heaters are located in the liquid and the supply is through a contactor. Heating is cut off either if there is a rise in vapour temperature due to condensing water failure or if the sump temperature exceeds 120°C owing to an accumulation of oil. |
| **Gas heating** | Gas is burned in an enclosed combustion chamber with flue pipe, which requires to be led to the atmosphere or to a point at which air is extracted from the workshop. Plants are available which can be heated by bottle gas (propane, butane or proprietary mixtures) or heated by natural gas. |

Most other types of work are best handled in baskets, which can rest on grille plates or work plates in the plant. As horizontal surfaces retain a film of liquid solvent on leaving the plant, baskets should be made of rod, covered with mesh if necessary, rather than angle section or perforated metal.

*After reaching vapour temperature, work should be removed from the plant slowly and steadily.* Small articles such as castings with blind holes or cup-shaped pressings, which will trap solvent, should be loaded into cylindrical baskets that can be turned over before they are removed from the vapour. Materials used for baskets are mild steel (galvanised), brass and stainless steel.

## SAFETY

Degreasing plants should be installed on a site that is well ventilated and free from direct draughts. *It is important that air from the vicinity of the plant should not be drawn into any naked flame, welding arc or other open source of heat.* The plant should not be installed in a fabrication shop in which *argon-arc welding* is carried out.

Exposure of the vapour to naked flames or red-hot surfaces such as welding arcs and open space-heaters may produce acidic gases.

Like most organic solvents, trichloroethylene will, if inhaled (or sniffed) in small quantities, cause drowsiness and giddiness. This will quickly pass off in fresh air. If inhaled in larger quantities, however, it may cause a feeling of euphoria, followed by unconsciousness and, in extreme cases, death. *If the fumes are drawn through the lighted end of a cigarette or a pipe (which is in effect a minature furnace), the danger is greatly increased, which is why 'No smoking' notices are displayed and should not be ignored!*

Contact of the solvent with the skin should be avoided, as it will remove the natural grease. Should contact accidentally occur, some benefit may be obtained by the application of barrier creams based on lanoline, which restores grease to the skin. However, should occasional contact with the hands be unavoidable, then PVC gloves should be worn.

## 6.28 Heat treatment of metals

The terms 'annealing', 'grain growth', 'recrystallisation', 'temper' and 'quench hardening', mentioned in this chapter, all refer to the many *heat treatment* processes that can have a marked effect upon the properties of metals.

The purposes underlying the heat treatment of steel are manifold, but include:

(1) increased hardness,
(2) increased strength,
(3) increased toughness.
(4) increased wear resistance,
(5) relief of internal strains,
(6) reduced brittleness,
(7) improved machinability.

Figure 6.30 summarises the methods by which the structure, and therefore the properties, of plain carbon steels can be changed by heat treatment processes.

The temperature to which the steel must be heated to wipe out the initial structure and properties will depend upon the carbon content of the steel. Figure 6.31 shows the temperatures from which plain carbon steels should be quench hardened. It will be seen that these are the same temperatures as for annealing. The only difference being in the rate of cooling (Figure 6.30). The degree of hardness the steel achieves is solely dependent upon:

1. The carbon content.
2. The rate of cooling.

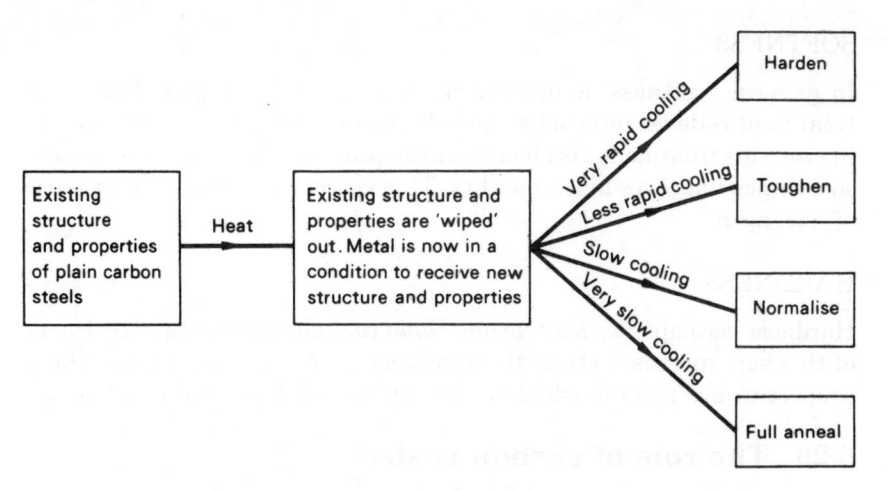

Fig. 6.30 Heat treatment of plain carbon steels

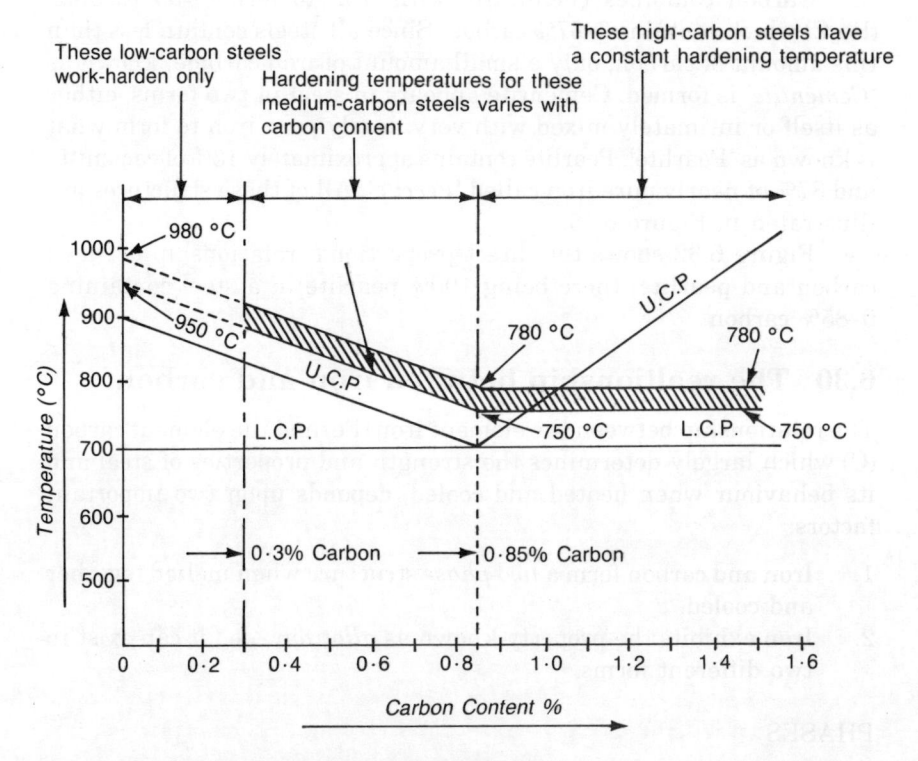

Fig. 6.31 Hardening of plain carbon steels

## SOFTNESS

In general, 'softness' is induced by *heating and slowly cooling*. Such treatment relieves internal strains. If grains have not increased greatly in size, this treatment also increases toughness, reduces brittleness, and sometimes improves machinability. This is accomplished at the expense of 'strength'.

## HARDNESS

Hardness accompanies *heating and rapid cooling* (quenching). Treatment of this sort increases strength, wear resistance and brittleness. These properties are accomplished at the expense of decreased toughness.

### 6.29 The role of carbon is steel

The essential metallurgy of steel is due to the particular relationship between the basic elements iron (Fe) and carbon (C), and the way in which this relationship is affected by changes of temperature.

Carbon combines chemically with iron to form 'iron carbide' ($Fe_3C$) which contains 6·67% carbon. Since all steels contain less than this amount of carbon, only a small amount of iron carbide, known as 'Cementite' is formed. Cementite appears in steel in two forms, either as itself or intimately mixed with very nearly pure iron to form what is known as 'Pearlite'. Pearlite contains approximately 13% of cementite and 87% of nearly pure iron called 'Ferrite'. All of these structures are illustrated in Figure 6.15.

Figure 6.32 shows the direct proportional relationship between carbon and pearlite, there being 100% pearlite in a steel containing 0·85% carbon.

### 6.30 The realtionship between iron and carbon

The relationship between the element iron (Fe) and the element carbon (C) which largely determines the strength and properties of steel and its behaviour when heated and cooled, depends upon two important factors:

1. Iron and carbon form a *two-phase* structure when melted together and cooled.
2. Iron exhibits the property known as *allotropy*, i.e., it can exist in two different forms.

## PHASES

In any system of solids and liquids there is the existence of *'phases'*,

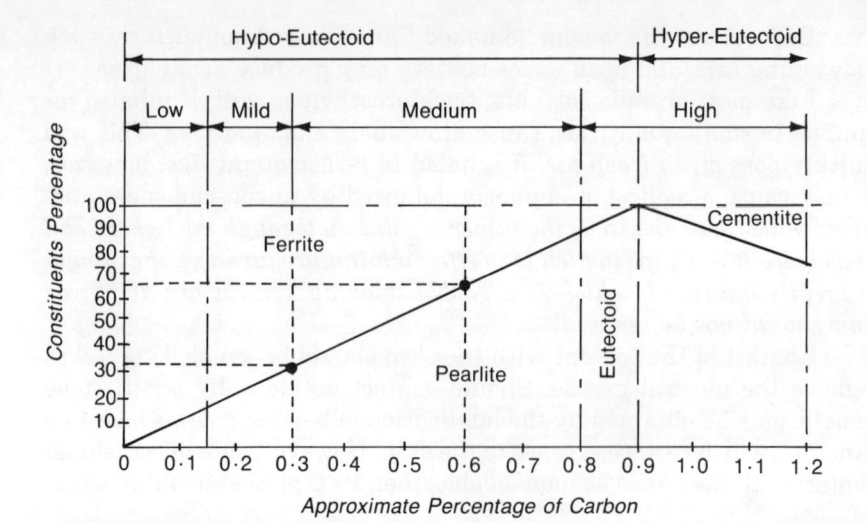

*Note: For simplicity the diagram shows an approximate percentage of carbon which produces 100% PEARLITE as 0·9%.*

Plain carbon steels which contain 0·85% carbon are called EUTECTOID steels. Those containing less than 0·85% C are termed HYPO-EUTECTOID steels, whilst above 0·85% C they are known as HYPER-EUTECTOID steels.

Hyper-eutectoid steels contain more *iron carbide* than is required to form PEARLITE and the excess carbide exists as CEMENTITE surrounding the grains of pearlite. Steels with free cementite when heat-treated are extremely hard and brittle. File steels, containing 1·3% carbon, are typical of the hardness and brittleness obtained.

Reference to the above diagram illustrates that a steel containing 0·3% carbon would consist of 33% pearlite and 67% Ferrite whilst a steel containing 0·6% carbon would have 66% pearlite and 33% Ferrite.

**Fig. 6.32 The relationship between carbon content and the constituents of slowly-cooled carbon steels**

that is parts of the whole which are separated from the other parts by definite detectable boundaries.

Tea, when poured straight from the pot, without milk, is a brown transparent liquid consisting of water and a little organic matter from the infused tea leaves. Although two substances have gone into the making of it, the resultant liquid is only a *one-phase system*. It is impossible to detect any boundary between colourless water and the brown organic matter, i.e., *it is a true solution*.

If a couple of spoonfuls of sugar are added it will not be dissolved at once, and grains of sugar will, for a time, exist separately in the tea, producing a *two-phase system* until all the sugar is completely dissolved, once again producing a clear liquid.

When the liquid tea is allowed to cool to below the freezing point of water by placing it in a refrigerator, the brown organic matter will remain in *solid solution* with the water in the form of brown ice. However, the sugar content will definitely start to 'crystallise out' as the temperature drops. *Water and tea in the solid state, therefore, remain as one phase, whereas water and sugar revert again to two phases.*

The above example has been used as an analogy to basically explain the behaviour of carbon and other alloying elements when dissolved in molten metal.

On cooling molten steel, elements such as silicon and nickel, for example, will remain in *solid solution* in the metal; they form a *single-phase system* like the tea does in ice. *They cannot be detected when solid steel is examined under a microscope.*

However, carbon resembles the sugar; it will not all remain in solid solution as the metal cools. Below a certain '*critical temperature*' it will '*crystallise out*' as a distinct second phase in solid steel.

ALLOTROPY — IRON

At ordinary temperatures the element iron is a soft, grey, ductile metal which exhibits the property of *magnetism*. However, if iron is *heated to a temperature above 910°C, it undergoes an atomic re-arrangement* which causes it to lose its magnetism. If cooled again, then it will change back into its original state and the magnetism will re-appear.

The low-temperature (magnetic) structure is termed a BODY-CENTRED CUBIC (B.C.C.), and the high-temperature (non-magnetic) structure a FACE-CENTRED CUBIC (F.C.C.) as illustrated in Figure 6.1.

When iron has a B.C.C. lattice it is referred to as an *alpha-iron* ($\alpha$) and when it has a F.C.C. lattice as a *gamma-iron* ($\gamma$).

Two very important facts about the allotropy exhibited by iron in relationship to the structure and heat-treatment of steel are:
1. The high-temperature gamma form of iron can hold carbon (as cementite) in solid solution while the low-temperature alpha form cannot. Thus, it is not absolutely correct to compare carbon in steel with sugar in tea, as we have done in a previous paragraph, because in fact when steel solidifies the carbon (cementite) is at first not precipitated but held in solid solution. It is only later, when the steel is already quite solid, that the precipitation of cementite takes place.
2. The existence of carbon up to a maximum content of 0·85% dissolved in gamma-iron has the effect of *depressing the temperature* at which the gamma to alpha change, and the resulting precipitation of cementite, commences. Figure 6.33(a) shows that as the level of carbon content rises from nothing to 0·85% so the temperature at which the change takes place drops from 910°C to 730°C. Above this critical carbon content, however, the temperature at which the change begins starts to rise again.

## 6.31 The iron—carbon equilibrium diagram

A horizontal line is drawn through the minimum transition temperature reached, and is labelled LOWER CRITICAL POINTS, whilst the existing line (Figure 6.33(a)) represents the UPPER CRITICAL POINTS. This simple version of what is known as an EQUILIBRIUM DIAGRAM graphically represents the critical temperatures at which important changes in the crystal structure of steel occur during heat-treatments.

The diagram has been completed in the manner shown in Figure 6.34 by labelling the various positions and areas according to what phase or phases apply to a particular point or range of temperatures and carbon contents.

Table 6.21 summarises the *effect of carbon phases* on the hardness, toughness and ductility of steel.

## 6.32 The effect of heat-treatment on the structure of steel

The structures and properties so far considered were produced under equilibrium conditions. The steel was heated to a certain critical temperature and allowed to cool sufficiently slowly for all the changes in the crystal structure in the metal to take place and establish a state of *chemical balance* or equilibrium. However, by altering the '*rate*' at which the temperature is lowered from above the UPPER CRITICAL POINTS, it is possible to produce a wide variety of different structures in steel even though the carbon content remains the same.

In the system of solid phases associated with steel the rate of diffusion of the different constituents is inevitably slow, and it is quite possible to speed up the rate of cooling so that by the time the metal is quite cold, and therefore diffusion has to all intents and purposes, come to a halt, the system has been '*frozen*' in a state of '*disequilibrium*'.

There are three principal single stage methods of heat-treatment of plain carbon steels, namely: *Annealing, Normalising* and *Hardening* (Quenching), all of which have an effect on the structure of the steel.

ANNEALING

Figure 6.35 shows the annealing temperatures for plain carbon steels. Reference to Figure 6.31 will indicate that the range of full annealing temperatures corresponds with that shown for hardening.

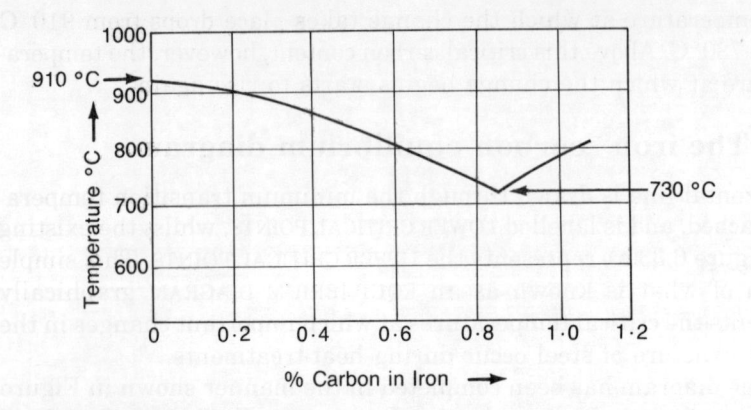

(a) The temperatures at which the structural changes take place and depress as the carbon content increases to a maximum of 0·85%

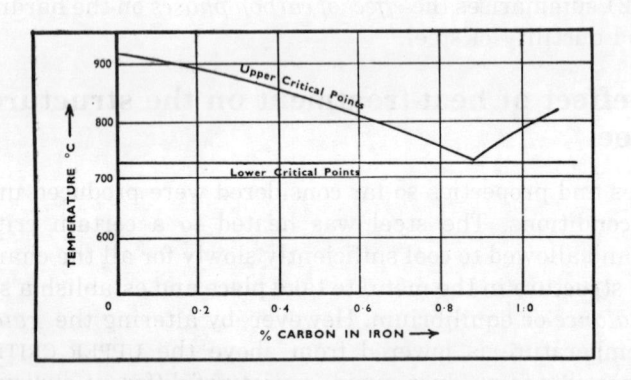

At all temperatures and carbon contents above the U.C.P., the carbon (cementite) is held in gamma-iron in solid solution, in a form known as AUSTENITE.

At a carbon content of 0·85%, i.e. the L.C.P., the change from gamma- to alpha-iron throws all the carbon out of solution at once to form an intimate mixture of iron and cementite called PEARLITE.

For carbon contents less than 0·85% the structural change takes place over the temperature range indicated by the difference between the upper and lower critical points. At first almost pure iron called FERRITE appears and goes on appearing until the remaining gamma-iron has been enriched to 0·85% carbon, when it changes at once to PEARLITE.

At carbon contents above 0·85% the change also takes place over a range of temperatures, but this time the first material to appear is pure CEMENTITE. This continues until the carbon content of the remaining gamma-iron is reduced to 0·85% when this again also changes at once to PEARLITE.

(b) Critical temperatures at which changes in structure of steel takes place

Fig. 6.33 Simple iron–carbon equilibrium diagram

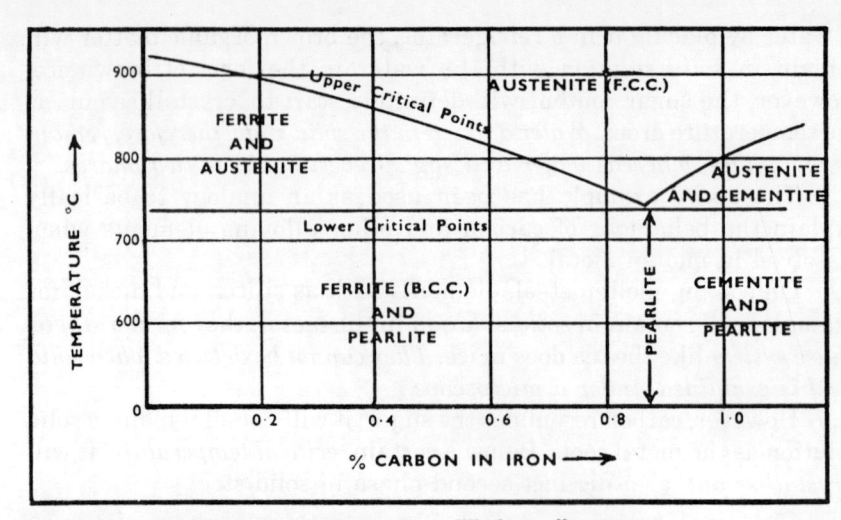

Fig. 6.34 Complete iron–carbon equilibrium diagram (plain carbon steels)

Table 6.21 Effect of carbon phases on mechanical properties of steel

| CARBON PHASE | REMARKS |
| --- | --- |
| Ferrite | Contains such a minute quantity of carbon in solid solution that like pure iron, it is a soft, tough, ductile phase. A steel containing a lot of ferrite will therefore be soft, with a low tensile strength, but will be tough and durable. |
| Pearlite | A laminated structure of ferrite and cementite. It is hard and brittle, mainly because the intimate mixing of ferrite and cementite ensures that any slip in a ferrite plate quickly meets the boundary between it and a plate of hard cementite. As the carbon content of the steel rises towards 0·85%, so the hardness of the metal as a whole increases. However, this is accompanied by a decrease in ductility and toughness, that is resistance to shock loads. |
| Cementite | Extremely hard and brittle. Steels containing significant quantities of free cementite find their greatest commercial use where extreme hardness is required, for example, for ball bearings, ball races and tools. These types of steels have to be very carefully heat-treated. |

*Note:* There is no 'free' carbon in plain carbon steels (free carbon only appears in cast irons), all the carbon being combined with some of the iron.

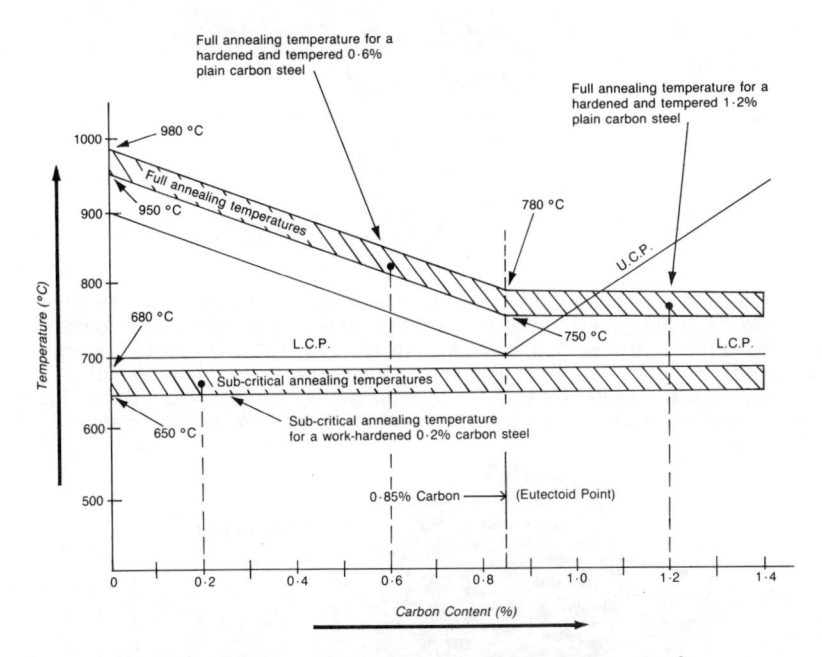

**Fig. 6.35 Annealing temperatures for plain carbon steels**

'*Process annealing*' is a form of annealing where steel or other metal is heated to a comparatively low temperature after cold-working in order to cause '*recrystallisation*' (see Section 6.4). In this process the metal is not generally heated to a temperature above the U.C.P., as this is not necessary. The recrystallisation temperature for mild steel lies within the band shown in Figure 6.35 as the '*sub-critical annealing temperatures*'. The process consists of heating '*work-hardened*' steel within this 550°C to 650°C band, followed by slow cooling.

Quench-hardened steels, however, also possess a distorted or stressed crystal structure, but such recrystallisation treatment would be ineffective.

'*Full annealing*' is the process used for quench-hardened steels and consists of heating the metal within the temperatures shown in Figure 6.35 for its particular carbon content, followed by very slow cooling. This band of '*full annealing temperatures*' lies just above the U.C.P.

Figure 6.36 illustrates the change in crystal structure when a medium carbon steel is annealed.

*Annealing is essentially a softening process of heat-treatment; its purpose is to relieve internal stresses, reduce brittleness and improve machinability.*

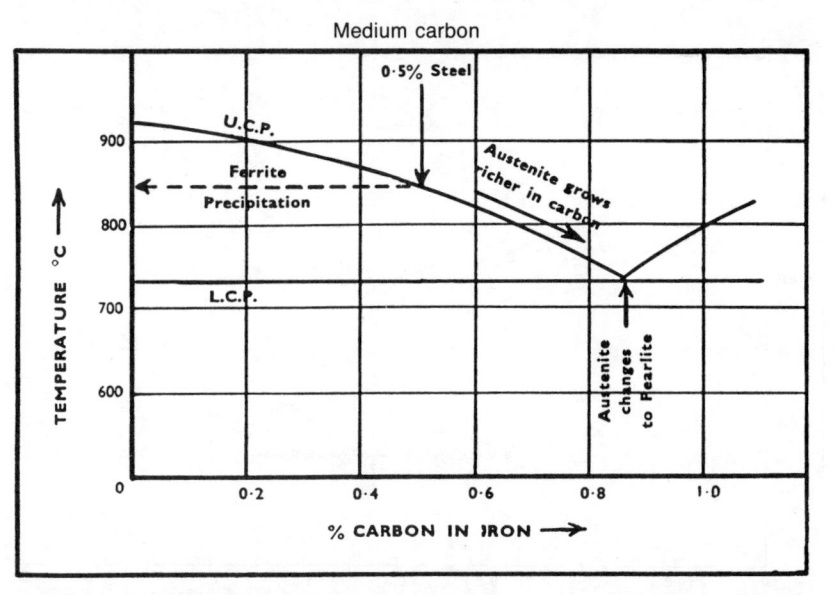

The steel is heated to a point just above the U.C.P., in this case approximately 830 °C to 860 °C, to ensure all carbon is absorbed into solid solution in gamma-iron as *Austenite*, and allowed to cool slowly in the furnace. As the metal cools though the U.C.P., so the gamma-iron will become unstable and start to change into the alpha form.

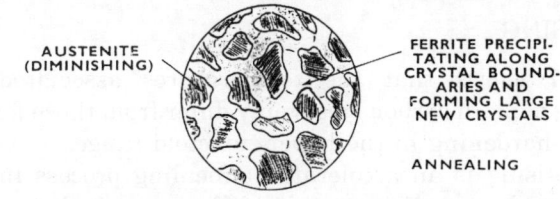

*Ferrite* starts to precipitate out as the temperature drops below the U.C.P. and new crystals of carbon-free metal begin to appear around the original crystal boundaries. As more and more carbon comes out it will form a series of large new crystals. Meanwhile the carbon diffuses into the centre of the remaining austenite crystals until they contain 0·85% carbon. When the temperature passes through the L.C.P. then the remaining austenite completely changes at once into *Pearlite*.

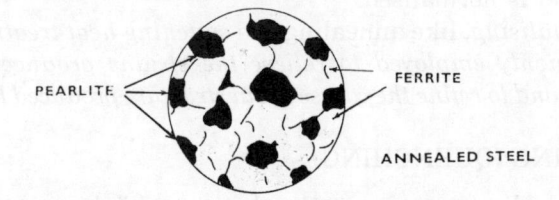

In *annealed* steel the *Ferrite* and *Pearlite* are divided up into separate areas or 'lakes', resulting in a comparatively coarse crystal structure.

**Fig. 6.36 Changes in crystal structure — annealing**

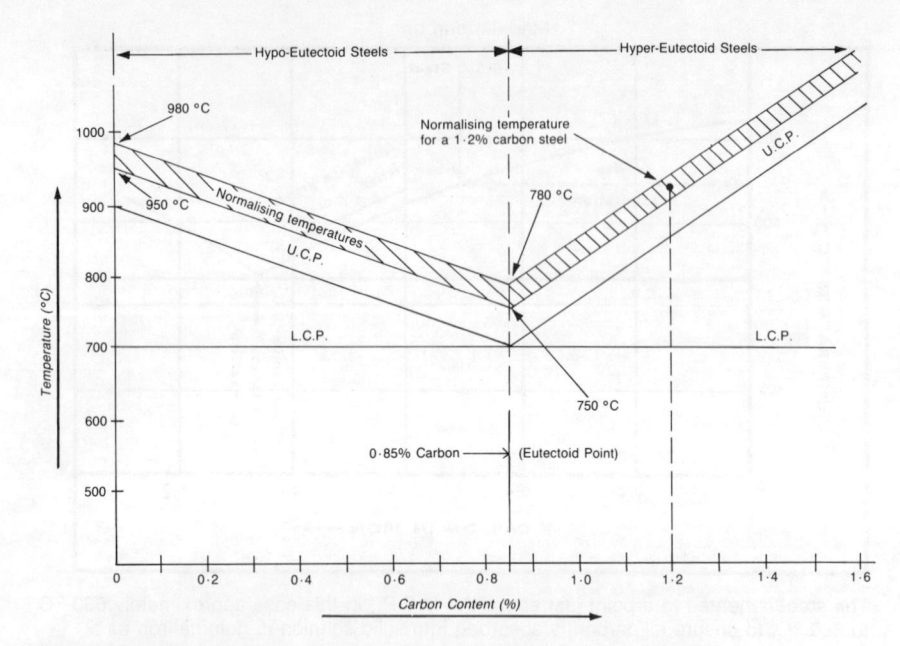

**Fig. 6.37 Normalising temperatures for plain carbon steels**

## NORMALISING

Figure 6.37 shows that the temperatures associated with the normalising of plain carbon steels only differ from those for annealing and quench-hardening in the hyper-eutectoid range.

Normalising is an accelerated annealing process in which the carbon does not have sufficient time to diffuse completely from the ferrite areas into the austenite. The process is conducted by removing the hot steel from the furnace and allowing it to cool in still air. Draughts may cause it to cool too quickly and should be avoided.

Figure 6.38 explains the change in structure when a medium carbon steel is normalised.

*Normalising, like annealing is a softening heat-treatment process. It is commonly employed to relieve the strains produced by forging operations and to refine the coarse grain structure produced by hot-rolling.*

## HARDENING (QUENCHING)

The rapid cooling necessary to harden a steel is known as *quenching*. The liquid into which the steel is immersed to promote this rapid cooling is called the QUENCHING BATH which contains either WATER or a SPECIAL QUENCHING OIL. This heat-treatment process consists of plunging the

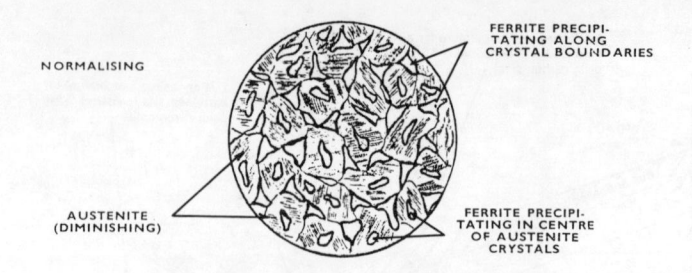

The steel is heated to just above the U.C.P. to ensure an *Austenitic structure* and then allowed to cool in a draught-free atmosphere. The relatively quick rate of cooling does not allow the carbon sufficient time to diffuse through the metal. This results, initially, in small crystals of *Ferrite* appearing both around the crystal boundaries and in the bodies of the original crystals. These areas of ferrite are unable to merge together and form large 'lakes' and neither is the separation into *Ferrite* and *Pearlite* complete.

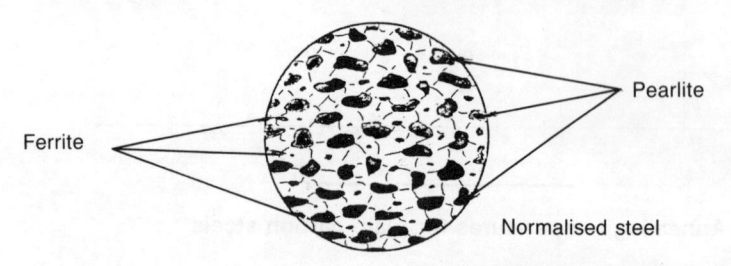

Instead of the resulting ferrite and pearlite areas being large and well separated, like the final coarse crystal structure produced by annealing, they are intimately blended together to form a much finer structure as much fo the ferrite has come out, not before but with the pearlite.

*Note*: The actual quantity of pearlite in a normalised steel is no different from that present in an annealed steel of the same carbon content. Normalising, however, spreads the pearlite over the mass of soft ferrite, thus *refining the crystal or grain structure*.

**Fig. 6.38 Changes in crystal structure — normalising**

steel, at a temperature just above its U.C.P., into the quenching medium which causes a drastic and rapid drop in temperature.

In quenching steel from an austenitic condition, the rate of cooling is so fast that not merely does the carbon fail to diffuse throughout the the mass and form large concentrations of ferrite, but its entire precipitation is suppressed altogether. Most of the carbon content is 'frozen' in a super-saturated state within the B.C.C. iron itself. This condition in which the carbon atoms are held in solid solution in a distorted B.C.C. lattice prevails throughout the entire structure of the steel. Under a microscope this type of steel appears in the form of needles, as shown in Figure 6.39(a), which is an *acicular structure called Martensite*.

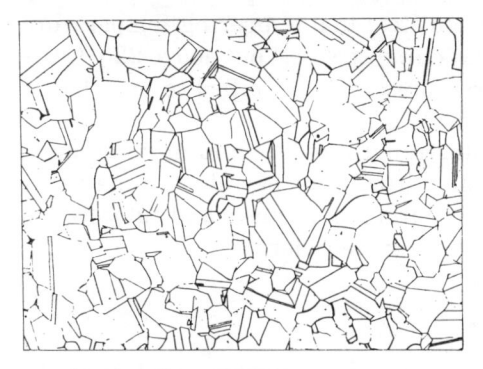

(a) **Austenite.** A solid solution of carbon in a 'Face-centred iron'. *This structure is not retained in quenched plain carbon steels.* However, with alloy steels containing a high percentage of MANGANESE or NICKEL, the structure will remain 'Austenitic' at normal temperatures.

18% Chromium — 8% nickel stainless steel microstructure. An 'Alpha' structure retained at room temperature due to alloying additions.

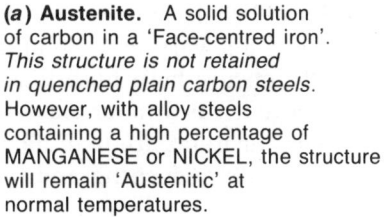

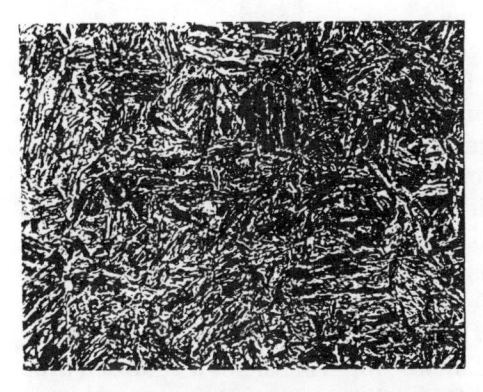

(b) **Martensite.** A fine needle-like structure in angular arrangement. This acicular structure is the hardest that can exist in plain carbon steels. Found in rapidly cooled steel. *The lower the carbon content, the more rapid must be the cooling in order to retain 'martensite'.*

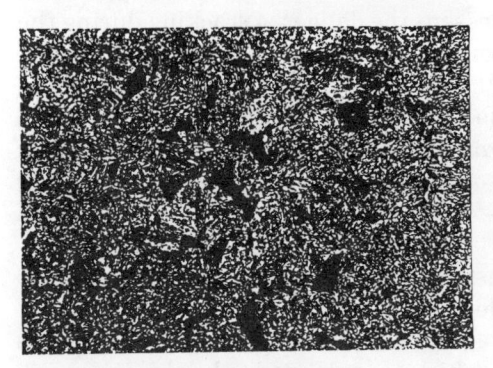

(c) * **Troosite.** Appears as a dark constituent in an acicular structure of 'martensite'. Obtained when the steel is cooled at a slightly less rapid rate than that required to produce a martensitic structure; or when a steel with a martensitic structure is re-heated. *This structure is much tougher but softer than martensite.*

Dark etching fine 'pearlite phase' (formerly known as 'troosite' in a martensitic matrix).

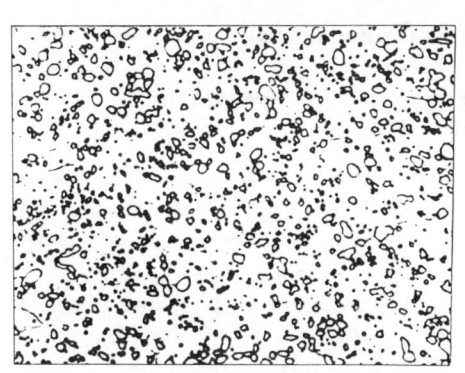

(d) * **Sorbite.** This structure is a mixture of 'cementite' and 'ferrite'. It is obtained when the steel is cooled at a slightly less rapid rate than that required to produce troosite; or as a result of tempering steel after hardening by quenching. *This structure is predominant in 'normalised steels'. It is not as hard as troosite, but is tougher with good all-round strength.*

Spherodised carbides in a ferrite matrix (formerly known as 'sorbite'). A soft structure resulting from prolonged sub-critical annealing.

\* *Note: These terms are no longer used by British Steel Technical.*

**Fig. 6.39 Microstructures — plain carbon steels (heat-treatable)**

Because the martensitic structure of a quenched steel is so intensely hard and brittle, in general workshop practice there are not a great many uses that can be found for it. However, *the principal purpose of quenching in the heat-treatment of steel is as a first operation in a two-stage treatment known as 'QUENCHING and TEMPERING'.*

## TEMPERING

Tempering consists of a gentle re-heating of a quench-hardened steel to a suitable temperature below the L.C.P. and again quenched in oil or water.

*The effect of this heat-treatment process is to greatly increase the toughness of the steel at the expense of hardness.*

Tempering temperatures below 200°C only relieve the hardening stresses, but above 200°C the hard and brittle martensitic structure can be modified. This relatively low temperature is sufficient to 'loosen' the severely distorted B.C.C. lattice structure of the quench-hardened steel and cause the carbon to be precipitated into a more stable condition. Some of the carbon is enabled to come out of solution, not in the normal laminated form of pearlite but in harder and more finely divided forms of pearlite known as '*Troosite*' and '*Sorbite*'.

Figure 6.39 illustrates some of the microstructures associated with heat-treated plain carbon steels.

Sometimes the two operations of quenching and tempering are

**Table 6.22 Tempering temperatures (carbon steels)**

| *COLOUR | EQUIVALENT TEMPERATURE °C | APPLICATION |
|---|---|---|
| Very light straw | 220 | Scrapers; lathe tools for brass |
| Light straw | 225 | Turning tools; steel-engraving tools |
| Pale straw | 230 | Hammer faces; light lathe tools |
| Straw | 235 | Razors; paper cutters; steel plane blades |
| Dark straw | 240 | Milling cutters; drills; wood-engraving tools |
| Dark yellow | 245 | Boring cutters; reamers; steel-cutting chisels |
| Very dark yellow | 250 | Taps; screw-cutting dies; rock drills |
| Yellow-brown | 255 | Chasers; penknives; hardwood-cutting tools |
| Yellowish brown | 260 | Punches and dies; shear blades; snaps |
| Reddish brown | 265 | Wood-boring tools; stone-cutting tools |
| Brown-purple | 270 | Twist drills |
| Light purple | 275 | Axes; hot setts; surgical instruments |
| Full purple | 280 | Cold chisels and setts |
| Dark purple | 285 | Cold chisels for cast iron |
| Very dark purple | 290 | Cold chisels for iron; needles |
| Full blue | 295 | Circular and band saws for metals; screwdrivers |
| Dark blue | 300 | Spiral springs; wood saws |
|  | 450 to 600 | Toughening (crankshafts) |

* At the temperatures associated with the tempering of ferrous metals, the colour of the oxide film that forms on the polished surface of the metal indicates the temperature of the metal. The higher the temperature, the thicker the oxide film and the darker it appears.

combined in one by rapidly cooling the steel to its tempering temperature and then holding at that temperature for some time to allow the precipitation of the carbon content and lessen the danger of cracking.

Table 6.22 gives the tempering temperatures for various applications of hardened plain carbon steels.

Re-heating a quench-hardened steel to a tempering temperature below 200°C only relieves the hardening stresses, but above 220°C the hard and brittle martensitic structure can be modified to one that is less hard but very much tougher.

To ensure both high tensile strength and maximum toughness, products such as armour plate, boiler drums, gun barrels, turbine rotors and similar items are manufactured from alloy steels and subjected to an elaborate quenching and tempering treatment.

### 6.33 The effect of alloys

Alloying elements, such as chromium, cobalt, manganese, molybdenum, nickel, tungsten and vanadium, do not all behave in the same way as carbon. *They may either be held in solid solution in the steel and therefore cannot be detected as a separate phase under a microscope, or they may wholly or partially form carbides, associated with the cementite in pearlite and not separately distinguishable.*

Alloying elements such as chromium, nickel and molybdenum are constituents of steels designed for quench-hardening and tempering. Their presence means that the initial quenching operation does not have to be so drastic as with plain carbon steels.

Plunging a red-hot piece of steel into water is likely to lead to serious cracking, especially if it is of complex section. The steel contracts and thin sections will cool and contract considerably faster than thicker ones, setting up internal stresses. If the steel has a high carbon content it will not be sufficiently ductile to absorb stresses; it will tear apart. By adding alloying elements to steel, which has the effect of making the change from F.C.C. to B.C.C. more sluggish, the now retarded drop in temperature from above the U.C.P. will have the effect of suppressing the carbon precipitation. This lessens the danger of cracking during the volume expansion which accompanies the F.C.C. to B.C.C. change.

*Manganese* and *nickel* will lower the critical points in steel whilst *chromium* has the opposite effect and raises the critical points. *When manganese is added in comparatively large quantities to steel it has the effect of stabilising austenite at ordinary temperatures so that no change takes place as the metal cools* — in other words the U.C.P. is depressed to room temperature.

One of the best known austenitic steels (which are non-magnetic) is the 13% manganeses alloy, used where a metal capable of withstanding intense wear is required. It is used extensively for points and crossings in railway tracks which are subjected to hard wear during service. The tough teeth fitted to the buckets of dredgers and mechanical diggers are cast in this alloy steel. These teeth consist basically of very tough austenite, and during service the constant abrasive and scraping action has the effect of changing the outside contact surfaces to a very hard martensite.

Another example of austenitic steel is the ductile and tough 18% chromium and 8% nickel alloy steel which contains less than 0.06% carbon, referred to as 18/8 stainless.

*Cobalt* is one of the few alloying elements that will reduce hardness in steel. It raises the tempering temperature which can be of great importance for certain types of high-speed tool steels.

*Molybdenum*, in solid solution, produces greater strength and toughness. It raises the temperature at which *grain growth* normally takes place in plain carbon steels.

*Tungsten* forms a strong hard carbide when alloyed with steel, resulting in a desirable strengthening of the metal over an extensive range of temperatures. It increases the ability of the steel to remain hard and to resist tempering when subjected to relatively high temperatures.

*Vanadium*, like tungsten, will combine with carbon to form carbides which have a beneficial effect on the mechanical properties of heat-treated steels. It raises the temperature at which grain growth would normally take place in carbon steels and is an active grain refiner.

## 6.34 Heat-treatment of non-ferrous metals and alloys

Cold-working of any metal by such processes as drawing, pressing, rolling, beating, spinning or stamping, increases its hardness and tensile strength but reduces its ductility, and if cold-working (strain-hardening) is carried too far, the metal may crack. There is a limit, therefore, to the amount of cold-working a metal will sustain, and if this is reached before the intended deformation is completed, the material must be softened again.

When a cold-worked metal is heated above a certain temperature (the RECRYSTALLISATION TEMPERATURE), softening occurs and the process is known as ANNEALING. Table 6.23 gives the temperatures at which some typical metals recrystallise after cold-working, and Figure 6.40 represents annealing stages.

### SOLUTION TREATMENT

Aluminium wrought alloys are divided into two main groups according to the means by which their maximum properties are obtained, namely by 'work-hardening' only or by a combination of working and heat-treatment.

**Table 6.23  Recrystallisation temperatures**

| METAL | TEMPERATURES (°C) |
|---|---|
| Aluminium | 500 to 550 |
| Cold-working brasses | 600 to 650 |
| Copper | 650 to 750 |
| Duralumin | 480 to 500[1] |
| Steel (cold-worked) | 650 to 680[2] |
| Zinc | 100 to 150[3] |

*Notes:*

1. This wrought aluminium alloy is susceptible to '*age hardening*', i.e., after annealing, it commences to re-harden by itself. After standing for five days it can no longer be cold-worked and requires re-annealing. Age hardening can be retarded by refrigerating the components at −10°C after annealing. See 'Solution Treatment'.
2. This is the '*sub-critical*' annealing temperature for all steels. It only applies if the steel has been *work-hardened*. If the steel has been hardened by heat treatment, then full annealing or normalising will be necessary (see Figures 6.35 and 6.37).
3. At a higher temperature of about 250°C, the metal becomes extremely brittle and can be hammered into powder form.

*Work-hardened alloys* are supplied in various '*tempers*' from the SOFT condition up to HARD; intermediate strengths being indicated by QUARTER-HARD, HALF-HARD and THREE-QUARTER-HARD. The degree of hardening is produced during manufacture by cold-working the metal a certain amount after annealing.

*Heat-treatable alloys* are not softened by annealing, nor are they quench-hardened. *The heat-treatment of the wrought light alloys is essentially a precision operation in which the temperature, and time at temperature, are critical.* It is imperative that both temperatures and minimum times be capable of exact adjustment under normal factory conditions, which means that the equipment used must provide uniform heat distribution and be fitted with automatic temperature control and recording instruments capable of working with precision. The forced air-circulation furnace and the molten salt bath are the most efficient of all the types of equipment used for light alloy heat-treatment. The salts used are a mixture of equal parts of SODIUM and POTASSIUM NITRATES, the proportion being chosen to give the lowest melting point, 220°C, for any mixture of these salts. However, since sodium nitrate melts at 310°C, and the salt bath is normally used for solution heat-treatments above 450°C, it may be used alone as the liquid heating medium.

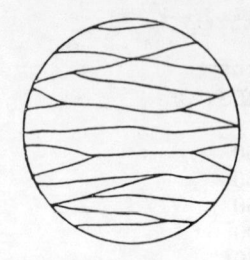

### (a) Cold-worked structure

Crystal structure distorted by cold-working and *work-hardening* occurs.

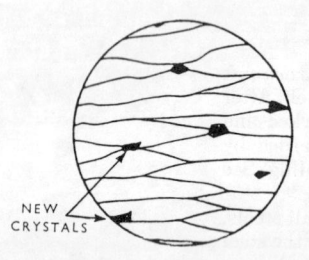

### (b) The commencement of recrystallisation

The temperature of the metal is raised above its recrystallisation point.
As the temperature of the cold-worked metal rises, the energy of the individual atoms increases, and at many points within the distorted crystal grains the atoms begin to rearrange themselves into regular strain-free equiaxed crystals.

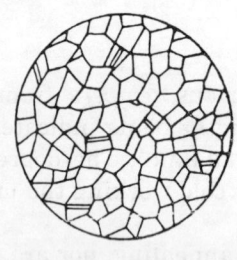

### (c) The annealed structure

An annealing continues the fresh crystals gradually enlarge, feeding on the old, distorted crystals until the whole workpiece once more consists of a regular equi-axed grain structure. The increased strength imparted to the metal by cold work is lost but ductility and malleability is regained. The metal is again in a suitable condition for further deformation.

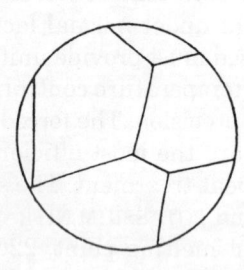

### (d) Development of large grains

If annealing is allowed to continue beyond its specified time, the crystals will continue to grow, but at each others expense, and an undesirable coarse grain structure will develop.
With work-hardening aluminium alloys, excessive grain growth may give an undesirable '*orange-peel*' effect on the surface when the material is subsequently worked.

**Fig. 6.40 Stages in the annealing process**

*The salt bath cannot be used for the heat-treatment of magnesium-base light alloys, as any attempt to do so will result in a reaction of explosive violence.*

The heat-treatment involves three distinct phases:

1. Heating the alloy at a specific temperature for the requisite time. The figures given in Table 6.24 provide a rough guide to the time necessary for material of different thicknesses.
2. Cooling the alloy rapidly from this temperature, usually by quenching in water. *Oil quenching should not be used for material heated in a salt bath, owing to the risk of fire when hot nitrate makes contact with oil.*
3. Ageing, either spontaneously at ordinary temperatures or as a result of low-temperature re-heating.

The first two operations together are known as SOLUTION TREATMENT. Solution treatment softens the material and consists of heating the alloy to a temperature ranging from 450°C to 540°C, depending upon its composition and form, to allow the alloying elements to go into solid solution in the aluminium. The alloy is then quenched to preserve the solid solution at room temperature and is now soft enough to cold work. This working may include shaping or forming of sheet, strip and section or the removal of distortion, such as by stretching, roll flattening, or straightening by hand or in a press.

**Table 6.24  Approximate heating times for solution-treated aluminium alloys**

|  | SECTION |  | TIME |
|---|---|---|---|
|  | s.w.g. | Thickness (mm) | (minutes) |
| **Sheet or strip** | up to 26 | up to 0·50 | 6 to 10 |
|  | 22 to 24 | 0·80 to 0·60 | 10 to 14 |
|  | 18 to 20 | 1·25 to 1·00 | 14 to 20 |
|  | 14 to 16 | 2·00 to 1·60 | 18 to 22 |
|  | 10 to 12 | 3·15 to 2·80 | 22 to 28 |
|  | 6 to 8 | 5·00 to 4·00 | 28 to 34 |
|  | 3 to 5 | 6·40 to 5·60 | 30 to 38 |
|  | Diameter (mm) |  |  |
| **Bars and extrusions** | up to 12·7 |  | 30 |
|  | 12·7 to 25 |  | 45 |
|  | For each additional 25 |  | 45 |

*Note*: The above times are for work immersed in a salt bath. For air furnaces an additional period must be allowed to bring the load up to the required temperature.

The alloying elements, such as copper, magnesium, silicon and zinc, are hardening constituents dissolved in solid solution with aluminium, and when the metal is rapidly cooled (quenched) much of the dissolved material remains in solid solution. However, if the alloy is allowed to stand, this artificially rich solid solution is not stable, and the excess elements or compounds tend to precipitate out of solution. *This precipitation causes straining of the crystals of the solid solution, and thus increases the hardness of the metal.* This spontaneous hardening of certain alloys is known as NATURAL AGE-HARDENING.

As age-hardening reduces ductility, any appreciable cold-working must be performed whilst the alloy is still soft and must be completed within 2 hours of quenching, or, for intricate pressings within 30 minutes. Once age-hardened, the metal must be re-solution-treated before further work is possible. In all age-hardening alloys, the strength begins to increase shortly after quenching, but some alloys harden more rapidly and to a greater extent than others. *Generally their maximum strength and hardness is reached after 4 or 5 days at normal shop temperature.*

Precipitation may take place spontaneously as in the natural age-hardening alloys, or it may only take place when the temperature is raised slightly, i.e., by precipitation treatment.

Precipitation treatment involves re-heating the quenched alloy to a temperature between 130°C and 200°C for a suitable period, according to the composition of the material. This relatively low temperature re-heating operation accelerates hardening and is commonly referred to as ARTIFICIAL AGEING or TEMPER-HARDENING.

Since solution-treated aluminium alloys only remain ductile for a maximum period of 2 hours after quenching, this could lead to a considerable 'bottleneck' during fabrication. This problem can be overcome by storing blanks, sheet or strip and rivets, immediately after solution-treatment, at sub-normal temperatures down to −20°C, thus temporarily retarding age-hardening. Table 6.25 compares storage times for rivets that have been solution-treated.

**Table 6.25 Storage periods — solution treated duralumin rivets**

| TEMPERATURE (°C) | MAXIMUM STORAGE PERIOD (HOURS) |
| --- | --- |
| Atmospheric temperature | 2 |
| 0 to −5 | 45 |
| −15 to −20 | 150 |

## 6.35 Annealing — workshop practice

Where annealing furnaces are not available or their use is uneconomical, or the components to be treated are relatively small in size and/or number, then the process may be successfully carried out with the aid of workshop indicators.

### WORKSHOP INDICATORS

The use of workshop indicators to estimate annealing temperatures are fairly reliable methods used in small workshops where hand methods of fabrication are employed. An estimation of temperature can be made by use of an indicator such as a smear of machine oil which 'smokes' and leaves a clear black mark at approxiately 350°C. This same temperature is also indicated when a piece of household soap or a stick of white wood leaves a black or dark brown mark when rubbed on the metal's surface.

The metal or workpiece to be annealed is placed on *clean firebricks* in a brazing hearth or a blacksmith's forge and the actual heating accomplished with the aid of a gas-air torch. The outer envelope of the flame should be allowed to cover as large an area as possible to ensure uniform heating. It is very important that any stray splashes of solder which may be adhering to the firebricks be removed before annealing is attempted. *Solder readily alloys with copper and brass at elevated temperatures.* It can penetrate the metal being annealed or cause hard spots to form on the surface thus lowering the melting point of the affected area with the consequent danger of local melting during the process.

The colour of a metal at elevated temperature may also serve to estimate its temperature for annealing purposes, but the brightness of the light must be taken into consideration when the observation is made. All metals glow and radiate visible light when heated above approximately 500°C. The radiant light changes colour and brightness with changes in temperature. Table 6.26 lists these changes.

Details of typical workshop methods of annealing are given in Table 6.27.

Some of the guess work can be taken out of the estimation of temperatures by the use of heat-sensitive paints as shown in Figure 6.41 and heat-sensitive crayons as shown in Figure 6.42. However, these crayons and paints only indicate that the specified temperature has been reached and not how much it has been exceeded.

**Table 6.26  Colours associated with temperature**

| COLOUR | TEMPERATURE °C |
| --- | --- |
| Dull red | 500 |
| Blood red | 650 |
| Cherry red | 750 |
| Bright red | 850 |
| Salmon | 900 |
| Orange | 950 |
| Yellow | 1050 |
| White | 1200 |

## 6.36 Hot and cold forging

The process of *forging* is the hammering of metal to shape. That is shaping the metal by *plastic flow*. Metals that are *worked* in this manner become *work hardened*. The temperatures at which metals re-crystallise after being cold-worked have already been discussed.

1. Forging the metal *below* the temperature of re-crystallisation is called COLD FORGING. Most rivet-heading operations are cold forging. The degree of flow-forming that is possible below the temperature of re-crystallisation is limited since the metal will work harden and eventually crack.
2. Forging the metal *above* the temperature of re-crystallisation is called HOT FORGING. The processes of hot forging are normally performed by a blacksmith. It is these processes that will be considered in the following sections of the chapter.

Metals whose internal structure has been modified by working (hot or cold) are called WROUGHT METALS.

## 6.37 The effect of forging on the properties of metals

The rolling and drawing processes to which metal bars are subject are themselves forming processes and effect the 'lay' of the grain in the bar as shown in Figure 6.43.

Figure 6.43(*a*) shows a cast bar with its random grain or crystal structure. Figure 6.43(*b*) shows a rolled or drawn bar with its grain directed along the length of the bar.

The component shown in Figure 6.44(*a*) may be machined from the bar or 'upset' forged (see Chapter 9). The differences in the lay of

**Table 6.27  Some typical workshop methods of annealing**

| METAL | PROCEDURE |
| --- | --- |
| **Aluminium** | The approximate annealing temperature (350°C to 400°C) is indicated by one of the following methods:<br>1. Heat the metal slowly until when stroked with a thin stick of white wood, a dark brown ('charred') stain appears on the metal's surface.<br>2. Lightly smear the metal's surface with machine oil and heat until the oil smokes and is burnt away.<br>3. Heat until the surface of the metal is stained black when rubbed with a piece of soap.<br>After heating the metal may be cooled slowly or quenched; the cooling rate is immaterial. |
| **Brass** | The cold-working brasses commonly used in sheet-metal work should not be heated above dull red (500°C). Those containing up to 63% copper must be cooled slowly while brasses containing more than 70% copper can be quenched in water. |
| **Copper** | The metal is heated between blood red (650°C) and cherry red (750°C) and may be quenched or slowly cooled. Quenching removes the black oxide scale formed on the metal's surface during heating. This scale, which remains when the metal is slowly cooled, can be removed by 'pickling'. |
| **Mild steel** | Heat to a bright red (approximately 900°C) and allow to cool as slowly as possible. Dry clean sand or ashes can be used to retard the cooling rate. |
| **Zinc** | Zinc must not be worked when completely cold because normal room temperature is below that at which this metal will remain in an annealed condition. Before folding or severe forming operations can be carried out, the metal should be heated to approximately 100°C, and this temperature should be maintained during forming. As soon as drops of water begin to hiss and rapidly evaporate when allowed to fall on the warm surface, the zinc is ready for forming. |

The general practice is to uniformly heat the metal to be annealed with a gas flame.

the grain between machining and forging the component are shown in Figure 6.44(*b*) and (*c*).

If the component shown in Figure 6.44 was to be used for a gear blank, this difference in grain orientation would be very important. Figure 6.44(*a*) shows that the grain of the machined blank is parallel

Heat-sensitive paint

Used for general heat-treating or heat-processing, particularly on areas not readily accessible for temperature indicating crayons; ideal for smooth and polished surfaces such as plastics and polished metal surfaces

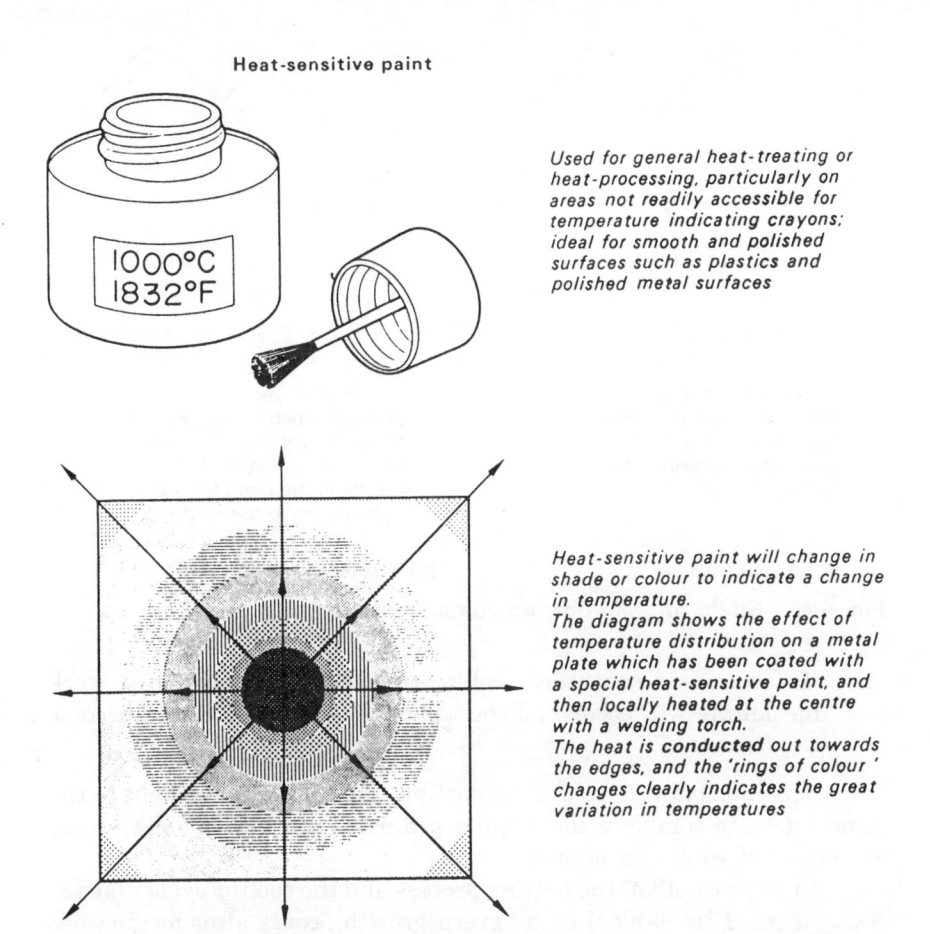

1000°C
1832°F

Heat-sensitive paint will change in shade or colour to indicate a change in temperature.
The diagram shows the effect of temperature distribution on a metal plate which has been coated with a special heat-sensitive paint, and then locally heated at the centre with a welding torch.
The heat is **conducted** out towards the edges, and the 'rings of colour' changes clearly indicates the great variation in temperatures

**Fig. 6.41 Heat spread and temperature variation indicated by use of heat-sensitive paints**

Temperature indicating crayon

330°C
625°F

1. Select 'crayon' for the required temperature

2. Mark the workpiece

3. Crayon mark melts when specified temperature is reached

This is a simple but accurate method of determining temperatures in soldering, brazing, welding and heat treating. It can be used for many other heat-dependent operations, for example when forming 'thermo-plastics'

**Fig. 6.42 The use of temperature indicating crayons**

to the teeth of the gear, causing planes of weakness. The teeth would break off relatively easily. Figure 6.45(b) shows that the grain of the forged blank would be at right angles to the teeth of the gear. This produces strong teeth. Remember that grain in metal behaves like grain in wood; metal breaks more easily *along* the grain than *across* the grain.

The forging process also breaks up and refines the crystal structure of the metal. The finer the crystals become, the tougher and stronger the metal becomes. The temperature at which the metal is forged is important. Figure 6.47 gives the upper and lower temperatures for forging various metals.

1. If the upper temperature is exceeded, burning and grain growth will occur.

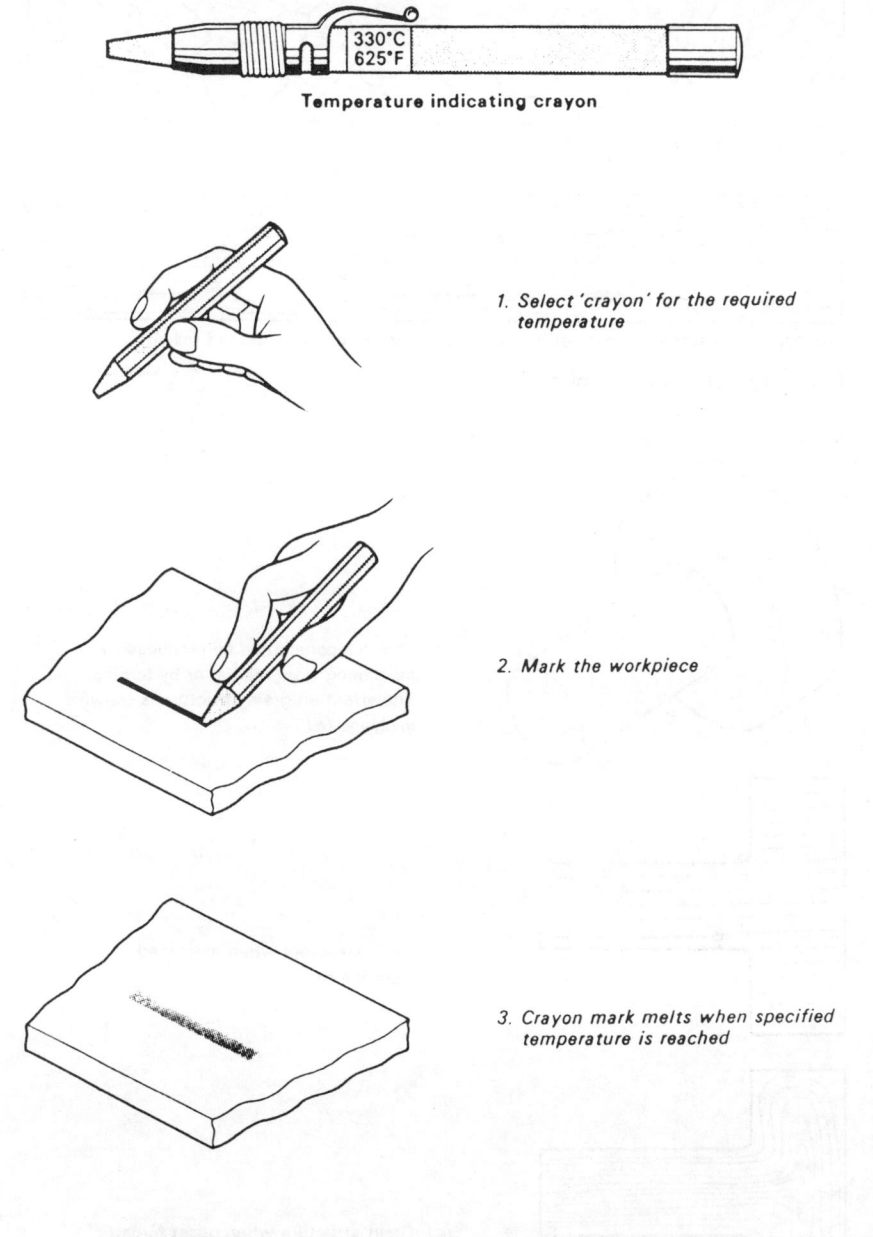

219

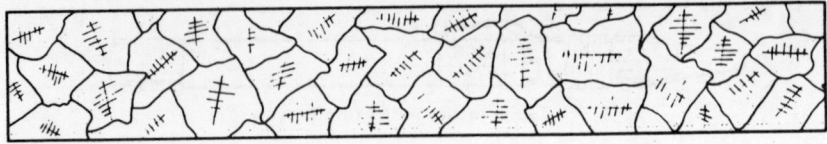

(a) Crystal structure as cast

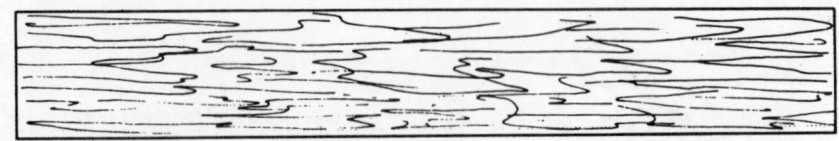

(b) Crystal structure after rolling (Crystals elongated in direction of rolling)

Fig. 6.43 Grain orientation

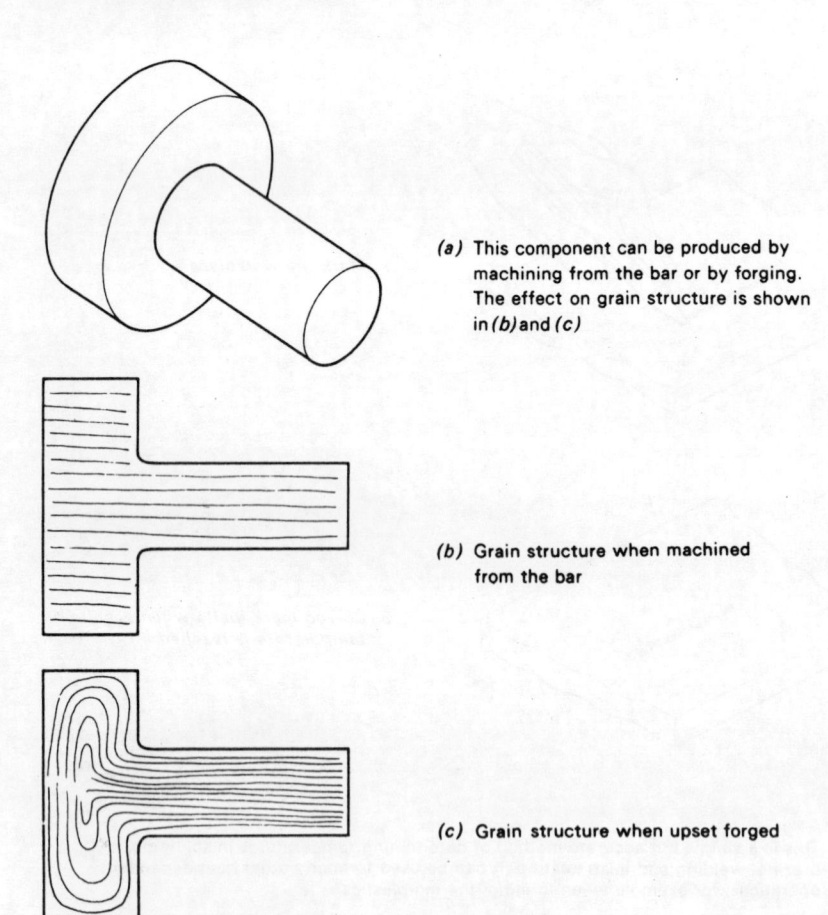

(a) This component can be produced by machining from the bar or by forging. The effect on grain structure is shown in (b) and (c)

(b) Grain structure when machined from the bar

(c) Grain structure when upset forged

Fig. 6.44 Comparison of machining and forging

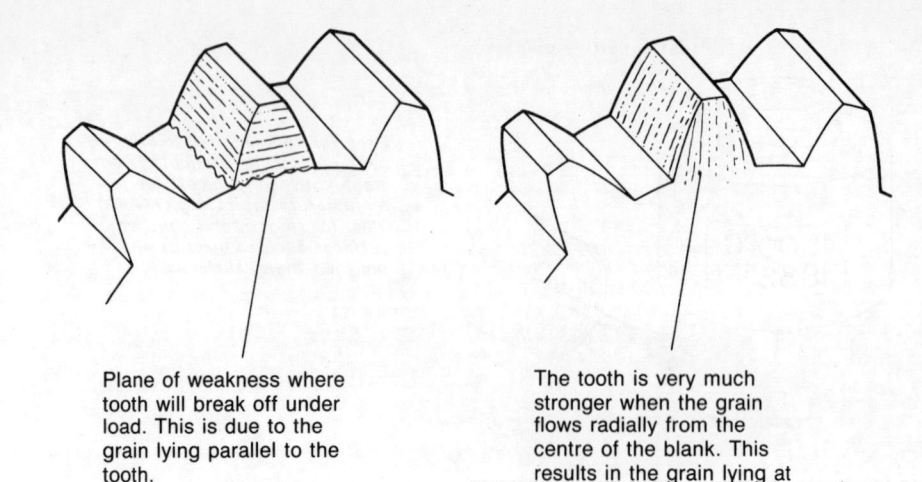

Plane of weakness where tooth will break off under load. This is due to the grain lying parallel to the tooth.

The tooth is very much stronger when the grain flows radially from the centre of the blank. This results in the grain lying at right angles to the tooth.

(a) Machined from bar          (b) Machined from forging

Fig. 6.45 Effect of grain flow on component strength

2. If forging is continued below the lower temperature, work hardening will occur and the component may crack and become weakened.

It is important to relate the cooling time of the component to the time taken to complete the forging process. Figure 6.46 shows some examples of what can happen.

In Figure 6.46(a) the forging process and the cooling cycle take the same time. This is ideal as no grain growth occurs after forging has finished, neither has any work hardening and cracking occurred.

In Figure 6.46(b) the forging time is very much shorter than the cooling time. Grain growth occurs after forging has finished and this reduces the strength of the component. To prevent this, either the component is not heated to the maximum forging temperature, or the forging is subjected to grain refinement heat treatment processes after it has cooled down.

In Figure 6.46(c) the forging time is longer than the cooling time and reheating is necessary. As in Figure 6.46(b) the maximum forging temperature of the reheat must be carefully judged, or grain refinement after cooling must be carried out.

The skill of the blacksmith lies in suiting the forging temperature to the process. However, in production forging, or where large components are being forged, grain refinement is carried out after cooling. Although an added expense, it is cheaper than scrapping a large

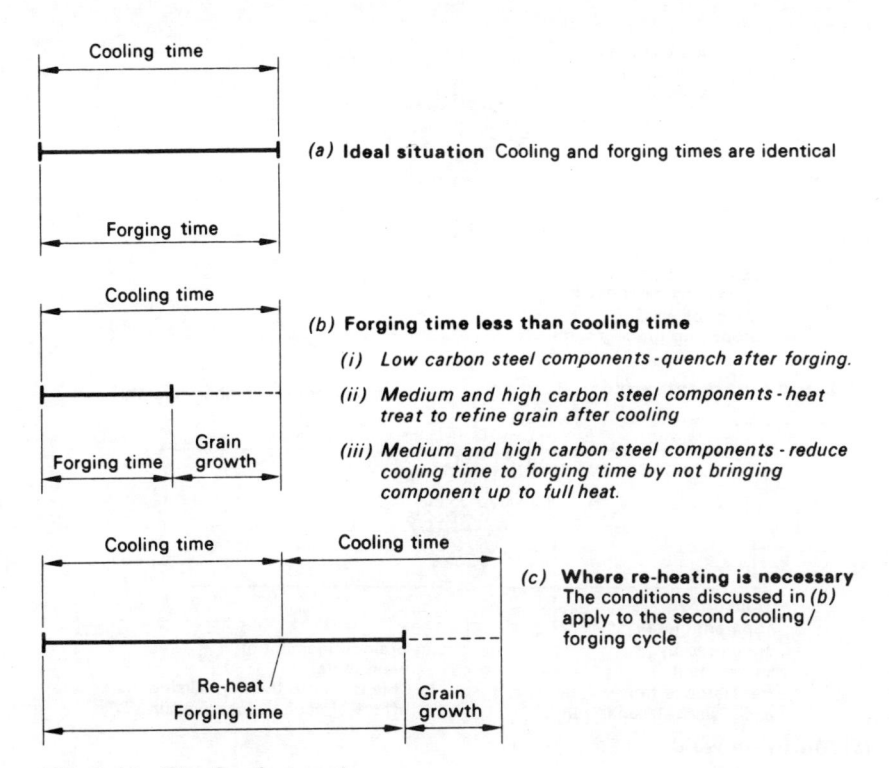

**Fig. 6.46 The forging cycle**

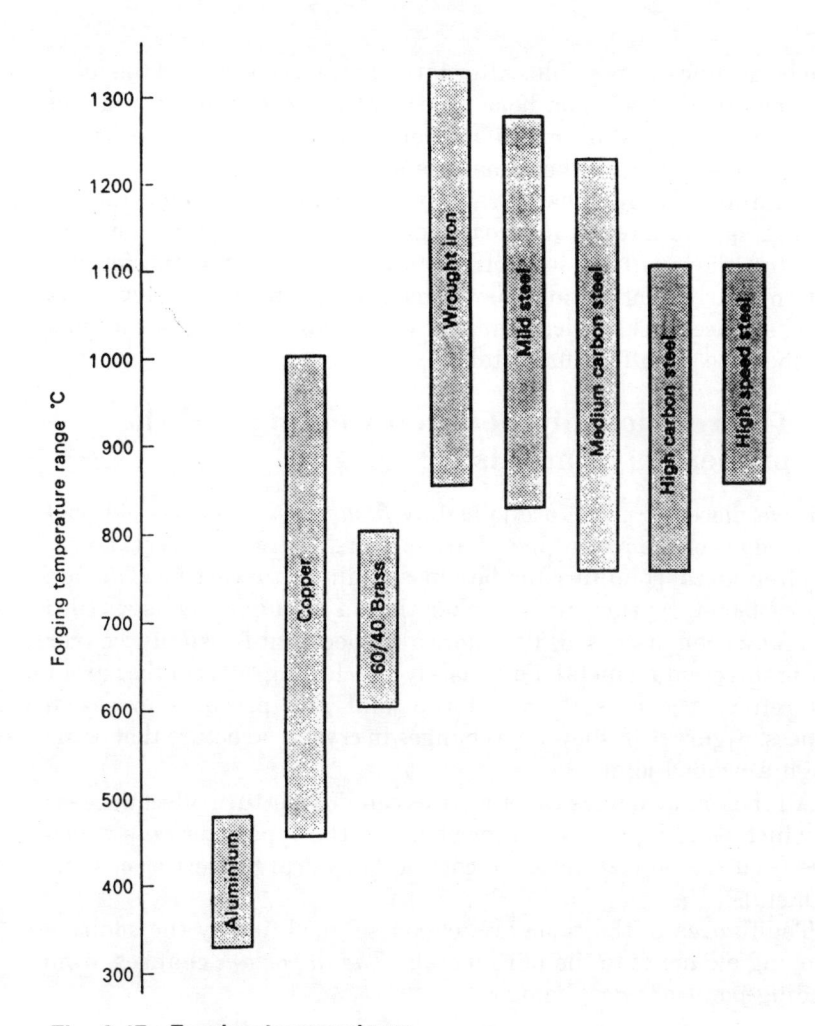

(a) **Ideal situation** Cooling and forging times are identical

(b) **Forging time less than cooling time**
   (i) Low carbon steel components - quench after forging.
   (ii) Medium and high carbon steel components - heat treat to refine grain after cooling
   (iii) Medium and high carbon steel components - reduce cooling time to forging time by not bringing component up to full heat.

(c) **Where re-heating is necessary** The conditions discussed in (b) apply to the second cooling/forging cycle

**Fig. 6.47 Forging temperatures**

and costly component or a large quantity of mass-produced components. The advantages of the forging process may be summarised as:

(a) *Economy in the use of raw materials.* Less time is used and less swarf is produced when machining from a forging, than when machining from the solid.
(b) *Increased strength* compared with the same component made from a casting or machined from the solid.

Examples of forging operations are to be found in Section 9.20.

## 6.38 Forging copper alloys and aluminium alloys

Forging has little or no effect on commercially-pure copper and aluminium.

1. Neither of these metals is susceptible to quench hardening, so the rate of cooling is unimportant.
2. Since forging is always carried out above the temperature of re-crystallisation both these metals will be in the annealed condition when they have cooled down.

However, most high duty copper and aluminium alloys are heat treatable. Therefore, their properties will be affected by:

(a) the forging temperature
(b) the time they are held at that temperature
(c) the rate of cooling.

Aluminium alloys are normally forged between 340°C and 450°C. Copper alloys are normally forged between 750°C and 810°C. Cooling should be as quick as possible. Alternatively the properties of the alloy can be modified by solution heat treatment after forging. If in doubt always consult the alloy manufacturer's technical literature before forging or heat treating these materials.

Aluminium alloys must be rapidly heated to forging temperature. Slow heating would result in a coarse grain structure. If the alloys are overheated they will be susceptible to cracking during the forging operation. Fairly light hammer blows must be used because heavy blows result in surface cracks which cannot be welded up by subsequent forging (as in the case with ferrous metals).

## 6.39 The relationship between welding and the properties of metals

When two pieces of metal are joined by *fusion welding* the weld pool and the edges of the parent metal are molten. As the joint cools down, the molten metal solidifies and becomes a miniature casting. The fact that a cast metal structure is weaker than a wrought steel structure has already been discussed. Therefore a welded joint is usually weaker than the surrounding metal. Fortunately the chilling effect of the parent metal refines the crystal size of the joint and prevents excessive weakness. Figure 6.48 shows the changes in crystal structure that occur through a welded joint.

All the common physical properties can in their turn affect the ease with which welding can be performed, and the type of process which can be used. Table 6.28 lists the common physical properties of some pure metals.

The figures in the table are, of course, modified by the addition of alloying elements to the pure metals. *The important* changes, from a welding point of view, being:

1. A reduction in the thermal and electrical conductivities
2. The appearance of 'melting ranges' in place of 'melting points'.

The first property of interest when welding is the MELTING POINT or MELTING RANGE. Whether the material to be welded has a melting point or a melting range is of very great importance.

The temperature at which a material melts has a considerable influence in determining methods by which it can be welded. As would be expected, ARC METHODS are more suited to the HIGHER MELTING POINT materials and GAS WELDING to those with LOWER MELTING POINTS. For example, the welding of lead by metal-arc processes is not conceivable, but it was one of the earliest metals to be welded by a low-temperature flame. Aluminium is frequently welded by oxy-coal gas or oxy-hydrogen

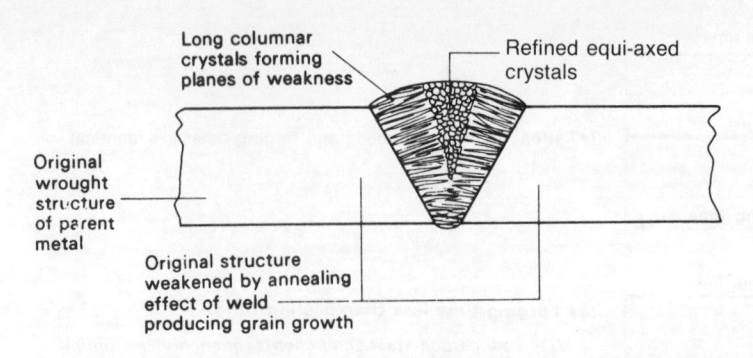

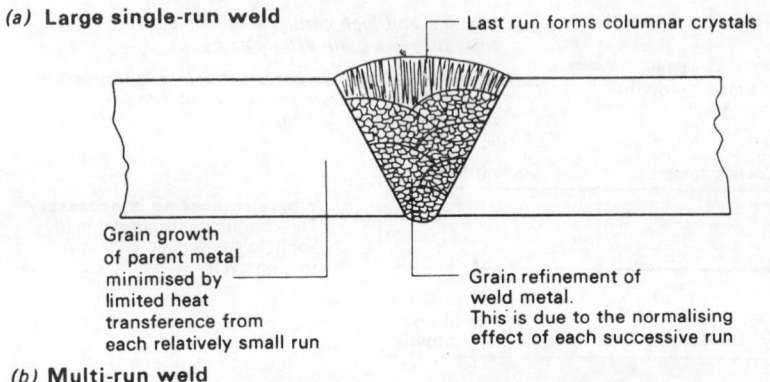

*(a)* **Large single-run weld**

*(b)* **Multi-run weld**

**Fig. 6.48 Structure of weld metal (mild steel)**

flames, whilst nickel with its higher melting point requires the hotter oxy-acetylene flame or arc welding methods.

*The need to maintain a correct relationship between the temperature of melting and that of the arc or flame is associated with the heat required to melt the metal*, but not in an obvious way. It is not just a question of the heat content of the molten metal, but is due rather to *the influence of the temperature on the heat losses during welding*. The theoretical amount of heat required to melt the higher melting point alloys is small in relation to the heat extracted from the weld at these temperatures by CONDUCTION, RADIATION and CONVECTION. For this reason, differences in LATENT HEATS and even SPECIFIC HEATS appear to play little part in determining differences in welding behaviour, whilst CONDUCTIVITY is very important in this respect.

COPPER *is outstanding among the more common NON-FERROUS metals in requiring a great deal of heat in welding*. This high rate of heat input and high rate of heat loss by conduction from the welding zone makes it more difficult than usual in maintaining the delicate THERMAL EQUILIBRIUM necessary just to melt the welding edges and not

**Table 6.28a  Common physical properties of some pure metals**

| METAL | MELTING POINT °C | BOILING POINT °C | DENSITY g/cm$^3$ | THERMAL CONDUCTIVITY J/cm$^2$/sec/°C/g |
|---|---|---|---|---|
| Aluminium | 658 | 2500 | 2·7 | 2·89 |
| Copper | 1083 | 2500 | 8·92 | 3·86 |
| Iron | 1535 | 2900 | 7·87 | 0·73 |
| Lead | 327 | 1750 | 11·34 | 0·34 |
| Magnesium | 650 | 1103 | 1·74 | 1·47 |
| Molybdenum | 2620 | 4700 | 10·4 | 1·45 |
| Nickel | 1455 | 3100 | 8·9 | 0·59 |
| Silver | 961 | 3180 | 10·5 | 4·45 |
| Zinc | 419 | 907 | 7·14 | 1·12 |

**Table 6.28b**

| METAL | LATENT HEAT J/g | SPECIFIC HEAT J/g/°C | COEFFICIENT OF EXPANSION $\times 10^{-3}$ |
|---|---|---|---|
| Aluminium | 390 | 0·885 | 23·6 |
| Copper | 210 | 0·386 | 16·4 |
| Iron | 251 | 0·458 | 12·3 |
| Lead | 24·4 | 0·130 | 29 |
| Magnesium | 361 | 1·030 | 26 |
| Molybdenum | — | 0·272 | 4·9 |
| Nickel | 315 | 0·458 | 13 |
| Silver | 105 | 0·235 | 19 |
| Zinc | 109 | 0·396 | 31·2 |

burn through. *The operator sees this as a difficulty in controlling penetration*; the molten metal drops right through the joint and leaves a hole or fails to fuse the edges. This applies more to metal-arc welding than to gas-welding, since in the former the heat input cannot be controlled independently of the filler addition. To some extent the effect of HIGH CONDUCTIVITY can be neutralised by PRE-HEATING.

The effect of reduced conductivity is very apparent with the alloys of copper, since some of these are welded relatively easily even with metallic-arc. The same comparison can be made between pure aluminium and aluminium alloys.

In general, the lower the conductivity the less the practical difficulty in welding. As a point of interest, in this respect, when welding PLASTICS, where conductivity is extremely low, it is very easy to char the top surface of the weld while the underside is still cold. This condition, fortunately, does not occur in the fusion welding of metals.

With COLD-WORKED metals, conductivity influences the width of the ANNEALED ZONE on either side of the weld. The higher the conductivity, the wider is this zone.

Electrical conductivity appears to be of little importance in fusion welding processes but seems to be a major consideration to resistance welding. For example, where nickel alloys may be satisfactorily 'spot-welded' using a low-power machine similar to that used for mild steel, aluminium requires a machine capable of delivering a very high current over a very short time, and pure copper is virtually unweldable by this process.

The high thermal expansion of copper is a considerable nuisance in all welding processes. It can cause movement of the joint edges which interferes with the proper execution of the weld and gives rise to distortion which must be subsequently removed. Contraction stresses are also responsible for producing cracked welds in those alloys which are 'hot short'. Even in those cases where thermal expansion and contraction do not appear to be an immediate nuisance they may indirectly lead to problems giving rise to RESIDUAL STRESSES. The problems of distortion and residual stresses have been considered in Chapter 4.

Where the components being joined operate at normal atmospheric temperatures they can benefit by being bonded by modern adhesives. The bonding processes are carried out below the temperatures at which changes in the grain structure occur thus preventing weakening of the parent metal. Adhesive bonding is now being used on motor-car body panels and even on such highly-stressed components as wheel studs.

## 6.40 The effects of welding on the parent metal

Fusion welding operations cause temperature gradients in the plates to be joined. The limits of the gradient are the melting point of the parent metal and room temperature. In the case of steel, a portion of the parent metal will be within the upper and lower critical temperature ranges. The effect that the welding heat exerts on the temperature of the plates or parts to be welded and the amount of heat conduction is illustrated in Figure 6.49.

Figure 6.50 shows the approximate temperatures of the parent metal when arc-welding. There is a great variation in the temperature of the parent metal during metallic-arc welding.

Point A (shaded area) indicates the metal immediately adjacent to the molten weld pool heated to a pasty stage.

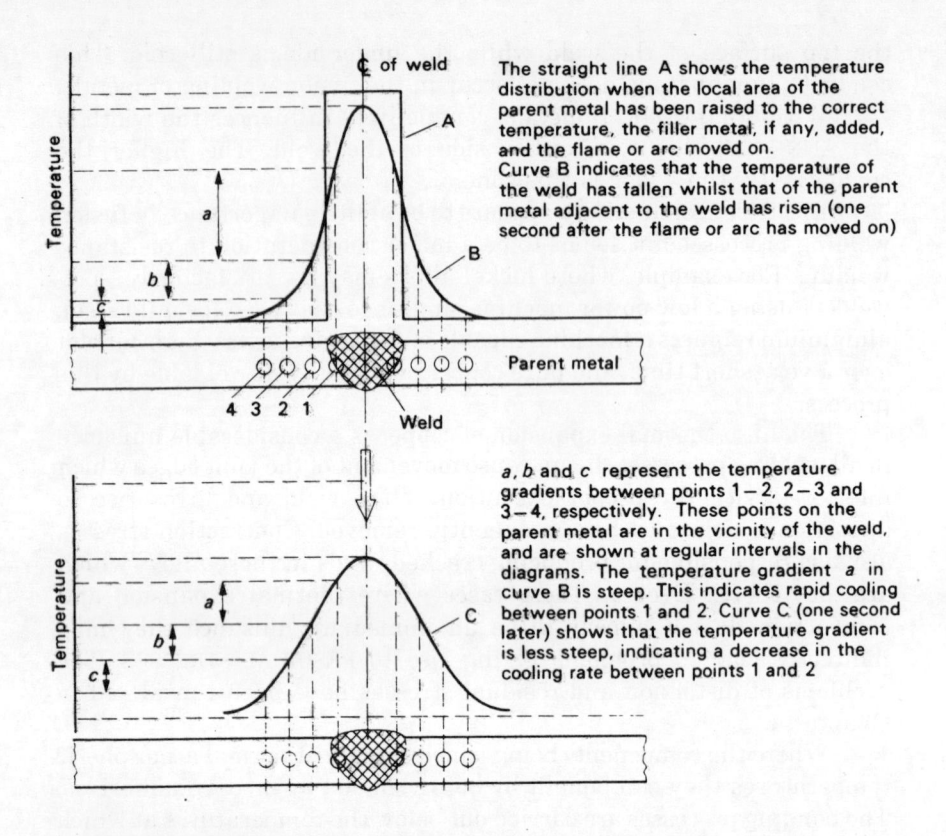

The straight line A shows the temperature distribution when the local area of the parent metal has been raised to the correct temperature, the filler metal, if any, added, and the flame or arc moved on.
Curve B indicates that the temperature of the weld has fallen whilst that of the parent metal adjacent to the weld has risen (one second after the flame or arc has moved on)

$a$, $b$ and $c$ represent the temperature gradients between points 1–2, 2–3 and 3–4, respectively. These points on the parent metal are in the vicinity of the weld, and are shown at regular intervals in the diagrams. The temperature gradient $a$ in curve B is steep. This indicates rapid cooling between points 1 and 2. Curve C (one second later) shows that the temperature gradient is less steep, indicating a decrease in the cooling rate between points 1 and 2

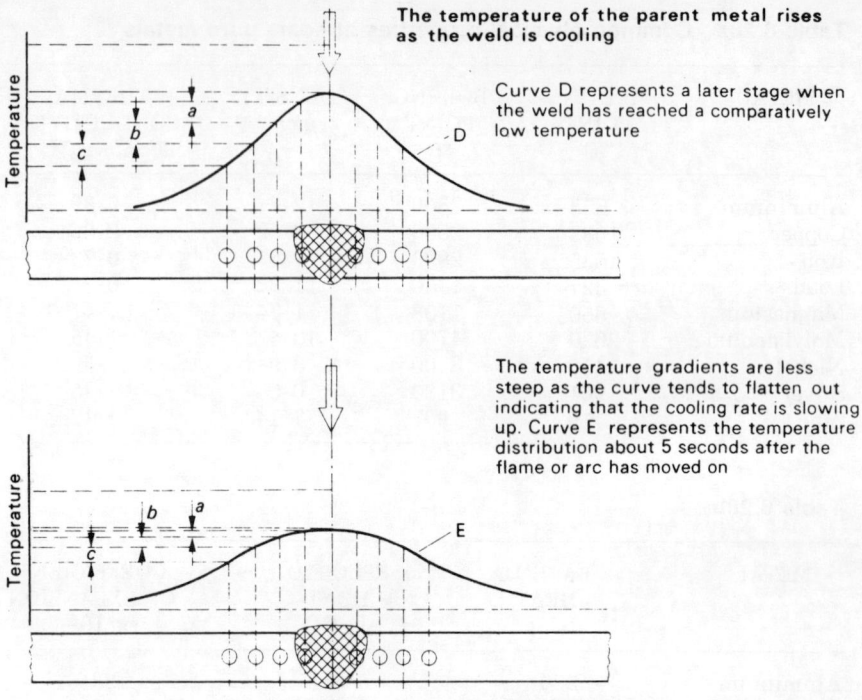

The temperature of the parent metal rises as the weld is cooling

Curve D represents a later stage when the weld has reached a comparatively low temperature

The temperature gradients are less steep as the curve tends to flatten out indicating that the cooling rate is slowing up. Curve E represents the temperature distribution about 5 seconds after the flame or arc has moved on

Parts of the parent metal remote from the weld area may rise in temperature, but only to a slight degree. The temperature at which the temperature distribution curve becomes horizontal will depend upon the size of the part. In the majority of welding applications the temperature does not become equalised throughout the part until it has reached room temperature

**Fig. 6.49 The temperature of all points in the vicinity of a weld at five different times (instantaneous heat source)**

Figure 6.51 shows the temperature distribution on a fairly thick steel plate. Table 6.29 summarises the effect of the temperature gradient on the properties of the metal in the vicinity of the weld.

A comparison of the heat spread associated with the leftward and rightward methods of gas-welding is made in Chapter 12, showing that less heat spread occurs in the plate with the rightward method. With the rightward method, the welding torch is so directed that the large bushy envelope of the flame is directed over the completed weld. *This produces an annealing effect of the weld metal itself.*

When a metal is heated up to its melting point and then allowed to cool suddenly the resulting structure is coarse, the metal is brittle and its mechanical properties are poor. If, however, the metal is allowed to cool down slowly in a special furnace much better properties are obtained. *The welding flame, in the case of rightward welding, has precisely this effect.* With leftward welding, the torch is directed away from the line of the finished weld and so the weld metal must cool abruptly in air.

Figure 6.52 illustrates the effects of welding on the parent metal.

## 6.41 Welding mild steel

This is one of the easiest metals that can be joined by fusion welding. With care the joint that is formed is almost as strong as the parent metal. The claim by some welders that their joints are stronger than the parent metal is nonsense. Such claims have grown up due to weakening of the parent metal adjacent to the weld by grain growth. This causes any subsequent fracture to occur at the side of the weld rather than through the weld itself. The effects of non-metallic inclusions in the weld and the effects of faulty techniques are considered in Chapter 12.

In arc welding mild steel, the HEAT AFFECTED ZONE will be much narrower than in gas-welding. This is because the heat of the arc is more localised and the temperature is raised more quickly.

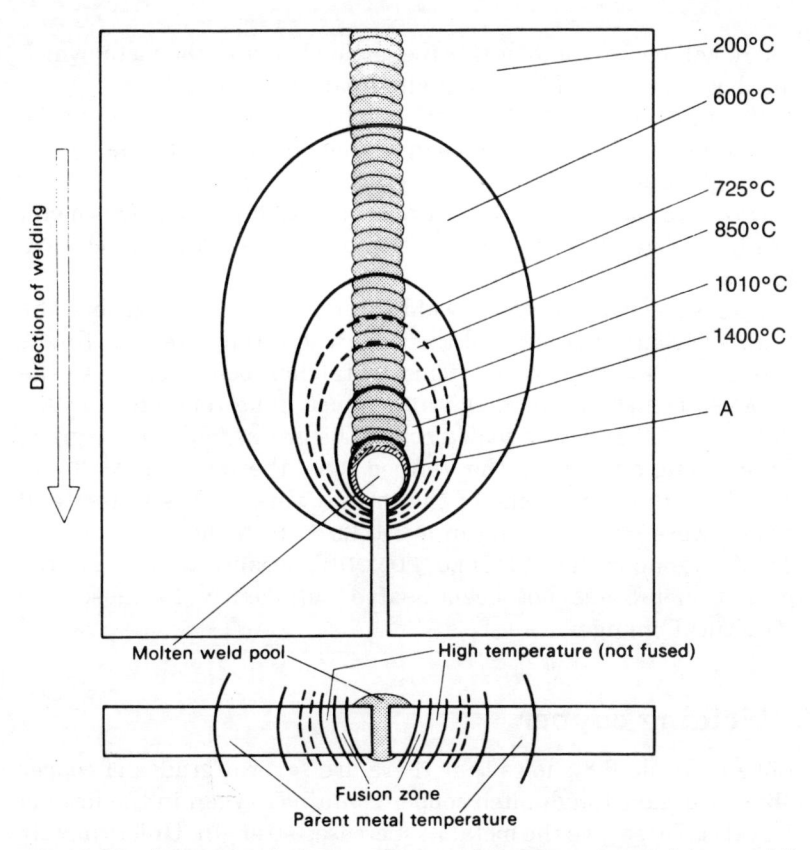

**Fig. 6.50 Approximate temperatures of the parent metal (arc welding)**

**Table 6.29 Effect of welding heat on the temperature of the plate**

| TEMPERATURE ZONES | REMARKS |
| --- | --- |
| Fusion zone | Temperature reaches melting point. The cooling rate is in the order of 350–400°C/min., which is the maximum quenching range. The weld is less hard than the adjacent area of the parent metal because of loss of useful elements (carbon, silicon and manganese). |
| Overheated zone | The temperature reaches 1100°C–1500°C. Cooling is extremely rapid in the order of 200°C–300°C/min. Some grain coarsening occurs. |
| Annealed zone | Here the temperature reaches slightly higher than 900°C. The parent metal has a refined normalised grain structure. The change is not complete because the cooling rate is still high, in the order of 170°–200°C/min. |
| Transformation zone | The temperature here is between 720°C and 910°C. These are the upper and lower critical temperatures between which the iron in steel transforms from a body-centred-cubic to a face-centred-cubic structure. The parent metal tends to recrystallise. |

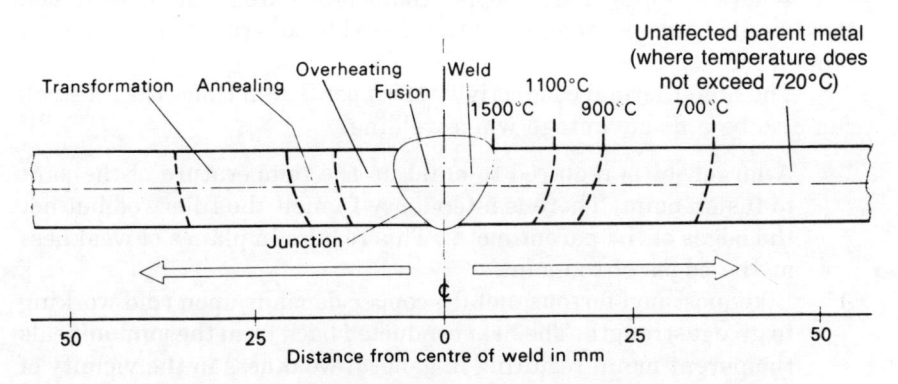

**Fig. 6.51 Temperature distribution during oxy-acetylene welding 10-mm-thick mild steel**

Key:
A — Weld metal
B — Heat-affected zone

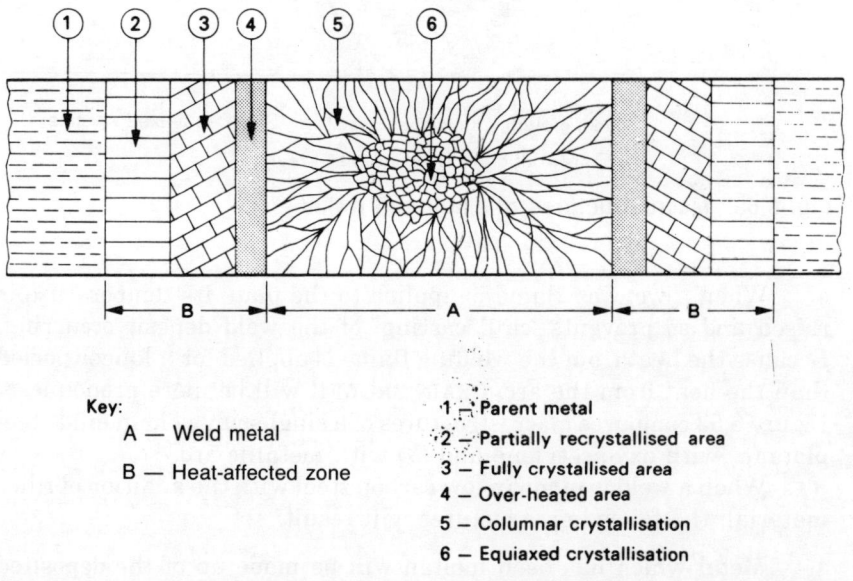

1 — Parent metal
2 — Partially recrystallised area
3 — Fully crystallised area
4 — Over-heated area
5 — Columnar crystallisation
6 — Equiaxed crystallisation

**Fig. 6.52 The effect of welding on the structure of a welded joint (schematic)**

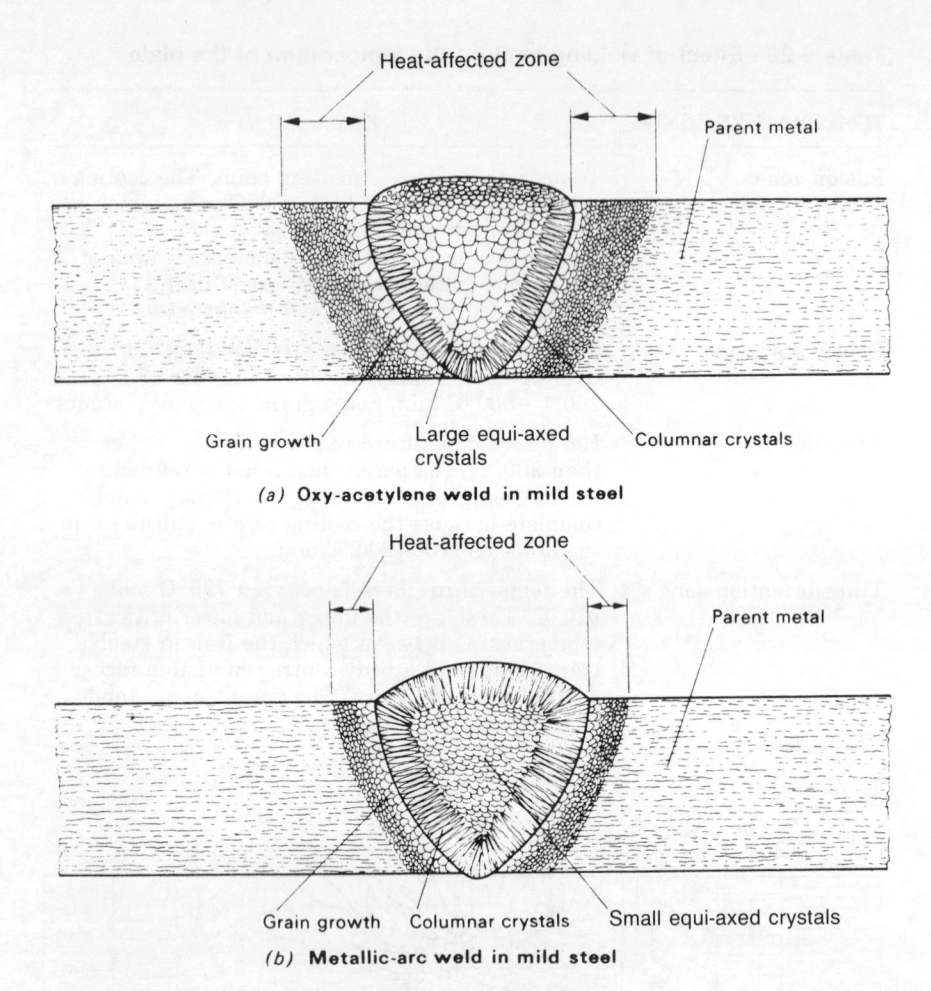

**Fig. 6.53 Macrostructures of single run welds**

(a) Oxy-acetylene weld in mild steel

(b) Metallic-arc weld in mild steel

When a welding flame is applied to the plate its temperature is raised and so prevents 'chill casting' of the weld deposit occurring. Because the heat from the welding flame is applied for a longer period than the heat from the arc, GRAIN GROWTH will be more pronounced. Figure 6.53 compares macrostructures of a single-run weld in mild-steel plate (a) with oxy-acetylene and (b) with metallic arc.

When a weld is made in low carbon steel with the addition of filler material, the following structures will result:

1. Metal which has been molten will be made up of the deposited metal mixed with the parent metal.
2. There will be a FUSION LINE at the juction between the metal which has been melted and the unmelted parent metal.
3. A HEAT-AFFECT ZONE which extends from the 'fusion line' to the parent metal which has not been heated hot enough to change the original structure.
4. Adjacent to the 'fusion line' is a zone of COARSE GRAINS. This metal has been heated nearly to its melting point and 'GRAIN GROWTH' has resulted.
5. Progressing away from the weld through the 'coarse grains' the grains become smaller, and the zone where they are very fine is called the 'REFINED ZONE'. This metal has been heated to the 'TRANSFORMATION TEMPERATURE' just long enough to 'RE-CRYSTALLISE' and then cooled before the grains had time to grow.
6. Progressing away from the 'refined zone' there is a 'TRANSITION ZONE'. In this zone some of the grains were re-crystallised and others were not, resulting in a 'MIXED STRUCTURE'.
7. The last zone is the 'UNAFFECTED ZONE', because this part of the parent metal has not been heated sufficiently to cause any structural changes.

## 6.42 Welding copper

Reference to Table 6.8 shows that there are several grades of copper available. Ordinary tough pitch copper contains oxygen in the form of copper oxides. These give the metal its increased strength. Unfortunately these oxides react with the welding flame to produce steam. This causes '*gas porosity*' in the weld leading to weakness. This can be reduced by using a slightly oxidising flame and a filler rod containing phosphorus.

Where it is known that copper components are going to be welded they should be made from one of the 'de-oxidised' grades and a neutral flame used.

The high thermal conductivity of copper (seven times that of steel) can also be a disadvantage when welding.

1. A larger jet is required to build up the temperature of the joint to fusion point. There is a tendency to melt the filler rod but not the edges of the parent metal. This results in planes of weakness at the edges of the joint.
2. Like most non-ferrous metals, copper depends upon cold working to give it strength. The heat conducted back from the joint anneals the parent metal resulting in general weakness in the vicinity of the component as shown in Figure 6.54. The structure of the weld metal can be improved by lightly hammering the metal whilst it

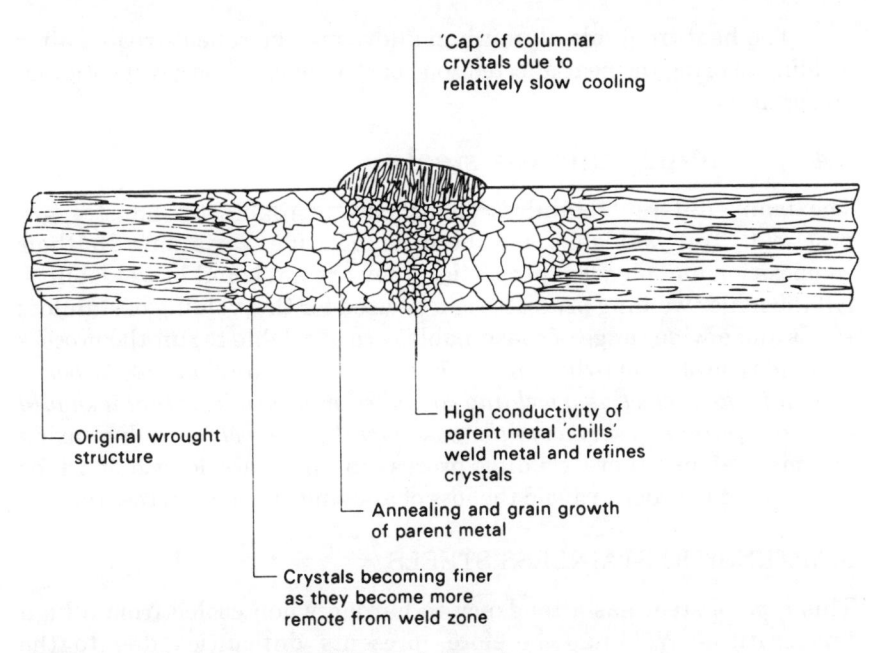

**Fig. 6.54 Structure of weld zone (copper)**

is still red hot, and continuing with slightly heavier blows as the metal cools. This refines the crystal structure as when forging.

## 6.43 Welding aluminium

Like copper, aluminium also has a high conductivity and depends upon cold working to improve its strength. Therefore the conditions shown in Figure 6.54 also apply to aluminium. Further, aluminium oxidises very easily and has to be protected from atmospheric oxygen by the use of fluxes and a reducing flame setting. Unfortunately aluminium absorbs hydrogen more readily than any other metal when in the molten state. The hydrogen comes from various sources such as the welding flame, fluxes and atmospheric moisture. As the weld cools, dissolved hydrogen is expelled resulting in 'gas porosity'. The fact that the conditions for preventing oxidation and gas porosity in aluminium conflict with each other is the reason why aluminium and its alloys are such difficult metals to weld.

The HEAT AFFECTED ZONE in joints made by metallic arc welding rarely extends beyond 50 mm from the centre line of the weld. Gas-welding usually affects a wider area. With regard to TENSILE properties, the weakest point of the joint would be immediately adjacent to the weld.

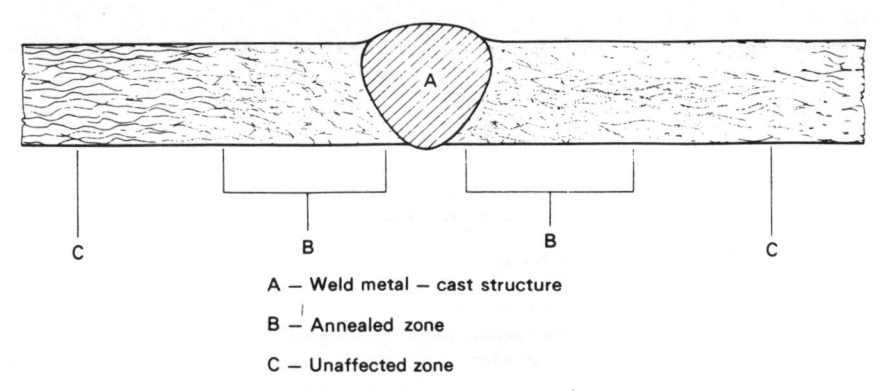

A — Weld metal — cast structure
B — Annealed zone
C — Unaffected zone

**Fig. 6.55 The various zones in a cross-section of a weld in a non-heat-treatable aluminium alloy**

In welded joints made in commercially pure aluminium and in NON-HEAT-TREATABLE aluminium alloys there are usually three zones:

1. Weld metal: the weld bead with its 'as-cast' structure where parent metal is alloyed with deposited weld metal.
2. Annealed zone: region where the heat from welding has caused RE-CRYSTALLISATION or annealing.
3. Unaffected zone: region where heating has not affected the structures.

Figure 6.55 shows the various zones in a cross-section of a weld in aluminium.

The HEAT-TREATABLE aluminium alloys contain alloying elements which exhibit a marked change in solubility with temperature change. These elements are very soluble in aluminium at elevated temperatures but have low solubility at room temperature. They tend to separate out as various microconstituents in the parent metal structure. The high strengths of the heat-treatable alloys are due to the controlled solution and precipitation of the alloying elements.

The alloying elements in heat-treatable aluminium alloys are dissolved in the aluminium at high temperature by a process commonly known as 'SOLUTION HEAT TREATMENT'. They are maintained in solid solution by rapidly quenching from this temperature. The solution of specific elements or compounds in aluminium governs the strength of these alloys in the 'as-quenched' condition. Additional increases in strength are affected by 'precipitation' of a portion of the soluble elements in finely divided form.

'Solution heat treatment', 'natural age hardening' and 'precipitation hardening' are discussed in Section 6.34.

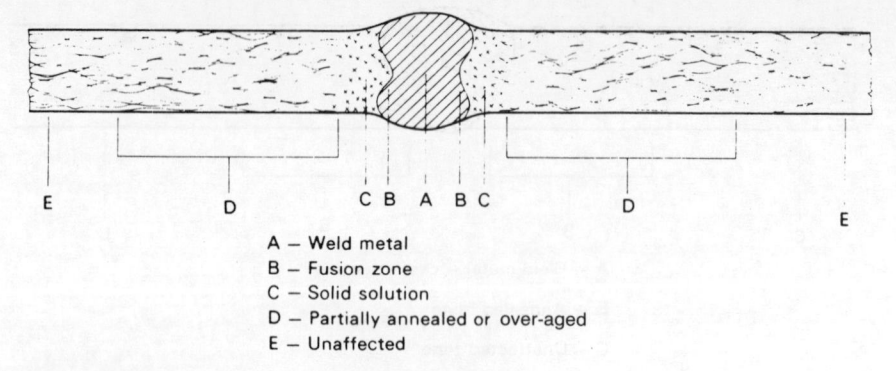

**Fig. 6.56** The various zones in a cross-section of a weld in a heat-treatable aluminium alloy

A — Weld metal
B — Fusion zone
C — Solid solution
D — Partially annealed or over-aged
E — Unaffected

Welds in heat-treatable alloys generally exhibit five microstructural zones:

1. **Weld metal** Weld bead with 'as-cast' structure where parent metal is alloyed with deposited weld metal.
2. **Fusion zone** Region where partial melting of the parent metal occurs, primarily at the grain boundaries.
3. **Solid solution zone** Region where the heat from welding is high enough to dissolve soluble constituents which are partially retained in solid solution if cooling is sufficiently rapid.
4. **Partially annealed or over-aged zone** Region where the heat from welding has caused precipitation and/or coalescence of particles of soluble constituents.
5. **Unaffected zone** Region where heating has not affected the structure.

Figure 6.56 shows the various zones in a cross-section of a weld in a heat-treatable aluminium alloy.

The five zones are generally quite evident in welds made in heat-treatable alloys in which copper and zinc are the major alloying elements.

Alloys of the MAGNESIUM–SILICON type exhibit microstructural changes in the HEAT-AFFECTED ZONE that are somewhat different, the principal heat effect is generally OVER-AGEING. The partially annealed and over-aged zones being of much greater widths.

The speed of welding has a marked effect upon the properties of welds in heat-treatable alloys. High welding rates not only decrease the width of the heat-affected zone, but they minimise effects such as grain boundary precipitation, over-ageing and grain growth.

The heat-treatable aluminium alloys may be reheat-treated after welding to bring the heat-affected zone back to approximately its original strength.

## 6.44 Welding stainless steels

The many different stainless steels now available to industry are practically all weldable, but because of wide variations in their composition and physical properties, they are not equal in weldability. All the major welding processes can be used for the welding of stainless steels and a wide range of consumables are available to suit the process and the type of steel to be welded. *The most important factor to be borne in mind when selecting a welding rod, wire or electrode, is that it should be of comparable analysis to the base metal to be welded.* When using the manual metal-arc welding process, a short arc length must be maintained in order to avoid the loss of alloying elements across the arc.

MARTENSITIC STAINLESS STEELS

This type of steel has a tendency to harden when cooled from a high temperature. Welding, therefore, presents difficulties due to the hardening effect that can be produced within the '*heat-affected zone*' of the parent metal and, to some extent, in the weld deposit. To overcome this problem and reduce the hardness and subsequent brittleness, precautions can be taken by employing pre-heating and post-heat treatment.

FERRITIC STAINLESS STEELS

These ductile alloys are sometimes referred to as '*stainless irons*'. Alloys in this group do not harden by heat-treatment due to their low carbon content. However, they are prone to brittleness which is caused by excessive '*grain growth*' at high temperatures, resulting in an 'as welded' deposit which can be brittle at ordinary temperatures but which may be quite tough at red heat. Ductility may be improved or modified through recrystallisation by pre-heating and post-heat treatment. Care must also be taken in the choice of filler material when welding steels in the ferritic group to compensate for '*dilution*'.

AUSTENITIC STAINLESS STEELS

In general, the austenitic stainless steels are more amenable to welding than the martensitic and ferritic types and for this reason most of the weldable stainless steels are from this group.

Austenitic steels have a greater coefficient of expansion and a considerably lower heat conduction than ferritic steels, therefore the tendency to distort is much greater.

DISTORTION

The problem of distortion can be controlled by the use of jigs and *'tacking'* wherever possible, particularly for material less than 3 mm in thickness. *Tack weld* should be made at intervals of half the pitch employed for mild steels. Also, in order to keep distortion to a minimum, weld deposit should be balanced and distributed by *'step-back'* techniques or similar sequences as described in Section 4.16.

WELD DECAY

If unstabilised austenitic steels are heated through a temperature range 500°C–800°C and allowed to cool slowly, they undergo a structural change that is detrimental to their inherent corrosion resistance. Such a problem can occur in the *'heat-affected zone'* of a weld and a band is formed in the adjacent parent metal parallel to the weld deposit where corrosion resistance is greatly reduced.

*This condition is due to the chromium in the grain boundary areas combining with carbon to form chromium carbides which subsequently precipitate out leaving a chromium depleted layer within the grain boundaries themselves.* This depletion of chromium is directly responsible for the drastic reduction in the metal's ability to resist corrosion. Consequently, when subjected to a corrosive medium under service conditions, the depleted areas are 'eaten out' and the grains of metal simply fall apart — hence the name *'weld decay'*.

Intergranular corrosion can be overcome in a number of ways. *Titanium or niobium have a greater affinity for carbon than chromium* and, because they form harmless titanium or niobium carbides, are added to stainless steels to act as *'stabilisers'*. In this way the grain boundaries do not suffer chromium depletion and retain their corrosion resistance.

Another and possibly better way of avoiding chromium carbide precipitation is to reduce the carbon content in the steel to less than 0·3% so that negligible carbide precipitation is possible at any temperature. Extra low carbon stainless steels are not subject to carbide precipitation during welding.

Unstabilised steels which have been welded may also have their corrosion resistance restored by heat-treatment in the form of heating to 1100°C followed by quenching.

Welding electrodes and other filler materials are available with either low carbon content or containing niobium to stabilise the higher carbon weld deposit against weld decay. *Titanium which is used to stabilise the wrought metal is not suitable for stabilising weld metal since much of it is oxidised, especially during transfer across the electric arc.*

## 6.45 Weld cracking

There are several reasons for the intercrystalline cracking of welds. These are:

1. Cracking due to contractional strains when a weld cools down whilst it is too rigidly restrained.
2. A coarse-grained deposit with large columnar crystals. These have a relatively small grain boundary area and are less able to resist the cooling strains even when the weld is not rigidly constrained. A multi-run weld helps to prevent this as shown in Figure 6.48.
3. The presence of grain boundary films with a low melting point. Iron sulphide in mild-steel welds gives rise to hot cracking. Steels for welding should have a high manganese and a low sulphur content to prevent iron sulphide forming.
4. Pure aluminium is not so susceptible to cracking as some aluminium alloys and it can be strengthened by reheating to approximately 500°C and quenching in water. This is solution treatment as described in Section 6.34. It refines and strengthens the grain structure.

## 6.46 The effects of flame cutting on the properties of metals

The effect on the structure of the metal is similar to that in the weld zone. In addition, hard oxide films are also formed along the *flame cut* edge of the *metal* and this can make machining difficult. Figure 6.57 shows a section through the edge of a flame cut component, whilst Figure 6.58 shows how it should be machined.

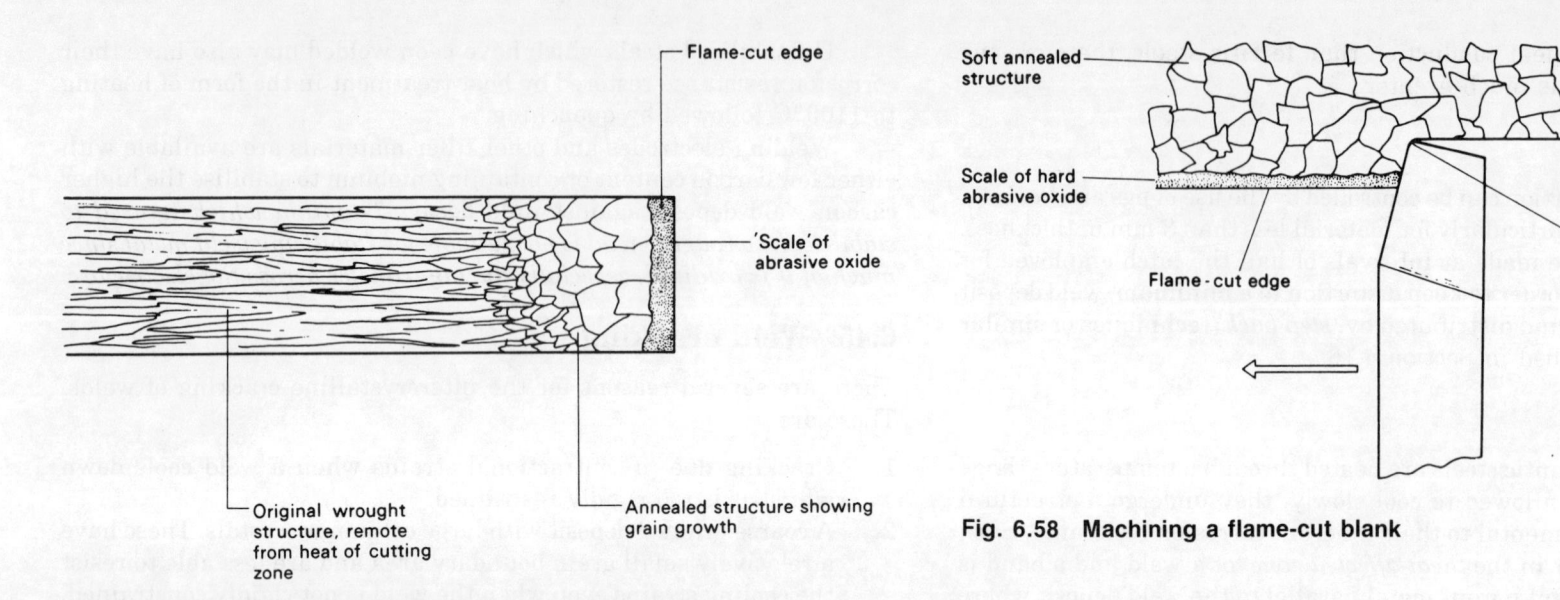

Fig. 6.57 Effect of flame-cutting on crystal structure

- Flame-cut edge
- 'Scale' of abrasive oxide
- Original wrought structure, remote from heat of cutting zone
- Annealed structure showing grain growth

Fig. 6.58 Machining a flame-cut blank

- Soft annealed structure
- Scale of hard abrasive oxide
- Flame-cut edge
- On first cut, nose of tool must operate in soft zone behind the hard, abrasive scale

230

# 7 Material removal

## 7.1 Metal cutting — basic principles

To ensure reasonable success, when cutting metals, certain established fundamental principles are incorporated in the tool design.

### 1. CLEARANCE ANGLE

When attempting to sharpen a pencil, we soon realise that, to be successful, the knife blade must be presented to the wood at a definite angle, as shown in Figure 7.1. This angle $\gamma$ (gamma), which is called the CLEARANCE ANGLE, allows the tool to 'bite' into the workpiece, reduces friction and, therefore prolongs the life of the tool.

The clearance angle is kept to the minimum required for the tool to cut (5–7°). If it is too small the tool will rub and wear out quickly, or even refuse to cut. If it is too large the tool is not only weakened but will tend to 'dig in' and 'chatter', producing a poor finish.

### 2. TOOL ANGLE

If an attempt is made to cut a piece of soft metal, such as brass, with a knife blade, it will be found that its cutting edge quickly becomes blunt. Examination under a magnifying glass would reveal that the cutting edge has crumbled away. For the blade to cut brass successfully, the cutting edge must be ground to a less acute angle to give greater strength, as shown in Figure 7.2.

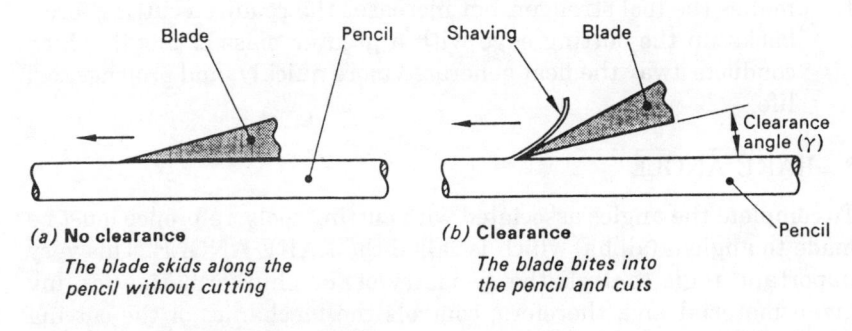

*(a)* **No clearance**
The blade skids along the pencil without cutting

*(b)* **Clearance**
The blade bites into the pencil and cuts

**Fig. 7.1** The clearance angle

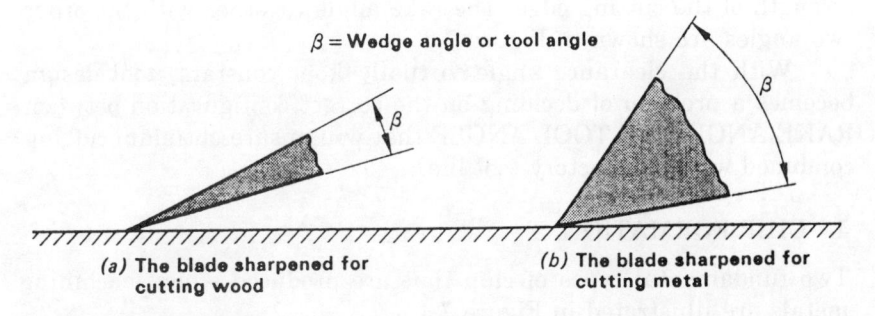

$\beta$ = Wedge angle or tool angle

*(a)* The blade sharpened for cutting wood

*(b)* The blade sharpened for cutting metal

**Fig. 7.2** Tool angle

231

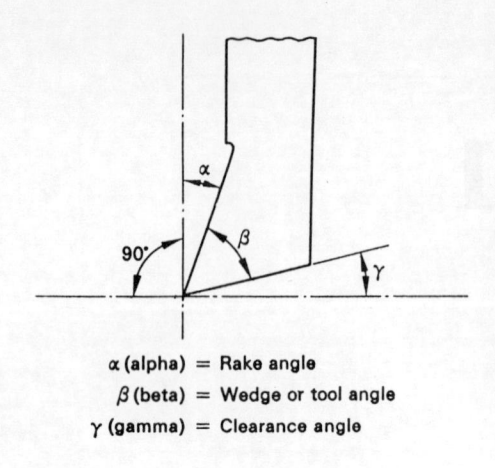

| Some typical rake angles for high-speed steel tools ||
| Material being cut | Rake |
| --- | --- |
| Cast iron | 0° |
| Free-cutting brass | 0° |
| Ductile brass | 14° |
| Tin bronze | 8° |
| Aluminium alloy | 30° |
| Mild steel | 25° |
| Medium carbon steel | 20° |
| High carbon steel | 12° |
| 'Tufnol' plastic | 0° |

$\alpha$ (alpha) = Rake angle
$\beta$ (beta) = Wedge or tool angle
$\gamma$ (gamma) = Clearance angle

**Fig. 7.3 Cutting tool angles**

The angle to which the blade is gound is $\beta$ (beta) and is called the TOOL ANGLE or WEDGE ANGLE. Increasing the tool angle:

1. makes the tool stronger, but increases the required cutting force;
2. backs up the cutting edge with a greater mass of metal, which conducts away the heat generated more quickly, and prolongs tool life.

### 3. RAKE ANGLE

To complete the angles associated with cutting tools, reference must be made to angle $\alpha$ (alpha) which is called the RAKE ANGLE. This very important angle controls the geometry of the chip formation for any given material and, therefore, controls the mechanics of the cutting action of the tool. Increasing the rake angle makes cutting easier when ductile and low strength materials are being cut, but reduces the strength of the cutting edge. The rake angle together with the other two angles are shown in Figure 7.3.

With the clearance angle virtually kept constant, tool design becomes a problem of deciding on the correct configuration between RAKE ANGLE and TOOL ANGLE that will ensure optimum cutting combined with satisfactory tool life.

### 4. CHIP FORMATION

Two fundamental types of chip that are produced when machining metals are illustrated in Figure 7.4.

A study of Figure 7.5 will help to explain how chips are produced

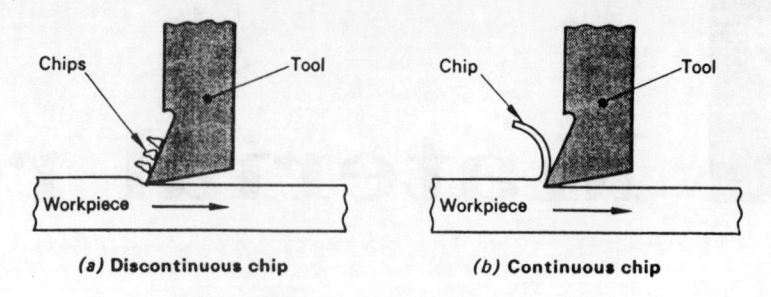

**Fig. 7.4 Types of chip**

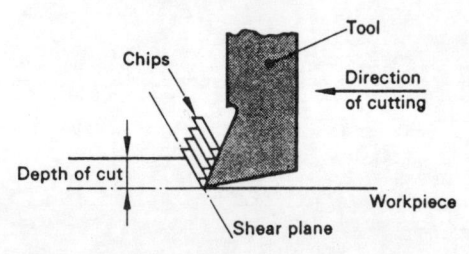

(a) Discontinuous chip

(b) Continuous chip

**Fig. 7.5 Chip formation**

by the SHEARING action of the tool and also help us to understand how an increase in rake angle makes cutting easier.

Any increase in the rake angle will have the following effects on cutting:

1. Consider the face of the tool as an inclined plane, then the greater the rake angle, the easier it will be for the chips to slide up the tool, because the slope will not be so steep;
2. The greater the rake angle, the shorter the shear plane becomes, and the shorter the shear plane, the less the force that is required to part the chip from the workpiece. This is easily proved by reducing the depth of cut which also shortens the shear plane and obviously reduces the cutting force.

Figure 7.6 shows that *shearing* occurs ahead of the cutting edge, leaving a rough torn surface on the work piece. This torn surface is 'shaved' smooth by the sharp cutting edge of the tool.

Some of the tools, machines and methods commonly employed in the fabrication industry will now be considered in detail.

## 7.2 The drill

The sole purpose of the drill is to remove the maximum volume of metal from a hole in a minimum period of time. It does not produce a precision

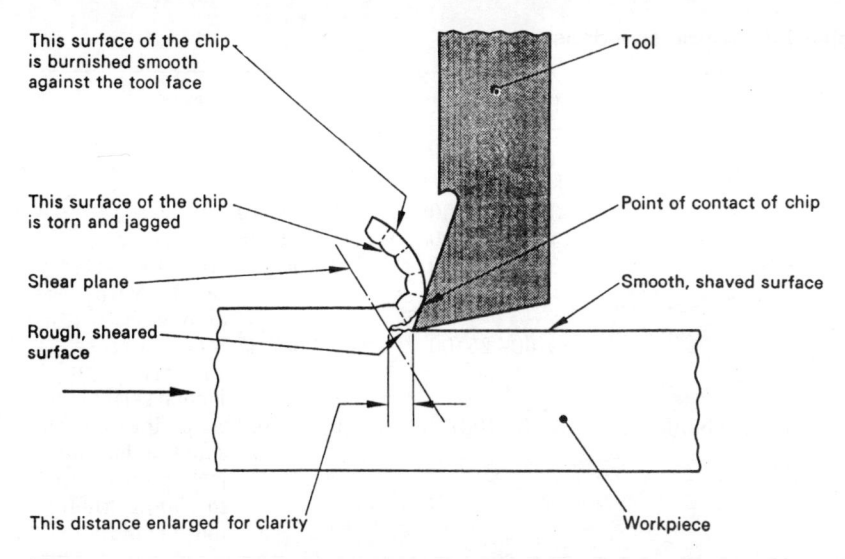

Note:
(a) A blunt tool will remove a chip but will not shave the surface smooth.
(b) A soft material will not shear cleanly, but will tear too deeply for even a sharp tool to shave smooth. This is why it is difficult to produce a good surface finish on soft, ductile metals such as copper.
(c) Soft metals such as copper can be machined to a high finish by cutting at very high speeds with diamond tipped tools. Under these conditions the metal behaves as though it were very much harder and stiffer at the point of cutting.

**Fig. 7.6 Cutting action of tool**

hole; however, this can be achieved by first drilling slightly undersize, followed by reaming to the correct size. The twist drill is the most common type of drill used. It cuts from solid materials and, under satisfactory conditions, can be employed to 'open up' existing holes. A typical twist drill is shown in Figure 7.7, together with its important features.

Five different standard types of twist drill are illustrated in Figure 7.8, whilst Table 7.1 gives the range of standard metric sizes that they are available in.

## 7.3 Twist drill cutting angles

As with any other cutting tool, the twist drill must be provided with the correct tool angles, and Figure 7.9 shows how these compare with the angles for a single-point lathe tool.

The CLEARANCE ANGLE of a twist drill can be adjusted during the grinding of the point.

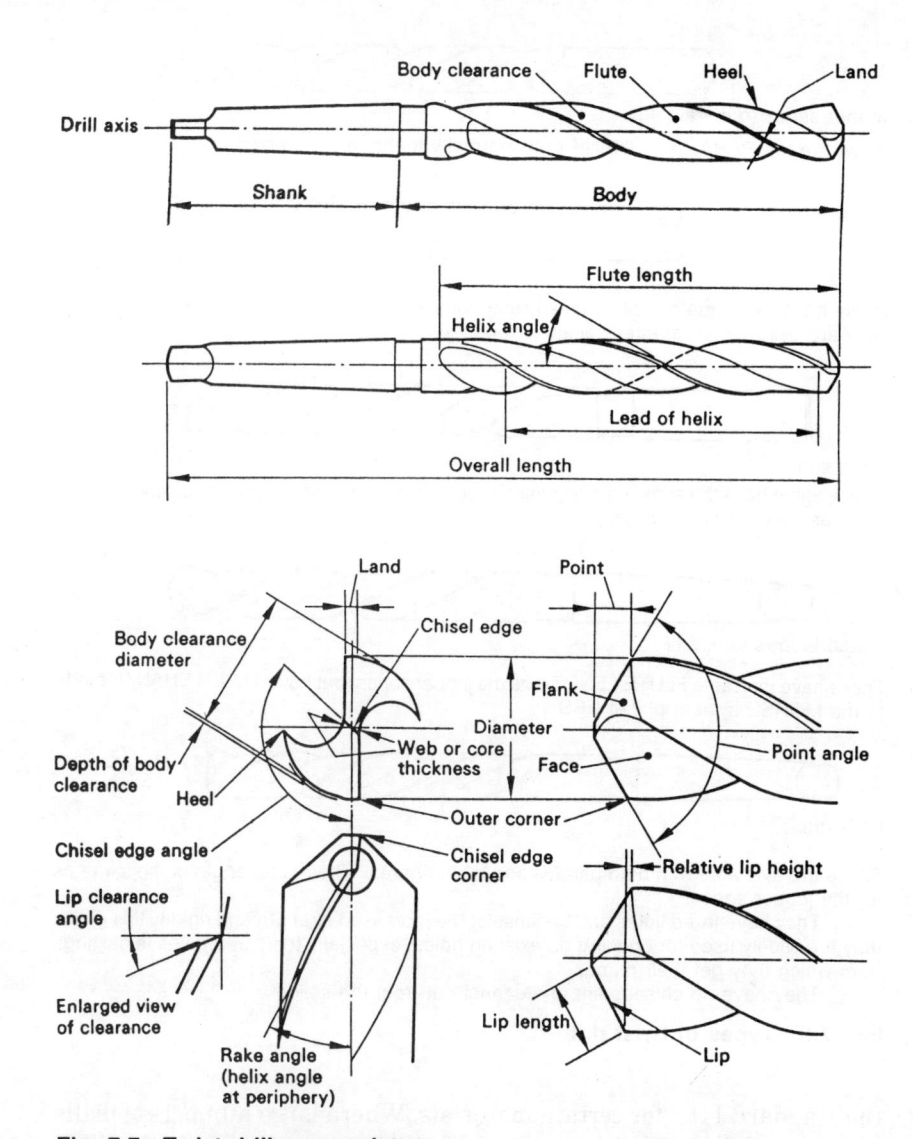

**Fig. 7.7 Twist drill nomenclature**

The RAKE ANGLE of a twist drill is not so easily altered as it is formed by the helix of the flutes during manufacture.

Some control of the rake angles is possible by choosing drills with the correct helix angle for the material being cut. Figure 7.10 shows the various types available.

As well as varying the basic cutting angles of the drill, its performance can also be improved by modifying the point angle from

233

Jobber series (general use)

These are the PARALLEL SHANK drills normally found in the workshop.

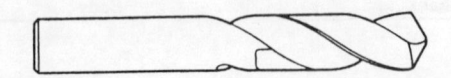

Stub drill

These have the same parallel shank proportions as the drill above but shorter FLUTES. They are used where greater rigidity is required.

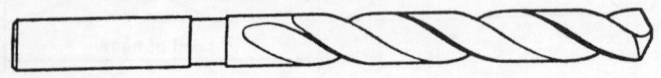

Long series

These again have the same parallel shank proportions as the jobber series but longer flutes. They are used for drilling deep holes.

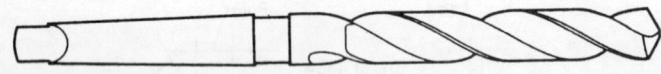

Morse taper shank drill

These have the same FLUTE LENGTH as the jobber series, but have TAPER SHANKS made to the MORSE system of TAPERS.

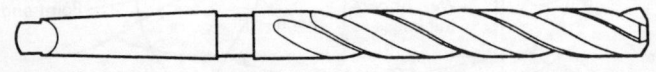

Core drill

These are available with both parallel and taper shanks. Their flute length is the same as for the jobber series.

They have three flutes and because of the increased strength and rigidity this gives, they are widely used for opening up existing holes, especially the cored holes in castings from which they get their name.

They have no chisel point and cannot cut from the solid.

**Fig. 7.8 Types of twist drill**

Table 7.1 Metric size drills (mm)

| TYPE | DIAMETER | STEPS | |
|---|---|---|---|
| Jobber series | 0·20–1·00 | 0·02/0·03 | |
| | 1·05–3·00 | 0·05 | |
| | 3·10–14·00 | 0·10 | |
| | 14·25–15·00 | 0·25 | |
| Stub drills | 0·50–25·00 | 0·20/1·00 | Steps increase rapidly with diameter |
| Long series | 2·00–25·00 | 0·05/0·25 | Steps increase similarly to the jobber series |
| Morse taper shank | 3·00–100·00 | 0·02/1·00 | Steps increase the same as for the jobber series up to 15 mm, then more rapidly |

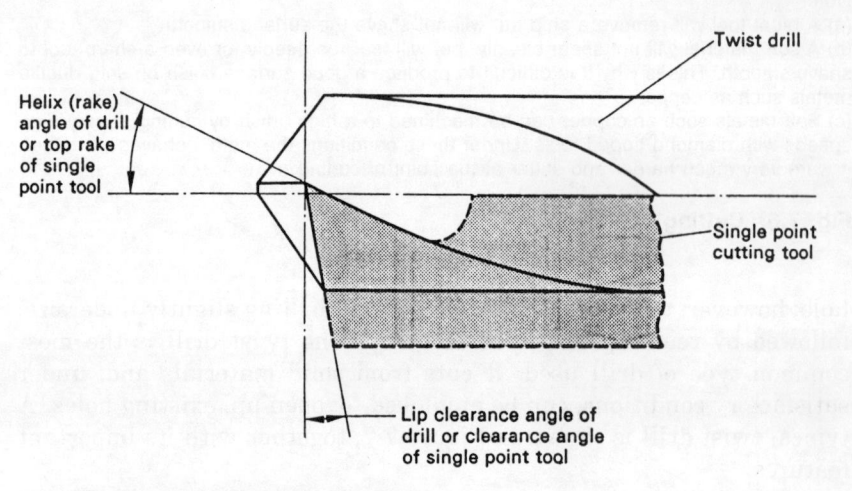

**Fig. 7.9 Twist drill cutting angles**

the standard 118° for certain materials. Where large numbers of drills of the same size are being purchased for long production runs, the WEB and LAND can also be varied with advantage. Figure 7.11 shows how the point angle, web and land can be varied for different materials.

## 7.4 Twist drill cutting speeds and feeds

For a drill to give satisfactory performance it must operate at the correct speed and correct rate of feed. For optimum results it is essential that:

1. The work is rigidly clamped;
2. The machine is in good condition;
3. A coolant is used if required;
4. The drill is correctly selected and ground to suit the material being cut.

A craftsman should be capable of grinding a twist drill by hand; however, one can never equal the twist drill point-grinding machine for consistent results. Whenever possible, the point should be ground on such a machine and checked on a point angle gauge.

Normal helix angle for drilling low and medium tensile materials

Reduced or 'slow' helix angle for high tensile materials

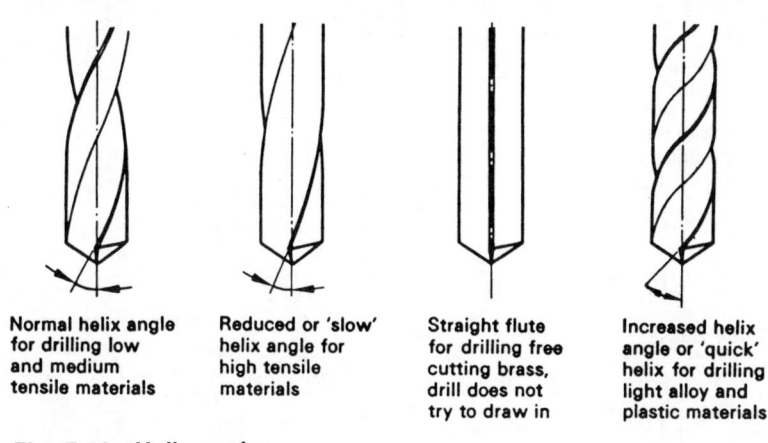

Straight flute for drilling free cutting brass, drill does not try to draw in

Increased helix angle or 'quick' helix for drilling light alloy and plastic materials

Fig. 7.10 Helix angles

Table 7.2 gives a range of cutting speeds suitable for jobbing work. The rates of feed and cutting speeds for twist drills are lower than for most other machining operations. This is because:

1. The drill is relatively weak compared with other cutting tools — the cutting forces are only resisted by the slender web;
2. In deep holes it is relatively difficult for the drill to eject the chips;
3. It is difficult to keep the point and cutting edge cool when they are enclosed in a hole.

A suitable range of feeds for jobbing work is given in Table 7.3.

## 7.5 Trepanning

It is not only foolhardy but dangerous to attempt to cut a large diameter hole in sheet metal with a twist drill. There is insufficient thickness of metal to guide the drill point and offer resistance to its cutting edge. The drill tends to 'snatch' and the resultant hole is anything but circular. This problem can be avoided by the use of a trepanning tool. Unlike the cutting action of a twist drill, where the metal is removed as swarf, the trepanning cutter (a minature single point lathe tool) removes a disc of metal as shown in Figure 7.12. This sturdy type of cutting can be safely used on a drilling machine provided the *metal sheet is securely clamped and the cutter operated at a very low speed and rate of feed.*

Where circular holes are required to be cut in sheet metal ducting or trunking, either the 'tank cutter' (simple type of trepanner) or a hole saw may be used. The latter is far superior where a number of holes of the same size are required. Both these types of cutters are illustrated

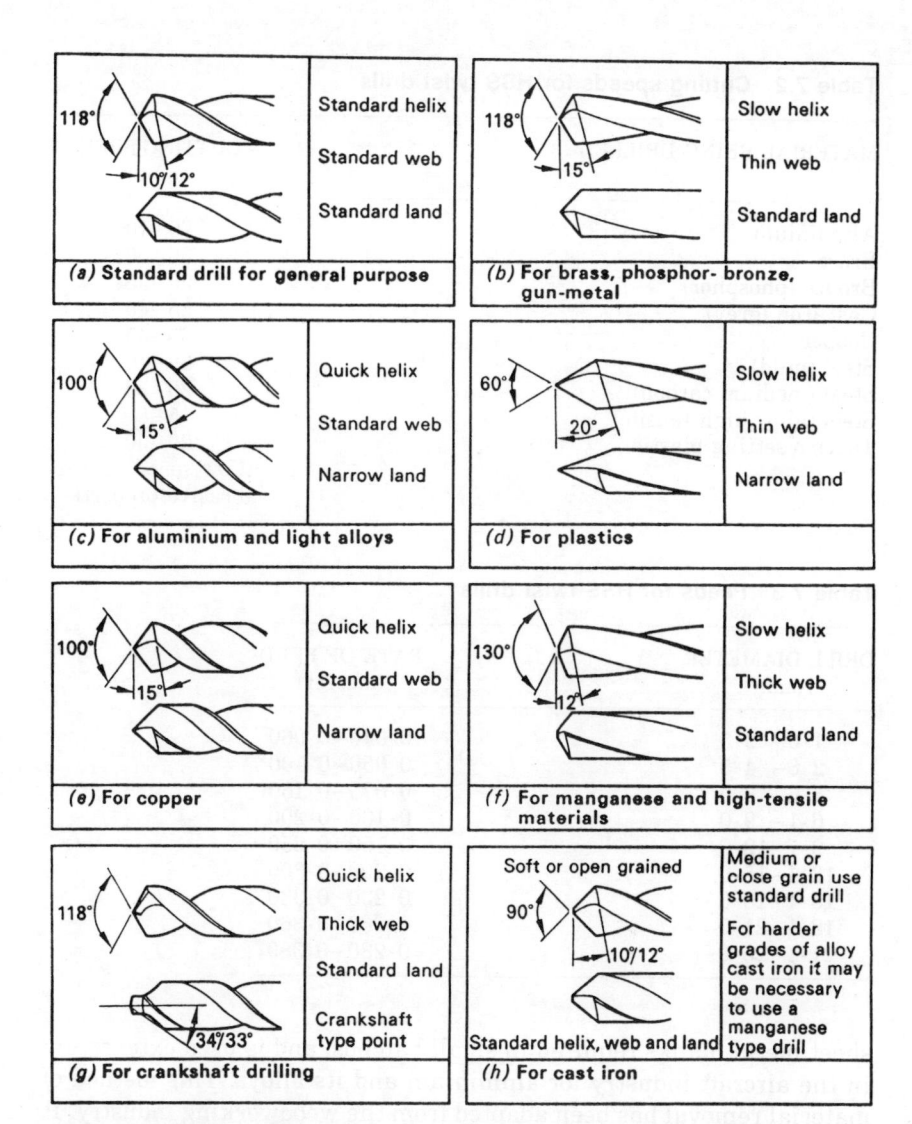

(a) Standard drill for general purpose — 118°, 10/12°, Standard helix, Standard web, Standard land

(b) For brass, phosphor-bronze, gun-metal — 118°, 15°, Slow helix, Thin web, Standard land

(c) For aluminium and light alloys — 100°, 15°, Quick helix, Standard web, Narrow land

(d) For plastics — 60°, 20°, Slow helix, Thin web, Narrow land

(e) For copper — 100°, 15°, Quick helix, Standard web, Narrow land

(f) For manganese and high-tensile materials — 130°, 12°, Slow helix, Thick web, Standard land

(g) For crankshaft drilling — 118°, 34/33°, Quick helix, Thick web, Standard land, Crankshaft type point

(h) For cast iron — 90°, 10/12°, Soft or open grained Standard helix, web and land. Medium or close grain use standard drill. For harder grades of alloy cast iron it may be necessary to use a manganese type drill

Fig. 7.11 Point angles

in Figure 7.13, and a comparison of their respective advantages and limitations is made in Table 7.4.

## 7.6 Routing

The choice of cutting out sheet metal blanks is governed by such factors as size and shape, quantity required and type of machine available.

Routing, possibly, provides the most economical method for cutting

235

**Table 7.2  Cutting speeds for HSS twist drills**

| MATERIAL BEING DRILLED | CUTTING SPEED m/min |
|---|---|
| Aluminium | 70–100 |
| Brass | 35–50 |
| Bronze (phosphor) | 20–35 |
| Cast iron (grey) | 25–40 |
| Copper | 35–45 |
| Steel (mild) | 30–40 |
| Steel (medium carbon) | 20–30 |
| Steel (alloy-high tensile) | 5–8 |
| Thermo-setting plastic | 20–30 (Low speed due to abrasive properties) |

**Table 7.3  Feeds for HSS twist drills**

| DRILL DIAMETER (mm) | RATE OF FEED (mm/rev) |
|---|---|
| 1·0– 2·5 | 0·040–0·060 |
| 2·6– 4·5 | 0·050–0·100 |
| 4·6– 6·0 | 0·075–0·150 |
| 6·1– 9·0 | 0·100–0·200 |
| 9·1–12·0 | 0·150–0·250 |
| 12·1–15·0 | 0·200–0·300 |
| 15·1–18·0 | 0·230–0·330 |
| 18·1–21·0 | 0·260–0·360 |
| 21·1–25·0 | 0·280–0·380 |

sheet metal blanks required in small batches, and is used extensively in the aircraft industry for aluminium and its alloys. This method of material removal has been adapted from the woodworking industry. It is extremely rapid and can be used for blanks up to approximately 1·2 metres wide by any length. The cutting tool used is very similar to an 'end mill' and revolves at very high speeds up to 10 000 r.p.m., cutting a number of sheets or blanks simultaneously.

Figure 7.14 illustrates the principles of material removal by three types of routing machine, two of which are the fixed-cutter type.

With reference to Figure 7.14:

*Type 'A'*: With this simple type of arrangement the cutter projects vertically through the centre of a horizontal table. At the lower

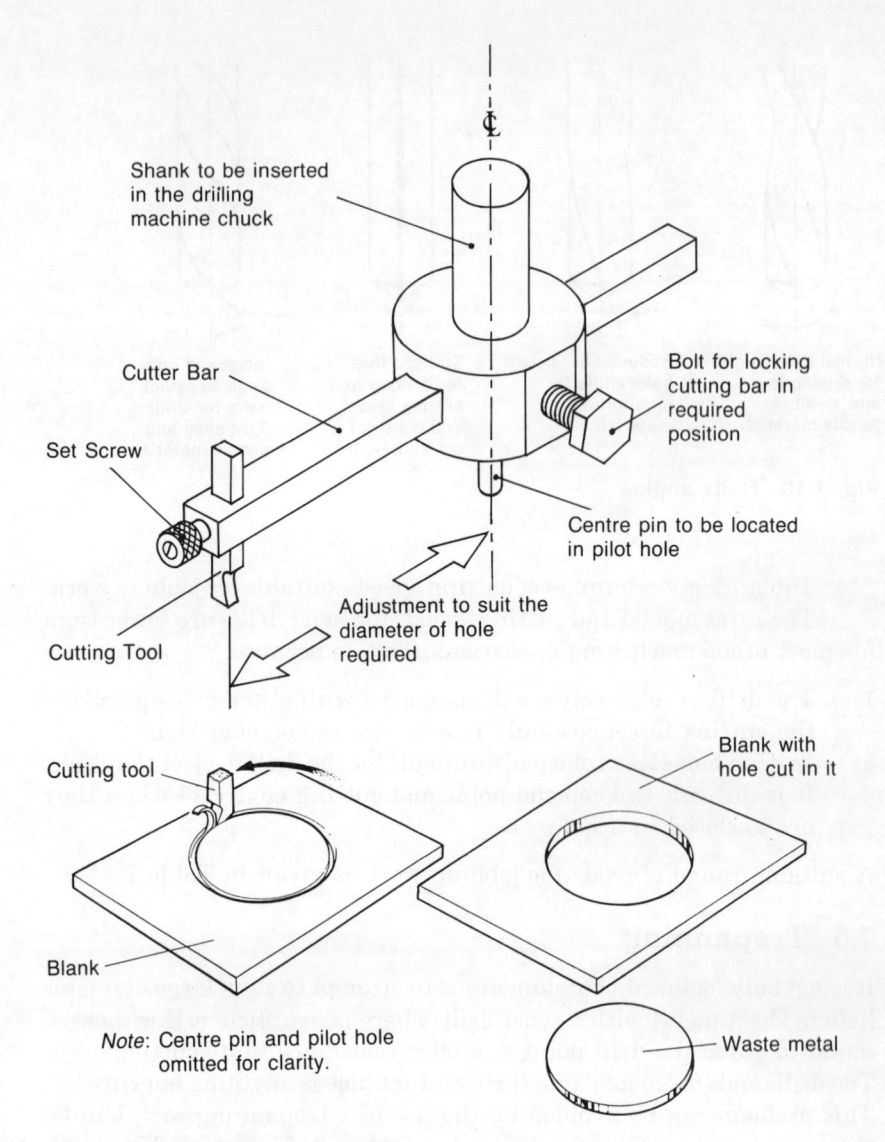

**Fig. 7.12  Principle of trepanning**

end of the cutter is a stationary collar of the same diameter. Some or all of the holes required in the finished component are pre-drilled through the stack of sheets, which are then secured on the top of a thick wood template by screws. The work is pressed against the cutter until the collar, under the cutter, encounters the edge of the template. Sideways pressure is maintained in order to keep the

**(a) Tank cutter**  **(b) Hole saw**

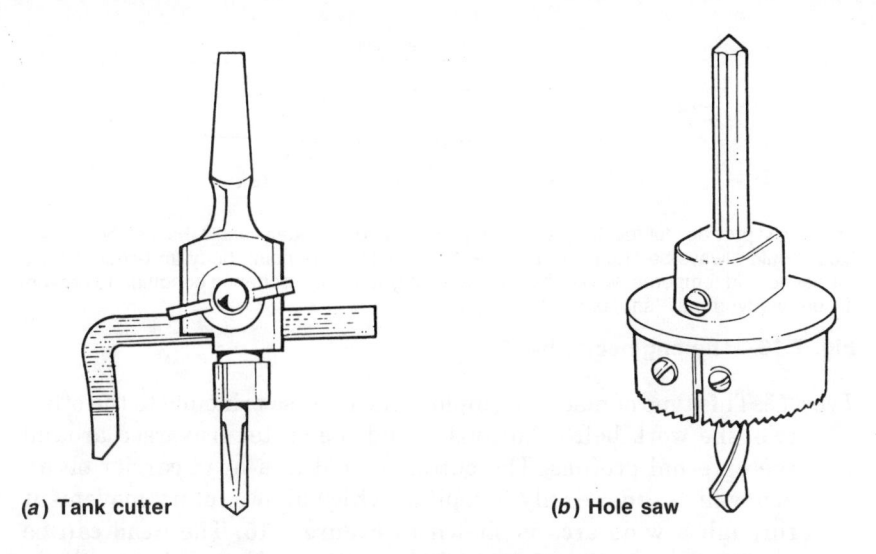

The tank cutter is used with a drill brace, the shank being made with a square taper portion for this purpose.

**Fig. 7.13  Large diameter hole cutters for sheet metal**

**Table 7.4  Comparison of tank cutters and hole saws**

TANK CUTTER

| ADVANTAGES | LIMITATIONS |
| --- | --- |
| 1. Simple<br>2. Robust<br>3. Cheap<br>4. Easy to re-grind<br>5. Adjustable for any size hole | 1. Cutting forces out of balance. This causes hole to wander and finish to be poor. In thin metal the pilot may tear the pilot hole, resulting in an oval hole being cut<br>2. Not suitable for power tools |

HOLE SAW

| ADVANTAGES | LIMITATIONS |
| --- | --- |
| 1. Simple<br>2. Cheap<br>3. Relatively robust<br>4. Cannot be re-ground but blades are easily replaceable<br>5. Leaves a round hole of good finish<br>6. Suitable for power tools | 1. Not adjustable for size, a separate cutter is required for each size of hole<br>2. Cannot be re-ground<br>3. Blades of low-tungsten steel have relatively short life |

(A) Fixed spindling-type cutter which projects vertically through the centre of a horizontal table.

With fixed cutter type machines the material is removed by manipulating the template around the cutter.

(B) Fixed-head vertical cutter which projects downwards over the centre of a horizontal table.

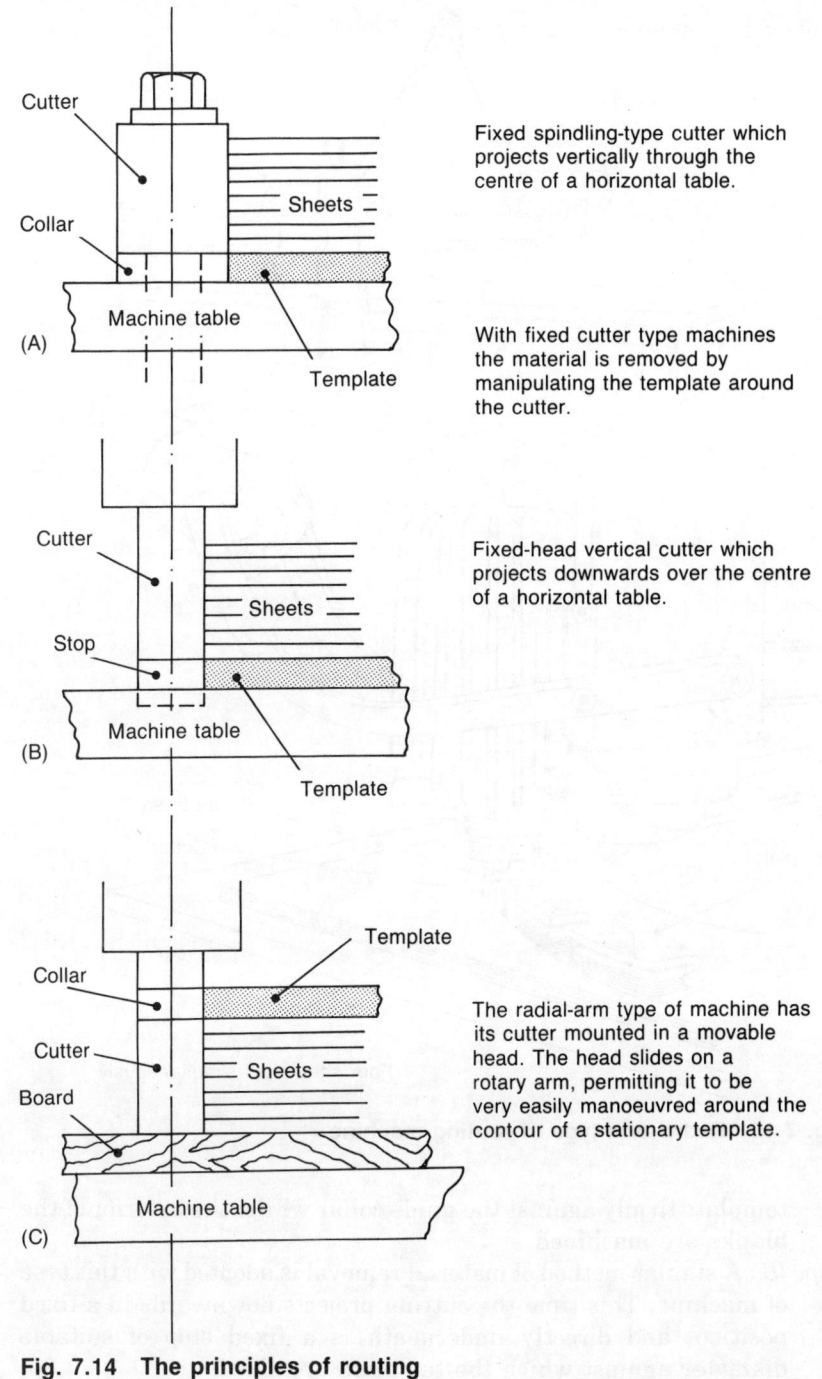

(C) The radial-arm type of machine has its cutter mounted in a movable head. The head slides on a rotary arm, permitting it to be very easily manoeuvred around the contour of a stationary template.

**Fig. 7.14  The principles of routing**

237

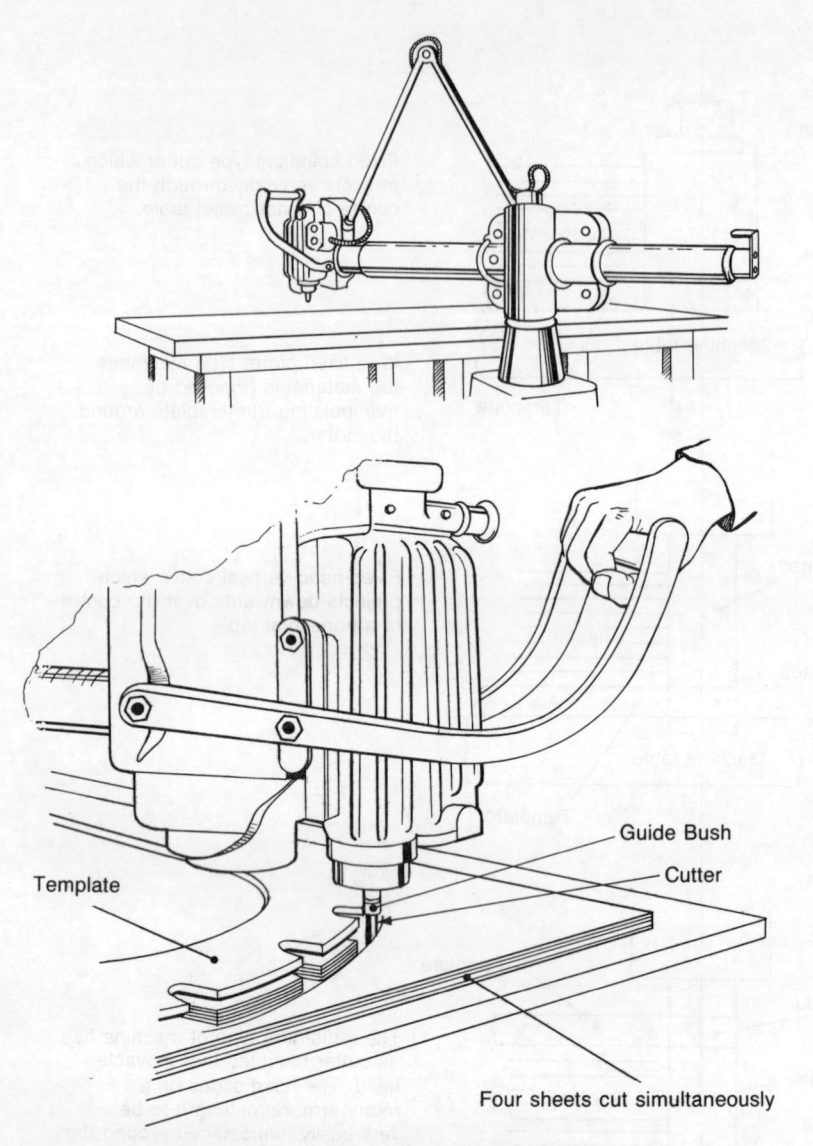

**Fig. 7.15 Radial-arm type of routing machine**

template firmly against the guide collar whilst the exterior of the blanks are machined.

*Type 'B'*: A similar method of material removal is adopted with this type of machine. This time the cutting projects downwards in a fixed position, and directly underneath is a fixed stop of suitable diameter against which the template is guided.

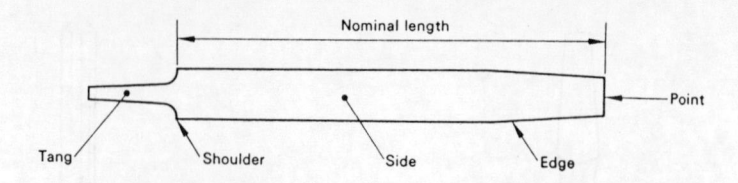

The tang is provided for the purpose of fitting the file into a wooden handle. For safety reasons, files should always be fitted with suitable file handles. The point, body and shoulder are hardened and tempered, whilst the tang is usually left in the softened condition to prevent it from inadvertently 'snapping off'.

**Fig. 7.16 The engineer's file**

*Type 'C'*: This type of machine employs the reverse principle to the other two, the work being stationary and the cutter traversed around the external profiles. The cutter is held in a head carrier on an arm which slides easily in a pillar which allows rotary movement through a wide arc, as shown in Figure 7.15. The head can be manipulated by hand around the work profiles with very little effort on the part of the operator. The cutter carries a stationary guide bush, or collar, at its upper end. The template is laid on top of a number of sheets and is secured by ordinary wood screws passing through into a sturdy board at the bottom.

## 7.7 The file

The file plays a very important part in material removal, and filing operations range from 'roughing down' to final finishing on flat and curved surfaces. The main features of a file are shown in Figure 7.16. The specifications for files are determined by *length, grade of cut and shape*. The grade or cut of a file depends upon its length: the shorter the file, the smaller will be the pitch of its teeth.

Table 7.5 shows the pitch range and applications of normal available files.

There is a very wide variety of files available for engineering purposes; the type or shape selected is generally governed by a particular application. Figure 7.17 shows a range of files and typical uses.

The file tooth, just like any other cutting tool, must have correctly applied tool angles, as shown in Figure 7.18. It will be seen from Figure 7.18(*b*), that unlike the cold chisel (Figure 7.26), the file tooth has *negative rake*.

*The single-cut file* has a series of parallel teeth formed at an angle of about 70° to the axis of the file.

*The double-cut file* has a second series of parallel teeth formed in the opposite direction and crossing the others at an angle of about 45°,

**Table 7.5 File grades**

| GRADE | PITCH mm | USE |
| --- | --- | --- |
| Rough | 1·8–1·3 | Soft metals and plastics |
| Bastard | 1·6–0·65 | General roughing out |
| Second cut | 1·4–0·60 | Roughing out tough materials Finishing soft materials |
| Smooth | 0·8–0·45 | General finishing and draw filing |
| Dead smooth | 0·5–0·25 | Not often used except on tough die steels where high accuracy and finish is required |

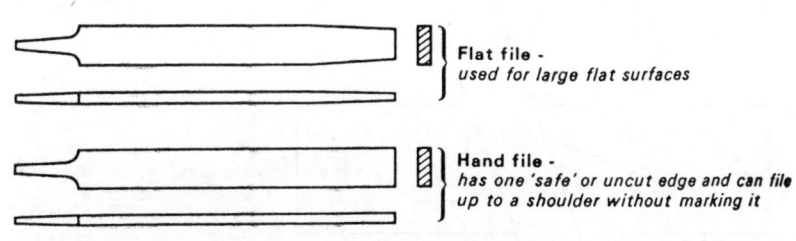

**Flat file** – used for large flat surfaces

**Hand file** – has one 'safe' or uncut edge and can file up to a shoulder without marking it

The pillar file – narrower and thicker than the hand file – is useful for filing die steels where greater pressure per unit area is required to make the file bite

The warding file – thinner than the flat file – is used for filing out narrow slots

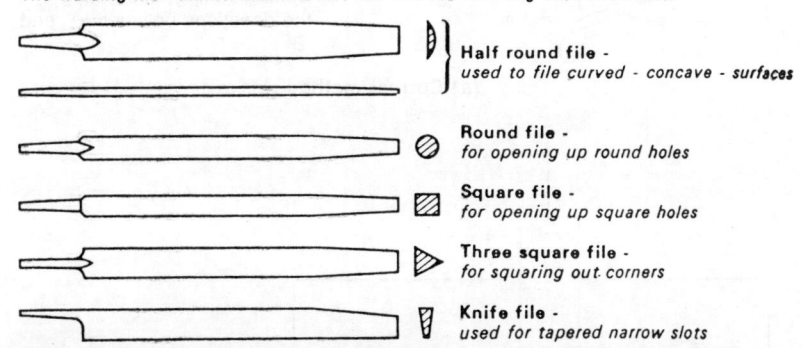

**Half round file** – used to file curved - concave - surfaces

**Round file** – for opening up round holes

**Square file** – for opening up square holes

**Three square file** – for squaring out corners

**Knife file** – used for tapered narrow slots

There are many other types of file, and for the full range consult B.S. 498: part 1 : 1960

**Fig. 7.17 Sections and uses of files**

and is usually made by over-cutting the single-cut file, as shown in Figure 7.18. The double-cut file offers a number of individual cutting teeth as opposed to a long single tooth across the width of the file, as is the case with the single-cut file. Since the filing effort is less than that required for a single-cut file, heavy-duty files are usually made with double-cut teeth.

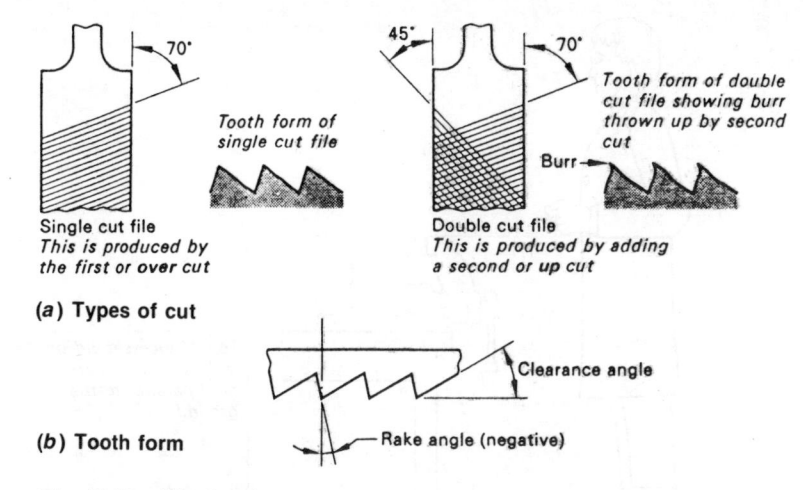

(a) Types of cut

(b) Tooth form

**Fig. 7.18 File teeth**

## 7.8 Filing technique

In order to generate a plane surface by filing, the file must be moved parallel to the plane of the required surface. This carefully controlled movement depends solely on the muscular co-ordination of the operator which involves certain essential basic principles.

HEIGHT OF VICE

For correct control, the top of the vice jaws should be level with the forearm held parallel to the ground as shown in Figure 7.19(a).

CORRECT STANCE

The file can only be controlled when the body is correctly balanced. Figure 7.19(b) illustrates how the feet should be placed relative to the vice to achieve this.

APPLICATION OF PRESSURE

An important point to remember when filing metal parts is that most, if not all, files are designed to cut on the forward direction. Therefore *filing pressure should be given only when the file is pushed away from the operator and released when the file is returned for the next forward movement.* When using a flat file, one hand applies the pressure whilst the other provides the surface movement of the file. Wherever possible, commence with the point of the file and move the file forward so as to finish with the shoulder portion. Using the whole length of the file

(a) Height of vice

*Top of vice should be in line with forearm held parallel to the ground*

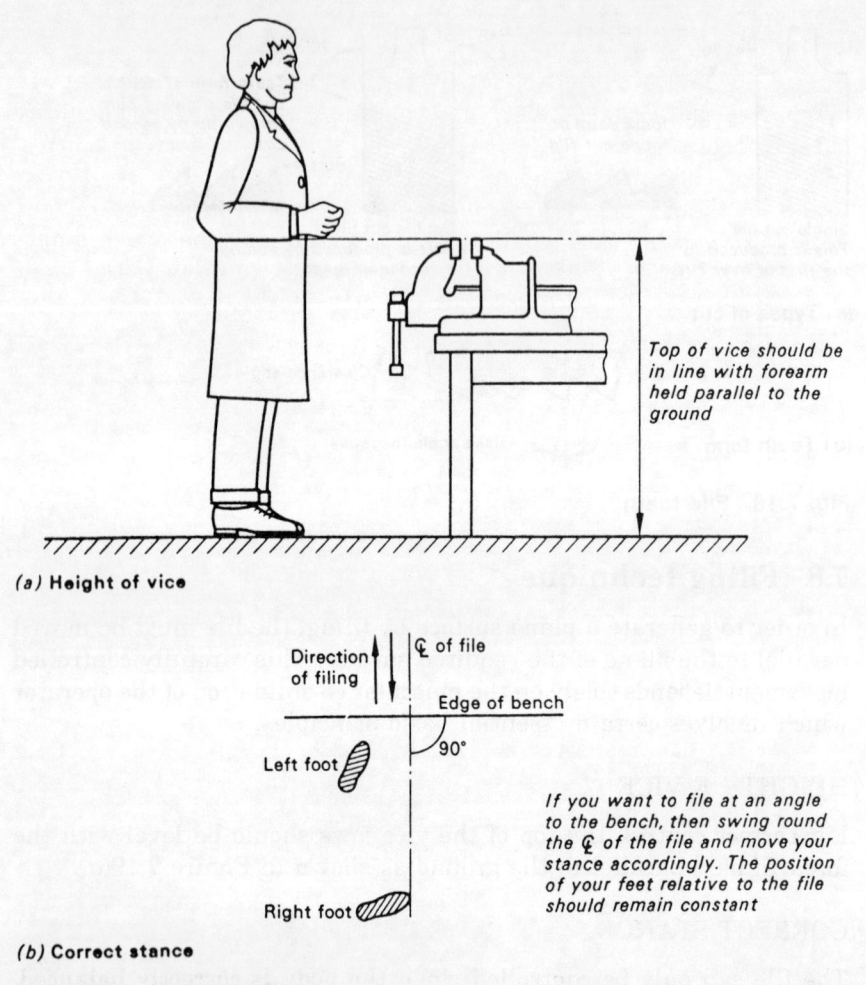

(b) Correct stance

*If you want to file at an angle to the bench, then swing round the ℄ of the file and move your stance accordingly. The position of your feet relative to the file should remain constant*

**Fig. 7.19 Use of the file**

prevents local wear. It also ensures a quicker and more efficient cutting action. The method of filing a flat surface or the edge of a metal strip is shown in Figure 7.20.

## CORRECT GRIP

The fingertips play an important part in keeping the file straight during the filing of a flat surface. Care must be exercised not to 'rock' the file during the forward movement, otherwise a convex, rather than the required flat surface, will be generated. Figure 7.21 shows how the file should be held for various operations.

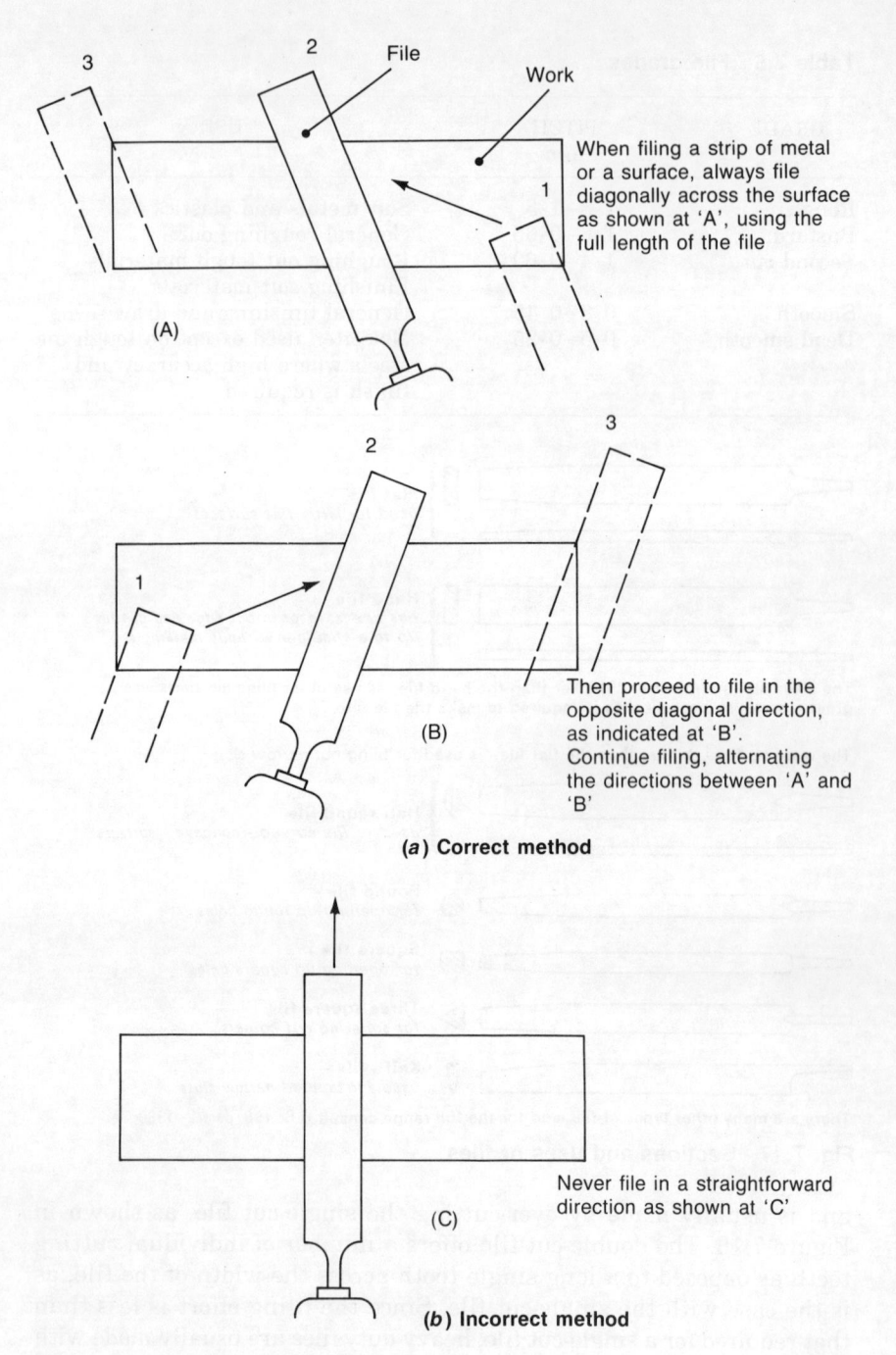

240

(A) *When filing a strip of metal or a surface, always file diagonally across the surface as shown at 'A', using the full length of the file*

(B) *Then proceed to file in the opposite diagonal direction, as indicated at 'B'. Continue filing, alternating the directions between 'A' and 'B'*

(a) Correct method

(C) *Never file in a straightforward direction as shown at 'C'*

(b) Incorrect method

**Fig. 7.20 Method of filing flat objects**

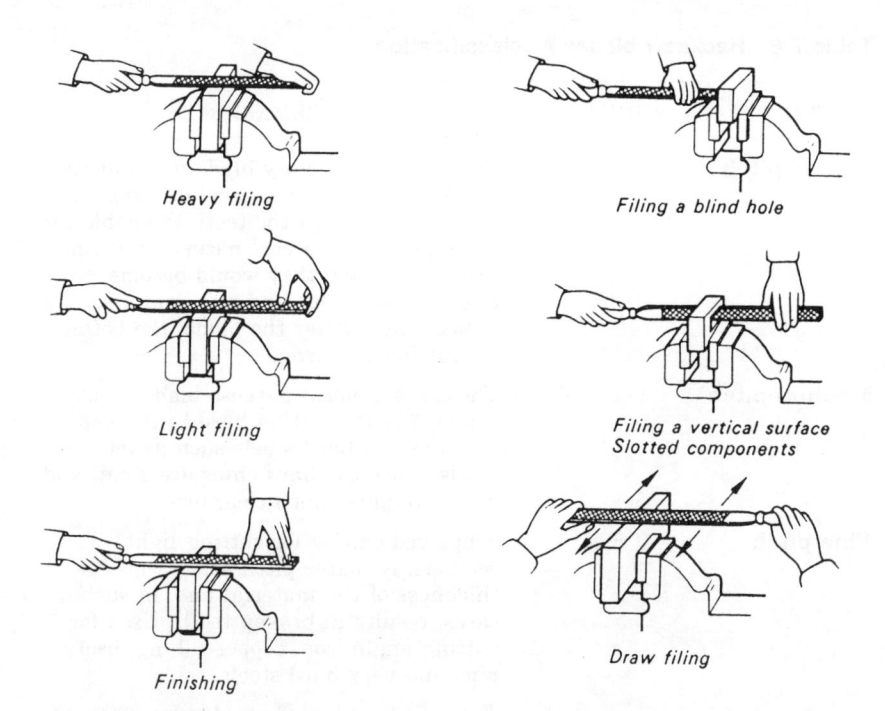

**Fig. 7.21  Correct grip**

Care must be taken to ensure 'flatness' of filing. Frequent inspection of the work, during the filing operation, using a straight edge or a surface table, will check for any unevenness in this respect. When using round or half-round files, it is good practice to give a certain amount of rotational action about their axes whilst moving them back and forth. This action produces a smoother and more regular concave surface than when they are merely moved in a fore-and-aft direction. *Remember that during each stroke of the file, the weight must be gradually transferred from the front hand to the back hand.*

### DRAW FILING

As draw filing is employed for finishing, the teeth of the file must be free from adhering metal particles, otherwise scratches will occur on the metal's surface. Regular use of the file card, in order to clear the teeth, is necessary during draw filing. *For final finishing it is advisable to rub over the teeth of the file with a block of chalk, so as to fill the spaces between the teeth.* This not only prevents scratching, but also gives a smoother finish than would otherwise be the case.

## 7.9  The hacksaw

Figure 7.22 shows a typical hacksaw with an adjustable frame that will accept a range of blade sizes. For optimum results, the blade should be carefully selected for the work in hand, it must be correctly fitted, tensioned, and correctly used.

*Like the file, the hacksaw blade cuts only in the forward direction.* Steady forward strokes should be applied, utilising the whole length of the blade, whilst exerting sufficient pressure to prevent the blade

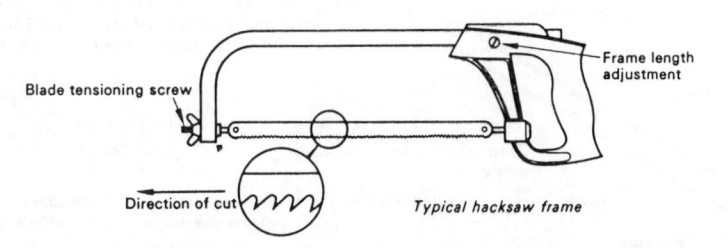

The hacksaw blade is held tightly in the frame by means of a tensioning screw and wing nut fitted to one end. A pin hole each end of the blade locate on the pins provided — one on the tensioning device, the other on the fixed part of the frame. The adjustable and fixed members of the frame which carry the locating pins are made of square section permitting them to be housed in the frame in one of two positions:

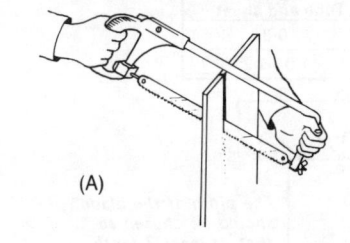

A — The normal position which enables the blade to be fitted parallel with the frame;

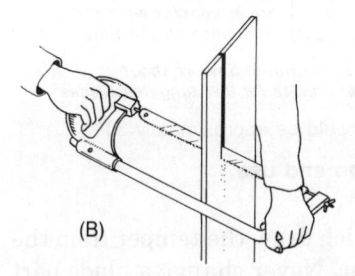

B — Locating pins turned through 90° enables the blade to be fitted with its plane at right angles to that of the frame.

With the blade fitted in the normal position, the depth or length of cut is restricted by the frame, as shown opposite.

With the blade fitted in the 90° position, the frame does not obstruct when making long cuts, or when cutting deep sections and similar parts.

**Fig. 7.22  The hacksaw**

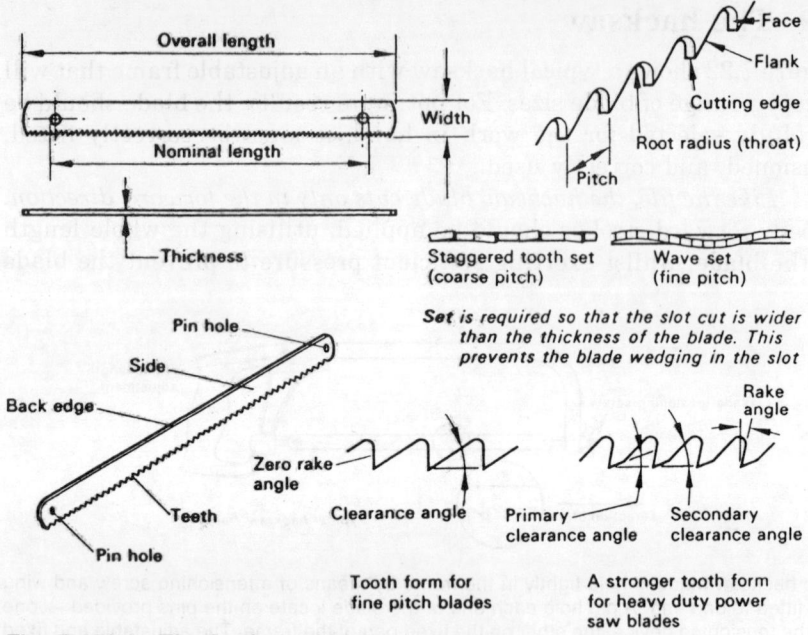

**Fig. 7.23(a)** The hacksaw blade — elements

**Fig. 7.23(b)** The hacksaw blade — selection and use

*When using the hacksaw the correct stance is the same as for filing. Saw at about 60 strokes per minute for H.S.S. blades, or 50 strokes per minute for low tungsten blades*

Where possible, the angle of the blade to the cut should be approximately 30°.

sliding. Avoid using short fast strokes which draw the temper from the teeth of the blade causing rapid local wear. Never change a blade part way through a cut. The greater 'set' of the new blade will cause it to jam in the slot already cut and break. If a change of blade is unavoidable, start a new cut to the side of the old one.

Figure 7.23 shows details of various hacksaw blades and the points

**Table 7.6 Hacksaw blades — classification**

| TYPE | PITCH (mm) | REMARKS |
|---|---|---|
| **Coarse pitch** | 1·8–2·2 | The most satisfactory blade for mild steel, since the coarse pitch provides proper clearance between the teeth to enable the chips to twist and curl naturally. If fine teeth were used they would become clogged with chips and may jam and break. The coarser the blade, the better for cutting cast iron. |
| **Medium pitch** | 1·4–1·6 | These are general purpose blades used principally for cutting hard brass, and medium and hard steels such as tool steels, since the hard chips are small and do not require much clearance. |
| **Fine pitch** | 0·8–1·3 | Employed chiefly for cutting light sections. A coarse pitch spans the thickness of the material and, in such cases, results in broken teeth. Used for cutting angle iron, copper tubing, iron pipe and very hard steels. *Extra fine pitch* (0·8mm) blades are particularly suitable for cutting thin material in order that as many teeth as possible may be in contact with the work. They are used for cutting thin-walled tubing, sheet metal and stainless steels. |

*Note*: A general classification is given in Figure 7.23(b).

to observe when selecting the correct blade. It also shows how the principles concerning cutting-tool angles also apply to hacksaw blades.

## CLASSIFICATION

A hacksaw blade is classified by the pitch of its teeth, length of the blade and the material of which it is made. Flexible low tungsten blades are hardened at the cutting edge, whilst high speed blades are hardened throughout. High speed blades are the strongest and best blades for cutting hard metals. There are different types of saw teeth for cutting various kinds of metal, as shown in Table 7.6.

The ordinary hacksaw blade is useless for profiling round a curved component. This sort of work requires the use of a *tension file*, which is a long thin circular file held in tension in the hacksaw frame. Figure

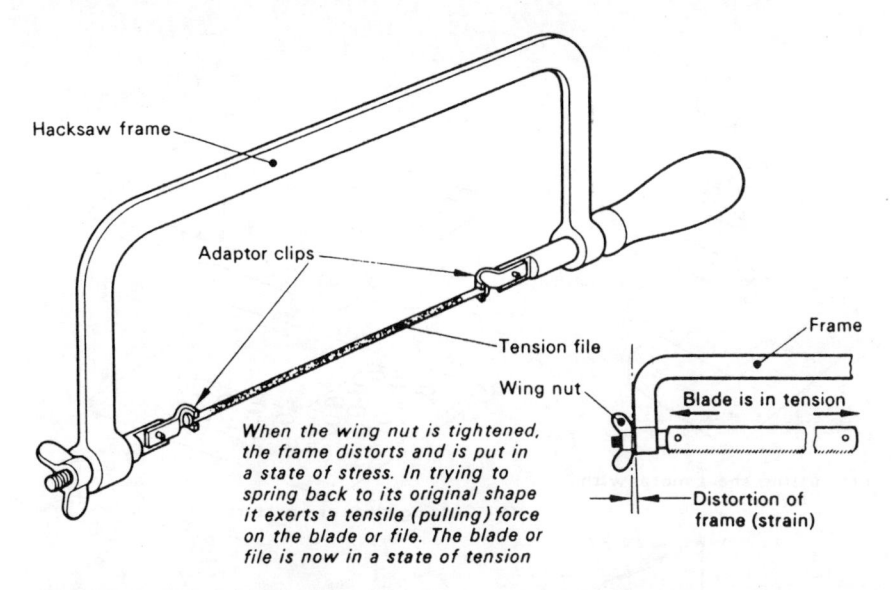

Fig. 7.24 The tension file

7.24 illustrates how the tension file is secured in a hacksaw frame with the aid of adaptor clips.

## 7.10 The chisel

Cold chisels are used for rapidly breaking down a surface. It is the quickest way to remove metal by hand, but the accuracy is low and the finish poor. Figure 7.25 shows a selection of chisels and the operations they are used for.

Since the amount of energy that is dissipated at the cutting edge is relatively small, very little heat is generated, so plain carbon steels are quite adequate for these tools. Although basically a roughing tool it is important that the principles regarding tool angles are correctly applied, as shown in Figure 7.26.

## 7.11 The use of the chisel for cutting sheet metal

Whe used for cutting sheet metal, the flat chisel must be held at a slight angle to the line of cut, as shown in Figure 7.27(a). The reasons for inclining the chisel are:

1. To provide a shearing angle.
2. To make the chisel move along the line of cut smoothly and continuously.

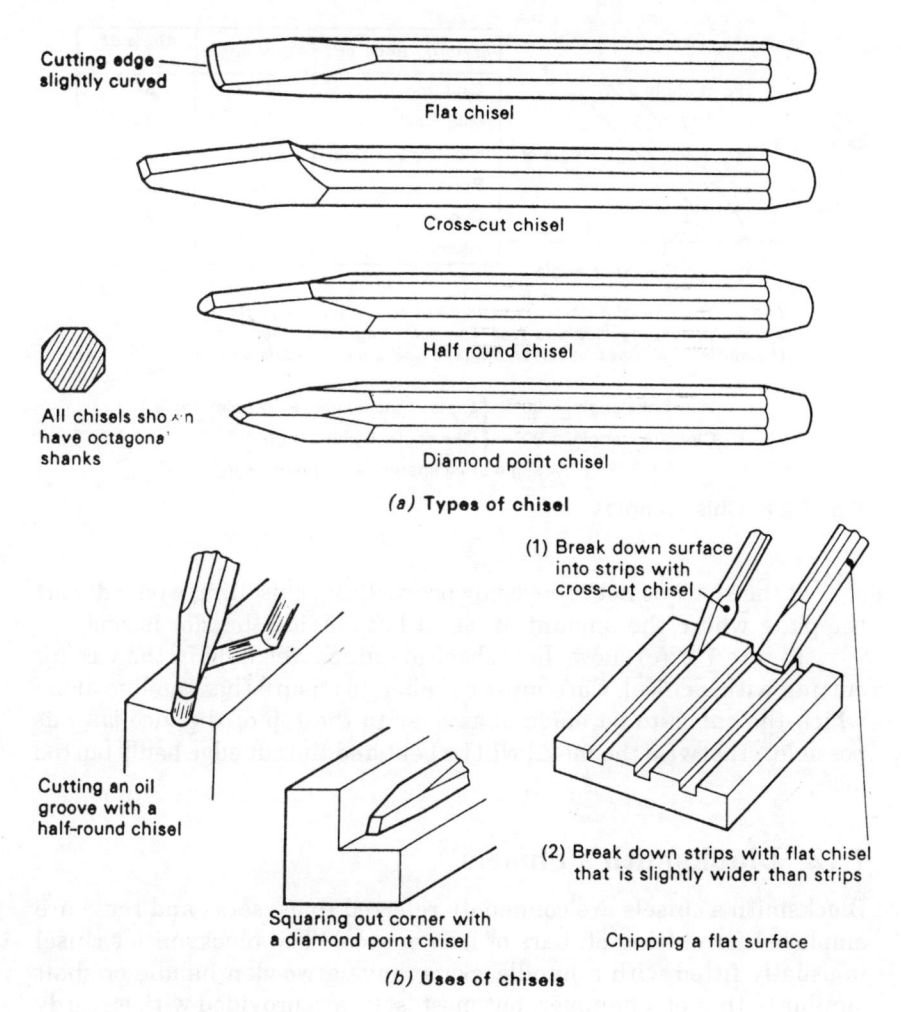

Fig. 7.25 Cold chisels

If the chisel is held vertically, a separate cut is made each time a hammer blow is delivered, and the 'line' becomes a series of irregular cuts. A block of soft iron is generally used to support the sheet metal whilst it is being cut.

The chisel is also inclined when cutting slots or apertures of various shapes or sizes. In this case the removal of the material is simplified by punching or drilling a series of holes, as near together as possible, before the chisel is used. The advantages of pre-drilling or pre-punching the sheet metal are shown in Figure 7.27(b).

243

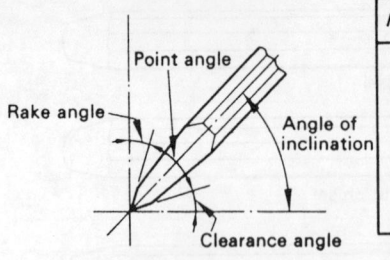

| Material to be cut | Point angle | Angle of inclination |
|---|---|---|
| Cast iron | 60° | 37° |
| Mild steel | 55° | 34½° |
| High carbon steel | 65° | 39½° |
| Brass | 50° | 32° |
| Copper | 45° | 29½° |
| Aluminium | 30° | 22° |

*The point angle of a chisel is equivalent to the wedge angle of a lathe or shaping machine tool. The point angle together with the angle of inclination forms the rake and clearance angles*

Rake angle = 90° − {angle of inclination + ½ point angle}
Clearance angle = 90° − {rake angle + point angle}
or = angle of inclination − ½ point angle

**Fig. 7.26 Chisel angles**

If the slot was not relieved by pre-drilling, chiselling would distort the plate where the amount of metal left outside the slot is small.

Figure 7.27(c) shows how sheet metal may be held in the vice for cutting with a chisel. Care must be taken to ensure that the line along which the cut is to be made is as near to the top of the vice jaws as possible, otherwise the metal will be bent and the cut edge badly burred over.

## 7.12 Blacksmith's chisels

Blacksmith's chisels are commonly referred to as 'sets', and these are employed for cutting off bars of iron or steel. The blacksmith's chisel is usually fitted with a handle. Some have a wooden handle or shaft similar to that of a hammer, but most 'sets' are provided with a sturdy twisted wire handle.

There are two types of blacksmith's chisel, the 'COLD SET' and the 'HOT SET'.

The 'hot set' is employed when cutting hot metal, and does not require to be hardened and tempered. Its cutting edge is keener than that of a 'cold set'. Hot sets are manufactured from a tough variety of steel in order that it may cut through relatively soft red-hot metal with ease. A hot set cannot be used on cold metal because the cutting edge would soon become blunted.

*The 'cold set', as the name implies, is employed when cutting cold metal, therefore it has to be correctly hardened and sharpened.*

Figure 7.28 illustrates the two types of blacksmith's chisel.

(a) Cutting sheet metal with a flat chisel

(b) Cutting slots or apertures

(c) Cutting sheet metal supported in a vice

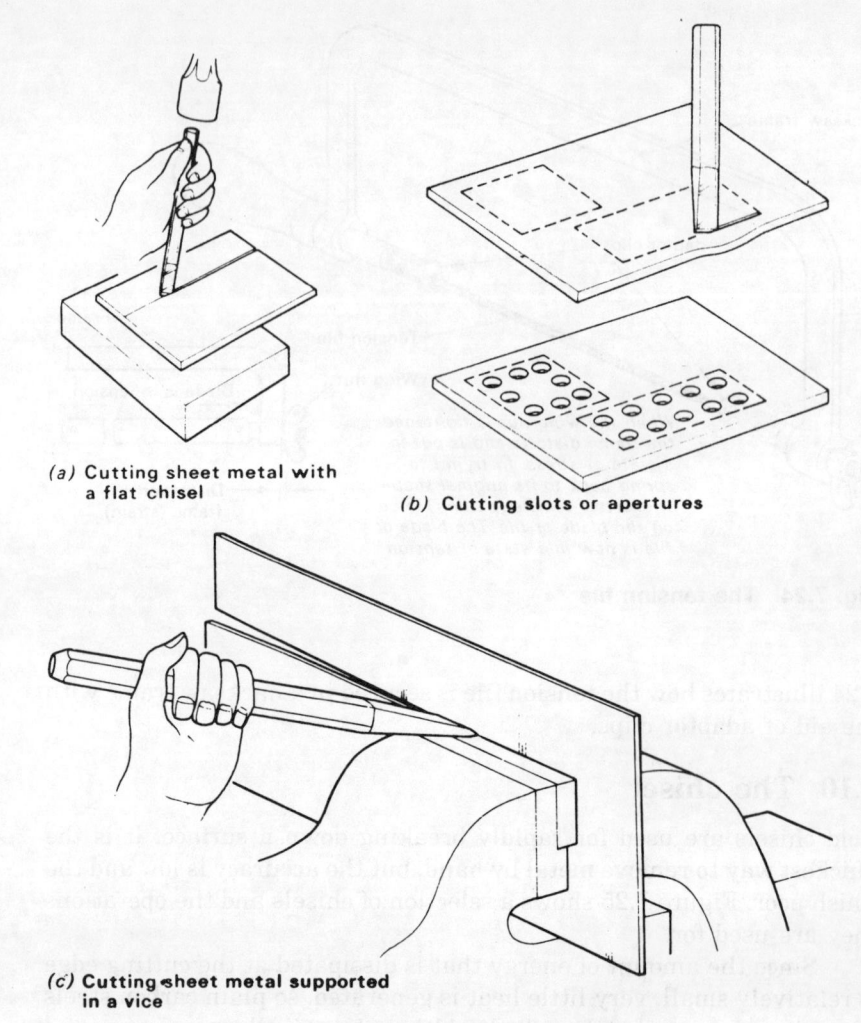

**Fig. 7.27 The use of a chisel for cutting thin plate**

When using cold sets, the work is supported on the soft table of the anvil *to avoid damaging the cutting edge as the chisel breaks through.*

Figure 7.28(c) illustrates cutting with a hot set.

HARDIES

The use of sets requires an extra pair of hands, therefore special chisels are available with a tapered square shank which fits into a square hole in the anvil. Those are called 'hardies' and allow the craftsman to work on his own.

(a) Cold sets

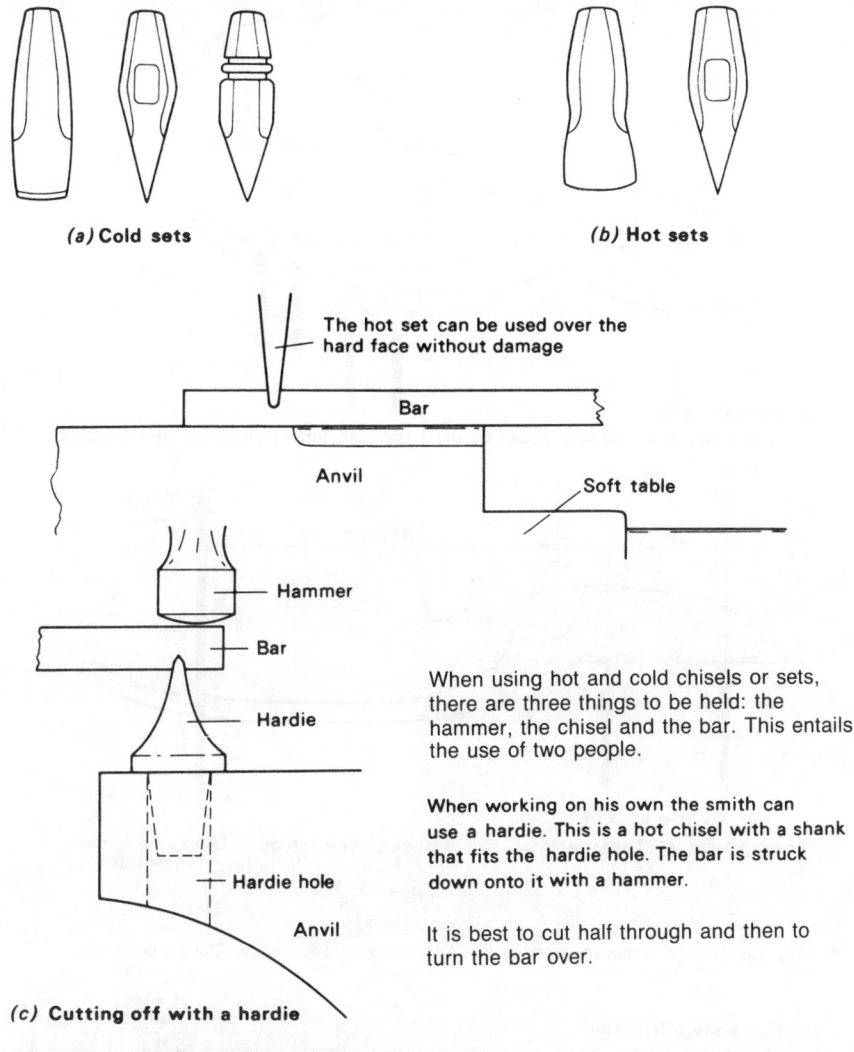

(b) Hot sets

(c) Cutting off with a hardie

When using hot and cold chisels or sets, there are three things to be held: the hammer, the chisel and the bar. This entails the use of two people.

When working on his own the smith can use a hardie. This is a hot chisel with a shank that fits the hardie hole. The bar is struck down onto it with a hammer.

It is best to cut half through and then to turn the bar over.

Fig. 7.28 Blacksmiths' chisels

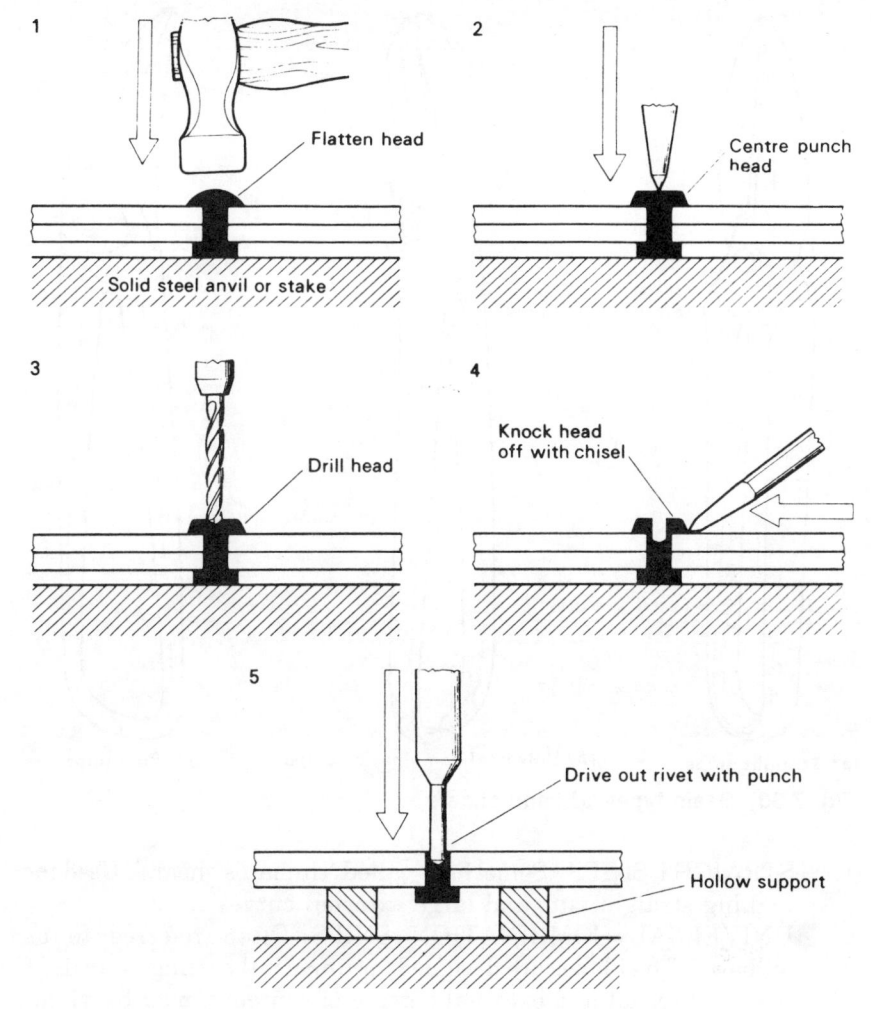

Fig. 7.29 Removing a rivet

The work is struck down on to the hardie as shown in Figure 7.28(c).

## 7.13 Rivet removal

Cold chisels are useful for removing defective rivets in sheet-metal work. Figure 7.29 illustrates the basic steps in removing a rivet. The removal of rivets in platework will be discussed in Section 7.37.

## 7.14 Snips and hand shears

Various types of snips and hand shears are available, the most common being:

SNIPS

These can be obtained in very many sizes and types up to about 400 mm in length. Figure 7.30 illustrates the basic types of snips together with their application.

245

(a) Straight snips     (b) Universal combination snips     (c) Pipe snips

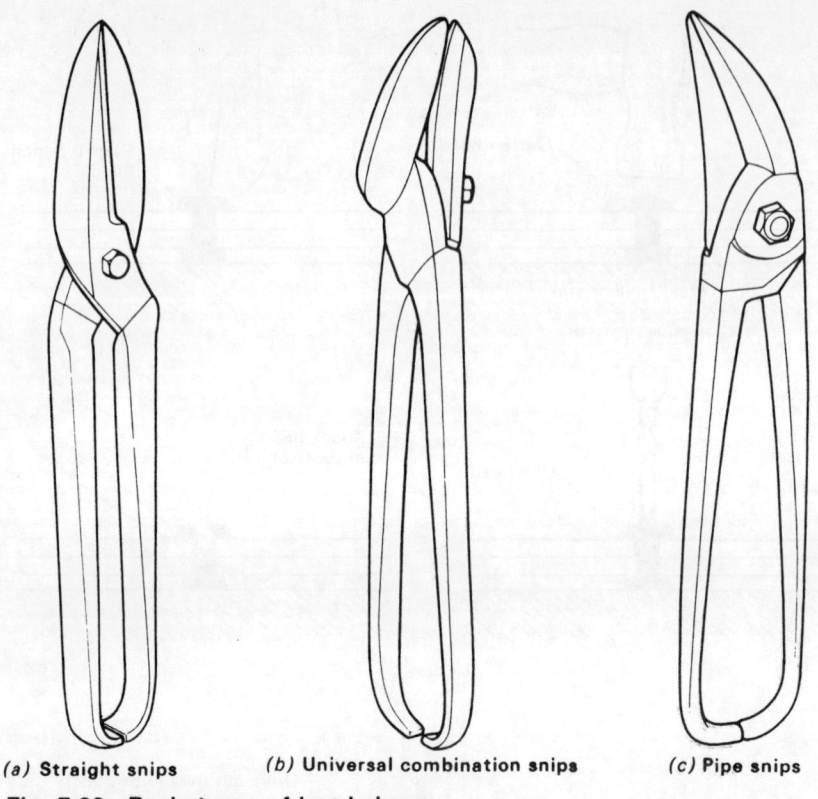

**Fig. 7.30   Basic types of hand shears**

1. STRAIGHT SNIPS   Sometimes called 'tinman's shears'. Used for making straight cuts and large external curves.
2. UNIVERSAL COMBINATION SNIPS   Often referred to as 'gilbows'. The blades are designed for universal cutting — straight line or internal and external cutting of contours. may be 'right-hand' or 'left-hand', *easily identifiable as the top blade is either on the right or the left.*
3. PIPE SNIPS   As the name implies used for trimming cylindrical or conical work in sheet metal. Have smaller thinner blades than universal snips.

Figure 7.31(*a*) shows the correct grip when using 'snips'. The scissors action is provided by the finger movement indicated by the arrows. To produce the maximum cutting force, the hand must be as far as possible from the pivot and the metal being cut must be kept close up to the pivot. Figure 7.31(*b*) shows how the *principle of moments* is exploited to magnify the effort of the craftman's hand (Section 4.27).

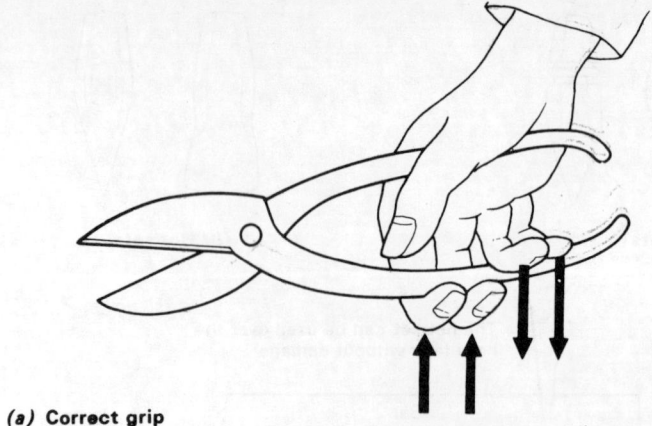

*(a)* **Correct grip**
The scissors action is provided by the finger movement indicated by the arrows

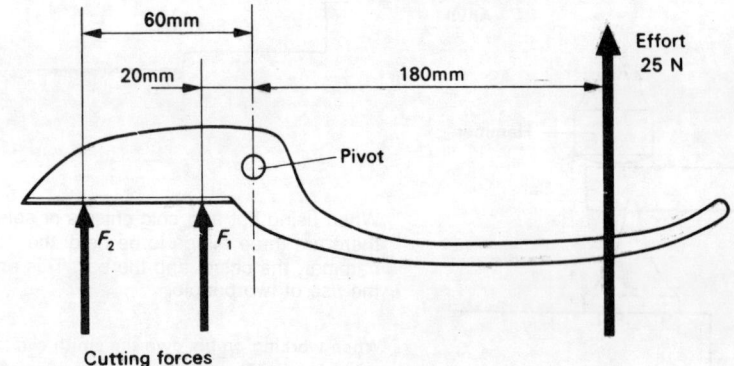

1. *Shearing force* $(F_1)$ *near to pivot*   Anti-clockwise forces = Clockwise forces
$F_1 \times 20$ = $25 \times 180$
$F_1 = \frac{25 \times 180}{20}$
$F_1 = \underline{225\ N}$

2. *Shearing force* $(F_2)$ *remote from pivot*   Anti-clockwise forces = Clockwise forces
$F_2 \times 60$ = $25 \times 180$
$F_2$ = $\frac{25 \times 180}{60}$
$F_2$ = $\underline{75\ N}$

*(b)* **Force magnification**

**Fig. 7.31   Principle of the hand shear**

## HAND-OPERATED SHEARS

These are operated by means of a lever (handle) and are generally for bench mounting. Various types of hand-operated shears are illustrated in Figure 7.32.

1. *Hand-lever bench shears*   Used for straight line cutting of sheet metal up to 3 mm thick.

*(a)* **Hand-lever bench shears**

*(b)* **Lever-operated throatless shears**

*(c)* **Hand-lever corrugated bench shears**

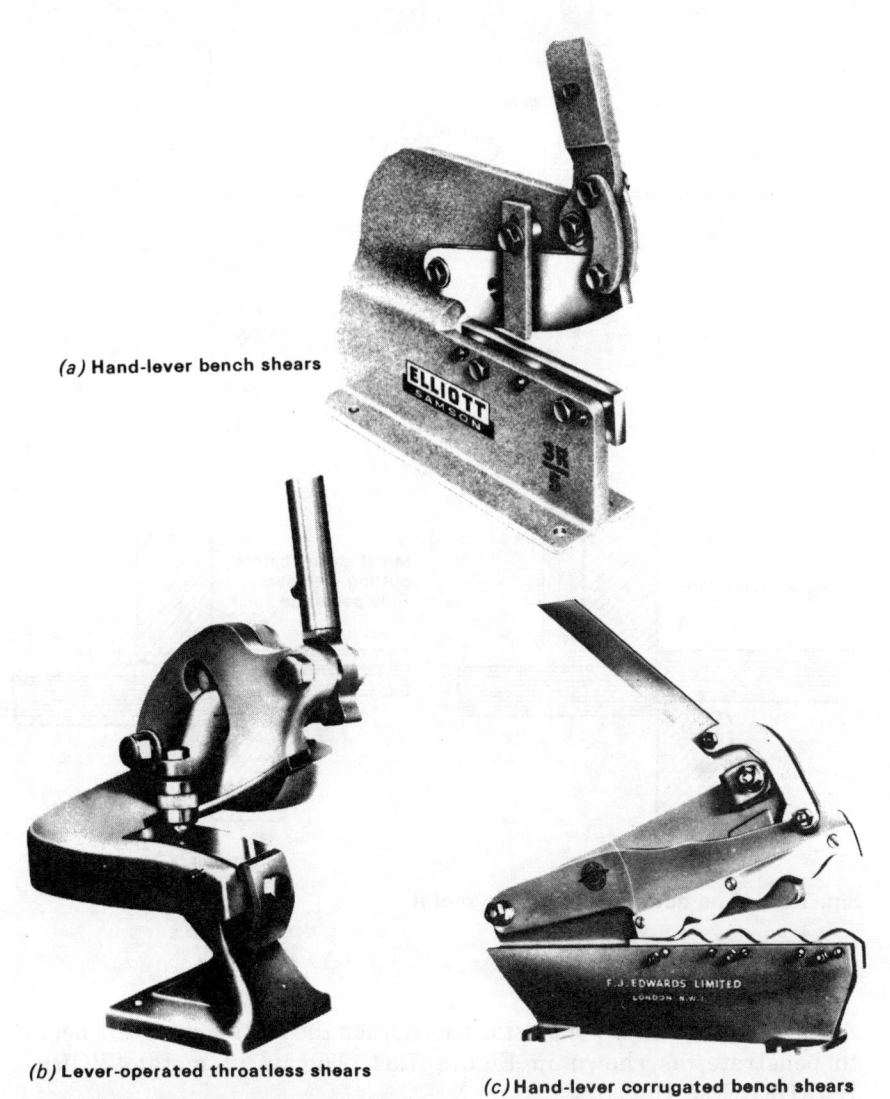

**Fig. 7.32  Typical hand lever shears**

Note: The movable top blade is curved to provide a constant SHEAR ANGLE at cutting position (may be used for CONVEX cutting).

2. *Lever-operated throatless bench shears* May be used for either CONVEX or CONCAVE cutting by suitable manipulation of the sheet. The heavier type of this machine is capable of cutting 4·7 mm thick mild steel.

3. *Hand-lever corrugated bench shears* Used for straight shearing or corrugated material — capacity 1·6 mm in mild steel.

## 7.15  Basic principles (shearing)

There is a wide range of machines for shearing sheet metal from the basic elementary snips to static and portable power machines. In each case the basic principle of metal cutting, whether the machine is operated by hand or by power, is the shearing action of a moving blade in relation to a fixed blade.

The standard type of bench shear and all guillotines are used for *straight-line cutting*. The basic principle of these machines is that one blade is fixed (bottom blade) and the moving blade (inclined to the fixed blade) is brought down to meet the fixed blade, as shown in Figure 7.33(*a*).

The moving cutting member of a shearing machine may be actuated by:

1. HAND-LEVER — bench shearing machines
2. FOOT TREADLE — treadle guillotines
3. ELECTRIC MOTOR or HYDRAULIC — power guillotines

If the cutting members of a guillotine or shearing machine were arranged parallel to each other, the area under shear would be the CROSS-SECTION of the material to be cut, i.e. 'length × thickness', as shown in Figure 7.33(*b*).

The top cutting member of a shearing machine is always inclined to the bottom member to give a 'SHEARING ANGLE' of approximately 5°. Figure 7.33 shows that with this arrangement of the blades, *the area under shear is greatly reduced and, consequently the force required to shear the material is considerably reduced.*

Figure 7.34 explains how the shearing action is used to cut metal. The shear blades are provided with a RAKE ANGLE of approximately 87° and there must be CLEARANCE between the cutting edges of the blades to assist in the cutting action. The importance of clearance will be discussed later in this section.

STAGE 1

As the top cutting member is moved downwards and brought to bear on the metal with continuing pressure, the top and bottom surfaces of the metal are deformed, as shown in Figure 7.34.

STAGE 2

As the pressure increases, the internal fibres of the metal are subjected to deformation. This is 'PLASTIC DEFORMATION' prior to 'SHEARING'.

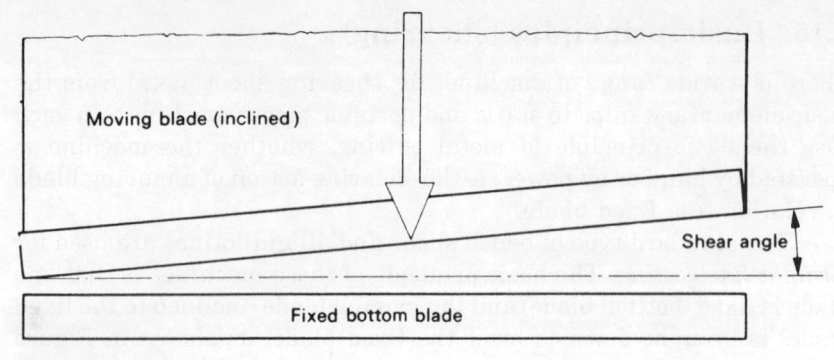

*(a) Shear blade movement*

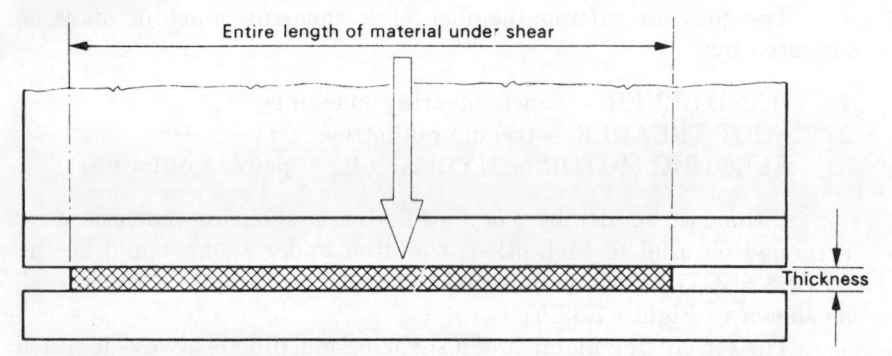

*(b) Parallel cutting blades*

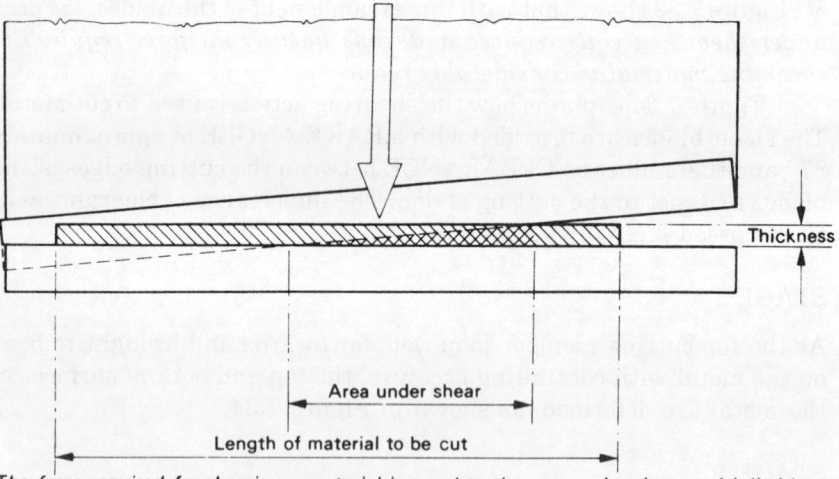

*(c) Top cutting blade inclined*

The force required for shearing a material is equal to the area under shear multiplied by the 'shear strength' of the material

**Fig. 7.33 The effect of shear angle (shearing machines)**

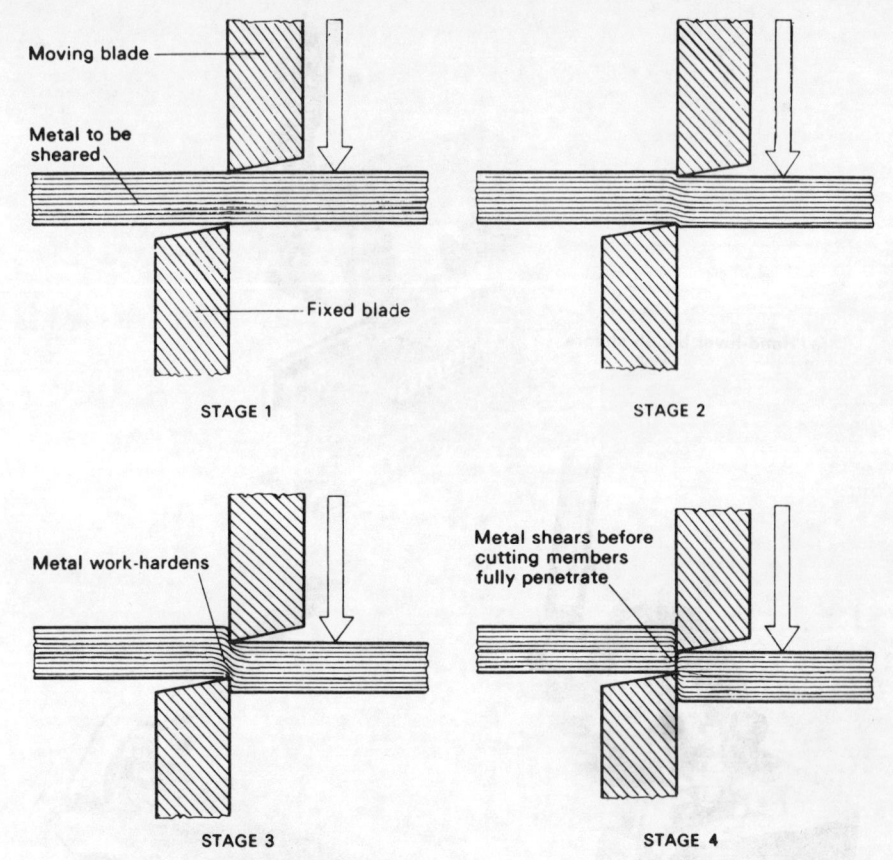

**Fig. 7.34 The action of shearing metal**

### STAGE 3

After a certain amount of plastic deformation the cutting members begin to penetrate, as shown in Figure 7.34. The uncut metal 'WORK-HARDENS' at the edges.

### STAGE 4

Fractures begin to run into the work-hardened metal from the points of contact of the cutting members. *When these fractures meet, the cutting members penetrate the whole of the metal thickness.*

*With all shearing machines a sufficient force must be applied to the moving blade to overcome the* shear strength *of the material and cause it to shear along the line of action.* In respect of snips and bench shears,

Mechanical advantage (force ratio) = $\frac{\text{Load}}{\text{Effort}}$

*The greater the force ratio, the greater will be the load that can be moved or produced for a given effort*

The load applied to the linkage by the first lever system:

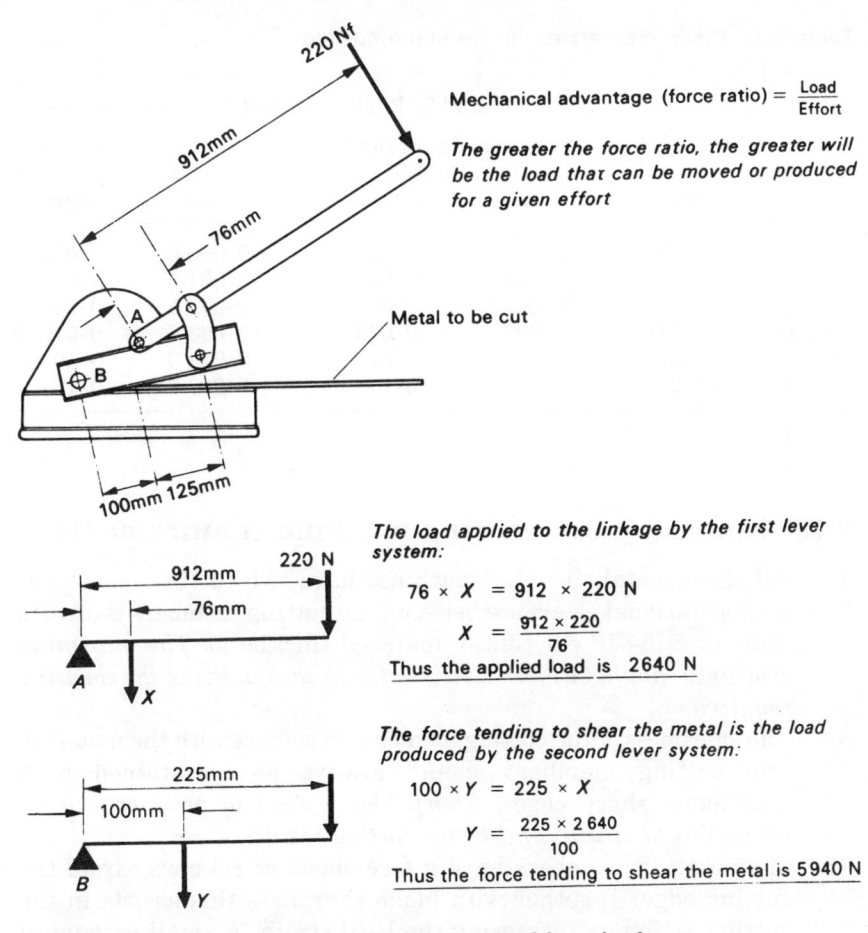

$$76 \times X = 912 \times 220 \text{ N}$$

$$X = \frac{912 \times 220}{76}$$

Thus the applied load is 2640 N

The force tending to shear the metal is the load produced by the second lever system:

$$100 \times Y = 225 \times X$$

$$Y = \frac{225 \times 2640}{100}$$

Thus the force tending to shear the metal is 5940 N

**Fig. 7.35 The lever system of hand-operated bench shears**

*Universal snips have short thick blades which will withstand the twisting of the snips when being used on irregular curved cuts*

**Universal snip**

*Straight snips have thin blades which are only strong on a vertical plane. They are therefore only suitable for straight cuts and external curves when surplus waste has be removed*

**Straight snips**

$\alpha$ = Rake
$\gamma$ = Front clearance

*The action of a pair of snips is to cut without forming chips*

*Cross-sectional view of the cutting blades of a pair of universal shears*

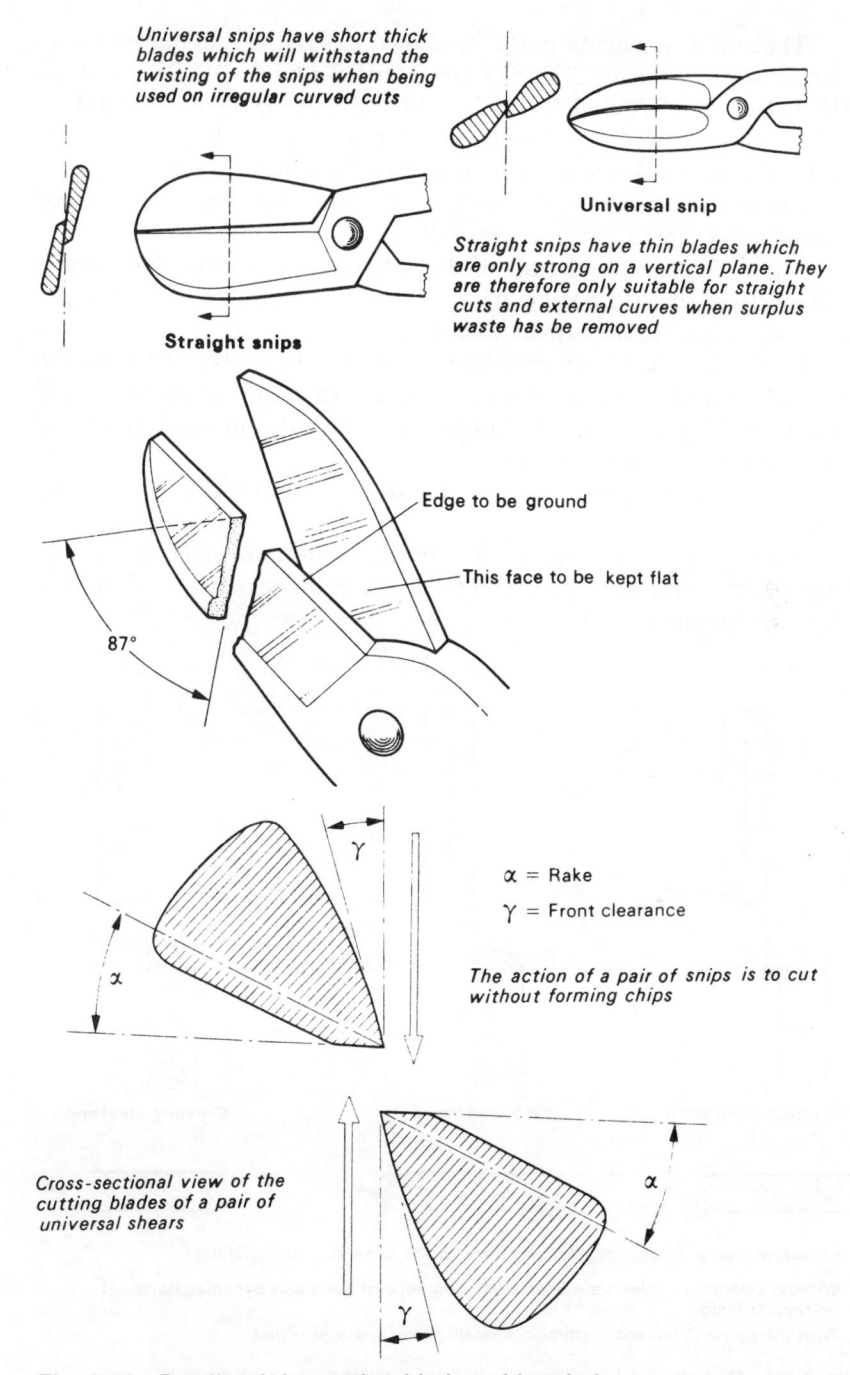

**Fig. 7.36 Details of the cutting blades of hand shears**

these are designed in such a manner that the fulcrums of the moving blades are carefully positioned to provide an optimum MECHANICAL ADVANTAGE to enable a relative small EFFORT (hand power) to be applied in order to cut the metal with comparative ease. Figure 7.35 shows a simplified diagram of the lever system of a hand-operated shearing machine.

A lever is a simple machine which applies the principle of moments to give mechanical advantage.

*By means of a lever, a comparatively small force, usually termed 'effort' applied at a relatively large distance from the fulcrum (pivot), will either overcome or balance a greater force or load at a small distance from the fulcrum.*

249

The simple shearing machine shown in Figure 7.35 makes use of a double application of 'THE SECOND ORDER OF LEVERS' in which THE LOAD IS BETWEEN THE FULCRUM AND THE EFFORT.

Never extend the arms of a pair of snips or a bench shear in order to cut a greater thickness of metal than that for which they are designed. *This careless action will only result in deflecting the cutting blades and bending or shearing the fulcrum bolt.*

Figure 7.36 shows details of the cutting blades of hand shears. It will be noticed that the design principles are the same as those employed in respect of guillotine shear blades.

Blade clearances are very important and should be set to suit the material being cut. An approximate rule is that the clearance should not exceed 10 per cent of the thickness to be cut and must be varied to suit the particular material.

Correct clearance is essential in order to obtain optimum shearing results.

Figure 7.37 shows the results of incorrect and correct setting of shear blades, and Table 7.7 gives details of clearance in relation to material thickness.

**Table 7.7  Blade clearances for optimum cutting**

| METAL THICKNESS ||  BLADE CLEARANCE (Tested by Feeler gauges) ||||
|  ||  Low Tensile Strength (e.g. BRASS) || High Tensile Strength (e.g. STEEL) ||
| in. | mm | in. | mm | in. | mm |
|---|---|---|---|---|---|
| 0·015 | 0·381 | 0·000 3 | 0·007 5 | 0·000 5 | 0·013 |
| 0·032 | 0·813 | 0·001 5 | 0·038 | 0·001 8 | 0·046 |
| 0·065 | 1·651 | 0·002 0 | 0·051 | 0·002 5 | 0·064 |
| 0·100 | 2·540 | 0·002 2 | 0·056 | 0·003 0 | 0·076 |
| 0·125 | 3·175 | 0·003 0 | 0·076 | 0·004 0 | 0·10 |
| 0·250 | 6·350 | 0·005 5 | 0·14 | 0·007 0 | 0·18 |

## 7.16  Summary of shearing machine requirements

1. All sheet metal shearing machines have two cutting members.
2. The operational clearance between the cutting members is always important (5–10 per cent of material thickness). *The maximum clearance should be given consistent with the quality of the shearing required.*
3. The sharpness of the cutting members in contact with the material. The cutting members should always be maintained with reasonably sharp edges. *Sharp blades tend to produce a more intense local strain, and better cutting results.*
4. Most cutting members have a face shear or rake-way from the cutting edge. Together with blade sharpness this assists in the cutting action by increasing the local strain. A small amount of face shear (about 2°) on the blades is useful where there is a risk of the blades or cutters being brought together almost to touch when shearing very light gauge material.

With regard to shearing machines in general, two principal basic requirements for a good machine (irrespective of type) are as follows:

1. The rigidity or robustness of the cutting member mountings, so that *deflection of the cutting members is kept to a minimum during the cutting action.* That is, once the cutting members are set in their correct position they will stay there.
2. A satisfactory means of adjustment should be provided between the cutting members. *This enables the cutting clearance to be adjusted and the position of the cutting members to be maintained with regard to the feed of the material.*

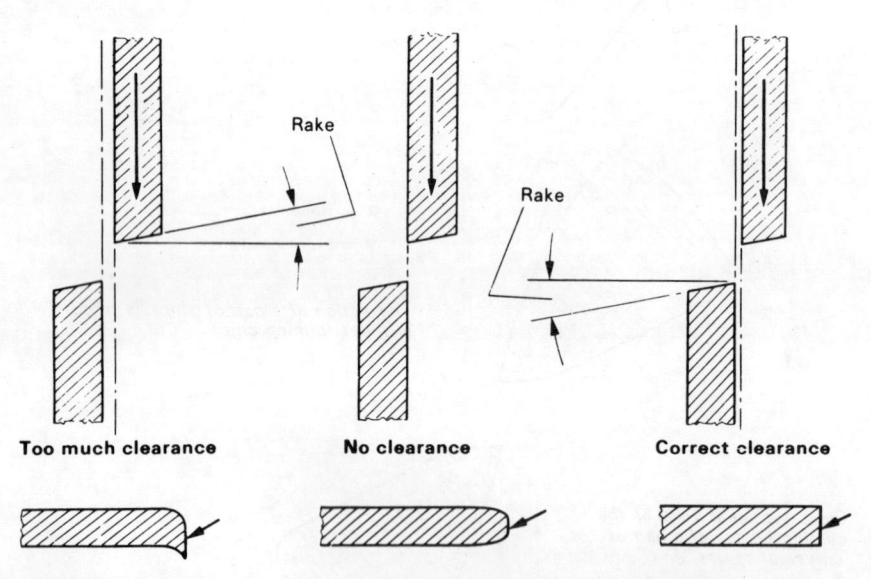

1. Excessive clearance causes a burr to form on the underside of the sheet
2. With no clearance, overstrain is caused, the edge of the sheet becomes flattened on the underside
3. With the correct clearance optimum shearing results are obtained

**Fig. 7.37  Results of incorrect and correct setting of shear blades**

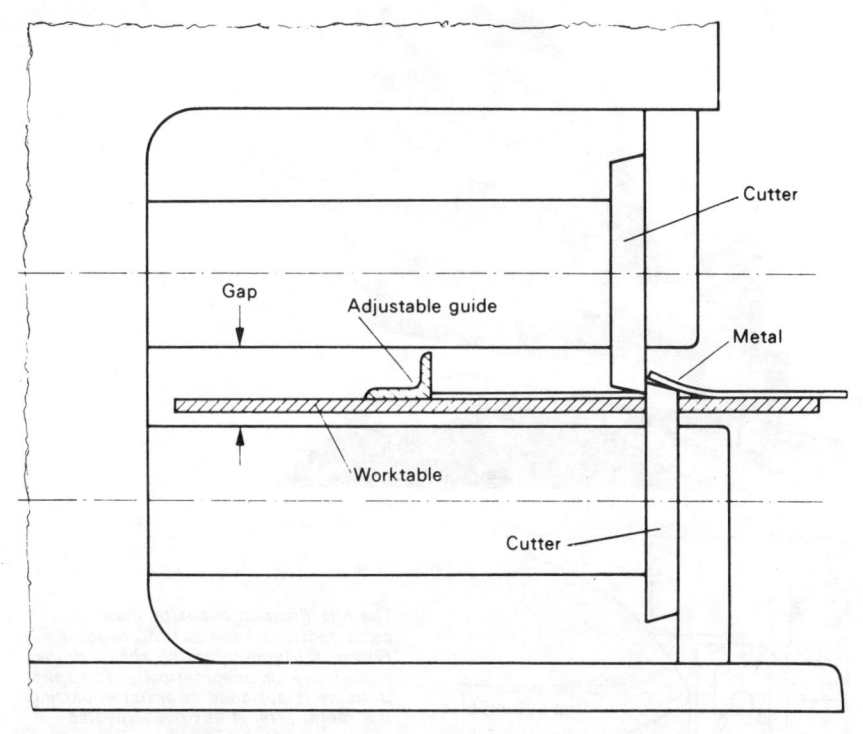

(a) Bench rotary machine for straight-line cutting (horizontal spindles)
**Fig. 7.38 Rotary shears**

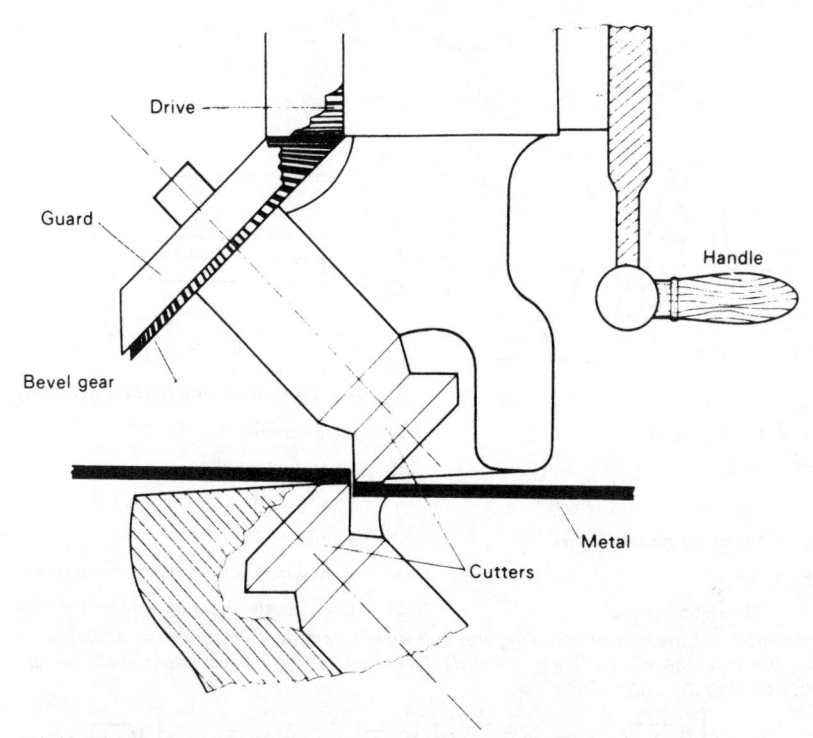

(b) Hand-operated rotary throatless shear (inclined spindles)

## 7.17 Rotary shears

With these types of machines a pair of ROTARY CUTTERS replaces the conventional flat blades used in those shearing machines which are essentially for straight-line cutting. *One advantage of rotary shears is that there is no restriction on the length of cut.*

The cutters rotate producing a continuous cutting action with very little distortion of the material. These machines may be hand or power driven and consist of two basic types, and these are illustrated in Figure 7.38.

SHEARING ESSENTIALS (ROTARY CUTTERS)

1. The edges of the cutters must overlap by the smallest amount consistent with clean cutting.
   Excessive overlap tends to distort the material.
   Insufficient overlap does not shear the material.
2. There must be clearance between the working edges of the cutters (see Section 7.15) and a means to adjust this.
3. Both cutters must run dead true, both on face and diameter.
4. The shafts on which the cutters are mounted must have no lateral movement. Neither the shafts nor their mountings should deflect under load.
5. The cutters must be kept reasonably sharp. In the case of machines which have no adjustment between the spindles, the cutters should be sharpened by grinding on the face only.

Figure 7.39 shows details of a circle cutting machine which operates on the rotary cutter principle.

## 7.18 Portable shears and nibblers

Portable power cutting tools are either electrically or pneumatically operated and may be used for straight-line cutting and for cutting irregular curves. *The pneumatic machines are much safer for work 'on site'*, the power source being a compressor.

There are two basic types of portable shearing machines — the 'shear type nibbler' and the 'punch type nibbler'.

251

**Circle cutting machine (hand operated)**

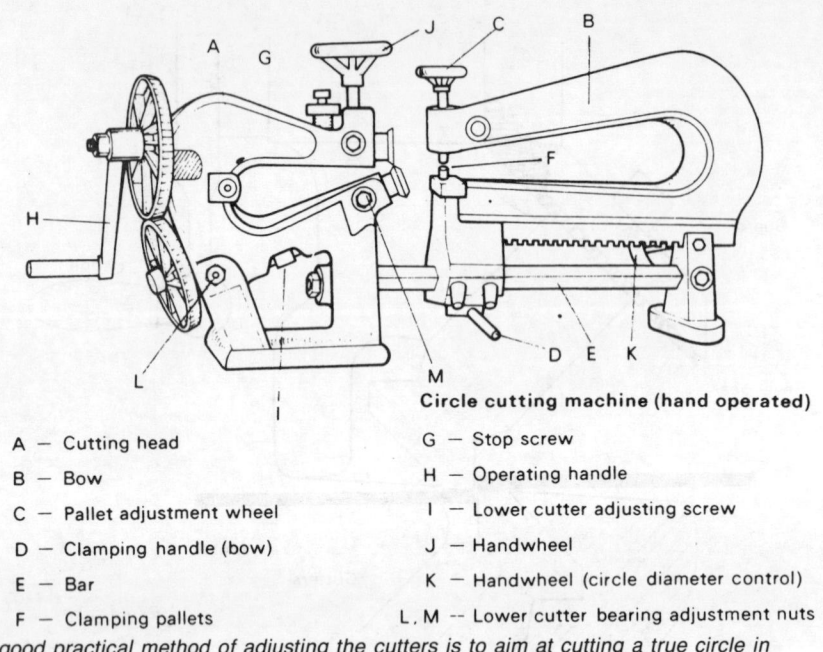

- A — Cutting head
- B — Bow
- C — Pallet adjustment wheel
- D — Clamping handle (bow)
- E — Bar
- F — Clamping pallets
- G — Stop screw
- H — Operating handle
- I — Lower cutter adjusting screw
- J — Handwheel
- K — Handwheel (circle diameter control)
- L, M — Lower cutter bearing adjustment nuts

*A good practical method of adjusting the cutters is to aim at cutting a true circle in paper. If a machine will do this in a satisfactory manner, then it will shear sheet metal without burring the edge of the disc*

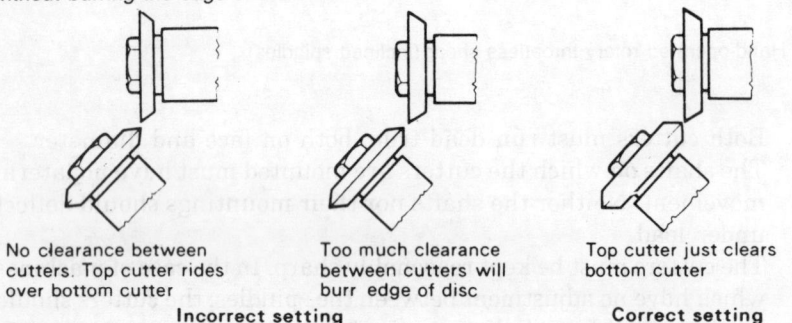

**Incorrect setting**
No clearance between cutters. Top cutter rides over bottom cutter

Too much clearance between cutters will burr edge of disc

**Correct setting**
Top cutter just clears bottom cutter

**Fig. 7.39 Circle cutting machine**

### THE SHEAR TYPE NIBBLER

This portable power tool is used for rapid and accurate straight-line or curved cutting of material up to 4·5 mm thickness. It is basically a short-stroke power shear fitted with a rapidly reciprocating cutting blade, so that each stroke makes a cut approximately 3 mm in length. The speed of the cutting blade is between 1200 and 1400 strokes per minute, and the linear cutting speed for material up to 1·62 mm thickness is approximately 10 metres per minute, and for 4·5 mm thickness the cutting speed is reduced to a maximum of approximately 4·6 metres per minute, and produces a cleanly cut edge.

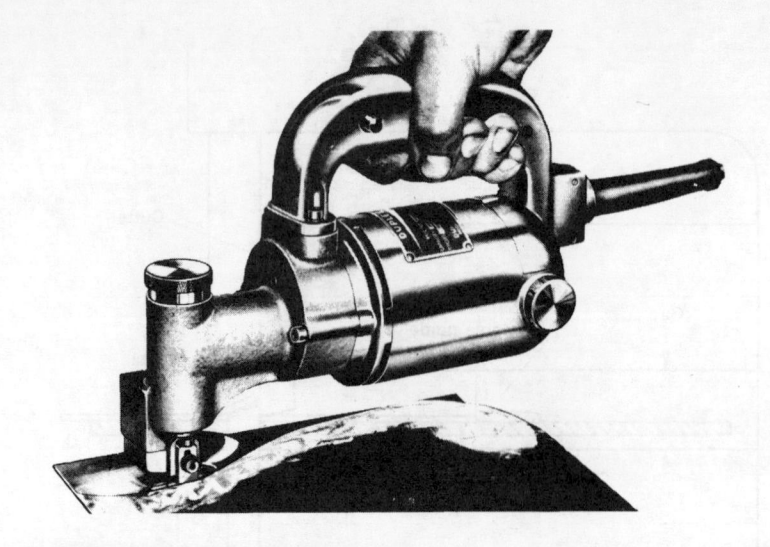

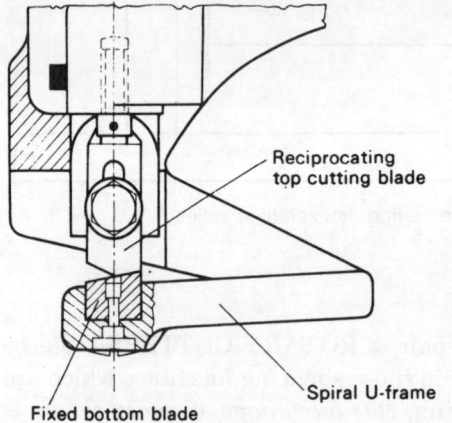

*The line diagram opposite shows a cross-section of the cutting head of a typical portable vibrating shears driven electrically or pneumatically. The spiral U-frame is designed to assist in parting the metal after it has been sheared*

Reciprocating top cutting blade

Spiral U-frame

Fixed bottom blade

*Basic details of cutting blades are given in the diagram below*

Moving blade (top)   12°   8°   12°

Fixed blade (bottom)   2°   90°

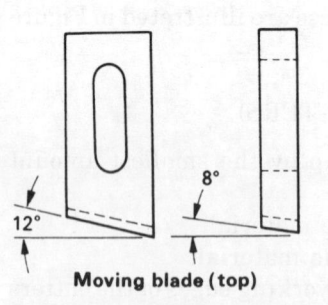

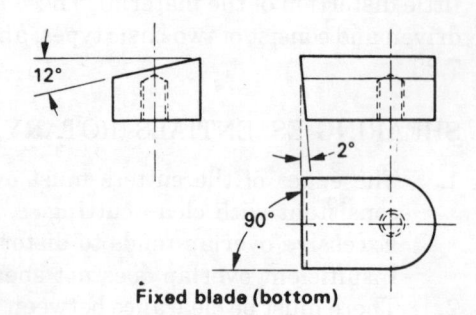

**Fig. 7.40 Portable shearing machine**

252

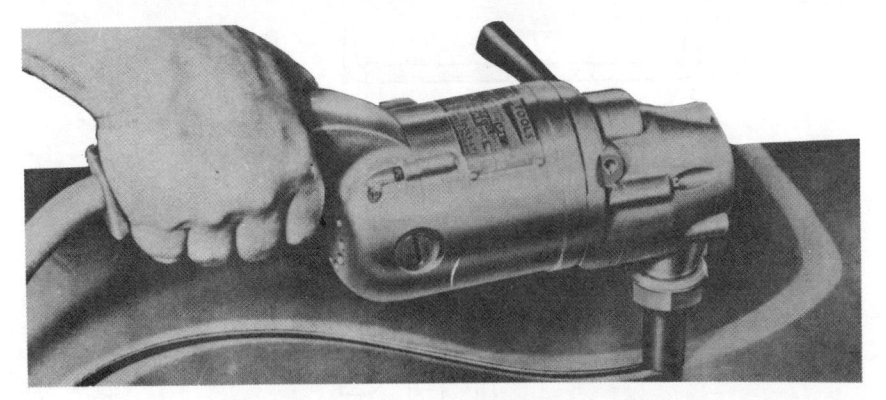

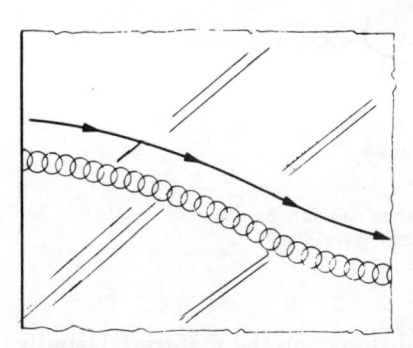

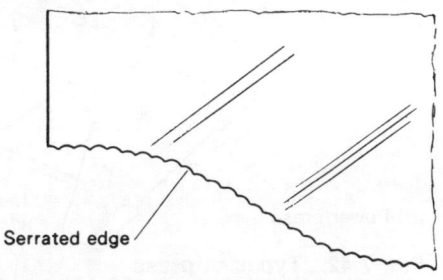

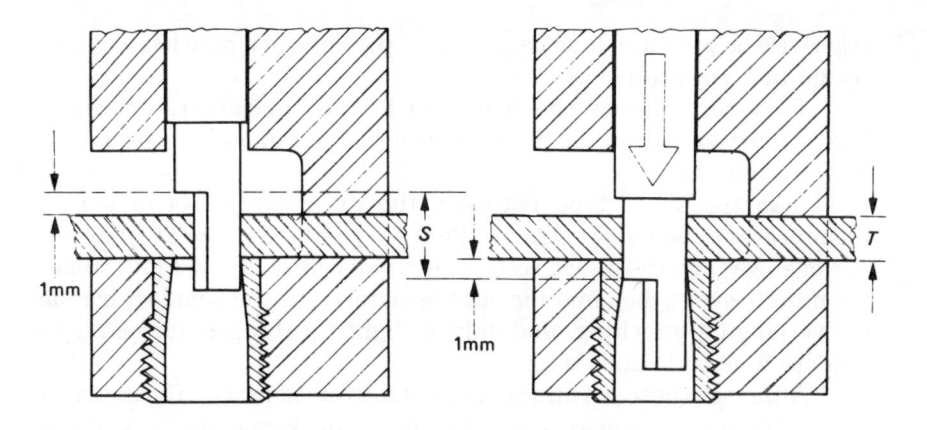

**Setting of nibbling punch**   $S$ = Length of stroke   $T$ = Metal thickness

The stroke is adjusted to give movement of approximately 1mm above the material and 1mm through the material, as shown in the diagram above

(b)

**Fig. 7.41   Portable nibbling machine**

The shear type nibbler is fitted with a pair of very narrow flat blades, one of which is usually fixed, and the other moving to and from the fixed blade at fairly high speeds. Generally these blades have a very pronounced RAKE to permit piercing of the material for internal cutting, and since the blades are so narrow, the sheet material can be easily manoeuvered during cutting.

The top blade is fixed to the moving member or ram, and the bottom blade on a spiral extension or 'U-frame'. This extension is shaped like the body of a *'throatless shear'*, to part the material after cutting.

The ligher machines have a minimum cutting radius of 16 mm, and the heavier ones about 50 mm.

There is usually provision for vertical adjustment to allow for re-sharpening of the blade by grinding, and an adjustment behind the bottom blade to allow for setting the cutting clearance.

Figure 7.40 shows details of the 'shear type nibbler'.

## THE PUNCH TYPE NIBBLER

This portable nibbling machine does not operate on the same principle as the shear type. *A punch and die is employed instead of shearing blades, and the nibbling principle is a special application of punching.*

The advantage of these machines is that they will effect certain operations that cannot be accomplished on other shearing machines. For example, they may be used to cut out apertures which could only

*The principle of nibbling is basically that of overlapping punching*

The width of the cut produced by nibbling machines is determined by the diameter of the punch in relationship with the thickness of the material to be cut.   For example:

**Capacity of machine – 2mm   Width of cut – 8mm   Approximate cutting speed – 1·8mm/min**
**Capacity of machine – 3·2mm  Width of cut  9·5mm  Approximate cutting speed – 1·5mm/min**

(a)

253

otherwise be produced by means of specially designed punches and dies set up in a powerful press.

These portable power tools are used for rapid and accurate straight-line or curved cutting of material from approximately 1·62 mm to 3·2 mm thickness.

Like the shear type machine, the top cutting tool (a punch) reciprocates at fast short strokes. Punch type nibblers are available in various sizes and the punch reciprocates at a rate of 350 to 1400 strokes per minute over a die, nibbling out the material by the simple principle of overlapping punching, and only a slight finishing is necessary to produce a smooth clean edge.

Although these machines are generally used for cutting material up to 3·2 mm thickness, there are heavy-duty machines available capable of cutting steel up to 6·35 mm thickness.

Standard punches of 4·8, 6·35 and 9·5 mm diameter are employed with different sizes of machines, and the maximum linear cutting speed is approximately 1·8 metres per minute.

One main advantage of nibbling over shearing is that there is less distortion of the work.

Figure 7.41 shows details of the punch type nibbler.

Some models are fitted with a controlled drip feed to LUBRICATE the punch.

Portable nibbling machines are also available with narrow rectangular punches. Rectangular punches produce a cut without any ragged edges such as would be experienced with a circular punch.

## 7.19 Principles of piercing and blanking

Hand and power presses similar to those shown in Figure 7.42 can also be used for cutting sheet metal and thin plate when fitted with suitable punches and dies (press tools).

The operation of cutting out the metal blank is called 'BLANKING'.

The operation of punching holes in the metal is called 'PIERCING'.

Figure 7.43 shows a typical piercing tool and the principle upon which it works.

If a piece of sheet metal is placed on the lower cutting member (the 'DIE') and the top cutting member (the 'PUNCH'), correctly located with regard to the bottom member, is brought to bear on the metal with continuing pressure, after a certain amount of deformation, the ELASTIC LIMIT of the metal is exceeded and the top cutting member penetrates the metal. The metal is sheared before the punch fully penetrates the whole of the metal thickness. The degree of penetration

(a) Fly press

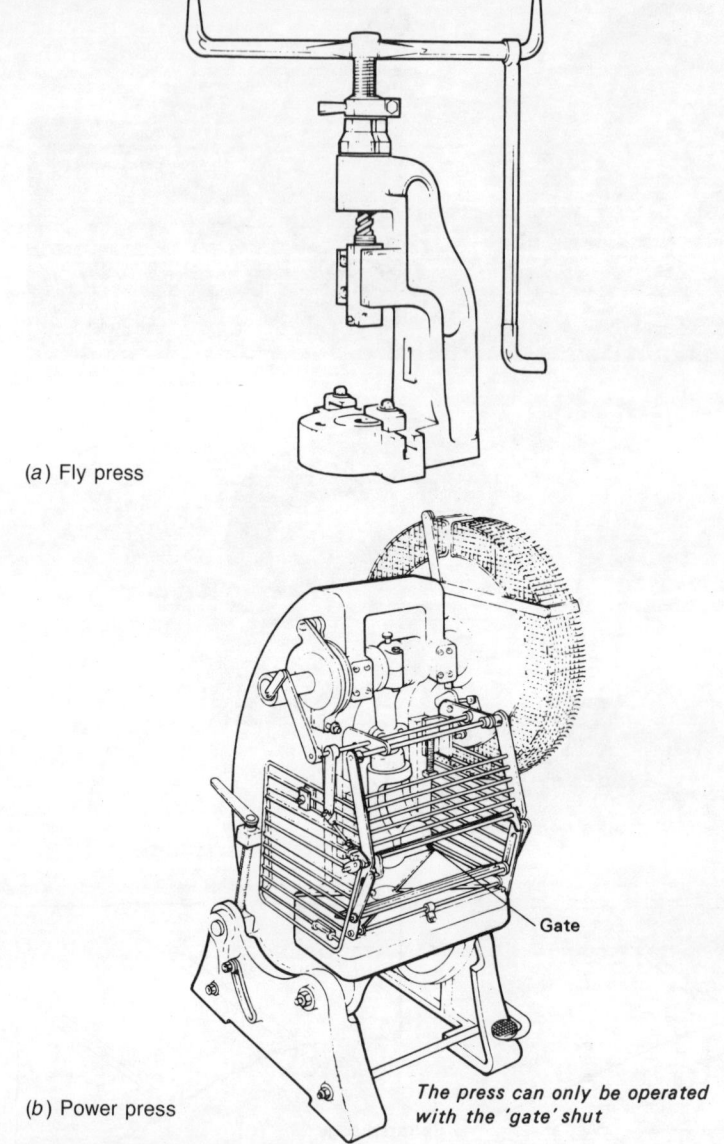

(b) Power press

The press can only be operated with the 'gate' shut

**Fig. 7.42 Types of press**

can vary widely with the cutting conditions and the material. Usually the thicker the material, the smaller the penetration required before fracture. The average degree of penetration is between 30 and 60 per cent.

The principles of punches and dies can be applied to the punching

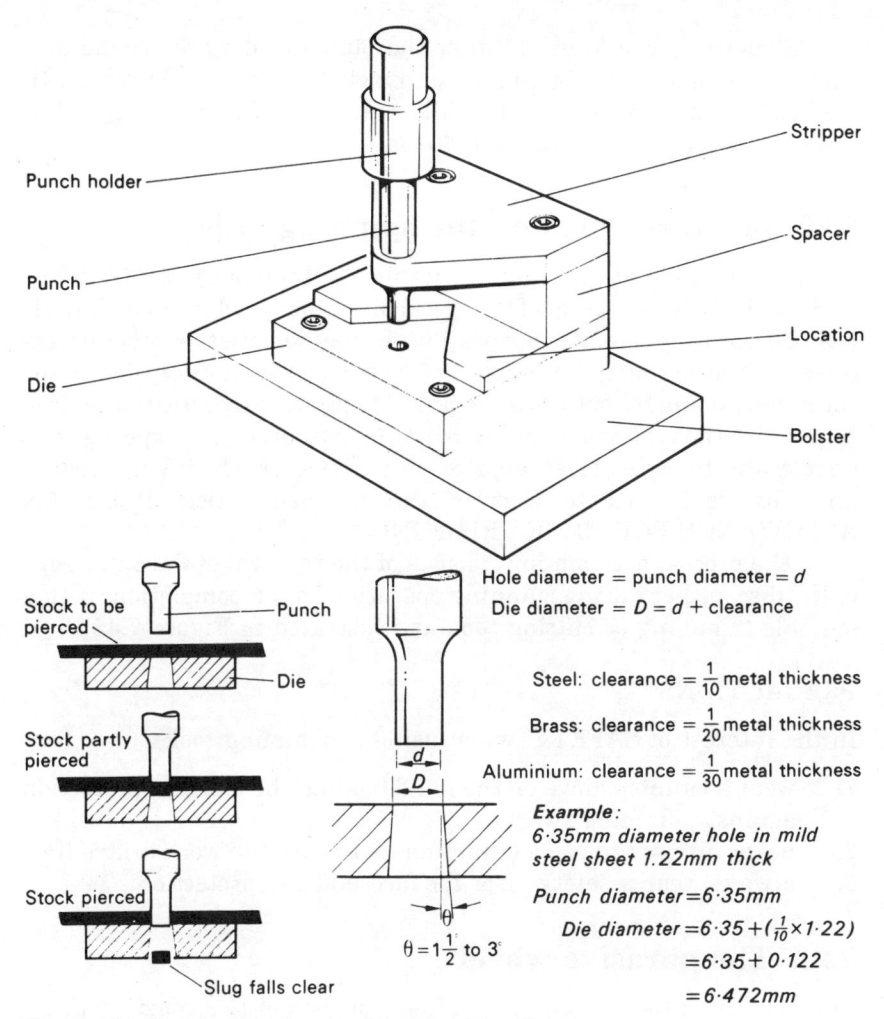

**Fig. 7.43 Simple piercing tool for use in a fly press**

Hole diameter = punch diameter = $d$
Die diameter = $D = d$ + clearance

Steel: clearance = $\frac{1}{10}$ metal thickness
Brass: clearance = $\frac{1}{20}$ metal thickness
Aluminium: clearance = $\frac{1}{30}$ metal thickness

*Example*:
6·35mm diameter hole in mild steel sheet 1.22mm thick

Punch diameter = 6·35mm
Die diameter = 6·35 + ($\frac{1}{10}$ × 1·22)
            = 6·35 + 0·122
            = 6·472mm

$\theta = 1\frac{1}{2}°$ to 3°

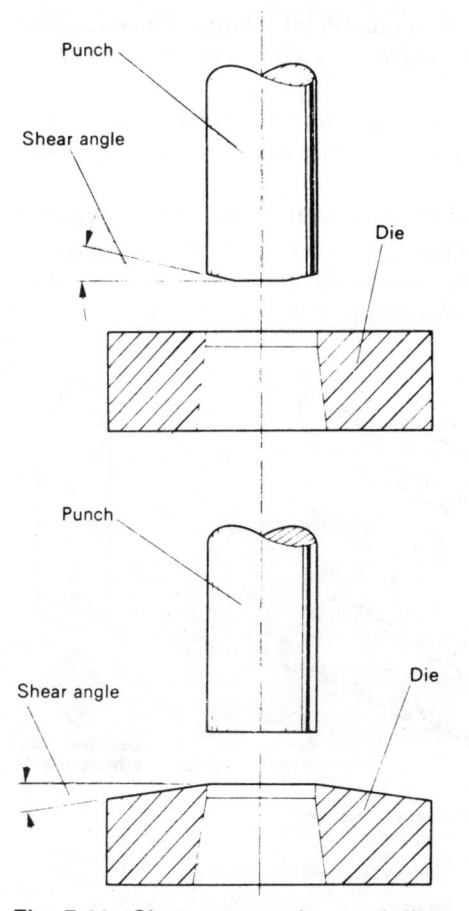

**Fig. 7.44 Shear on punches and dies**

When it is required to leave the metal sheet or strip flat, the DIE FACE IS GROUND FLAT and the PUNCH IS GIVEN THE NECESSARY 'SHEAR' as shown opposite

If the punched metal or blank is required to be left flat, the PUNCH IS GROUND FLAT and the NECESSARY 'SHEAR' IS PUT ON THE DIE as shown opposite

of any shape, and the general arrangement is basic whether a hole is required to be round, square, rectangular or some irregular figure.

The chief difference between piercing and blanking is determined by the clearance between the punch and die, as follows:

1. In the case of piercing, the size of the hole is governed by the size of the punch, and the correct cutting clearance must be provided in the die.
2. In the case of blanking, the size of the blank is governed by the die, and the correct cutting clearance must be provided on the punch.

## CLEARANCE OF PUNCHES AND DIES

'Clearance' (see Figure 7.43) *is the distance that determines the quality of the cut and the load on the machine being used.*

When piercing or blanking hard materials, the clearance between the punch and die will be greater than that required for soft materials, since a harder material has greater resistance and stiffness and shows less tendency to form burrs.

*If the clearance is too small* then the top cutting member has to penetrate much deeper into the material thickness before severance is obtained, and *the load on the machine is increased.*

*If the clearance is too great* for the particular material being cut, then the fractures from the two cutting members will not meet, and a certain amount of tearing apart of the material results. The cut edges will be very rough but, *providing the clearance is not too great, the load on the machine is reduced.*

There is only one correct clearance allowance for any particular material, and this is dependent upon its thickness and its properties, particularly its DUCTILITY.

In general clearance varies between 5 to 10 per cent of the material thickness — this is best tested for quality of cut. *The greater the clearance, the less the load required, and the thicker the material to be pierced or blanked, the greater will be the clearance.*

Sometimes the faces of either the punch or die are ground at a slight angle in order to obtain greater punching efficiency. This SHEAR ANGLE (approximately 10°) is illustrated in Figure 7.44, and is essential when blanking thick material.

## 7.20 Metal removal on the spinning lathe

The process of metal spinning inevitably results in a COMPRESSION or STRETCHING (or both) of the metal disc from which the spun article is produced. Therefore, it is not possible, by mathematical or other means (except when spinning the simplest of contours), to accurately determine the correct diameter of the disc required to produce a particular article. In practice the craftsman relies, to a great extent, on his experience to enable him to estimate an approximate diameter which will produce an equivalent area to that of the finished article PLUS AN ALLOWANCE FOR EDGE TRIMMING.

The process of trimming consists of the removal of the outer edge of the disc, either during spinning operations or on completion, with a suitable trimming or cutting tool, as illustrated in Figure 7.45.

PRECAUTIONS

In the interest of SAFETY, when using a trimming tool:

1. wear a suitable glove on the hand holding the tool as a protection against trimming swarf;
2. never attempt to use a trimming tool above the work centre line;
3. always wear a safety visor for face and eye protection.

## 7.21 The abrasive wheel

The abrasive wheel or grinding wheel consists of two constituents:

1. The *abrasive* that does the cutting.
2. The *bond* that holds the abrasive particles together.

The specification of a grinding wheel gives a clue as to its construction and suitability for a particular operation. For example a wheel carrying the marking:

38A60—J5VBE

would indicate that the wheel has an aluminium oxide abrasive; that the abrasive grit is medium to fine in grain size; that the grade is soft; that the structure shows medium spacing; that a vitrified bond is used. How the code marked on the wheel can indicate all this information

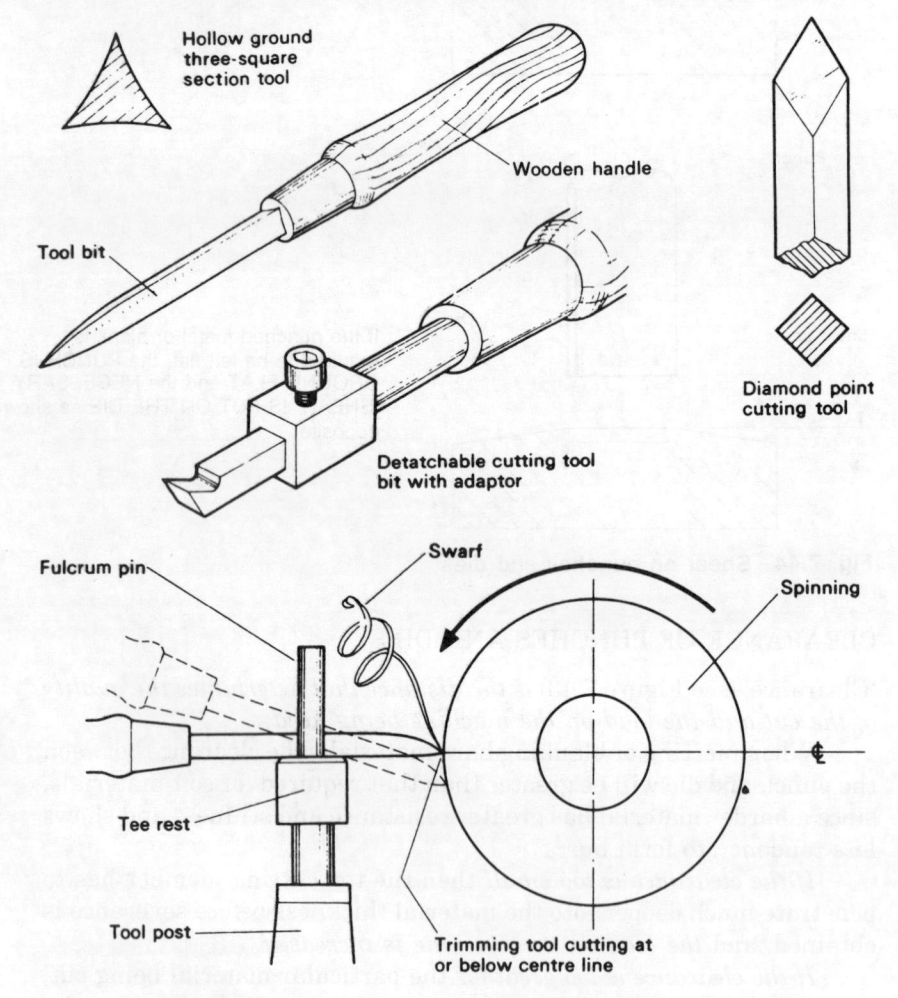

**Fig. 7.45 Trimming and cutting tools (metal spinning)**

required in selecting the correct wheel for a particular job will now be examined in some detail.

## 7.22 Abrasive

This must be chosen depending upon the material being cut.

1. 'Brown' aluminium oxide for general purpose grinding of tough materials.
2. 'White' aluminium oxide for grinding hard die steels.
3. 'Green' silicon carbide for very hard materials with low tensile strength such as cemented carbides.

| MANUFACTURER'S TYPE CODE | BS CODE | ABRASIVE | APPLICATION |
| --- | --- | --- | --- |
| — | A | Aluminium oxide | A high strength abrasive for hard, tough materials |
| 32 | A | Aluminium oxide | Cool; fast cutting, for rapid stock removal |
| 38 | A | Aluminium oxide | Light grinding of very hard steels |
| 19 | A | Aluminium oxide | A milder abrasive than 38A used for cylindrical grinding |
| 37 | C | Silicon carbide | For hard, brittle materials of high density such as cast iron |
| 39 | C | Silicon carbide (green) | For very hard, brittle materials such as tungsten carbide |

Note: In the above examples, the manufacturer's type code is based upon Norton abrasives.

## 7.23 Grit size

The number indicating the size of the grit represents the number of openings per linear 25 mm in the sieve used to size the grains. The larger the grit size number, the finer the grit.

| COARSE | MEDIUM | FINE | VERY FINE |
| --- | --- | --- | --- |
| 10 | 30 | 70 | 220 |
| 12 | 36 | 80 | 240 |
| 14 | 40 | 90 | 280 |
| 16 | 46 | 100 | 320 |
| 20 | 54 | 120 | 400 |
| 24 | 60 | 150 | 500 |
|  |  | 180 | 600 |

## 7.24 Grade

This indicates the strength of the bond and therefore the 'hardness' of the wheel. In a hard wheel the bond is strong and securely anchors the grid in place and therefore reduces the rate of wear. In a soft wheel the bond is weak and the grit is easily detached resulting in a high rate of wear.

The bond must be carefully related to the use of the wheel. If it is too hard the wheel will glaze and become blunt, if it is too soft it will wear away too quickly. This would be uneconomical and also cause loss of accuracy.

| VERY SOFT | SOFT | MEDIUM | HARD | VERY HARD |
| --- | --- | --- | --- | --- |
| EFG | HIJK | LMNO | PQRS | TUWZ |

## 7.25 Structure

This indicates the amount of bond present between the individual abrasive grains and the closeness of the individual grains to each other. An open structure wheel will cut more freely. That is, it will remove more metal in a given time and produce less heat. It will not produce such a good finish as a closer structured wheel.

| CLOSE SPACING | MEDIUM SPACING | WIDE SPACING |
| --- | --- | --- |
| 0 1 2 3 | 4 5 6 | 7 8 9 10 11 12 |

## 7.26 Bond

1. *Vitrified bond* This is the most widely used bond and is similar to glass in composition. It has a high porosity and strength giving a wheel suitable for high rates of stock removal. It is not adversely affected by water, acid, oils or ordinary temperature conditions.

2. *Rubber bond* This is used where a small degree of flexibility is required in the wheel, as in cutting off wheels and centreless grinding control wheels.

3. *Resinoid bond* This is used for high-speed wheels. Such wheels are used in foundries for dressing castings. Resinoid bond wheels are also used for cutting off. They are strong enough to withstand considerable abuse. They are cool cutting even when metal is being removed rapidly.

4. *Shellac bond* This is used for heavy duty, large diameter wheels where a fine finish is required. For example, the grinding of mill rolls.

| TYPE OF BOND | B.S. CODE |
|---|---|
| Vitrified bond | V |
| Resinoid bond | B |
| Rubber bond | R |
| Shellac bond | E |

## 7.27 Wheel selection

the correct selection of a grinding wheel depends upon many factors.

1. MATERIAL TO BE GROUND

(a) An aluminium oxide abrasive should be used on materials with a high tensile strength.
(b) A silicon carbide abrasive should be used on materials with a low tensile strength.
(c) A fine grain wheel can be used on hard brittle materials.
(d) A coarser grain wheel should be used on soft ductile materials.
(e) *Grade*. Use a hard wheel for soft materials and a soft wheel for hard materials.
(f) *Structure*. A close structured wheel can be used on hard, brittle materials. An open structured wheel should be used on soft, ductile materials.
(g) The bond is seldom influenced by the material being ground.

2. ARC OF CONTACT

(a) Figure 7.46 shows what is meant by 'arc of contact'.
(b) *Grain size*. For a small arc of contact a fine grain can be used. For a large arc of contact a coarse grain should be used.
(c) *Grade*. For a small arc of contact a 'hard' wheel can be used. For a large arc of contact a 'soft' wheel should be used.

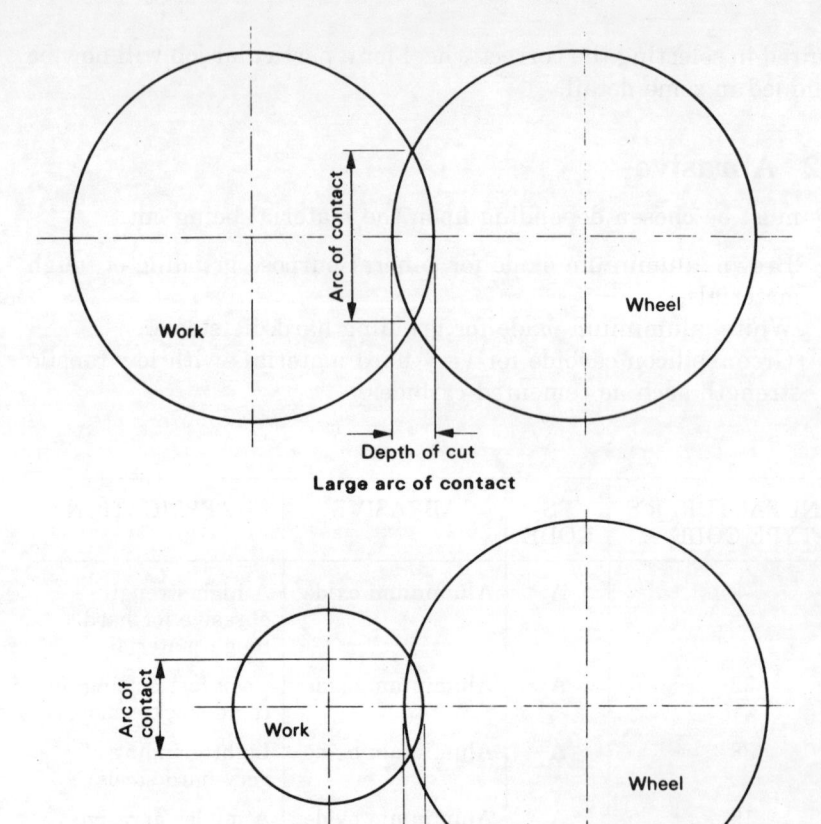

Fig. 7.46 Arc of contact

(d) *Structure*. For a small arc of contact a close grained wheel can be used. For a large arc of contact an open structured wheel should be used.

3. TYPE OF GRINDING MACHINE

A heavy, rigidly constructed machine can produce accurate work using softer grade wheels. This reduces the possibility of 'burning' the workpiece. Also broader wheels can be used increasing the rate of metal removal.

Lightly constructed machines or machines with worn wheel spindle bearings require a harder wheel.

## 4. WHEEL SPEED

(a) *Grade.* Increasing the surface speed of the wheel has the same effect as increasing the hardness of the wheel.

(b) *Bond.* The strength of the bond should be sufficient to resist the bursting effect caused by rotating the wheel at high speed. Do not exceed the safe working speed marked on the wheel.

## 5. PROCESS

As well as being used for tool sharpening and precision grinding, the abrasive wheel is also used for the following processes:

(a) Dressing the weld and grinding the weld bead flush.
(b) Edge preparation.
(c) Cutting off awkward sections.

Abrasive wheels used for these purposes are subject to far more abuse than those used for precision grinding. They are also required to remove metal more quickly, but they do not have to leave such a good finish or work so accurately as precision grinding wheels.

## 7.28 Weld grinding

1. A suitable straight wheel for heavy-duty dressing using a portable hand grinder (7.31) could be specified as:

A  16  R  4  B7

Reference to the earlier sections of this chapter will show that such a wheel would be free cutting and resistant to impact.

- A   Heavy duty aluminium oxide.
- 16  Coarse grain for rapid metal removal.
- R   Hard structure for long life.
- 4   Medium spacing for free cutting.
- B7  Tough, shock resistant, cool-cutting resinoid bond.

2. A suitable reinforced, dished wheel for fine finishing on light gauge sheet metal-work similar to automobile body building could be specified as:

A  241  S  10  BD

Reference to the earlier sections of this chapter will show that such a wheel would give a good finish and be cool cutting.

- A    General purpose aluminium oxide.
- 241  Very fine grain for fine finish and close control.
- S    Hard structure for long life.
- 10   Wide spacing for cool cutting.
- BD  Reinforced resinoid bond for strength and cool cutting.

## 7.29 Cutting off

A suitable cutting-off wheel for mild steel would be:

A  24  Q  8  B or R

Reference to the earlier sections of this chapter will show that such

Fig. 7.47 Abrasive wheel cutting-off machine

a wheel would have cool, free cutting characteristics and the ability to withstand the deflections inherent in cutting off processes.

- A   Heavy duty, general purpose aluminium oxide.
- 24  Fairly coarse grain for rapid metal removal.
- Q   Hard structure to resist excessive wear to which such a thin wheel is subjected.
- 8   Fairly wide spacing for cool cutting.
- B   Resinoid bond for strength and cool cutting. (wheels over 1 mm wide)
- R   Rubber bond to be unaffected by deflections of the wheel when less than 1 mm wide. It is also used to give less burr and a better finish in wider wheels.

### 7.30 Abrasive wheel cutting-off machines

A typical abrasive wheel cutting-off machine is illustrated in Figure 7.47.

These machines are driven by high power electric motors, the abrasive wheel revolving at speeds in excess of 3000 rev/min.

An adjustable stop is fitted to the counterbalanced head to compensate for wheel wear. A 'vee' belt drives the cutting head from the motor. The diameter of the abrasive wheel is generally about 400 mm and its thickness approximately 3 mm. The section to be cut is clamped in a screw vice, which has a swivelling back-plate for mitre cutting. The head is brought down by hand to sever the material. The high-speed abrasive wheel cuts through steel sections like a knife cutting through butter. Table 7.8 shows typical cutting speeds.

Table 7.8  Typical cutting times (abrasive cutting-off machine)

|  | MATERIAL SECTION | | CUTTING TIME (in seconds) |
|---|---|---|---|
|  | in. | mm |  |
| Tube | $1\frac{1}{2}$ O.D. | 38 O.D. | 2·5 |
| Angle | $3 \times 3 \times \frac{1}{4}$ | $76 \times 76 \times 6·35$ | 5 |
| R.S.J. | $3 \times 1\frac{1}{2} \times \frac{3}{16}$ | $76 \times 38 \times 4·76$ | 5 |
| Flat bar | $2\frac{1}{4} \times \frac{3}{8}$ | $57·2 \times 9·53$ | 2 |
| Round bar | 1 Diameter | 25·4 Diameter | 2·5 |
| Channel | $4 \times 2 \times \frac{1}{4}$ | $101·6 \times 50·8 \times 6·35$ | 9 |

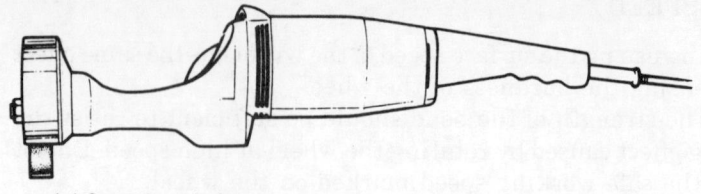

(a) Straight grinder

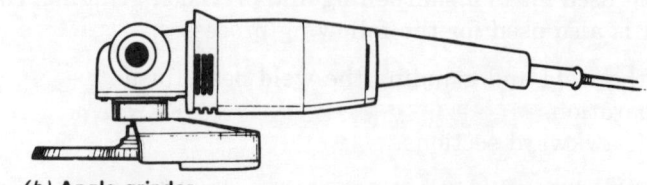

(b) Angle grinder

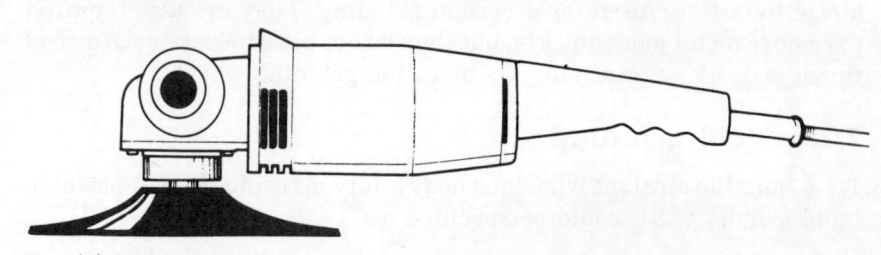

(c) Angle sander

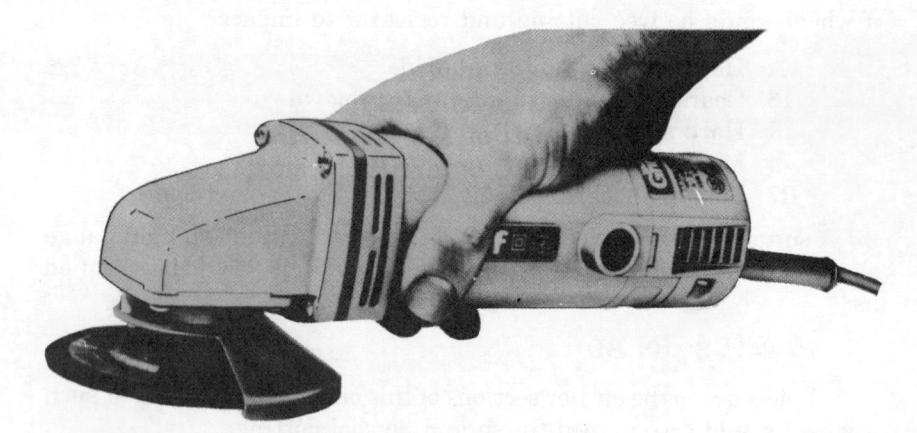

(d) Lightweight general duty high-speed 'grinderette'

Fig. 7.48  Portable electric grinding machines

## 7.31 Portable grinding machines

Portable grinding machines are often used for smoothing down welded joints and seams and generally do much of the fabrication workshop jobs which would otherwise be done by the laborious methods of chiselling and filing.

These portable tools are basically of two types, ELECTRIC and PNEUMATIC, and the three most commonly used variations of portable grinders are:

1. THE STRAIGHT GRINDER
2. THE ANGLE GRINDER
3. THE SANDER GRINDER

Figure 7.48 illustrates electrically-powered portable grinders.

Figure 7.48(a) shows a 'straight' portable grinder. This uses an ordinary grinding wheel, cutting on its periphery as in tool grinding.

Usually two sizes are available:

1. Grinding wheel diameter 102 mm
   Grinding wheel thickness 19 mm
   Spindle speed (running light) 8400 rev/min
2. Grinding wheel diameter 127 mm
   Grinding wheel thickness 19 mm
   Spindle speed (running light) 6 600 rev/min

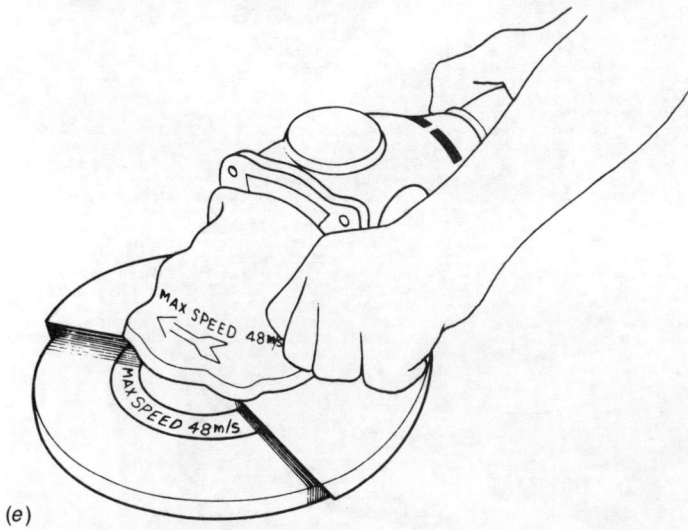

(e)

*Portable grinders and all abrasive wheels larger than 55mm diameter must be marked with the maximum speed specified by the manufacturer*

**Fig. 7.48 Portable electric grinding machines (continued)**

Both these machines have the same nett mass (less wheel and guard) of 5·6 kg.

When grinding materials of high tensile strength, such as alloy, carbon and high grade steels, an ALUMINOUS OXIDE grinding wheel is used. For grinding materials of low tensile strength, such as cast and chilled irons, brass, soft bronze and cemented carbides, a SILICON CARBIDE wheel must be used.

Do not allow the wheels to wear more than 25 mm below their **normal diameter**. To exceed this amount of wear lowers the peripheral speed of the wheel below its efficient cutting action.

Figure 7.48(b) shows an 'angle' portable grinder using a depressed centre, reinforced wheel. It will be seen that such wheels are pierced with a honeycomb of small holes to ventilate both wheels and work for cool cutting.

These machines have been designed for use with depressed centre reinforced high-speed abrasive discs. They are available in three sizes.

1. Grinding disc capacity 127 mm
   Spindle speed (running light) 11 500 rev/min
2. Grinding disc capacity 178 mm
   Spindle speed (running light) 8500 rev/min
3. Grinding disc capacity 228 mm
   Spindle speed (running light) 6500 rev/min

The grinding disc thickness may be 5, 6 or 10 mm, and the nett mass of these machines (less disc and guard) is 5·2 kg.

Only slight pressure, sufficient to keep the disc cutting, should be applied to the machine. *When working on a flat surface tilt the machine slightly — just sufficient to present a cutting angle to the edge of the disc.*

Figure 7.48(c) shows the 'angle sander'. These are designed for use with resin bonded fibre-backed SANDING DISCS on a flexible backing pad made of virtually indestructible POLYURETHANE.

The sanding disc capacity is 178 mm.
Spindle speed (running light) 4200 rev/min.

The abrasive sanding discs are available with various coarseness or fineness of abrasive grits, as follows:

0/80, $\frac{1}{2}$/60, 1/50, 2/36, 3/24 and 5/16.

As a rough guide for selecting grit sizes, 0/80 is a fine abrasive disc suitable for finishing operations on metal surfaces, and will remove scratches left by a medium abrasive disc; $\frac{1}{2}$/60 is a medium abrasive disc, generally used for removing weld reinforcements on light fabrications. For rough grinding of metal surfaces, abrasive discs 2/36, 3/24 and 5/16 are most suitable. Abrasive discs may be 'CLOSE

*(a)* Grinding welds with angle grinder

'COATED' or 'OPEN COATED'. The latter with medium to coarse grits are used for fast stripping down of old paint and cutting down solder, where the NON-CLOG characteristics of the coating are advantageous.

Figure 7.49 shows typical applications of portable electric grinding machines.

SAFETY

Hand-powered tools with exposed rotating heads must be switched off and have stopped revolving before being laid down. Otherwise, they can spin themselves off scaffolding, for instance, causing damage and injury.

Power grinders and cutting-off wheels must have guards for protection and to prevent oversize wheels being used.

ALWAYS USE THE CORRECT SIZE AND TYPE OF WHEEL FOR THE JOB — if it is too hard or too fine it becomes glazed. The

*(b)* Angle grinder used for bevel grinding

Preparation of internal surfaces of large diameter pipe ends for welding

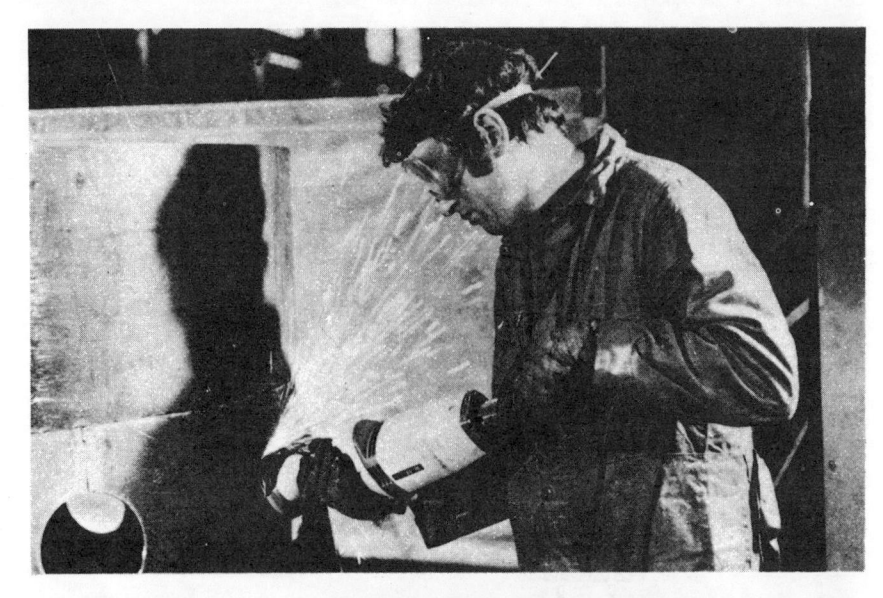

*(c) Grinding edges of flame-cut apertures with a straight grinder*

*(d) Use of portable sander for metal surface finishing*

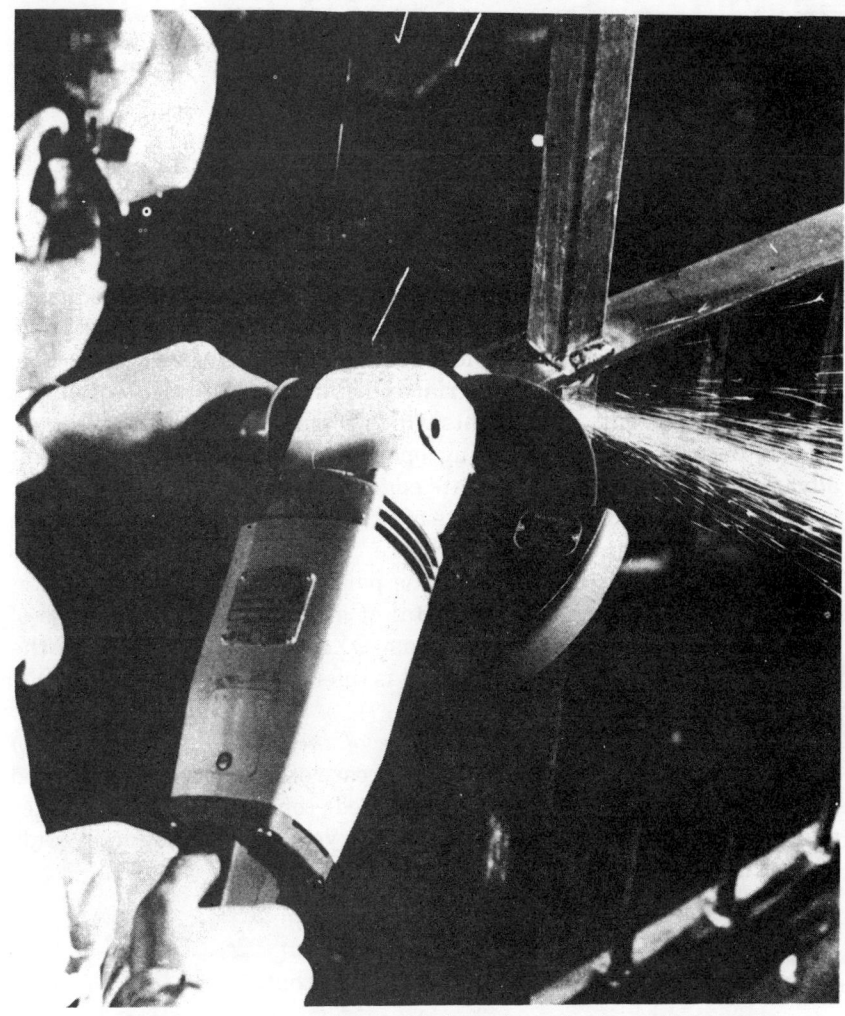

*(e) Use of angle grinder — preparing steel balustrade section for welding on site*

**Fig. 7.49 Applications of portable grinding machines**

operator must then use excessive pressure resulting in more breakages and reduced productivity.

Pneumatic grinders must have a mechanical governor to prevent the spindle exceeding its maximum speed.

## 7.32 Thermal-cutting processes

In the fabrication industry the main types of thermal-cutting processes employed are:

1. Flame cutting : (a) Oxygen  (b) Powder
2. Electric arc : (a) Oxygen  (b) Air/carbon  (c) Plasma
3. (a) Electron beam  (b) Laser

In this chapter only FLAME CUTTING by oxygen/fuel-gas processes will be considered.

## 7.33 Flame cutting

Of all the methods used for material removal in the fabrication industry, the flame-cutting process plays a prominent part in the preparation of mild-steel plate material for welded fabrications. It is readily applicable to very large thicknesses of material and allows a multiplicity of shapes or contours to be cut — two points which restrict the use of guillotines.

It is faster than machining operations, which is an important advantage in connection with plate edge preparations for welded joints.

Most metals oxidise. The rate of oxidation in air depends upon the type of material and the temperature. The properties of the oxides formed are different from that of the parent metal.

Oxygen combining with a metal at a *slow rate*, as in the case of the *rusting of iron*, is referred to as 'OXIDATION', whereas if the formation of the oxide is very rapid, it is referred to as 'COMBUSTION' or 'BURNING'.

Generally, *a rise in temperature of the metal has the effect of accelerating the rate of oxidation*. In the case of mild steel when heated at a temperature of 890°C (bright cherry red), complete combustion takes place if it is in any atmosphere of pure oxygen, and a magnetic oxide of iron is formed.

### THE PROCESS

Flame cutting is made possible by the fact that:

1. IRON BURNS IN PURE OXYGEN WHEN HEATING TO ITS 'IGNITION' TEMPERATURE (890°C).
2. HEAT IS GIVEN OUT BY THE EXOTHERMIC REACTION BETWEEN THE IRON AND OXYGEN.

   Iron + Oxygen → Iron Oxide + Heat

The process consists of creating a local hot spot on the surface of the steel with a flame to ignition temperature and directing a high pressure jet of pure oxygen on to this PRE-HEATED spot. *A vigorous and rapid chemical reaction takes place*, the steel burns and oxides are formed. *The exothermic reaction produces a great deal more* heat. This heat is sufficient to melt the oxides formed on the metal, and also to

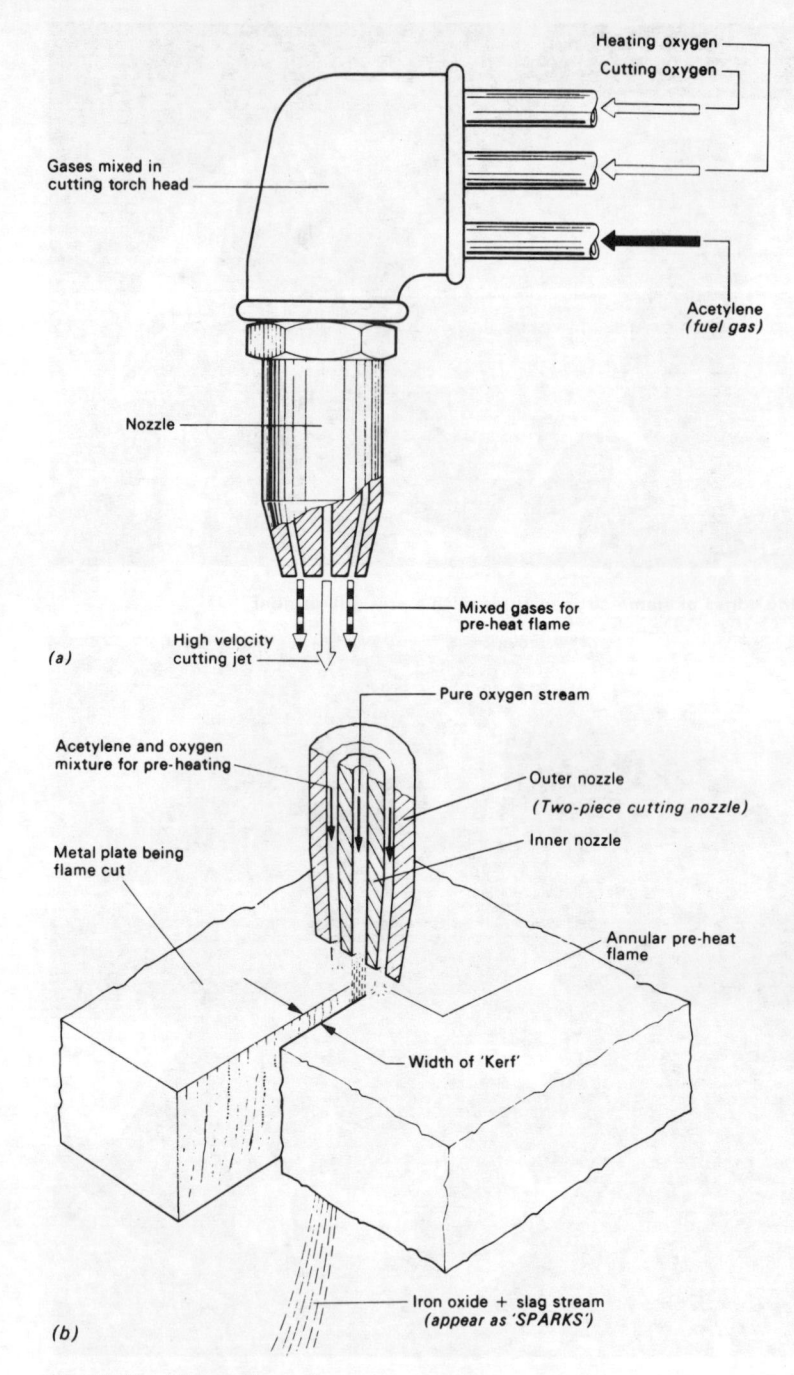

**Fig. 7.50** The action of oxygen cutting

heat and melt the metal itself over a very localised area. The melted oxides are fluid and are blown away by the force of the oxygen stream, so that more metal is exposed.

The action of 'oxygen cutting' is shown in Figure 7.50.

## 7.34 Flame-cutting equipment

Since the oxygen-cutting process involves directing a high-pressure jet of oxygen continuously on to an area of steel that has been previously heated to ignition temperature, the basic equipment combines a pre-heating flame and a pure oxygen source. Both these requirements are provided by a specially designed cutting torch and nozzle combination which is connected to a FUEL GAS and an OXYGEN supply, in the same manner as a gas-welding torch. The cutting torch and nozzle combinations are of varying design, depending on the specific type of cutting to be employed.

The equipment required for gas welding has been described in Part A of Chapter 12, and oxy-gas cutting equipment is basically similar except for the following items:

### PRESSURE REGULATORS

*Oxygen*

An oxygen pressure regulator, complete with gauges giving a higher outlet pressure (up to 6·5 bar) than for welding is required. *The standard acetylene pressure regulator and gauges, giving outlet pressures up to 0·28 bar, used for welding is also used for oxy-acetylene cutting.*

*Propane*

If propane is the fuel gas used for cutting, a special pressure regulator must be used giving outlet pressures up to 0·6 bar.

*Natural gas*

When natural gas is used for cutting torches, no pressure regulator is required (mains pressure) *but a non-return valve must be fitted in the supply line to prevent flash-back.*

Both OXY-PROPANE and OXY-NATURAL GAS cutting processes use the same type of OXYGEN pressure regulator as the OXY-ACETYLENE cutting process.

### CUTTING TORCH

The cutting torch may be either a HIGH-PRESSURE or LOW-

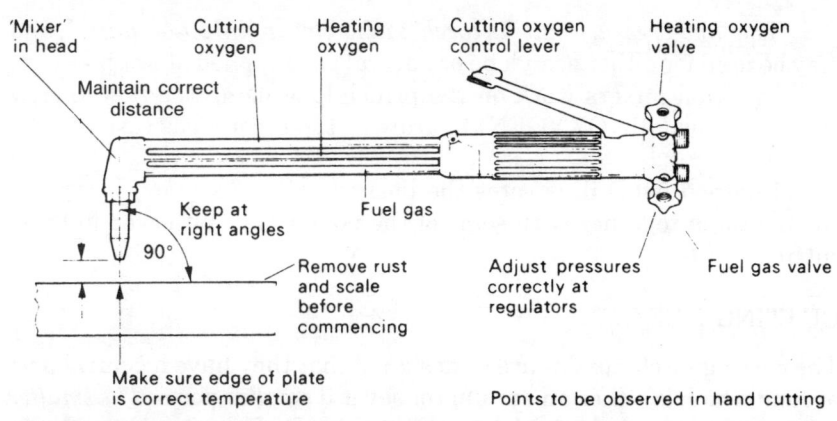

(a) **The cutting torch**
The size of the torch used depends upon whether it is for light duty or heavy continuous cutting and the volume of oxygen used is much greater than that of fuel gas (measured in LITRES PER HOUR)

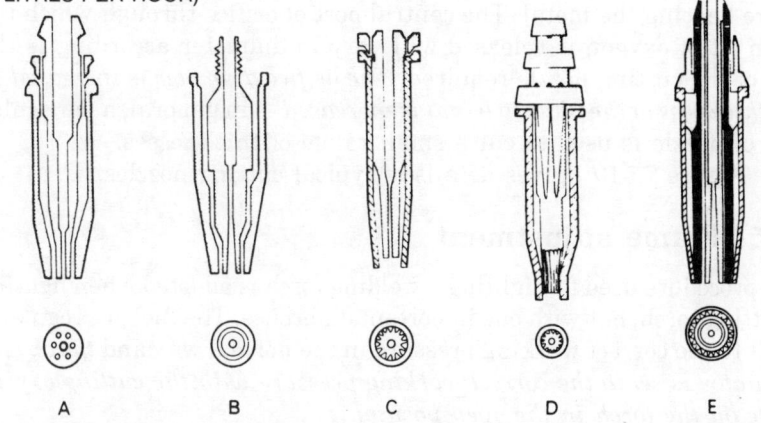

(b) **Cutting nozzle design feature**

A  One-piece ACETYLENE cutting nozzle - parallel bore, 3 - 9 pre-heat holes, no skirt.
B  Two-piece ACETYLENE cutting nozzle - venturi bore, pre-heat annulus, no skirt
C  Two-piece NATURAL GAS nozzle-venturi bore, pre-heat flutes, long skirt
D  Two-piece PROPANE nozzle - parallel bore, pre-heat slots, long skirt
E  Two-piece PROPANE nozzle - parallel bore, pre-heat flutes, long skirt, oxygen curtain

**Fig. 7.51 Oxygen cutting-torch details**

PRESSURE type. It is similar in construction to that employed for welding.

The '*high-pressure cutting torch*' uses acetylene or propane as the fuel gas, and these are supplied from cylinders. It has a 'Gas Mixer' which is usually incorporated in the torch head.

The '*low-pressure cutting torch*' embodies an '*injection mixer*', and may be used for either acetylene or natural gas supplied at low pressure.

Injection mixers work on the principle of using the flow energy of one gas (usually OXYGEN) to draw in the second gas (usually the fuel gas).

Figure 7.51(a) illustrates the basic details of any oxy-acetylene cutting torch together with some of the points to be observed in hand cutting.

### CUTTING NOZZLES

The cutting torch nozzles are so designed that they have a central port around which is either an annulus or several smaller ports. *The smaller ports are circular holes in the case of acetylene cutting nozzles, and take the form of annular slots for propane.* Through the smaller ports is fed the fuel gas mixed with the correct proportion of oxygen for the purpose of pre-heating the metal. The central port of orifice through which the main jet of oxygen is released will vary in diameter according to the size of the cutting nozzle required. *The orifice diameter is increased as the thickness of the plate to be cut is increased* — in general, a particular size of nozzle is used to cut a small range of thicknesses.

Figure 7.51(b) gives details of typical cutting nozzles.

## 7.35 Flame adjustment

The procedure used for lighting a welding torch is adopted when lighting a cutting torch, but with one important difference. The fuel gas regulator is set to the correct working pressure in the normal way and *the oxygen regulator is set to the correct working pressure with the cutting oxygen valve on the torch in the open position.*

The fuel gas is lit and the flame adjusted, in the same manner as for a welding torch, until it ceases to smoke. The heating oxygen valve is then opened and adjusted (similar to a neutral flame setting) until there is a series of nicely defined white inner cones in the flame (in the case of the multi-port nozzle) or a short white conical ring, if the nozzle is of the annular port type.

At this stage the cutting oxygen valve is opened and the flame readjusted to a neutral condition. *The oxygen cutting valve is then closed and the torch is ready for use.*

Figure 7.52 illustrates flame adjustment for cutting.

When OXY-PROPANE is used for cutting, the correctly adjusted pre-heating flame will be indicated by a small non-luminous central cone with a pale blue envelope.

In the case of OXY-NATURAL GAS the flame is adjusted until

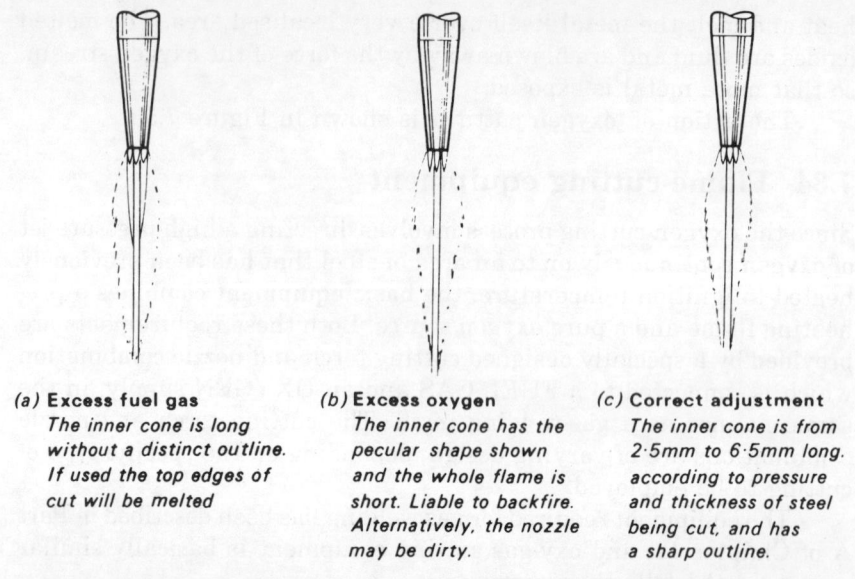

*(a)* **Excess fuel gas**
The inner cone is long without a distinct outline. If used the top edges of cut will be melted

*(b)* **Excess oxygen**
The inner cone has the peculiar shape shown and the whole flame is short. Liable to backfire. Alternatively, the nozzle may be dirty.

*(c)* **Correct adjustment**
The inner cone is from 2·5mm to 6·5mm long. according to pressure and thickness of steel being cut, and has a sharp outline.

Fig. 7.52 Flame conditions (set with oxygen cutting valve open)

the luminous inner cone has a clear definite shape, usually up to 8–10 mm in length.

The correct procedure to extinguish the flame is as follows:

1. Turn off the cutting oxygen.
2. Close the fuel gas control valve.
3. Close the heating oxygen control valve.

### GAS PRESSURES

Table 7.9 lists the approximate gas pressures for cutting mild-steel plate.

## 7.36 Factors influencing the quality of the cut

The success of the flame-cutting operation depends upon:

1. Selecting the correct size of cutting nozzle for the thickness of the material being cut.
2. Operating the cutting torch at the correct oxygen pressure.
3. Moving the cutting torch at the correct cutting speed.
4. Maintaining the nozzle at the correct distance from the plate surface.

If the torch is adjusted and manipulated correctly, a smooth narrow cut, termed the 'KERF' (see Section 7.39), is produced.

**Table 7.9 Approximate pressures for hand cutting steel plate**

| PLATE THICKNESS | | NOZZLE SIZE | | GAS PRESSURES | | | |
|---|---|---|---|---|---|---|---|
| | | | | Acetylene | | Oxygen | |
| in. | mm | in. | mm | lbf/in² | bar | lbf/in² | bar |

Acetylene

| 1/8 | 3.2 | 1/32 | 0.75 | 2 | 0.14 | 15 | 1.05 |
| 1/4 | 6.4 | 1/32 | 0.75 | 2 | 0.14 | 25 | 1.8 |
| 1/2 | 12.5 | 3/64 | 1.0 | 2 | 0.14 | 30 | 2.1 |
| 1 | 25.4 | 1/16 | 1.5 | 2 | 0.14 | 35 | 2.5 |
| 2 | 51 | 1/16 | 1.5 | 2 | 0.14 | 45 | 3.2 |
| 3 | 76 | 1/16 | 1.5 | 2 | 0.14 | 50 | 3.5 |
| 4 | 100 | 5/64 | 2.0 | 2 | 0.14 | 60 | 4.2 |
| 6 | 150 | 5/64 | 2.0 | 2 | 0.14 | 75 | 5.3 |

Propane

| 1/8 | 3.2 | 1/32 | 0.75 | 3 | 0.21 | 25 | 1.8 |
| 1/4 | 6.4 | 1/32 | 0.75 | 3 | 0.21 | 25 | 1.8 |
| 1/2 | 12.5 | 3/64 | 1.0 | 3 | 0.21 | 40 | 2.8 |
| 1 | 25.4 | 1/16 | 1.5 | 3 | 0.21 | 45 | 3.2 |
| 2 | 51 | 1/16 | 1.5 | 3 | 0.21 | 50 | 3.5 |
| 3 | 76 | 1/16 | 1.5 | 3 | 0.21 | 60 | 4.2 |
| 4 | 100 | 5/64 | 2.0 | 4 | 0.28 | 70 | 4.9 |
| 6 | 150 | 5/64 | 2.0 | 4 | 0.28 | 80 | 5.6 |

Natural gas

| 1/8 | 3.2 | 1/32 | 0.75 | Mains | — | 25 | 1.8 |
| 1/4 | 6.4 | 1/32 | 0.75 | Mains | — | 25 | 1.8 |
| 1/2 | 12.5 | 3/64 | 1.0 | Mains | — | 30 | 2.1 |
| 1 | 25.4 | 1/16 | 1.5 | Mains | — | 45 | 3.2 |
| 2 | 51 | 1/16 | 1.5 | Mains | — | 55 | 3.9 |
| 3 | 76 | 5/64 | 2.0 | Mains | — | 60 | 4.2 |
| 4 | 100 | 5/64 | 2.0 | Mains | — | 65 | 4.6 |
| 6 | 150 | 3/32 | 2.5 | Mains | — | 70 | 4.9 |

*Note*: The above figures are given only as a guide since the actual requirements may vary according to the nature of the work.
Some pressure regulators are fitted with gauges which are calibrated in kg/cm². *1 kg/cm² is approximately equal to 1 bar.*

The PRE-HEATING FLAME has two specific functions:

1. *To transmit sufficient heat to the surface of the steel to compensate for the heat loss due to* thermal conductivity.

In this respect ACETYLENE, because of its high flame temperature, has the following advantages over PROPANE and NATURAL GAS:

(a) Faster pre-heat starts.
(b) More able to penetrate priming, scale and rust.
(c) Provides for faster cutting speeds on thinner plate where the HEAT SPREAD across the plate is the controlling factor.

*The disadvantage of the high-temperature acetylene preheat flame is the tendency for excessive melting of the top edges of the cut.*

This is generally associated with the use of too large a pre-heating flame once the cut has started, and this problem may be eliminated by correct flame control and distance from the workpiece.

2. *To provide sufficient heat to raise a small local area of the surface to 'ignition temperature'.*

The pre-heating flame must be of sufficient intensity in order to break up surface scale and to maintain the steel at ignition temperature irrespective of surface irregularities.

*Only those materials whose combustion or ignition temperature is below their melting point can be flame cut.* Otherwise the material would melt away before OXIDATION could take place, making it impossible to obtain a cleanly cut edge.

NON-FERROUS METALS CANNOT NORMALLY BE FLAME CUT. Cast iron and stainless steel require special procedures and even then a 'flame-cut edge' of the same quality as with plain carbon steel is difficult to obtain.

Figure 7.53 illustrates the effects of variation in the flame-cutting procedure.

## 7.37 Applications of flame cutting (oxy-fuel gas)

HAND CUTTING

Before commencing to cut thick plate, it is necessary for the edge to be well heated. The procedure for starting the cut is illustrated in Figure 7.54(a).

When starting a cut away from an edge, a hole should either be drilled, or produced with the cutting torch. Figure 7.54(b) shows the procedure for piercing a hole with the cutting torch.

When round bar is to be flame cut, it is advisable to make a nick with a cold chisel at the point where the cut is to start. This enables the cut to be started much more easily.

Once the cut is commenced, the cutting torch should be moved steadily and at a uniform speed, with the small cone of the pre-heating

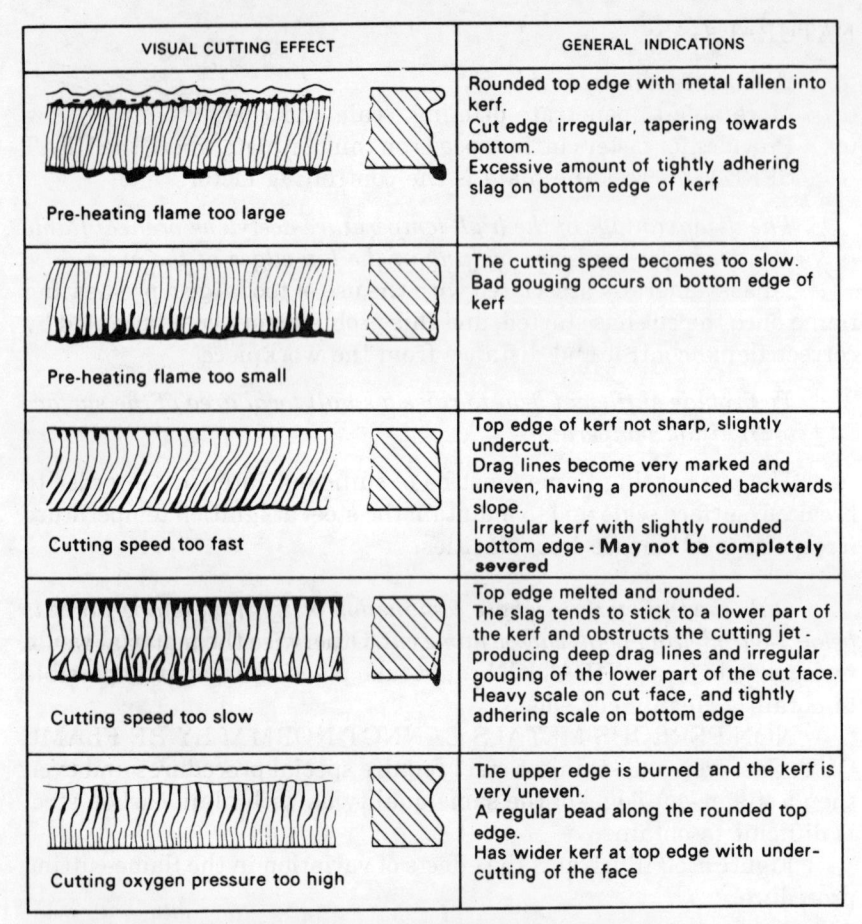

| VISUAL CUTTING EFFECT | GENERAL INDICATIONS |
|---|---|
| Pre-heating flame too large | Rounded top edge with metal fallen into kerf. Cut edge irregular, tapering towards bottom. Excessive amount of tightly adhering slag on bottom edge of kerf |
| Pre-heating flame too small | The cutting speed becomes too slow. Bad gouging occurs on bottom edge of kerf |
| Cutting speed too fast | Top edge of kerf not sharp, slightly undercut. Drag lines become very marked and uneven, having a pronounced backwards slope. Irregular kerf with slightly rounded bottom edge - **May not be completely severed** |
| Cutting speed too slow | Top edge melted and rounded. The slag tends to stick to a lower part of the kerf and obstructs the cutting jet - producing deep drag lines and irregular gouging of the lower part of the cut face. Heavy scale on cut face, and tightly adhering scale on bottom edge |
| Cutting oxygen pressure too high | The upper edge is burned and the kerf is very uneven. A regular bead along the rounded top edge. Has wider kerf at top edge with undercutting of the face |

**Fig. 7.53 Effects of variations in flame-cutting procedure**

flame just clear of the work surface. *There must be no vibration of the cutting head as such movement will result in a ragged cut and, in some cases, the cut being halted.*

Various attachments are available which, when fitted to the hand-cutting torch, ensure a steady rate of travel and enable the operator to execute straight lines, bevels or circles with much greater ease. These very useful devices (Figure 7.55) are generally attached to the cutting nozzle. The following are some of the devices in common use:

### SINGLE CUTTING SUPPORT

This simple device may either be a 'spade support' or a single 'roller guide' which can be adjusted vertically for 'stand off'. Figure 7.55(a) shows a single cutting support used in conjunction with a 'straight edge'.

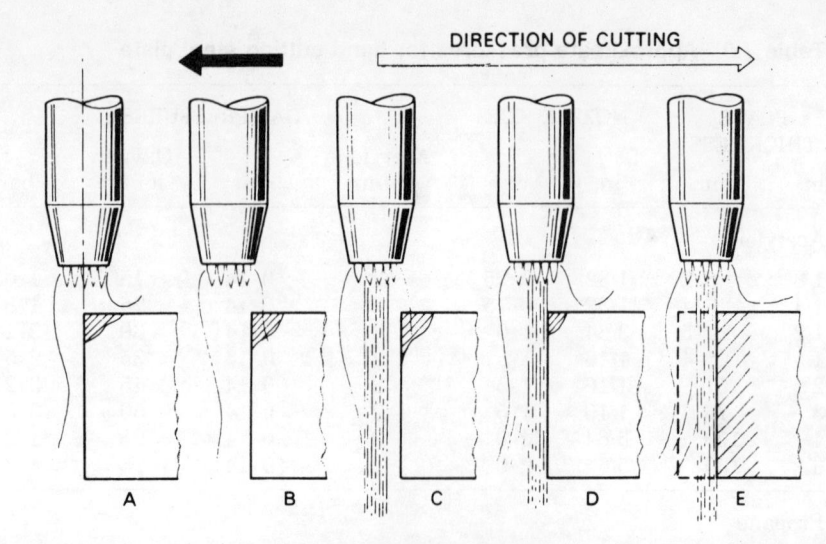

DIRECTION OF CUTTING

A  Pre-heat to ignition temperature
B  Move the cutting torch backwards, just clear of edge
C  Open the cutting oxygen valve
D  Commence cutting
E  Continue cutting

*(a)* **Starting a cut on the edge of heavy plate**

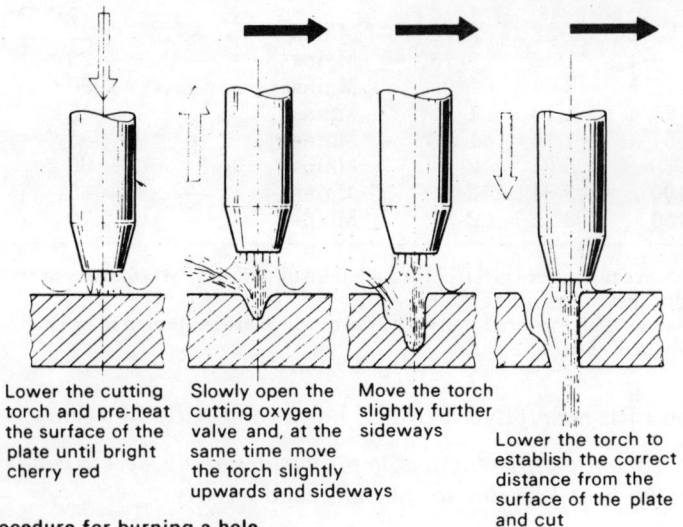

Lower the cutting torch and pre-heat the surface of the plate until bright cherry red

Slowly open the cutting oxygen valve and, at the same time move the torch slightly upwards and sideways

Move the torch slightly further sideways

Lower the torch to establish the correct distance from the surface of the plate and cut

*(b)* **Procedure for burning a hole**

**Fig. 7.54 Cutting techniques**

*(a) Cutting with straight-edge and single support*

*(b) Small circle cutting*

*(c) Large circle cutting*

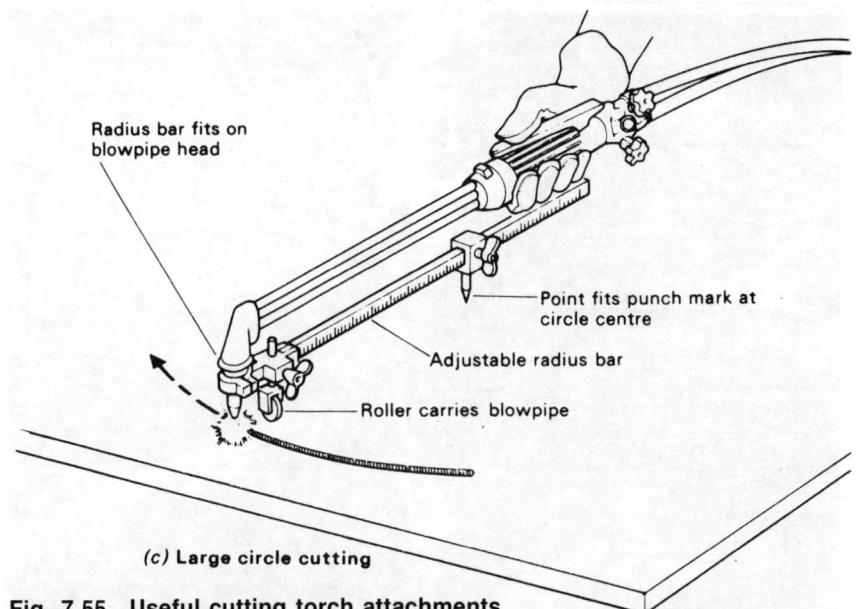

**Fig. 7.55 Useful cutting torch attachments**

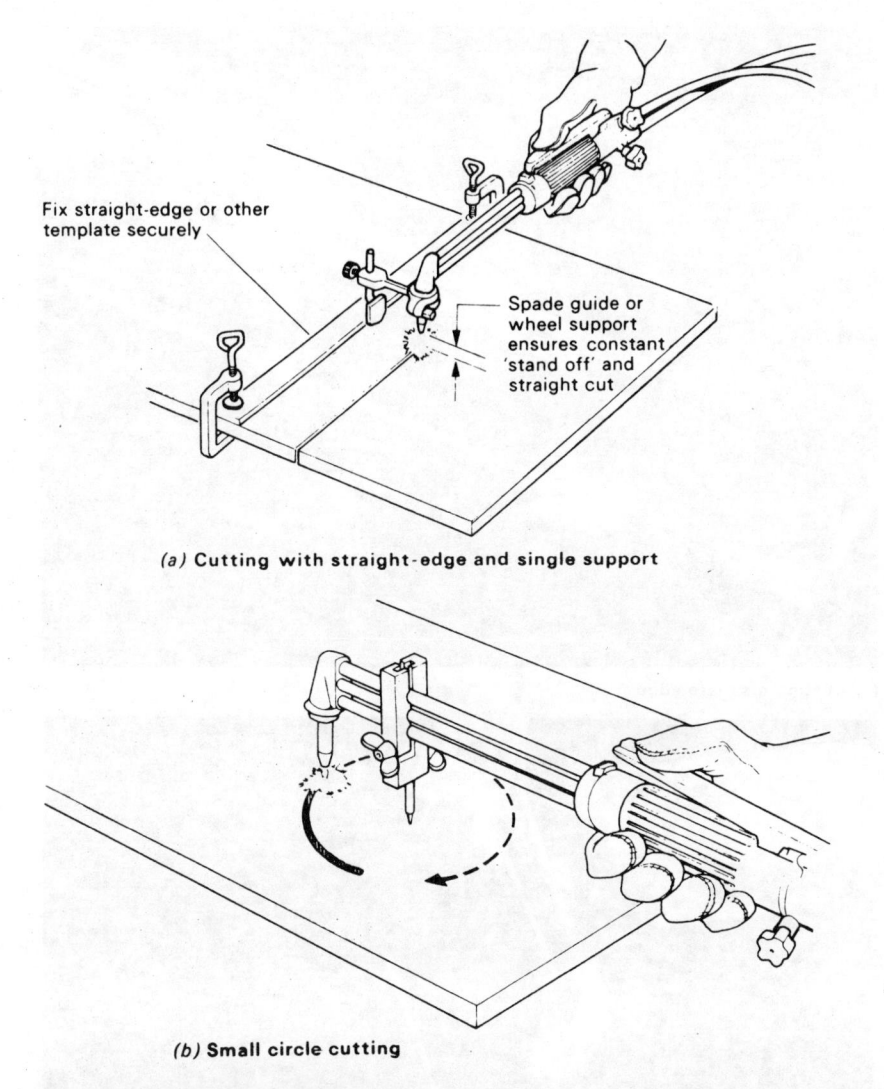

## CIRCLE-CUTTING DEVICE

Basically there are three types of devices which may be used for radial cuts and circle cutting.

1. *Small circle-cutting attachment* This is simply a pivot which is attached to the shank of the torch at a particular distance from the nozzle according to the radius required. Such a device is illustrated in Figure 7.55(*b*).

2. *Large circle-cutting attachment* This extremely useful device is illustrated in Figure 7.55(*c*).

3. *Special cutting device for small holes* Some cutting torches are supplied in kit form and include a specially designed interchangeable shank in which the cutting head is 'in line' instead of being at the normal 90°. This special head is used in conjunction with a pivot for cutting small holes, as illustrated in Figure 7.56.

## DOUBLE CUTTING SUPPORT

This device consists of a U-frame and two rollers which makes it easier to move the cutting torch along the workpiece at the correct distance from the plate. Like the single support, it has provision for vertical adjustment. Certain types of double cutting supports are designed for square and bevel cutting. These are commonly referred to as 'bevel-cutting attachments'. Figure 7.57 illustrates the versatility of a bevel-cutting attachment.

## RIVET REMOVAL

When carrying out repair work on riveted structures it is often necessary to remove rivets. They may be quickly removed with the aid of a cutting torch by following the procedure illustrated in Figure 7.58.

## FLAME GOUGING

This process is very similar to flame cutting except that instead of

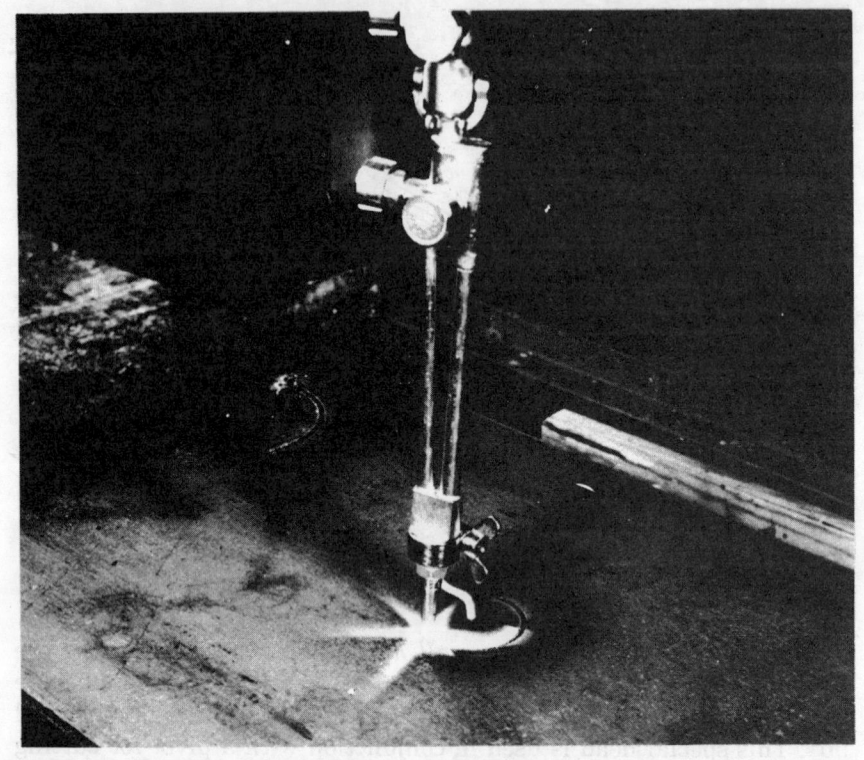

Fig. 7.56 Special cutting device for small holes

severing the metal, a groove is gouged out of the surface of the plate. The principle of operation is the same as that used in oxy-fuel gas-cutting processes, except that a special type of nozzle is used in a standard cutting torch. A pre-heating flame is used to bring the metal to ignition temperature, the cutting oxygen is switched on, OXIDATION occurs and cutting is commenced. It is a very useful process for the removal of defective welds and defects in the parent metal. Various details of the flame gouging process are illustrated in Figure 7.59.

### 7.38 Cutting machines (oxy-fuel gas)

Cutting machines basically consist of one or more cutting torches and a means for supporting and propelling them in the required directions with high precision.

There are many machines available, ranging from simple lightweight portable types to static types which are large and versatile with several cutting heads.

A typical portable cutting machine is illustrated in Figure 7.60(a).

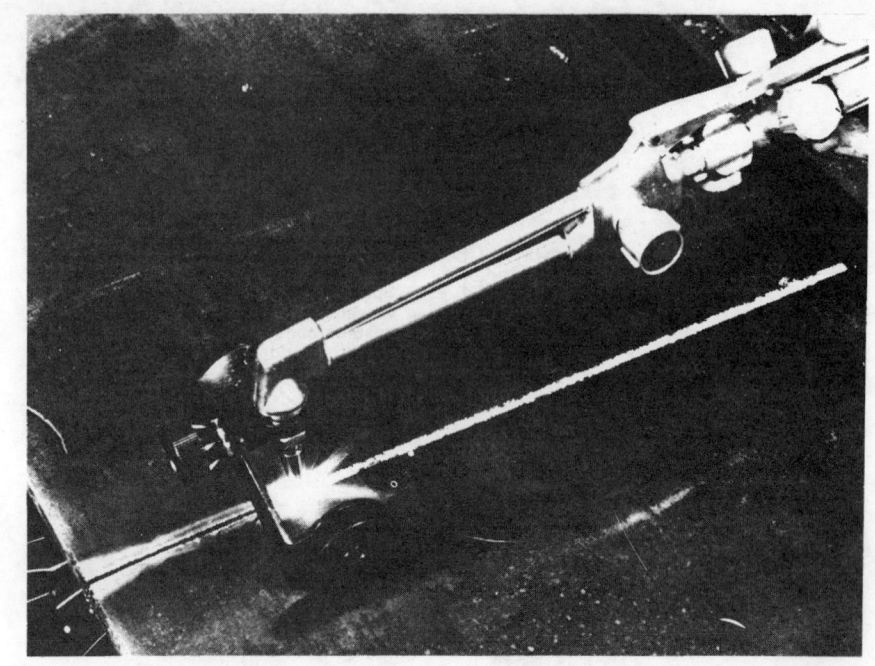

(a) Cutting a square edge

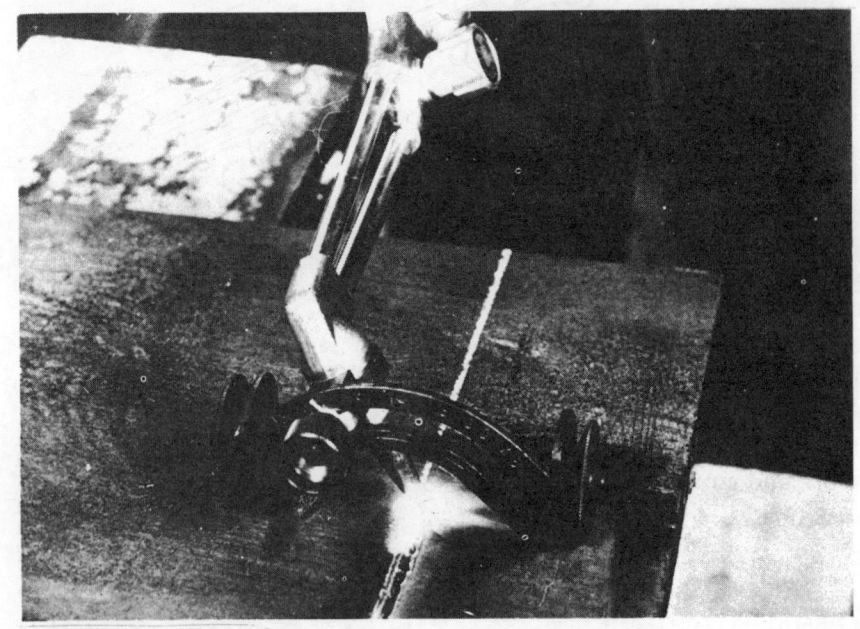

(b) Cutting a bevel edge

Fig. 7.57 Use of a double cutting support

**Rivet to be removed**
**Cutting torch**
**Steel plates**

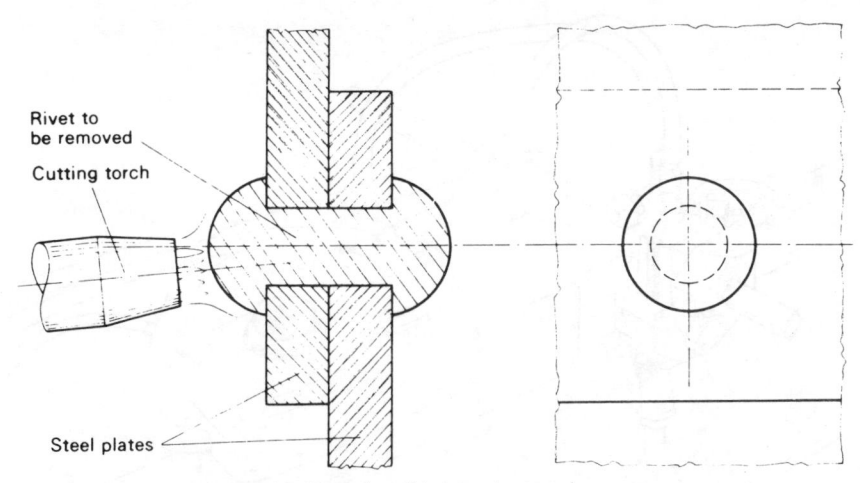

1. The centre of the rivet head is heated until bright cherry red, and the edge of the hole becomes clearly visible (A)
2. The cutting oxygen jet is turned on, and with the torch slightly inclined towards the centre, a cut is carefully made around the edge of the hole, without damaging it (B)
3. On reaching the halfway point the cut is made towards the centre (C)
4. With the cutting torch slightly inclined inwards complete the cut around the edge of the hole (D)
5. The rivet is easily removed by knocking out with a sharp hammer blow on a solid steel punch, as indicated in (E)

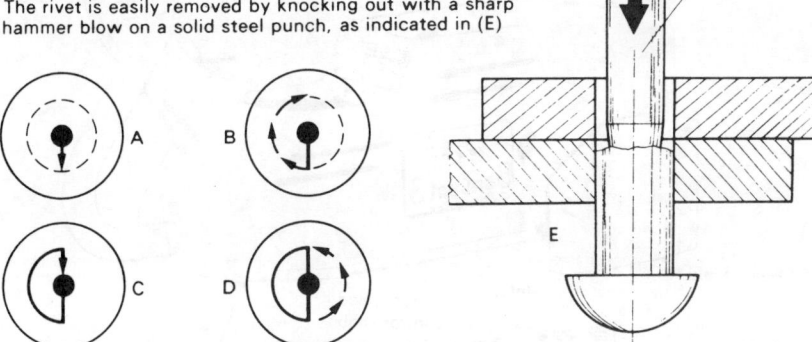

Solid steel punch

**Fig. 7.58 Rivet removal by flame cutting**

It is basically a 'SELF-PROPELLED LIGHTWEIGHT TRACTOR' which can be adapted to carry a range of cutting or welding equipment. The machine details are also shown in Figure 7.60.

The machine will make runs of any length and can be controlled for straight-line work by the use of, either an extruded aluminium track, or by guiding the tractor along a straight line. One or two cutting heads may be used. Contours may be followed by hand steering the tractor along a drawn line and radial or circle-cutting by using a radius bar attachment. *When using the machine for contour or radial cutting, the castor wheel must be free to swivel and all three wheels should be in contact with the work.*

This machine is very useful for plate edge preparation for welding, as bevel cutting may be carried out by setting the swivel cutting head to the required angle. Two cutting heads may be fitted, one slightly in advance of the other, to enable 'nose and bevel cutting' to be done, and it is important that:

Preheating edge for start of gouging 25°-35°

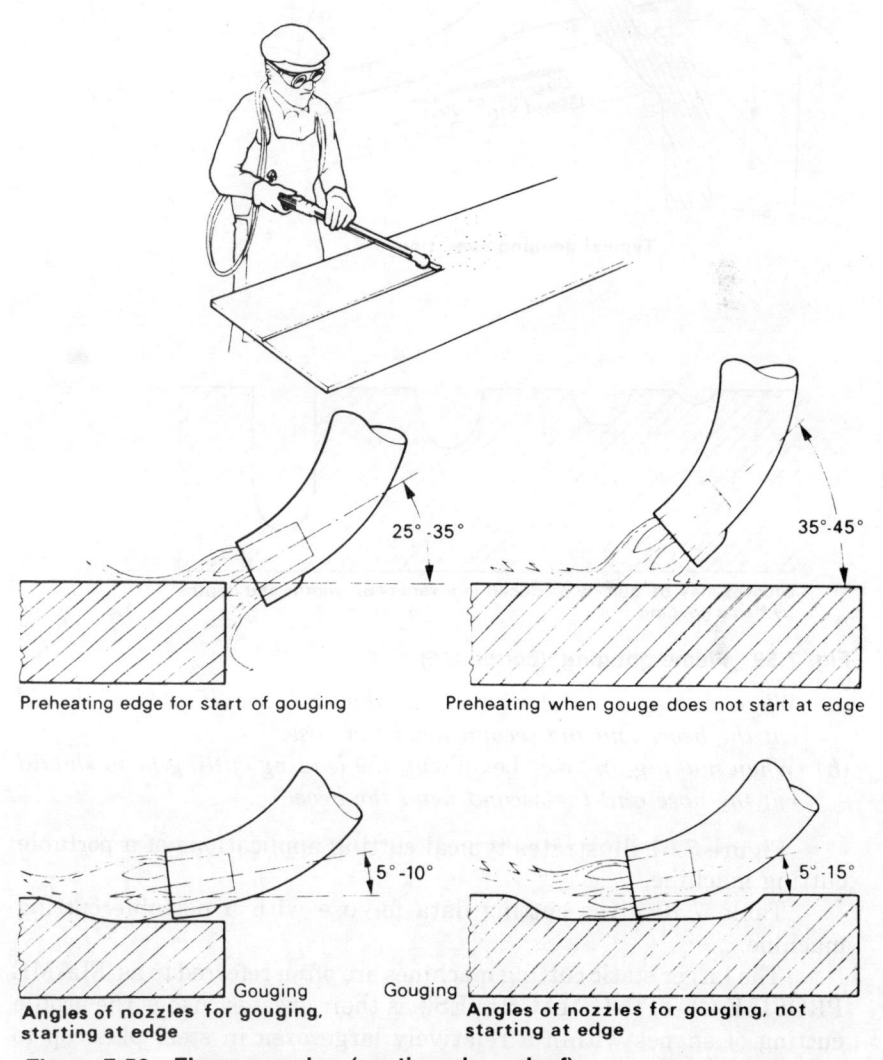

Preheating when gouge does not start at edge 35°-45°

Angles of nozzles for gouging, starting at edge 5°-10°

Angles of nozzles for gouging, not starting at edge 5°-15°

**Figure 7.59 Flame gouging (continued overleaf)**

271

**Fig. 7.59 Flame gouging (continued)**

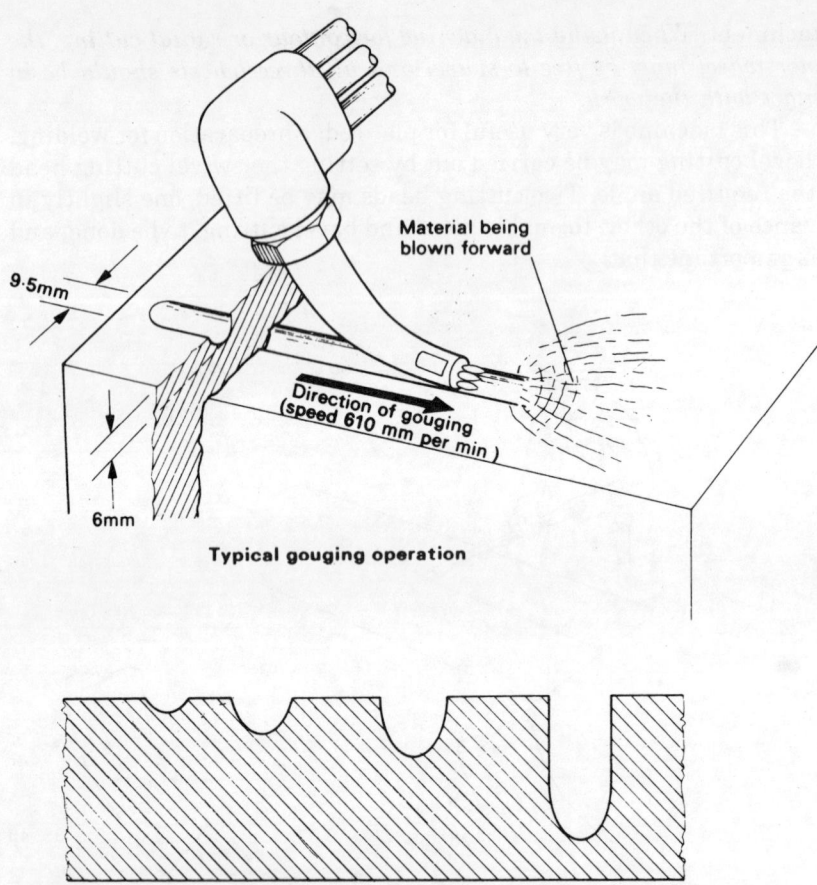

(a) *When making an 'under bevel' cut, the leading cutting head should cut the bevel and the second head the nose.*
(b) *When making an 'over bevel' cut, the leading cutting head should cut the nose and the second head the bevel.*

Figure 7.61 illustrates typical cutting applications of a portable cutting machine.

Table 7.10 gives cutting data for use with a portable cutting machine.

The larger static cutting machines are often referred to as 'FLAME PROFILE CUTTING MACHINES', as their greatest use is the profile cutting of shapes within a relatively large area in steel plate up to considerable thickness.

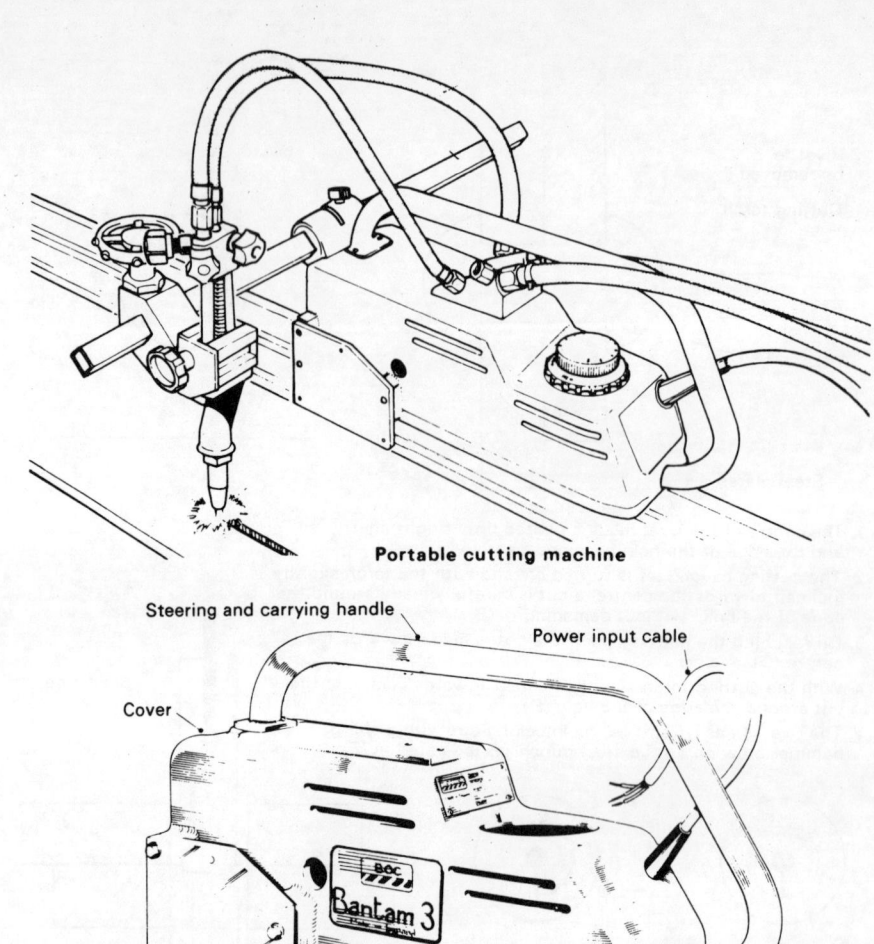

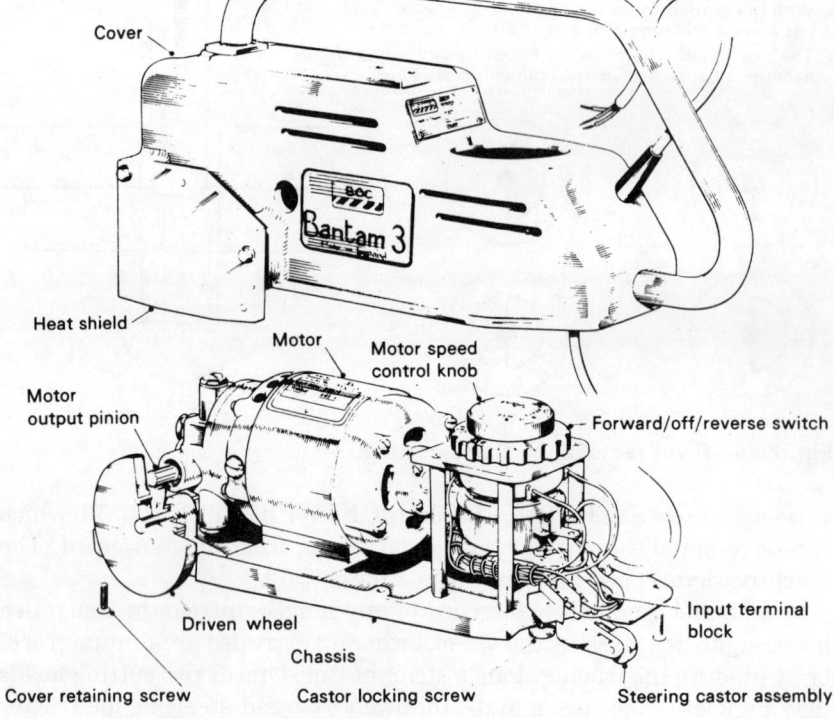

**Fig. 7.60 Portable type oxygen-cutting machine**

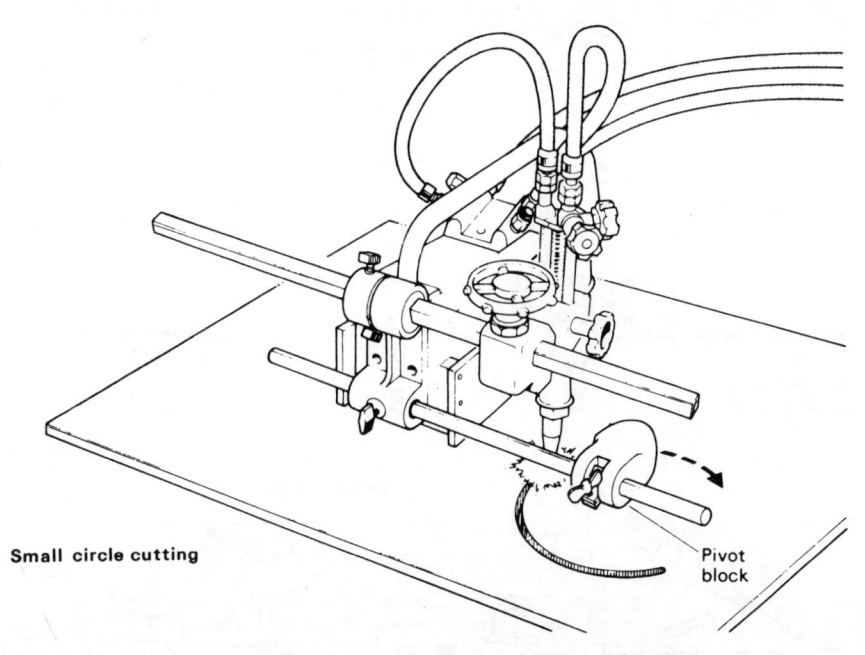

**Fig. 7.61** Typical cutting applications (portable machines)

The machine illustrated in Figure 7.62 is a cross-carriage profile cutting machine capable of flame-cutting shapes from steel plate up to a maximum thickness of 150 mm and, if desired, two cutting heads may be mounted and operated simultaneously. This machine can profile cut shapes with a plate area of 1·2 m × 1·2 m and the cutter heads are so mounted as to permit cutting up to 45° on either side of the vertical.

Three interchangeable tracing heads may be utilised with the machine illustrated in Figure 7.62, depending upon the type of TEMPLATE used for profile cutting, as follows:

1. *A wheel tracing head* Used for tracing direct from an outline drawing. This method of tracing is most economical for 'one-off' production. It consists of a fine-toothed tracking wheel, steered by hand along the line to be cut. It is driven through a vertical spindle assembly and bevel gear linkage contained within the main body. The body screwed connection remains static to the drive unit while the main body, complete with tracing wheel, is free to turn through 360°. A pointer to assist the operator in guiding the tracing wheel along its path is also incorporated.

Provision is made on the head for attaching a circle-cutting radius bar, capable of cutting circles and flanges up to 350 mm diameter.

2. *A spindle tracing head* Used with either wooden, hardboard or non-ferrous templates. This is a revolving vertical spindle attached to the drive unit. It has a detachable knurled roller that rotates round either the inner or outer profile of a template. Generally there are two diameters of roller available, either 3 mm or 6 mm, and the larger roller should be used where possible except when the template corners require the smaller one. The spindle rotates in a stationary outer sleeve whose lower end is knurled to provide a non-slip hand grip to enable the roller to be held against the edge of the template in order to follow the profile.

3. *A magnetic head* Used for steel templates only and *has the*

**Table 7.10  Operating data for portable machines**

| **English** | | Acetylene | | | | | | Propane | | | | |
|---|---|---|---|---|---|---|---|---|---|---|---|---|
| Plate thickness (in) | | 1/4 | 1/2 | 1 | 2 | 4 | 6 | 1/4 | 1/2 | 1 | 2 | 4 | 6 |
| Nozzle size (in) | | 1/32 | 3/64 | 1/16 | 1/16 | 5/64 | 5/64 | 1/32 | 3/64 | 1/16 | 1/16 | 5/64 | 3/32 |
| Speed in ins/min | | 21 | 14 | 11 | 8 | 7 | 6 | 18 | 13 | 10 | 7 | 6 | 5 |
| Regulator | Fuel Gas | 2 | 2 | 2 | 2 | 2 | 2 | 2 | 2 | 4 | 4 | 4 | 4 |
| Pressure | Heating Oxygen | 5 | 5 | 6 | 6 | 8 | 8 | 9 | 9 | 10 | 10 | 12 | 12 |
| lbf/in² | Cutting Oxygen | 20 | 30 | 35 | 40 | 55 | 65 | 20 | 30 | 35 | 40 | 55 | 65 |

Speed Control Settings: 1  2  3  4  5  6  7  8  9  10

Roller Diameter in inches — Cutting Speed (in/min). These columns correspond with the above settings in relation to the appropriate size roller

| | | 1 | 2 | 3 | 4 | 5 | 6 | 7 | 8 | 9 | 10 |
|---|---|---|---|---|---|---|---|---|---|---|---|
| Spindle | 1/8in | 3 | 6 | 9 | 13 | 16 | 19 | 22 | 25 | 28 | 31 |
| Roller | 1/4in | 6 | 13 | 19 | 25 | 31 | 38 | 44 | 50 | 57 | 63 |
| Magnet | 3/8in | 2 | 5 | 7 | 9 | 12 | 14 | 16 | 19 | 21 | 24 |
| Roller | 1/2in | 3 | 6 | 10 | 13 | 16 | 19 | 22 | 25 | 29 | 32 |
| Wheel tracing | | 4 | 9 | 13 | 18 | 22 | 26 | 31 | 36 | 40 | 44 |

| **Metric** | | Acetylene Nozzle | | | | | | Propane Nozzle | | | | |
|---|---|---|---|---|---|---|---|---|---|---|---|---|
| Plate Thickness (mm) | | 6 | 12 | 25 | 50 | 100 | 150 | 6 | 12 | 25 | 50 | 100 | 150 |
| Nozzle | | 1/32 | 3/64 | 1/16 | 1/16 | 5/64 | 5/64 | 1/32 | 3/64 | 1/16 | 1/16 | 5/64 | 3/32 |
| Cutting Speed (cm/min) | | 53 | 36 | 28 | 20 | 18 | 15 | 46 | 33 | 25 | 18 | 15 | 13 |
| Regulator | Fuel Gas | 0·14 | 0·14 | 0·14 | 0·14 | 0·14 | 0·14 | 0·14 | 0·14 | 0·28 | 0·28 | 0·28 | 0·28 |
| Pressures | Heating O₂ | 0·35 | 0·35 | 0·42 | 0·42 | 0·56 | 0·56 | 0·63 | 0·63 | 0·7 | 0·8 | 0·8 | 0·8 |
| Bar (b) | Cutting O₂ | 1·4 | 2·1 | 2·5 | 2·8 | 3·9 | 4·6 | 1·4 | 2·1 | 2·5 | 2·8 | 3·9 | 4·6 |

Speed Control Setting: 1  2  3  4  5  6  7  8  9  10

Roller Dia (mm) — Cutting Speed (m/min) (Corresponding with the related Speed Control Setting)

| | | 1 | 2 | 3 | 4 | 5 | 6 | 7 | 8 | 9 | 10 |
|---|---|---|---|---|---|---|---|---|---|---|---|
| Spindle | 3·175 | 0·076 | 0·15 | 0·23 | 0·33 | 0·41 | 0·48 | 0·56 | 0·41 | 0·71 | 0·79 |
| Roller | 6·35 | 0·15 | 0·33 | 0·48 | 0·64 | 0·78 | 0·97 | 1·12 | 1·27 | 1·42 | 1·60 |
| Magnet | 9·525 | 0·05 | 0·13 | 0·18 | 0·23 | 0·30 | 0·36 | 0·41 | 0·48 | 0·53 | 0·61 |
| Roller | 12·7 | 0·076 | 0·15 | 0·23 | 0·33 | 0·41 | 0·48 | 0·56 | 0·64 | 0·74 | 0·81 |
| Wheel Tracing | | 0·10 | 0·23 | 0·33 | 0·46 | 0·56 | 0·66 | 0·78 | 0·91 | 1·02 | 1·12 |

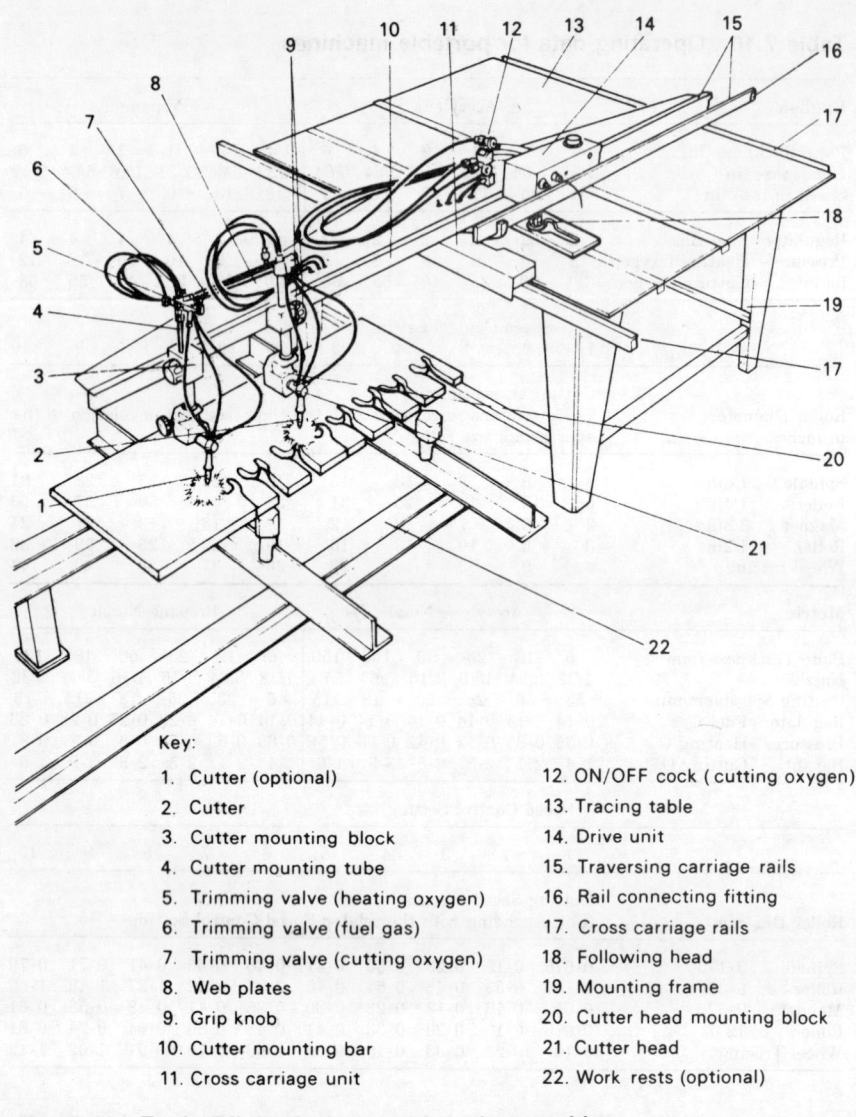

Key:
1. Cutter (optional)
2. Cutter
3. Cutter mounting block
4. Cutter mounting tube
5. Trimming valve (heating oxygen)
6. Trimming valve (fuel gas)
7. Trimming valve (cutting oxygen)
8. Web plates
9. Grip knob
10. Cutter mounting bar
11. Cross carriage unit
12. ON/OFF cock (cutting oxygen)
13. Tracing table
14. Drive unit
15. Traversing carriage rails
16. Rail connecting fitting
17. Cross carriage rails
18. Following head
19. Mounting frame
20. Cutter head mounting block
21. Cutter head
22. Work rests (optional)

**Fig. 7.62 Typical light-duty oxygen cutting machine**

*advantage over the other tracing heads in that it automatically follows the template profile.* The vertical driving spindle is magnetised by a solenoid encased in the outer shell of the main body of the driving head. The head is generally supplied with two interchangeable knurled steel rollers, usually 9 mm and 12 mm in diameter.

Both the 'spindle tracing head' and the 'magnetic head' find their greatest use in repetition work, and these are illustrated in Figure 7.63.

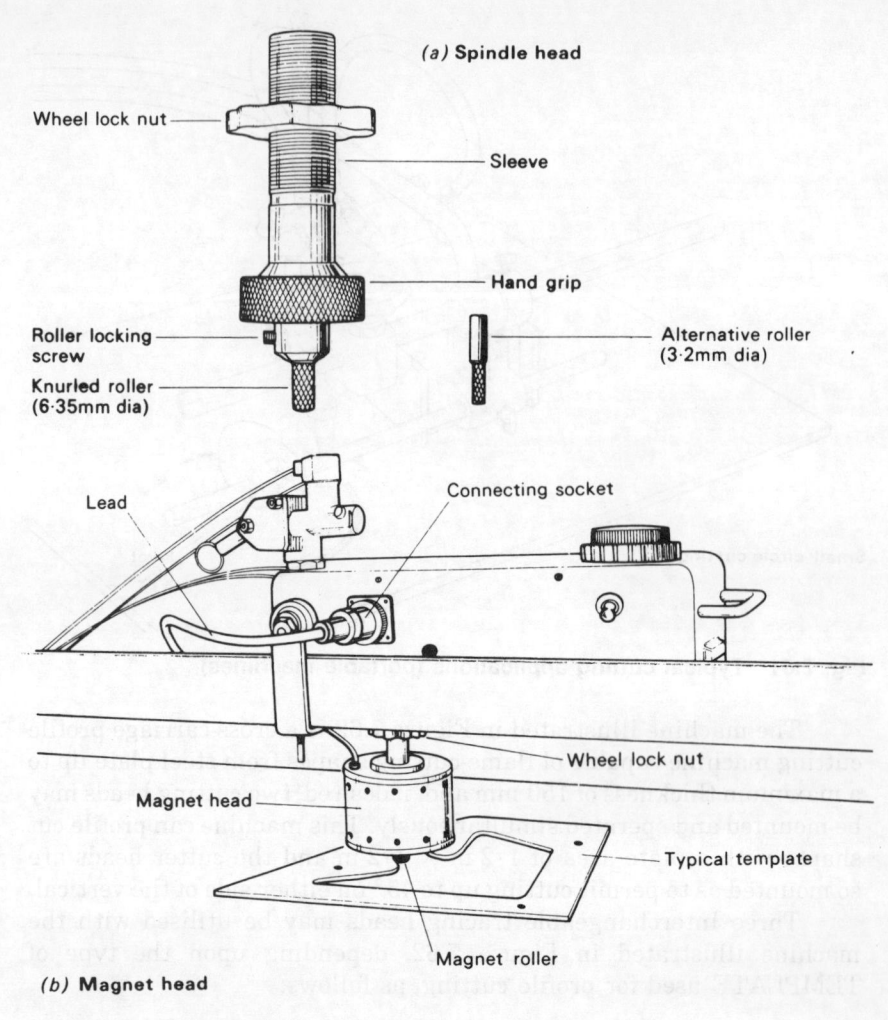

**Fig. 7.63 Profile cutting machine tracing heads**

Static flame profiling machines require three pressure regulators:

1. FUEL GAS PRESSURE REGULATOR (acetylene or propane)
2. HEATING OXYGEN PRESSURE REGULATOR
3. CUTTING OXYGEN PRESSURE REGULATOR

Table 7.11 gives useful cutting data when using 'mixer type' nozzles for flame cutting with the machine illustrated in Figure 7.62.

## 7.39 Template and allowance for profile cutting

Suitable templates may be manufactured from steel or light alloy sheet,

## Table 7.11 Cutting data for profile cutting machines

### Fuel Gas — Acetylene

Vertical cutting

| PLATE THICKNESS | | NOZZLE | | GAS PRESSURES | | | | NOZZLE HEIGHT ABOVE PLATE | | CUTTING SPEED APPROX. | |
|---|---|---|---|---|---|---|---|---|---|---|---|
| | | | | ACETYLENE | | OXYGEN | | | | | |
| in | mm | mm | in | lbf/in² | Bar (6) | lbf/in² | Bar (b) | in | mm | in/min | m/min |
| 1/8 | 3·2 | 0·75 | 1/32 | 2 | 0·14 | 20 | 1·41 | 1/4 | 6 | 28 | 0·71 |
| 1/4 | 6·4 | 0·75 | 1/32 | 2 | 0·14 | 30 | 2·11 | 1/4 | 6 | 21 | 0·53 |
| 1/2 | 12·5 | 1·0 | 3/64 | 2 | 0·14 | 30 | 2·11 | 1/4 | 6 | 16 | 0·41 |
| 1 | 25·4 | 1·5 | 1/16 | 2 | 0·14 | 40 | 2·80 | 5/16 | 8 | 15 | 0·38 |
| 2 | 51 | 1·5 | 1/16 | 2 | 0·14 | 55 | 3·80 | 5/16 | 8 | 11 | 0·28 |
| 4 | 100 | 2·0 | 5/64 | 3 | 0·21 | 70 | 4·9 | 3/8 | 9·5 | 6·7 | 0·15–0·18 |
| 30° BEVEL CUTTING | | | | | | | | | | | |
| 1/8 | 3·2 | 0·75 | 1/32 | 2 | 0·14 | 25 | 1·75 | 1/4 | 6 | 20 | 0·51 |
| 1/4 | 6·4 | 0·75 | 1/32 | 2 | 0·14 | 30 | 2·11 | 1/4 | 6 | 18 | 0·46 |
| 1/2 | 12·5 | 1·0 | 3/64 | 2 | 0·14 | 50 | 3·50 | 1/4 | 6 | 16 | 0·41 |
| 1 | 25·4 | 1·5 | 1/16 | 2 | 0·14 | 50 | 3·50 | 5/16 | 8 | 13 | 0·33 |
| 2 | 51 | 1·5 | 1/16 | 2 | 0·14 | 70 | 4·92 | 5/16 | 8 | 9 | 0·23 |

### Fuel Gas — Propane

Vertical cutting

| PLATE THICKNESS | | NOZZLE | | GAS PRESSURES | | | | NOZZLE HEIGHT ABOVE PLATE | | CUTTING SPEED APPROX. | |
|---|---|---|---|---|---|---|---|---|---|---|---|
| | | | | PROPANE | | OXYGEN | | | | | |
| in | mm | mm | in | lbf/in² | Bar (6) | lbf/in² | Bar (b) | in | mm | in/min | m/min |
| 1/8 | 3·2 | 9·75 | 1/32 | 3 | 0·21 | 25 | 1·76 | 1/4 | 6 | 20 | 0·51 |
| 1/4 | 6·4 | 0·75 | 1/32 | 3 | 0·21 | 25 | 1·76 | 1/4 | 6 | 19 | 0·48 |
| 1/2 | 12·5 | 1·0 | 3/64 | 3 | 0·21 | 40 | 2·81 | 1/4 | 6 | 16 | 0·41 |
| 1 | 25·4 | 1·5 | 1/16 | 3 | 0·21 | 45 | 3·17 | 5/16 | 8 | 14 | 0·36 |
| 2 | 51 | 1·5 | 1/16 | 3 | 0·21 | 50 | 3·52 | 5/16 | 8 | 10 | 0·25 |
| 4 | 100 | 2·0 | 5/64 | 4 | 0·28 | 75 | 5·3 | 3/8 | 9·5 | 6·7 | 0·15–0·18 |
| 30° BEVEL CUTTING | | | | | | | | | | | |
| 1/8 | 3·2 | 1·0 | 3/64 | 3 | 0·21 | 25 | 1·76 | 1/4 | 6 | 16 | 0·41 |
| 1/4 | 6·4 | 1·0 | 3/64 | 3 | 0·21 | 30 | 2·11 | 1/4 | 6 | 11 | 0·28 |
| 1/2 | 12·5 | 1·5 | 1/16 | 3 | 0·21 | 50 | 3·52 | 1/4 | 6 | 11 | 0·28 |
| 1 | 25·4 | 1·5 | 1/16 | 3 | 0·21 | 55 | 3·87 | 5/16 | 8 | 10 | 0·25 |
| 2 | 51 | 1·5 | 1/16 | 3 | 0·21 | 70 | 4·92 | 5/16 | 8 | 7 | 0·18 |

*Note*: The above figures are given only as a guide since the actual requirements may vary according to the nature of the work.

plywood or hardboard, depending on the type of spindle head to be used.

In general, an EXTERNAL TEMPLATE is used when the piece to be cut from the plate is the component and an INTERNAL TEMPLATE when the piece cut from the plate is not required for the component.

Composite templates may be used where the component to be cut has both an external and an internal profile, and also includes a hole or holes.

To obtain the best results for accurate cutting, templates should conform to the following basic requirements:

1. Minimum thickness 3 mm.
2. Edges must be square. Plywood or hardwood templates should have their edges prepared with a coarse sandpaper finish, and metal template edges should have a good file finish but not too smooth, *in order to provide sufficient frictional grip for the knurled steel rollers.*
3. When the inside corner of the component to be cut is radiused, *the corner radius on the template must be greater than that of the roller.*
4. Correct allowances must be made in respect of the width of the KERF and the diameter of the tracing roller.

## ALLOWANCES FOR FLAME CUTTING

These will vary according to the width of the kerf, the diameter of the tracing roller and whether an internal or external template is used.

*The kerf This is a term used to define the width of the metal consumed in the cutting process. It may vary between $1\frac{1}{2}$ and 2 times the diameter of the cutting oxygen orifice of the cutting nozzle used; for example, a 1·6 mm diameter nozzle will produce a kerf of between 2·4 mm and 3·2 mm in width.*

Allowances (to compensate for the kerf and the diameter of tracing roller) must be made on the size of the template and these will differ PLUS or MINUS depending whether an internal or external template is used.

Figure 7.64 illustrates template allowance; the component profile is shown in heavy outline with its applicable template shown hatched.

For future use it is advisable to mark templates with the following information:

1. Nozzle type and size.
2. Fuel gas used.
3. Tracing roller diameter.
4. Thickness of plate cut.
5. Speed of cut.
6. Part number (if applicable) of component.

When using a WHEEL TRACING HEAD, allowances must be made on the drawing dimensions for the kerf width. As a general guide, allow the diameter of the cutting nozzle orifice per side, plus or minus for external and internal cuts respectively.

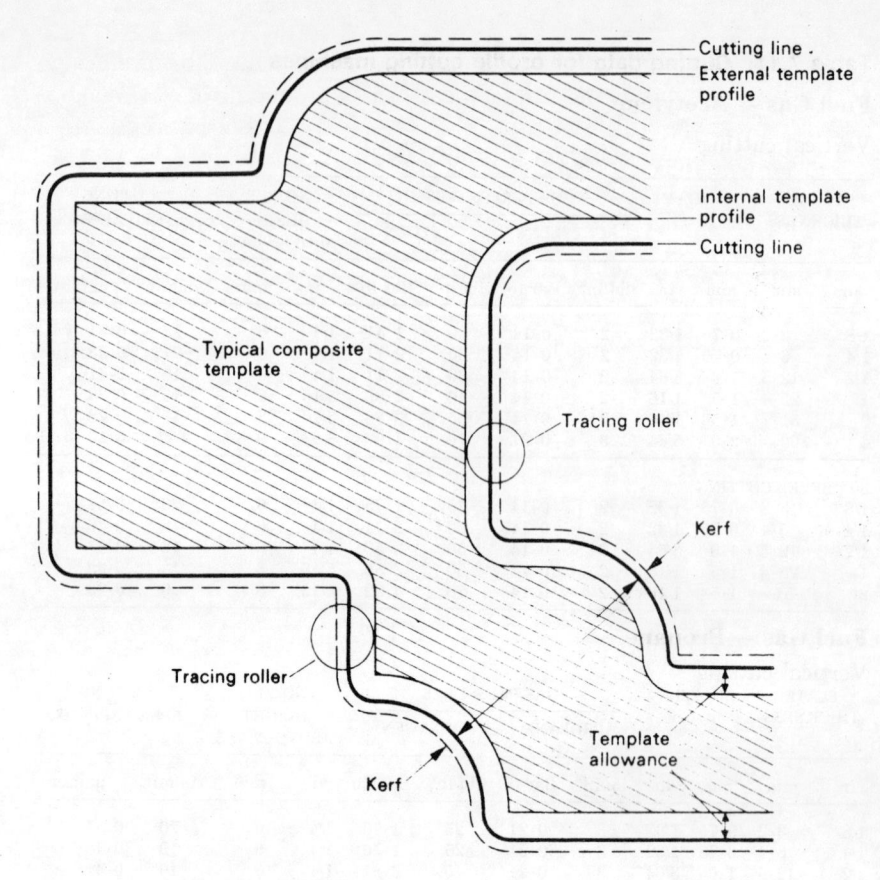

When preparing EXTERNAL TEMPLATES, DEDUCT from the OUTSIDE DIMENSIONS of the required component:

$$\frac{\text{DIAMETER OF ROLLER}}{2} + \frac{\text{WIDTH OF KERF}}{2}$$

The KERF can be effected by the distance of the tip of the nozzle from the upper surface of the material. *The tendency is for wider kerfs to result when this distance is increased.*

For INTERNAL TEMPLATES, ADD to the required INSIDE DIMENSIONS of the required component:

$$\frac{\text{DIAMETER OF ROLLER}}{2} + \frac{\text{WIDTH OF KERF}}{2}$$

Widths of kerf vary considerably with:
1. Operating pressure
2. Speed of cutting
3. Size of oxygen-cutting orifice employed

**Fig. 7.64 Template allowances**

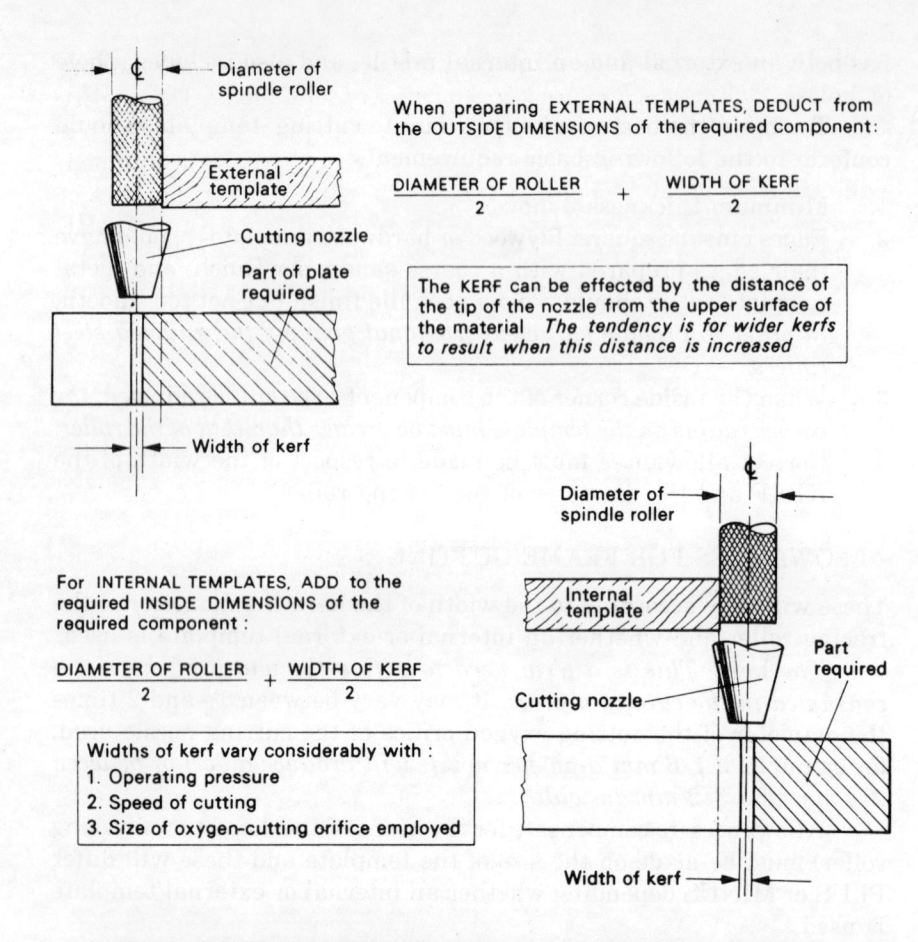

When CIRCLE-CUTTING with a RADIUS BAR, allowances are the same as those for wheel tracing.

## 7.40 Comparison of flame cutting with shearing

Since the introduction of the first cutting torch at the beginning of the twentieth century, oxygen cutting has brought about far-reaching changes in the industrial practices of cutting and shaping of steel.

*The flame-cutting process is more versatile than any other known cutting or shearing process.* A multiplicity of components in a wide range of sizes and thicknesses can easily be shaped by oxygen cutting which could otherwise only be performed with great difficulty by other material removal processes. *Because of its versatility, one oxygen-cutting machine can replace several mechanical types of cutting machine.*

The main disadvantage of shearing machines is that they are of very limited capacity with regard to the thickness of the material being cut. *Guillotines can be used for straight-line cutting only, and the length of cut is governed by the capacity of the machines.*

However, shearing machines have the advantage over oxygen cutting in respect of material up to 4 mm thick because it is difficult to produce a clean flame cut edge below this thickness. The pre-heat flame tends to melt the top edges of the cut causing them to fuse together.

Oxygen cutting is faster than sawing and it *cuts greater thicknesses than the shear or the guillotine*. For example, 25 mm thick plate can be cut at the rate of 15·6 m/hour, 100 mm thick at 7·8 m/hour. The cutting of steel 600 mm thick is frequently undertaken with a high degree of accuracy.

Bevel-edge cutting is no problem with the oxygen-cutting process, whereas, with the exception of certain specially designed portable shears, shearing machines only produce square cut edges.

Portable and static nibbling machines may be used in conjunction with a profile template but are only capable of producing one component at a time. Flame profile cutting machines are capable of producing a number of components from a template simultaneously.

Some large profile machines are capable of performing multi-torch operations in which as many identical shapes can be cut simultaneously as there are torches in use. Ten or more cutting torches may be used in normal operations.

In the ship-building industry and in some large plate fabrication shops, large automatic machines are used for cutting several large plates simultaneously. This is usually done from reduced 1:10 or 1:100 scale TEMPLATE DRAWINGS.

The thinnest steel which can be cut satisfactorily by oxygen cutting is 3·2 mm, but thinner sheets can be successfully cut by using the 'STACK-CUTTING METHOD'. *In this way as many as 100 sheets may be cut simultaneously.* The advantage and economy of this method for the production of regular or irregular shaped thin plates in repetition is obvious. *The stack-cutting method involves piling a number of thin plates one on top of the other, clamping them tight together, and making the cut as if the clamped plates were one piece of solid metal.* The quality and accuracy of the cut will depend upon the stack thickness. With a stack not exceeding 50 mm a cut edge tolerance of 0·8 mm is obtained, while stacks from 50 to 100 mm can be cut with an edge tolerance of 1·6 mm. THE EDGES OF THE STACK-CUT SHEETS ARE SQUARE AND FULL WITH NO BURRS SUCH AS ARE OFTEN PRESENT ON SHEETS PRE-FABRICATED BY SHEARING METHODS. *This advantage is of particular importance on sheet metal discs that are subsequently to be flanged or dished.*

The versatility of oxygen cutting cannot be over emphasised, and by adopting machine-cutting processes, small workshops in particular, can extend their activities to a much greater variety of work on material considerably heavier and more intricate, thus increasing their capacity with regard to material removal.

The author is indebted to The British Oxygen Company for much of the information offered on oxygen cutting in this chapter.

# 8 Restraint and location

## 8.1 Location and restraint (introduction)

A body in space, free of all restraints, is able to move as shown in Figure 8.1. It may be moved back and forth; from side to side; upwards and downwards. It may rotate clockwise or anti-clockwise about any of three axes (plural of axis). That is, the body has SIX DEGREES OF FREEDOM.

In order that the body may be worked upon it must be LOCATED in a given position by RESTRAINING its freedom of movement. Figure 8.2 shows how these restraints may be built up.

## 8.2 The fitter's vice

The fitter's vice, or bench vice as it is also called, is widely used for holding work whilst it is being operated on by hand (Figure 8.3). Figure 8.4 shows how the jaws of the vice provide the restraint necessary to prevent the work from moving.

For heavy work the vice jaws are serrated (rough) so that they will bite into the component being held. This would damage a finished surface and for fine work either smooth, ground jaws or fibre shoes, as shown in Figure 8.5 are used.

To hold more delicate or awkwardly shaped work, special vice jaws as shown in Figure 8.6 are used.

## 8.3 Use of the bench vice

The work held in the vice must be positioned so that it is given adequate support both against the cutting force (Figure 8.7) and the clamping force of the vice itself (Figure 8.8).

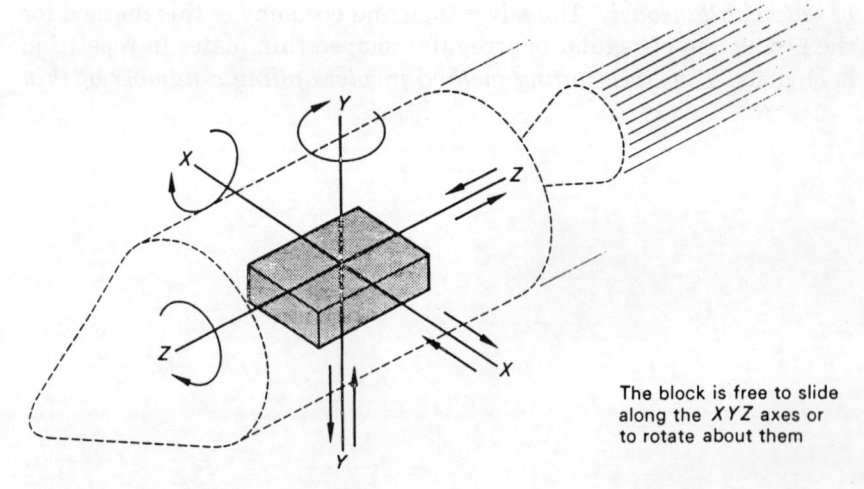

The block is free to slide along the $XYZ$ axes or to rotate about them

**Fig. 8.1 Six degrees of freedom**

## 8.4 The Vee block

Figure 8.9 shows a circular disc of metal being located by various methods. It will be seen that the *Vee location* is the most successful in the range of work it will accept.

Since the circular disc can be considered as a very short cylinder, it can be seen that for components of cylindrical shape, the Vee block is the best method of location. Figure 8.10 shows the restraint given to a cylinder located in a Vee block.

The Vee block is also used in conjunction with parallel jaw vices

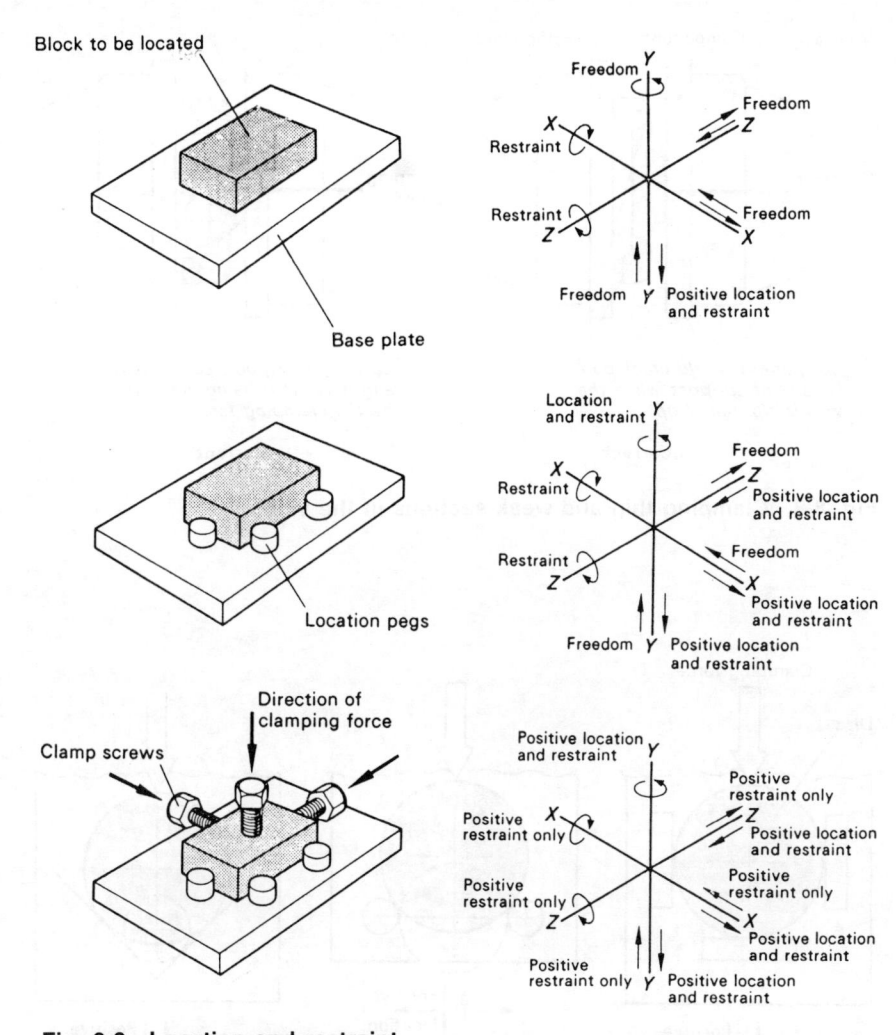

**Fig. 8.2** Location and restraint

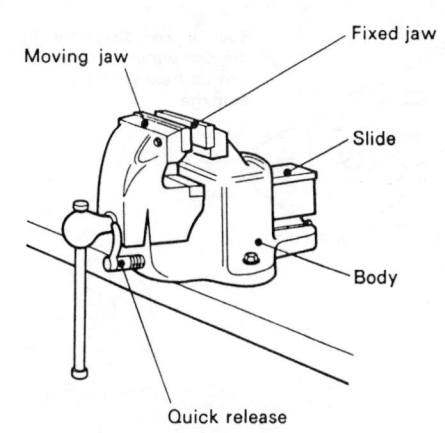

**Fig. 8.3** The fitter's vice

For accurate work the vice must be kept in good condition.
1. Oil the screw regularly.
2. Ensure that the vice is heavy enough for the job in hand.
3. Heavy hammering and bending should be done on the anvil.
4. When chipping, the thrust of the chisel should be against the fixed jaw.
5. Never hammer on the top surface of the slide.

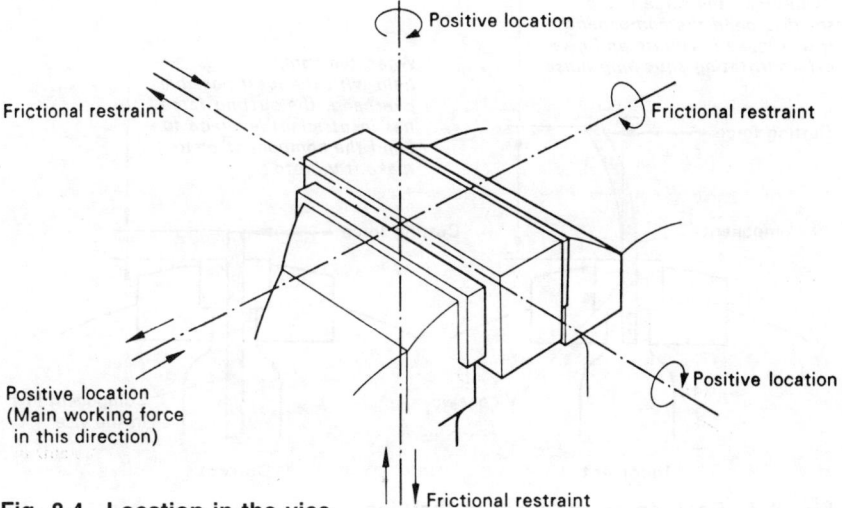

**Fig. 8.4** Location in the vice

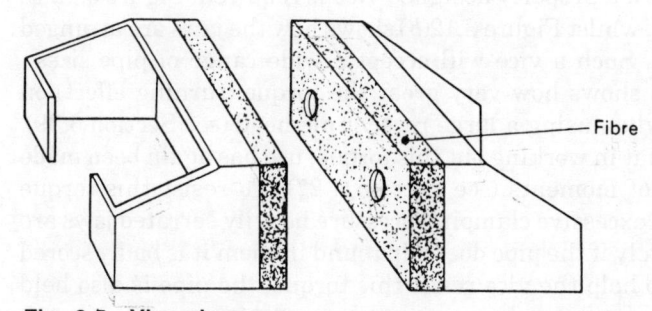

**Fig. 8.5** Vice shoes

— for example, the fitter's vice and the machine vice — to hold cylindrical work as shown in Figure 8.11.

## 8.5 The pipe vice

The fitter's vice is not convenient for holding pipe whilst cutting it off, threading it, or tightening fittings, as the clamping force required would flatten the pipe. The use of a Vee block as shown in Figure 8.11 is not the complete answer either.

279

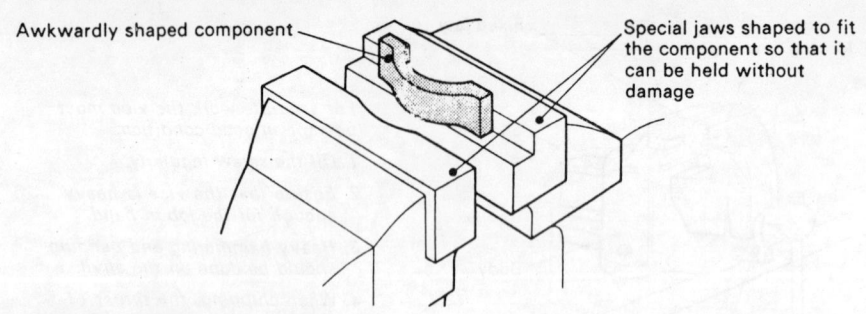

**Fig. 8.6  Shaped vice jaws**

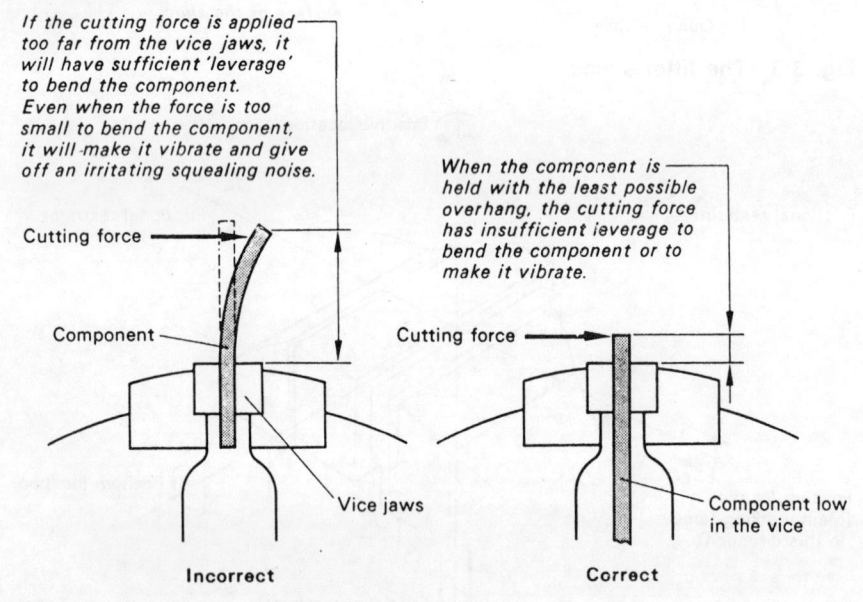

**Fig. 8.7  Positioning the work in the vice**

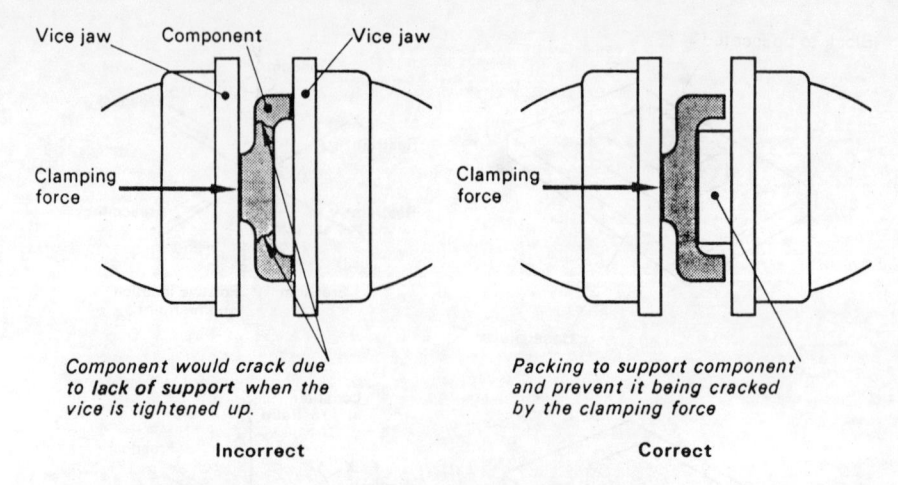

**Fig. 8.8  Clamping thin and weak sections in the vice**

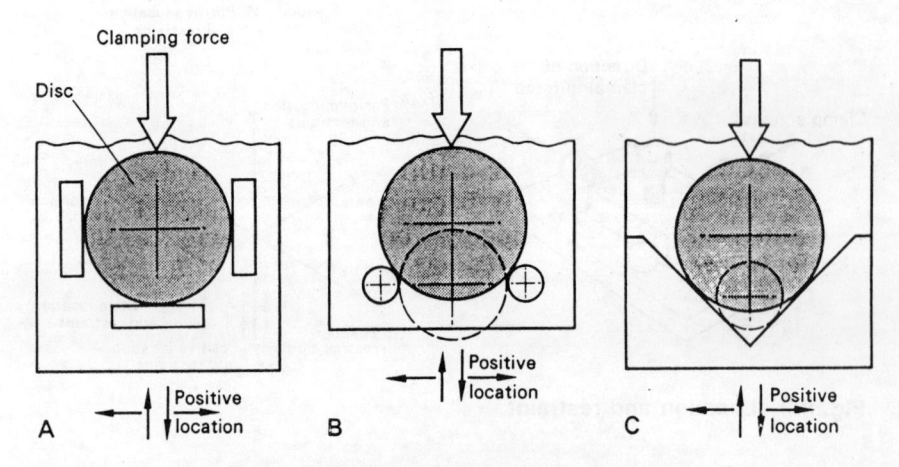

**Fig. 8.9  The location of circular sections**

It can be seen that A, B and C all locate the disc, but with varying degrees of success

A will only locate a correct size disc. Slightly larger and it fouls the side locations, slightly smaller and location is lost.

B will locate a small range of discs. It can be seen that the disc does not have to decrease in size very much before it tends to slip between the location pegs.

C it can be seen that a Vee location can accommodate and locate a large range of disc diameters.

For pipe work a properly designed vice is required. Figure 8.12(*a*) shows such a vice, whilst Figure 8.12(*b*) shows how the jaws are arranged in greater detail. Such a vice will accept a wide range of pipe sizes.

Figure 8.13 shows how very great the torque (turning effect) on the pipe can be when using a large receder die-head (see Section 3.39). It will be noted that in working out this torque, use has again been made of the principle of moments (see Section 4.27). To resist this torque without using an excessive clamping pressure heavily serrated jaws are used. Unfortunately if the pipe does slip round in them it is badly scored and damaged. To help the vice resist this torque, the pipe is also held by the fitter's mate using chain grips (Figure 3.44).

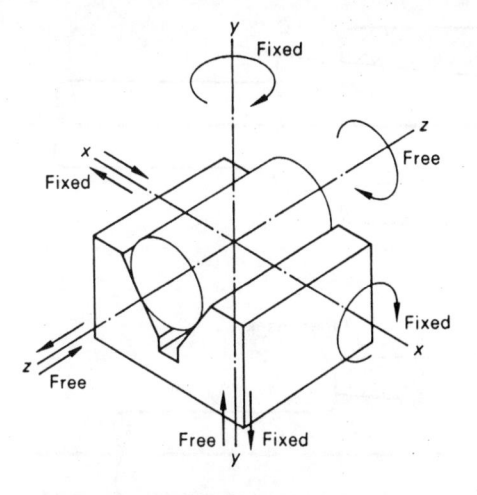

**Fig. 8.10 Location in a Vee block**

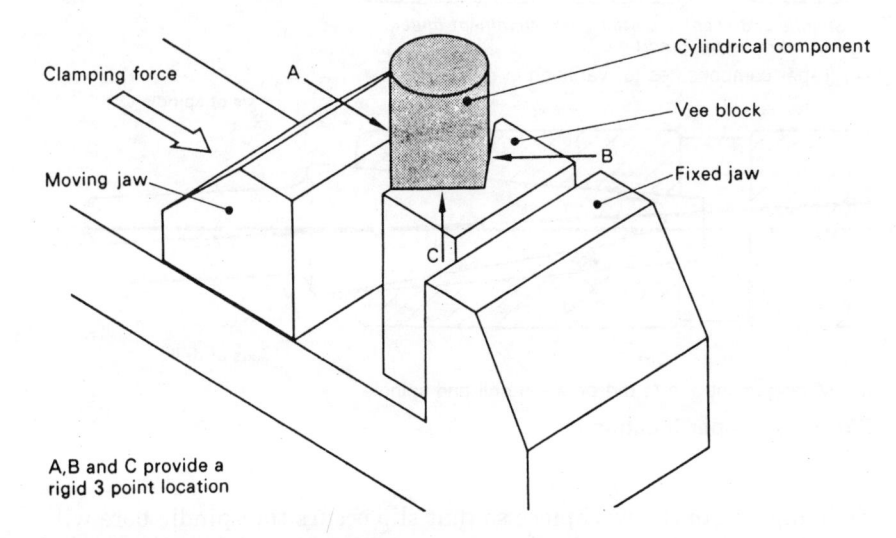

**Fig. 8.11 A use of the Vee block**

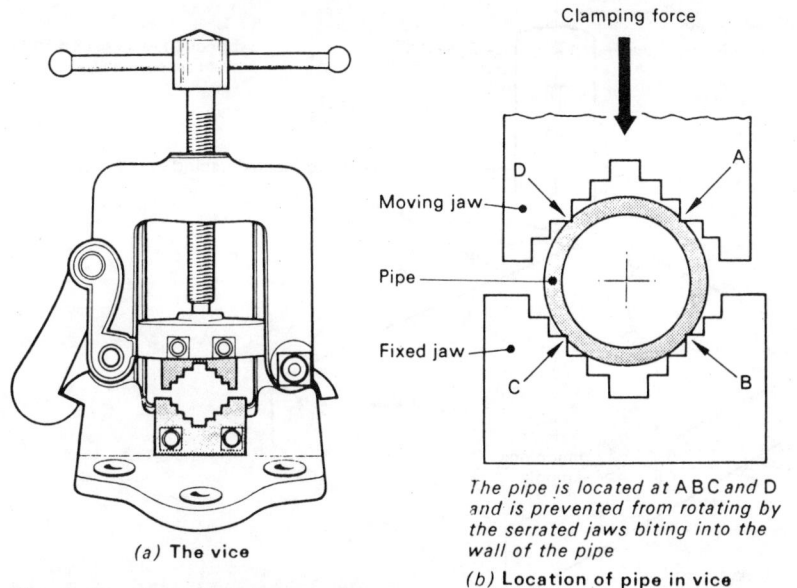

A, B and C provide a rigid 3 point location

**Fig. 8.12 The pipe vice**

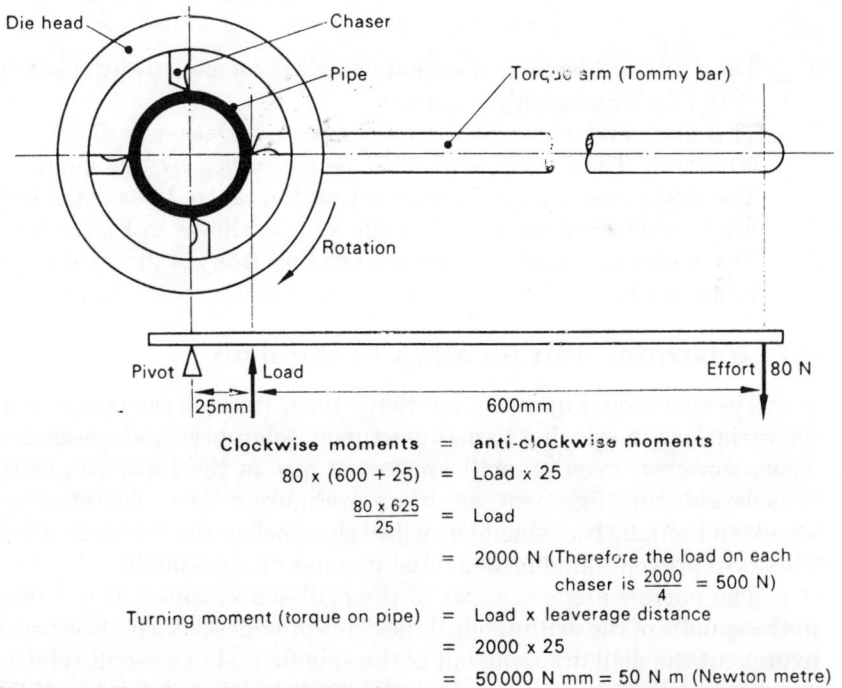

(a) The vice

(b) Location of pipe in vice

The pipe is located at A B C and D and is prevented from rotating by the serrated jaws biting into the wall of the pipe

**Fig. 8.13 Turning effect on the pipe**

Clockwise moments = anti-clockwise moments
80 × (600 + 25) = Load × 25
$\frac{80 \times 625}{25}$ = Load
= 2000 N (Therefore the load on each chaser is $\frac{2000}{4}$ = 500 N)

Turning moment (torque on pipe) = Load × leaverage distance
= 2000 × 25
= 50000 N mm = <u>50 N m</u> (Newton metre)

This chapter sets out to consider restraint and location as it applies to components being drilled and fabricated in greater detail.

## 8.6 Drilling process

To successfully drill a hole in a component so that it is correctly positioned, four basic conditions must be satisfied.

281

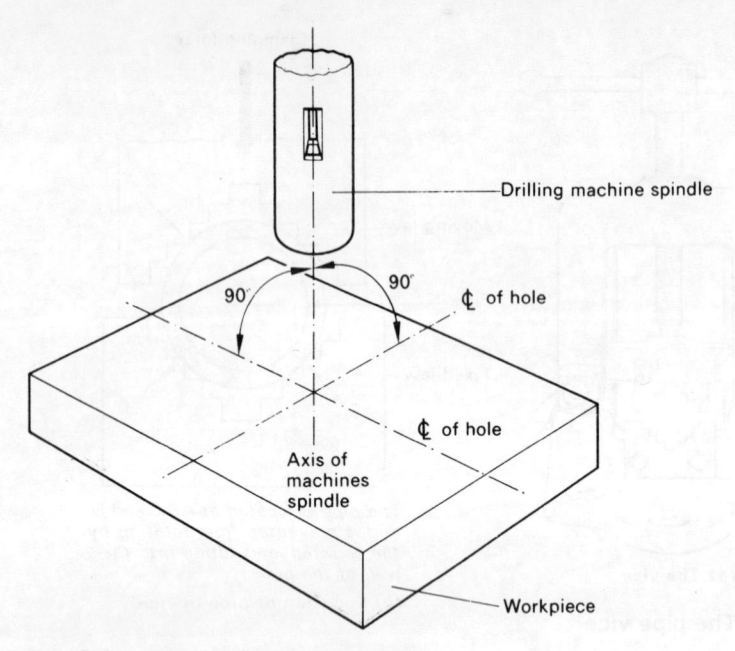

**Fig. 8.14  Basic drilling alignments**

1. The drill must be *located* so that its axis is coincident with the axis of the drilling machine spindle.
2. The drill and the spindle must rotate together without slip occurring. That is the drill must be *restrained* by the spindle.
3. The workpiece must be *located* so that the centre lines of the hole are in alignment with the spindle axis as shown in Figure 8.14.
4. The workpiece must be *restrained* so that it is not dragged round by the drill.

## 8.7  Restraint and location of the drill

It will be seen from Figure 8.15(a) that a taper location can compensate for variations in size due to manufacturing tolerances and reasonable wear. However, even a small amount of dirt in the taper can cause considerable misalignment, as shown in Figure 8.15(b). Therefore the shank and spindle bore should be wiped clean before the drill is inserted. This also prevents undue wear and damage to the spindle.

The narrow angle of taper of the drill shank causes it to wedge in the spindle of the drilling machine. This provides sufficient restraint to prevent the drill dropping out of the spindle and to prevent relative movement (slip) between the drill and the spindle as it rotates. If the

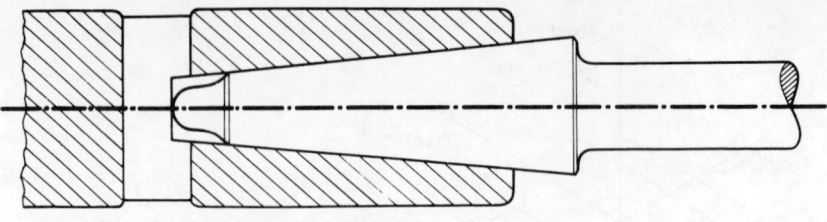

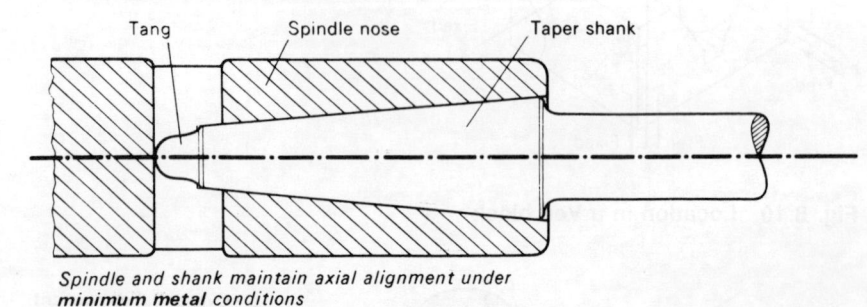

(a) Taper compensates for variation in drill size

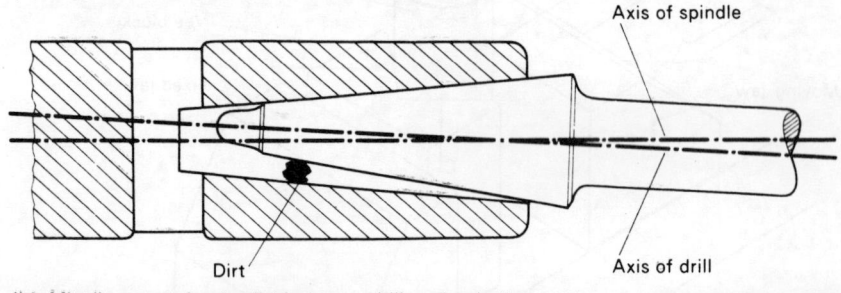

(b) Misalignment due to dirt between drill and spindle

**Fig. 8.15  Taper location**

drill 'digs in' to the workpiece so that slip occurs the spindle bore will be quickly scored up and become useless for locating and retraining the drill.

Straight-shank drills and other small tools are often held in self-centring chucks. The principle of the drill chuck as a device for locating and restraining small drills is explained in Figure 8.16.

## 8.8  Restraint and location of a drilled component

1. Figure 8.17(a) shows the *restraints* acting on a simple component held in a machine vice. The geometric alignments necessary to

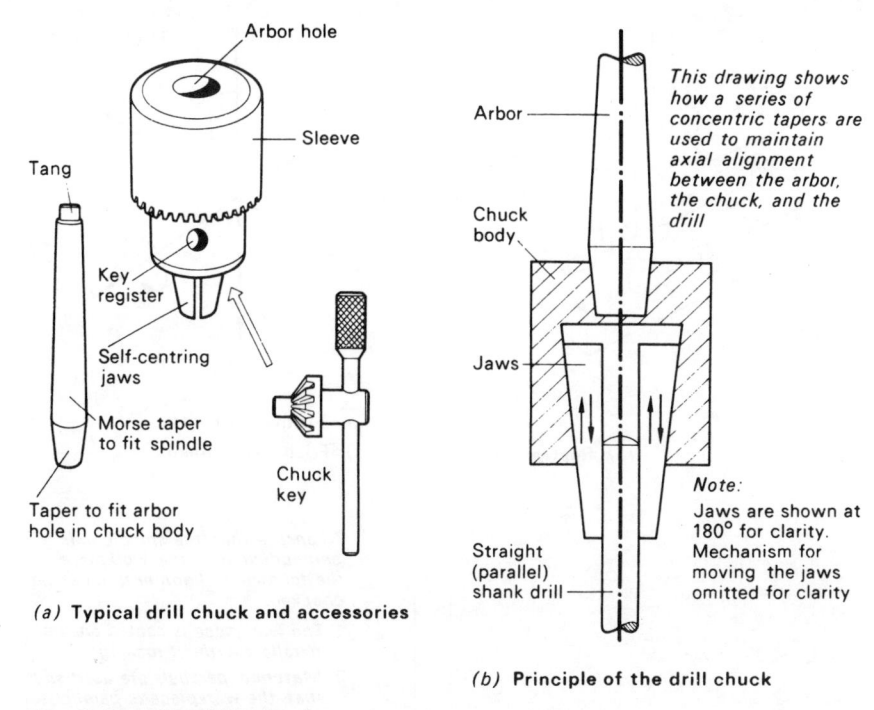

**Fig. 8.16 The drill chuck**

*locate* the component correctly relative to the axis of the drilling machine spindle are shown in Figure 8.17(b).

2. Figure 8.18(a) shows the *restraints* acting on a cylindrical component held in a machine vice. The geometric alignments necessary to *locate* the component correctly, relative to the axis of the drilling machine spindle, are shown in Figure 8.18(b).

3. Figure 8.19(a) shows the *restraints* acting on a simple component clamped directly to the drilling machine table. The geometric alignments necessary to *locate* the component correctly, relative to the axis of the drilling machine spindle, are shown in Figure 8.19(b).

4. Figure 8.20(a) shows the *restraints* acting on a cylindrical component supported on Vee blocks. The geometric alignments necessary to *locate* the component correctly, relative to the axis of the drilling machine spindle, are shown in Figure 8.20(b).

5. The relatively chunky precision engineering components shown in Figures 8.17 to 8.20 inclusive are mounted on parallels, or other datum surfaces to ensure that the hole axis is perpendicular to the drilling machine table. However, sheet metal components are not

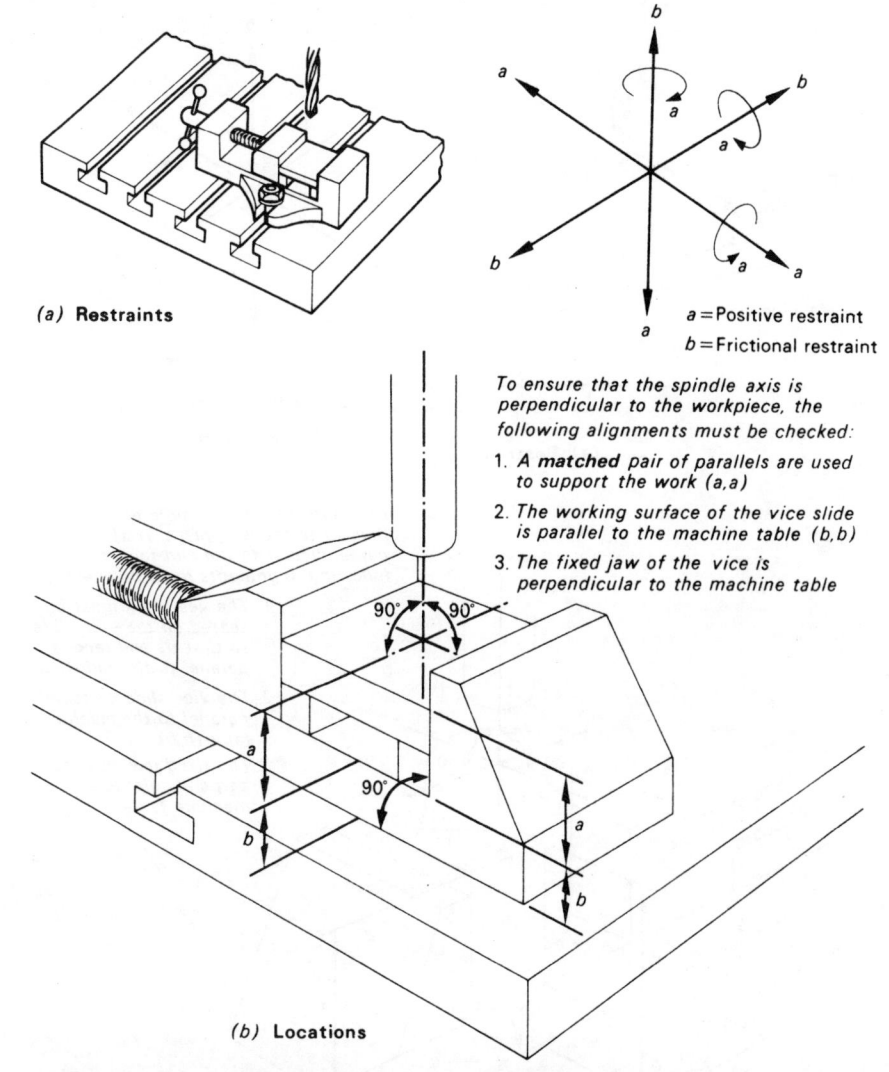

**Fig. 8.17 Workholding in the vice — restraints and location**

sufficiently rigid to locate in this manner and are better mounted on *flat* wood to ensure support right up to the edge of the hole as shown in Figure 8.21.

## 8.9 Clamping work for folding

For simple bending or folding of thin sheet metal the work may be held in '*bending bars*' or in a pair of '*angle bars*' which, in turn, are usually

283

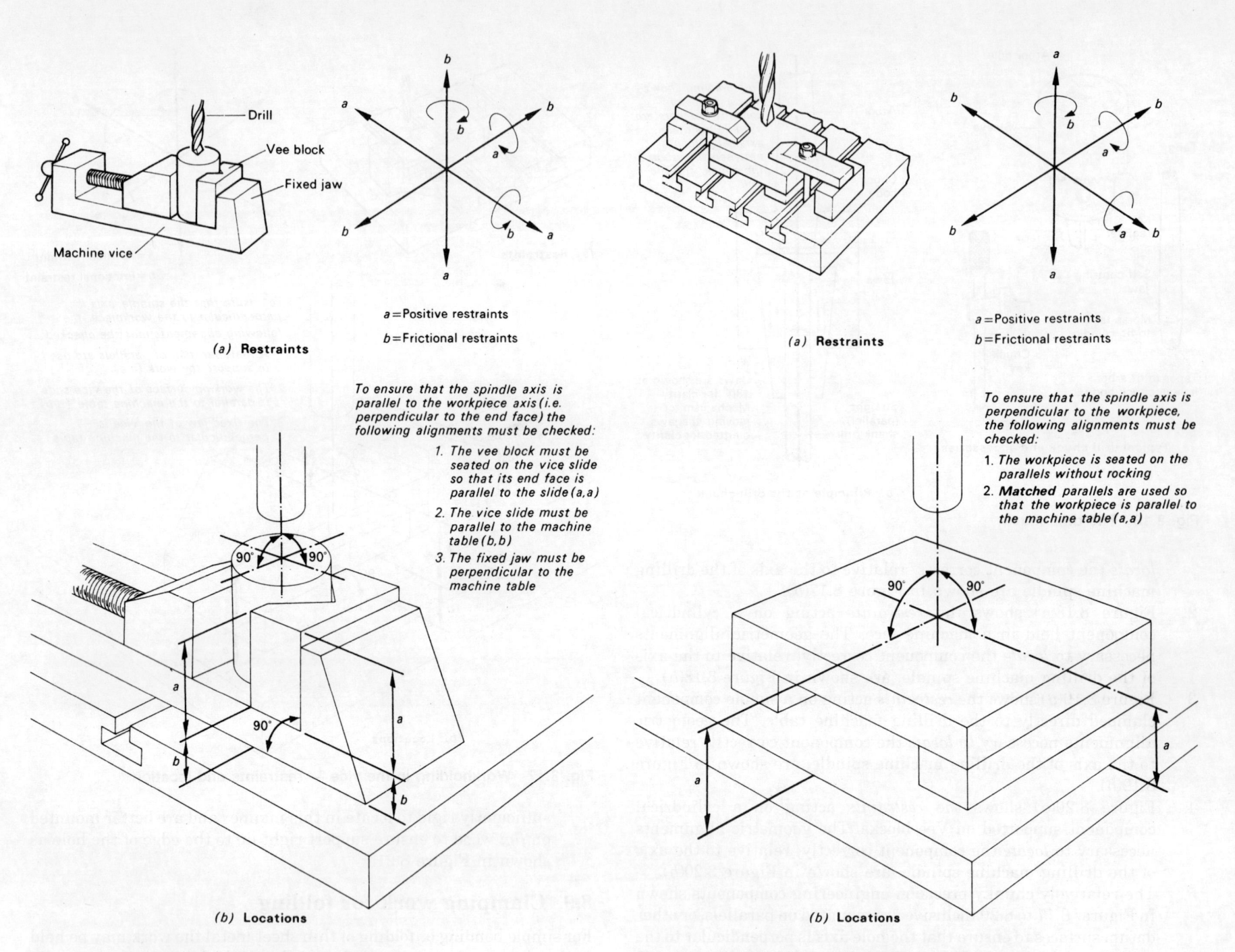

### Fig. 8.18 Workholding cylindrical work in the vice

(a) **Restraints**

$a$ = Positive restraints
$b$ = Frictional restraints

*To ensure that the spindle axis is parallel to the workpiece axis (i.e. perpendicular to the end face) the following alignments must be checked:*

1. The vee block must be seated on the vice slide so that its end face is parallel to the slide ($a, a$)
2. The vice slide must be parallel to the machine table ($b, b$)
3. The fixed jaw must be perpendicular to the machine table

(b) **Locations**

### Fig. 8.19 Workholding on the drilling machine table

(a) **Restraints**

$a$ = Positive restraints
$b$ = Frictional restraints

*To ensure that the spindle axis is perpendicular to the workpiece, the following alignments must be checked:*

1. The workpiece is seated on the parallels without rocking
2. **Matched** parallels are used so that the workpiece is parallel to the machine table ($a, a$)

(b) **Locations**

*a* = Positive restraint
*b* = Frictional restraint

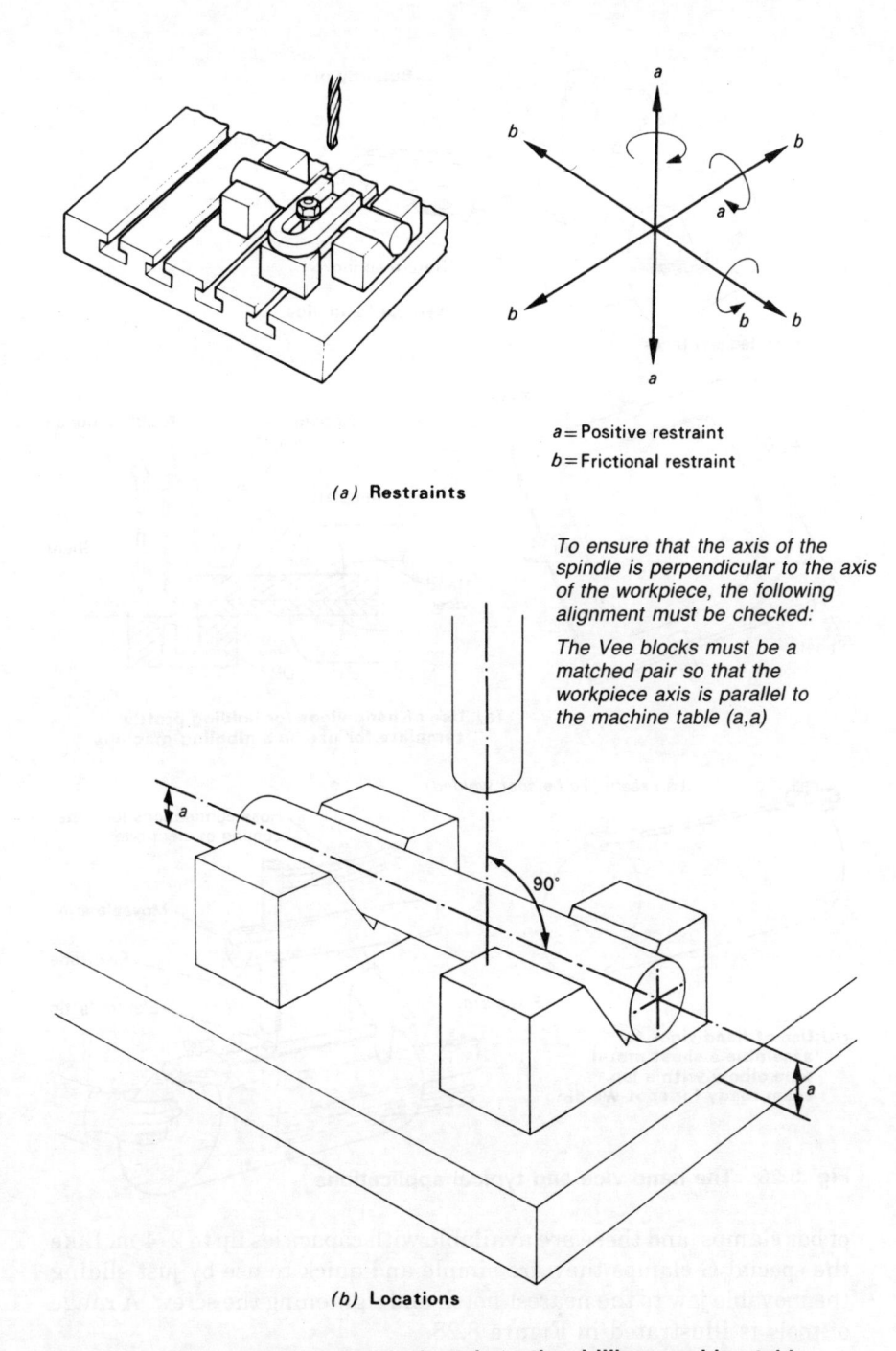

**(a) Restraints**

To ensure that the axis of the spindle is perpendicular to the axis of the workpiece, the following alignment must be checked:

The Vee blocks must be a matched pair so that the workpiece axis is parallel to the machine table (a,a)

**(b) Locations**

**Fig. 8.20 Workholding cylindrical work on the drilling machine table**

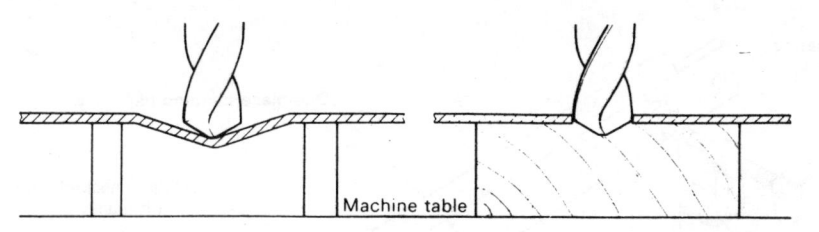

*(a)* Sheet metal inadequately supported on parallels will deflect under the thrust of the drill

*(b)* Sheet metal supported right up to the cutting edge of the drill on a planed wood block

**Fig. 8.21 Drilling sheet metal**

held in an engineer's vice, as illustrated in Figure 8.22. When bending in a *folding machine* the work is held by the clamping beam as explained in Chapter 10.

A pair of angle sections are useful for holding sheet metal when making short bends by hand.

The metal is positioned and clamped between the angle sections by means of a G clamp each end. The whole assembly is then held in the vice and the edge or flange knocked over with a mallet, as shown in Figure 8.22.

## 8.10 Hand tools used for clamping (general)

One of the most common tools used for clamping or holding work is the *engineer's hand vice*. Hand vices are an invaluable aid in fabrication work, and no craftsman should be without them. Figure 8.23 shows a hand vice together with typical applications.

Another type of hand vice depends for its gripping power upon its 'toggle action' (Figure 8.24). With an ordinary type of hand vice, the friction force available to grip the component is somewhat limited. Hand vices with a toggle action tend to grip more securely, and unlike the ordinary hand vice, provide a means of quick release. Figure 8.24 illustrates various types of toggle-action hand vices.

Considerable use is made of G clamps, and these are available in a range of sizes. *When using G clamps care should be taken to select the correct size of clamp for the job.* For example, when large clamps are used where a much smaller size is called for, there is a tendency to put too great a strain on the screw. This results in a permanent misalignment of the two gripping surfaces, as indicated in Figure 8.25(*a*).

Applications of the use of G clamps are illustrated in Figure 8.25(*b*).

285

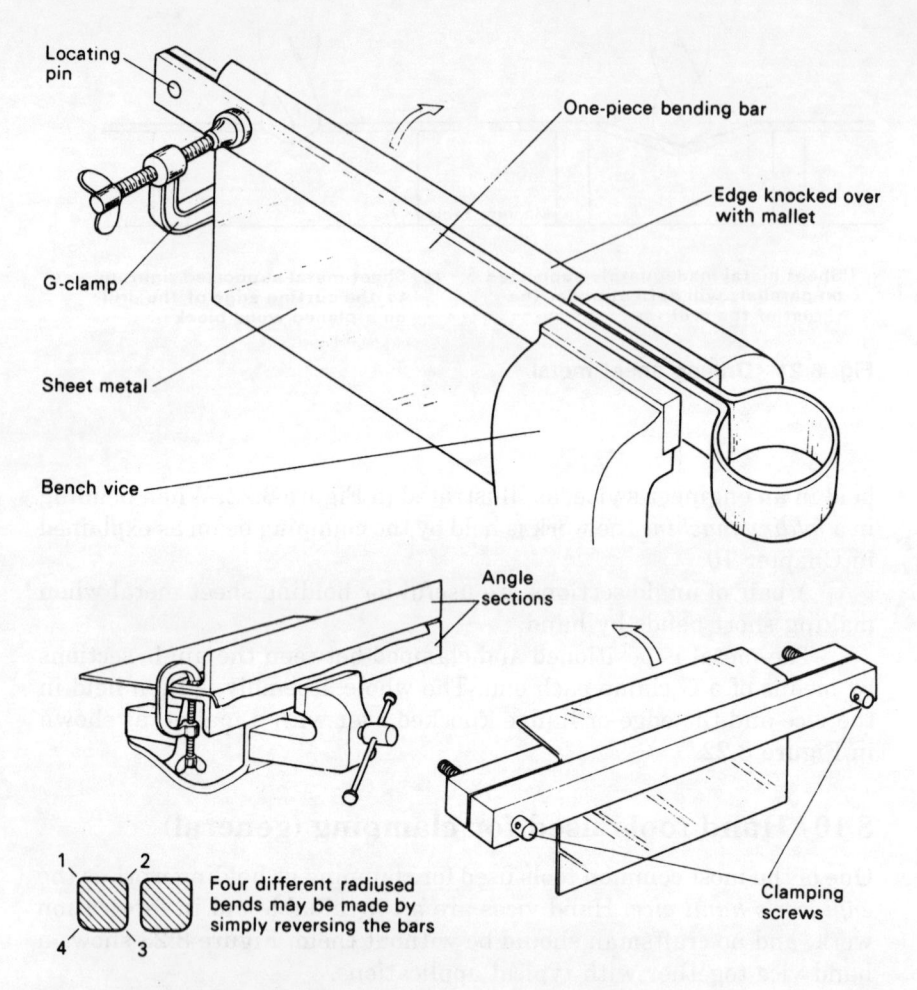

**Fig. 8.22 Clamping sheet metal for folding by hand**

## 8.11 Hand tools used for clamping (specific)

Special G clamps are available which are a great improvement on the ordinary G clamp. These special clamps are capable of performing many clamping operations more efficiently and quickly than by other means. They are fitted with a short, shielded screw which will outlast many ordinary clamp screws, the fine-pitched thread provides a very powerful grip. *Unlike the ordinary G clamp, there is no long screw to bend or damage.* Details of these clamps are given in Figure 8.26, whilst typical applications are shown in Figure 8.27.

For all types of work where a large capacity is required, use is made

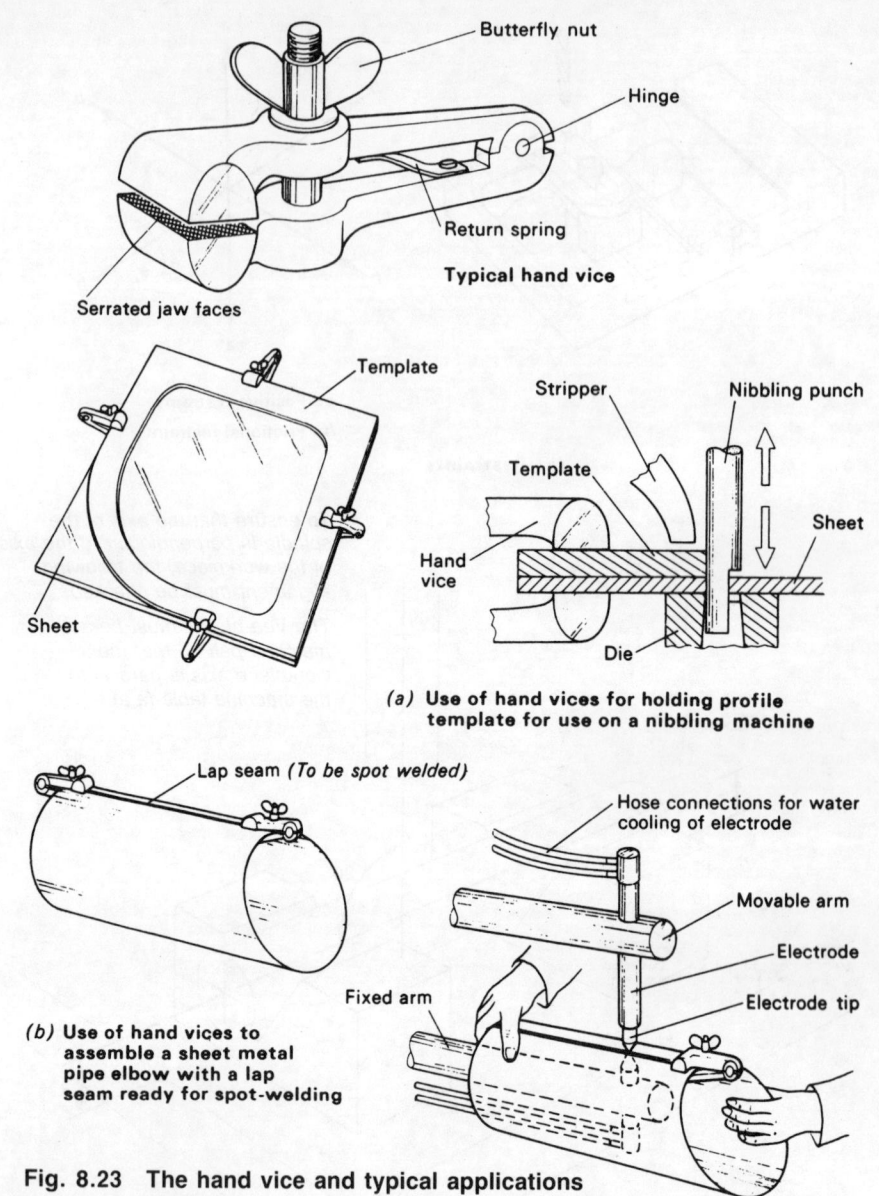

**Fig. 8.23 The hand vice and typical applications**

of bar clamps, and these are available with capacities up to 2·4 m. Like the special G clamps they are simple and quick to use by just sliding the movable jaw to the nearest notch and tightening the screw. A range of tools is illustrated in Figure 8.28.

Tee-slot clamps are used for clamping the work on all Tee-slotted

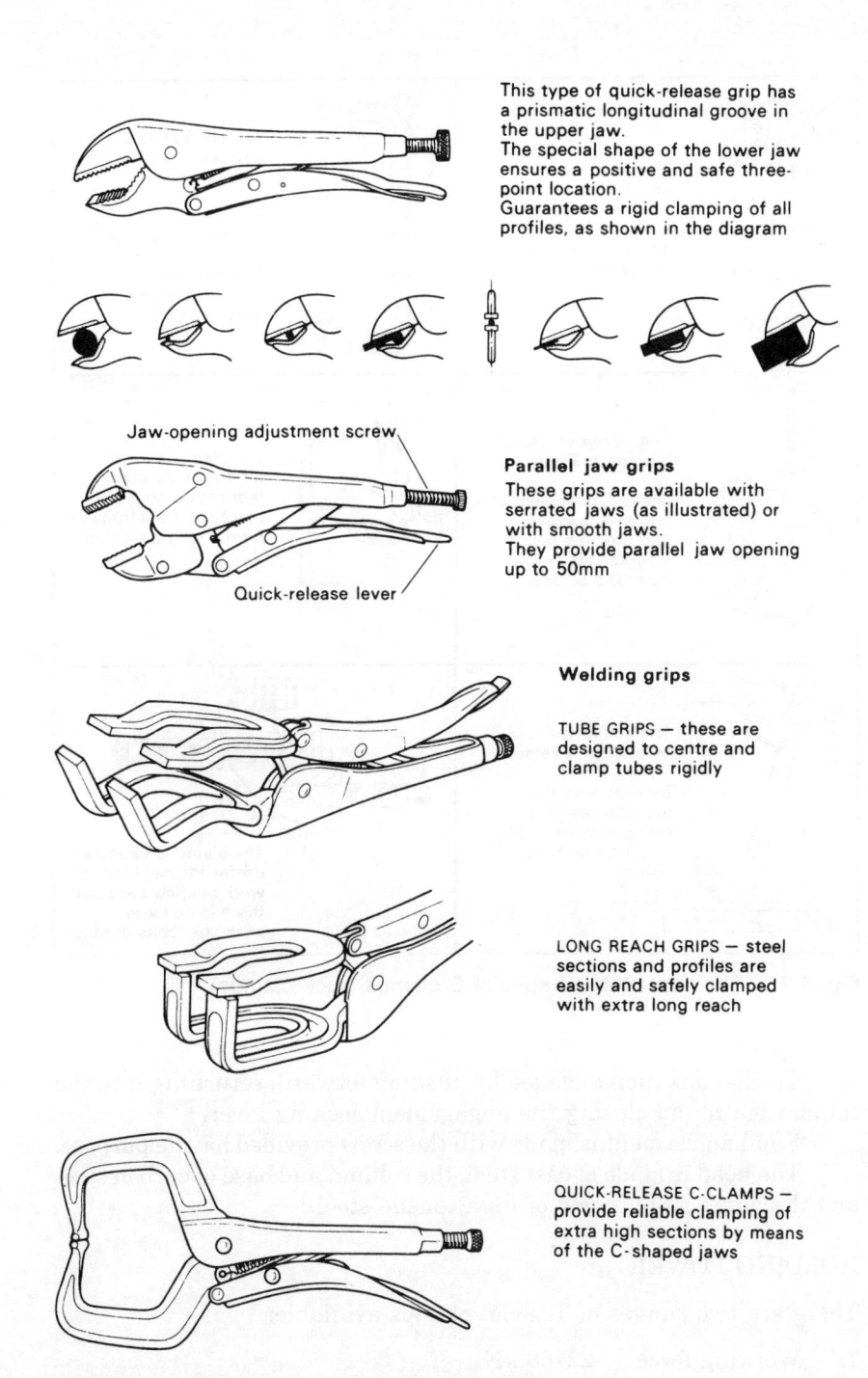

This type of quick-release grip has a prismatic longitudinal groove in the upper jaw.
The special shape of the lower jaw ensures a positive and safe three-point location.
Guarantees a rigid clamping of all profiles, as shown in the diagram

**Parallel jaw grips**

These grips are available with serrated jaws (as illustrated) or with smooth jaws.
They provide parallel jaw opening up to 50mm

**Welding grips**

TUBE GRIPS — these are designed to centre and clamp tubes rigidly

LONG REACH GRIPS — steel sections and profiles are easily and safely clamped with extra long reach

QUICK-RELEASE C-CLAMPS — provide reliable clamping of extra high sections by means of the C-shaped jaws

**Fig. 8.24 Toggle-action hand vices or grips**

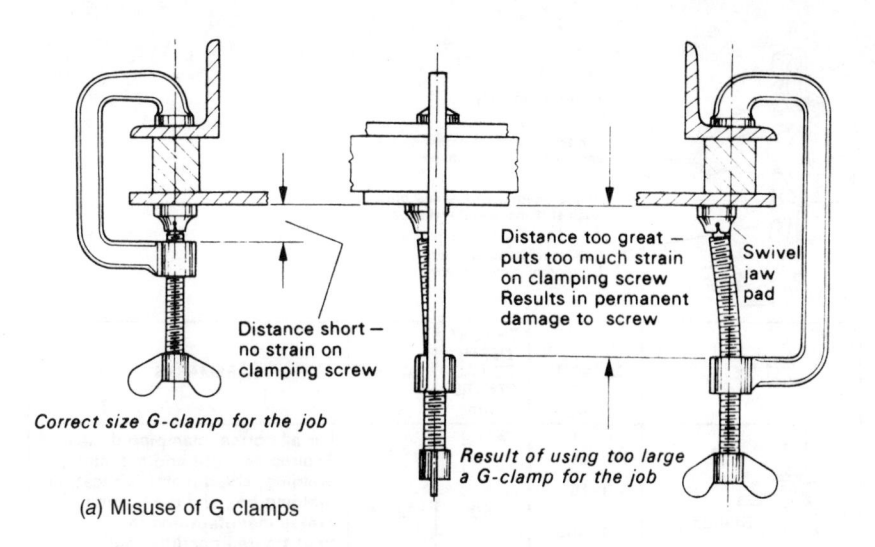

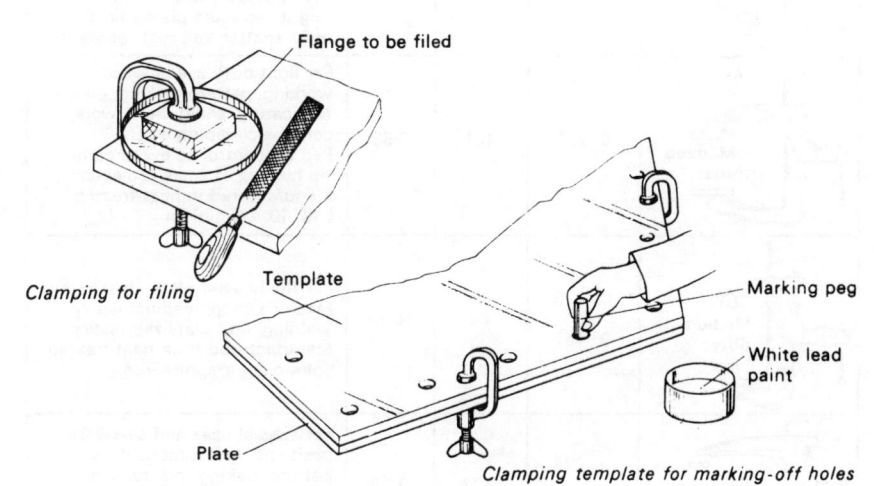

*Correct size G-clamp for the job* — Distance short — no strain on clamping screw

*Result of using too large a G-clamp for the job* — Distance too great — puts too much strain on clamping screw. Results in permanent damage to screw. Swivel jaw pad

(a) Misuse of G clamps

Flange to be filed
Clamping for filing
Template
Marking peg
White lead paint
Plate
Clamping template for marking-off holes

(b) Applications of use of G clamps from workholding

**Fig. 8.25 Use of G clamps**

machine tables, or welding positioners. The basic principle of a very efficient type of Tee-slot clamp is shown in Figure 8.29.

## METHOD OF OPERATION

The required height is obtained by releasing the locking lever, pulling the head assembly rearwards out of engagement with the column teeth, and raising or lowering as required.

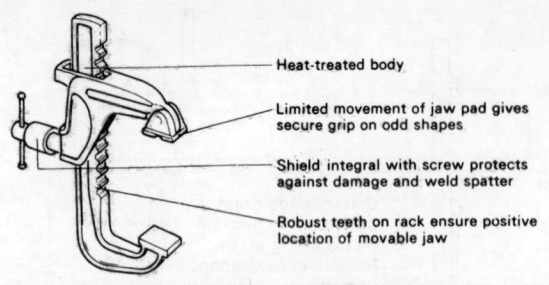

| TYPE | CAPACITY (mm) | THROAT DEPTH TO PAD CENTRE (mm) | RATING (kN) | REMARKS |
|---|---|---|---|---|
| Standard duty | 0-152<br>0-305 | 60 | 11·34 | For all normal clamping duties. Toolrooms, light and medium welding, sheet metal fabrication, welding box and die holding. Frame manufactured in heat-treated pearlite iron; bright cadmium-plated finish; weld spatter and rust resistant |
| L.M. deep throat | 0-228 | 121 | 3·63 | For light duty use in wood-working, pattern making, plastic fabricating, sheet metal work, corebox clamping etc. Pad tilt control screw to permit tip holding of small projections. Manufactured in heat-treated L.M. 10 Aluminium |
| Medium duty | 0-228<br>0-457 | 89 | 18·15 | For the heavier work, deeper jaw. Machine shop, medium heavy welding, and plate fabrication. Manufactured from heat-treated spheroidal graphite iron |
| Deep throat medium | 0-228<br>0-457 | 203 | 9·08 | For special uses and awkward positions. Metal fabrication, pattern making, plastic and fibre-glass moulding. Manufactured from heat-treated spheroidal graphite iron |
| Heavy duty | 0-152<br>0-305<br>0-457<br>0-609<br>0-914<br>0-1372 | 114 | 27·23 | The ultimate in hand applied clamping where extreme pressure or length is required. Frame forged steel, heat treated |

Fig. 8.26 Special G clamps and applications

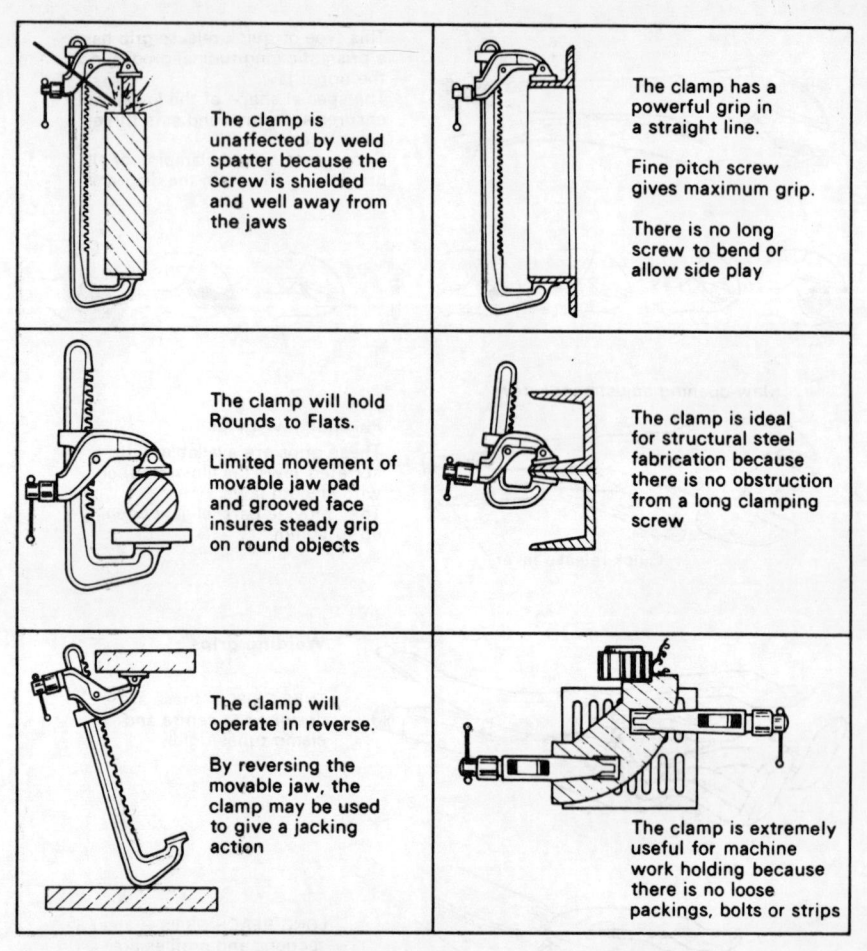

Fig. 8.27 The versatility of special G clamps (rack clamps)

The head is then engaged by pushing forward, returning it to the column teeth and closing the engagement locking lever.

Final adjustment is made with the screw provided for the purpose.

The head is made of cast steel, the column and base of carbon steel and the adjustment screw of high tensile steel.

## HOLDING POWER

There are two ranges of Tee-slot clamps available:

1. Working force     22·68 kN
   Ultimate force     36·3 kN

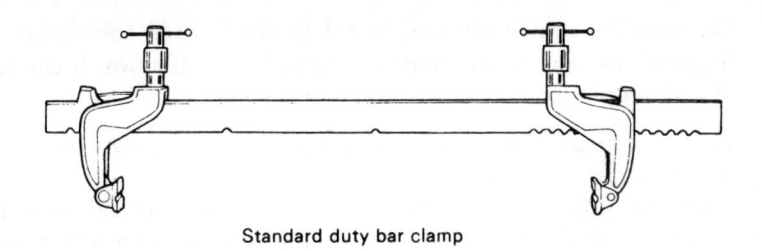

| TYPE | MAXIMUM CLAMPING LENGTH (mm) | CAPACITY (mm) | RATING (kN) | REMARKS |
|---|---|---|---|---|
| 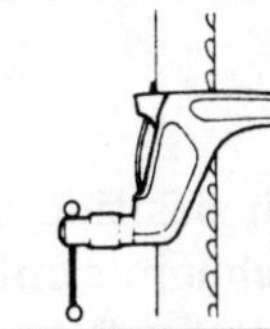 Standard duty | 457, 609, 914, 1219, 1524, 1829 | 140 minimum to indicated length | 9·1 | For all types of work where a large capacity is required. The bar is manufactured from HIGH TENSILE STEEL |
| 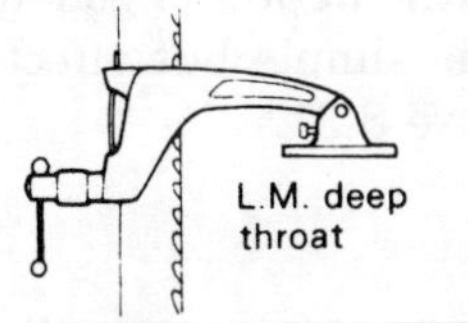 L.M. deep throat | 457, 609, 914, 1219, 1524, 1829 | 152 minimum to indicated length | 3·6 | With all bar clamps, multiples of jaws may be used on a bar to give positive component location |
| 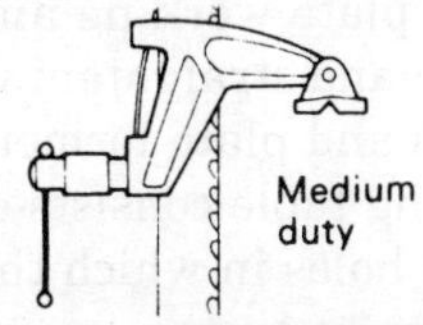 Medium duty | 609, 914, 1219, 1524, 1829 | 216 minimum to indicated length | 18·2 | Principally for use in medium weight fabrication shops. The bar is manufactured from HEAT-TREATED HIGH TENSILE STEEL |
| 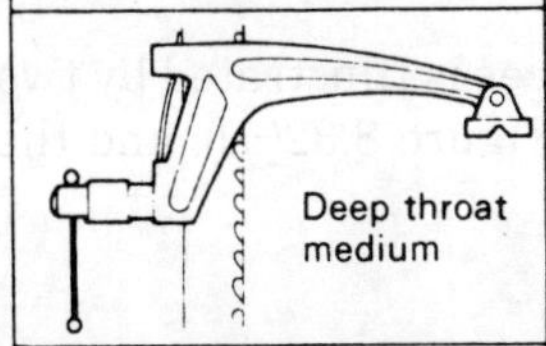 Deep throat medium | 609, 914, 1219, 1524, 1829 | 254 minimum to indicated length | 9·1 | |
| 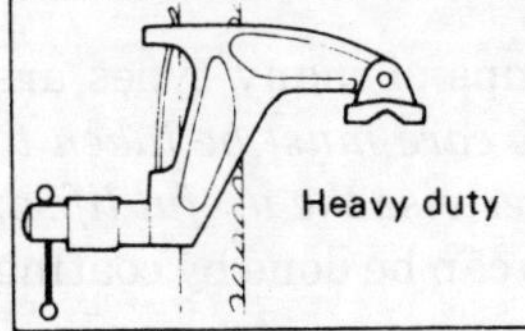 Heavy duty | 1829, 2134, 2438 | 305 minimum to indicated length | 27·25 | The ultimate in heavy duty clamping. HEAT-TREATED HIGH TENSILE BAR The jaws on all bar clamps may be reversed to give opening action |
|  | **Bar connector** Standard duty, medium duty and heavy duty connection pieces may be used to couple any two bars for greater overall length | | | |

Fig. 8.28 Range of bar clamps and applications

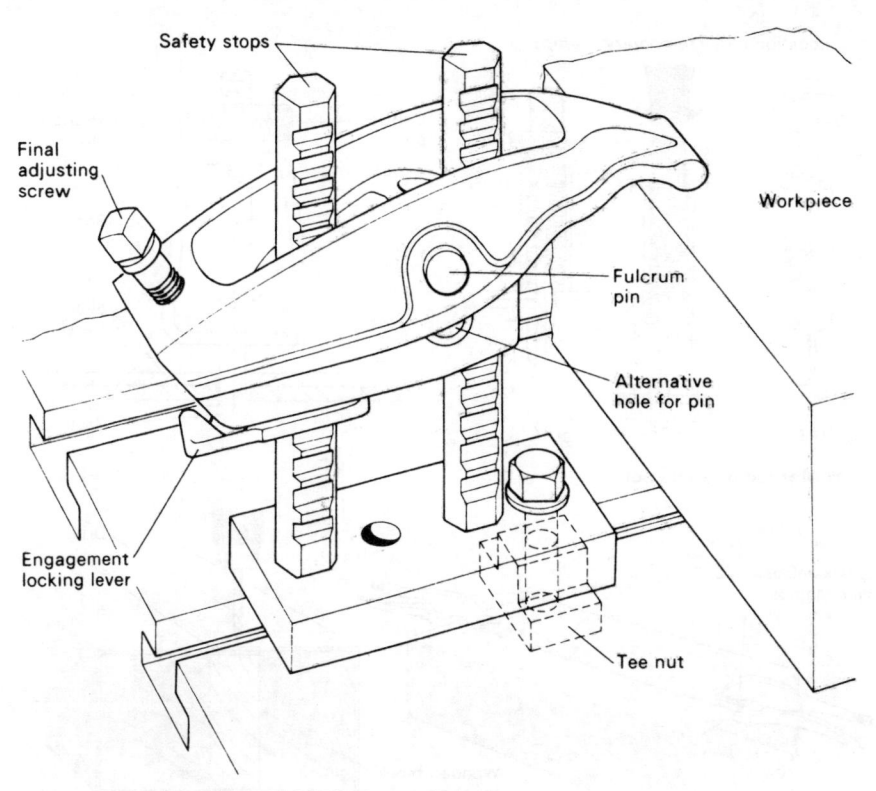

Fig. 8.29 Basic details of Tee-slot clamps

2. Working force    36·3 kN
   Ultimate force    113·4 kN

The small type are available in three heights, 102 mm, 152 mm and 203 mm.

The larger type is made in eight heights from 152 mm to 1219 mm in increments of 152 mm.

## 8.12 Use of clamping devices for riveting

The toggle-action hand vices described in Section 8.10 are an invaluable aid for holding work for riveting, and are used in conjunction with special 'locating pins' or 'skin pins'. A typical application is shown in Figure 8.30.

METHOD OF OPERATION

1. Unscrew the knurled nut.

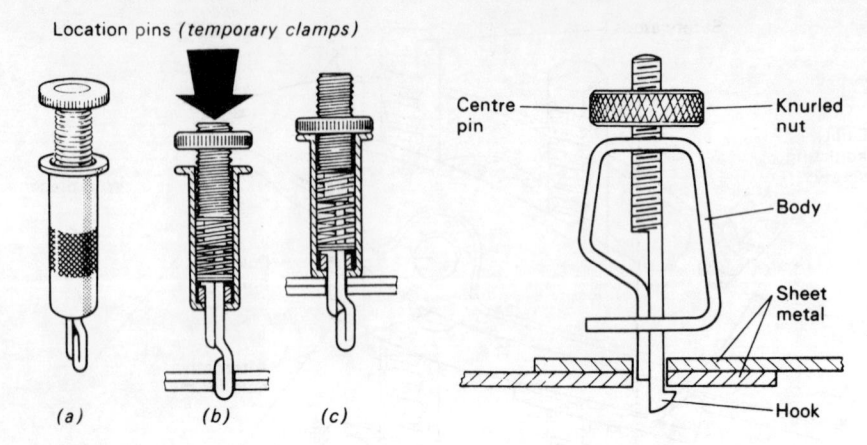

**(a) Method of operation**

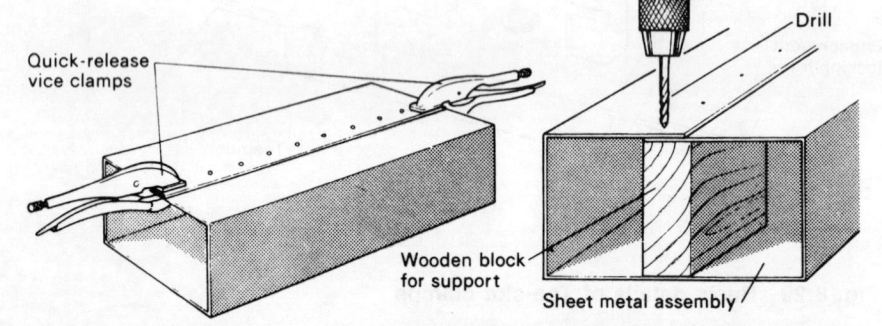

**(b) Work clamped for drilling**

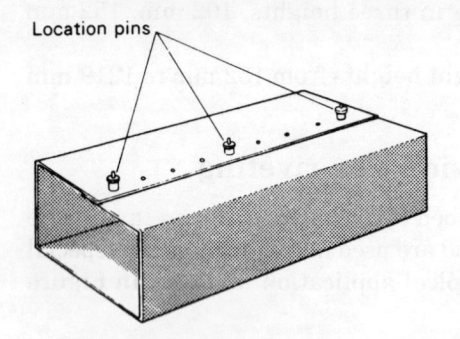

*Work of this nature must be supported to prevent sagging under the drill pressure which would result in misalignment of the rivet holes*

**(c) Work clamped for riveting**

*After drilling the holes are deburred and the assembly clamped with the aid of location pins ready for riveting*

Fig. 8.30 Workholding for riveting

2. Depress the centre pin and insert in the hole in the sheets.
3. Tighten the knurled nut and draw the centre pin upwards clamping the sheets between the body and the hook.

*Location pins are removed by simply unscrewing the knurled nut and depressing the centre pin.*

Location pins are an essential aid for sheet metal assembly, particularly in the aircraft industry. The pins give accurate location, positive grip, no distortion of the holes, rapid one-hand operation, and additional clamping pressure by means of the knurled nut if required. They can be removed as rapidly as they are inserted, they are suitable for holes of 2·4 mm, 3·2 mm and 4 mm diameter, and cut assembly time considerably in sheet metal assembly work.

## 8.13 Use of soft iron wire and spring pegs

Many soldered or brazed lap joints require to be held in position to prevent the mating surfaces from moving when heated. Such joints only require light clamping, and this is achieved with the aid of soft iron wire or simple spring pegs. Applications of this simple but effective method of work holding are illustrated in Figure 8.31.

## 8.14 Use of plate dogs and pins

'Plate dogs' and 'pins' are used extensively in the plate working and steel fabrication workshops for a variety of forming and straightening operations. They are used to clamp plates, sections and plate formers on to the bending floor. The bending floor or bending table consists of a large cast iron slab which has numerous location holes in which the 'dogs' and 'pins' are inserted.

The basic principle of the use of these devices is illustrated by two practical applications: bevelling steel section (Figure 8.32(a)); and the hot bending of angle section (Figure 8.32(c)).

## 8.15 Clamping devices for welding

In the assembly of parts prior to welding, clamps of many types are needed. G clamps are most commonly used, but *care must be taken to prevent 'spatter' from damaging the threads, otherwise the useful life of an ordinary G clamp will be extremely short*. This can be done by coating the threads with 'anti-spatter compound'.

Because of the damage which can be caused by 'spatter', the special G clamps (Figure 8.26) have a very definite advantage over ordinary G clamps by the fact that they are fitted with a shield which protects the screw against damage and spatter.

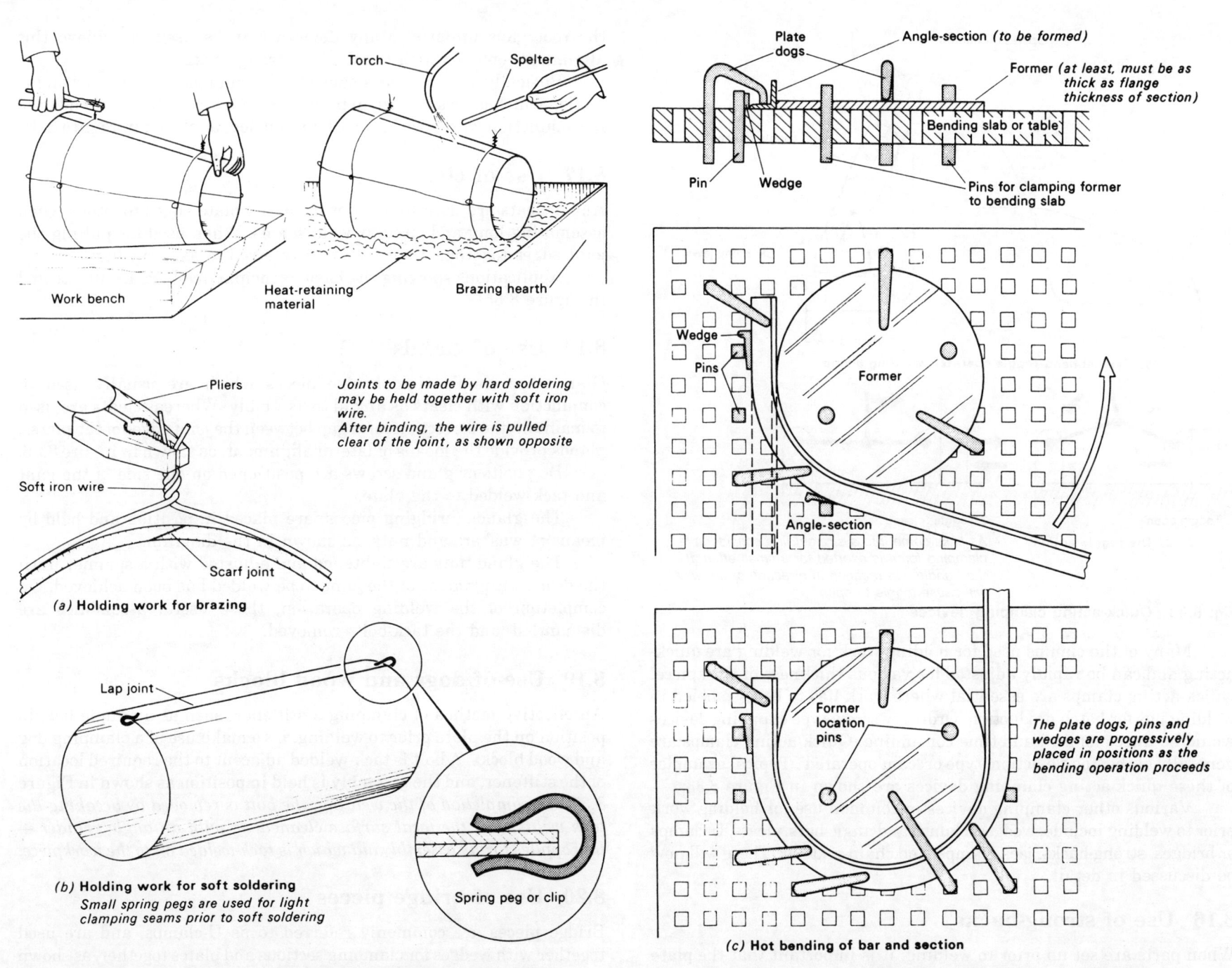

Fig. 8.31 Use of soft iron wire and spring pegs for holding work

(a) Holding work for brazing

(b) Holding work for soft soldering
Small spring pegs are used for light clamping seams prior to soft soldering

Joints to be made by hard soldering may be held together with soft iron wire.
After binding, the wire is pulled clear of the joint, as shown opposite

Fig. 8.32 Applications of plate-dogs and pins for workholding

(c) Hot bending of bar and section

The plate dogs, pins and wedges are progressively placed in positions as the bending operation proceeds

291

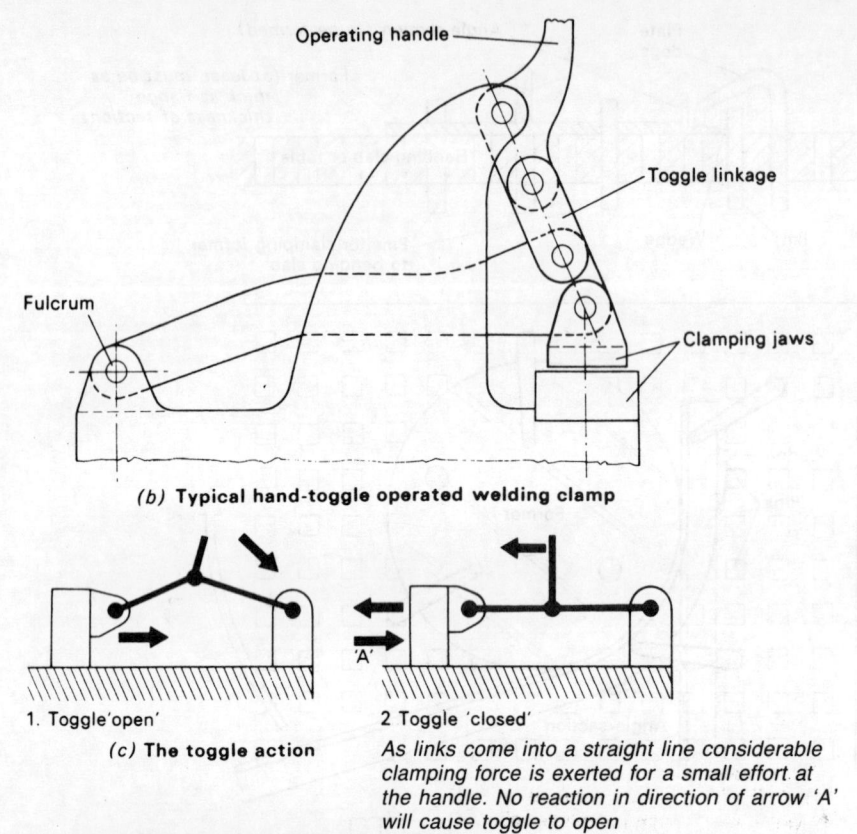

**Fig. 8.33 Quick-acting clamping devices**

Many of the clamps used for holding work for welding are quick-acting and can be rapidly adjusted for various thickness of workpiece. Quick-acting clamps are essential where work has to be assembled in welding jigs for batch production. Ordinary screw-type clamping devices would be uneconomical and time consuming. Quick-acting clamps are generally of the 'toggle-action' type or 'cam-operated'. Typical examples of these quick-acting clamping devices are shown in Figure 8.33.

Various other clamping devices in common use for holding work prior to welding include, wedges, clamps and angle bars, cleats, U-clamps or bridges, strong-backs, jack-clamps, and chain-and-bar. These will now be discussed in detail.

## 8.16 Use of strong-backs

When parts are set up prior to welding, it is important that the plate edges should be correctly aligned. The root faces should be aligned and the root gaps uniform. Many devices may be used to achieve this alignment, one of which is called a 'strong-back'.

Basically, a strong-back consists of a rigid piece of plate or angle-section which is tack-welded in position on one side of the joint and used in conjunction with a wedge or a bolt and dog, as shown in Figure 8.34.

## 8.17 Use of cleats

Angle cleats are used to push or draw the plate edges together when assembling long and circular weld seams. When used for pulling the joint edges together, they are often referred to as 'draw-lugs'.

Applications showing the basic principle involved are illustrated in Figure 8.35.

## 8.18 Use of glands

Glands are simple rigid bridge pieces which are usually used in conjunction with cleats as an aid to assembly. Whereas cleats are used to maintain the correct uniform gap between the joint edges of the plates, glands provide the means of lateral alignment, as shown in Figure 8.36.

Dog bolts or gland screws are positioned on one side of the joint and tack-welded to the plate.

The glands (bridging pieces) are placed in position and held by means of washers and nuts, as shown in the diagram.

The gland nuts are tightened and adjusted with a spanner until the desired alignment of the joint to be welded has been achieved. On completion of the welding operation, these useful assemblies are dismantled, and the tack-bolts removed.

## 8.19 Use of dogs and wood blocks

An effective method of clamping a stiffener, such as an angle bar, in position on the plate prior to welding, is to make use of a clamping dog and wood blocks. A bolt is tack-welded adjacent to the required location of the stiffener, and the assembly is held in position as shown in Figure 8.37. *On completion of the welding, the bolt is removed by breaking the tack welds, and the local surface cleaned up with an angle grinder — this applies to any assembly aid which is tack-welded on to the workpiece.*

## 8.20 Use of bridge pieces

Bridge pieces are commonly referred to as U-clamps, and are used together with wedges for clamping sections and plates together, as shown in Figure 8.38.

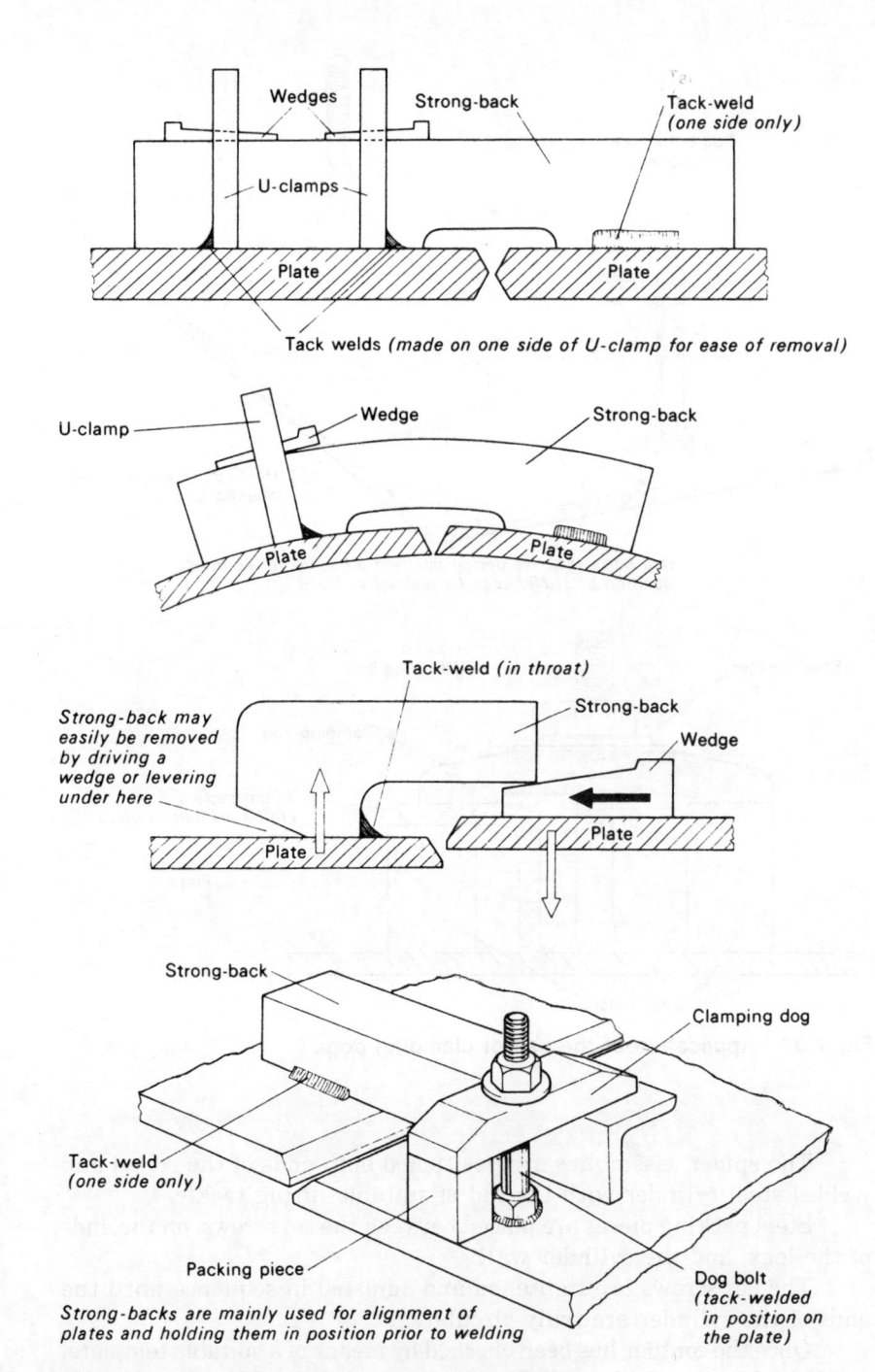

**Fig. 8.34** The use of strong-backs for welding

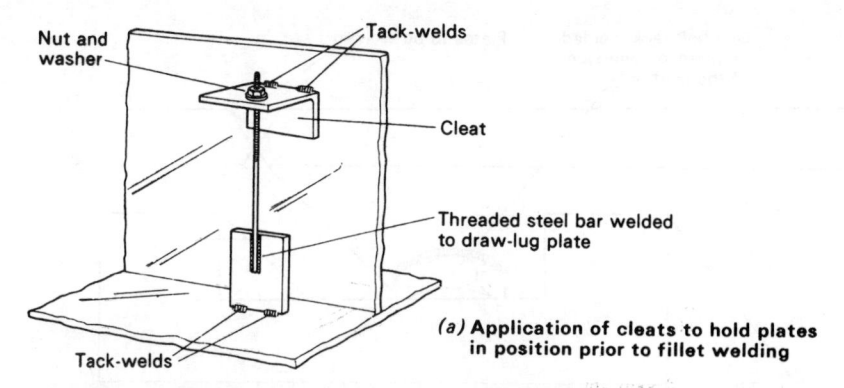

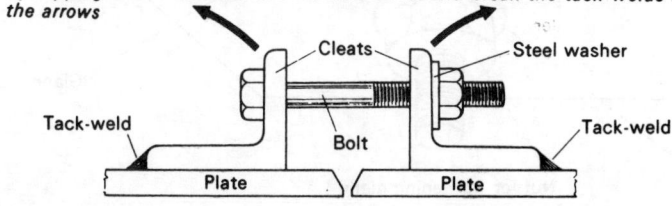

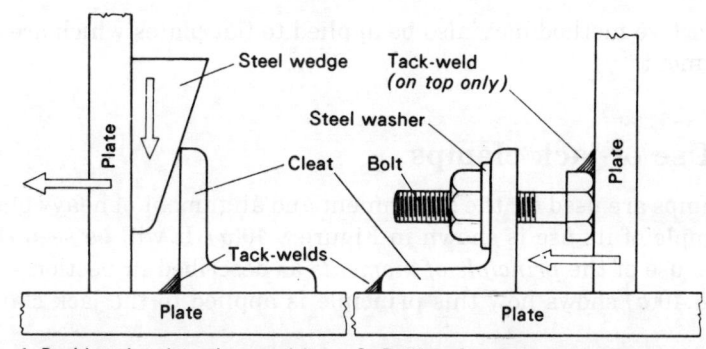

**Fig. 8.35** The use of cleats for holding work for welding

## 8.21 Use of chain and bar

Difficulty is often experienced when welding longitudinal seams on cylindrical vessels which have been incorrectly rolled, causing considerable misalignment of the joint edges. This difficulty may be overcome by the use of a chain and bar, as illustrated in Figure 8.39.

293

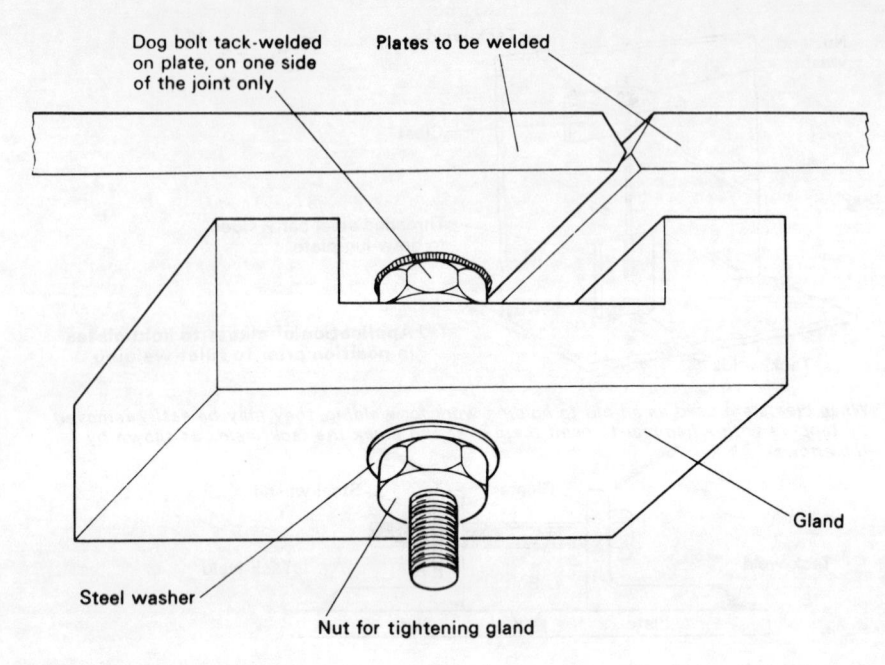

**Fig. 8.36 Use of glands**

This effective method may also be applied to flat plates which are out of alignment.

## 8.22 Use of jack clamps

Jack clamps are used for the attachment and alignment of heavy plate. An example of its use is shown in Figure 8.40(*a*). It will be seen that it makes use of the *principle of moments* as described in Section 4.27. Figure 8.40(*b*) shows how this principle is applied to the jack clamp.

## 8.23 Use of spiders

Large cylindrical vessels are often made up of two or more sections of rolled plate. The sections are then welded together circumferentially. It is essential that each section is truly circular to ensure correct alignment of the joint to be welded.

In order to provide correct alignment of the circumferential joints, 'spiders' are fitted at both ends of each section to ensure concentricity. Figure 8.41 illustrates the use of spiders for this purpose.

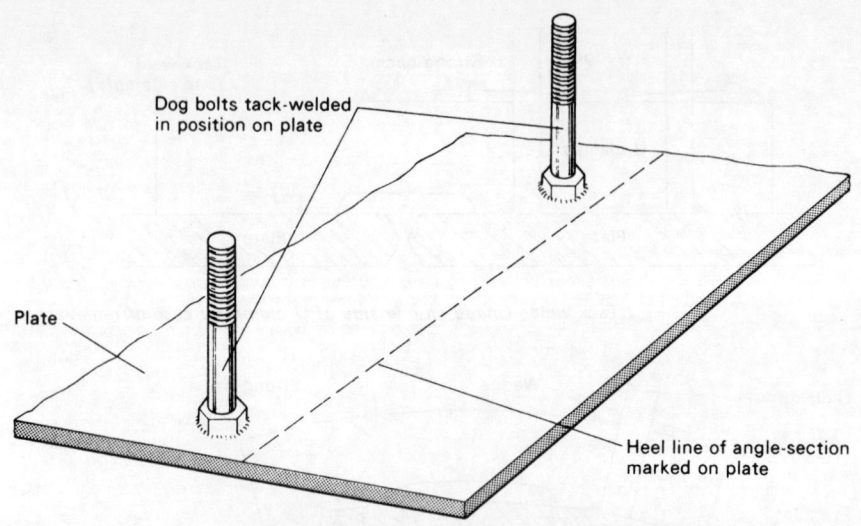

*Dog bolts tack-welded in position adjacent to required position of stiffener to be welded to plate*

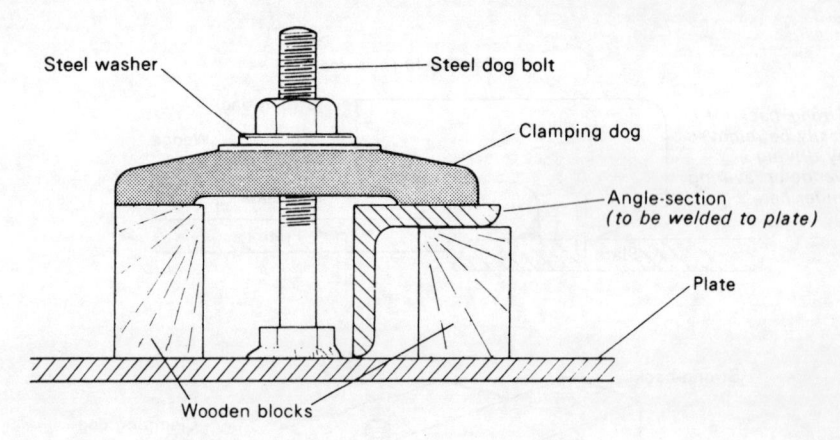

**Fig. 8.37 Application of the use of clamping dogs**

The spider assemblies are positioned both ends of the rolled and welded steel cylinder with the aid of suitable lifting tackle.

Steel packing pieces are placed between the set screws, on the ends of the legs, and the cylinder wall.

The set screws are tightened and adjusted in sequence until the ends of the cylinder are truly circular.

Once the contour has been checked by means of a suitable template, location plates are tack-welded to the legs and the inside cylinder wall.

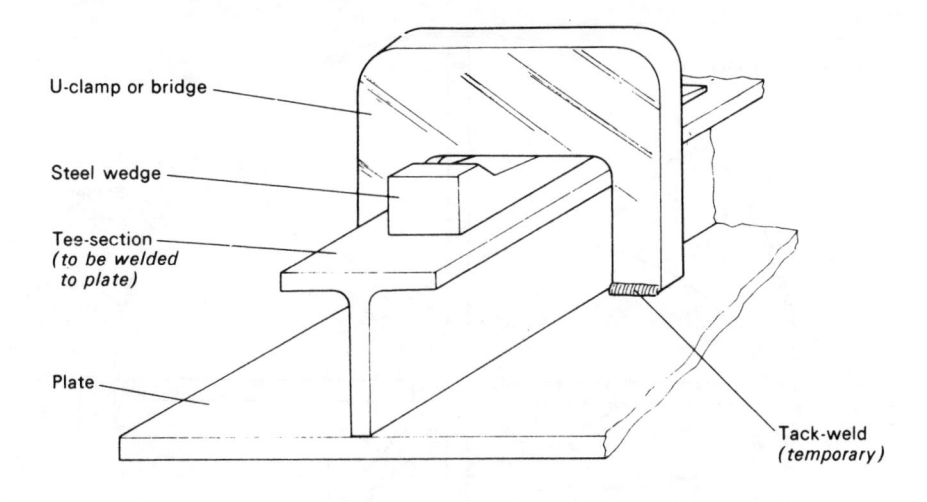

(a) Application of clamping bridge

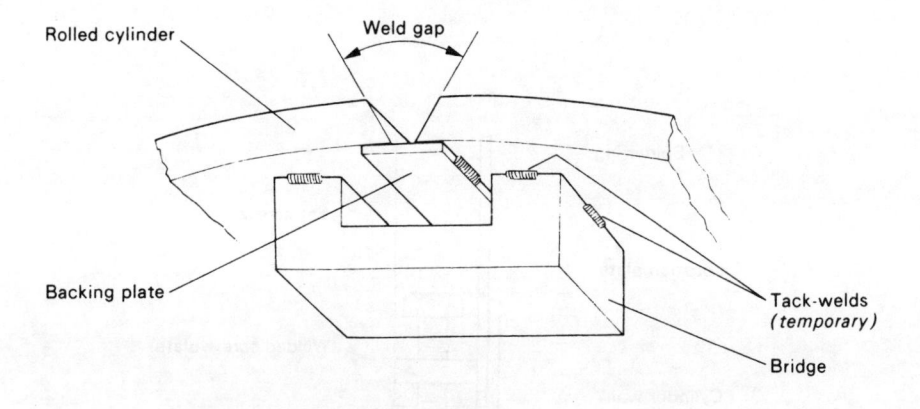

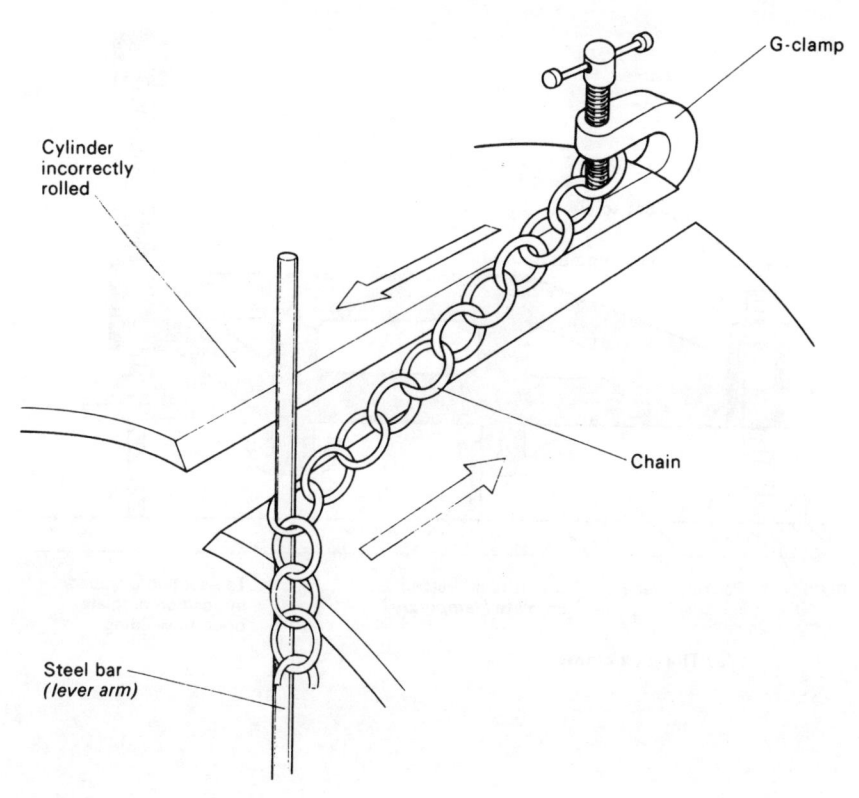

Fig. 8.39 Use of chain and bar

(b) Application of assembly bridge

*Assembly bridges are a useful holding aid for maintaining the weld gap during welding.*

*They are used mainly on cylindrical work where backing straps are required*

Fig. 8.38 Use of bridge pieces

## 8.24 General information on the use of clamping aids (welding)

When tack-welding clamping aids such as cleats, bridges, strong-backs and dogs, the tack welds should be small, but sufficient to perform their temporary function. They should be made so that they can be easily removed after they have served their purpose.

*These clamping devices should not be used on hardenable steels because there is a tendency for local hard spots to develop as a result of the small tack welds.*

For shop assemblies, a large, flat, rigid base plate with Tee-slots is generally used for bolting-down purposes. Such tables or plates are ideal for the use of Tee-slot clamps. It is common practice to use angle plates and simple framework for holding components in position prior to welding.

295

(a) The jack clamp

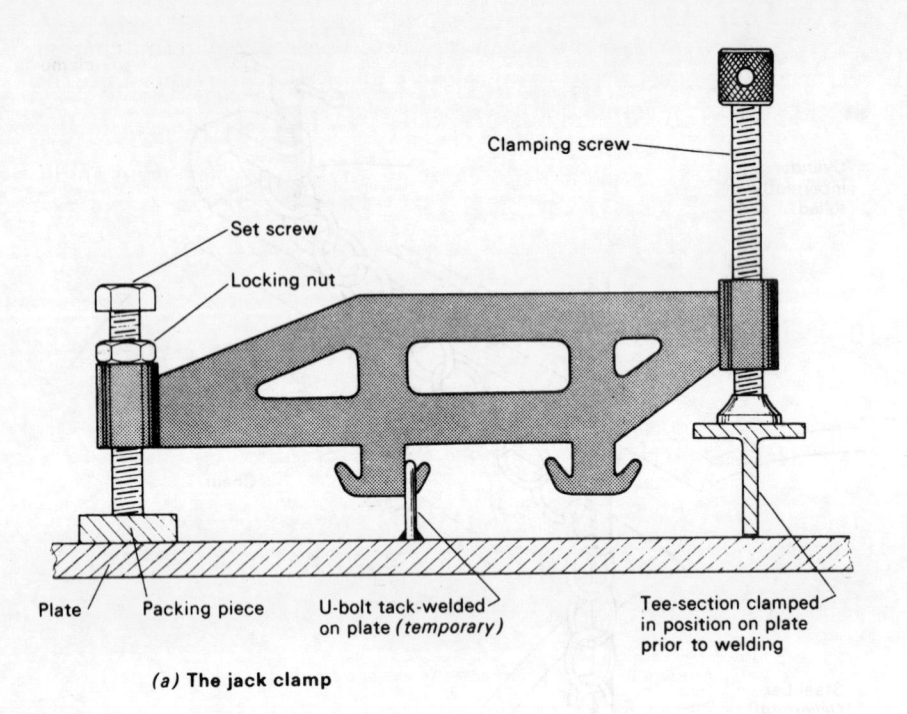

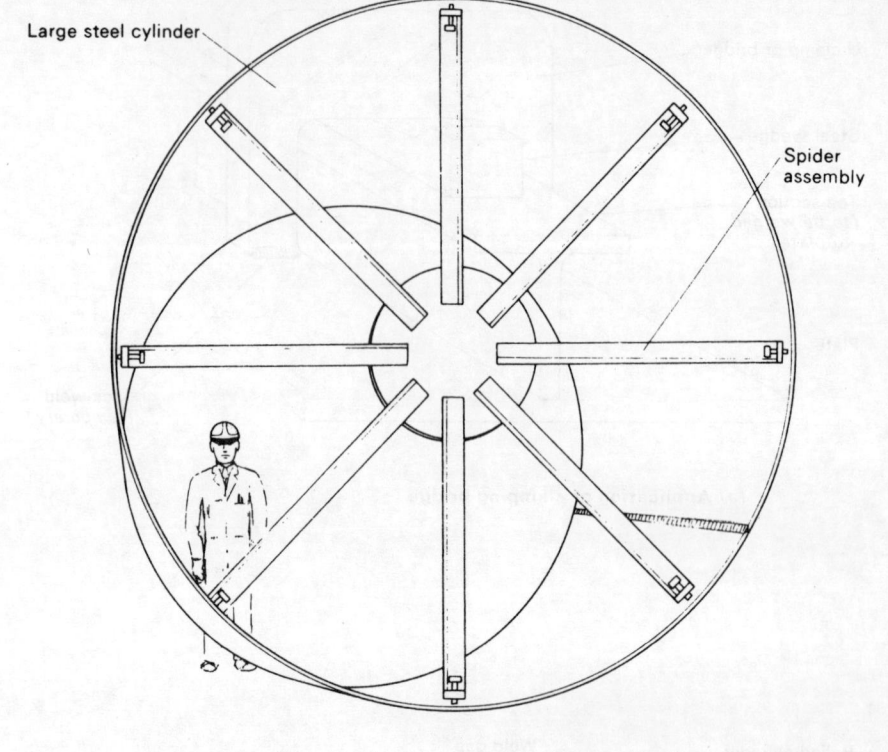

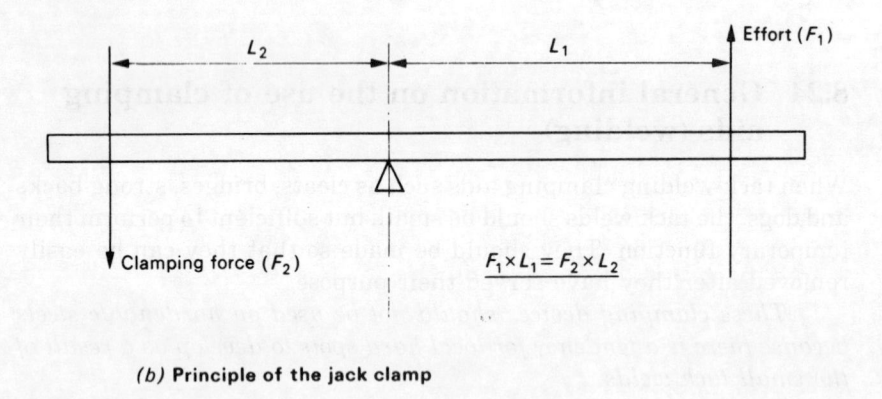

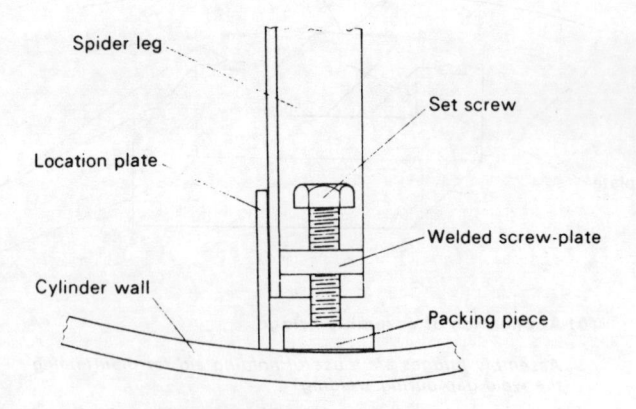

$F_1 \times L_1 = F_2 \times L_2$

(b) Principle of the jack clamp

**Fig. 8.40 Use of jack clamps for holding work**

**Fig. 8.41 Use of spider assemblies for holding work**

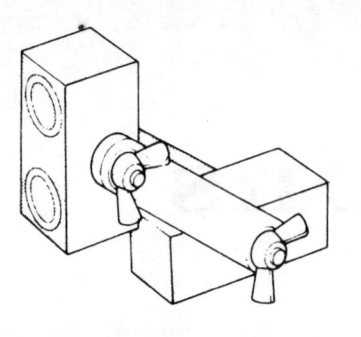

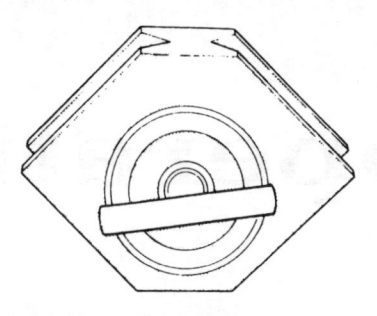

(b) Adjustable magnetic links    (a) Magnetic holder

**Fig. 8.42 Magnetic clamps for welding**

On site, a number of the clamping devices discussed in this chapter are extremely useful for manipulating plates and sections into the correct position and holding them there for welding.

Caliper clamps and magnetic clamps, as shown in Figure 8.42, are also useful where components have to be welded together at an angle.

# 9 Workshop operations

## 9.1 Basic fabrication procedure

Fabrication engineering involves a multiplicity of operations and the use of a range of tools and machines.

One can appreciate the immense variety and volume of work carried out in the fabrication industry by considering the very wide range of materials with which the fabricator works. A section of the market forms of supply of materials used in the industry is listed in Table 9.1. Tables 9.2 and 9.3 list the classifications of hot- and cold-rolled materials respectively.

The production of fabricated components or structures involves five principal operational stages:

1. MEASURING AND MARKING-OUT (see Chapter 5);
2. CUTTING TO SIZE (see Chapter 7);
3. FORMING TO REQUIRED DIMENSIONS (see Chapter 10);
4. JOINING AND ASSEMBLY (see Chapter 3);
5. SURFACE FINISHING FOR PROTECTION AND DECORATIVE PURPOSES (see Chapter 6).

In this chapter a number of examples of fabrication work will be used to illustrate a combination of basic fabrication procedures.

## 9.2 The need for planning operations

A sheet-metal article may be made from one or from several pieces of metal. If the article is made from more than one piece it has to be fabricated to required dimensions by means of hand tools, by machine operations, or by a combination of both. Most components and assemblies involve a sequence of operations which must be planned in order to produce the finished article as economically as possible. *This can be best achieved by ensuring that each operation is performed, where possible, with equipment designed specifically for the purpose.*

The sequence of operations may vary slightly with the individual craftsman and the equipment available in his fabrication workshop.

If a sheet-metal article requires a number of joints to be made during the course of its fabrication, the sequence of operations should be planned to ensure that access is possible.

Generally, operations on sheet and plate commence with the guillotine. It is considered good practice to make a 'trim cut' on standard sized sheets or plates to ensure one datum edge from which other cuts are made. This reduces the size of the flat material, usually in length, and this must be taken into consideration when marking out blank sizes for cutting in order to obtain a maximum yield. Intelligent marking-out on the flat material is essential in order to avoid considerable scrap or 'off-cut' material. Economy in the use of materials should be the prime consideration of all craftsmen engaged in fabrication. The economic arrangement of patterns has been discussed in Chapter 5 and serves to illustrate that with planned marking-out and by careful cutting, a maximum number of blanks can be obtained. This means:

1. A REDUCED NUMBER OF STANDARD SHEETS ARE REQUIRED;
2. OFF-CUT MATERIAL IS KEPT TO A MINIMUM — resulting in a reduction in material cost per part.

## Table 9.1 Market forms of supply

| MARKET FORM OF SUPPLY | REMARKS |
|---|---|
| Sheet metal | Produced in widths up to 1·83 m, in thicknesses ranging from 0·762 mm to 3·25 mm in FERROUS and NON-FERROUS materials, and in a range of qualities and finishes. may be coated or uncoated. |
| Plates | These are HOT-ROLLED in a wide range of sizes and thicknesses, and are usually supplied with square edges. Thicknesses in common use range between 3·25 mm and 51 mm. Heavy plates are available in widths up to 3·66 m and thicknesses from 9·5 mm to 51 mm or more, and light plates in thicknesses from 3·25 mm to 9·5 mm. Steel plate is the flat material used for ship's hulls, bridge building, large tanks and vessels such as gas holders. It is also used to make structural members and pipes of large diameter. |
| Flats | Narrow plates up to 508 mm width and 25 mm thickness, with rolled edges. |
| Slabs | Thick plates with rolled edges and are commonly used for the base plates of heavy stanchions. |
| Sections | In general, 'sections' are structural shapes used for making framework and bridge building. They include CHANNELS and ANGLES, TEES, UNIVERSAL BEAMS and COLUMNS, ROLLED STEEL JOISTS, RAILS, BARS (are usually straight lengths, which are produced in a wide variety of cross-sections, such as round, square, flat, hexagonal, etc., and in a wide range of sizes). Considerable use is made of ROUND, SQUARE and RECTANGULAR HOLLOW TUBES. |
| Wire rod and wire | Wire rods range from 6·35 to 12·7 mm in diameter. Wire is produced from wire rod by wire drawing. *Some grades of wire can be reduced by over 90% of the starting cross-section to produce over 10 times the original length of the rod.* |

## Table 9.2 Classification of cold-rolled sheets

### CLASSIFICATION OF COLD-REDUCED STEEL SHEETS

| MATERIAL TYPE | CLASSIFICATION | REMARKS |
|---|---|---|
| *En2A/1* | CR1 | A non-ageing (stabilised) steel for exceptionally severe draws and/or pressings where it is necessary to avoid the appearance of stretcher strains and loss of ductility due to strain ageing. |
| *En2A/1* | CR2 | For severely drawn parts. |
| *En2A* | CR3 | For general deep drawing. |
| *En2* | CR4 | For moderate drawing and simple bending. |

### FINISHES FOR COLD-REDUCED SHEETS

| FINISH | CLASSIFICATION | DESCRIPTION |
|---|---|---|
| *Cold-reduced general purpose* | CR/GP | A commercially flat sheet, whilst reasonably free from surface defects, should not be used where finish is of prime importance. |
| *Cold-reduced full finish* | CR/FF | A commercially flat 'skin passed' high grade sheet with a minimum of surface blemishes suitable where a high standard of surface finish is desirable. The surface usually has a 'matt finish' and should be used on outside or visible parts and is suitable for any kind of high grade painting. |
| *Cold-reduced vitreous enamelling* | CR/VE | A commercially flat sheet of special quality low metalloid steel with a surface specially prepared for vitreous enamelling. |

*The above table is based on BS1449: Part 1B.*

## 9.3 Forming a 'Pittsburgh lock' (use of folding machine)

The 'Pittsburgh lock' or 'lock seam' is used extensively in ductwork shops. These seams are normally produced by special 'lock-forming machines'. They can, however, be successfully produced on a folding machine provided the correct sequence of operations is performed. The need for adopting a planned sequence of operations and the necessary skill required to make a Pittsburgh lock can be appreciated by practising with scrap metal or off-cuts. Figure 9.1 shows the lock and the layout for forming the lock.

**Table 9.3 Classification of hot-rolled materials**

### CLASSIFICATION OF HOT-ROLLED SHEETS

| MATERIAL TYPE | CLASSIFICATION | REMARKS |
|---|---|---|
| *En2A/1* | HR1 | For severely drawn parts. |
| *En2A/1* | HR2 | For severely drawn parts. |
| *En2A* | HR3 | For general deep drawing. |
| *En2* | HR4 | For moderate drawing and simple bending. |

### FINISHES FOR HOT-ROLLED SHEETS

| FINISH | CLASSIFICATION | DESCRIPTION |
|---|---|---|
| *Hot rolled* | HR | A commerically flat mill finish sheet. |
| *Hot rolled pickled* | HRP | A commercially flat pickled sheet significantly free from scale. |
| *Hot rolled vitreous enamelled* | HR/VE | A commercially flat sheet of special quality, low metalloid steel with a surface specially prepared for vitreous enamelling. |

*The above information is based on BS 1449: Part 1B.*

### QUALITIES OF STEEL

| MATERIAL TYPE | DUCTILITY OR TENSILE STRENGTH | FINISH HOT ROLLED | FINISH HOT ROLLED PICKLED |
|---|---|---|---|
| *En2A/1* | Extra deep drawing quality (killed steel) | HR11 | HRP11 |
| *En21/1* | Extra deep drawing quality. | HR12 | HRP12 |
| *En2A* | Deep drawing quality. | HR13 | HRP13 |
| *En2A* | Flanging or drawing quality. | HR14 | HRP14 |
| *En2* | Commercial quality. | HR15 | HRP15 |

### CLASSIFICATION OF HOT-ROLLED PLATES

| MATERIAL TYPE | CLASSIFICATION | REMARKS |
|---|---|---|
| *En2A/1* | HR11 | For severely drawn parts. |
| *En2A/1* | HR12 | For severely drawn parts. |
| *En2A* | HR13 | For moderately drawn parts. |
| *En2A* | HR14 | Flanging or drawing quality steel. |
| *En2* | HR15 | Commercial quality steel. |

*The above information is based on BS 1449: Part 1A.*

*A pencil should be used for marking the fold lines because a scriber would cut through the protective zinc coating on the surface of the sheet.*

Figure 9.2 shows the sequence of operations which may be adopted for making the lock, using two off-cuts of 24 s.w.g. (0·56 mm) galvanised steel sheet. The first operation is to cut two pieces of sheet metal to the required sizes. The fold lines are then marked in accordance with the dimensions provided on the drawing (Figure 9.1). *The width of the flanged edge is normally made slightly less than the depth of the pocket.* The allowance for the pocket is equal to twice the width of the pocket plus an allowance for knocking over. In the example given, this allowance is $W + W + 6·35$ mm. Lock seams are used as longitudinal corner seams for various shaped ducts, and the necessary allowances have to be added to the pattern or layout.

## 9.4 Transferring paper or metal patterns on to sheet metal

It is most important when using patterns or templates, to position them in the proper manner on the metal to be used in order to avoid unnecessary scrap. Templates must be held in position in order to restrain any tendency to move during the marking out operation. These two factors for successful marking out with patterns or templates are illustrated in Figure 9.3.

## 9.5 The use of notched corners (sheet-metal work)

A common requirement in light sheet-metal work is the notched corner, as in the manufacture of simple folded boxes, trays or electronic chassis. Notching is an essential operation where light-gauge fabrications

## Assembling the lock seam

Flanged edge *(check with try-square after folding)*

Flange knocked over and flattened

Pocket

W

W + 6·35mm

Flange inserted in pocket *(may have to be forced in by striking lightly with a hammer)*

Note: This fold line is marked on the underside

Fold lines

Material —24 s.w.g. galvanised steel

(Pocket section)

152mm

(Flanged section)

9·5mm | 16mm | 6·35mm
152mm | | 76mm

Layout for forming lock

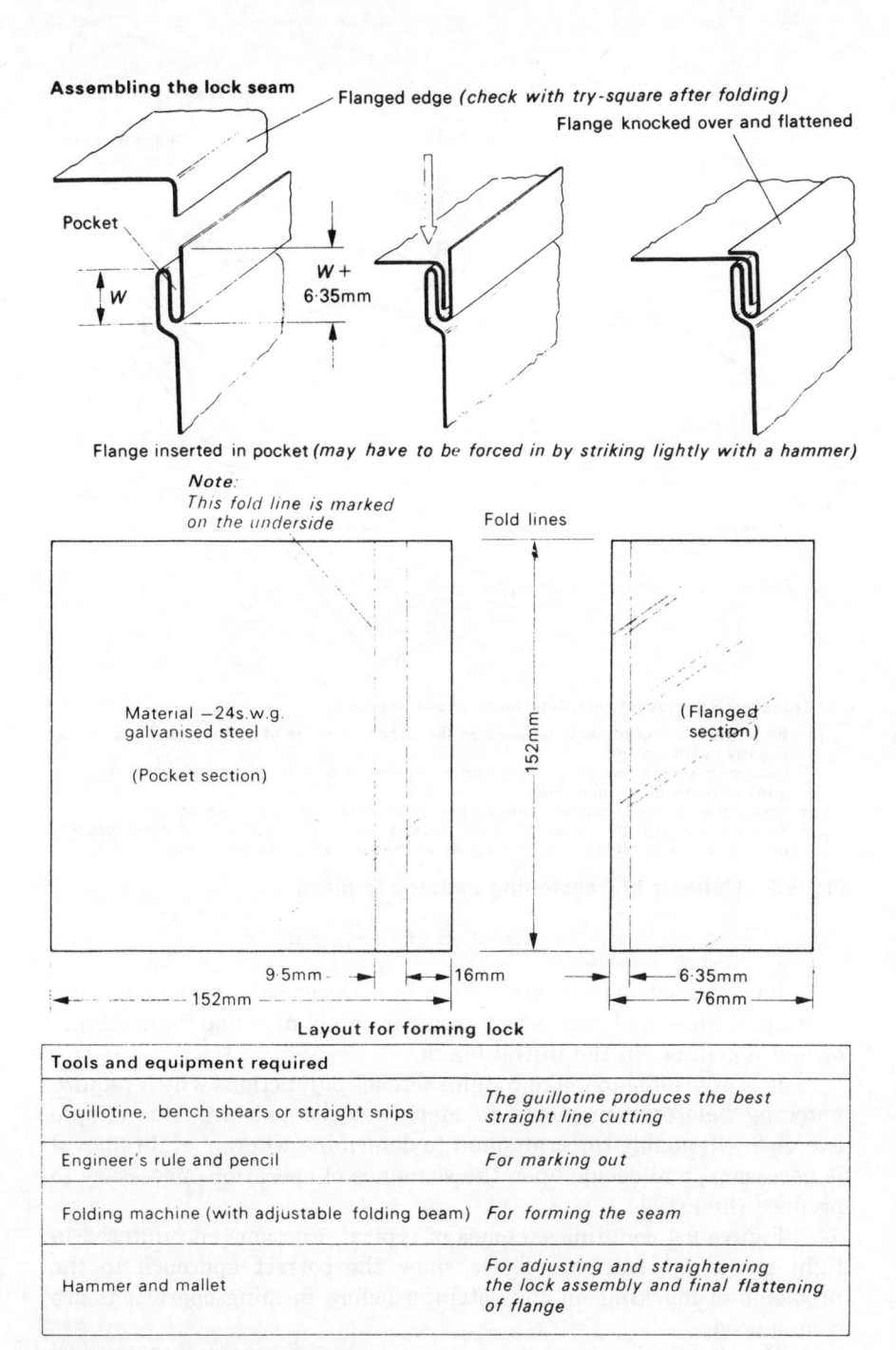

| Tools and equipment required | |
|---|---|
| Guillotine, bench shears or straight snips | The guillotine produces the best straight line cutting |
| Engineer's rule and pencil | For marking out |
| Folding machine (with adjustable folding beam) | For forming the seam |
| Hammer and mallet | For adjusting and straightening the lock assembly and final flattening of flange |

**Fig. 9.1 The Pittsburgh lock**

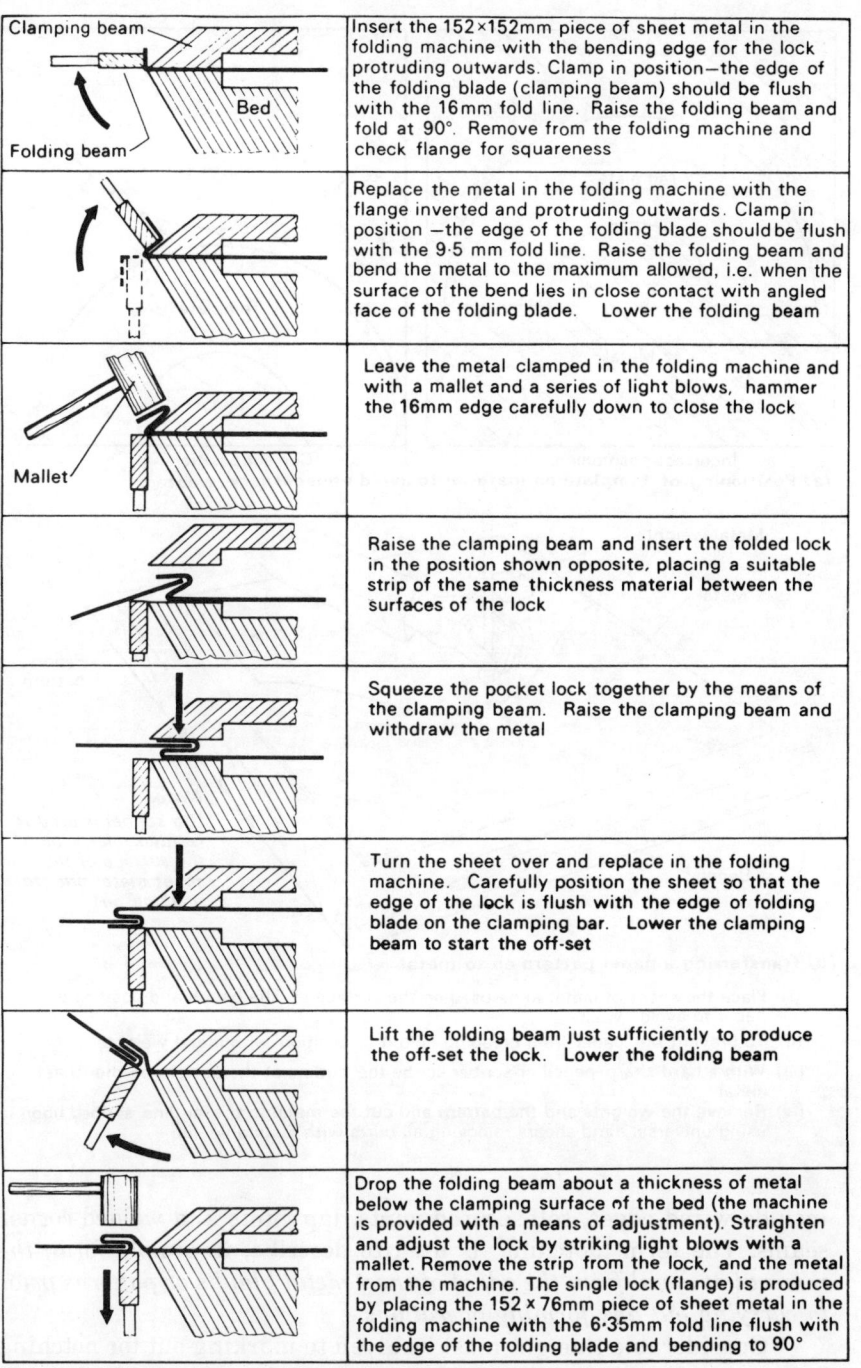

---

| | |
|---|---|
| Clamping beam / Bed / Folding beam | Insert the 152×152mm piece of sheet metal in the folding machine with the bending edge for the lock protruding outwards. Clamp in position—the edge of the folding blade (clamping beam) should be flush with the 16mm fold line. Raise the folding beam and fold at 90°. Remove from the folding machine and check flange for squareness |
| | Replace the metal in the folding machine with the flange inverted and protruding outwards. Clamp in position—the edge of the folding blade should be flush with the 9·5 mm fold line. Raise the folding beam and bend the metal to the maximum allowed, i.e. when the surface of the bend lies in close contact with angled face of the folding blade. Lower the folding beam |
| Mallet | Leave the metal clamped in the folding machine and with a mallet and a series of light blows, hammer the 16mm edge carefully down to close the lock |
| | Raise the clamping beam and insert the folded lock in the position shown opposite, placing a suitable strip of the same thickness material between the surfaces of the lock |
| | Squeeze the pocket lock together by the means of the clamping beam. Raise the clamping beam and withdraw the metal |
| | Turn the sheet over and replace in the folding machine. Carefully position the sheet so that the edge of the lock is flush with the edge of folding blade on the clamping bar. Lower the clamping beam to start the off-set |
| | Lift the folding beam just sufficiently to produce the off-set the lock. Lower the folding beam |
| | Drop the folding beam about a thickness of metal below the clamping surface of the bed (the machine is provided with a means of adjustment). Straighten and adjust the lock by striking light blows with a mallet. Remove the strip from the lock, and the metal from the machine. The single lock (flange) is produced by placing the 152 x 76mm piece of sheet metal in the folding machine with the 6·35mm fold line flush with the edge of the folding blade and bending to 90° |

**Fig. 9.2 Sequence of folding operations (Pittsburgh lock)**

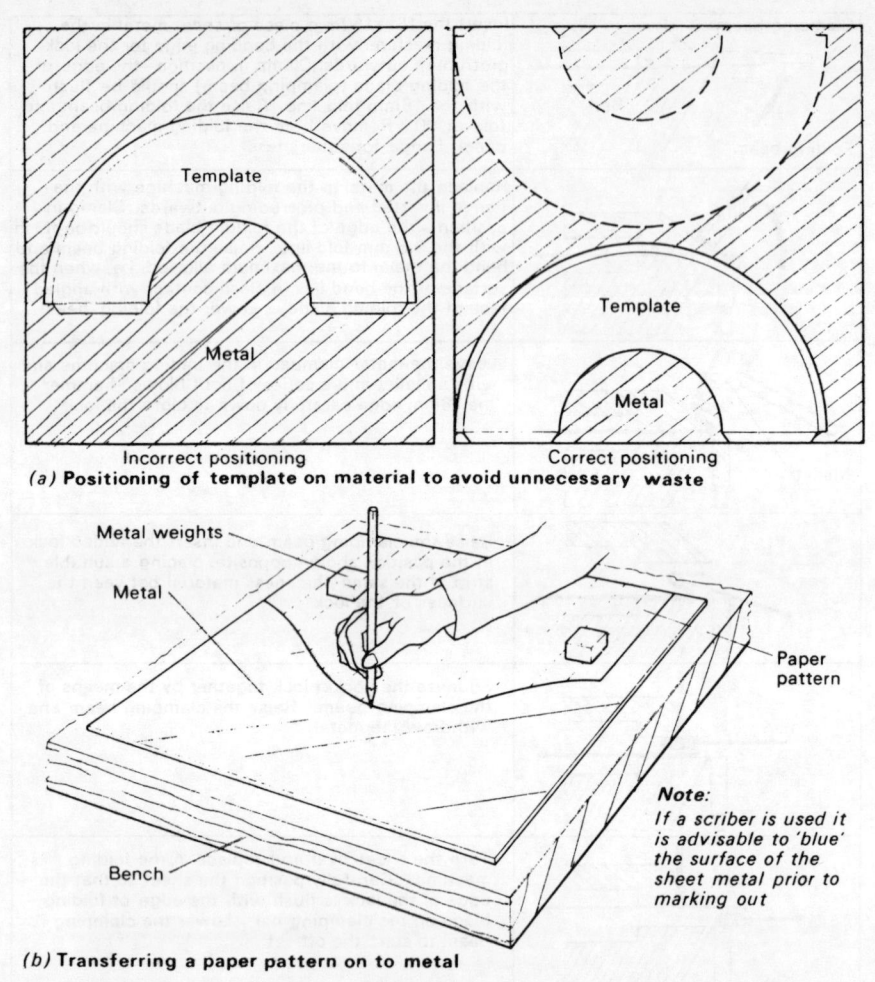

*(a)* **Positioning of template on material to avoid unnecessary waste**

*(b)* **Transferring a paper pattern on to metal**

(i) Place the sheet of metal to be used on the surface of the bench and position the paper to avoid waste
(ii) To prevent the paper from creeping hold it in position with metal weights
(iii) With a hard sharp pencil or scriber scribe the outline of the pattern on the sheet metal
(iv) Remove the weights and the pattern and cut the metal to the outline scribed upon it using universal hand shears removing all burrs with a suitable file

*Note:* If a scriber is used it is advisable to 'blue' the surface of the sheet metal prior to marking out

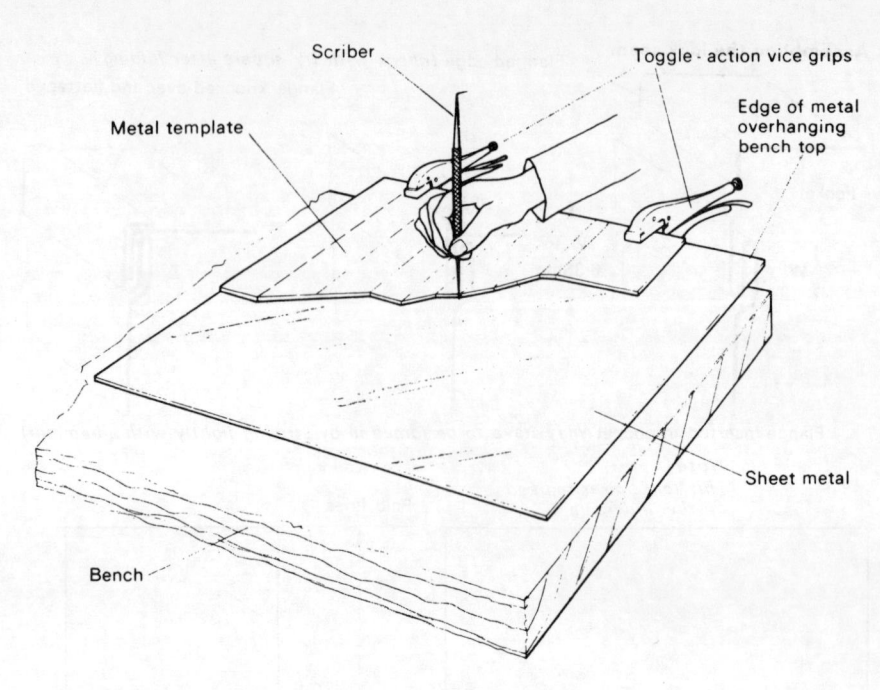

*(c)* **Transferring a metal template on to sheet metal**

(i) Place the sheet of metal to be used on the bench with one of its squared sides slightly overhanging
(ii) Position the metal template in position as shown, and clamp it securely with vice grips to restrain any movement
(iii) Scribe the outline of the template on the sheet metal, using a sharp scriber
(iv) Release vice grips to remove template and cut the sheet metal to the outline scribed upon it with a suitable pair of hand shears. Remove all burrs with a file

**Fig. 9.3 Methods of transferring patterns to metal**

include wired edges, self-secured joints, lap joints and welded corner seams. The term 'notching' is used to describe, in simple form, *the removal of metal from the edges of sheet-metal blanks or patterns prior to carrying out any forming operations.*

Considerable thought must be given to marking-out for notching because good notching is of prime importance where the finished article is to have a neat appearance. There is nothing more unsightly than overlaps, bulges and gaps which result from not allowing for notching, or bad notching, on the initial blank.

It is advisable to make rough sketches of junctions which require notching before commencing to mark out the pattern. This simple exercise will enable the craftsman to determine where metal removal is necessary, and decide upon the sequence of operations necessary to produce the article.

Figure 9.4 contains sketches of typical junctions encountered in light gauge fabrications. These show the correct approach to the problems of marking-out for notching before forming operations are commenced.

Usually, the notching of corners on sheet-metal blanks to be formed

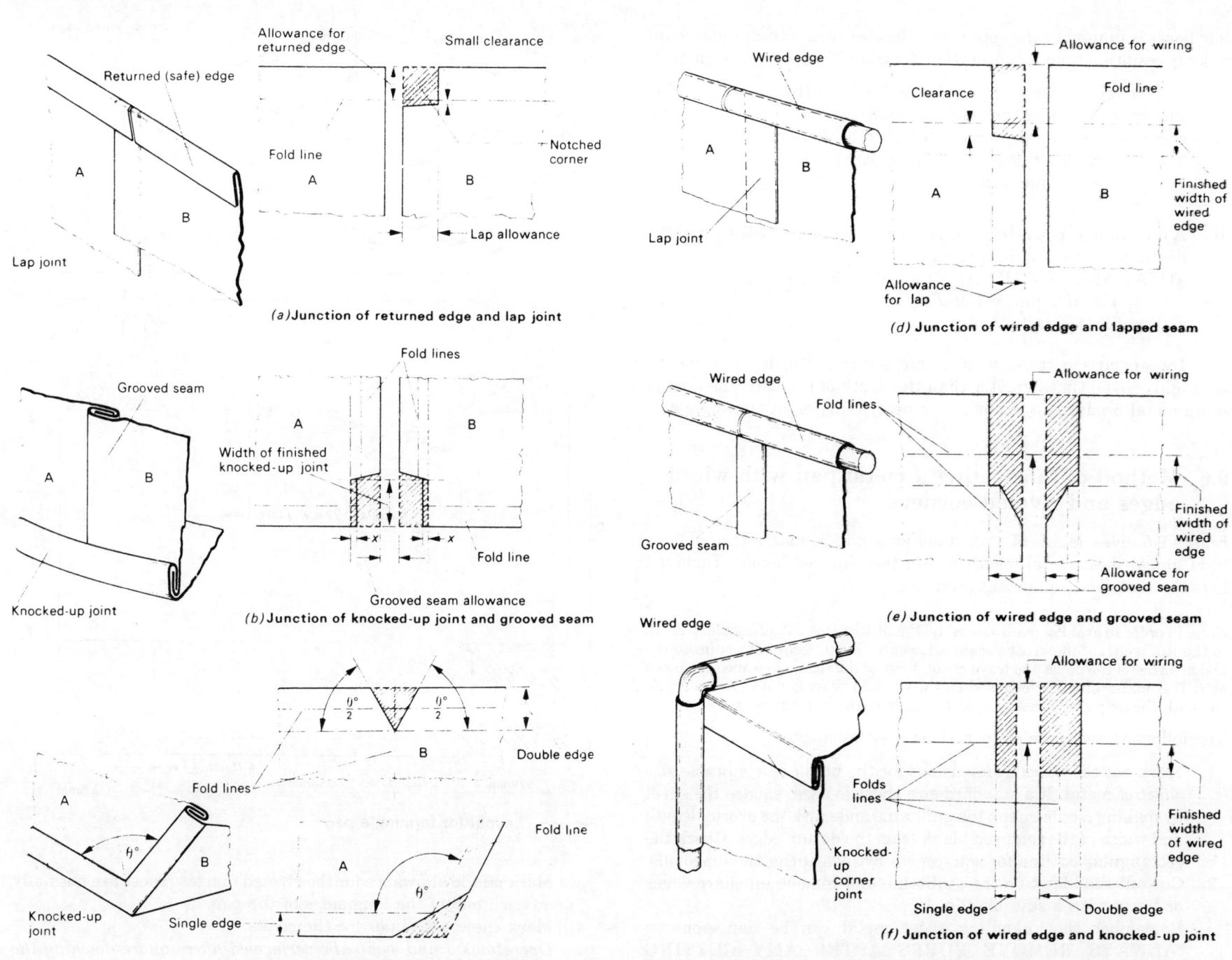

**Fig. 9.4 Junctions which require notching prior to forming**

(a) Junction of returned edge and lap joint
(b) Junction of knocked-up joint and grooved seam
(c) Junction of knocked-up joints (mitred corner)
(d) Junction of wired edge and lapped seam
(e) Junction of wired edge and grooved seam
(f) Junction of wired edge and knocked-up joint

303

into boxes is limited by the depth of the box because of the large amount of scrap resulting from this operation. Consider the following examples:

(a) A rectangular tray is required to be fabricated by square notching the corners, folding the sides up square and welding the vertical corner seams, the depth of notch is to be 20 mm.
TOTAL AREA OF METAL REMOVED BY NOTCHING
= 4 × 20 mm × 20 mm
= $\underline{\underline{1600 \text{ mm}^2}}$

(b) A rectangular box is to be fabricated by the same method, and the depth of notch is to be 152 mm.
TOTAL AREA OF METAL REMOVED BY NOTCHING
= 4 × 152 mm × 152 mm
= $\underline{\underline{92\,416 \text{ mm}^2}}$

By comparison, it can be seen that although the depth of the box is roughly seven times greater than the depth of the tray, the area of scrap metal produced by notching is nearly fifty-eight times greater.

## 9.6 Method of fabricating a metal pan with wired edges and riveted corners

Figure 9.5 gives details of a rectangular pan to be manufactured from light gauge sheet metal together with the required layout. Table 9.4 lists the tools and equipment required.

Note: In order to avoid gaps on assembly, Fig. 9.4(b), it is not advisable to notch to the full width of the groove seam allowance. Make a cut with snips to the inside corner of the notch and leave about 3 mm of the seam allowance, as shown at x. This material will overlap when the grooved seam is made, and a single cut with the snips will ensure a perfect butt of the notched edges.

The following sequence of operations may be adopted:

1. Mark out the overall dimensions for the blank on a suitable size sheet of metal. If a standard size sheet is used, square the sheet by making a trim-cut on the guillotine and mark the overall length and width of the required blank from the datum edges. Check the rectangular outline for squareness by measuirng the diagonals.
2. Cut out the blank on the guillotine and remove all sharp edges or burrs with a suitable file.
Remember that handling sheet metal can be dangerous so 'ALWAYS REMOVE BURRS AFTER ANY CUTTING OPERATION'.

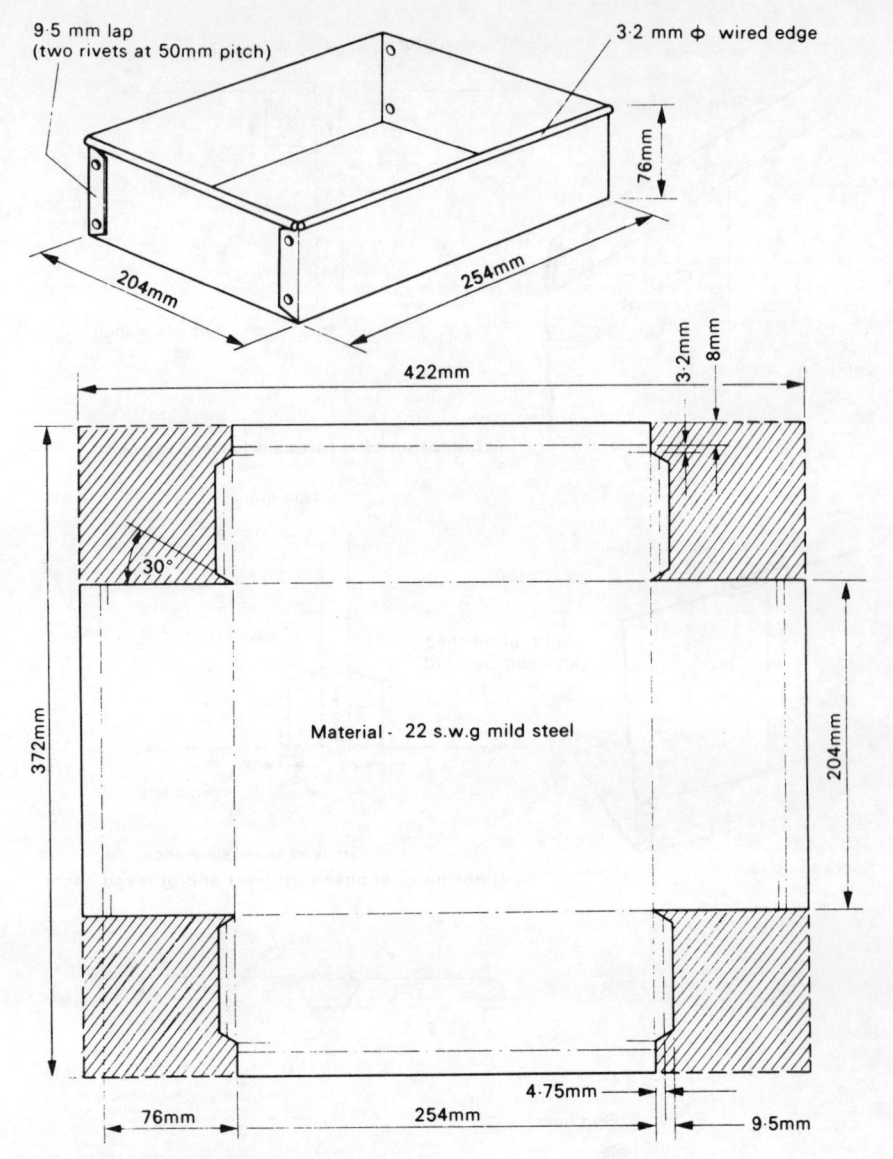

**Fig. 9.5 Layout for forming a pan**

3. Mark off the allowance for the riveted flanges. These are normally on each end of the long sides of the pan.
4. Mark the centre lines for the rivets.
*Operations 3 and 4 may be performed after square notching the corners.*

Table 9.4 Tools and equipment required to make a rectangular pan

| TOOLS AND EQUIPMENT REQUIRED | REMARKS |
| --- | --- |
| Steel rule<br>Straight edge<br>Flat steel square<br>Dividers<br>Jenny odd-legs<br>Bevel<br>Scriber | These are used for marking-out the blank. |
| Centre punch<br>Nipple punch<br>Hammer | Required for marking the positions of the rivet holes. The hammer is also used for riveting. |
| Guillotine<br>Universal snips | For cutting out the blank and notching the corners. |
| Rivet set | For the riveting operation. |
| Folding machine | For bending up the sides |
| Bench stakes<br>Mallet | These are required for completing the bending operations by hand and for wiring the edge. |
| Cutting pliers | For cutting the wire and holding it in position during the wiring operation. |
| Bench vice | For bending the wire if a frame is used, and for holding the hatchet stake when throwing the flange off for the wired edge. |
| Tinman's hand-lever punch | May be used for punching the rivet holes in the corner flanges on the blank. |
| Drilling machine or portable drill | For drilling the rivet holes on assembly. |

5. Notch the corners. Care must be taken, when cutting with the snips, not to 'over-cut' in the corners. The notching of corners can be performed much quicker and more efficiently on a notching machine. Most machines are capable of making square notches up to 102 mm in depth in sheet metal up to 1·6 mm thickness.
6. Mark out the clearance notches at each end of the rivet flanges. These are normally made at an angle of 30° as shown in Figure 9.4, and may be marked with the aid of a bevel.
7. Mark the positions for the rivets on the flanges and centre punch.
8. Drill or punch the rivet holes. *If a drill is used the blank must be supported on a block of wood and the whole assembly securely clamped to the drilling table.*

*Figure 9.6 illustrates a 'tinman's hand-lever punch' which may be employed for punching holes in sheet metal.*

9. Bend the long sides up first in the folding machine to an angle of about 45° and flatten back as shown in Figure 9.7(a). This provides a crease line for final bending by hand over a hatchet stake after the other two sides have been folded up.
10. Bend up the ends in the folding machine to 90°. This operation will also bend up the rivet flanges as shown in Figure 9.7(b).
11. Complete the bending operations over a hatchet stake as shown in Figure 9.7(c).
12. During operation 13, the flanges have to be knocked back slightly to accommodate the ends. This is rectified by bending the laps over a suitable stake with a mallet as shown in Figure 9.7(d).
13. Support each corner, in turn, on a suitable anvil or bench stake and mark the centres of the rivet holes through the holes in the flange with a nipple punch.
14. Drill the top hole in each corner first and insert a rivet to maintain the correct location before drilling the bottom holes. During this drilling operation the job must be properly supported on a block of wood and securely clamped.

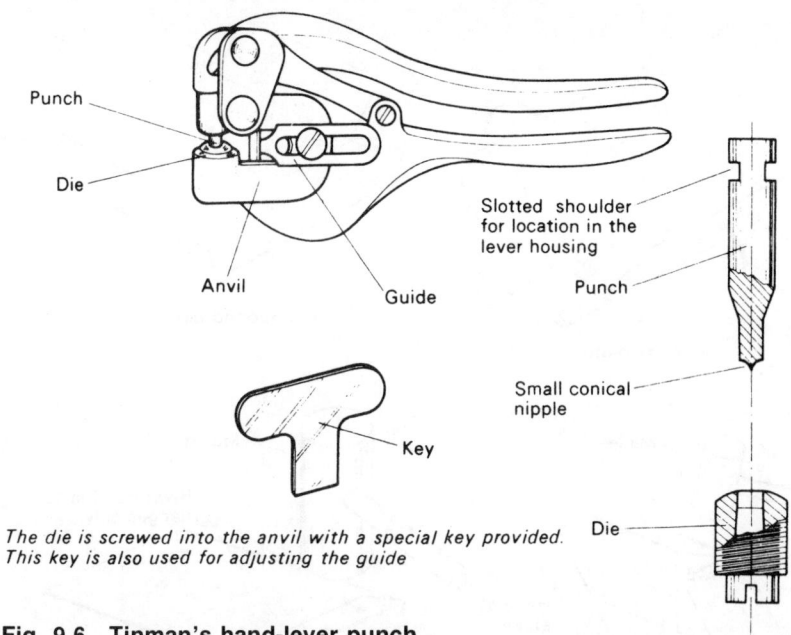

The die is screwed into the anvil with a special key provided. This key is also used for adjusting the guide

**Fig. 9.6 Tinman's hand-lever punch**

*(a)* **First bend**

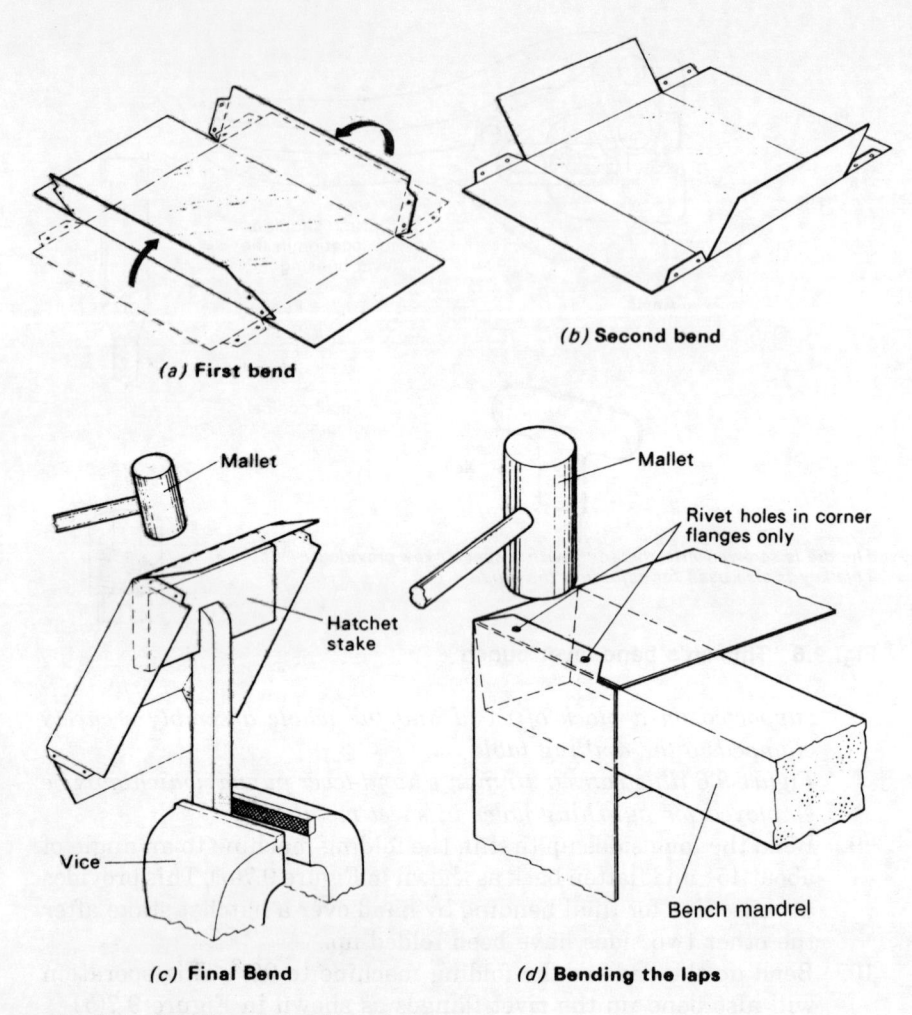

*(b)* **Second bend**

*(c)* **Final Bend**

*(d)* **Bending the laps**

**Fig. 9.7 Bending the sides of the pan**

15. Deburr the holes. The assembly is now ready for riveting.
16. Rivet the corners. The method of riveting is shown in Figure 9.8.
17. Bend the edges for the wire over a hatchet stake, using a mallet as shown in Figure 9.9(*a*).
18. Cut the wire to the required length and make a frame as shown in Figure 9.9(*b*). The frame may be bent to shape in the bench vice. A great many craftsmen prefer to apply the wire direct to the job, thus dispensing with the operation of forming the frame. *In either case the ends of the wire should terminate about the middle of one of the short sides of the pan.*

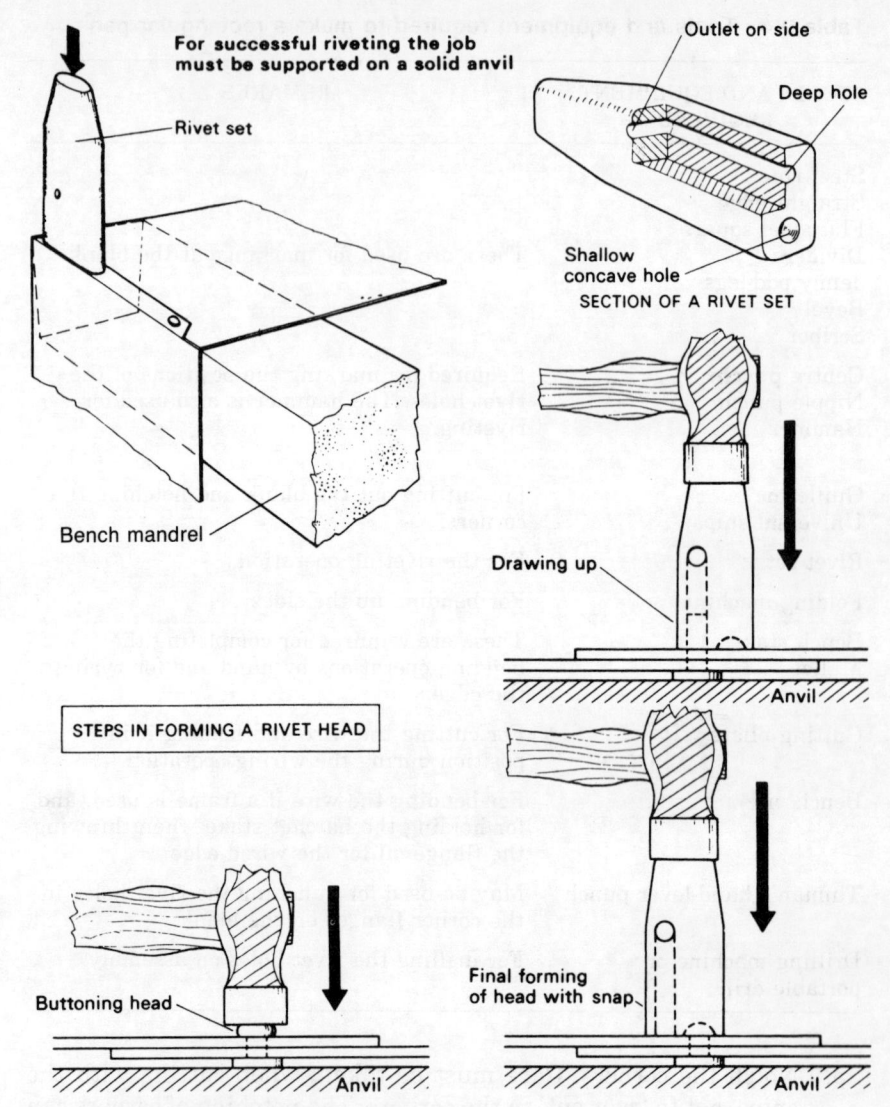

For successful riveting the job must be supported on a solid anvil

SECTION OF A RIVET SET

STEPS IN FORMING A RIVET HEAD

**Fig. 9.8 Method of riveting**

Note: Hand-lever punches are supplied with a range of punches and dies suitable for punching small diameter holes in sheet metal. No distortion of the sheet occurs, and the holes are practically free from burrs — one advantage over drilling. The punch is located exactly in centre punch marks on the surface of the metal by 'feel' rather than by 'sighting'. The punch is designed with a small conical nipple for this purpose.

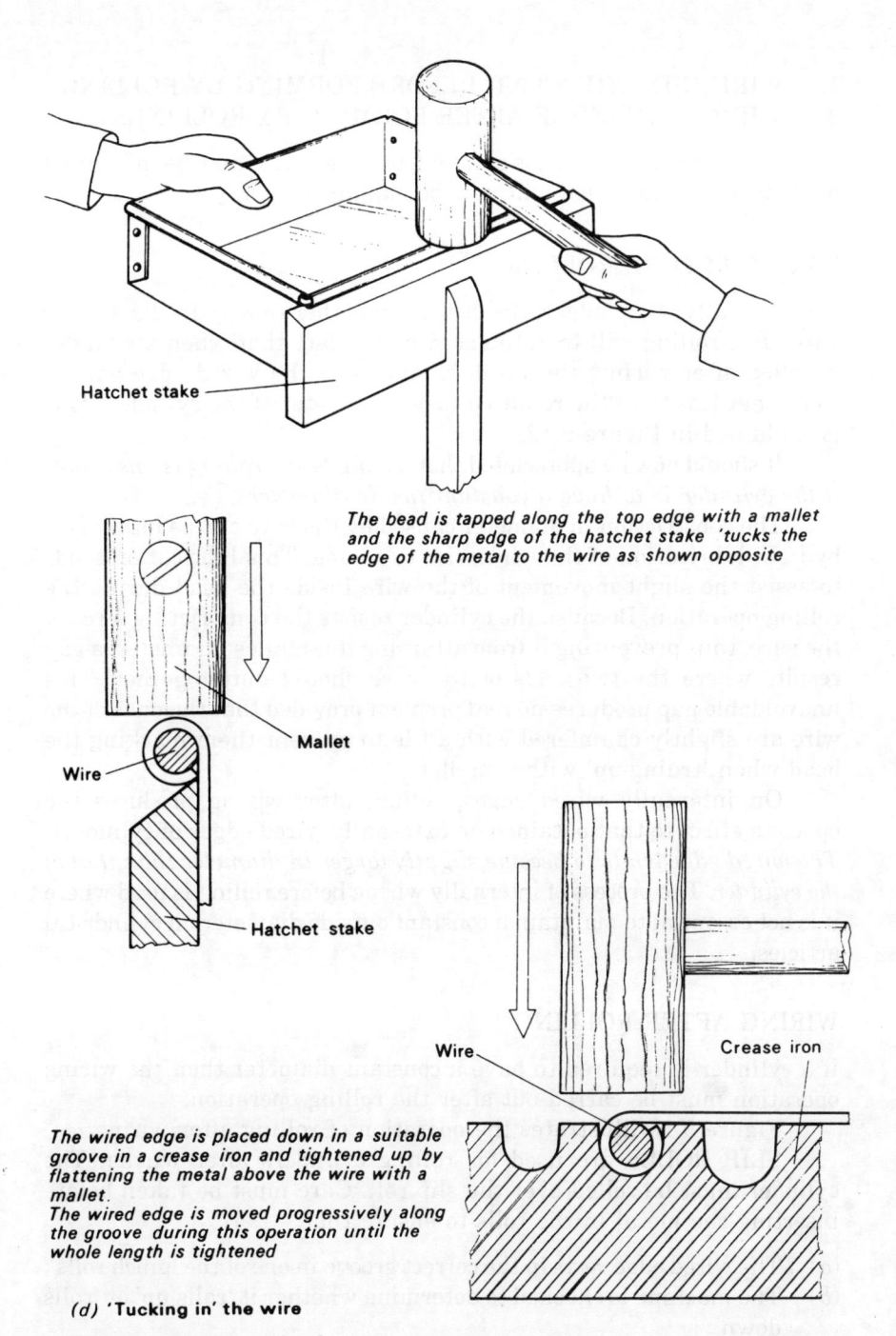

*The bead is tapped along the top edge with a mallet and the sharp edge of the hatchet stake 'tucks' the edge of the metal to the wire as shown opposite*

*The wired edge is placed down in a suitable groove in a crease iron and tightened up by flattening the metal above the wire with a mallet.*
*The wired edge is moved progressively along the groove during this operation until the whole length is tightened*

(d) 'Tucking in' the wire

**Fig. 9.9 Wiring the edge of the pan**

**FAULTS WHEN WIRING AN EDGE**

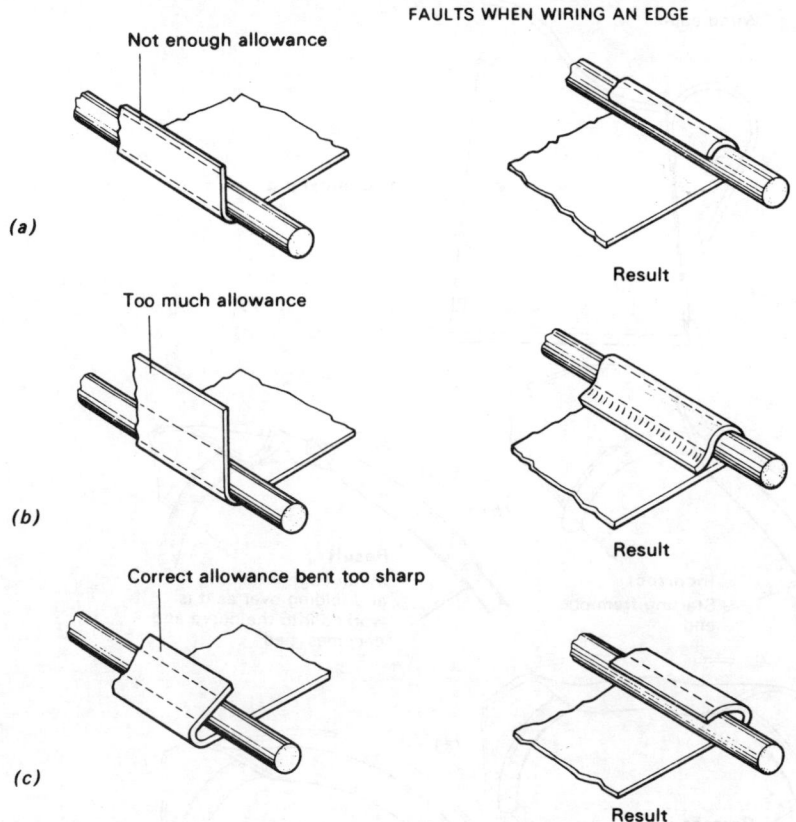

(a) Not enough allowance — Result

(b) Too much allowance — Result

(c) Correct allowance bent too sharp — Result

**Fig. 9.10 Typical faults when wiring a straight edge**

19. Insert the wire and form the metal over the wire using a mallet and a suitable bench stake as shown in Figure 9.9(c).
20. Complete the wiring by 'tucking-in' over a hatchet stake using a mallet as shown in Figure 9.9(d).

## 9.7 Faults when wiring an edge

In most fabrication shops wiring is essentially a hand operation. As a general rule the allowance for a wired edge is $2\frac{1}{2}$ times the diameter of the wire. The flange or edge is folded up just over 90° and knocked over the wire with a mallet. A skilled craftsman can control the operation of forming the metal over the wire by pulling over or knocking back the metal with the mallet. Figure 9.10 illustrates common faults which may occur when wiring a straight edge.

307

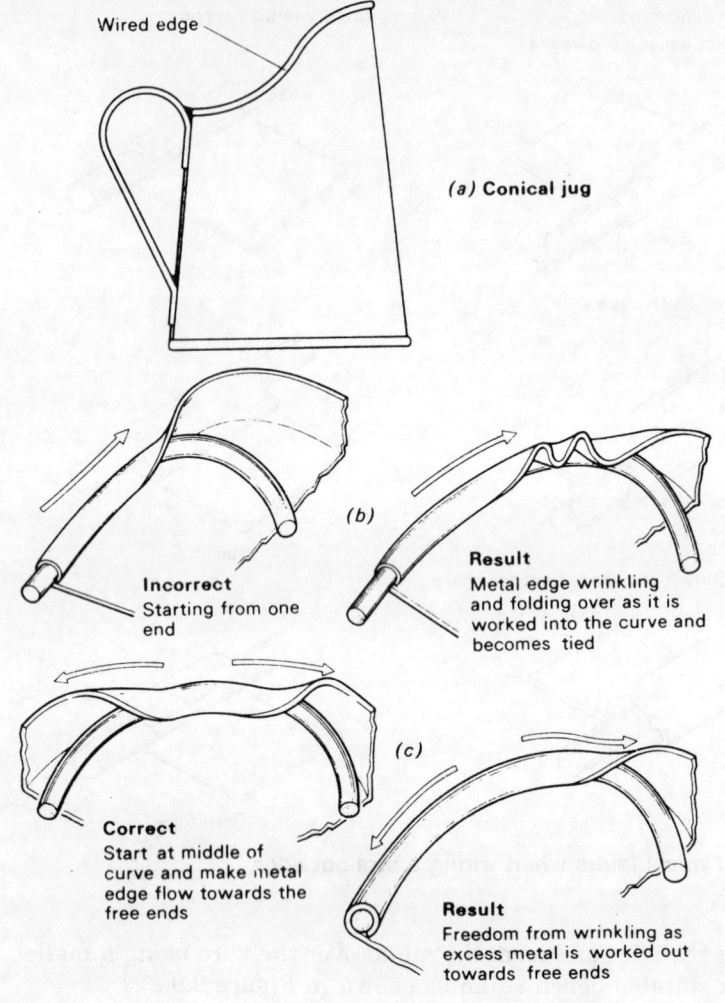

**Fig. 9.11 Typical faults when wiring a curved edge**

Figure 9.11 shows the faults that may occur when wiring a curved edge. The conical jug shown in Fig. 9.11(a) has a wired edge. This edge, like all contoured edges must be wired in the flat. This type of wiring operation can prove difficult if the correct technique is not adopted. Figures 9.11(b) and (c) illustrate the incorrect and correct methods of performing the operation of wiring a curved surface in the flat.

## 9.8 Methods of wiring cylinders and cones

The two methods of producing a wired edge on curved surfaces are:

1. WIRING IN THE 'FLAT' BEFORE FORMING BY ROLLING;
2. WIRING THE EDGE AFTER FORMING BY ROLLING.

The method employed is optional, but the choice may be influenced by certain factors which will now be explained.

### WIRING BEFORE ROLLING

On externally wired edges, the decision whether to wire in the 'flat' or wire after rolling will be influenced by the fact that when a cylinder is rolled after wiring, the inside diameter of the wired edge will be somewhat less than the required inside diameter of the cylinder. This is explained in Figure 9.12.

It should now be appreciated that *wiring before rolling is unsuitable if the cylinder is to have a constant inside diameter*.

The constriction in diameter caused by the wire can be minimised by light lubrication of the wire prior to wiring. This lubrication tends to assist the slight movement of the wire inside the bead during the rolling operation. Because the cylinder resists the constricting force of the wire, thus preventing it from attaining its smallest diameter, a gap results where the two ends of the wire should butt together. This unavoidable gap produces no real problem provided that the ends of the wire are slightly chamfered with a file to prevent them marking the bead when 'truing-up' with a mallet.

On internally wired edges, rolling after wiring produces the opposite effect to that obtained on externally wired edges of cylinders. *The wired edge tends to become slightly larger in diameter than that of the cylinder*. The process of internally wiring before rolling is used where it is not essential to maintain a constant outside diameter on cylindrical articles.

### WIRING AFTER ROLLING

If a cylinder is required to have a constant diameter then the wiring operation must be carred out after the rolling operation.

Figure 9.13 illustrates the operation of rolling after wiring.

SLIP ROLLS are used for rolling cylinders after wiring. The cylinder must be rolled over the slip roll. Care must be taken before inserting the metal in the rolls to ensure that:

(a) The wired edge rests in the correct groove in one of the 'pinch rolls';
(b) The machine is checked to determine whether it 'rolls up' or 'rolls down'.

The slip roll enables the cylinder to be withdrawn easily from the

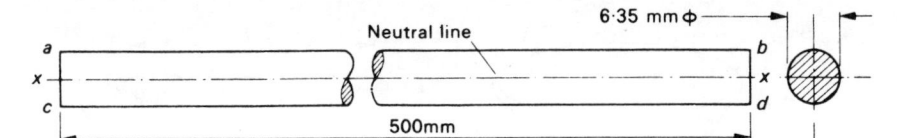

If a piece of 6·35mm diameter wire 500mm in length is rolled to form a ring it will have a mean diameter of 159mm. Likewise, when sheet of 1·6mm thick metal of the same length is rolled to form a cylinder, it will have a mean diameter of 159mm.

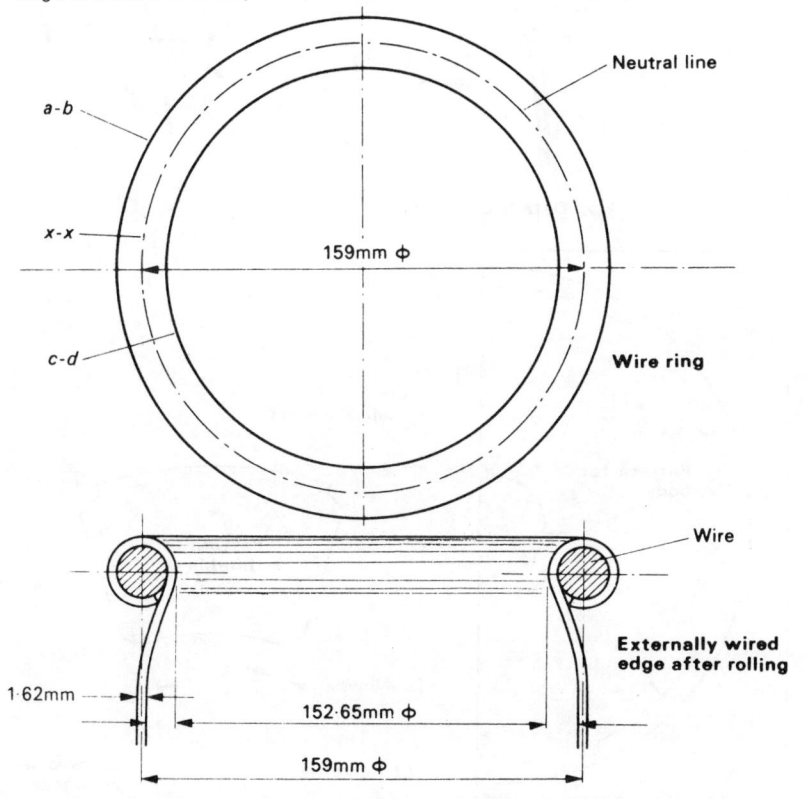

If the sheet is wired along on edge (i.e. the 500mm length) and then rolled to form a cylinder, the above effect will result. It will be seen that both sheet metal and wire forms assume a common **mean diameter** thus causing a considerable constriction as illustrated above

**Fig. 9.12** The effect of wiring before rolling

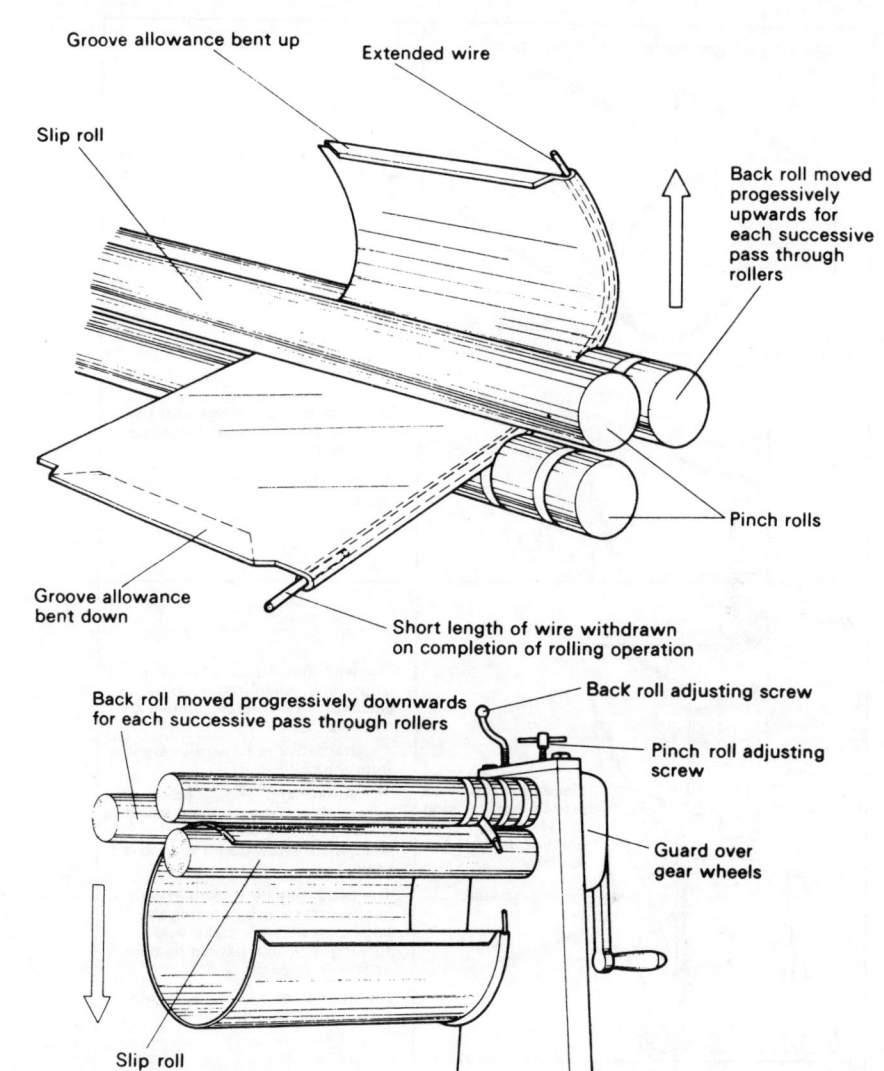

**Fig. 9.13** Rolling after wiring

machine on completion of the rolling. The sequence of operations for wiring after rolling is shown in Figure 9.14.

## 9.9 Method of fabricating a domestic funnel (tinplate)

Figure 9.15 gives details of a domestic funnel which is to be fabricated from tinplate. A funnel has been selected for the purpose of showing the sequence of operations and the combined use of hand tools and machines necessary for the construction of a tapering article with a grooved seam and a wired edge.

When making a funnel, the normal procedure, as is the case with all tapering articles, is to wire the edge after it has been formed to the required shape and fastened together.

309

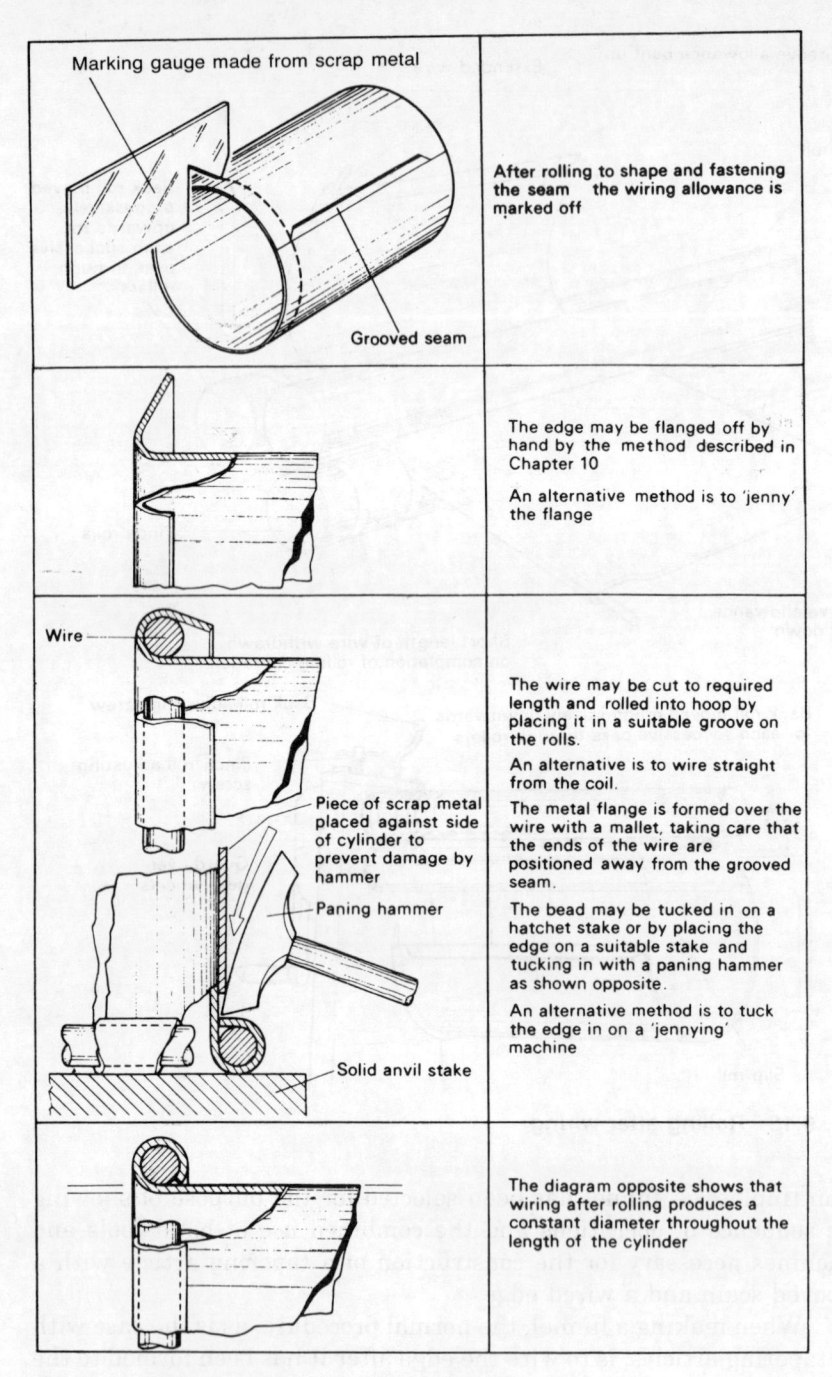

**Fig. 9.14 Wiring a cylinder after rolling**

*Marking gauge made from scrap metal*

After rolling to shape and fastening the seam the wiring allowance is marked off

*Grooved seam*

The edge may be flanged off by hand by the method described in Chapter 10

An alternative method is to 'jenny' the flange

*Wire* — *Piece of scrap metal placed against side of cylinder to prevent damage by hammer* — *Paning hammer* — *Solid anvil stake*

The wire may be cut to required length and rolled into hoop by placing it in a suitable groove on the rolls.

An alternative is to wire straight from the coil.

The metal flange is formed over the wire with a mallet, taking care that the ends of the wire are positioned away from the grooved seam.

The bead may be tucked in on a hatchet stake or by placing the edge on a suitable stake and tucking in with a paning hammer as shown opposite.

An alternative method is to tuck the edge in on a 'jennying' machine

The diagram opposite shows that wiring after rolling produces a constant diameter throughout the length of the cylinder

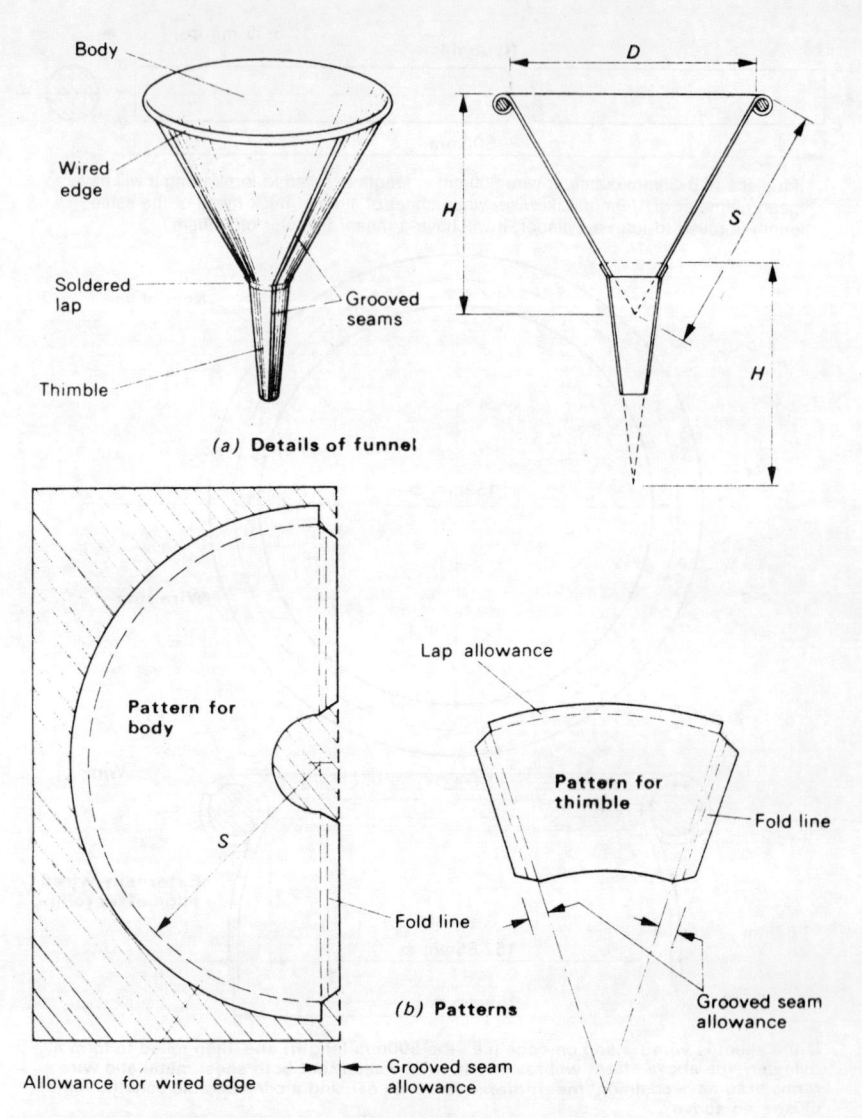

310

**Fig. 9.15 Details and patterns for fabricating a domestic funnel**

*(a) Details of funnel*

Body, Wired edge, Soldered lap, Thimble, Grooved seams

*(b) Patterns*

Pattern for body, Pattern for thimble, Fold line, Grooved seam allowance, Allowance for wired edge, Grooved seam allowance

The funnel shown in Figure 9.15(a) is made in two parts, a BODY and a THIMBLE. Each are frustums of a right cone, the patterns for which are developed by the RADIAL LINE METHOD which is explained in Chapter 2. By making the diameter equal to the slant height S the development of the body will be a quadrant, as shown in Figure 9.15(b). It is common practice to make the distance H from the base of the cones to the apex constant. The patterns may be marked out on paper and transferred on to the tinplate, or marked out directly on to the sheet. The appropriate allowances for the seams and for notching are made in the usual manner.

## 9.10 Breaking the grain of the metal

*The first operation when working with tinplate should be to 'break the grain' of the metal* to prevent ridges forming on its surface. This consists of rolling the piece of tinplate backwards and forwards a few times in the rolls, reversing the bending each time. This process will ensure that the article to be formed by rolling will have a smooth surface free from kinks or ridges. If this 'breaking' operation is not carried out, a cylindrical or conical article, with only one rolling, develops ridges around the shaped body which, if not seen, may be more easily felt by passing the palm of the hand over the rolled surface.

It is sound practice always to break the grain, especially on metals which have been 'cold reduced' before commencing forming operations by rolling.

Once the breaking operation is completed the pattern or blank should be rolled out flat in readiness for forming operations.

Note: It is common practice to insert a narrow strip of metal between the hooked flanges for the grooved seam to prevent them from being flattened during the rolling operation. *The material being rolled should be lightly gripped between the pinch rolls.*

## 9.11 Forming the funnel

Figure 9.16 shows typical workshop operations for forming the body and thimble for the funnel, and Table 9.5 lists the tools and equipment required.

The tapered flange on the thimble for the lap can be produced by a light stretching of the edge on a suitable stake using a stretching hammer.

1. Fold the locks for the grooved seam one up and one down as shown at A and B in Figure 9.16(a) on both the body and thimble. A suitable machine for this operation is the BENCH ROLL TYPE FOLDER.

    The edge to be folded is pushed under the forming blade until it butts against the back guide plate. The sliding back gauge plate is adjusted backwards or forwards until the fold line is aligned with the edge of the forming blade. The edge is folded right over by pulling the operating handle upwards. In this case the metal blank is positioned at one end of the machine, as shown in the diagram. One edge is folded and then the blank is turned over and inserted in the same position, thus ensuring that one edge is turned up and the other one down.

2. The body may be formed to shape over a funnel stake. This operation consists of bending the body of the funnel by hand with a sliding motion over the stake. A mallet is used for final forming to shape after grooving the seams (Figure 9.16(b)).

**Table 9.5 Tools and equipment required to make a funnel**

| TOOLS AND EQUIPMENT REQUIRED | REMARKS |
| --- | --- |
| Steel rule<br>Scriber<br>Dividers | These are used for marking out. |
| Guillotine<br>Bench shears | For cutting out the blanks and developed shapes. |
| Universal snips | Straight snips may be used for straight line and external curve cutting, but bent pattern snips will be required for cutting internal curves. Universal snips will perform both these operations. |
| Rolling machine | For BREAKING-IN the TINPLATE. |
| Flat bed folding machine | For forming the locks for the grooved seams. An alternative method for performing this operation is to throw the edges over a hatchet stake using a mallet. |
| Funnel stake | Used for forming the body of the funnel. |
| Bick iron | Used for forming the thimble. |
| Mallet<br>Stretching hammer | For stretching lap circumference of thimble. |
| Grooving tool<br>Hammer | For fastening the seams. |
| Soldering stove<br>Soldering iron | For soft soldering the inside of the seams and the lap joint between body and thimble. |
| Cutting pliers | For cutting wire for the wired edge. |
| Jennying machine | Two operations are required:<br>1. Turning the wire allowance.<br>2. 'Tucking' the edge over the wire. |

3. The thimble is formed in a similar manner over a long tapering bick iron. It is formed roughly to shape, the grooved seams are interlocked and the thimble is driven on to the bick iron to hold the interlock tight before completing the grooved seam (Figure 9.16(c)).
4. The body and thimble are joined by soft soldering. A bick iron is used to flange the joint as shown in Figure 9.16(d).

An alternative method of attaching the thimble to the body is to make the allowance for the soldered lap on the thimble. The patterns must be adjusted accordingly. In this case the small end of the body for the funnel is reamed out on the bick iron as shown above. This operation will set the lap edge to the correct angle to allow the thimble to be forced in.

Note: Once the body and the thimble have been formed to the required shape, the seams are grooved with a grooving tool and soft soldered on the inside. When forming the thimble, it is advisable to over bend with the locks over-lapping. Then when the thimble is forced on to the tapering bick iron the interlocking edge will be held securely in position ready for the grooving operation.

The next operation is to cut the required length of wire and roll it into a hoop on the rolls. The wire allowance is flanged up on the jenny, the wire inserted and the metal closed over the wire with a suitable pair of rolls on the jenny, making sure that the butt joint on the hoop is positioned away from the grooved seam on the body.

## 9.12 Forming operations on a universal jennying machine

Figure 9.17 shows the types of wheels or rolls which are used on a universal jennying machine together with the various operations which they perform.

Figure 9.18 illustrates a number of operations for preparing edges with the universal jenny machine. In Figure 9.18(a) the working faces of the rolls are radiused and are used for forming narrow edges on circular and irregular articles. In Figure 9.18(b) the upper roll has a sharp edge and is used for turning single edges on curved work and discs. These rolls are also used for tucking in wired edges.

The 'V'-shaped rolls shown in Figure 9.18(c) are ideal for turning up a double edge on elbows for paned-down or knocked-up seams.

Precaution: When using sharp edged rolls for flanging operations, do not over-tighten the top roll. If the top roll is too tight the metal will be sheared. The top roll should be adjusted so as to afford a light grip on the metal between the rolls.

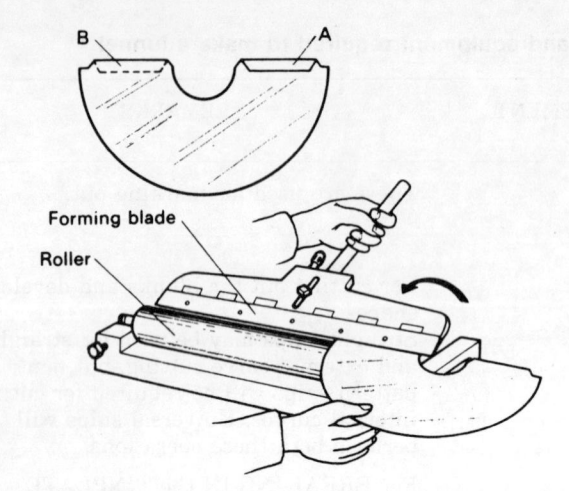

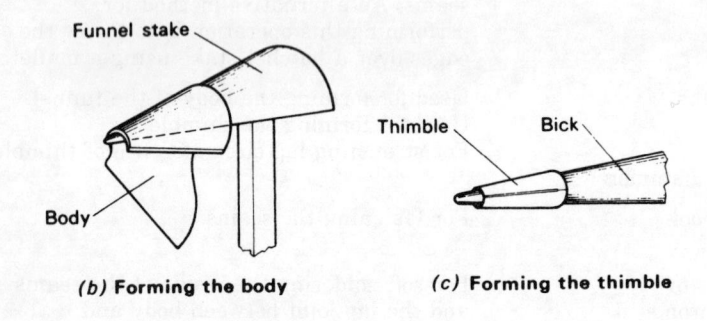

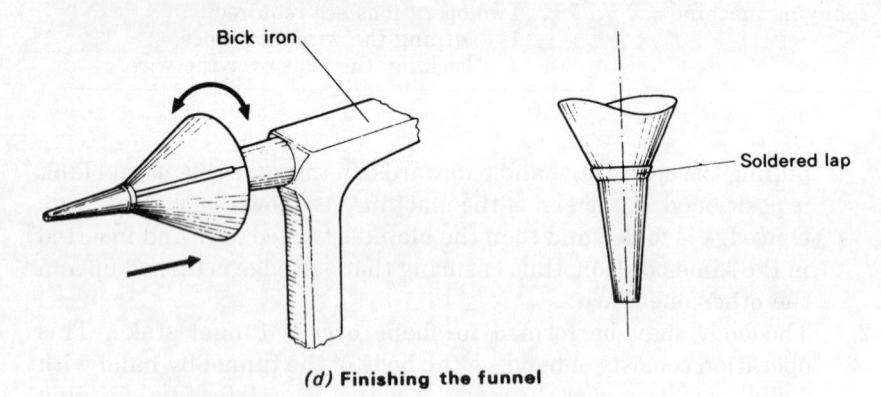

**Fig. 9.16 Operations for forming a funnel**

(a) Folding the locks
(b) Forming the body
(c) Forming the thimble
(d) Finishing the funnel

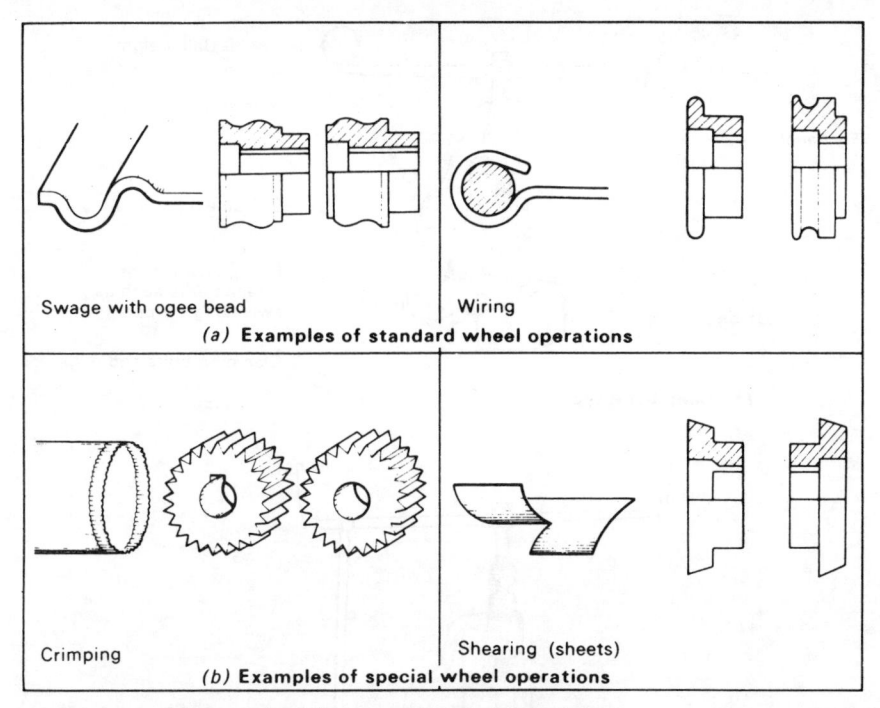

**Fig. 9.17 Wheels and rollers for jennying machines**

For a wired edge the allowance must be equal to $2\frac{1}{2}$ times the diameter of the wire. The measurement is taken from the face of the gauge to the centre of the upper roll as shown in Figure 9.18(d). The gauge is adjusted, usually by a knurled nut at the side of the machine. After setting to the required measurement the gauge is locked in position.

When flanging a disc, a small piece of metal folded (as shown in Figure 9.19) should be used to prevent injury to the operator's hand. As an extra precaution remove all burrs from the metal disc before commencing the operation.

## 9.13 Fly press operations

A fly press is basically a hand press where the ram is worked in the frame by the operation of a screw which is rotated by the operator turning a handle or 'fly'.

The types of fly press usually found in a fabrication shop are:

1. *Standard or C-frame* The main dimensions are the bed to guides, centre to back, and the screw diameter.

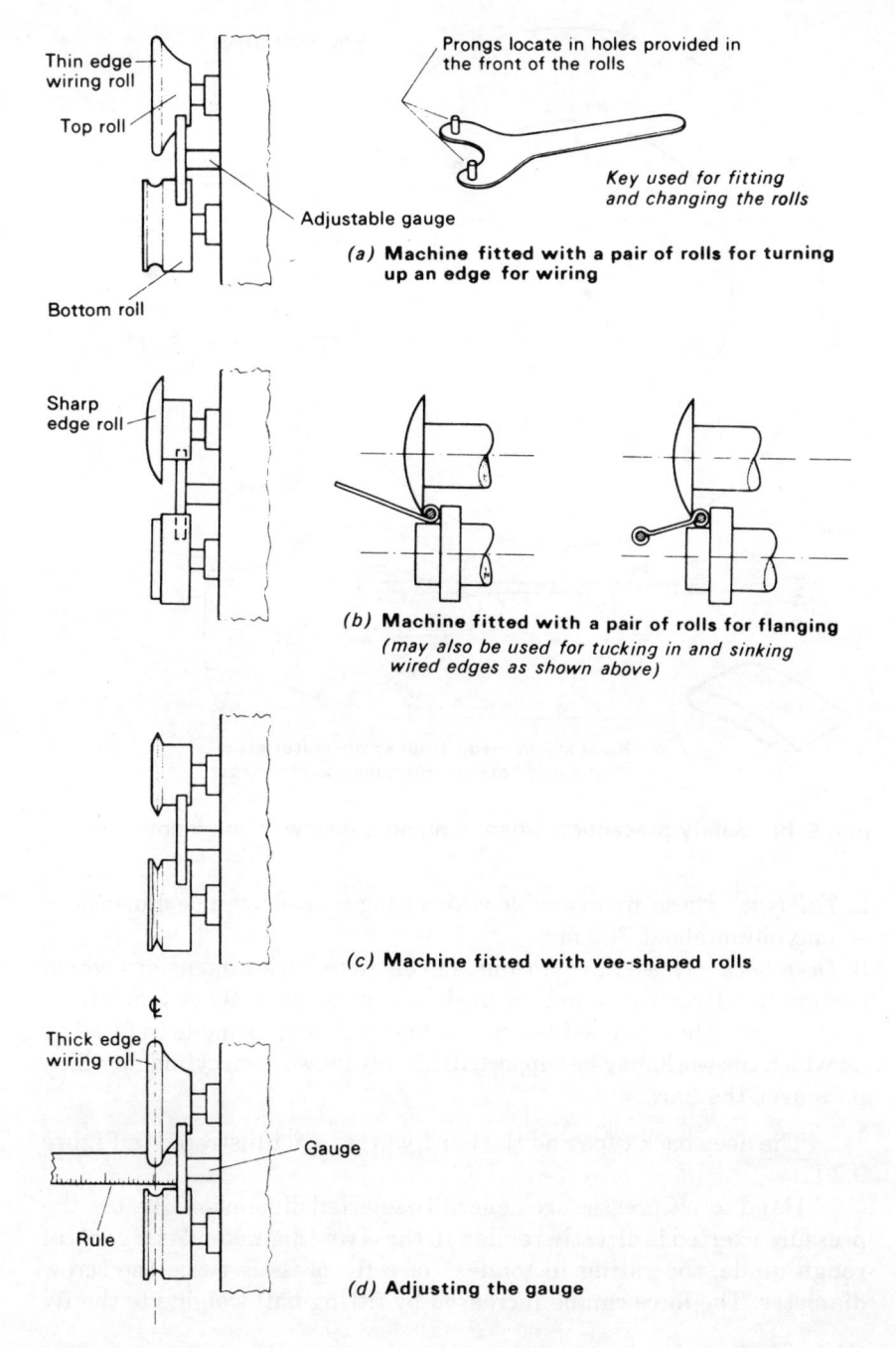

**Fig. 9.18 Edge preparation on a jennying machine**

313

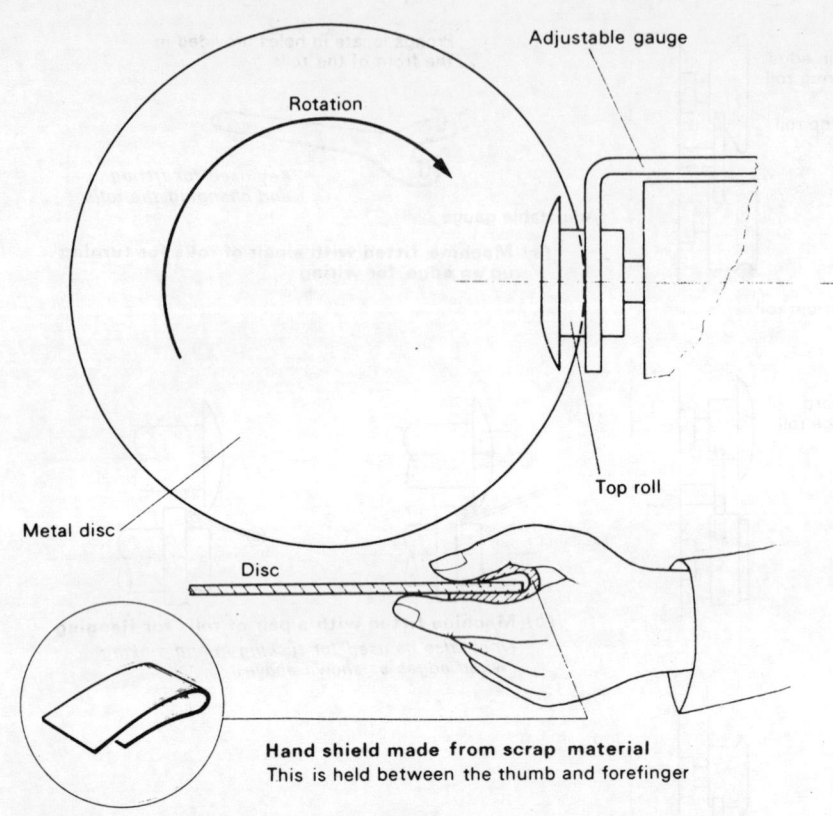

**Fig. 9.19** Safety precaution when flanging a disc with the jenny

2. *Tall type* These are available with a range of bed to guide dimensions — maximum about 360 mm.
3. *Deep back* These have a range of centre to back dimensions which are greater than the standard models — maximum about 306 mm.
4. *Bar type* The solid bed is omitted and provision is made to fit a bar on which the work may be supported. Mainly for work on cylinders which pass over the bar.

The deep back type and the bar fly press are illustrated in Figure 9.20.

Hand screw presses are generally selected dimensionally, but the pressure exerted is directly related to the screw diameter. As a general rough guide, the rating in tonnes* of a fly press is twice the screw diameter. The force can be increased by fitting ball weights to the fly

*Note: The force, in kilo-newtons exerted by the press on the workpiece is equal to ten times the rating in tonnes mass.

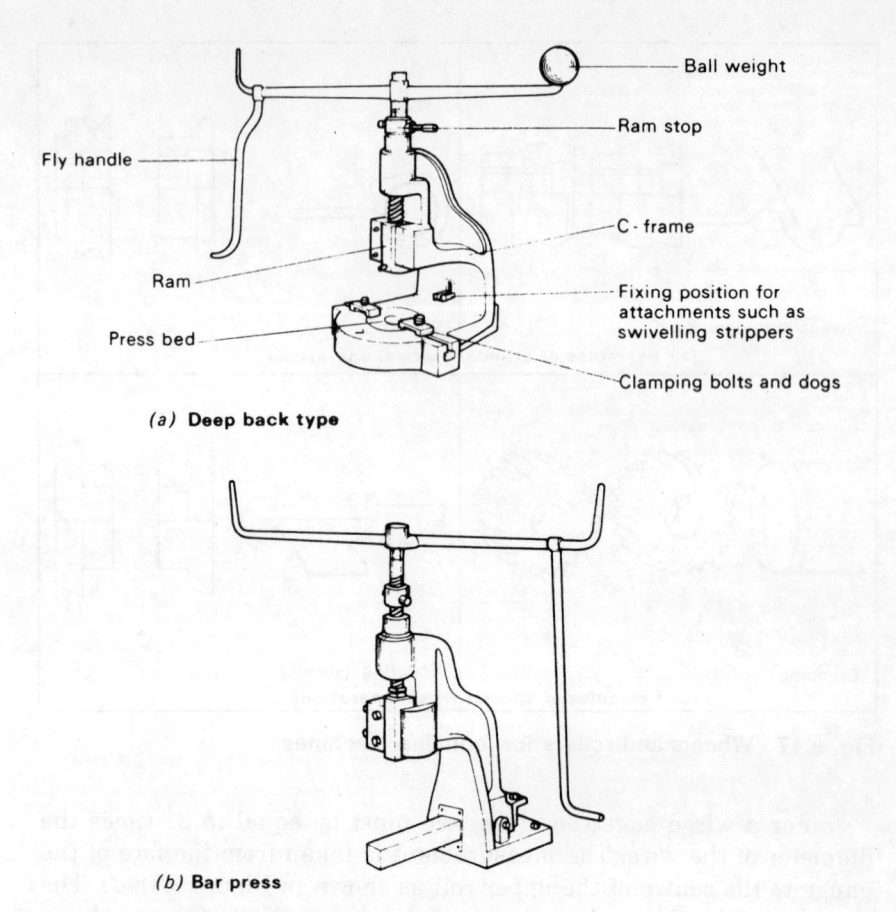

**Fig. 9.20** Types of fly press in common use

arms. *In the interest of safety, take care when operating a fly press because the fly arm and the weights can cause injury, not only to the operator but to those in close proximity.*

Notching, piercing and punching, light pressing, bending and riveting are all operations which can be performed on a fly press.

Figure 9.21 shows details of a typical sub-chassis in 18 s.w.g. mild steel sheet. All the operations necessary to form this chassis from the blank can be easily performed on a fly press fitted with appropriate tooling. This type of sheet-metal work is normally done in batches, which means that the fly press is set up for one specific operation and the batch is run through before resetting for the next operation. Alternatively more than one fly press may be set up to reduce waiting time due to resetting operations.

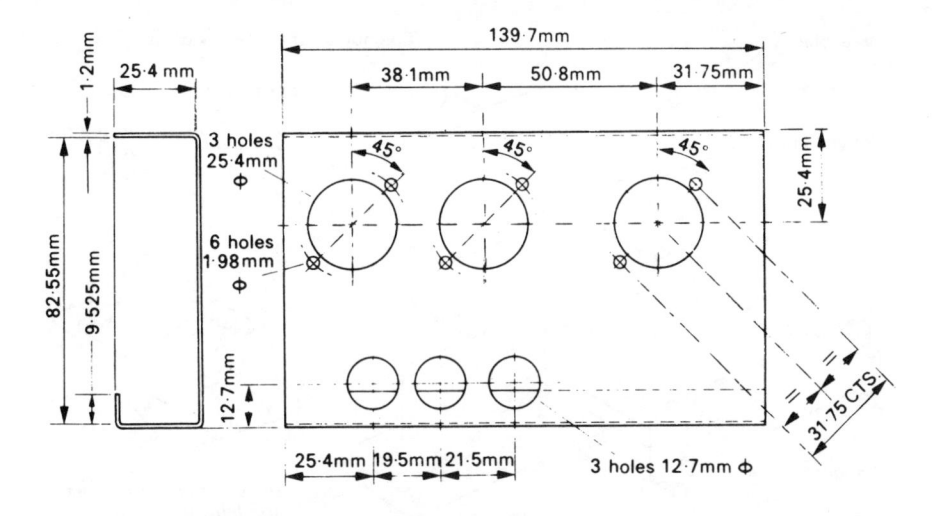

**Fig. 9.21 Details of a typical sub-chassis**

The sequence of operations may be as follows:

1. A template is used to mark the positions of all the holes. Such a template is usually marked out on 10 s.w.g. mild steel plate on a surface table using a vernier height gauge and an angle plate. Small pilot holes are drilled, and once the template has been passed by inspection these are opened out with the correct size drill to suit the diameter of a nipple punch. The template is provided with location buttons to give an accurate location for the blanks. Figure 9.22 shows the template positioned over the blank ready for transferring the hole positions with a nipple punch. The use of such a template is a foolproof system which not only provides identical hole positions on each blank, but dispenses with the use of guides and locations having to be set up on the press.

2. The fly press is set up for punching the holes and the ram-stop is set to give optimum punching conditions for ease of stripping. A universal bolster outfit is used extensively on fly-press work of this nature. The bolster frame is a heavy casting which has a central hole bored to receive the dies and standard holders for the smaller dies. It is also provided with two pairs of holes drilled and bored for adjustable gauges through bosses cast on the sides. In addition four tapped holes are provided in the top face of the bolster for securing special attachments in place of the gauges mentioned. Two types of stripper are available with the bolster outfit.

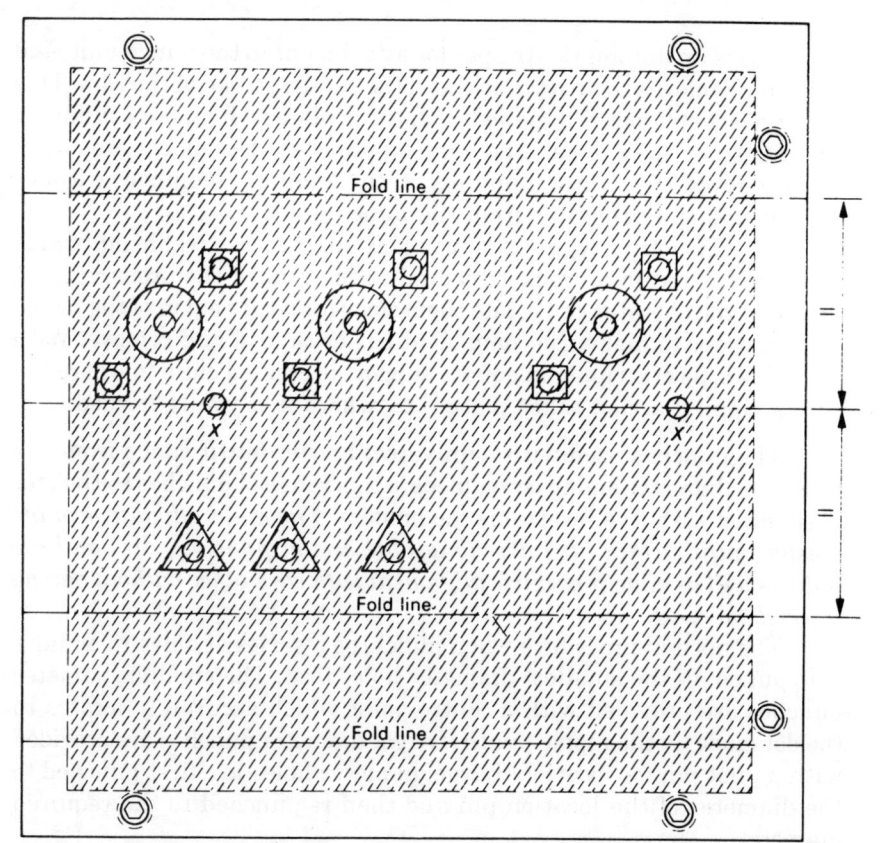

The master template is placed over the blank (shown shaded) and is accurately located by means of location buttons. The master template may be used for transferring hole positions to the blank using a nipple punch or, where large diameter holes are indicated, the pilot holes are punched through the template and the blank in one operation. Location holes (sometimes termed 'tooling holes') as shown at *X* may be provided on the template for more accurate location for bending

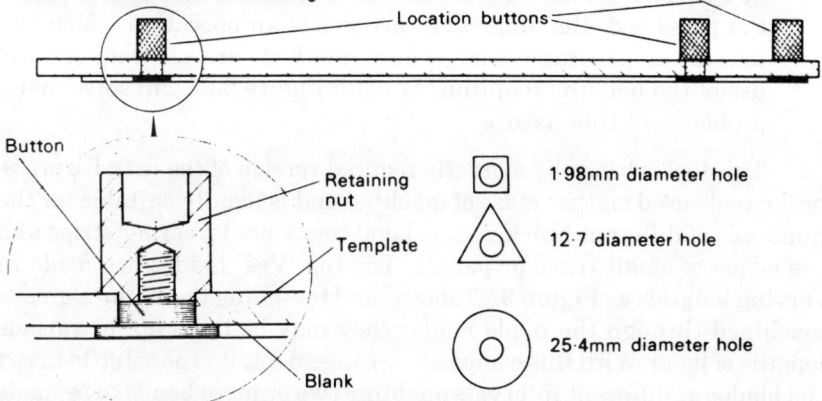

**Fig. 9.22 Marking hole centres (master template)**

315

One is an automatic stripper for attachment to the punch unit, and this removes the part quickly on the return stroke of the ram. This type of stripper is supplied with three plates which have different aperture sizes to suit the varying punches. The plates are easily removed for interchanging with another member without dismantling the unit from the punch holder.

The other type of stripper consists of two swivel heads which are located in a bracket secured to the rear of the press column, and the operator can rotate them clear from the region of the punch in order to make the component easier to load and remove. With this type of stripper adequate support is applied close to where the stripping action takes place, i.e., where it is really needed.

The punches used with a universal bolster outfit range from 2·4 to 95 mm diameter and to accommodate this range of punches the bolster is supplied with five insertable die holders. *The main advantage of the bolster outfit is that all interchangeable sets of punches, dies and die holders will assume concentric alignment with each other without having to reset the unit.*

The normal procedure when punching a series of holes in a blank is to punch all the smallest diameter holes first. The smaller diameter punches are provided with a centre point which accurately locates in the dot made by the nipple punch. Larger diameter punches are provided with a pilot pin for location. Large diameter holes are first punched to the diameter of the location pin and then re-punched to the required diameter.

3. The bolster is removed from the flypress by releasing the clamping arrangement which is usually 'Tee'-bolts and dogs. This is replaced by a standard bending tool consisting of a 'Vee' block mounted on the press bed. Bending small articles is an operation which can take up considerable time if hand methods are utilised, and the use of the bending tool illustrated in Figure 9.23 can solve many problems of this nature.

This tool is basically a greatly reduced version of the massive press-brake tools used on that class of machine and is ideally suitable for the hundreds of different brackets and bend components, corner strips and the edges of small trays or panels. The top 'Vee' blades are made in varying lengths, as Figure 9.23 shows, and by sliding them into a groove machined through the blade holder they may be built up for various lengths of bend. With this composite arrangement, it is possible to insert the blades at different intervals enabling two or more bends to be made along a sheet of material.

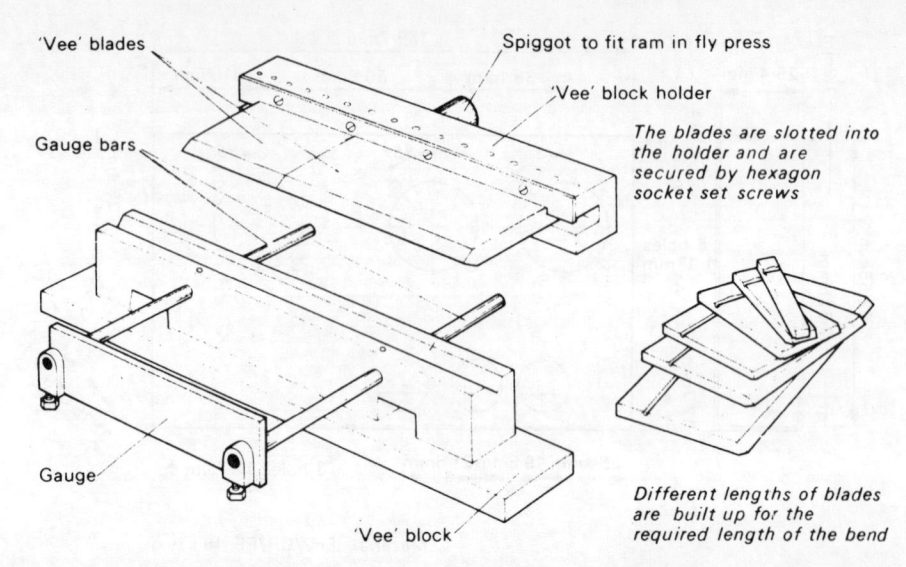

**Fig. 9.23 Bending tool for small sheet-metal components**

4. The ram-stop is set to control the downwards movement of the 'Vee' blade, thus ensuring a constant bending angle, in this case 90°. A back stop mounted on the 'Vee' block itself is set to provide the location for a specific bend. The blank is pushed up to this stop with one hand and the press is operated with the other. Provision is made on the top of the ram screw for positioning the fly handle at a suitable station for this two-handed operation. Figure 9.24 shows details of the bending operation:
(a) with the use of a back stop;
(b) with a pair of tooling holes in the blank which are placed on a centre-line parallel to the fold required.

Every sheet-metal workshop encounters examples where it becomes necessary to cut and bend sheets to the form of boxes or shallow trays, and in order to obtain this shape the blank needs notching at each corner prior to undertaking the bending operations. Figure 9.25 illustrates the following useful cutting tools which can be used with a universal bolster on the fly press:

*Corner notching tool* The die is inserted in the bolster frame and the punch is aligned by the dovetail leg operating in a corresponding groove — this useful feature makes incorrect setting impossible;

*Corner radius tool* Which can remove four different radii according to the set of the tools. The four legs on the punch extend into the die

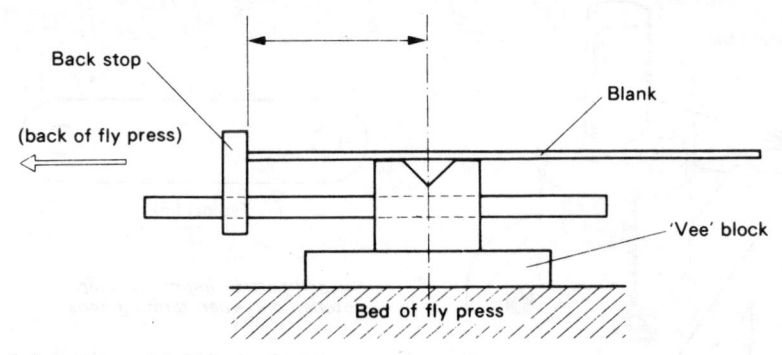

(a) Position of fold by back stop

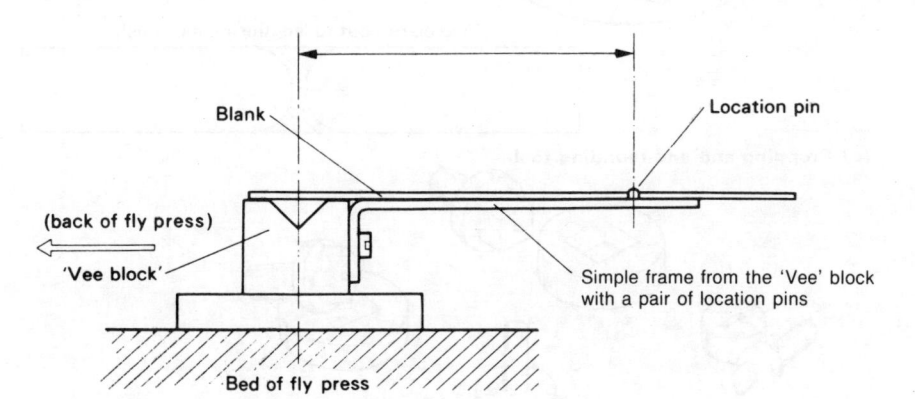

(b) Position of fold by tooling hole and pin

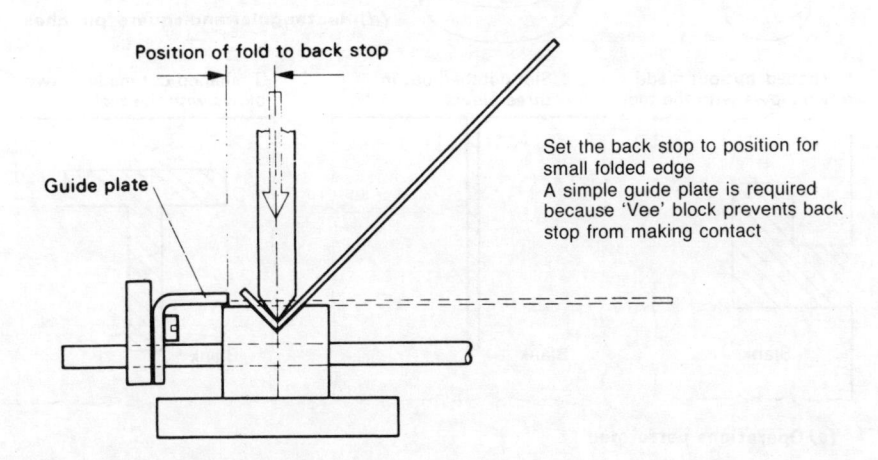

(c) Make the small fold first

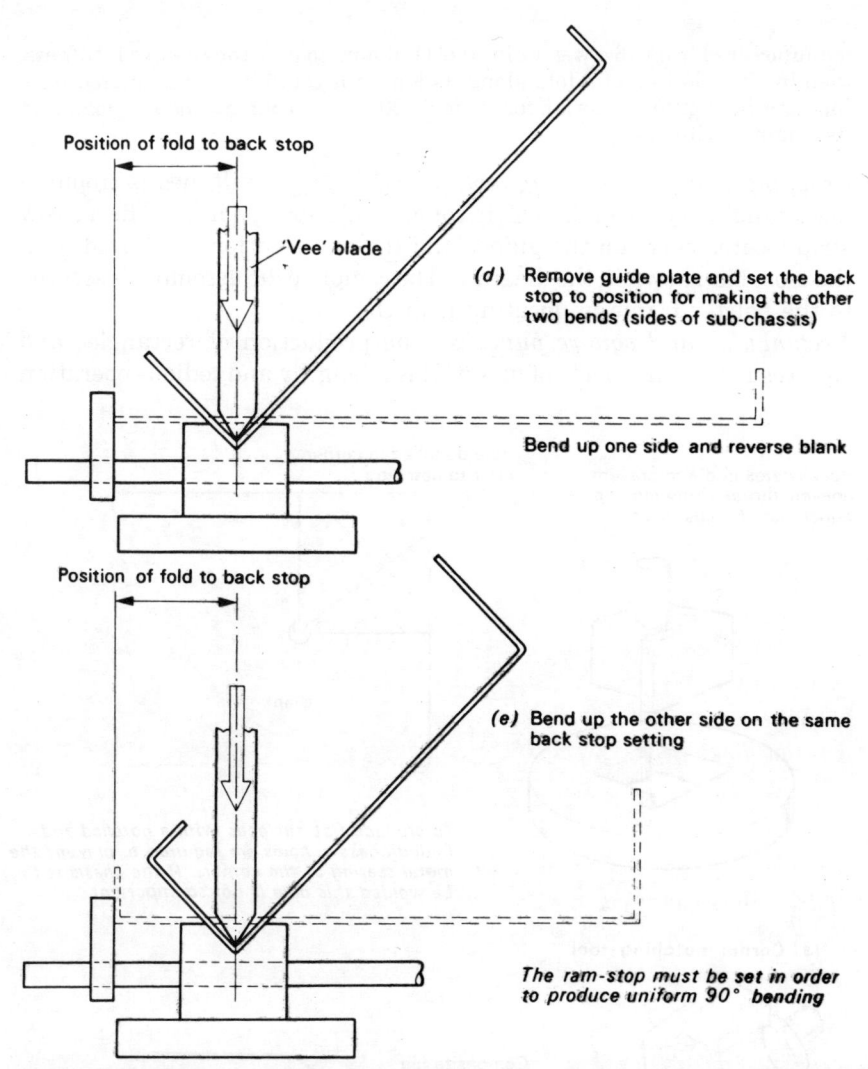

Set the back stop to position for small folded edge
A simple guide plate is required because 'Vee' block prevents back stop from making contact

(d) Remove guide plate and set the back stop to position for making the other two bends (sides of sub-chassis)

Bend up one side and reverse blank

(e) Bend up the other side on the same back stop setting

*The ram-stop must be set in order to produce uniform 90° bending*

**The general allowance on the blank for fly press bending is inside dimensions plus half a thickness of material for each bend**

**Fig. 9.24 Bending the sub-chassis on the fly press**

aperture and so ensure proper alignment. The die is attached to the bolster frame by screws which locate in the four tapped holes provided on its top face. This tool is used primarily for cutting the corners of sheets and panels;

Note: Producing a fold in a blank stiffens the part considerably. If the component has been designed to support a certain weight over a span and has to be

317

manufactured from 16 s.w.g. mild steel (1·6 mm) to give the required stiffness, then by the addition of a fold along its length it could be manufactured from considerably lighter material, for example 20 s.w.g. (0·9 mm) and still maintain its original stiffness.

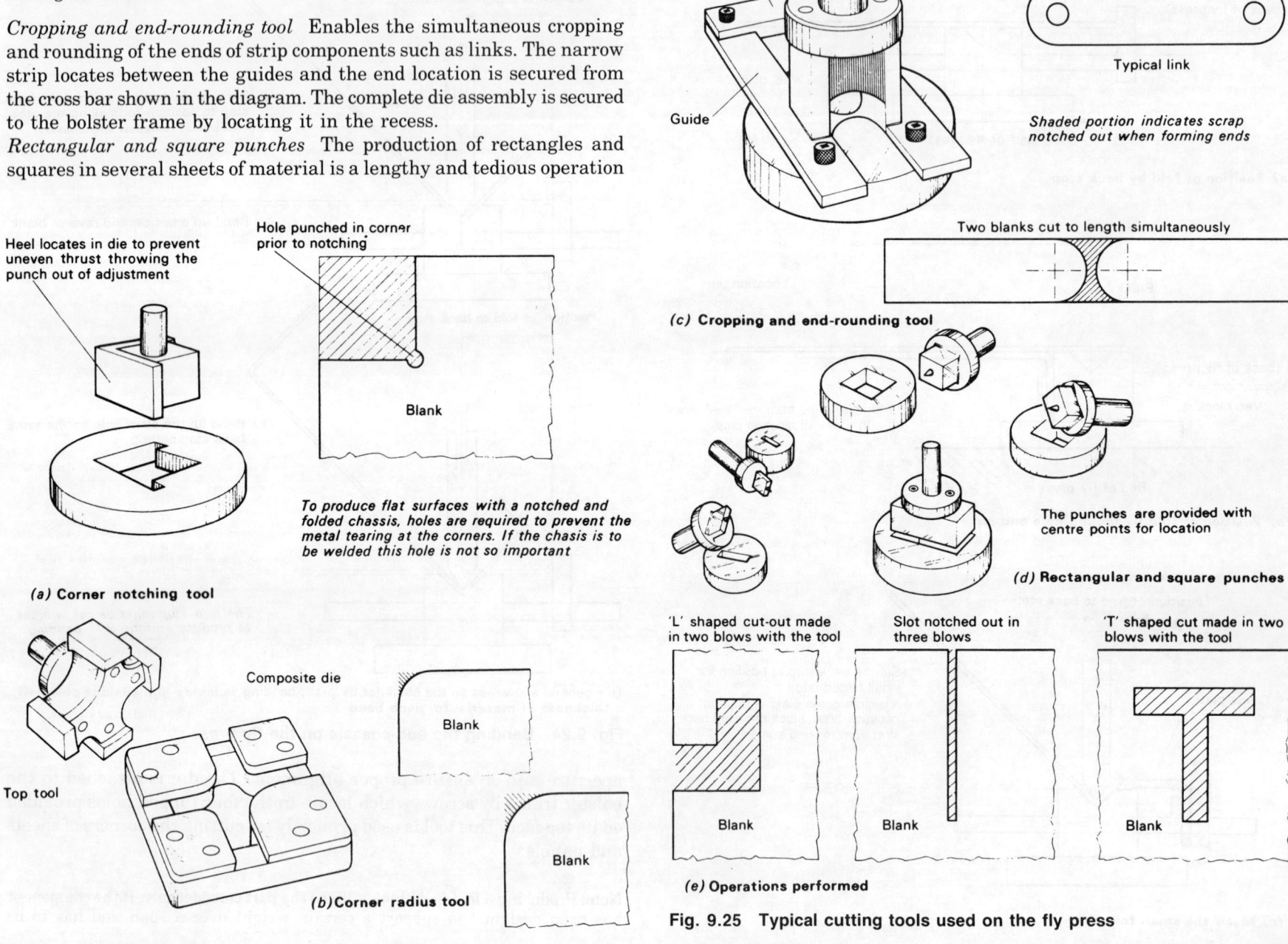

*Cropping and end-rounding tool* Enables the simultaneous cropping and rounding of the ends of strip components such as links. The narrow strip locates between the guides and the end location is secured from the cross bar shown in the diagram. The complete die assembly is secured to the bolster frame by locating it in the recess.

*Rectangular and square punches* The production of rectangles and squares in several sheets of material is a lengthy and tedious operation

*(a) Corner notching tool*

Heel locates in die to prevent uneven thrust throwing the punch out of adjustment

Hole punched in corner prior to notching

Blank

*To produce flat surfaces with a notched and folded chassis, holes are required to prevent the metal tearing at the corners. If the chasis is to be welded this hole is not so important*

*(b) Corner radius tool*

Top tool

Composite die

Blank

Blank

End location bar

Guide

Typical link

*Shaded portion indicates scrap notched out when forming ends*

Two blanks cut to length simultaneously

*(c) Cropping and end-rounding tool*

The punches are provided with centre points for location

*(d) Rectangular and square punches*

'L' shaped cut-out made in two blows with the tool

Slot notched out in three blows

'T' shaped cut made in two blows with the tool

Blank

Blank

Blank

*(e) Operations performed*

**Fig. 9.25 Typical cutting tools used on the fly press**

if press tools are not employed for the task. A range of punches and dies may be used with the bolster unit, and may be set to produce single holes, several appertures in different positions in the component, notch the sheet edges, or produce 'L' of 'T' shaped cuts.

## 9.14 Setting the guillotine

Treadle and power guillotines are fitted with FRONT and SIDE GAUGES. These usually consist of a FIXED SIDE GAUGE, sometimes referred to as the 'SQUARING GUIDE', and a FLAT BAR FRONT GAUGE.

On some machines the side gauge can be extended for wide sheets, and may be graduated in millimetres.

The front gauge is adjustable across the bed or table, and further along extension bars (arms fitted to the front of the machine). Figure 9.26(a) shows a plan view of the side and front gauges. Figures 9.26(b) and 9.26(c) show how the gauges are set.

The bulk of cutting performed on the guillotine is when the sheet or plate is located against a BACK GAUGE, and there are several types of these. The simplest, and usual standard type, consists of an ANGLE GAUGE BAR. Back gauges are mounted on an attachment fixed to the movable cutting beam and move up and down with it. Figure 9.27 gives basic details of gauges and guides used on guillotines.

Where precision cutting is required, the simple back gauge will take a while to set in order to achieve the exact size register.

On batch or bulk cutting operations where the back gauge has to be moved constantly a more elaborate type of back gauge is necessary to enable the settings to be carried out more quickly, reducing both the actual time taken to move the gauge and then the time to set it to a dead size.

Even the simple type of back gauge can have an attachment for final fine adjustment, as shown in Figure 9.27. The guide is set in an approximate position and locking lever B tightened. The final setting to correct register is achieved by slackening lever A, and turning the hand wheel on the fine adjustment screw. The back gauge is then locked in position by tightening locking lever A.

## 9.15 Operations on the guillotine

The fixed side gauge is used for positioning the material. To square off two adjacent sides of a sheet or plate, a trim-cut of approximately 6 mm is made on one edge. The second edge is then sheared at 90° to the first by holding the trimmed edge firmly against the side gauge which is normally located on the left-hand side of the table.

(a) Plan view of guillotine table showing front gauges

1. Make sure machine is switched off
2. Place rule between blades
3. Position front gauge against end of rule, and sight as shown opposite

Set each end of bar, and lightly tighten bolts. Check each end then fully tighten in position

(b) Setting the front gauge - using a rule

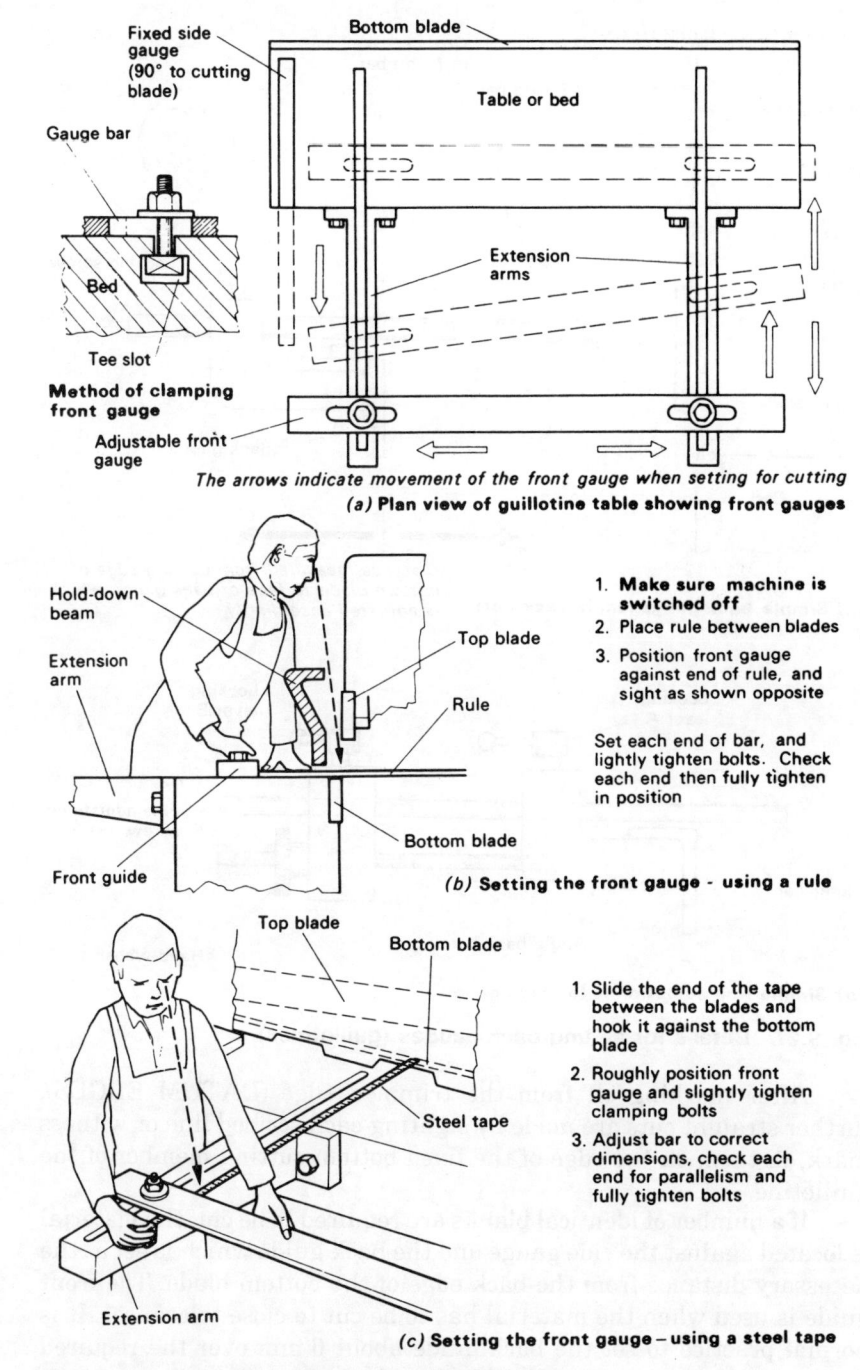

1. Slide the end of the tape between the blades and hook it against the bottom blade
2. Roughly position front gauge and slightly tighten clamping bolts
3. Adjust bar to correct dimensions, check each end for parallelism and fully tighten bolts

(c) Setting the front gauge – using a steel tape

Fig. 9.26 Front and side gauges (guillotines)

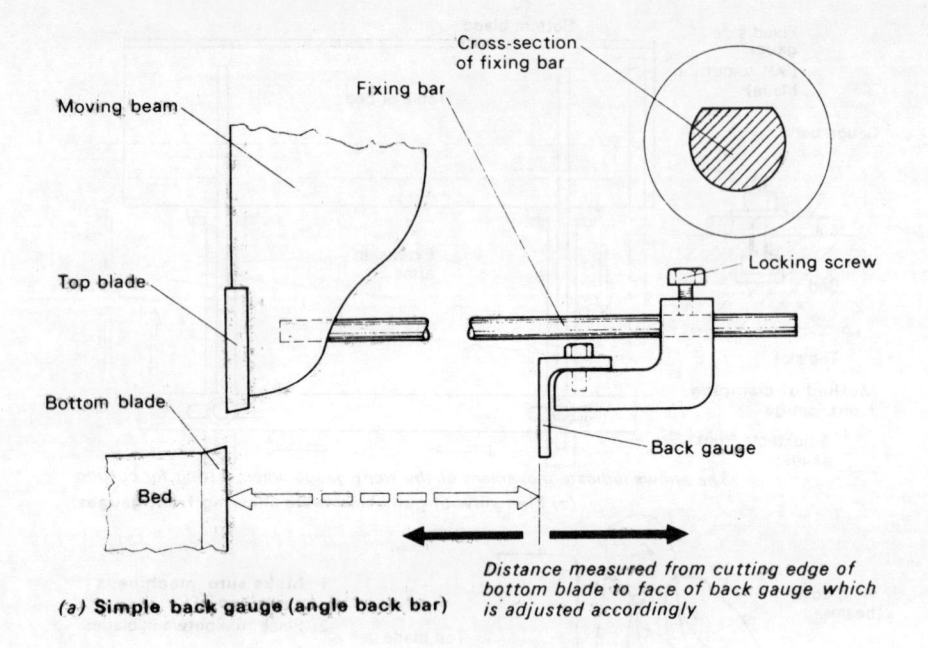

**(a) Simple back gauge (angle back bar)**

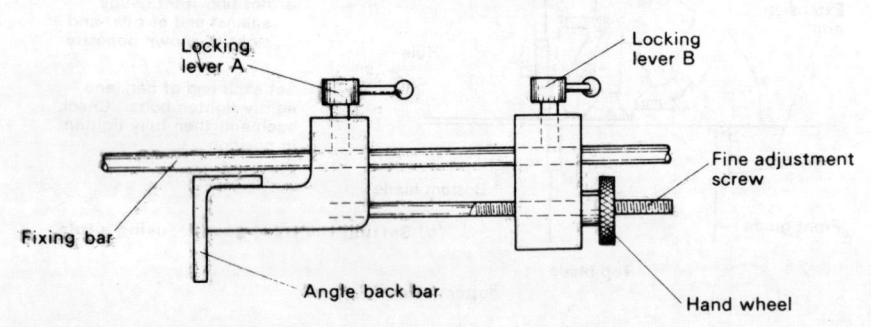

Distance measured from cutting edge of bottom blade to face of back gauge which is adjusted accordingly

**(b) Simple fine adjustment for back gauge**

**Fig. 9.27 Details for setting back gauges (guillotines)**

After marking off from the trimmed edge (DATUM EDGES), further straight cuts are made by sighting each scribed line or witness mark, in turn, on the edge of the fixed bottom cutting member of the guillotine.

If a number of identical blanks are required to be cut, the material is located against the side gauge and the back guide which is set at the necessary distance from the back edge of the bottom blade. The front guide is used when the material has to be cut to close tolerances. It is normal practice to set the back guide about 6 mm over the required dimension for the cut, and the front guide to the exact measurement.

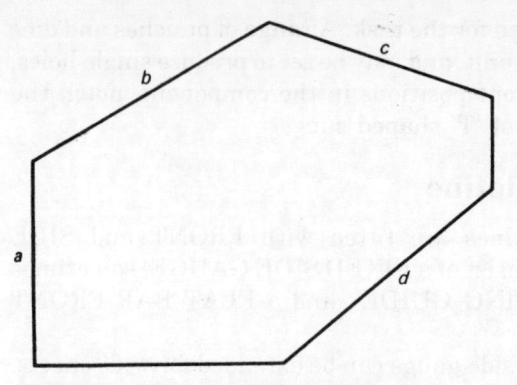

**Fig. 9.28 Typical gusset plate**

The material is passed over the front guide, through the gap between the blades to locate on the back guide, and the initial cuts made off the back guide. The cut blanks are then located on the front guide and finally cut to required dimension. Taper cutting is performed by setting the front guide at an angle.

Simple guide plates may be made up and clamped in position on the table by means of 'Tee'-bolts. These are very useful improvised attachments which enable many bevel and mitre cutting operations to be easily performed on the guillotine. Figure 9.28 shows a typical gusset plate that can be produced on the guillotine shear.

The sequence of operations for cutting out the gusset plate is explained below and shown in Figure 9.28. (The required gusset plate is shown shaded for all the sequence of operations.)

OPERATION 1

Set the back guide to cut the edge marked $a$. The length of flat is located against the side gauge and the back gauge, as shown in Figure 9.29(a).

OPERATION 2

Cut the number of blanks required.

Notes:
(a) If a suitable width of flat is not available, suitable widths may be cut from a standard plate. The back gauge would be set for this purpose.
(b) A simple bevelling plate may be flame cut and into it a series of location holes, as shown in D, E, F and G, drilled for clamping purposes. It will be seen that a pair of holes is used for each setting. The first blank is marked out and used as the pattern for all the settings. The bevelling plate is set for cutting edge $b$ and is securely clamped to the table using the

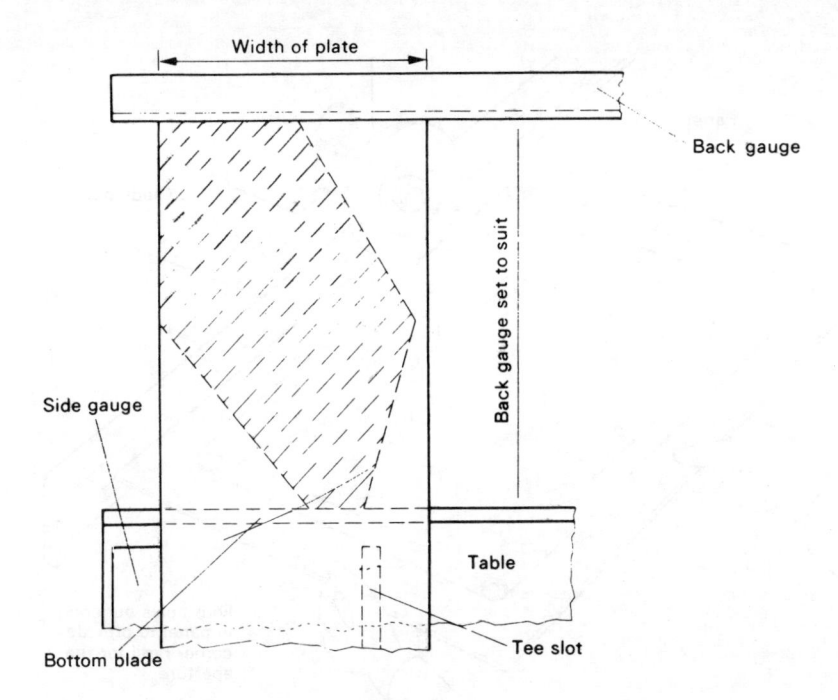

(a) Operation 1 and 2

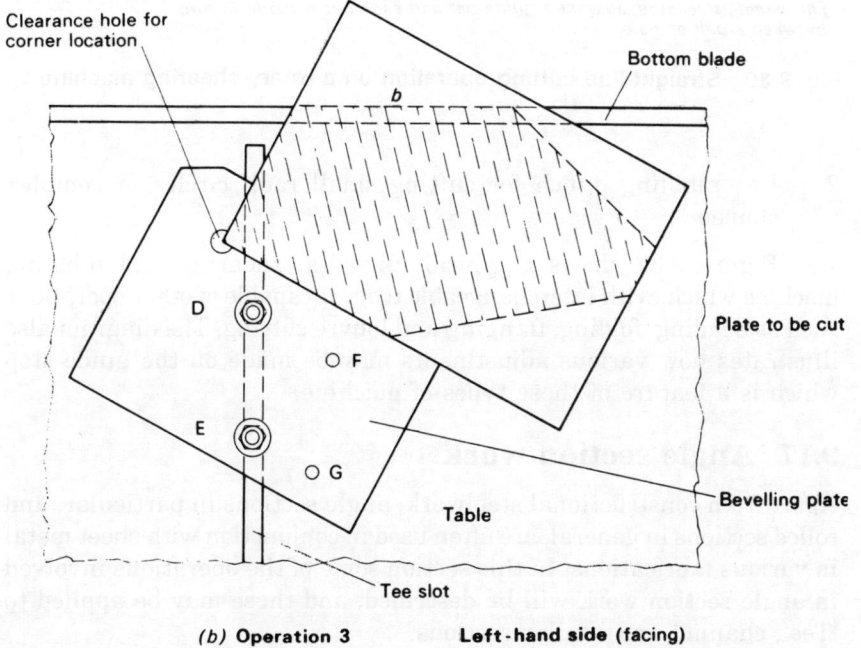

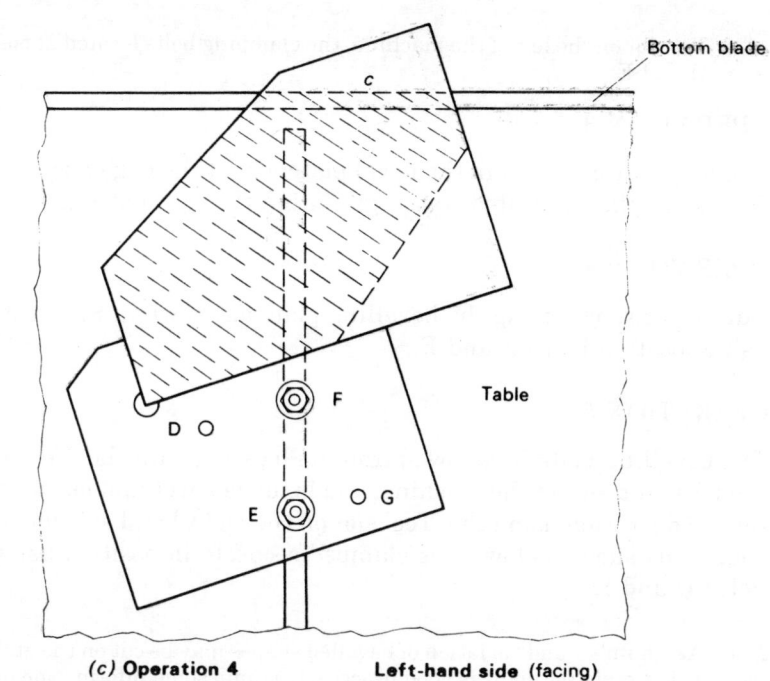

(b) Operation 3   Left-hand side (facing)

(c) Operation 4   Left-hand side (facing)

(d) Operation 5   Right-hand side (facing)

Fig. 9.29 Sequence of operations for cutting gussets

321

Tee slot on the left of the machine, the clamping bolts located at positions D and E.

OPERATION 3

The blanks are located in the bevel plate whilst the edge *b* is trimmed. The bevel plate is bolted to the guillotine through holes D and E.

OPERATION 4

Edge *c* is cut by setting the bevelling plate as shown in Figure 9.29(*c*), using location holes F and E.

OPERATION 5

The bevelling plate is removed from its location with the 'Tee' slot on the left-hand side of the machine, and is turned over and located in the required position using the 'Tee' slot on the right-hand side for cutting edge *d*, as shown below. It is clamped securely in position using bolt holes D and E.

Note: Any number and variation of bevelled shapes may be cut on the guillotine using this simple set up. It is good practice to stamp some information on such bevelling plates. They can then be stored and used time and time again when a similar cutting operation is called for. On completion of the cutting, the gusset plates are marked off for the hole centres and punched or drilled.

## 9.16 Cutting apertures and circles (shearing machines)

Rotary shears or combined nibbler-shearing machines are used for straight or irregular cutting of metal sheets and plate.

The basic principle of these machines has been explained in Chapter 7.

Rotary shears may be used for cutting out apertures in panels on which the outline has been marked, and this operation requires considerable skill. Figure 9.30 shows a method of cutting out a rectangular aperture with radiused corners with the aid of a suitable guide. This operation requires less skill.

Circular flanges and discs can be easily produced on shearing machines, and this operation is explained in Figure 9.31. As a general guide, when performing cutting operations on a combination shearing and nibbling machine:

1. Use shearing blades for straight line or large radius curves;

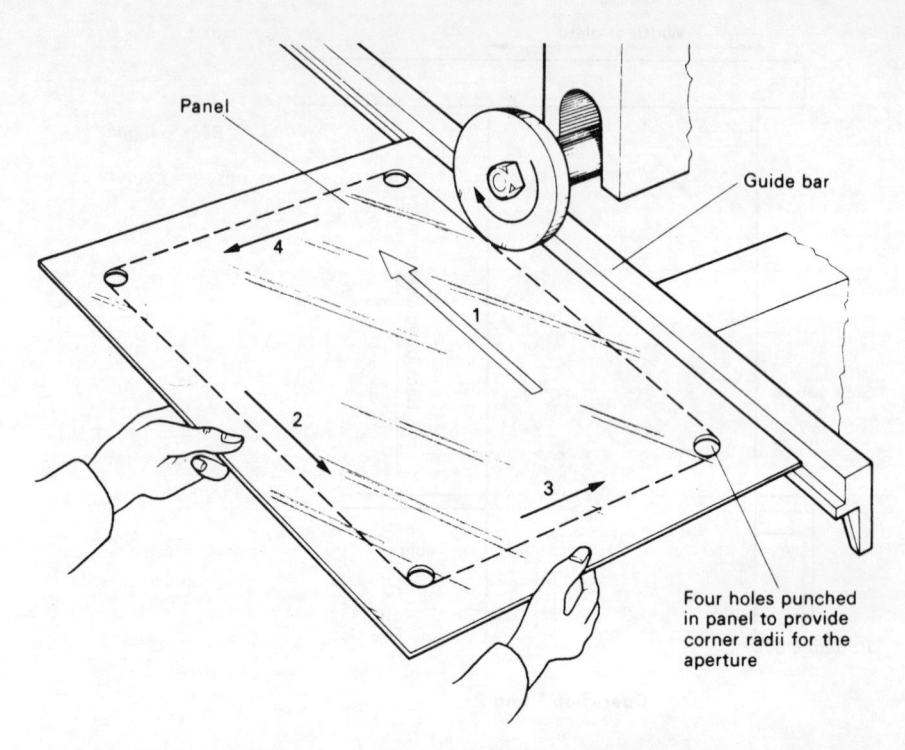

*The panel is located against a guide bar and each cut is made in turn between a pair of holes*

**Fig. 9.30 Straight-line cutting operation on a rotary shearing machine**

2. Use nibbling punch for cutting small radii circles or complex shapes.

Figure 9.32 shows a typical universal shearing and nibbling machine which, with interchangeable tools, is capable of other operations such as beading, folding, flanging and louvre cutting. The diagram also illustrates how various adjustments may be made on the guide stop which is a feature of these types of machine.

## 9.17 Angle section work

Apart from constructional steelwork, angle sections in particular, and rolled sections in general, are often used in conjunction with sheet metal in various fabrications. In this section some of the operations involved in angle section work will be described, and these may be applied to 'Tee', channel and similar sections.

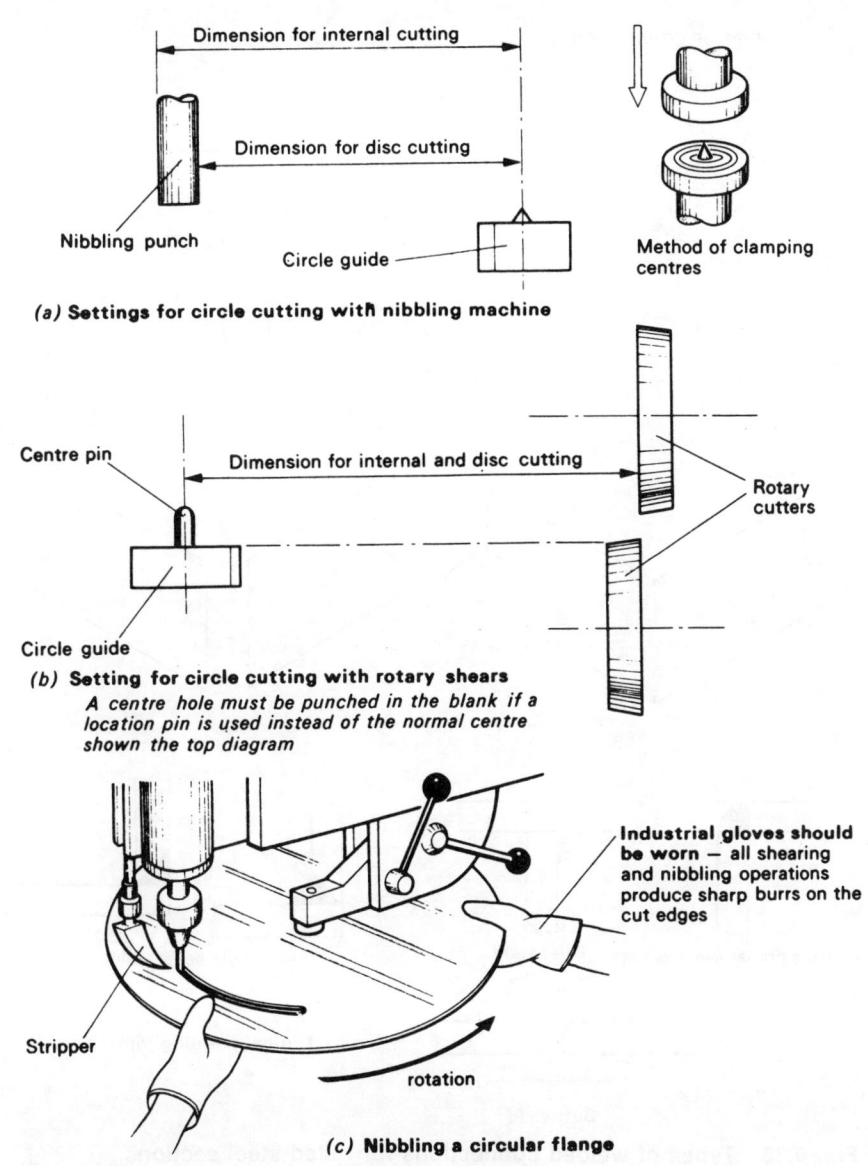

(a) Settings for circle cutting with nibbling machine

(b) Setting for circle cutting with rotary shears
*A centre hole must be punched in the blank if a location pin is used instead of the normal centre shown the top diagram*

**Industrial gloves should be worn** — all shearing and nibbling operations produce sharp burrs on the cut edges

(c) Nibbling a circular flange

Fig. 9.31 Circle cutting on shearing machines

## CUTTING

The methods generally used for cutting angle sections are:

1. Shearing.
2. Use of abrasive wheel machines (see Chapter 7).

(a) Typical curve, cutting-out and nibbling shears

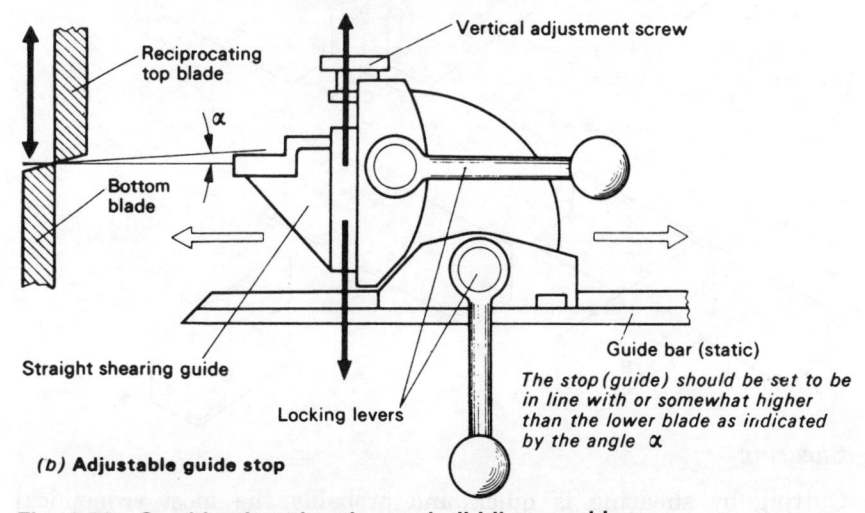

(b) Adjustable guide stop

*The stop (guide) should be set to be in line with or somewhat higher than the lower blade as indicated by the angle α*

Fig. 9.32 Combination shearing and nibbling machine

3. Cold sawing.
4. Notching — see under 'shearing' below.
5. Flame cutting. This process has been fully described in Chapter 7.

Figure 9.33 illustrates types of welded connections for rolled steel sections, all of which have to be prepared by cutting operations.

323

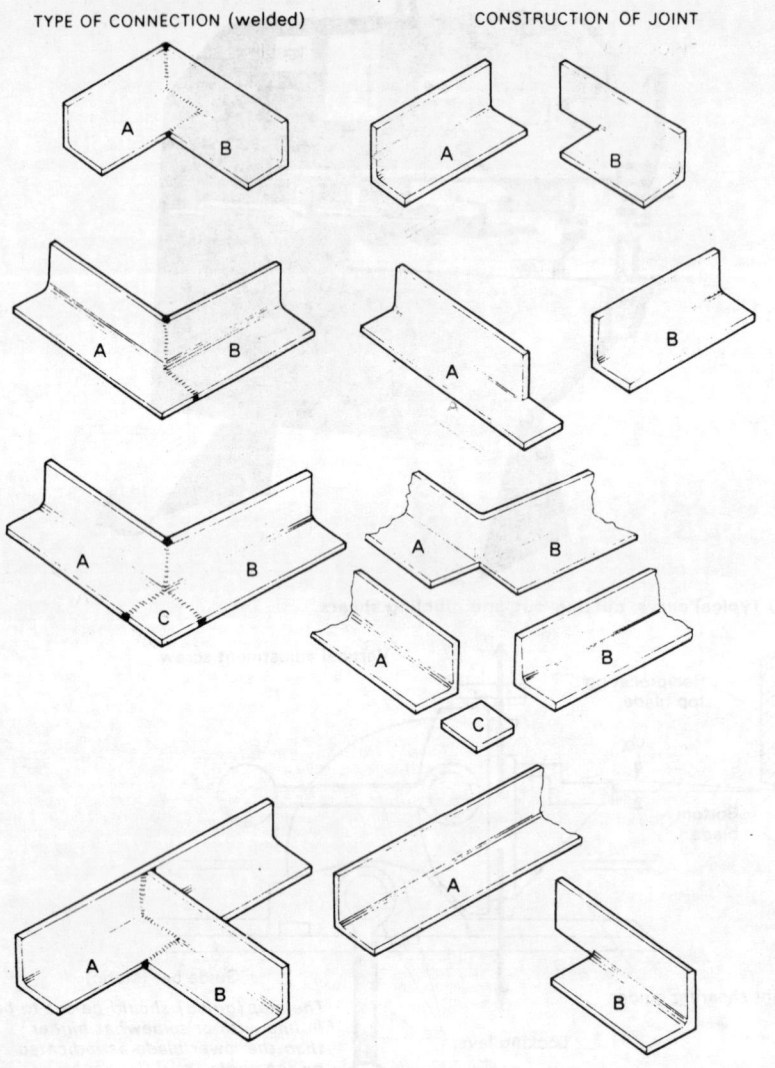

TYPE OF CONNECTION (welded)   CONSTRUCTION OF JOINT         TYPE OF CONNECTION           CONSTRUCTION

*Shearing*

Cutting by shearing is quick and probably the most economical production method. The shearing of rolled steel sections is performed in dies designed to suit the section. The dies are mounted in a special shearing machine. This operation is commonly referred to as CROPPING.

*With all cropping operations there is always some risk of distortion on one of the cut edges.* This problem may be easily overcome by reinserting the portion with the distorted edge in the opposite direction

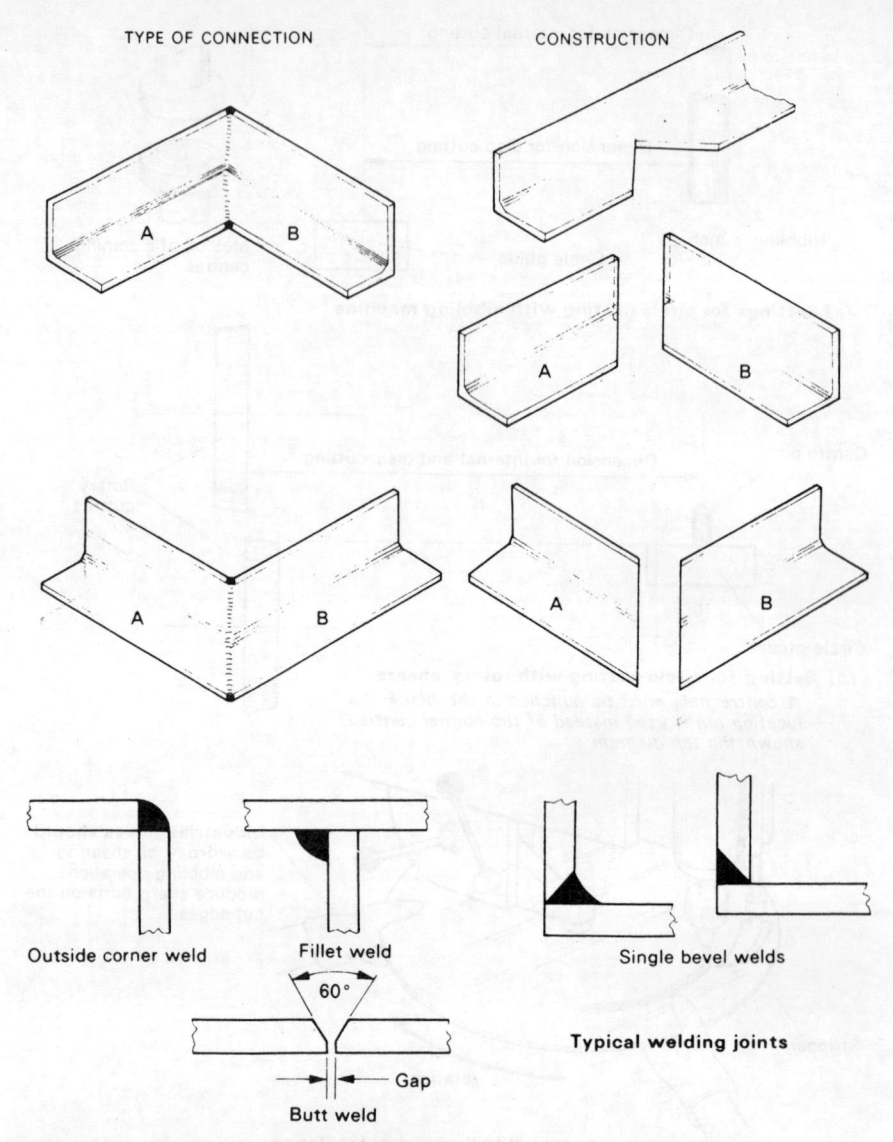

**Fig. 9.33 Types of welded connections for rolled steel sections**

and making a trim-out. Some allowance must be made for this in the initial marking-off or setting-up prior to cutting.

Hand-operated angle shearing machines are generally limited to sections up to $76 \times 76 \times 9.5$ mm for, as will be appreciated, a considerable amount of manual effort is required to shear, for example, $51 \times 51 \times 6.35$ mm section.

324

**Main components**

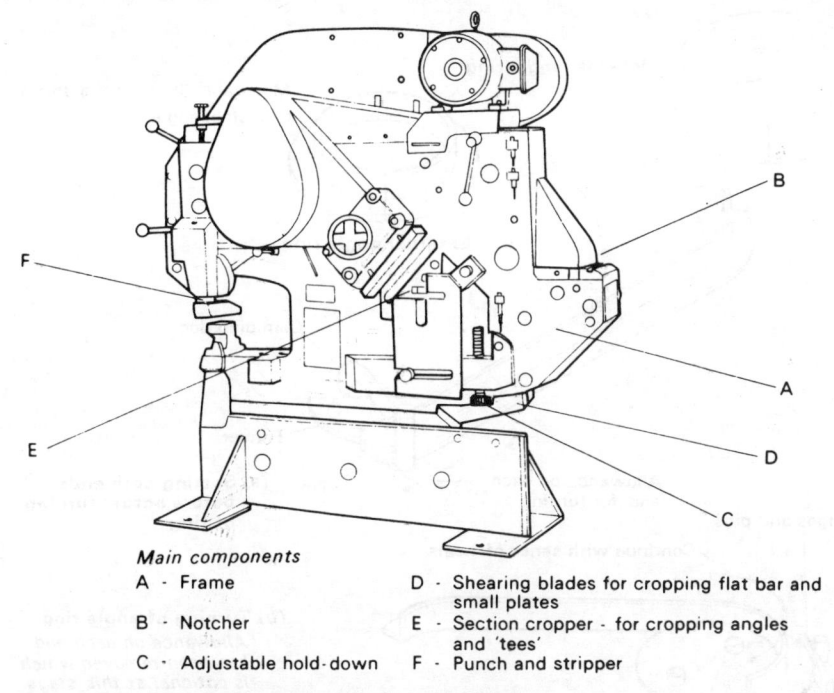

A - Frame
B - Notcher
C - Adjustable hold-down
D - Shearing blades for cropping flat bar and small plates
E - Section cropper - for cropping angles and 'tees'
F - Punch and stripper

**Fig. 9.34 Universal steel shearing machine**

In most fabrication shops, cutting operations on rolled steel sections are carried out on power machines. Machines are available which perform a combination of cutting operations, such as punching, shearing and notching, the shearing operations including not only section shearing, but round and square bar cropping, and plate shearing. Angle section has to be notched in order to permit it to be bent, and most of the notches are of the 'Vee'-notch or the square-notch type. Figure 9.34 shows a typical universal shearing machine, whilst Figure 9.35 shows some of the operations performed on such a machine.

*Abrasive wheel cutting*

In general, cutting operations using an abrasive wheel machine are confined to the lighter sections. Cutting may be carried out 'wet' or 'dry'. Wet cutting gives a rougher cut, but does give longer wheel life. Data for dry cutting is given in Table 9.6.

*Cold sawing*

For cutting operations on the heavier sections, the cold saw is preferred to the abrasive cutting wheel.

(a) Shearing plate

(b) Provision for holding plate

(c) Notching Tee-section

(d) Mitring an angle flange with the notching tool

(e) Typical mitre cutting operations performed on section cropper

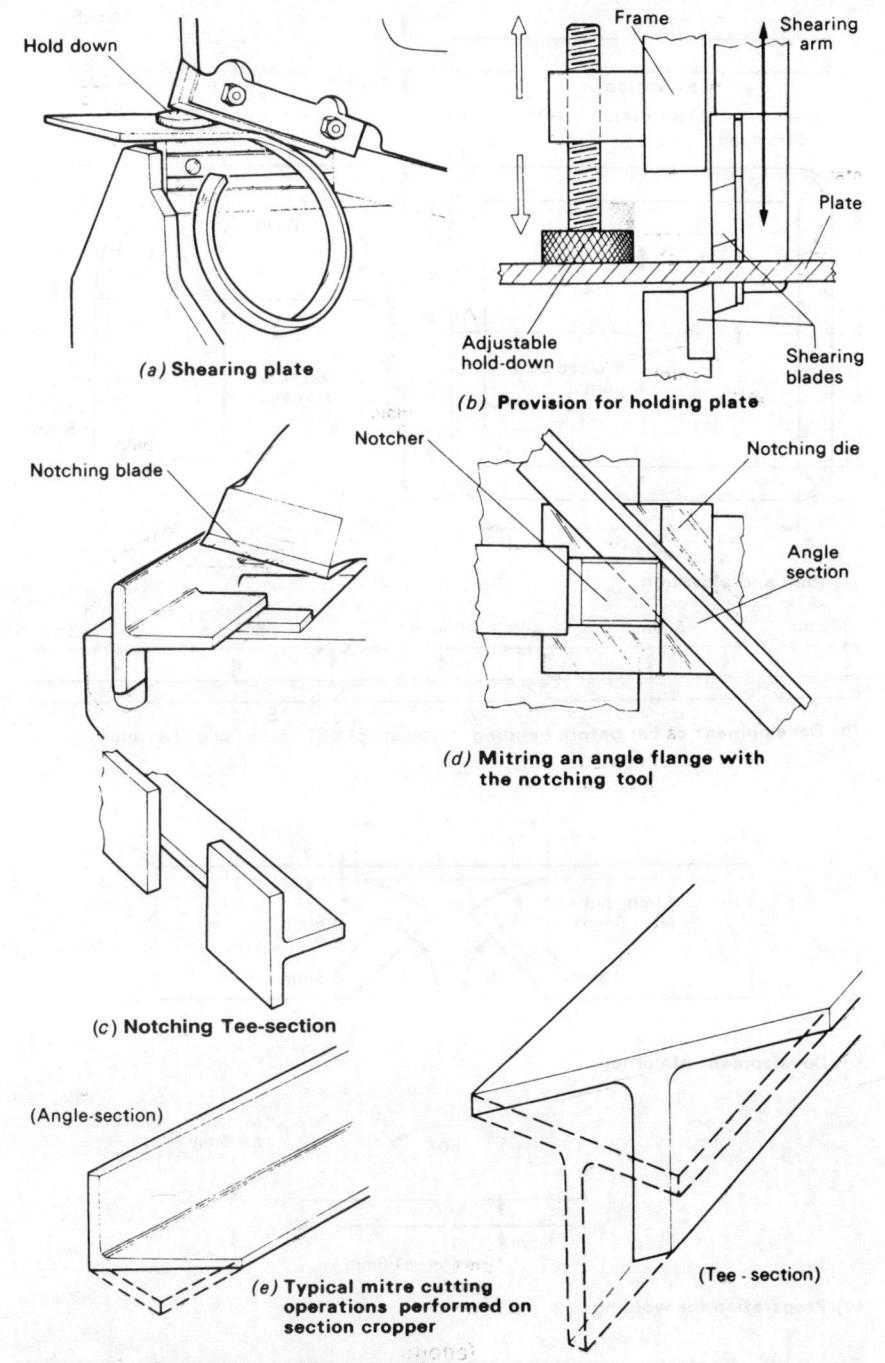

**Fig. 9.35 Cutting operations (sections)**

**Elevation**

51mm

610mm overall

51mm rad

Length $X = 2\left(R - \dfrac{t}{2}\right)\dfrac{\pi}{4}$

A

Ç

Welded butt joint

C

508mm overall

C

B

**Plan**

Weld

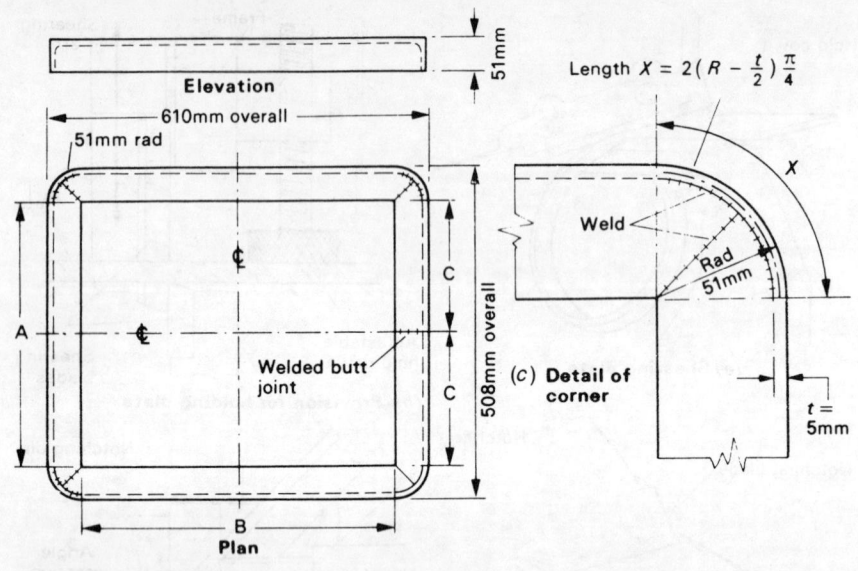

Rad 51mm

$t = 5\text{mm}$

(c) **Detail of corner**

(a) **Plan and elevation**

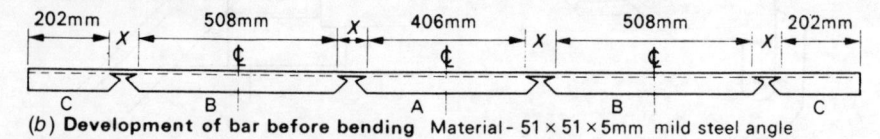

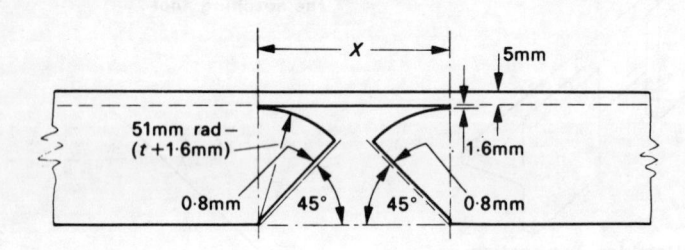

| 202mm | 508mm | 406mm | 508mm | 202mm |
|---|---|---|---|---|
| X | X | X | X | X |
| C | B | A | B | C |

(b) **Development of bar before bending** Material – 51 × 51 × 5mm mild steel angle

X

5mm

51mm rad – $(t + 1.6\text{mm})$

1.6mm

0.8mm   45°   45°   0.8mm

(d) **Development of corner**

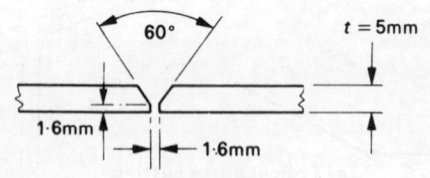

60°

$t = 5\text{mm}$

1.6mm   1.6mm

(e) **Preparation for welding**

Fig. 9.36 Angle frame with rounded corners

**Internal angle ring**

Material – 38 × 38 × 6.35mm mild steel angle

**External angle ring**

Clamping dog

Former

Allowance on each end for turning

Pin

(a) **Setting both ends before actual turning of ring**

Wedges and pins

Continue with series of heats

(b) **Turning of angle ring**
*Allowance on each end has been removed which is optional at this stage*

Present heat

(c) **Turning of angle ring**
*Commencing from opposite end and complete as in (b)*

Last heat

Lever tool

(1) Position of last heat

Incorrect position of last heat

$D_1$

(2)

Horizontal flange

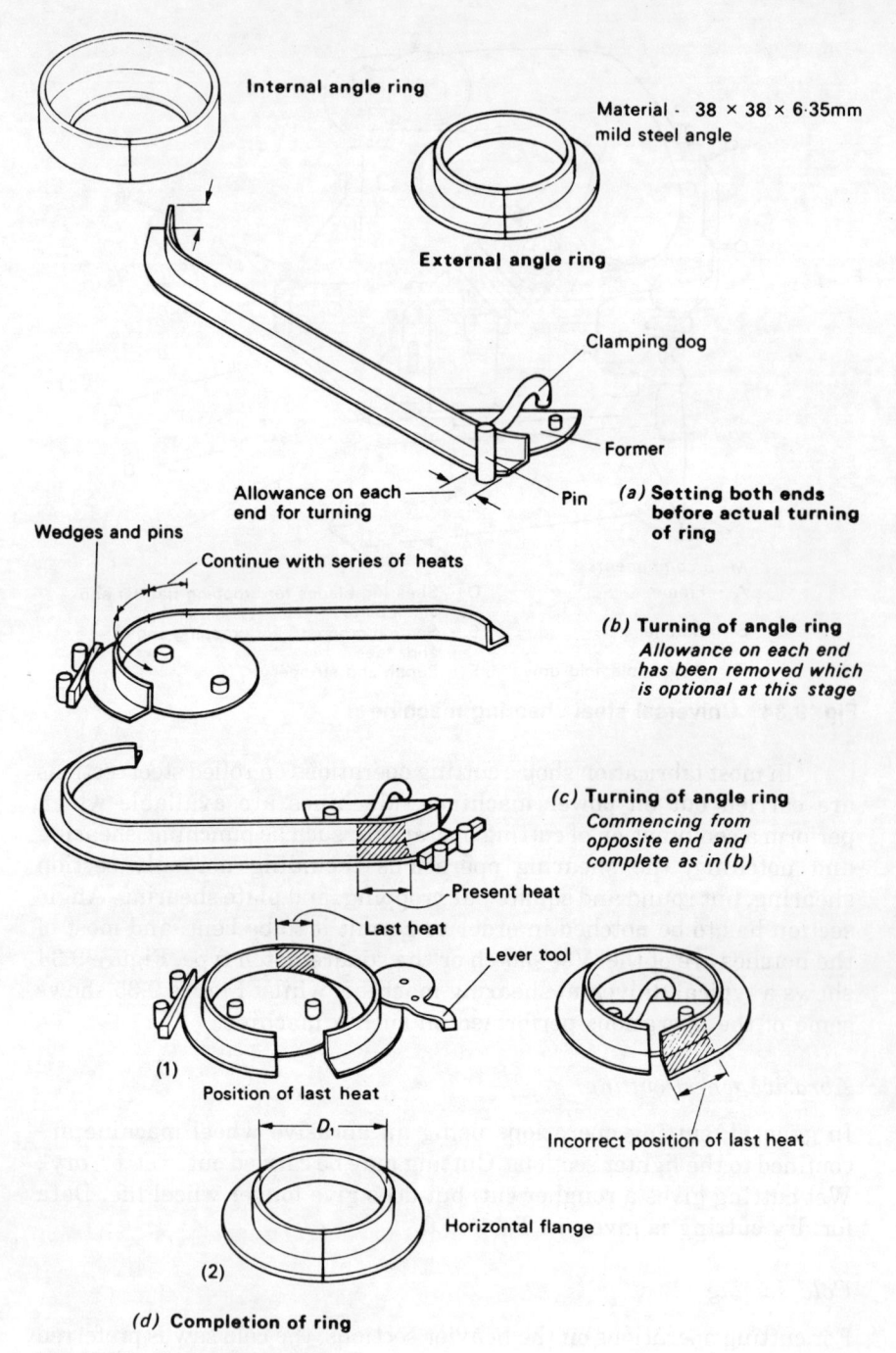

(d) **Completion of ring**

Fig. 9.37 External angle ring

Table 9.6 Abrasive cutting data for some common sizes of angle section

| ANGLE SECTION (Dimensions in mm) | APPROXIMATE CUTTING TIME IN MINUTES FROM FLOOR TO FLOOR | APPROXIMATE NUMBER OF CUTS PER WHEEL |
|---|---|---|
| 76 × 76 × 9·5 | 2 | 110 |
| 51 × 51 × 6·35 | 1 | 300 |
| 38 × 38 × 6·35 | 0·5 | 400 |

*Note:* The above figures are typical for DRY CUTTING. Times for WET CUTTING will be slightly slower, but would give increased wheel life.

Three types of high-speed cold sawing machines in general use in the fabrication industry are:

1. *Table type sawing machine* This is especially useful for cutting long bevels through the webs of joists and channels, in addition to making normal square cuts. The machine is similar to a woodworker's circular saw in that the blade centre is below the level of the work table, but the travel of the blade is provided for in a slot through the length of the table. The material to be cut must be securely clamped to the table during the cutting operation.

2. *Upright type sawing machine* With this type of machine, the circular blade is fed in a horizontal direction. Joists or other sections are placed with the widest part of the section upright. When bevels are required to be cut through the flanges of channels or joists, this machine may be used on a turntable and is fitted with a 'Vee'-shaped side-block device which is specially designed for this purpose.

3. *Vertical-lift type sawing machine* This circular blade is fed vertically downwards into the section being cut. This machine is normally fitted with self-centring vices for square cutting, and the section to be cut is fixed with the narrowest part in the vertical position.

*Power- or hand-operated holding clamps are fitted to cold sawing machines to secure the steel sections against the powerful thrust of the blade while they are being cut.*

The blades used on cold saws consist basically of a body with segments made from hard tool steel rebated over and riveted to the body, the centre of which is reinforced and provided with a suitable hole for fixing to the machine. To ensure that the saw teeth are kept free from steel chips during the sawing operation, chip removers are fitted.

*The blade must be adequately lubricated*, and provision is made for a continuous flow of a soluble oil and water mixture on to the blade during the cutting operation.

BENDING

For bending angle section to various angles after notching, the standard type of bar bender (see Chapter 10) is used.

Three examples of angle bending together with a sequence of operations are illustrated in Figures 9.36, 9.37 and 9.38.

## 9.18 Locating hole centres on large diameter flanges

The method of marking out hole centres on circular flanges has been explained in Chapter 5. Using a centre-punch mark for datum centre, this operation is relatively easy when marking out on a blank before the flange is cut. However, very often in the fabrication industry the craftsman has to mark out hole centres for drilling on large diameter blank flanges. Figure 9.39 explains how this problem is overcome by the use of drilling jigs.

## 9.19 Forging equipment

The principles of forging have already been introduced in Sections 6.9 and 6.10. Some of the more practical aspects of this forging process will now be considered. The equipment required for hand forging is as follows:

1. *The hearth* Since the forging operations performed in the average workshop are rarely more ambitious than 'hot bending', the hearths are usually gas- or oil-fired. These have the advantage of being clean in use and do not throw off abrasive dust that could settle on the slideways of nearby machines, causing excessive wear.

However, if the full range of forging operations are to be performed, then the traditional open-fire hearth, burning coke breeze or anthracite peas, is essential. Figure 9.40 shows a typical hearth and its blowing equipment.

In skilled hands this type of hearth can be fired to produce a wide range of temperatures from a dull red heat for bending to white heat for forge welding. The area of the fire can also be controlled. Unfortunately, this type of hearth produces a lot of abrasive dust. Therefore, it is best kept in a separate shop, away from machine tools.

2. *The anvil* The anvil is used to support the work while it is hammered. Figure 9.41 shows a typical SINGLE BICK LONDON PATTERN anvil.

The anvil itself is made from a number of forgings that are forge welded together. The workface is usually made of a medium carbon steel and hardened.

Fig. 9.38 Internal angle ring

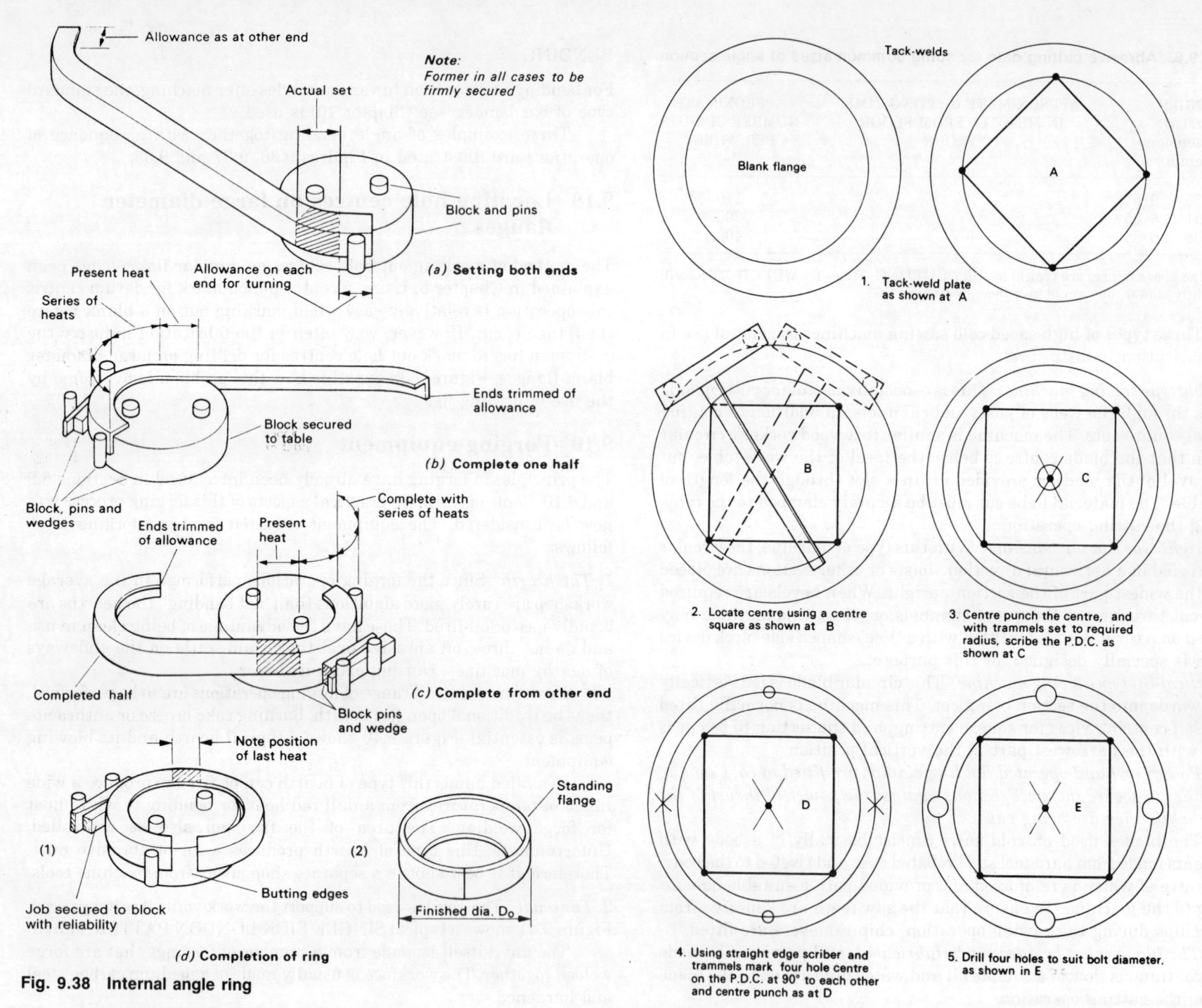

1. Tack-weld plate as shown at A
2. Locate centre using a centre square as shown at B
3. Centre punch the centre, and with trammels set to required radius, scribe the P.D.C. as shown at C
4. Using straight edge scriber and trammels mark four hole centre on the P.D.C. at 90° to each other and centre punch as at D
5. Drill four holes to suit bolt diameter, as shown in E

328

**Cross-section of drilling jig**

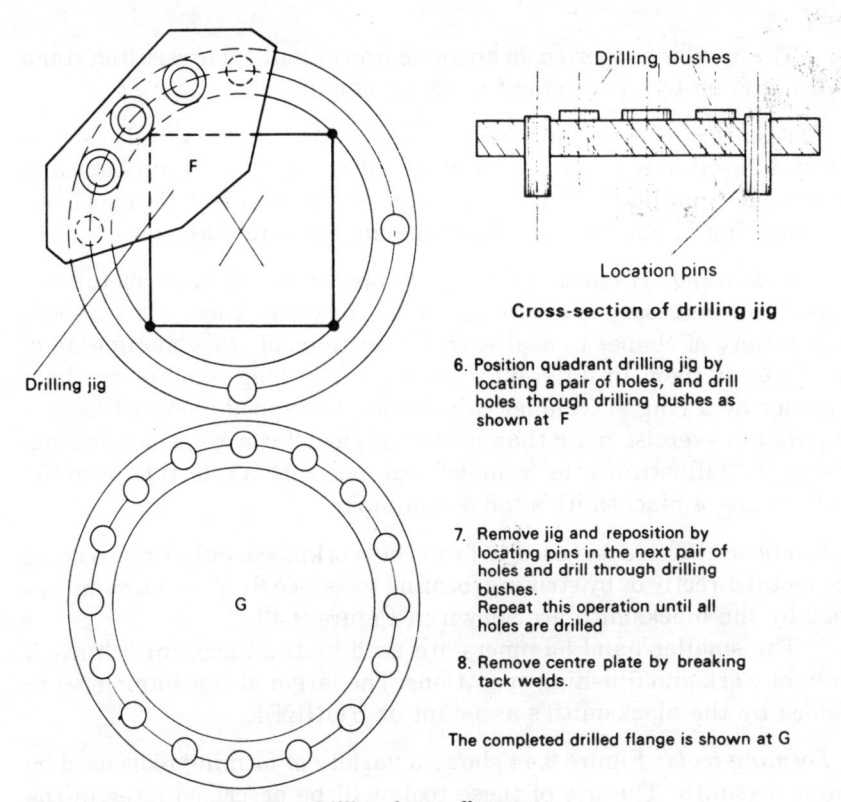

6. Position quadrant drilling jig by locating a pair of holes, and drill holes through drilling bushes as shown at F
7. Remove jig and reposition by locating pins in the next pair of holes and drill through drilling bushes. Repeat this operation until all holes are drilled
8. Remove centre plate by breaking tack welds.

The completed drilled flange is shown at G

**Fig. 9.39 Use of jig for drilling large flanges**

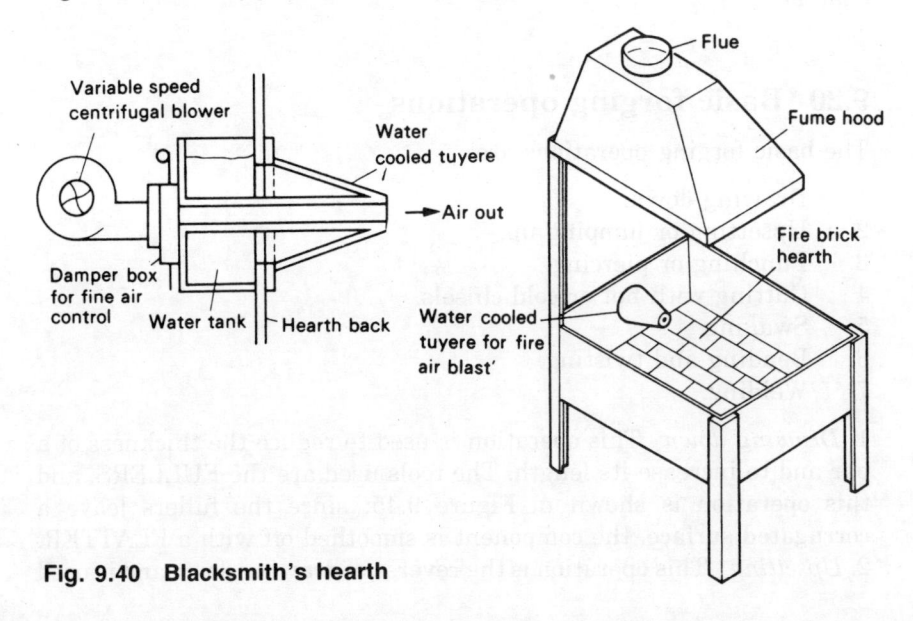

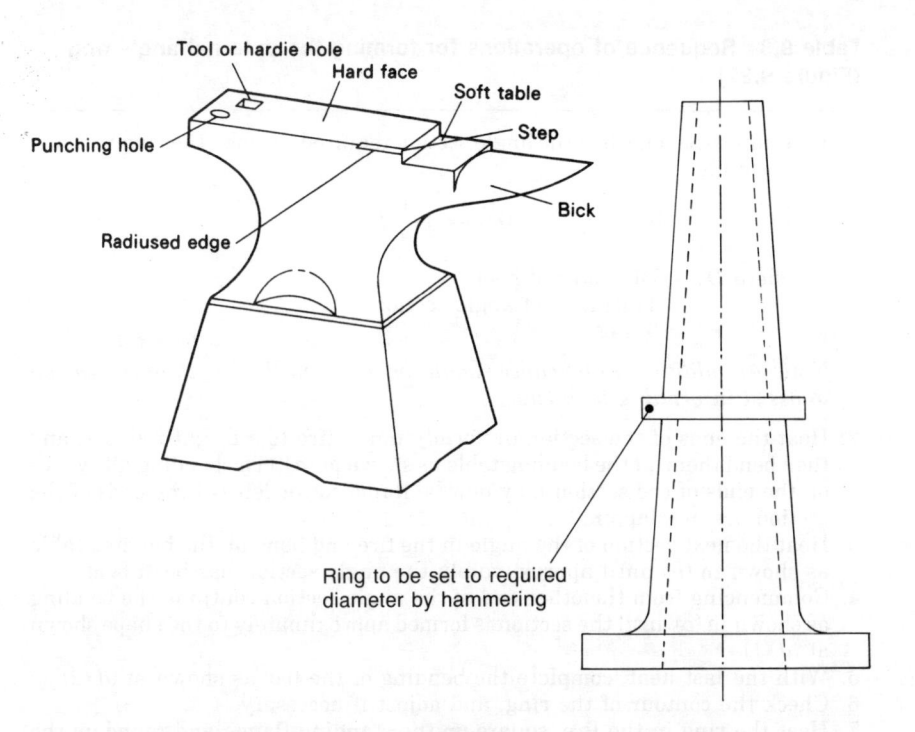

**Fig. 9.40 Blacksmith's hearth**

**Fig. 9.41** (*a*) The anvil and (*b*) the Smith's cone or mandrel

**Table 9.7 Sequence of operations for forming the frame (Figure 9.36)**

1. Draw a plan view of one corner showing the joint preparation as shown at (*a*)
2. Calculate the corner length, marked '*X*', using the formula:

$$\text{LENGTH } X = 2\left(R - \frac{t}{2}\right)\frac{\pi}{4}$$

3. Calculate total length of angle section required.
4. On a suitable length of angle section, scribe a line at 90° to the heel in the centre of the section as shown in (*b*).
5. On each side of this centre line mark-off the notched corners as shown at (*b*), using a sheet metal template for the development of the corners as shown at (*c*). With a centre punch and hammer 'witness-mark' the lines to be cut.
6. Flame cut and prepare the joints for welding.
7. Bend the angle section around a suitable jig to form the corners.
8. Check the frame for dimensions and shape. Adjust if necessary.
9. Tack-weld mild steel straps diagonally to corners of frame, or clamp frame to suitable plate (jig) to prevent undue warping during welding.
10. Weld using metallic arc welding equipment.
    Allow to cool and check frame for size and adjust if necessary.

**Table 9.8  Sequence of operations for forming the external angle ring (Figure 9.37)**

1. Determine the length of the angle section required to make the ring by using the formula

$$(D_1 + T + \frac{1}{3}T)\pi$$

where $D_1$ = Diameter of heel,
$T$ = Thickness of angle section,
$\pi$ = 3·142.

Note: *An additional allowance should be made on the length of the section to assist in bending both ends.*

2. Heat the ends of the section uniformly in the fire to a bright red heat and then bend them on the bending table as shown at (*a*). (The bending allowance on the ends of the section may now be flame cut or left till the ends of the section are overlapping).
3. Heat the next section of the angle in the fire and bend on the bending table as shown in (*b*) until approximately the angle section has been bent.
4. Commencing from the other end of the angle section continue the bending as shown in (*c*) until the section is formed approximately to the shape shown at (*d*)(1).
5. With the last heat, complete the bending of the bar as shown at (*d*)(2)
6. Check the contour of the ring, and adjust if necessary.
7. Heat the ring in the fire, square-up the standing flange and round-up the angle section to correct diameter.
8. Flame cut the joint and prepare for tack-welding. Tack-weld and completely weld the joint.

Precautions:
1. Do not overheat the angle section.
2. When rounding-up and flattening the section use the 'fuller' and 'flattener' in preference to the hammer to avoid damaging the angle section.

**Table 9.9  Sequence of operations for forming the internal angle ring (Figure 9.38)**

The same sequence of operations may be adopted as shown in Figure 9.3, but in this case determine the length of angle section required to form the ring by using the formula

$$(D_o - T - \frac{1}{3}T)\pi$$

where $D_o$ = Diameter of the ring,
$T$ = Thickness of angle section.
$\pi$ = 3·142.

The methods used for clamping have been explained in Chapter 8.

The anvil is supported on an angle iron or malleable cast iron stand to bring it up to a convenient working height.

3. *The Smith's cone or mandrel*  Figure 9.41(*b*) shows a Smith's cone which is used to set and correct rings after they have been bent and force welded on the anvil. It is supported on the floor by a flanged base, and the ring is hammered around the appropriate diameter.

4. *Work-holding*  Because of the high temperatures at which metals are forged, it is necessary to use tongs to hold the work. These are available in a variety of shapes to deal with the sections of stock manipulated, and present them in different directions. The long handles are held together by a ring or 'coupler' which may be tapped along when it is required to exercise more than a short period of control when forging. Figure 9.42 illustrates the 'coupler' together with a selection from the wide range of blacksmith's tongs available.

5. *Hammers*  These are used to form the workpiece, either by striking the metal directly or by striking forming tools (see 6). Various hammers used by the blacksmith are shown in Figure 9.43.

The smaller hand hammers are used by the blacksmith himself on light work and finishing operations. The larger sledge hammers are welded by the blacksmith's assistant or STRIKER.

6. *Forming tools*  Figure 9.44 shows a variety of forming tools used by the blacksmith. The use of these tools will be described later in the chapter.

## 9.20  Basic forging operations

The basic forging operations are:

1. Drawing down.
2. Upsetting or jumping up.
3. Punching or piercing.
4. Cutting with hot or cold chisels.
5. Swaging.
6. Bending and twisting.
7. Welding.

1. *Drawing down*  This operation is used to reduce the thickness of a bar and to increase its length. The tools used are the FULLERS, and this operation is shown in Figure 9.45; since the fullers leave a corrugated surface, the component is smoothed off with a FLATTER.

2. *Upsetting*  This operation is the reverse of drawing down and is used

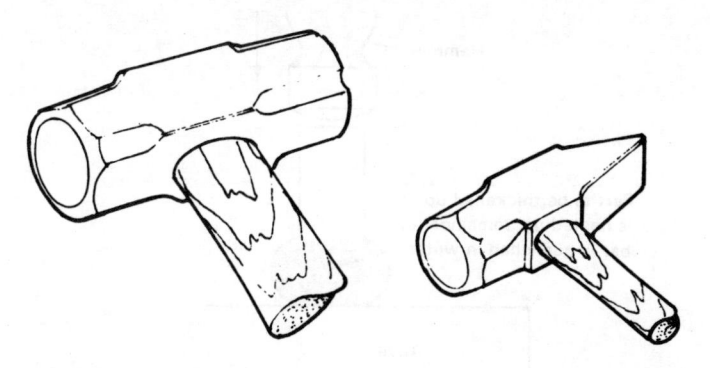

**Fig. 9.43** Hammers

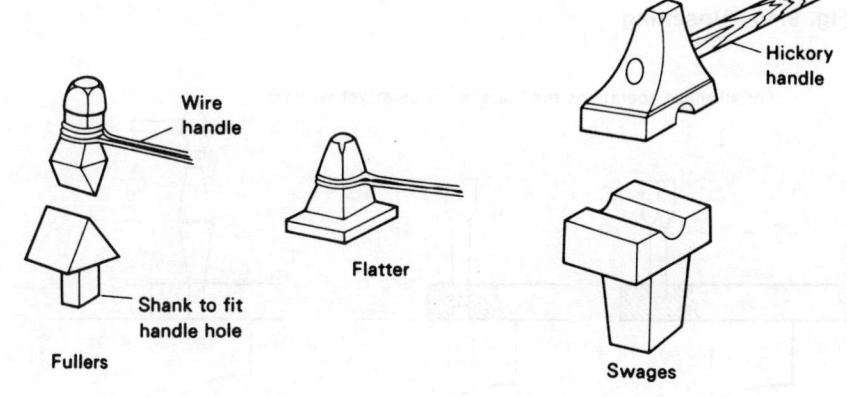

**Fig. 9.44** Forming tools

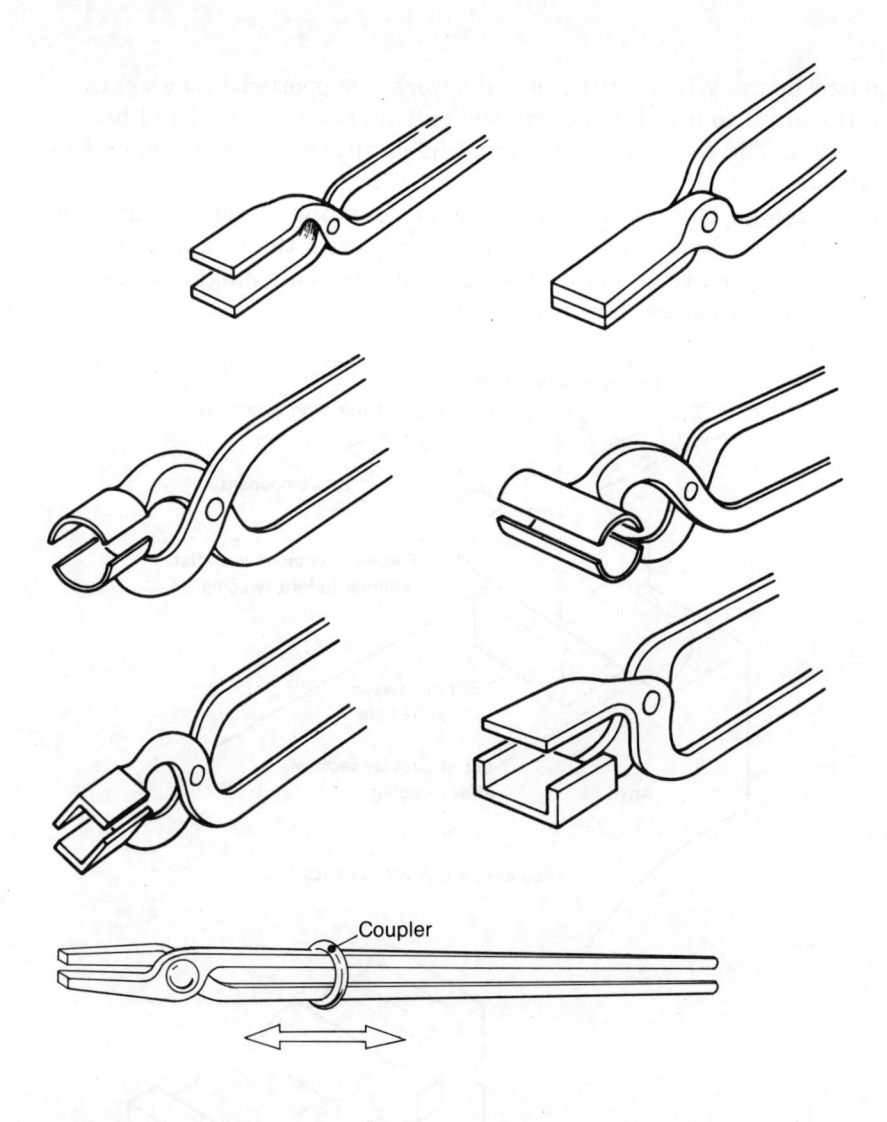

**Fig. 9.42** Tongs

to increase the thickness of a bar and to reduce its length. Generally, the increase in thickness is only local as when forming a bolt head. Figure 9.46 shows the operation of upsetting and it will be seen that only the hammer is used.

3. *Punching* Holes are started by being punched out (round or square) at a size convenient for the piercing or slug to pass through the punch hole to the anvil.

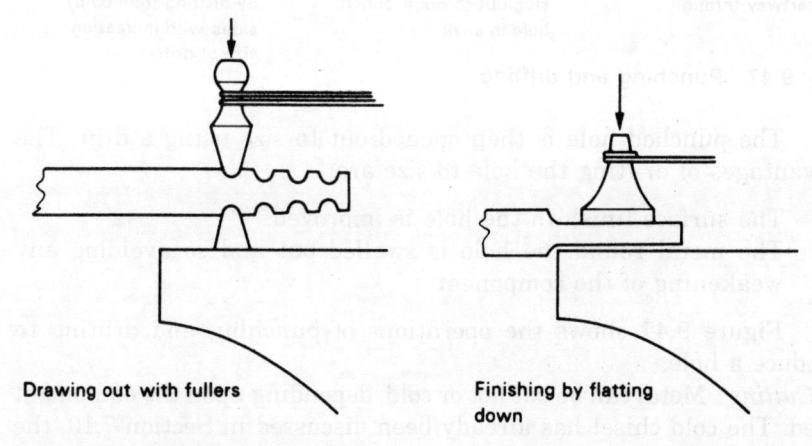

**Fig. 9.45** Drawing down

331

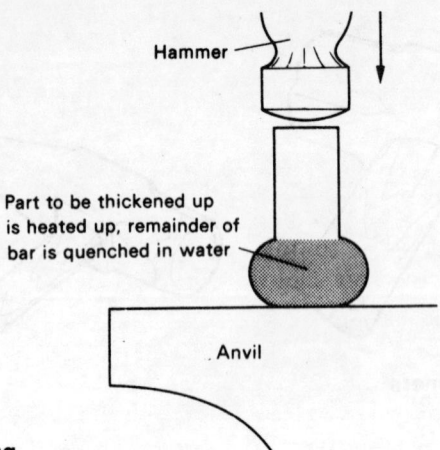

**Fig. 9.46 Upsetting**

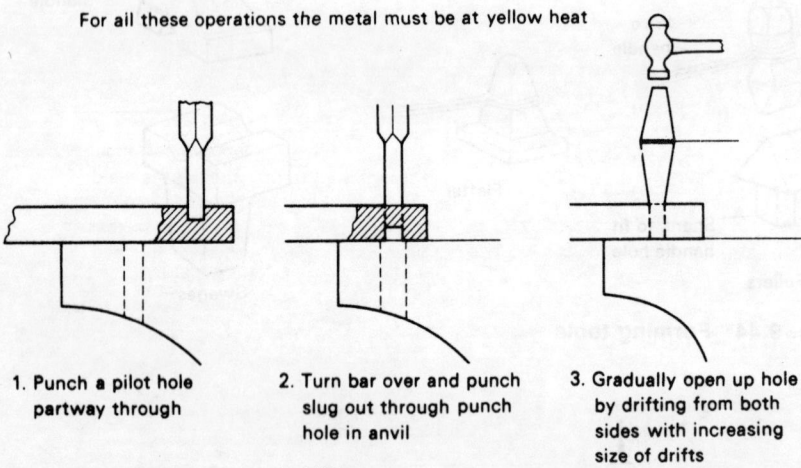

**Fig. 9.47 Punching and drifting**

The punched hole is then opened out to size using a drift. The advantages of drifting the hole to size are:

1. The surface finish in the hole is improved.
2. The metal round the hole is swelled out and so avoiding any weakening of the component.

Figure 9.47 shows the operations of punching and drifting to produce a hole.

4. *Cutting* Metal can be cut hot or cold, depending upon the equipment used. The cold chisel has already been discussed in Section 7.10. the hot chisel is more slender so that it can knife its way through the hot, plastic metal. When cutting cold, the work is supported on the soft table of the anvil to avoid damaging the cutting edge as the chisel breaks through. The techniques of cutting have already been introduced in Chapter 7.

5. *Swaging* This is the operation of rounding a component. The component is broken down into an octagon with the flatter, and then rounded off between a pair of swages or between a single swage and the swage block, as shown in Figure 9.48.

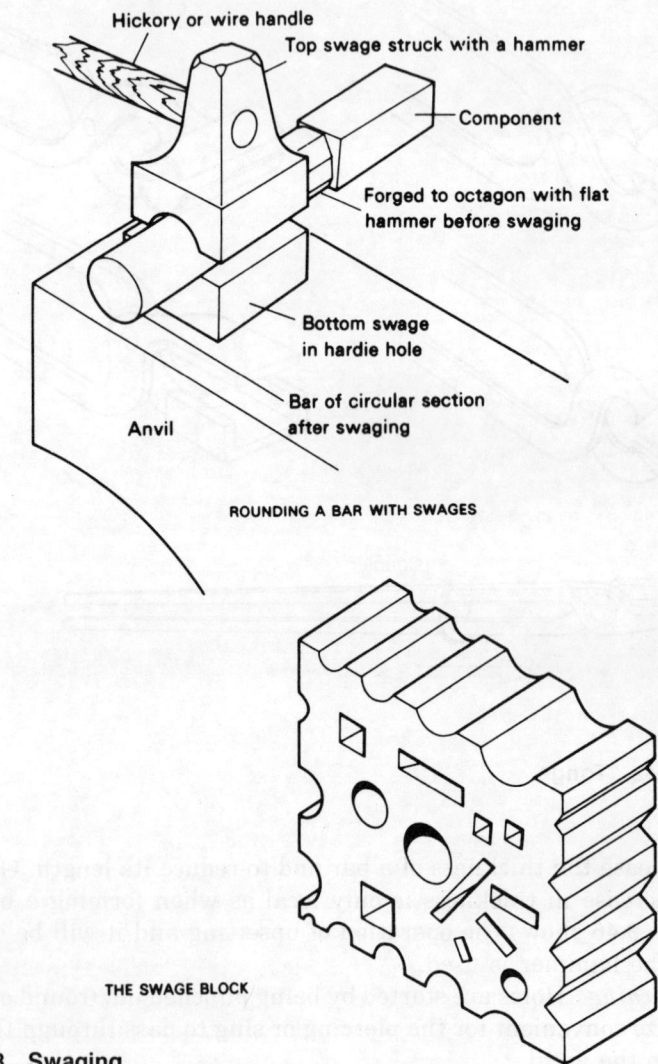

**Fig. 9.48 Swaging**

**Fig. 9.49 Bending**

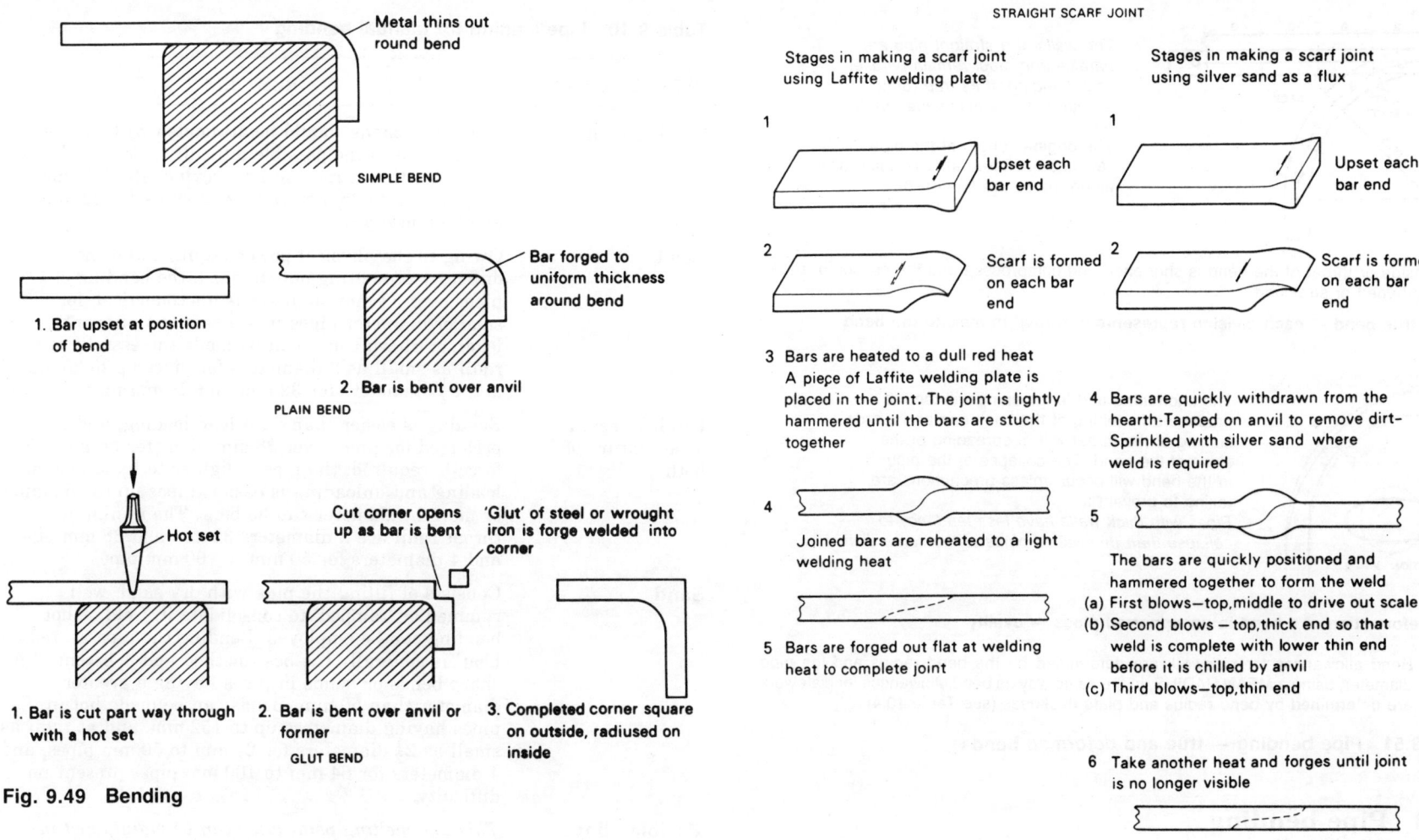

6. *Bending* This operation is shown in Figure 9.49. The point to note when bending is that there is a tendency for the metal to thin out round the bend, causing weakness. This can be overcome by either upsetting the bar prior to bending, or forge welding an additional piece of metal, or GLUT, into the bend. The latter method forms a very strong corner.

7. *Welding* This is a difficult operation requiring great skill. When done correctly the resulting joint is superior in strength to either arc or oxy-acetylene welding. Forge welding is restricted to wrought iron and mild steel. A flux is necessary, and this may be a proprietary brand such as Laffite, or a natural material such as silver sand. The former is the easier to use. There are various types of weld, and two examples are given in Figure 9.50.

**Fig. 9.50 Forge welding**

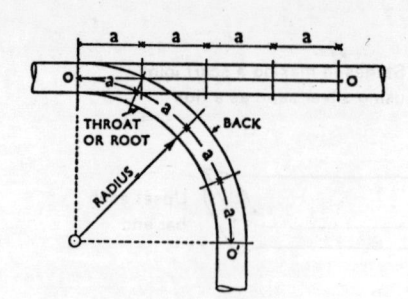

*The walls of a straight pipe are parallel and must remain parallel after bending, if its true round section is to be maintained in the bend.*
*The original length of the pipe (0–0) remains unaltered after bending only along the centre line (0–0).*

The inside or throat of the bend is shortened and compressed and the outside or back is lengthened or stretched.

**(a) A true bend — each division represents a 'throw' in making the bend**

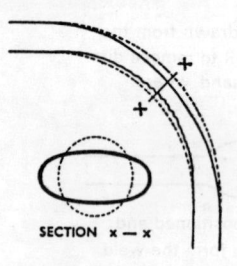

The shortening and lengthening will tend to produce a flattening of the back and an inwards kinking of the throat with a spreading of the sides of the bend. The collapse of the pipe in the bend will occur unless precautions are taking to prevent it.
*Pipes with thick walls have less tendency to collapse than thin-walled pipes of the same diameter.*

**(b) Deformation of a bend in unsupported pipes — ovality**

*Note*: Bend allowances for pipework are determined by the bend radius and the pipe diameter, using a MEAN RADIUS, in the same way as bend allowances for platework are determined by bend radius and plate thickness (see Table 10.4).

**Fig. 9.51 Pipe bending — true and deformed bends**

## 9.21 Pipe bending

For a proper understanding of the various techniques and tools employed in pipe bending, it is essential that one clearly understands what deformations occur during the process. This deformation is simply explained in Figure 9.51.

Owing to their relatively thin walls, light gauge pipes need greater care if satisfactory bends are to be made by manual methods. In order to prevent 'ovality' or collapse, it is necessary to fill or 'load' the pipe with some material that will support the pipe walls.

The materials commonly used for 'loading' are detailed in Table 9.10.

Before loading with lead, pitch, resin, or low melting point alloy, the pipe must be sealed at one end with a wooden plug. When sand is used, the pipe must be plugged at both ends in order to retain the compacted loading.

**Table 9.10 Pipe loading for manual bending**

| TYPE OF LOADING | REMARKS |
|---|---|
| **Steel springs** | Only easy bends should be attempted with spring loading due to the possible difficulty of removal. The minimum throat radii are approximately 3 diameters for pipes up to 25 mm and 4 diameters for 32 mm and 50 mm pipes. |
| **Lead** | Owing to the physical power required and the difficulty of melting out the lead, the bending of pipes over 38 mm diameter is uneconomical; for smaller diameter pipes it is a safe and efficient loading. Sharp bends can be made successfully with radii as small as 2 diameters for pipes up to 25 mm and $2\frac{1}{2}$ diameters for 32 mm and 38 mm pipes. |
| **Pitch** or **resin** or a mixture of both | Bending is easier than with lead loading, and is preferred for pipes over 38 mm diameter because less force is required, the pipe is lighter to handle, and loading and unloading is easier. Pipes up to 152 mm or more in diameter can be bent. The minimum throat radii are 2 diameters 32 mm and 38 mm pipes, and 4 diameters for 50 mm to 100 mm pipes. |
| **Sand** | Consists of filling the pipe with dry sand, well rammed or 'tamped' to consolidate it. Cold or hot bending can be employed. Bending the pipe at 'red heat' is generally the best method and is essential for sharp bends or bends in pipes having a greater diameter than 50 mm. Bends can be made hot in pipes having diameters up to 152 mm, and at radii as small as $2\frac{1}{2}$ diameters for 25 mm to 50 mm pipes, and 4 diameters for 64 mm to 100 mm pipes present no difficulty. |
| **Fusible alloy** | *This low melting point alloy can be maintained in its molten state at temperatures lower than the boiling point of water, and quickly solidifies at room temperature.* After bending, the filler will quickly run out when the pipe is dipped into a tank of boiling water, leaving the interior of the pipe perfectly clean. |

## 9.22 Use of flexible bending springs

The use of flexible steel springs when pipe bending is probably the most practical method employed when forming thin-walled copper pipes by

**Fig. 9.52  Use of springs for bending**

*The effect of bending thin wall tube without a spring* — Tube collapses

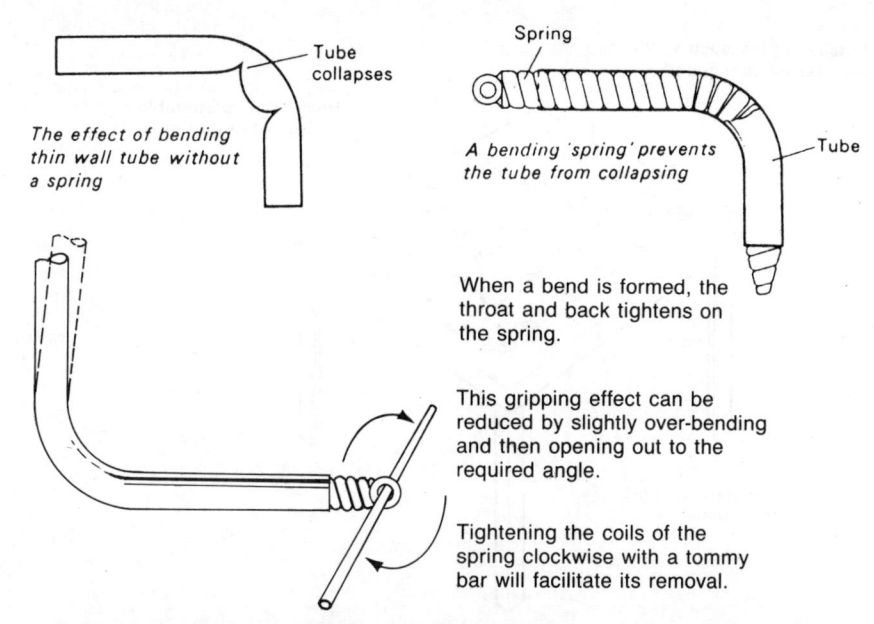

*A bending 'spring' prevents the tube from collapsing* — Spring, Tube

When a bend is formed, the throat and back tightens on the spring.

This gripping effect can be reduced by slightly over-bending and then opening out to the required angle.

Tightening the coils of the spring clockwise with a tommy bar will facilitate its removal.

Use of an extension rod hooked into the eyelet of the spring will enable it to be inserted and withdrawn when a bend is made in the middle of a length of pipe.

hand. It is recommended that bending springs should be lightly greased before use. This will facilitate ease of removal after bending.

Figure 9.52 illustrates the use of bending springs.

## 9.23  Safety — pipe loading

### LEAD LOADING

The pipe should be securely fixed or held in tongs, since it will become very hot when the molten lead is poured into it. The pipe should be warmed before pouring commences.

*Great care must be taken to ensure that the pipe is thoroughly dry, especially if it has been quenched after annealing.* If a pipe is at all damp, steam will be generated and its pressure will drive out the molten lead, with very grave risk to the operator.

### PITCH OR RESIN LOADING

Pitch or resin must be slowly heated until it is liquid, taking care that it does not ignite. If this does occur, quickly smother the mouth of the vessel containing it with a wet sack (one should be kept handy in case of such an accident).

*Attempts should never be made to anneal a pipe already loaded, because of the high risk of a dangerous explosion.*

When the bend is completed and the loading has to be melted out, care should be taken to start heating at an end. If the flame is started away from the end, gases will form to drive out a semi-molten plug of resin or pitch with great force and risk to anyone in the near vicinity. It is possible to cause an explosion and split the pipe.

### SAND LOADING

Before use, sand should be thoroughly dry, any moisture in the sand may generate steam when the pipe is heated, and thus may cause an explosion. Sand used for loading can be conveniently dried on a red-hot iron plate.

## 9.24  Pipe bending by hand

One of the simplest devices for hand bending small diameter pipes can be made by boring holes equal in diameter to the outer diameter of the pipe in a wood block or plank 76 mm to 100 mm in thickness. These are then gouged out to form easy bend channels in the thickness of the block. The pipe is gradually drawn through a suitable hole and forced downwards progressively in the channel, according to the sharpness of the bend required. Only easy cold bends on loaded pipes should be attempted using this method.

Another simple device for making cold bends by hand is shown in Figure 9.53.

Light gauge copper pipes are usually supplied in 'half-hard' temper, which preserves sufficient rigidity and straightness when installed, and minimises damage in transit. During the final stage of manufacture, soft or annealed copper pipe is converted to 'half-hard' temper by the cold work imposed upon it when drawn through a die to final size. The work of cold-forming copper pipe by hand or machine methods can be made considerably easier by annealing. The cold work imposed on the pipe by the bending operation has the effect of restoring the original 'half-hard' temper to the annealed length.

Heavy gauge copper pipes can be forge-bent while red-hot, in a similar manner to that employed when bending thick-walled mild steel and wrought iron pipes. This is a highly skilled operation and the basic principles are shown in Figure 9.54.

Forge-bending, correctly executed, produces a bend having a

**Fig. 9.53 Pipe bending — simple bench former**

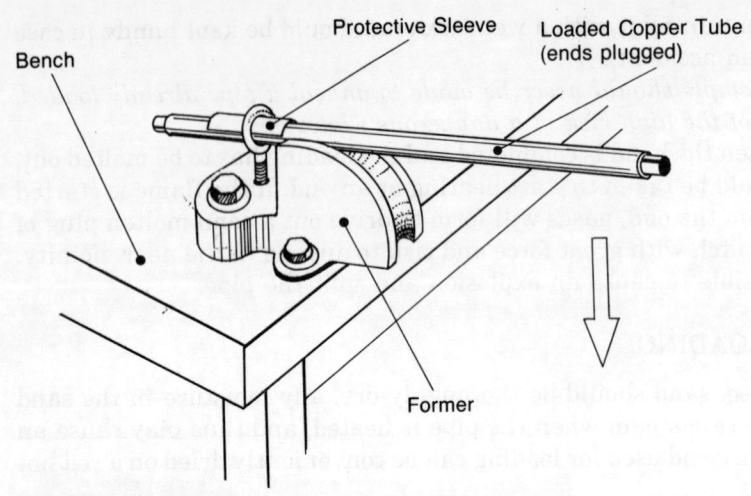

A former with a curvature to correspond with the required radius of the bend is bolted to the bench.
One end of the loaded pipe is passed through the screwed eyelet provided, and the other end is forced downwards around the grooved former.
Pipes up to 38 mm diameter may be formed in this manner.

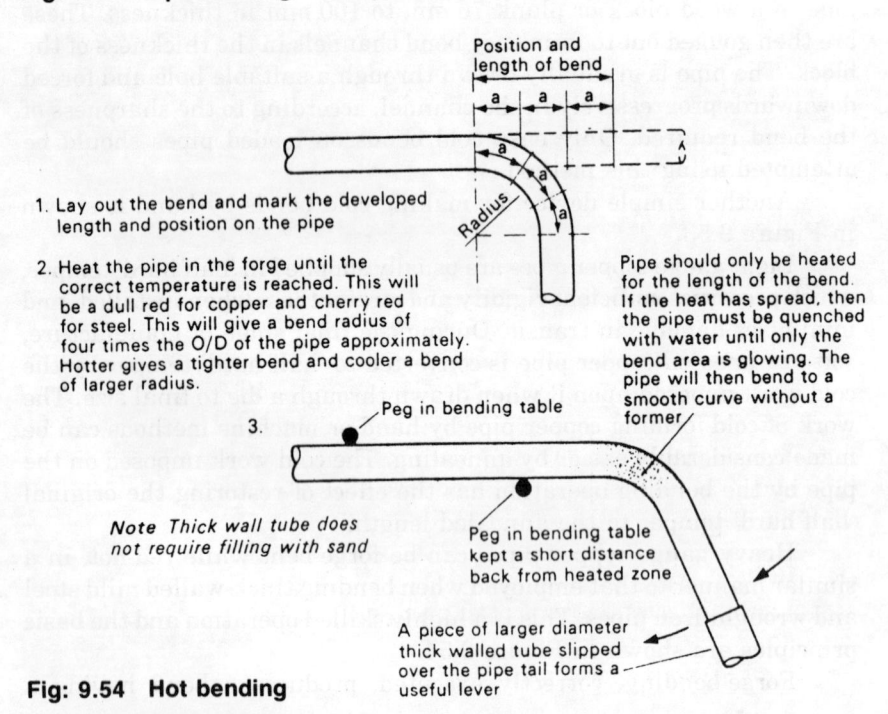

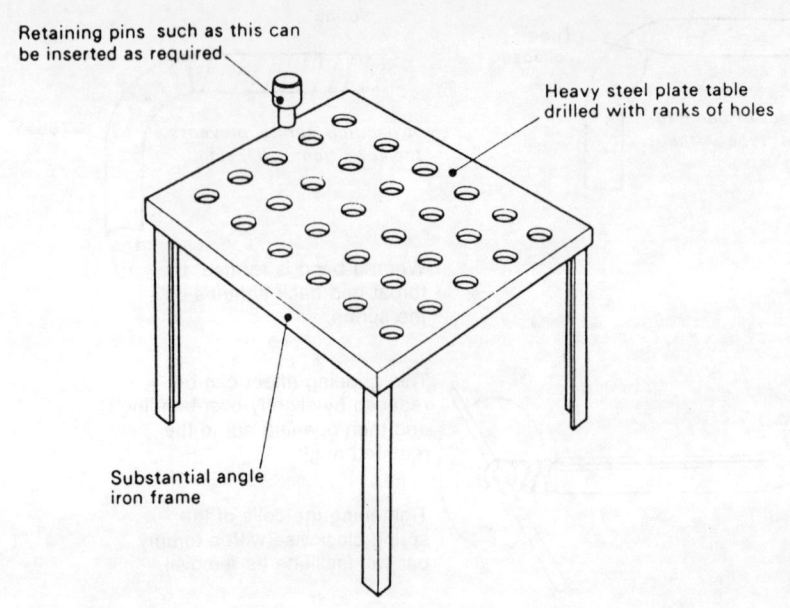

1. Lay out the bend and mark the developed length and position on the pipe

2. Heat the pipe in the forge until the correct temperature is reached. This will be a dull red for copper and cherry red for steel. This will give a bend radius of four times the O/D of the pipe approximately. Hotter gives a tighter bend and cooler a bend of larger radius.

Pipe should only be heated for the length of the bend. If the heat has spread, then the pipe must be quenched with water until only the bend area is glowing. The pipe will then bend to a smooth curve without a former.

3. 

*Note* Thick wall tube does not require filling with sand

A piece of larger diameter thick walled tube slipped over the pipe tail forms a useful lever

**Fig: 9.54 Hot bending**

**Fig. 9.55 Bending table**

smooth curve, with a minimum of thinning and ovality. The pipes are filled with compacted sand to prevent them collapsing during bending. When a pipe has been raised to forging heat at the location of the bend, it is transferred to a bending table where it is located by retaining pins (Figure 9.55) and pulled round by an iron sling or lever. Very large diameter pipes are heated in a specially designed furnace and a winch is employed when forming the bend. The real skill lies in the local quenching of the pipe so that it only 'gives' at the required point.

## 9.25 Pipe bending by machine

Machines of various types and sizes, worked by direct hand power, are capable of bending pipes up to 50 mm diameter. Some are so small, compact and light, that they are practically hand tools which can be easily transported and used on site work. Ratchet action or geared machines are used for large size pipework. For batch production, draw-bar machines are available. These machines have an adjustable mandrel that is pivoted at the outer end, supporting the walls of the pipe at the bending point, and is always tangential to the bend at that point whatever the radius of the former. Leverage is applied to the former, to which the pipe is attached, and the former itself revolves while the bend is being made. Some draw-bar machines are machine driven.

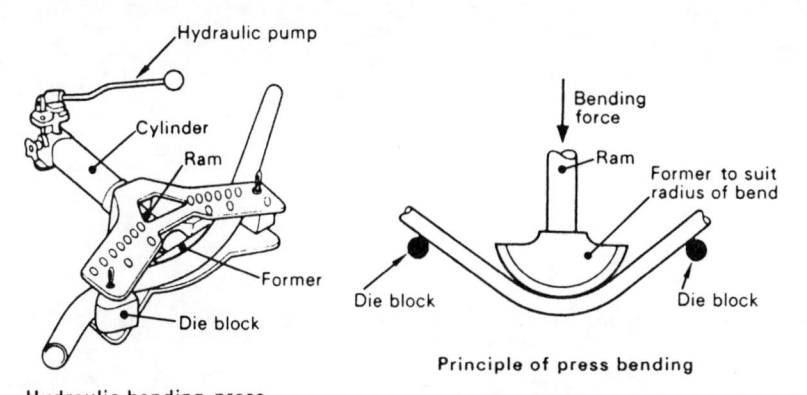

**Fig. 9.56 Bending press**

Figure 9.56 illustrates the principle of a typical hydraulic machine used for press-bending thick-walled steel pipes.

Rotary type machines have two principle components: a *quadrant former* and a *back guide*, as shown in Figure 9.57(a), which enable pipes to be bent without loading. The former supports the throat of the pipe while the back guide rotates around the bend as pressure is applied by the roller. The sectional view in Figure 9.57(b) illustrates how these components support the pipe during the process, thus enabling the bend to be made without thickening of the throat metal which would reduce the bore of the pipe.

It is advisable, when using hand pull machines for heavy gauge pipes, to anneal the section to be bent and so save labour and undue stress on the machine.

Some machines are fitted with a device on the lever arm which, by proper adjustment, determines automatically the exact point at which pressure by the roller is best applied to the pipe. This device is the 'pointer'. Figure 9.57(c) shows the correct adjustment for bending. *The pointer is parallel with the pipe when the roller is making tight contact with the back of the guide.*

Figure 9.58 illustrates how incorrect positioning of the roller will result in defects in the bend.

Setting out and measurements for making machine bends are considered in Chapter 5.

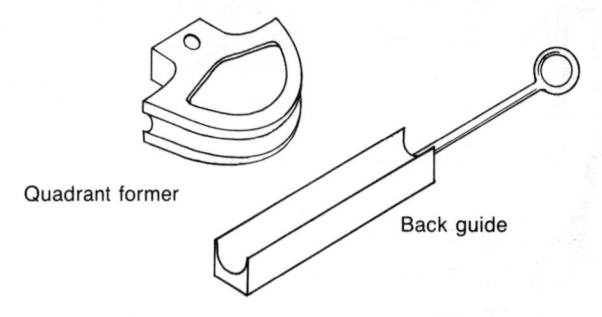

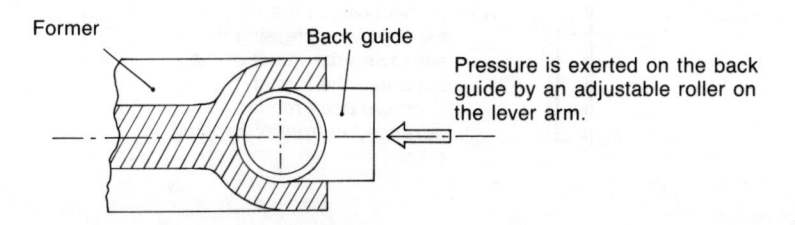

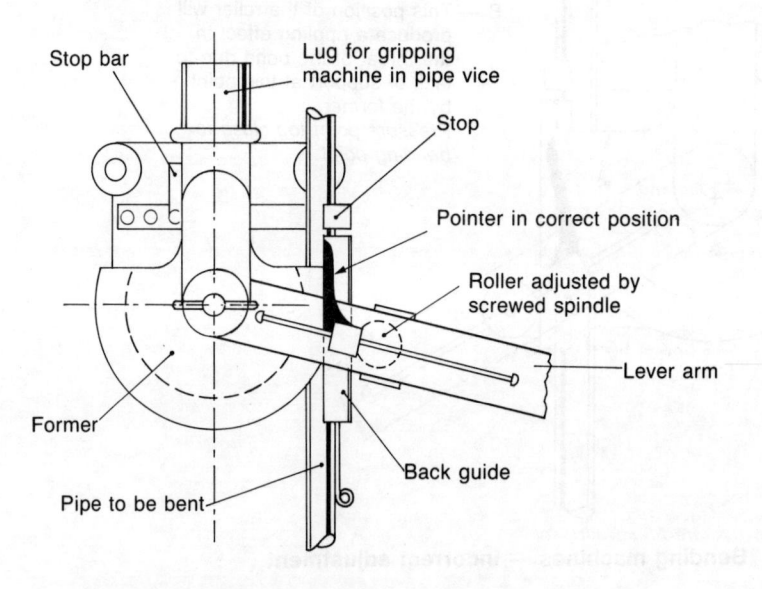

**Fig. 9.57 Rotary bending machines**

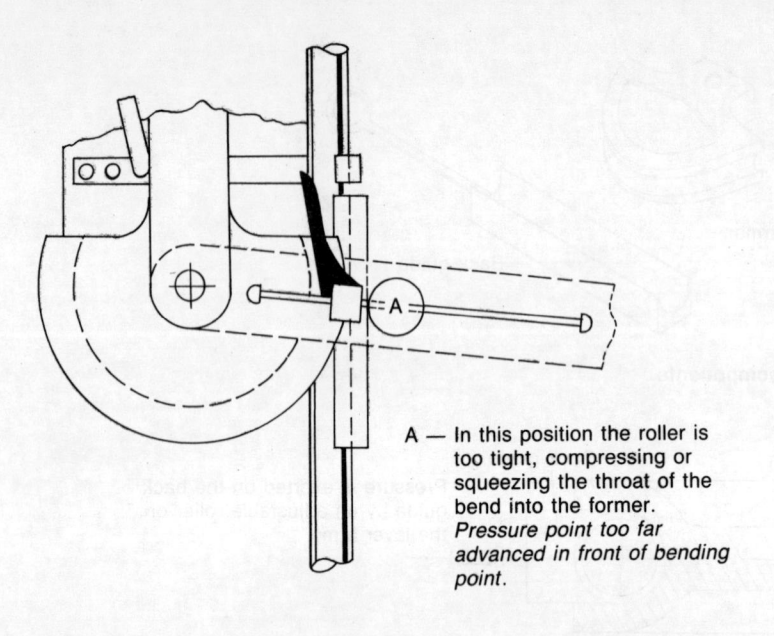

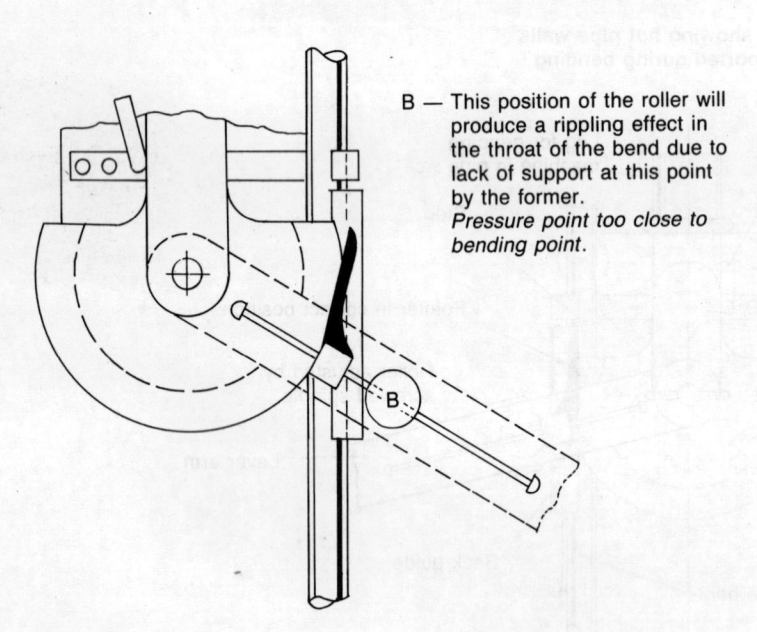

A — In this position the roller is too tight, compressing or squeezing the throat of the bend into the former.
*Pressure point too far advanced in front of bending point.*

B — This position of the roller will produce a rippling effect in the throat of the bend due to lack of support at this point by the former.
*Pressure point too close to bending point.*

Fig. 9.58 Bending machines — incorrect adjustment

# 10 Fabrication processes

This chapter is concerned with the plastic manipulation of sheet metal at room temperature. Figure 10.1 shows some typical cold-forming operations.

## 10.1 Folding or bending

The terms '*folding*' and '*bending*' are rather loosely used in industry because they are so similar. The difference between them is so slight that they are both carried out with the same purpose in view, which is to deflect the metal from one flat plane to another so that it stays there permanently.

If the deflection is sharp, and the radius small, the metal is said to be *folded*.

Should the curvature be large and the deflection cover a large area, it is called *bending*. In this respect the rolling of a hollow body, such as a cylinder, is called bending.

Folding or bending involves the deformation of a material along a straight line in two dimensions only.

## 10.2 The mechanics of bending

When a bending force is gradually applied to a workpiece under free bending conditions, the *first stage* of bending is elastic in character. This is because the TENSILE AND COMPRESSIVE stresses that are developed on opposite faces of the material are not sufficiently high to exceed the YIELD STRENGTH of the material. The movement or STRAIN which takes place as a result of this initial bending force is elastic only, and upon removal of the force the workpiece returns to its original shape.

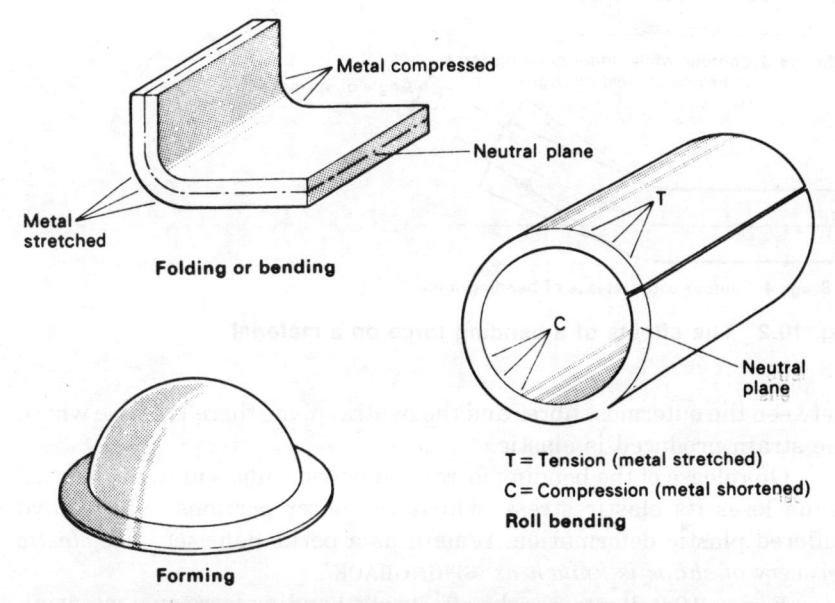

Fig. 10.1 Comparison of common cold-forming applications

As the bending force is continued and gradually increased, the stress produced in the outermost fibres (on both the compression and tension sides) of the material eventually exceeds the yield strength.

Once the yield strength of the material has been exceeded, the movement (strain) which occurs is PLASTIC. This permanent strain occurs only in the outermost regions furthest from the NEUTRAL PLANE.

339

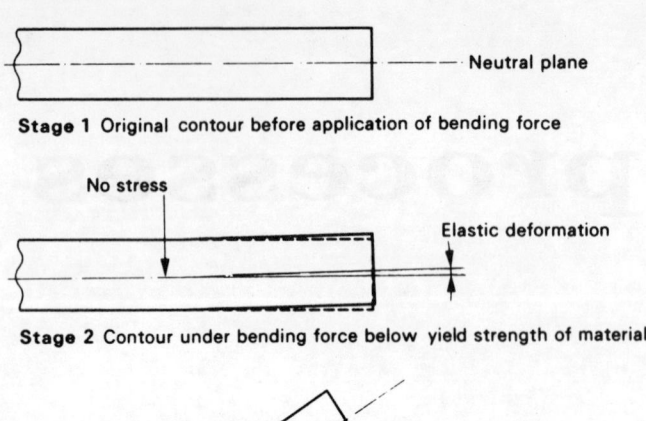

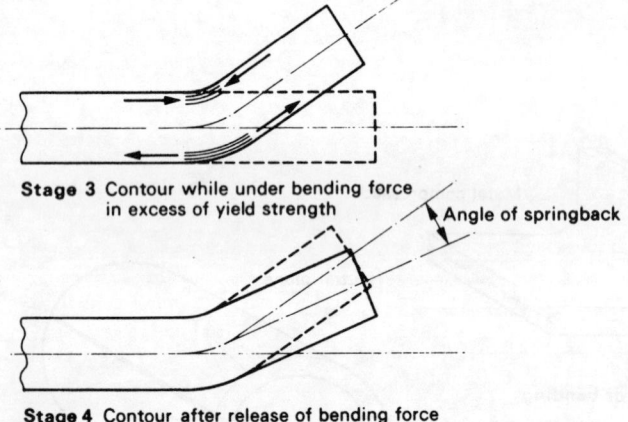

**Fig. 10.2  The effects of a bending force on a material**

Between the outermost fibres and the neutral plane there is a zone where the strain produced is elastic.

On release of the bending force, that portion adjacent to the neutral plane loses its elastic stress, whilst the outer portions, which have suffered plastic deformation, remain as a permanent set. *The elastic recovery of shape is known as* 'SPRINGBACK'.

Figure 10.2 illustrates the effects of a bending force on a material.

When bending sheet metal to an angle, the *inner fibres* of the bend are *compressed* and given COMPRESSION STRESSES. The *outer fibres* of the bend are *stretched* and given TENSILE STRESSES.

Between the two stressed zones, which are in opposition to each other, lies a boundary which is a NEUTRAL PLANE. This boundary is termed the 'NEUTRAL AXIS' or 'NEUTRAL LINE' (see Figure 10.3). See Section 10.27 for a fuller explanation of neutral line or axis.

The position of the neutral line will vary in different metals

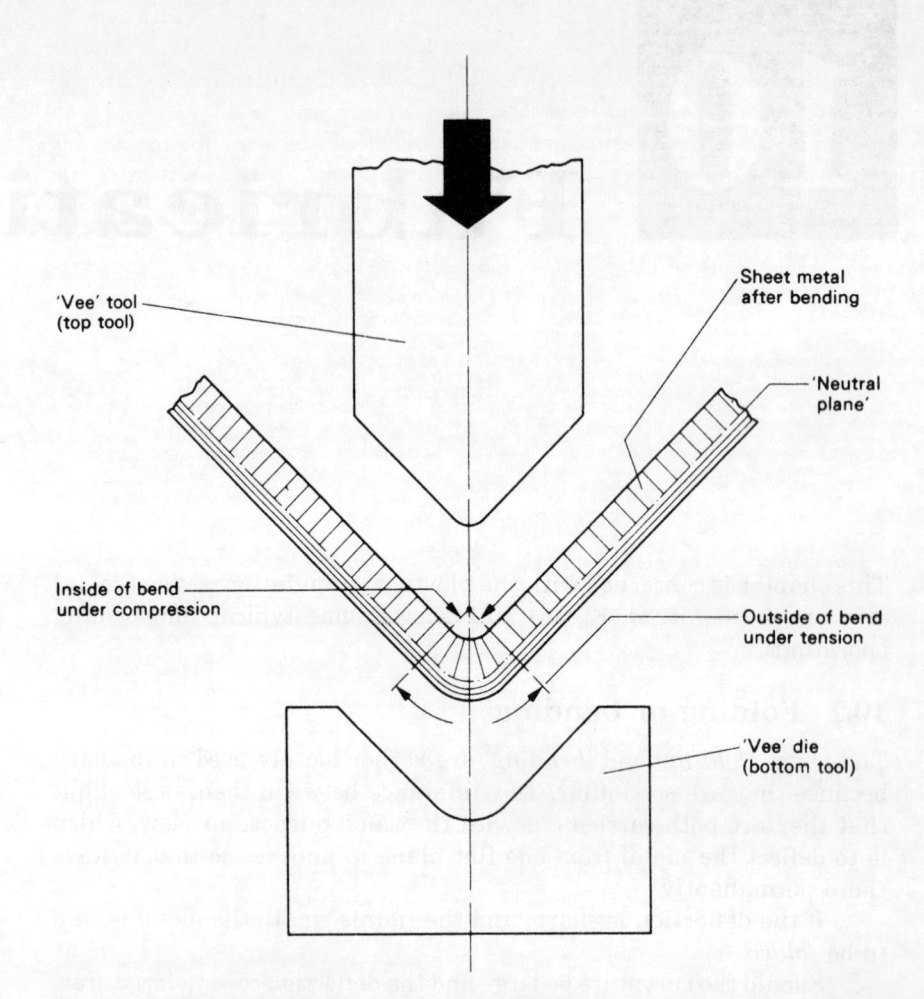

**Fig. 10.3  Bending action — pressure bending**

because of their differing properties, and also vary due to the thickness of the material and its physical condition. It is important to establish the position of the neutral line as it is required, in practice, for the purpose of calculating bend allowances (see Section 10.27).

## 10.3  Springback

During the bending of a material an unbalanced system of varying stresses is produced in the region of the bend. When the bending force is removed (on completion of the bending operation) this unbalanced system tends to bring itself to equilibrium. The bend tends to spring back, and any part of the elastic stress which remains in the material becomes RESIDUAL STRESS in the bend zone.

The amount of springback action to be expected will obviously vary because of the differing compositions and mechanical properties of the materials used in fabrication engineering. Some materials, because of their composition, can undergo more severe cold-working than others.

The severity of bending a specific material depends on two basic factors:

1. The radius of the bend.
2. The thickness of the material.

A 'tight' (small radius) bend causes greater cold deformation than a more generous bend in a material of the same thickness.

A thicker material develops more STRAIN HARDENING than is experienced in thinner material bent to the same inside radius.

The '*condition*' of the material upon which bending operations are to be performed, has an influence on the amount of springback likely to result. For example, using the same bend radius, a COLD-ROLLED NON-HEAT-TREATABLE aluminium alloy in the 'HALF-HARD' temper, or condition, will exhibit greater springback than the same alloy of equal thickness when in the 'FULLY ANNEALED' condition.

The limit to which free bending can be carried out is determined by:

1. The extent to which the material will stretch (ELONGATE) on the *tension side* (outside of bend).
2. The failure due to such COMPRESSIVE EFFECTS as *buckling*, *wrinkling* or *collapse* on the inside of the bend in respect of hollow sections (pipe bending).

## 10.4 Methods of compensating for springback

The clamping beam on a folding machine is specially designed to compensate for springback. This is illustrated in Figure 10.4.

There are two methods of reducing springback when using a press brake, or a 'Vee' tool in a fly press. These are illustrated in Figure 10.5.

### AIR BENDING

This allows partial bending and various angles to be bent by THREE POINT LOADING. The three points are the two edges of the VEE DIE (bottom tool) and the nose of the VEE PUNCH (top tool).

During air bending, the sheet or plate retains its ELASTICITY. In this case the bending angle must be over-closed to compensate for the springback of the material after removal.

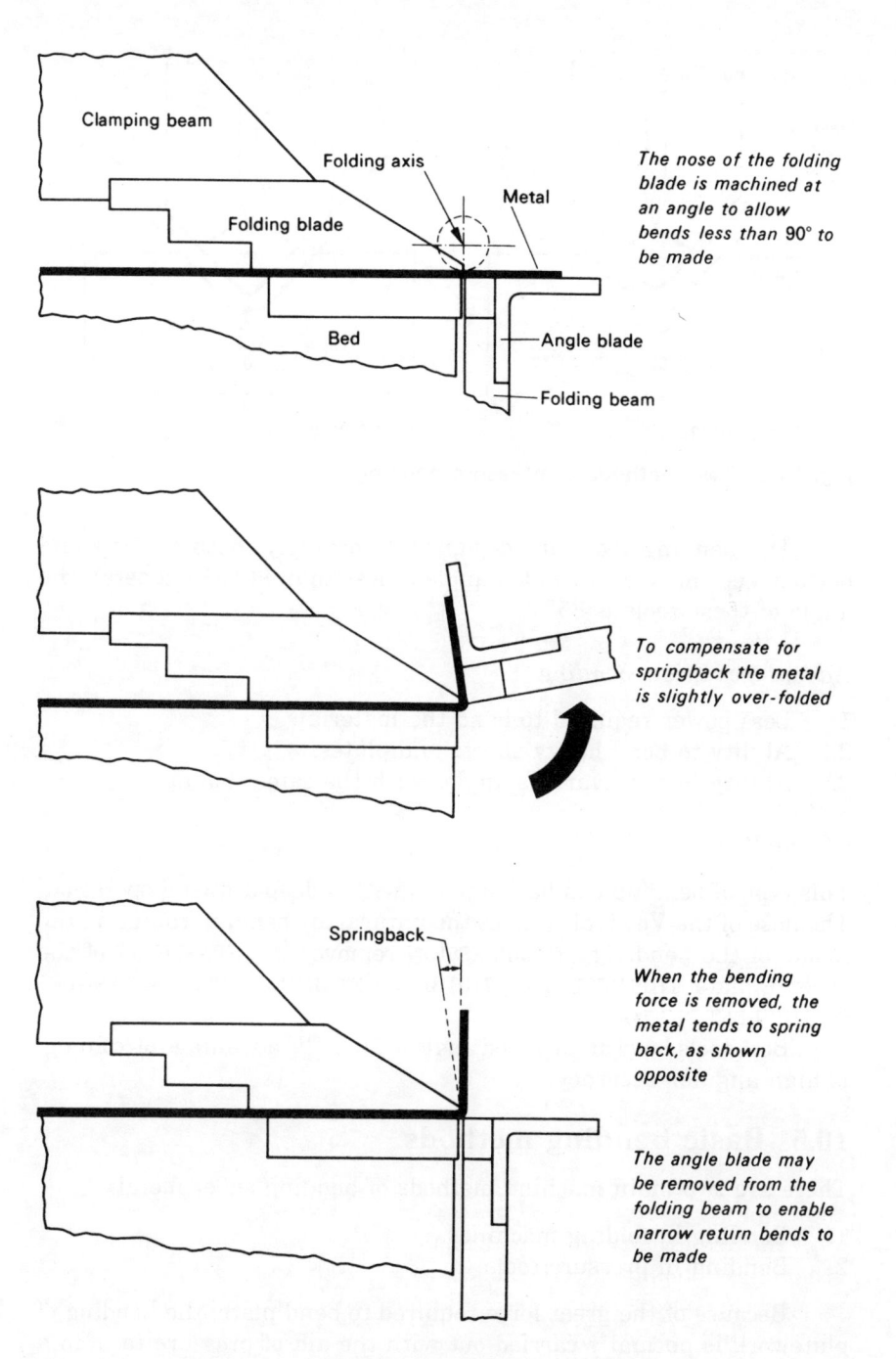

Fig. 10.4 Allowing for springback on a folding machine

(a) Air bending  (b) Coining

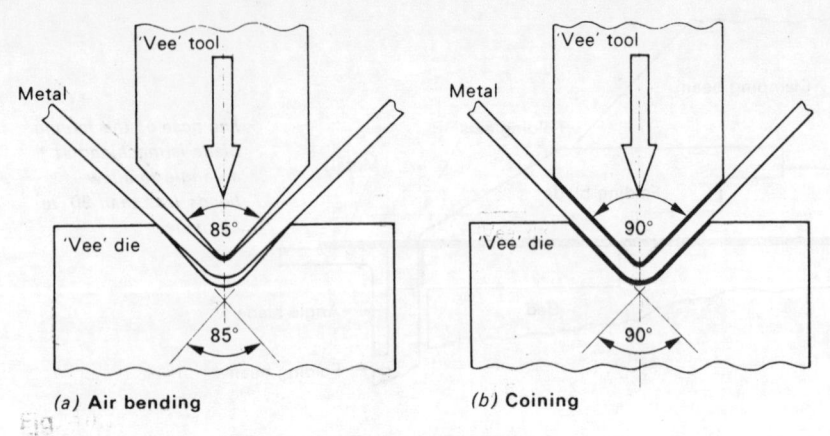

**Fig. 10.5 Two methods of pressure bending**

The bending tools are designed accordingly, both the top and bottom Vees have an included angle of less than 90°. In general, the angle of these tools is 85°.

*Advantages in air bending*

1. Less power required to bend the material.
2. Ability to bend heavy sheets and plates.
3. Ability to form various angles with the same tooling.

COINING

This type of bending can be compared with a deep-drawing operation. The nose of the Vee tool crushes the natural air-bending radius on the inside of the bend. The COMPRESSION removes the ELASTICITY of the sheet or plate. THIS RESULTS IN THE BEND RETAINING THE EXACT ANGLES OF THE BENDING TOOLS.

Both tools have an included angle of 90°. The advantage of 'coining' is high angular accuracy.

## 10.5 Basic bending methods

There are two main machine methods of bending sheet metal:

1. Bending in folding machines.
2. Bending in pressure tools.

Because of the great force required to bend plate, the bending of platework is normally carried out with the aid of pressure tools in a press brake.

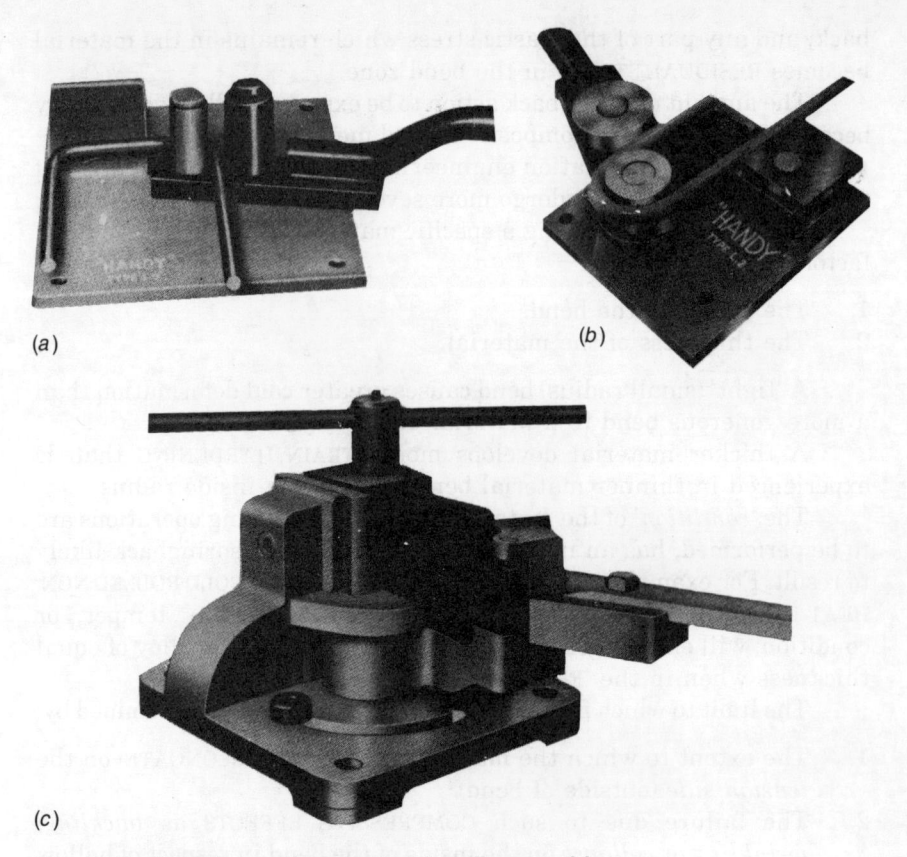

**Fig. 10.6 Simple bench mounted bending machines**

The bending of bar, angle and flat section may be carried out with simple bench-mounted bending machines, as shown in Figure 10.6.

The simple bar-bending machines shown in Figure 10.6(a) have the following capacities:

Bends 1 rod of 9·53 mm dia.
Bends 2 rods simultaneously of 7·94 mm dia.
Bends 3 rods simultaneously of 6·35 mm dia.
Bends 4 rods simulateously of 4·76 mm dia.
The diameter of the centre spindle is 25·4 mm.

The heavier machine shown in Figure 10.6(b) has replaceable formers for bending the bar round. This enables bends of different radii to be made on the machine, which has the following capacities:

Bends rounds 12·7 to 19 mm dia. (supplied with 3 bending formers)

Bends rounds 12·7 to 25·4 mm dia. (supplied with 5 bending formers)

Bends rounds 12·7 to 31·75 mm dia. (supplied with 6 bending formers).

The angle and flat bar bending machine shown in Figure 10.6(c) enables the material to be bent hot or cold.

The machine shown is supplied with a locking bar for the angle stop and swivel blade, an operating lever and double-edged folding blade. It is used for making bends in notched angle, flats, rounds and squares. Its capacity is increased when making hot bends.

## 10.6 Folding machines

The main specifications of folding machines are as follows:

1. The maximum length and thickness to be bent. For example, the capacity of the machine may be 1·5 m times 1·62 mm. This means that the machine is capable of folding a sheet of metal 1·5 m wide and of 1·62 mm (16 s.w.g.) thickness.
2. The lift and shape of the clamping beam.

The smallest width of bend is 8 to 10 times the metal thickness. The minimum inside corner radius of the bend is $1\frac{1}{2}$ times the metal thickness.

The THREE MAIN STEPS in folding work are:

1. *Clamping.* In clamping, the amount of lift of the clamping beam is important. It should be sufficient to allow the fitting and use of special clamping blades, or to give adequate clearance for previous folds.
2. *Folding.* Care must be taken to see that the folding beam will clear the work, particularly when making second or third folds. Some folding machines are designed to fold radii above the minimum, either by the fitting of a radius bar or by adjustment of the folding beam.
3. *Removal of the work.* Care must be taken in folding to ensure that the work may be easily removed on completion of the final bend. The sequence of folding must be carefully studied. The lift of the clamping beam is important here. Some folding machines known as 'UNIVERSAL FOLDERS' have a swing beam. The work may be completely folded around this beam, which is then swung out to allow removal of the work.

Some of the above points are illustrated in Figure 10.7.

Figure 10.7(a) shows a section of a 'box and pan' folding machine.

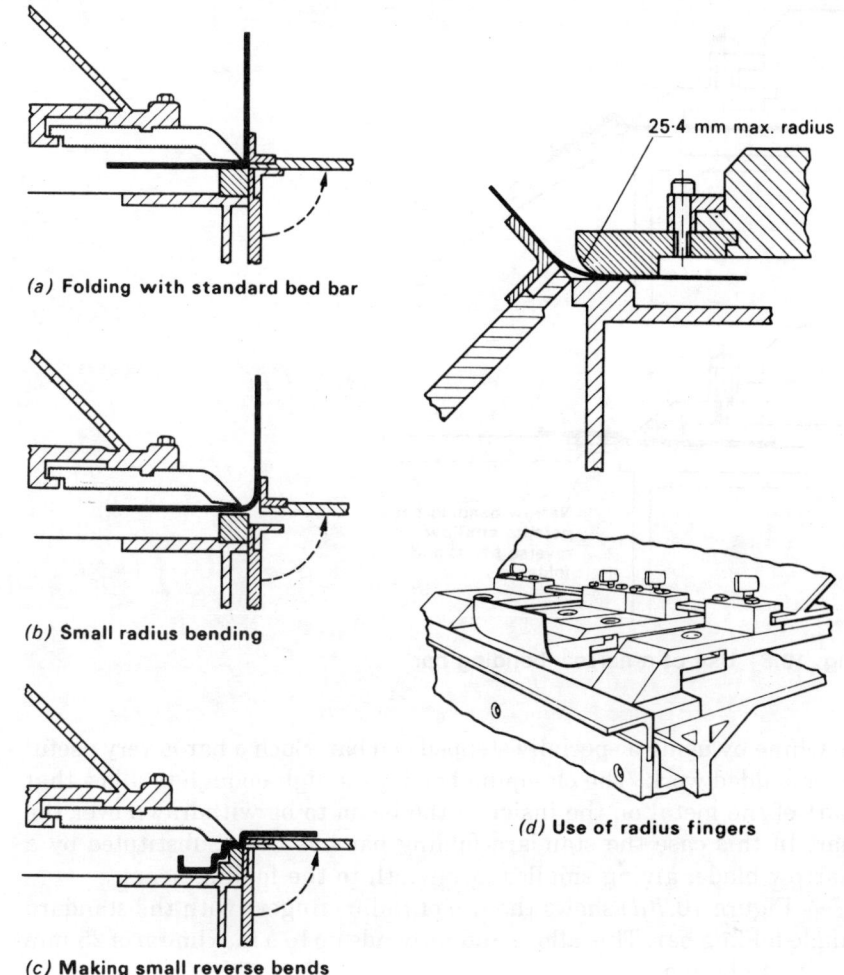

*(a)* Folding with standard bed bar

*(b)* Small radius bending

*(c)* Making small reverse bends

*(d)* Use of radius fingers

**Fig. 10.7 The use of folding machines**

It is fitted with a standard bed bar and fingers. The sheet metal is shown in position after completion of a right-angle bend, using a standard angle folding bar.

Figure 10.7(b) shows small radius bends being made. The folding beam is lowered, and the metal is clamped in the normal way.

When the folding beam is raised, the gap between the nose of the folding blade and the face of the folding bar allows a larger radius to be made.

Figure 10.7(c) shows small return bends being made on this

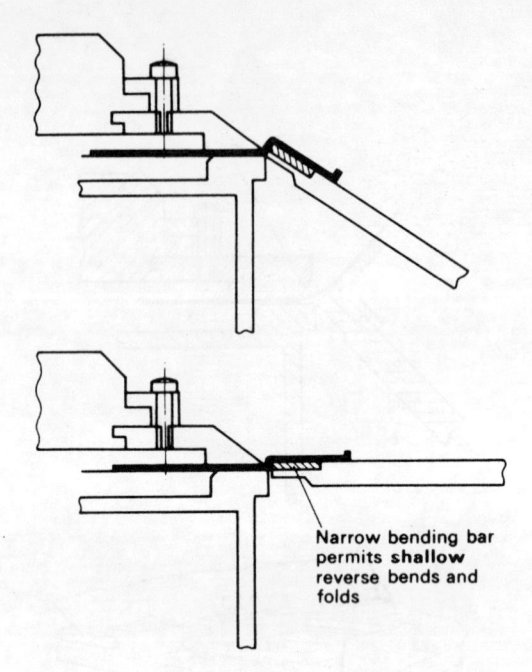

**Fig. 10.8** Use of a narrow bending bar

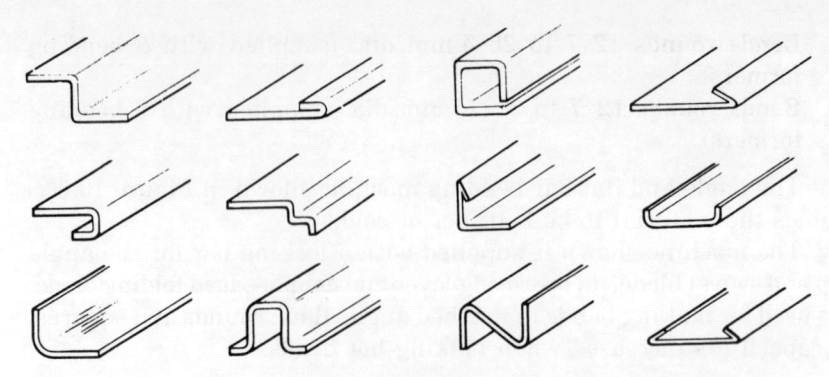

**Fig. 10.9** Examples of work produced on a folding machine

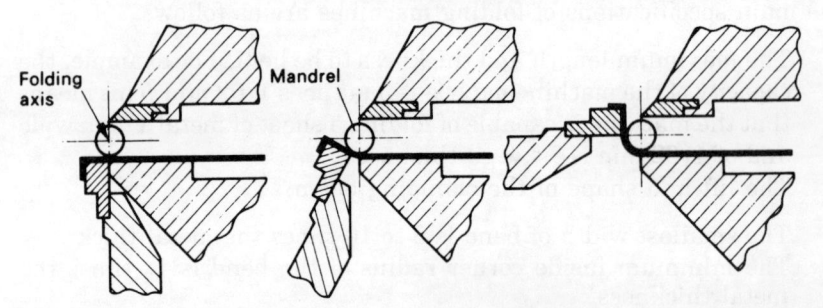

**Fig. 10.10** Use of a mandrel in a folding machine

machine by using a specially stepped bed bar. Such a bar is very useful for moulded work. The clamping beam lifts high enough to allow that part of the metal on the inside of the beam to be withdrawn over the bar. In this case the standard folding bar has been substituted by a narrow blade, giving smaller face width to the folding beam.

Figure 10.7(d) shows the use of radius fingers with the standard angle folding bar. This allows radius bends up to a maximum of 25 mm radius to be made.

The fingers may be positioned where required on the clamping beam to allow short lengths to be folded.

By fitting a special narrow bar to the folding beam, it is possible to form reverse bends narrower than the face of the standard angle bar. This is shown in Figure 10.8. This, of course, reduces the maximum gauge of sheet metal which would be normally folded with the standard blade.

The variety of bends and combinations of bends that can be made on the folding machine are shown in Figure 10.9.

Some folding machines have provision for inserting a round mandrel in trunnion arms on the machine. Such is the case with UNIVERSAL SWING-BEAM FOLDERS, as shown in Figure 10.10.

The amount of lift in the clamping beam is very important. It governs the maximum size of the mandrel used. A machine with a clamping beam lift of between 175 and 200 mm will allow a mandrel of 152 mm diameter to be used.

The folding of shallow depth boxes and pans may be successfully performed on a universal folding machine, provided there is sufficient lift of the clamping beam. An angle clamping blade is attached to the clamping beam as shown in Figure 10.11.

Figure 10.12 shows three types of sheet-metal folding machines in common use in the light fabrication industry.

## 10.7 Bending in pressure tools

The simplest types of machines which may be used for pressure bending are:

1. A fly press fitted with a 'Vee' punch and a 'Vee' die. This arrangement is ideal for bending small sheet metal components such as brackets or electronic chassis.

2. A hand-operated angle bender. This is limited to short lengths of sheet metal angle or 'Zed' section. An angle bender is shown in Figure 10.13.

By far the greatest amount of pressure bending is done on a press brake. The larger capacity machines are capable of bending heavy plate.

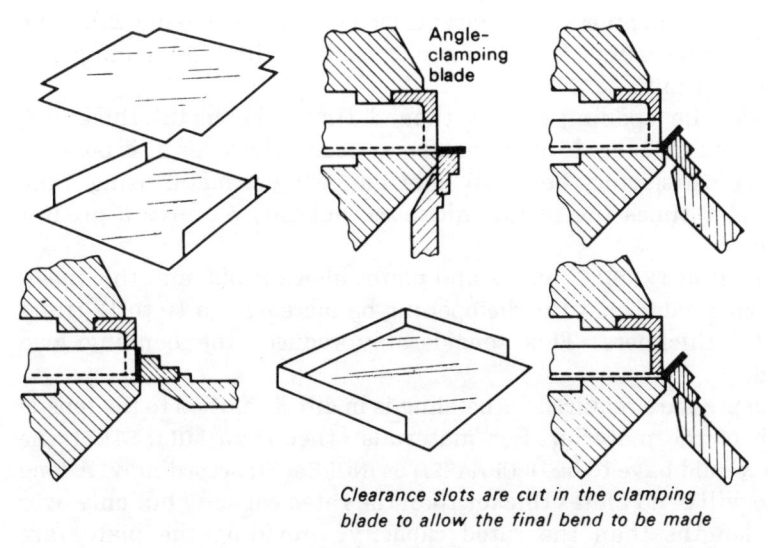

*Clearance slots are cut in the clamping blade to allow the final bend to be made*

**Fig. 10.11 Use of an angle-clamping blade in a folding machine**

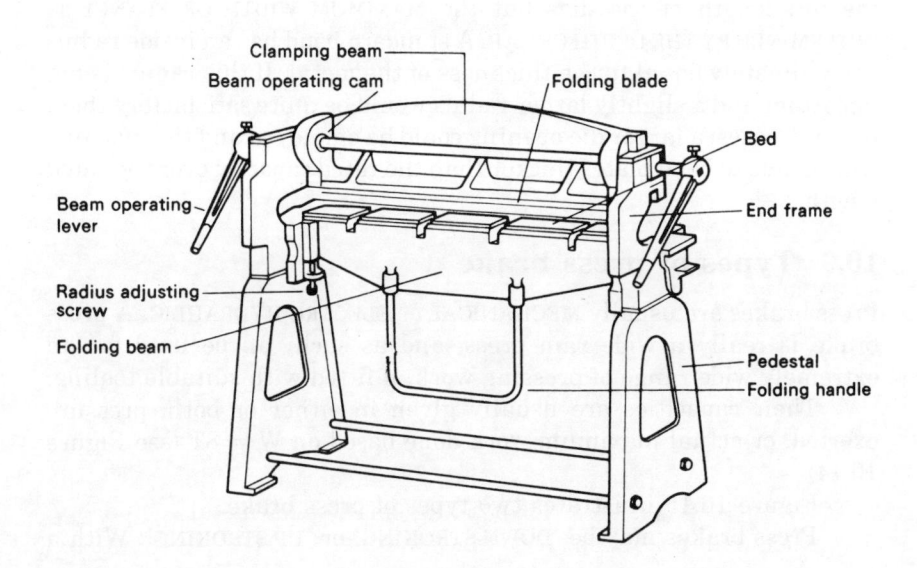

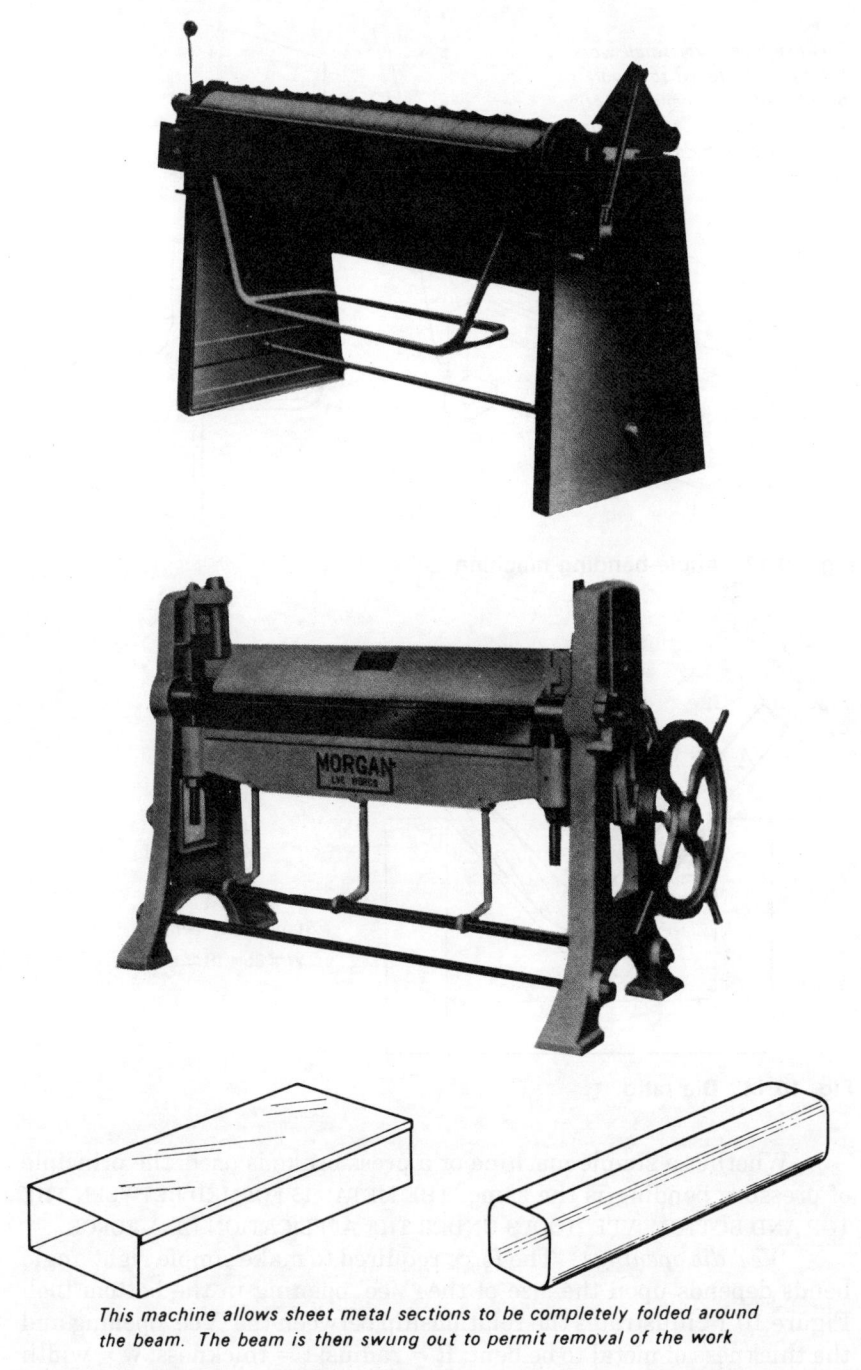

*This machine allows sheet metal sections to be completely folded around the beam. The beam is then swung out to permit removal of the work*

**Fig. 10.12 Types of folding machine**

345

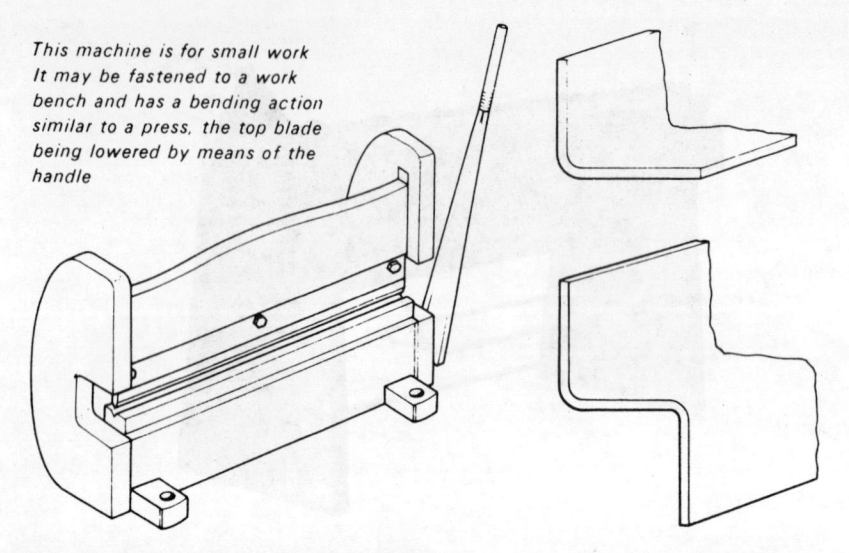

*This machine is for small work It may be fastened to a work bench and has a bending action similar to a press, the top blade being lowered by means of the handle*

**Fig. 10.13  Angle-bending machine**

Standard air bend
W = 8t • R ≃ t

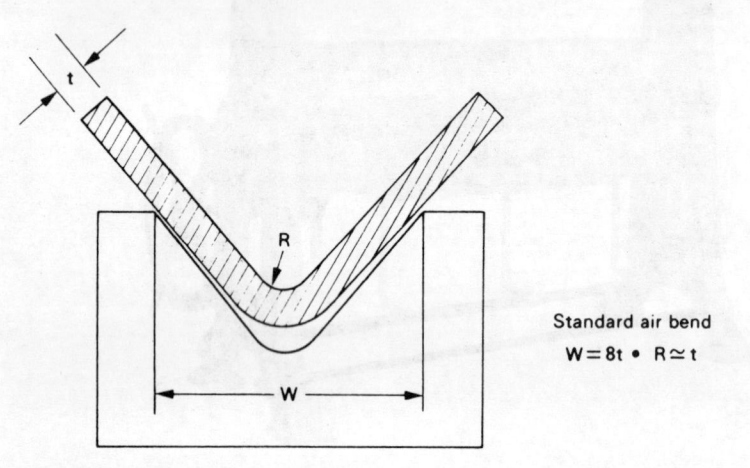

**Fig. 10.14  Die ratio**

Whether a simple machine or a press brake is used, the principle of pressure bending is the same. THE METAL IS FORMED BETWEEN THE TOP AND BOTTOM 'VEE' TOOLS UNDER THE APPLICATION OF A FORCE.

*'Vee' die openings.* The force required to make simple right-angle bends depends upon the size of the 'Vee' opening in the bottom tool. Figure 10.14 illustrates the relationship between the 'Vee' opening and the thickness of metal to be bent: R = radius; t = thickness; w = width at die opening.

## 10.8  Principle of the press brake

Press brakes are designed to bend to a rated CAPACITY based on a '*die ratio*' of 8:1 which is accepted as ideal conditions. Figure 10.14 shows the meaning of *Die ratio*. This is recommended for use with a standard 'Vee' die for 90° 'air bends', and gives an INSIDE RADIUS approximately equal to the THICKNESS of the metal.

Different thicknesses of plate formed over the same die will have the same inside radius, but the force or load required for bending will vary considerably.

If the die opening is less than 8 times the metal thickness, fracturing on the outside of the bend may occur. However, it is possible to produce satisfactory bends in light gauge sheet metal using a die opening of 6 times the metal thickness, but this requires a greater pressure.

For 'HIGH TENSILE' plates and plates above 9·525 mm thickness, it is recommended that the die opening be increased to 10 to 12 times the metal thickness. This considerably reduces the bending load required.

The pressure required for bending is in direct relation to the tensile strength of the material. For materials other than MILD STEEL the capacity would have to be DECREASED or INCREASED accordingly. A long machine will bend plates thicker than the rated capacity but only over shorter lengths than the rated capacity, providing the plates are appreciably shorter than the maximum tool or die length.

Thinner plates than the rated capacity can usually be bent over the full length of the dies but the MAXIMUM WIDTH OF FLANGE IS DETERMINED BY THE DEPTH OF GAP. A standard bend has an inside radius approximately equal to the thickness of the metal. If this radius is not important and a slightly larger radius would be quite satisfactory then, in many cases, a larger die opening could be employed and the machine will be able to bend plates thicker than the rated capacity over the rated length.

## 10.9  Types of press brake

Press brakes are usually MECHANICAL or ELECTRO-HYDRAULIC. A press brake is really a wide ram press, and as such can be used for an extremely wide range of pressing work, if fitted with suitable tooling.

Their capacities are usually given in either or both, pressure exerted, or actual maximum work done based on W = 8T (see Figure 10.14).

Figure 10.15 illustrates two types of press brake.

Press brakes may be 'DOWN-STROKING' or 'UP-STROKING'. With a

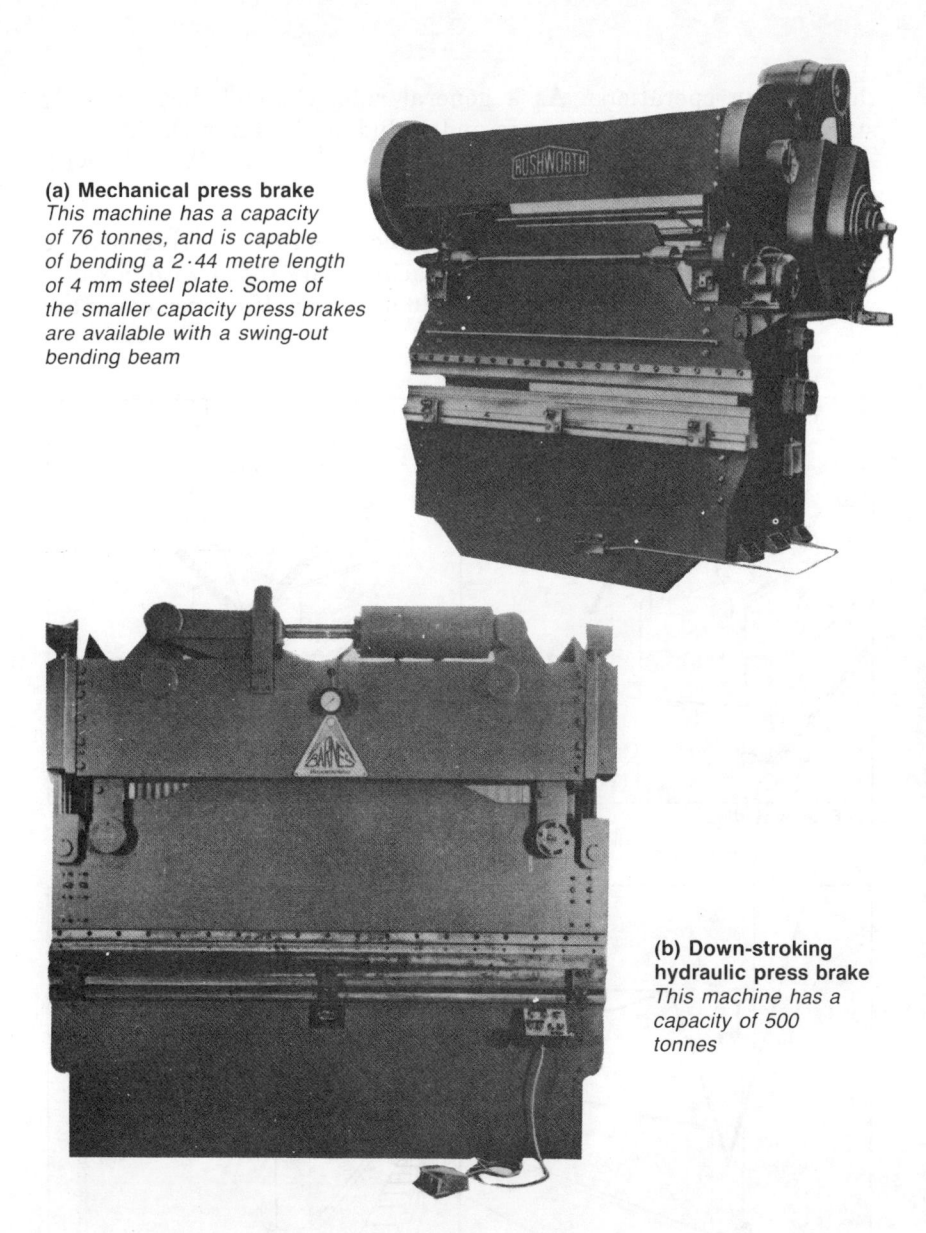

**(a) Mechanical press brake**
*This machine has a capacity of 76 tonnes, and is capable of bending a 2·44 metre length of 4 mm steel plate. Some of the smaller capacity press brakes are available with a swing-out bending beam*

**(b) Down-stroking hydraulic press brake**
*This machine has a capacity of 500 tonnes*

**(c) Details of the dead stop mechanism on a down-stroking hydraulic press brake**

**Fig. 10.15 Press brakes**

down-stroking press the ram brings the top tool down to the bottom fixed tool. An up-stroking press brake is one in which the ram pushes the bottom tool up to the fixed top tool. Many hydraulic press brakes are up-stroking.

Some smaller press brakes are available with a swing-out bending beam. Light-duty machines are rated between 25 tonnes and 75 tonnes,

347

medium-duty machines between 75 tonnes and 150 tonnes, and heavy-duty machines between 150 tonnes and 500 tonnes. Some of the very large machines have table length of 5·5 m. Figure 10.16(*a–h*) illustrates the versatility of the press brake. Note that the tonne is a unit of mass equal to 1000 kg. Therefore the load on the ram of a 25 tonne machine would be equivalent to a mass of 25 × 1000 = 250 000 kg. Such a load would exert a force on the work piece of 25 000 × 9·81 newtons. That is 245 250 N or 2·5 MN. Similarly, a 152 tonne machine would exert a maximum force on the workpiece of 15·2 MN.

*Interchangeable four-way dies* — Figure 10.16(*a*). The interchangeable female dies are used for bending medium and heavy plate. They are provided with 85° openings on each of the four faces.

Male punches for use with four-way dies are usually made with a 60° angle.

*Acute angle dies* — Figure 10.16(*b*). Acute angle dies have many uses and, if used in conjunction with flattening dies, a variety of seams and hems may be produced on sheet metal.

These tools are available for any angle, but if the female die is less than 35° the sheet tends to stick to the die.

Acute angle dies may be set to bend 90° by adjusting the height of the ram.

*The Goose-neck punch* — Figure 10.16(*c*). When making a number of bends on the same component, clearance for previous bends has to be considered. Goose-neck punches are specially designed for the above purpose. These tools are very versatile, enabling a variety of sheet-metal sections to be formed.

The bending force for MILD STEEL is given in Table 10.1, whilst the bending force for other materials is given in Table 10.2.

*Flattening (planishing) tools* — Figure 10.16(*d*). Flattening tools of various forms may either be used in pairs for flattening a returned edge, or hem, on the edge of sheet metal or in conjunction with a formed male or female die (as illustrated for closing a countersunk grooved seam in sheet-metal work).

*Radius bending* — Figure 10.16(*e*). A radius bend is best formed in a pair of suitable tools. The radius on the male punch is usually slightly less than that required to allow for 'springback' in the material. A large radius can be produced by simply adjusting the height of the ram and progressively feeding the sheet through the tools.

*Channel dies* — Figure 10.16(*f*). Channel dies are made with 'pressure pads' so that the metal is held against the face of the male die during the forming operations. As a general rule, channel dies are only successful on sheet metal up to and including 2·64 mm thickness.

A channel in heavy gauge metal is best made in a 'Vee' die with a 'Goose-neck' type of male punch.

*Boxmaking* — Figure 10.16(*g*). Male punches for box making must be as deep as possible. Most standard machines are fitted with box dies which will form a sheet-metal box 170 mm deep. If deeper boxes are

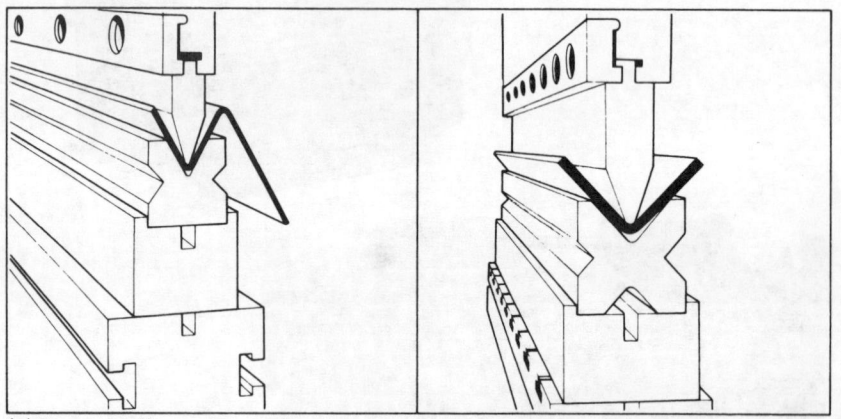

*(a)* Four-way dies

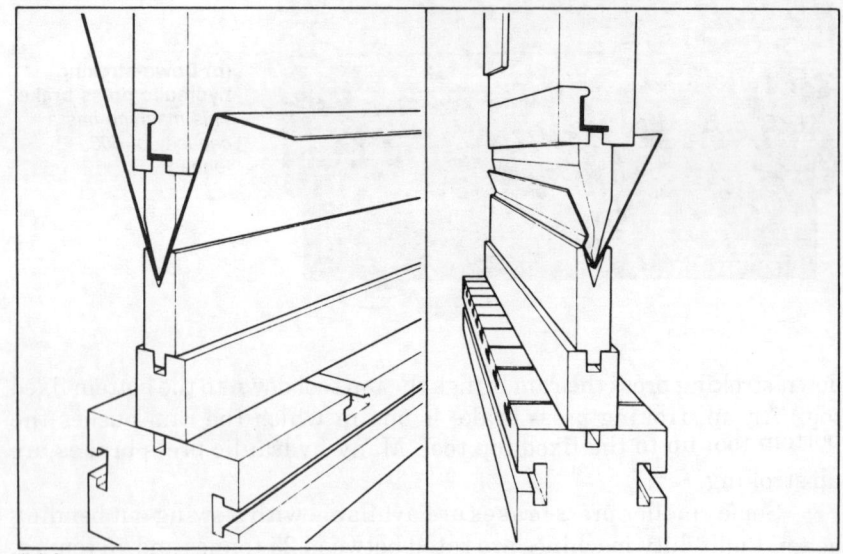

*(b)* Acute angle tools

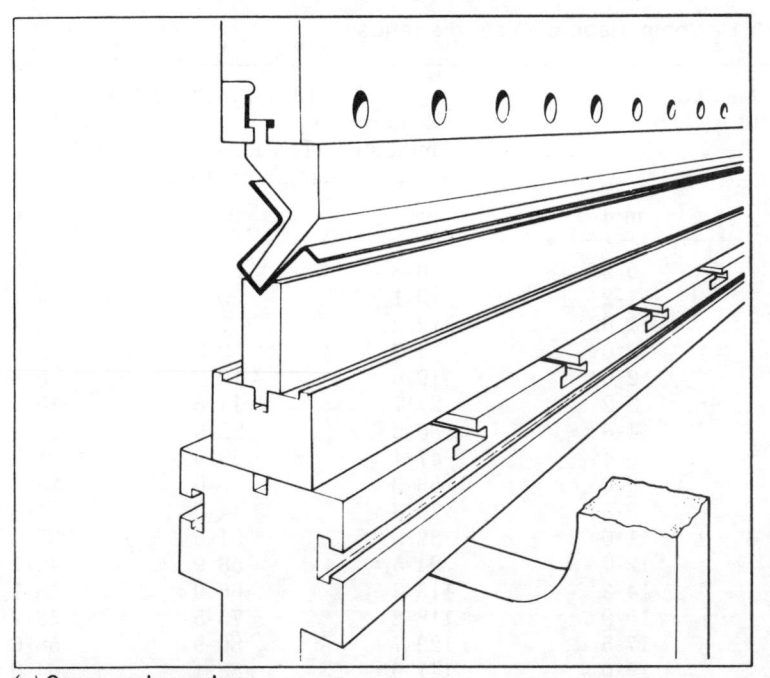

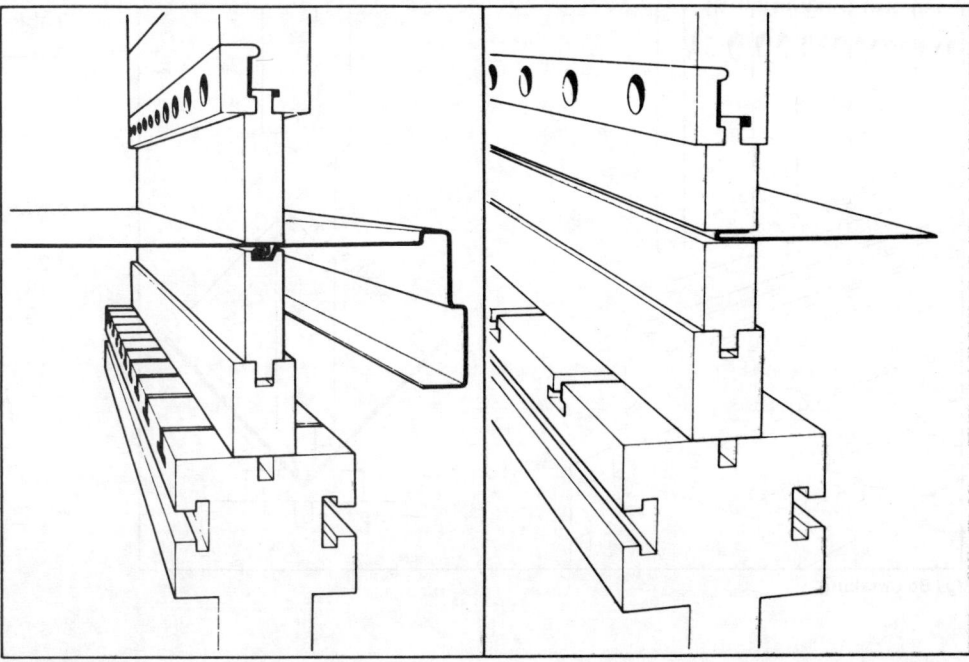

*(c)* **Goose-neck punches**

*(d)* **Flattening**

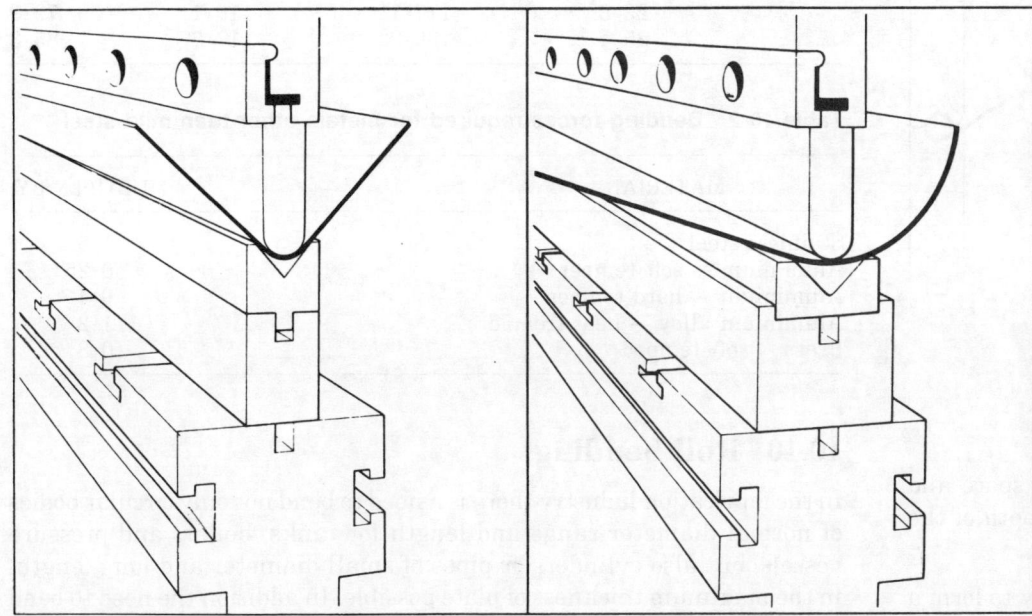

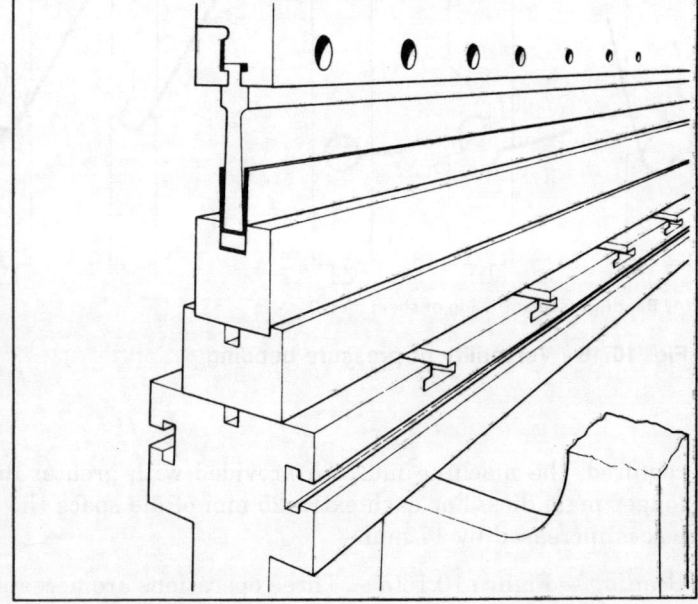

*(e)* **Radius bending**

*(f)* **Channel forming**

349

(g) Box making

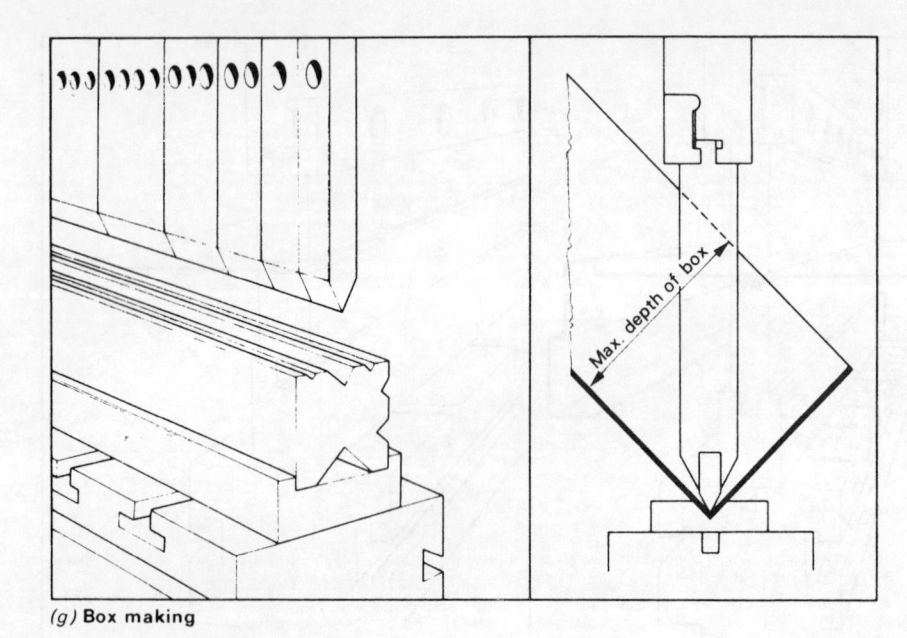

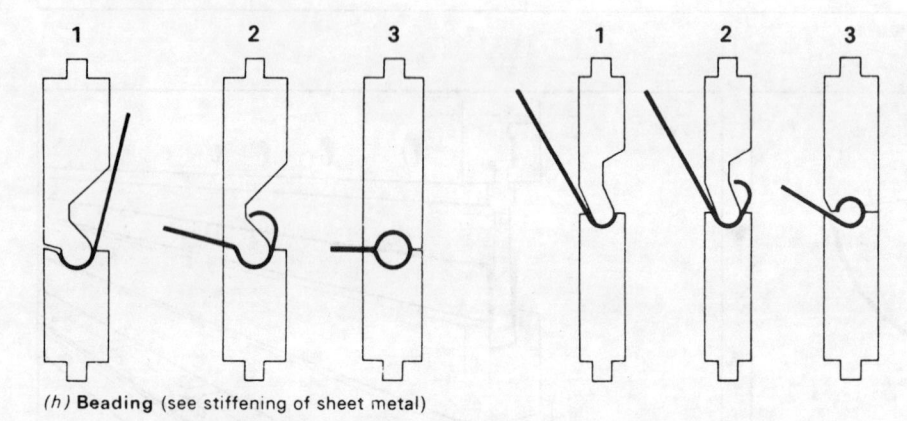

(h) Beading (see stiffening of sheet metal)

Fig. 10.16  Versatility of pressure bending

Table 10.1  Comparison of 'Vee' die ratios

| METAL THICKNESS | | FORCE TONNES/METRE Required to produce 90° 'air bends' in mild steel (Tensile strength 450 N/m² using a die ratio of:) | | |
|---|---|---|---|---|
| s.w.g. | mm. | 8:1 | 2:1 | 16:1 |
| 20 | 0.9 | 6.8 | 4.1 | 3.0 |
| 18 | 1.2 | 9.1 | 5.8 | 4.1 |
| 16 | 1.62 | 12.2 | 7.5 | 5.4 |
| 14 | 2.0 | 14.9 | 9.5 | 6.8 |
| 12 | 2.64 | 19.6 | 12.2 | 8.8 |
| | 3.2 | 23.7 | 14.6 | 10.5 |
| | 4.8 | 35.2 | 22.0 | 15.9 |
| | 6.4 | 47.4 | 29.5 | 21.3 |
| | 8.0 | 58.9 | 36.6 | 26.8 |
| | 9.5 | 70.8 | 44.0 | 31.8 |
| | 11.0 | 82.6 | 51.5 | 37.3 |
| | 12.7 | 94.5 | 58.9 | 42.7 |
| | 14.3 | 115.3 | 66.0 | 48.1 |
| | 15.9 | 118.2 | 73.5 | 53.2 |
| | 17.5 | 129.7 | 80.9 | 58.6 |
| | 19.0 | 141.9 | 88.1 | 64.0 |
| | 20.4 | 153.5 | 95.5 | 69.1 |
| | 22.2 | 165.2 | 102.6 | 74.5 |
| | 23.8 | 177.1 | 110.1 | 79.9 |
| | 25.4 | 189.3 | 117.5 | 85.3 |

Table 10.2  Bending forces required for metals other than mild steel

| MATERIAL | MULTIPLY BY: |
|---|---|
| Stainless steel | 1.5 |
| Aluminium — soft temper | 0.25 |
| Aluminium — hard temper | 0.4 |
| Aluminium alloy — heat treated | 1.2 |
| Brass — soft temper | 0.8 |

## 10.10  Roll bending

In the fabrication industry there is a need to bend not only circular bodies of normal diameter range and length for tanks, boilers and pressure vessels, but also cylinders or pipes of small diameter and long length, in the maximum thickness of plate possible. In addition the need to bend conical sections has also to be catered for.

required, the machine must be provided with greater die space and longer male dies. For each extra 25 mm of die space the depth of the box is increased by 17 mm.

*Beading* — Figure 10.16(*h*).  Three operations are necessary to form a bead on the edge of sheet metal.

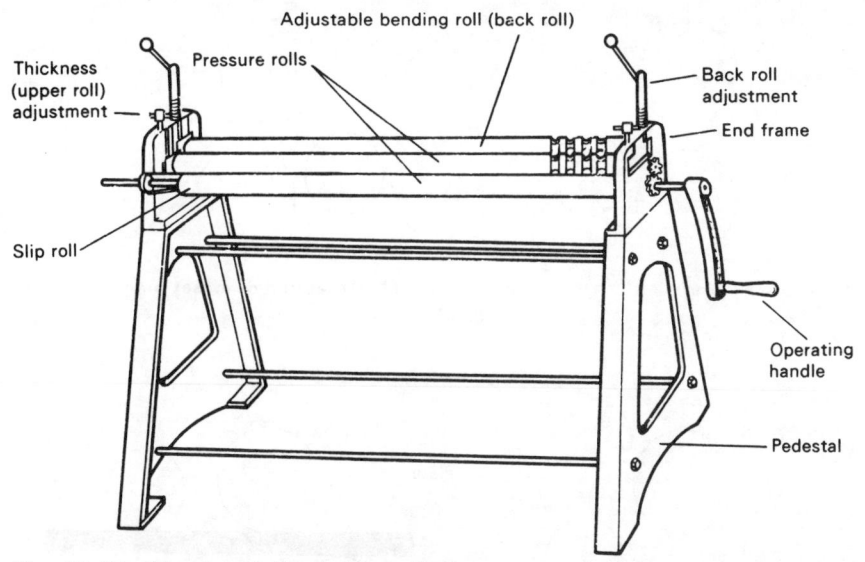

**Fig. 10.17** Sheet-metal bending rolls

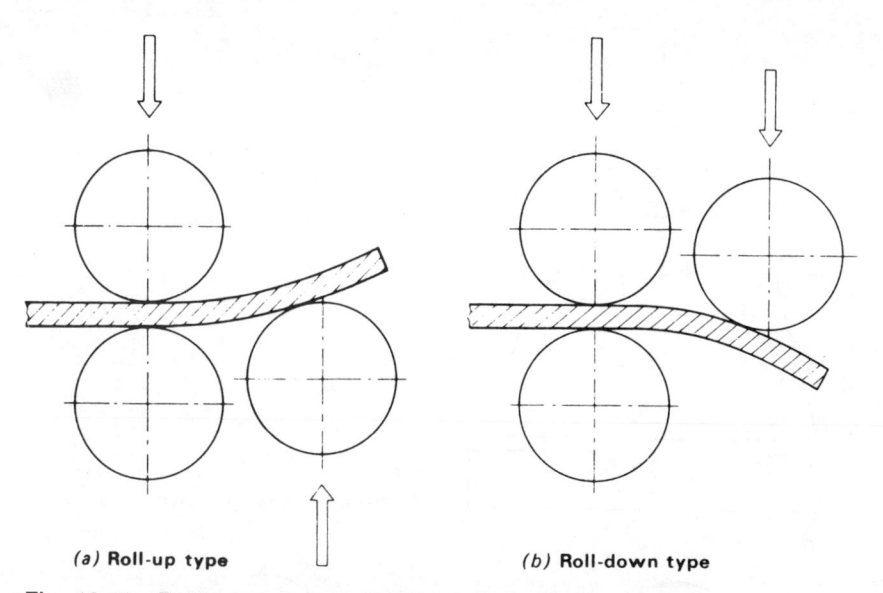

**Fig. 10.18** Basic arrangement of the rolls in pinch-type rolls

Machines used to bend sheet metal or plate into cylindrical or part cylindrical forms with either parallel or conical sides are called 'BENDING ROLLS'.

Bending rolls for sheet metal are made in various sizes from the bench type for tinplate work to the larger pedestal types which are suitable for general sheet-metal work.

Figure 10.17 shows the main features of a bending rolls suitable for bending sheet metal.

The basic type of rolls used in sheet-metal work is known as the 'PINCH-TYPE ROLLERS'. These machines have two front rollers which lightly grip (PINCH) and draw the sheet through, and a 'free roller' at the rear to 'set' the metal to the desired radius.

There are two kinds of pinch-type rollers, basic details of which are illustrated in Figure 10.18.

*Roll-up type* — Figure 10.18(*a*). These machines have adjustment in a vertical direction on the top or bottom pinch roll, and in an upward direction on the back roller. This type will roll any size of curvature above the size of the top roll.

*Roll-down type* — Figure 10.18(*b*). These machines have adjustment in a vertical direction on top or bottom pinch roll, and in a downward direction on the back roller. This type will not roll more curvature than will pass beneath the pedestal frame of the machine.

As a general rule, the minimum diameter which can be rolled on a rolling machine is in the order of $1\frac{1}{2}$ to 2 times the diameter of the roll round which it is being rolled.

Most machines roll the metal in an upward direction because this does not restrict the size of the cylinder or curve to be rolled.

The IDENTIFICATION of either kind of pinch-type rollers can easily be determined by visual inspection, as follows:

Where pinch-type rollers have wiring or beading grooves, if these grooves are in the top and back rolls, the machine is designed to roll down. When the grooves are in the bottom and back rolls, then the machine is designed to roll up.

## 10.11 The rolling of plate

The majority of machines in general use for the roll bending of plate are of the horizontal type, although considerable use is made of vertical plate rolls.

Rolls used for platework are more robust than those used for sheet-metal work, and are normally power operated. Whereas pinch-type rolls are suitable for bending light plate, 'PYRAMID-TYPE ROLLERS' are used for medium or heavy plate bending. Pyramid-type rolls, as the name suggests, have three rolls arranged in pyramid fashion as shown in Figure 10.19.

*(a)* **Pyramid-type rolls** (standard design)

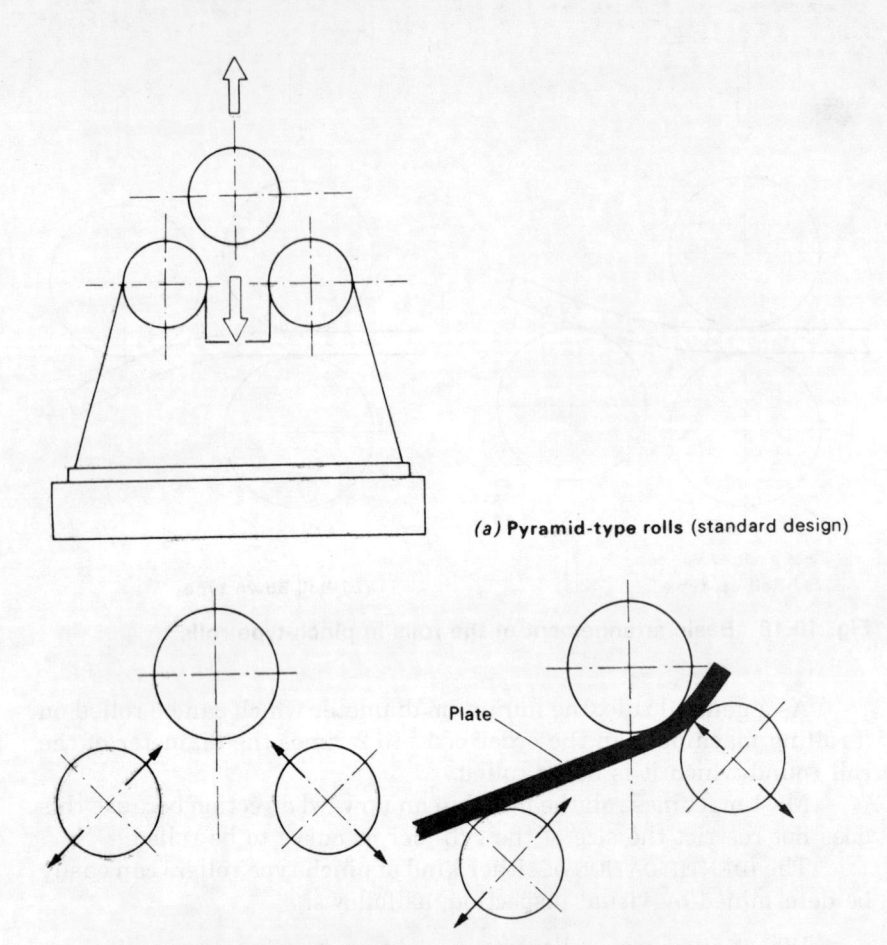

*(b)* **Pyramid-type rolls with adjustable bottom rollers**

**Fig. 10.19 Pyramid-type rolling machines**

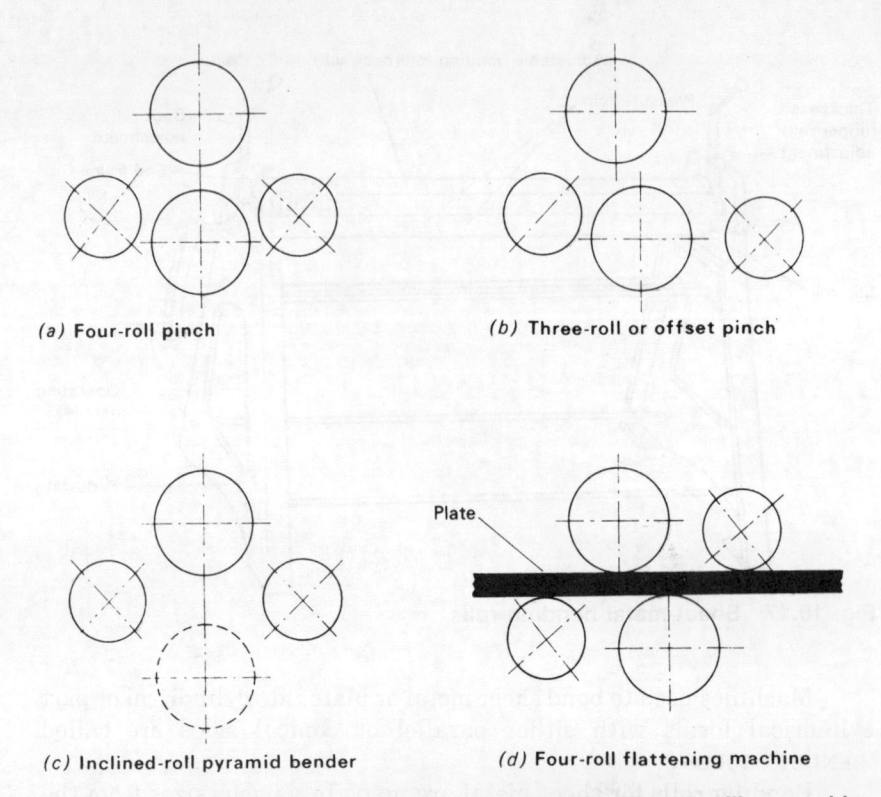

*(a)* Four-roll pinch

*(b)* Three-roll or offset pinch

*(c)* Inclined-roll pyramid bender

*(d)* Four-roll flattening machine

**Fig. 10.20 The four-in-one universal pyramid/pinch-type rolling machine showing alternative settings of the rollers**

Most plate-rolling machines are provided with longitudinal grooves along the lower rolls to assist in gripping the plate. These grooves are useful for initial alignment of the plate.

Figure 10.19(a) shows the basic arrangement of the rollers in a standard pyramid-type rolls. The top roll is adjustable up or down, and may be 'slipped' to allow removal of the work when completely rolled around it, as in the case with cylinders (see Section 10.12).

Figure 10.19(b) shows the rolls set for 'pinch bending'. One great advantage of this type of machine is that for heavy plate the bottom roll centres are wide, and the load on the top roller consequently reduced. Being mounted in inclined slideways, the bottom roller centres are automatically reduced as the rollers are adjusted up to the work on thin plate or to small diameters, thereby providing a slightly increased capacity at the top and bottom ends of plate range.

Figure 10.20 shows a 'four-in-one' universal pyramid/pinch-type rolling machine. These machines are capable of performing all the roll bending operations normally carried out in fabrication workshops. In the hands of a capable operator it is a universal machine for all types of roll bending in light or heavy platework.

## 10.12 Slip rolls

When rolling complete cylinders, the finished cylinder is left round the roll, so provision has to be made for its removal. Rolls with this provision are referred to as 'SLIP ROLLS'. With most sheet-metal rolling machines, the roll around which the cylinder formed is made to slip out sideways. The slip roll on the powered plate-rolling machines usually slips upwards for the removal of the cylinder. Figure 10.21 shows a plate-rolling machine with a slip roll.

**Fig. 10.21** Bending a pipe-line section on a plate-rolling machine

(b) Sequence of operations for rolling a steel pipe line section of the four-roller twin initial pinch hydraulic bending machine illustrated in figure (a).

**These machines are available in the following sizes:**
Plate width 380mm to 4·6m
Plate thickness 1·62mm to 76mm (cold)

## 10.13 Cone rolling

A certain amount of cone rolling may be carried out on both hand- and power-operated rolls. To enable this to be done provision must be made to adjust the curving roll to a suitable angle in the horizontal plane to the pinch rolls, or in the pyramid rolls to the other two.

## 10.14 Angle ring-bending rolls

These machines may be hand-operated by suitable gearing, or power-operated. They are used for the cold-bending of channel, angle and Tee bar rings. The rollers may either be horizontal or vertical, as shown in Figure 10.22.

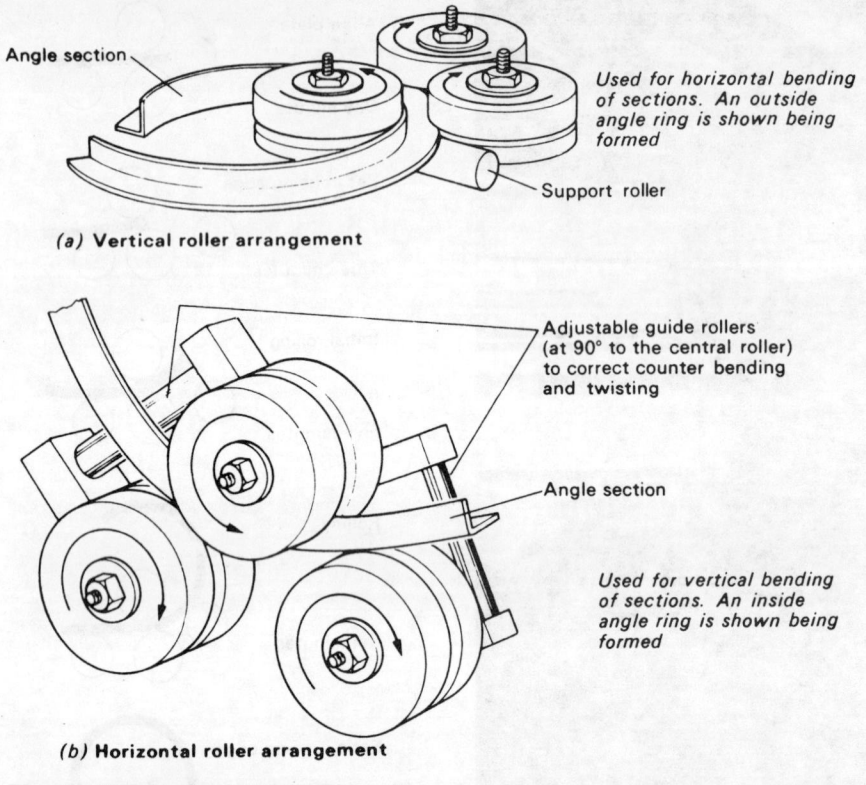

**Fig. 10.22 Angle ring-bending rolls**

An angle ring-bending roll consists of three rollers arranged in triangular formation. Each roller can be split into two sections to take the flat flanges of angles or channels as they are being bent.

When bending an outside ring, the flat flange of the angle is adjusted in the slots of the two rollers, and for an inside ring the flat flange is adjusted in the slot of the single central roller. 'Tee' sections may also be formed on these machines. Pressure is exerted, during rolling, by a screw arrangement which moves the single central roller towards the gap between the other two rollers.

## 10.15 Forming sheet metal

The expression 'forming metal' usually means shaping or bending the material in three dimensions. This is much more difficult than simple bending in two dimensions, for some part of the metal must be STRETCHED or SHRUNK, or both.

Consider a FLANGE to be 'thrown' on a curved surface, for example

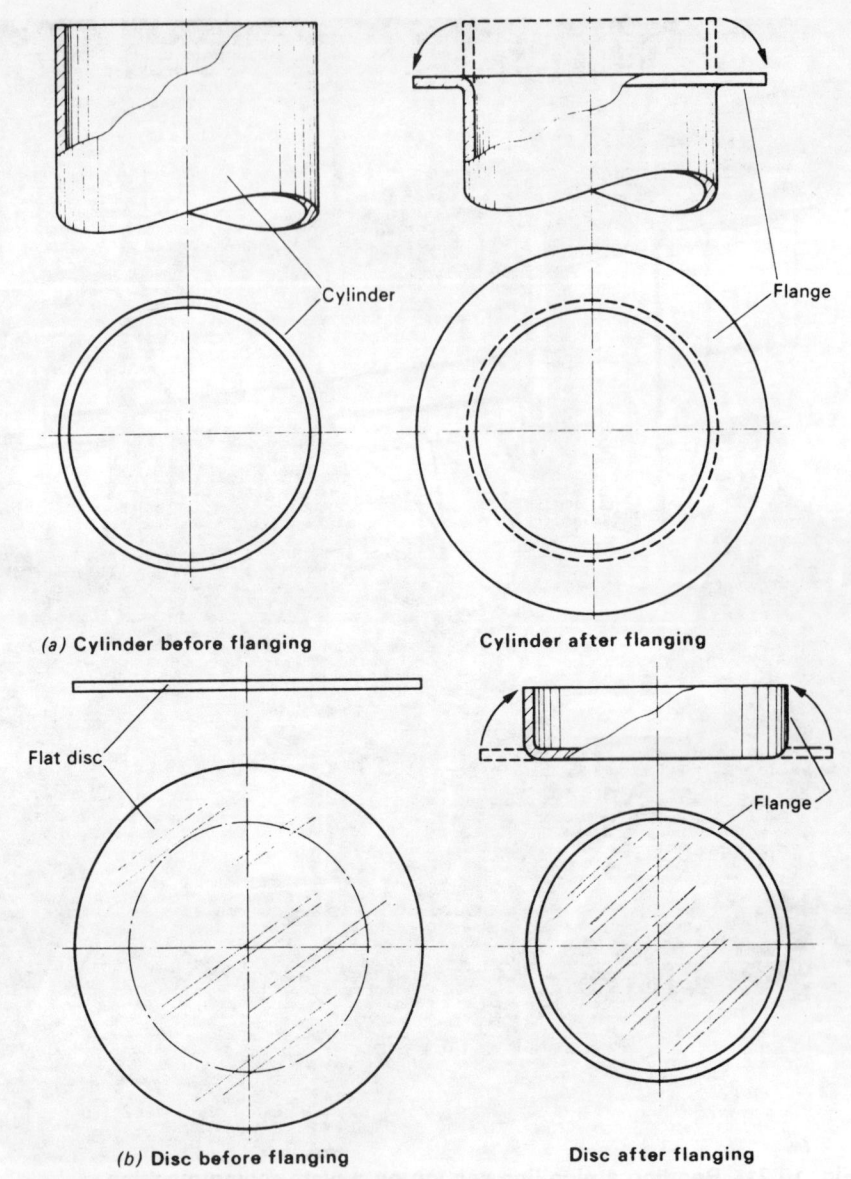

**Fig. 10.23 Comparison of flanging methods**

a cylinder, as illustrated in Figure 10.23(a). It will be apparent that the edge of the cylinder, after externally flanging, has a greater circumference than it had before the flange was thrown.

Now consider a FLANGE to be worked up around the edge of a flat metal disc, as shown in Figure 10.23(b). Here it will be appreciated that

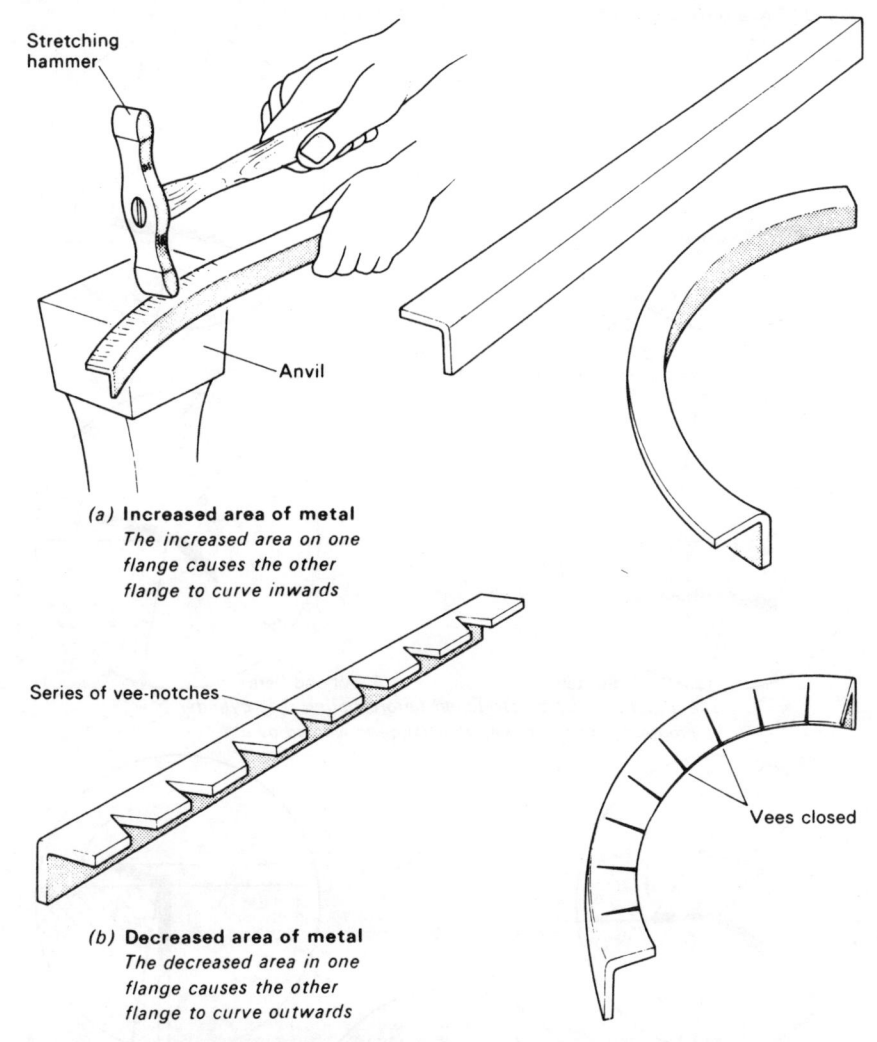

**(a) Increased area of metal**
*The increased area on one flange causes the other flange to curve inwards*

**(b) Decreased area of metal**
*The decreased area in one flange causes the other flange to curve outwards*

**Fig. 10.24  The effects of increasing and decreasing the surface area of one flange on an angle strip**

the edge of the disc, after flanging, will have a smaller circumference than it had before flanging.

In case (*a*) the metal has been STRETCHED, whilst in case (*b*) the metal has been SHRUNK or COMPRESSED.

By way of contrast, the effects of increasing or decreasing the surface area of one flange of an angle strip are illustrated in Figure 10.24.

In practice metal is not generally removed by the simple expedient of cutting 'Vee' slots. The surface area is reduced by compressing or shrinking the surplus metal. It is much more difficult to produce an internally curved flange than an externally curved one, because it is much easier to stretch metal than to shrink it.

## 10.16  The main aspects of forming sheet metal by hand

It is essential for the craftsman to possess a thorough basic knowledge of the properties of the materials which he has to use.

*This enables him to understand, and even forecast the behaviour of materials under applied forces, and thus be in control of the desired direction of their flow during a particular forming process.*

Failure to understand these points will, inevitably, result in much valuable material, time, and effort being spent producing faulty shapes, stressed areas, splitting and other undesirable features.

The methods employed for shaping work by hand are similar for most materials, the main difference being concerned with such factors as:

(*a*)  The FORCE with which the blow strikes the metal;
(*b*)  The DIRECTION in which the FORCE or blow is applied.

It should be appreciated that if, for example, a piece of aluminium and a piece of mild steel sheet of equivalent thicknesses are struck with blows of equal force, the aluminium — being softer and more malleable and ductile — will be deformed to a much greater extent than the mild steel.

Since the shaping of sheet metals by hand is essentially a HAMMERING PROCESS, it is important to consider the types of blow which can be struck on sheet metal, and that each has its own field of application for particular purposes.

### TYPES OF BLOW USED ON SHEET METAL

1. *A solid blow*  Where the metal is struck solidly over a solid steel head or anvil. a solid blow will stretch the metal. Figures 10.25 and 10.26 illustrate a typical application of the solid blow.
2. *An elastic blow*  Where either the head or the tool (or both) are made of a resilient material such as wood. An elastic blow will form sheet metal without unduly stretching it, and can be used to advantage to thicken the metal when shrinking it. The use of an elastic blow is shown in Figure 10.27.
3. *A floating blow*  Where the head or the anvil is not directly under

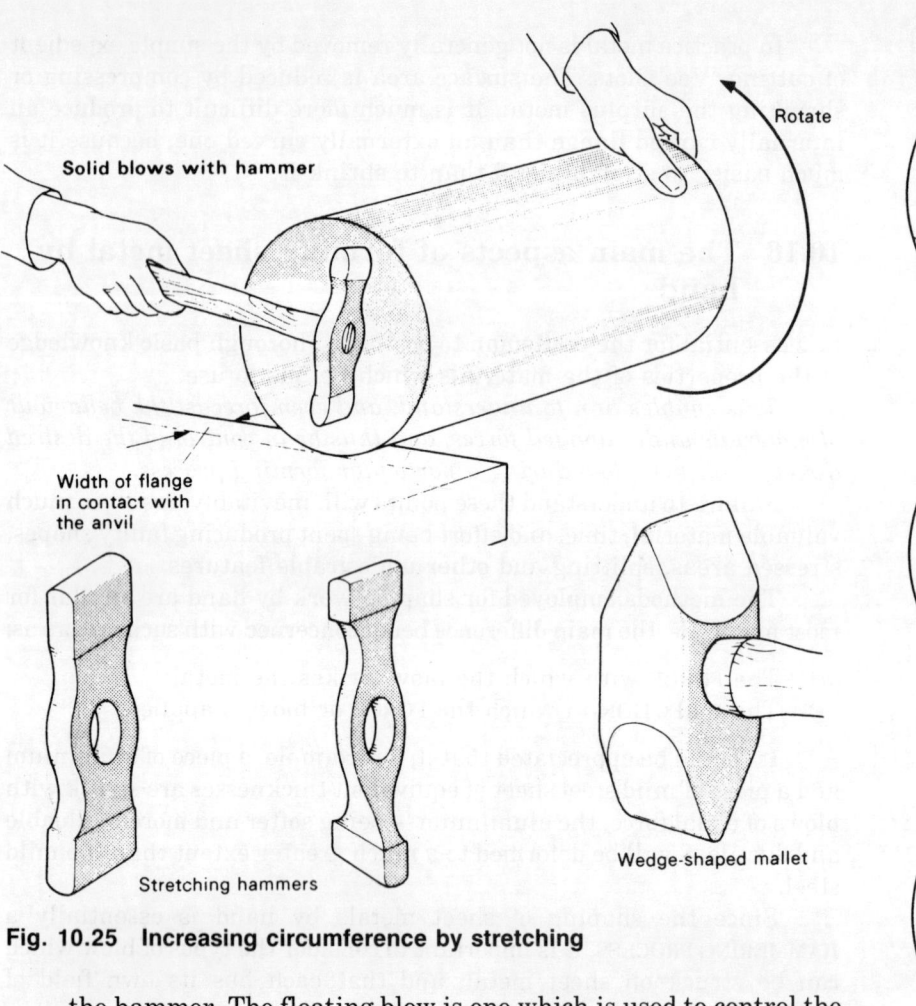

**Fig. 10.25 Increasing circumference by stretching**

the hammer. The floating blow is one which is used to control the direction in which the metal is required to flow during the forming process. It is delivered while the metal is held over a suitable head or stake, hitting it 'off the solid', thus forming an indentation at the point of impact. Figure 10.28 illustrates the use of floating blows.

## 10.17 The use of bench tools for forming sheet metal

The craftsman often finds it necessary, when suitable machines are not available, to resort to the use of various types of metal anvils when bending or forming sheet-metal articles.

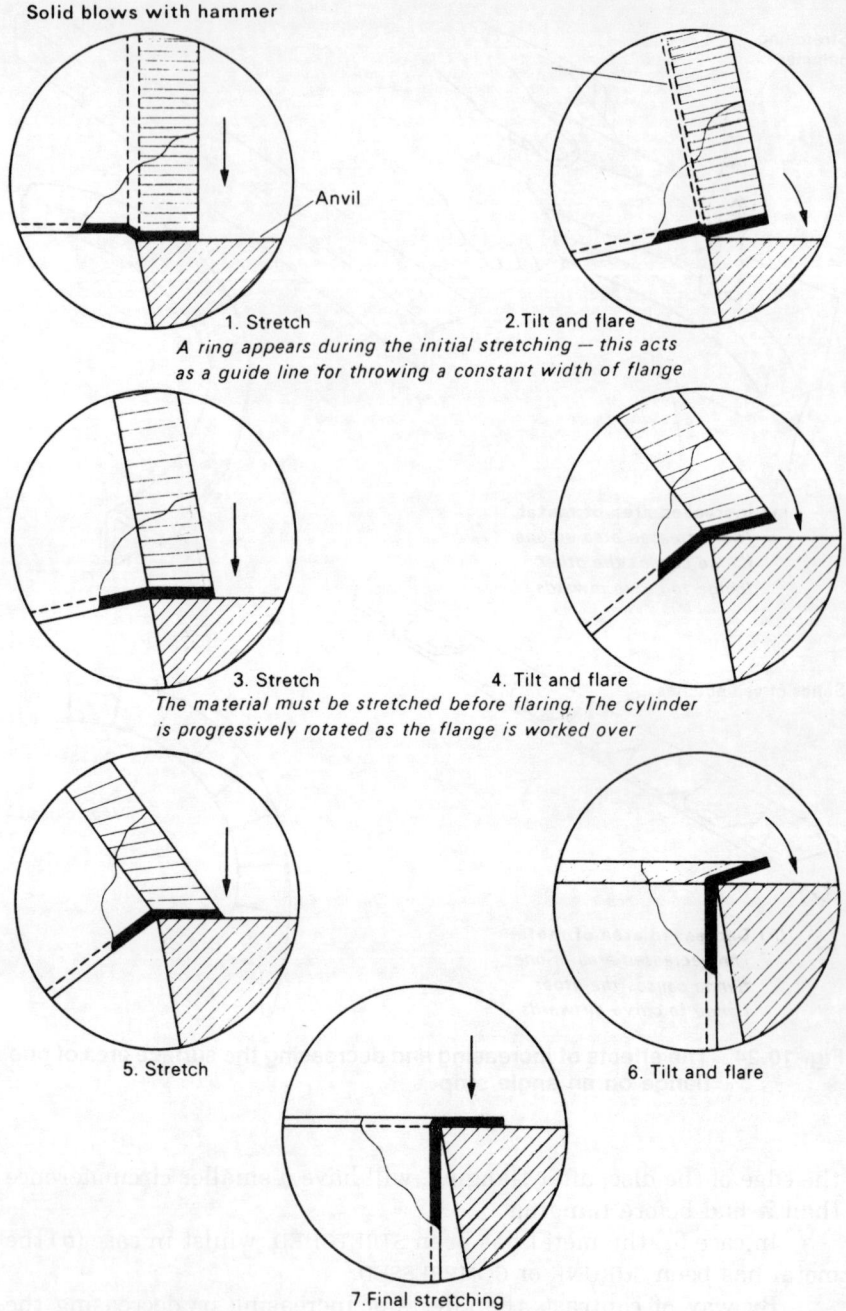

1. Stretch
*A ring appears during the initial stretching — this acts as a guide line for throwing a constant width of flange*

2. Tilt and flare

3. Stretch

4. Tilt and flare
*The material must be stretched before flaring. The cylinder is progressively rotated as the flange is worked over*

5. Stretch

6. Tilt and flare

7. Final stretching

**Fig. 10.26 Sequence of operations for producing an external flange on a cylindrical body**

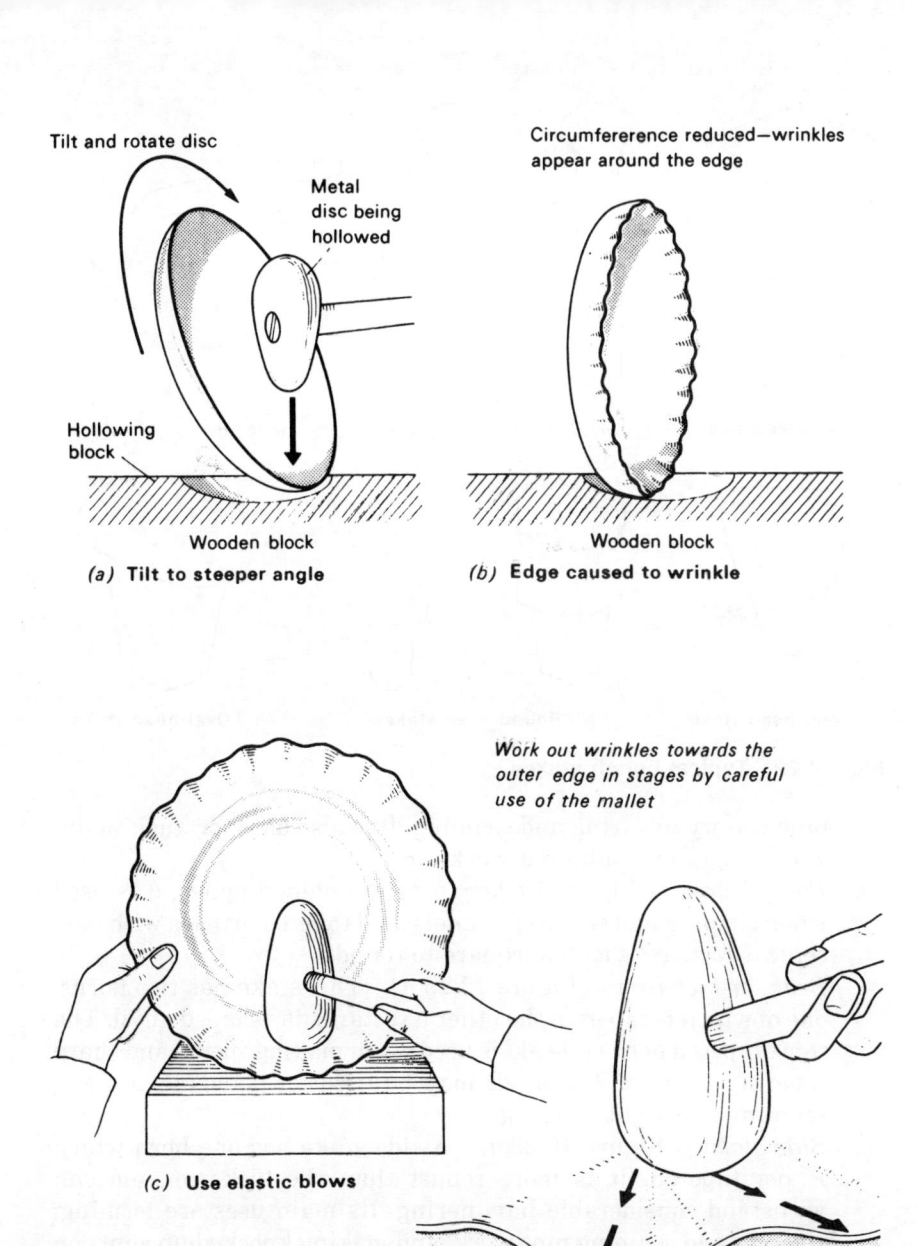

**Fig. 10.27 Shrinking the edge when hollowing**

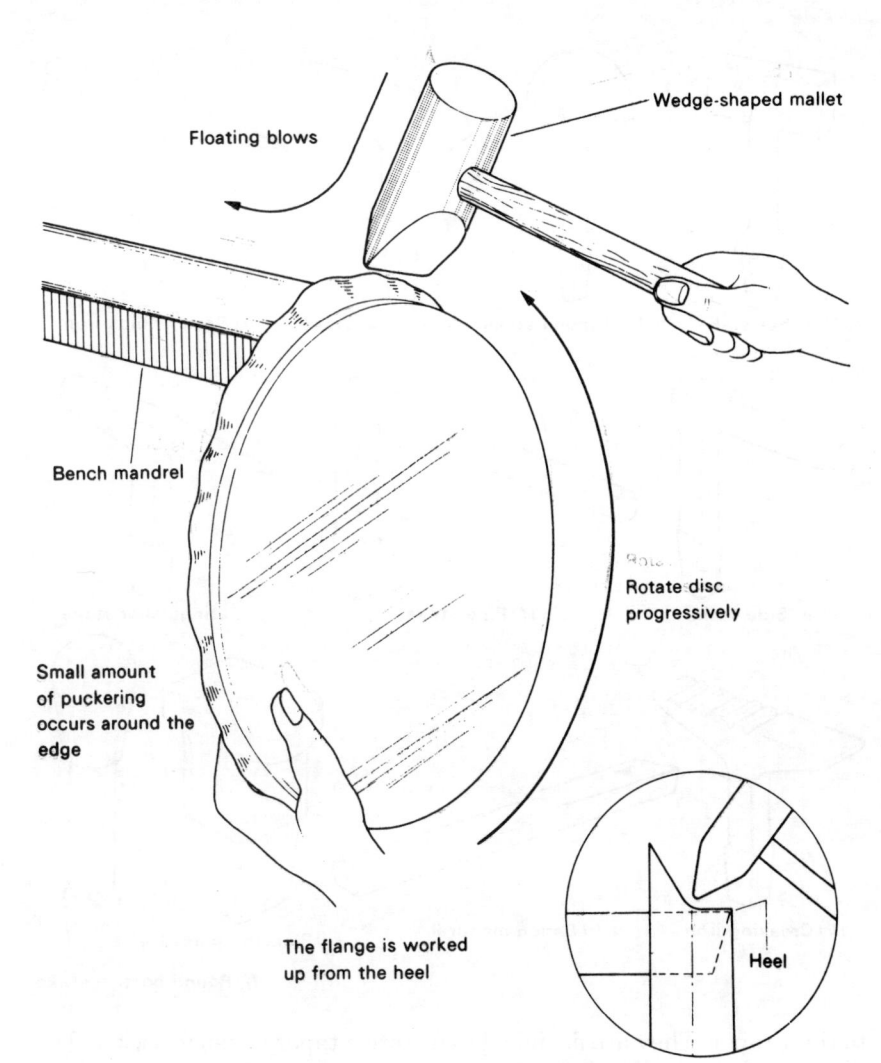

(a) Tilt to steeper angle
- Tilt and rotate disc
- Metal disc being hollowed
- Hollowing block
- Wooden block

(b) Edge caused to wrinkle
- Circumference reduced—wrinkles appear around the edge
- Wooden block

(c) Use elastic blows
- Work out wrinkles towards the outer edge in stages by careful use of the mallet

**Fig. 10.28 Raising a flange on the edge of a metal disc**

- Wedge-shaped mallet
- Floating blows
- Bench mandrel
- Rotate disc progressively
- Small amount of puckering occurs around the edge
- The flange is worked up from the heel
- Heel

These anvils are commonly referred to as 'STAKES' and are designed to perform many types of operations for which machines are not readily available or readily adaptable. Some stakes are made of forged mild steel faced with cast steel. The better-class stakes are made either of cast iron or cast steel. All are sold by weight because of their variety of size and shape.

A stake used for sheet-metal work consists basically of a shank and a head or horn. The shanks are generally standard with respect

357

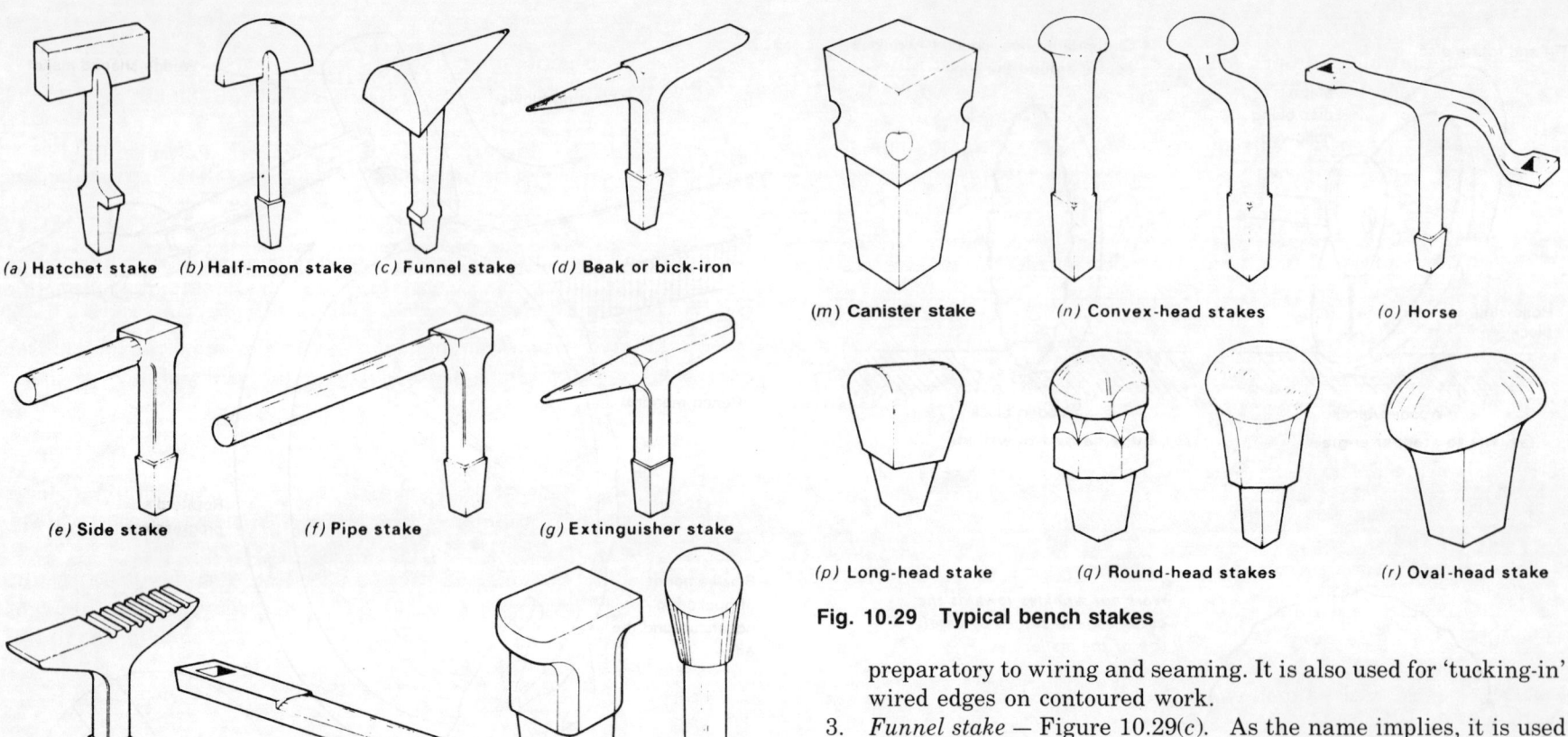

(a) Hatchet stake    (b) Half-moon stake    (c) Funnel stake    (d) Beak or bick-iron

(e) Side stake    (f) Pipe stake    (g) Extinguisher stake

(h) Creasing iron    (j) Bench mandrel    (k) Planishing anvil    (l) Round bottom stake

(m) Canister stake    (n) Convex-head stakes    (o) Horse

(p) Long-head stake    (q) Round-head stakes    (r) Oval-head stake

**Fig. 10.29 Typical bench stakes**

to their taper. They are designed to fit into a tapered bench socket. The heads and horns are available in a great variety of shapes and sizes. their working faces are machined or ground to shape.

The more common stakes and some of their uses are illustrated in Figure 10.29(a–r).

1. *Hatchet stake* – Figure 10.29(a). The hatchet stake has a sharp, straight edge, bevelled along one side. It is very useful for making sharp bends, folding the edges of sheet metal, forming boxes and pans by hand, 'tucking-in' wired edges, and seaming.
2. *Half-moon stake* – Figure 10.29(b). This stake has a sharp edge in the form of an arc of a circle, bevelled along one side. It is used for throwing up flanges on metal discs or contoured blanks, preparatory to wiring and seaming. It is also used for 'tucking-in' wired edges on contoured work.
3. *Funnel stake* – Figure 10.29(c). As the name implies, it is used when shaping and seaming funnels, and tapered articles with part conical corners, such as 'square-to-rounds'.
4. *Beak- or bick-iron* – Figure 10.29(d). This stake has two horns, one of which is tapered, the other a rectangular shaped anvil. The thick tapered horn or 'beak' is used when making spouts and sharp tapering articles. The anvil may be used for squaring corners, seaming and light riveting.
5. *Side stake* – Figure 10.29(e). A side stake has one horn which is not tapered. It is more robust than the bick-iron and can withstand considerable hammering. Its main uses are forming, riveting and seaming pipe work, and making knocked-up joints on small cylindrical articles. It may also be used when forming tapered work of short proportions.
6. *Pipe stake* – Figure 10.29(f). This is a much longer version of the side stake and, as the name implies, is useful when forming and seaming pipes.
7. *Extinguisher stake* – Figure 10.29(g). Is very similar to the bick-iron, in that it has a round and tapered horn at one end and a

rectangular shaped horn at the other. Some contain a number of grooving slots on the working surface of the rectangular horn. These are useful when creasing metal and bending wire. The tapered horn is used for forming, riveting, or seaming small tapered articles. It is also useful when forming wrinkles or puckers prior to 'raising'.

8. *Creasing iron* — Figure 10.29(h). This has two rectangular shaped horns, one of which is plain. The other horn contains a series of grooving slots of various sizes. The grooves are used when 'sinking' a bead on a straight edge of a flat sheet, i.e., reversing wired edges. These irons are also used when making small diameter tubes with thin-gauge metal.
9. *Bench mandrel* — Figure 10.29(j). This is firmly fixed to the bench by means of strap clamps which may be quickly released allowing the mandrel to be reversed or adjusted for length of overhang. It is double ended — the rounded end is used for riveting and seaming pipes, while the flat end is used when seaming corners of pans, boxes, square or rectangular ducting, and riveting. It has a square tapered hole in the flat end for receiving the shanks of other stakes and heads. Bench mandrels are available in four sizes ranging from 20 to 114 kg in mass.
10. *'Planishing anvils'* — Figure 10.29(k). Used for planishing all types of flat and shaped work, they are highly polished on their working surfaces. The one illustrated is called a 'TINSMITH'S ANVIL', and is used when planishing flat surfaces.
11. *Round bottom stake* — Figure 10.29(l). These stakes are available in various diameters and have flat working surfaces. They are used when forming the bases of cylindrical work, and for squaring knocked-up seams.
12. *Canister stake* — Figure 10.29(m). This has a square and flat working surface. Its main use is for working in corners, and squaring-up seams when working with square or rectangular articles.
13. *Convex-head stakes* — Figure 10.29(n). Used when forming or shaping double-contoured and spherical work. It is usually available in two patterns — with a straight shank or with an offset or cranked shank.
14. *Horse* — Figure 10.29(o). This adaptable stake is really a double-ended support. At the end of each arm — one of which is usually cranked downwards for clearance purposes — there is a square hole for the reception of a wide variety of heads. Four of these heads are illustrated:

   *A long head* — Figure 10.29(p). This is used when making knocked-up joints on cylindrical articles, and for flanging;
   Two types of *round head* — Figure 10.29(q) which are used when raising;
   An *oval head* — Figure 10.29(r) which is oval in shape and has a slightly convex working surface. It sometimes has a straight edge at one end.

The condition of the stake has much to do with the workmanship of the finished article. Therefore great care must be taken when using them.

If a stake has been roughened by centre-punch marks or is chisel marked, such marks will be impressed upon the surface of the workpiece and spoil its appearance.

A stake should never be used to back up work directly when centre-punching or cutting with a cold chisel.

A mallet should be used wherever possible when shaping sheet metal, and when a hammer is used care must be exercised to avoid making 'half-moons' on the surface of the stake.

Bench tools which have been abused and damaged as a result, should be reconditioned immediately. Regular maintenance will avoid the risk of marking the surface of the workpiece, as such marks in the metal cannot be removed.

## 10.18 'Hollowing' and 'raising'

Hollowing or raising are both methods employed for the purpose of forming sheet metal into hollow or double-curvature work by hand. As a general guide with regard to the choice of method employed:

HOLLOWING is employed where the desired shape or contour is to be only slightly domed or hollowed.

RAISING is always employed where shapes or contours of much greater depth are required.

Figures 10.30 to 10.35 inclusive show the comparison between the two processes.

*Hollowing* is a process whereby sheet metal is beaten into a small indentation. The metal being formed is stretched and has its original thickness reduced.

*Raising* is a process whereby sheet metal is beaten and induced to flow into the required shape by application of carefully controlled 'floating' blows struck whilst the metal is slightly off the head or former being used. The metal being formed is compressed upon itself and has its original thickness increased.

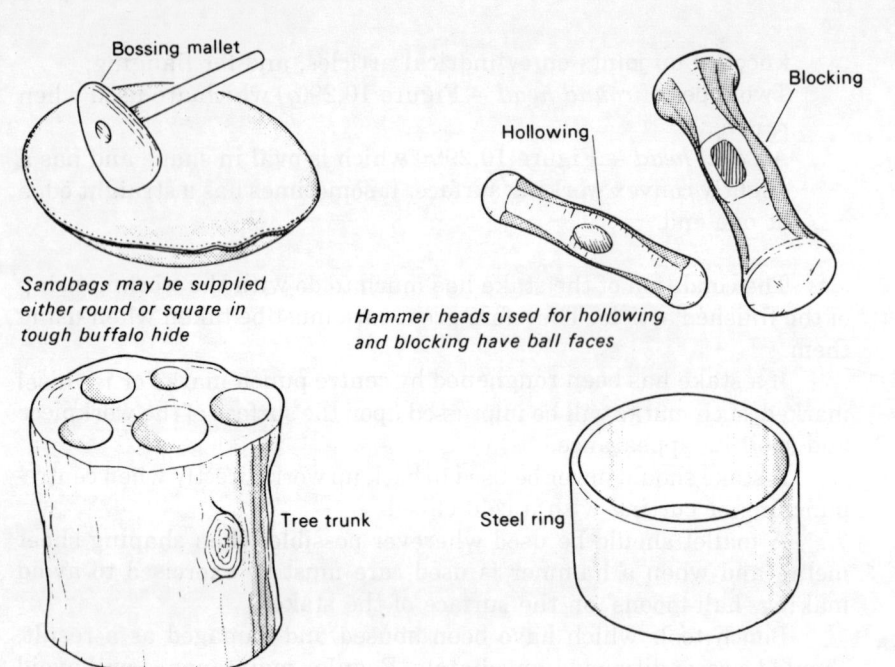

**Fig. 10.30** The basic tools for hollowing

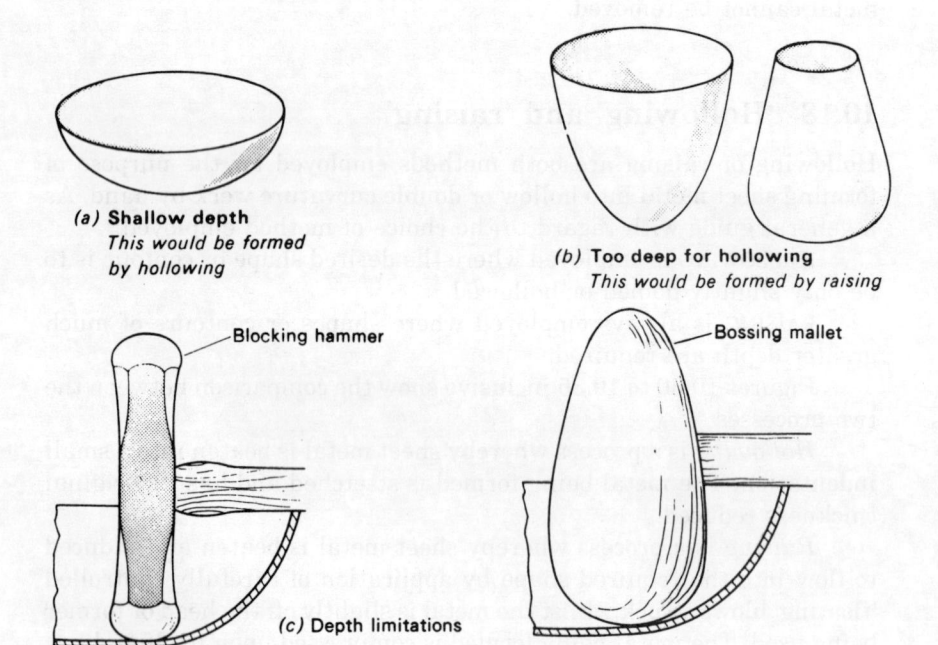

**Fig. 10.31** Comparison of hollowing and raising

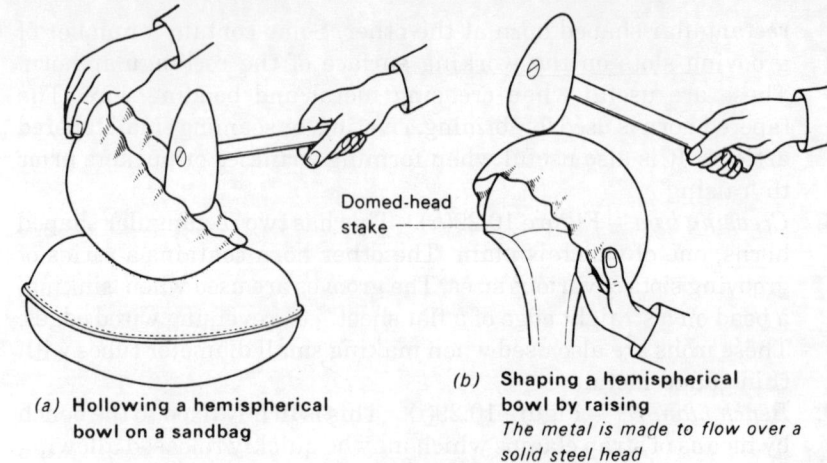

**Fig. 10.32** The two basic methods of forming a bowl

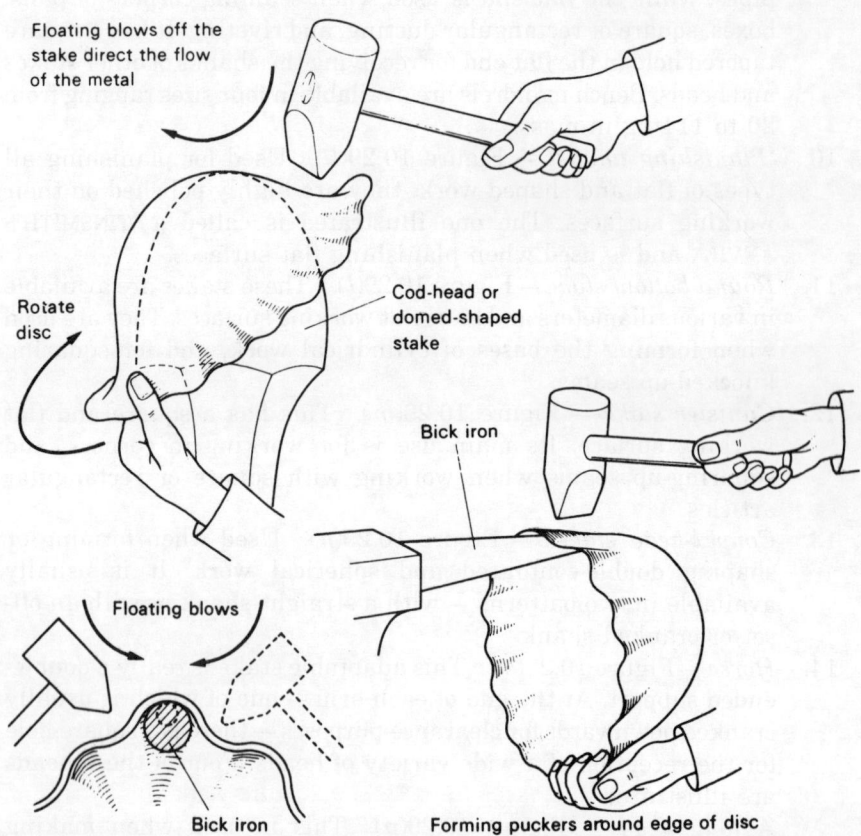

360

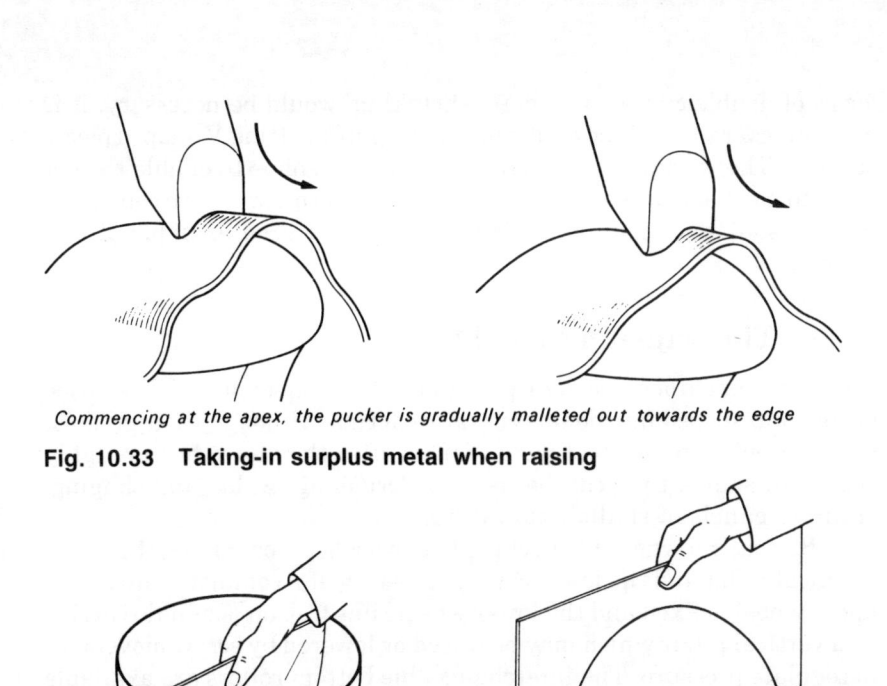

*Commencing at the apex, the pucker is gradually malleted out towards the edge*

**Fig. 10.33 Taking-in surplus metal when raising**

*(a)* **Checking the contour**
*Checked with the aid of inside and outside templates*

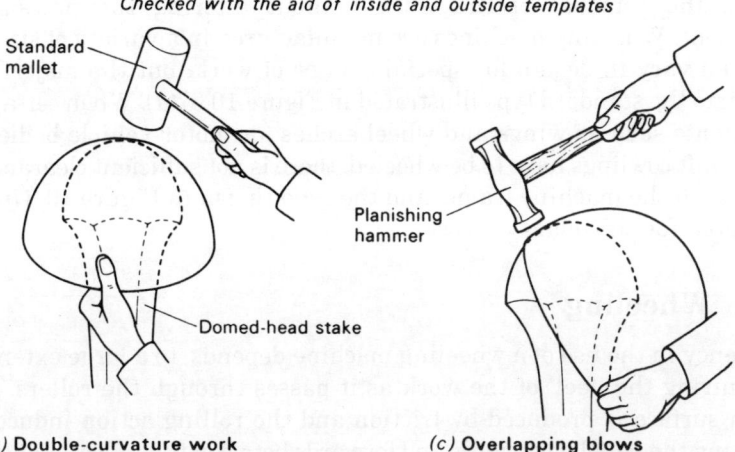

*(b)* **Double-curvature work**
*Produced by either hollowing or raising is placed on a suitable stake and lightly malleted all over in order to remove any high spots on the surface*

*(c)* **Overlapping blows**
*These are made with the planishing hammer directly over the point of contact of the workpiece with the surface of the stake*

Standard mallet — Domed-head stake — Planishing hammer

**Fig. 10.34 Finishing processes for double curvature work**

*(a)* **Pipe bend**
*Made from two cheeks — shaped, planished and welded together*

*(b)* **Working up the throat of one cheek**

*Shaping speeded up by means of puckers*

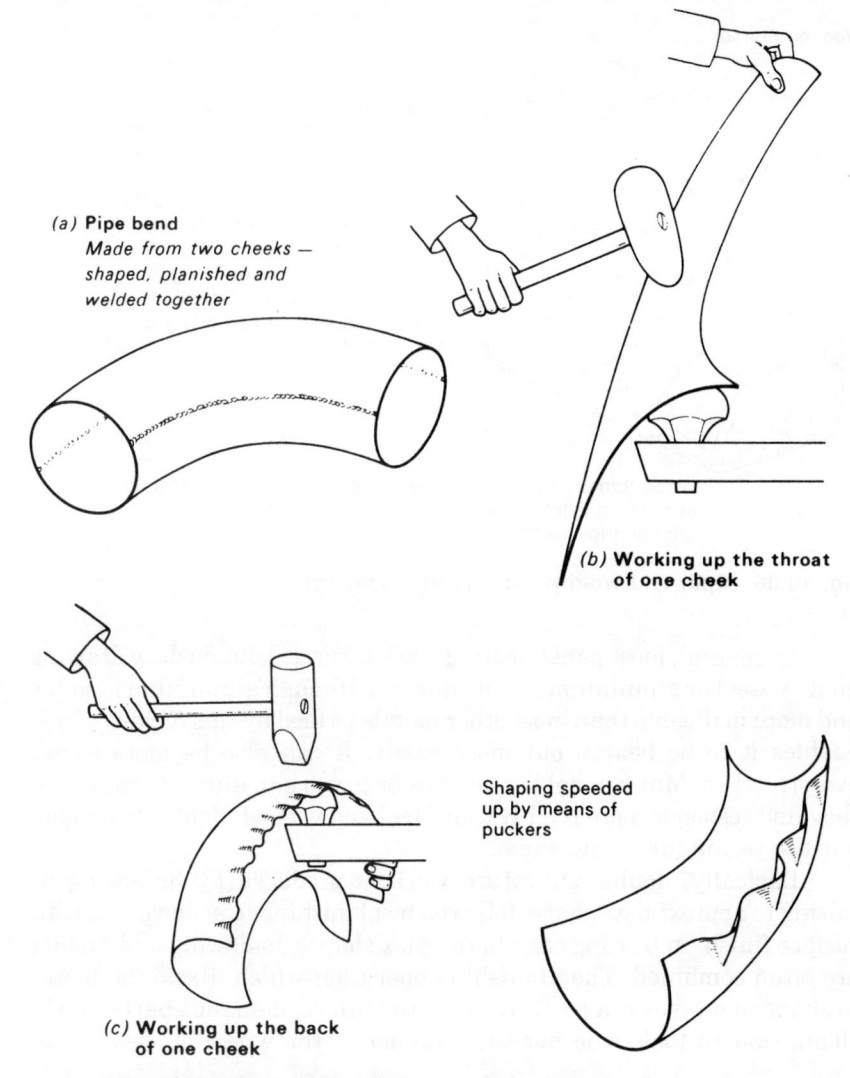

*(c)* **Working up the back of one cheek**

**Fig. 10.35 Pipe bend fabricated from sheet metal**

## 10.19 Panel beating

In coach building, in particular, the panel beater is employed for the fabrication, by hand, of wings, wheel arches, roof domes and similar double curvature work. The skilled panel beater is a craftsman, who, to a great extent, relies on a good eye for line and form. It is a specialised skill which can only be cultivated by years of experience and application combined with dexterity in the use of hand tools.

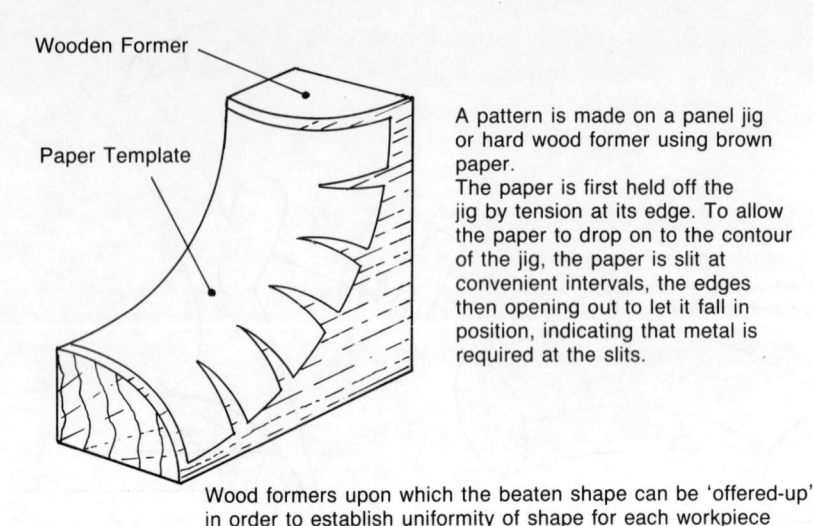

**Fig. 10.36 Split and weld pattern (panel beating)**

A pattern is made on a panel jig or hard wood former using brown paper.
The paper is first held off the jig by tension at its edge. To allow the paper to drop on to the contour of the jig, the paper is slit at convenient intervals, the edges then opening out to let it fall in position, indicating that metal is required at the slits.

Wood formers upon which the beaten shape can be 'offered-up' in order to establish uniformity of shape for each workpiece are used for many jobs.

In general, most panel-beating work is carried out on deep drawing quality steel or aluminium and its alloys. Although aluminium is softer and more malleable than most other metals formed by this means, which enables it to be beaten out more easily, it can also be more easily overstretched. Many panel beaters prefer to work in aluminium, which they find responds more readily than steel and is much lighter to handle, but the technique is the same.

Basically, double curvature work is produced by hollowing or raising to approximate shape, followed by planishing to achieve a smooth surface finish. In beating certain complex shapes, hollowing and raising are often combined. The planishing operation which 'fixes' the metal to shape and gives it a perfectly smooth surface, demands particularly clean, smooth tools whether the hammer or the wheel is used.

'SPLIT AND WELD' METHOD

This process, often used because it is simpler, less laborious and much quicker than other methods of panel beating, is best explained by reference to Figure 10.36 which shows the procedure for making a paper pattern on a panel jig.

It is obvious that material is required where the slits open out. Stretching the metal at these points, until enough is obtained to meet requirements, is time consuming. However, the process can be speeded up by welding in V-shaped pieces of metal. Conversely, in beating certain forms of double curvature where 'shrinking' would be necessary, it is much quicker to 'lose' the excess metal by simply cutting V-shaped pieces out of it. The workpiece is roughly beaten into shape to enable the cut edges to meet and be welded together. Further forming is then completed in the usual way. Figure 10.39(a) illustrates a good example of this method.

## 10.20 The wheeling machine

Wheeling machines are indispensable where sheet metal has to be beaten and shaped into double curvature forms. Although their greatest use is for smoothing sheet-metal panels after they have been roughly beaten to shape, they can be used for 'crushing' welds and shaping, entirely, panels of shallow curvature.

Basically a wheeling machine has two wheels or rollers, the upper practically flat and the lower convex, meeting at a common centre. The upper wheel is fixed and the lower wheel runs free on a spindle carried by a vertical pillar which may be raised or lowered by screw movement to regulate pressure. The interchangeable bottom rollers are available in a number of shapes to suit various curvatures. The rollers are made of steel, their centres being bored out and the ends recessed to house ball races. Through the centre is fixed a steel spindle which, projecting from either side, is used to suspend the roller, in a carrier, parallel to the fixed top roller or wheel. A quick release rise-and-fall movement permits the removal of the work without altering the pressure adjustment. Wheeling machines are manufactured in a variety of sizes, and often vary in design for special classes of work, but the majority conform to the standard type illustrated in Figure 10.37(a). When certain components such as wings and wheel-arches for motor vehicle bodies, and aircraft cowlings have to be wheeled, there is not sufficient clearance underneath the machine frame, and the type shown in Figure 10.37(b) is suitable for this class of work.

## 10.21 Wheeling

Proficiency in the use of a wheeling machine depends, to a large extent, on acquiring the 'feel' of the work as it passes through the rollers. A smooth surface is produced by friction and the rolling action induced by passing the panel backwards and forwards between the rollers. Panels of moderate curvature can be produced by wheeling alone. The sheet is placed between the two rollers at one edge, or in the middle, and pressure of the required intensity, depending upon the gauge and temper of the material, is applied. By repetition of the passing movement

Ball bearing bottom rollers

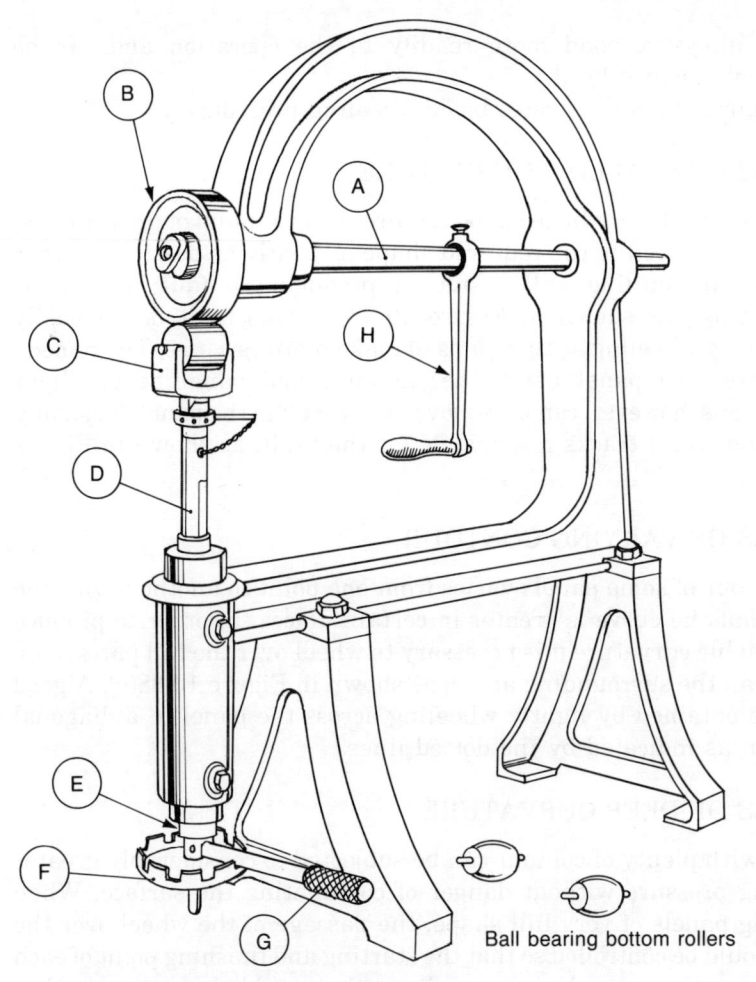

A — Main steel shaft running on ball bearings.
B — Flanged top roller (fixed).
C — Detachable bottom roller supported in carrier that has quick release action.
D — Perpendicular pillar supporting bottom roller.
E — Screwed shaft for pillar adjustment.
F — Flanged turret wheel enabling bottom roller assembly to be raised or lowered, during operation, by turning the wheel with the foot lever 'G'.
H — Handle freely rotating on shaft, can, when necessary, be tightened and used to rotate top roller when wheeling work of small dimensions.

**Fig. 10.37(a)   Standard type wheeling machine**

through the rollers, both longitudinally and transversely, as shown in Figure 10.38(a), the metal stretches and takes on a convex curvature. It is very important that each backward and forward movement should be accompanied by alteration of direction so that the rollers make contact

This type of wheeling machine is more versatile than the standard type, but the principle is unchanged. It can be arranged in pairs, back-to-back on a common bed, thus saving considerable floor space.

**Fig. 10.37(b)   Wheeling machine with additional clearance**

with the panel in a different place at each pass, at the same time ensuring complete coverage of the entire surface. The panel should not be gripped too tightly and, by 'feel', the hands should allow it to follow its own shape during wheeling. The movement is varied until the desired shape is obtained, those parts of the panel which are required to be of light curvature receiving less wheeling than other more curved portions.

Wheeling pressure is applied as required by raising the lower roller. If the contour of the panel requires raising, the bottom roller is raised, increasing the pressure on the work between the rollers, and after a few passes, the necessary raising will be accomplished.

*When working with soft metals such as aluminium, which is very easily scratched, it is essential that the surfaces of the tools, rollers and workpiece are perfectly clean.* Care should be taken not to put too much pressure on the workpiece because up to three times more 'lift' is obtainable than that experienced with the harder metals such as steel. The craftsman used to wheeling harder metals will find that aluminium and its alloys respond more readily to the operation and can be extensively shaped by it.

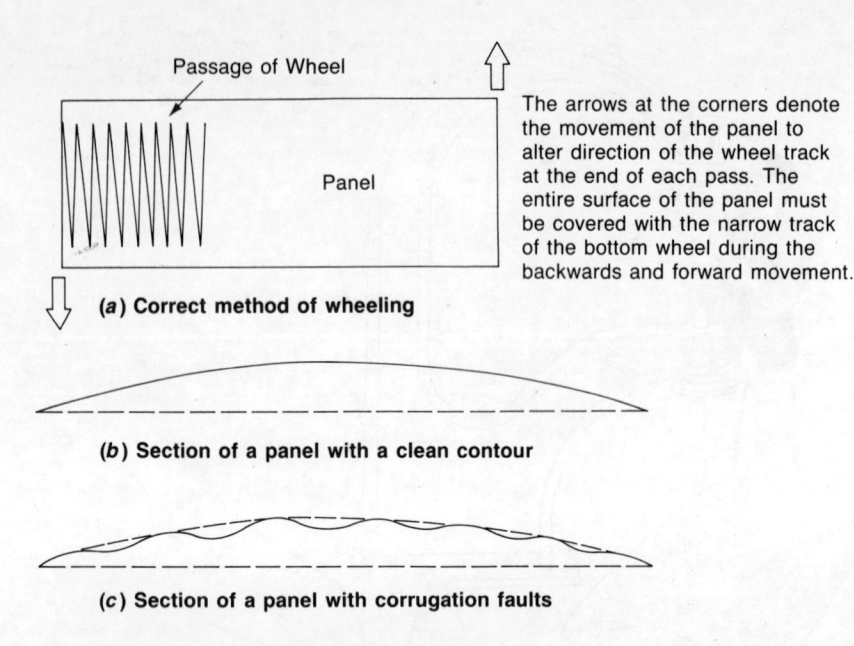

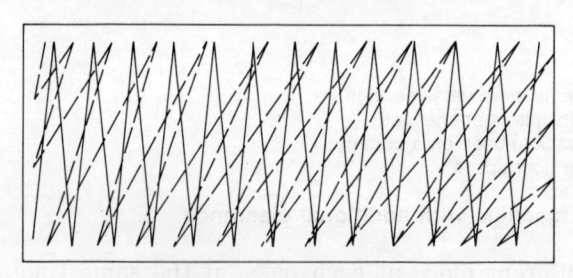

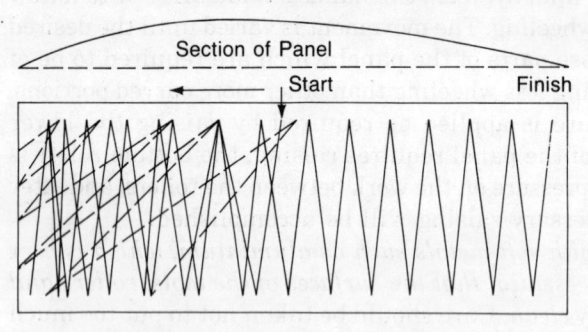

Figure 10.38 illustrates basic wheeling procedures.

## PANELS OF SHALLOW CURVATURE

Flat panels with very little shape require the minimum of pressure and should be wheeled to the required shape as gently as possibly. Undue pushing and pulling will result in producing a fault known as 'corrugating', as shown in Figure 10.38(c). This fault is not easily corrected by wheeling as this tends to make matters worse. The remedy is to reset the panel using the hammer and panel head. Slight corrugations, however, can be removed by wheeling the panel diagonally across the wheel tracks responsible for the fault, as shown in Figure 10.38(d).

## PANELS OF VARYING CONTOUR

The contour of some panels varies from one point to another, with the result that the curve is greater in certain places. In order to produce this variable curvature it is necessary to wheel over the full parts more often than the surrounding areas, as shown in Figure 10.38(e). A good finish is obtained by lightly wheeling across the panel in a diagonal direction as indicated by the dotted lines.

## PANELS OF DEEP CURVATURE

Panels with plenty of contour can be subjected to considerably greater wheeling pressure without danger of corrugating the surface. When wheeling panels of very full shape, the passage of the wheel over the work should be controlled so that the starting and finishing point of each pass does not occur in the same position; the panel being moved so that the raising of the shape is performed evenly and not by a series of lumps. Sometimes too much shape is wheeled into a panel. This undesirable shape can be corrected by turning the panel upside down and wheeling the outside edges. With panels of very slight curvature this reverse wheeling can be carried through the entire panel.

## LARGE PANELS

An assistant may be required to hold one side when wheeling large panels. *For successful wheeling when two operators are working on one panel it is important that each person should do their own pulling.* Since no two operators 'pull' alike, they change sides mid-way through to avoid giving the panel uneven curvature. On no account should either operator

Fig. 10.38 The technique of wheeling

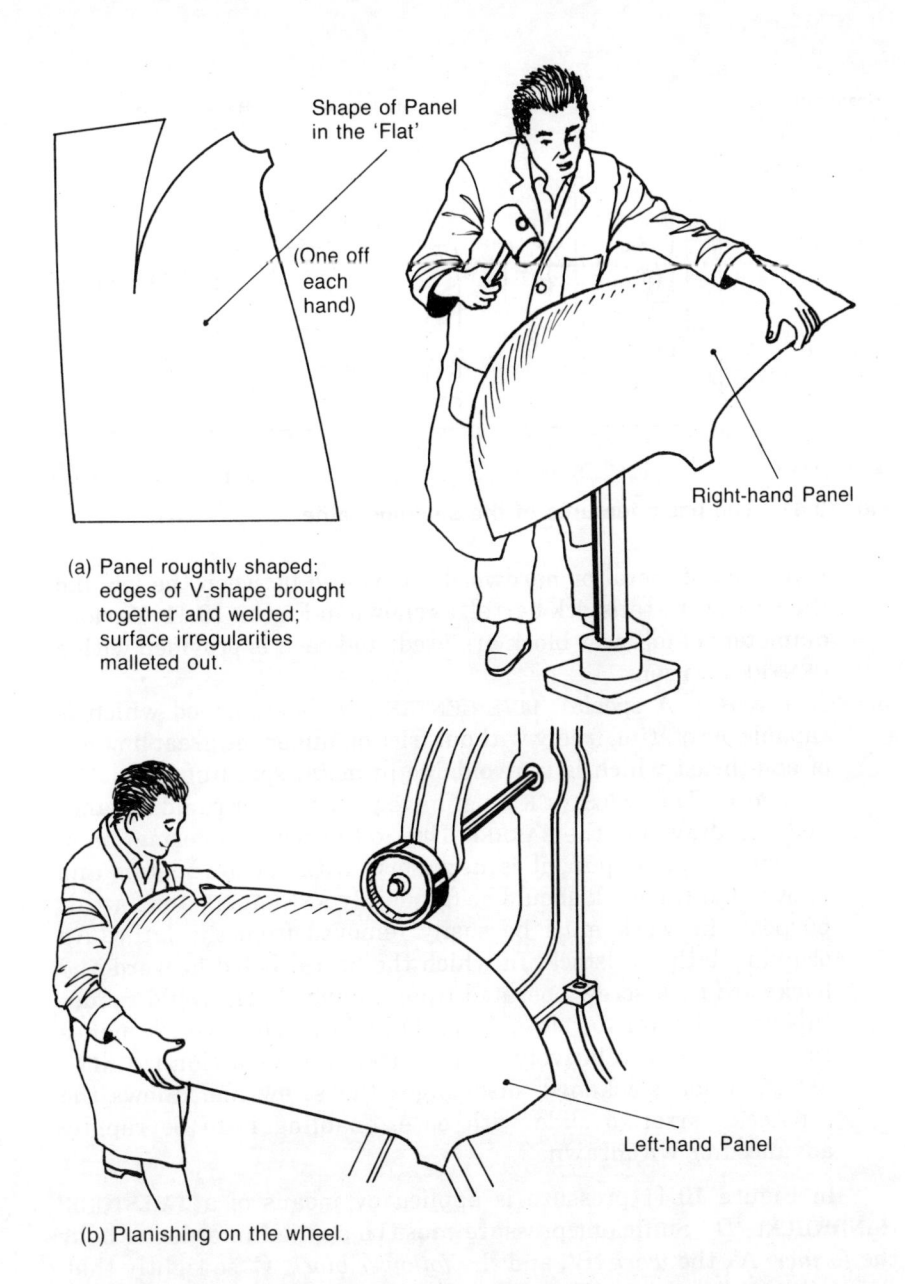

Shape of Panel in the 'Flat'

(One off each hand)

Right-hand Panel

(a) Panel roughtly shaped; edges of V-shape brought together and welded; surface irregularities malleted out.

(b) Planishing on the wheel.

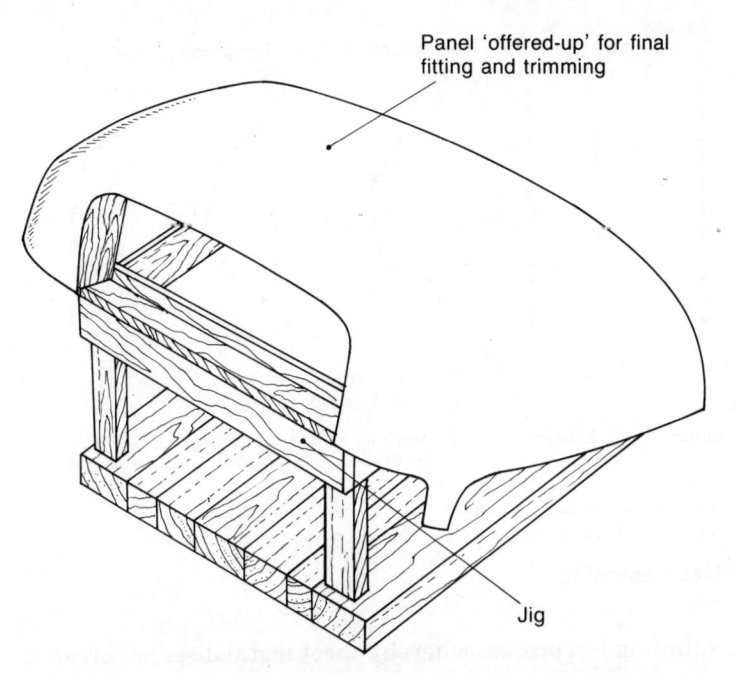

Left-hand Panel

push the panel during the reverse passage through the wheels. Any roughness when pulling or pushing will cause corrugations and unevenness of shape.

As previously stated, wheeling is widely used for finishing panels formed by the 'cut-and-weld' method. The very rough shape produced

Panel 'offered-up' for final fitting and trimming

Jig

(c) Frequent use of jigs during the fabrication of the component panels, and for final fitting and trimming, ensure the symmetry of the final composite panel.
The composite panel is fabricated by welding. It consists basically of a central panel and two end panels. The central panel is shaped entirely by wheeling.

**Fig. 10.39 Stages in forming large panels (coach building)**

by welding the edges together can be brought to the desired contour on the wheel, after preliminary malleting over a suitable panel head. Wheeling not only crushes and smooths over the weld, which will disappear altogether after dressing with a file or portable sander, but also partially restores, by work-hardening, the temper of the metal adjacent to the weld which has been softened by the heat of welding. Figure 10.39 illustrates stages in forming a large roof panel for a coach.

## 10.22 Introduction to metal spinning

Most sheet metals lend themselves readily to cold-forming by spinning. There will, of course, be some variations in performance, and this is due to the individual characteristic mechanical properties and work-hardening behaviours. However, the *basic principles* of spinning are identical regardless of the particular metal being formed.

**Fig. 10.40 Metal spinning**

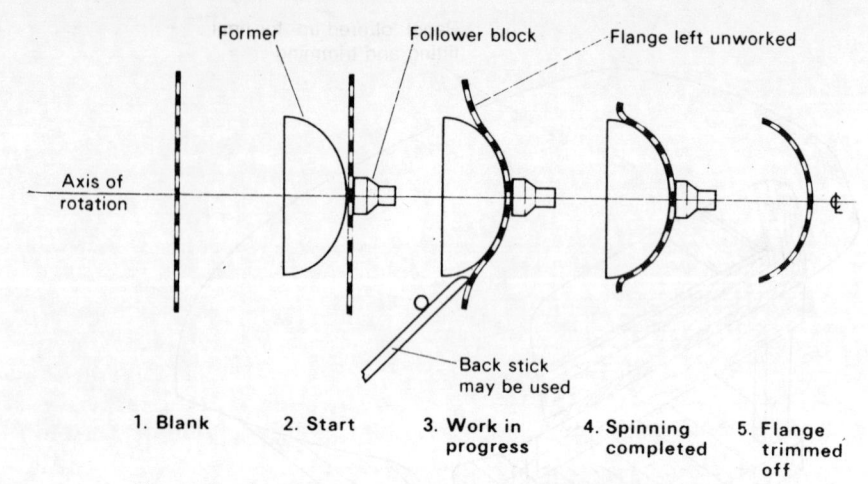

*The successive stages in manual metal spinning showing the spreading or drawing action which is characteristic of the process*

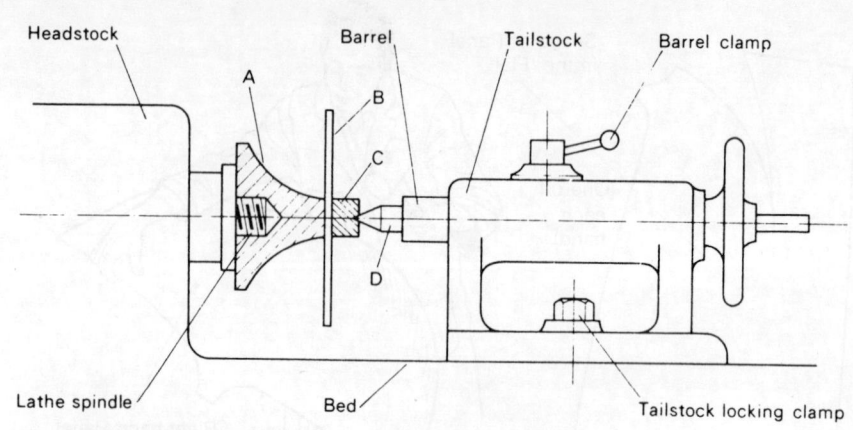

**Fig. 10.41 The basic features of the spinning lathe**

Metal spinning is a process whereby sheet metal discs are pressed or rolled to specific shapes by forcing the metal to flow over a suitable mandrel or chuck of the required form, usually by hand tools whilst chuck and metal are rotating together.

Causing flat sheet metal to alter its shape by PLASTIC FLOW under pressure manually applied through the LEVERAGE of spinning tools, calls for skill that can only be acquired through training and experience. The manner in which such shape-changing is accomplished is illustrated in Figure 10.40.

## 10.23 Spinning lathes

The general design of a spinning lathe (Figure 10.41) follows closely that of an ordinary machine-shop lathe. There are, however, several necessary and important differences:

1. *Headstock and spindle* The HEADSTOCK and SPINDLE must be sturdily built to be able to withstand the considerable force which is applied during the spinning operation.

    The HEADSTOCK is unlike that of a machinist's lathe in that the sheet metal disc ($B$) from which the article is to be shaped is not held in a CHUCK, but is held by FRICTION between a FORMER and a FOLLOWER.

2. *Former and follower* The FORMER ($A$) is solidly fixed to the headstock spindle and turns with it (see below). The FOLLOWER ($C$) is a block of metal or hardwood introduced between the job and the nose of the tailstock barrel, as shown in Figure 10.41. Various diameters of follower block are used, and each is provided with a CENTRE location.

3. *Live centre* A special 'LIVE CENTRE' ($D$) is employed which is capable of rotating freely without friction under the great amount of end-thrust which is unavoidable in metal spinning.

4. *Tailstock* The TAILSTOCK must have provision for rapid advance and withdrawal of the BARREL. Due to the severe deformation of the metal being spun, it is necessary, with certain metals and alloys, that the work should be frequently ANNEALED, and for this purpose the work must be easily removed from the lathe. An ordinary lathe tailstock, in which the barrel is fed forward and backward by a screw operated from a handwheel, would render this operation far too slow. A specially designed tailstock makes this operation a relatively simple task. A quick-action LOCKING LEVER, when slackened, disengages the screw and allows the tailstock barrel to slide with ease, enabling it to be rapidly advanced or withdrawn.

In Figure 10.41 pressure is applied by means of a TAILSTOCK HANDWHEEL 'D'. Sufficient pressure must be applied in order to hold the *former* 'A', the *work* 'B', and the *follower block* 'C' so tightly that they will revolve as one. The special 'LIVE CENTRE' is shown at 'E'.

### FORMERS

Formers, upon which the final contour or shape of the spun article is dependent, are usually made of hardwood (for short runs), steel or cast

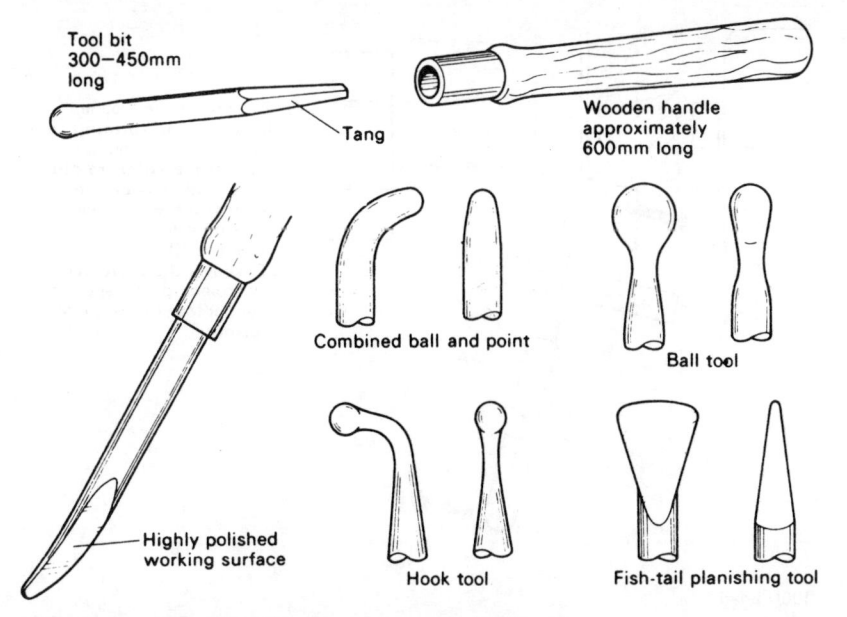

**Fig. 10.42 Hand-forming tools for metal spinning**

iron (for long runs). Wooden formers are generally made from MAHOGANY or LIGNUM VITAE. These hardwoods have *high strength* and *resistance to wear*. Small formers are made in one piece, i.e. from a single section of hardwood. For economy, and also to lessen the effects of shrinkage, some large formers are made from laminated hardwood blocks.

## 10.24 The spinning process

The hand-spinning process which is the most commonly used, except when considerable quantities of components are mass produced, is performed with the aid of a number of uniquely designed tools which have their actual working surfaces of special shapes according to the nature of the work, or part being spun. Some of these hand-spinning tools are illustrated in Figure 10.42.

Hand-spinning tools are not standardised and may have a wide variety of shapes — some craftsmen shape their own forming tools.

A hand-forming tool consists of two parts:

(a) TOOL BIT. This is approximately 300–450 mm in length and usually forged to shape from high-speed steel (round bar) and hardened. It has a 'tang' which fits into a handle.

(b) A WOODEN HANDLE. This is approximately 600 mm in length.

The 'bit', when securely fitted, projects from the handle for a distance of 200 mm.

THE AVERAGE OVERALL LENGTH OF A HAND-FORMING TOOL IS BETWEEN 750 AND 850 mm.

The most common forming tool used is the COMBINATION BALL AND POINT. Its range of usefulness is large on account of the variety of shapes that may be utilised merely by rotating the tool in different direction.

A BALL TOOL is used for finishing curves.

The HOOK TOOL is shaped for use on inside work.

The bulk of spinning operations involve starting the work and bringing it approximately to the shape of the former, after which 'smoothing' or 'planishing' tools are used to remove the spinning marks and produce a smooth finish.

The FISH-TAIL planishing tool is one which is commonly used for finishing the work. It is also a very useful tool for sharpening any radii in the contour.

These hand tools are used in conjunction with a TEE-REST and FULCRUM PIN. The manner in which the fulcrum pin is advanced as the spinning operation progresses is extremely simple, as will be seen by studying Figures 10.43 to 10.46. The Tee-rest is accommodated in the adjustable TOOL-REST HOLDER which is clamped to the bed of the lathe. The tool-rest provides a wide range of adjustment in six directions, and a further fine adjustment can be made by releasing a clamp bolt and swivelling the Tee-rest. All these features are illustrated in Figure 10.43. The action of the spinning tool is shown in Figure 10.44 (see Section 4.27).

When commencing spinning operations, the initial strokes are made outwards towards the edge of the disc being spun. In order to speed up the operation and also to avoid possible thinning of the metal, strokes are also made in the opposite direction, i.e. inward strokes.

In either case it is important not to dwell in any one position on the workpiece so as to cause excessive WORK-HARDENING. The Tee-rest and the position of the fulcrum pin are reset as the work progresses, and both the forming tool and the outer surface of the metal disc are LUBRICATED frequently.

*The process of spinning* HARDENS THE METAL *by reason of the* COMPRESSION AND STRETCHING *which it undergoes whilst being shaped to the contour of the former.*

Figure 10.45 illustrates the metal-spinning process.

When spinning fairly large diameter discs, particularly when the shape of the former does not lend any support, the use of a 'BACK-STICK' is necessary.

Back-sticks, as their name implies, are always positioned at the

'B' represents the working portion of the fulcrum pin. This varies in length between 75 and 100mm.
'C' represents the portion of the pin which is a free fit in the holes provided in the tee-rest.
It is smaller diameter than 'B' and is approximately 40 to 50mm in length

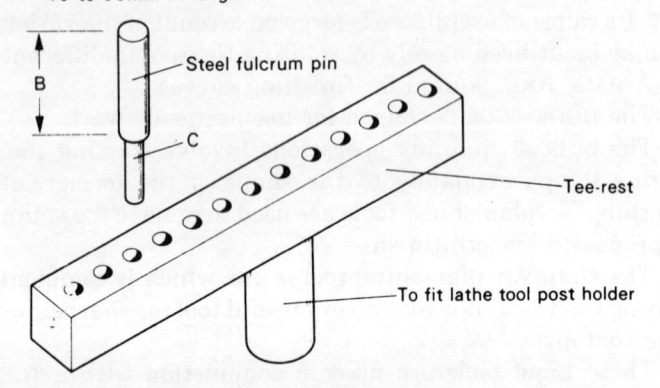

Steel fulcrum pin
Tee-rest
To fit lathe tool post holder

Simple tee-rest and fulcrum pin

(a) By moving the wooden handle of the **forming tool** 'A' in the direction 'B', against the **fulcrum pin** 'C' a force may be exerted against the workpiece causing it to flow in the direction 'D'.
'E' shows the position of workpiece partly spun to the contour of the simple **former**

100N Load
150mm | 750 mm
Fulcrum
Effort 20N

(b) Moments acting on the spinning tool

Former
Metal disc
Follower block
Forming tool
Tee-rest
Fulcrum pin

View in the direction of the arrow

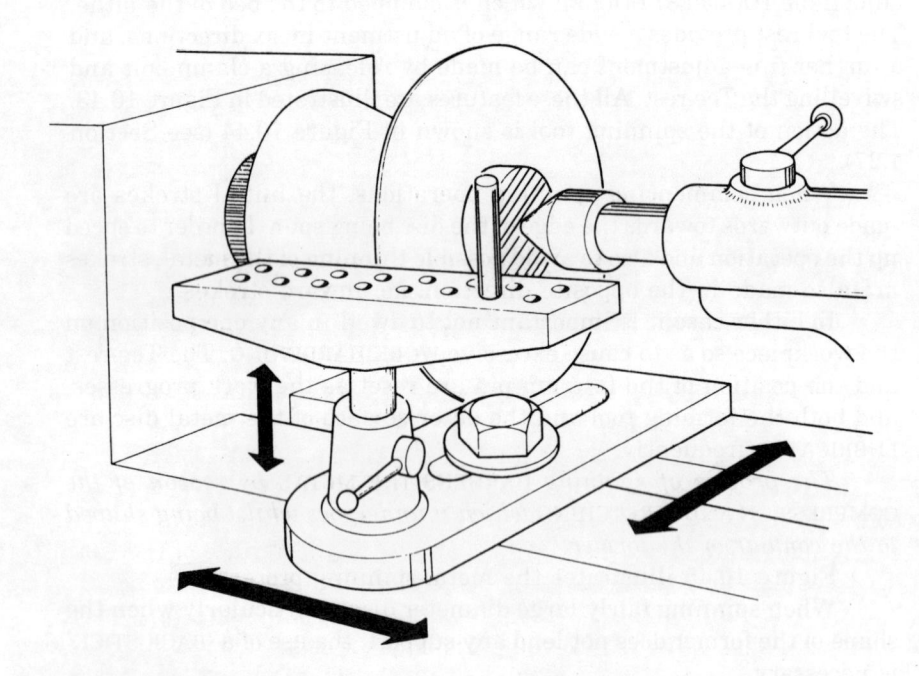

(c) Three dimensional movement of the forming tool

**Fig. 10.43 Adjustment for variations of fulcrum pin position**

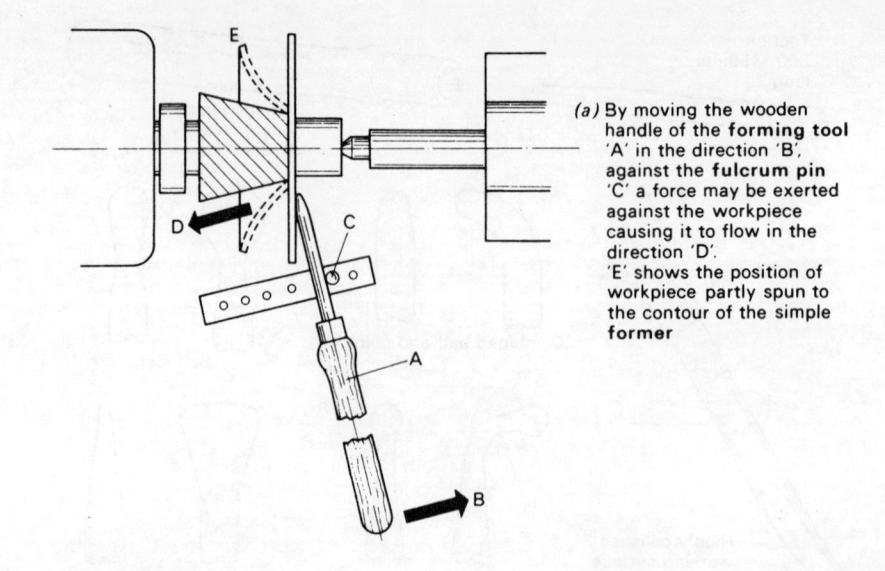

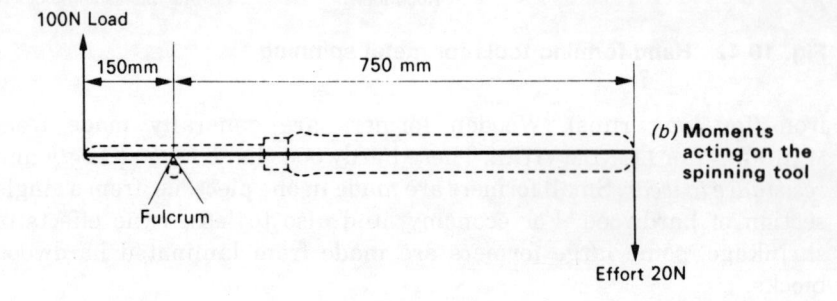

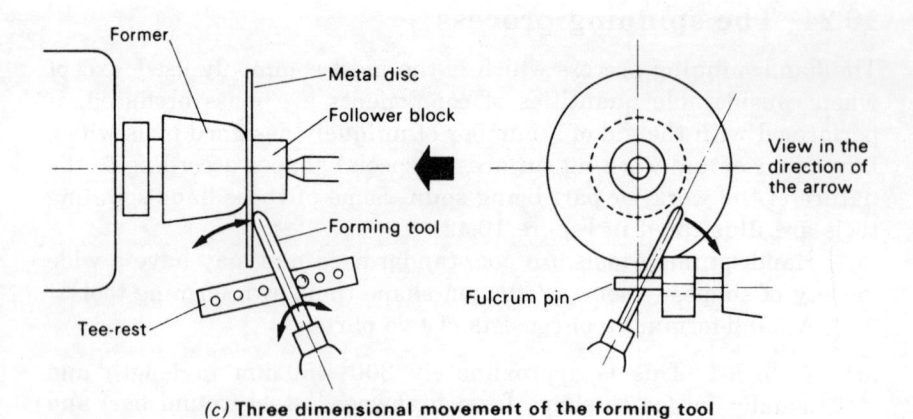

**Fig. 10.44 The action of hand spinning**

368

Final shape / Initial stroke will dish or set the metal disc

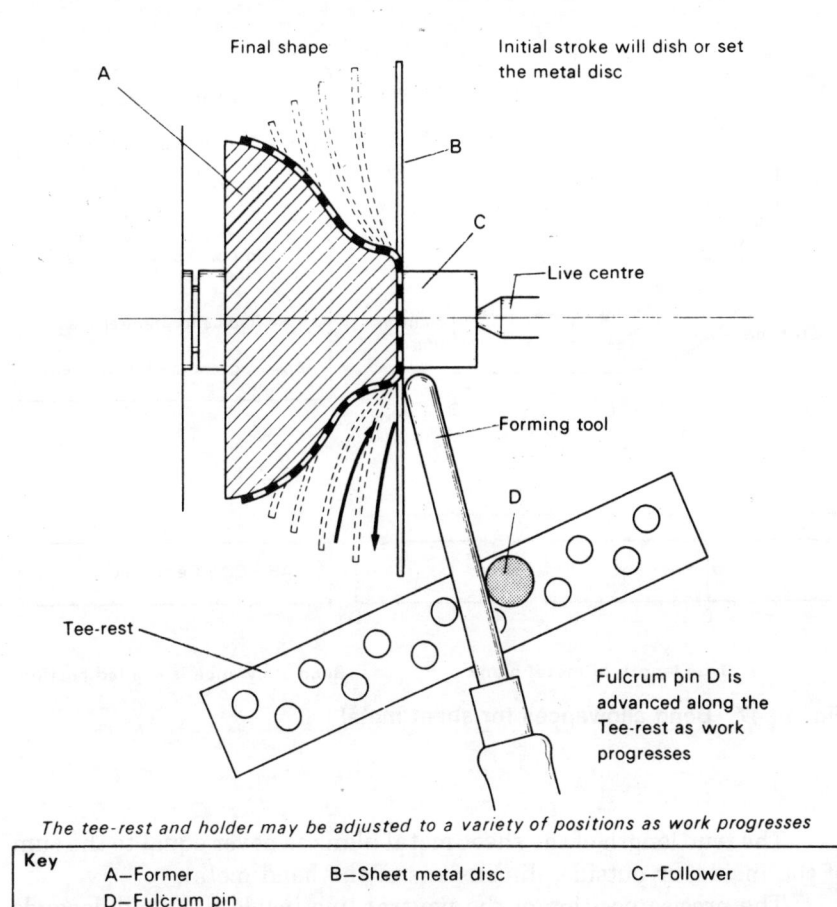

Key: A—Former  B—Sheet metal disc  C—Follower  D—Fulcrum pin

The tee-rest and holder may be adjusted to a variety of positions as work progresses

**Fig. 10.45 The metal spinning process**

Back-sticks are made from hardwood and are approximately 300 mm long

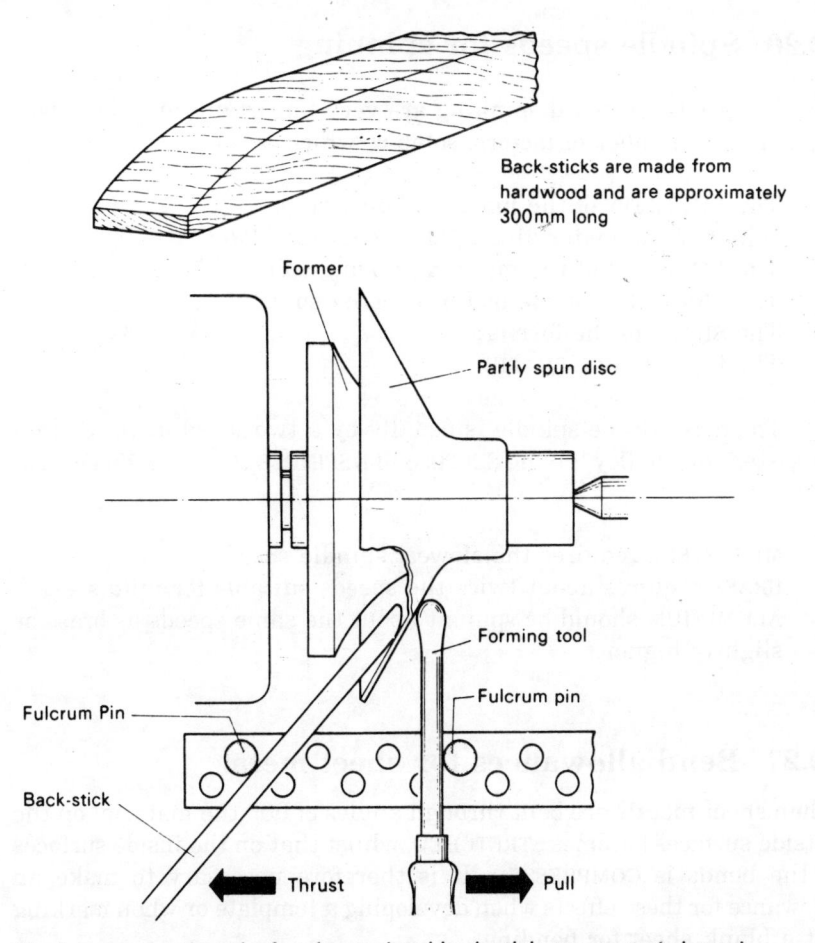

The forces exerted on the forming tool and back-stick are counterbalanced

**Fig. 10.46 The use of the back-stick**

back of the metal disc immediately opposite the forming tool. They are used to prevent wrinkles forming and, in the case of thin gauge metal, to hold and stiffen the edge and prevent it from collapsing. Figure 10.46 shows that a back-stick is used in the same manner as a forming tool, but on the opposite side of the disc. Pressure is applied in the direction indicated, and the work revolves between the back-stick and the forming tool, two fulcrum pins being used.

## 10.25 Lubrication for spinning

It is necessary that the FRICTION between the nose of the forming tool and the work be reduced to a minimum in order to prevent: excessive HEAT being developed; scratching or cutting of the metal surface; and possible damage to the tool.

LUBRICANTS are used for this purpose. **ANY LACK OF LUBRICANT WILL RESULT IN DAMAGE TO THE WORK-PIECE AND THE TOOL.**

Lubricants must be frequently applied to the surface of the work and the tools during the spinning operations. It is important to use the correct type of lubricant. It must be sufficiently ADHESIVE to cling to the metal disc when it is revolving at high speeds.

In hand spinning, TALLOW and INDUSTRIAL SOAP, or a mixture of both, are generally used as lubricants.

## 10.26 Spindle speeds for spinning

Spindle speeds for metal spinning are very important, and these will depend on a number of factors, some of which are:

(a) The DUCTILITY of the material being spun;
(b) Whether the material is FERROUS or NON-FERROUS;
(c) The DIAMETER of the metal disc being spun;
(d) The THICKNESS of the material being spun;
(e) The SHAPE of the former;
(f) The type of TOOL SHAPE.

The drive to the spindle is usually by a two-speed motor on to a three-step cone pulley, giving SIX POSSIBLE SPINDLE SPEEDS. As a general rule:

MILD STEEL requires the slowest spindle speeds.
BRASS requires about twice the speeds suitable for mild steel.
ALUMINIUM should be spun at about the same speeds as brass or slightly higher.

## 10.27 Bend allowances for sheet metal

When sheet metals are bent through angles of 90° the material on the outside surfaces becomes STRETCHED, whilst that on the inside surfaces of the bends is COMPRESSED. It is therefore necessary to make an allowance for these effects when developing a template or when marking out a blank sheet for bending.

Figure 10.47 illustrates the importance of the 'NEUTRAL LINE.'
An enlarged cross-section of a 90° bend in sheet metal is shown in Figure 10.47.

THE NEUTRAL LINE IS AN IMAGINERY CURVE SOMEWHERE INSIDE THE METAL IN THE BEND. ITS POSITION DOES NOT ALTER FROM THE ORIGINAL FLAT LENGTH DURING BENDING.

Because there is a slight difference between the amount of COMPRESSIVE STRAIN and the amount of TENSILE STRAIN, the NEUTRAL LINE lies in a position nearer the inside of the bend.

For the purpose of calculating the allowance for a bend in sheet metal, the neutral line curve is regarded as an arc of a circle whose radius is equal to the sum of inside bend radius and the distance of the neutral line in from the inside of the bend.

$T$ = Thickness of metal
$R$ = Outside radius of bend
$r$ = Inside radius of bend
$x$ = Distance of neutral line in from compression or inner side of bend
$r+x$ = Radius for neutral line
$bc$ = Arc length of neutral line

$L$ = AB + CD + Calculated length $bc$

True length of metal blank                 Bend allowance is shaded portion

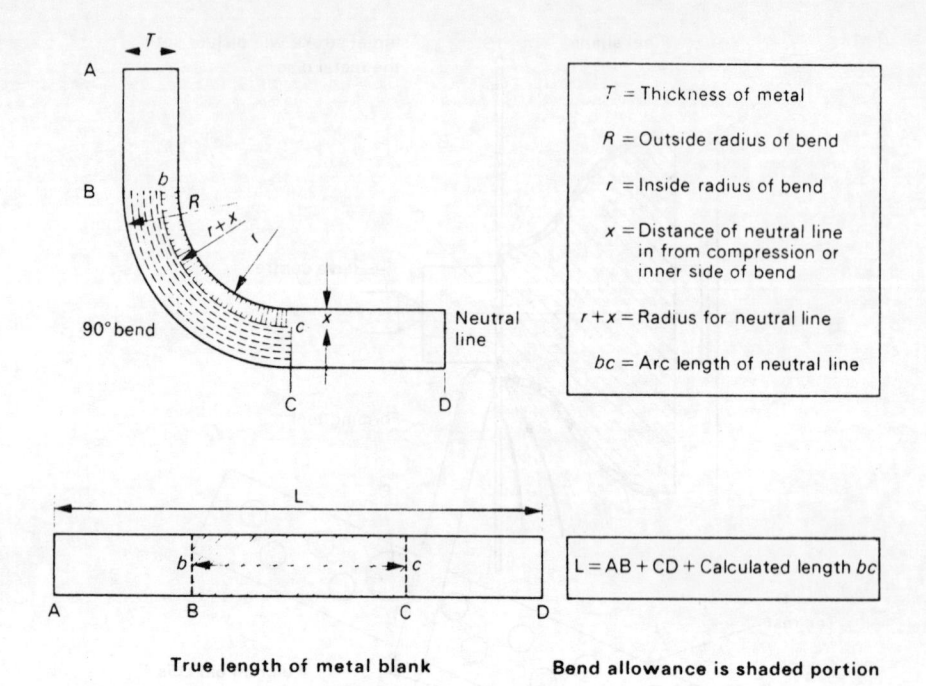

**Fig. 10.47  Bend allowances for sheet metal**

The true length of the sheet-metal blank is never equal to the sum of the inside, or outside, dimensions of the bend metal.

The precise position of the neutral line inside the bend depends upon a number of factors which include:

(a) The properties of the materials;
(b) The thickness of the material;
(c) The inside radius of the bend.

Table 10.3 lists the approximate positions of the neutral line for some materials.

### DEFINITIONS

In this section the following British Standard (BS 1649) definitions apply:

Neutral line. The boundary line between the area under COMPRESSION and the area under TENSION in any angle bend.

**Table 10.3  Neutral line data for bending sheet metal**

| MATERIAL (20 to 14 s.w.g.) | AVERAGE VALUE OF RATIO $\frac{x}{T}$ |
|---|---|
| Mild steel | 0·433 |
| Half-hard aluminium | 0·442 |
| Heat-treatable aluminium alloys | 0·348 |
| Stainless steel | 0·360 |

In general the position of the neutral line is 0·4 times the thickness of the material in from the inside of the bend.

This means that the radius used for calculating the bend allowance is equal to the sum of the inside bend radius and 0·4 times the thickness of the metal. The inside bend radius is rarely less than twice the thickness of the material or more than four times.

For the purpose of calculating the required length of blank when forming cylindrical or part-cylindrical work a mean circumference is used — i.e. the neutral line is assumed to be the central axis of the metal thickness.

For general sheet-metal work the following values for the radius of the neutral line may be used (where precision is unimportant):

| THICKNESS OF MATERIAL s.w.g. | mm | APPROXIMATE VALUE OF NEUTRAL LINE RADIUS |
|---|---|---|
| 30 to 19 | 0·315 to 1·016 | One-third metal thickness plus inside bend radius |
| 18 to 11 | 1·219 to 2·346 | Two-fifths metal thickness plus inside bend radius |
| 10 to 1 | 3·251 to 7·620 | One-half metal thickness plus inside bend radius |

<u>Radius of bend.</u> The radius of the inside of the bend.

<u>Outside radius of the bend.</u> The inside radius of the bend plus the metal thickness.

<u>Bend allowance.</u> The length of the metal required to produce the radius portion only of the bend.

## 10.28  Applications of bending allowances

As previously stated the length of the neutral line is represented by an ARC of a CIRCLE.

Arc lengths are dependent upon their SECTOR ANGLES, and can be determined by calculation as follows:

SECTOR ANGLE DIVIDED BY 360° AND MULTIPLIED BY THE CIRCUMFERENCE.

For example, consider an arc of RADIUS 100 mm whose subtended angle is 90°. Then its length will be:

$$\frac{90}{360} \times 2\pi R$$

$$= \frac{1}{4} \times 2 \times 3\cdot142 \times 100 \text{ mm}$$

$$= 50 \times 3\cdot142 = \underline{157\cdot1 \text{ mm}}$$

Alternatively, by inspection the ratio $\frac{2\pi}{360}$ is a constant which may be used for all bend allowance calculations:

$$\frac{2\pi}{360} = \frac{2 \times 3\cdot142}{360}$$

$$= \frac{3\cdot142}{180} = 0\cdot0175$$

Thus the length of the arc will be:

$$0\cdot0175 \times R \times 90$$
$$= 0\cdot0175 \times 100 \times 90 \text{ mm}$$
$$= 1\cdot75 \times 90 = \underline{157\cdot5 \text{ mm}}$$

From the above it will be seen that a formula may be derived for calculating bend allowances. It is as follows:

BEND ALLOWANCE
= 0·0175 multiplied by inside radius to the neutral line multiplied by the subtended angle of the bend.

This can be expressed as follows:

$$A = \theta \times R \times 0\cdot0175$$

Where: $\theta$ = (180 − angle given)  $A$ = Bend allowance
$R$ = Inside radius to the NEUTRAL LINE

Table 10.4 shows two worked examples for bend allowances using the 'MEAN LINE' method.

Tables 10.5, 10.6 and 10.7 show worked examples using the approximate values for the neutral line from Table 10.3.

Tables 10.8, 10.9 and 10.10 show worked examples for 'Precision Sheet-Metal Work' using the ratio $x/T = 0\cdot4$.

### Table 10.4(a)  Calculation – centre line bend allowance

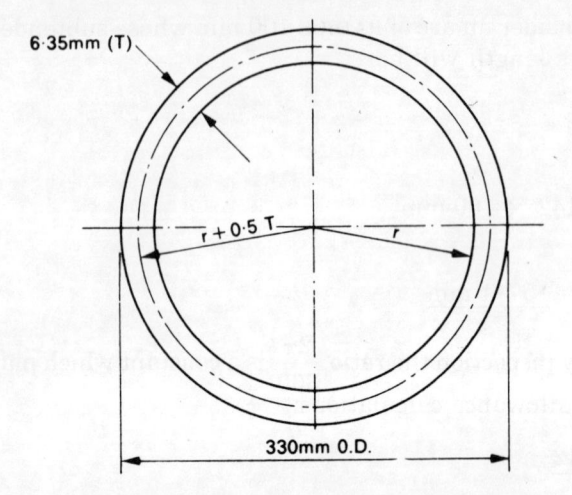

Calculate the length of the blank required to roll the cylinder shown opposite.
The position of the neutral line = $0.5T$, where the thickness of the plate, $T = 6.35$ mm.

**Solution:**

The length of the blank required is equal to the MEAN CIRCUMFERENCE.
*The MEAN RADIUS 'R' is equal to the INSIDE RADIUS 'r' plus half the thickness 'T' of the plate.*

| The outside diameter | = | 330 mm |
| The inside diameter | = | $330 - 2T$ |
| | = | $330 - (2 \times 6.35)$ |
| | = | $330 - 12.7$ |
| | = | 317.3 mm |

The inside radius

$$r = \frac{317.3}{2} = 158.65 \text{ mm}$$

From which $R = 158.6 + (0.5 \times 6.35)$
$\phantom{From which R} = 158.6 + 3.175$
$\therefore R = 161.825$ mm

The mean circumference $= 2\pi R$
$= 2 \times 3.142 \times 161.825$
$= 6.284 \times 161.825$

| No. | Log |
|---|---|
| 6.284 | 0.7983 |
| 161.825 | 2.2089 |
| 1 016 | 3.0072 |

**Length of blank required = 1016 mm**

### Table 10.4(b)  Calculation – centre line bend allowance

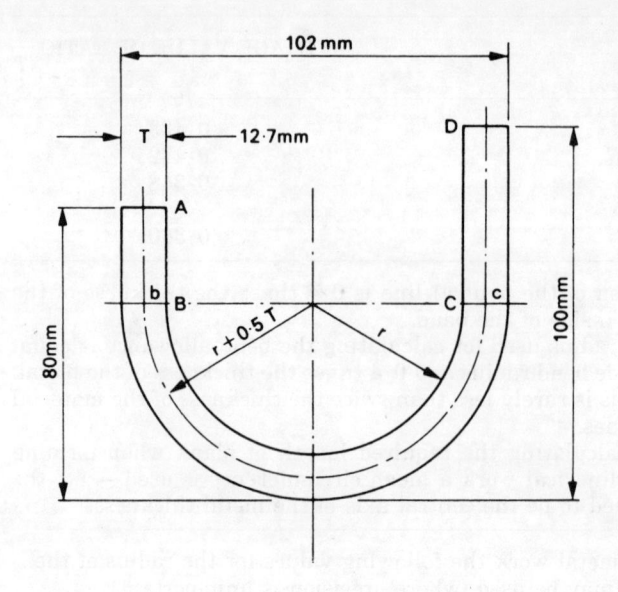

Calculate the length of the blank required to form the 'U-bracket' shown opposite.
The position of the neutral line = $0.5T$, where the thickness of the plate, $T = 12.7$ mm.

**Solution:**
The length of the blank required is equal to the sum of the flats, 'A B' and 'C D', plus the length of the mean line, '$b\,c$'.
Thus $L = AB + CD + `b\,c`$.
Now $b\,c$ represents a semi-circular arc whose mean radius $R$ is equal to the INSIDE RADIUS $r$ plus half the thickness $T$ of the plate.

| The outside diameter of the semi-circle | = | 102 mm |
| The inside diameter of the semi-circle | = | $102 - (2T)$ |
| | = | $102 - 25.4$ |
| | = | 76.6 mm |

From which the inside radius $r = \dfrac{76.6}{2} = 38.3$ mm

The mean radius $R = 38.3 + (0.5 \times 12.7)$
$\phantom{The mean radius R} = 38.3 + 6.35$
$\therefore R = 44.65$ mm

**Length of flats**

$AB = 80 - \dfrac{102}{2}$
$\phantom{AB} = 80 - 51 \quad = 29$ mm

$CD = 100 - \dfrac{102}{2}$
$\phantom{CD} = 100 - 51 \quad = 49$ mm

**Total length of flats = 78 mm**

Bend allowance for 'b c'
'Bend Allowance' = $\pi R$ (semi-circle)
'Bend Allowance' = 3·142 × 44·65
= 140·3 mm

| No. | Log |
|---|---|
| 3·142 | 0·4971 |
| 44·65 | 1·6498 |
| 1403 | 2·1469 |

**Length of blank**
$L = AB + CD + bc$
= 78 + 140·3

Length of blank required = 218·3 mm

Tables 10.5, 10.6 and 10.7 show worked examples using the approximate values for the neutral line from Table 10.3.

### Table 10.5  Calculation — neutral line bend allowance

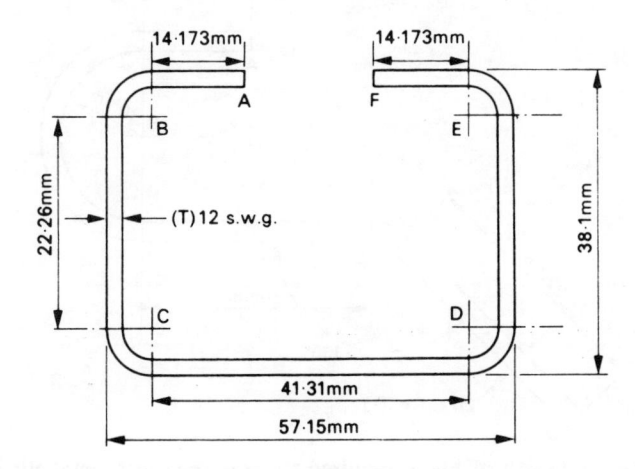

Calculate the length of the blank required to form the four-bend channel section shown in the diagram. Use the approximate value for the neutral line given in Table 10.3. The thickness of the sheet metal, $T = 2·64$ mm (12 swg) and the inside radius for the bends $r = 2T$.

**Solution:**
From the Table 10.3, the approximate value for the neutral line when bending 12 swg

is $\frac{2}{5}T$    $\frac{2}{5}T = 0·4T$

The inside bend radius, $r$, may be found as follows:

**(1)** Subtract inside dimensions from outside dimensions

(a) 57·15 − 41·31
 = 15·84 mm

(b) 38·1 − 22·26
 = 15·84 mm

**(2)** Subtract two thicknesses of metal from (1) and divide by two (2 bends)

15·84 − (2 × 2·64)
= 15·84 − 5·28
= 10·56 mm

Divide by 2
$\frac{10·56}{2}$
∴ $r = 5·28$ mm

It will be seen that $r = 2T$.

**Length of flats**

| | | |
|---|---|---|
| A B | = | 14·173 |
| B C | = | 22·260 |
| C D | = | 41·310 |
| D E | = | 22·260 |
| E F | = | 14·173 |

Total length of flats = 114·176 mm.

**Bend allowances for four quadrant corners**
The four quadrant corners, B + C + D + E represent a circle for which the bend allowance radius

$R = r + 0·4T$
= $2T + 0·4T$
= $2·4T$
= 2·4 × 2·64
∴ $R = 6·336$ mm

| No. | Log |
|---|---|
| 2·4 | 0·3802 |
| 2·64 | 0·4216 |
| 6336 | 0·8018 |

**Total bend allowance**
'Bend Allowance' = $2\pi R$
= 2 × 3·142 × 6·336
= 6·284 × 6·336
∴ Total bend allowance = 39·82 mm

| No. | Log |
|---|---|
| 6·284 | 0·7983 |
| 6·336 | 0·8018 |
| 3982 | 1·6001 |

**Length of blank**
= Total length of flats + total bend allowance
= 114·176 + 39·82

Length of blank required = 153·996 mm

## Table 10.6 Calculation – neutral line bend allowance

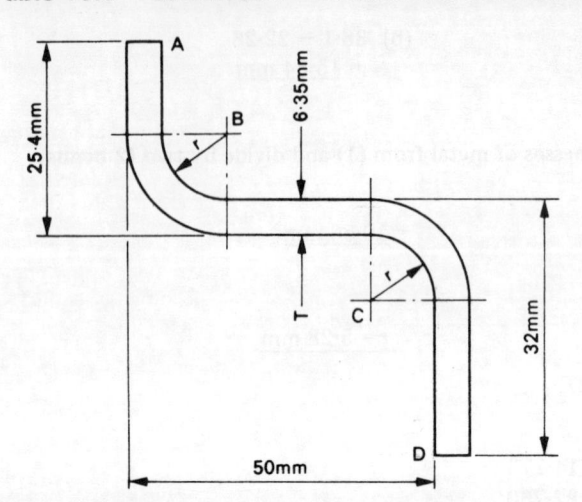

Calculate the length of blank required to form the support bracket shown opposite.
Use the approximate value for the neutral line from Table 10.3. The thickness of the steel plate, $T = 6.35$ mm and the inside bend radius $r = 2T$.

**Solution:**
From the Table 10.3, the approximate value for the neutral line when bending plate of 6.35 mm thickness is given as $0.5T$.

**Length of flats**

$$\begin{aligned}
A B &= 25.4 - (r + T) \\
&= 25.4 - (2T + T) \\
&= 25.4 - 3T \\
&= 25.4 - (3 \times 6.35) \\
&= 25.4 - 19.05 \quad = 6.35 \text{ mm}
\end{aligned}$$

Similarly $\quad CD = 32 - 3T$
$\quad\quad\quad\quad\quad = 32 - 19.05 \quad = 12.95$ mm

But $\quad BC = 50 - (r + T + r)$
$\quad\quad\quad\quad = 50 - (2r + T)$
$\quad\quad\quad\quad = 50 - (4T + T)$
$\quad\quad\quad\quad = 50 - 5T$
$\quad\quad\quad\quad = 50 - (5 \times 6.35)$
$\quad\quad\quad\quad = 50 - 31.75 \quad = \underline{18.25 \text{ mm}}$

∴ Total length of flats $\quad = 37.55$ mm

**The bend allowance radius**

$$\begin{aligned}
R &= r + 0.5T \\
&= 2T + 0.5T \\
&= 2.5T \\
&= 2.5 \times 6.35
\end{aligned}$$

∴ $R = \underline{15.88 \text{ mm}}$

| No. | Log |
|---|---|
| 2·5 | 0·3979 |
| 6·35 | 0·8028 |
| 1588 | 1·2007 |

**Bend allowance for bends B and C**
Total bend allowance for 2 bends
'Bend Allowance' $= 2 (\theta \times R \times 0.0175)$
(where $\theta = 180° - 90° = 90°$)
$\quad\quad\quad\quad = 2 \times 90 \times 15.88 \times 0.0175$
$\quad\quad\quad\quad = 180 \times 15.88 \times 0.0175$

Total bend allowance $= 50$ mm

**Length of blank**
$\quad =$ Total length of flats + total bend allowance
$\quad = 37.55 + 50$

Length of blank required $= \underline{87.55 \text{ mm}}$

| No. | Log |
|---|---|
| 180 | 2·2553 |
| 15·88 | 1·2007 |
| 0·0175 | 2·2430 |
| 5000 | 1·6990 |

## Table 10.7 Calculation – neutral line bend allowance

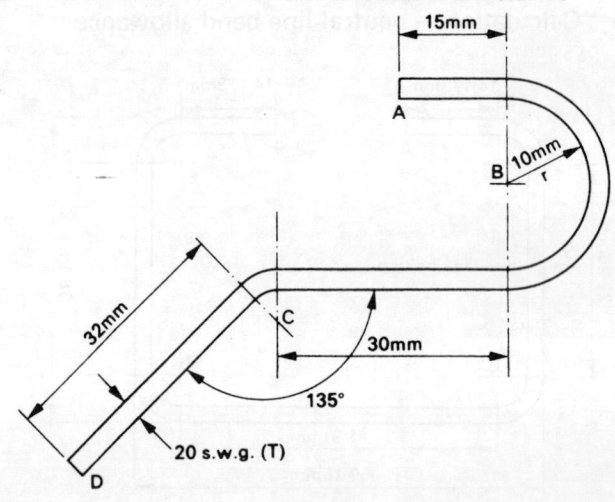

Calculate the length of blank required to form the sheet metal clip shown in the diagram. Use the approximate value for the neutral line from Table 10.3 for bending sheet metal whose thickness $T = 0.914$ (20 swg). The inside bend radius, $r = 2T$ for the bend at C.

**Solution:**
From the Table 10.3 the value of the neutral line for 20 swg $= \frac{1}{3}T$.

**Length of flats**

$$\begin{aligned}
A B &= 15 \text{ mm} \\
B C &= 30 \text{ mm} \\
C D &= \underline{32 \text{ mm}}
\end{aligned}$$

Total length of flats $= 77$ mm

374

The inside bend radius for bend B
$r = 10$ mm

∴ The bend allowance radius for bend B

$$R = r + \frac{1}{3}T$$
$$= 10 + \frac{0.914}{3}$$
$$= 10 + 0.305$$
$$\underline{R = 10.305 \text{ mm}}$$

**Bend allowance for bend B**

The bend at B is a semi-circle

| | | No. | Log |
|---|---|---|---|
| Bend allowance = | $\pi R$ | 3·142 | 0·4971 |
| | = 3·142 × 10·305 | 10·305 | 1·0132 |
| Bend allowance for B | = 32·38 mm | 3238 | 1·5103 |

The inside bend radius for bend C
$r = 2T$

Bend allowance radius for bend C

$$R = r + \frac{1}{3}T$$
$$= 2T + \frac{1}{3}T$$
$$= 2\frac{1}{3}T$$
$$= \frac{7}{3} \times 0.914$$
$$= 7 \times 0.305$$
$$\therefore R = 2.135 \text{ mm}$$

**Bend allowance for bend C**

Bend allowance = $\theta \times R \times 0.0175$
(where $\theta = 180° - 135° = 45°$)
= 45 × 2·135 × 0·0175

Bend allowance for C = 1·681 mm

| **Length of blank** | | No. | Log |
|---|---|---|---|
| = Total length of flats + total bend allowance | | 45 | 1·6532 |
| = 77 + 32·38 + 1·681 | | 2·135 | 0·3294 |
| | | 0·0175 | 2·2430 |
| **Length of blank required = 111·061 mm** | | 1681 | 0·2256 |

Tables 10.8, 10.9 and 10.10 show worked examples for 'Precision Sheet Metal Work' using the ratio $\dfrac{x}{T} = 0.4$

**Table 10.8(a)   Calculation — precision bend allowance**

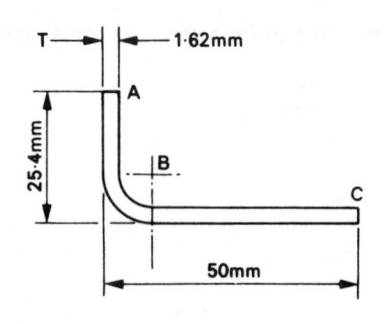

Calculate the length of blank required to make the simple sheet metal bracket shown opposite. The value for the neutral line = $0.4T$. The thickness of the sheet metal, $T = 1.62$ mm. The inside bend radius $r = 2T$.

**Solution:**
**Length of flats**

$$\begin{aligned}
AB &= 25.4 - (r + T) \\
&= 25.4 - (2T + T) \\
&= 25.4 - 3T \\
&= 25.4 - (3 \times 1.62) \\
&= 25.4 - 4.86 \qquad = 20.54 \text{ mm}
\end{aligned}$$

Similarly   $BC = 50 - 3T$
$= 50 - 4.86$   $\underline{= 45.14 \text{ mm}}$

Total length of flats   $= 65.68$ mm

The inside bend radius
$r = 2T$

∴ The bend allowance radius for bend B

| | | No. | Log |
|---|---|---|---|
| $R$ | = $2T + 0.4T$ | | |
| | = $2.4T$ | 2·4 | 0·3802 |
| | = 2·4 × 1·62 | 1·62 | 0·2095 |
| ∴ $\underline{R}$ | = 3·888 mm | 3888 | 0·5897 |

**Bend allowance for bend at B**

'Bend allowance' = $\theta \times R \times 0.0175$
(where $\theta = 180° - 90° = 90°$)
= 90 × 3·888 × 0·0175

| | No. | Log |
|---|---|---|
| | 90 | 1·9542 |
| Bend allowance for B = 6·122 mm | 3·888 | 0·5897 |
| | 0·0175 | 2·2430 |
| | 6122 | 0·7869 |

**Length of blank**
= Total length of flats + bend allowance
= 65·68 + 6·122

**Length of blank required = 71·802 mm**

375

### Table 10.8(b)  Calculation – precision bend allowance

Calculate the length of blank required to form the sheet metal detail shown in the diagram.

The value for the neutral line = $0.4T$.
The thickness of the metal, $T = 2.642$ mm
The inside bend radius, $r = 2T$.

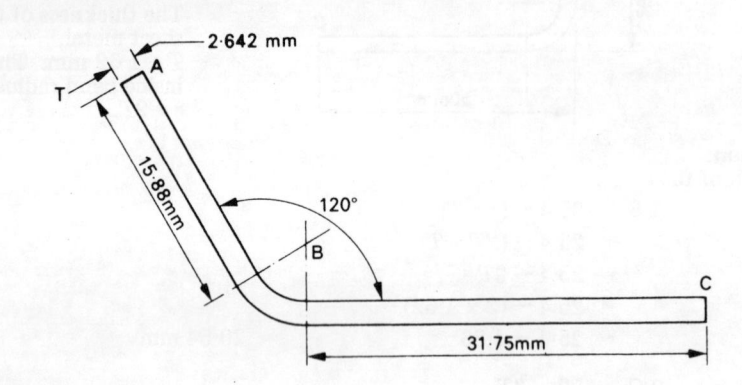

**Solution:**
**Length of flats**

$$AB = 15.88 \text{ mm}$$
$$BC = \underline{31.75 \text{ mm}}$$
Total length of flats = $\underline{47.63 \text{ mm}}$

The inside bend radius
$$r = 2T$$
∴ The bend allowance radius
$$R = 2T + 0.4T$$
$$= 2.4T$$
$$= 2.4 \times 2.642$$
∴ $\underline{R = 6.34 \text{ mm}}$

| | No. | Log |
|---|---|---|
| | 2.4 | 0.3802 |
| | 2.642 | 0.4219 |
| | 6340 | 0.8021 |

**Bend allowance for bend at B**
'Bend allowance' = $\theta \times R \times 0.0175$
(where $\theta = 180° - 120° = 60°$)
= $60 \times 6.34 \times 0.0175$
Bend allowance for B = $\underline{6.658 \text{ mm}}$

| | No. | Log |
|---|---|---|
| | 60 | 1.7782 |
| | 6.34 | 0.8021 |
| | 0.0175 | 2.2430 |
| | 6658 | 0.8233 |

**Length of blank**
= Total length of flats + bend allowance
= $47.63 + 6.658$
Length of blank required = $\underline{54.288 \text{ mm}}$

### Table 10.9  Calculation – precision bend allowance

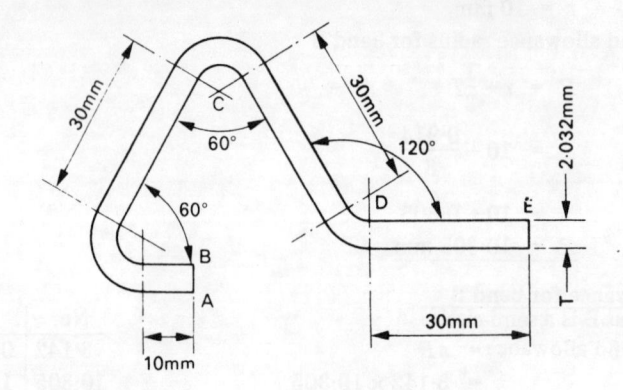

Calculate the length of blank required to form the sheet metal section shown in the diagram.

The value for the neutral line = $0.4T$.
The thickness of the metal, $T = 2.032$ mm
The inside bend radius, $r = T$

**Solution:**
**Length of flats**

$$AB = 10 \text{ mm}$$
$$BC = 30 \text{ mm}$$
$$CD = 30 \text{ mm}$$
$$DE = \underline{30 \text{ mm}}$$
Total length of flats = $\underline{100 \text{ mm}}$

The inside bend radius
$$r = T$$
∴ The bend allowance radius
$$R = T + 0.4T$$
$$= 1.4T$$
$$= 1.4 \times 2.032$$
∴ $\underline{R = 2.844 \text{ mm}}$

| | No. | Log |
|---|---|---|
| | 1.4 | 0.1461 |
| | 2.032 | 0.3079 |
| | 2844 | 0.4540 |

For bend B   $\theta = 180° - 60° = 120°$
Similarly for bend C   $\theta = 120°$
For bend D   $\theta = 180° - 120° = \underline{60°}$
Total sum of bending angles = $300°$

**Total bending allowance**
B.A. = $\theta \times R \times 0.0175$
(where $\theta = 300°$)
= $300 \times 2.844 \times 0.0175$
Total bend allowance = $\underline{14.93 \text{ mm}}$

| | No. | Log |
|---|---|---|
| | 300 | 2.4771 |
| | 2.844 | 0.4540 |
| | 0.0175 | 2.2430 |
| | 1493 | 1.1741 |

Length of blank
= Total length of flats + total bend allowance
= 100 + 14·93
Length of blank required = 114·93 mm

## Table 10.10  Calculation — precision bend allowance

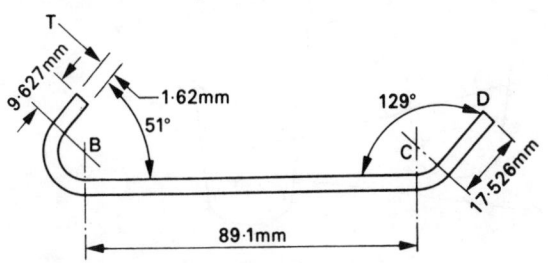

Calculate the length of the blank required to form the sheet metal section shown in the diagram.

The value for the neutral line = 0·4T
The thickness of the metal, T = 1·62 mm
The inside bend radius, r = 2T

**Solution:**
**Length of flats**

$$A B = 9·627 \text{ mm}$$
$$B C = 89·100 \text{ mm}$$
$$C D = 17·526 \text{ mm}$$

Total length of flats = 116·253 mm

The inside bend radius
$$r = 2T$$

∴ The bend allowance radius
$$R = 2T + 0·4T$$
$$= 2·4T$$
$$= 2·4 \times 1·62$$
∴ $R = 3·888$ mm

| No.  | Log    |
|------|--------|
| 2·4  | 0·3802 |
| 1·62 | 0·2095 |
| 3888 | 0·5897 |

For bend B  θ = 180°−51° = 129°
For bend C  θ = 180°−129° = 51°
Total sum of bending angles = 180°

**Total bending allowance**
B.A. = θ × R × 0·0175
(where θ = 180°)
= 180 × 3·888 × 0·0175
∴ Total bend allowance = 12·25 mm

| No.    | Log    |
|--------|--------|
| 180    | 2·2553 |
| 3·888  | 0·5897 |
| 0·0175 | 2·2430 |
| 1225   | 1·0880 |

Length of blank
= Total length of flats + total bend allowance
= 116·253 + 12·25
Length of blank required = 128·503 mm

## 10.29  Deep drawing and pressing

Deep drawing and pressing are among the most common methods of forming sheet metal.

Metals upon which drawing and pressing operations are to be carried out must be ductile.

The simplest of drawing operations is that of 'CUPPING'. In this operation sheet-metal blanks are formed into cup shapes by the application of a force.

The basic principle of cupping operations is very simple and the essential tools consist of a die, a blankholder or pressure plate and a punch. These are set up in a suitable power press.

The function of the pressure plate is to prevent 'wrinkling' of the metal.

It is possible, with some materials, to produce cups of diameter somewhat less than half that of the original blanks.

Figure 10.48 shows how an article may be formed by drawing.

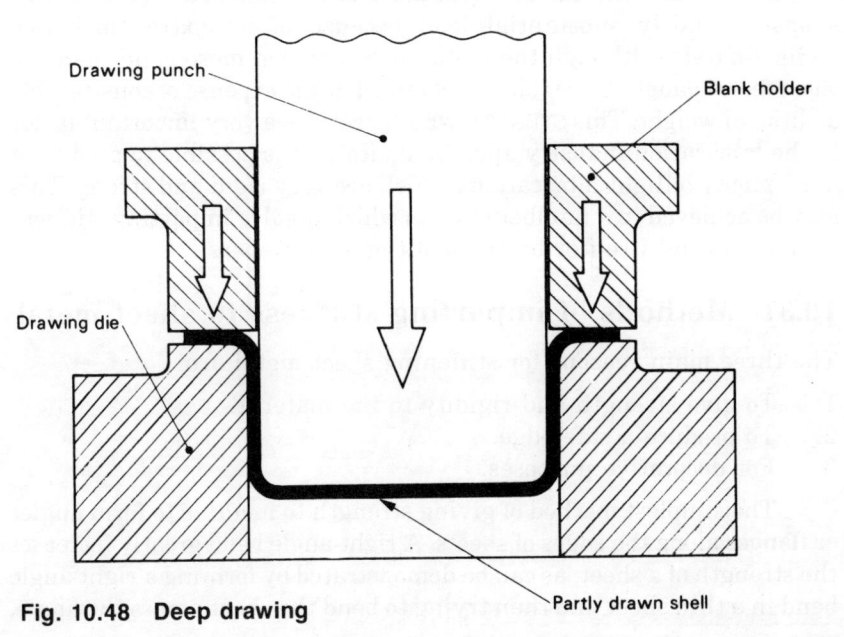

**Fig. 10.48  Deep drawing**

The stresses involved during the drawing operation are:

(a) COMPRESSIVE at the periphery, here the metal tends to thicken;
(b) TENSILE in the walls because of the natural resistance to the action of drawing.

The metal blank retains its original dimensions at the base whilst the rest of the blank is compressed into shape with parallel sides as the metal is drawn from the flange into the walls.

With metals which WORK-HARDEN, the drawing process is generally commenced with the material in the fully annealed (soft) condition.

Some of the most common materials which are suitable for deep drawing are: Mild steel; Austenitic stainless steel; Aluminium and aluminium alloys; Copper and copper alloys (particularly cartridge brass).

## 10.30 Stiffening of fabricated material (introduction)

The basic principle of stiffening may be illustrated by a popular party trick — that of supporting a tumbler of water on a piece of note paper bridging two other tumblers. This simple trick is explained in Figure 10.49.

A sheet-metal panel will not support a very great load due to the thinness of the material. A metal plate of the same surface area will support a fairly substantial load because of its extra thickness. Unfortunately, although the metal plate is much more rigid than the sheet-metal panel, this rigidity is obtained at the expense of considerable additional weight. This STENGTH/WEIGHT ratio is a very important factor in the fabrication industry and, fortunately, it is possible to produce a multiplicity of light fabrications which are very rigid and strong. This may be achieved in a number of ways which involve imparting stiffness to the material itself or by the addition of stiffeners.

## 10.31 Methods of imparting stiffness to sheet metal

The three main reasons for stiffening sheet metal are:

1. To give strength and rigidity to the material.
2. To produce a safe edge.
3. For decorative purposes.

The simplest method of giving strength to metal is to form angles or flanges along the edges of sheets. A right-angle bend greatly increases the strength of a sheet, as can be demonstrated by forming a right-angle bend in a thin sheet and then trying to bend the sheet across the angle.

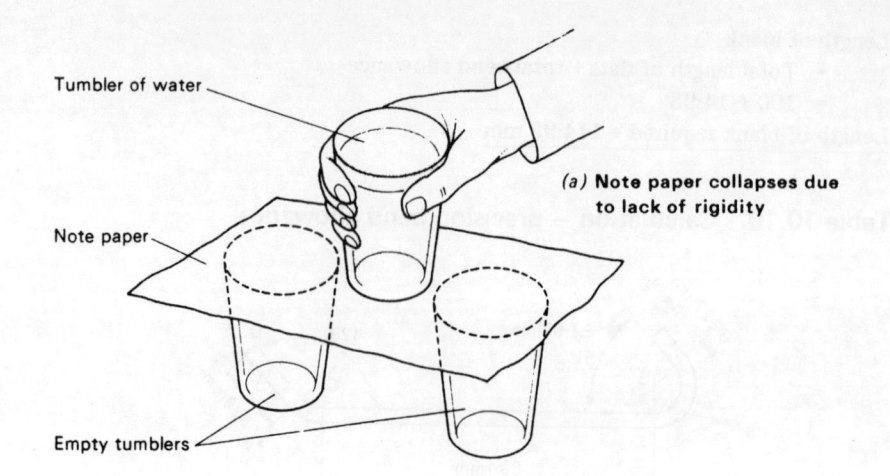

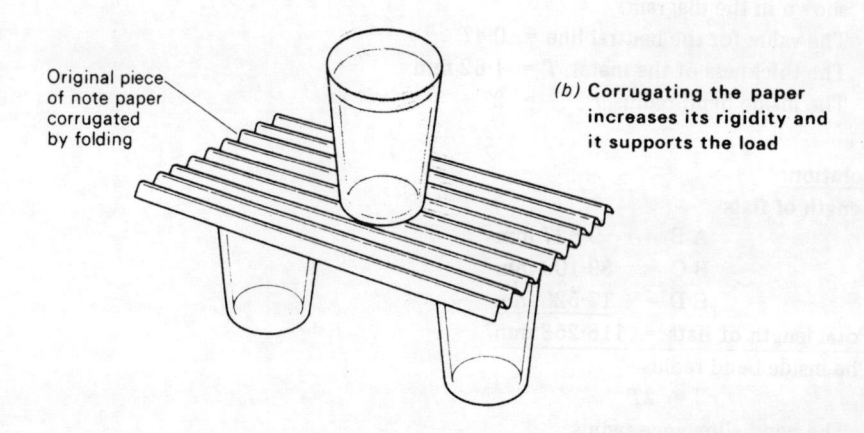

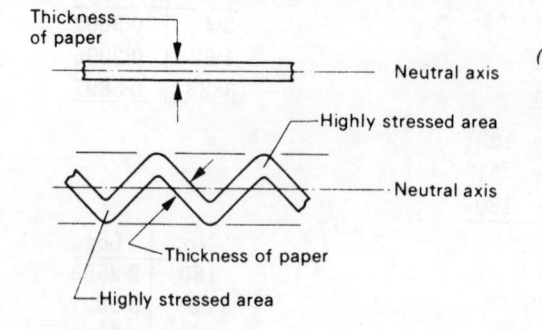

**Fig. 10.49 The basic principle of stiffening**

Curving the surface of sheet metal by rolling or pressing will increase its strength and rigidity. These methods are used to great advantage in the manufacture of car bodies. A curved surface is much stronger than a flat sheet. This is because the metal on the outer surface of the bend is under TENSION whilst the metal on the inner surface of the bend is in COMPRESSION, and these two stressed zones balance one another and will resist any force tending to change the shape or contour of the panel.

A surface with very little curve such as on some panels, bonnets or the top surface of roof panels on car bodies is springy, whilst surfaces with a lot of contour or shape, such as wings or the edges of roof panels, are very resistant to an applied force.

The raw edges of thin sheet metal can be extremely dangerous. Because of their sharpness they can cause nasty cuts when being handled. Components made out of thin sheet metal should have their raw edges made safe. One method of providing a safe edge is to make a return fold or hem. A 'double-returned edge' imparts greater rigidity to the metal than a single hem. Figure 10.50 shows various methods of stiffening metal components.

## 10.32 Wiring the edges of sheet metal

Wiring sheet metal not only makes the edge rigid and safe, but also provides a pleasing and decorative appearance.

*(a)* **Little rigidity**
*A flat sheet metal panel possesses little rigidity*

*(b)* **Strength and rigidity**
*This can be imparted to the panel by making right-angled folds along the two longest edges*

*(c)* **Greater strength and rigidity**
*Given to the panel by folding all four edges.
Greater strength has been given to the longest sides of a double fold*

Single fold    Double return fold

*(d)* **Dished ends**
*The tops and bottoms of copper domestic hot water cylinders are generally stiffened by means of 'dished-ends'*

*(e)* **Return fold or single hem**
*The raw edge of a sheet metal article may be stiffened, and at the same time made safe, by means of a returned fold or single hem*

*(f)* **Double hem**
*Greater stiffness may be achieved by folding a double hem*

Single hem    Double hem
Safe edges

*(g)* **Lightening holes**
*Lightening holes in sheet metal support brackets are stiffened by means of a flange*

*Usually the holes are punched and flanged in one operation, using a specially designed punch and die*

Cross-section of lightening hole

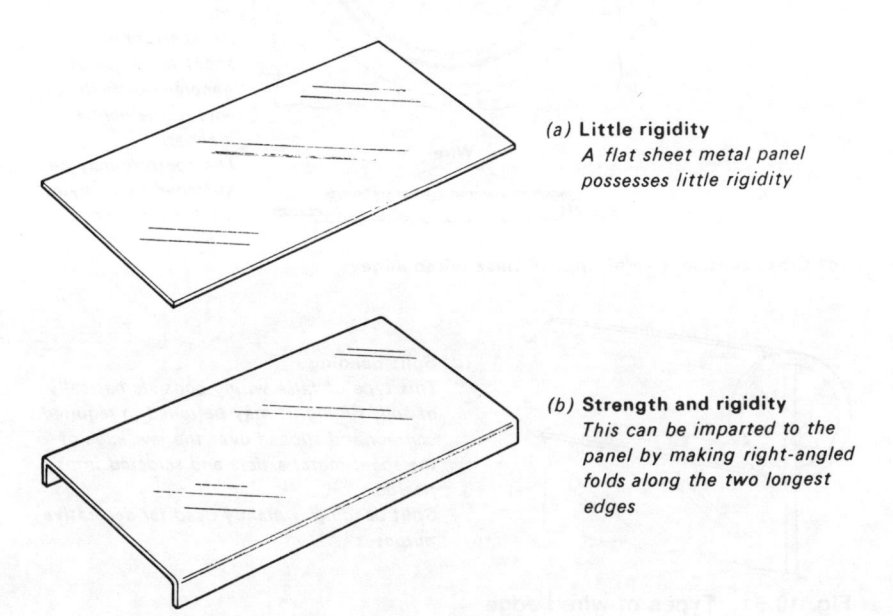

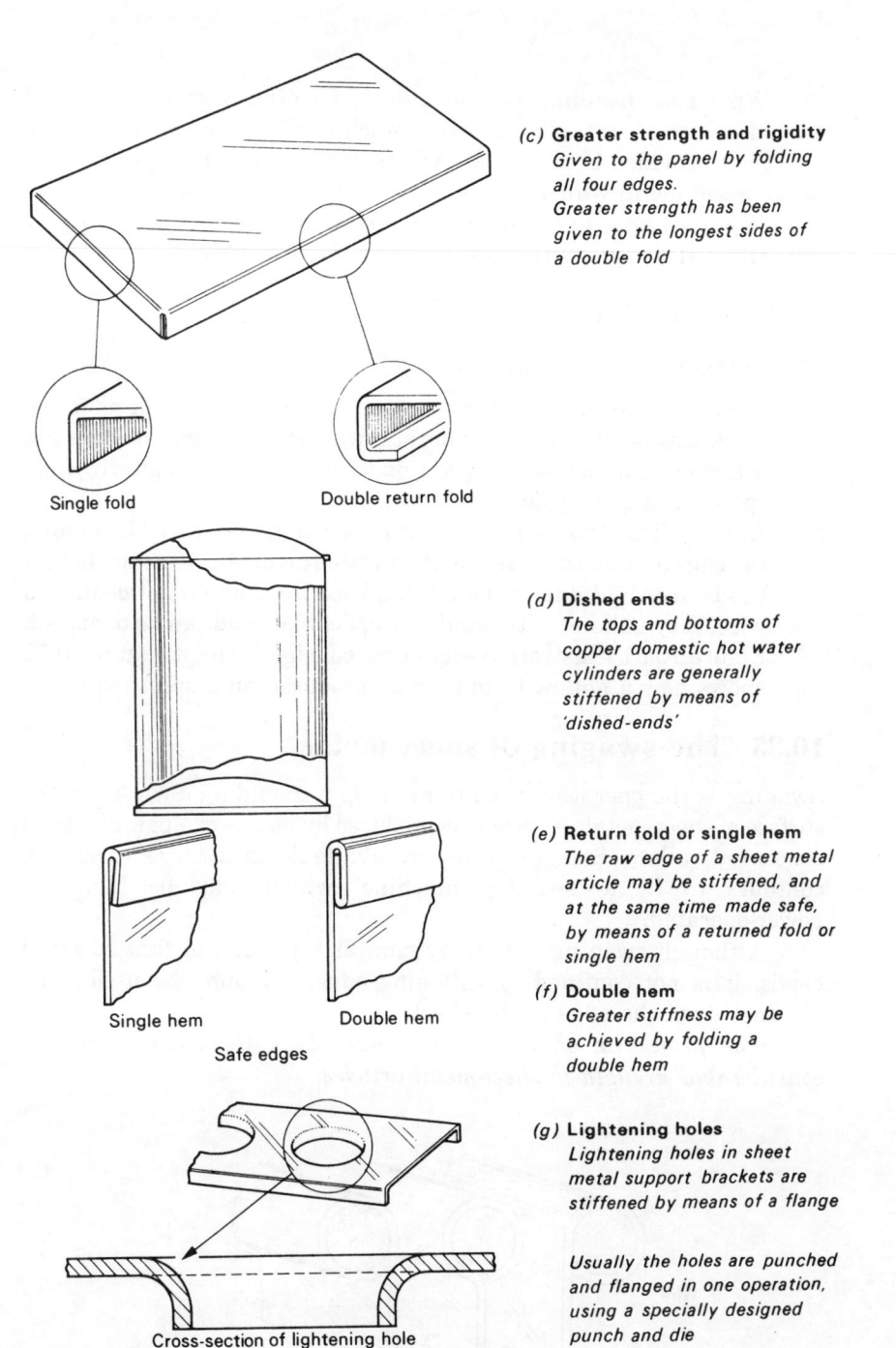

**Fig. 10.50 Imparting stiffness to sheet metal**

Wiring or 'beading' is a process of forming a sheet-metal fold around a wire of suitable diameter. Much of the strength of this type of edge is provided by the wire. Additional strength is obtained from the stressed metal which closely follows the exact contour of the wire.

*The allowance to be added to the sheet metal edge to be wired is* TWO AND A HALF TIMES THE DIAMETER *of the wire used.*

### FALSE WIRED EDGES

These may be one of two types:

1. *Applied* The APPLIED type is used when the position or metal thickness is unsuitable for normal wiring. These wired edges are attached and fastened in position by a returned flange, riveting, spot-welding or soldering.
2. *Hollow* This type of HOLLOW bead is usually produced by folding the edge around the wire and then withdrawing it. Although hollow beads are rigid due to their form, because they do not contain a wire, they will not withstand an impact blow and can be damaged. Figure 10.51 illustrates types of wired edges, whilst Figure 10.52 shows how a hollow bead may be produced on a spinning lathe.

## 10.33 The swaging of sheet metal

Swaging is the operation used to raise up a moulding (SWAGE) on the surface of sheet metal. A 'swage' is produced by means of a pair of special contoured rollers. Swaging rolls are available in a large variety of contours to fit a 'swaging machine' which may be hand- or power-operated.

Although swaging has many similar functions to that of wired beads, it is not confined to stiffening edges but may be used some distance from the edge of the sheet.

*The projecting shape of the swage above the surface imparts considerable strength to sheet-metal articles.*

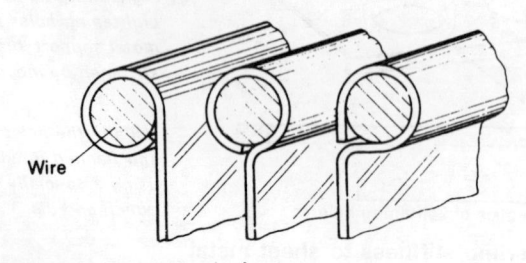

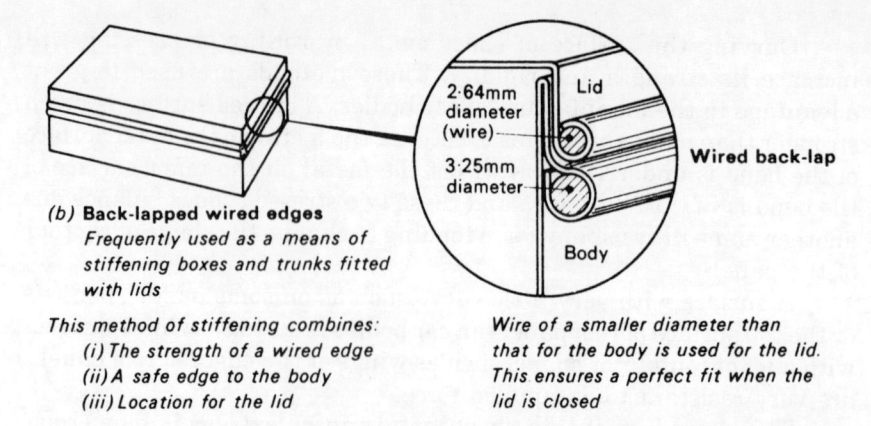

*(b) Back-lapped wired edges*
Frequently used as a means of stiffening boxes and trunks fitted with lids

This method of stiffening combines:
(i) The strength of a wired edge
(ii) A safe edge to the body
(iii) Location for the lid

Wire of a smaller diameter than that for the body is used for the lid. This ensures a perfect fit when the lid is closed

The edge of the metal is hidden on the inside in the case of the 'back-lapped wired edge'

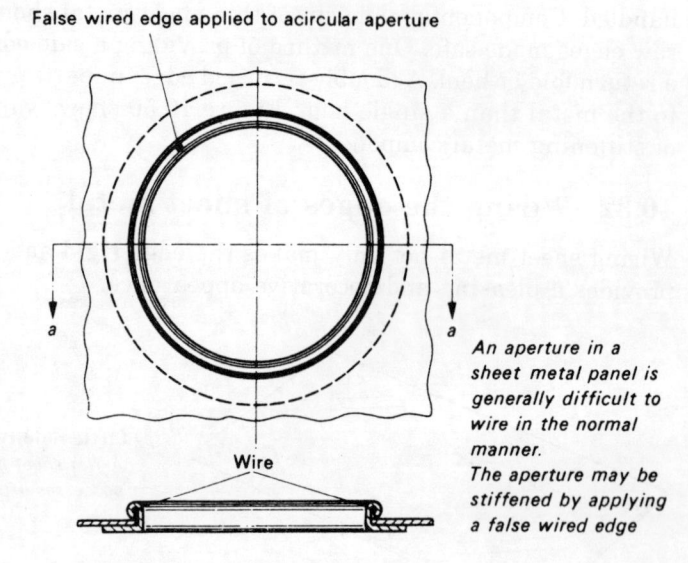

False wired edge applied to a circular aperture

An aperture in a sheet metal panel is generally difficult to wire in the normal manner.
The aperture may be stiffened by applying a false wired edge

*(c) Cross-section a–a of applied false wired edge*

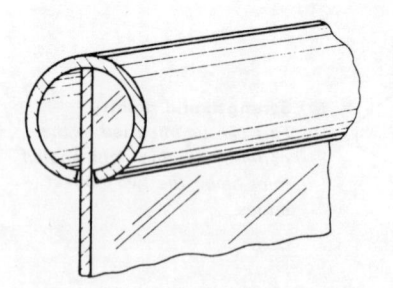

*(d) Split beading*
This type of false wiring consists basically of split tubing. It may be rolled to required contour and slipped over the raw edge of the sheet-metal article and soldered into position.
Split beading is mainly used for decorative purposes

*(a) Three common types of wired edges*

**Fig. 10.51** Types of wired edge

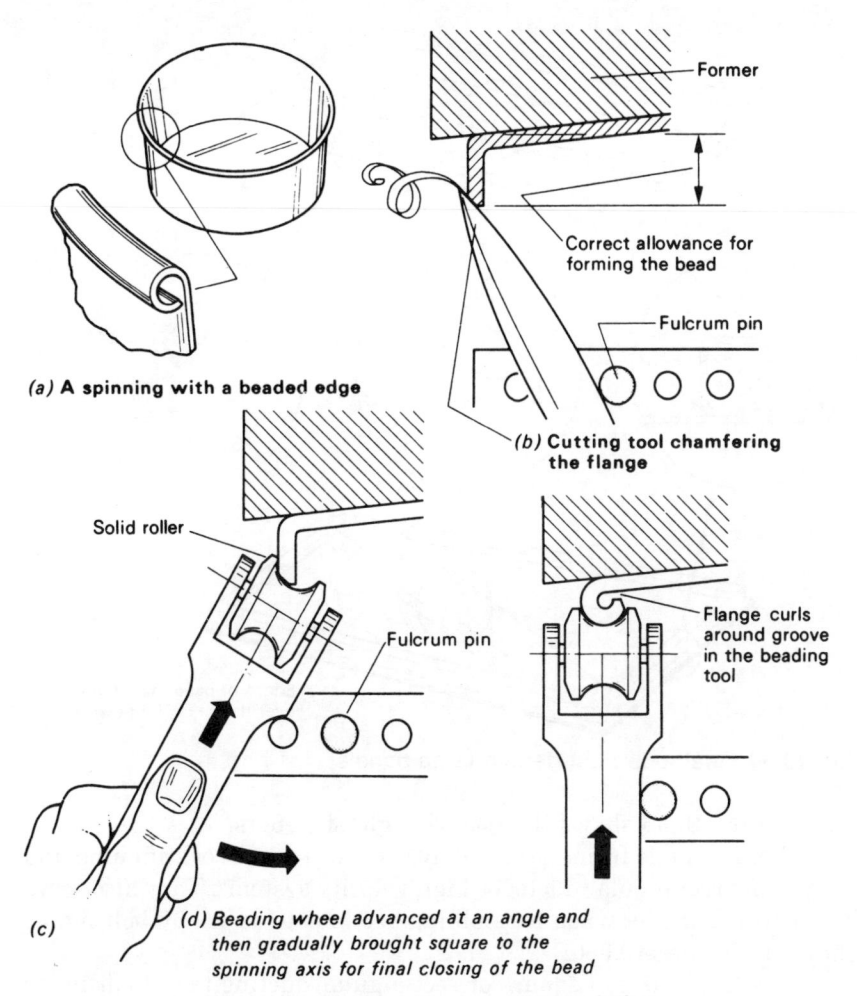

**Fig. 10.52 Forming a hollow bead on a spinning lathe**

The 'OGEE', or return curve swage, is frequently used to strengthen the centre portions of cylindrical containers or drums because of its high resistance to internally or externally applied forces in service.

Examples of the combination of strength with decoration associated with swaging are to be found in the design of circular sheet-metal articles such as drums, dustbins, waste-paper bins, buckets and water tanks.

The swage is a very important part of the decoration of a motor car body, and a great deal of attention is usually given to it. In motor body production the panels are formed to the correct contour on very large power presses and the swage is formed as an integral part of the panel. Figure 10.53 shows some aspects of swaging.

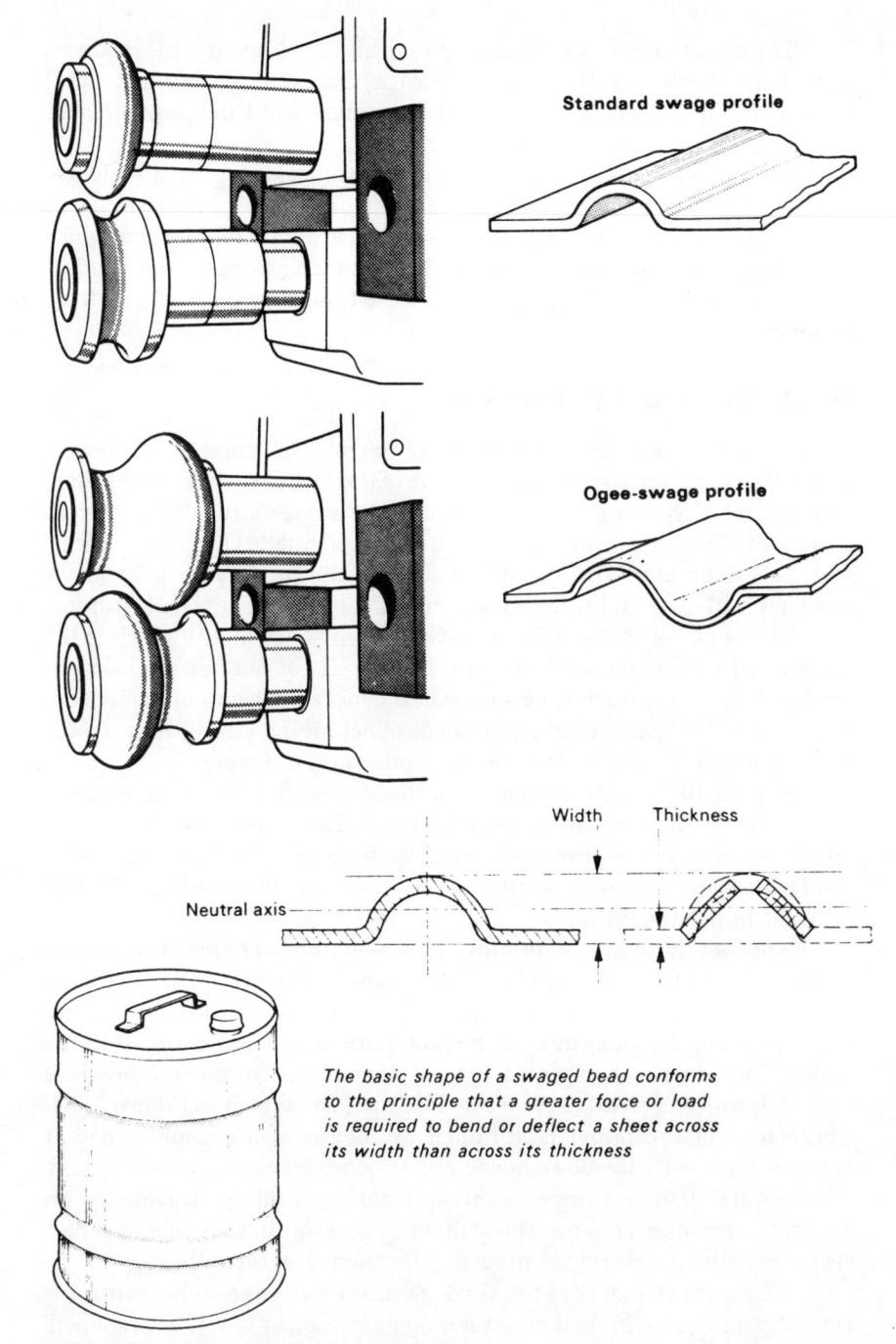

*The basic shape of a swaged bead conforms to the principle that a greater force or load is required to bend or deflect a sheet across its width than across its thickness*

**Fig. 10.53 The swaging of sheet metal**

381

The metal in the swaged bead is very highly stressed. This produces a much greater STRENGTH/THICKNESS ratio than that of the sheet metal with which it is formed. In general the maximum thickness of sheet metal which may be swaged is 1·62 mm.

A container or drum made of thin sheet metal will be of little use if when full of liquid or other substance it distorts and imposes extra load on its seams. Such drums are swaged around the circumference of the body as shown in Figure 10.53. This adds strength and rigidity to the walls of the drum enabling it to withstand severe rough handling in service.

## 10.34 The use of stiffeners

Large panels may be reinforced by means of applied stiffeners. Generally, panels are stiffened by virtue of the fact that they are fastened to some sort of framework. These frameworks are usually fabricated from sheet-metal sections which are strong and rigid due to their form. Sheet-metal sections may also be roll-formed for the purpose of internal or external stiffening of large components of cylindrical or circular shape.

The edges of fabrications constructed in sheet metal which is too thick to wire or hem can be stiffened by the use of flat bar or D-shaped section. It may be attached by spot-welding, brazing, welding or riveting.

One of the most common methods of achieving strength by means of attached stiffeners is the use of 'angle section frames'.

Figure 10.54 shows various methods of stiffening large panels.

A large sheet-metal panel may be stiffened as shown in Figure 10.54. All four edges are made rigid by folding. 'TOP-HAT SECTION' is used to stiffen the centre section of the panel and is usually secured in position by spot-welding.

Another method of stiffening large sheet-metal panels is to attach them to a rigid framework. The welded frame is fabricated from lengths of 'P-section' which has a very high STRENGTH/WEIGHT ratio for a sheet-metal section. All four edges of the panel are folded at 90° to a suitable width. The panel is then placed in position over the frame and the edges 'paned-down' over the flange on the 'P-section', as shown in Figure 10.54. The centre of the panel is stiffened by means of a diagonal top-hat section. Figure 10.55 shows how circular components may be stiffened.

Figure 10.55(a) illustrates an application of internal stiffening on a panel of circular shapes. The stiffening section, in this case 'top-hat' sections rolled to correct contour, are attached externally.

When sheet metal is too thick to allow the edge to be wired the raw edge may be stiffened by attaching either a flat bar or a 'D-shaped' bar as shown in Figure 10.55(c).

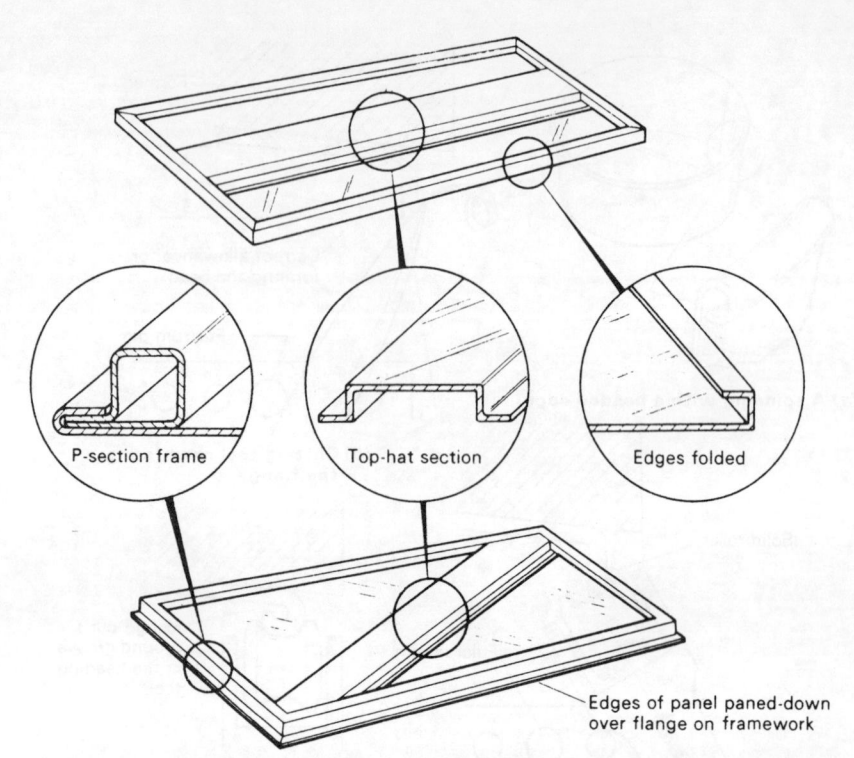

**Fig. 10.54 Methods of stiffening large panels**

Figure 10.56 shows the use of angle stiffeners.

Welded angle frames are widely used as a means of stiffening and supporting rectangular ducts for high-velocity systems. They also serve as a jointing media when assembling sections together by bolting, as shown in Figure 10.56(a).

The large sizes of square or rectangular ducting tend to drum as the air pressure passing through them varies. To overcome this drumming it is necessary to provide adequate stiffening to the walls of the duct. This may be achieved by use of swaging, but often a 'diamond-break' is used, as shown in Figure 10.56(b).

Simple angle frames of welded construction may be used as a means of supporting and stiffening the open ends of tanks or bins fabricated from sheet metal. Two methods of attaching the angle frame are shown in Figure 10.56(c).

## 10.35 Fabricated structural members

The material most commonly used in constructional steelwork is Mild steel to BS 4360, which is produced by steel rolling mills in a variety

(a) Internal stiffening

(b) External stiffening

(c) Use of flat bar    Use of D-shaped bar

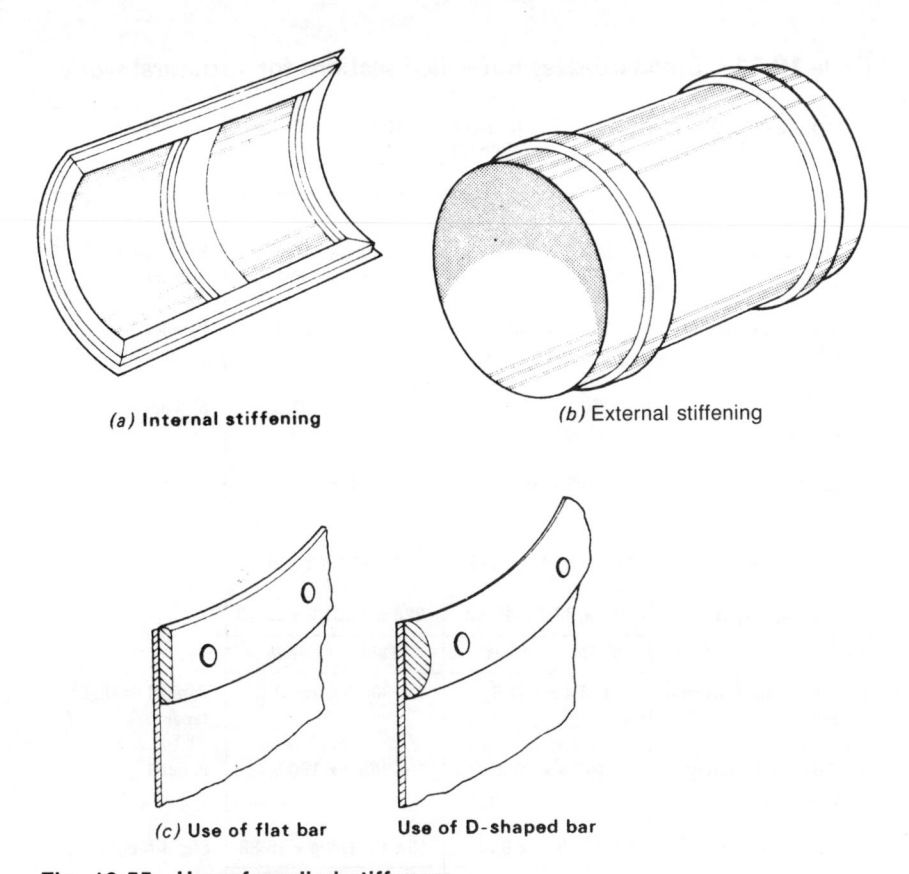

**Fig. 10.55** Use of applied stiffeners

of shapes and sizes, some of which are illustrated in Figure 10.57 and set out in Table 10.11.

For efficiency a structural member should use as little material as possible to carry as much load as possible.

The I-section, extensively used in steelwork, is designed to give an efficient STRENGTH/WEIGHT ratio, and the 'universal sections' provide even better efficiency (see Section 4.32).

## 10.36 Typical structural steel connections and assemblies

In constructional steel work the connections of the members are made by one of the following four methods:

1. Use of BLACK BOLTS,
2. Use of HIGH-STRENGTH FRICTION-GRIP BOLTS,

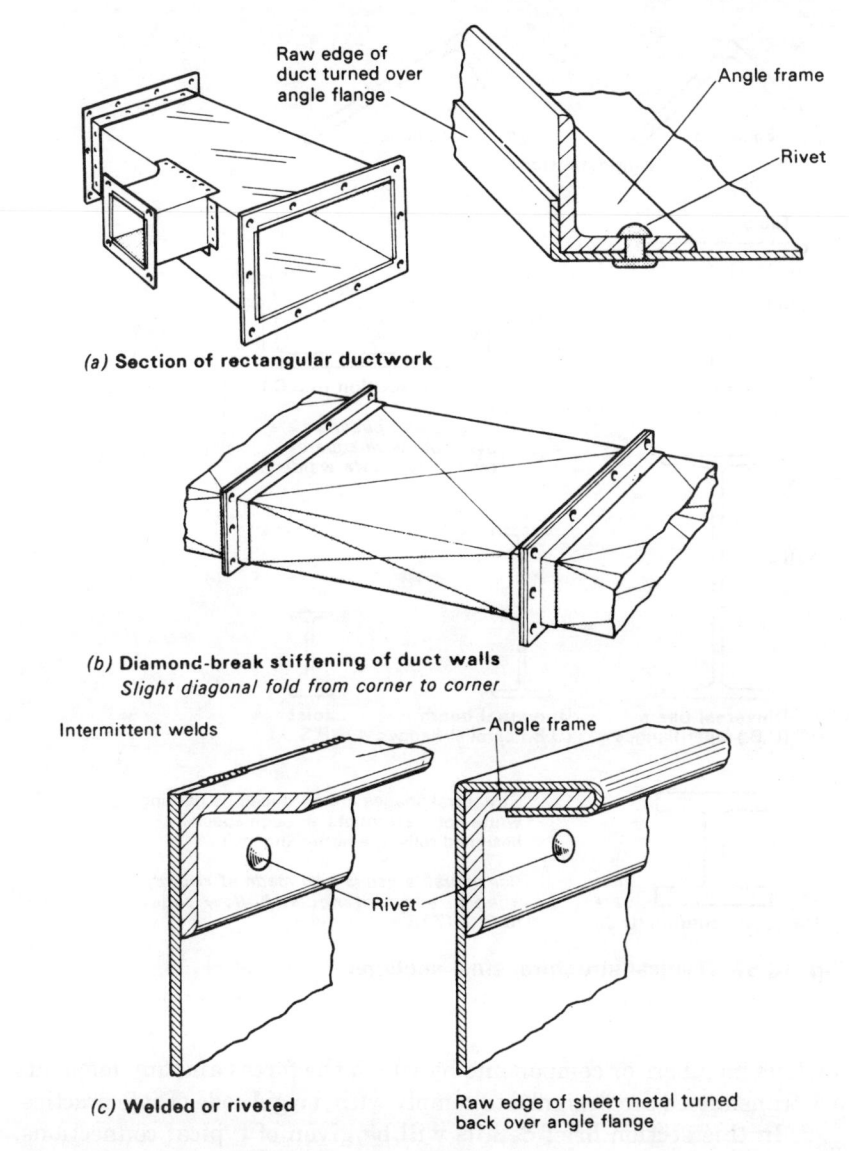

(a) Section of rectangular ductwork

(b) Diamond-break stiffening of duct walls
*Slight diagonal fold from corner to corner*

(c) Welded or riveted    Raw edge of sheet metal turned back over angle flange

**Fig. 10.56** The use of angle stiffeners

3. Riveting,
4. Welding.

Basically a steel structure is so designed that the sizes of the various components are determined by their ability to withstand the effects of the applied loads in service. The connections between the

383

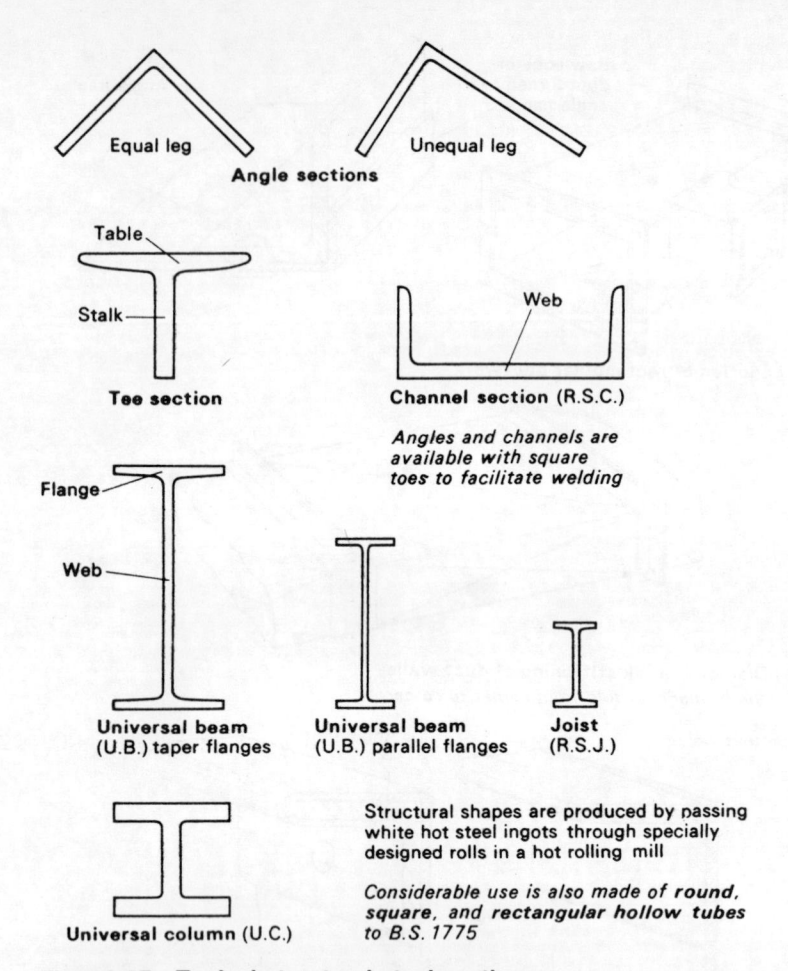

**Fig. 10.57 Typical structural steel sections**

*Structural shapes are produced by passing white hot steel ingots through specially designed rolls in a hot rolling mill*

*Angles and channels are available with square toes to facilitate welding*

*Considerable use is also made of round, square, and rectangular hollow tubes to B.S. 1775*

**Table 10.11  Standard sizes hot-rolled sections for structural work**

| ROLLED STEEL SECTION | RANGE OF SIZES (Dimensions in mm) From: | RANGE OF SIZES (Dimensions in mm) To: | FLANGES |
|---|---|---|---|
| | Web    Flange | Web    Flange | |
| Universal beams (U.B.) | 609·6 x 304·8<br>203·2 x 133·4 | 920·5 x 420·5<br>617   x 230·1 | Slight taper (2° 52')<br>Parallel |
| Universal columns (U.C.) | 152·4 x 152·4 | 474·7 x 424·1 | Parallel |
| Joists (R.S.J.) | 76·2 x 50·8 | 203·2 x 101·6 | 5° Taper |
| Channels (R.S.C.) | 76·2 x 38·1 | 431·8 x 101·6 | 5° Taper |
| Equal angles | 25·4 x 25·4 x 3·175 | 203·2 x 203·2 x 25·4 | |
| Unequal angles | 50·8 x 38·1 x 4·762 | 228·6 x 101·6 x 22·13 | |
| | Table    Stalk | Table    Stalk | |
| Tees (from Universal beams) | 133·4 x 101·6 | 305·5 x 459·2 | Parallel or slight taper |
| Tees (from Universal columns) | 152·4 x 76·2 | 395·0 x 190·5 | Parallel |
| Tees (short stalk) | 38·1 x 38·1 x 6·35 | 152·4 x 152·4 x 15·88 | Slight taper (½°) |
| | Table    Stalk | Table    Stalk | |
| Tees (long stalk) | 25·4 x 76·2 | 127 x 254 | 8° Taper |

All Universal beams and Universal columns are specified by their serial size and mass per metre

---

various members or components by which the forces and any moments are transmitted are designed to comply with a rigid code of safe practice.

In this section brief details will be given of typical connections, all of which combine strength, appearance and safety.

## STANCHION BASES

There are two types of stanchion base:

1. Bases for stanchions transmitting DIRECT LOAD only to the foundations,
2. Bases required where the stanchion has to transmit to the foundations a considerable bending moment in addition to the vertical loading.

These are illustrated in Figure 10.58.

Generally the top face only of a slab base is machined to give close contact with the stanchion.

A slab under 50 mm thickness does not normally require machining. The foot of the stanchion must be machined square on an 'end machine'. This ensures that the load is carried direct on to the base plate or slab and into the foundation (footing).

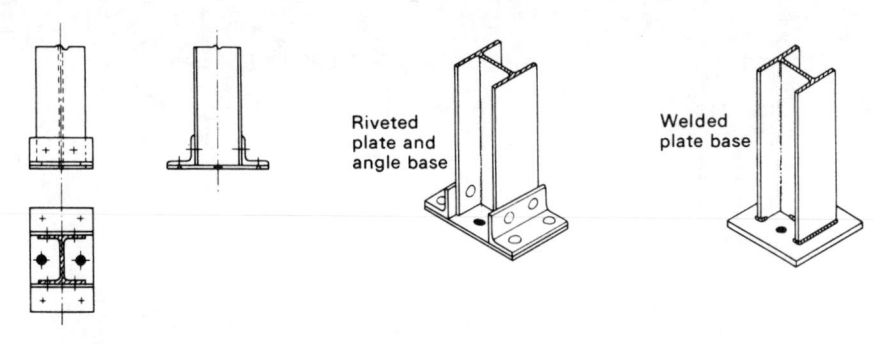

(a) Bases for small stanchions

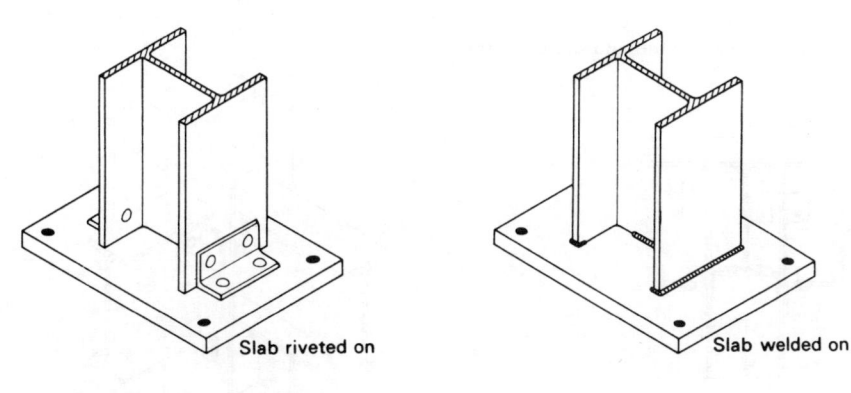

(b) Stanchions with slab bases

If the foot of the stanchion were not square to the base plate, the total load on the stanchion would be carried through the bolt shanks or rivets, in the case of bolted or riveted assemblies.

## STANCHION SPLICES

Splices in stanchions or columns should be arranged at a position above the adjacent floor level so that the joint, including any splice plates are well clear of any beam connection. Splices should never be made on a connection otherwise the bolts making the joint will be subjected to double loading. Figure 10.59 illustrates stanchion splices.

## CONNECTIONS OF BEAMS TO STANCHIONS

Figure 10.60 illustrates simple beam to stanchion connections. It will be noticed that all the load is transmitted from the beam to the stanchion through the seating cleat and its fixings.

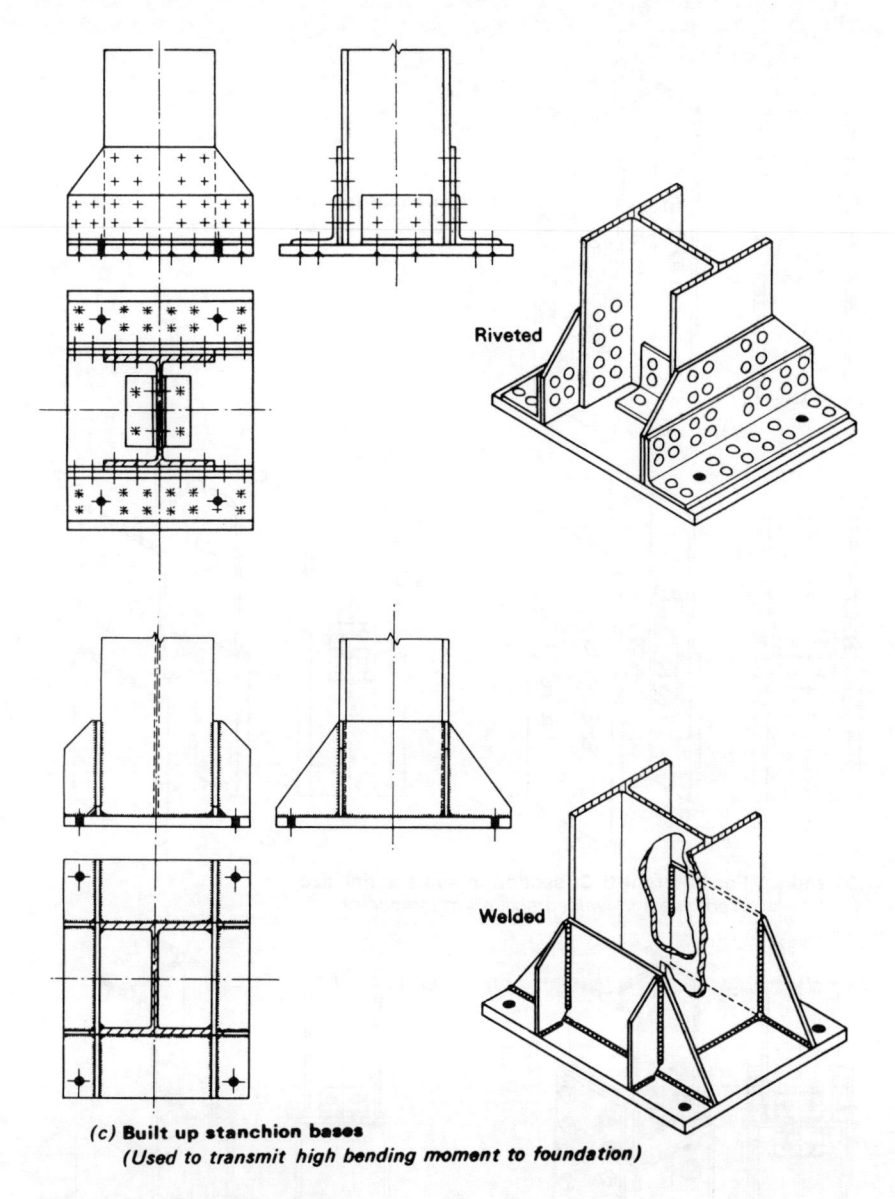

(c) Built up stanchion bases
(Used to transmit high bending moment to foundation)

**Fig. 10.58 Stanchion bases**

## BEAM TO BEAM CONNECTIONS

Where beam to beam connections are made, web cleats or an end plate are used to transmit loads from the secondary to the main beams.

Figure 10.61 illustrates this type of connection.

385

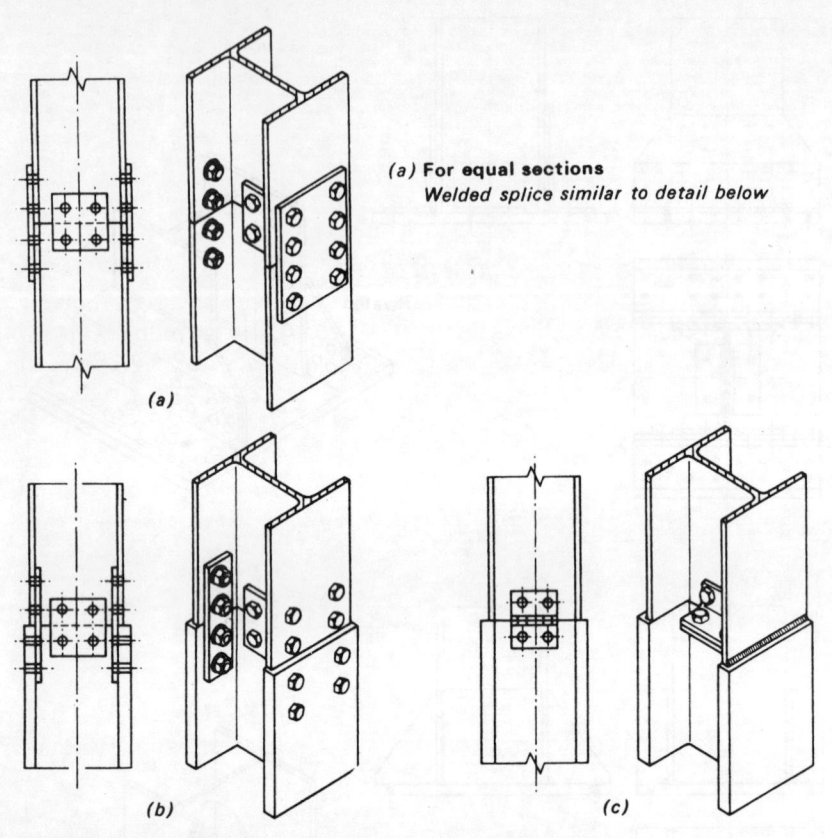

(a) For equal sections
Welded splice similar to detail below

(a)

(b) and (c) For different U.C. section in same serial size
Web cleats in welded splice are temporary

(d) and (e) For different sections
Web cleats in welded splice are temporary

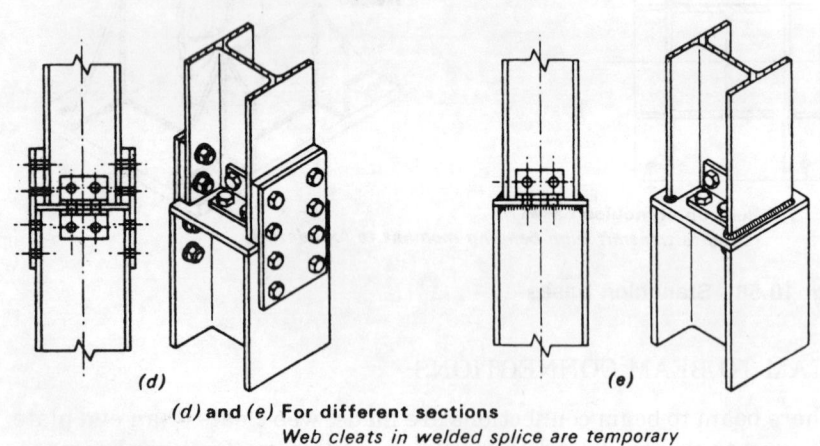

Fig. 10.59  Stanchion splices

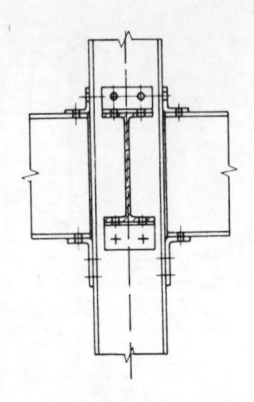

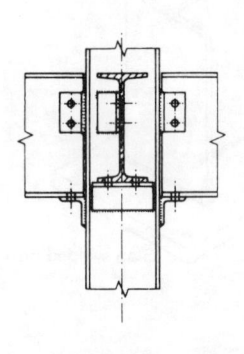

(a) Simple riveted and bolted detail

(b) Equivalent welded detail

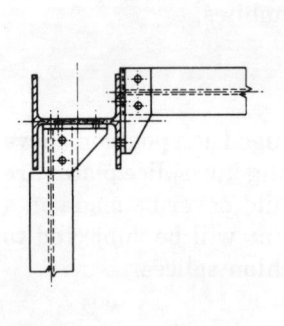

(c) Simple connections of beams to corner stanchion

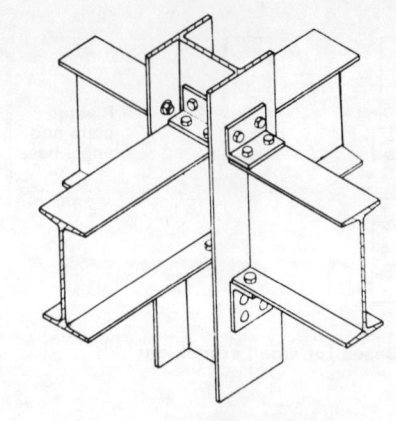

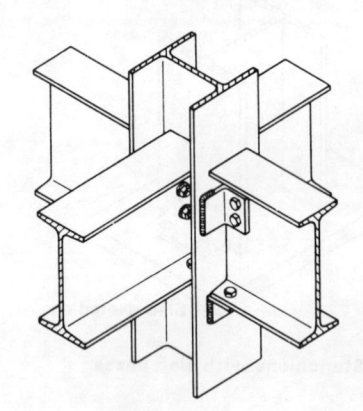

Fig. 10.60  Connection of beams to stanchions

386

## TRUSSES AND OTHER LATTICE FRAMES

Roof trusses are plane frames consisting of sloping rafters which meet at the ridge. A main tie connects the feet of the rafters, and the internal bracing members. They are used to support the roof covering in conjunction with purlins. Purlins are secondary members laid longitudinally across the rafters to which the roof covering is attached.

Figure 10.62 gives details of both riveted and welded roof trusses.

Lattice girders, also sometimes called 'trusses', are plane frames of open-web construction. They usually have parallel chords or booms which are connected with internal web bracing members. There are two basic types of lattice girder, the 'N' type and the 'Warren' type as shown in Figure 10.63.

As with roof trusses, the framing of a lattice girder should be triangulated, taking into account the span and the spacings of the applied loads. The booms are divided into panels of equal length and, as far as possible, the panel points are arranged to coincide with the applied loads. Lattice girders may be used in flat roof construction or in conjunction with trusses or trussed rafters.

Because of the greater forces usually borne by members of lattice girders, shop connections are generally riveted or welded. Site connections are normally made with HIGH-STRENGTH FRICTION-GRIP BOLTS (which have largely replaced rivets for site work) or by welding. Figure 10.64 shows how lattice girders are fabricated from standard sections.

## 10.37 Web stiffeners

Web stiffeners are required when a beam or plated structure is subjected to a twisting force (TORSION) or a sideways thrust.

The need for web stiffeners, or gussets, increases as the depth of the beam increases, as shown in Figure 10.65.

Web stiffeners may be welded or riveted. When fabricating stiffeners which are to be welded in position, it is important that the stiffener is an exact fit on the beam. The slope of the tapered flanges should be copied faithfully. With triangular-shaped gussets 'feather edges' must be avoided, the sharp corners should be cut off, for otherwise the strength of the assembly may be reduced rather than increased. Another important reason for cutting off the corners of webs or gussets is to provide ample clearance from fillet welds or bending radii. This allows for ease of assembly and permits welding through the gap with the web, or gusset, in position. It also avoids costly fitting operations at the corners.

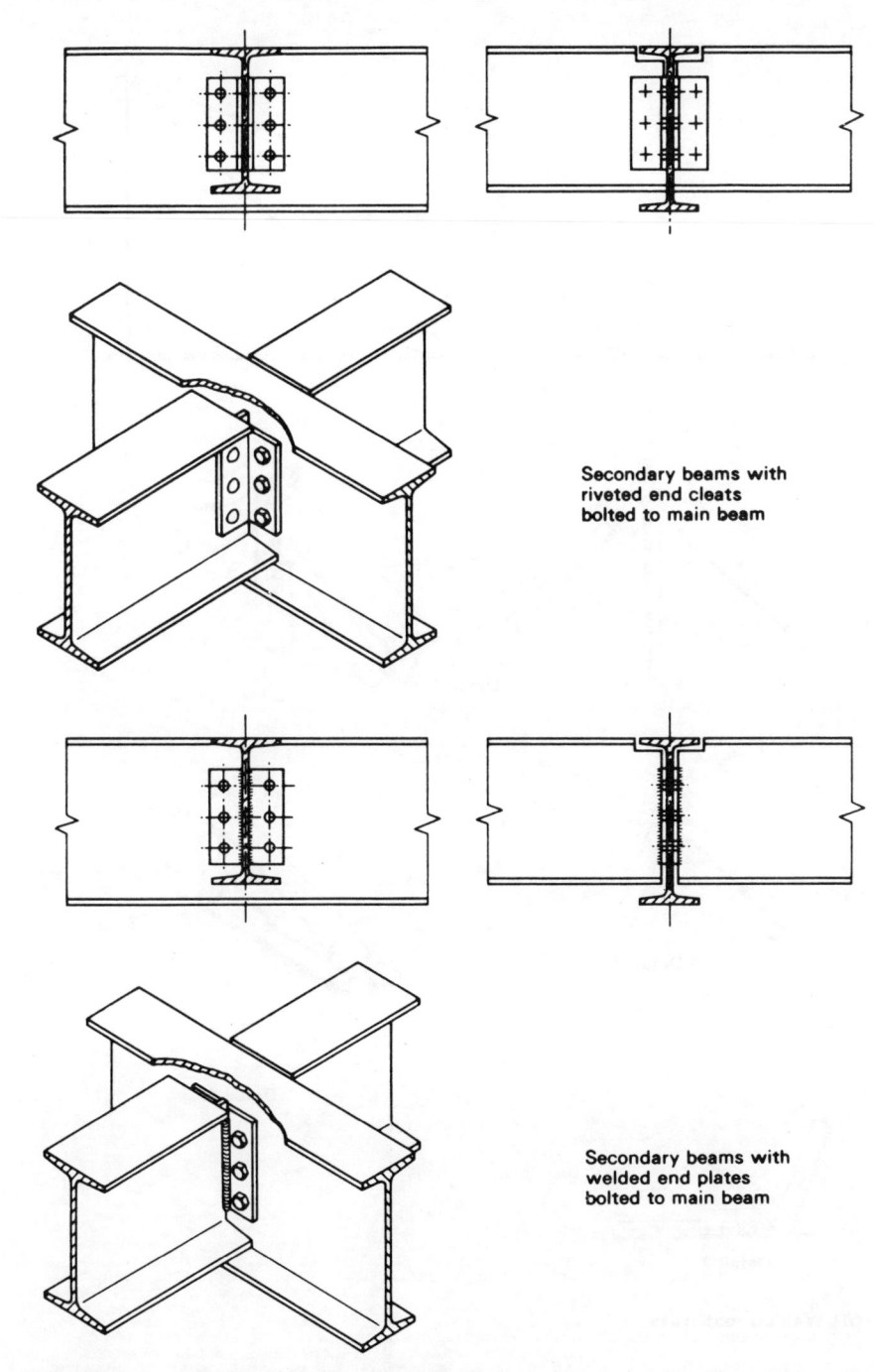

**Fig. 10.61 Beam to beam connections**

Secondary beams with riveted end cleats bolted to main beam

Secondary beams with welded end plates bolted to main beam

*(a)* **Riveted roof truss**

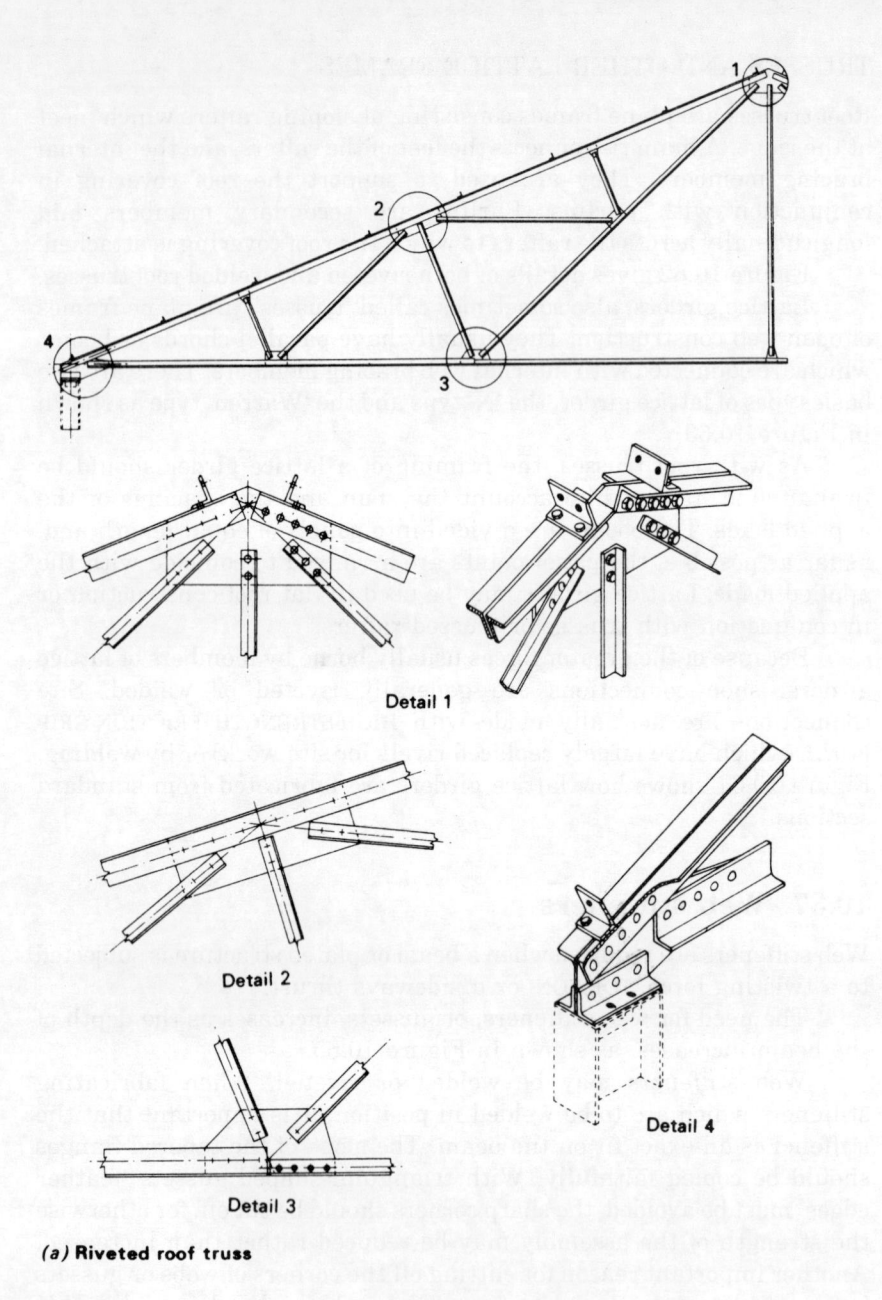

*The gusset plate are designed to withstand the forces applied by the members which they connect together*

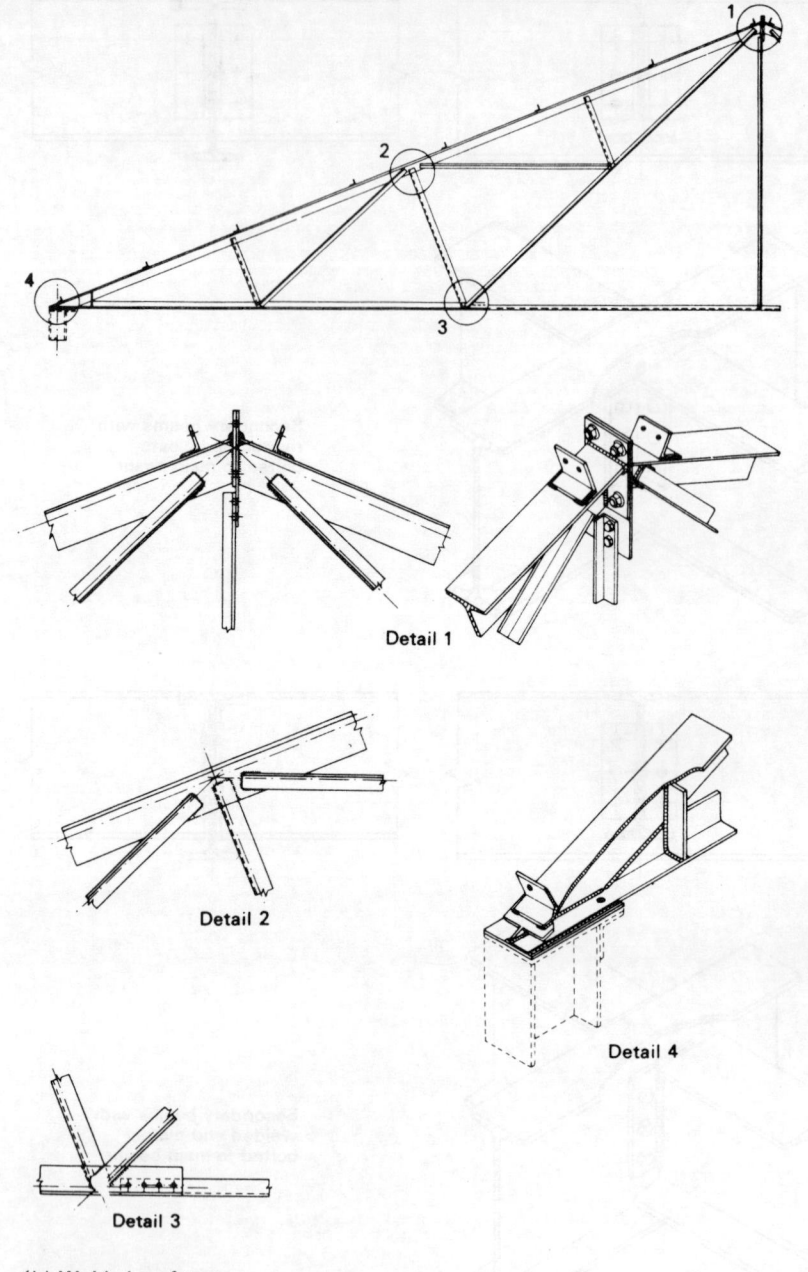

*(b)* **Welded roof truss**

Fig. 10.62 Roof trusses

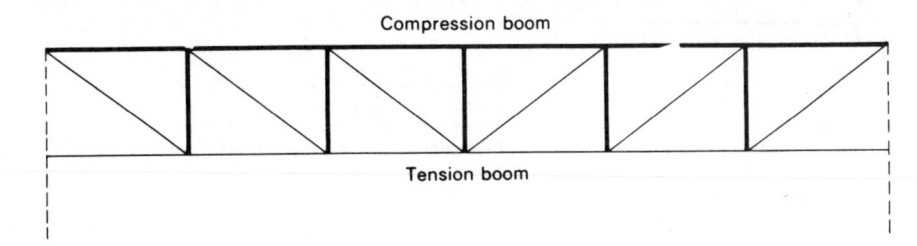

#### The 'N-type' lattice girder

The diagonal bracing members are arranged so that they act as ties. If reversed they would become struts and the shorter vertical members would be ties

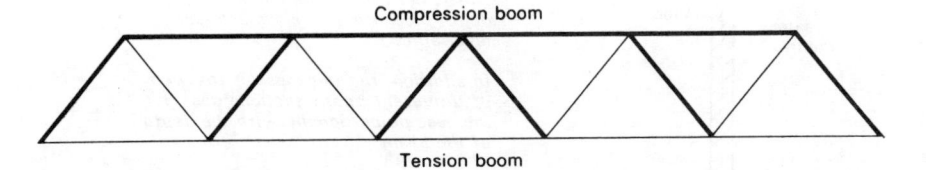

#### The 'Warren-type' lattice girder

*Note:* The thick lines in the diagrams represent 'struts'

**Fig. 10.63 Lattice girders**

Detail 1

Detail 2

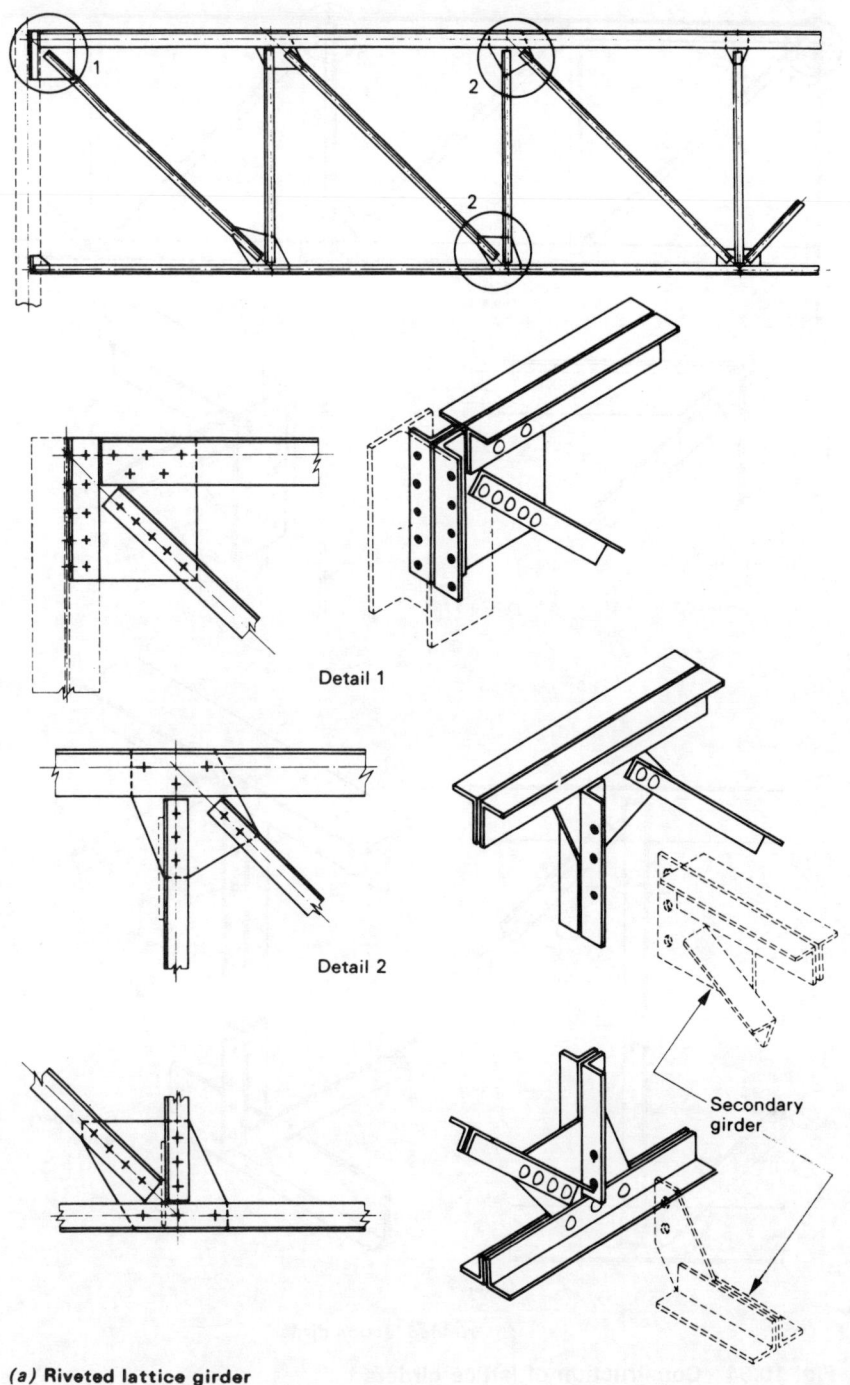

*(a)* Riveted lattice girder

Various applications of web stiffeners and gussets are illustrated in Figures 10.66 and 10.67.

## 10.38 Fabricated lightweight beams

*Bar frames or open web beams* are of ultra-light construction and are particularly suitable for use as long-span roof purlins. The flanges are usually formed of angle sections and the web diagonals of round or square sections, bent to shape and welded to the flanges, as shown in Figure 10.68.

Detail 1

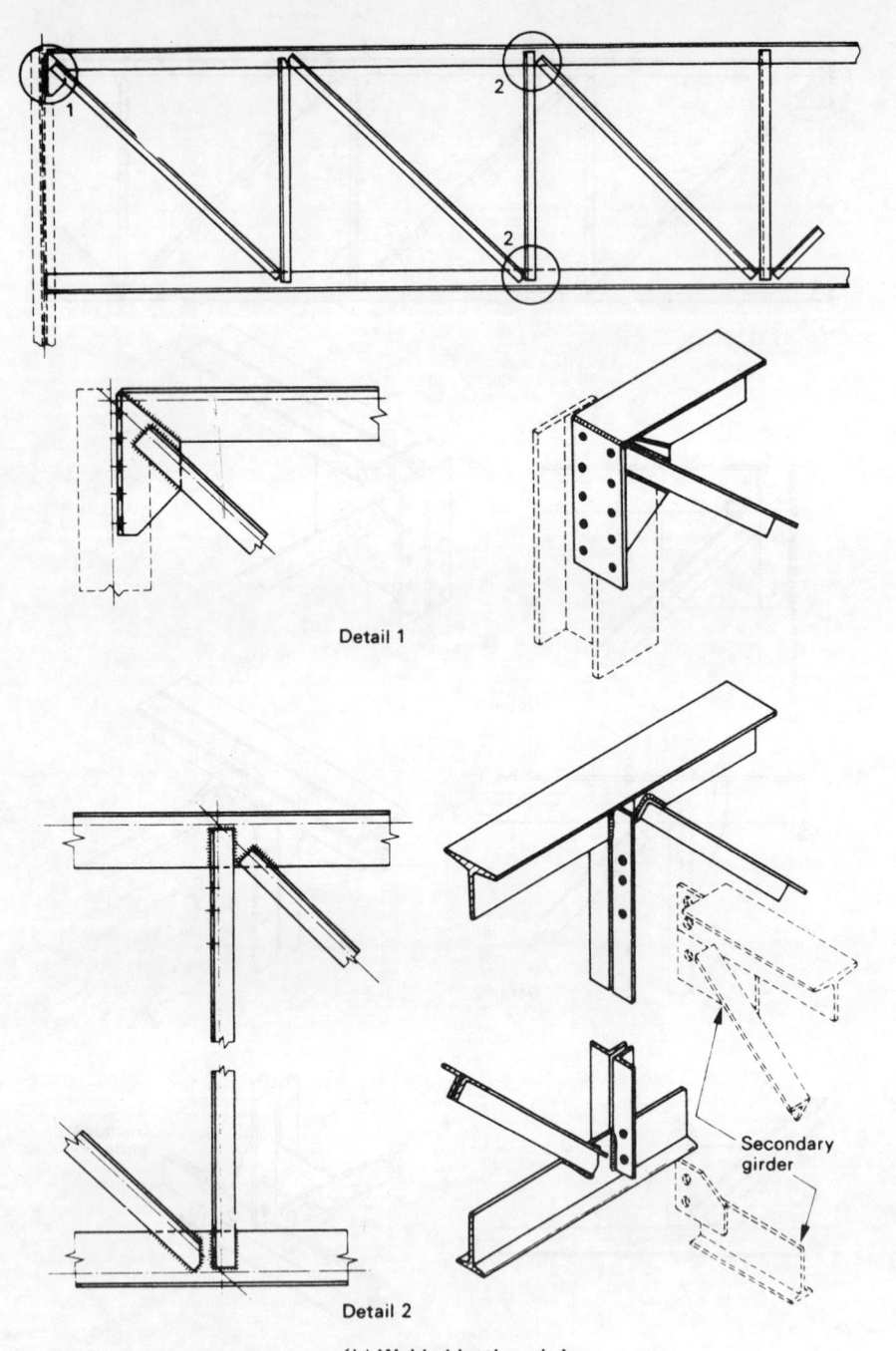

Detail 2

*(b)* **Welded lattice girder**

**Fig. 10.64  Construction of lattice girders**

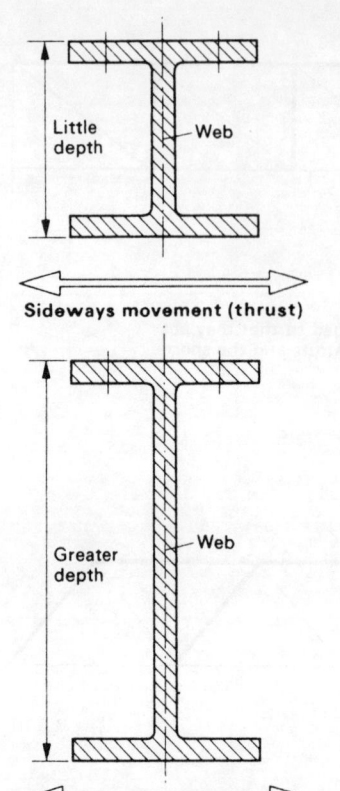

*The smaller sizes of **universal beam** sections will resist a twisting force within its web which has relatively little depth, presenting a short lever arm*

*The larger sizes of **universal beam** sections are unable to resist a twisting force within its web as effectively as the smaller sizes*

*This is because of the much greater depth presenting a very much longer lever arm*

*In addition, the thickness of the web of **universal beam** sections does not increase proportionally with the depth of the beam*

**Fig. 10.65  The need for web stiffeners**

## CASTELLATED BEAMS

The castellation system of construction is a very simple, yet ingenious, method of increasing the strength of a section without increasing its weight and without waste. Figure 10.69 explains the method of construction. By flame cutting and arc welding the depth of any beam can be increased by nearly 50%, increasing its load capacity for large spans. Tapered beams and ridged beams can be satisfactorily produced by this method.

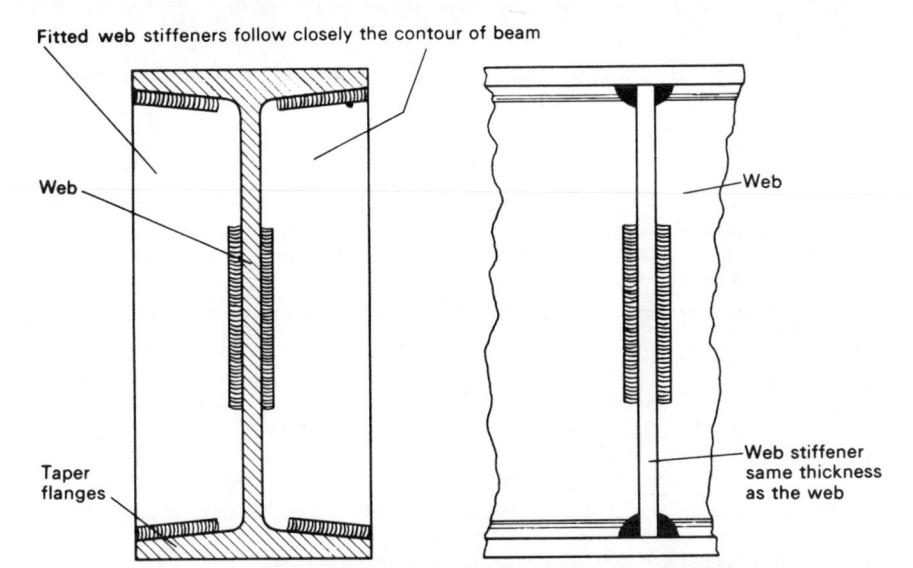

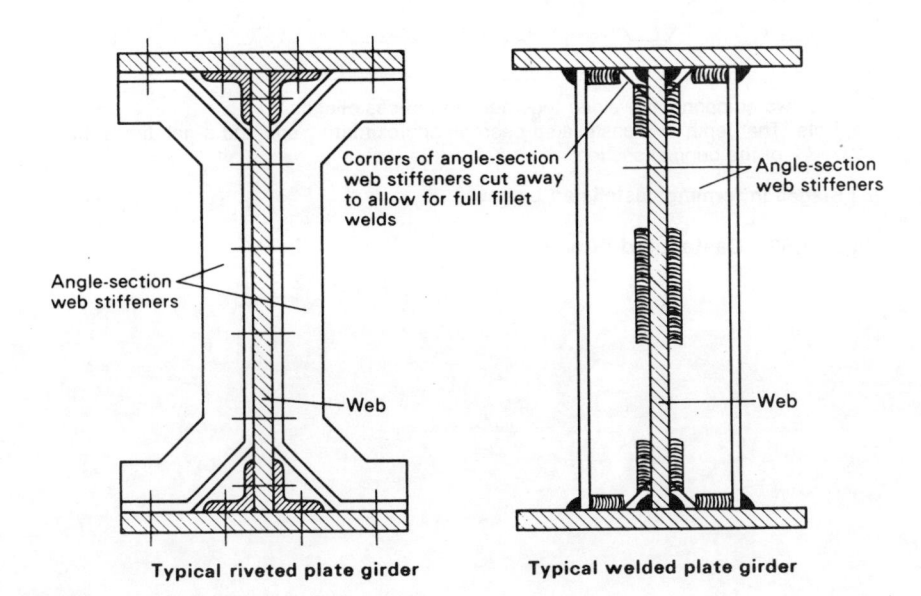

Fig. 10.66  Web stiffeners

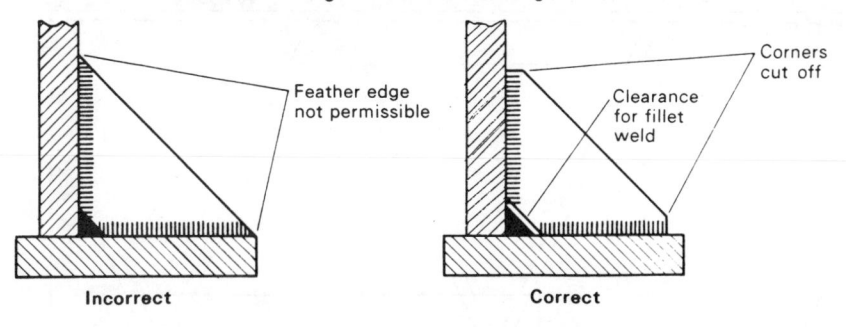

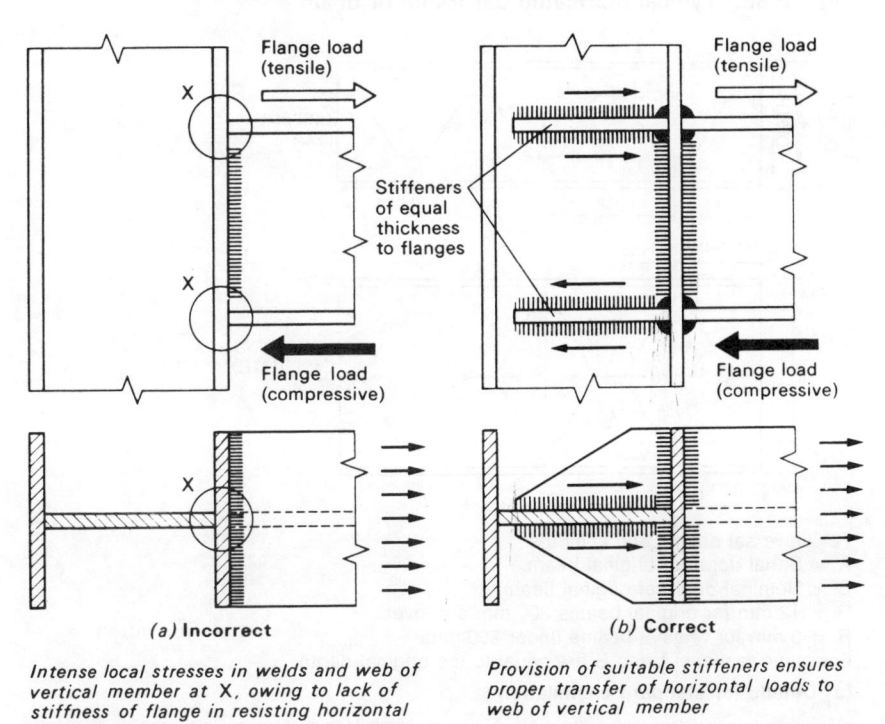

Fig. 10.67  Applications of stiffeners

391

View in direction 'A.A'    (All dimensions in mm)

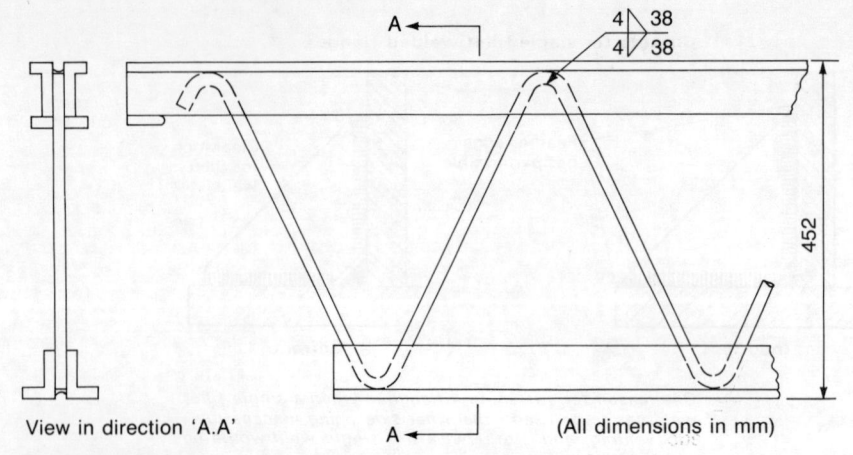

This simple lightweight beam is fabricated by welding together 64 mm × 45 mm × 8 mm M.S. angle sections and 16 mm diameter M.S. bar.
The web diagonals are produced by ordinary workshop methods such as localised heat at the bends and pulling the bars round pins.

**Fig. 10.68  Typical fabricated bar frame or beam**

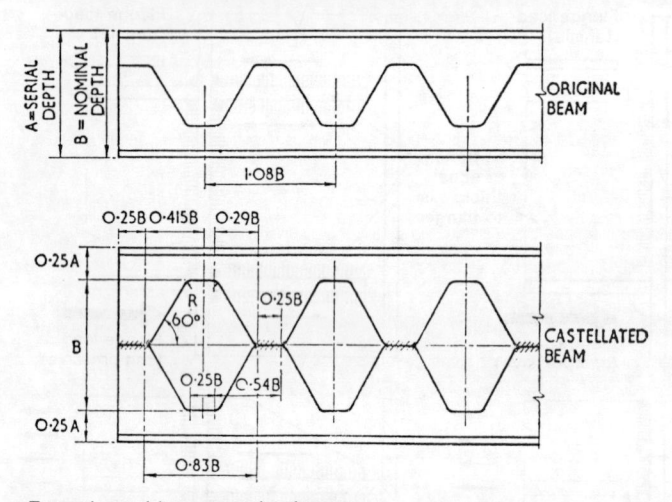

For universal beams and columns:
A = Serial depth of original beam.
B = Nominal depth of original beam.
R = 12 mm for original beams 200 mm and over.
R = 9 mm for original beams under 200 mm.
For other sections A and B are equal to the original depth.

(*a*) **Details for standard castellation**

*Note*: A castellated beam will inherit a greater STRENGTH/WEIGHT RATIO than the original section from which it is fabricated.
To enable suitable connections to be made, it is sometimes necessary to fill some holes by welding in shaped plates.

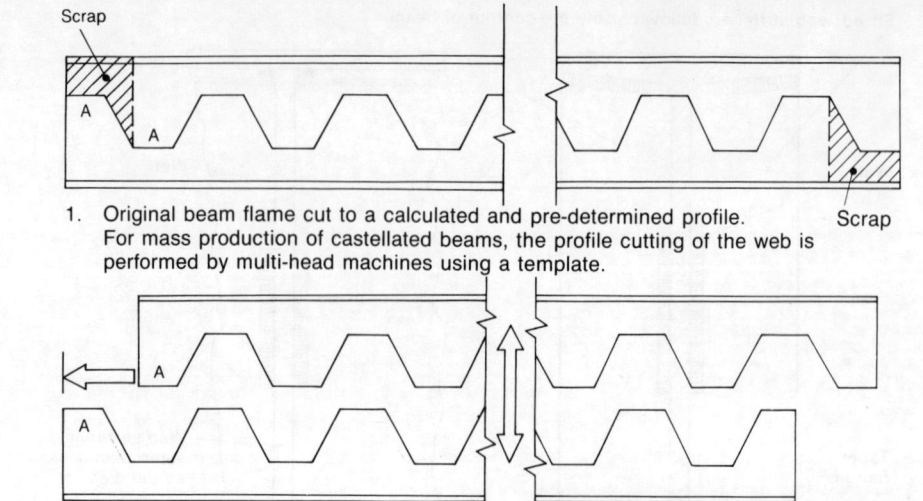

1. Original beam flame cut to a calculated and pre-determined profile.
   For mass production of castellated beams, the profile cutting of the web is performed by multi-head machines using a template.

2. Cut beam separated into two components for fitting A to A.

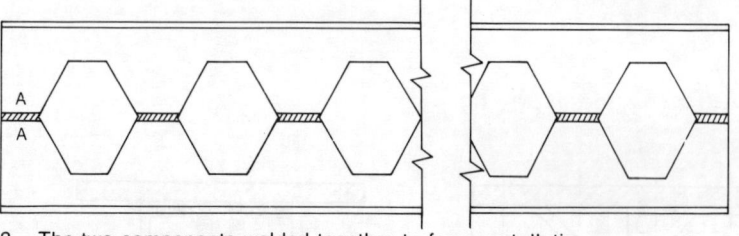

3. The two components welded together to form castellation.
   Note: The depth of a castellated beam is approximately one and a half times that of the original section.

(*b*) **Stages in forming castellated beams**

**Fig. 10.69  Castellated beams**

# 11 Metal joining —
## thermal and adhesive processes

## 11.1 Joining materials by thermal processes

Thermal processes in common use for the joining of metals and their alloys include:

1. Soft soldering.
2. Hard soldering or brazing.
3. Braze welding.
4. Full fusion welding.
5. Resistance welding.

One common factor in all the above processes is that each of them is dependent upon a suitable source of HEAT ENERGY. The application or choice of a particular thermal joining process may vary according to the TEMPERATURES used to produce an efficient joint and, of course, according to the materials being joined.

## 11.2 Soft soldering

Soft soldering is a **low temperature thermal process** in which the actual metals being joined are not melted. The process involves the use of a suitable **low-melting-temperature alloy** of **tin** and **lead** which is 'bonded', by the application of heat, to an unmelted parent material.

**Soft solders**, therefore, **must have a lower melting point than the metals they join.**

It is an essential feature of a soft-soldered joint that each of the joined surfaces is 'tinned' by a film of solder and that these two films of solder are made to 'fuse' with the solder filling the space between them.

Basically the action of a soft solder when applied to a prepared metal joint which has been heated to the required temperature is to:

(a) Flow between the parent metal surfaces which remain unmelted;
(b) Completely fill the space between the surfaces;
(c) Adhere thoroughly to the surfaces;
(d) Solidify.

In a correctly soldered joint, examination under a microscope has shown that in the action of tinning metals, such as BRASS, COPPER and STEEL, a definite CHEMICAL REACTION takes place. *The metal surface and the tin in the solder react together to form an 'intermetallic compound'* which, to some extent, becomes part of the parent metal, as illustrated in Figure 11.1.

In general there are two stages in making a soft-soldered joint:

1. Tinning the metal surface.
2. Filling the space between the tinned surfaces with solder.

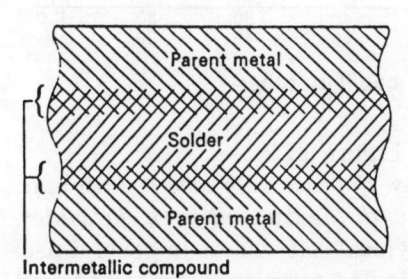

Fig. 11.1 A section of a soft-soldered joint

*The intermetallic compound layer will continue to grow in thickness the longer the joint is kept at soldering temperature.*
*Solder cannot be completely wiped or drained off when in the molten state. The metal surface, therefore, remains permanently wetted or 'tinned' by a film of solder.*
*A film of solder cannot be mechanically prised off, leaving the surface of the parent metal bare in its original state, unlike lead which has been cast upon a metal surface.*

*The metal to be soldered is supported on a wooden block (heat insulator) to prevent unnecessary heat loss by conduction*

(a) Tinning the metal surface

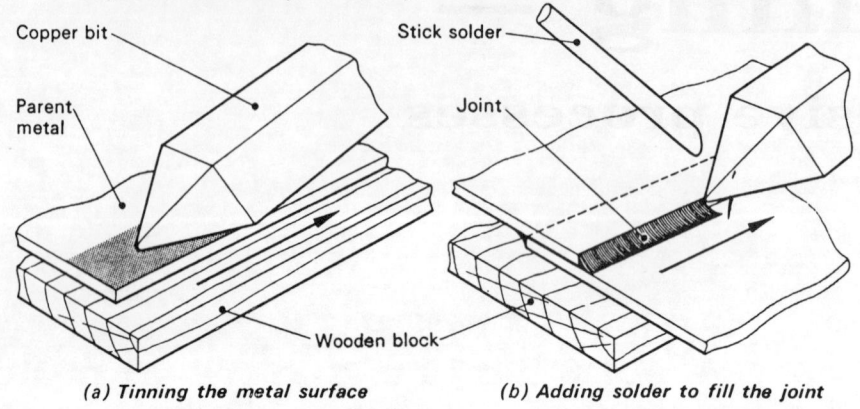

(b) Adding solder to fill the joint

**Fig. 11.2 Basic stages in soft soldering**

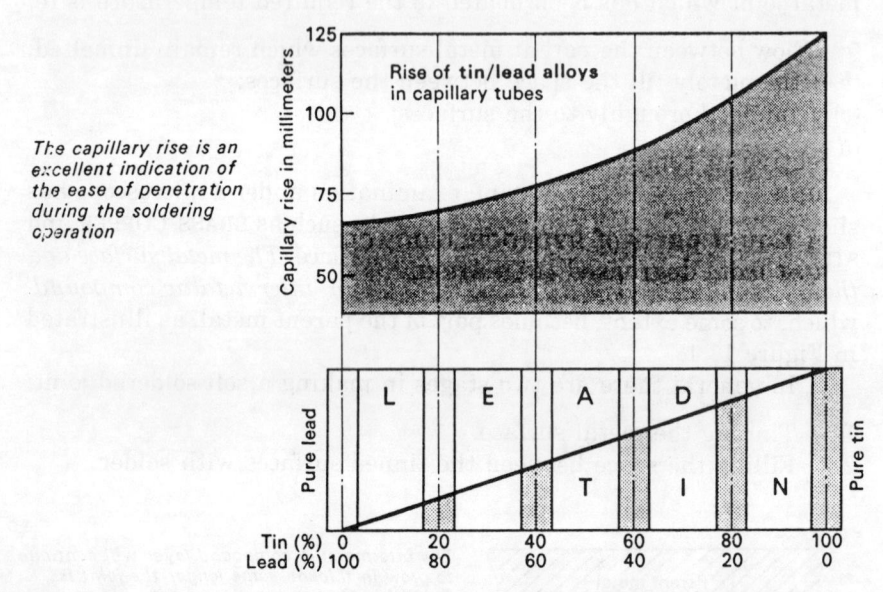

*The capillary rise is an excellent indication of the ease of penetration during the soldering operation*

**Fig. 11.3 The rise of tin/lead alloys in capillary tubes**

These stages are illustrated in Figure 11.2.

Figure 11.3 clearly indicates how *the ease with which a* **tin/lead** *solder penetrates the joint, increases with the* **tin** *content of the solder*.

Table 11.1 gives a selection of typical soft solders in general use, together with their properties and uses. The complete list may be found in BS 219:1959.

**Table 11.1  Types of solder**

| B.S. SOLDER | COMPOSITION % Tin | Lead | Antimony | MELTING RANGE (°C) | REMARKS |
|---|---|---|---|---|---|
| A | 65 | 34·4 | 0·6 | 183–185 | Free-running solder ideal for soldering electronic and instrument assemblies. Commonly referred to as **electricians's solder**. |
| K | 60 | 39·5 | 0·5 | 183–188 | Used for high-class tin-smith's work, and is known as **tinman's solder**. |
| F | 50 | 49·5 | 0·5 | 183·212 | Used for general soldering work in coppersmithing and sheet-metal work. |
| G | 40 | 59·6 | 0·4 | 183–234 | **Blow-pipe solder.** This is supplied in strip form with a D cross-section 0·3 mm wide. |
| J | 30 | 69·7 | 0·3 | 183–255 | **Plumber's solder.** Because of its wide melting range this solder becomes 'pasty' and can be moulded and wiped. |

## 11.3 Preparing the joint

*The 'tinning action' of solder cannot take place unless the two surfaces to be joined are* **chemically clean**. Not only must the mating surfaces be free from dirt and grease, but also of any OXIDE FILM.

CLEANING FOR SOLDERING

Prior to any soft soldering, brazing or welding operations it is necessary to expose bare metal at the mating or joining surfaces. *Dirt, grease, oil or oxides are themselves unsolderable and would act as barriers between the molten solder and the metal surfaces to be joined*. Table 11.2 lists some of the common methods of preparing a bright surface:

When a metal is exposed to the air at 'room temperature' it will

**Table 11.2 Common methods of cleaning metal surfaces**

MECHANICAL (ABRASIVE) CLEANING

**Methods**
1. Rubbing stained or corroded surfaces with EMERY CLOTH.
2. Use of FILES.
3. Use of SCRAPERS.
4. Scouring with the aid of WIRE WOOL.
5. SAND-BLASTING.

**Remarks**
Particular attention should be paid to any pitting or small depressions which may be in the surfaces of the metal to be soldered. These depressions tend to harbour dirt and OXIDES which, if not effectively removed, will render any application of soldering flux ineffective in such local areas.

Sand-blasting can reach into fine recesses not normally accessible by other means.

---

**METAL SURFACES SHOULD LOOK BRIGHT AND BE FREE FROM STAINS, BUT NEED NOT BE HIGHLY POLISHED, IN PRACTICE IT HAS BEEN FOUND THAT VERY SMALL SCRATCHES ON A METAL SURFACE ACTUALLY PROMOTE THE SPREADING OF THE MOLTEN SOLDER.**

---

CHEMICAL CLEANING

**Methods**
1. Application of a suitable SOLVENT. For 'one-off' or small batches, swabbing with either cotton waste or rag dipped in WHITE SPIRIT or CARBON TETRACHLORIDE is usually adequate. For 'production lines', a suitable TRICHLOROETHYLENE degreasing apparatus is essential.
2. Use of HOT ALKALINE SOLUTIONS.
3. Use of ALKALI DETERGENTS of the soapless type.
4. PICKLING in acids.

**Remarks**
Coated metals such as TINPLATE, TERNEPLATE and GALVANISED IRON or STEEL do not require mechanical cleaning, but it is essential that any OIL or GREASE on their surfaces should be removed.

Trichloroethylene is effective for removing most oils or greases in common use.

Two chambers are used in the degreasing apparatus, the article first being dipped in liquid solvent, followed by immersion in a condensing vapour of the boiling solvent.

Alkaline degreasing is effective in removing the last traces of oil or grease, whether animal or vegetable in origin.

A cold solution containing **equal parts** of HYDROCHLORIC ACID and WATER is used for the effective removal of RUST from degreased MILD STEEL.

---

**AFTER CHEMICAL CLEANING BY MEANS OF ALKALI OR ACID SOLUTIONS, ARTICLES SHOULD BE RINSED THOROUGHLY IN SOFT OR DISTILLED WATER AND ADEQUATELY DRIED.**

---

acquire a thin OXIDE COATING within a few minutes. *With few exceptions almost all metals on exposure to air at* **low temperatures** *become covered with a* **surface film of oxide**.

**Oxygen combines with substances much more rapidly at elevated temperatures**. Therefore, it is very important that metal surfaces must be prevented from OXIDISING whilst they are being HEATED to the soldering temperature. A suitable SOLDERING FLUX is used for this purpose.

## 11.4 Soldering fluxes

The requirements of a good soldering flux are:

1. It must remain liquid at soldering temperature.
2. In its liquid state it must act as a cover over the joint and EXCLUDE THE AIR.
3. It must dissolve any OXIDE FILM present on the surfaces being joined.
4. It should be readily displaced from the metal surfaces by molten solder.

Figure 11.4 illustrates the essential functions of a soldering flux, how it removes the oxide, protects the surface from atmospheric attack and then gives way for the 'wetting action' of the molten solder.

Fluxes used for soft-soldering operations may be classified as ACTIVE or INACTIVE (passive).

*A molten solder is said to 'wet' when it leaves a continuous permanent film on the surface of the parent metal instead of rolling over it.*

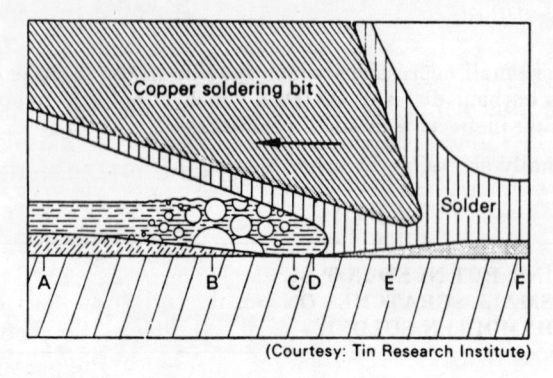

(Courtesy: Tin Research Institute)

Diagrammatic representation of the displacement of flux by molten solder.
A Flux solution lying above oxidised metal surface.
B Boiling flux solution removing the film of oxide (e.g. as chloride).
C Bare metal in contact with fused flux.
D Liquid solder displacing fused flux.
E Tin reacting with the basis metal to form compound.
F Solder solidifying.

**Fig. 11.4 The essential functions of a soldering flux**

*The active fluxes will quickly dissolve the* **oxide film** *on a metal and, at the same time act as a barrier to prevent further oxidation.*

**All active fluxes have a corrosive action.** Although any active flux residue which is present along the edges of the joint after soldering *may at first appear to be dry and harmless, it must be removed as soon as the soldering has been completed.* **These apparently dry flux residues which remain after soldering absorb water readily from the atmosphere.** *Within a few hours any active flux residue, if not removed, will become wet spots at which CORROSION will take place rapidly.* Corrosive flux residues may be removed by washing in hot water containing a few drops of HYDROCHLORIC ACID, rinsing in a hot dilute SODA SOLUTION, and finally in hot water to remove the salts formed. If only hot water is available, the soldered part should be thoroughly rinsed.

*The inactive fluxes protect a previously cleaned metal surface and have little effect on oxides.*

**Because of the difficulty of avoiding corrosion the use of active fluxes is banned in electrical and other work that cannot be washed effectively.**

Soldering fluxes in general use are listed in Table 11.3.

## Table 11.3 Soldering fluxes in general use

Soldering fluxes are of two basic types:
(i) Substances which yield an acid only when heated — these are classed as 'non-corrosive' because the flux residue is non-active and even protective;
(ii) Acids — these are classed as 'corrosive'.

NON-CORROSIVE TYPES

| FLUX | REMARKS |
| --- | --- |
| **RESIN or ROSIN** | In its natural form is the gum extracted from the bark of pine trees. It is an amber-coloured substance which is solid at rook temperature and does not cause corrosion, but it reacts mildly at soldering temperatures. It is used mainly for electrical work, but because it is non-toxic it is safe to use on food containers. |
| **TALLOW** | This is a product of animal fat. It is virtually inactive at room temperature, and like resin is only slightly active at soldering temperatures. This flux is used extensively with 'plumber's solder' for jointing lead sheets and pipes, and with 'body solder' on previously tinned steel for motor-vehicle repair work. |
| **OLIVE OIL** | This is a natural vegetable oil. It forms a weak vegetable acid at soldering temperatures. A useful flux when soldering pewter. |

CORROSIVE TYPES

| FLUX | REMARKS |
| --- | --- |
| **ZINC CHLORIDE** | Commonly called 'Killed Spirits of Salts' and forms the base for most commercially produced fluxes. A good general flux suitable for MILD STEEL, BRASS, COPPER, TERNPLATE and TINPLATE. |
| **AMMONIUM CHLORIDE** | Commonly called 'Sal Ammoniac'. This substance in a paste form is used as an ELECTROLYTE in the cells of dry batteries. As a soldering flux it is generally used in liquid form when tinning CAST IRON, BRASS or COPPER. It can be obtained in the form of a solid block (this is useful for the tinning of a soldering bit) or in the form of crystals. |
| **HYDROCHLORIC ACID** | This is known as 'Raw Spirits of Salts' and is extremely corrosive. It is used in dilute form when soldering ZINC and GALVANISED IRON or STEEL. *Zinc chloride (killed spirits of salts) is produced by the chemical action of hydrochloric acid on zinc.* |

PHOSPHORIC ACID — This flux is very effective when soldering STEEL, COPPER and BRASS. It tends to leave a 'glassy' residue.

**A mixture of equal parts of zinc chloride and hydrochloric acid** provides a very active flux for soldering STAINLESS STEELS. *This flux is dangerous because it gives off irritant fumes.*

**A solution of zinc chloride and 10% ammonium chloride** has a lower operating temperature than ZINC CHLORIDE (262°C) and is more suitable for use with low-melting-point solders.

PASTE FLUXES
Both the corrosive and non-corrosive types of soldering fluxes are obtainable in paste form. There are a number of paste-forming ingredients which may be added to fluxes, both of the active and non-active types. These include PETROLEUM JELLY (Vaseline), TALLOW, LANOLINE and GLYCERINE. An example of both types of paste flux are:

NON-CORROSIVE TYPE
METHYLATED SPIRITS or INDUSTRIAL SPIRITS are used to digest crushed RESIN in the proportion of 1 part resin with 2 to 4 parts 'spirits' by weight. Resin paste is used as the flux in single and multi-cored solders.

CORROSIVE TYPE
PETROLEUM JELLY is the main paste forming ingredient and is mixed with a solution of ZINC CHLORIDE, AMMONIUM CHLORIDE and WATER in approximately the following proportions:

| | |
|---|---|
| Petroleum Jelly | 65 % |
| Zinc Chloride | 25 % |
| Ammonium Chloride | 3·5% |
| Water | 6·5% |

**The main advantage of paste fluxes is that unlike liquid fluxes they do not tend to drain off the surface and run over other parts of the work where flux would be harmful.**

## 11.5 Heat sources for soft soldering

There are many different heat sources used in soft soldering, most of which are beyond the scope of this book. The following are a selection of the practical soft soldering methods in common use in industry:

1. **Soldering with the copper bit.**
2. **Soldering by flame.**
3. **Hot-plate soldering** — The principle is that clean and fluxed parts are either assembled with solder paste applied to their adjoining surfaces, or with preplaced foil or fine wire of solder, and placed on a tray or hotplate in an oven and there heated to soldering temperature.
4. **Induction heating** — Prepared articles are placed within a coil of copper tube and a high frequency electric current is passed through the coil which is suitably cooled. The part to be soldered is usually located at the strongest part of an INDUCED MAGNETIC FIELD where the electrical eddy currents produced raise the temperature of the component sufficiently to melt the solder after only a few seconds exposure.
5. **Resistance soldering** — The principle is similar to 'resistance welding' in that the heat required for melting the solder is generated by the resistance of the joint surfaces to the passage of a large current of electricity.
6. **Dip soldering** — This method of soldering is particularly suitable for sheet-metal work where the joint clearances are of consistent close tolerance. Previously cleaned and fluxed articles are immersed in molten solder contained in a heated bath. Baths are made from fine-grained non-porous cast iron — *copper cannot be used because it is rapidly dissolved by the tin in soft solders.*

## 11.6 Soldering with a copper bit

The traditional 'hand soldering iron' consists of a copper bit, an iron shank or yoke and a wooden handle.

COPPER is exclusively used for the 'head' or 'bit' for three basic reasons:

1. **Heat capacity** — a soldering bit must be capable of storing and carrying a large quantity of heat. Although both IRON and NICKEL have a higher 'Heat Capacity' than COPPER, they are not suitable for other reasons.
2. **Thermal conductivity** — a soldering bit must be capable of delivering heat to the job, and *copper has the highest 'thermal' or 'heat' conductivity of the common metals used in industry.*
3. **Reservoir of molten solder** — Copper is readily 'wetted' or 'tinned'. As a tool for manipulating molten solder the facets of the bit must be completely 'wet' with molten solder. *The larger the area of facets at the tip of the bit, the more molten solder it will hold.*

Various types of soldering irons in general use are illustrated in Figure 11.5.

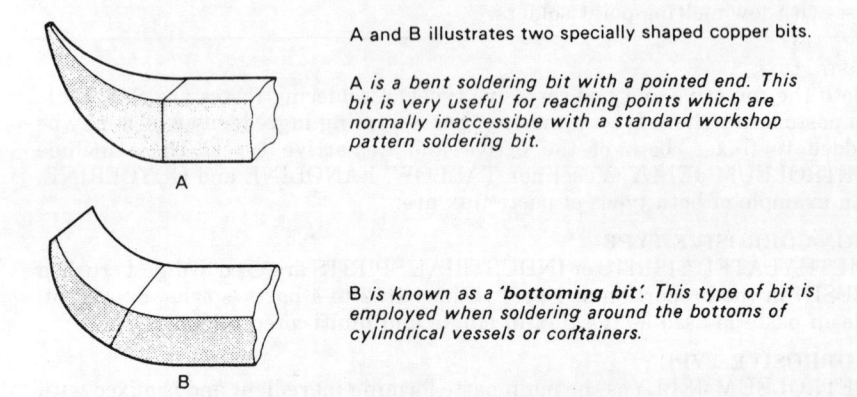

**Fig. 11.5  Typical soldering irons in common use**

## 11.7 Methods of heating the copper bit

One method which can be employed for heating the bit is to direct a petrol or paraffin blow-torch flame on to the bit close to the tinned facets (Figure 11.6). If the flame starts to show a green tinge, this indicates that the bit will become too hot if not withdrawn. With this method of heating, although very useful on site, there is a risk of burning the bit.

Where a supply of town gas is available, the bit is more likely to be heated in a soldering stove. Soldering stoves are usually made of cast iron, and the more expensive types have a fire-clay or suitable fire-brick lining and are heated by gas (Figure 11.7). Stoves which can take more than one soldering iron are extremely useful where repetition soldering operations are involved.

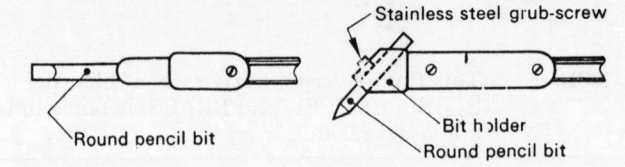

Considerable use is made of GAS-HEATED and ELECTRICALLY-HEATED soldering irons; these are illustrated in Figure 11.5.

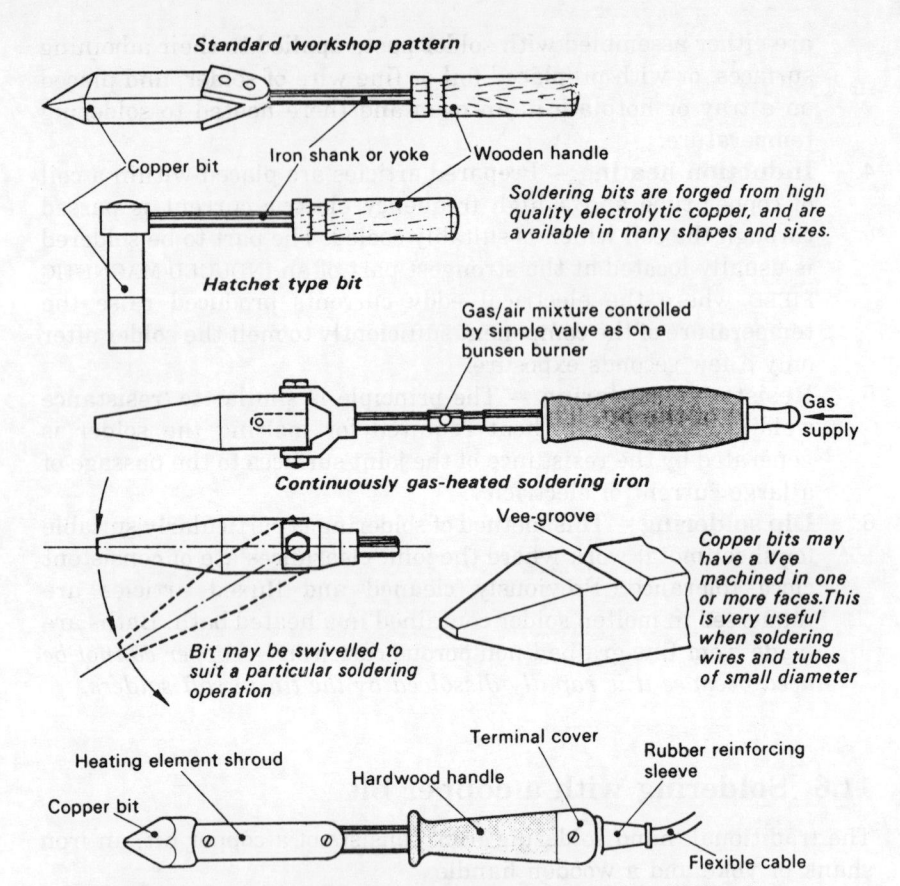

With gas-heated soldering irons a bunsen-type jet is arranged to play upon the back of the copper bit, with some internally-heated gas soldering irons the flame passes through the centre of the bit. The gas is supplied through an armoured flexible hose and may be regulated on the handle by a suitable valve.

Electrically-heated soldering irons are heated by an element in the same manner as the domestic electric iron. This method of heating is ideal for the smaller size of bits which are in general use on electric and electronic assembly work, since the amount of solder required for each joint is quite small and little heat is lost from the bit. **For safety reasons electric soldering irons may use a low-voltage heater element and are connected to the mains supply through a low voltage transformer.** Because of the time wasted in waiting for the bit to regain working temperature due to CONVECTION and CONDUCTION

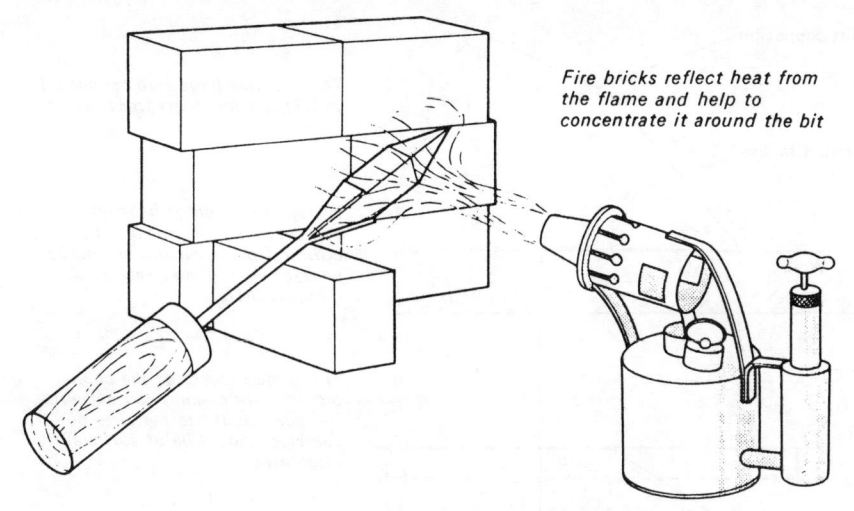

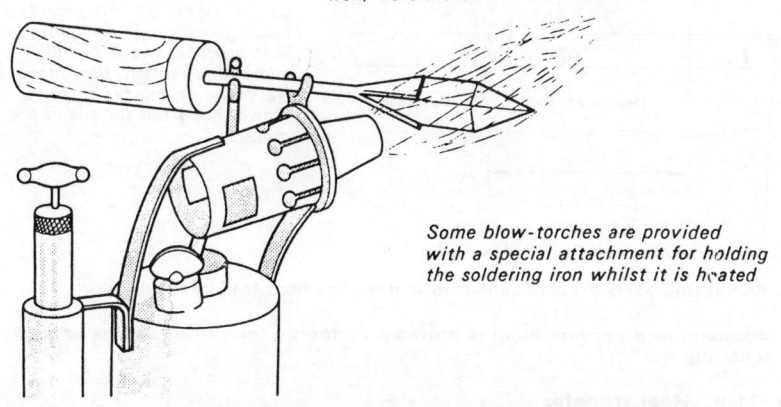

**Fig. 11.6 Heating the bit with a blow-torch flame**

*Fire bricks reflect heat from the flame and help to concentrate it around the bit*

*Adaptor for supporting the soldering iron; note the bit is in the flame*

*Some blow-torches are provided with a special attachment for holding the soldering iron whilst it is heated*

**Petrol and paraffin blow torches are gradually being superseded by gas/air torches which employ bottled butane or propane as fuel gases.**

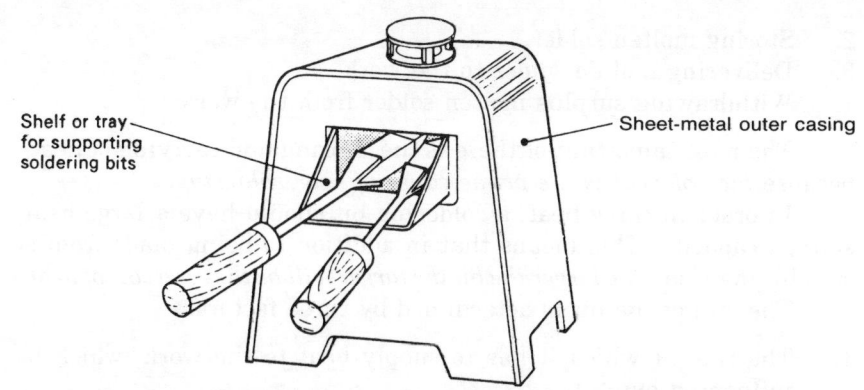

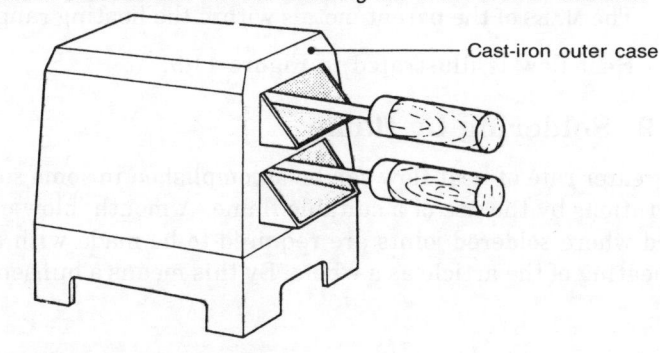

**Fig. 11.7 Gas-fired soldering stoves**

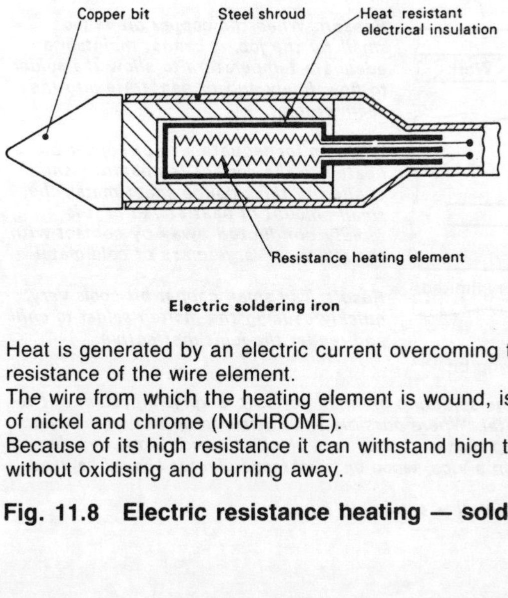

*A gas ring or row of burner jets is incorporated below the chamber. The flames do not come in direct contact with the soldering bits in the oven or chamber.*

Shelf or tray for supporting soldering bits — Sheet-metal outer casing

Cast-iron outer case

**Electric soldering iron**

Copper bit — Steel shroud — Heat resistant electrical insulation — Resistance heating element

Heat is generated by an electric current overcoming the resistance of the wire element.
The wire from which the heating element is wound, is an alloy of nickel and chrome (NICHROME).
Because of its high resistance it can withstand high temperatures without oxidising and burning away.

**Fig. 11.8 Electric resistance heating — soldering iron**

losses, electrically-heated bits are not suitable for work on heavy sheet-metal seams.

The basic construction of an electric soldering iron is illustrated in Figure 11.8.

## 11.8 Heat transfer

The main functions of a soldering bit are as follows:

1. Storing and carrying HEAT from the source to the work.

399

2. Storing molten solder.
3. Delivering molten solder to the work.
4. Withdrawing surplus molten solder from the work.

The most important of these is the 'storing and carrying of heat' because *lack of heat is the prime cause of poor soldering.*

In order to carry heat, a soldering bit should have a large heat storage capacity. This means that in addition to being made from a suitable material, *the larger the bit the larger will be the reservoir of heat.*

The size of the bit is determined by three factors:

1. The rate at which it has to supply heat to the work, which is influenced by:
2. The CONDUCTIVITY of the parent metals to be soldered; and
3. The MASS of the parent metals within the heating range of the bit.

Heat flow is illustrated in Figure 11.9.

## 11.9 Soldering by flame

A greater rate of heat flow can be accomplished in some soft-soldering operations by the use of a suitable flame. A mouth 'blow-pipe' may be used where soldered joints are required to be made with a minimum of heating of the article as a whole. By this means a bunsen-type flame of exactly the right size can be directed where the heat is required, solder being applied when the locally heated part has reached soldering temperature. Special 'blow-pipe solder' is available for the purpose, or solder can be applied from fine flux-cored wire or in the form of a paste in which the flux and solder are incorporated.

Large gas flames or 'blow-pipes' are extensively used where a more rapid and wider spread of heat is required, for example:

1. Filling in dents or sunken welded joints on motor vehicle bodies.
2. Sweating pre-tinned joints together. For example, pipe joints or long seams where the heat tends to be rapidly conducted away from the joint.

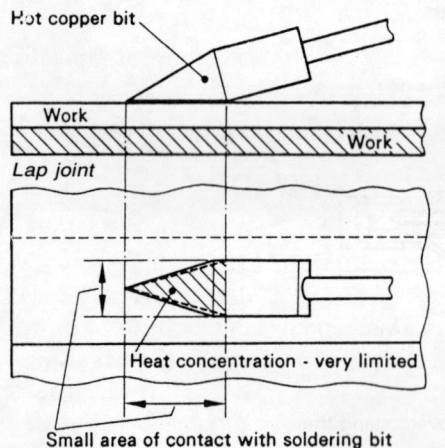

*A small soldering bit is not suitable for soldering any heavy or comparatively large-size sheet-metal products.*

*Reason: When the copper bit is too small for the job, it cannot maintain an adequate temperature to allow the solder to flow freely and to penetrate into the seam or joint.*

*When an inadequate size of copper bit is heated to the correct temperature and applied to thick gauge sheet metal, the small amount of heat stored in it is rapidly conducted away by contact with the relatively large mass of cold metal.*

*Result: The small copper bit cools very quickly causing the molten solder to chill and render the joint ineffective.*

When using a small copper bit for the soldering of small parts, it is very important not to rest the parts on a cold mass of metal. Where possible small parts to be soldered should be placed on some sort of insulating material such as a block of wood. Similarly when parts to be soldered are to be held in a vice, wood or fibre faced clamps should be used.

*(a) Lack of heat when using too small a soldering iron*

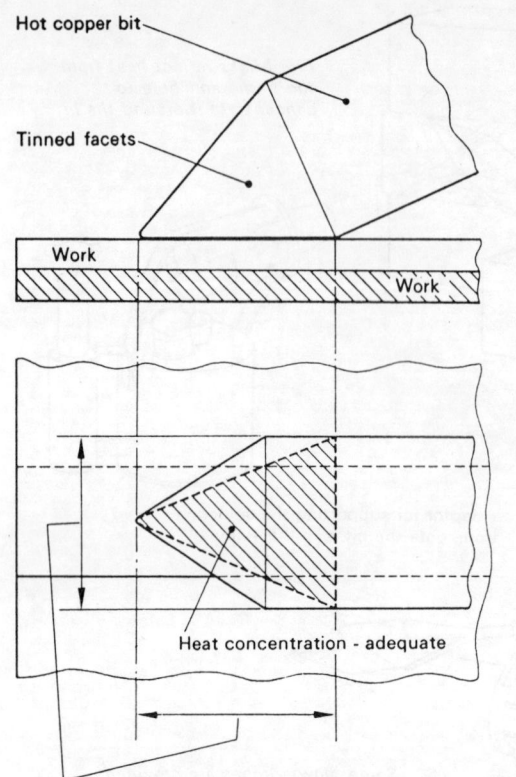

*The use of a large size copper bit provides a much greater heat source.*

*A large reservoir of heat will compensate any heat loss by conduction when the bit makes contact with a relatively cold mass of metal.*

*The tinned facets of the copper bit actually making contact with the surface of the metal to be soldered should be of sufficiently large area.*

*Reason: A large area, tinned face on the copper bit will provide a much greater heat concentration. A large concentration of heat will allow time for it to flow into the work while the bit is slowly drawn along the surface of the metal.*

**Preheating parts prior to soldering will reduce heat loss by conduction**

*(b) Adequate heat concentration is provided by the by the use of a large enough soldering iron*

**Fig. 11.9 Heat transfer**

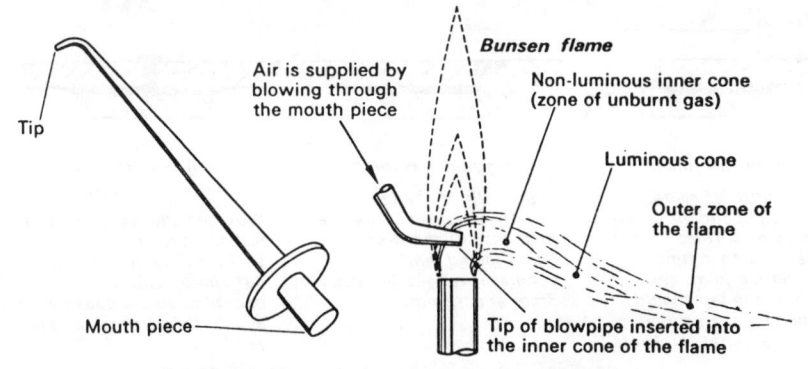

*(a) Mouth blow pipe – heat source for soldering*

Solder paint applied to tube ends when cold – i.e. at room temperature. This method has an advantage over conventional tinning because the components are not hot to handle when being assembled.

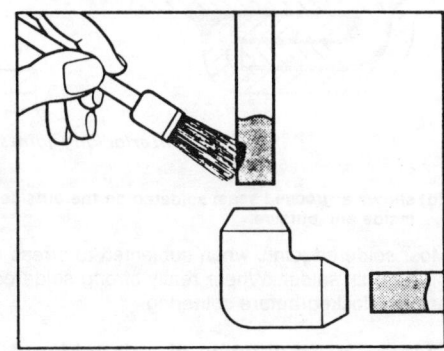

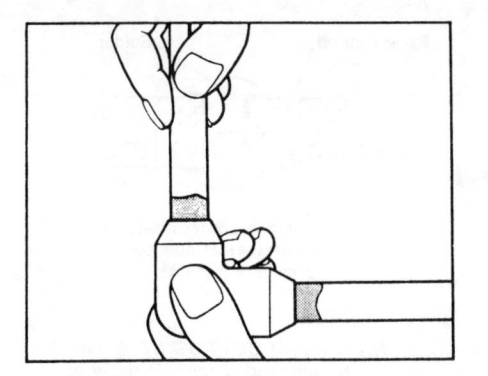

Assembling the tubes and elbow capillary fitting prior to the 'sweating' operation.

Sweat soldering the assembled pipe joint using a blow-torch flame. The final 'fillet' around the shoulder of the fitting may be made simply by touching the joint with blowpipe solder whilst heat is still being applied.

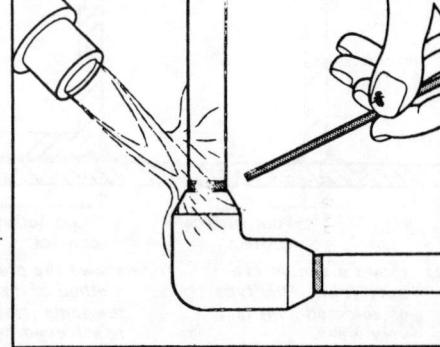

*(b) Soft soldering a pipe joint using a blow-torch flame*

**Fig. 11.10 Flame soldering**

3. The tinning of castings and some metal bearings.

Figure 11.10 illustrates flame soldering.

## 11.10 Types of soft-soldered joints

Soft soldering, as a joining method, relies almost entirely on ADHESION for its strength.

**Soft-soldered joints are strongest at room temperature, their mechanical strength decreasing rapidly as the temperature increases.**

The strength of a soft-soldered joint is determined by three basic factors:

1. The strength of the solder itself — this is governed by its composition (i.e. ratio of TIN/LEAD); **surplus solder does not add any strength to the joint**.
2. The strength of the bond between the bulk of the solder and the surfaces which it 'tins' (i.e. the interfaces);
3. The design of the joint. Where strong joints are required, the joint edges may be interlocked prior to soldering.

Factors influencing joint design for soft soldering are illustrated in Figure 11.11 and Figure 11.12.

As illustrated in Figure 11.11(a), a soldered joint is strongest 'in shear' (42–49 MN/m$^2$) and weakest 'in tension' (0·28–0·35 MN/m$^2$).

A straight butt joint under tension places the bulk solder between the interfaces under TENSILE STRESS (see Figure 11.11(b)). This means that the stress-carrying capacity of this type of joint is limited to the tensile strength and cross-sectional area of the solder itself. **Soft solder has a very low tensile strength**.

*Joints designed for 'shear loading' can be strengthened by increasing the lap area (**surface contact area**) as required, and by reducing the thickness of the solder deposit (Figure 11.11(e)).*

Soldered lap joints made with 'close clearances' derive the greater proportion of their overall strength from the 'bonded zones' (interfaces) where the solder is in intimate contact with the surfaces of the parent

401

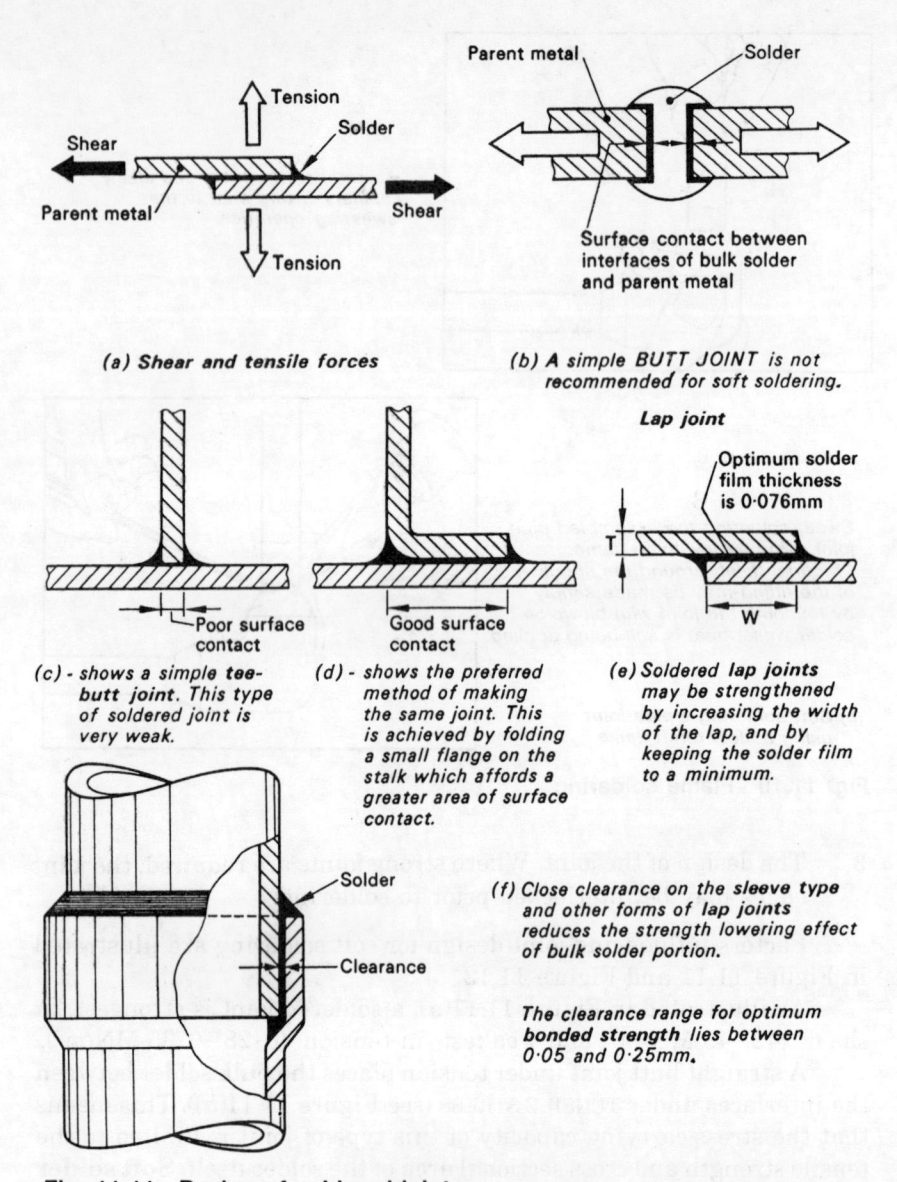

### (a) Shear and tensile forces

### (b) A simple BUTT JOINT is not recommended for soft soldering.

Parent metal — Solder — Surface contact between interfaces of bulk solder and parent metal

### Lap joint

(c) - shows a simple tee-butt joint. This type of soldered joint is very weak. (Poor surface contact)

(d) - shows the preferred method of making the same joint. This is achieved by folding a small flange on the stalk which affords a greater area of surface contact. (Good surface contact)

(e) Soldered lap joints may be strengthened by increasing the width of the lap, and by keeping the solder film to a minimum. Optimum solder film thickness is 0·076mm

(f) Close clearance on the sleeve type and other forms of lap joints reduces the strength lowering effect of bulk solder portion.

The clearance range for optimum bonded strength lies between 0·05 and 0·25mm.

**Fig. 11.11 Design of soldered joints**

(a) *Simple lap joint.* The strength of soldered joints is dependent upon the area of surface contact. With a simple soldered lap joint, the width of the lap should not be less than three times the metal thickness.

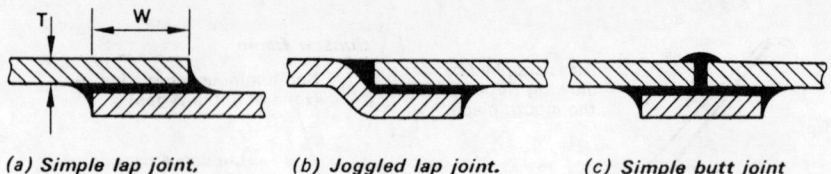

(b) *Joggled lap joint.* The width of the lap is usually determined by the thickness of the metal being joined, and, in general it ranges between 3mm and 12mm.

(c) *Simple butt joint with cover strap.* Provided the joint space is full and that there is sufficient solder to blend out sharp corners, an excess of solder does not add to the strength of the joint.

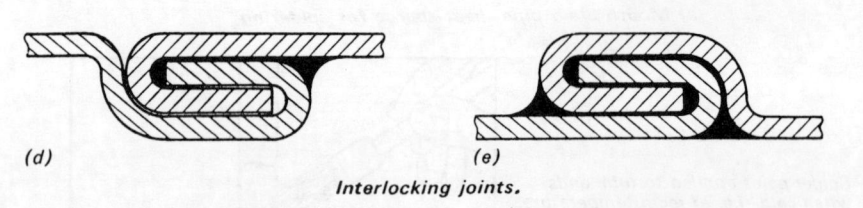

(d) *Interlocking joints.* (e)

(d) shows a grooved seam soldered on the outside only, whilst (e) is soldered both inside and outside.

Most soldered joints when subjected to stress have to rely on the strength of the bulk solder. Where really strong soldered joints are required the surfaces are interlocked before soldering.

**Fig. 11.12 Various types of soft-soldered joints**

metal. *Any tendency to yield under load is opposed by the force required to displace the solder bond.*

Figures 11.13, 11.14 and 11.15 illustrate three common soft-soldering operations: tacking, sweating, and floating.

TACKING (Figure 11.13). When making long soldered seams it is considered good practice to 'tack' the surfaces together using the following sequence.

First at the middle, then at each end, and at regular intervals in between.

*Reason:* This tacking sequence maintains a **uniform spread of temperature** and so reduces any tendency by the parent metal being tacked to warp or buckle due to uneven **expansion** and **contraction**.

Figure 11.12 shows various types of joint design used in sheet-metal work.

The process of tacking (Figure 11.13(b)) is simply carried out by methodically placing small 'blobs' of solder at regular intervals along the seam in order to hold it in position prior to the main soldering operation.

The tip of the hot copper bit is used and the local heating effect causes the small amount of solder placed on it to penetrate the joint at the point of contact.

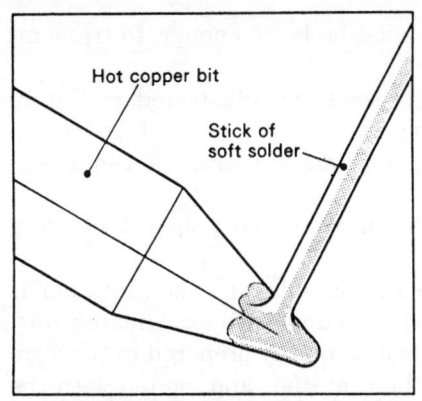

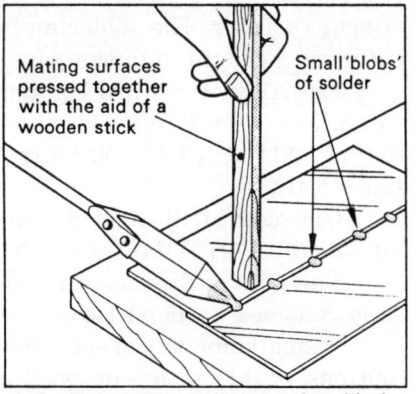

(a) Loading the copper bit with soft solder

(b) Tacking a seam prior to soft soldering

A little solder is picked up on the point of the hot soldering bit by touching it with a stick of solder. Any excess solder may be removed by simply shaking or flicking the bit.

**Fig. 11.13 Soft-soldering processes — tacking**

During tacking, the mating surfaces of the seam are pressed tightly together by means of a wooden stick or the 'tang' of a file.

The second common soft-soldering process is illustrated in Figure 11.14(a), (b) and (c).

(a) A thin coat of soldering flux is applied to the mating surfaces which have been previously cleaned.

It is advisable to arrange the seam to be soldered on some suitable support such as a block of wood or a piece of solid asbestos sheet.

**Avoid placing the seam to be soldered over a metal support.**

*Reason: A cold mass of metal, when used as a support, will absorb too much of the heat by* CONDUCTION *and make soldering difficult.*

(b) A uniform coating of solder is applied to each of the surfaces to be joined. This operation is called TINNING.

**Solder flows more readily between surfaces that have been pre-tinned.**

The advantage of tinning the mating surfaces is that the seam or joint can be made very much quicker and at a lower temperature. *One other advantage of a pre-tinned surface is that a* **corrosive flux** *is not required.*

(c) The joint edges are placed together with their tinned surfaces in contact and are held by pressing downwards with a hold-down stick or the tang of an old file. A heated soldering iron is placed on one end of

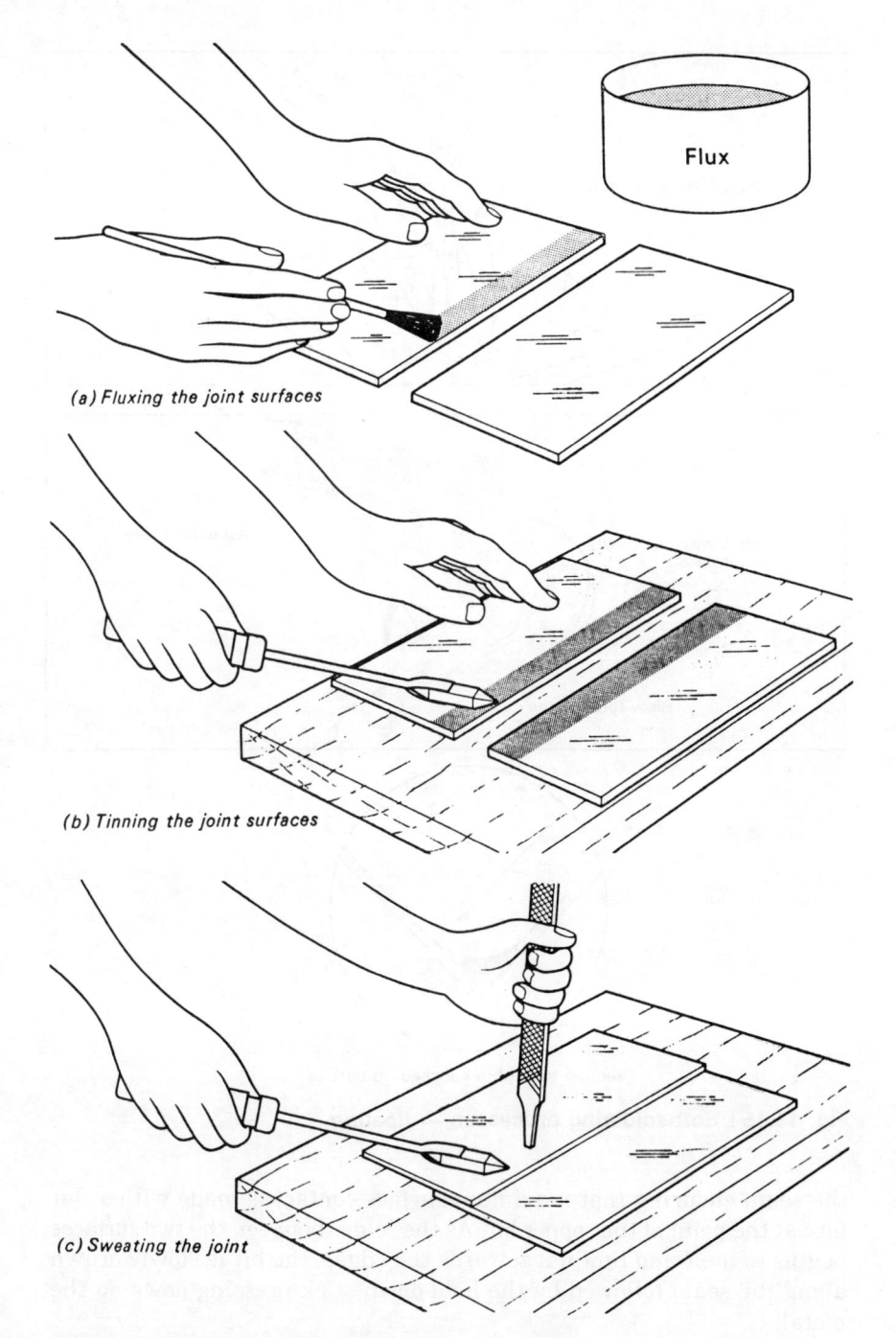

(a) Fluxing the joint surfaces

(b) Tinning the joint surfaces

(c) Sweating the joint

**Fig. 11.14 Soft-soldering processes — sweating**

403

**Fig. 11.15 Soft-soldering processes — floating**

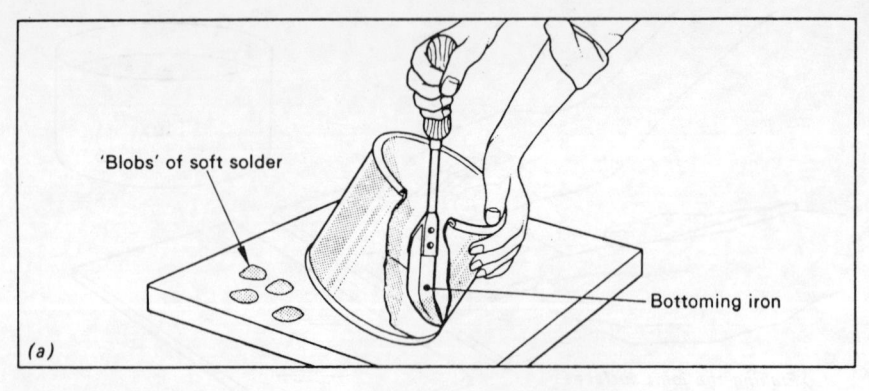

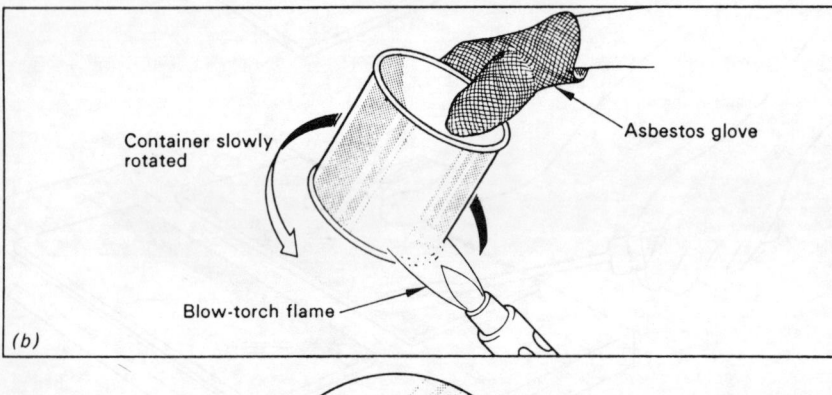

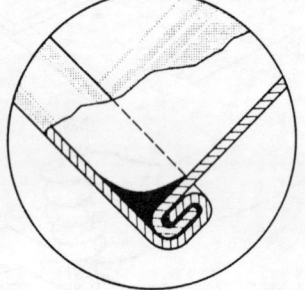

the seam ensuring that maximum surface contact is made with a flat face at the point of the copper bit. As the solder between the two surfaces begins to melt and flow out towards the edges, the bit is slowly drawn along the seam followed by the hold-down stick pressing down on the metal.

THE SUCCESS OF SWEATING DEPENDS UPON HAVING A CONSTANT SUPPLY OF HEAT. The soldering bit must be large enough to transmit sufficient heat through the joint.

FLOATING, the third common process and illustrated in Figure 11.15, is a useful soft-soldering operation which is employed when sealing SELF-SECURED JOINTS in order to make vessels or containers liquid-tight.

One method, illustrated in (a), is to use a specially shaped soldering bit called a BOTTOMING IRON.

The joint is fluxed both inside and outside, and the container is supported on a wooden base and tilted at an angle of approximately 45°.

A small quantity of soft solder, conveniently prepared in the form 'buttons' or 'blobs', are dropped in the container and melted with the heated and tinned bottoming iron.

This molten solder is then carefully flooded along the inside of the joint as the container is slowly rotated. The quantity of solder is uniformly controlled by steady manipulation of the soldering iron, adding blobs of solder if and when required.

If a bottoming iron is not available, an alternative method is to use a flame as the heat source. This method is shown in Figure 11.15(b).

An asbestos glove should be worn as a protection from the flame and the heat radiated from it. The flame is carefully applied to the outside of the joint, and the same procedure is adopted.

With either method the solder not only penetrates the interlocking surfaces but presents a smooth and nicely rounded corner on the inside of the joint.

A floating of solder not only prevents liquid foodstuffs from penetrating the joint and becoming rancid, but also provides a radiused corner which can easily be kept clean by simply wiping with a damp cloth.

## 11.11 Hard soldering or brazing

Hard soldering is a general term used to cover 'brazing' and 'silver soldering' which are very similar thermal joining processes.

*In this process, as in soft soldering, melting or fusion of the parent metals to be joined does not take place, which means that a suitable filler material which has a lower melting point than that of the parent metal is employed.*

There are a number of alloys other than the familiar TIN/LEAD alloys which can be used as SOLDER. They do not possess the low melting point of the soft solders, but they have other properties such as HIGHER STRENGTH, which makes them preferable for certain jobs.

For many years bicycle frames have been made by joining alloy steel tubes to the brackets by a BRAZING operation. In this case the solder is a grade of BRASS (hence the name BRAZING) which usually consists of 60% ZINC and 40% COPPER — this alloy melts at about 850°C, which is much higher than the soft-soldering temperatures. Such high melting point solders are called HARD SOLDERS, and hard soldered or brazed joints are much stronger than ordinary soft soldered ones.

Brazing is defined as '*a process of joining metals in which molten filler metal is drawn by capillary attraction into the space between closely adjacent surfaces of the parts to be joined.* In general the melting point of the filler metal is above 500°C.' In this respect, and broadly in its application, brazing lies between soft soldering and fusion welding.

## 11.12 Basic principles of brazing

The success of all brazing operations depends on the following general conditions:

1. Selection of a suitable brazing alloy which has a melting range appreciably lower than that of the parent metals to be joined;
2. Thorough cleanliness of the surfaces to be brazed;
3. Complete removal of the oxide film from the surfaces of the parent metals and the brazing alloy by a suitable flux;
4. Complete 'wetting' of the mating surfaces by the molten brazing alloy.

*The above principles also apply to the process of* **silver soldering**.

When a surface is 'wetted' by a liquid, a continuous film of the liquid remains on the surface after draining. This condition is essential for brazing and silver soldering, for the flux having removed the oxide film completely 'wets' the surfaces of the joint faces. This wetting action by the flux assists spreading and feeding of the filler material to the CAPILLARY SPACES, leading to the production of completely filled joints.

AN IMPORTANT FEATURE OF HARD SOLDERING IS THAT THE FILLER MATERIAL IS DRAWN INTO THE JOINT AREA BY CAPILLARY ATTRACTION. For any one combination of solid and liquid, *the smaller the gap the deeper is the capillary penetration.* The principles of capillary flow, illustrated in Figure 11.16 are explained below.

Consider a clean piece of glass plate with a small droplet of water lying on it as shown in Figure 11.16(*a*).

If another piece of clean glass plate is laid on the first piece and made to slide until its edge is brought into contact with the water droplet, *some of the water will immediately flow into the very narrow gap between the mating surfaces, for an appreciable distance, by* **capillary attraction**.

Exactly the same effect occurs when a joint is hard soldered, except

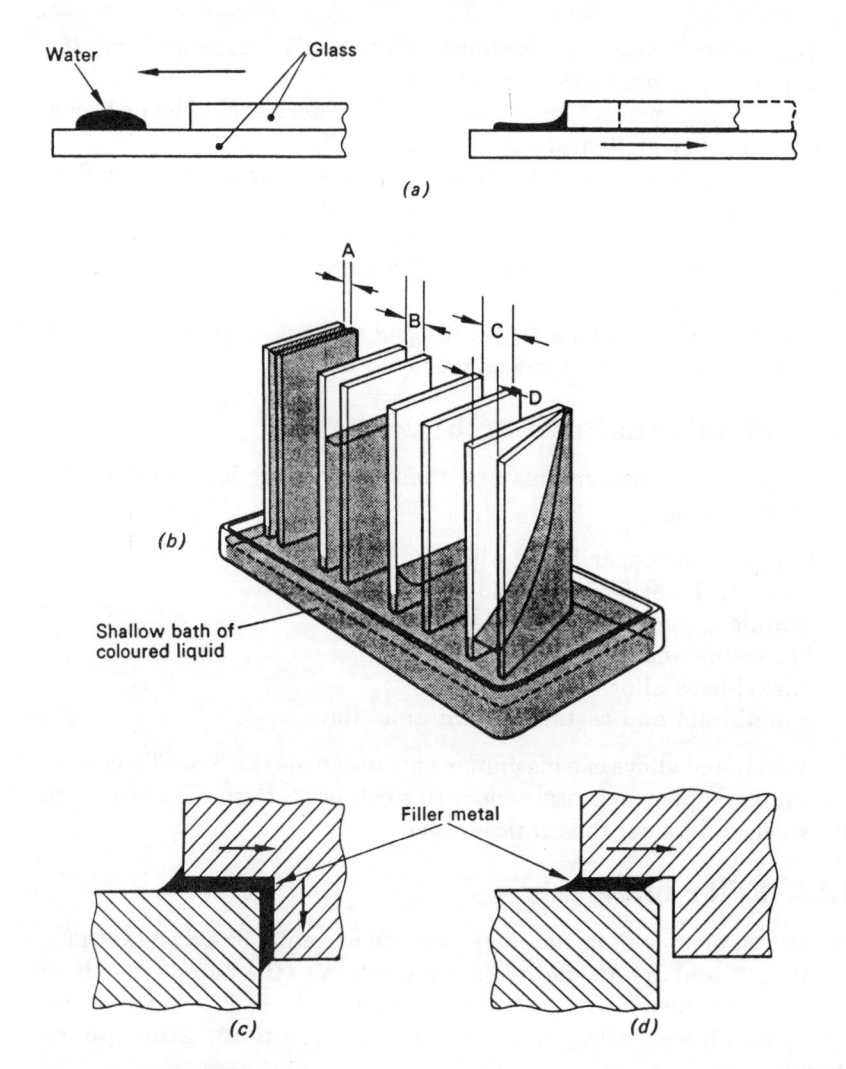

**Fig. 11.16 Capillary flow**

that instead of glass plates there are the mating surfaces of the components to be joined and in place of water, molten filler material.

The effects of 'capillary attraction' are shown schematically in Figure 11.16(*b*). Four pairs of glass plates are arranged with increasing gaps A, B, C, and D standing in a shallow bath of coloured liquid. *The liquid has risen the highest between the pair with the smallest gap*, A. In the right-hand pair liquid has risen to the top of the back edge but only slightly at the front, wide gap D.

*The effectiveness of capillary attraction is governed by the maintenance of appropriate joint clearances.*

The mating surfaces should be parallel. There should be no break in the uniformity of clearance.

*If a break occurs due to a widening or closing, then capillary flow will stop in that vicinity and may not go beyond it.*

As shown in Figure 11.16(c) CAPILLARY ATTRACTION CAN BE EXPECTED TO DRAW THE FILLER METAL AROUND A CORNER WHEN THE CONDITIONS ARE CORRECT.

Capillary flow may easily be stopped and cut off from the rest of the joint by a chamfered corner, as shown at (d).

## 11.13 Metals that can be brazed

The following common metals and their alloys may be joined by the process of brazing:

Copper and copper based alloys;
Mild steel, carbon steel and alloy steels;
Stainless steels and stainless irons;
Malleable and wrought iron;
Nickel-base alloys;
Aluminium and certain aluminium alloys.

Metals and alloys of a dissimilar nature can also be brazed together, for example: copper to brass; copper to steel; brass to steel; cast iron to mild steel; mild steel to stainless steel.

## 11.14 Filler materials

For most metals the brazing alloys used are normally based on COPPER–ZINC ALLOYS and are dissimilar in composition to the parent metals to which they are applied. COPPER is commonly used as a brazing material for the flux-free brazing of MILD STEEL in **reducing atmosphere furnaces**.

Brazing alloys may be classified into three main types:

1. Silver-bearing brazing alloys or 'silver solders';
2. Brazing alloys containing phosphorus;
3. Brazing brasses or 'spelters'.

### SILVER SOLDERS

These are more expensive than the normal brazing alloys because they contain a high percentage of SILVER, but they offer the advantages of producing very strong and ductile joints of much lower temperatures.

**Table 11.4 Composition of silver solders**

| BRITISH STANDARD 1845 Type | Silver min. | Silver max. | Copper min. | Copper max. | Zinc min. | Zinc max. | Cadmium min. | Cadmium max. | APPROXIMATE MELTING RANGE (°C) |
|---|---|---|---|---|---|---|---|---|---|
| 3 | 49 to 51 | | 14 to 16 | | 15 to 17 | | 18 to 20 | | 620–640 |
| 4 | 60 to 62 | | 27·5 to 28·5 | | 9 to 11 | | — | | 690–735 |
| 5 | 42 to 44 | | 36 to 38 | | 18·5 to 20·5 | | — | | 700–775 |

Type 4 possesses a high conductivity and is, therefore, very suitable for making electrical joints. It is the most expensive because of its high silver content. Type 3 is extremely fluid at brazing temperatures which makes it ideal when brazing dissimilar metals. *A low melting point alloy.*
Type 5 is a general purpose solver solder which can be employed at much higher brazing temperatures.

Silver solders are very free-flowing at brazing temperature. Their tensile strength is in the region of 500 MN/m$^2$, and *because of the low brazing temperatures required, have very little heat-effect upon the properties of the parent metals.*

By using silver solders it is possible to increase the speed of brazing and to eliminate or to limit 'finishing' operations. *One of its main applications is for delicate work in which small neat joints are essential.*
**Borax-type fluxes are not suitable for silver soldering**. Because of their low melting temperatures silver solders should not be used with high-melting brazing fluxes based on BORAX or BORIC ACID. These fluxes are not sufficiently fluid for silver soldering temperatures below 760°C. POTASSIUM FLUOBORATE can be used as a suitable flux, it is very active and completely molten at 580°C.

*An efficient flux should melt at a temperature at least 50°C lower than the melting point of the brazing material and retain its activity at a temperature at least 50°C above the melting temperature of brazing alloy being used.* Because of this factor it is advisable to use the proprietary fluxes which are available on the market. Table 11.4 gives the compositions of three silver solders together with their melting ranges.

### BRAZING ALLOYS CONTAINING PHOSPHORUS

Filler materials which contain PHOSPHORUS are usually referred to as 'self-fluxing brazing alloys'. These alloys contain SILVER, PHOSPHORUS and COPPER or, COPPER and PHOSPHORUS, the former possessing a lower

Table 11.5 Composition of brazing alloys containing phosphorus

| BRITISH STANDARD 1845 Type | Silver min. | Silver max. | Phosphorus min. | Phosphorus max. | Copper | APPROXIMATE MELTING RANGE (°C) |
|---|---|---|---|---|---|---|
| 6 | 13 to 15 | | 4 to 6 | | Balance | 625–780 |
| 7 | — | | 7 to 7·5 | | Balance | 705–800 |

Table 11.6 Composition of brazing spelters

| BRITISH STANDARD 1845 Type | Copper min. max. | Zinc | APPROXIMATE MELTING RANGE (°C) |
|---|---|---|---|
| 8 | 49 to 51 | Balance | 860–870 |
| 9 | 53 to 55 | Balance | 870–880 |
| 10 | 59 to 61 | Balance | 885–890 |

melting range. *The outstanding feature of these alloys is their ability to braze* **copper** *in air without the use of a flux.* When melted in air the products of OXIDATION form a fluid compound which acts as an efficient flux. *A suitable flux is required when brazing copper-based alloys.*

Brazing alloys containing phosphorus are only effective if melted in an **oxidising atmosphere** and should only be used when brazing COPPER and all COPPER-BASED alloys excepting those containing more than 10% NICKEL.

Nickel, nickel-base alloys and ferrous metals should not be brazed with these phosphorus-containing alloys. Although they will 'wet' and flow on such materials, they form brittle compounds which weaken the joint.

By comparison, the TENSILE strength of the silver/phosphorus/copper brazing alloys is about 450 MN/m$^2$ whilst the higher melting point copper/phosphorus alloys is much less at 350 MN/m$^2$.

These brazing alloys find their greatest application in resistance brazing operations, in refrigerator manufacture, electrical assemblies (electric motor armatures), for brazing seams and fittings in domestic copper hot-water cylinders, and in plumbing. Table 11.5 gives the composition and melting ranges of two common brazing alloys containing phosphorus.

BRAZING BRASSES

The oldest and best known method of brazing involves the use of brazing brasses or 'brazing spelters', using BORAX as a flux.

These alloys melt at much higher temperatures than the silver solders and the phosphorus-containing brazing alloys, but produce sound joints having TENSILE strengths between 400 MN/m$^2$ and 480 MN/m$^2$. The composition and melting ranges of three common brazing spelters are shown in Table 11.6 in which it can be seen that *increasing the zinc content decreases the melting range*. This makes it possible to make a joint in 60/40 brass using a 50/50 brass as the brazing alloy. Conversely,

This group of copper alloys tends to lose zinc by vapourisation and oxidation when the parent metal is heated above 400°C. This loss of zinc produces relatively higher tensile strength. The brazing alloys containing a high percentage of zinc, therefore, produce joints of the lowest strength.
Type 8 is used for medium strength joints, whilst the strongest joints can be produced by using type 10.

it is important that a brass to be joined by brazing should have a high copper content compared with the brazing alloy used.

## 11.15 Aluminium brazing

There is a distinction between the brazing of ALUMINIUM and the brazing of other metals. For aluminium, the filler material is one of the ALUMINIUM ALLOYS having a melting point below that of the parent metal.

The various grades of commercially pure aluminium alloys containing $1\frac{1}{4}$% MANGANESE, and certain ALUMINIUM/MAGNESIUM/SILICON alloys can be successfully joined by brazing.

Aluminium/magnesium alloys containing more than 2% magnesium are difficult to braze, as the oxide film is tenacious and difficult to dissolve with ordinary brazing fluxes (see Section 4.7).

*Borax-base fluxes are unsuitable for brazing aluminium and its alloys*, but many proprietary fluxes are available, and these are basically mixtures of the ALKALI METAL CHLORIDES and FLUORIDES. A standard aluminium brazing flux contains essentially chlorides of SODIUM, POTASSIUM and LITHIUM.

These fluxes readily absorb moisture when exposed to the atmosphere, i.e. are 'HYGROSCOPIC'. It is important, therefore, that all lids or covers be firmly replaced on flux containers, when not in use, to prevent rapid deterioration of their contents. Care should be taken when handling aluminium brazing fluxes containing fluorides, as personal contact may result in skin complaints, and toxic effects may follow the inhalation of fumes from these compounds.

Table 11.7  Composition of filler alloys for aluminium brazing

| BRITISH STANDARD 1942 Type | COMPOSITION PERCENTAGE |  |  |  |  | APPROXIMATE MELTING RANGE (°C) | BRAZING RANGE (°C) |
|---|---|---|---|---|---|---|---|
|  | Aluminium | Copper min. | Copper max. | Silicon min. | Silicon max. |  |  |
| 1 | Balance | 2 to 5 |  | 10 to 13 |  | 550–570 | 570–640 |
| 2 | Balance | — |  | 10 to 13 |  | 565–575 | 585–640 |
| 3 | Balance | — |  | 7 to 8 |  | 565–600 | 605–615 |
| 4 | Balance | — |  | 4·5 to 6·5 |  | 565–625 | 620–640 |

**It is impossible to braze certain aluminium alloys whose melting points are below those of the available brazing alloys.**

The melting points of the parent metals which can be **aluminium brazed** range between 590°C, and 660°C. Reference to the Table 11.7 will clearly indicate that the temperature margin between the melting ranges of the parent metals and those of the brazing alloys employed is very close.

Extreme care must be taken when **aluminium flame brazing** because of the very small margin in temperature which permits the joint to be made without melting and possible collapse of the parent metal during the operation. Other methods of heating for aluminium brazing are critically controlled, providing a uniform brazing temperature.

## 11.16 Aluminium flame brazing — procedure

Aluminium flame or torch brazing technique is simple, but *of the brazing processes it is the one most dependent on the skill of the operator, who by manipulation must maintain the correct brazing temperature without over-heating.*

Aluminium flame brazing can be successfully accomplished with relative ease, provided the following procedure is adopted:

1. The flame is adjusted to a slightly CARBURISING condition. This is essential to keep the joint in a REDUCING ATMOSPHERE to overcome the great affinity of aluminium for oxygen.

    The oxy-acetylene flame is lit and the oxygen valve slowly opened until the inner cone is well defined, but with a slight haze or 'feather' around its point. This feather should be as small as possible as shown in Figure 11.17(*a*).
2. The joint assembly is pre-heated using the envelope of the flame.
3. The filler rod is heated at one end with the flame and when dipped into flux a 'tuft' of flux will adhere to it.

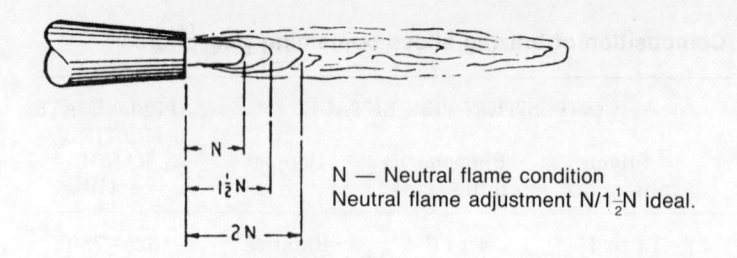

N — Neutral flame condition
Neutral flame adjustment N/1½N ideal.

(*a*) Essential flame condition

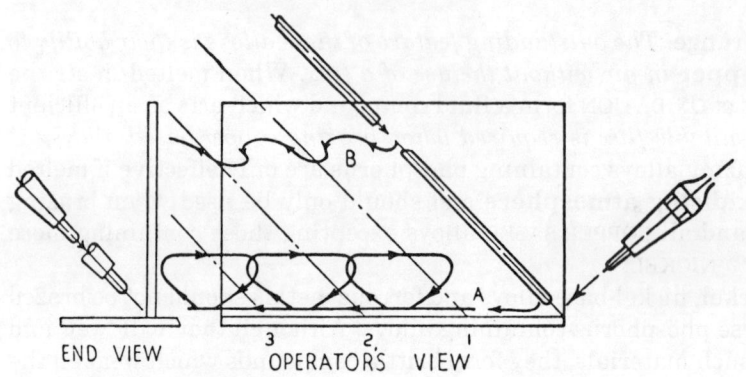

A — Movement of Torch.
B — Movement of Filler Rod.
1, 2 and 3 — Points of application of Filler Rod.

(*b*) Filler rod and torch manipulation

Fig. 11.17  Aluminium flame brazing

4. The joint is heated with the inner flame of the torch held about 50 mm away, until the flux commences to flow freely along it; the flux being applied from the tuft on the end of the filler rod.
5. As soon as the flux begins to flow rapidly and smoothly along the metal at the joint (due to capillary action and the driving power of the flame), a small quantity of filler rod is added and the rod withdrawn. *In good practice the brazing alloy is melted by the heat of the assembly rather than directly by the torch flame*:
6. As the filler metal begins to flow, so is the torch brought forward towards the joint until it is almost in the normal welding position. At the same time it is moved forward and away in a sweeping or punching action. More filler rod is added progressively along the joint as the movement of the torch is maintained in order to drive the filler metal along where the joint has been 'wetted' by the flux.

The above technique is illustrated in Figure 11.17(*b*).

Table 11.8 Methods of flux removal by chemical cleaning

| STAGE | METHOD A SOLUTION | TEMP. | TIME (min.) | METHOD B SOLUTION | TEMP. | TIME (min.) |
|---|---|---|---|---|---|---|
| 1 | Concentrated nitric acid | Room | 5–15 | 10% nitric acid 0·23% hydrofluoric acid | Room | 5–10 |
| 2 | Water rinse | Room | — | Water rinse | Room | — |
| 3 | 10% nitric acid 5–10% sodium dichromate | Room | 5–10 | — | — | — |
| 4 | Water rinse | Hot | — | Water rinse | Hot | — |
| 5 | Dry using warm air | — | — | Dry using warm air | — | — |

Method A is the most common, and method B gives a good uniform appearance to the joint surfaces, as it counteracts the darkish appearance in the joint caused by the silicon content of the filler metal.

## 11.17 Aluminium flame brazing – flux removal

FLUX REMOVAL IS AN ESSENTIAL PART OF THE BRAZING PROCESS, and it is achieved by scrubbing the joint and surrounding metal, immediately after brazing, with a stiff bristle or wire brush and *hot soapy soft water*. This is followed by rinsing with cold soft water and drying. Hard water tends to leave a thin deposit of lime on the metal surface.

Where flux residue is difficult to remove, for example from crevices, CHEMICAL CLEANING may also be used in addition to water cleaning. The methods of flux removal by chemical means are given in Table 11.8.

## 11.18 Heating methods for brazing

The choice of brazing method obviously depends on workshop conditions, equipment available and quantities required, the skills of the operator and the nature of the part to be brazed, as well as those factors involved in the economic manufacture of the product.

Several methods of heating are used to produce brazed joints. These are listed in Table 11.9 which also indicates the sections in this chapter where the particular heating technique is dealt with in detail.

Table 11.9 Heat sources for brazing

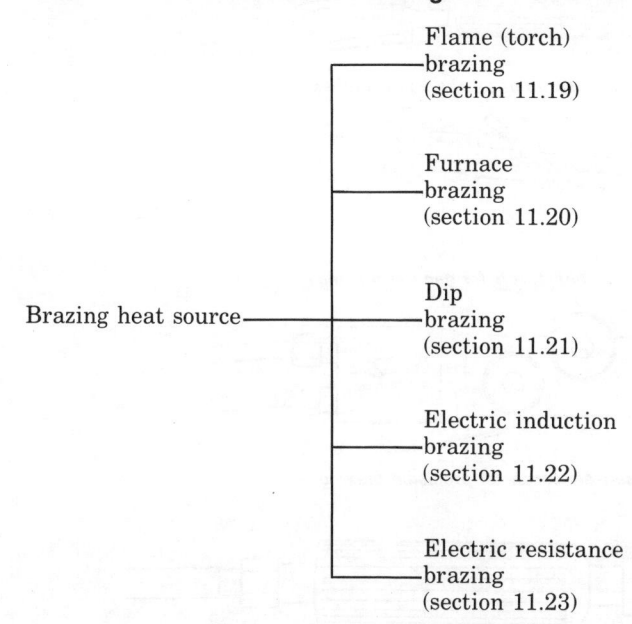

Brazing heat source
- Flame (torch) brazing (section 11.19)
- Furnace brazing (section 11.20)
- Dip brazing (section 11.21)
- Electric induction brazing (section 11.22)
- Electric resistance brazing (section 11.23)

## 11.19 Flame or torch brazing

Flame brazing may be used to fabricate almost any assembly, *and gives particular advantage where the joint area is small in relation to the bulk of the assembly*. The torch is the most common method of heating.

The brazing spelter may be applied in one of two ways:

(a) by carefully loading the joint with granulated spelter which has been 'calcined' by boiling it in a borax solution. The calcined granules are mixed with borax paste. As heat is applied from the torch and the joint brought uniformly up to brazing temperature the spelter melts. Borax is usually sprinkled on the joint which is tapped with an iron spatula in order to assist the flow of the molten spelter throughout the joint.

(b) in wire or strip form. The end of the spelter is heated and dipped into the flux which adheres to it. When the joint has been brought up to temperature the fluxed end of the filler metal is applied and flows into the joint.

A heating flame is produced by a suitably designed torch supplied with an OXYGEN/FUEL GAS or an AIR/FUEL GAS mixture, and *the filler metal is fused by the heat CONDUCTED from the hot component parts*.

A wide variety of gas mixtures are available for torch brazing and

*Oxy-acetylene, the most versatile of all the hand torches*

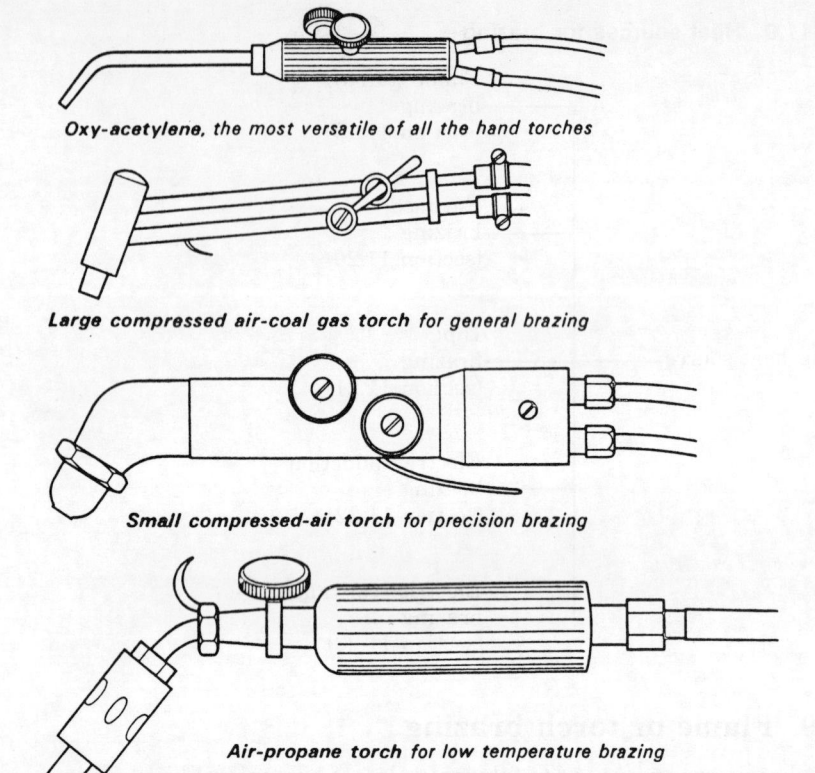

*Large compressed air-coal gas torch for general brazing*

*Small compressed-air torch for precision brazing*

*Air-propane torch for low temperature brazing*

**Fig. 11.18 Typical hand torches used for brazing**

the most useful of these, in approximate order of decreasing heating power are:

1. Oxy-acetylene
2. Oxy-hydrogen
3. Oxy-propane
4. Oxy-coal gas
5. Compressed air-coal gas
6. Air-propane: Air-butane: Air-methane.

Of these the two mixtures the most commonly used are oxy-acetylene and compressed air-coal gas. A wide range of heat control is obtained with the use of oxy-acetylene torches. These torches have been developed to a high degree of perfection and, if properly used, offer the most flexible and versatile method of torch heating.

Some aspects of torch brazing are illustrated in Figures 11.18 and 11.19.

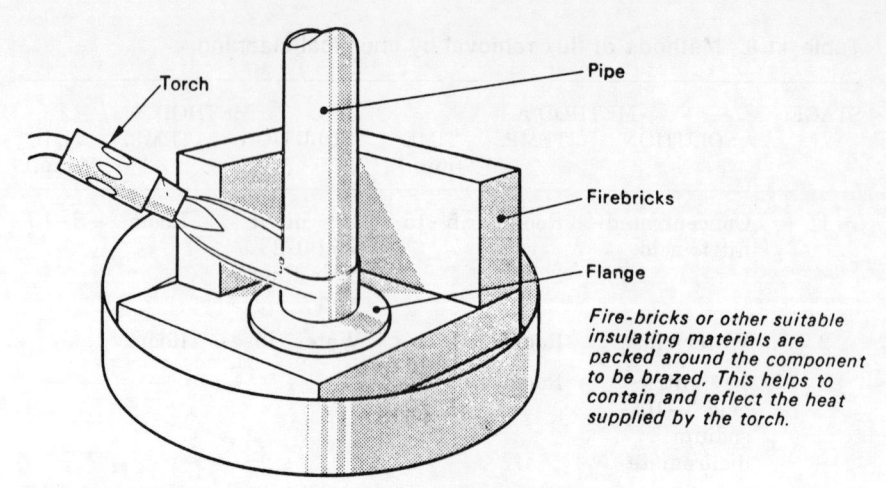

*Fire-bricks or other suitable insulating materials are packed around the component to be brazed. This helps to contain and reflect the heat supplied by the torch.*

**Fig. 11.19 Hand torch in use with brazing hearth**

## 11.20 Furnace brazing

Furnace brazing is used extensively when:

(i) the parts to be brazed can be pre-assembled or jigged to hold them in position;
(ii) the brazing filler material can be pre-placed;
(iii) when a controlled atmosphere is required.

The method of heating varies according to the application, and in general **muffle furnaces** are used in order that the flame does not impinge directly on the parts being brazed. Some furnaces are heated by gas or oil, but the majority are electrically heated.

The furnaces can be classified into types:

1. The BATCH TYPE with either an air or controlled atmosphere;
2. The CONVEYOR TYPE with a controlled atmosphere.

*The air-atmosphere furnaces may be used for brazing parts which can be protected by the use of a suitable flux or when the joint is to be made with phosphorus-containing brazing alloys.* They are not suitable for brazing when pure copper brazing filler metal is used.

Controlled-atmosphere furnaces are usually electric-resistance heated.

The conveyor type furnace has a mesh belt passing completely through it on which the parts are loaded. Alternatively, it may be of the 'pusher type' in which trays of jigged assemblies are loaded at the entrance end and pushed through the furnace by succeeding parts.

Figure 11.20 illustrates some aspects of furnace brazing.

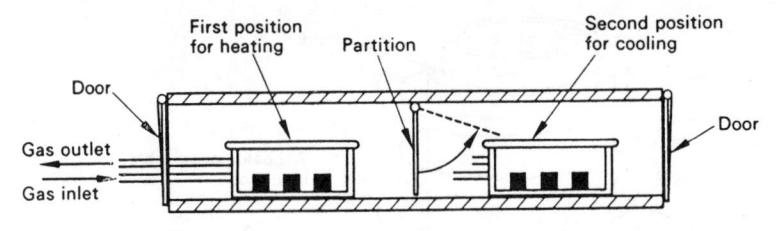

**Schematic layout of batch brazing in a sealed container**

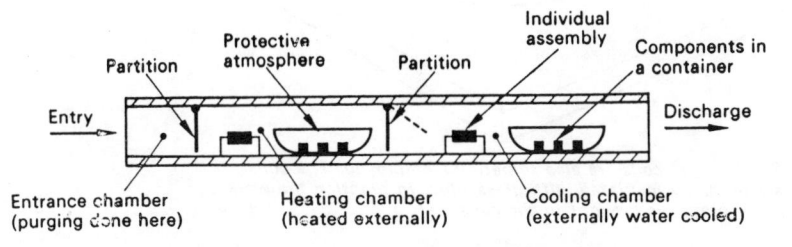

**Schematic layout of continuous brazing furnace**

*Individual assemblies or components for batch brazing (in trays) are passed through the furnace on a conveyor system. Inert or reducing atmospheres can be used to protect the work from oxidation.*

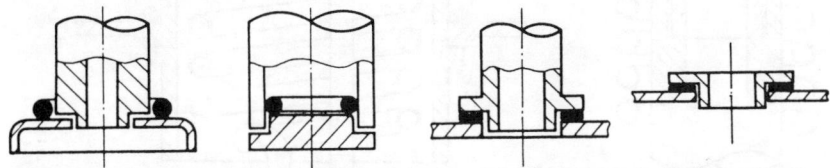

*Filler metal for furnace brazing is used in the form of pre-placements. The components to be brazed are assembled with the filler alloy in the required position and fluxed if necessary.*
*Pre-placed brazing alloy inserts are available in a variety of forms, such as wire rings, bent wire shapes, washers and foils.*

**Fig. 11.20 Furnace brazing**

## 11.21 Dip brazing

Two methods of heating are employed:

1. Molten filler metal dip heating
2. Molten chemical flux bath dip heating

Instead of a furnace, the parts to be brazed may be submerged in a bath of molten filler metal. A crucible, usually made of graphite is heated externally to the required temperature to maintain the brazing alloy in fluid form. A cover of flux is maintained over the molten brazing alloy, and in practice the assembled parts to be joined should be raised and lowered two or three times in order that flux and filler metal may flow into the joint clearance. The size of molten bath and the heating method must be such that the immersion of the parts to be brazed will not lower the temperature of the bath below that necessary for brazing. *Large articles are generally* pre-heated before being placed into the bath.

The most common application of this process is the brazing of STEEL or MALLEABLE IRON articles, using a COPPER/ZINC alloy as the filler metal.

**Salt bath brazing** is a process developed mainly for the brazing of ALUMINIUM, and for brazing MILD STEEL with COPPER/ZINC filler metal. *In this process the molten chemical salts contained in the bath supplies the heat required to raise the parent metal to brazing temperature.* The chemical salts are contained in a metal or ceramic container (the bath) which may be heated externally by gas or oil, or heated internally by an electric low-voltage alternating current passing between two submerged CARBON ELECTRODES. The temperature of the bath is critically controlled. *Salt bath brazing makes it possible to heat the parts to the brazing temperature more quickly than in the controlled-atmosphere furnace because of the greater* **heat conductivity** *of a liquid medium.*

The constituents of the salt bath are carefully chosen in order that they have some fluxing action upon the parent metals and the filler material.

The main use of the salt bath as a heat source for brazing is on ALUMINIUM and certain of its alloys, for which NITRATE SALTS are in common use. When a salt bath is used for other applications such as low-temperature silver soldering, the salts are usually composed of SODIUM CYANIDE (**extremely poisonous**) and SODIUM CARBONATE. Where higher temperature salt bath brazing is used the salts will be based on SODIUM CARBONATE.

The filler material is pre-placed on jigged assemblies and *the main advantage of salt baths is that very critical brazing temperatures, where the difference in melting points between parent metal and filler material are slight, can be precisely controlled.*

A comparison between dip-brazing methods is illustrated in Figure 11.21.

## 11.22 Electric induction brazing

Induction heating is used extensively on parts which are self-jigging and where very rapid heating is required. The component to be brazed is placed in the magnetic field of a coil carrying high-frequency electric current. **Induction coils provide the most rapid source of heat for brazing.** Parts may be assembled in jigs, provided effective heating is

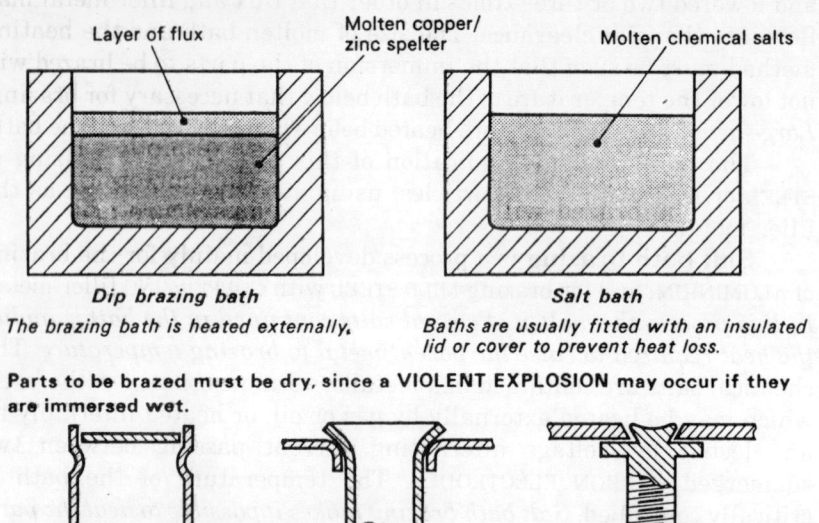

**Fig. 11.21 Two methods of dip brazing**

*Dip brazing bath* — Layer of flux; Molten copper/zinc spelter. The brazing bath is heated externally.

*Salt bath* — Molten chemical salts. Baths are usually fitted with an insulated lid or cover to prevent heat loss.

Parts to be brazed must be dry, since a VIOLENT EXPLOSION may occur if they are immersed wet.

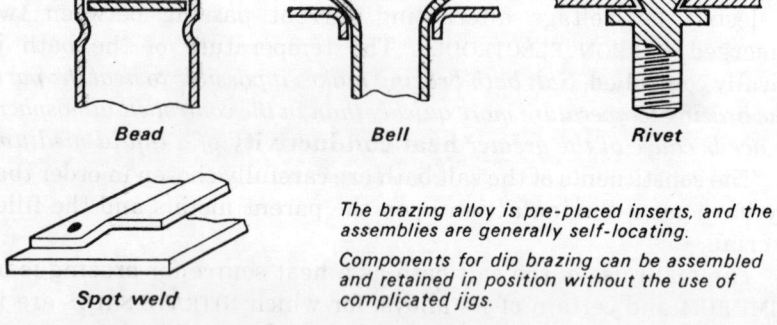

*Bead*, *Bell*, *Rivet*, *Spot weld*

The brazing alloy is pre-placed inserts, and the assemblies are generally self-locating.

Components for dip brazing can be assembled and retained in position without the use of complicated jigs.

not reduced by their use. In practice the operator loads one jig whilst another assembly is being brazed.

In this process, which is normally operated continuously, *the heat required for brazing is generated within the material itself from a* **high-frequency** *electric current passing through a high-conductivity water-cooled copper tube coil encircling the joint*. Most of the intense heat generated by this method is relatively near the surface. The interior is heated by **thermal conduction** from the hot surface. As a rule, *the higher the frequency, the shallower the heating*.

Internal and external coils may be used, as shown in Figure 11.22, and several different coil designs may be used.

A paste or powder flux is commonly used, and the filler material is pre-placed. Silver solders find an extensive use of this process.

## 11.23 Electric resistance brazing

In this process the heat required for brazing is developed by:

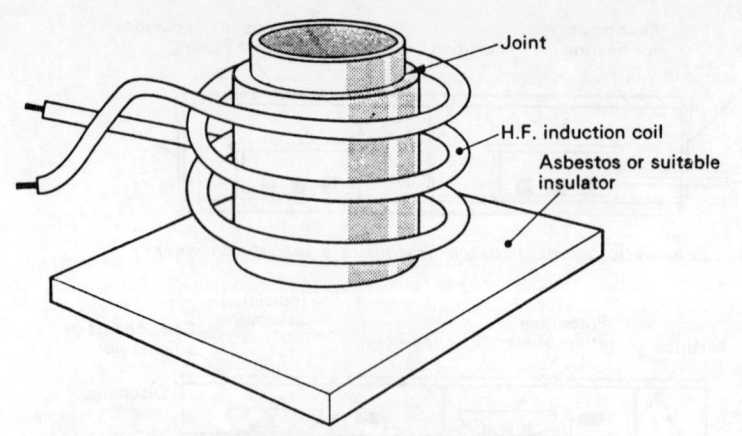

**Fig. 11.22 Electric induction brazing**

Joint; H.F. induction coil; Asbestos or suitable insulator.

Solid copper coils are also sometimes employed. The coils are normally insulated with glass fibre. In practice the work to be brazed is brought to the coils.

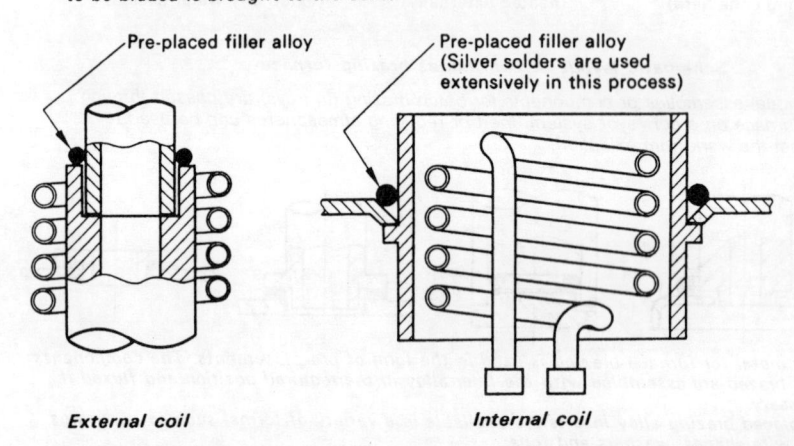

*External coil* — Pre-placed filler alloy

*Internal coil* — Pre-placed filler alloy (Silver solders are used extensively in this process)

It is usual for induction coils to be designed to surround the joint, but internal coils can be used for certain applications.

(i) **Resistance at the joint interface**, as in 'resistance welding';
(ii) **Resistance in carbon electrodes** which conduct heat to the joint faces.

**The basic principle of resistance heating is that a high electrical current at low voltage is passed through a resistive circuit in such a way that a heat source is generated which is suitable for brazing operations.**

Heating can be precisely localised and this ensures no general loss of mechanical properties throughout the parent metal.

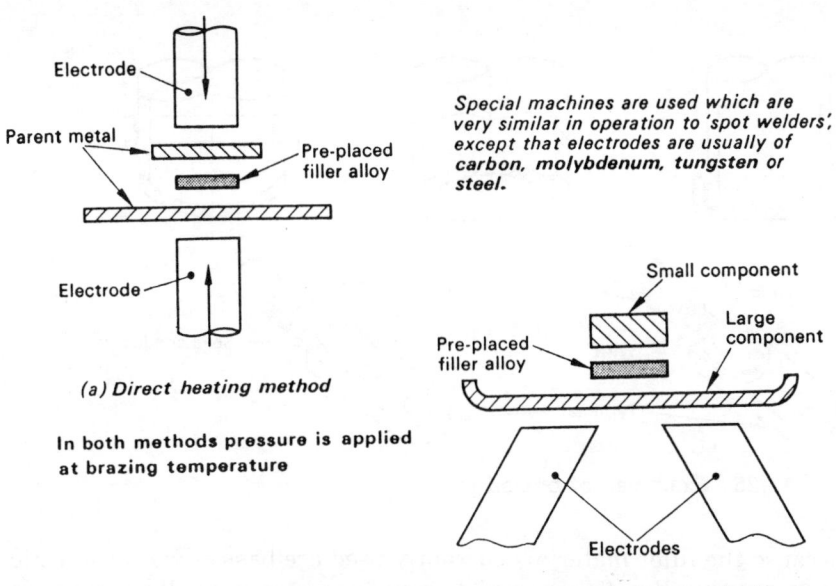

Fig. 11.23 Electric resistance brazing

Fig. 11.24 The two basic types of joint used in brazing

One main advantage of resistance brazing is that normally no jigs are required, for the electrodes themselves act as jigs and hold the components in correct relationship.

There are two methods of heating, either DIRECT or INDIRECT, and these are shown schematically in Figure 11.23.

With direct heating (Figure 11.23(*a*)), the assembly is placed between the electrodes (as in spot welding) so that the current passes from one electrode through the parent metals and pre-placed brazing insert to the other electrode.

In the indirect heating method (Figure 11.23(*b*)), the electrodes are so placed that the current does not pass through the joint, but *passes through only* one component of the assembly. This method finds its greatest use where it is required to braze a relatively small component to a large one.

In practice, it is recommended that only carbon electrodes be used, and the current is passed through the large or backing component, from which the heat is CONDUCTED to the filler material and the other component.

## 11.24 Types of brazed joints

There are basically two types of joints used in brazing operations. They are the LAP and BUTT joints, and these are illustrated in Figure 11.24.

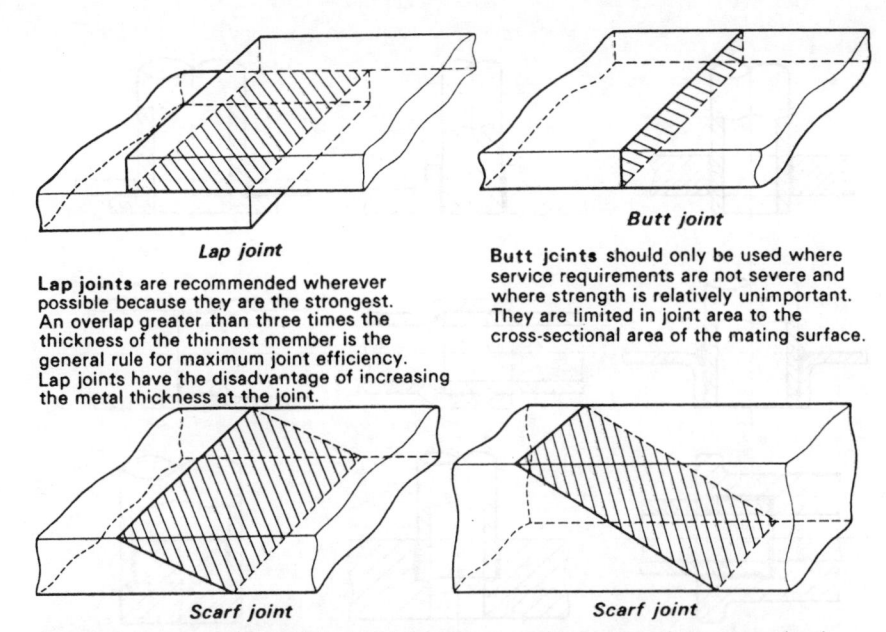

There are, however, many combinations and variations of these joints designed to meet the requirements of specific conditions.

The factors affecting the strength of hard soldered joints are the same as those which apply to soft soldering, *but since the mechanical strength of the brazing alloys used is very much greater than that of soft solders, the overall strnegth of the joint is higher.* A selection of the many joint designs for brazing are shown in Figure 11.25.

## 11.25 Braze welding (bronze welding)

Like brazing processes, **braze welding** is used to join either similar or dissimilar metals by using a filler material of lower melting point than either of the components to be joined.

Before continuing, it is necessary to clearly distinguish between the two terms **brazing** and **braze welding**.

When the parts to be joined are tightly fitted together, so that *the filler alloy flows through the joint by* **capillary attraction**, the process is known as **brazing**.

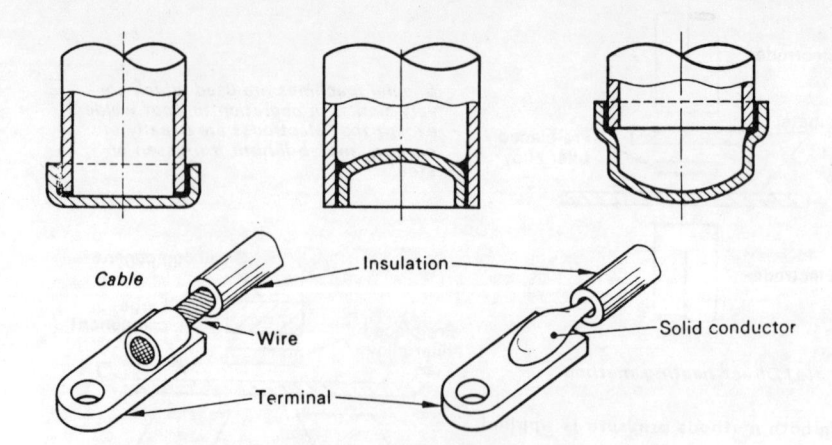

Fig. 11.25 Examples of brazed joints

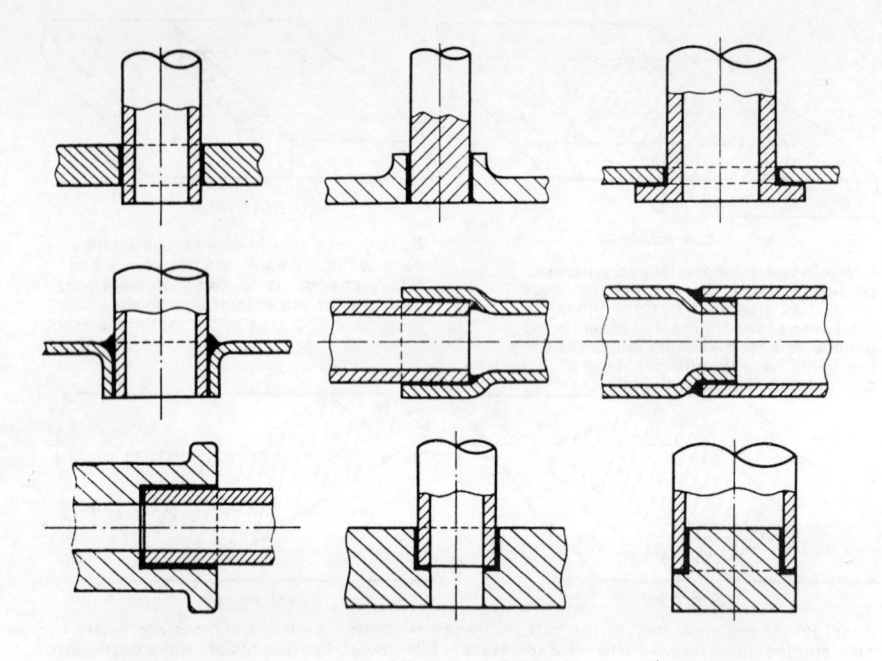

If the parts are not closely fitted, so that a relatively large quantity of filler alloy is used, the process is called **braze welding**.

Braze welding has been termed for many years **bronze welding**. This was because the process resembled welding and, originally, the filler material used was a COPPER/PHOSPHORUS/TIN alloy, in other words a BRONZE. However, the term bronze welding is now a misnomer because the filler materials currently used are basically a 60/40 ratio COPPER/ZINC alloy (brass) which may incorporate small amounts of SILICON, MANGANESE, and/or NICKEL depending upon the application. Thus in general terms, modern filler alloys used for braze welding are **brasses**. A list of commonly used filler materials are given in Table 11.10.

## 11.26 Basic principles of braze welding

British Standard 1742 gives the following definition of braze welding:

**Table 11.10 Composition of copper alloy filler rods for braze welding**

| BRITISH STANDARD 1453 Type | Copper min. max. | Zinc min. max. | Silicon min. max. | Tin max. | Manganese min. max. | Nickel min. max. | MELTING POINT °C |
|---|---|---|---|---|---|---|---|
| C2 | 57 to 63 | 36 to 42 | 0.2 to 0.5 | 0.5 | — | — | 875 |
| C4 | 57 to 63 | 36 to 42 | 0.15 to 0.3 | 0.5 | 0.05 to 0.25 | — | 895 |
| C5 | 45 to 53 | 34 to 42 | 0.15 to 0.5 | 0.5 | 0.5 | 8 to 11 | 910 |
| C6 | 41 to 45 | 37 to 41 | 0.2 to 0.5 | 1.0 | 0.2 | 14 to 16 | |

Type C2 is termed a SILICON-BRONZE filler rod and is specially recommended for the braze welding of BRASS and COPPER sheet and tubes as are used for sanitary and hot water installations. It is also suitable for the braze welding of MILD STEEL and GALVANISED STEEL.

Type C4 is termed a MANGANESE-BRONZE filler rod, it has a higher melting point and is especially suitable for braze welding CAST or MALLEABLE IRON, and also for building up worn parts such as gear teeth.

Types C5 and C6 are NICKEL-BRONZE rods which are recommended for braze welding STEEL or MALLEABLE IRON where the **highest mechanical strength** is required. These are the high melting point welding rods and have a valued application in the reclaiming and building up of wearing surfaces.

*A method of joining materials with a gas-welding technique by means of the deposition of molten, copper-rich filler metal on the parts to be joined. The bonding of the joint results from the wetting of the unmelted surfaces and interdiffusion of the filler metal and the parent metal.*

In braze welding, as distinct from **fusion welding**, the melting point of the filler material is lower than that of the materials to be joined. This means that *the process requires a lower temperature than fusion welding, and troubles such as distortion and oxidation of the parent metal, which are generally associated with fusion welding, are reduced considerably.*

One important general principle relating to the preparation of joints is that *the workpiece must be thoroughly clean and that scale, oil, grease, paint or stains must be removed before attempting the braze welding operation.*

*The main advantage of braze welding over fusion welding is that the parent metal is only 'tinned' and* **never melted**.

Fluxes of the BORAX type should be used. A paste or a powdered flux wetted to make a paste can be more thoroughly and evenly distributed than a flux in powder form. If a powder is used the rod should be heated at one end and dipped into it so that a 'tuft' of flux will adhere to the filler metal.

The parent metal is preheated with the flame, care being taken not to cause it to melt, especially if it is copper. The main function of the flame is to heat the parent metal's surface locally to about the melting point of the filler metal. This temperature, in excess of 900°C, is not sufficient to cause the parent metal to melt, but will permit alloying between it and the filler metal.

A globule of filler metal is first melted off and deposited in the joint. Circular motion of the welding flame will cause the deposit of filler metal to tin the surface. When local tinning of the joint has taken place, the filler metal should be reheated to the fluid condition and additional filler rod applied on to the weld deposit already in the joint and not in front of it.

This fresh globule of filler metal is manipulated with the flame and pushed forward to form an advanced 'wet' edge in the joint. This operation is then repeated until the joint is complete.

*The strength of the deposited filler metal will be found equal to, and in some cases in excess of the strength of the parent metal.* The overall strength of a braze welded joint is, therefore, entirely dependent on the soundness of the bond between the filler and parent metal.

The operator should watch that the filler metal is actually tinning the surface immediately in front of the welding flame. This will indicate that the joint is at the correct temperature, and that it has been properly prepared. Figure 11.26 illustrates the 'wetting' action in braze welding.

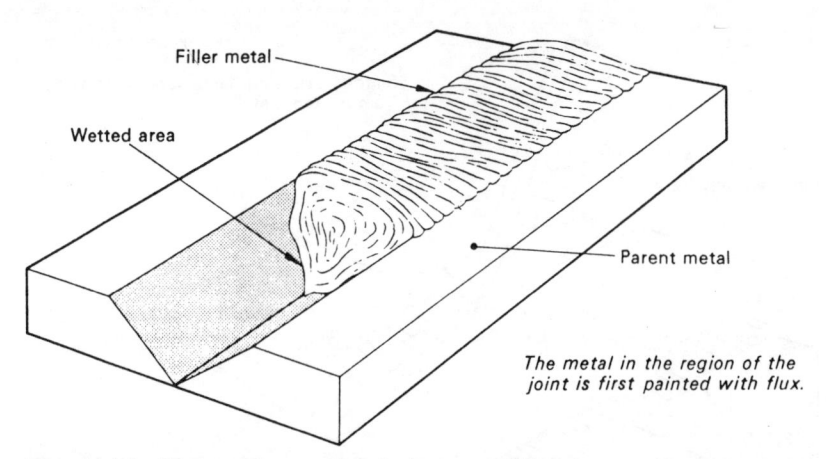

**Fig. 11.26  The wetting action in braze welding**

Braze welding utilises the concentrated head of the **oxy-acetylene flame**. By careful control of the torch this heat can be localised to bring any area of the joint up to braze welding temperature, and enable perfect control of the molten filler metal to be achieved, without capillary attraction.

The **leftward method** of welding is employed, the torch nozzle should be about two sizes smaller than for butt welds on metals of the same thickness, and the flame condition should be slightly OXIDISING. Figure 11.27 illustrates the angles of rod and welding torch for the braze welding of CAST IRON.

## 11.27  Metals that can be braze welded

The following metals may be joined by braze welding:

1. Copper;
2. Mild steel;
3. Galvanised mild steel;
4. Malleable iron;
5. Cast iron;
6. A combination of any two of the above.

## 11.28  Types of braze-welded joints

Braze welding depends for its overall strength on the high SHEAR STRENGTH of the bond between the filler alloy and the parent metal.

The edges of the parent metal should, where possible, be bevelled

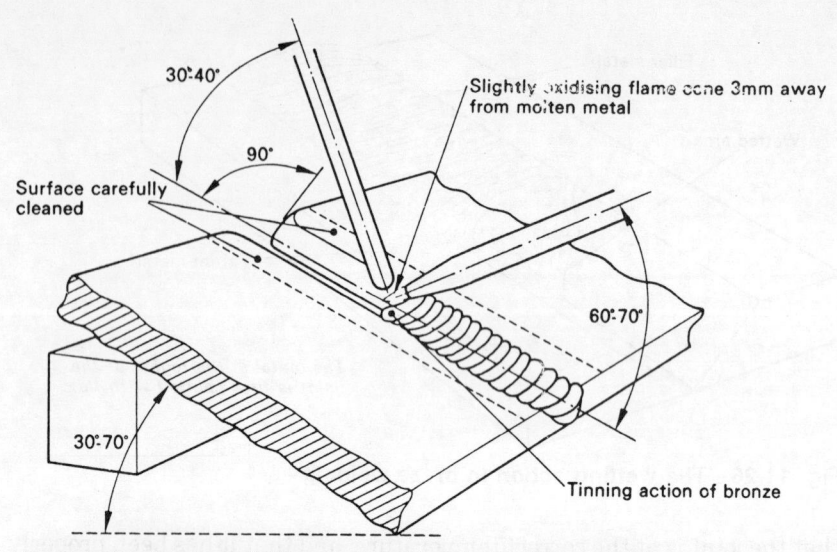

**Fig. 11.27** Angles of filler rod and torch for braze welding

to give a 'Vee' of 90° by filling, grinding or chipping. The upper and lower surfaces of the metal should be cleaned by wire brushing for a short distance on either side of the 'Vee'. Where bevelling is not possible, the width of the filler metal deposit at the top should be at least twice the thickness of the parent metal, and the deposit should penetrate just over the edge of the underside. Types of joints suitable for braze welding are shown in Figures 11.28 and 11.29.

## 11.29 Full fusion welding

**Fusion welding**, as the name implies, is a thermal joining process in which the parent metal to be joined is melted and caused to fuse together with or without the use of filler metal.

The principle of fusion welding is illustrated by two typical joint arrangements — one requiring a separate filler metal, the other self-supplying — in Figure 11.30.

## 11.30 Heat sources for welding

The two basic types of welding are commonly referred to as GAS WELDING and ARC WELDING which will now be compared.

### OXY-ACETYLENE WELDING

In general all fusion welding processes involve a high temperature heat source, especially when joining high-melting-temperature materials. The

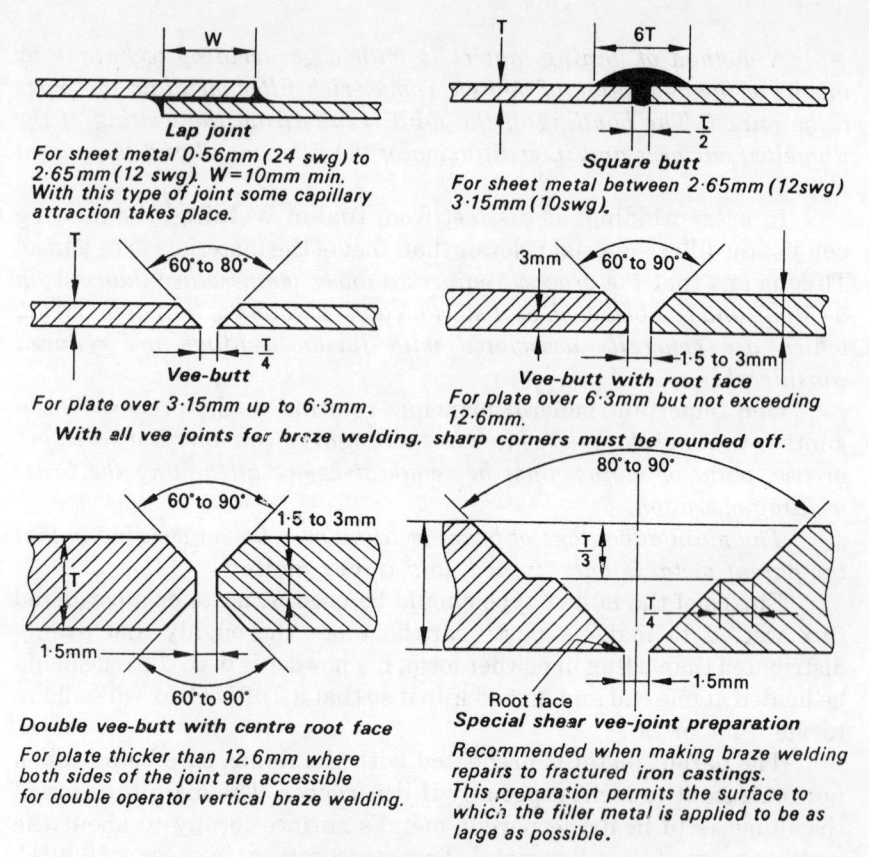

*Lap joint*
For sheet metal 0·56mm (24 swg) to 2·65mm (12 swg). W=10mm min. With this type of joint some capillary attraction takes place.

*Square butt*
For sheet metal between 2·65mm (12swg) 3·15mm (10swg).

*Vee-butt*
For plate over 3·15mm up to 6·3mm.

*Vee-butt with root face*
For plate over 6·3mm but not exceeding 12·6mm.

With all vee joints for braze welding, sharp corners must be rounded off.

*Double vee-butt with centre root face*
For plate thicker than 12·6mm where both sides of the joint are accessible for double operator vertical braze welding.

*Special shear vee-joint preparation*
Recommended when making braze welding repairs to fractured iron castings. This preparation permits the surface to which the filler metal is applied to be as large as possible.

**Fig. 11.28** Forms of braze-welded joints

oxy-acetylene flame provides the heat required to melt most of the metals and alloys in commercial use.

Other OXY-FUEL GAS mixtures can be used, but the vast majority of gas-welding processes employ ACETYLENE as the fuel gas. This is because acetylene is the most economical gas to use in conjunction with relatively pure OXYGEN supplied from high-pressure cylinders, to give a flame with a maximum temperature of 3200°C.

Figure 11.31 shows the main characteristics of the oxy-acetylene welding flame, and gives a comparison of flame temperatures obtained with other fuel gas mixtures.

### METALLIC-ARC WELDING

By contrast the heat required for welding in arc-welding processes is not obtained by the combustion of a fuel gas with oxygen.

Examples of 'Tee fillet' joints in mild steel.

Sleeve joint for mild steel pipes
The mating surfaces of the pipes should be accurately machined. The sleeve and the pipes should be a close fit.

Tube plate joint for mild steel assemblies

(a) Bell type butt joint.

(b) Diminishing joint
The end of the smaller pipe is belled out to fit into the large pipe.

(c) Branch tee joint
Suitable for small pipes of equal and unequal diameters.

(d) Bell type tee joint
Suitable for all diameters and thickness of pipes.

(e) Stub branch joint.

(f) Short bell branch joint.

Branch cut at angle

Braze welded joints in copper pipes

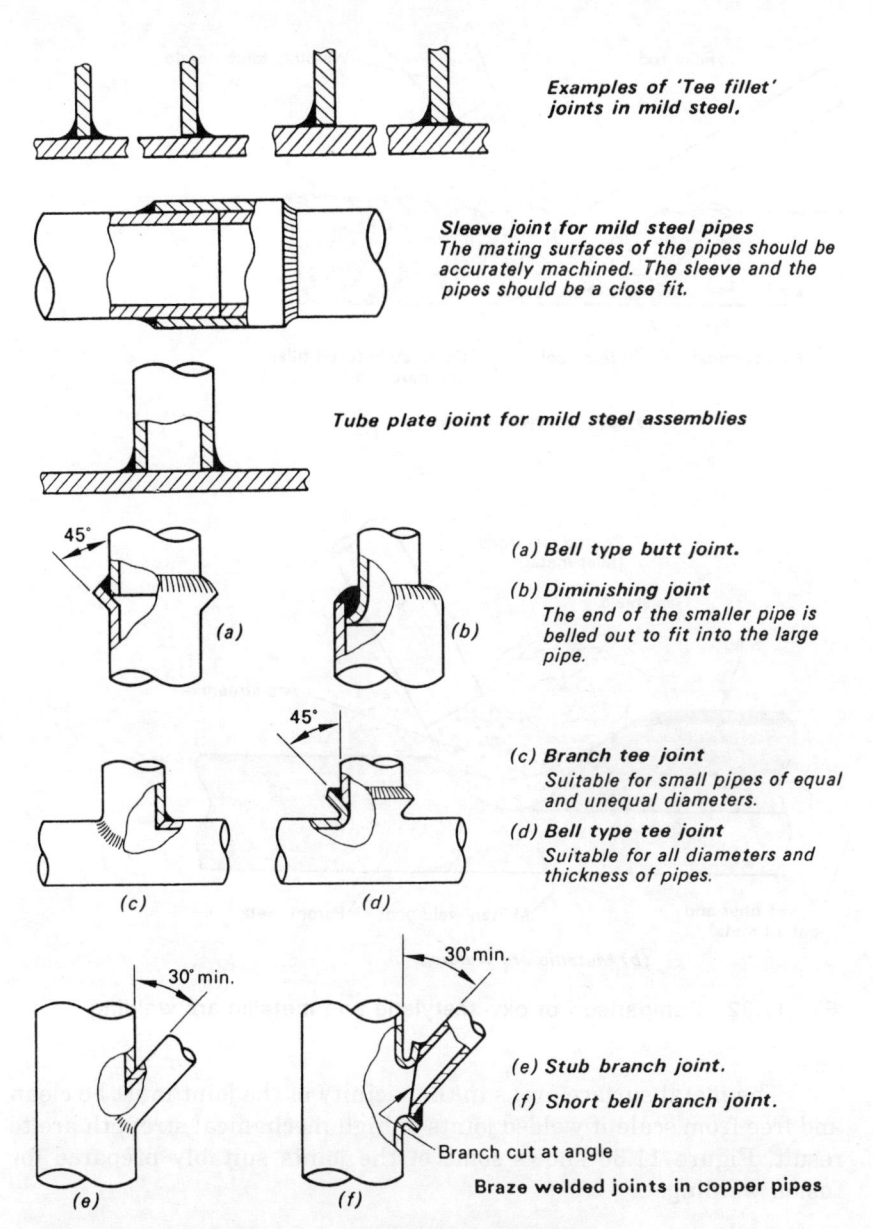

**Fig. 11.29 Applications of braze-welded joints**

SINGLE VEE BUTT requires extra metal.

The edges of vee are melted and fused together with the molten filler metal.

SQUARE BUTT preparation used when welding thin sheet metal. The small folded edges provide their own filler metal.

The edges are melted and fuse together to form the welded joint *without the addition of filler metal*.

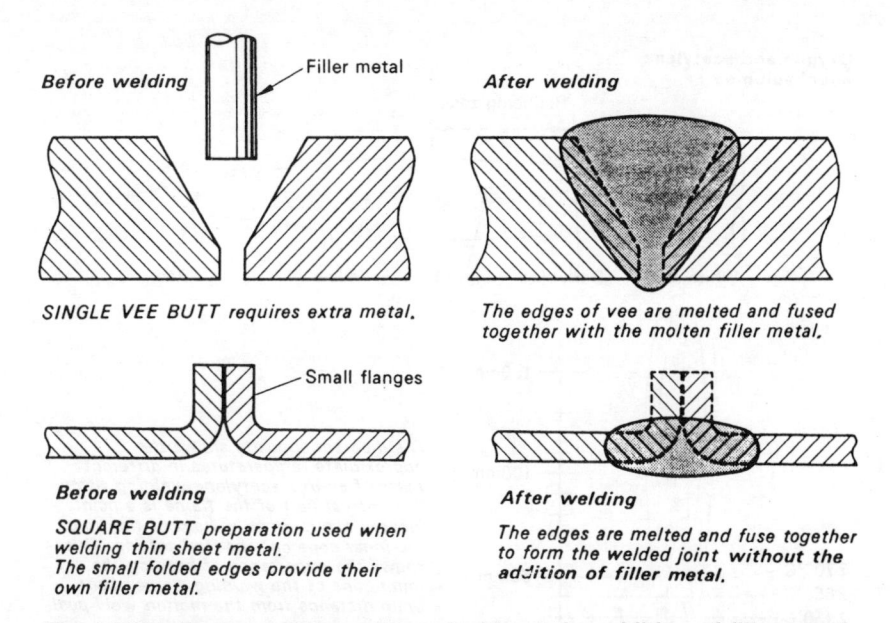

**Fig. 11.30 Fusion welding with and without the addition of filler metal**

An ELECTRIC ARC is simply a prolonged spark between two terminals of an electric circuit, which in the case of arc welding is the workpiece and the electrode.

The arc is extremely hot (approximate average temperature in the region of 6000°C). This very high temperature, being concentrated in a relatively small area, is sufficient to melt instantly the surface of the workpiece and the end of a metal electrode. In metallic-arc welding the electrode not only provides a means of melting the surface of the workpiece, but actually provides metal to fill the joint. Figure 11.32 illustrates the basic differences between the oxy-acetylene and manual metallic-arc welding processes.

## 11.31 Joint preparation

Although when joining metals by the braze-welding method basic welding joint preparations may be used, joints which are fusion welded are very much stronger. This is because their overall strength is not dependent upon the strength of the filler metal — as in a brazing and braze welding. *Whether filler metal is used or not, when the joint edges are melted and fused together, the resultant weld is as strong as the parent metal.*

In order to achieve good penetration and full fusion of both the joint edges and the filler metal, edges of the parent metal to be joined must

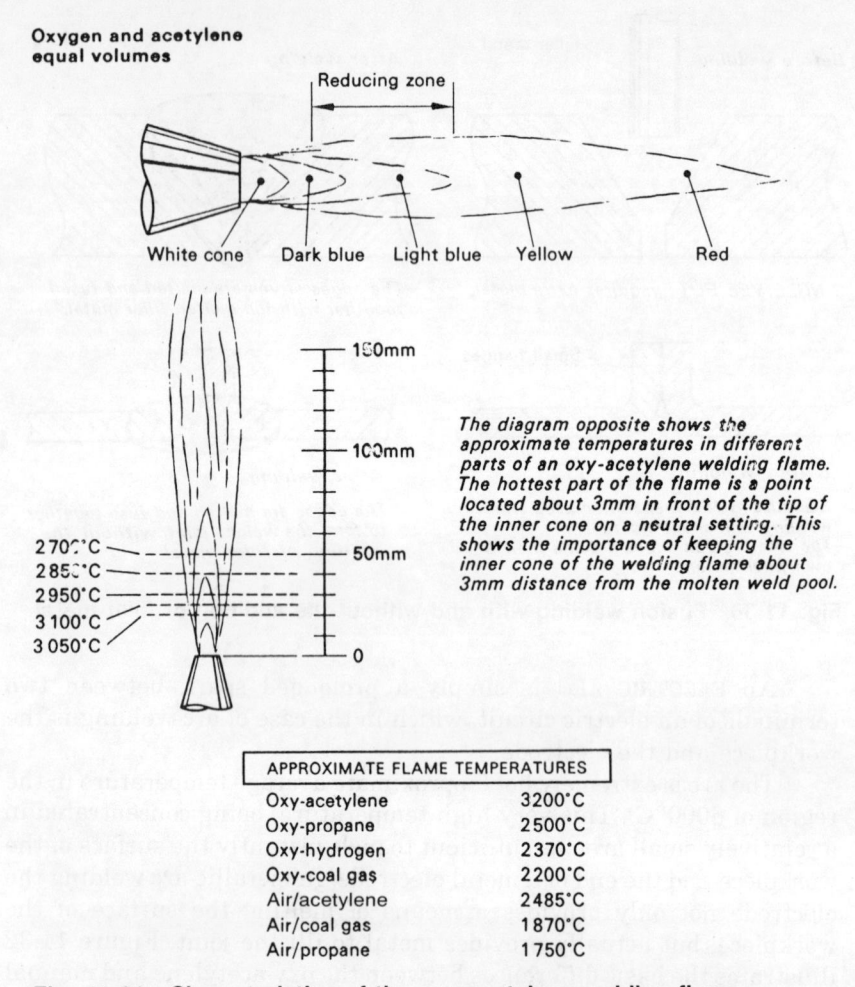

**Fig. 11.31 Characteristics of the oxy-acetylene welding flame**

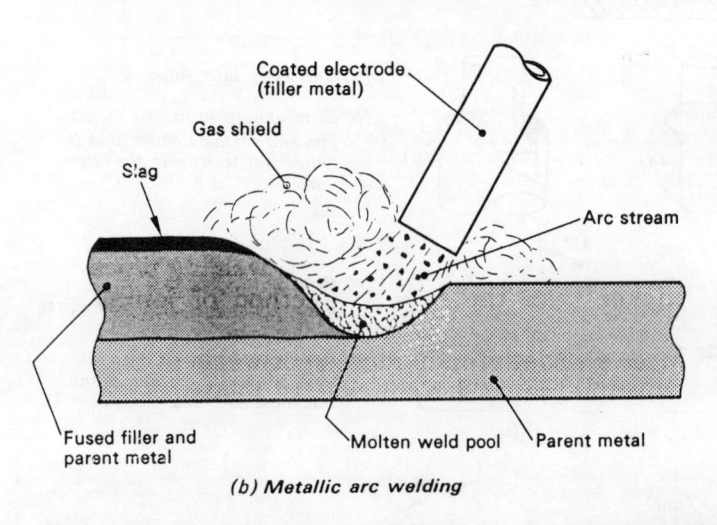

| APPROXIMATE FLAME TEMPERATURES | |
|---|---|
| Oxy-acetylene | 3200°C |
| Oxy-propane | 2500°C |
| Oxy-hydrogen | 2370°C |
| Oxy-coal gas | 2200°C |
| Air/acetylene | 2485°C |
| Air/coal gas | 1870°C |
| Air/propane | 1750°C |

**Fig. 11.32 Comparison of oxy-acetylene and metallic arc welding**

be carefully prepared and accurately aligned. With the welding of plate, edge preparation can become very costly, therefore, consideration has to be given to the most economical method to be adopted. The following are some of the ways in which plate edges are prepared for welding:

1. By use of guillotines — limited to straight line cuts;
2. By machining — useful when bevelling circumferential edges;
3. By grinding — very time consuming;
4. By flame cutting — this method is widely used because of its versatility — can produce straight and contoured edges suitable for either 90° butt or 'Vee' butt joints.

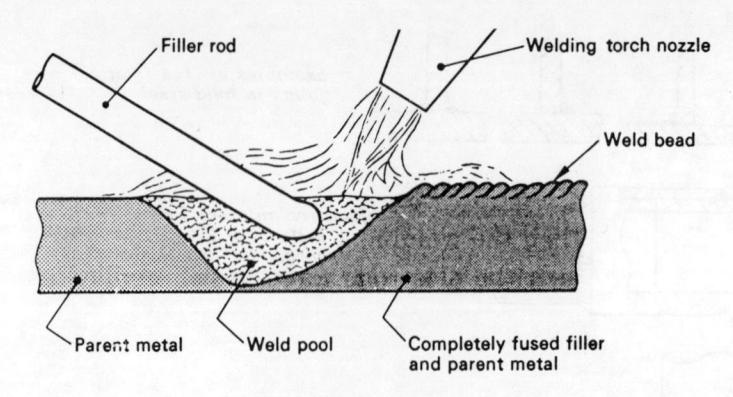

The metal surface areas in the vicinity of the joint must be clean and free from scale if welded joints of high mechanical strength are to result. Figure 11.33 shows some of the joints suitably prepared for fusion welding.

## 11.32 Resistance welding

*A number of welding processes are based upon the principle that when an electric current is sent through a resistance, heat is developed.*

Basically, with resistance welding, the heat required at the joints to be welded is generated by the resistance offered through the workparts

Dimensions in millimetres

| THICKNESS OF METAL | DIAMETER OF WELDING ROD | EDGE PREPARATION |
|---|---|---|
| Less than 1·0 | 1·2 to 1·6 | |
| 1·0 to 3·2 | 1·6 to 3·2 | 0·8-3·2 |
| 3·2 to 4·8 | 3·2 to 5·0 | 80° V, 1·6-3·2 |
| 4·8 to 8·0 | 3·2 to 5·0 | 3·2-4·0 |
| 8·0 to 16·0 | 5·0 to 6·0 | 60° V, 3·2-4·0 |
| 16·0 and over | 6·0 | Top V 60°, Bottom V 80°, 3·2-4·0 |

**Fig. 11.33 Edge preparation**

**Fig. 11.34 Principles of resistance welding**
(a) Electric spot-welding machine (Schematic diagram)
(b) Spot welding
(c) Seam welding
(d) Projection welding

to the relatively short time flow of electricity at **low potential and high current density**.

Resistance welding is also termed **pressure welding** because **force is always applied** before, during, and after the application of the current to ensure a continuous electrical circuit and to FORGE the heated parts together.

As shown in Figure 11.34(a), the electrode tips are held by two copper arms through which water is circulated to keep the tips relatively cool. The lower electrode tip is stationary, but the upper one is movable which allows an adequate gap for work entry. Pressure on a pedal brings the electrodes together and clamps the workpiece in position. The machine is fitted with a device which controls the current and the timing. Further pressure on the pedal activates a switch and an electric current is passed through the workpiece until the welding cycle is completed. There are four stages of time in the spot-welding cycle:

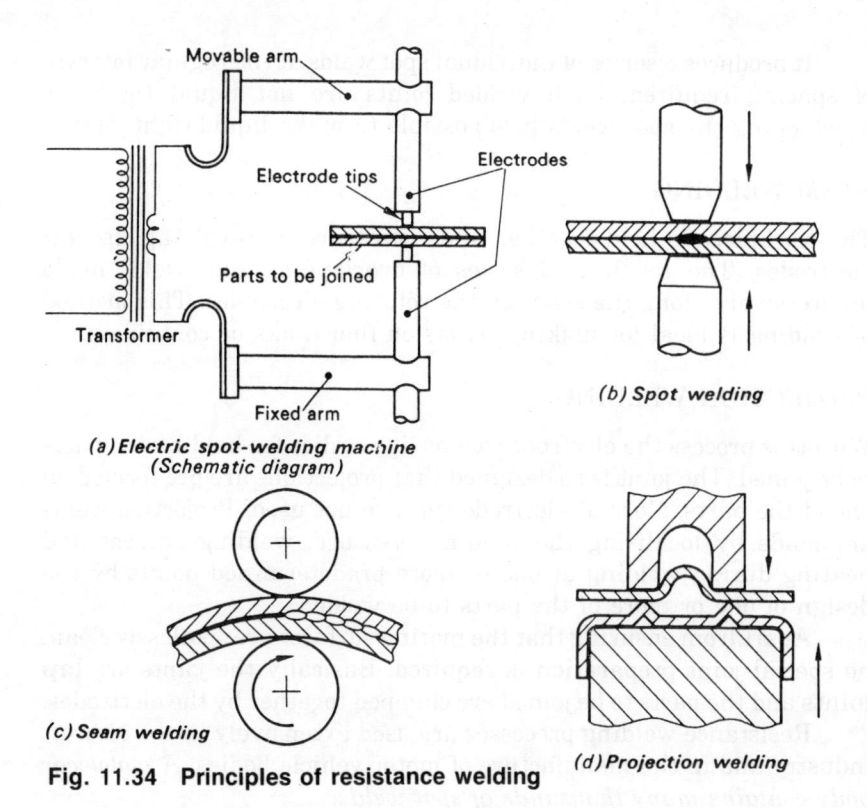

(i) **Squeeze time** — time between the first application of the electrode force and the first application of the welding current.
(ii) **Weld time** — when the welding current flows.
(iii) **Hold time** — when the electrode force is still applied after the current has stopped flowing.
(iv) **Off time** — time during which the electrodes are not in contact with the work.

The maximum temperature achieved is ordinarily above the melting point of the parent metal. Unlike other thermal jointing processes no filler metal is required to make a resistance welded joint.

In this chapter only three types of resistance welding will be considered, namely SPOT, SEAM and PROJECTION welding processes and are illustrated in Figure 11.34(b), (c), (d).

## SPOT WELDING

This is the most common of the resistance welding processes. It is much quicker than the riveting method of joining sheet metal parts.

It produces a series of individual spot welds at the regular interval of spacing required. Such welded joints are not liquid tight. By overlapping the spot welds it is possible to make liquid-tight joints.

## SEAM WELDING

The workpiece is held together under pressure by revolving circular electrodes. The result is a series of overlapping spot welds made progressively along the seam by the rotating electrodes. This method of welding is ideal for making seams on fuel tanks or containers.

## PROJECTION WELDING

With this process the electrodes act as jigs or dies for holding the parts to be joined. The joint is so designed that projections are pre-formed on one of the parts. Normal electrode tips are not used. Projection welds are made by localising the welding pressure, welding current and heating during welding at one or more predetermined points by the design of one or more of the parts to be welded.

Apart from ensuring that the mating surfaces are spotlessly clean, no special joint preparation is required. Basically the joints are **lap joints** and the parts to be joined are clamped together by the electrodes.

Resistance welding processes are used extensively in the aircraft industry and in the manufacture of motor vehicle bodies. *A motor car body contains many thousands of spot welds.*

## 11.33 Thermal joining processes — temperature relationships

The temperatures relating to the melting points of common workshop materials and thermal joining processes are shown in Figure 11.35.

Table 11.11 gives the melting temperature range for a number of alloys.

Figure 11.36 shows the temperature range for fusible alloys in greater detail. A knowledge of these melting points is important if the correct thermal joining process is to be selected.

## 11.34 Use of adhesives

Traditionally, adhesives fell into two categories:

1. **Glues**. These were made from the bones, hooves and horns of animals and the bones of fishes. Derivatives of milk and blood were also used. Glues were largely used for joining wood and were

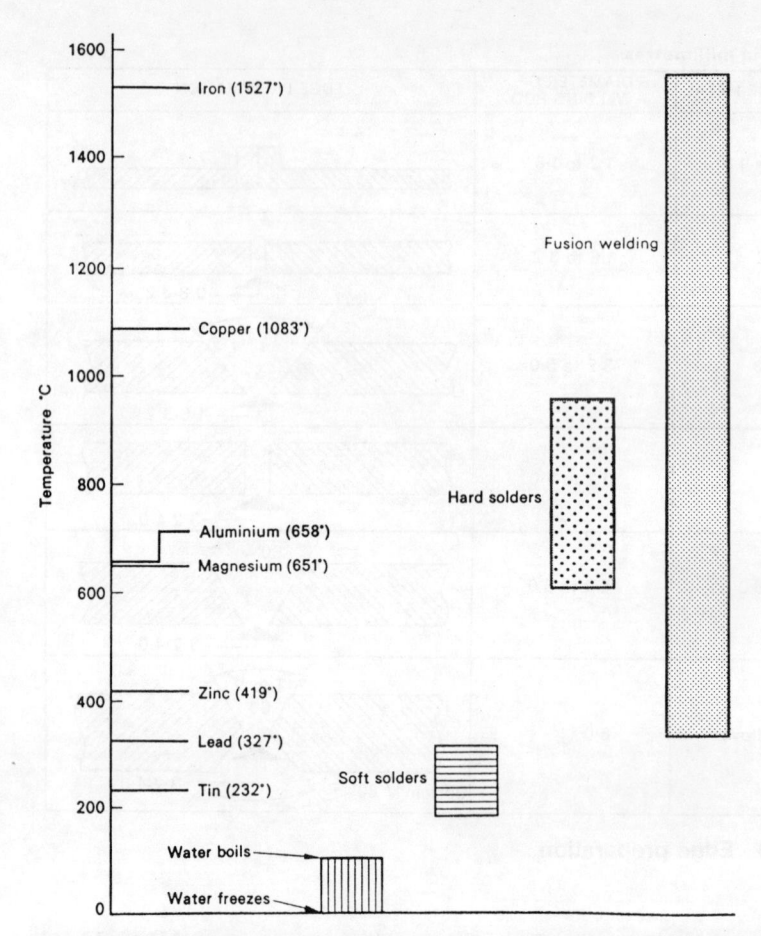

**Fig. 11.35 Temperature relationship — melting points and thermal jointing processes**

particularly important in the funiture and toy manufacturing industries.

2. **Gums**. These were made from vegetable matter. Resin and rubber being extracted from trees, and starches being extracted from the by-products of flour milling.

Although natural glues and gums are still widely used for relatively low strength applications, they are being increasingly supplanted by high strength synthetic adhesives.

Shortly before the Second World War, the growth of the high-polymer (plastics) industry led to the development of many new synthetic materials suitable for high strength adhesives. These new adhesives are a considerable advance on traditional, natural glues and gums and are

## Table 11.11 Melting ranges of alloys

| ALLOY | °C |
|---|---|
| Fusible alloys | 70– 180 |
| Soft solders | 183– 310 |
| Silver solders | 620– 740 |
| Brazing and bronze-welding alloys | 860– 950* |
| Brasses | 850–1000 |
| Gunmetal | (approx) 995 |
| Monel | 1300–1350 |
| Inconel | 1395–1425 |
| Stainless steels | 1430–1450 |

*Note:* Special 'bronzes' have been developed for the brazing of TUNGSTEN CARBIDE inserts into 'rock drills' where high temperature HEAT TREATMENT of the rock drill shank must be carried out at temperatures between 800 and 1000°C. Some of these have melting ranges of 980–1030°C, and even 1090–1100°C.

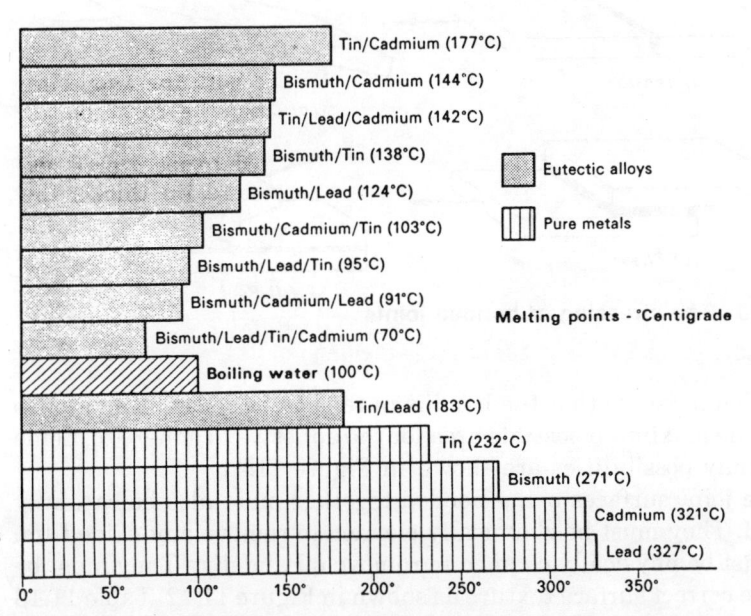

Fig. 11.36 Composition and melting points of fusible alloys

used extensively for bonding together materials in a wide variety of industries. Table 11.12 lists some of the more important advantages and limitations of adhesive bonding as compared with the mechanical and thermal joining processes discussed so far.

## Table 11.12 Advantages and limitations of bonded joints

**Advantages**
1. The ability to joint dissimilar materials, and materials of widely different thicknesses.
2. The ability to join components of difficult shape that would restrict the application of welding or riveting equipment.
3. Smooth finish to the joint which will be free from voids and protrusions such as weld beads, rivet and bolt heads, etc.
4. Uniform distribution of stress over entire area of joint. This reduces the chances of the joint failing in fatigue.
5. Elastic properties of many adhesives allow for flexibility in the joint and give it vibration damping characteristics.
6. The ability to electrically insulate the adherends and prevent corrosion due to galvanic action between dissimilar metals.
7. The joint will be sealed against moisture and gases.
8. Heat sensitive materials can be joined.

**Limitations**
1. The bonding process is more complex than mechanical and thermal processes, i.e., the need for surface preparation, temperature and humidity control of the working atmosphere, ventilation and health problems caused by the adhesives and their solvents. The length of time that the assembly must be jigged up whilst setting (curing) takes place.
2. Inspection of the joint is difficult.
3. Joint design is more critical than for many mechanical and thermal processes.
4. Incompatibility with the adherends. The adhesive itself may corrode the materials it is joining.
5. Degradation of the joint when subject to high and low temperatures, chemical atmospheres, etc.
6. Creep under sustained loads.

## 11.35 The adhesive bond

Figure 11.37 shows a typical bonded joint and explains the terminology used for the various features of the joint.

The strength of the bond depends upon two factors:

(*a*)  Adhesion
(*b*)  Cohesion

*Adhesion* is the ability of the bonding material (adhesive) to stick (adhere) to the materials being joined (adherends). There are two ways in which this bond can occur and these are shown in Figure 11.38.

*Cohesion* is the ability of the adhesive to resist the applied forces within itself.

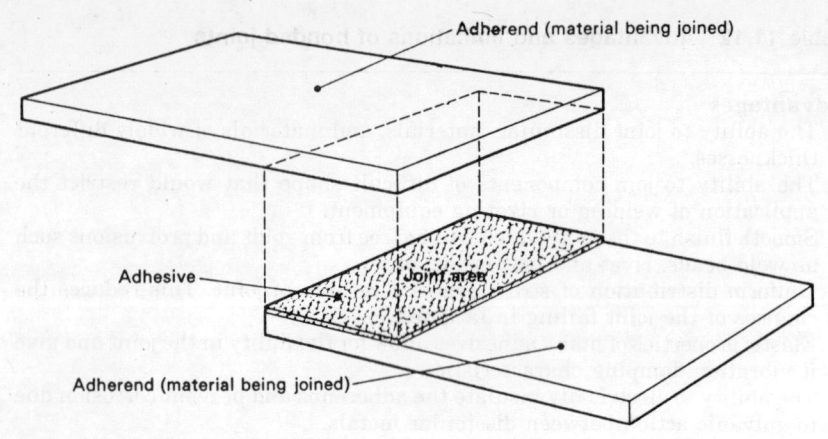

**Fig. 11.37 The bonded joint**

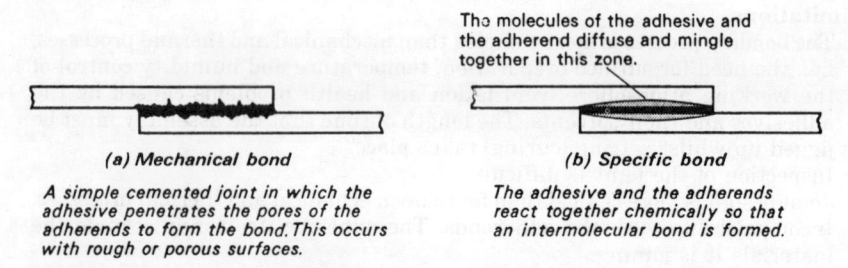

**Fig. 11.38 Types of bond**

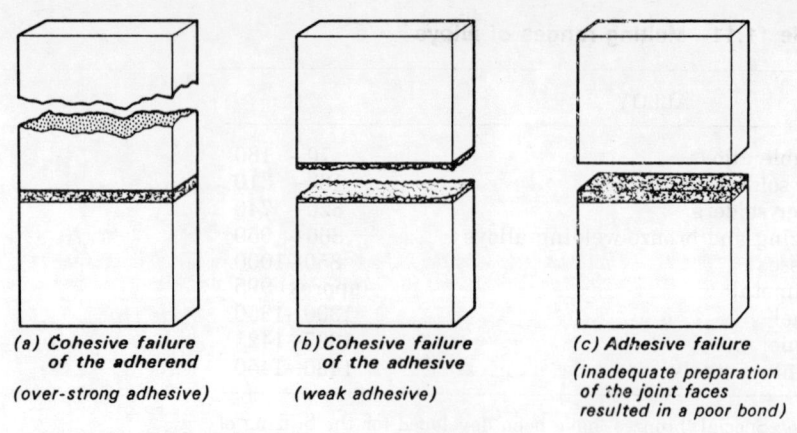

(a) Mechanical bond — A simple cemented joint in which the adhesive penetrates the pores of the adherends to form the bond. This occurs with rough or porous surfaces.

(b) Specific bond — The adhesive and the adherends react together chemically so that an intermolecular bond is formed.

**Fig. 11.39 Adhesive and cohesive failure**

(a) Cohesive failure of the adherend (over-strong adhesive)
(b) Cohesive failure of the adhesive (weak adhesive)
(c) Adhesive failure (inadequate preparation of the joint faces resulted in a poor bond)

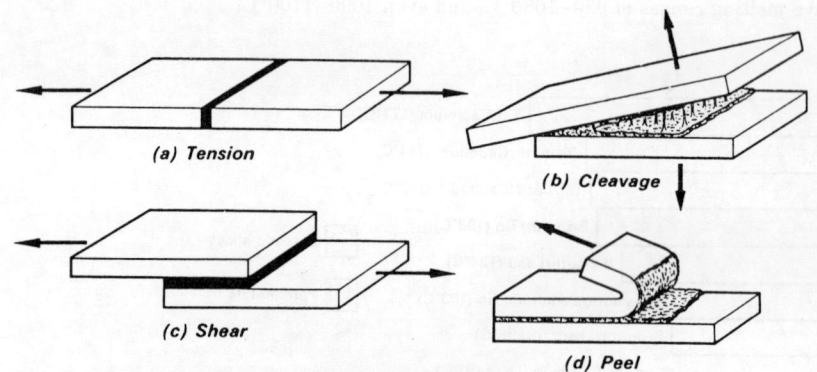

**Fig. 11.40 The stressing of bonded joints**

(a) Tension  (b) Cleavage  (c) Shear  (d) Peel

## 11.36 Failure in adhesive bonds

Figure 11.39 shows three ways in which a bonded joint can fail. These failures can be prevented by careful design of the joint, correct selection of the adhesive, careful preparation of the joint surfaces and control of the working environment (cleanliness, temperature and humidity).

## 11.37 The strength of bonded joints

No matter how effective the adhesive is and how carefully it is applied, the joint will be a failure if it is not correctly designed and executed. It is bad practice to apply an adhesive to a joint that was originally proportioned for bolting, riveting or welding. The joint must be proportioned to exploit the properties of adhesives.

Most adhesives are relatively strong in tension and shear; and weak in cleavage and peel. These terms are explained in Figure 11.40.

The shape of the joint is also important. Not only must it restrict the applied forces so that the bond is in tension or shear, it must also present the maximum possible area for the adhesive to adhere to. Some of the many possibilities are shown in Figure 11.41.

The joint surfaces being bonded together must also be carefully prepared. They must be chemically cleaned to remove any oxide film. They must be physically cleaned to remove oil and dirt. They must be given the correct surface texture as shown in Figure 11.42. Table 11.13 lists the recommended procedures for preparing the joints of several groups of materials.

The adhesive must 'wet' the joint surfaces throrougly, otherwise voids will occur and the bonded area will be considerably less than the theoretical maximum. This will weaken the joint considerably. Figure 11.43 shows the effect of 'wetting' on the formation of the joint.

## Figures (left column)

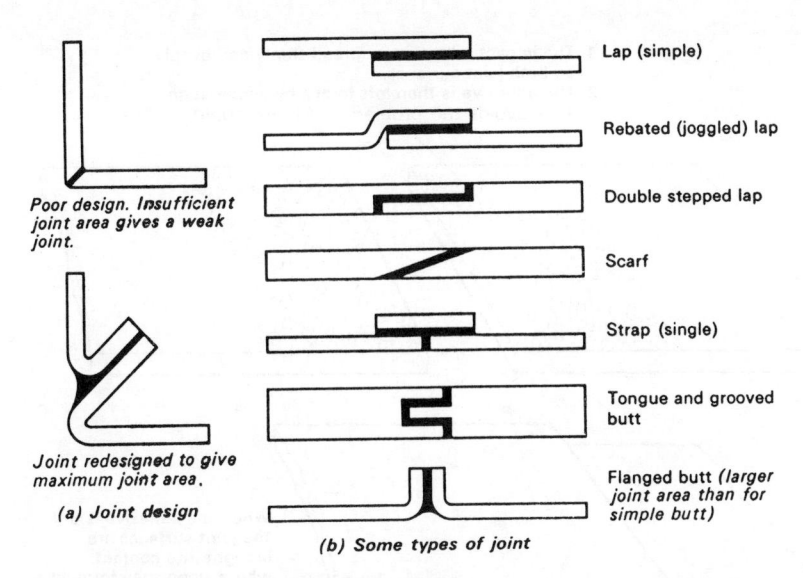

**Fig. 11.41 Some typical bonded joints**

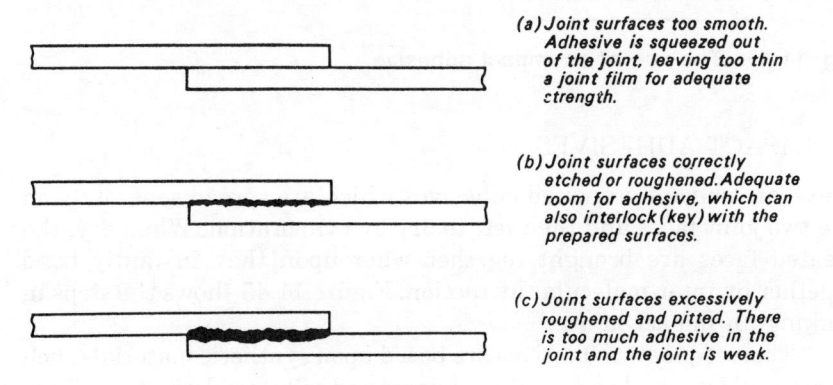

(a) Joint design
- Poor design. Insufficient joint area gives a weak joint.
- Joint redesigned to give maximum joint area.

(b) Some types of joint
- Lap (simple)
- Rebated (joggled) lap
- Double stepped lap
- Scarf
- Strap (single)
- Tongue and grooved butt
- Flanged butt (larger joint area than for simple butt)

**Fig. 11.42 Joint surface preparation**

(a) Joint surfaces too smooth. Adhesive is squeezed out of the joint, leaving too thin a joint film for adequate strength.

(b) Joint surfaces correctly etched or roughened. Adequate room for adhesive, which can also interlock (key) with the prepared surfaces.

(c) Joint surfaces excessively roughened and pitted. There is too much adhesive in the joint and the joint is weak.

## 11.38 Thermo-plastic adhesives

These are materials that soften when they are heated and harden again when cooled. They may be applied to the joint in three ways:

1. HEAT ACTIVATED

The adhesive is softened by heating until it is fluid enough to spread freely over the whole joint and adhere to the materials being joined. Upon cooling to room temperature a bond is achieved.

## Table 11.13 Preparation of joint surfaces

**Aluminium**
1. Vapour blast or rub with 100 grit emery cloth to give correct surface texture.
2. Degrease in trichlorethylene vapour. (Do not use alkaline degreasing agents for aluminium and its alloys.)

This technique is suitable for general purpose bonding.

For precise control of highly stressed joints, as in aircraft work, the surface texture is given by chemical or electro-chemical etching. This is followed by rinsing in distilled water and drying in warm air.

**Copper, Brass, Bronze**
1. Sand blast, wire brush or rub with 100 grit emery cloth to give correct surface texture.
2. Degrease in trichloroethylene vapour.

This technique is for general purpose bonding.

For precise control, the surface texture is given by chemical etching, followed by a neutralising rinse, finally washing in distilled water and drying in warm air. For maximum bond strength when using epoxy resins, an X83/65 primer should be used.

**Non-absorbent, non-metallic materials** (*plastics, glass, ceramics, etc.*)
1. Light abrasion to give surface texture.
2. Remove dust.
3. Degrease using a suitable solvent so as not to attack the adherend.
4. Rinse in distilled water and dry in warm air.

This technique is for the general purpose bonding of materials in popular use. For maximum strength joints in all these materials, follow the maker's recommendations most carefully, no matter how complex.

**Steels**
1. Grit, vapour blast or rub with emery cloth to give correct surface texture.
2. Degrease with moisture-free solvents to prevent re-oxidation.
3. Bond immediately before rusting sets in.

This technique is for general purpose bonding. For maximum strength joints, prepare the surface with chemical etchants and check that the degreasing agents are compatible with the adhesive as washing is not possible.

**Wood**
1. Sand or plane smooth. This not only gives the correct surface texture but ensures minimum glue line.
2. Remove surface dust from the pores of the wood by brushing and vacuum cleaning.
3. Reduce wood to correct moisture content.
4. To prevent starving the glue line when joining highly absorbent timbers the joints should first be primed. Note: this does NOT imply the use of paint primer.
5. Bond joint immediately after preparation, as the freshly prepared joint rapidly absorbs moisture and other contaminants.

*Note:* These instructions are very brief and only intended as an indication of the problems involved. For consistent results, always follow the manufacturers' instructions to the letter.

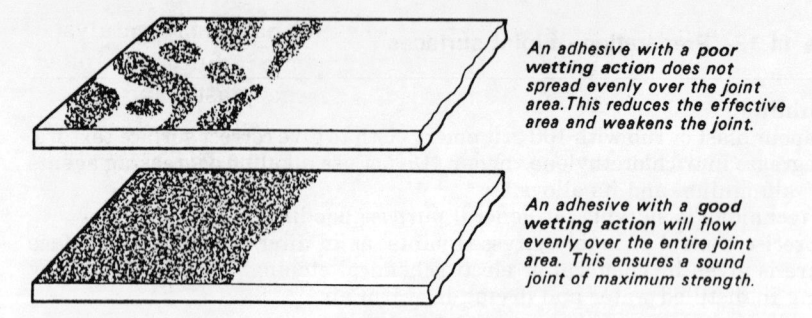

**Fig. 11.43  Wetting capacity of an adhesive**

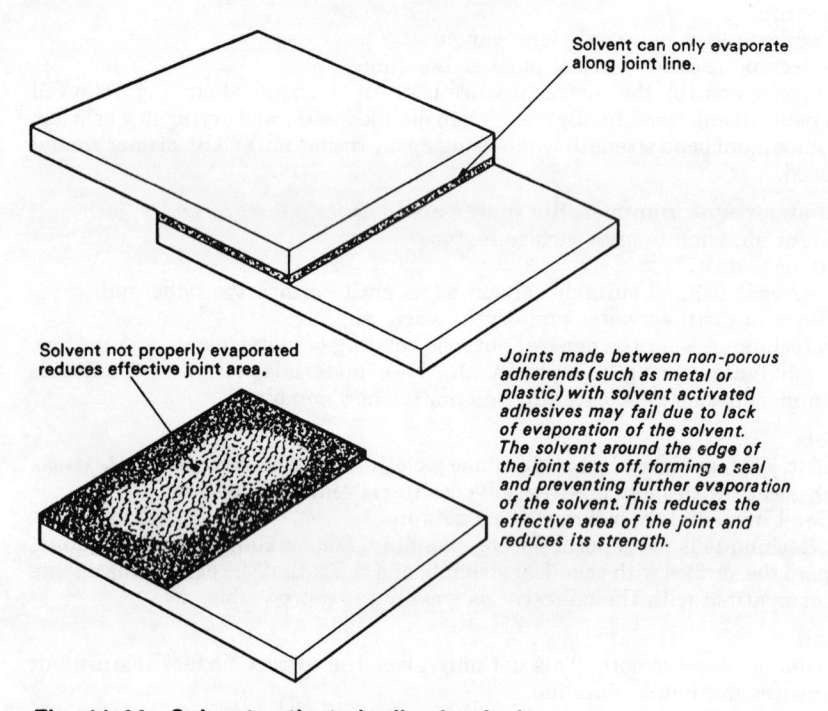

*An adhesive with a poor wetting action does not spread evenly over the joint area. This reduces the effective area and weakens the joint.*

*An adhesive with a good wetting action will flow evenly over the entire joint area. This ensures a sound joint of maximum strength.*

Solvent can only evaporate along joint line.

Solvent not properly evaporated reduces effective joint area.

*Joints made between non-porous adherends (such as metal or plastic) with solvent activated adhesives may fail due to lack of evaporation of the solvent. The solvent around the edge of the joint sets off, forming a seal and preventing further evaporation of the solvent. This reduces the effective area of the joint and reduces its strength.*

**Fig. 11.44  Solvent activated adhesive fault**

### 2. SOLVENT ACTIVATED

The adhesive is softened by a suitable solvent and the bond is achieved by the solvent evaporating. Because evaporation is essential to the setting of the adhesive, a sound bond is almost impossible to achieve at the centre of a large joint area as shown in Figure 11.44. This is particularly the case when joining non-absorbent materials.

1. The impact adhesive is spread thinly and evenly on both joint surfaces.
2. The adhesive is then left to dry by evaporation. This avoids the problem in Figure 10.40

3. When the adhesive is dry, the joint surfaces are brought into contact, where upon they form an immediate intermolecular bond.

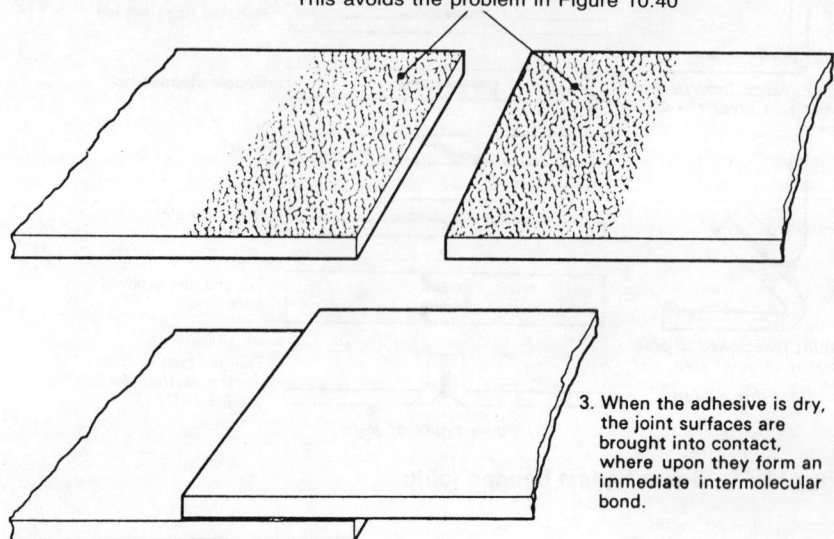

**Fig. 11.45  The use of an impact adhesive**

### 3. IMPACT ADHESIVES

These are solvent activated adhesives which are spread separately on the two joint faces and then left to dry by evaporation. When dry, the treated faces are brought together whereupon they instantly bond together by inter-molecular attraction. Figure 11.45 shows the steps in making an impact joint.

Thermo-plastic adhesives are based upon synthetic materials such as polyamides, vinyl and acrylic polymers and cellulose derivatives. They are also based on such natural materials as rosin, shellac, mineral waxes and rubber. They are not so strong as thermo-setting plastics (Section 11.39) but, being more flexible, are most suitable for joining non-rigid materials.

## 11.39  Thermo-setting plastics

These are materials which depend upon heat to make them set. The setting (curing) process causes chemical changes to take place within the adhesive. Once cured they cannot be softened again by the re-application of heat. This makes them less heat sensitive than thermo-plastics.

The heat necessary to cure the adhesive can be applied externally as when phenolic resins are used, or internally by adding a 'hardener' as when epoxy resins are used. The hardener is a chemical that reacts with the adhesive to generate heat (exothermic reaction).

Since the setting process is a chemical reaction and not dependent upon evaporation, the area of the joint is unimportant.

Thermo-setting adhesives are extremely strong and are used for making structural joints in high strength materials such as metals. The body shells of motor cars and stressed members of aircraft are increasingly dependent upon these adhesives for their joints in place of spot welding and riveting. The stresses are more uniformly distributed and the joints are sealed against corrosion.

Thermo-setting adhesives tend to be brittle when cured and, therefore, are not suitable for flexible (non-rigid) materials.

## 11.40 Safety

One great advantage of natural gums and glues is that they are non-toxic. Therefore, they are widely used in the labelling and packaging of foodstuffs.

Most synthetic adhesives and their solvents, hardeners, catalysts, etc., are highly toxic and must be used under carefully controlled conditions. In addition, the solvents used in thermo-plastic and impact adhesives are HIGHLY FLAMMABLE. They must be stored and used in well-ventilated conditions and the working area declared a NO SMOKING zone.

The health hazards presented by these materials range from dermatitis and sensitisation of the skin to permanent damage to the internal organs of the body if inhaled or accidentally swallowed.

### PRECAUTIONS

1. Use only in well-ventilated areas.
2. Wear protective clothing appropriate to the process no matter how inconvenient.
3. Use a barrier cream.
4. After use, wash thoroughly in soap and water. Do not use solvents.

# 12 Welding

## Part A  Oxy-acetylene welding

The principles of *gas welding* have already been introduced in Sections 11.27 and 11.28. These basic principles will now be examined in greater depth.

## 12.1  Equipment

In this chapter reference will only be made to the components of 'HIGH PRESSURE' oxy-acetylene welding systems.

The assembled, basic welding equipment is illustrated in Figure 12.1.

## 12.2  Oxygen and acetylene cylinders

The differences between an oxygen cylinder and an acetylene cylinder are clearly illustrated in Figure 12.2.

### OXYGEN

This is supplied to the welding torch from a solid drawn steel cylinder where it is contained in compressed form. Oxygen cylinders are usually supplied in capacities of $3 \cdot 4$ m$^3$, $5$ m$^3$, and $6 \cdot 8$ m$^3$. Mild steel cylinders are charged to a pressure of 13 660 kN/m$^2$ (136·6 bar) and alloy steel cylinders to 17 240 kN/m$^2$ (172 bar).

*The oxygen volume in a cylinder is directly proportional to its pressure*, and the consumption for a welding job can, therefore, be found by noting the pressure drop during the welding operation. For example, if it were noted that the original pressure of a full oxygen cylinder had dropped 5% during a welding operation, then 1/20 of the cylinder contents would have been consumed.

THE VALVE OUTLET ON AN OXYGEN CYLINDER HAS A RIGHT-HAND SCREW THREAD.

### ACETYLENE

For high-pressure welding, acetylene is supplied in solid drawn steel cylinders as shown in Figure 12.2. *High-pressure acetylene is not stable and for this reason it is dissolved in* acetone, *which has the ability to absorb a large volume of the gas and release it as the pressure falls*. One volume of acetone at atmospheric pressure and at a temperature of 15°C is capable of dissolving about 25 volumes of acetylene. *This dissolving capacity can be increased in proportion to the pressure*. Acetylene cylinders are charged to a pressure of 1152 kN/m$^2$ (15·5 bar), and as one atmosphere is equivalent to one bar it follows that, at the same temperature, one volume of acetone absorbs approximately $25 \times 15 = 375$ times its own volume of acetylene at a pressure of 15 bar.

Because of the danger of explosions to which compressed acetylene is susceptible, the steel cylinder is filled with a porous substance, and the construction of the cylinder, its filling and testing, are all strictly controlled by the manufacturers in the interests of safety. The pores in the filling material divide the space into a large number of very small compartments which are completely filled with 'dissolved acetylene'. *These small compartments prevent the sudden decomposition of the*

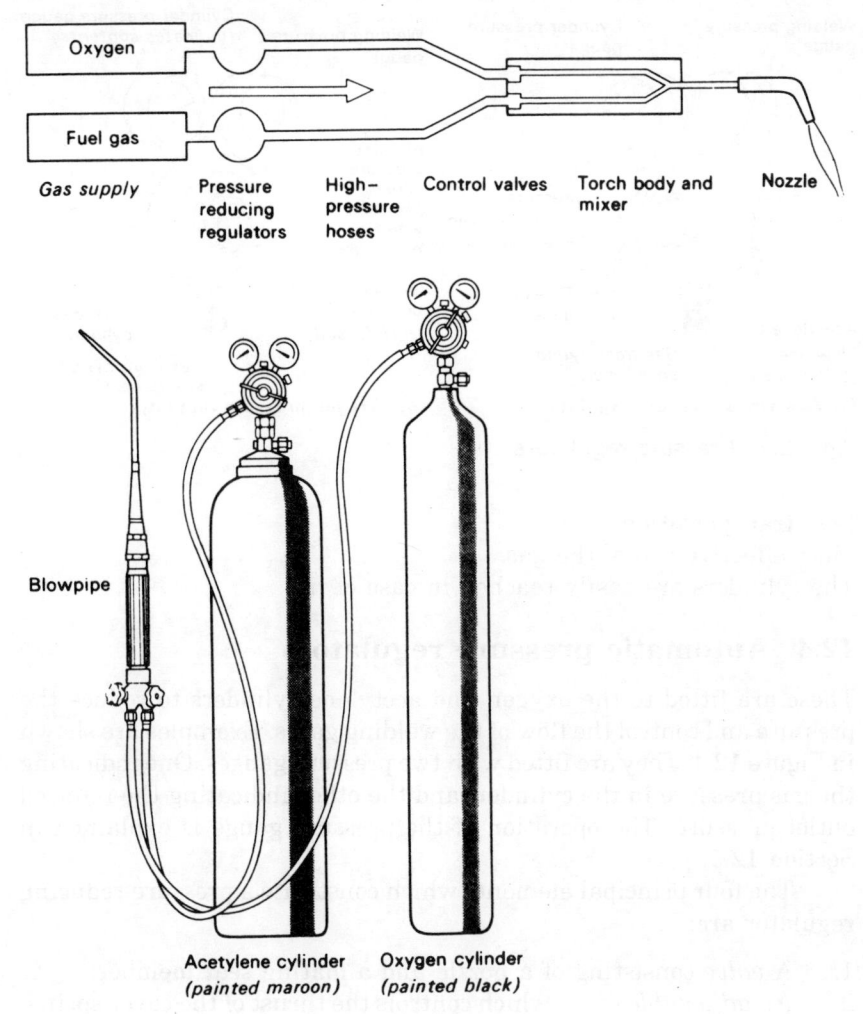

**Fig. 12.1 Basic gas-welding equipment**

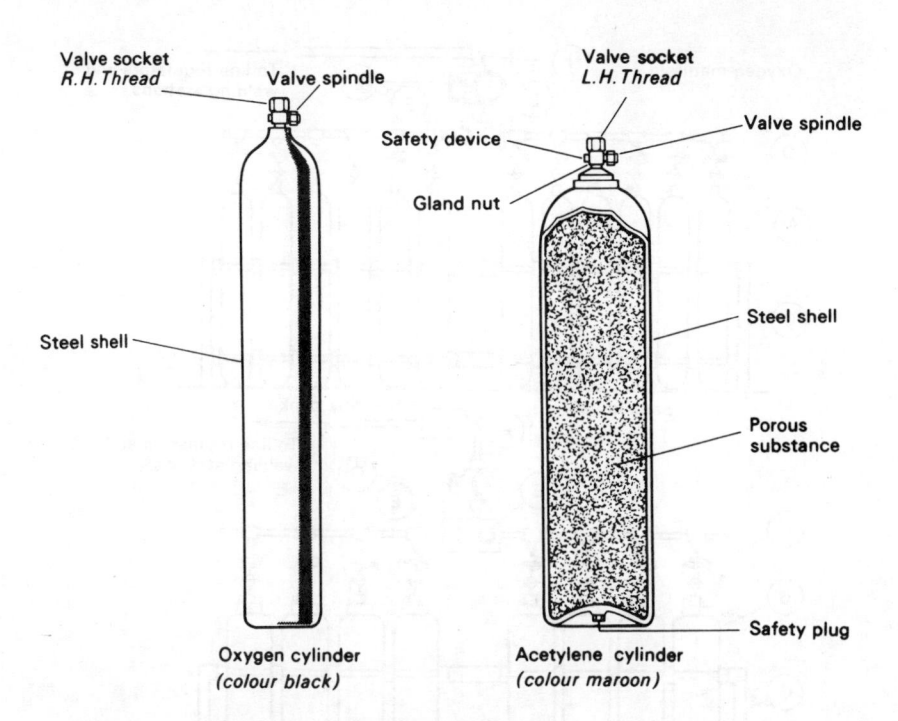

**Fig. 12.2 Acetylene and oxygen cylinders**

*acetylene throughout the mass, should it be started by local heating or other causes.*

THE VALVE OUTLET ON ACETYLENE CYLINDERS IS FITTED WITH A LEFT-HAND SCREW THREAD.

## 12.3 Discharge rate

Should OXYGEN be withdrawn from a cylinder at too great a rate of consumption, *a rapid drop in pressure will occur, with the result that the cylinder valve may freeze.* When flame cutting heavy cast-iron sections, which involves a high rate of gas consumption, it is advisable to couple together sufficient oxygen cylinders to complete the work — this also applies when repairing large castings which have to be preheated.

The rate of acetylene consumption also has to be kept below a certain limit. Acetylene cylinders should never be discharged at a rate which will empty them in less than 5 hours. This means that the *discharge rate must be limited to less than 20% of the total cylinder content per hour.* This is not because of freezing, as with oxygen, but because should this limit be exceeded acetone will be drawn off and mixed with the acetylene. Acetylene which contains small quantities of acetone vapour will lower the flame temperature. Unlike oxygen, the volume and pressure of acetylene are not in linear proportion which means that the gas consumption cannot be reliably found from the loss in pressure in the cylinder during a welding operation.

The sizes of acetylene cylinders in general use are $2 \cdot 8$ m$^3$ and $5 \cdot 7$ m$^3$.

In workshops, where welding gases are needed in several places, or at high rates of consumption, it is of considerable advantage to use

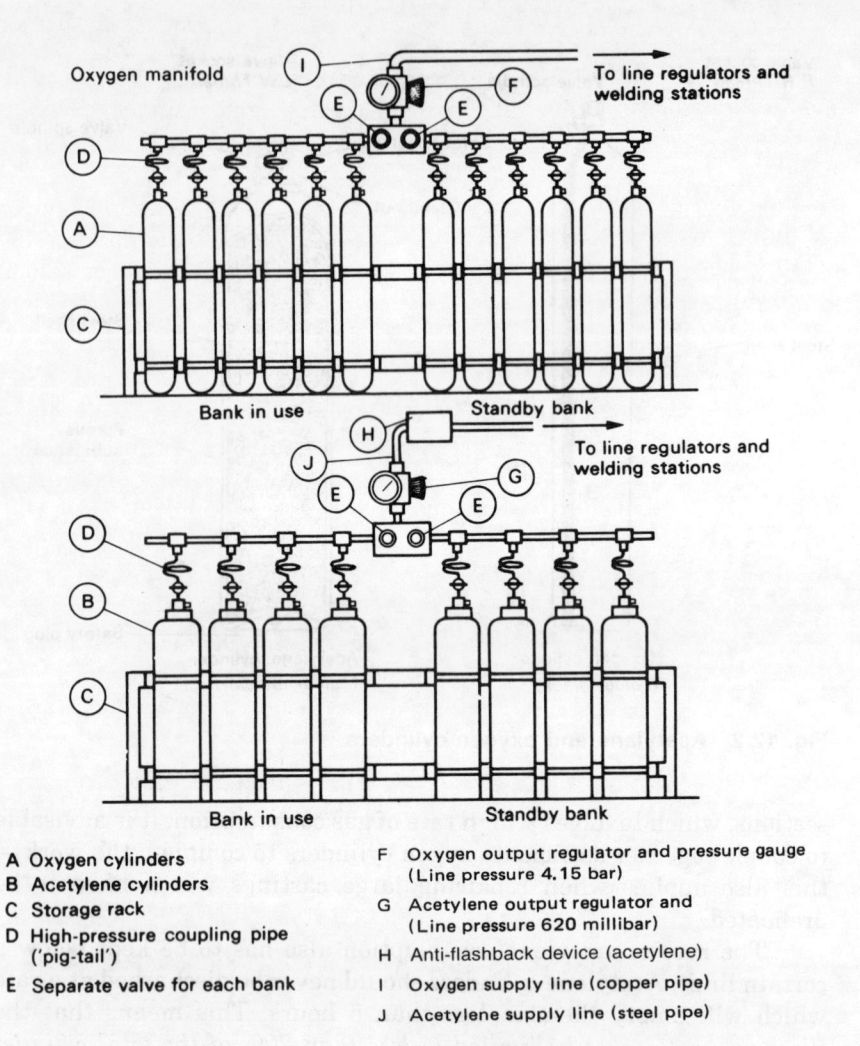

Fig. 12.3 The manifold system for gas welding

- A  Oxygen cylinders
- B  Acetylene cylinders
- C  Storage rack
- D  High-pressure coupling pipe ('pig-tail')
- E  Separate valve for each bank
- F  Oxygen output regulator and pressure gauge (Line pressure 4.15 bar)
- G  Acetylene output regulator and (Line pressure 620 millibar)
- H  Anti-flashback device (acetylene)
- I  Oxygen supply line (copper pipe)
- J  Acetylene supply line (steel pipe)

a manifold system. Instead of having cylinders at each place of work, they are assembled in one centralised position in specially designed racks and connected by a manifold, as shown in Figure 12.3. These gases are then distributed by means of a pipeline to the different work-places.

ADVANTAGES OF MANIFOLDS:

More space available at work-place.
No replacement of cylinders inside the workshop.

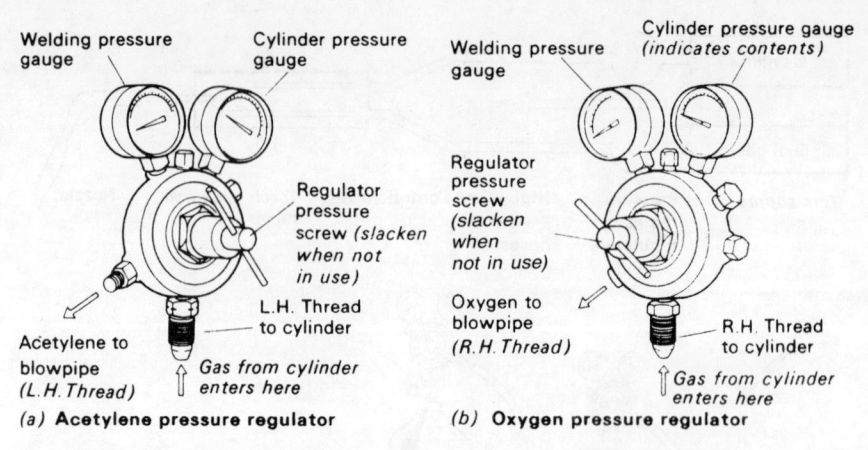

Fig. 12.4 Pressure regulators

Less transportation.
More effective use of the gas.
The cylinders are easily reached in case of fire.

## 12.4 Automatic pressure regulators

These are fitted to the oxygen and acetylene cylinders to reduce the pressure and control the flow of the welding gases. Examples are shown in Figure 12.4. They are fitted with two pressure gauges. One indicating the gas pressure in the cylinder, and the other indicating the reduced outlet pressure. The operation of the pressure gauge is explained in Section 12.5.

The four principal elements which constitute a pressure-reducing regulator are:

1. A *valve* consisting of a nozzle and a mating seat member.
2. An *adjustable screw* which controls the thrust of the cover spring.
3. A cover spring which transmits to a diaphragm the thrust created by the adjusting screw.
4. A diaphragm connected with the mating seat member.

Figure 12.5 illustrates two basic types of *'single-stage' regulator*.

In the 'needle type' regulator the inlet pressure tends to close the seat member against the nozzle. The outlet pressure on this type of regulator has a tendency to increase somewhat as the inlet pressure decreases. *This increase is caused by a decrease of the force produced by the gas pressure against the seating area as the inlet pressure decreases.* This type of single-stage regulator is sometimes referred to as 'inverse' or 'negative' type.

When the inlet pressure decreases, its force against the seat member decreases, allowing the cover spring force to move the seat member away from the nozzle. Thus more gas pressure is allowed to build up to re-establish the balanced condition.

A smaller outlet pressure on the underside of the diaphragm is all that is necessary to close the seat member against the nozzle. The opening between the seat members and the nozzle is reduced, which results in less gas flow.

## 12.5 The pressure gauge

Inside a pressure gauge there is a BOURDON TUBE. This is a copper-alloy tube of oval section, bent in a circular arc. One end of the tube is sealed shut and attached by light linkage to a mechanism which operates a pointer. The other end is fixed, and is open for the application of the pressure which is to be measured. The internal pressure tends to change the section of the tube from oval to circular and this causes it to straighten out slightly. The resultant movement of the tube causes the pointer to move over a suitably calibrated scale. An example of a bourdon tube pressure gauge is shown in Figure 12.6.

## 12.6 Welding torches

The gases having been reduced in pressure by the gas regulators are fed through suitable hoses to a welding torch.

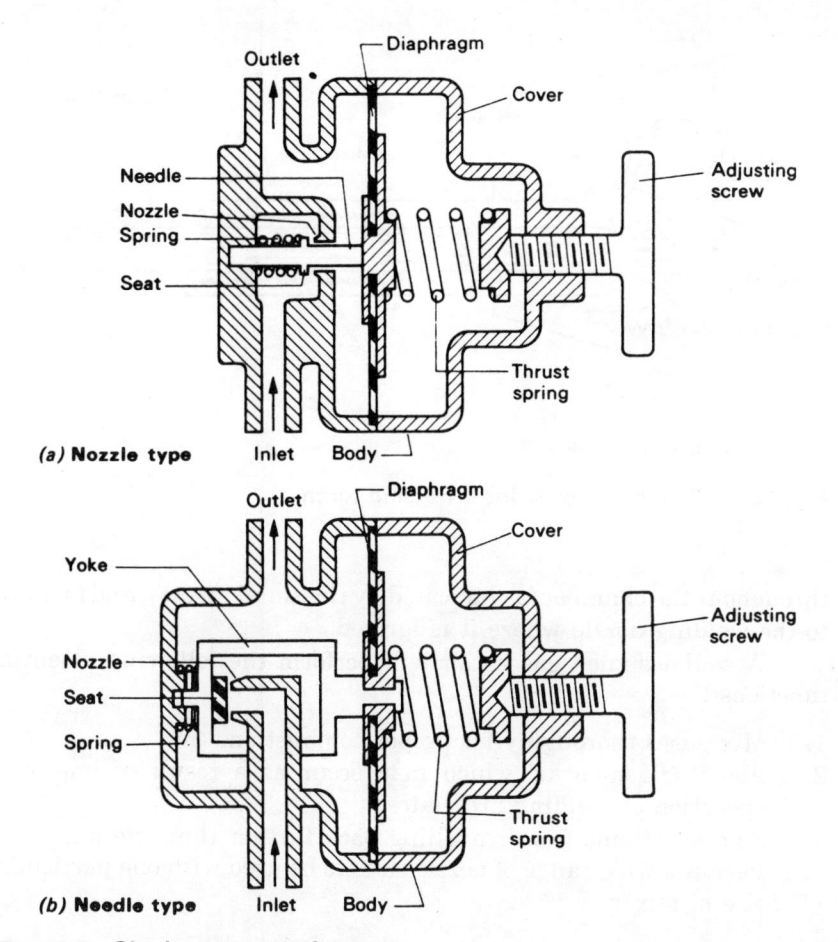

**Fig. 12.5 Single-stage regulators**

In the *'nozzle type' regulator* the inlet pressure tends to move the seat member away from the nozzle, therefore the outlet pressure decreases somewhat as the inlet pressure decreases. *This is because the force tending to move the seat member away from the nozzle is reduced as the inlet pressure decreases.* This type of single-stage regulator is referred to as 'direct acting' or 'positive' type.

THE GAS OUTLET PRESSURE FOR ANY PARTICULAR SETTING OF THE ADJUSTING SCREW IS REGULATED BY A BALANCE OF FORCES. This balance is between the cover spring thrust and the opposing forces created by a combination of the outlet pressure against the underside of the diaphragm and the inlet pressure against the seating area.

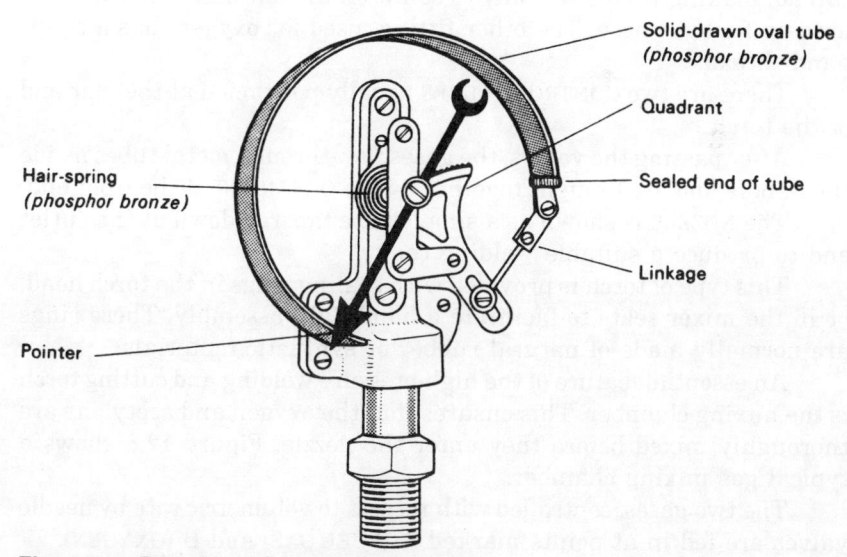

**Fig. 12.6 Bourdon tube pressure gauge**

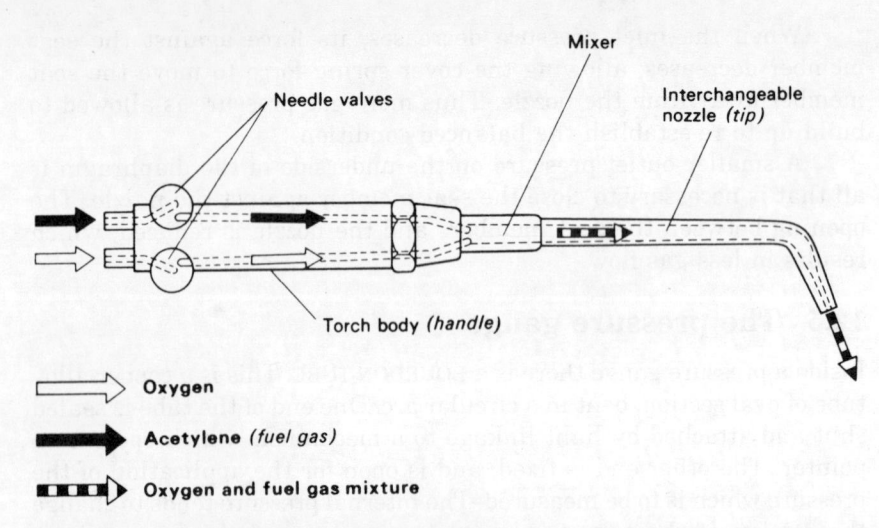

**Fig. 12.7** High-pressure welding torch

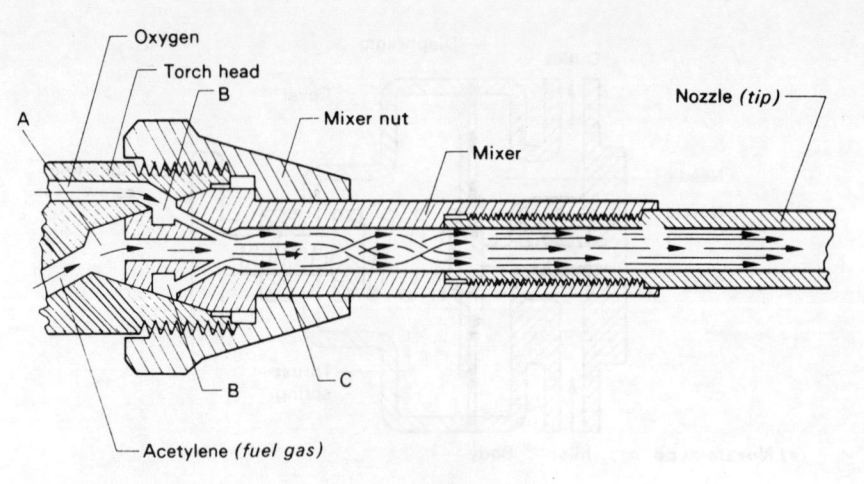

**Fig. 12.8** Mixing chamber for a welding torch

*The WELDING TORCH is a specially designed piece of equipment used for mixing and controlling the flow of gases to the WELDING NOZZLE or TIP.* The torch provides a means of holding and directing the welding nozzle. The basic elements of a high-pressure welding torch are shown in the simplified drawing, Figure 12.7.

The fuel gas hose fitting on all welding torches has a left-hand thread, making it possible only to screw on the left-hand grooved nuts used on fuel gas hose. The other fitting, used for oxygen, has a right-hand thread.

There are two CONTROL VALVES usually positioned at the rear end of the torch.

After passing the valves, the gases flow through metal tubes inside the handle and are brought together by the GAS MIXER at the front end.

The NOZZLE is shown as a simple tube tapered down at the outlet end to produce a suitable welding cone.

This type of torch is provided with sealing rings in the torch head, or in the mixer seats to facilitate a hand-tight assembly. These rings are normally made of natural rubber or synthetic materials.

An essential feature of the high-pressure welding and cutting torch is the mixing chamber. This ensures that the oxygen and acetylene are thoroughly mixed before they enter the nozzle. Figure 12.8 shows a typical gas mixing chamber.

The two gases, controlled with respect to volumetric rate by needle valves are fed in at points marked A (FUEL GAS) and B (OXYGEN).

The mixing of the gases commences at point C, and continues throughout the chamber (as indicated by the small arrows) and forward to the welding nozzle where it is ignited.

A well-designed gas mixer will perform the following essential functions:

1. Mix gases thoroughly for proper combustion.
2. Arrest 'flash-backs' which may occur as a result of improper operation or welding procedure.
3. Stop any flame from travelling back further than the mixer.
4. Permit a wide range of nozzle sizes to be used with one particular size of mixer.

## 12.7 Welding nozzles

The welding nozzle or tip is that portion of the torch through which the gases pass prior to their ignition and combustion. A welding nozzle enables the operator to guide the flame and direct it with the maximum ease and efficiency.

Nozzles are made from a NON-FERROUS metal such as COPPER or a COPPER ALLOY. *These materials possess a HIGH THERMAL CONDUCTIVITY and their use greatly reduces the danger of burning the nozzle at high temperatures.*

Welding nozzles are available in a variety of sizes, shapes, and construction. However, there are two main classes of *nozzle and mixer combinations* in general use:

1. A SEPARATE NOZZLE AND MIXER UNIT — each size of nozzle has the

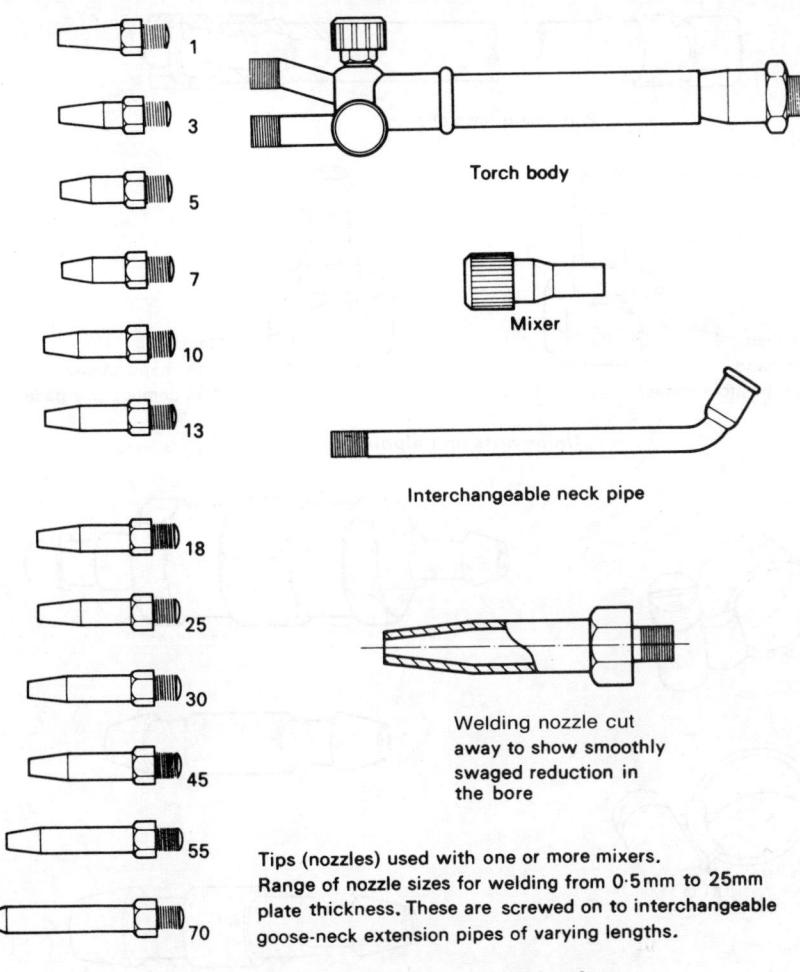

**Fig. 12.9 Welding torch and nozzle/mixer combinations**

proper mixer which provides efficient combustion of the gases and optimum heating efficiency.

2. ONE OR MORE MIXERS USED FOR THE ENTIRE RANGE OF NOZZLE SIZES — The nozzle is screwed into the appropriate mixer, and each size of mixer has a particular thread size to ensure the correct combination of the proper nozzle and mixer.

Figure 12.9 shows a welding torch combination in this category. In this case use is made of the interchangeable 'goose-neck' extensions which are screwed into the mixer portion of the torch body. A wide range of nozzles or tips may be fitted to these extensions.

HEAT RADIATION *of the flame produced with the larger sizes of nozzles is great, therefore it is advisable to have the torch body at a comfortable distance from the flame.* In this respect, nozzle mixer units or interchangeable neck pipes will vary in length according to the thickness of plate being welded.

CARE AND MAINTENANCE OF NOZZLES

Since nozzles are made from materials which are relatively soft, care must be taken to guard them against damage. The following simple precautions are recommended:

1. Make sure that the nozzle seat and threads are absolutely free from foreign matter in order to prevent any scoring when tightening on assembly.
2. Nozzles should only be cleaned with tip cleaners, which are specially designed for this purpose. A special *nozzle cleaning compound* is available for dirty nozzles. The correct strength of cleaning solution is obtained by using approximately 50 grammes of compound to 1 litre of water. Dirty nozzles should be immersed in this solution for a period of *at least* two hours.
3. Nozzles should never be used for moving or holding the work.

## 12.8 Gas welding hose

For welding and cutting, hose should be used which is specially manufactured for the purpose. *Other types of hose may cause considerable gas losses and accidents.*

Welding hose has a seamless lining which is manufactured from rubber (or a rubber compound) which is reinforced with canvas or wrapped cotton plies. It is resistant to the action of gases normally used in welding and cutting. The outer casing is made of tough abrasion-resistant rubber. The hose is very robust and capable of withstanding high pressure; it is available in two colours, black and red.

BLACK HOSE is used for OXYGEN and other 'non-combustible' gases.

RED HOSE is used for ACETYLENE and other 'combustible' or fuel gases.

Table 12.1 lists the sizes of welding hose in general use:
*Care must be taken not to damage welding hose by dropping heavy weights where it crosses the workshop floor, playing the torch flame on to it, or allowing it to come in contact with hot or molten metal.*

*Hose clips* For all welding and cutting operations the hose 'union nipples and nuts' must be secured with the aid of reliable hose clips.

Table 12.1 Sizes of welding hose in general use

| INTERNAL DIAMETER | | APPROXIMATE OUTSIDE DIAMETER | | USED FOR OXYGEN AND ACETYLENE SUPPLIES WITH: |
|---|---|---|---|---|
| mm | (in) | mm | (in) | |
| 4·8 | (3/16) | 13·5 | (17/32) | Light-duty high-pressure torches |
| 6·3 | (1/4) | 15·1 | (19/32) | |
| 8·0 | (5/16) | 16·7 | (21/32) | Heavy-duty high-pressure torches |
| 9·5 | (3/8) | 18·3 | (23/32) | Large cutting torches and where *low-pressure generated acetylene* is used |

Standard lengths of hose are available on which the coupling nipples are permanently attached by means of special clips, as shown in Figure 12.10.

*Hose couplers* These are used when joining welding hoses of equal or unequal sizes (Figure 12.10).

## 12.9 Miscellaneous accessories (equipment)

Safety goggles and protective clothing have been fully described in Chapter 1.

In addition to the equipment previously described there are numerous pieces of auxiliary equipment which may be used in gas welding and cutting processes. For example, cylinder trucks, jigs and fixtures, devices for introducing liquid flux into the acetylene stream for brazing operations, tip cleaners, friction lighters, and gas economisers for automatically shutting off the torch flame. Of these only the last two items will be described in this chapter.

1. *Gas lighters* A spark lighter is a very useful accessory which provides a convenient and inexpensive means of lighting the torch. *From a safety point of view, it is advisable to use such friction lighters rather than matches to ignite the gas.*
2. *Gas economiser* This unit provides an efficient means of conserving gas to the user who frequently needs to interrupt the welding or cutting operation. The unit may be mounted on a light metal stand, or fastened to the welding table, in some convenient position.

The basic elements and function of a gas economiser are shown in Figure 12.11.

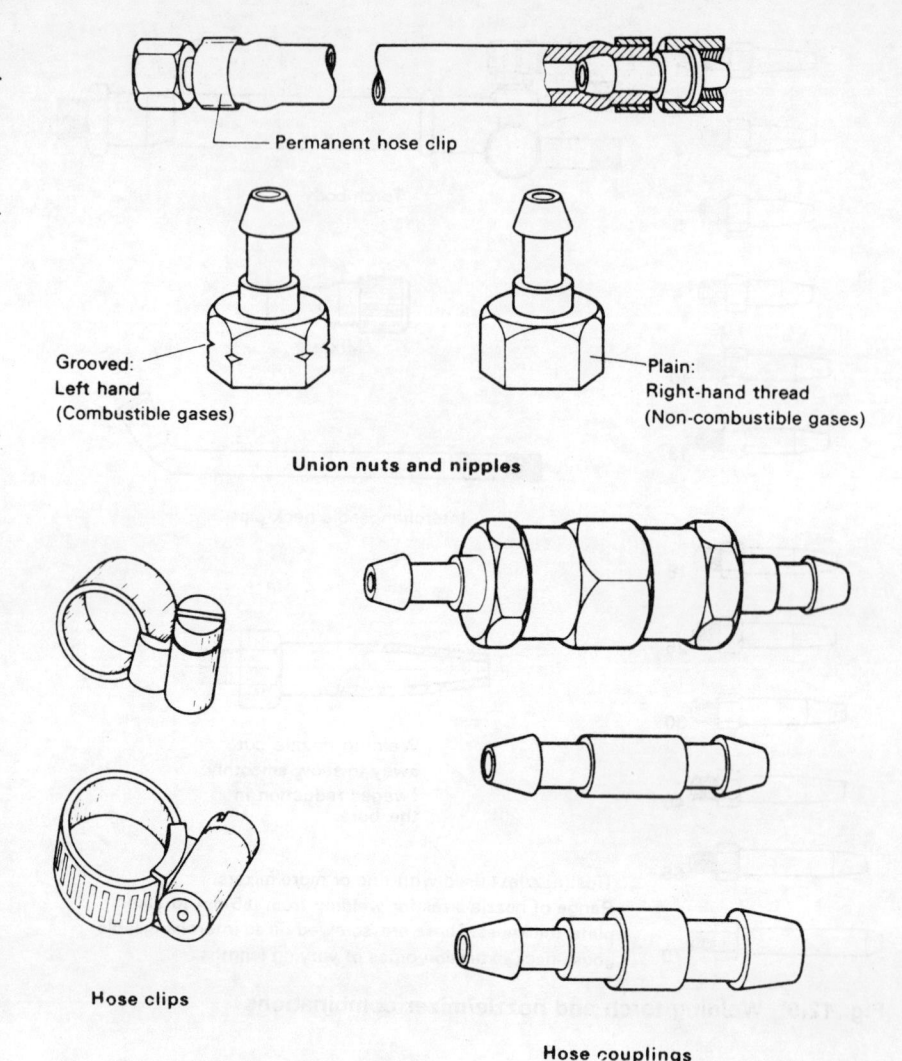

Fig. 12.10 Welding hose fittings

The two lengths of hose from the regulators are attached to the inlet connections. The outlets are connected to the torch with additional lengths of hose. The torch is ignited, and the flame set in the normal manner, but at any interruption of the welding operation, or after completing a weld, it is simply hooked to the lever arm control of the economiser.

The weight of the torch depresses the lever, causing the valve to operate and cut off the supply of gas to the torch and the flame is

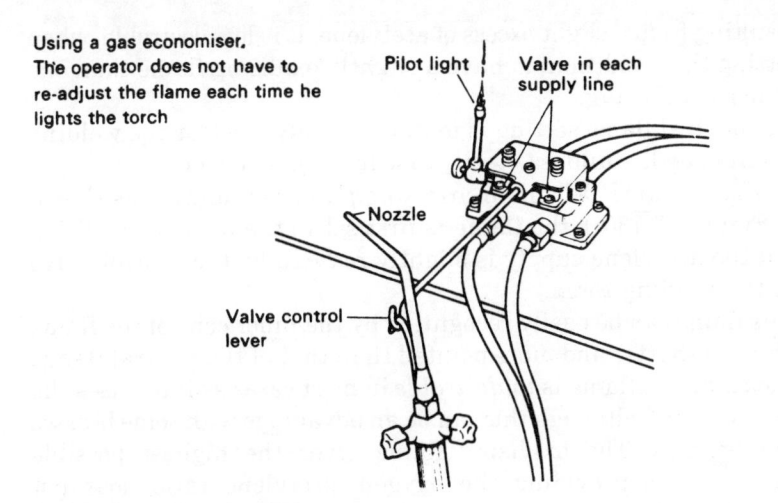

Using a gas economiser. The operator does not have to re-adjust the flame each time he lights the torch

**Fig. 12.11 Gas economiser unit**

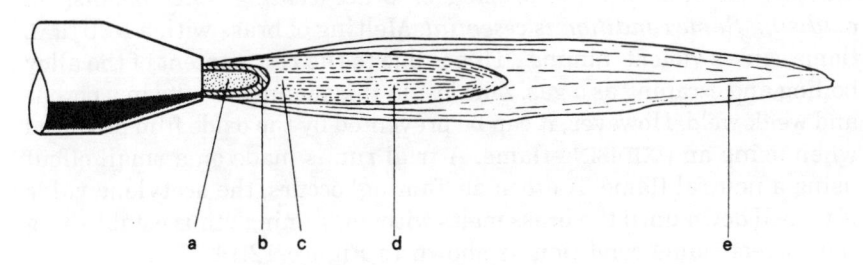

**Fig. 12.12 The structure of the oxy-acetylene welding frame**

extinguished. When the torch is required again, it is lifted from the lever arm and can be ignited by a small pilot flame on the economiser.

## 12.10 Structure of the oxy-acetylene flame

The welding flame is produced by burning approximately equal volumes of the oxygen and acetylene which are supplied to the torch. Figure 12.12 illustrates the basic structure of the oxy-acetylene welding flame.

The oxy-acetylene flame, like most oxy-gas flames used for welding, is characterised by the following zones:

(a) The innermost cone of MIXED UNBURNT GASES leaving the nozzle. It appears intensely white and clearly defined with the correct NEUTRAL setting. *Any small excess of* ACETYLENE *is indicated by a white 'feather' around the edge of this cone. With excess* OXYGEN *the cone becomes shorter, more pointed, and tinged with blue.*

(b) A very narrow stationary zone wherein the chemical reaction of the FIRST STAGE OF COMBUSTION takes place *producing a sudden rise in temperature.*

(c) The REDUCING ZONE, appearing dark to light blue, in which the PRIMARY COMBUSTION PRODUCTS are concentrated. *The nature of these products determines the* CHEMICAL NATURE *of the flame, i.e., whether* 'NEUTRAL', 'OXIDISING' *or* 'CARBURISING'.

(d) Within the region of PRIMARY COMBUSTION and approximately 3 mm from the tip of the nicely defined cone of unburnt gases there is a REGION OF MAXIMUM TEMPERATURE, this is the zone used for welding.

(e) The yellow to pinkish outer zone or 'plume' around and beyond the previous zones represents the *chemical reaction* of the SECOND STAGE OF COMBUSTION. *Here the two* COMBUSTIBLE *gases* CARBON MONOXIDE *and* HYDROGEN, *which are the products of primary combustion, combine with* OXYGEN FROM THE SURROUNDING ATMOSPHERE. This part of the flame is always OXIDISING and contains large amounts of NITROGEN. *The presence of a high percentage of nitrogen in the* OXIDISING OUTER ZONE *is not surprising because the* ATMOSPHERE *contains approximately 80%* NITROGEN *and 20%* OXYGEN *by volume.*

For the oxy-acetylene welding flame the CHEMICAL REACTIONS can be summarised as follows:

*Primary combustion* Produces the REDUCING ZONE consisting of approximately 60 per cent CARBON MONOXIDE and 40 per cent HYDROGEN:

ACETYLENE + OXYGEN = CARBON MONOXIDE + HYDROGEN

*Secondary combustion* The products of the primary reaction *react with* OXYGEN *in the surrounding air* producing the outer envelope or plume of the flame:

1. CARBON MONOXIDE + AIR (oxygen and nitrogen) = CARBON DIOXIDE + NITROGEN

2. HYDROGEN + AIR (oxygen and nitrogen) = WATER VAPOUR + NITROGEN

*Thus the products of complete combustion can be said to be* CARBON DIOXIDE *and* WATER VAPOUR.

With the oxy-acetylene welding it is essential to maintain a NEUTRAL FLAME for, except in special cases, an OXIDISING FLAME or a CARBURISING FLAME will result in welds with unsatisfactory mechanical

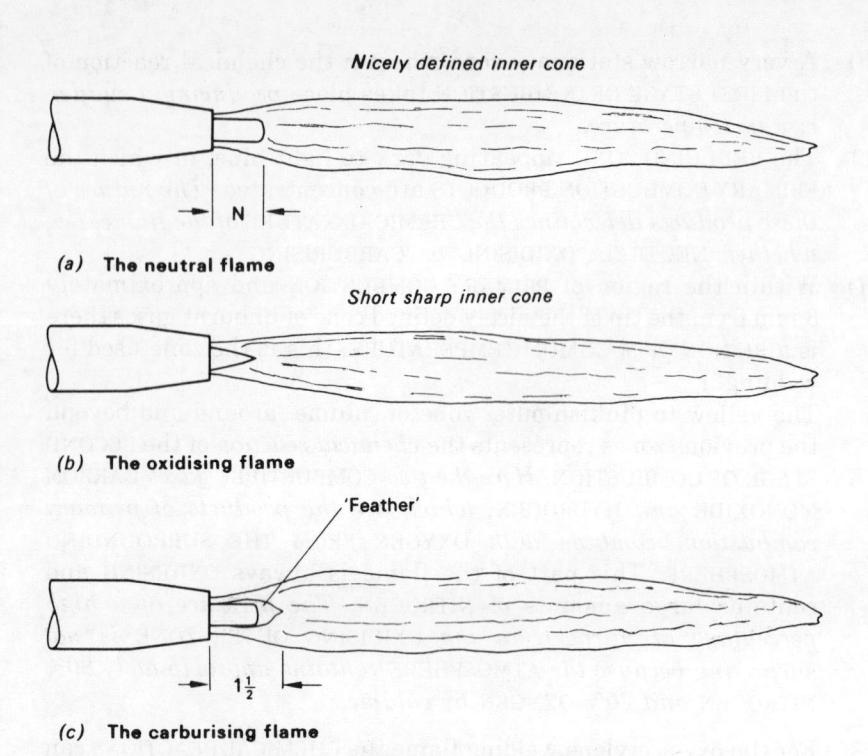

**Fig. 12.13 Oxy-acetylene welding flame conditions**

(a) The neutral flame — Nicely defined inner cone
(b) The oxidising flame — Short sharp inner cone
(c) The carburising flame — 'Feather'

properties. *The presence of even the smallest quanity of an* OXIDISING AGENT, *such as* CARBON DIOXIDE, WATER VAPOUR, *or* OXYGEN, *will rapidly destroy the* REDUCING PROPERTIES *of the flame.*

## 12.11 The three oxy-acetylene flame conditions

The correct type of flame is essential for the production of satisfactory welds, and the characteristics of the three flames are shown in Figure 12.13.

1. *The neutral flame* For most applications the *neutral flame* condition is used, as shown in Figure 12.13(*a*). This is produced when approximately equal volumes of oxygen and acetylene are mixed in the welding torch. It is termed NEUTRAL because it effects no chemical change on the molten metal and therefore will not oxidise or carburise the metal.

   It is easily recognised by its characteristic clearly defined white inner cone at the tip of the torch nozzle. Correct adjustment is indicated by a slight white flicker (feather) on the end of this cone resulting from a slight excess of acetylene. It is less desirable, when setting this condition, to have a slightly oxidising flame than one which is reducing.

   The neutral flame setting is most commonly used for the welding of mild steel, stainless steels, cast iron, and copper.

2. *Oxidising flame* When an *oxidising flame* is required, as shown in Figure 12.13(*b*), the flame is first set to the neutral condition and the acetylene supply is slightly *reduced* by the control valve on the welding torch.

   This flame can be easily recognised by the inner cone of the flame which is shorter and more pointed than that of the neutral flame. The oxidising flame is *undesirable* in most cases as it oxidises the molten metal, although this can be an advantage with some brasses and bronzes. The oxidising flame gives the highest possible temperatures providing the oxygen : acetylene ratio does not exceed 1·5:1.

   *Whether welding brasses, brazing or braze-welding other metals, an oxidising flame condition is essential.* Melting of brass with a NEUTRAL flame causes violent 'fuming'. This is due to the ZINC content of the alloy boiling and escaping as a gas. Zinc lost in this manner results in a porous and weak weld. However, it can be prevented by the oxide film produced when using an OXIDISING flame. A trial run is made on a small offcut using a neutral flame. As soon as 'fuming' occurs, the acetylene valve is turned down until the brass melts without 'fuming', thus establishing the correct flame condition as shown in Figure 12.14.

3. *Reducing flame* When a *reducing flame* is required, as shown in Figure 12.13(*c*), the flame is first set to the neutral condition and the acetylene supply is slightly *increased* by the control valve on the welding torch.

   This flame is recognised by the 'feather' of incandescent carbon particles between the inner cone and outer envelope.

   The reducing flame is used for carburising (surface hardening) and for the 'flame brazing' of aluminium.

## 12.12 Nozzle size

For a given welding torch, the NOZZLE OUTLET SIZE has a much greater influence on governing the flame size than changing the gas pressures or adjusting the control valves.

The manufacturers of gas welding equipment have adopted various methods of indicating nozzle sizes, such as:

1. By the approximate consumption of each gas per hour.

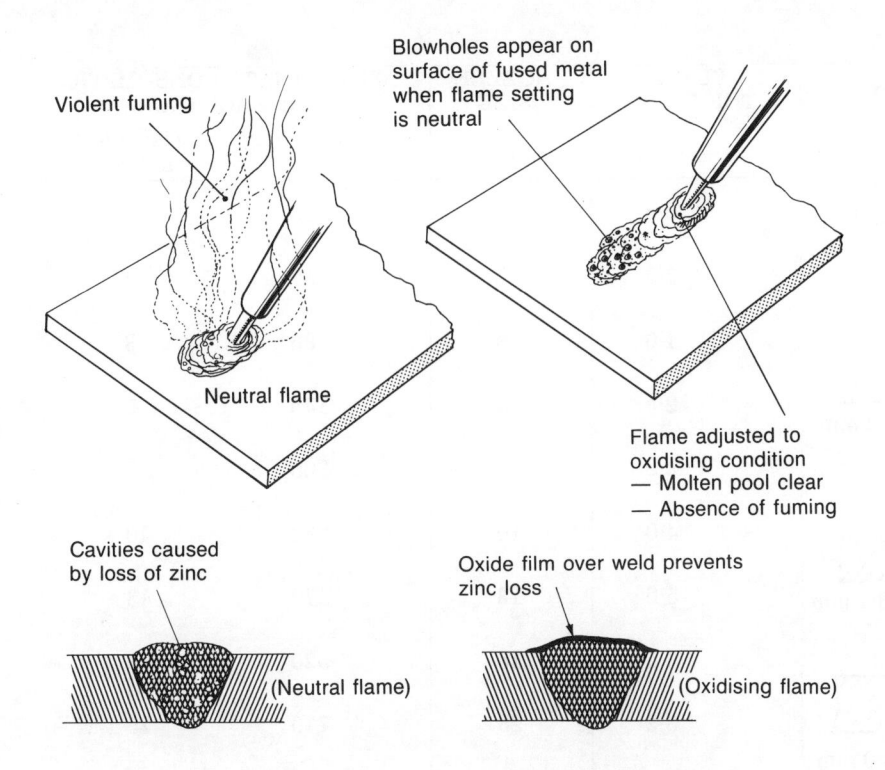

**Fig. 12.14 The importance of an oxidising flame — brass welding**

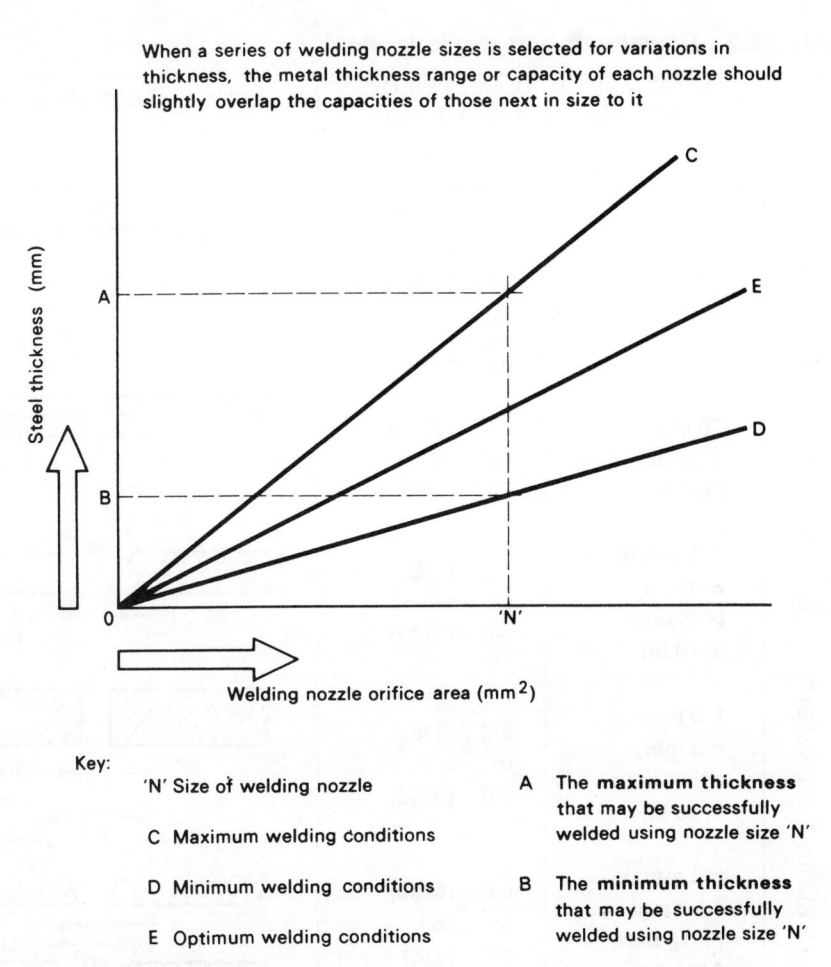

*Note*: Brazing filler rods usually contain about 0·25% SILICON, which, during the brazing operation, produces a 'silica film' on the surface of the liquid metal which prevents 'fuming', and thus produces a sounder deposit.

2. By the nozzle outlet bore size (orifice diameter).
3. By a reference number corresponding to a metal thickness range which may be welded with a specific nozzle.

Whatever the method employed for indicating nozzle sizes *there is a definite relationship between the sizes of welding nozzles and the metal thicknesses.* Figure 12.15 indicates this relationship where the thickness of steel plate is plotted against the orifice diameter of the welding nozzle.

Manufacturer's recommendations should always be followed with regard to nozzle sizes and gas pressures for a particular application.

Table 12.2 shows the nozzle sizes, gas consumption, and working pressures required for welding various thicknesses of mild steel using a neutral flame. The values given are for butt welds made in the downhand position.

Key:
'N' Size of welding nozzle
C Maximum welding conditions
D Minimum welding conditions
E Optimum welding conditions

A The **maximum thickness** that may be successfully welded using nozzle size 'N'
B The **minimum thickness** that may be successfully welded using nozzle size 'N'

**Fig. 12.15 Relationships between nozzle size and metal thickness**

## 12.13 Gas velocities

Flame cones are produced by the velocity gradient which exists across the circular orifice of the nozzle when the high-pressure gases are flowing. Since the gas velocity is greatest in the centre of the stream it follows that the flame length at the centre is similarly the greatest. Likewise, since the gas velocity is lowest at the periphery of the nozzle orifice the flame around the edge is shortest. This is because *friction* to the gas stream is greatest near the walls of the bore. The velocity of the gas increases towards the centre of the stream producing a nicely defined flame cone, as illustrated in Figure 12.16(*a*). The effect of nozzle size on the cone profile is shown in Figure 12.16(*b*).

**Table 12.2 Downhand butt welds in steel**

| | THICKNESS OF METAL | DIAMETER OF FILLER ROD mm (in) | JOINT EDGE PREPARATION | PLATE THICKNESS | NOZZLE SIZE | APPROXIMATE CONSUMPTION OF EACH GAS: Litres per hour | Cubic feet per hour |
|---|---|---|---|---|---|---|---|
| LEFTWARD WELDING TECHNIQUE | Less than 20 swg | 1·2 (3/64) to 1·6 (1/16) for square each preparation | No filler rod | 0·8 | 1 | 28 | 1 |
| | | | | 1·2 | 2 | 57 | 2 |
| | | | | 1·6 | 3 | 86 | 3 |
| | 20 swg to 3·2 mm (1/8 in) | 1·6 (1/16) to 3·2 (1/8) | 0·8–3·2 mm Gap | 2·4 | 5 | 140 | 5 |
| | | | | 3·2 | 7 | 200 | 7 |
| RIGHTWARD WELDING TECHNIQUE | 3·2 mm to 5·0 mm (1/8 in to 3/16 in) | 3·2 (1/8) to 4·0 (5/32) | 80° 1·6–3·2 mm Gap | 4·0 | 10 | 280 | 10 |
| | | | | 5·0 | 13 | 370 | 13 |
| | | | | 6·5 | 18 | 520 | 18 |
| | 5·0 mm to 8·2 mm (3/16 in to 5/16 in) | 3·2 (1/8) to 4·0 (5/32) | 3·2–4·0 mm Gap | 8·2 | 25 | 710 | 25 |
| | | | | 10·0 | 35 | 1000 | 35 |
| | 8·2 mm to 16·2 mm (5/16 in to 5/8 in) | 4·0 (5/32) to 6·5 (1/4) | 60° 3·2–4·0 mm Gap | 13·0 | 45 | 1300 | 45 |
| | | | | 16·2 | 55 | 1600 | 55 |
| | 16·2 mm and over | 6·5 (1/4) | 60° 80° 3·2–4·0 mm Gap | 19·0 | 60 | 1700 | 60 |
| | | | | 25·0 | 70 | 2000 | 70 |
| | | | | Over 25·0 | 90 | 2500 | 90 |

In practice, the efficiency of a welding nozzle can be determined by observing the action of the flame on the metal. If the gas velocity is too high a violent flame results (easily recognised because it burns with a harsh hissing sound) which has the effect of tending to blow the metal out of the weld pool. A harsh flame condition can readily be corrected by simply reducing the volumteric rates of oxygen and acetylene to a point where a satisfactory weld can be performed. THIS POINT WOULD INDICATE THE MAXIMUM VOLUMETRIC RATE WHICH WOULD GIVE OPTIMUM EFFICIENCY FOR A SPECIFIC SIZE OF NOZZLE.

## 12.14 Leftward welding

In this method of gas welding the flame is directed away from the finished weld, i.e., towards the unwelded part of the joint. Filler rod, when used, is directed towards the welded part of the joint.

(a) **Effect of gas velocities**

The variation in the laminar flow of the mixed gases as they leave the nozzle orifice determines the shape of the welding cone

The cones produced by a small nozzle will vary from a pointed to a semi-pointed shape. Cones from medium-sized nozzles will vary from a semi-pointed to a medium shape, and cones produced with a large-size nozzle will vary from a semi-blunt to a blunt shape. All these variations in cone shapes are determined by gas velocities for a particular size

(b) **Welding flame-cone shapes**

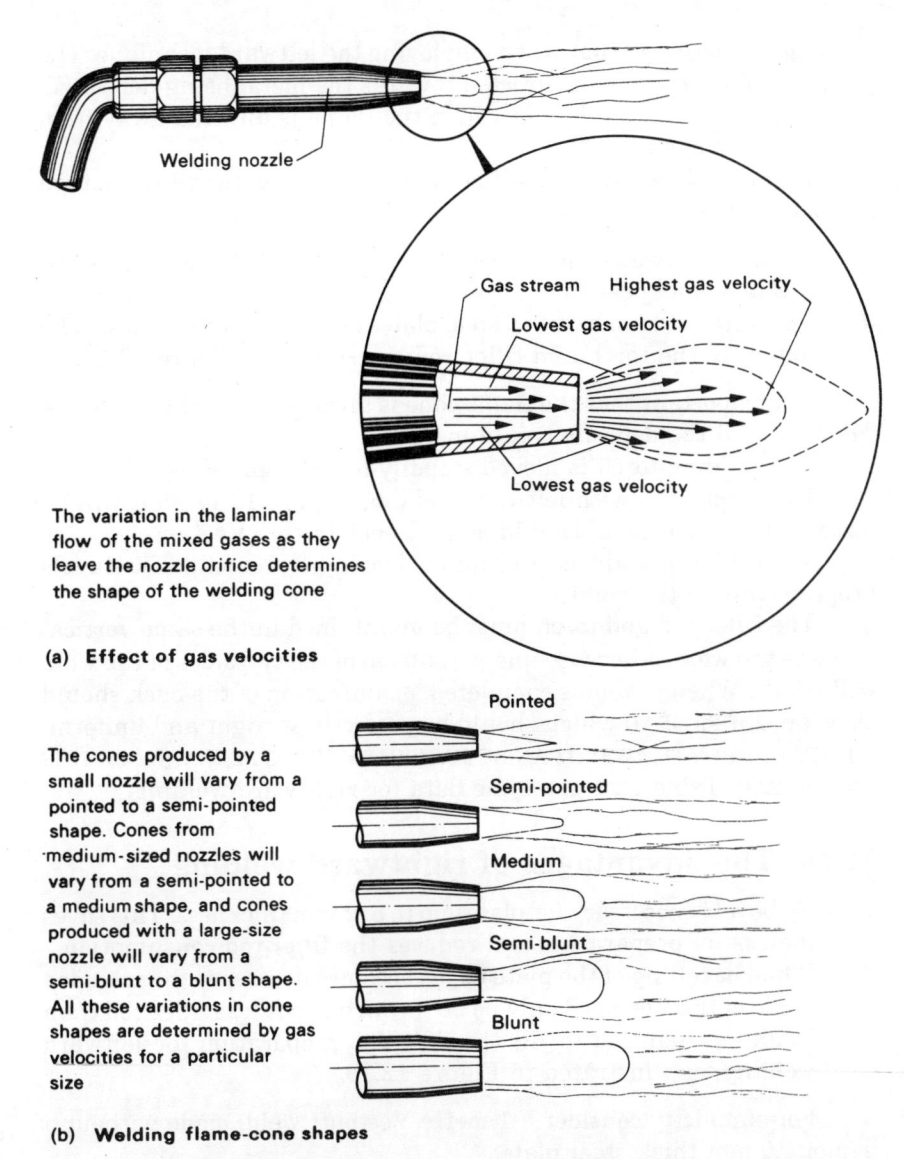

**Fig. 12.16 Factors affecting the weld flame-cone profile**

Although this technique is termed 'leftward welding', it is not confined to right-handed operators. Normally, with right-handed persons, the welding torch is held in the right hand and welding proceeds from right to left, i.e., LEFTWARDS. With a left-handed operator the torch is held in the left hand and welding proceeds from left to right.

Leftward welding is used on normal low carbon steels for the following:

(a) Flanged edge welds for thicknesses less than 20 s.w.g. These welds are made without the addition of a filler rod.
(b) Square-butt welds on unbevelled steel plates up to 3·2 mm thicknesses.
(c) Vee-butt welds on bevelled steel plates over 3·2 mm and up to 5 mm thickness.

The leftward method of welding is not considered to be economical for thicknesses above 5 mm thickness for reasons which will be explained later.

The angles of the torch and filler rod are clearly illustrated in Figure 12.17.

*Flanged-edge weld* (Figure 12.17(a)) No filler rod required. The flame should be manipulated with steady semi-circular sideways movement and progressively forwards only as fast as the edges of the sheet metal are melted. *The tip of the flame cone must be kept about 3 mm from the weld pool.*

*Square-butt weld* (Figure 12.17(b)) As the filler rod is melted it should be fed forward to build up the molten pool of weld metal and then retracted slightly. The flame is guided progressively forwards with small sideways movements as fast as the plate edges are melted.

A pear-shaped melted area (often referred to as an 'onion') should be maintained ahead of the weld pool. Care must be exercised to ensure uniform melting of both plate edges.

*Vee-butt weld* (Figure 12.17(c)) The tip of the flame cone should never touch the weld metal or the filler rod.

As the filler rod is melted it should be fed forward into the molten weld pool in order to build it up. The rod is then retracted slightly to enable the heat to fuse the bottom edges of the 'Vee'. *This ensures full penetration of the weld to the bottom of the plate edges.*

The side-to-side movement of the welding torch should only be sufficient to melt the sides of the 'Vee'.

This weld deposit is built up as the torch and filler rod move progressively forwards filling the 'Vee' to a level slightly higher than the edges of the plate. Table 12.3 gives the data for leftward welding.

## 12.15 Rightward welding

With this technique the weld is commenced at the left-hand end of the joint and the welding torch moves towards the right. The direction of

437

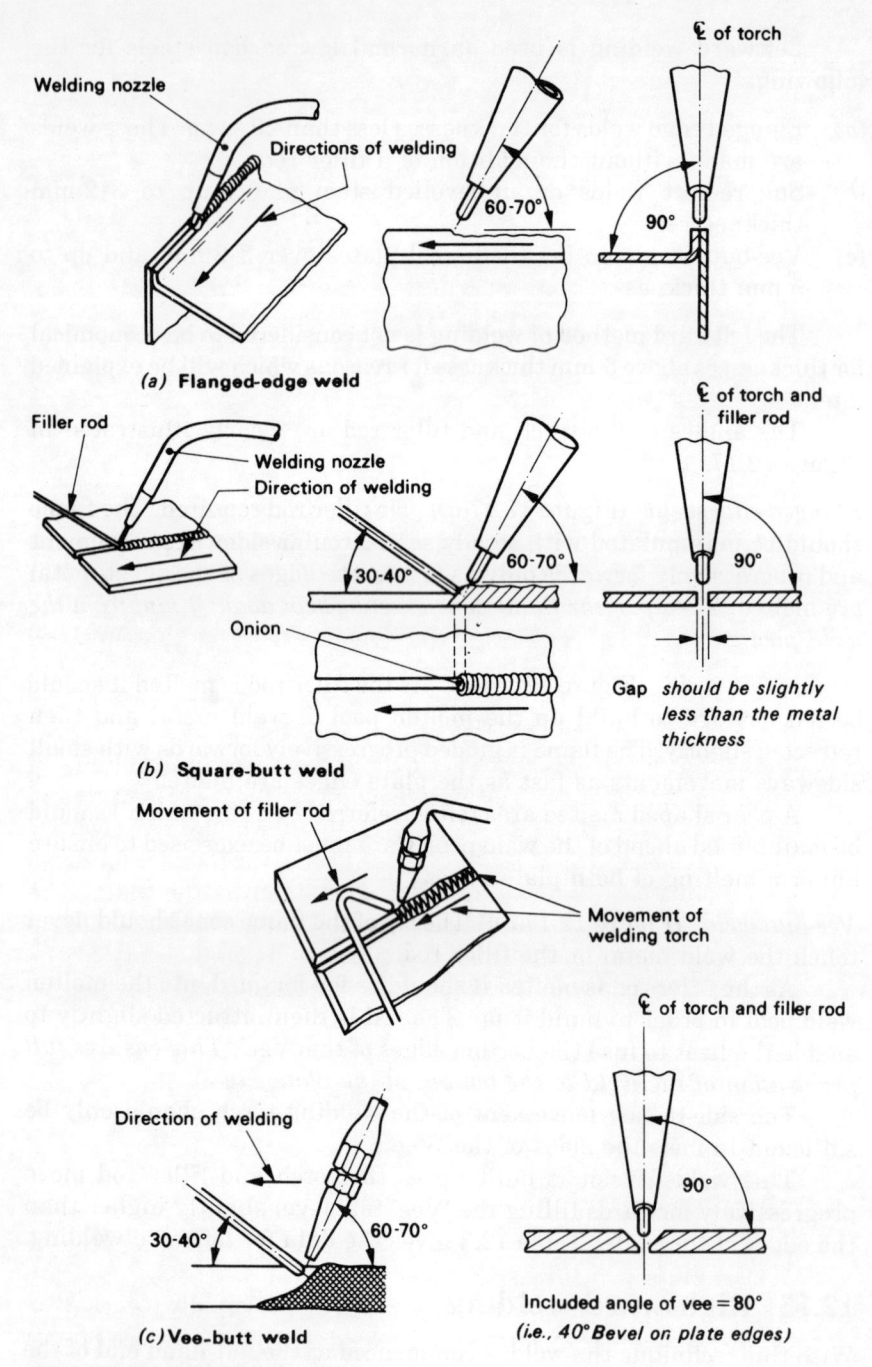

**Fig. 12.17 Leftward welding**

(a) Flanged-edge weld
(b) Square-butt weld
(c) Vee-butt weld

welding is opposite to that when employing the leftward technique. The torch flame in this case is directed towards the metal being deposited, unlike the leftward method in which the flame is directed away from the deposited metal.

Rightward welding should be used for steel plates which exceed 5 mm thickness as follows:

(a) Square-butt welds on unbevelled steel plates between 5 mm and 8·2 mm thickness (inclusive).
(b) Vee-butt welds on bevelled steel plates over 8·2 mm thickness. The angles of the torch and filler rod are shown in Figure 12.18.

It is important that the flame cone is always kept just clear of the filler rod and the deposited weld metal.

The welding torch is moved steadily to the right along the joint.

By comparison with leftward welding (Figure 12.17(c)) it will be noticed that the cone of the flame is deeper in the 'Vee'.

The filler rod is given an elliptical looping movement as it travels progressively to the right.

The filler rod and torch must be maintained in the same vertical plane as the weld, otherwise unequal fusion of the two sides of the weld will result. When a weld is completed, examination of the back should show an UNDER BEAD which should be perfectly straight and uniform. THE QUALITY OF THE WELD CAN BE JUDGED BY THE APPEARANCE OF THIS UNDERBEAD. Table 12.4 gives the data for rightward welding.

## 12.16 The advantages of rightward welding

1. No bevel is necessary for plates up to 8·2 mm thickness. This saves the cost of preparation and reduces the filler-rod consumption.
2. When bevelling of the plate edges becomes necessary the included angle of the Vee need only be 60°, which needs less filler rod than would be required to fill the 80° Vee preparation for leftward welding, as illustrated in Figure 12.19.

For simplicity consider a 1 metre Vee-butt weld, made without a gap on 10 mm thick steel plate:

For LEFTWARD WELDING the included angle of the Vee will be 80° (although in practice 90° is often used), whilst for RIGHTWARD WELDING the included angle will be 60°. *In each case the volume of metal deposited can be calculated by multiplying the cross-sectional area by the length.*

**Leftward**

$L \times T \times T \tan 40°$
$1000 \times 10 \times 10 \times 0.8391 = \underline{83\,910 \text{ mm}^3}$

**Rightward**

$L \times T \times T \tan 60°$
$1000 \times 10 \times 10 \times 0.5774 = \underline{57\,740 \text{ mm}^3}$

Table 12.3 Welding speeds and data for leftward welding

| THICKNESS OF METAL | | EDGE PREPARATION | GAP | | DIAMETER OF FILLER ROD | | POWER OF TORCH | | RATE OF WELDING | | FILLER ROD USED | |
|---|---|---|---|---|---|---|---|---|---|---|---|---|
| | | | | | | | (Gas consumption) | | | | per metre | per foot |
| mm | (in) | | mm | (in) | mm | (in) | Litres/hr | Cubic ft/hr | Metres/hr | Feet/hr | (m) | (ft) |
| 0·8 | 1/32 | Square | — | — | 1·6 | 1/16 | 28 — 57 | 1 — 2 | 6·0 — 7·6 | 20 — 25 | 0·3 | 1 |
| 1·6 | 1/16 | Square | 1·6 | 1/16 | 1·6 | 1/16 | 57 — 86 | 2 — 3 | 7·6 — 9·0 | 25 — 30 | 0·53 | 1·75 |
| 2·4 | 3/32 | Square | 2·4 | 3/32 | 1·6 | 1/16 | 86 — 140 | 3 — 5 | 6·0 — 7·6 | 20 — 25 | 0·84 | 2·75 |
| 3·2 | 1/8 | Square | 3·2 | 1/8 | 2·4 | 3/32 | 140 — 200 | 5 — 7 | 5·4 — 6·0 | 18 — 20 | 0·50 | 1·65 |
| 4·0 | 5/32 | 80° Vee | 3·2 | 1/8 | 3·2 | 1/8 | 200 — 280 | 7 — 10 | 4·6 — 5·4 | 15 — 18 | 0·64 | 2·1 |
| 5·0 | 3/16 | 80° Vee | 3·2 | 1/8 | 4·0 | 5/32 | 280 — 370 | 10 — 13 | 3·6 — 4·6 | 12 — 15 | 1·40 | 4·8 |

*Rightward welding is sometimes termed 'backward' or 'back-hand' welding*

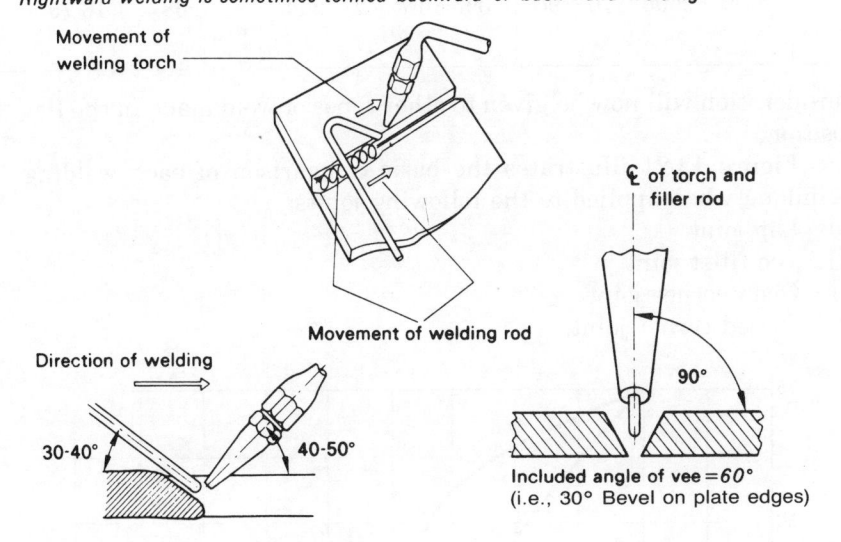

**Fig. 12.18 Rightward welding**

*The above results indicate that in the case of rightward welding the amount of filler rod is less than 69 per cent of that required for leftward welding.*

3. Larger welding nozzles must be used which results in higher welding speeds. With the use of more powerful nozzles the force of the flame 'holds back' the molten metal in the weld pool and allows more metal to be deposited so that welds in plates up to 10 mm thickness can be completed in one pass.

4. The operator's view of the weld pool and the sides and bottom of the Vee is unobstructed, enabling him to control the molten metal. This ensures that full fusion of the plate edges, particularly of the bottom edges, is always maintained, and results in an adequate and continuous bead of penetration.

5. The quality and appearance of the weld is better than that obtained with the leftward technique. This is due, to a large extent, to the fact that the deposited metal is protected by the envelope of the flame which retards the rate of cooling.

6. Compared with leftward welding, the smaller total volume of deposited metal with rightward welding reduces shrinkage and distortion. The graph in Figure 12.20 shows clearly how, by adopting the rightward technique, the heat of the flame is localised in the joint and is not allowed to spread across the plate. The blowpipe movement associated with leftward welding causes a greater spread of heat. The amount of distortion depends on the amount of heat put in, therefore it is important that this heat be confined as far as possible to the weld-seam itself.

The graph shows the maximum temperatures attained at various distances from welds made in 10 mm thick steel plate by the rightward and leftward techniques in the downhand position. A study of the graph will clearly indicate that in the case of
RIGHTWARD WELDING:
(a) The temperature of the plate is considerably lower, even 12·5 mm from the weld seam, than the corresponding area on the plate which was leftward welded.
(b) The temperature falls away much more rapidly, so that at approximately 28 mm from the weld the plate is almost 300°C cooler than the corresponding area for leftward welding.

**Table 12.4 Welding speeds and data for rightward welding**

| THICKNESS OF METAL | | EDGE PREPARATION | GAP | | DIAMETER OF FILLER ROD | | POWER OF TORCH (Gas consumption) | | RATE OF WELDING | | FILLER ROD USED | |
|---|---|---|---|---|---|---|---|---|---|---|---|---|
| mm | (in) | | mm | (in) | mm | (in) | Litres/hr | Cubic ft/hr | Metres/hr | Feet/hr | per metre (m) | per foot (ft) |
| 5·0 | 3/16 | Square | 2·4 | 3/32 | 2·4 | 3/32 | 370 — 520 | 13 — 18 | 3·6 — 4·6 | 12 — 15 | 1·03 | 3·4 |
| 6·5 | 1/4 | Square | 3·2 | 1/8 | 3·2 | 1/8 | 520 — 570 | 18 — 20 | 3·0 — 3·6 | 10 — 12 | 1·03 | 3·4 |
| 8·2 | 5/16 | Square | 4·0 | 5/32 | 4·0 | 5/32 | 710 — 860 | 25 — 30 | 2·1 — 2·4 | 7 — 8 | 1·03 | 3·4 |
| 10·0 | 3/8 | 60° Vee | 3·2 | 1/8 | 5·0 | 3/16 | 1000 — 1300 | 35 — 45 | 1·8 — 2·1 | 6 — 7 | 1·22 | 4·6 |
| 13·0 | 1/2 | 60° Vee | 3·2 | 1/8 | 6·5 | 1/4 | 1300 — 1400 | 45 — 50 | 1·3 — 1·5 | 4·5 — 5 | 1·69 | 4·75 |
| 16·2 | 5/8 | 60° Vee | 3·2 | 1/8 | 6·5 | 1/4 | 1600 — 1700 | 55 — 60 | 1·1 — 1·3 | 3·75 — 4·25 | 2·05 | 6·75 |
| 19·0 | 3/4 | 60° Vee (Top) 80° Vee (Bottom) | 4·0 | 5/32 | 6·5 | 1/4 | 1700 — 2000 | 60 — 70 | 0·9 — 1·0 | 3 — 3·25 | 2·96 | 9·75 |
| 25·0 | 1 | 60° Vee (Top) 80° Vee (Bottom) | 4·0 | 5/32 | 6·5 | 1/4 | 2000 — 2500 | 70 — 90 | 0·6 — 0·7 | 2 — 2·25 | 5·08 | 16·75 |

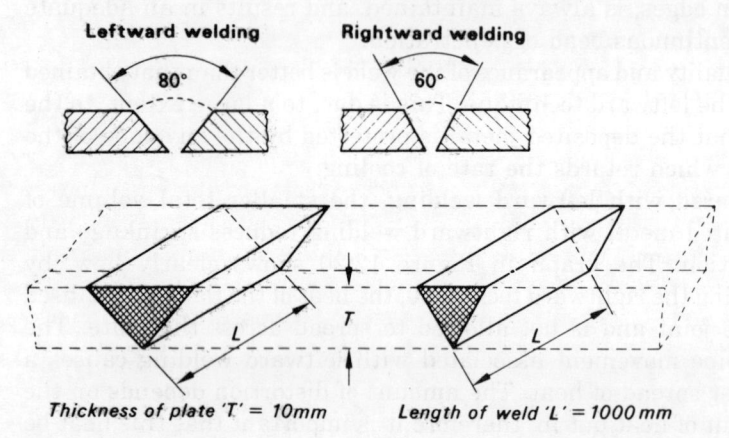

**Fig. 12.19 Amounts of deposited metal**

7. The cost of welding is lower than the leftward technique, despite the use of more powerful welding nozzles. The greatest economy is in the consumption of filler rods owing to the smaller amount of metal deposited with the rightward technique.

## 12.17 Comparison of welding techniques

A basic comparison of the leftward and rightward welding techniques has been made by illustrating butt welds in the flat or downhand position (Figures 12.17 and 12.18). In order to make a much broader comparison, consideration will now be given to other types of weld made in the flat position.

Figure 12.21 illustrates the basic comparison of each welding technique when applied to the following joints:

(a) Lap joint
(b) Tee fillet joint
(c) Open corner joint
(d) Closed corner joint.

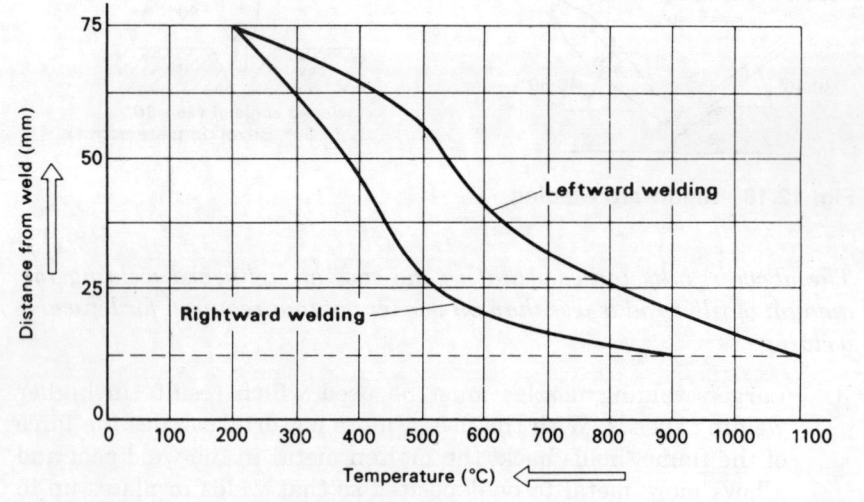

**Fig. 12.20 Comparison of heat spread**

(a) Lap joint

Length of lap 'L' = 3 times metal thickness 'T'

Single lap

Double lap

(b) Fillet joint

(c) Open corner joint

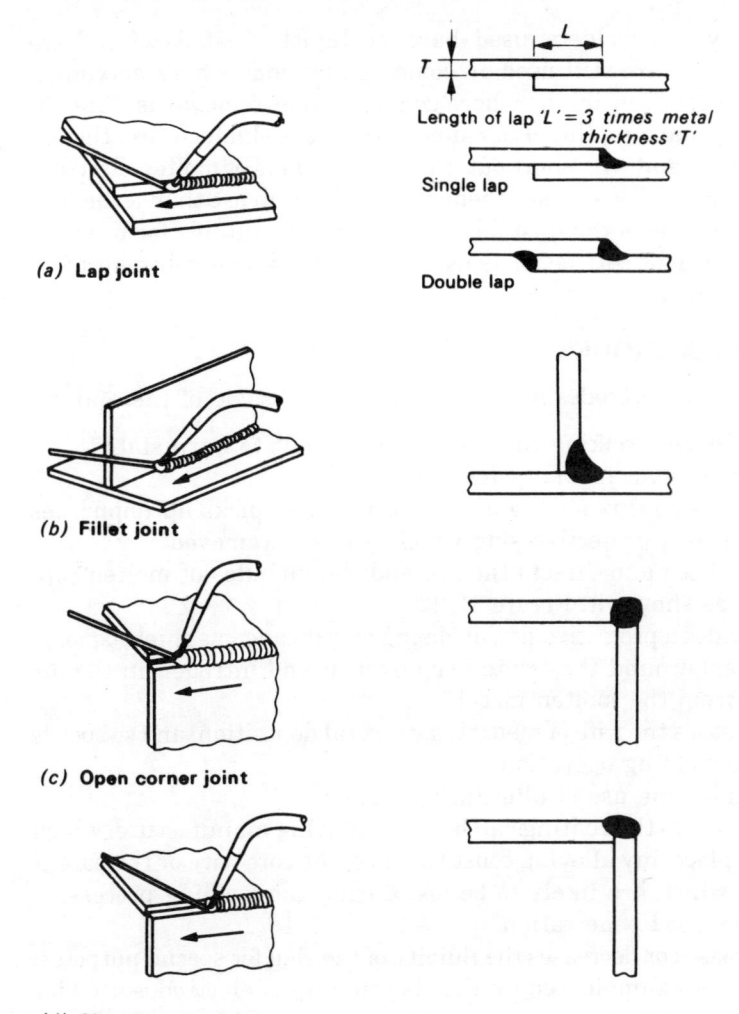

(d) Closed corner joint

*Note: All the joints in the above examples are being produced by the leftward technique. The rightward technique can also be used for all these joints*

Fig. 12.21 Types of joint

# Part B  Metal-arc welding

The principles of 'metal-arc' welding have already been introduced in Section 11.28. These basic principles will now be examined in greater depth. Examples of typical arc-welding plant have been illustrated in Chapter 1.

## 12.18  The metal-arc process

The arc is produced by a low-voltage, high-amperage electric current jumping an air gap between the electrode and the joint to be welded. The heat of the electric arc is concentrated on the edges of two pieces of metal to be joined. This causes the metal edges to melt. While these edges are still molten additional molten metal, transferred across the arc from a suitable electrode, is added. This molten mass of metal cools and solidifies into one solid piece.

As soon as the arc is struck, the tip of the electrode begins to melt, thus increasing the gap between electrode and work. Therefore it is necessary to cultivate a continuous downward movement with the electrode holder in order to maintain a constant arc length of 3 mm during the welding operation. The electrode is moved at a uniform rate along the joint to be welded, melting the metal as it moves.

## 12.19  The electrode

The greatest bulk of electrodes used with manual arc welding are *coated* electrodes. A coated electrode consists mainly of a core wire of closely controlled composition having a concentric covering of flux and/or other material, which will melt uniformly, with the core wire forming a partly vaporised and partly molten screen around the arc stream. This shield protects the arc from contamination by atmospheric gases.

The liquid slag produced performs three important functions:

1. Protects the solidifying weld metal from further contamination from the atmosphere.
2. Prevents rapid cooling of the weld metal.
3. Controls the contour of the completed weld.

## 12.20  The 'arc stream'

The function of an electrode is more than simply to carry the current to the arc. The core wire melts in the arc and tiny globules of molten metal shoot across the arc into the molten pool (arc crater in parent metal) during welding. These tiny globules are explosively forced through the arc stream. They are not transferred across the arc by the

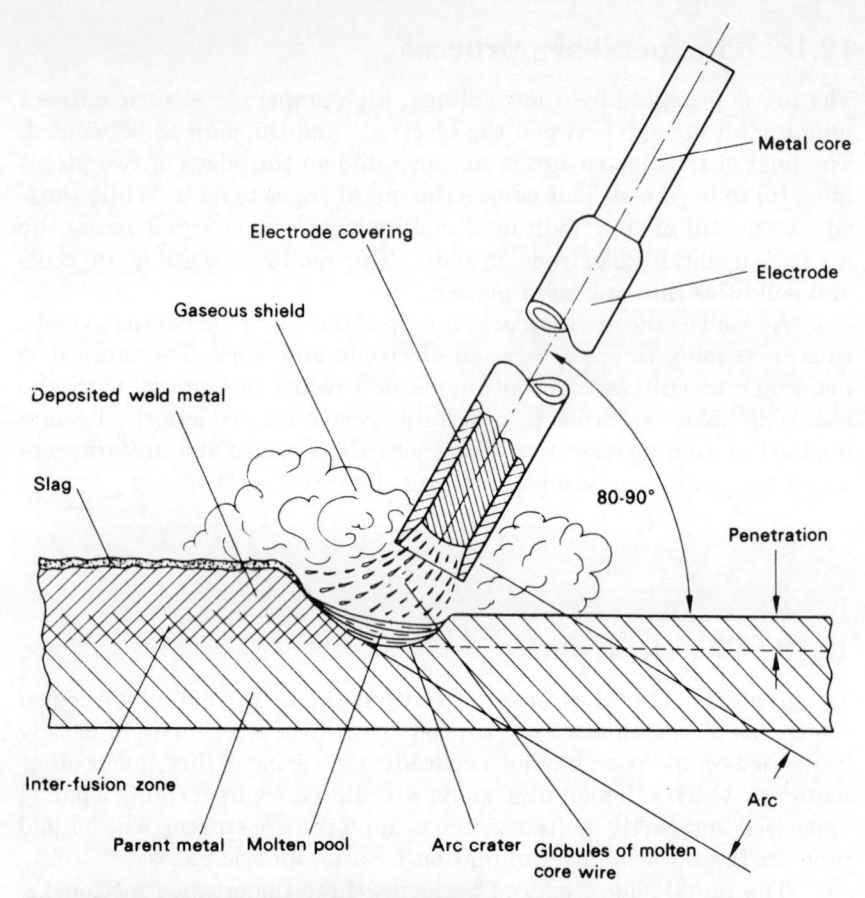

**Fig. 12.22 Basic features of arc-welding**

force of gravity, otherwise it would not be possible to use the manual arc welding process for overhead welding.

The chemical coating surrounding the core wire melts or burns in the arc. It melts at a slightly higher temperature than the metal core and therefore extends a little beyond the core and directs the arc. This extension also prevents sideways arcing when welding in deep grooves.

The arc stream and other basic features of manual *Gas-shielded metal-arc welding* are illustrated in Figure 12.22.

## 12.21 Functions of the electrode coating

### UNCOATED ELECTRODES

Arc-welding has not always been as easy as one finds it today using the shielded electrode. Before coated electrodes were produced commercially, arc-welders used bare or lightly coated wire. Bare (uncoated) wires are still used on some applications where maximum strength is not essential, or where complete slag removal is difficult. Because they have no flux and cannot produce a shielded arc, the arc is less powerful and will 'short out' easily as droplets of molten electrode bridge the arc gap. This makes them more difficult to use than a shielded electrode. Because of the unstable arc they are unsuitable for use with a.c., but they can be satisfactorily used on d.c. equipment using electrode negative polarity.

### COATED ELECTRODES

The coating on electrodes has several functions, some of these are:

(a) It facilitates striking the arc and enables it to burn stably.
(b) It serves as an insulator for the core wire.
(c) It provides a flux for the molten pool, which picks up impurities and forms a protective slag which is easily removed.
(d) It stabilises and directs the arc and the globules of molten core metal as shown in Figure 12.22.
(e) It provides a protective non-oxidising or reducing gas shield (smoke-like gas) around the arc to keep oxygen and nitrogen in the air away from the molten metal.
(f) It increases the rate of melting (i.e., metal deposition) and so speeds up the welding operation.
(g) It enables the use of alternating current.
(h) Additions to the coating can be made (during manufacture) which will replace any alloying constituents of the core wire or the parent metal which are likely to be lost during the welding process.
(i) It gives good penetration.
(j) It increases or decreases the fluidity of the slag for special purposes. It can, for example, reduce the slag fluidity of electrodes used for overhead welding.

## 12.22 Welding with direct current

The basic equipment used for welding with d.c. operates on an 'OPEN CIRCUIT' VOLTAGE which is much lower than that of an a.c. plant.

*With* DIRECT CURRENT *the electron current always flows in the same direction*. The POSITIVE pole therefore becomes somewhat more heated than the NEGATIVE. From a practical point of view this is very important to the welder because *the* HEAT *generated by d.c. is not distributed equally between the two poles (i.e., the electrode and the work)*. TWO-THIRDS OF THE HEAT IS GENERATED AT THE POSITIVE POLE AND ONLY ONE-THIRD AT THE NEGATIVE POLE. This has considerable influence on

**Connections to heavy-gauge work pieces**
Electrode NEGATIVE
Workpiece POSITIVE to concentrate the heat into the parent metal

**Connections to light-gauge work pieces**
Electrode POSITIVE
Workpiece NEGATIVE to prevent burning a hole

*Note: Some coated electrodes can only be connected to the positive pole*

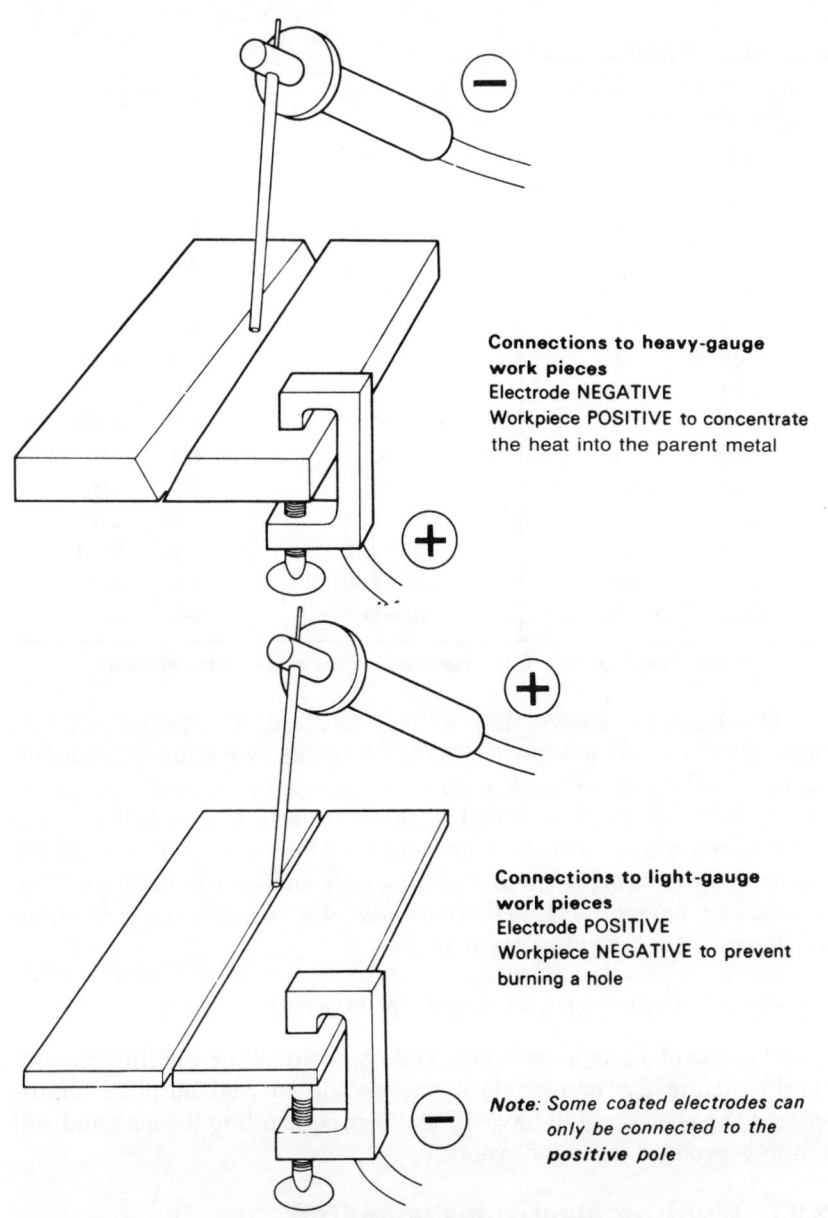

**Fig. 12.23 Connections for direct-current arc-welding**

welding procedure. By connecting the electrode to the appropriate pole the heat input may be increased or reduced. This variation may be used to advantage as shown in Figure 12.23.

The polarity of the electrode is most important when welding with d.c. and the manufacturer's instructions should be rigidly adhered to. In order to prevent any misuse, electrodes to be connected to the positive pole are termed 'ELECTRODE POSITIVE'.

## 12.23 Welding with alternating current

*With* ALTERNATING CURRENT *the electron current is changing its magnitude and direction 50 times per second*, and therefore there are no positive and negative poles in the ordinary sense. THE SAME AMOUNT OF HEAT IS PRODUCED AT THE ELECTRODE AND THE WORK.

Thus, while manufacturers specify that a particular electrode should be connected either to the negative or the positive pole of a d.c. supply, no polarity is quoted for electrodes to be used with a.c.

The arc is extinguished each time the current changes direction, but re-establishes itself because the arc atmosphere is heavily ionised. Figure 12.24 shows the comparison between the two types of electric current used for arc welding.

## 12.24 Welding current values

All electrodes are designed for use with a specific welding current, and for this reason the values quoted by the manufacturers must always be closely adhered to. Any alteration will affect the behaviour of the electrode, the appearance of the weld, and possibly the properties of the deposited metal. Variations in current values and their resultant effects will be discussed later in this chapter.

The approximate currents to be used when welding mild steel plates are shown in Table 12.5.

## 12.25 Open circuit and welding voltages

In order to produce an arc which is suitable for welding it is necessary to have an ELECTRO-MOTIVE FORCE (VOLTAGE) to drive the CURRENT through the circuit. Depending on the type of welding set used (i.e., a.c. or d.c.) a voltage between 60 and 100 V is necessary to start the arc. Once the arc has been started, the arc voltage drops and only 20 to 45 V is necessary to maintain it.

The OPEN CIRCUIT VOLTAGE is often referred to as the 'STRIKING VOLTAGE' and is the e.m.f. required when striking the arc. Most a.c. welding transformers operate at an open circuit voltage of 90 V. Welding sets providing d.c. operate with an open circuit voltage of between 60 and 70 V.

The ARC VOLTAGE is the e.m.f. required to maintain the arc during welding and usually varies between 20 and 45 V.

**Arc-welding with direct current**
When welding with d.c. the electron current flow is always in the same direction. The current and voltage have constant values for the same resistance. More heat is generated at the POSITIVE POLE than at the NEGATIVE POLE

**Arc-welding with alternating current**
When welding with a.c. the electron current flow is constantly changing its MAGNITUDE and DIRECTION. Therefore the CURRENT and VOLTAGE are constantly changing in magnitude and direction. There is no negative and positive pole in the general sense, therefore the same amount of heat is generated at the electrode and the parent metal

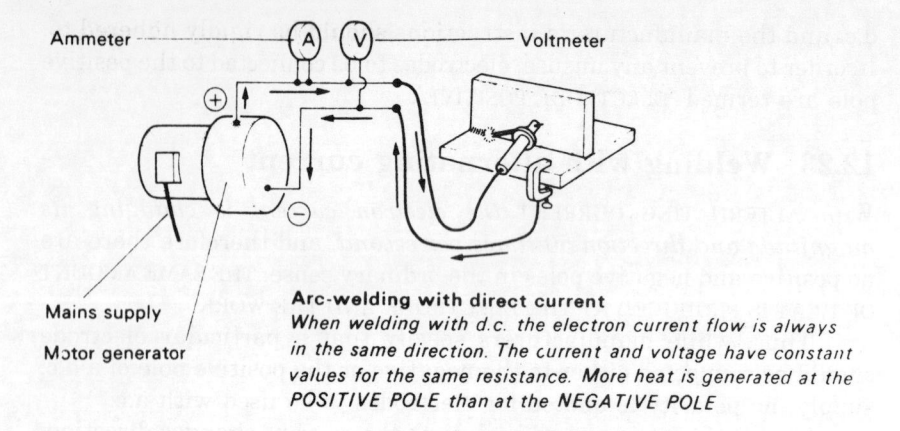

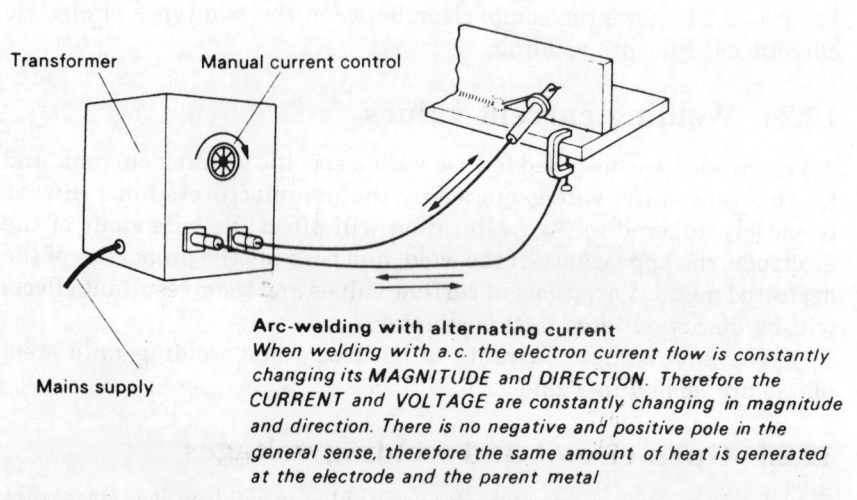

**Fig. 12.24 Comparison of alternating-current and direct-current arc-welding**

Figure 12.25 illustrates the changes which occur in producing the arc.

## 12.26 Welding current stability

### DIRECT CURRENT FROM A GENERATOR

As the arc voltage depends upon the size and type of electrode being used, the generator is designed so that the arc voltage can adjust itself to these conditions. This is done by use of a d.c. generator having a 'drooping voltage' characteristic, as shown in Figure 12.26.

**Table 12.5 Welding currents**

| MINIMUM THICKNESS (MILD STEEL PLATE) | | WELDING CURRENT VALUE | DIAMETER OF ELECTRODE | |
|---|---|---|---|---|
| mm | (s.w.g.) | Amps. | mm | (in) |
| 1·62 | 16 | 40—60 | 1·6 | 1/16 |
| 2·03 | 14 | 60—80 | 2·4 | 3/32 |
| 2·64 | 12 | 100 | 3·2 | 1/8 |
| 3·18 | 1/8 (in) | 125 | 3·2 | 1/8 |
| 3·25 | 10 | 125 | 3·2 | 1/8 |
| 4·06 | 8 | 160 | 4·8 | 3/16 |
| 4·76 | 3/16 (in) | 190 | 4·8 | 3/16 |
| 4·88 | 6 | 190 | 4·8 | 3/16 |
| 5·89 | 4 | 230 | 6·4 | 1/4 |
| 6·35 | 1/4 (in) | 250 | 6·4 | 1/4 |
| 7·01 | 2 | 275—300 | 7·9 | 5/16 |
| 8·23 | 0 | 300—400 | 7·9 | 5/16 |
| 8·84 | 00 | 400—600 | 9·5 | 3/8 |
| 9·53 | 3/8 (in) | 400—600 | 9·5 | 3/8 |

Note: The diameter of the electrode is the size of the core wire.

The diagram shows that RELATIVELY LARGE CHANGES IN ARC VOLTAGE WILL ONLY RESULT IN SMALL CHANGES IN WELDING CURRENT. Changes in the arc voltgage occur with any variation in the length of the arc. With manual arc-welding this is due to the inability of the welder to maintain a constant arc length. *These curves (Figure 12.26) show that the welding arc will remain stable and easy to hold, and that the welding current will remain practically constant despite slight variations in the length of the arc.*

### ALTERNATING CURRENT FROM A TRANSFORMER

As in the case of d.c. equipment arc voltages control the welding current. Whether using a.c. or d.c. plant, unless the current output remains constant the operator will have difficulty in controlling the slag and will therefore produce unsound welds.

## 12.27 Welding electrodes (specific)

During manufacture, coated electrodes for welding are subjected to a very high standard of quality control at all stages of the manufacturing process. Most electrodes are made by solid extrusion, whereby the flux covering is applied direct to the core wire under high pressure. The covering size is controlled to very precise limits of concentricity. If the

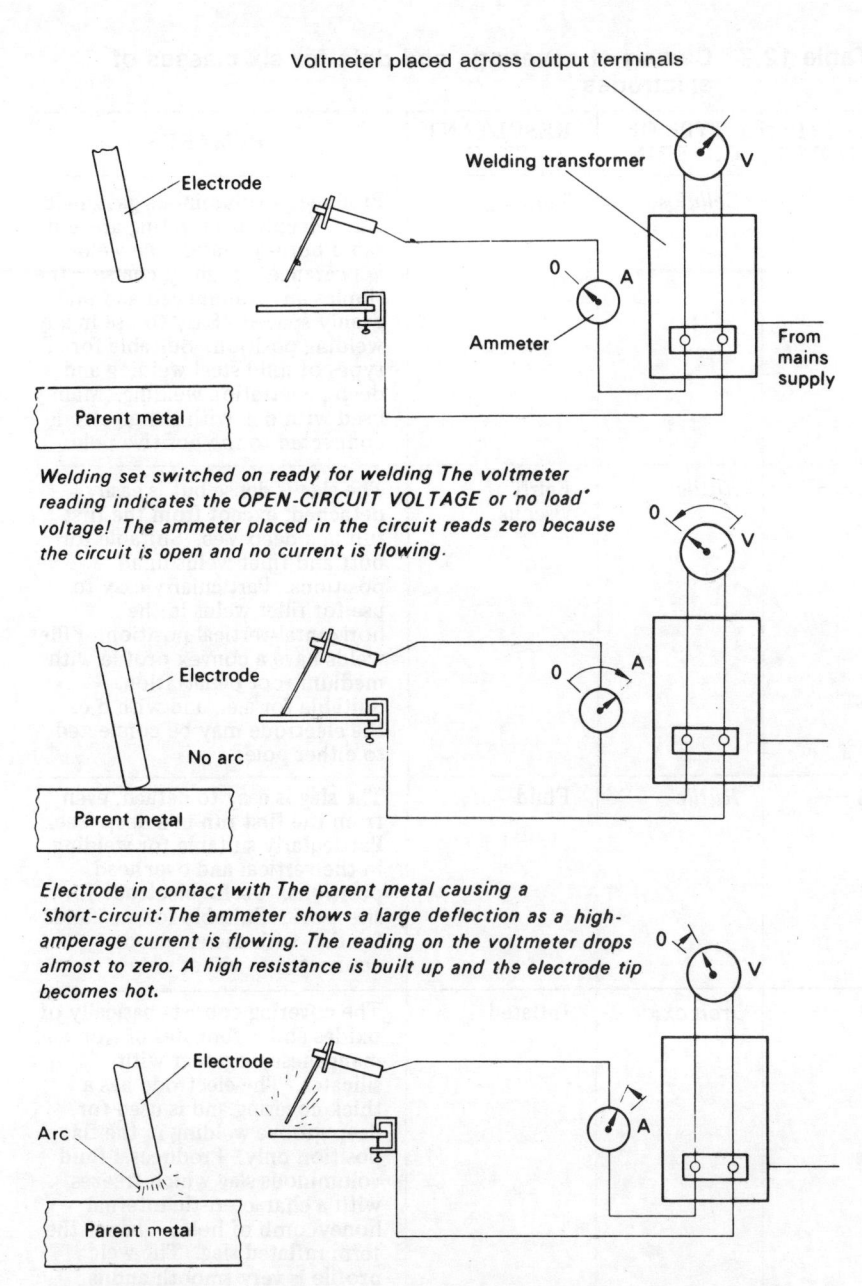

**Fig. 12.25 Effects on current and voltage when striking the arc**

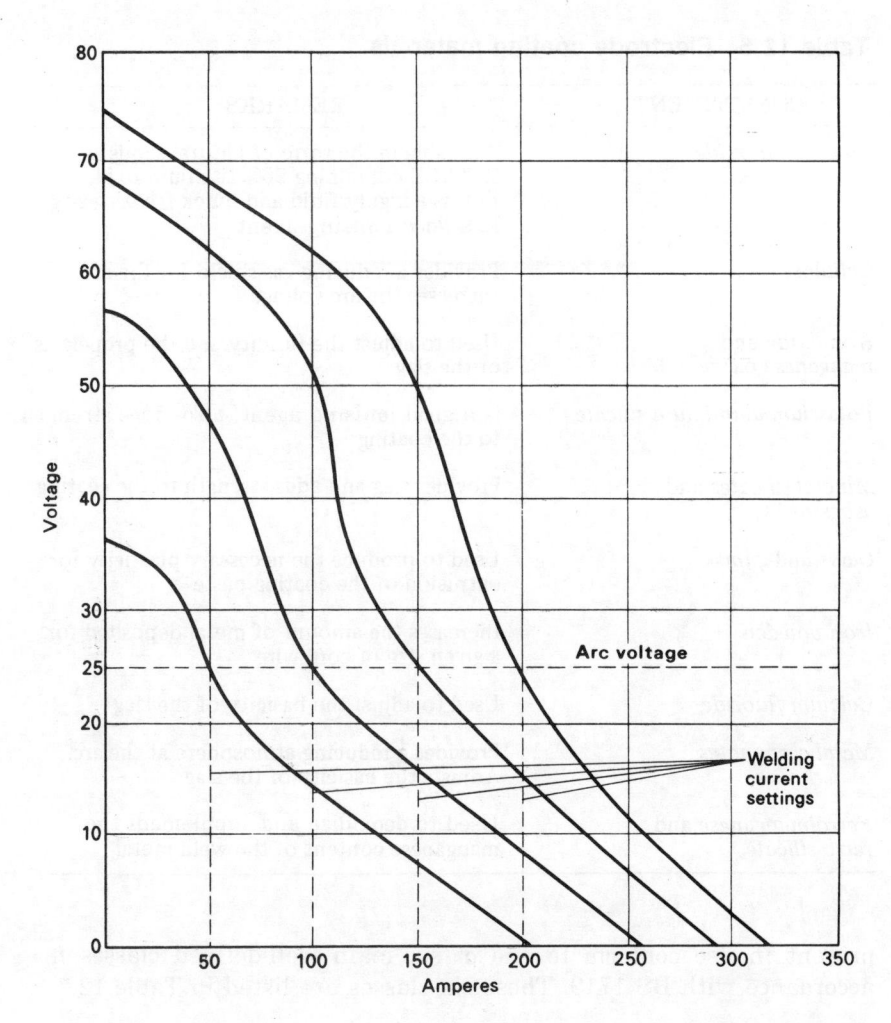

*Welding set switched on ready for welding The voltmeter reading indicates the OPEN-CIRCUIT VOLTAGE or 'no load' voltage! The ammeter placed in the circuit reads zero because the circuit is open and no current is flowing.*

*Electrode in contact with the parent metal causing a 'short-circuit'. The ammeter shows a large deflection as a high-amperage current is flowing. The reading on the voltmeter drops almost to zero. A high resistance is built up and the electrode tip becomes hot.*

*Electrode withdrawn slightly to form an air gap. The current jumps the gap and produces an arc. The voltmeter reads between 20 and 45 volts and the current drops to the WELDING CURRENT set on the transformer.*

**Fig. 12.26 Static characteristic curves of a direct-current welding generator**

covering were not concentric to the core wire an unstable arc would result.

*Covering constituents* The covering of the electrodes for the welding of mild and high tensile steels consists of various mixtures of the substances shown in Table 12.6.

*Type of flux covering* By using different properties and combinations of the constituents listed in Table 12.6 it is possible to produce an infinite variety of electrode types. In practice, practically all the electrodes at

445

**Table 12.6  Electrode coating materials**

| CONSTITUENT | REMARKS |
|---|---|
| *Titanium dioxide* | Available in the form of natural sands as RUTILE containing 96% titanium oxide. Forms a highly fluid and quick freezing slag. Is a good ionising agent |
| *Cellulose* | Provides a reducing gas shield for the arc. Increases the arc voltage |
| *Iron oxide* and *manganese oxide* | Used to adjust the fluidity and the properties of the slag |
| *Potassium aluminium silicate* | Is a good ionising agent, also gives strength to the coating |
| *Mineral silicates* and *asbestos* | Provides slag and adds strength to the coating |
| *Clays* and *gums* | Used to produce the necessary plasticity for extrusion of the coating paste |
| *Iron powder* | Increases the amount of metal deposited for a given size of core wire |
| *Calcium fluoride* | Used to adjust the basicity of the slag |
| *Metal carbonates* | Provides a reducing atmosphere at the arc. Adjusts the basicity of the slag |
| *Ferro-manganese* and *ferro-silicate* | Used to deoxidise and supplement the manganese content of the weld metal |

present in use conform to one of six main well-defined classes in accordance with BS 1719. These six classes are listed in Table 12.7.

*Storage of electrodes*  Because electrode coatings tend to absorb moisture from the atmosphere it is essential that they be kept dry. *If the electrode coating becomes damp, steam will be generated during the welding process causing excessive spatter, and porosity in the weld.* Low hydrogen electrodes should be dried in an oven prior to their use. Specially designed electric ovens are available which can be thermostatically controlled over a temperature range of 38 to 275°C.

## 12.28  Effect of variable factors in metal-arc welding

The variable factors in metal-arc welding are:

1. *The welding current*  If the welding current is set *too low* the

**Table 12.7  Coding of coverings and data for six classes of electrodes**

| BS 1719 CODING | TYPE OF COVERING | RESULTANT SLAG | REMARKS |
|---|---|---|---|
| 1 — — | Cellulose | Thin | Produces a voluminous gas shield. Has a deeply penetrating arc and rapid burn-off rate. The weld appearance is slightly coarse — the ripples are pronounced and unevenly spaced. Easy to use in any welding position. Suitable for all types of mild steel welding and deep penetration welding. Mainly used with d.c. with the electrode connected to the positive pole. |
| 2 — — | Rutile | Fairly viscous | The slag is dense but is easily detached, except from the first run in a deep Vee. Suitable for butt and fillet welds in all positions. Particularly easy to use for fillet welds in the horizontal-vertical position. Fillet welds have a convex profile with medium root penetration. Suitable for a.c., and with d.c. the electrode may be connected to either pole. |
| 3 — — | Rutile | Fluid | The slag is easy to detach, even from the first run in a deep Vee. Particularly suitable for welding in the vertical and overhead positions. Useful electrode for site work. Suitable for a.c., and for d.c. with the electrode connected to either pole. |
| 4 — — | Iron oxide | Inflated | The covering consists basically of oxides and carbonates of iron and manganese, together with silicates. The electrode has a thick covering and is used for deep groove welding in the flat position only. Produces a fluid voluminous slag which freezes with a characteristic internal honeycomb of holes — hence the term inflated slag. The weld profile is very smooth and is concave. The electrode is suitable for use with d.c. with the electrode connected to the positive pole, but may also be used with a.c. |

| 5 — — | *Iron oxide* | Solid | The covering is thick and consists basically of iron oxides with or without oxides of manganese. A heavy solid slag is produced which is sometimes self-detaching. Used for single-run fillet welds where smooth contour is more important than high mechanical properties in the weld metal. Covering melts with a distinctive 'cupped' effect, enabling the electrode to be used touching the work — touch welding. The weld has a smooth concave profile. These electrodes are often referred to as dead soft electrodes because the weld metal has a low carbon and manganese content. Suitable for a.c. and on d.c. the electrode may be connected to either pole. |
| 6 — — | *Lime fluorospar* | Basic | This class of electrode is often termed low hydrogen or basic type. Suitable for welding in all positions. The slag is fairly fluid. The weld deposit is usually convex to flat in profile. Is less sensitive to variation in plate quality than any other class of electrode. Used for welding of heavy sections and highly restrained joints. Can be used with a.c. but with some types d.c. is preferred with electrode positive. |

electrode core wire is not melted sufficiently. This results in poor penetration due to the lack of heat necessary to complete fusion. The molten weld metal tends to pile up instead of flowing smoothly, producing a weld bead which is irregular in width and contour, as shown in Figure 12.27(a). There is a tendency for slag to become entrapped within the weld bead itself.

The weld crater is not very well defined. The arc is unsteady and burns with an irregular spluttering sound.

An *exessive current* can cause the weld metal to become too fluid and difficult to control. *There is considerable spatter and unnecessary wastage of electrode*, as shown in Figure 12.27(b). When the welding current is set too high the electrode tends to become red hot and the arc burns with a regular explosive sound.

Too high a current can often result in blow holes being formed in the parent metal accompanied by excessive penetration and some undercutting along the edge of the weld.

The slag produced is usually difficult to remove. Table 12.9 gives the British Standard Coding for welding currents.

2. *The arc length* Too short an arc length causes irregular build up of the molten metal. The ripples on the weld bead are not uniform with regard to both width and height, as shown in Figure 12.28(a). With too short an arc length it is difficult to control and maintain the arc, and there is a tendency for the electrode to 'freeze' or stick to the weld pool.

**Table 12.8 Definitions of welding positions**

| POSITION | BS 1719 SYMBOL | LIMITS OF SLOPE | LIMITS OF ROTATION |
|---|---|---|---|
| *Flat* | F | 0° to 5° | 0° to 10° — 0° to 10° |
| *Horizontal-Vertical* | H | 0° to 5° | 30° to 90° / 30° to 90° |
| *Vertical-Up* | V | 80° to 90° | 0° to 180° — 0° to 180° |
| *Vertical-Down* | D | 80° to 90° | |
| *Overhead* | O | 0° to 10° | 115° to 180° / 115° to 180° |

Any intermediate position not specified above is undefined but the general term 'inclined' is used.

**Table 12.8 cont.**

DIAGRAMMATIC REPRESENTATION OF PRINCIPLE WELDING POSITIONS

*Key:*
| | |
|---|---|
| A – Horizontal-vertical | F – Inclined fillet |
| B – Vertical fillet | G – Flat butt |
| C – Vertical butt | H – Overhead butt |
| D – Flat fillet | I – Overhead fillet |
| E – Inclined butt | J – Horizontal-vertical fillet |

| BS 1719 CODING | WELDING POSITIONS |
|---|---|
| –0– | F, H, V, D, O |
| –1– | F, H, V, O |
| –2– | F, H |
| –3– | F |
| –4– | F, Hf (fillet welding only) |

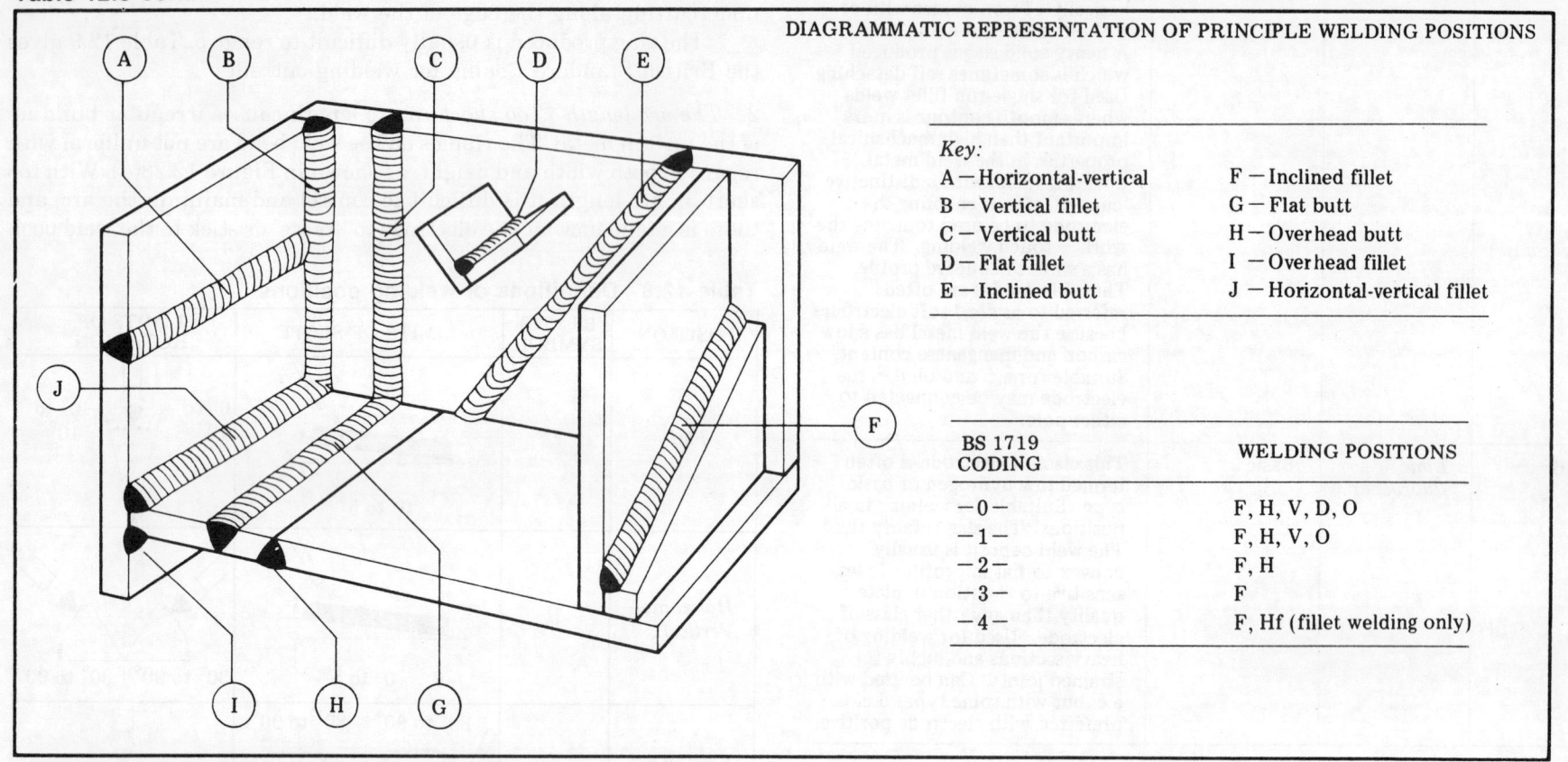

Table 12.8 defines the standard welding positions as recommended in BS 1719.

Note  A normal arc length should be slightly less than the diameter of the electrode, and is generally considered to be between 1·6 and 3·2 mm.

*Too long* an arc causes an appreciable increase in spatter, and penetration is poor. The weld bead is wide and of poor appearance because the core wire metal is deposited in large globules instead of a steady stream of fine particles. The arc crater is flat and blistered as shown in Figure 12.28(b).

Note  To ensure complete penetration the arc length should be kept as short as possible to enable the heat from the arc to melt the parent metal and the electrode core-wire simultaneously. If the length is too great a considerable amount of this heat is lost to the air.

3. *Speed of travel*  When the speed of travel is too fast a narrow and thin weld deposit, longer than normal, is produced. Penetration is poor, the weld crater being small and rather well defined, as shown in Figure 12.28(a). The surface of the weld bead has elongated ripples and there is 'undercutting'. *The reduction in bead size and amount of undercutting depends on the ratio of speed and current.*

Note  Most manufacturers specify the length of run which may be obtained with electrodes of different, types sizes, and lengths, used under their recommended current conditions. Thus, if the length of deposit per electrode is smaller or greater than is specified, the speed of travel is either slower or faster than that intended by the manufacturer.

*As the rate of travel decreases the width and thickness of the deposit increase*, as shown in Figure 12.29(b). The weld deposit is much shorter

*(a)* Welding current too low    *(b)* Welding current too high

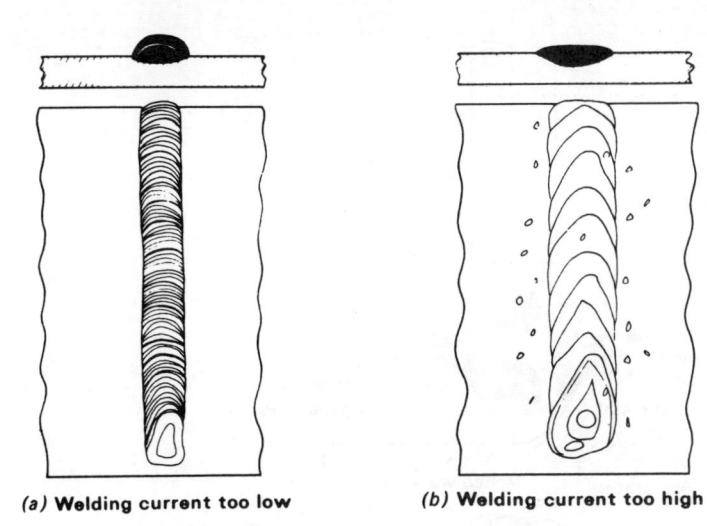

Fig. 12.27 Effect of variation in welding current

*(a)* Welding with too short an arc    *(b)* Welding with too long an arc

Fig. 12.28 Effect of variation in arc length

Table 12.9 Coding for welding current conditions

*The welding current conditions as defined in BS 1719 are as follows:*

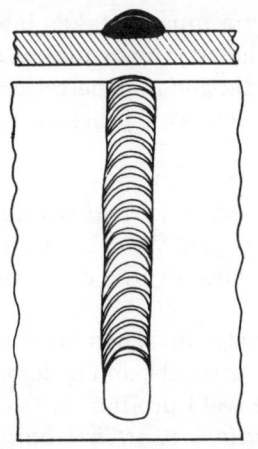

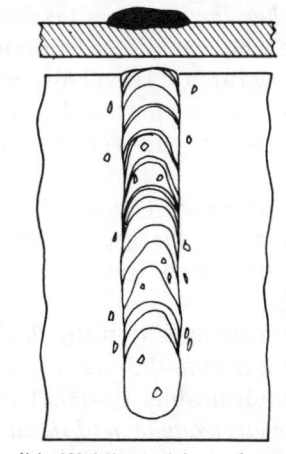

| WELDING CURRENT CONDITIONS | SYMBOL |
|---|---|
| D.C. with electrode positive | D + |
| D.C. with electrode negative | D − |
| D.C. with electrode positive or negative | D ± |
| A.C. with an open-circuit voltage not less than 90 volts | $A_{90}$ |
| A.C. with an open-circuit voltage not less than 70 volts | $A_{70}$ |
| A.C. with an open-circuit voltage not less than 50 volts | $A_{50}$ |

| BS 1719 CODING | WELDING CURRENT | BS 1719 CODING | WELDING CURRENT |
|---|---|---|---|
| − − 0 | D + | − − 4 | D + $A_{70}$ |
| − − 1 | D + $A_{90}$ | − − 5 | D ± $A_{90}$ |
| − − 2 | D − $A_{70}$ | − − 6 | D ± $A_{70}$ |
| − − 3 | D − $A_{50}$ | − − 7 | D ± $A_{50}$ |

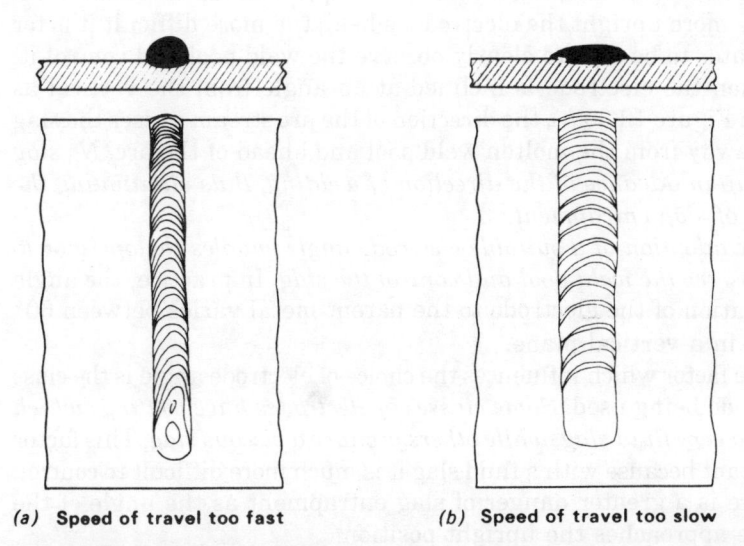

*(a)* Speed of travel too fast    *(b)* Speed of travel too slow

Fig. 12.29 Effect of variation in rate of travel

than the normal length. The weld crater is flat. The surface of the weld bead has coarse evenly spaced ripples. With a slower rate of travel the molten weld metal will tend to pile up and cause excessive overlap of the weld bead.

*Slower speeds cause the parent metal and the weld bead to become*

449

*hotter than when the movement is faster.* This factor must be taken into consideration when welding certain metals, or when trying to minimise distortion. Slow rates of travel also result in a reduction in penetration, and there is a tendency for the slag to flood the molten weld pool making it difficult to control the weld deposit.

4. *Angle of electrode* The more upright the electrode is held relative to the parent metal the greater the depth of 'PENETRATION'. This is because the full force of the arc is directed and concentrated on to the parent metal.

In the vertical position the force of the arc stream tends to drive the molten metal from the edge of the weld pool outwards, and in doing so produces undesirable 'UNDERCUTTING' of the weld profile.

*When an electrode is held in an upright position, 'SLAG' will build up and surround the weld metal deposit.* This will cause a problem as soon as the electrode is moved in the directions shown by the arrows.

These effects are shown in Figure 12.30(*a*), *and the diagram clearly indicates that some slag must become entrapped in the resultant weld.*

The more upright the electrode is held, the more difficult it is for the operator to be able to clearly observe the weld pool and control it.

When the electrode is inclined at an angle from the vertical as shown in Figure 12.30(*b*), the direction of the arc stream causes the slag to form away from the molten weld pool and ahead of the arc. *No slag is built up in advance of the direction of welding, thus eliminating the problem of slag entrapment.*

*The adoption of a suitable electrode angle enables the operator to easily observe the weld pool and control the slag.* In practice, the angle of inclination of the electrode to the parent metal varies between 60° and 90° in a vertical plane.

One factor which influences the choice of electrode angle is the class of electrode being used. *Some classes of electrodes have coatings which produce a very fluid slag, while others produce a viscous slag.* This factor is important because with a fluid slag it is much more difficult to control, and there is a greater danger of slag entrapment as the angle of the electrode approaches the upright position.

Between the flat and the upright positions there is one which provides the optimum welding conditions, with the following result:

(*a*) Adequate penetration.
(*b*) Correct weld profile.
(*c*) Correct width of weld bead.
(*d*) Minimum spatter.
(*e*) Minimum difficulty in controlling the slag.

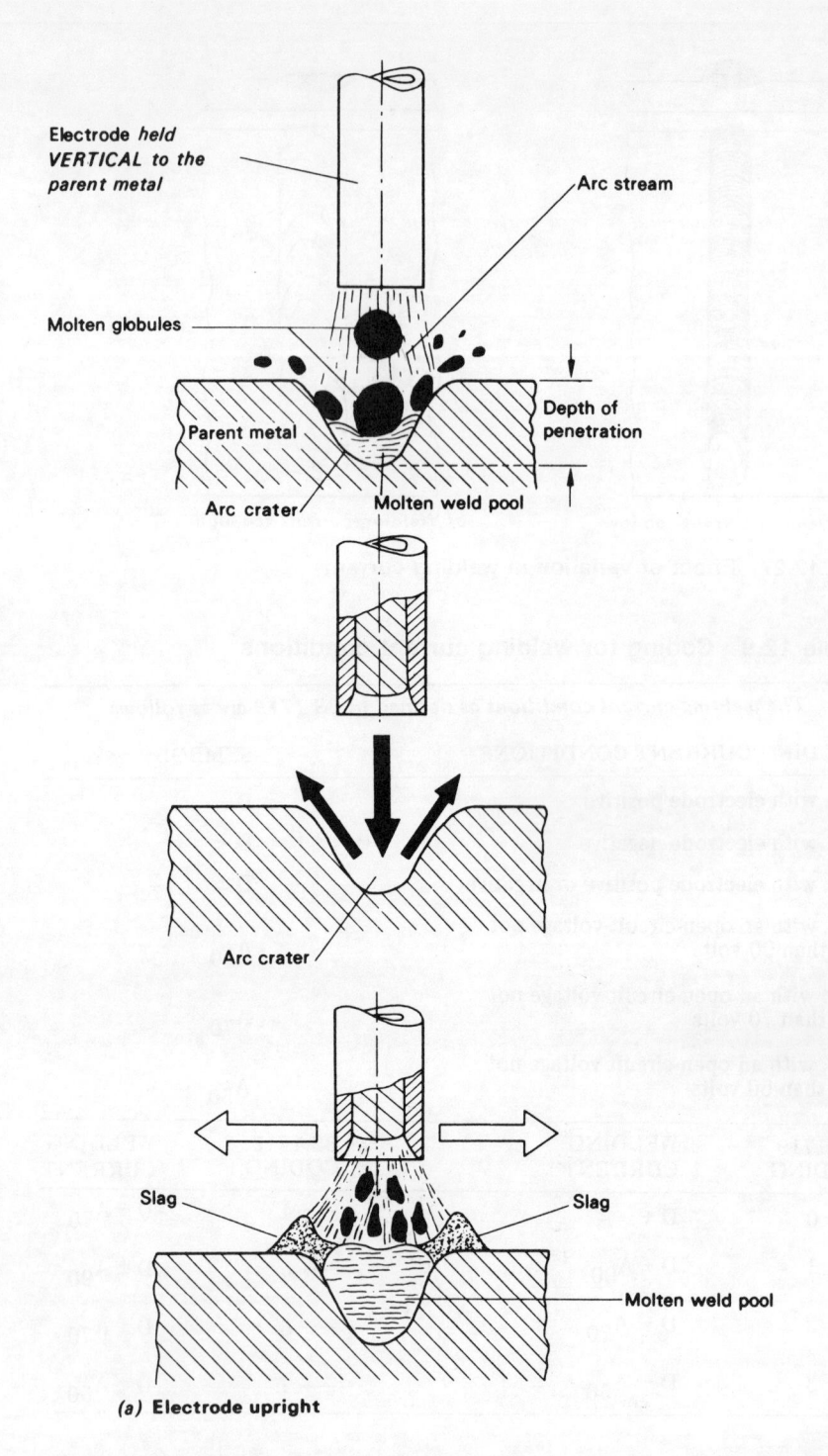

(*a*) Electrode upright

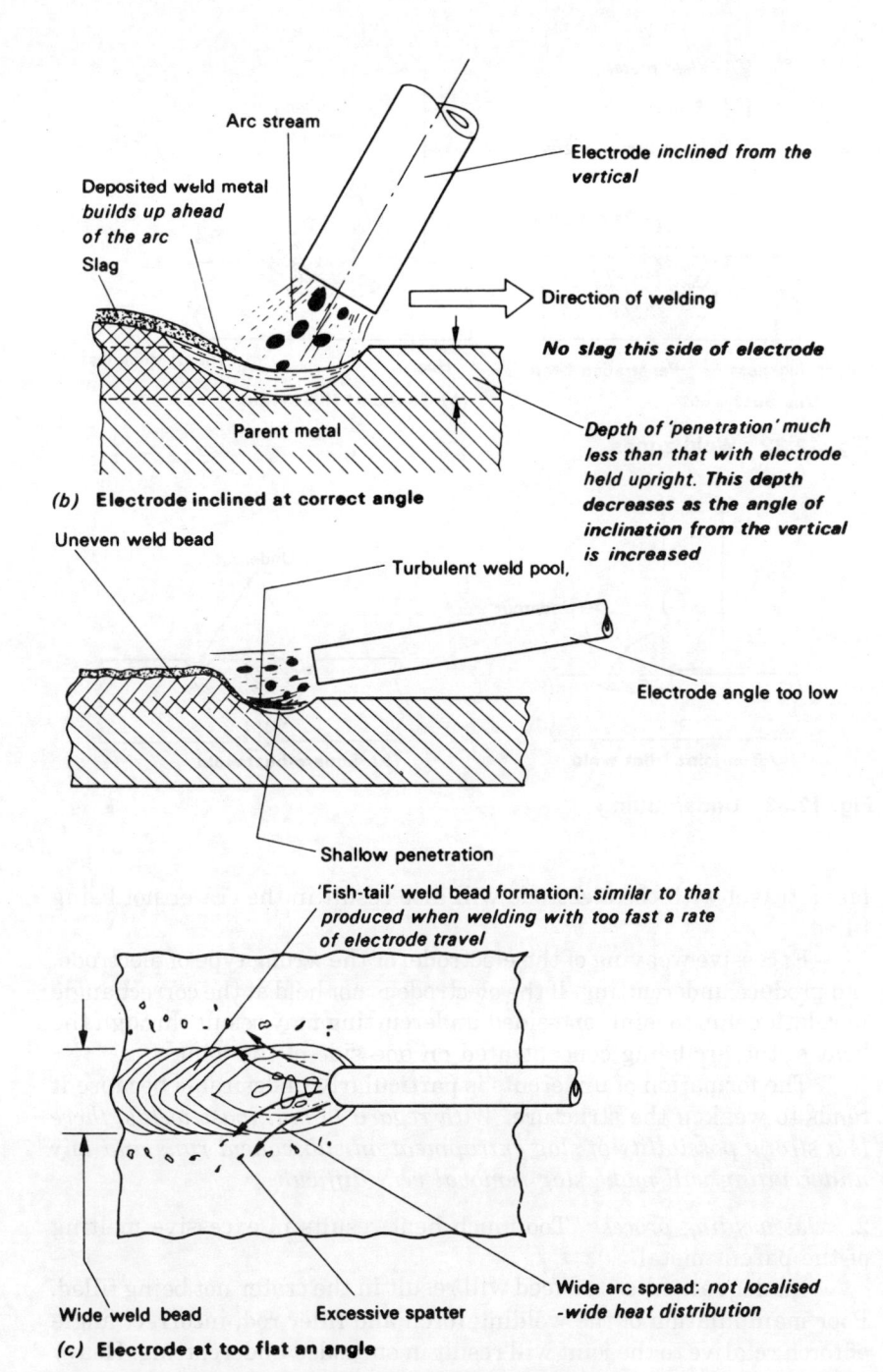

Fig. 12.30 Effect of variation in electrode angle

# Part C  Welding defects

## 12.29  British Standard 1295 (1959)

British Standard 1295 (1959) *Tests for use in the training of welders* is subtitled 'manual, metal-arc, and oxy-acetylene welding of mild steel'. Referring to this standard, some of the main factors to be considered when making an assessment of weld quality are:

(a) Shape of profile.
(b) Uniformity of surface.
(c) Degree of undercut.
(d) Smoothness of join where weld is recommended.
(e) Freedom from surface defects.
(f) Penetration bead.
(g) Degree of fusion.
(h) Degree of root penetration.
(i) Non-metallic inclusions and gas cavities.

Figure 12.31 illustrates desirable and undesirable weld profiles.

The weld metal and the parent metal affected by the welding process can be divided into THREE DISTINCT ZONES which can be clearly seen in a polished and etched section through the weld. These are as follows:

1. ACTUAL WELD METAL ZONE
2. WELD PENETRATION ZONE — consisting of parent metal which has been fused during the welding process.
3. HEAT-AFFECTED ZONE — where the parent metal, although not fused has been affected by welding heat (see Chapter 6). Figure 12.32 shows these zones diagrammatically.

## 12.30  Welding defects

### UNDERCUTTING

This is a term used to denote either the burning away of the side walls of the joint recess, or the reduction in parent metal thickness at the line where the weld bead is joined to the surface, as shown in Figure 12.33.

*Causes of undercutting*

1. *Manual metal-arc welding process*  Use of excessive welding currents will melt a relatively large amount of parent metal and cause it to sag under its own weight which results in the crater not being filled. Too

451

**(A) DESIRABLE BUTT-WELD PROFILE**

(a) Undercut
(b) Too abrupt reinforcement
(c) Incompletely filled groove
(d) Overlap
(e) Notch at edge of reinforcement due to incompletely filled groove
(f) Excessively high reinforcement

**(B) UNDESIRABLE BUTT-WELD PROFILES**

(a) Concave, equal leg lengths
(b) Convex, equal leg lengths

**(C) DESIRABLE FILLET-WELD PROFILES**

(a) Imperfect root penetration, Notch
(b) Undercut, Notch

**(D) UNDESIRABLE FILLET-WELD PROFILES**

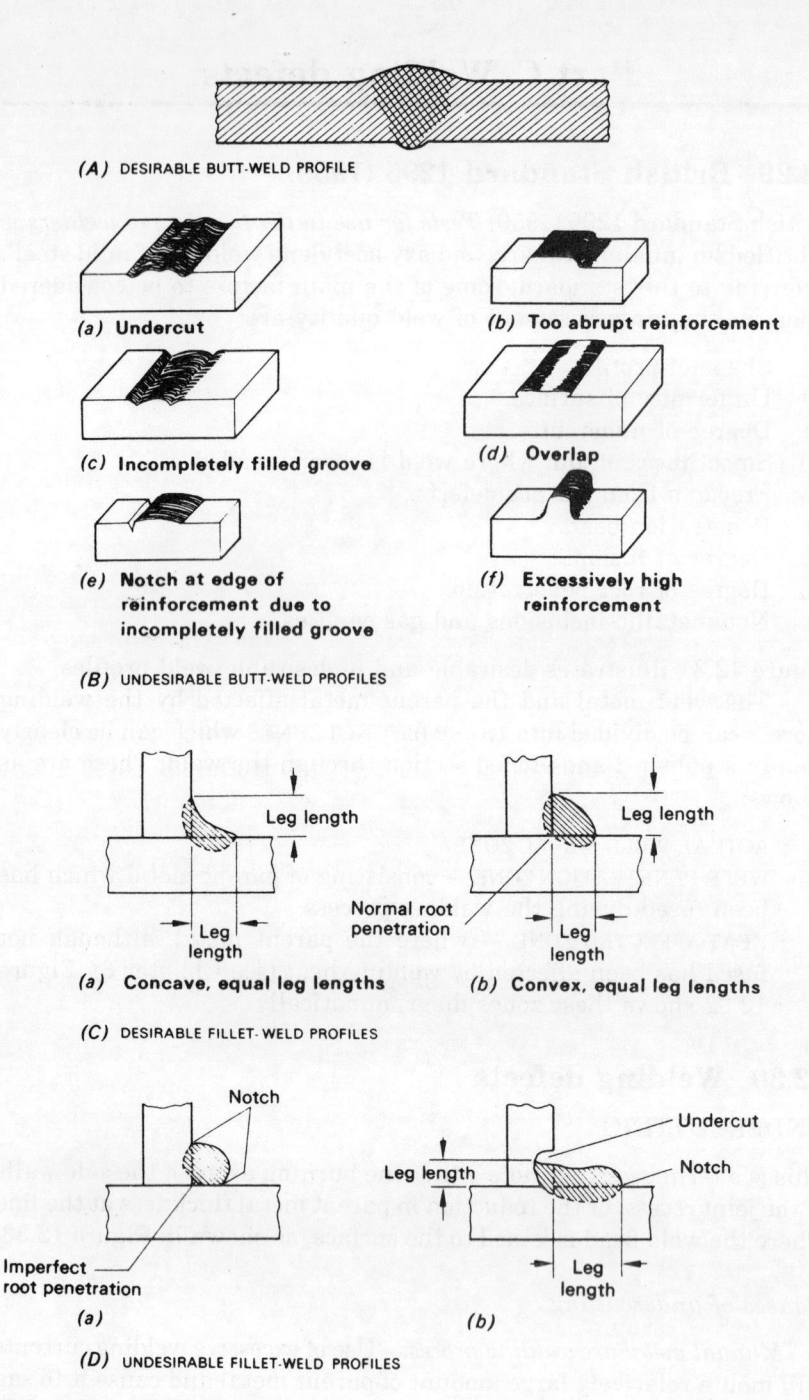

**Fig. 12.31 Weld profiles**

Weld metal
Penetration
Heat-affected zone

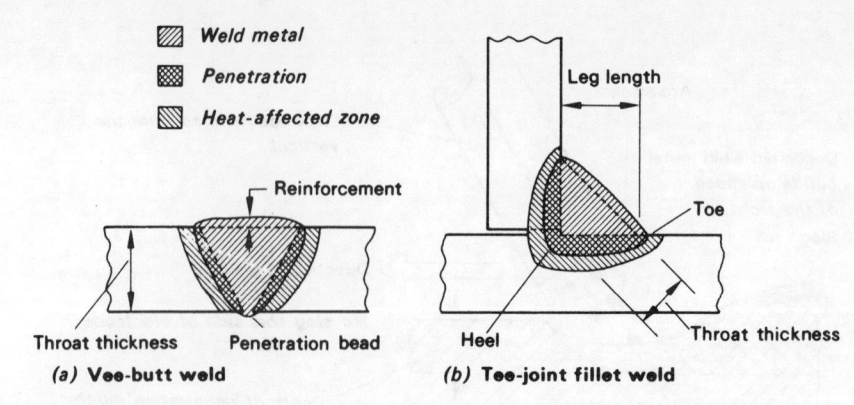

(a) Vee-butt weld — Reinforcement, Throat thickness, Penetration bead
(b) Tee-joint fillet weld — Leg length, Toe, Heel, Throat thickness

**Fig. 12.32 Weld zones**

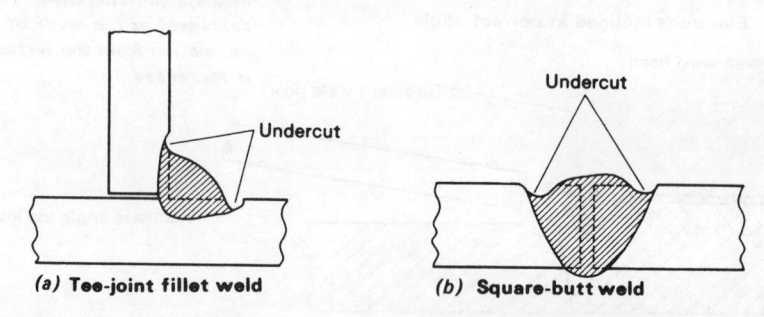

(a) Tee-joint fillet weld — Undercut
(b) Square-butt weld — Undercut

**Fig. 12.33 Undercutting**

fast a travel with the electrode will also result in the crater not being filled.

Excessive weaving of the electrode, or the wrong type of electrode, can produce undercutting. If the electrode is not held at the correct angle in relationship to joint, one-sided undercutting may result through the heat of the arc being concentrated on one side of the joint.

The formation of undercuts is particularly undesirable because it tends to weaken the structure. *With regard to multipass welds, there is a strong possibility of slag entrapment on subsequent runs and any undercutting will make slag removal very difficult.*

2. *Gas welding process* Too much heat results in excessive melting of the parent metal.

Inadequate filler-rod feed will result in the crater not being filled. Poor manipulation of the welding torch and filler rod, incorrect angle of torch relative to the joint will result in one-sided undercutting. Heavy mill-scale on the parent metal is also a possible cause of undercutting.

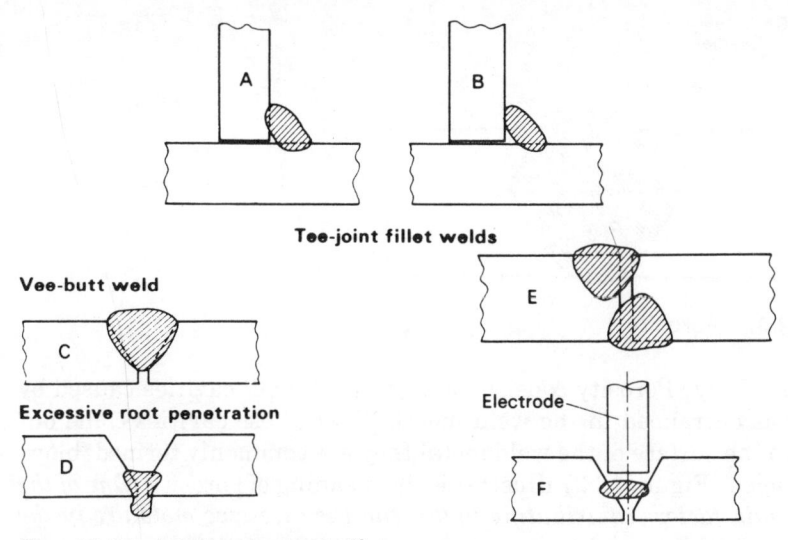

**Fig. 12.34  Incorrect penetration**

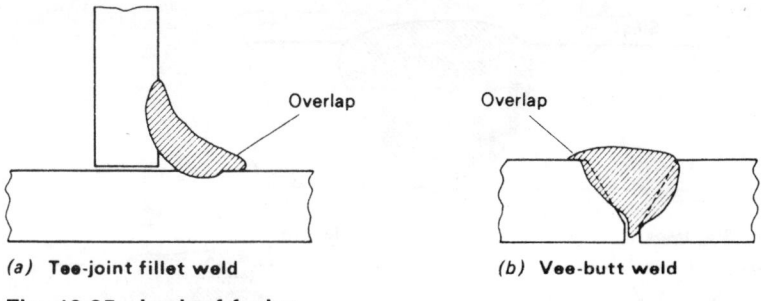

**Fig. 12.35  Lack of fusion**

## SMOOTHNESS OF JOIN WHERE WELD IS RECOMMENCED

Whenever a welding run has to be interrupted and then restarted it is very important that where the weld is recommenced the join should be as smooth as possible. The join should show no pronounced hump or crater in the weld surface. The joins at the ends of the weld runs are liable to have poor strength. This is caused by crater cracks producing stress concentrations. The welded joint may also be weakened by overlap and lack of fusion.

## SURFACE DEFECTS

Surface defects in welds are generally due to the use of unsuitable materials and/or incorrect techniques. The surface should be free from porosity, cavities, and either burnt-on scale (in the case of gas welding) or trapped slag (in the case of metal-arc welding). Surface cavities may be caused by lack of fusion, gas bubbles, or trapped slag.

## PENETRATION

One of the more common causes of faulty welding is the *lack of penetration*, often termed 'incomplete penetration' or 'lack of fusion'. When viewed from the underside, a sound weld should have a slight penetration bead. The size of this penetration bead will be influenced by the welding process used, the type of joint preparation, and the skill of the welder. Figure 12.34 shows lack of penetration.

With manual *metal-arc welding* lack of penetration can be attributed to the welding current being set too low for the particular type of electrode. With *gas welding* the use of too large a diameter filler rod or incorrect angles of the welding torch. Use of the wrong polarity when *arc welding* with DIRECT CURRENT will result in lack of penetration.

Excessive penetration in the root of the joint will result with too high a welding current or concentration of heat in the case of gas welding, and with the use of an unsuitable edge preparation.

Unsatisfactory penetration may result if the welds are not located directly opposite each other, as shown in the double-side square-butt weld 'E'.

Lack of penetration can also be caused by using an electrode for the first run which is not small enough in diameter to reach down to the ROOT of the weld, as shown at 'F'.

*Incorrect joint design, including the type of preparation, can be the cause of incomplete penetration.* The joint must allow entry of the electrode and permit unrestricted manipulation. Too small a diameter of electrode in a root run will result in inadequate fusion or penetration.

Other causes of inadequate penetration with the manual metal-arc welding process which can usually be attributed to the welder and must be avoided are:

(a) *Bad incorporation of tack welds* (This also applies to gas welding).
(b) *Inadequate removal of the slag.*
(c) *The use of too low an arc* — the arc must always be kept as short as possible.

## LACK OF FUSION

Lack of fusion can be defined as the failure to fuse together adjacent layers of the weld metal or adjacent weld metal and parent metal as shown in Figure 12.35. This condition may be caused by failure to raise

453

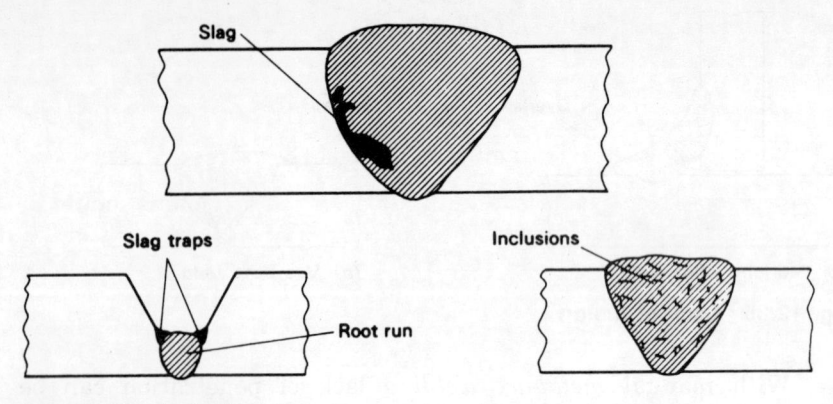

**Fig. 12.36 Inclusions**

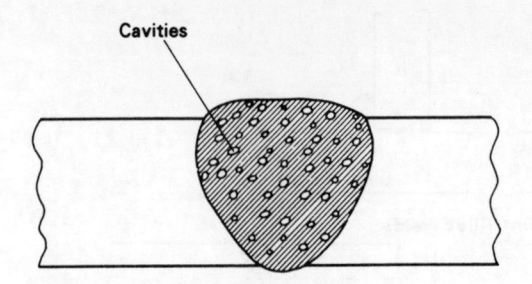

**Fig. 12.37 Porosity**

the temperature of the parent metal to its melting point. Another cause is the failure to remove oxides or other foreign matter.

## NON-METALLIC INCLUSIONS AND GAS CAVITIES

1. *Slag inclusions* The term 'slag inclusions', or merely 'inclusion', refers to slag or other non-metallic foreign matter entrapped in the weld metal. The usual source of inclusions is the slag formed by the electrode covering, although other substances may be present due to welding on dirty surfaces. Figure 12.36 illustrates the problem of slag inclusion.

   Inclusions may also be caused by contamination from the atmosphere by oxidising conditions, faulty manipulation of the welding torch or filler rod, or fluxes used in welding operations. With gas welding they are usually caused by dirty parent metal surfaces or millscale.

   Slag inclusions in manual metal-arc welding, are generally caused by using either too high or too low a welding current. Other factors include incorporation of bad tack welds, too long an arc or too high a speed of travel, lack of penetration, or the use of too large a diameter electrode which is also a relevant factor.

   Carelessness in slag removal prior to putting down another run of weld metal when producing multi-run weld may be the cause, or the inaccessibility of the weld not allowing the removal of slag resulting in 'slag traps'.

   IN GENERAL, THE WELDING CHARACTERISTICS OF AN ELECTRODE ARE SUCH THAT THE MOLTEN SLAG FLOATS FREELY TO THE SURFACE OF THE MOLTEN WELD METAL AND IS EASILY REMOVED WITH THE AID OF A CHIPPING HAMMER.

2. *Porosity* Porosity consists of a group of small cavities caused by gas entrapped in the weld metal. When these cavities come out to the surface of the weld metal they are commonly termed 'blowholes'. Figure 12.37 illustrates the meaning of porosity. *One of the main factors contributing to this fault is excessive moisture in the electrode or in the joint.* ALWAYS USE DRY ELECTRODES AND DRY MATERIALS.

   Porosity may be scattered uniformly throughout the entire weld, isolated in small groups, or concentrated at the root. Other causes of porosity are:

| *Cause:* | *Remedy:* |
| --- | --- |
| 1. High rate of weld freezing. | Increase the heat input. |
| 2. Oil, paint, or rust on the surface of the parent metal. | Clean the joint surfaces. |
| 3. Improper arc-length, current, or manipulation. | Use proper arc-length, (within recommended voltage range), control welding technique. |
| 4. Heavy galvanised coatings. | Remove sufficient zinc on both sides of the joint. |

## 12.31 Workshop testing of welds

The correct procedure for preparing test welds is fully explained in BS1295, and for the purpose of bend tests two specimens are cut from the test weld. *One specimen is tested with the face of the weld in tension*

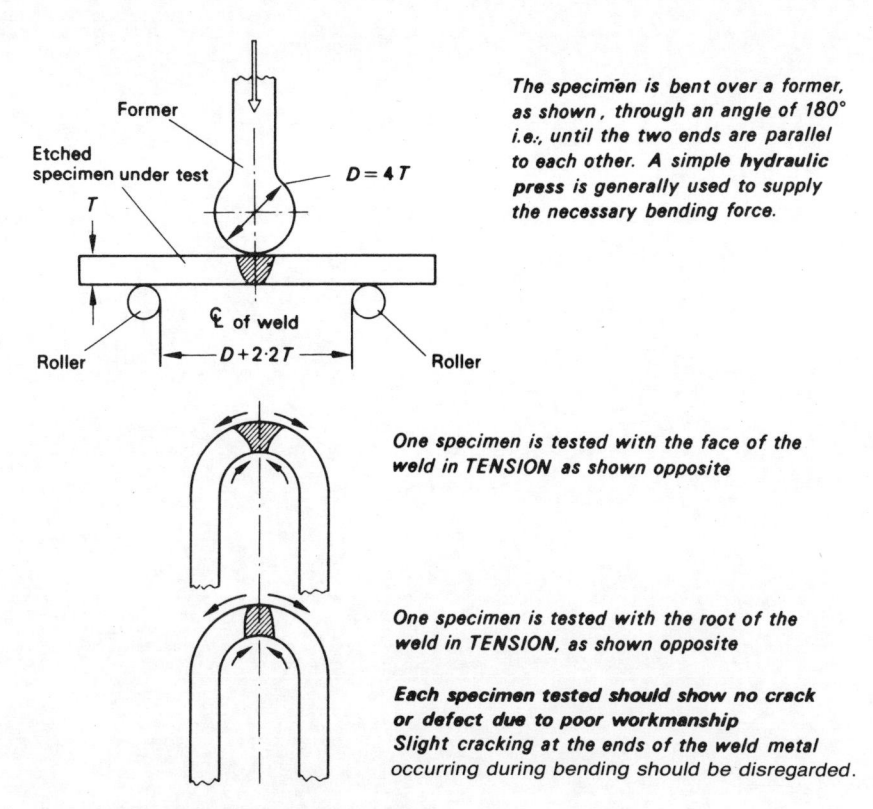

**Fig. 12.38 Bend-testing of welds**

*The specimen is bent over a former, as shown, through an angle of 180° i.e., until the two ends are parallel to each other. A simple hydraulic press is generally used to supply the necessary bending force.*

One specimen is tested with the face of the weld in TENSION as shown opposite

One specimen is tested with the root of the weld in TENSION, as shown opposite

**Each specimen tested should show no crack or defect due to poor workmanship** Slight cracking at the ends of the weld metal occurring during bending should be disregarded.

*and the other with the root in tension.* The method of bend-testing is simply explained by Figure 12.38.

The upper and lower surfaces of the weld may be filed, ground, or machined level with the surface. The direction of machining should be along the specimen and across the weld. The sharp corners of the test piece should be rounded to a radius not exceeding one-tenth of the thickness. The centre line of the weld must be accurately aligned with the centre line of the former, as shown. The specimen is bent over a former through an angle of 180° — i.e., until the two ends are parallel to each other. A simple HYDRAULIC PRESS is generally used to supply the necessary bending force.

## 12.32 Inspection of etched sections and bend specimens

Etched sections and weld fractures will give information of the following defects if present:

1. Incorrect profile.
2. Undercutting.
3. Slag inclusions.
4. Porosity and cracks.
5. Poor root penetration and lack of fusion.

It is recommended that the examination of etched specimens should be made with the aid of, at least, a hand magnifying glass.

## 12.33 Method of preparing etched specimens

1. *Preparation of surface for etching* The surface should first be filed using a coarse file until all the deep marks are removed. A smooth file is then used to file the surface at right angles to the initial coarse file marks. This smooth filed surface should then be polished with successively finer grades of emery paper, the direction of polishing being at right angles to the marks made by the previous grade in each case. *It is important to polish with each grade of emery paper until the scratches made by the previous paper have been removed before proceeding to the next finer grade.* This will ensure a first-class finish to the surface to be etched.

2. ETCHING FOR MACRO-EXAMINATION A suitable etching solution is as follows:

   10–15 ml NITRIC ACID (specific gravity 1·42)
   90 ml ALCOHOL (industrial spirit)

   The specimen may be immersed in the etching solution until a well-defined macro-structure is obtained, or the surface of the specimen may be swabbed with cotton wool saturated with the solution. In either case, the specimen should be thoroughly washed in hot water, followed by rinsing with acetone or industrial spirit and drying in a current of air. It is possible to preserve etched specimens by a coating of clear lacquer.

   Examples of macro-structures of weld specimens are shown in Appendix 1.

# Appendix 1

## Weld defects

(b) The defect magnified

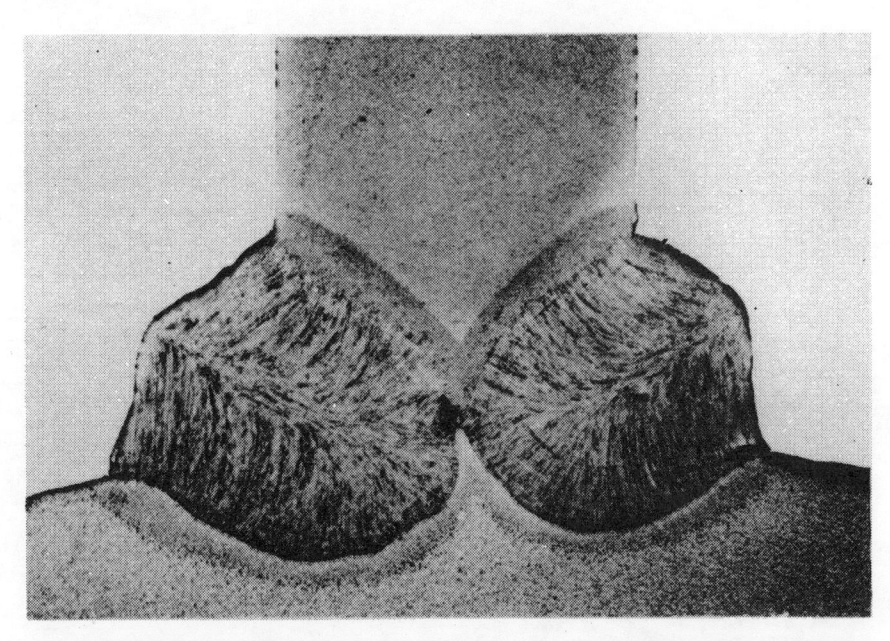

(a) Fillet welds in low alloy steel, showing slag inclusion at the root of two runs

A manual metal arc single 'V' butt weld, showing lack of penetration and misalignment of the plate edges

A manual metal arc single 'V' butt weld, showing undercut and misalignment of the plate edges

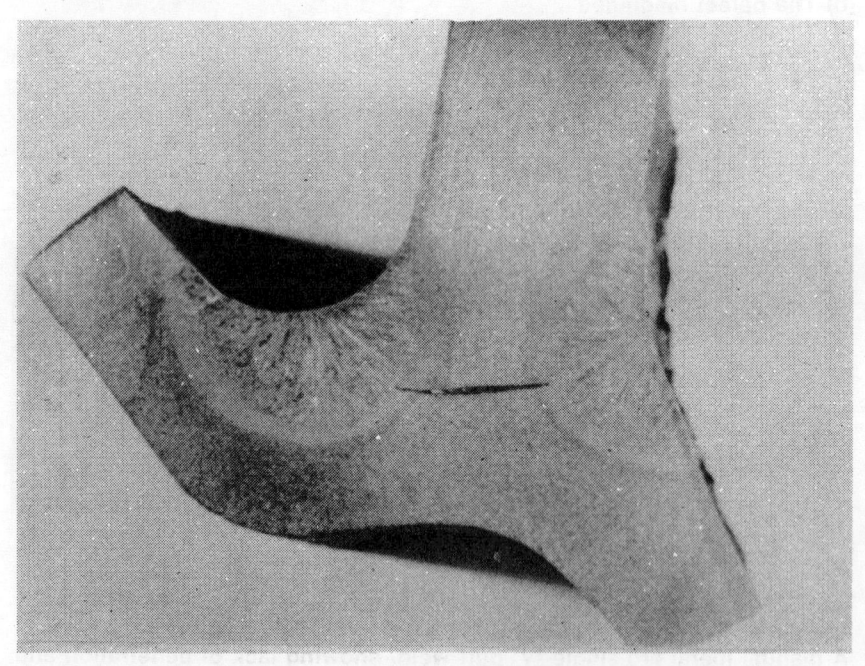

Submerged arc welds, showing lack of penetration

# Appendix 2 — Applications

A batch of completed flame-brazed units

An unusual application of Aluminium Flame Brazing (see Sections 11.15, 11.16, and 11.17)

Brazing a ventilator sleeve inserted into a large corrugated sheet — both sleeve and sheet being of 0·71 mm (22 s.w.g.) aluminium

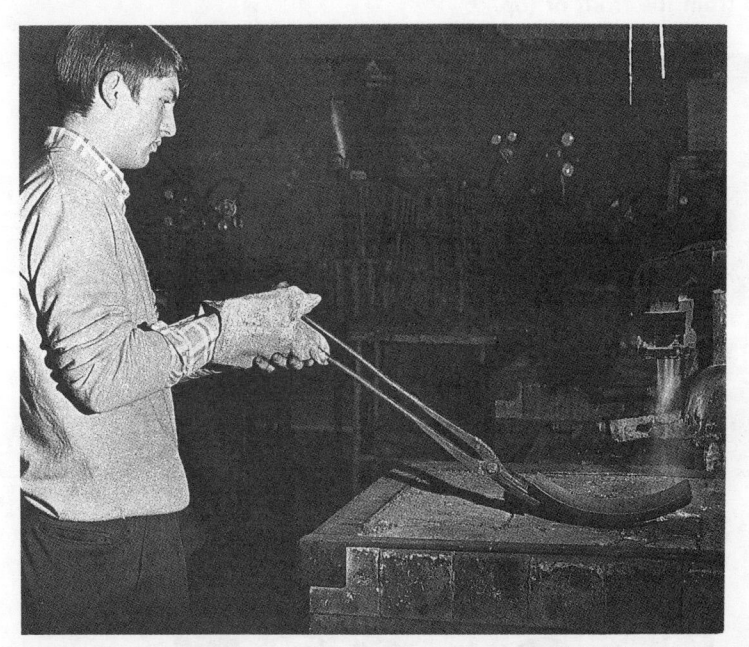

Intermittent annealing on a typical workshop gas-fired hearth (Section 6.35) to relieve a work-hardened material during the fabrication of a 90° pipe bend.

A gas-fired furnace specially designed for bending pipes (Section 9.24) up to 762 mm outside diameter over a length of up to 5·34 m. The furnace can be loaded from the front or top.

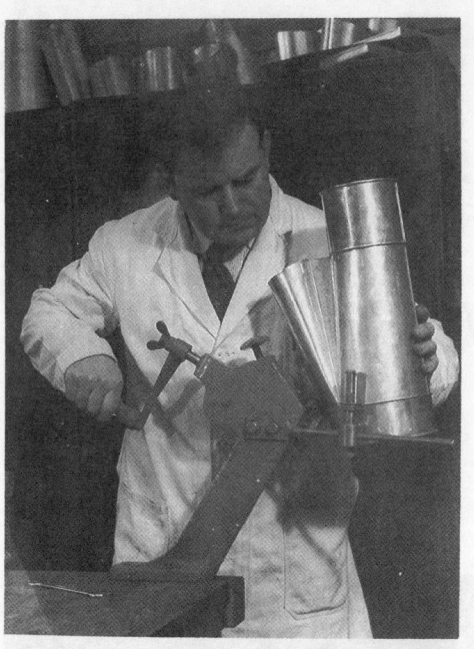

Use of a hand-operated machine to complete a paned-down joint (Section 3.12)

A power-driven mandril-type pipe bending machine (Section 9.25)

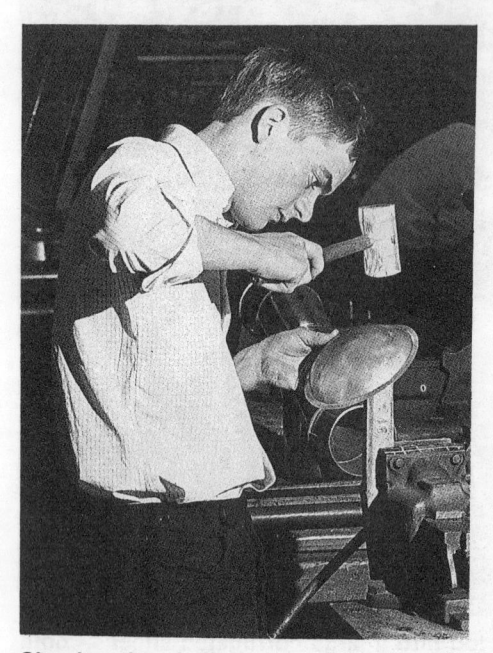

Shaping the shallow domed top for a tinplate kettle (Figure 10.34)

Use of a wheeling machine to planish the hemispherical surface of an aluminium dome of welded construction (Sections 10.20 and 10.21)

Squaring the flanges during the fabrication of a transition-piece for joining two rectangular section ducts in a plane at right angles to each other. All four apparently twisted surfaces are produced by straightforward rolling (Section 10.10) to the desired contour and then joined together by open-corner fillet welding (Figure 12.21(c)).

Wiring (Sections 9.8 and 10.32) a 90° pipe bend fabricated from aluminium by hollowing, raising (Section 10.18) and welding

461

# Index

Abrasives, *see* material removal
Adhesives, *see* bonded joints
Age hardening, *see* heat treatment
Allotropy (iron), *see* materials
Alloy steels, *see* materials
Aluminium, *see* materials
  alloys, *see* materials
  bronze alloys, *see* materials
Angle section work, *see* bending, fabrication
Annealing, *see* heat treatment
Arc welding, *see* metal arc welding
Austenite, 210–13

Beams
  bar frames, 389, 392
  castellated, 390, 392
  deflection of, 131, 132
  holes and notches, 132
Bending
  air, 341, 342
  allowances for sheet metal, 370–77
  angle section, rolls for, 353, 354
  basic methods, 342, 343
  forging, 333
  machines, 342, 343
  mechanics of, 339, 340
  pipe work, *see* pipe bending
  pressure tools, 344, 346
  press brake
    operations, 348–50
    principles of, 346
    types of, 346–8
  pyramid types roll bending machines, 351, 352
  roll bending machines, 350, 353
    operations, 350 *et seq.*
    plate, 351, 352
    spring back, 340–43
Beryllium, *see* materials
Bevelled plates, *see* dihedral angle
Blanking, principles of, *see* press work
Bolted joints, *see* mechanical fastening
Bended joints
  adhesives
    impact, 424
    thermo-plastic, 423
    thermo-setting, 424, 425
    use of, 420
  adhesive bond, 421 422
  failures, 422
  preparation, 422
  strength of, 422, 423
Brass alloys, *see* materials
Brazing
  aluminium, 407–9
  filler materials, 406–8
  hard soldering, 404, 405
  heat sources, 409
  dip brazing, 411
  electric induction, 411, 412
  electric resistance, 412, 413
  flame, 409, 410
  furnace, 410, 411
  principles, 405, 406
  spelter, *see* filler materials
  types of joint, 413, 414
Braze welding
  filler rods, 414
  principles, 413–15
  types of joint, 415–17

Calipers, *see* measurement
Cast iron, *see* materials
Cementite, 208–10
Chemical changes in metal, 97, 98
  *see also* 103–5
Chemical reactions during welding, 99, 100
Chisels
  blacksmiths', 244, 245
  cold, 243, 244
Chromium, *see* materials
Cladding (Alclad), 190, 191
Cobalt, *see* materials
Colour coding, compressed gas cylinders, 46, 47
Compressed air, *see* pneumatics
Compressive strength, *see* materials
Connections, *see* fabrication
Constructions, *see* geometry
Copper, *see* materials
Copper-nickel (Cupro-nickel) alloys, *see* materials
Corrosion, *see* materials
Crystal structure, 177–81
Cutting tools
  chip formation, 232
    types of, 232
  chisel, 243–5
  clearance angle, 231
  file, 238–41
  hacksaw, 241–43
  tool angle, 231–233
  twist drill, 232 *et seq.*
    speeds and feeds, 234, 235

Datum edge, line, surface, *see* marking out
Dead mild steel, *see* materials
Dendrites, 178–80
Developments, *see* geometry
Dimensioning, 39–44
  selection of, 44–6
Dividers, *see* marking out
Dihedral angle, *see* marking out
Drawing
  division of lines, 58, 59
  orthographic
    first and third angle, 33–6
    pictorial, 36–9
    true lengths, 65–7
    use of square and isometric paper, 39–42
Drilling process, *see* restraint and location
Ductility, *see* materials
Duralumin, *see* materials

Elasticity, *see* materials

Electric arc, 97, 98
Exothermic reaction, 264
  (*see also* Thermit process)

Fabrication
  angle section work, 322 *et seq.*
  applications of, *see* appendix 2
  beam to beam connections, 385, 387
  cone rolling, 353
  lattice frames, 387, 389
  slip rolls, 351–3
  stanchion
    bases, 384, 385
    splices, 385, 386
    to beam connections, 385, 386
  structural connections and assemblies, 383 *et seq.*
  trusses, 387–90
  web stiffeners, 387, 390, 391
Ferrite, 208, 210–12
Ferrous metals, *see* materials
Files and use of, *see* cutting tools
First angle projection, *see* drawing
Flame cutting, *see* material removal
Flanges
  locating hole centres, 327–29
  forming, *see* sheet metal work
Fluxes, *see* soldering
  use of, gas welding, 102, 103
Fly press, *see* press work
Folding machine operations, 343–46
Force
  gravity, 116
    centre of, 117 *et seq.*
    controid, 117 *et seq.*
  mass, 116, 117
  moment of, 125–7
  moments, principles of, 126–9
  movement 122, 124
  turning effect, 125, 127
  vector quantity, 122, 124, 125
  weight, 116 *et seq.*
Forging
  aluminium alloys, 221, 222
  anvil, 327, 329, 330
  bending, 330, 333
  copper alloys, 221, 222
  drawing down, 330, 331
  forming tools, 331
  hammers, 331
  hearth, 329
  hot and cold, 218
  material properties, effect on, 218–21
  punching and drifting, 331, 332
  Smith's cone, use of, 329, 330
  swaging, 332
  temperatures, 221

tongs, 331
up-setting, 330, 331
welding (pressure), 333
work holding, 330, 331
Forming, see sheet metal work
Fusibility, see materials

Gravity, force of, see force
Geometry
angles, setting out,
use of compasses, 50, 51
use of set squares, 50
constructions
miscellaneous, 58 et seq.
ovals and ellipses, 54–8
plane figures, 51–4
development
of surfaces, 62
parallel line, 62–4
radial line, 64, 65
triangulation, 65–7
right and oblique cones, comparison of 67, 68
Grinding and grinding machines, see material removal

Hacksaw, see cutting tools
Hank bushes, see mechanical fastening
Hardness, see materials
Hard soldering, see brazing
Heat affected zone, see welding (mild steel)
Heat
energy sources, 97
Heat treatment
age-hardening, 217
alloys, effect of, 214, 215
annealing, 209, 211, see also 216
workshop practice, 217, 218
carbon content, effect of, 208–10
critical temperatures, 209 et seq.
hardening, 212, 213
iron-carbon equilibrium diagram, 209, 210
non-ferrous metals and alloys, 215–17
normalising, 212
precipitation treatment, 217
purpose of, 207
quenching, 212–14
recrystallisation, 181, see also 211, 215
solution treatment, 215, 216
structure of steel, effect on, 209 et seq.
tempering, 213, 214
Heat straightening, 113–15
High carbon steel, see materials
High-speed steel, see materials
Hole saw, see material removal

Hollowing, see sheet metal work
Hydraulics, see pressure
Hydraulic press, see pressure

Inconel, see materials
Iron-carbon equilibrium diagram, see heat treatment
Iron, see materials

Lathe, spinning, see spinning sheet metal
Lead, see materials
Linear expansion 105, 106
in welding processes, 106
Locking devices, 69, 71
Location and restraint, see work holding
Lubrication, spinning, see spinning sheet metal

Magnesium and alloys, see materials
Manganese, see materials
Marking out
angle sections, holes in, 165, 166
beam compasses, see trammels
bevel, use of the, 155, 156
centre punch, use of the, 156, 157
channel sections, holes in, 165–7
columns and beams, holes in, 167, 168
cumulative error, 146, 148
datum
edges, 146, 149
lines, 146
surfaces, 147, 150
dihedral angle, 176
dividers, 145, 147
flanges, holes in, 154, 155
hermaphrodite calipers, 146, 147
large plates, 151–54
pipe bends, setting out for, 171 et seq.
plumb line, use of, 156, 157
precision sheet metal work, 169–71
bend allowances, 169–71
ordinates, use of, 169, 170
rule, use of, 139, 140
scratch gauge, 155, 156
scribing, 141, 145, 146
scribing block (universal surface gauge) 146, 148
setting scribing instruments, 146, 148
steel tape, 139 see also 153
surface preparation for, 145
'tee' sections, holes in, 167
templates
box type, 164
large, 149–54
making, 158–61
plain and bushed, 165
templates, use of, 161–64

templates for inspection and checking, 161, 162
trammels, 146, 147, see also 150–54
vernier height gauge, 144, 146
wing compasses, 146, 147
Martensite, 212, 213
Materials
allotropy (iron), 209
alloy steels, 186, 187
aluminium, 188, 189
alloys, 189, 194, 195
aluminium bronze alloys, 192, 194, 195
beryllium, 189
brass alloys, 192, 193
cast iron (grey), 184–6
chromium 187 et seq.
cobalt, 188, 215
cold reduced sheets, 299
copper, 188, 190
copper and nickel alloys (cupro-nickels) 192, 193
corrosion 198 et seq.
metals that resist, 200–2
prevention, 202, 203
resistance, 200–2
season cracking, 200
surface preparation, 203, 204
duralumin, 190, 191
ferrous, 182 et seq.
hot and cold working, 196–8
hot rolled plates and sheets, 300
inconel, 194
lead, 188, see also 195, 197
low carbon steels, 182, 185
magnesium, 188 et seq.
alloys, 190–2
manganese, 191, 194, see also 214
market forms of supply, 299
medium carbon steels, 185, 186
metallic crystals, 177
mild steels, 185
molybdenum, 214, 215
monel, 196
nickel, 188, see also 201, 202, 214
nickel-chromium alloys, 193–5
'nimonic series', 194, 195
niobium, 189
non-ferrous alloys, 189 et seq.
non-ferrous metals, 187–9
physical changes, 197
plain carbon steels, 185, 186
properties of,
compressive strength, 182
conductivity, 183
ductility, 183
elasticity, 183
fusibility, 183

hardness, 183
malleability, 183
plasticity, 183
tensile strength (tenacity), 182, 184
toughness, 182
stainless steels, 201 see also 228, 229
slip planes, effect of, 180, 181
tantalum, 189
tellurium, 189
tin-bronze alloys, 192, 194
tin-lead alloys (soft solders), 195, 196
titanium, 188, 189
tungsten, 187 see also 215
vanadium, 187, 188
wrought iron, 185
zinc, 202
zirconium, 189
Material removal (see also cutting tools)
abrasives, 256 et seq.
abrasive wheel cutting off machine, 259, 260
flame cutting,
applications of, 267–74
distortion caused by, 113, 114
effect on metal properties, 229, 230
equipment, 265, 266
factors influencing quality of cut, 266, 267
machines, 270 et seq.
principles, 264 et seq.
grinding machines, portable, 260–3
routing, 235–8
thermal cutting processes, 264 et seq.
trepanning, 235–7
hole saw, 235, 237
tank cutter, 235, 237
weld grinding, 259
Measurement (see also marking out)
direct eye, 139–41
micrometer caliper, 142
micrometer scale (metric), 142, 148
tension wire, 156–8
vernier caliper, 143, 144
vernier scale (metric), 144, 145
Mechanical fastenings
alignment of holes, 90, 91
defects in riveted and bolted holes, 84–6
hank bushes, 71
locking devices, 69, 71
'pop' rivets, 87, 88
riveted joints, 79 et seq.
screw thread
elements, 69, 71
forms, 69–72
systems, 72, 73
screwed, types of, 69, see also 88–90
self-secured, 74 et seq.

463

self-tapping screws, 74
taps and dies, 73
types of, 69
washers, 90
Metal arc welding
   alternating current, use of, 443, 444
   angle of electrode, effect of, 450, 451
   arc length, effect of, 447–9
   arc stream, 441, 442
   current stability, 444
   direct current, use of, 442
   electrode coating, function of, 442
   electrodes, 441, see also 444–6
   excess current, effect of, 447–9
   open circuit voltage, 443, 444
   positions, definitions of, 447, 448
   shielding gases, 101
   speed of travel, effect of, 448–50
   variable factors, effect of, 446–51
   welding current values, 443, 444
   welding voltage, 443–5
Metal degreasing, 205–7
Metal inert gas (MIG) welding, see welding
Metallic crystals, see materials
Micrometer caliper, see measurement
Moments of a force, see force
Movement, freedom of, see restraint and location

Nibbling machines, 251–4
Nickel, see materials
Niobium, see materials
Normalising, see heat treatment

Oblique drawing, see pictorial drawing
Orthographic drawing, see drawing
Oxy-fuel gas welding
   acetylene cylinders, 426
   discharge rate, cylinders, 427
   economiser, gas, 432, 433
   equipment, welding, 426, 427
   flame conditions, 434
   flame structure of, 433, 434
   hoses, 431, 432
   leftward welding, 436, 437
   lighter, gas, 432
   manifold, 428
   mixing chamber, 430
   nozzle size, 434–6
   nozzle, welding torch, 430, 431
   oxygen cylinders, 426, 427
   pressure gauge, 429
   pressure regulator (automatic), 428, 429
   rightward welding, 437, 438
      advantages of, 438–40
   torch welding, 429, 430

velocity, gas, 435–7
welding techniques, camparison of, 440
Oxidation of welds, 100
   fluxes, use of, 102, 103

Panel beating, see sheet metal work
Patterns, use of, see sheet metal work
Planning, need for, see sheet metal work
Pearlite, 210–12
Pictorial drawing
   isometric
     construction, 37–9
     paper, use of, 39
     scale, 39
   oblique construction, 36
   squared paper, use of, 39
Piercing, see press work
Pipe bending
   by hand, 335, 336
   by machine, 336–8
   loading for, 334
   setting out for, see marking out
   use of flexible springs, 334, 335
Pipe fitting
   colour coding, 95, 96
   compression joints, 94
   flanged joints, 93, 94
   screwed joints, 93
   screwing equipment, 94, 95
   types of pipe, 92
Plain carbon steels, see materials
Plastic materials, see bonding joints
Plumb line, see marking out
'Pop' riveting, see mechanical fastenings
Power press, see press work
Precipitation treatment, see heat treatment
Pressure
   atomspheric, 133, 134
   fluid (hydraulic), 135, 136
   hydraulic
     bending press, 136–8, see also 337
     jack, principle of, 137
   pneumatic, 135
   portable tools, 135
Press work
   blanking, principles of, 254, 255
   corner notching tool, 316, 318
   corner radius tool, 316, 318
   cropping tool, 318
   deep drawing, 377, 378
   fly press, 254, see also 314
   master template, use of, 315
   piercing, principles of, 254–6
   power press 254
   sheet metal operations, 313 et seq.

'vee' bend tool, 316, 317, see also 340 et seq.
'vee' die, 340–2, 346
Proof stress, 184
Properties of materials, see materials
Protective clothing, see safety

Raising, see sheet metal work
Rake angle, 232
Recrystallisation, see materials
Relative density, 121, 122
Restraint and location, see also work holding
   drilled components, 282
     machine table, 283–5
     machine vice, 282–4
     vee blocks, 278, 280, 281
   drilling process, 281, 282
     sheet metal, 283, 285
   bench vice, use of, 278, 280
   fitter's vice, 278–80
   movement, freedom of, 278
   pipe vice, 279–81
Riveting, see sheet metal work
Roll bending machines, see bending
Routing, see material removal

Safety
   adhesives, use of 425
Safety (Cranes), precautions when using, 22–4
Safety (Hazards)
   health
     lead poisoning, 31, 32
     metal degreasing, 206, 207
     zinc fumes, 30, 31
   mains operated arc welding equipment, 8, 9
Safety (Forging), 20
Safety (Lifting tackle), 27–30
   safe working loads (SWL)
     chain slings, 26, 27
     wire ropes, 24
     wire rope slings, 24–6
Safety (Metal arc welding)
   body protection, 13, 14
   cable connectors, 9, 10
   circuit diagram, 8, 9
   electrode holder, 12
   external welding circuit, 9, 10
   eye and head protection, 13
   fire hazards, 15
   mains operated equipment, 8
   mobile welding plant, 12, 13
   return current clamps, 10
   screens, 14, 15
   ventilation (fume extraction) 15

welding cables, 9–11
Safety (Oxy-fuel gas welding and cutting)
   explosion risks, 17, 18
   fire hazards, 17
   gas cylinders, use of, 19
   goggles, 16
   personal protection, 16 et seq.
   testing for leaks, 18, 19
   welding in confined spaces, 19, 20
Safety (pipe work)
   colour coding, 95, 96
   pipe fitting, 91, 92
   pipe loading, 334
Safety (pressure vessels)
   testing, 20, 21
   welded repairs, 21
Safety (Site)
   ladders, care and use of, 6–8
   protective clothing, 1–3
   protective footwear, 4
   protective gloves, 3
   protective helmets, 3, 4
   safety blocks
     self-contained, 5
     static, 5
   working aloft, 5, 6
Salt bath, 215, 216
Scratch gauge, see marking out
Screwed fastenings, see mechanical fastenings
Screwed joints, see pipe fitting
Screw threads, see mechanical fastenings
Scribing, see marking out
Scribing block, see marking out
Season cracking, see corrosion
Self-secured joints, 74 et seq.
Setting out
   pipe bends, see marking out
   roof trusses, 168, 169
Shears, bench, 246, 247
   circle cutting, 322, 323
   guillotine
     operations on the, 319 et seq.
     setting the, 319, 320
   hand snips, 245 et seq.
   portable (hand), 251–4
   rotary, 251, 252
   summary of requirements, 250
   universal steel machines, 325
Shearing, principles of, 247–50
Sheet metal work
   applied stiffeners, 382, 383
   bench tools, use of, 356 et seq.
   bend allowances, see bending
   breaking the grain, 311
   cold forming operations, comparison of, 339

flanging methods, 354, 355
fly press operations, *see* press work
forming operations, 354 *et seq.*
hand lever punch, 305
hatchet stake, use of, 306 *see also* 358
hollow bead (false wiring), 380, 381
hollowing, 359, 360
jennying machine, use of, 312–14
notched corners, 300, 302–4
panel beating, 361, 362
patterns, use of, 300, 302
pittsburgh lock, 299–301
planning, need for, 298
raising, 359–61
riveting, methods of, 306
rivets, removal of, 245
spinning, *see* spinning sheet metal
spring back, *see* bending
stiffening
  introduction to, 378
  methods of, 378 *et seq.*
swaging, 380–2
wheeling, 362–5
wire edges, cylinders and cones, 308–10
wired edge, faults, 307, 308
wiring straight edges, 307, *see also* 379, 380
Sheet metal work examples
  funnel, tin plate domestic, 309 *et seq.*
  simple box (metal pan), 304 *et seq.*
Silver soldering, *see* hard soldering
Skip welding, 110, 113
Solder
  soft, types of, 394
Soldering
  hard, *see* brazing
  soft
    fluxes, 395–7
    heat source
      copper bit, 397–400
      flame, 400, 401
    operations, 402–4
    preparation of surfaces, 394, 395
    principles of, 393–5
    types of joint, 401, 402
Sorbite, 213
Spinning sheet metal
  back stick, use of, 367, 369
  hollow bend, forming a, 380, 381
  hand tools for, 367
  introduction to, 365, 366
  lathes for, 366
  lubrication for, 369
  metal removal, 256
  process of, 367–9
  spindle speeds for, 370
Stainless steels, *see* materials

Standards of measurement, 139
Steel rule, *see* marking out
Steel tape, *see* marking out
Stiffening, *see* sheet metal work
Stress
  and strain, 127, 128
  practical effects, 131, 132
  types of, 128, 130
Stress/Strain curve (mild steel) 184


Tantulum, *see* materials
Temperature
  estimation of, 217, 218
  forging, 213, *see also* 220–2
  indicators
    heat sensitive crayons, 217, 219
    heat sensitive paints, 217, 219
    melting points, fusible alloys, 420, 421
    melting ranges, alloys, 420, 421
    relationship, thermal joining processes, 420, 421
Tempering, *see* heat treatment
Tempering colours, 217, 218
Templates, *see* marking out
Tensile strength, *see* materials (properties)
Tellurium, *see* materials
Thermal cutting processes, *see* material removal
Thermit process, *see* welding
Third angle projection, *see* drawing
Tin, *see* materials
Tin-bronze alloys, *see* materials
Tin-lead alloys, *see* materials
Titanium, *see* materials
Toughness, *see* materials (properties)
Trammels, *see* marking out
Trepanning, *see* material removal
Troosite, 213
Tungsten, *see* materials
Tungston inert gas (TIG) welding, *see* welding
Twist drills nomenclature (BS 328) 233


Undercutting, *see* welding defects
Universal surface gauge (scribing block), *see* marking out
Up-setting, *see* forging


Vanadium, *see* materials
Vector quantities, *see* force
Vernier calper, *see* measurement
Vernier height gauge, *see* marking out
Vernier scale, *see* measurement
Vice (bench, fitter's, pipe), *see* restraint and location

Welding
  aluminium, 227, 228
  brass, 434, 435
  copper, 226, 227
  distortion
    causes of, 108, 109
    minimising, 109 *et seq.*
    types of, 108, 109
    effects on parent metal, 223–5
  forge (pressure), *see* forging
  fusion
    heat source
      metallic arc, 416, 417
      oxy-acetylene, 416
    joint preparation, 417, 418
    principle, 416, 417
  grinding, *see* material removal
  manufacturing stresses, 108
  metal inert gas (MIG), 101
  mild steel, 224–6
  processes, *see* oxy-fuel gas, metal arc, forging
  relationship with material properties, 222, 223
  resistance
    principles, 418–20
    projection, 420
    seam, 420
    spot, 419, 420
  stainless steel, 228, 229
  symbols, 46 *et seq*
  thermit process, 98, 99
  tungston inert gas (TIG), 101
Weld decay, 229
Welding defects
  cracking, 229
  etched specimens
    inspection of, 455
    preparation of, 455
  fusion, lack of, 453
  illustrations of (Appendix 1)

inclusions, (slag), 454
penetration, lack of, 453
porosity, 454
profiles of, 451, 452
surface, 453
undercutting, 451, 452
workshop testing for, 454, 455
Wheeling, *see* sheet metal work
Wheeling machine, 362
Wire ropes, 24–6
Work holding (*see also* restraint and location)
  bending bars, sheet metal, 283, 285, 286
  bridge pieces, 292, 295
  chain and bar, 293, 295
  clamps, sheet metal, 283, 285, 286
  cleats, 292, 293
  dogs and blocks, 292, 294
  G-clamps
    types and applications, 286 *et seq.*
  glands, 292, 294
  hand vice, 285, 286
  jack clamps, 294, 296
  machine clamps, 289
  magnetic clamps, 297
  plate and dog pins, 290, 291
  riveting, whilst, 289, 290
  skin pins, 289, 290
  soft iron wire, use of, 290, 291
  spiders, 294, 296
  spring clips, use of 290, 291
  strong backs, 292, 293
  Tee-slot clamps, 289
  toggle action grips, 287
  welding clamps, 287 *et seq.*
Wrought iron, *see* materials


Zinc, *see* materials
Zinc fumes, *see* health hazards (safety)
Zirconium, *see* materials

**465**